UNITEXT for Physics

UNITEXT for Physics series, formerly UNITEXT Collana di Fisica e Astronomia, publishes textbooks and monographs in Physics and Astronomy, mainly in English language, characterized of a didactic style and comprehensiveness. The books published in UNITEXT for Physics series are addressed to graduate and advanced graduate students, but also to scientists and researchers as important resources for their education, knowledge and teaching.

More information about this series at http://www.springer.com/series/13351

Giovanni Giusfredi

Physical Optics

Concepts, Optical Elements, and Techniques

Springer

Giovanni Giusfredi
European Laboratory for Non-Linear
Spectroscopy (LENS)
Istituto Nazionale di Ottica—Consiglio
Nazionale delle Ricerche (INO-CNR)
Sesto Fiorentino, Italy

ISSN 2198-7882 ISSN 2198-7890 (electronic)
UNITEXT for Physics
ISBN 978-3-030-25281-6 ISBN 978-3-030-25279-3 (eBook)
https://doi.org/10.1007/978-3-030-25279-3

This Springer imprint is published by the registered company Springer Nature Switzerland AG
The registered company address is: Gewerbestrasse 11, 6330 Cham, Switzerland

1 In the beginning was the Lògos:
the Lògos was with God
and the Lògos was God.
2 He was with God in the beginning.
3 Through him all things came into being,
not one thing came into being
except through him.
4 What has come into being in him was life,
life that was the light of men;
5 and light shines in darkness,
and darkness could not overpower it.
…
9 The Lògos was the real light
that gives light to everyone;
he was coming into the world.
…
14 The Lògos became flesh,
he lived among us,
and we saw his glory,
the glory that he has from the Father
as only Son of the Father,
full of grace and truth.
…
18 No one has ever seen God.
The only Son, God,
who is close to the Father's heart,
has revealed him.

From the *Prologue, Gospel of John*
John 1:1-5, 9, 14, 18

Foreword

Optics is everywhere in our daily lives. Just imagine what the Earth would be without light... Light is the primary source of energy for life on our planet. It is also our primary channel of communication, not only through vision, but also in transmissions of optical data through sub-terrestrial or submarine fiber-optic networks and "invisible-light" radio communications in the atmosphere. Light is also one of the primary investigation tools in science, from microscopy in biological sciences to Universe-scale astronomical observations in astrophysics. Mastering the study of light and its interaction with matter enabled the scientists of the XX century to unveil the fundamental laws of our physical world and put ultimate limits on their validity.

Theodor W. Hänsch, recipient of the Nobel Prize 2005 in Physics for the invention of the "optical frequency comb", a revolutionary technique for the measurement of optical frequencies with unprecedented accuracy, entitled his Nobel lecture "A passion for precision". The motivation for that title resided in the thrilling precision that the extreme developments in the manipulation of light had brought into the field of atomic and molecular spectroscopy, and, ultimately, in our knowledge of the quantum world.

Giovanni Giusfredi is a true master of optics. I had the privilege of knowing him since 50 years: in 1969 we met in Pisa at the Scuola Normale Superiore. Then, I shared with him many exciting scientific adventures, mostly in the context of those high-precision spectroscopic studies that Hänsch had pioneered. During those years I witnessed the sincere personal "passion for precision" that moves Giovanni Giusfredi in all the aspects of his work, from theory to experiments. In his person the acute intellect of the scientist, always striving to grasp the finest details of the problems, coexists with an exceptional talent as "craftsman" of refined experimental devices that he kept building personally during his whole scientific life.

This book is a true gem. It is an essential resource for any researcher approaching the study of optics or wishing to consolidate her/his knowledge. Here students, scholars and professors, can find basically everything they need to know of classical optics. Geometrical optics and wave optics are covered in a comprehensive and original way: not only rigorous demonstrations of laws and equations, but also the perspective and the "tricks" of an experienced, talented experimentalist, with very interesting historical notes complementing the scientific content.

Optics is everywhere in our daily lives. And this truly "enlightened" book merits to be on the shelf of every scientist that makes use of optics in her/his research.

Florence, July 31, 2019 Massimo Inguscio

Preface to the Italian edition

In 1989, I was asked to teach the course in General Optics of the postgraduate School of Optics at the University of Florence. I gladly accepted this assignment, with a little fear. Indeed, I knew I was only just a beginner, and thus I set a rule for myself that, before each lesson, I would study the subject to be treated in depth. To establish order, I started writing these notes, choosing every word carefully and, through use of the various books available at the splendid Library of the *Istituto Nazionale di Ottica*, determining the best teaching method for exposing the arguments and demonstrations of Physical Optics. This text has thus grown little by little, lengthening and thickening gradually every time I learned something new. The goal that I set for myself was to give my students a solid foundation in the study of optics, leaving the deeper study of specific topics, such as optical design or holography, to other courses at the School.

As a starting point, I chose Maxwell's equations of classical electromagnetism: this is the subject of Part I, Chapter 1, which serves as both a refresher and as a support for the rest of the book. In particular, plane waves are described, including their polarization, the laws of refraction and reflection by dielectric and metallic media, absorption and dispersion.

In Part II, Chapter 2, the main concepts of Geometrical Optics are exposed. This approach reduces the waves to *rays of light* by neglecting the transverse and oscillating nature of the fields, devoting itself to the intensity and then to the *incoherent* superposition of light beams. On the other hand, it allows us to handle many of the optical phenomena of everyday in a truly excellent way, often involving inhomogeneous and heterogeneous media. For the practical aspect, elementary optical systems such as lenses, mirrors and prisms and their aberrations are treated.

In Part III, I treat the core of Physical Optics, namely, Interference and Diffraction. Chapter 3, which is on Interference, addresses the difficult problem of the superposition of monochromatic or quasi-monochromatic waves by the undulatory point of view. Limiting myself to the case of plane and spherical waves in homogeneous media, I describe the classic experiments in the division of wavefront and amplitude, various interferometers, of which the plane mirrors Fabry-Perot more extensively, and finally, the dielectric multilayers with various application examples. Lastly, in Chapter 4 on the Diffraction, I consider the propagation of monochromatic waves of any shape, particularly with respect to objects or diffracting apertures that are on the wave path. Even here, we will only deal with this phenomenon through a scalar and paraxial theory, discussing the Kirchhoff integral and those of Rayleigh - Sommerfeld, and then the Fresnel and Fraunhofer approximations, accompanied by several examples.

By contrast, in Part IV, Chapter 5, I expose what is called Fourier Optics, in which diffraction is studied through the plane wave spectrum and the numerical methods based on the *Fast Fourier Transform*. These methods are then used for the analysis of the quality of the optical systems. In addition, with the study of spatial and temporal coherence, we leave the restricted ambit of the plane and monochromatic waves. Finally, I present a review of spatial filtering methods and describe the diffraction gratings.

In Part V, I return to discussing wave propagation in isotropic and anisotropic linear media. In particular, in Chapter 6, I describe the monochromatic waves, those of laser beams, waves no longer planar but of limited transverse extension, which are called *Gaussian beams*. Also, the technique employed in their description contains a simplification, that of considering waves that mainly propagate around an axis, which is called *Gaussian* or *paraxial approximation*. This approximation is, in turn, borrowed from Paraxial Optics, whose laws can thus be reinterpreted, allowing us to easily manipulate, even in a conceptual sense, laser beams with lens systems. I also discuss the *Bessel waves* and the *Bessel-Gauss beams*, which are modern topics that I picked up directly from magazine articles. Finally, I present a brief discussion of the resonant cavity. In Chapter 7, I detail the propagation in anisotropic media, beginning with a summary on crystallography and the punctual and spatial symmetry groups. In addition, on the basis of relativity, I discuss the propagation in *bianisotropic* media to explain phenomena such as optical activity, the Faraday effect and the Fresnel - Fizeau effect. This discussion was also obtained from magazine articles. Finally, I describe the main manipulation devices of polarization and their use as spectral filters.

There came a point in my learning when I realized that something was missing, that is, a historical memory of the optical knowledge and the hard work that our predecessors had done to obtain it. I subsequently dedicated myself to the study of the History of Optics, consulting many books, from which I finally created a summary that I divided into interludes that serve as historical introductions for the respective Parts. Without any claim to completeness and accuracy, I hope that the news reported there be an incentive for a more detailed historical study and that readers can draw passion and fun from them.

I ask readers to forgive the mistakes that are surely present in the text and I gratefully welcome any corrections that they may wish to send my way.

Finally I want to thank my wife Marcela and my son Michele for the love and patience that they have shown me in the many years of writing this book.

Campi Bisenzio, December 19, 2014 Giovanni Giusfredi

Index

Part III Physical Optics

Abbreviations and used symbols

(illustrative list of symbols with local exceptions)

Abbreviations

A.D.	*Angular Deviation*
APSF	*Amplitude Point Spread Function*
CTF	*Coherent Transfer Function*
CW	*Continuous Wave*
FFT	*Fast Fourier Transform*
FSR	*Free Spectral Range*
GDD	*Group Delay Dispersion*
GVD	*Group Velocity Dispersion*
MTF	*Modulation Transfer Function*
NA	*Numerical Aperture*
OPD	*Optical Path Difference*
OPL	*Optical Path Length*
OTF	*Optical Transfer Function*
PVD	*Phase Velocity Dispersion*
PSF	*Point Spread Function*
TE, TM	*Transverse wave to the plane of incidence, Electrical, Magnetic*
TEM	*Transverse Electromagnetic Mode*

Real fields

$\mathbf{E}$	electric field
$\mathbf{B}$	magnetic induction field
$\mathbf{D}$	electric displacement field
$\mathbf{H}$	magnetic (displacement) field
$\mathbf{P}$	electric polarization
$\mathbf{M}$	magnetization
$\mathbf{A}$	vector potential
Φ	scalar potential
$\mathbf{S}$	Poynting vector

Complex amplitudes of the fields (vectors)

$\boldsymbol{E}, \boldsymbol{E}_0$	electric field
$\boldsymbol{B}, \boldsymbol{B}_0$	magnetic induction field
$\boldsymbol{D}, \boldsymbol{D}_0$	electric displacement field
$\boldsymbol{H}, \boldsymbol{H}_0$	magnetic field
$\boldsymbol{P}$	electric polarization
$\boldsymbol{M}$	magnetization
$\mathcal{E}, \mathcal{B}, \mathcal{D}, \mathcal{H}, \mathcal{P}, \mathcal{M}$	their analytical functions

Physical constants

$\mathfrak{c}$	light velocity
$\mathfrak{e}$	electron charge
$\hbar$	Plank constant/(2π)
$\mathfrak{m}_e$	electron mass
ε_0	electric permittivity of vacuum
μ_0	magnetic permeability of vacuum

Geometric entities

A, S, V, ...	surfaces, volumes
C, ...	lines
P, V, O, ...	points

Mathematical sets

$\mathbb{N}$, $\mathbb{Z}$, $\mathbb{R}$, $\mathbb{C}$	set of natural numbers, integers, real, complex
$\mathbf{L}^1$, $\mathbf{L}^2$, $\mathbf{L}^\infty$	Lebesgue spaces

Matrices and tensors

$\mathfrak{A}$, $\mathfrak{B}$, $\mathfrak{C}$, $\mathfrak{D}$	submatrices of an optical system
$\mathbf{C}$	surface curvature matrix
$\mathfrak{F}$	electromagnetic tensor (**E** and **B**)
$\mathfrak{G}$	displacement tensor (**D** ed **H**)
$\mathfrak{J}$	coherency matrix of polarization
$\mathfrak{Q}$	wavefront curvature matrix

(the $\mathfrak{Q}$ matrices have the opposite sign with respect to the **C** matrices)

Mathematical operators

$\mathcal{P}$	principal part
∇	vector differential operator, nabla
∇f	gradient of f
$\nabla\bullet$	divergence
$\nabla\times$	rotor
Δ, ∇^2	Laplacian
$\square$	d'Alembertian
Ξ	optional minus sign (sign convention in Geometrical Optics)

Transformation operators

$\mathscr{L}$	Lorentz transformation
$\mathscr{R}$	rotation

Properties of the medium

n	refractive index
$\tilde{n}$	complex refractive index
n', n''	eigenvalues of the refractive indices in an anisotropic medium
n_1, n_2, n_3	principal refractive indices in a biaxial medium
n_e, n_o	principal indices of an uniaxial medium, extraordinary and ordinary

α	polarizability
ε, μ	permittivity and permeability of the medium
κ	extinction coefficient
σ	conductivity
χ_e, χ_m	electrical and magnetic susceptibility

Cartesian reference frames

$\mathfrak{L}$, $\mathbb{S}$, $\mathfrak{R}$, ℓ	of the laboratory, of the surface, of the ray, local

Source terms

J	current density
ρ	charge density

Variables

a, *b*, *c*, ...	scalar variables
j, *k*, *l*, ...	indices
x, *y*, *z*, *t*	spatial and temporal coordinates
α, β, γ, ϑ, φ, ...	angles, phases, ...

Variables of the waves

f_x, f_y, f_z	spatial frequencies
$\boldsymbol{k}$	wave vector
k_x, k_y, k_z	components of the wave vector, often k_z abbreviated with *k*, when the other two components are null
k	module of the wave vector
L	eikonal
s	optical path along a ray
v, *u*	phase velocity, group velocity
λ	wavelength in the medium
λ_o	wavelength in vacuum
ν	frequency
ω	angular frequency

Particular variables in optics

f	focal length
$\mathcal{F}$	finesse of a resonant cavity
$\mathcal{P}$	power of an optical system
$\mathcal{R}$	resolving power
$\mathcal{V}$	visibility of the fringes

Versors

e, **b**, **d**, **h**	versors of the fields
n	versor of the wave vector
s	versor of the Poynting vector
$\mathbf{P}_x$, $\mathbf{P}_y$, **L**, **R**	versors of polarization
S, **T**, **N**, ...	versors in Geometrical Optics: sagittal, tangential, normal, ...

About the Author

Giovanni Giusfredi is currently a senior associate at INO-CNR, having retired in 2016 from INO, where he had been a lead researcher for almost 30 years. He remains actively involved in research and collaborates with colleagues at the European Laboratory for Non-linear Spectroscopy. He is also a founding member of CNR ppqSense. His research interests are wide ranging and he has coauthored more than 220 titles, including 71 articles in international journals. He is skilled and experienced in the teaching of Physical Optics and the History of Optics.

Introduction to Optics

The study of the emission, propagation and perception of electromagnetic waves, in particular those corresponding to visible radiation, is a science, *optics*, whose origin dates back to classical antiquity. Optics has made crucial contributions to culture and technology and has been and still is today the testing ground of many physical and philosophical theories. Its laws have been refined over time, thanks to the contribution of many scholars along a difficult, long, and uncertain path: what seems obvious to us was most assuredly not so for our ancestors, who came up with some rather strange conceptions, far removed from our way of reasoning. We need to ponder this, as we too could suffer the same fate if, as we hope, civilization keeps progressing. For the sake of conciseness, the historical labor necessary to attain knowledge often goes unmentioned, causing us to forget how rich in perspective, and yet terribly fragile, human culture and civilization are. At the same time, the history of optics is a paradigm for the evolution of scientific thought and the way in which Science works.

The center of attention for the ancient Greek philosophers was Man himself, and light and his perception of it were integral parts of the sense of sight. In this context, the foundations of Geometrical Optics and Perspective were laid. In the considerable time since this initial fact was perceived, several partitions have occurred: Physical Optics focused on the physical agent of light, the *lumen*, while Physiological and Psychological Optics were given the task of exploring the mechanisms of perception and interpretation of the visual information, the *lux*. Physical Optics was born as a geometric theory of light rays and then developed as a wave theory, up to the modern quantum theories. It branched out into many disciplines, mainly dealing with three aspects of the visual process: sources, propagation and detectors. In the pages of this book, we will mainly deal with the propagation, using Maxwell's equations of classical electromagnetism as a starting point. They look very elegant, however, the wave equations are not easy to solve, except for plane waves in simple media. Therefore, we often use more practical conceptual tools, sometimes rather ancient ones, even though they necessarily introduce some degree of approximation. There are various tools that can be more or less ordered on a scale of increasing simplicity and practicality, as well as inaccuracy. As a rule, we just need to adopt the simplest technique that is still capable of providing the accuracy required for our goals. As a matter of fact, several ways of handling wave propagation have been developed, each one fit for some particular application.

Now, we have a large variety of optical instruments and techniques for endless applications. We have the optics of classical instruments that deals with the production of images, such as eyeglass lenses, camera lenses, microscopes for observing the physical and biological microscopic world, telescopes for exploring the Universe, theodolites and tacheometers for trigonometric measurements of the territory. Various optical methods have been developed for non-invasive measurements in the absence of contact with the object to be measured. Thus, we have interferometry, for precision measurements of distances and for assessing the quality of optical components, and Fourier Optics, on the basis of diffraction, for filtering and optical pattern recognition. With the use of infrared- or ultraviolet-sensitive detectors, we can explore paintings in depth so as to discover the techniques of the artists who produced them and facilitate restoration. Recently, the optics of solar energy collection has been in development. As direct sources, lamps for lighting, spectral lamps, semiconductors emitters, and other luminescent materials have been developed. Huge progress then occurred in the field of Quantum Optics and the invention of laser sources. These have revolutionized traditional optical investigation techniques. In particular, the spatial and temporal coherence properties of the lasers have made holography possible, that is, the recording of three-dimensional objects. The laser has also led to the development of the optical fibers on which modern telephony and data transmission are based. Special interest is now being devoted to quantum cryptography. The semiconductor lasers allow for the recording of information on optical discs, with which we can enjoy music, watch a movie, read a book, and store data and computer programs. Micro-optical components are being developed for many applications, in particular, to be used with optical fibers. The significant intensity of the radiation supplied by lasers has led to the study and development of non-linear optical materials and devices, capable of generating coherent radiation with sum or difference frequencies from the pump laser beams. Lasers also allow for phase conjugation, with which it is possible to reconstruct a clean image from aberrations in transmissions through an inhomogeneous medium. Powerful lasers also provide the capability for various invasive techniques, such as the cutting and heat treatment of metals, the shaping of tridimensional objects and microelectronic components, and even surgical operations. Besides the eye, which is the object of Physiological Optics, other detectors worthy of being mentioned are photographic and holographic materials, semiconductor detectors for a great variety of spectral bands, from ultraviolet to infrared, photo tubes, image intensifier, and CCD arrays.

Optics is also a vital component of many other sciences, such as Spectroscopy and Astronomy. Biology and Medicine also utilize optical techniques, for example, many clinical and biological analyses are conducted with spectrophotometers and microscopes. The bond granted by technical applications such as photography, television and numerical image processing (and faking) should also be remembered. Physiological Optics has provided the basis of visual perception for lighting and for color synthesis through the trichromy. The eye, in its various forms, and the functioning of the brain's visual cortex are considered a model for the perception of the world for use in artificial intelligence devices. Optics, both physiological and psychological, is thus essential for the production and enjoyment of works of art, since sight is, in most cases, the only one of our senses that permits it.

Much more can be written on optics, which, by the pervasiveness of its object of study in all disciplines, represents one of the most fruitful sciences in regard to ideas and applications.

Part I
Electromagnetism

Historical notes: The first discoveries about magnetism and electricity

1 In the beginning, when God created the
heavens and the earth,
2 the earth was a formless wasteland,
and darkness covered the abyss,
while the spirit of God swept over the waters.
3 Then God said, “Let there be light”,
and there was light.
4 God saw how good the light was.
God then separated the light from the darkness.
5 God called the light “day”,
and the darkness he called “night”.
Thus evening came, and morning followed:
the first day.

Genesis I, 1-5.

The dawning

The magnetic properties of some minerals and the electric ones of rubbed amber have been known since ancient times: amber is a fossil resin that the Greeks used for their experiments and which they called ηλεκτρον, while, at Magnesia, a city in Asia Minor, Phoenician merchants were selling magnetic stones. These materials have in common the unique ability to attract other bodies, so their electric or magnetic properties were assimilated into the magical conceptions of the ancients [Bernal 1969]. They had fed the theories of affinity and attraction at a distance and the idea of a “virtue” (as it was called by Aristotélēs), held by certain substances, that could be transferred to others by simple contact. Think of the myth of King Midas and the philosopher’s stone. In turn, the concept of the field comes from a long tradition that attributes dynamic or energy properties to space, as in the Aristotelian idea of aether. However, the study of magnetism became a science only after the magnetic virtue was associated with a practical purpose, as with the compass: it was discovered in China around the first century BC and, perhaps in the twelfth century, traveled to the West in a manner that remains unknown. The first Western scientific study on magnetism was carried out by Pierre le Pèlerin de Maricourt (Petrus Peregrinus), who, in his *Epistola de magnete* (1269) , gives instructions on how to build a magnetic needle and describes for the first time the simple properties of magnets, including the law that opposite poles attract each other. This study was followed, after a considerable time, by *The new attractive* (1581), written by the navigator and manufacturer of compasses Robert

Norman (m. ~1590), who discovered the magnetic declination (on a vertical plane) of terrestrial magnetism, associated with the latitude, and by *De Magnete, Magneticisque Corporibus, et de Magno Magnete Tellure* (1600), written by the physician William Gilbert (1544–1603). In this book, Gilbert described his experiments on magnetic bodies and electrical attraction, distinguishing it from the magnetic. In fact, he noted that a magnet does not require any stimulus to show its effects, while glass or amber need to be rubbed. In addition, when a magnet is broken into two fragments, they always retain two poles; indeed, it is not possible to find a fragment with a single pole. From his experiments, he derived new concepts and general principles, including the idea that the Earth is a great magnet and the fact that it is the same magnetic property of *attraction* that keeps the planets tied to their pattern of motion. His book also became a classic of scientific literature for its method of investigation: indeed, in it, Gilbert supports the need to verify the traditions passed down over time through the execution of appropriate and detailed experiments. His work was appreciated by Galileo, who dedicated several pages to it in the third day of the *Dialoghi*, and it paved the way for Newton's general theory of gravitation. Gilbert addressed not only magnets, but also electrical phenomena. He found that various materials, besides amber, could be electrified through rubbing and theorized that this would cause the emanation of certain *effluvia* that, returning to the rubbed body, would drag along the surrounding light bodies. Furthermore, he invented the first electric instrument, the balanced needle, or *versorium*, and was the first to use the terms *electric attraction*, *electric force*, and *magnetic pole*.

From 1600 onwards, there have been major new discoveries about electricity and magnetism, from which began the electromagnetic revolution of human society that is still dramatically changing the conditions of life on Earth. At first, electricity was considered a mere intellectual pastime without serious applications, such as the little cloud sighted by Noah before the Great Flood.

Thus, Niccolò Cabeo (1586–1650) described, in his *Philosophia Magnetica* of 1629, the discovery of electrostatic repulsion. Francesco Maria Grimaldi (1618 - 1663) claims, in the frontispiece of his *De Lumine*, posthumously published in 1665, that the magnetic effluvium pervades all bodies. In 1655, Otto von Guericke (1602 - 1686), inventor of the vacuum pump, built the first machine for producing electricity, consisting of a rotating sulphur globe that emitted sparks when it was rubbed by hand. Subsequently, other researchers used a fast rotating glass globe instead of the sulphur one: with this machine, electricity became reproducible and susceptible to accurate experimentations.

In 1675, Jean Picard (1620–1682) noted that a barometer he had shaken in the dark had emitted green light (produced by the mercury): in the early 1700s, this phenomenon aroused the interest of Francis Hauksbee (1660 – 1713), who showed that, through rubbing, we could create light effects in vacuum. In particular, if the glass globe of the electric machine was partially emptied of air, the rubbing caused intense luminous phenomena in its interior.

At that time, bodies were distinguished as being either *electric* or *non-electric*.

To the first group belonged only amber, glass and some precious stones that could be electrified by rubbing. It was believed that these were abundantly supplied with a particular electrical substance that emanated in response to rubbing. This phenomenon is called *triboelectricity*. The other non electric bodies, indifferent to rubbing, were deemed to be without this substance.

In 1729, Stephen Gray (~1666–1736) discovered that the electrical virtue can be transmitted from an excited electric body to a non-electrical one by contact and conduction without any apparent material movement, like an imponderable fluid: a piece of metal touched by an electrified glass tube acquired the property of attracting the test corpuscles. Moreover, he found that, in some cases, one could have an electrification for induced charges. In particular, he noted that electricity can be generated and stored in certain materials that he called *electris*, while it flows in metals and wet bodies, which he called *non electris*; a few years later, the first would come to be called insulators and the seconds conductors.[1] Indeed, a metal piece was electrified only when it was supported by special materials, such as silk threads, but not by metallic wires. In addition, by suspending a conductive strip with silk threads, he found that he could transmit the electricity to more than 300 feet.

These studies were extended by the superintendent of the royal gardens of the King of France, Charles François Dufay (1698–1739), who emphasized the need to place the bodies to be electrified on insulating supports: this precaution was called *Dufay's rule*. In 1733, he discovered that there were two kinds of electricity, one deriving from the rubbing of the glass and another by the rubbing of resins such as amber. These two electricities were associated with two kinds of electric force, one attractive and one repulsive: indeed, he noticed that two bodies both electrified by resin, or by glass, repel each other, while there is attraction between a body electrified by resin and one electrified by glass.

Around 1740, the production of electricity with a globe machine was improved. In place of the hands, the globe was rubbed with leather pads or other materials, while the electricity was collected by a chain or by a metal bar, called the *prime conductor*, suspended with insulating cords. Dufay's rule suggested that the person who rubs the globe should stand on an insulating base, but, surprisingly, it was discovered that the machine gave off more electricity if the rubber was in good contact with the ground, violating the rule. This contradiction has been resolved by imagining that the earth, and, in general, the non-electric bodies, were actually rich in electrical substance, while the electric ones were poor in it, reversing the previous hypotheses. An attempt was then made to enclose this electric fluid within glass containers filled with water, which was electrified through a wire with one end immersed in the liquid and the other end brought into contact with the prime conductor. In this way, three amateur electrologists discovered a new

---

[1] It should be noted that, for the small amount of charge and the high voltages generated, both the volume and surface resistance of the insulators must be particularly high: indeed, it just needs a bit of moisture to disperse the charges and convert them into conductors.

shocking violation of Dufay's rule. To these researchers is credited the invention of the *Leyden jar*, which was a primitive form of an electrical capacitor. In 1745, Jürgen von Kleist, dean of the Cathedral of Kammin in Pomerania, took an "electric shock" while wielding an electrified bottle. Independently, a few months later, in the Netherlands, a professor at the University of Leiden and manufacturer of scientific instruments, Pieter van Musschenbroek (1692–1761), gave notice of a similar experiment. His report of January 1746 is particularly dramatic: his laboratory was frequented by Andreas Cunaeus (1712–1788), who, neglecting Dufay's prescription that the bottle be kept on an insulating support, gripped it by the belly, bringing the wire into contact with the prime conductor. During this operation, he accidentally touched the first conductor with the other hand, receiving a shock that stunned him and frightened him to death. Informed of the incident, Musschenbroek could also *personally* verify the painful effects of the electric shock. Contrary to expectations, the charging of the bottle was much stronger than what was obtained by following Dufay's rule.

In the wake of these findings, interest in electrical phenomena began to spread with the construction of curious contrivances, which were exhibited in public for a fee, offering the thrill of trying out electric shocks.

Some order in the understanding of these phenomena began to appear when Benjamin Franklin (1706–1790) intrigued by the burgeoning scientific field, got himself some electrical equipment and, with various experiments, succeeded in founding the basics of electrical theory. Born in Boston into the family of a poor candle merchant, in the course of his intense life, he was an apprentice typographer at age 12, the inventor of the lightning rod and the rocking chair, a philosopher, politician and diplomat; he was one of the three drafters of the *Declaration of Independence* and was the first famous physicist of the New World. He interpreted the «electric fire» as an immaterial fluid that existed in every body, imperceptible under normal conditions, but this fluid could be pumped from one body to another through rubbing. The body that receives the electricity is charged in the *positive* sense, the other in the *negative*; with this, Franklin enunciated the law of conservation of electric charge and produced a new interpretation of the two types of electricity discovered by Dufay using a pneumatic-attractive model. The electrical attraction was therefore due to the tendency towards the equilibrium condition, which could be re-established by a spark. Through various experiments, he explained the functioning of the Leyden jar, concluding that a *superficial* excess of electrical substance is formed on one face of the glass wall and an equivalent depletion on the opposite face. Franklin discovered that the bottle works better by coating the outer and inner surfaces with thin metal sheets. In addition, the shape of the bottle was inessential, and therefore he used simple glass slabs, armed with metallic sheets on both surfaces; this device was called *Franklin's square*. If a thin surface layer is sufficient to produce such great effects, the *electris* must also be rich in electrical substance within, the expansive power of which is counterbalanced by the "power of attraction and condensation" that the matter of the substance exerts on the electric fire.

Franklin captured the popular imagination, especially when, in 1753, he was able to prove that lightning is an electrical phenomenon similar to the spark, and the damage from this dramatic natural effect could be avoided with his lightning rod.

Also in 1753, John Canton (1718–1772), inspired Franklin's theories, discovered a new induction phenomenon: a conductor, placed near an electrified body without touching it, accumulates charges of the opposite sign on the side near to this body and charges of the equal sign on the far side.

Unlike Franklin, the German astronomer Franz Aepinus (1724–1802) made use of a gravitational analogy, instead of a mechanical one, to explain the electrostatic effects and the functioning of the Franklin squares. Assuming that the bodies are provided with matter and electric fire, he considered the various forces of repulsion and attraction between the elements that constitute two distinct neutral bodies A and B, separated from each other. Therefore, between the two electrical fires of A and B, there is the repulsion r, while the attraction a is between the fire of A on the matter of B and the attraction a' is between the fire of B on the matter of A. Since there is no attraction or electric repulsion between neutral bodies, these forces should balance each other, for which we have $r+a+a' = 0$. On the other hand, there is no attraction or repulsion, not even between the electric fire of the body A and the neutral body B, so it must also be valid that $r+a = 0$, from which we obtain the absurd result of $a' = 0$. Aepinus concluded that this contradiction could be resolved assuming a fourth force, *repulsive*, between the matter of the two bodies. He came to this realization to his own horror, since this fact contradicted the idea of an attractive force (gravitational) among the matter, but remained convinced of the validity of his proof. Aepinus also worked on magnetism and, in 1759, published his studies in *Tentamen theoriae electricitatis et magnetismi*, which was considered harsh owing to its strong mathematical content, and was thus neglected by electrologists of the time.

The laws of electromagnetism

In 1729, Gray had noted that two cubes of oak, one hollow and one solid, equally charged, show the same electrical effects, suggesting that the charges are distributed only on their surface. A similar experiment was done in 1766 by Joseph Priestley (1733–1804): he found that, inside a metallic spherical shell, there is no electrical force. Priestly deduced that the electrical attraction is subject to the same law of gravitational attraction, proportional to the inverse square of the distance. This observation was repeated with more precision by Henry Cavendish (1731–1810) in 1771, but his research long remained unpublished. Already in 1750, John Michell (1724–1793) had found, with a simple torsion balance, that the attraction or repulsion between magnetic poles also follows the same law. In 1785, Charles-Augustin Coulomb (1738 –1806) came to the same conclusions: he suspended the needle of a compass with a thin wire and used it to measure the strength of magnetic poles; in a similar way, Coulomb used a very sensitive torsion balance, the first prototype of an electric instrument, to measure the force between electric charges.

One authentic prophet of modern physics was the Jesuit Giuseppe Ruggero Boscovich (1711–1787), who, ahead of his time, had developed a non-mechanical theory of matter in 1758, introducing concepts such as punctiform elementary particles and the unity of all physical forces. He published his studies in the treaty *Theoria philosophiae naturalis redacta ad unicam legem virium in natura existentium*. Dissatisfied with Newton's principle of attraction, Boscovich introduced the idea that matter is composed of identical physical points, without extension, with a sphere of action alternately attractive and repulsive depending on the distance, and with asymptotic behavior at extreme distances. Repulsive and diverging to infinity at distances tending to zero, the material would otherwise collapse at a point; it would be attractive with decreasing intensity for distances tending to infinity, to account for the gravity. Thus, ordinary matter can exist, since the overlap of the forces that originate from the various centers of force gives rise to positions of equilibrium in which these centers can stay, forming stable agglomerates.

Important contributions came from Alessandro Volta (1745–1827), who became occupied with electricity from an early age. Inspired by the work of Newton and Franklin, in his study of electrical motions, he preferred the non-mechanical way of attraction to the mechanical principles developed by such famous electrologists as the abbe Jean Antoine Nollet (1700 –1770) and father Giambattista Beccaria (1716 –1781), to whom he wrote about this conviction in 1763. In the earliest years of his activity, Volta studied triboelectricity, which was still the only known way to produce electricity. Thus, in a letter of 1765 sent to Beccaria, Volta classified the various material substances according to a scale in which glass was the poorest of electrical fire and sulfur was the richest of it, while other substances were in intermediate positions depending on their reciprocal properties of granting or acquiring electric fire. This scheme, however, was frustrated by some experiments by Beccaria, in which, for example, the rough glass behaved in the opposite manner as the polished.

Beccaria described these experiments in the first of two papers that he sent to the *Royal Society* of London in 1766. There, he also describes an electrometer consisting of two silver wires left free to diverge for electric repulsion when they are put into contact with a charged cylindrical conductor. Following the pneumatic analogy, Beccaria attributed a pressure to the charge possessed by the conductor body and, with his electrometer, verified that this pressure was halved when the cylinder was placed in contact with another equal but discharged cylinder. In the second memory, Beccaria described some observable effects with the Franklin's square; however, for these effects and for the triboelectricity, he postponed the explanation, referring to vague principles that still had yet to be discovered.

Stimulated by this incompleteness, Volta then proposed an electric theory based on two principles. The first was still the pneumatic repulsion of the electric fire, while the second was the existence of a non-mechanical force of attraction between the matter and the electric fire. Volta was influenced by the work of Boscovich and resumed the saturation concept that he had used in his theory of heat: the condition of equilibrium between two bodies of different substances is reached when the force of attraction that these exert on the substance of the heat balance out the expansive power of the pneumatic repulsion. So, the different substances become "sated" with an amount of heat proportional to their different forces of attraction. In an analogous way, Volta could coherently explain both the electrical attraction or repulsion motions between electrified bodies and the effects of electrostatic induction. Indeed, two plates, charged in the opposite way and approached at a small distance without touching, lose their electric signs, while, when their distance is increased, the plates regain such signs, as happens for the armatures of Franklin's square. In addition, he could explain the triboelectricity by appealing, in this case, to the microscopic level of Boscovich's theory on matter. In 1769, only 24 years old, Volta wrote a paper entitled *De vi attractiva ignis electrici ac phaenomenis independentibus*, which he sent to the major electrologist of his time, although he got a rather cold answer.

Volta did not lose heart and, on the suggestion of Lazzaro Spallanzani, devoted himself to numerous experimental tests. In 1775, he announced the invention of the "*perpetual electrophorus*", by itself similar to Franklin's square, but consisting of a metal plate on which was deposited a layer of resin, much more capable of retaining electricity than the glass, and a metallic shield with rounded edges and an insulating handle. The resin was initially electrified. Then, the shield was placed over the resin, short-circuited with the dish, and finally raised, producing a strong spark. The sequence could be repeated many times, causing the instrument to seem to generate a nearly unlimited quantity of electricity. This produced a sensation in Europe similar to that of the invention of the Leyden jar; however, almost no one understood the operation, and Volta's appeal to reconsider his *De vi actractiva* was not yet accepted. As a result of the experiments carried out with long conducting wires, he resumed the pneumatic concept of *tension* used by Beccaria to indicate the "pressure" of the electric fluid in a charged body and introduced the concept of *capacity* to indicate the wire's attitude toward containing the electric fluid. Using an electrometer for voltage measurements and counting the number of revolutions of the electric machine as an estimate of the charge Q that was necessary to load a conductor with a capacity C up to the voltage V, Volta was thus able to formulate the law of proportionality $Q \propto VC$. In 1780, Volta realized the *condenser*, so named for its ability to store significant amounts of charge. It was similar to the electrophorous, but the resin was replaced with a thin insulating layer. In this way, once the plates were loaded one over the other, by separating them, one could produce a much higher voltage, even one a hundred times higher. This made it possible to measure voltages that were otherwise too weak to be detected by electroscopes. Upon the recommendation of Volta, between 1781 and 1782, Pierre Simon de Laplace (1749 –1827) and Antoine Laurent de Lavoisier could find electric signs from the evaporation of water and opened the study of electrochemical potentials.

Volta subsequently realized a version of a leaf electrometer, equipped with various contrivances that increased its precision, and he devoted himself to the problem of defining a unit of tension, reproducible unambiguously by all the experimenters. For this purpose, he described an electric balance, equipped with two metal plates at one side, one fixed and connected to the ground, the other mobile, suspended from insulating ropes and held in equilibrium by a counterweight on the other side of the balance. As a result of this invention, the unit of potential difference has been dedicated to him with the name of *volt*.

Afterwards, the great French physicists and mathematicians dealt with the theory of electricity, developing a general formulation of the phenomena governed by force laws of the type $1/r^2$, in the context of the doctrine of action at a distance. In particular, they considered the effects of a mass or a charge distributed in the bodies. Coulomb had already examined the case of the sphere. Laplace, in his studies of celestial mechanics, was concerned with establishing the gravitational force acting on a point P of mass m by an extended body of a given geometric shape. In particular, he considered a spherical or ellipsoidal body, in which its mass M was distributed in a continuous manner, and introduced a function of the spatial coordinates of P, which is known today as the *potential function*, with which to express this force in the differential equations of motion.

From 1811, Siméon Denis Poisson (1781–1840), assuming the existence of both positive and negative electric fluids, developed the theory of electrostatics, which he published in 1812 in his *Mémoire sur la distribution de l'électricité à la surface des corps conducteurs*, laying the foundation for an analytical formulation of the law of attraction of Coulomb and Cavendish. In this paper, he applied the Laplacian function of the potential, but, in the following year, he showed that it was true for the charged bodies only externally and gave a more general expression, known as the Poisson equation, which is also valid within them. Poisson also treated the magnetism, enriching it with the concept of magnetic moment.

So far, the phenomena studied were only of the static type: a new trail of investigation relating to electrical currents was opened around 1780 by Luigi Galvani (1737 –1798), a professor of anatomy at the University of Bologna who was also interested in electricity and had bought some electric instruments for his laboratory. He discovered that, when the crural nerve of the leg of a dissected frog is touched with a bistouri, while a spark is emitted from a nearby electric machine, the muscles of the leg shrink violently. Later, he discovered that the same phenomenon happens when two conductors of different metals are put into contact, one of which touches a nerve and the other a muscle. Galvani deduced that an electrical fluid was transported from the nerve to the muscle through the metal and that, in living things, a peculiar animal electricity was generated.

In 1792, a year after the publication of Galvani, a physical explanation of this phenomenon was given by Volta, who found that there was passage of electricity even in a circuit consisting of two different metals and any wet body. He deduced that the source of electricity was in the joint between the two metals and that the frog leg was only a delicate electrometer for detecting the passage of the current. However, Galvani had shown that even the transmission of signals in a living being is done through electricity.

In 1800, Volta built thc first electrical power source not of an electrostatic type, namely, a *pile* constituted by a series, repeated many times over, of a copper disk, a zinc disc and a wet cloth disk, with which Volta was able to demonstrate the same electrical effects that were obtained with the rubbing machines. Soon, chemists began to realize the importance of this discovery and they used large and expensive batteries for the separation and identification of the chemical elements. Among them, Humphry Davy (1778–1829), director of the *Royal Institution* founded in London in 1799 by Benjamin Thompson (Count Rumford), used the pile to decompose many metallic salts. His personal assistant was Michael Faraday (1791 –1867), who began his research investigating the boundary between chemistry and physics, subsequently succeeding in enunciating the laws of electrochemical decomposition. From these, by the way, one could foresee the existence of the elementary unit of electric charge, as it was later noted by George Johnstone Stoney (1826–1911) in 1874 and by Hermann von Helmholtz (1821–1894) in 1881. The electron was finally discovered in 1897 by Joseph John Thomson (1856–1940) [Peruzzi 1997].

Thus, electricity became a bridge between physics and chemistry, corroborating the idea of unity of nature. Contrary to the dominant idea that the electric and magnetic phenomena were associated with completely different fluids, German philosophers argued instead that light, electricity, magnetism and chemical forces were different aspects of the same reality, and the various natural phenomena were brought back to the "conflict" between the primordial forces opposing each other. Thus, Hans Christian Oersted (1777–1851), professor of Physics in Copenhagen, inspired by this *Naturphilosophiae*, for years sought a relationship between electricity and magnetism, but the connection was hard to find. Finally, at the end of 1819, while preparing a teaching experiment on the thermal effects of electric current in a copper wire, he discovered that the needle of a compass that had been placed nearby by a lucky chance was diverted in the presence of current in the wire. He then deepened the study of this phenomenon and found that the sense of this deviation depends on the direction of the current in the wire and the position of the compass relative to the wire. He thus deduced from this the existence of an "electric conflict" that operates in a circle around a straight wire carrying a current, i.e., with the Faraday expression, that there are circular lines of magnetic force around the wire. Oersted said, among other things, that this electric conflict was able to cross the non-magnetic materials and deduced from this that it "can act only on the magnetic particles of matter". On July 21, 1820, he sent his results to the Danish Academy of Sciences, which published them under the title: *Experimenta circa effectum conflictus electrici in acum magneticam*. In a subsequent article, Oersted showed that the effect was mutual: if a magnetic needle is held near a movable wire, this wire rotates in the opposite direction when a current passes through it. Therefore, in addition to finding an unexpected connection between magnetism and electricity, he had demonstrated the existence of a shocking *non-central force* acting at a distance.

The news of this discovery was immediately taken to Paris by Jean Dominique François Arago, who repeated the experiment in front of the members of the French Academy, which included his friend André Marie Ampère (1775 –1836). Ampere realized that it was appropriate to cancel the Earth's magnetic field around the compass to determine the direction of the magnetic field generated by currents. Thanks to this measure, he created a tool called a galvanometer: with it, he could define the current path in a closed circuit and discovered that, within the pile, it flows the same current that goes through the rest of the circuit.

Ampère then performed a series of experiments in which he found that, between two parallel conductors, a force of attraction or repulsion is developed, depending on the mutual orientation in which the currents flow, and that a solenoid behaves in a way similar to a magnet. From here, he concluded that the same magnetism of the magnets was generated by internal currents. In this way, Ampère could bring the magnetic force back to a central force due to currents, understood as a result of the integration on closed circuits, or extended indefinitely [Jackson 1974, Cap. 5]. Faraday challenged this conclusion in an anonymous publication of 1821, citing the counterexample of one of his experiments, in which he noted that the magnetic field of a magnetized iron tube reverses its direction, by going from outside to inside of the tube along its axis, while this does not happen for the solenoid. On the suggestion of his friend Augustin Fresnel, Ampère was able to solve the question by assuming, then, that the currents of a permanent magnet circulate around the individual molecules that compose it, instead of uniformly in a circle around its magnetic axis, i.e., the axis of the tube, contrary to what occurs in the solenoid. Ampère finally succeeded in giving a mathematical expression for the force that acts between any two conductors by vectorially integrating the elementary forces produced by the current in the individual volume elements of the conductors. These forces resulted dependent on both the distance, as $1/r^2$, and the current direction. Contributions to the laws that determine the magnetic effects as a function of the currents, distances and directions also came from Jean-Baptiste Biot (1774–1862), Felix Savart (1791–1841), and, in particular, Pierre-Simon de Laplace. The latter by using infinitesimal calculus, determined the magnetic intensity inside a solenoid and at the center of a circular loop where a current flows.

A few years later, in 1823, these discoveries already had a practical application with the invention of the electromagnet by William Sturgeon (1783 –1850), which was later improved by the American Joseph Henry (1799–1878), opening the way for the invention of the telegraph and the electric motor. Georg Simon Ohm (1789–1854), a teacher of physics in Cologne, investigated the way in which the current flows in the conductors: he found that this depends on the metal used and that it was proportional to the cross-section of the wire through which the current passes. Adapting the concepts of a fluid flowing in a tube to electricity, as it had been done for the heat conduction with the *caloric*, Ohm defined the concepts of resistance and potential difference or electromotive force, and linked them to the current intensity with his homonymous law of 1826.

After the discovery of magnetic effects by electric currents, Michael Faraday, (1791–1867) faced the task of investigating the inverse relationship. He tried for years to produce electricity from magnetism and, finally, in 1831, found that this was possible, but only dynamically, as a result of a relative movement between the magnet and electric circuit. Moreover, since a magnetic field can be generated by a current, he also found that, by wrapping two separate coils around an iron core (a ring transformer) and sending a current pulse through the first, an electric current is induced in the second. Faraday called this effect *magneto electric induction* and presented the results of his experiments to the Royal Society at the end of 1831, subsequently publishing them in a paper of 1832. This phenomenon has made it possible to generate electric currents mechanically and, reciprocally, operate machines with electric currents, but it took another 50 years before it came to its full implementation. In the meantime, Henry, after conducting experiments with an electromagnet, also tried to produce electricity from magnetism: he succeeded before Faraday, but was unable to take advantage of the discovery until after the publication of Faraday's memory in 1832. That same year, Henry also discovered and came to understand the phenomenon of self-induction, ultimately publishing his work in the *American Journal of Science*.

Faraday came to the discovery of induction by placing his attention to the "overall process", understood by means of conceptual geometric tools, rather than with the analytical methods of the French physicists (such as Ampere) which integrated the contributions of each element of conductors and magnets. He, in fact, rejected the atomistic conception, which divides matter into atoms and empty space, since he considered it absurd to separate matter from the actions it carries, which extend everywhere in space. Instead, he resumed Boscovich's atomic theory, assuming that matter was made up of centers radiating *lines of force* that fill the entire space, generating the various electrical, magnetic, gravitational phenomena and the radiation itself. In particular, in the case of a magnet, these lines of force «are indicated in a general manner by the disposition of iron filings». Faraday regarded this as an experimental proof of their existence, while there was none whatsoever for the aether that he considered the result of speculation and not observable. Faraday believed that matter was a continuous entity distributed everywhere and concluded that «the powers around the centers give these centers the properties of atoms of matter», as he wrote in the issue of Philosophical Magazine from 1844, page 142.

Following these insights, Faraday devoted himself to the study of electrical induction through dielectric materials, finding that they modify it, and therefore there are charges "spread inside" of these substances. In addition, driven by the attempt to unify the forces of nature, he went in search of the effects that an electric or magnetic field may have on light. In 1845, he found that a block of glass subjected to the action of a strong magnet is able to rotate the plane of polarization of a beam of light that passes through it. There are not, therefore, only a few materials, such as iron, that are affected by magnetism. Then, a few weeks later, he performed a series of experiments on various substances traditionally considered inert to magnetism; as he expected, he found that these had an influence on the induction. He also discovered that some substances hinder the flow of the lines of force, while others facilitate it, and distinguished them as *diamagnetic* and *paramagnetic*.

In conclusion, if the medium changes the induction, the concept of *action at a distance* loses validity; therefore, even the possibility of an instantaneous propagation of physical actions was being questioned, as Faraday noted in a letter entitled *Thoughts on Ray-Vibrations* published in the issue of Philosophical Magazine from 1846. Instead, he introduced the concept of a field that was effectively represented with *electric or magnetic lines of force*. In particular, he interpreted the radiative processes such as the vibration of the lines of force, so we could «dismiss the aether». Furthermore, since «the propagation of light, and therefore probably of all radiant action, occupies time», he concluded: «that a vibration of the line of force [from one extreme to another] should account for the phaenomena of radiation, it is necessary that such vibration should occupy time also».

Meanwhile, a group of German physicists and mathematicians were continuing the researches in the tradition of action at a distance. Wilhelm Eduard Weber (1804–1891), between 1846 and 1848, developed the first theory of electric and magnetic phenomena based on the idea of discrete electric charge and of electric current as the motion of charges, succeeding in giving an analytical expression for the force between two charges in relative motion to each other. In addition to the term Coulomb's force, a term appeared (of relativistic correction) proportional to $(\upsilon/\mathrm{c})^2$, where υ is their relative speed, as well as a (radiative) term, proportional to the acceleration and inversely proportional to the distance. In both terms, there was then a coefficient c, which would later prove equal to the speed of light. Besides being a non-radial force, it was no longer dependent on the distance alone: a possible solution was that this force was not instantaneous, but delayed, according to a conjecture that Carl Friedrich Gauss (1777–1855) had proposed to Weber in a letter of 1845.

The collaboration between Gauss and Weber began in 1831. Their first joint success was the construction of the first electromagnetic telegraph, which was operative from 1833 to 1845 between the Observatory and the physics laboratory at the University of Göttingen. To study the Earth's magnetism, a magnetic observatory was built in 1833 in Göttingen, where the measurements performed throughout Europe were coordinated and collected. Subsequently, the Society for Magnetism was born, under the direction of Gauss and Weber. Other magnetic observatories were built in various places on Earth. At the conclusion of the work, around 1842, Gauss and Weber, in collaboration with Carl Goldschmidt, had produced a detailed mapping of the magnetic field of the whole Earth. To determine the field strength, in 1837, they devised a new tool that used the rotation from the equilibrium position of a magnet suspended between two wires. In this way, the intensity of the field was traced back to the classical measures of mass, length, and time. The unit of measurement of the magnetic field introduced by Gauss was, appropriately enough, the *gauss*, which was inserted between the CGS units by the Paris International Congress of 1881. Following these studies, Gauss demonstrated his important theorem on the flow, which is a fundamental part of the electromagnetic laws.

Among thc few who appreciated Faraday's work was William Thomson (Lord Kelvin, 1824 - 1908), who, in 1841, at only 17 years old, had already found an analogy between the propagation of heat and the phenomena of electrical conduction. In a later work, he emphasized the superiority of the Faraday method in the explanation of the influence of dielectric media on electrostatic forces. In those same years, George Gabriel Stokes (1819–1903) had obtained a number of differential equations to express the equilibrium conditions of an elastic body subject to tensions. In a publication from 1847, Thomson noted that some solutions to Stokes' equations manifested the same properties of the electrostatic, magnetostatic and electrodynamic magnitudes [Peruzzi 1998]. Furthermore, with the affirmation of the wave theory of light, various researchers, among them George Green (1793–1841) and Stokes, tried to formulate models of elastic luminiferous aether consistent with observations. In particular, the magneto-optics effect discovered by Faraday had suggested to Thomson, in 1856, the existence of a coupling between the luminiferous aether and the electromagnetic one.

The discovery of electromagnetic waves

In the year 1800, William Herschel (1738–1822), examining the warming produced by different portions of the solar spectrum with a thermometer, found that it extends to "infrared", beyond the visible spectrum. Still, the nature of these *infrared rays* was not clear, that is to say, if they were a particular caloric radiation or an invisible extension of the lumen. This second hypothesis became more credible when the physical chemists Johann Wilhelm Ritter (1776–1810), in 1801, and, independently, William Hyde Wollaston (1766–1828) discovered a similar extension on the other end of the spectrum: these *ultraviolet rays* were revealed by their chemical activity in the blackening of silver chloride. Therefore, it became natural to think that light represents only a visible portion in the spectrum of a wave radiation, orderable according to a wavelength parameter, which, thanks to Young and Fresnel, could be measured with interference experiments.

The decisive step in establishing the connection between light and electromagnetism was accomplished by the Scottish physicist James Clerk Maxwell (1831 –1879). He was particularly influenced by the concept of lines of force and developed the theory of the electromagnetic field, translating into precise and quantitative equations the insights of Faraday, with whom he had been in constant personal contact during his time at the King's College of London, from 1860 to 1865. Maxwell expressed his debt to Faraday in all of his memories, but he went further, succeeding in determining the first great unification of the laws of nature, between the phenomena of electricity, magnetism and light.

Continuing the work of William Thomson, in his first major contribution to electromagnetism published under the title *On Faraday's Lines of Force* in 1856, Maxwell used the analogy of the motion of an incompressible fluid to illustrate the concept of a line of force. In this way, he could represent, in addition to the direction, the intensity of the force by means of the speed of the imaginary fluid. This model was well suited to explaining the electrical conduction, the electrostatic and magnetostatic phenomena, but not those of induction, whereby a magnetic flux variation gives rise to an electromotive force in a concatenated circuit. Faraday had assumed that this depended on the variation in an *electrotonic state* of the circuit. Maxwell devoted the second part of the work to this effect, published in 1856 with the title *On Faraday's "Electro-tonic state"*: instead of referring to mechanical models, here, he used the results of the theory of potential, understood, however, as an energy density distributed in space, and he applied the general concept of conservation of energy. In this way, he introduced the «electrotonic function», which carries out the function of potential for the magnetic field and which is today known as the *vector potential*, and he obtained a series of differential equations for the current, the electromotive force, the induction **B** and the magnetic field **H**.

In his second work on electromagnetism, published between 1861 (Parts I and II) and 1862 (Parts III and IV) with the title *On the Physical Lines of Force*, Maxwell begins with a complex mechanical model to illustrate the physical characteristics of a medium suitable for transmitting electromagnetic actions. He imagines the space as being filled with «innumerable vortices» whose rotation axes are aligned with the direction of the magnetic field. These vortices are separated into cells whose walls are made out of the «matter of electricity», which, like the steel spheres of the ball bearings, is able to transmit the rotary movement between adjacent cells, and is also free to move in the conductors, while it is blocked in dielectrics, including the empty space. In this way, Maxwell could explain the electrotonic function as an angular momentum, associated with the vortices, whose time derivative contributed to electromotive force.

In 1861, Maxwell, inspired by the trend of researches on luminiferous elastic aether and stimulated by Thomson's work of 1856, introduced elastic properties into his own model. This allowed him to consider the cases of polarization of dielectrics as a particular case of electrical conduction: the applied tension is stored as potential energy in the form of elastic energy associated with the deformation of the molecular vortices. The matter of electricity would move from its equilibrium position, while remaining bound to the molecules, which would be positively polarized on one side and negatively on the other. Overall, such matter would shift in a given direction, and Maxwell writes «The effect of this action on the whole dielectric mass is to produce a general displacement of the electricity in a certain direction. This displacement does not amount to a current, because when it has attained a certain value it remains constant, but it is the commencement of a current, and its variations constitute currents in the positive or negative direction, according as the displacement is increasing or diminishing». This interpretation allowed Maxwell to equate the "closed" circuit to the "open" circuit, including the Leyden jar: a conductor behaves like a porous medium that can offer a given resistance to the passage of the current, while an insulator behaves like an impermeable elastic membrane, but it transmits its pressure. This was the crucial breakthrough that allowed Maxwell to complete the Ampere equation with the contribution of what is still known today as *displacement currents*. Ampere's law was therefore modified, with the addition of a reciprocal term, to Faraday's law of induction, that is, one could obtain the generation of a magnetic field by a variable electric field, in addition to that produced by a current.

With this hypothesis, it appeared immediately clear to him that the propagation of electromagnetic waves had been made possible, with a speed suggestively equal to that of light, bringing optics and, in general, radiant energy back into Maxwell's theory. This coincidence in the speeds was, in fact, found in 1856 by Weber and Rudolph Kohlrausch (1809–1858), who had determined the ratio between the units of measure for the electrical charge in the magnetostatic and electrostatic CGS systems [Jackson 1974] that has the dimensions of a velocity, obtaining 310740 km/s [Weber and Kohlrausch 1856; Kirchner 1957]. They had succeeded in this venture by charging a Leyden jar to about 30000V. Putting the bottle in contact with a conductive sphere 13 inches in diameter and, immediately afterward, putting this sphere in contact with a smaller one of a torsion balance, they determined, for proportion, the charge in electrostatic units that remained in the bottle. Discharging the bottle, they obtained a pulse of current that was measured for its magnetic effects in a galvanometer. Finally, through the oscillation of the galvanometer, they obtained the charge in electromagnetic units. Maxwell considered the correspondence between the two speeds as evidence that light is composed of electromagnetic waves and expounded this view in the article *A Dynamical Theory of the electromagnetic Field*, divided into seven parts, published in 1865 in the Philosophical Transactions of the *Royal Society*, and then ultimately in his monumental *Treatise on Electricity and Magnetism* of 1873. In 1868, he also published an experimental work [Maxwell 1868] carried out with the help of Charles Hockin, in which they repeated the measures of Weber and Kohlrausch by using a resistor instead of a capacitor and obtaining 288000 km/s.

In *Dynamical Theory*, Maxwell perfects the electromagnetic field theory, in which the field becomes a dynamic entity distributed in space, a theory founded on the essential concepts of mechanics, such as energy and momentum, and on the use of the Lagrangian mathematical methods. At the same time, he avoids recourse to mechanical models, of which he had made use before, considering the propositions related to these models «as illustrative and not as explanatory». But he writes, «in speaking of the Energy of the field, however, I wish to be understood literally». Starting from the potential function, Maxwell can express the laws of induction, and from the energy conservation principle, he derives the expressions for the electromagnetic forces. In the third part of the work, he sums up his theory with a set of 20 differential equations for the 20 variables used to describe the actions of the electromagnetic field. Among these, in addition to the vector potential, there is even present a term to take account for magnetic monopoles, if they existed. These equations were later simplified in the 1890s by Heinrich Hertz (1857–1894) and Oliver Heaviside (1850–1925) in the version we know today as *Maxwell's equations*. Finally, in the sixth part of the work, with the title of *Electromagnetic Theory of Light*, Maxwell derives the first wave equations for the propagation of the potential function and the magnetic field, which indicates that they are transverse waves traveling at the speed of light.

In *Treatise*, Maxwell collects and mathematically refines the job done, trying to clarify the meaning of the electromagnetic field theory. His equations also required new mathematical tools: so, Maxwell emphasizes the role of the vectors to geometrically represent the physical quantities and introduces the differential operators *rotor* and *convergence* (which we now call *divergence*). Although Maxwell still illustrates these equations with the idea of elastic actions on aether, responding to the views of his era, they are, in fact, independent of an aether modeling, and thus have withstood the test of time. Incidentally, Maxwell even derived the existence of a radiation pressure, which had already been invoked by Kepler to explain the shape of comets; the experimental evidence of this pressure was obtained in 1899 by Pyotr N. Lebedev (1866–1912) [Landsberg 1979].

Maxwell's theory, however, contained a problematic aspect: he considered the electric charge as an "electric displacement" produced by the field, unlike the continental physicists, such as Weber, who supposedly considered the charge as a primitive notion. If, indeed, the idea of displacement had favored the completion of Ampere's equation, at the same time, it prevented consideration of the radiative sources, such as those constituted by oscillating charges, hampering the realization of experiments that could prove the validity of his theory. This obstacle was finally overcome by Hertz, who accepted the substantialist notion of the electric charge.

Thanks to Heaviside, and also to Hertz, who, however, recognized the priority of Heaviside, we have the modern formulation of the Maxwell equations and the use of the concise notation used in the vector calculation, which Maxwell had instead decomposed into its individual components. Moreover, Heaviside took care of telegraphy and telephony and found a way to compensate for the dispersion in the transmission lines, making it possible to transmit signals at acoustic frequencies, even over great distances. Now, this compensation enables us to receive even video signals on old telephone pairs. He also postulated the existence of the ionosphere and foresaw the superluminal radiation that occurs when a charged particle passes through a medium at a speed greater than that of the light in it.

Hertz was initiated into the studies on electromagnetism by Helmholtz, to whom he was a student and then an assistant. From theoretical studies of Maxwell's theory, the existence of electromagnetic waves appeared evident to him. For the waves, the frequency exerts the dominant role: Hertz strove then as to how to obtain high-frequency oscillations. In 1855, the Danish physicist Berend Wilhelm Feddersen (1832–1918), with the use of rotating mirrors, had discovered that the discharge of an electric spark current has a very rapid oscillation: this fact opened up the possibility of generating electromagnetic waves with the means available at that time. In 1888, Hertz experimentally verified the existence of these waves in the band of centimeter waves: he triggered high frequency discharges in simple electrical circuits (*Hertz oscillators*), including an inductance, and succeeded in picking them up using resonant circuits where the wave reception was again revealed by a discharge. Moreover, he demonstrated that these waves exhibit the same phenomena as light, such as reflection, refraction, diffraction, interference, and polarization. These studies confirmed the electromagnetic theory of light expoused by Maxwell.

Suggested reading

These historical notes are derived from the following texts:

V.A. *Encyclopaedia Britannica*, at entries “Electricity”, “Gilbert”.

V.A. *Enciclopedia della scienza e della tecnica*, Mondadori Ed., at the entry “Elettricità”.

Bernal John D., *Storia della Scienza*. Editori Riuniti, Roma, III ed. (1969). This book describes the mutual relationship between science and society and allows us to frame the progress of optics and electromagnetism in the general context of scientific development and civilization.

Dragoni Giorgio, *Le origini del campo elettromagnetico*. Le Scienze n° 356, insert *I grandi esperimenti* (April 1998). In addition to discussing the history of the electromagnetic field, it gives many details of the experiments of Oersted, Arago, Ampere and Faraday.

Fregonese Lucio, *Volta: teorie ed esperimenti di un filosofo naturale*. I grandi della scienza n° 11 (1999), ed. Le Scienze S.p.A., Milano.

Fuchs Walter R., *La fisica moderna illustrata*. Rizzoli ed., Milano (1967). This describes the path of physics of the last three centuries and addresses the details of individual discoveries, analyzing the conceptual and philosophical value, with particular emphasis on the difficult relationship between model and reality.

Hall A.R. e Boas Hall M., *Storia della Scienza*. Ed. il Mulino, Bologna (1979). This tells the history of science through the men who, through their thinking and their work, have left a fundamental imprint.

Jackson J.D., *Classical Electrodynamics*, John Wiley & Sons, New York (1974).

Kirchner F., *Determination of the velocity of light from electromagnetic measurements according to W. Weber and R. Kohlrausch,* Am. J. Phys. 25, 623 (1957).

Landsberg G.S., *Ottica*, MIR, Mosca (1979).

Nahin P.J., *Oliver Heaviside*, Le Scienze **264**, p.74 (agosto 1990).

Maxwell J. Clerk, *On a method of making a direct comparison of electrostatic with electromagnetic force; with a note on the electromagnetic theory of light*, Phil. Trans. Roy. Soc. Lond., **158**, 643-657 (1868).

Pearce Williams, L., *André Marie Ampère*. Le Scienze n° 247, p. 90 (marzo 1989).

Peruzzi Giulio, Le Scienze **351**, p. 48 (November 1997). *Maxwell: dai campi elettromagnetici ai costituenti ultimi della materia*. I grandi della scienza n° 5 (1998), ed. Le Scienze S.p.A., Milano. A beautiful text on the life and discoveries of Maxwell, with an accurate account of the “present state of electrical science” of his time, including the

contributions of Faraday and other researchers.
Segrè E., *From falling bodies to radio waves*, W.H. Freeman and Company ed. New York (1984).
Tazzioli R., *Gauss, principe dei matematici e scienziato poliedrico*, I grandi della scienza n° 28 (2002), ed. Le Scienze S.p.A., Milano.
Weber W. and Kohlrausch R., *Ueber die Elektricitätsmenge, welche bei galvanischen Strömen durch den Querschnitt der Kette fliesst*, Annalen der Physik **175**, 10-25 (1856). Extract in Il nuovo Cimento **5**, 280-284 (1857).

Chapter 1
Recollections of electromagnetism

The bases of electromagnetic radiation

Introduction

The starting point of this text is given by the equations of electromagnetism, which are assumed to be a postulate on which to develop the whole of Optical Physics. In this chapter, we will start by addressing the microscopic equations for the fields and for the matter, in which all of the elementary charge and current densities must be made explicit, and the relativistic formulation of these equations. Then, we will move on to the macroscopic equations, in which the properties of the material media are collected in suitable auxiliary vectors by means of an operation of averaging over small distances compared to the wavelength, but large relative to atomic dimensions. This step is crucial for understanding the concepts of index of refraction and dispersion. In fact, the response of the matter to the field generally can be very complex, but here, we will limit ourselves to the cases in which this response is linear, although delayed in time. In this situation, mathematics comes to our aid, by introducing a complex representation of the fields by which the dispersion can be treated more easily. Maxwell's equations are generally defined for media with continuous properties, however, in optics, the cases in which the medium changes abruptly (but not too unevenly) in the space, on much smaller distances than the wavelength, are fundamental. Therefore, we will see the boundary conditions that dictate the way in which the field is transformed from side to side of an interface between two media. Finally, we will see how we can obtain the equations of the electromagnetic waves and their properties in terms of transport of energy, momentum and angular momentum from Maxwell's equations. In particular, we will study their transverse nature, that is, their polarization. Lastly, we will discuss two key aspects of the propagation of electromagnetic waves: their reflection and transmission at interfaces and their speed of propagation in dispersive media.

1.1 The fundamentals

1.1.1 Charges and fields

The concept of *field* was born to describe the interaction at a distance between material particles. In Classical Mechanics, it was merely a way of representing the forces; afterwards, Relativistic Mechanics showed that the speed of propagation of

G. Giusfredi, *Physical Optics*, UNITEXT for Physics,
https://doi.org/10.1007/978-3-030-25279-3_1

interactions is finite, whereby the force acting on a particle is affected by the variations of the position of the other particles only with a certain delay. This means that the field is an intrinsic physical reality: we cannot speak of direct interaction between particles, but we will talk about the interaction of a particle with the field and the consecutive interaction of the field with another particle [Landau and Lifchitz (VIII) 1966b].

An operational definition of the *electromagnetic field* is usually given in terms of the Lorentz force induced on a small charged particle (e.g., an electron), which moves slower than the speed of light:

$$\boldsymbol{F} = m\boldsymbol{a} = e\left(\mathbf{E} + \boldsymbol{v} \times \mathbf{B}\right), \tag{1.1.1}$$

where $\boldsymbol{v}$ is the velocity of the particle, of mass m and charge e. Thus, the *electric field* $\mathbf{E}$ is determined with the charge at rest (that is, with $\boldsymbol{v} = 0$), while the *magnetic field* $\mathbf{B}$ contributes, and then is obtained with the additional term, proportional to the speed. The exact equation for Relativity has $\dot{\boldsymbol{p}}$ instead of $m\boldsymbol{a}$, with

$$\boldsymbol{p} = m\boldsymbol{v} \Big/ \sqrt{1 - v^2/c^2},$$

where c is the light speed. In SI, the dimensions of $\mathbf{E}$ and $\mathbf{B}$ are:

$$[\mathbf{E}] = \frac{\mathrm{N}}{\mathrm{C}} = \frac{\mathrm{V}}{\mathrm{m}}, \quad [\mathbf{B}] = \frac{\mathrm{N \cdot sec}}{\mathrm{C \cdot m}} = \frac{\mathrm{V \cdot sec}}{\mathrm{m}^2} = \frac{\mathrm{Wb}}{\mathrm{m}^2} = 1\ \mathrm{Tesla} = 10000\ \mathrm{Gauss}\,.$$

In turn, these fields, determined by the charges and their motion, evolve according to the splendid equations of Maxwell, here transcribed in their pure differential form, that is, in empty space:

$$\begin{cases} \nabla \cdot \mathbf{E} & = \rho/\varepsilon_0\,, \\ \nabla \cdot \mathbf{B} & = 0, \\ \nabla \times \mathbf{E} & = -\partial \mathbf{B}/\partial t\,, \\ \nabla \times \mathbf{B} & = \mu_0 \mathbf{J} + c^{-2}\,\partial \mathbf{E}/\partial t\,, \end{cases} \tag{1.1.2}$$

where ρ is the *charge density* ($\mathrm{C/m^3}$) and $\mathbf{J}$ is the *current density* ($\mathrm{A/m^2}$), including the contribution of the spins. These quantities are introduced to avoid singularities associated with punctiform charges and the currents they generate. Moreover, in SI, we have the exact definitions for vacuum:

$\varepsilon_0 = (\mu_0 c^2)^{-1}$ ($\cong 8.86 \cdot 10^{-12}$ F/m) is the *electric permittivity,*

$\mu_0 = 4\pi\ 10^{-7}$ H/m ($\cong 1.236 \cdot 10^{-6}$ H/m) is the *magnetic permeability,*

c = 299 792 458 m/sec is the *light velocity,*
(from the definition of *meter* in SI of the 17ª CGPM, 1983).

Maxwell's equations also implicitly contain the *continuity equation* for the charge and current densities:

$$\nabla \bullet \mathbf{J} + \partial\rho/\partial t = 0 \,. \qquad (1.1.3)$$

This can be derived by combining the time derivative of the first equation with the divergence of the fourth equation.

In addition, these equations are linear in all variables, and since the sum of more solutions is still their solution, they include the superposition principle, for which the fields are added linearly and depend linearly on the charges and currents. Nonlinearities will then arise because of the dynamic response of the same charges and currents to the field.

Eqs. (1.1.2) represent, with a modern and concise notation, various fundamental laws of electromagnetism discovered at the turn between the eighteenth and nineteenth centuries. The first equation is the differential form of *Gauss's law*:

$$\oint_S \mathbf{E} \bullet \mathbf{n} dS = \frac{1}{\varepsilon_o} \int_V \rho dV \,,$$

where S is a closed surface containing the volume V and **n** is a versor normal to S at each point. It states that the net flow of the electric field (or, according to Faraday's conception, the flow of *lines of force*) crossing S is independent of the size of the surface and is proportional to the net charge contained therein. This law is equivalent to *Coulomb's law*

$$F = \frac{q_1 q_2}{4\pi\varepsilon_o r^2} \qquad (1.1.4)$$

of the force F between two charges q_1 and q_2 at a distance r. This law is valid for at least 24 orders of magnitude in the scale of lengths, from 10^{-17} m upward and, in turn, it corresponds to taking the rest mass of the photon as equal to zero [Jackson 1974, p. 5-9].

The second equation is analogous to the first, but with **B** instead of **E**,

$$\oint_S \mathbf{B} \bullet \mathbf{n} dS = 0 \,.$$

Having a zero as the source term, it responds to the absence of magnetic monopoles, which have not yet been found. On the other hand, P.A.M. Dirac showed that the mere existence of at least *one* magnetic monopole in the Universe would stand as reason why the charge of elementary particles is quantized! [Jackson 1974, p. 251-260].

The third equation is the differential form of *Faraday's law of induction*, whereby a magnetic field that varies in time generates an electric field:

$$\oint_C \mathbf{E}\bullet dl = \int_{S'} \frac{\partial \mathbf{B}}{\partial t}\bullet \mathbf{n} dS',$$

where C is a closed circuit that encloses the surface S′ and **n** is the versor perpendicular to that surface at each point.

Finally, the last equation contains *Ampère's law*, for which a current generates a magnetic field, but with the important addition made by James Clerk Maxwell in 1865 of a term similar to that of the Faraday's Law, whereby the temporal variations of the electric field also contributes to the magnetic field:

$$\oint_C \mathbf{B}\bullet dl = \int_{S'} \mu_0 \left(\mathbf{J} + \varepsilon_0 \frac{\partial \mathbf{E}}{\partial t} \right) \bullet \mathbf{n} dS'.$$

It was not (just) an aesthetic consideration of symmetry that guided Maxwell. In modern words, we can state that, without that contribution, the equations would be inconsistent with the principle of conservation of the charge, which is expressed in differential form just by the equation of continuity. The "faulty" equation was Ampère's law

$$\nabla \times \mathbf{B} = \mu_0 \mathbf{J},$$

which, as Coulomb's law, was demonstrated in stationary conditions. Indeed, taking the divergence of both sides of this equation, we obtain $\nabla\bullet\mathbf{J} = 0$, which is valid only in problems at steady state. Maxwell thus converted Eq. (1.1.3) into an expression with zero divergence exploiting Coulomb's law:

$$\nabla\bullet\mathbf{J} + \partial\rho/\partial t = \nabla\bullet\left(\mathbf{J} + \varepsilon_0 \partial\mathbf{E}/\partial t\right) = 0,$$

and he obtained Eq. (1.1.2.d), generalizing **J** in the Ampère law with the substitution $\mathbf{J} \rightarrow \mathbf{J} + \varepsilon_0\partial\mathbf{E}/\partial t$. The term added by Maxwell finally revealed the reciprocity of the effects that exists between **E** and **B**, which makes possible the electromagnetic waves, and therefore the light, and opened the study of new and yet unknown phenomena. The new equation was thus the fundamental contribution of Maxwell to electromagnetism and optics.

1.1.2 The scalar and vector potentials

Maxwell's equations consist of a set of first-order partial differential equations for the electric and magnetic fields. They can be turned into a smaller number of second-order equations by introducing the potential of these fields, as is done, for example, in electronics, where we prefer to work with the tensions instead of the fields. In particular, since the second Maxwell's equation states that the divergence of **B** is zero, the magnetic field can be defined in terms of a *vector potential* **A**:

$$\mathbf{B} = \nabla \times \mathbf{A} \,. \tag{1.1.5}$$

In this way, Faraday's law can be written as

$$\nabla \times \left(\mathbf{E} + \frac{\partial \mathbf{A}}{\partial t} \right) = 0 \,. \tag{1.1.6}$$

This means that the term in brackets in Eq. (1.1.6), the curl of which is a zero vector, can be written as the gradient of a scalar function, that is, a *scalar potential* Φ, for which the electric field can now also be redefined as

$$\mathbf{E} = -\nabla\Phi - \frac{\partial \mathbf{A}}{\partial t} \,. \tag{1.1.7}$$

The definition of **B** and **E** in terms of the potentials **A** and Φ according to Eqs. (1.1.5) and (1.1.7) identically satisfies the two homogeneous Maxwell's equations. Finally, the two non-homogeneous Eqs. (1.1.2a) and (1.1.2d), expressed in terms of the potentials, become

$$\begin{cases} -\nabla^2\Phi - \dfrac{\partial}{\partial t}(\nabla \cdot \mathbf{A}) = \dfrac{\rho}{\varepsilon_0} \,, \\ \dfrac{1}{c^2}\dfrac{\partial^2 \mathbf{A}}{\partial t^2} - \nabla^2 \mathbf{A} + \nabla\left(\nabla \cdot \mathbf{A} + \dfrac{1}{c^2}\dfrac{\partial \Phi}{\partial t} \right) = \mu_0 \mathbf{J}. \end{cases} \tag{1.1.8}$$

We note now that, for Eq. (1.1.5), there is a certain arbitrariness in the choice of **A**. This last can indeed be changed by the addition of the gradient of any scalar function Λ without changing the value of **B**. Since, for Eq. (1.1.7), **A** also then determines part of **E**, we can always simultaneously change **A** and Φ with the transformation

$$\begin{cases} \mathbf{A} \to \mathbf{A}' = \mathbf{A} + \nabla\Lambda \,, \\ \Phi \to \Phi' = \Phi - \dfrac{\partial \Lambda}{\partial t} \,, \end{cases} \tag{1.1.9}$$

without affecting either **B** or **E**, that is, without changing physics. This change is called *gauge transformation*, and the invariance of the fields under this transformation is called *gauge invariance*.

This freedom allows us to simplify Eqs. (1.1.8) by choosing a suitable value for $\nabla\bullet\mathbf{A}$. One of the allowed choices is to take

$$\nabla\bullet\mathbf{A}+\frac{1}{c^2}\frac{\partial\Phi}{\partial t}=0\,. \tag{1.1.10}$$

This decouples Eqs. (1.1.8), and leads to two elegant non-homogeneous wave equations, one for Φ and one for $\mathbf{A}$:

$$\frac{1}{c^2}\frac{\partial^2\Phi}{\partial t^2}-\nabla^2\Phi=\frac{\rho}{\varepsilon_0},\quad \frac{1}{c^2}\frac{\partial^2\mathbf{A}}{\partial t^2}-\nabla^2\mathbf{A}=\mu_0\mathbf{J}. \tag{1.1.11}$$

The set of Eqs. (1.1.10) and (1.1.11) is equivalent throughout the Maxwell's equations (1.1.2). These new equations are the natural form in which to treat the Electromagnetism in Special Relativity. Moreover, note that Eq. (1.1.10) has the form of an equation of continuity for the potential.

Since it leads to relativistic invariant equations, the condition imposed by Eq. (1.1.10) is called the *Lorentz condition*, and the solutions for the potentials obtained with this choice are said to belong to the *Lorentz gauge*.

Another possible choice for the potential is what is called the *Coulomb gauge*, for which we take

$$\nabla\bullet\mathbf{A}=0\,, \tag{1.1.12}$$

and, with this choice, Eqs. (1.1.8) become

$$\begin{cases} \nabla^2\Phi=-\dfrac{\rho}{\varepsilon_o}\,, \\ \nabla^2\mathbf{A}-\dfrac{1}{c^2}\dfrac{\partial^2\mathbf{A}}{\partial t^2}=-\mu_0\mathbf{J}+\dfrac{1}{c^2}\nabla\dfrac{\partial\Phi}{\partial t}\,. \end{cases} \tag{1.1.13}$$

The first equation is simply the Poisson's equation of electrostatics. Therefore, in this gauge, the scalar potential is just the *instantaneous* Coulomb potential generated by the charge density. From this originates the name of this gauge, which is especially useful in Quantum Mechanics. We can even say that one of the differences between Relativity and Quantum Mechanics resides in choosing the best gauge to be used in the resolution of electromagnetic potentials.

1.1.3 Relativistic formulation of fields and Lorentz transformations

In Relativity, the physical space to consider is the space-time with four dimensions, the Lorentz transformations replacing those of Galileo between reference frames in motion with different constant velocities and without rotation. In the relativistic notation [Jackson 1974], a point in this space is described by the *contravariant four-vector*

$$x^{\iota} \equiv \left(x^{0}, x^{1}, x^{2}, x^{3}\right) = \left(ct, x, y, z\right). \tag{1.1.14}$$

The general form of the transformation between two different reference frames is given by $x'^{\iota} \equiv x'^{\iota}\left(x^{0}, x^{1}, x^{2}, x^{3}\right)$, meaning that x'^{ι} is a function of x^{κ}, and a generic vector is said to be *contravariant* or *covariant* if it is transformed, respectively, as

$$a'^{\kappa} = \frac{\partial x'^{\kappa}}{\partial x^{\iota}} a^{\iota} \quad \text{or} \quad a'_{\kappa} = \frac{\partial x^{\iota}}{\partial x'^{\kappa}} a_{\iota}, \tag{1.1.15}$$

where, as is customary, it is understood as the sum over repeated indices. The index is placed up for the contravariant vectors and tensors, while it is placed down for those that are covariant.[1]

The Lorentz transformations of Special Relativity are derived from the invariance of the *interval* between two points a and b of space-time whose square is defined as

$$s_{\mathrm{ab}}{}^{2} = \left(x_{\mathrm{a}}^{0} - x_{\mathrm{b}}^{0}\right)^{2} - \left(x_{\mathrm{a}}^{1} - x_{\mathrm{b}}^{1}\right)^{2} - \left(x_{\mathrm{a}}^{2} - x_{\mathrm{b}}^{2}\right)^{2} - \left(x_{\mathrm{a}}^{3} - x_{\mathrm{b}}^{3}\right)^{2}.$$

In turn, this invariance is derived from the postulate of light speed constancy proposed for the first time by Einstein. The transformations that leave $s_{\mathrm{ab}}{}^{2}$ invariant belong to a group that is called the *non-homogeneous Lorentz group* or *Poincaré group*. It includes transformations between inertial reference systems in motion with each other, including spatial translations and rotations and inversions both spatial and temporal. In other words, the *norm* of the space-time is here defined by the *metric tensor* $g_{\iota\kappa}$, which, in our case, is diagonal; it is equal to $g^{\iota\kappa}$, and its non-zero elements are [2]

$$g_{00} = 1, \quad g_{11} = g_{22} = g_{33} = -1, \tag{1.1.16}$$

for which the scalar product is defined as

---

[1] For the four-dimensional space-time, the coordinates of the indices are denoted here by Greek letters and take the values *0, 1, 2, 3,* written in italics so as to distinguish them, as far as possible, from the operation of exponentiation, while the indices for the sole spatial coordinates are denoted by Latin letters.

[2] According to Jackson's convention [Jackson 1974]: in this way, the interval is real when the point b lies in the *cone of light* centered at a. Instead, Landau and Lifchitz (1966a) take the opposite signs.

$$a \bullet b = g_{\iota\kappa} a^{\iota} b^{\kappa} . \tag{1.1.17}$$

The metric tensor also allows for changing any index from contravariant to covariant, and vice versa:

$$a_{\iota} = g_{\iota\kappa} a^{\kappa}, \quad a^{\iota} = g^{\iota\kappa} a_{\kappa} . \tag{1.1.18}$$

Namely, if $a^{\iota} = \left(a^0, a^1, a^2, a^3\right)$, then $a_{\iota} = \left(a^0, -a^1, -a^2, -a^3\right)$. In particular, from Eq. (1.1.14), we have $x_{\iota} = \left(ct, -x, -y, -z\right)$. Therefore, the dot product of two four-vectors is explicitly defined as

$$a \bullet b = a_{\iota} b^{\iota} = a^{\iota} b_{\iota} = a^0 b^0 - \boldsymbol{a} \bullet \boldsymbol{b} . \tag{1.1.19}$$

The differential four-dimensional operator of space-time is given by

$$\nabla^{\iota} = \left(\frac{1}{c} \frac{\partial}{\partial t}, -\frac{\partial}{\partial x}, -\frac{\partial}{\partial y}, -\frac{\partial}{\partial z} \right). \tag{1.1.20}$$

Finally, we can define, in analogy to the Laplacian operator, a new differential operator called the *d'Alembertian*:

$$\Box = \nabla_{\iota} \nabla^{\iota} = \frac{1}{c^2} \frac{\partial^2}{\partial t^2} - \nabla^2 . \tag{1.1.21}$$

For vectors and tensors of rank 2, the Lorentz transformations between two inertial reference frames in motion between them can be expressed as, respectively,

$$a'^{\kappa} = \mathscr{L}^{\kappa}{}_{\iota}\, a^{\iota}, \quad b'^{\iota\kappa} = \mathscr{L}^{\iota}{}_{\lambda}\, b^{\lambda\mu}\, \mathscr{L}_{\mu}{}^{\kappa}, \tag{1.1.22}$$

where

$$\mathscr{L}^{\kappa}{}_{\iota} = \frac{\partial x'^{\kappa}}{\partial x^{\iota}}, \quad \mathscr{L}_{\kappa}{}^{\iota} = \frac{\partial x^{\iota}}{\partial x'^{\kappa}}, \tag{1.1.23}$$

whose ι, κ indexes are also ordered horizontally to recall, respectively, row and column in the representation of $\mathscr{L}$ as a matrix. The invariance of the scalar product requires, in particular, that

$$\mathscr{L}_{\lambda}{}^{\iota} g_{\iota\kappa} \mathscr{L}^{\kappa}{}_{\mu} = g_{\lambda\mu} . \tag{1.1.24}$$

Since the determinant of a product of matrices is equal to the product of the determinants of the matrices, we have that

$$\det(\mathcal{L}) = \pm 1\,. \tag{1.1.25}$$

The two matrices $\mathcal{L}^{\iota}{}_{\kappa}$ and $\mathcal{L}_{\iota}{}^{\kappa}$ are different from each other, and we find that

$$\nabla'_{\iota} x'^{\kappa} = \delta^{\kappa}_{\iota} \Rightarrow \mathcal{L}_{\iota}{}^{\mu} \nabla_{\mu} \mathcal{L}^{\kappa}{}_{\lambda} x^{\lambda} = \delta^{\kappa}_{\iota} \Rightarrow \mathcal{L}_{\iota}{}^{\lambda} \mathcal{L}^{\kappa}{}_{\lambda} = \delta^{\kappa}_{\iota} \Rightarrow \mathcal{L}^{\iota}{}_{\kappa} = \left(\mathcal{L}_{\iota}{}^{\kappa}\right)^{-1}, \tag{1.1.26}$$

In the case in which the two reference frames are parallel and the second is moving with respect to the first along the x-axis (for both frames) at speed $\boldsymbol{\upsilon}$, we have

$$\mathcal{L}^{\kappa}{}_{\iota} = \begin{pmatrix} \gamma & -\gamma\beta & 0 & 0 \\ -\gamma\beta & \gamma & 0 & 0 \\ 0 & 0 & 1 & 0 \\ 0 & 0 & 0 & 1 \end{pmatrix} \quad \text{where } \beta = \frac{\upsilon}{c}, \quad \gamma = \frac{1}{\sqrt{1-\upsilon^2/c^2}}, \tag{1.1.27}$$

For example, for the position in the space-time, the transformation is

$$ct' = \gamma(ct - \beta x), \quad x' = \gamma(x - \beta ct), \quad y' = y, \quad z' = z\,. \tag{1.1.28}$$

In the case of any direction of the relative speed of translation, the $\mathcal{L}$ representation becomes more complicated. If the two reference systems are still taken as being "parallel" to each other, in the sense that the components of $\boldsymbol{\upsilon}$ are the same for both (remember anyway that $\boldsymbol{\upsilon}$ is defined as the speed of the second frame with respect to the first for both frames), we have [Jackson 1974]

$$\mathcal{L}^{\kappa}{}_{\iota} = \begin{pmatrix} \gamma & -\gamma\beta_x & -\gamma\beta_y & -\gamma\beta_z \\ -\gamma\beta_x & 1+\alpha{\beta_x}^2 & \alpha\beta_y\beta_x & \alpha\beta_z\beta_x \\ -\gamma\beta_y & \alpha\beta_x\beta_y & 1+\alpha{\beta_y}^2 & \alpha\beta_z\beta_y \\ -\gamma\beta_z & \alpha\beta_x\beta_z & \alpha\beta_y\beta_z & 1+\alpha{\beta_z}^2 \end{pmatrix}, \text{ with } \alpha = \frac{(\gamma-1)}{\beta^2}, \; \boldsymbol{\beta} = \frac{\boldsymbol{\upsilon}}{c}, \tag{1.1.29}$$

The space-time coordinates are now explicitly transformed as

$$\begin{aligned} ct' &= \gamma(ct - \boldsymbol{\beta}\cdot\boldsymbol{x}), \\ \boldsymbol{x}' &= \boldsymbol{x} + \frac{\gamma-1}{\beta^2}(\boldsymbol{\beta}\cdot\boldsymbol{x})\boldsymbol{\beta} - \gamma ct\boldsymbol{\beta}. \end{aligned} \tag{1.1.30}$$

Before the formulation of Relativity, it was assumed that the equations of electromagnetism were valid in a special reference system, one in which the *aether* is at rest. On the other hand, assuming for a postulate, experimentally confirmed, the invariance of the electric charge from the reference system, these equations have the peculiar property of being *invariant in form* according to the Lorentz transformations, but not those of Galileo. This invariance of the form implies the *covariance* of the terms of equations. To make this evident, it is necessary that we express the variables of electromagnetism as scalar, four-vectors, four-tensor, etc., in the physical space-time of Relativity. In particular, the charge and current densities and the scalar and vector potentials can be put in the form of four vectors:

$$J^\iota = \left(c\rho, J_x, J_y, J_z\right) \quad \text{and} \quad A^\iota = \left(\Phi, cA_x, cA_y, cA_z\right). \tag{1.1.31}$$

The continuity Eq. (1.1.3) thus takes the simple form

$$\nabla_\iota J^\iota = 0\,. \tag{1.1.32}$$

Furthermore, the Lorentz gauge (1.1.10) and the wave Eqs. (1.1.11), respectively, become the following elegant equations:

$$\nabla_\iota A^\iota = 0, \text{ and} \tag{1.1.33}$$

$$\Box A^\iota = c\mu_o J^\iota\,. \tag{1.1.34}$$

These are the microscopic Maxwell's equations in Relativity!

If we express the electric and magnetic fields using Eqs. (1.1.5) and (1.1.7) and with definition (1.1.31.b), we find that they do not transform independently, but together form an antisymmetric tensor given by

$$\mathfrak{F}^{\iota\kappa} = \nabla^\iota A^\kappa - \nabla^\kappa A^\iota = \begin{pmatrix} 0 & -E_x & -E_y & -E_z \\ E_x & 0 & -cB_z & cB_y \\ E_y & cB_z & 0 & -cB_x \\ E_z & -cB_y & cB_x & 0 \end{pmatrix}, \tag{1.1.35}$$

which is called the *electromagnetic field tensor*. With this tensor and with the four-current J^ι, the two homogeneous microscopic Maxwell's equations (1.1.2.b) and (1.1.2.c) and the two non-homogeneous ones (1.1.2.a) and (1.1.2.d) become, respectively,

$$\begin{cases} \nabla^\iota \mathfrak{F}^{\kappa\lambda} + \nabla^\kappa \mathfrak{F}^{\lambda\iota} + \nabla^\lambda \mathfrak{F}^{\iota\kappa} = 0, \\ \nabla_\iota \mathfrak{F}^{\iota\kappa} = \dfrac{1}{c\varepsilon_o} J^\kappa, \end{cases} \tag{1.1.36}$$

where ι, κ, λ assume the values 0, 1, 2, 3 and they must be taken as all being different from each other. In total, we still have 8 equations. Finally, the transformation of the electric and magnetic fields, from a reference frame K to a reference frame K′, with velocity $\boldsymbol{\upsilon}$ relative to K, is given explicitly by

$$\begin{aligned}\mathbf{E}' &= \gamma\left(\mathbf{E}+\boldsymbol{\upsilon}\times\mathbf{B}\right)-\frac{\gamma^2}{\gamma+1}\frac{1}{c^2}\boldsymbol{\upsilon}\left(\boldsymbol{\upsilon}\bullet\mathbf{E}\right),\\ \mathbf{B}' &= \gamma\left(\mathbf{B}-\frac{1}{c^2}\boldsymbol{\upsilon}\times\mathbf{E}\right)-\frac{\gamma^2}{\gamma+1}\frac{1}{c^2}\boldsymbol{\upsilon}\left(\boldsymbol{\upsilon}\bullet\mathbf{B}\right).\end{aligned} \tag{1.1.37}$$

1.2 Maxwell's equations in macroscopic media

1.2.1 Development of equations

Eqs. (1.1.2) are formulated for the empty space and, for a real medium, they are therefore referred to the actual microscopic fields and to all of the elementary charges and currents in the medium itself. Thus, for the purposes that concern us here, it is more appropriate to use *macroscopic* entities, i.e., fields and density averaged on many atomic particles contained in small volume elements. In the case of our optics studies, this means to mediate at least within small distances compared to the wavelengths of the radiation in the examination process. Following the procedure described, for example, by Jackson (1974) [for a broad discussion, see also Van Kranendonk and Sipe (1977)], and indicating the microscopic fields and variables with an index μ, the averaged, macroscopic, field is obtained by integrating on space at a certain assigned instant of time:

$$\mathbf{E}(\boldsymbol{r},t)=\left\langle \mathbf{E}_\mu(\boldsymbol{r},t)\right\rangle=\iiint f(\boldsymbol{r}')\,\mathbf{E}_\mu(\boldsymbol{r}-\boldsymbol{r}',t)\,d\boldsymbol{r}',$$

where $f(\boldsymbol{r})$ constitutes an appropriate positive real continuous function, normalized to unity over the entire space, but different from zero only in a small volume around $\boldsymbol{r} = 0$. This function must also decrease smoothly to zero with respect to molecular size so as to avoid the difficulties related to the discrete nature of the particles and their motion. The exactly same average process also needs to be done for the other quantities. In particular, the charge density can be conveniently decomposed in terms of *free and bound charges*:

$$\rho_\mu=\rho_{\mu\ \mathrm{free}}+\rho_{\mu\ \mathrm{bound}}.$$

The integration on the bound charges of each molecule can, in turn, be expressed in terms of multipole expansion around the respective center of mass of the molecule. The corresponding average charge density is thus given by a series:

$$\rho = \rho_m - \nabla \cdot \mathbf{P} + \dots, \tag{1.2.1.a}$$

where ρ_m represents the *macroscopic charge density*, formed by both the free charges and the balance of the bound charges, while **P** is the *macroscopic polarization*. The following terms are generally negligible and, from here on, we will omit them. However, it is to be mentioned that the first of these is given by a sum over the second derivatives of the *macroscopic quadrupole density*, which is a tensor of rank two. Similarly, but with considerably more complex calculations, for the *average* density of the current, one gets:

$$\mathbf{J} = \mathbf{J}_m + \nabla \times \mathbf{M} + \partial \mathbf{P}/\partial t + \dots, \tag{1.2.1.b}$$

where $\mathbf{J}_m$ is the *macroscopic current density* (to which both free charges and bound charges contribute), and **M** is the *macroscopic magnetization*, which is given by the magnetic moments of the motion of bound electrons and by the magnetic moment of spin of all particles. Here, too, the dots indicate terms that are normally negligible [Jackson 1974] and that we will omit in the following.

The new vectors of *polarization* **P** and of *magnetization* **M** enclose the properties of the medium, and in SI they have the dimensions:

$$[\mathbf{P}] = \mathrm{C/m^2}, \quad [\mathbf{M}] = \mathrm{A/m}\,.$$

Concerning the dimension of **M**, it should be noted that in the literature, there are different definitions of it. The one used here could be the most suitable.

For a medium in motion with respect to an observer, we have further contributions to the average densities of charge and current. For small velocities, in the non-relativistic limit, Eqs. (1.2.1) become

$$\begin{aligned} \rho &= \rho_m - \nabla \cdot \mathbf{P} + \nabla \cdot (\boldsymbol{v} \times \mathbf{M}) + \dots, \\ \mathbf{J} &= \mathbf{J}_m + \nabla \times \mathbf{M} + \partial \mathbf{P}/\partial t + \nabla \times (\mathbf{P} \times \boldsymbol{v}) + \dots, \end{aligned} \tag{1.2.2}$$

where $\boldsymbol{v}$ is the translation speed of the medium. The terms in $\boldsymbol{v} \times \mathbf{M}$ and $\mathbf{P} \times \boldsymbol{v}$ represent the mutual contribution between polarization and magnetization, but while the latter is derived directly from the averaging process on the microscopic particles [Jackson 1974], the first only follows from relativistic considerations [Becker and Sauter 1964]. Here, $\mathbf{J}_m$ is the macroscopic current density in the observer reference, which therefore also includes the product of the macroscopic charge density for the velocity of the medium.

Finally, the macroscopic Maxwell's equations are:

$$\begin{cases} \nabla\cdot\mathbf{E} = \dfrac{1}{\varepsilon_0}\rho_m - \dfrac{1}{\varepsilon_0}\nabla\cdot\mathbf{P} + \dfrac{1}{\varepsilon_o}\nabla\cdot(\boldsymbol{\upsilon}\times\mathbf{M}) + \dots, \\ \nabla\cdot\mathbf{B} = 0, \\ \nabla\times\mathbf{E} = -\dfrac{\partial\mathbf{B}}{\partial t}, \\ \nabla\times\mathbf{B} = \dfrac{1}{c^2}\dfrac{\partial\mathbf{E}}{\partial t} + \mu_0\mathbf{J}_m + \mu_0\nabla\times\mathbf{M} + \mu_0\dfrac{\partial\mathbf{P}}{\partial t} + \mu_0\nabla\times(\mathbf{P}\times\boldsymbol{\upsilon}) + \dots. \end{cases} \tag{1.2.3}$$

Together with **P** and **M**, we use two other auxiliary fields, **D** and **H**, called *displacement fields*:

$$\mathbf{D} = \varepsilon_0\mathbf{E} + \mathbf{P} - \boldsymbol{\upsilon}\times\mathbf{M} + \cdots, \quad [\mathbf{D}] = \mathrm{C/m^2}, \tag{1.2.4}$$

$$\mathbf{H} = \frac{1}{\mu_0}\mathbf{B} - \mathbf{M} - \mathbf{P}\times\boldsymbol{\upsilon} + \cdots, \quad [\mathbf{H}] = \mathrm{A/m}. \tag{1.2.5}$$

With these new vectors, the macroscopic Maxwell's equations become:

$$\begin{cases} \nabla\cdot\mathbf{D} = \rho_m, \\ \nabla\cdot\mathbf{B} = 0, \\ \nabla\times\mathbf{E} = -\partial\mathbf{B}/\partial t, \\ \nabla\times\mathbf{H} = \mathbf{J}_m + \partial\mathbf{D}/\partial t. \end{cases} \tag{1.2.6}$$

Finally, the continuity equation for the macroscopic case is:

$$\nabla\cdot\mathbf{J}_m + \partial\rho_m/\partial t = 0. \tag{1.2.7}$$

We must be very grateful to Oliver Heaviside for the concise and elegant form of these equations. In fact, after a really intense period of study and with the use of vector fields and operators that he himself propagandized, he enormously simplified the even 20 equations in 20 variables of the ponderous treatise by Maxwell, who had rightly left it «cumbered with the *débris* of his brilliant lines of assault, of his entrenched camps, of his battles»[3] [Nahin 1990]!

During the years when the laws of electromagnetism were formulated, the modern concepts of atomic physics had not yet been developed, and then, as now, the auxiliary fields **D** and **H** helped to gloss over the microscopic origins of the

[3] The phrase in quotes is by F. G. Fitzgerald from a eulogy for Heaviside who had finally eliminated these débris.

charge and current densities. Very often, for historical reasons, **B** is called the *magnetic induction* and **H** is said to be the *magnetic field*. However, although it is often more convenient to write equations with **H** instead of **B**, the true average magnetic field is **B** [Landau and Lifchitz (VIII) 1966b, page 154].

The electromagnetic tensor corresponding to the macroscopic **E** and **B** fields is formally the same as that given for the microscopic case in Eq. (1.1.35). Similarly, the Lorentz transformations for the macroscopic fields **E** and **B** are still given by Eqs. (1.1.37). Also, each of the pairs **D** and **H**, **P** and **M** form a tensor that is transformed as the electromagnetic tensor of **E** and **B**. In particular, for the **D** and **H** fields, we have

$$\mathfrak{G}^{\iota\kappa} = \begin{pmatrix} 0 & -cD_x & -cD_y & -cD_z \\ cD_x & 0 & -H_z & H_y \\ cD_y & H_z & 0 & -H_x \\ cD_z & -H_y & H_x & 0 \end{pmatrix}, \tag{1.2.8}$$

and with the source four-vector

$$J_m^\iota = \left(c\rho_m, J_{xm}, J_{ym}, J_{zm}\right), \tag{1.2.9}$$

the (2+2×3) Maxwell's Eqs. (1.2.6) become the (2×4) Minkowski's equations:

$$\begin{cases} \nabla^\iota \mathfrak{F}^{\kappa\lambda} + \nabla^\kappa \mathfrak{F}^{\lambda\iota} + \nabla^\lambda \mathfrak{F}^{\iota\kappa} = 0, \\ \nabla_\iota \mathfrak{G}^{\iota\kappa} = J^\kappa, \end{cases} \tag{1.2.10}$$

where we still have $\iota, \kappa, \lambda = 0, 1, 2, 3$ and they must all be taken different.

1.2.2 Electromagnetic properties of media

The relationships that exist between polarization, magnetization, current and charge densities and the electric and magnetic fields, and finally the state of motion, characterize the properties of the medium understood as a macroscopic entity. Thus, an insulating material, i.e., a material with a very low electrical conductivity, is defined as *dielectric*, while it is a *conductor* material, typically a metal, is one that has high conductivity. There are also intermediate situations. In our study of optics, we are primarily interested in the propagation in dielectric media, transparent enough in some spectral range. For the dielectrics, being insulators, we can usually write $\mathbf{J}_m = 0$. However, if they have an appreciable absorption, this effect can be formally associated with a conductivity, and therefore a current density that is not negligible. The non-transparency can also be due to diffusion when the medium is *not homogeneous*. The dielectric may also be electrified; however, a macroscopic electric charge is generally irrelevant for the visible radiation.

Thus, if we consider the case of a *transparent dielectric* medium in which $\mathbf{J}_m = 0$, $\rho_m = 0$ and that is *not in motion*, so $\boldsymbol{v} = 0$, the Eqs. (1.2.3) are simplified to

$$\begin{cases} \nabla\cdot\mathbf{E} &= -\dfrac{1}{\varepsilon_0}\nabla\cdot\mathbf{P}, \\ \nabla\cdot\mathbf{B} &= 0, \\ \nabla\times\mathbf{E} &= -\dfrac{\partial\mathbf{B}}{\partial t}, \\ \nabla\times\mathbf{B} &= \dfrac{1}{c^2}\dfrac{\partial}{\partial t}\left(\mathbf{E}+\dfrac{1}{\varepsilon_0}\mathbf{P}\right)+\mu_0\nabla\times\mathbf{M}. \end{cases} \tag{1.2.11}$$

Thus, the terms **P** and **M** are the source terms for the fields **E** and **B**, and their knowledge is necessary to continue in the calculations. In most of the transparent materials for the radiation that interests us, **M** is negligible. Instead, **P** is not generally negligible in optics, but it can often be considered as a more or less complicated function of **E**, and it is a function that is also affected by the past history. For example, to know **P** in a vapor of atoms illuminated by a beam of light resonant with some atomic transition, we need to solve the evolution of the atomic variables, which, in turn, are subjected to coupling with the electromagnetic wave field. In practice, as we will see, all of the charm and the heart of optics lies in this interdependence between **E** and **P**, no offense to **B**, **M**, ρ_m and $\mathbf{J}_m$, as, well, they too are sometimes important.

However, depending on the case, one can make reasonable approximations for **P** and **M**. One of the strongest is that the medium is:

linear, i.e., it responds linearly to the field,
isotropic, i.e., the coefficients are simple scalars,
local, i.e., it depends locally from the field,
non-dispersive, i.e., it responds instantly to the field.

With these conditions, we have

$$\mathbf{P} = \varepsilon_0\chi_e\mathbf{E}, \quad \mathbf{M} = \chi_m\mathbf{H} = \frac{\chi_m}{1+\chi_m}\frac{1}{\mu_0}\mathbf{B}, \tag{1.2.12}$$

where χ_e and χ_m are the *electric* and *magnetic susceptibility,* respectively. Thus, with the help of two new parameters of permittivity and permeability,

$$\varepsilon = (1+\chi_e)\varepsilon_0, \quad \mu = (1+\chi_m)\mu_0, \tag{1.2.13}$$

the **D** and **H** fields simply become proportional to **E** and **B**:

$$\mathbf{D} = \varepsilon\mathbf{E}, \qquad \mathbf{B} = \mu\mathbf{H}, \tag{1.2.14}$$

and the Maxwell's equations *for this particular medium* can be put in the following mixed form:

$$\begin{cases} \nabla\cdot(\varepsilon\mathbf{E}) = 0, & \nabla\times\mathbf{E} = -\dfrac{\partial}{\partial t}(\mu\mathbf{H}), \\ \nabla\cdot(\mu\mathbf{H}) = 0, & \nabla\times\mathbf{H} = \dfrac{\partial}{\partial t}(\varepsilon\mathbf{E}). \end{cases} \tag{1.2.15}$$

Finally, if the medium is also spatially *homogeneous*, i.e., of uniform electromagnetic characteristics, and *constant* in time, ε and μ can be brought out of the derivations.

Maxwell's equations will soon lead us to the study of electromagnetic waves, but before examining them, we see now how we can extend the treatment in the presence of discontinuities and for dispersive media.

1.2.3 Boundary conditions at interfaces between transparent media

The Maxwell equations above are expressed for regions of space in which the electromagnetic properties of the medium are continuous. However, in optics, we frequently have situations in which these properties (for example, those characterized by ε and μ) change abruptly through one or more surfaces, where, therefore, we can expect that the fields **E**, **D**, **B** and **H** will also become discontinuous. An appropriate way to solve this problem is to use the integral form of Maxwell's equations:

$$\begin{aligned} &\text{a}: \oint_{\text{S}} \mathbf{D}\cdot\mathbf{n}\,d\text{S} = \int_{\text{V}} \rho_{\text{m}}\,d\text{V}, && \text{b}: \oint_{\text{S}} \mathbf{B}\cdot\mathbf{n}\,d\text{S} = 0, \\ &\text{c}: \oint_{\text{C}} \mathbf{E}\cdot d\mathbf{l} = \int_{\text{S}'} \frac{\partial\mathbf{B}}{\partial t}\cdot\mathbf{n}\,d\text{S}', && \text{d}: \oint_{\text{C}} \mathbf{H}\cdot d\mathbf{l} = \int_{\text{S}'}\left(\mathbf{J}_{\text{m}} + \frac{\partial\mathbf{D}}{\partial t}\right)\cdot\mathbf{n}\,d\text{S}'. \end{aligned} \tag{1.2.16}$$

where S is the closed surface that encloses the volume V, **n** is the unit vector normal to its surface element *d*S, and, ultimately, ***dl*** indicates the path elements of a closed circuit C that delimits the surface S′. These equations are also valid in the presence of discontinuity, and are equivalent to Eqs. (1.2.6) if there is continuity. In the case that interests us here, the local properties of the fields can be found by considering small volumes or surfaces, places at the turn of the T interface under consideration (Fig. 1.1), and adequately ensuring that their size tends to zero.

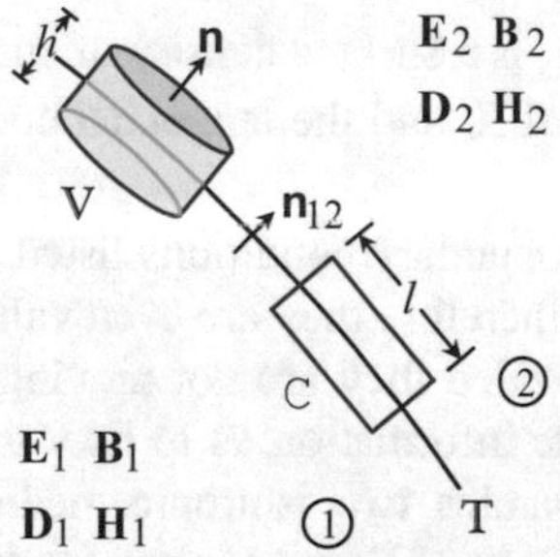

Fig. 1.1
Boundary conditions

Consider, for example, the field **B** and take a small cylindrical volume V of height *h*, delimited by a lateral wall perpendicular to T with the base and the roof parallel to T. We can now apply Eq. (1.2.16.b) for the flux of **B** through the surface S of V. By letting *h* go to zero, the contribution of the sidewall disappears. If, then, the two surfaces of the base and the roof of the cylinder are small enough, the field **B** on them can be approximated by two constant values $\mathbf{B}_1$ and $\mathbf{B}_2$ at the two sides of T. In conclusion, we get:

$$(\mathbf{B}_2 - \mathbf{B}_1)\cdot\mathbf{n}_{12} = 0\,, \tag{1.2.17}$$

where $\mathbf{n}_{12}$ is a unit vector perpendicular to T pointing from the 1st to the 2nd medium. In other words, Eq. (1.2.17) states that the component of **B** normal to T is continuous through T.

A similar reasoning can be applied to **D** through Eq. (1.2.16.a). It is found that

$$(\mathbf{D}_2 - \mathbf{D}_1)\cdot\mathbf{n}_{12} = \rho_s\,, \tag{1.2.18}$$

where ρ_s is a surface charge density on T, to express the presence of a possible net charge in V that does not vanish for $h \to 0$. If, in particular, $\rho_s = 0$, the normal component of **D** is also continuous across the interface T.

The case of the **E** and **H** fields can be treated in a similar manner using, respectively, Eqs. (1.2.16.c) and (1.2.16.d) applied to a small rectangular area placed perpendicular to T with height *h* and length *l*. Making sure the height *h* tends to zero, only the two sides parallel to T contribute to circulation, while the surface integrals tend to zero in the absence of singularities. Therefore, in the limit $h = 0$, the circulation is zero too. For small enough length *l*, the electric field can be taken with constant values $\mathbf{E}_1$ and $\mathbf{E}_2$ on the two sides of the interface. Then, the circulation integral is $0 = (\mathbf{E}_1 - \mathbf{E}_2)\cdot\boldsymbol{l} = (\mathbf{E}_1 - \mathbf{E}_2)\cdot(\mathbf{n} \times \mathbf{n}_{12})\,l = \mathbf{n}\cdot[\mathbf{n}_{12} \times (\mathbf{E}_1 - \mathbf{E}_2)]\,l$, where $\mathbf{n}$ is perpendicular to the rectangle's surface. However, the direction of $\mathbf{n}$ is arbitrary; therefore, the argument in the square bracket must be zero. The result for the **H** field is obtained with a similar procedure. In conclusion, we find that:

$$\mathbf{n}_{12} \times (\mathbf{E}_2 - \mathbf{E}_1) = 0, \quad \mathbf{n}_{12} \times (\mathbf{H}_2 - \mathbf{H}_1) = \mathbf{J}_s, \tag{1.2.19}$$

where $\mathbf{J}_S$ is a surface density of current on T. So, we have that the tangential component of **E**, and the tangential component of **H** if $\mathbf{J}_S = 0$, are continuous through T.

The boundary conditions listed above were obtained with very general premises, and therefore they are even valid for anisotropic, dispersive, and nonlinear media, provided they are not moving. However, they do not by themselves provide complete information as to how to transform the field through an interface T. If this separates two isotropic media, knowing the field on one side, the normal components of **E** and **H** must be determined by their respective relationships with the components of **D** and **B** normal to T. The same applies to the tangential components of **D** and **B**, which must be obtained from the tangential components of **E** and **H**. A more accurate analysis of these conditions, including the case of moving interface, is in Jackson (1974).

1.3 Complex field representation

In optics, we are essentially interested in oscillating fields and the propagation of *waves*, that is, optical *signals*, for which appropriate mathematical tools are required. Among these, the concept of an *analytical signal*, or a *complex analytic function*, was introduced in 1946 by D. Gabor in communications theory. It finds wide application in optics in which the real fields are replaced with complex variables, with a formalism that becomes essential in dealing with dispersive media and non-monochromatic waves.

A complex function $g(z)$ of the complex variable $z = x + iy$ is called *holomorphic* or *analytic* in z_0 if its derivative exists in a finite neighborhood of z_0, including the same point z_0. The two terms, analytical and holomorphic, are synonymous. If the function is holomorphic in the neighborhood of z_0, but not in z_0, such a point is called *singular* or a *singularity of* $g(z)$. If a given point z_0 is not singular, then it can be shown that, at this point, *all* derivatives of $g(z)$ exist. The analyticity of $g(z)$ imposes stringent conditions between its real and imaginary parts. Writing g as

$$g(z) = u(x,y) + iw(x,y), \tag{1.3.1}$$

where u and w are real functions, the necessary and sufficient condition so that $g(z)$ is analytic is that it fulfills the relations of *Cauchy-Riemann*

$$\begin{gathered}\frac{\partial g(x+iy)}{\partial x} = -\frac{\partial g^*(x+iy)}{\partial y}, \quad \text{that is,} \\ \frac{\partial u(x,y)}{\partial x} = \frac{\partial w(x,y)}{\partial y}, \quad \frac{\partial u(x,y)}{\partial y} = -\frac{\partial w(x,y)}{\partial x}.\end{gathered} \tag{1.3.2}$$

This condition also ensures that $u(x,y)$ and $w(x,y)$ are *harmonic* functions, i.e., that they satisfy the two-dimensional equations of Laplace

$$\nabla^2 u(x,y) = \nabla^2 w(x,y) = 0\,. \tag{1.3.3}$$

In many applications, we have to work with real functions that oscillate in time at a single frequency ν, for example, the real fields that represent monochromatic waves. It is convenient to represent them mathematically with the real part of a complex exponential

$$\mathrm{g}(t) = \mathrm{A}\cos(2\pi\nu t + \varphi) = \Re e\{A\exp(2\pi i\nu t + i\varphi)\}\,, \tag{1.3.4}$$

where A is a constant. More often, we have to consider more general functions, where, for example, their amplitude A is a complex function of time $A(t)$. Then, wanting to express g(t) as the real part of a complex function $g(t)$, in principle, we can simply write

$$g(t) = \mathrm{g}(t) + i\,\mathrm{h}(t)\,, \tag{1.3.5}$$

where h(t) is a real arbitrary function. Here, I use regular characters to indicate that the function g (and h, too) is real, so as to distinguish it from the complex function g, shown in italics. The best way to choose h(t) is to make $g(t)$ analytic when the time t is replaced with the complex variable $\tau = t + it'$. This choice eliminates the arbitrariness. Let us see how.

Consider a "good" real function g(t): its Fourier transform is defined as[4]

$$G(\omega) = \int_{-\infty}^{\infty} \mathrm{g}(t)\,e^{i\omega t}dt\,. \tag{1.3.6}$$

Because g(t) is real here, we can easily prove that

$$G(-\omega) = G^*(\omega)\,. \tag{1.3.7}$$

The inverse transform is

$$\mathrm{g}(t) = \frac{1}{2\pi}\int_{-\infty}^{\infty} G(\omega)\,e^{-i\omega t}d\omega\,. \tag{1.3.8}$$

---

[4] Here, I assume the property and the conditions of the existence of the Fourier transforms to be known. They will be recalled more fully in Chap. 5, which is dedicated to Fourier Optics. Moreover, I only consider "good" functions here, that is, functions sufficiently regular to be continuous, differentiable, limited, and delimited. Finally, compared to the standard definition, the exponential argument is taken with the opposite sign (see below).

We now define two new complex functions

$$\mathcal{g}(t) = \frac{1}{\pi}\int_0^{\infty} G(\omega) e^{-i\omega t} d\omega \quad \text{and} \quad \mathcal{g}_-(t) = \frac{1}{\pi}\int_{-\infty}^{0} G(\omega) e^{-i\omega t} d\omega . \qquad (1.3.9)$$

Their sum returns the double of the real function g. Replacing ω with −ω in the second integral and applying Eq. (1.3.7), we find that

$$\mathcal{g}(t) = \mathcal{g}_-^*(t). \qquad (1.3.10)$$

If, therefore, in $\mathcal{g}(t)$ and the integrals of Eqs. (1.3.9), we replace t with the complex variable $\tau = t + it'$, we can easily verify that $\mathcal{g}(\tau)$ satisfies the Cauchy-Riemann relations. Therefore, $\mathcal{g}$ is the sought *analytic function*, corresponding to g. We can then write

$$\mathrm{g}(t) = \frac{1}{2}\mathcal{g}(t) + \frac{1}{2}\mathcal{g}^*(t) \qquad (1.3.11)$$

and replace g with $\mathcal{g}$ in Eq. (1.3.5): with this choice, the function h that represents the imaginary part of $\mathcal{g}$ turns out to be the *Hilbert transform* of its real part g. Indeed, using the sign function sgn, Eq. (1.3.9.a) can be rewritten as

$$\mathcal{g}(t) = \frac{1}{2\pi}\int_{-\infty}^{\infty} G(\omega)\left[1 + \mathrm{sgn}(\omega)\right] e^{-i\omega t} d\omega . \qquad (1.3.12)$$

Therefore, the Fourier transform of $\mathcal{g}$ is given by $G(\omega) + G(\omega)\,\mathrm{sgn}(\omega)$. For the convolution theorem (see Chap. 5), the second term is the Fourier transform of the convolution integral between g(t) and the function

$$-\frac{i}{\pi}\mathcal{P}\left(\frac{1}{t}\right),$$

where the symbol $\mathcal{P}$ indicates the principal part. The transform of this function is sgn(ω). Thus,

$$\mathcal{g}(t) = \mathrm{g}(t) - i\frac{1}{\pi}\mathcal{P}\int_{-\infty}^{\infty} \frac{\mathrm{g}(t')}{t - t'} dt' = \mathrm{g}(t) + i\mathrm{h}(t) . \qquad (1.3.13)$$

A real function has an important property:

The analytic function of the derivative of a real function is equal to the derivative of the analytic function of the real function.

Indeed, integrating by parts, we find that the Fourier transform of the derivative of a "good" real function g (t) is

$$\int_{-\infty}^{\infty} \frac{\partial g(t)}{\partial t} e^{i\omega t} dt = i\omega \int_{-\infty}^{\infty} g(t) e^{i\omega t} dt = -i\omega G(\omega).$$

Therefore, the analytical function corresponding to $\partial g/\partial t$ is

$$\frac{1}{\pi}\int_{0}^{\infty} G(\omega)(-i\omega) e^{-i\omega t} d\omega = \frac{\partial}{\partial t}\frac{1}{\pi}\int_{0}^{\infty} G(\omega) e^{-i\omega t} d\omega = \frac{\partial \mathcal{g}(t)}{\partial t}.$$

This rule extends to all derivation orders. It follows, in particular, that, if a real function g($\boldsymbol{r}$,t) satisfies the wave equation

$$\nabla^2 g(\boldsymbol{r},t) - \frac{1}{\upsilon^2}\frac{\partial^2}{\partial t^2} g(\boldsymbol{r},t) = 0, \tag{1.3.14}$$

where ∇^2 is the Laplace operator on the coordinates of $\boldsymbol{r}$, and υ is a velocity, then the same wave equation is valid for its analytical function

$$\nabla^2 \mathcal{g}(\boldsymbol{r},t) - \frac{1}{\upsilon^2}\frac{\partial^2}{\partial t^2} \mathcal{g}(\boldsymbol{r},t) = 0. \tag{1.3.15}$$

These equivalences lead us to a particularly useful representation of the various (real) electromagnetic entities in terms of complex amplitudes. Indeed, by applying Eq. (1.3.11) to the electric field, we get that it can be decomposed as the sum of two complex conjugate vectors:

$$\mathbf{E}(\boldsymbol{r},t) = \frac{1}{2}\mathcal{E}(\boldsymbol{r},t) + \frac{1}{2}\mathcal{E}^*(\boldsymbol{r},t) = \Re e\left[\mathcal{E}(\boldsymbol{r},t)\right], \tag{1.3.16}$$

where $\mathcal{E}$ is the analytic function associated with **E** and that is

$$\mathcal{E}(\boldsymbol{r},t) = \frac{1}{\pi}\int_{0}^{\infty} \hat{E}(\boldsymbol{r},\omega) e^{-i\omega t} d\omega. \tag{1.3.17}$$

A similar decomposition into two complex vectors (conjugated to each other) can be done for the other real variables. This decomposition allows for simplifying the solution of Maxwell's equations, and a wide use is made of it, to the point of the complex vectors not being distinguished from the real ones. To avoid confusion, in this text, the vectors for real electromagnetic variables are denoted by regular Roman letters, while the corresponding complex vectors are indicated by *italic* letters.

The Maxwell's equations (1.2.6) can then be written in terms of analytic functions of the fields and the source densities

$$\begin{cases} \nabla \cdot \mathscr{D}(\boldsymbol{r},t) &= \tilde{\rho}_{\mathrm{m}}(\boldsymbol{r},t), \\ \nabla \cdot \mathscr{B}(\boldsymbol{r},t) &= 0, \\ \nabla \times \mathscr{E}(\boldsymbol{r},t) &= -\partial \mathscr{B}(\boldsymbol{r},t)/\partial t, \\ \nabla \times \mathscr{H}(\boldsymbol{r},t) &= \mathscr{J}_{\mathrm{m}}(\boldsymbol{r},t) + \partial \mathscr{D}(\boldsymbol{r},t)/\partial t. \end{cases} \tag{1.3.18}$$

where $\mathscr{E}$, $\mathscr{B}$, $\mathscr{D}$, $\mathscr{H}$, $\tilde{\rho}_{\mathrm{m}}$ and $\mathscr{J}_{\mathrm{m}}$ denote the analytical functions of **E**, **B**, **D**, **H**, ρ_{m} and $\mathbf{J}_{\mathrm{m}}$, respectively. These equations are general, although, to resolve them, we also need the constitutive relations between $\mathscr{E}$, $\mathscr{B}$ and $\mathscr{D}$, $\mathscr{H}$, $\tilde{\rho}_{\mathrm{m}}$, $\mathscr{J}_{\mathrm{m}}$, which are generally given by additional differential or integral equations in space and time.

Since the Maxwell equations are linear, one of their alternative forms is to express them in the space of the spatial angular frequency, $\boldsymbol{k}$, and temporal ones, ω, through the Fourier transform of the fields and of the real sources, defined here as

$$G(\boldsymbol{k},\omega) = \int_{-\infty}^{\infty}\int_{-\infty}^{\infty}\int_{-\infty}^{\infty}\int_{-\infty}^{\infty} g(\boldsymbol{r},t)\, e^{i(\boldsymbol{k}\cdot\boldsymbol{r}-\omega t)} d\boldsymbol{r}\, dt\,. \tag{1.3.19}$$

In this expression, a certain convention is established in the signs of the terms in the exponential: it is noteworthy that, while the sign of $\boldsymbol{k}\cdot\boldsymbol{r}$ agrees with the standard definition as a Fourier transform, the sign of ωt is instead opposite.[5]

In frequency space, the differential operators are replaced by multiplications

$$\nabla\cdot \to i\boldsymbol{k}\cdot, \quad \nabla\times \to i\boldsymbol{k}\times, \quad \frac{\partial}{\partial t} \to -i\omega \tag{1.3.20}$$

and the Maxwell's equations become

$$\begin{cases} i\boldsymbol{k}\cdot\hat{\boldsymbol{D}}(\boldsymbol{k},\omega) = \hat{\rho}_{\mathrm{m}}(\boldsymbol{k},\omega), \\ i\boldsymbol{k}\cdot\hat{\boldsymbol{B}}(\boldsymbol{k},\omega) = 0, \\ i\boldsymbol{k}\times\hat{\boldsymbol{E}}(\boldsymbol{k},\omega) = i\omega\hat{\boldsymbol{B}}(\boldsymbol{k},\omega), \\ i\boldsymbol{k}\times\hat{\boldsymbol{H}}(\boldsymbol{k},\omega) = \hat{\boldsymbol{J}}_{\mathrm{m}}(\boldsymbol{k},\omega) - i\omega\hat{\boldsymbol{D}}(\boldsymbol{k},\omega), \end{cases} \tag{1.3.21}$$

where the ^ symbol indicates the transform.

Very often, in treating the waves, we consider the case of *monochromatic waves*, with angular frequency ω_{o} assigned. In this case, in place of Eq. (1.3.17),

[5] The convenience of this opposition of signs will be apparent hereinafter. This choice is the one most commonly followed in optics, but be aware that some authors reverse both signs.

we can write

$$\mathscr{E}(\boldsymbol{r},t) = \tilde{E}(\boldsymbol{r})e^{-i\omega_0 t}, \tag{1.3.22}$$

where $\tilde{E}(\boldsymbol{r})$ is a complex vector that carries with it all of the spatial dependence, but is constant in time. It represents the *amplitude* of the field, while the exponential represents the *temporal carrier*. When the wave has the definite direction, $\boldsymbol{k}_0$, of a plane wave, the analytical function of the field can be factored as

$$\mathscr{E}(\boldsymbol{r},t) = \boldsymbol{E}_0\, e^{i(\boldsymbol{k}_0 \cdot \boldsymbol{r} - \omega_0 t)}. \tag{1.3.23}$$

On the other hand, a monochromatic wave constitutes a strong idealization, while, in general, the field is composed of a band of frequencies, for which, in principle, we should return to use its representation in the frequency-space. However, if the spectrum of the field **E** is sufficiently narrow around the angular frequency ω_0, the analytic function of the field is usually factored as

$$\mathscr{E}(\boldsymbol{r},t) = \tilde{E}(\boldsymbol{r},t)e^{-i\omega_0 t}, \tag{1.3.24}$$

where $\tilde{E}(\boldsymbol{r},t)$ is a slowly varying amplitude in time. This factorization can be generalized to include the dependence on the angular spatial frequency $\boldsymbol{k}_0$ in the carrier as well. The conditions in which we have a quasi-monochromatic wave will be specified in the next section.

The same type of expressions applies to all of the other variables.

1.3.1 Generalization of susceptibility for dispersive media

In the most general case, the polarization **P** does not instantly follow the field **E**, nor the magnetization **M** the field **B**, as would be suggested by Eqs. (1.2.12). In principle, any medium requires a certain response time to adapt to the field and retains some memory of the past. Therefore, remaining in the ambit of *linear* media, a generalization of Eq. (1.2.12.a) can be given by an integration on past times, for example,

$$\mathbf{P}(\boldsymbol{r},t) = \varepsilon_0 \int_{-\infty}^{t} \breve{\chi}_e(\boldsymbol{r},t-t')\,\mathbf{E}(\boldsymbol{r},t')\,dt', \tag{1.3.25}$$

where $\boldsymbol{r}$ is the position in space. Here, $\breve{\chi}_e(\boldsymbol{r},t) = \widehat{\chi}_e(\boldsymbol{r},-t)$ is a function of a time difference: it literally expresses the memory of the medium, and the upper limit of the integral corresponds to the *principle of causality* for which one *cannot* have a memory of the future! This expression is only valid in the case in which the interaction is *local*, while, in the case in which **P** is also influenced by the field in other positions in the space, the integration must be extended even to the variable $\boldsymbol{r}$ on the *whole* space.

The equation for **P** is just the temporal *correlation* of $\breve{\chi}_e(\boldsymbol{r},t)$ with $\mathbf{E}(\boldsymbol{r},t)$, and it is also the temporal *convolution* of $\widehat{\chi}_e(\boldsymbol{r},t)$ with $\mathbf{E}(\boldsymbol{r},t)$. Eq. (1.3.25) can be treated more conveniently by means of the (temporal) Fourier transform, in the space of the angular frequencies ω (in place of t). If the interaction is *non-local*, the spatial transform is also required, so as to express the field in the space of *wave vectors* $\boldsymbol{k}$ (in place of $\boldsymbol{r}$). For example, the overall space-time transform of $\mathbf{E}(\boldsymbol{r},t)$ is commonly formulated by

$$\hat{\boldsymbol{E}}(\boldsymbol{k},\omega)=\iiint d\boldsymbol{r}\int dt\,\mathbf{E}(\boldsymbol{r},t)\,e^{-i(\boldsymbol{k}\cdot\boldsymbol{r}-\omega t)},$$

where the ^ sign distinguishes the transform from the original function. Ordinary media are, however, generally *local*, and it is good to limit ourselves here to considering the sole temporal transform (in particular, this allows us easily to treat the ordinary inhomogeneous media). For example, for **E**, this is expressed by the Fourier integral:

$$\hat{\boldsymbol{E}}(\boldsymbol{r},\omega)=\int_{-\infty}^{\infty}\mathbf{E}(\boldsymbol{r},t)\,e^{i\omega t}dt, \tag{1.3.26}$$

where, we should remember that $\hat{\boldsymbol{E}}(\boldsymbol{r},-\omega)=\hat{\boldsymbol{E}}^*(\boldsymbol{r},\omega)$, since **E** is real.

If we apply this transform to Eq. (1.3.25), for the convolution theorem, we find that

$$\hat{\boldsymbol{P}}(\boldsymbol{r},\omega)=\varepsilon_0\chi_e(\omega)\,\hat{\boldsymbol{E}}(\boldsymbol{r},\omega), \tag{1.3.27}$$

where $\chi_e(\omega)$ is the transform of $\widehat{\chi}_e(t)$, that is, the integral:

$$\chi_e(\omega)=\int_{-\infty}^{\infty}\widehat{\chi}_e(t)\,e^{i\omega t}dt, \tag{1.3.28}$$

where we take $\widehat{\chi}_e(t)=0$ for $t<0$ (Fig. 1.2). Eq. (1.3.27) has a very similar form to Eq. (1.2.12.a) and, in fact, the instantaneous response implied by that equation corresponds to $\chi_e(\omega)$ = constant, as is evident from Eq. (1.3.28), for which a constant χ_e corresponds to a Dirac delta function centered at $t = 0$.

A completely analogous reasoning can also be applied to the magnetic susceptibility, for which we have

$$\hat{\boldsymbol{M}}(\boldsymbol{r},\omega)=\mu_0\chi_m(\omega)\,\hat{\boldsymbol{H}}(\boldsymbol{r},\omega), \tag{1.3.29}$$

and, in parallel with Eqs. (1.2.13), we can now define the electric permittivity ε

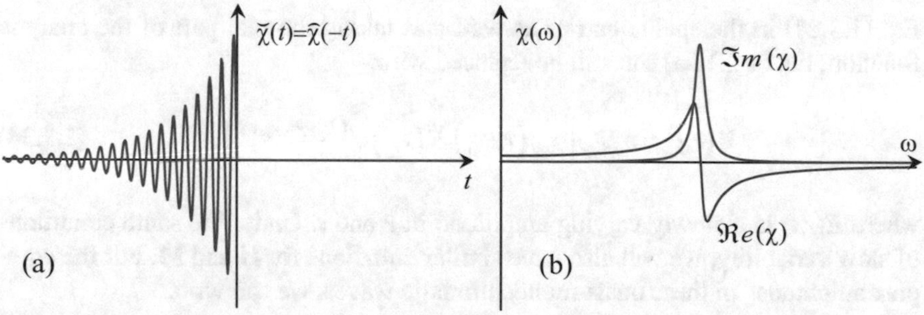

Fig. 1.2 (a) memory function of the medium in the presence of a resonance, (b) real and imaginary parts of the corresponding susceptibility

and the magnetic permeability μ as the complex quantities:

$$\varepsilon(\omega)=\left[1+\chi_e(\omega)\right]\varepsilon_0,\quad \mu(\omega)=\left[1+\chi_m(\omega)\right]\mu_0, \tag{1.3.30}$$

for which

$$\hat{\boldsymbol{D}}(\boldsymbol{r},\omega)=\varepsilon(\omega)\hat{\boldsymbol{E}}(\boldsymbol{r},\omega),\quad \hat{\boldsymbol{B}}(\boldsymbol{r},\omega)=\mu(\omega)\hat{\boldsymbol{H}}(\boldsymbol{r},\omega). \tag{1.3.31}$$

Then, if the medium is *linear* and *local*, but not necessarily non-dispersive, in the case of a *monochromatic* wave with angular frequency ω_o, represented, for the electric field, by Eq. (1.3.22), Eqs. (1.3.27) and (1.3.29) can be reversed, now taking the χ as assigned coefficients (constant over time). In these conditions, the polarization and magnetization have the complex representation:

$$\begin{aligned}&\mathcal{P}(\boldsymbol{r},t)=\varepsilon_0\chi_e(\boldsymbol{r},\omega_o)\mathcal{E}(\boldsymbol{r},t),\\&\mathcal{M}(\boldsymbol{r},t)=\chi_m(\boldsymbol{r},\omega_o)\mathcal{H}(\boldsymbol{r},t)=\frac{\chi_m(\boldsymbol{r},\omega_o)}{1+\chi_m(\boldsymbol{r},\omega_o)}\frac{1}{\mu_0}\mathcal{B}(\boldsymbol{r},t),\end{aligned} \tag{1.3.32}$$

in analogy to Eqs. (1.2.12). For our *monochromatic* wave in a *dielectric* medium, which is *linear* and *not in motion*, Eqs. (1.3.18) thus become:

$$\begin{cases}\nabla\cdot\mathcal{E} &= -\nabla\cdot(\chi_e\mathcal{E}),\\ \nabla\cdot\mathcal{B} &= 0,\\ \nabla\times\mathcal{E} &= -\partial\mathcal{B}/\partial t,\\ \nabla\times\mathcal{B} &= \dfrac{1}{c^2}(1+\chi_e)\dfrac{\partial}{\partial t}\mathcal{E}+\nabla\times\left(\dfrac{\chi_m}{1+\chi_m}\mathcal{B}\right).\end{cases} \tag{1.3.33}$$

In order for the concept of a *quasi-monochromatic* wave to make sense, it is necessary that the variations of χ_e and χ_m be small in its spectral band. Developing

Eq. (1.3.24) in the spatial carrier as well and taking the real part of the analytic function, Eq. (1.2.12.a) can still be replaced with:

$$\mathbf{P}(\boldsymbol{r},t) = \varepsilon_0 \Re e\left[\chi_e(\boldsymbol{r},\omega_0)\boldsymbol{E}(\boldsymbol{r},t)e^{i(\boldsymbol{k}_0\cdot\boldsymbol{r}-\omega_0 t)}\right], \tag{1.3.34}$$

where $\boldsymbol{E}(\boldsymbol{r},t)$ is a slowly varying amplitude in $\boldsymbol{r}$ and t. Under the same conditions of slow variability, we will also have similar equations for **H** and **M**. For the complex amplitudes of these quasi-monochromatic waves, we can write:

$$\begin{aligned} \boldsymbol{P}(\boldsymbol{r},t) &= \varepsilon_0 \chi_e(\boldsymbol{r},\omega_0)\boldsymbol{E}(\boldsymbol{r},t), \\ \boldsymbol{M}(\boldsymbol{r},t) &= \mu_0 \chi_m(\boldsymbol{r},\omega_0)\boldsymbol{H}(\boldsymbol{r},t). \end{aligned} \tag{1.3.35}$$

In conclusion, both $\chi_e(\omega)$ and $\chi_m(\omega)$ fully correspond to the susceptibility χ_e and χ_m introduced by Eqs. (1.2.12), taking care to use them in conjunction with the complex representation of the fields, as in Eq. (1.3.34), of which we will make extensive use later, dealing, in particular, with the laser beams.

Since the $\chi(\omega)$ are, in general, complex, they express two values, the first of which (the real part) is associated with the index of refraction of the medium, while the second (the imaginary part) is associated with the absorption, or the amplification by the medium itself. A more detailed discussion of the properties of $\chi(\omega)$ will be given later in §1.8 of this chapter.

1.3.2 Temporal average of the product of variables

The complex representation of the fields does not pose any problem as long as we consider linear operations like sum, differentiation and integration. However, when we have to treat non-linear operations such as the product of vector fields, we must return to the real form of the physical quantities. In turn, the time average of a product between quantities oscillating sinusoidally at the same frequency can be directly evaluated by the complex amplitudes. Indeed, if

$$\mathrm{U}(t) = \Re e\left[Ue^{-i\omega t}\right] \quad \text{and} \quad \mathrm{V}(t) = \Re e\left[Ve^{-i\omega t}\right],$$

the time average of the product of U to V is obtained simply by integrating on a time interval τ equal to one period of oscillation. This is given by

$$\begin{aligned} \langle \mathrm{U}(t)\mathrm{V}(t)\rangle &= \frac{1}{\tau}\int_0^\tau \mathrm{U}(t)\mathrm{V}(t)dt \\ &= \frac{1}{4\tau}\int_0^\tau \left(Ue^{-i\omega t} + U^* e^{i\omega t}\right)\left(Ve^{-i\omega t} + V^* e^{i\omega t}\right)dt = \frac{1}{2}\Re e\left(UV^*\right). \end{aligned} \tag{1.3.36}$$

A similar expression is also obtained in the case of non-monochromatic waves. Consider two real variables U(t) and V(t) that each represents any component of the electromagnetic fields of a statistically stationary wave and their complex analytic functions $\mathscr{U}(t)$ and $\mathscr{V}(t)$, respectively. It can be shown that, for time intervals T, much longer than the characteristic times of the wave to be measured, one has

$$\frac{1}{T}\int_{-T/2}^{T/2} \mathrm{U}(t)\mathrm{V}(t)\,dt = \frac{1}{4T}\int_{-T/2}^{T/2}\left[\mathscr{U}(t)\mathscr{V}^*(t)+\mathscr{U}^*(t)\mathscr{V}(t)\right]dt\,. \qquad (1.3.37)$$

For the definition of analytic function (1.3.9.a), U and V contain only positive frequencies, and therefore the terms $\mathscr{V}\,\mathscr{U}$ and $\mathscr{V}^*\mathscr{U}^*$ do not appear, as they have only oscillating contributions that are canceled by the integration

The same relationship exists both for the scalar product and for the vector product between vector quantities.

1.4 Electromagnetic waves

1.4.1 Wave equations

For dielectric media *not in motion*, the most general wave equations can be derived from Eqs. (1.2.11). With a few steps, we find that

$$\nabla\times(\nabla\times\mathbf{E})+\frac{1}{c^2}\frac{\partial^2}{\partial t^2}\mathbf{E} = -\frac{1}{\varepsilon_0 c^2}\frac{\partial^2}{\partial t^2}\mathbf{P}-\mu_0\nabla\times\frac{\partial}{\partial t}\mathbf{M}\,,$$

that is,

$$\nabla^2\mathbf{E}-\nabla(\nabla\cdot\mathbf{E})-\frac{1}{c^2}\frac{\partial^2}{\partial t^2}\mathbf{E} = \frac{1}{\varepsilon_0 c^2}\frac{\partial^2}{\partial t^2}\mathbf{P}+\mu_0\nabla\times\frac{\partial}{\partial t}\mathbf{M}\,. \qquad (1.4.1)$$

For the field **H**, we have instead

$$\nabla^2\mathbf{H}-\nabla(\nabla\cdot\mathbf{H})-\frac{1}{c^2}\frac{\partial^2}{\partial t^2}\mathbf{H} = \frac{1}{c^2}\frac{\partial^2}{\partial t^2}\mathbf{M}-\nabla\times\frac{\partial}{\partial t}\mathbf{P}\,. \qquad (1.4.2)$$

In many cases of interest, the optical medium can be magnetically considered *homogeneous*, *linear*, *local*, *isotropic* and *non-dispersive*. In this case, Eq. (1.4.1), through Eqs. (1.2.13.b), (1.2.12.b) and (1.2.6.c), simplifies to

$$\nabla^2\mathbf{E}-\nabla(\nabla\cdot\mathbf{E})-\mu\varepsilon_0\frac{\partial^2}{\partial t^2}\mathbf{E} = \mu\frac{\partial^2}{\partial t^2}\mathbf{P}\,. \qquad (1.4.3)$$

We can now do some simple considerations about the terms $\nabla\bullet\mathbf{E}$ and $\nabla\bullet\mathbf{H}$ appearing in Eqs. (1.4.1) and (1.4.2). If the medium is *not homogeneous* but linear, for Eqs. (1.2.11.a), (1.2.11.b) and (1.2.12), we have

$$\nabla\bullet\mathbf{H} = \nabla\bullet\left(\frac{1}{\mu}\mathbf{B}\right) = \mathbf{B}\bullet\nabla\frac{1}{\mu} = -\mathbf{H}\bullet\frac{\nabla\mu}{\mu} \tag{1.4.4}$$

and

$$\nabla\bullet\mathbf{E} = -\frac{1}{\varepsilon_o}\nabla\bullet(\varepsilon_o\chi_e\mathbf{E}) = -\chi_e\nabla\bullet\mathbf{E} - \mathbf{E}\bullet\nabla\chi_e ,$$

whence

$$\nabla\bullet\mathbf{E} = -\mathbf{E}\bullet\frac{\nabla\varepsilon}{\varepsilon} . \tag{1.4.5}$$

Thus, for a transparent dielectric medium that is *linear*, *local*, *isotropic* and *non-dispersive*, but *not homogeneous*, the wave equations can be derived from Eqs. (1.2.15) with the help of Eqs. (1.4.4) and (1.4.5):

$$\begin{aligned} \nabla^2\mathbf{E} - \varepsilon\mu\frac{\partial^2}{\partial t^2}\mathbf{E} &= -\frac{\nabla\mu}{\mu}\times(\nabla\times\mathbf{E}) - \nabla\left(\mathbf{E}\bullet\frac{\nabla\varepsilon}{\varepsilon}\right), \\ \nabla^2\mathbf{H} - \varepsilon\mu\frac{\partial^2}{\partial t^2}\mathbf{H} &= -\frac{\nabla\varepsilon}{\varepsilon}\times(\nabla\times\mathbf{H}) - \nabla\left(\mathbf{H}\bullet\frac{\nabla\mu}{\mu}\right). \end{aligned} \tag{1.4.6}$$

These equations are commonly used as a starting point to justify Geometrical Optics in electromagnetic terms. It should be noted that the propagation in inhomogeneous media is solvable with relative ease only when we can apply the Geometrical Optics.

Finally, if the medium is also *homogeneous*, Eqs. (1.4.6) become the most familiar equations

$$\nabla^2\mathbf{E} - \varepsilon\mu\frac{\partial^2}{\partial t^2}\mathbf{E} = 0, \quad \nabla^2\mathbf{H} - \varepsilon\mu\frac{\partial^2}{\partial t^2}\mathbf{H} = 0 , \tag{1.4.7}$$

which describe an electromagnetic wave that propagates with speed

$$\upsilon = \frac{c}{n} = \frac{1}{\sqrt{\varepsilon\mu}} , \tag{1.4.8}$$

where n is called the *index of refraction*.

1.4.2 Homogeneous plane waves

For *non-dispersive linear* media, where n does not depend on the frequency and it is real (zero absorption), *homogeneous* and *isotropic* (for which $\nabla \cdot \mathbf{E} = 0$), the form of Eqs. (1.4.7) holds for each component of both **E** and **B**:

$$\nabla^2 u - \frac{1}{\upsilon^2}\frac{\partial^2}{\partial t^2}u = 0\,, \tag{1.4.9}$$

with $\upsilon = c/n =$ constant. If we consider waves that propagate in only one direction (say, z), this equation has a general solution of the form:

$$u(z,t) = f(z - \upsilon t) + g(z + \upsilon t)\,,$$

where $f(z)$ and $g(z)$ are arbitrary functions. The meaning of this expression is that any waveform, composed of any spectral band, is propagated in space without change. Instead, in dispersive media, such a solution is not valid, nor does the same Eq. (1.4.9) apply and we must return, for example, to Eq. (1.4.3). In other words, the individual spectral components propagate with different speeds, and therefore their composition changes continuously. Thus, the Fourier expansion into monochromatic plane waves is commonly used, that is, the space of the transformed variables $\boldsymbol{k}$ and ω introduced in the previous section is used. Thus, let us try to apply a *monochromatic plane wave* to the Maxwell's Eqs. (1.2.11), expressed by the equations

$$\begin{aligned} \mathbf{E}(\boldsymbol{r},t) &= \Re e\left[\mathcal{E}(\boldsymbol{r},t)\right] = \Re e\left[\boldsymbol{E}_0 e^{i(\boldsymbol{k}\cdot\boldsymbol{r}-\omega t)}\right], \\ \mathbf{B}(\boldsymbol{r},t) &= \Re e\left[\mathcal{B}(\boldsymbol{r},t)\right] = \Re e\left[\boldsymbol{B}_0 e^{i(\boldsymbol{k}\cdot\boldsymbol{r}-\omega t)}\right], \end{aligned} \tag{1.4.10}$$

where $\boldsymbol{E}_0$, $\boldsymbol{B}_0$, $\boldsymbol{k}$ are complex vectors, constant in space and time. The complex fields $\mathcal{E}$ and $\mathcal{B}$ are now the main object of our study, bearing in mind that the "true" physical fields are obtainable simply by taking their real part. In the following, we will often use the complex notation for the fields that utilize the analytic functions or the complex amplitudes. In this second case, omitting, for brevity, the oscillation in time, it should be understood that the time dependence is $\exp(-i\omega t)$.

One might think that Eqs. (1.4.10) leave the imaginary part of field free to behave as it wishes, however, from the exponential of the second equality, we see, for example, that $\Im m[\mathcal{E}(t)] = \Re e[\mathcal{E}(t+\pi/2\omega)]$, namely, that the trend of the imaginary part of $\mathcal{E}$ is equal to that of its real part with an anticipation of 1/4 period. This happens thanks to the fact that $\mathcal{E}$ is an analytic function, so that its imaginary part is closely linked to the real one, as we already noted in §1.2.1. Thus, $\Im m(\mathcal{E})$

must also satisfy the same equations that $\Re e(\mathscr{E})$ must fulfill, and the same goes for the other fields.

Consider now the case of a medium that is linear and isotropic, but not necessarily non-dispersive. Therefore, for our monochromatic wave, the polarization and the magnetization have the complex shape already indicated by Eqs. (1.3.32) and the Maxwell equations for the complex vectors are reduced to Eqs. (1.3.33). Finally, if the medium is also *homogeneous* and *constant*, for which χ_e and χ_m are independent from the position in space and constant in time, these equations can be further simplified in the following mixed form:

$$\begin{cases} \nabla\cdot\mathscr{E} = 0, & \nabla\cdot\mathscr{H} = 0, \\ \nabla\times\mathscr{E} = -\mu\dfrac{\partial\mathscr{H}}{\partial t}, & \nabla\times\mathscr{H} = \varepsilon\dfrac{\partial\mathscr{E}}{\partial t}, \end{cases} \tag{1.4.11}$$

where $\mathscr{H} = \mathscr{B}/[\mu_0(1+\chi_m)] = \mathscr{B}/\mu$. The coefficients ε and μ are still given by Eqs. (1.2.13), but now they need to be interpreted as complex quantities.

Eqs. (1.4.11) are very similar to Eqs. (1.2.15) and hence, from them, we can immediately derive the wave equations corresponding to Eqs. (1.4.7):

$$\begin{aligned} \nabla^2\mathscr{E} - \varepsilon\mu\frac{\partial^2}{\partial t^2}\mathscr{E} &= 0, \\ \nabla^2\mathscr{H} - \varepsilon\mu\frac{\partial^2}{\partial t^2}\mathscr{H} &= 0. \end{aligned} \tag{1.4.12}$$

For our monochromatic wave, we have thus obtained the same equation (equal but now complex) to those that would occur for a generic wave if the medium were *also* non-dispersive. Let us now make explicit the space-time dependence of $\mathscr{E}$ and $\mathscr{H}$, as in the second equality of Eqs. (1.4.10). The operators ∇ and δ/δ*t* applied to exp($i\boldsymbol{k}\cdot\boldsymbol{r} - i\omega t$) correspond to simple multiplications by the terms, respectively

$$\begin{aligned} \nabla &\to i\boldsymbol{k}, & \nabla^2 &\to -\boldsymbol{k}\cdot\boldsymbol{k}, \\ \frac{\partial}{\partial t} &\to -i\omega, & \frac{\partial^2}{\partial t^2} &\to -\omega^2. \end{aligned} \tag{1.4.13}$$

Eqs. (4.12) thus pose a constraint between $\boldsymbol{k}$ and εμ:

$$\boldsymbol{k}\cdot\boldsymbol{k} - \varepsilon\mu\omega^2 = 0\,. \tag{1.4.14}$$

We will now examine the simple but important case in which εμ is real and positive. With this condition, $\boldsymbol{k}$ *can* also be real, for which Eq. (1.4.14) leads us to

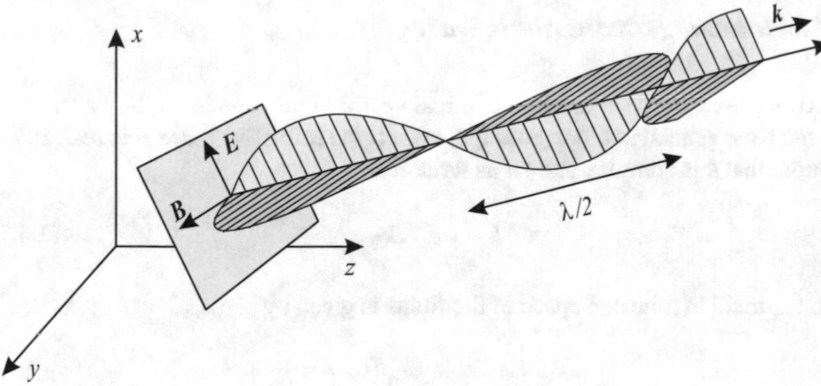

Fig. 1.3 Homogeneous plane wave (right-hand rule)

$$|\boldsymbol{k}| = k = \frac{n}{c}\omega\,, \tag{1.4.15}$$

where n (real) is the refractive index of the medium:

$$n = \sqrt{\varepsilon\mu/\varepsilon_0\mu_0}\,. \tag{1.4.16}$$

Maxwell's equations then lead to other constraints that indicate the relative orientation of the various vectors. For example, by the first two Eqs. (1.4.11) for the amplitudes, we have

$$\boldsymbol{k}\bullet\boldsymbol{E}_0 = 0, \quad \boldsymbol{k}\bullet\boldsymbol{H}_o = 0\,, \tag{1.4.17}$$

and this is to say that the amplitudes $\boldsymbol{E}_0$ and $\boldsymbol{H}_o$ are perpendicular to the *wave vector* $\boldsymbol{k}$, which is oriented [see Eqs. (1.4.10)] in the direction of wave propagation. For this reason, the electromagnetic waves are called *transverse waves*. The other two Maxwell's equations provide a further constraint:

$$\begin{aligned} \nabla\times\mathscr{E} &= -\mu\frac{\partial\mathscr{H}}{\partial t} &\rightarrow\quad i\boldsymbol{k}\times\boldsymbol{E}_0 &= i\omega\boldsymbol{B}_o, \\ \nabla\times\mathscr{H} &= \varepsilon\frac{\partial\mathscr{E}}{\partial t} &\rightarrow\quad i\boldsymbol{k}\times\boldsymbol{B}_o &= -i\omega\varepsilon\mu\,\boldsymbol{E}_0, \end{aligned} \tag{1.4.18}$$

thus $\mathscr{E}$ and $\mathscr{B}$ must also be perpendicular to each other. In particular, if $\boldsymbol{k}$ is real, $\mathscr{E}$ and $\mathscr{B}$ oscillate in phase. The whole plane electromagnetic wave *appears* as in Fig. 1.3, which is just a draft: nobody knows what the *true* field is. In the graphic representations of electromagnetic waves, it is customary to show only the trend of the electric field, and I will keep to this style in the following.

1.4.3 Inhomogeneous plane waves

Above, we have seen the case of a real vector of propagation $\boldsymbol{k}$, however, this is not the most general solution possible, even if the refractive index n is real. In fact, suppose that $\boldsymbol{k}$ is complex and let us write it as

$$\boldsymbol{k} = \boldsymbol{k}_{\Re} + i\boldsymbol{k}_{\Im}, \tag{1.4.19}$$

then the trend in time and space of the fields is given by

$$e^{i\boldsymbol{k}\cdot\boldsymbol{r}-i\omega t} = e^{-\boldsymbol{k}_{\Im}\cdot\boldsymbol{r}} e^{i\left(\boldsymbol{k}_{\Re}\cdot\boldsymbol{r}-\omega t\right)}.$$

This expression can be regarded as the representation of a wave with an amplitude decreasing exponentially in some direction, that of $\boldsymbol{k}_{\Im}$. This type of wave is called an *inhomogeneous plane wave*, and we can locate constant amplitude surfaces there ($\perp$ to $\boldsymbol{k}_{\Im}$), besides those of constant phase ($\perp$ to $\boldsymbol{k}_{\Re}$), but they are not necessarily parallel; indeed, as we shall see now, for a non-absorbent medium, they are perpendicular. In fact, for a medium with a *real* refractive index, condition (1.4.14) leads to an equation for the real part:

$$\boldsymbol{k}_{\Re}^2 - \boldsymbol{k}_{\Im}^2 = \left(\frac{n}{c}\omega\right)^2, \tag{1.4.20.a}$$

and one for the imaginary part:

$$\boldsymbol{k}_{\Re}\cdot\boldsymbol{k}_{\Im} = 0, \tag{1.4.20.b}$$

from which it is seen that $\boldsymbol{k}_{\Re}$ and $\boldsymbol{k}_{\Im}$ are orthogonal. Then, for n real, Eq. (1.4.14) can be satisfied generally by writing

$$\boldsymbol{k} = \frac{n}{c}\omega\left(\mathbf{e}_3 \cosh\vartheta + i\mathbf{e}_2 \sinh\vartheta\right), \tag{1.4.21}$$

where ϑ is a real parameter that allows us to modulate the relative amplitude between $\boldsymbol{k}_{\Re}$ and $\boldsymbol{k}_{\Im}$, while $\mathbf{e}_3$, $\mathbf{e}_2$ are two orthogonal versors, the first in the direction of $\boldsymbol{k}_{\Re}$ and the second in the direction of $\boldsymbol{k}_{\Im}$. The most general electric field vector that satisfies the condition $\boldsymbol{k}\cdot\mathbf{E} = 0$ is now

$$\boldsymbol{E}_o = \left(i\mathbf{e}_3 \sinh\vartheta - \mathbf{e}_2 \cosh\vartheta\right) A + \mathbf{e}_1 A', \tag{1.4.22}$$

where A, A' are complex amplitudes and $\mathbf{e}_1$ is a versor orthogonal to $\mathbf{e}_3$ and $\mathbf{e}_2$. The corresponding magnetic field $\boldsymbol{H}_o$ of our wave is obtained with Eqs. (1.4.18.a) and (1.4.20):

$$\boldsymbol{H}_o = \frac{n}{\mu c}\left[\left(-i\mathbf{e}_3 \sinh\vartheta + \mathbf{e}_2 \cosh\vartheta\right)A' + \mathbf{e}_1 A\right]. \tag{1.4.23}$$

In the direction of $-\boldsymbol{k}_{\mathfrak{I}}$, the wave amplitude increases indefinitely, and therefore, in an infinite medium, these waves cannot exist. However, within certain conditions, they can appear in proximity to the separation surface between two media, as in the case of an external *evanescent* wave, which, as we shall see later, is generated in conjunction with a total internal reflection. Other examples of inhomogeneous wave exist in the problem of diffraction [Born and Wolf 1980] and in the case of refraction and reflection in conductor (absorbent) media, for which we can define a complex refractive index in which the directions of $\boldsymbol{k}_{\mathfrak{R}}$ and $\boldsymbol{k}_{\mathfrak{I}}$ are not orthogonal (see §1.8).

1.4.4 Spherical waves

In many situations in optics it is necessary to consider spherical waves emitted by a point source. Let us look at a short outline here, considering the case of a simple medium, starting from Eq. (1.4.9). So, let us see if a wave that has a spherical symmetry, and therefore is of the type $\psi = \psi(r,t)$, where r is the distance from an assigned origin, can satisfy that equation. Expressing the ∇ operator in spherical coordinates, with a little calculation, it is found that

$$\nabla^2\psi = \frac{\partial^2}{\partial r^2}\psi + \frac{2}{r}\frac{\partial}{\partial r}\psi = \frac{1}{r}\frac{\partial^2}{\partial r^2}(r\psi), \tag{1.4.24}$$

thus Eq. (1.4.9) for our function ψ (and for $r > 0$) becomes:

$$\frac{1}{r}\frac{\partial^2}{\partial r^2}(r\psi) - \frac{1}{\upsilon^2}\frac{\partial^2}{\partial t^2}\psi = 0\,.$$

This equation can be simplified by multiplying it by r:

$$\frac{\partial^2}{\partial r^2}(r\psi) - \frac{1}{\upsilon^2}\frac{\partial^2}{\partial t^2}(r\psi) = 0\,. \tag{1.4.25}$$

Thus, we find that $r\psi$ must satisfy a one-dimensional wave equation, and therefore the solutions for ψ are of the type

$$\psi(r,t) = \frac{f(r-\upsilon t)}{r}, \tag{1.4.26}$$

where f is an arbitrary continuous function. This equation expresses a wave that expands and whose amplitude decays as $1/r$. A mathematically permissible solution is also that of a wave converging to a point, so, in place of the minus sign in Eq. (1.4.26), there is a plus sign. We also note that, just as we calculate ψ with Eq. (1.4.26), it diverges to infinity for $r = 0$, and actually, if we want a wave with amplitude different from zero, it is necessary that there is a source in the origin.

In obtaining this spherical wave, we started from a type of wave equation, Eq. (1.4.9), that the individual components of the electric and magnetic fields must satisfy. Therefore, we might think that the individual components of the electromagnetic field can evolve as a wave with spherical symmetry. However, this is *not* correct. In fact, the wave equations, even if they were derived from Maxwell's equations, do not contain all of their constraints. In particular, a vector field constructed with three Cartesian components with a space-time dependency of the kind in Eq. (1.4.26) would not have zero divergence, as is instead required for the electric and magnetic fields.

Therefore, in the presence of a punctiform source, that is, one that is very small compared to the wavelength of the emitted radiation, the admitted solutions for the field must contain a dependence on the angular coordinates, as well as from the radial one, such as is obtained with a multi-pole development of the field of an oscillating charge.

Nevertheless, in optics, we often use a scalar wave, as in Eq. (1.4.26), within appropriate approximations that we will study later.

1.5 Energy flow and density of momentum and energy

1.5.1 Derivation and general concepts

Some important results of Electromagnetism are related to the electromagnetic energy density u and the electromagnetic energy flow density called the *Poynting vector* $\boldsymbol{S}$. With just a few steps from the Maxwell's Eqs. (1.2.6), we obtain:

$$\operatorname{div}(\mathbf{E}\times\mathbf{H})+\mathbf{E}\cdot\frac{\partial\mathbf{D}}{\partial t}+\mathbf{H}\cdot\frac{\partial\mathbf{B}}{\partial t}+\mathbf{J}_{\mathrm{m}}\cdot\mathbf{E}=0\,. \tag{1.5.1}$$

Integrating on a volume V, and by making use of the divergence theorem, we get:

$$\oint_{\mathrm{A}}(\mathbf{E}\times\mathbf{H})\cdot\mathbf{n}\,d\mathrm{A}+\int_{\mathrm{V}}\left(\mathbf{E}\cdot\frac{\partial\mathbf{D}}{\partial t}+\mathbf{H}\cdot\frac{\partial\mathbf{B}}{\partial t}\right)d\mathrm{V}+\int_{\mathrm{V}}\mathbf{J}_{\mathrm{m}}\cdot\mathbf{E}\,d\mathrm{V}=0\,, \tag{1.5.2}$$

where A is the closed surface enclosing V and $\mathbf{n}$ is the versor perpendicular to the surface dA. Each term has the dimensions of a power. The first is interpreted as the flow of energy that passes through the area A, the second as the change in energy of the field and the third as the work done on the charges, per unit of time. However, great care is necessary to define the electromagnetic energy U, the corresponding density of momentum **g**, and the energy flow density **S**, because that which can be considered either electromagnetic or mechanical is somewhat arbitrary [Jackson 1974, §6-9]. All scholars adhere to the definition:

$$\mathbf{S} = \mathbf{E} \times \mathbf{H} \tag{1.5.3}$$

for the Poynting vector.

For the other quantities, Jackson (1974) writes that a correct resolution follows with a statistical mechanical treatment of the system of matter plus the (external) field, in which the electromagnetic quantities are defined as the difference between the quantities of the combined system and those for the matter alone at the same equilibrium temperature T and density d, in zero field. With this definition, the energy flux density is given by Eq. (1.5.3), and one can establish that the second term of Eq. (1.5.2) is the electromagnetic contribution to the *free energy F* of the system. This energy can be calculated once we know **E** as a function of **D**, and **H** as a function of **B**. In particular, it is necessary that $\mathbf{E} \bullet \partial \mathbf{D}/\partial t + \mathbf{H} \bullet \partial \mathbf{B}/\partial t$ is an exact differential form. For example, for a *linear*, *isotropic* and *non-dispersive* medium, with $\mathbf{D} = \varepsilon \mathbf{E}$ and $\mathbf{B} = \mu \mathbf{H}$, the free energy is

$$F = \frac{1}{2}\int_{\mathrm{V}} (\mathbf{E} \bullet \mathbf{D} + \mathbf{H} \bullet \mathbf{B})\, d\mathrm{V}\,, \tag{1.5.4}$$

while the electromagnetic energy U=F+TS, where T is the temperature and S the entropy, is

$$U = \frac{1}{2}\int_{\mathrm{V}} \left\{ \mathbf{E}^2 \left[\varepsilon + T\left(\frac{\partial \varepsilon}{\partial T}\right)_d \right] + \mathbf{H}^2 \left[\mu + T\left(\frac{\partial \mu}{\partial T}\right)_d \right] \right\} d\mathrm{V}\,. \tag{1.5.5}$$

For more details, see Jackson (1974), Chap.6, and Panofsky and Phillips (1966), §6.3.

An electromagnetic wave also carries a momentum. This can be guessed considering, for example, a monochromatic wave orthogonally incident on a reflective surface: the electrons of the material are placed in oscillation in the same direction and with the same frequency as the electric field, with some phase delay, so that they acquire a speed (oscillating with a certain phase relative to the field) parallel to the surface. This speed combined with the magnetic field of the wave, for the term dependent on **B** of the Lorentz force, constitutes a net force perpendicular to the surface. In the general case, the calculation is complicated, and, for the momentum density, the favor of scholars oscillated between the solutions of Hermann

Minkowski (1908):

$$\mathbf{g}_M = \mathbf{D} \times \mathbf{B}\,, \tag{1.5.6.a}$$

and that of Max Abraham (1914):

$$\mathbf{g}_A = \frac{1}{c^2}\mathbf{E} \times \mathbf{H} = \frac{1}{c^2}\mathbf{S}\,. \tag{1.5.6.b}$$

Still others were added later. The Minkowski solution comes from a direct application of the macroscopic equations to the conservation of momentum and is consistent with a hypothesis resulting from Quantum Mechanics. The Abraham solution is instead consistent with the principles of Relativity. With the above considerations, Jackson takes a position in favor of Abraham's solution. However, the dispute continued over the intervening years. According to some, for example, R. Peierls, Abraham's result gives only the momentum that resides in the electromagnetic field [Peierls 1976], while, for J.P. Gordon, the Minkowski expression represents a pseudo-moment, experimentally observable in many situations, such as the radiation pressure on a reflective surface immersed in a fluid or the recoil of the atoms in a gas [Gordon 1973]. Subsequent experiments are in agreement with the expression of Minkowski [Gibson et al 1980]. Others, in which the acceleration of a dielectric is measured, agree instead with the momentum of Abraham. Recently, S.M. Barnett proposed a resolution of the dilemma by noting that, while the total momentum of matter and radiation is unique, there are two ways to split it between matter and radiation within the medium [Barnett 2010]. Therefore, for the radiation, *two* distinct electromagnetic momenta should be considered, the *kinetic momentum* and the *canonical momentum*, already discussed by J.C. Garrison and R.Y. Chiao [Garrison and Chiao 2004]. For a photon, the first is $\hbar\omega/(c n_g)$ and the second, associated with the wavelength, is $\hbar\omega n/c$, where n_g and n are, respectively, the refractive indices for the group velocity and the phase velocity. The kinetic momentum is equal to that of Abraham, and the canonical momentum to that of Minkowski. Which of these two momenta is revealed depends on the experiment itself.

1.5.2 Flow of energy in simple media

1.5.2.1 Intensity of a monochromatic wave

Let us now see the case of a monochromatic wave in a simple dielectric medium, which is *linear*, *local*, *isotropic*, *homogeneous*, *transparent*, and *non-dispersive*. Making explicit the temporal part in the complex representation of the fields, we have

$$\begin{aligned}\mathbf{E}(\boldsymbol{r},t) &= \Re e\left[\boldsymbol{E}(\boldsymbol{r})e^{-i\omega t}\right],\\ \mathbf{H}(\boldsymbol{r},t) &= \Re e\left[\boldsymbol{H}(\boldsymbol{r})e^{-i\omega t}\right].\end{aligned} \tag{1.5.7}$$

Given that **E** and **H** are oscillating, **S** = **E**×**H** oscillates with a double frequency. It is therefore useful to consider its mean value calculated over a period. Such value, as is known from Eq. (1.3.36), is given by

$$\overline{\mathbf{S}} = \frac{1}{2}\Re e\left(\boldsymbol{E}\times\boldsymbol{H}^*\right). \tag{1.5.8}$$

To evaluate this vector, in the simplest cases, we can locally approximate our wave with a homogeneous plane wave:

$$\boldsymbol{E}(\boldsymbol{r}) = \boldsymbol{E}_r e^{i\boldsymbol{k}_r\cdot\boldsymbol{r}},\ \ \boldsymbol{H}(\boldsymbol{r}) = \boldsymbol{H}_r e^{i\boldsymbol{k}_r\cdot\boldsymbol{r}}, \tag{1.5.9}$$

where the amplitudes $\boldsymbol{E}_r$ and $\boldsymbol{H}_r$ and the wave vector $\boldsymbol{k}_r$ are slowly varying functions of $\boldsymbol{r}$. Thanks to this assumption, by applying Eqs. (1.5.9) to the Maxwell equations and to the wave equation, the contribution of the terms containing spatial derivatives of $\boldsymbol{E}_r$, $\boldsymbol{H}_r$ and $\boldsymbol{k}_r$ are negligible compared to those with the derivative of the carrier. Therefore, it is still found that $\boldsymbol{k}_r$ is subject to the constraints [see Eq. (1.4.15)]:

$$|\boldsymbol{k}_r| = \sqrt{\varepsilon\mu}\,\omega = \frac{n}{c}\omega = n\frac{2\pi}{\lambda_o}. \tag{1.5.10}$$

In addition, it is oriented perpendicularly to $\boldsymbol{E}_r$ and $\boldsymbol{H}_r$, as is clear from Eqs. (1.4.17). In addition, from Eqs. (1.4.18), we get:

$$\boldsymbol{k}_r \times \boldsymbol{E}_r = \omega\mu\boldsymbol{H}_r, \tag{1.5.11}$$

so that for the modules of $\boldsymbol{E}_r$ and $\boldsymbol{H}_r$, we have

$$\sqrt{\mu}\,H_r = \sqrt{\varepsilon}\,E_r. \tag{1.5.12}$$

Finally, we can calculate the time average of the Poynting vector, substituting Eq. (1.5.11) and then Eq. (1.4.17.a) in Eq. (1.5.8):

$$\overline{\mathbf{S}} = \Re e\left[\frac{1}{2\mu}\boldsymbol{E}_r\times\left(\frac{\boldsymbol{k}_r}{\omega}\times\boldsymbol{E}_r^*\right)\right] = \frac{1}{2\mu}|\boldsymbol{E}_r|^2\frac{\boldsymbol{k}_r}{\omega}.$$

Then, for Eq. (1.5.10), we can write

$$\overline{\mathbf{S}} = \frac{1}{2}\sqrt{\frac{\varepsilon}{\mu}}\,|\boldsymbol{E_r}|^2\,\mathbf{s}\,, \tag{1.5.13.a}$$

that is,

$$\overline{\mathbf{S}} = \frac{n}{2c\mu}|\boldsymbol{E_r}|^2\,\mathbf{s}, \quad \text{or also} \quad \overline{\mathbf{S}} = \frac{c}{2n}\varepsilon|\boldsymbol{E_r}|^2\,\mathbf{s}\,, \tag{1.5.13.b,c}$$

where $\mathbf{s}$ is the versor of $\boldsymbol{k_r}$. In particular, for a non-magnetic medium, with $\mu = \mu_0$, Eq. (1.5.13.a) becomes:

$$\overline{\mathbf{S}} = \frac{nc}{2}\varepsilon_0|\boldsymbol{E_r}|^2\,\mathbf{s}\,. \tag{1.5.14}$$

Let us now examine the time average of the energy density $\overline{w}$ of our wave in the medium under examination. From Eq. (1.5.4), we have:

$$\overline{w} = \overline{w}_e + \overline{w}_m = \tfrac{1}{4}\Re e\left(\boldsymbol{E_r}\cdot\boldsymbol{D_r}^* + \boldsymbol{H_r}\cdot\boldsymbol{B_r}^*\right) = \frac{1}{4}\left(\varepsilon|\boldsymbol{E_r}|^2 + \mu|\boldsymbol{H_r}|^2\right), \tag{1.5.15}$$

from which we see, with Eq. (1.5.12), that the electromagnetic energy is equally divided between electric and magnetic fields:

$$\overline{w}_e = \overline{w}_m\,, \tag{1.5.16}$$

and therefore Eq. (1.5.15) can be simplified to the expression

$$\overline{w} = \frac{1}{2}\varepsilon|\boldsymbol{E_r}|^2\,. \tag{1.5.17}$$

The intensity I of the wave is the modulus of $\overline{\mathbf{S}}$, and so it can finally be put in relation with the energy density $\overline{w}$:

$$I = \frac{c}{2n}\varepsilon|\boldsymbol{E_r}|^2 = \frac{c}{n}\overline{w}\,, \tag{1.5.18}$$

which expresses the fact that the energy is moving at a speed $\upsilon = c/n$.

An important result is obtained by applying the fields given by Eqs. (1.5.7) to the law of energy conservation: for a *non-conductive* medium ($\sigma = 0 \rightarrow \boldsymbol{J} = 0$) where no mechanical work is done (on the charges), and therefore non-absorbent, by taking the temporal average of Eq. (1.5.2), we find that

$$\operatorname{div}\overline{\mathbf{S}} = 0\,. \tag{1.5.19}$$

By integrating this equation on an arbitrary volume, which does not contain, however, sources or absorbers, and applying the Gauss theorem, it results that

$$\oint \overline{\mathbf{S}} \cdot \mathbf{n} \, dA = 0 , \tag{1.5.20}$$

where $\mathbf{n}$ is here the versor normal to the element dA of the integration surface. Thus, the total flow of energy through such a closed surface is zero.

The results of this section and those of the next section are obtained in the case of a homogeneous medium; however, they can also be extended to inhomogeneous media, with the approximations and the limits that we will see later.

1.5.2.2 Intensity of a quasi-monochromatic wave

In the case of a non-monochromatic wave, things get a little more complicated, because, for this type of waves, the field is subject to more or less irregular fluctuations. On the other hand, each intensity measurement involves an integration over some time interval, and, in general, subsequent measurements may give different results. Here, for simplicity, we shall consider only the case in which the wave is *statistically stationary*, in which by this we mean that it has statistical characteristics constant in time, for which the measures do not depend on the choice of the initial time to perform them. Furthermore, we limit ourselves to dealing with the case of a *non-dispersive linear* medium, or, anyway, a wave with a limited spectral band, that is, a *quasi-monochromatic* wave, such that the variations of ε and μ in this band are negligible.

The application of Eq. (1.3.37) to the average value of the Poynting vector leads us to write that the intensity I is given by

$$I = \frac{1}{4} \left\langle \mathscr{E} \times \mathscr{H}^* + \mathscr{E}^* \times \mathscr{H} \right\rangle , \tag{1.5.21}$$

where the brackets $\langle \rangle$ imply the time average. On the other hand (see §1.3.1), also for a quasi-monochromatic wave, we can follow the calculation done in the previous paragraph and write an expression similar to Eq. (1.5.12):

$$\sqrt{\mu} \left| \mathscr{H} \right| = \sqrt{\varepsilon} \left| \mathscr{E} \right| , \tag{1.5.22}$$

and, similarly to Eq. (1.5.13.c), the intensity of our quasi-monochromatic wave is given by:

$$I = \frac{c\varepsilon}{2n} \left\langle \mathscr{E} \cdot \mathscr{E}^* \right\rangle . \tag{1.5.23}$$

1.5.2.3 Momentum of photons and radiation pressure

We still have to make some considerations as to the momentum density **g** carried by an electromagnetic wave. In vacuum (at least) Eq. (1.5.6.b) establishes a relationship between **g** and the Poynting vector **S**, which is derived from the general principles of relativistic dynamics: you can read the nice discussion that Feynman (1969) gives about it, § 27-6. For our plane wave, with Eq. (1.5.18), the two possible choices of $\overline{\mathbf{g}}$, for Eqs. (1.5.6), become:

$$\overline{\mathbf{g}} = a\frac{\overline{w}}{c}\mathbf{s}\,, \tag{1.5.24}$$

where **s** is the versor of ***k***, while a is $1/n_g$ for the solution of Abraham or n for that of Minkowski. In other words, if a given energy E is transported per m^2 and sec, then there is also a corresponding momentum aE/c transported per m^2 and sec. In Quantum Mechanics, we depict the electromagnetic energy divided into quanta (*photons*), each with energy $h\nu$, where h is the Planck's constant. So, we can expect that each photon also transports a momentum $p = a\,h\nu/c = (a/n)h/\lambda$, where λ is the wavelength in the medium considered. In the vacuum, the two solutions coincide, and we have:

$$\boldsymbol{p} = \frac{h}{\lambda}\mathbf{s} = \hbar\boldsymbol{k}\,, \tag{1.5.25}$$

where, as usual, $\hbar = h/2\pi$. I recall here that h/λ also expresses the momentum of a particle whose de Broglie's wavelength is λ.

To the wave momentum density corresponds a *radiation pressure*. Consider, for example, the case of a light beam of intensity I that impinges orthogonally on an absorbent surface. The force exerted by the wave on this surface is equal to the momentum absorbed per unit of time, and therefore the pressure is $P_r = |\overline{\mathbf{g}}|\,c/n$. According to that which has been experimentally verified, the value of **g** to be used is that of Minkowski, and therefore

$$P_r = \frac{n}{c}I = \overline{w}\,, \tag{1.5.26}$$

thus the pressure is equal to the energy density of the beam.[6] If, instead of an absorber, we use a mirror, the impulse exchanged in reflection (for normal incidence) is also doubled, and therefore the pressure is doubled.

The radiation pressure is generally very weak, for example, for a beam with intensity 1 W/mm^2, from Eq. (1.5.26), the pressure on a mirror, on which the beam

[6] Although P_r and $\overline{w}$ express different concepts, their dimensions are equal.

is incident perpendicularly, is about 1.7×10^{-3} pascal. Despite its weak entity, the radiation pressure has important applications, such as the isotope separation, the laser cooling and confinement of neutral atoms, and the levitation of small objects.

1.6 Polarization

1.6.1 Jones vectors for a plane wave

Let us consider, for simplicity, the case of a monochromatic plane wave that propagates along the *z*-axis, that is, with the wave vector oriented in that direction. In this way, for Eq. (1.4.10.a), the wave is characterized by an electric field of the form:

$$\mathbf{E}(z,t) = \Re e\left[\boldsymbol{E}_0 e^{i(kz-\omega t)}\right], \qquad (1.6.1)$$

where k may be positive or negative, indicating the two opposite directions of propagation, but here, we will consider only the positive case. The amplitude $\boldsymbol{E}_o$ of the field is a complex vector oriented in the *xy*-plane (since it must be $\perp$ to $\boldsymbol{k}$):

$$\boldsymbol{E}_0 = \begin{pmatrix} E_{ox} \\ E_{oy} \end{pmatrix}. \qquad (1.6.2)$$

Such a vector, with this way of representing it in a column with (only) two components, is called a *Jones' vector* and expresses the state of *polarization* of the wave. Indeed, the modulus and relative phase of its components are arbitrary and, depending on their value, we will have a linearly, circularly or, in general, elliptically polarized wave, where, as we will see shortly, these terms express the motion type of the "tip" of the vector of the electric field in time. The Jones vectors can only be applied to beams of polarized and coherent light, but allow for easy handling of the simpler cases of light polarization by introducing a very concise technique, powerful and well suited to laser beams. A more general method, but one less easy to handle, for representing the polarization is to make use of the *Stokes' vectors*, which consist of four components, with which we can also treat light from natural sources, non-polarized or partially polarized, that we will study later.

1.6.2 Linear polarization

A normalized basis for the Jones vectors is given by the vectors

$$\mathbf{P}_x = \begin{pmatrix} 1 \\ 0 \end{pmatrix} \text{ and } \mathbf{P}_y = \begin{pmatrix} 0 \\ 1 \end{pmatrix}. \tag{1.6.3}$$

Any polarization state characterized by a complex amplitude $\boldsymbol{E}_0$ of the electric field can be obtained by a linear combination of these two entities, but it is important to emphasize that the coefficients are generally complex.

The vector $\begin{pmatrix} 1 \\ 0 \end{pmatrix}$ means an oscillating electric field oriented along the x-axis:

$\begin{pmatrix} 1 \\ 0 \end{pmatrix} \rightarrow$ 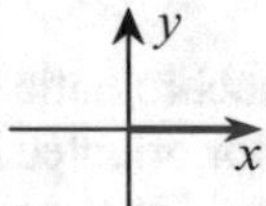

and similarly: $\begin{pmatrix} 0 \\ 1 \end{pmatrix} \rightarrow$ 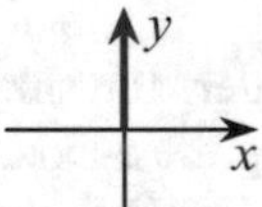

Of course, any other base formed by any two mutually perpendicular directions is a useful orthonormal basis. If you want to use two or more bases at the same time, it is good to differentiate the vectors with some index, as you like, for example:

$\begin{pmatrix} \\ \end{pmatrix}_\ell$ for 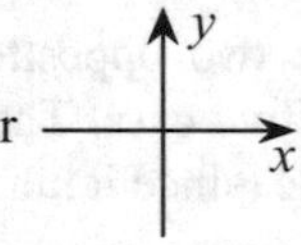

and $\begin{pmatrix} \\ \end{pmatrix}_L$ for 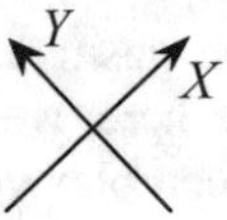

Well, $\begin{pmatrix} 1 \\ 0 \end{pmatrix}$ and $\begin{pmatrix} 0 \\ 1 \end{pmatrix}$ are two examples of *linear polarization*, that is, because of the term $\exp[i(kz - \omega t)]$, their electric field oscillates, respectively, on the zx-plane and yz-plane:

$\begin{pmatrix} 1 \\ 0 \end{pmatrix} \rightarrow$ 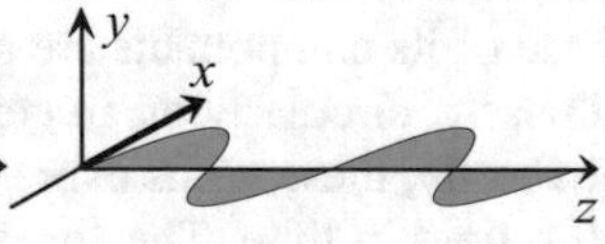

$\begin{pmatrix} 0 \\ 1 \end{pmatrix} \rightarrow$ 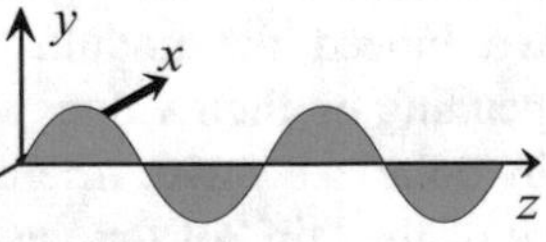

A generic linear polarization with the field **E** oscillating in a direction α is simply represented by the vector:

$\begin{pmatrix} \cos\alpha \\ \sin\alpha \end{pmatrix} \rightarrow$ 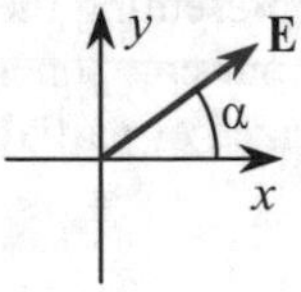

Even this last vector is normalized, i.e., it has a unit norm; indeed, the norm of $\begin{pmatrix} a \\ b \end{pmatrix}$ is defined by the square modulus of the two generally complex elements a

and b:

$$aa^* + bb^*,$$

thus, the norm of $\begin{pmatrix}\cos\alpha\\ \sin\alpha\end{pmatrix}$ is $\cos^2\alpha + \sin^2\alpha = 1$. It should be noted that this rule for the Jones vectors is equivalent to calculation of the wave intensity:

$$I = \frac{c\varepsilon}{2n}\left|\boldsymbol{E}_o\right|^2 = \frac{c\varepsilon}{2n}\left(E_{ox}^* E_{ox} + E_{oy}^* E_{oy}\right) = \frac{c\varepsilon}{2n}\left(E_{ox}^*, E_{oy}^*\right)\begin{pmatrix}E_{ox}\\ E_{oy}\end{pmatrix}. \tag{1.6.4}$$

While the vectors are indicated in bold, their moduli, as usual, will be indicated with regular characters.

1.6.3 Circular polarization

In general, E_{ox} and E_{oy} are independent complex numbers. Among all of the possible combinations, there are two particularly interesting complex vectors, which constitute a new basis:

$$\mathbf{L} = \frac{1}{\sqrt{2}}\begin{pmatrix}1\\ i\end{pmatrix} \text{ and } \mathbf{R} = \frac{1}{\sqrt{2}}\begin{pmatrix}1\\ -i\end{pmatrix}, \tag{1.6.5}$$

where $\sqrt{2}$ needs to normalize these vectors.

These new versors describe the *circular polarizations*. Let us see how they work. The real field **E** corresponding to $\frac{E_o}{\sqrt{2}}\begin{pmatrix}1\\ i\end{pmatrix}$ has components:

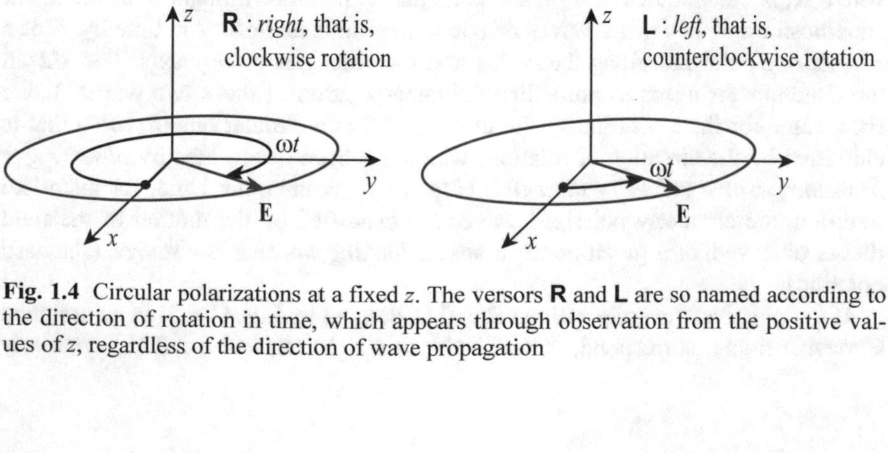

Fig. 1.4 Circular polarizations at a fixed z. The versors **R** and **L** are so named according to the direction of rotation in time, which appears through observation from the positive values of z, regardless of the direction of wave propagation

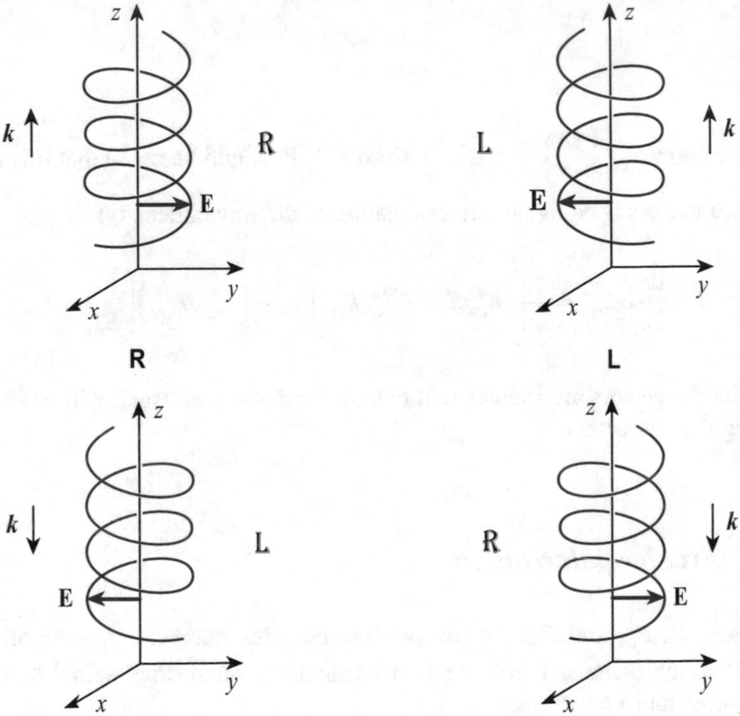

Fig. 1.5 Circular polarizations at a fixed time t for the two cases of Fig. 1.4. The names of the **R** and **L** waves correspond to the helicity of the curve plotted in space by the field vector

$$\begin{aligned} \mathrm{E}_x &= \frac{\mathrm{E_o}}{\sqrt{2}}\frac{1}{2}\left[e^{i(kz-\omega t+\phi)}+e^{-i(kz-\omega t+\phi)}\right] = \frac{\mathrm{E_o}}{\sqrt{2}}\cos\left(kz-\omega t+\phi\right), \\ \mathrm{E}_y &= \frac{\mathrm{E_o}}{\sqrt{2}}\frac{1}{2}\left[i\,e^{i(kz-\omega t+\phi)}-i\,e^{-i(kz-\omega t+\phi)}\right] = -\frac{\mathrm{E_o}}{\sqrt{2}}\sin\left(kz-\omega t+\phi\right), \end{aligned} \tag{1.6.6}$$

where $\mathrm{E_o}$ is the modulus of E_o and ϕ is its phase. It can be thought of as the linear superposition of two plane waves of equal amplitude and offset in time by ¼ of a period, one polarized along the x-axis and the other along the y-axis. The $\sqrt{2}$ in the denominator needs to normalize the superposition of these two waves. If we fix a value for the z coordinate, the motion of **E** is a circular one (*in time*) that is classified by the direction of rotation, which has been determined by *observing it from the positive values of the z-axis* (Fig. 1.4). On the other hand, for historical tradition, the circularly polarized waves are classified by the motion of the field that is observed, at a given point in space, *looking towards the source* (standard notation).

If k is positive, the polarizations **R** and **L**, defined by Eqs. (1.6.5) in an assigned Cartesian frame, correspond, respectively, to a right (**R**) and a left (**L**) circularly

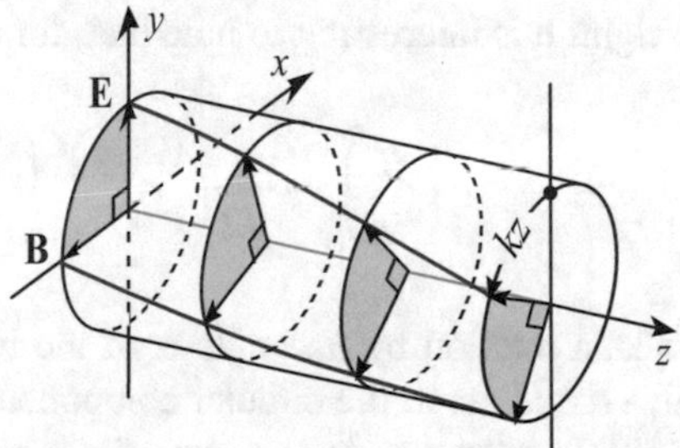

Fig. 1.6
Orientation of the fields along the z axis for a circularly polarized wave, at a given time

polarized waves, while, if k is negative, that is, if the wave proceeds in the opposite direction to the z-axis, it has the opposite correspondence.

Instead, if t is a fixed time, **E** appears to rotate *in space* following a helix that wraps around the z-axis, as in Fig. 1.5, whose *helicity* is indicated by the name according to the standard classification. In particular, the helix generated by a right circularly polarized wave, that is, an **R**-wave, corresponds to an ordinary screw that is screwed turning clockwise when viewed from the head.

The magnetic field, since it must always remain perpendicular to the electric field, rotates in the same direction. For example, an **R**-wave, at a fixed time, has the shape indicated by Fig. 1.6.

In conclusion, the versors **L** and **R** represent, at $z = 0$, two opposite circular motions of radius $1/\sqrt{2}$, as shown in Fig. 1.7 (a). They are orthogonal to each other, and therefore can be used as a new base for the Jones vectors. The vectors expressed in this base can be identified, for example, with an index "c", and those in the linear base with an index "ℓ". The transformation from the linear to the circular base is representable with

$$\Big(\;\Big)_\ell \to \Big(\;\Big)_c \; : \; \begin{pmatrix} a_L \\ a_R \end{pmatrix}_c = \frac{1}{\sqrt{2}} \begin{pmatrix} 1 & -i \\ 1 & i \end{pmatrix}_c^\ell \begin{pmatrix} a_x \\ a_y \end{pmatrix}_\ell , \tag{1.6.7}$$

as we can easily verify by replacing **L** or **R** of Eqs. (1.6.5) in the Jones vector at

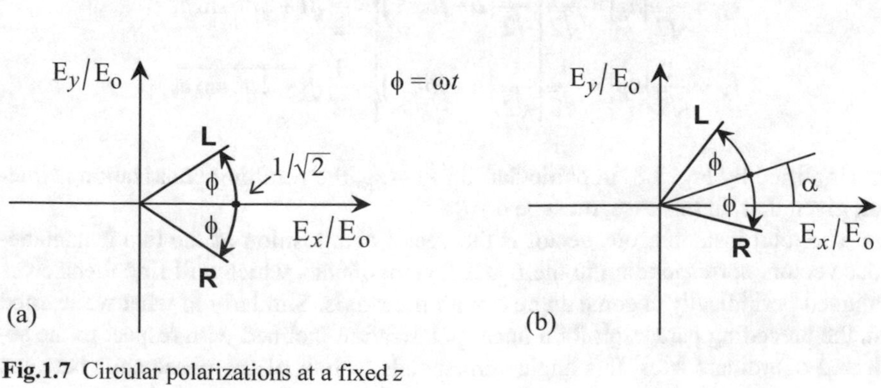

Fig.1.7 Circular polarizations at a fixed z

the right. It is interesting to note that, for a generic linear polarization, we have:

$$\begin{pmatrix} a_L \\ a_R \end{pmatrix}_c = \frac{1}{\sqrt{2}} \begin{pmatrix} 1 & -i \\ 1 & i \end{pmatrix}_c^{\ell} \begin{pmatrix} \cos\alpha \\ \sin\alpha \end{pmatrix}_{\ell} = \frac{1}{\sqrt{2}} \begin{pmatrix} e^{-i\alpha} \\ e^{+i\alpha} \end{pmatrix}_c , \tag{1.6.8}$$

that is, a rotation by an angle α of the linear polarization corresponds to a phase shift $-\alpha$ and $+\alpha$ in the circular components L and R, respectively, as shown in Fig. 1.7 (b). In other words, the sum of two opposite circular polarizations, of the same amplitude and a phase shift 2α, gives rise to a linear polarization oriented at the angle α.

1.6.4 Elliptical polarization

Since a common phase factor to both components a_x, a_y is irrelevant to the polarization, a generic normalized Jones vector can be written in the form:

$$\begin{pmatrix} a \\ be^{i\delta} \end{pmatrix}_{\ell} , \quad \text{with } a^2 + b^2 = 1 \,; \quad a, b \in \mathbb{R} \,.$$

Now, we want to figure out which state of motion it matches for the field **E**, at $z = 0$. To do this, it is better to transform it in the circular base!

$$\begin{pmatrix} a_L \\ a_R \end{pmatrix}_c = \frac{1}{\sqrt{2}} \begin{pmatrix} 1 & -i \\ 1 & i \end{pmatrix}_c^{\ell} \begin{pmatrix} a \\ be^{i\delta} \end{pmatrix}_{\ell} = \frac{1}{\sqrt{2}} \begin{pmatrix} a - ibe^{i\delta} \\ a + ibe^{i\delta} \end{pmatrix}_c .$$

This motion is thus given by the composition of two opposite circular motions, of radius

$$r_L = \frac{1}{\sqrt{2}} |a_L| = \frac{1}{\sqrt{2}} \left| \frac{1}{\sqrt{2}} \left(a - ibe^{i\delta} \right) \right| = \frac{1}{2} \sqrt{1 + 2ab\sin\delta},$$

$$r_R = \frac{1}{\sqrt{2}} |a_R| = \frac{1}{\sqrt{2}} \left| \frac{1}{\sqrt{2}} \left(a + ibe^{i\delta} \right) \right| = \frac{1}{2} \sqrt{1 - 2ab\sin\delta},$$

exemplified by Fig. 1.8. In particular, for $r_L = r_R$, the resulting polarization is linear, given that, in this case, must be $\delta = 0$.

The total instantaneous vector is the vector composition of the two instantaneous vectors corresponding to the L and R components, which will find themselves aligned periodically at some angle α with the x-axis. Similarly to what we learned in the preceding paragraph for a linear polarization inclined with respect to the selected coordinate axes, this angle corresponds to half of the phase shift between

Fig. 1.8
Analysis of an elliptical polarization

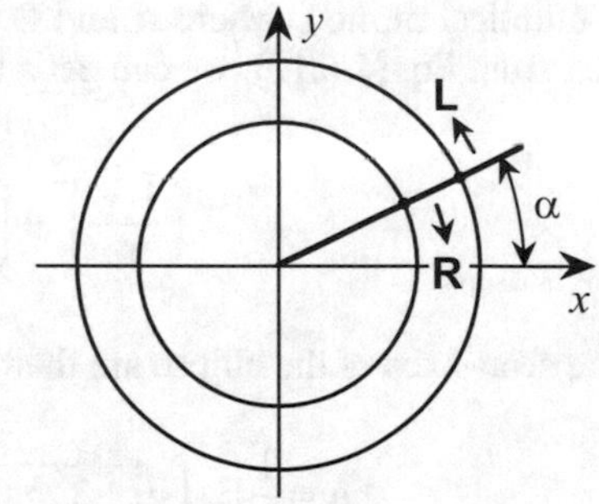

the circular components. Since

$$e^{i\phi_1}\left(e^{i\phi_2}\right)^* = e^{i(\phi_1-\phi_2)},$$

this phase shift can be immediately obtained from the product of the amplitudes a_R with the conjugate of a_L:

$$\frac{1}{2}\left(a+ibe^{i\delta}\right)\left(a+ibe^{-i\delta}\right)=\frac{1}{2}\left(a^2-b^2+i2ab\cos\delta\right),$$

whence

$$\tan 2\alpha = \frac{2ab\cos\delta}{a^2-b^2}. \tag{1.6.9}$$

Let us now consider a system of Cartesian axes (*X*, *Y*) rotated by this angle α. There is no change in the radius of the two circular motions: their composition is then given by

$$\begin{aligned} E_X &= E_o\left(r_L+r_R\right)\cos\omega t' = A\cos\omega t', \\ E_Y &= E_o\left(r_L-r_R\right)\sin\omega t' = B\sin\omega t', \end{aligned} \tag{1.6.10}$$

where t' is the time taken from the instant of alignment in the direction of α. This

Fig. 1.9
Elliptical polarization

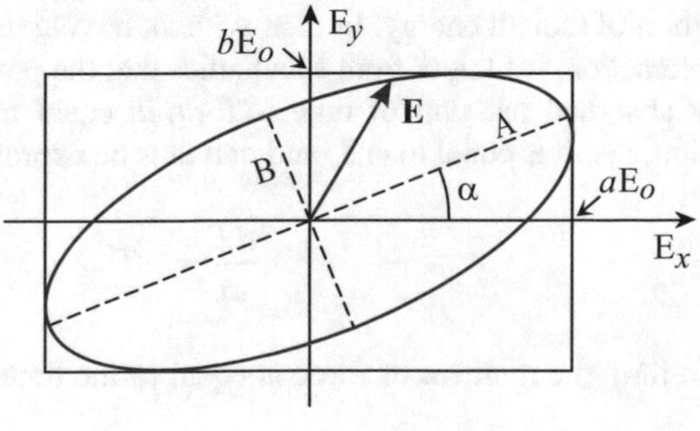

is an elliptical motion, where A and B are the lengths of the semi-axes (Fig. 1.9), In fact, from Eq. (1.6.10), we can get a bond between E_X and E_Y:

$$\left(\frac{E_X}{A}\right)^2+\left(\frac{E_Y}{B}\right)^2=1.$$

The semi-axes of the ellipse are then:

$$\begin{aligned} A&=\frac{1}{2}E_o\left(\sqrt{1+2ab\sin\delta}+\sqrt{1-2ab\sin\delta}\right),\\ B&=\frac{1}{2}E_o\left(\sqrt{1+2ab\sin\delta}-\sqrt{1-2ab\sin\delta}\right). \end{aligned} \tag{1.6.11}$$

It immediately occurs that $A^2 + B^2 = E_o^2$. Finally, the degree of ellipticity is expressed by the ratio B/A.

1.6.5 Angular momentum of radiation

We have already seen that, when the radiation is absorbed by a body, it gives it energy and momentum. With simple heuristic reasoning, we can see that, in general, there is also an angular momentum transferred from the field to the absorbent body, in other words, the wave can also carry an angular momentum. As an example, let us consider the case of a circularly polarized monochromatic wave: we can expect that the charged particles of the material (the electrons in particular) are put into a circular motion in response to the Lorentz force induced by the rotating field (and to which is added, to be more exact, the interaction with the spin). This motion has the same angular frequency ω of the wave and occurs in the same direction of rotation as the field, both for the negative charges and for the positive ones (although with a mutual phase shift). Therefore, overall, the action on the material (with this circular polarization) during a period of rotation, in addition to the thrust exerted by the impulse exchange, is constituted by the torsion, or the moment of force, $\mathcal{T}$, which involves a transfer of angular momentum associated with the absorption of radiant energy. In fact, without having to get into the details of dynamic interaction, we know from Mechanics that the power transferred, that is, the energy absorbed per unit of time, $d\mathcal{E}/dt$, is equal to the power generated by the torsion, i.e., it is equal to $\omega\mathcal{T}$, and can thus be expressed in the formula

$$\frac{d\mathcal{E}}{dt}=\omega\mathcal{T}.$$

In turn, the moment of force is equal to the time derivative of the angular momentum $\mathcal{L}$ transferred to the body, so that, averaged upon one rotation, we have

$$\frac{d\mathcal{E}}{dt} = \omega \frac{d\mathcal{L}}{dt},$$

from which it can be said that a charge that absorbs a given energy $\mathcal{E}$, simultaneously absorbs an angular momentum $\mathcal{L}$ such that

$$\mathcal{L} = \frac{\mathcal{E}}{\omega}. \tag{1.6.12}$$

As a vector, this angular momentum is oriented towards the direction of propagation for an **L**-wave and in the opposite direction for an **R**-wave.

As we have already mentioned in §1.5.2.3, according to the quantum description, the transfer of energy between matter and field occurs in quanta (photons), each with energy $\hbar$, so that, for Eq. (1.6.12), each photon has an intrinsic angular momentum, that is, a spin equal to $+\hbar$ for the **L**-waves and $-\hbar$ for the **R**-waves, taking the *direction of propagation* as *positive*. It is noteworthy that, contrary to what happens for the impulse, the angular momentum of a photon is *independent* of its frequency.

So far, we have considered the case of a circularly polarized wave. What happens instead for a linearly or elliptically polarized coherent wave? We have seen earlier that any polarization can be decomposed into the sum of two opposite circularly polarized waves. We are therefore tempted to say that a linearly polarized wave is composed of an equal number of photons with spin $+\hbar$ and $-\hbar$. However, from considerations of Quantum Mechanics (think, for example, of the photon emitted from an atom in a π transition, that is, with the angular momentum change $\Delta J = 0$), we must say, instead, that the state of each photon is the superposition with equal amplitudes of the two states of opposite spin.

Finally, let us give consideration to the names used to distinguish the various types of wave. From the point of view of an atom, in particular, in electric dipole transitions, the propagation verse of the wave matters little (that little is the impulse exchanged). What is important is the angular momentum of the photons in the atom reference system, in particular, the one established by selecting a quantization axis for the atom, which is possibly the most suitable for understanding what happens in quantum terms. Well, in this case, it is generally preferable to distinguish the waves according to their effect, that is, according to the type of transition that they induce, especially when there are several of them with different directions of propagation, and if the transition is uniquely determined by a selection rule on the angular momentum.

In particular, the waves that are related to transitions involving a change in absorption of $+\hbar$ or $-\hbar$ of the atomic angular momentum will be called waves σ_+ or σ_-, respectively. An example: in the case of an **L**-polarized beam that illuminates an atomic system (whose quantization axis is chosen parallel to the beam direction) and is then reflected back by a mirror, the reflection changes its helicity passing from **L** to **R**. However, returning to the same atomic system, the beam will continue to induce the same transition, for example, a transition σ_+, and therefore it can be initialed σ_+ for both directions.

It is not always possible to make an unambiguous assignment of a beam to a particular transition: if the quantization axis has not been chosen parallel to the direction of propagation of a given beam, this may induce, in general, both transitions σ and π. In this case, some caution is required in designating the beam.

1.6.6 Natural light

Now, we want to leave the narrow context of the monochromatic waves. Basically, the wave equations allow for the propagation of waves of arbitrary shape (although changing because of the dispersion). This arbitrariness is also present for the direction of oscillation of the field, while still complying with the transverse nature of the waves. In particular, by *natural light*, we mean a set of light waves, which present all of the possible vibration directions of the vector **E**. These directions occur rapidly in a disorderly manner, in such a way that the set is statistically symmetric around the direction of propagation.

Let us see how this light can be generated. An ordinary "natural" source consists of a very large number of atoms that emit without a mutual relationship of phase and direction. The radiation of each atom, in favorable conditions, such as in a rarefied gas, can keep the initial phase and orientation of the vibrations up to a typical time of 10^{-8} sec constant. The individual emissions overlap to give total wave that has an "instantaneous" polarization, which persists only for a time similar to that of the individual emission, and therefore is generally too short to be discriminated. In this way, what we have is the collection of all of the possible orientations for **E** and **H**, always orthogonal to each other however, which follow each other rapidly and which thus correspond to the natural light. Sometimes, there is some asymmetry, which requires the atoms to emit with some preferred direction of polarization, whereby the natural sources may also generate radiation with a certain degree of polarization. We can realize this by observing the blue light of the sky with Polaroid glasses on a beautiful, clear day. Other times, it is the crossed medium that imposes some polarization on the radiation, as with our Polaroid or the reflection on a surface.

In general, the radiation will be partially polarized, and the cases of natural and completely polarized light represent the extremes. Mathematically, a wave of this type can still be represented as the superposition of two polarized waves orthogonal to each other, but the two components are, in general, partially uncorrelated between them. In particular, for natural light, they have the same intensity and are completely incoherent. At the other extreme, it is important to remember that a strictly monochromatic wave is always polarized.

1.6.7 Coherency matrix and Stokes' parameters

Let us now quantitatively characterize the degree of polarization of a non-monochromatic plane wave that propagates along a z-axis. In particular, we will here examine the parameters introduced by G.G. Stokes in 1852, which take his name and are a function only of easily measurable quantities. For this purpose, we allow ourselves three experimental devices. The first is a *phase advancer* or *compensator*, an optical device that is able to introduce a variable phase delay between two orthogonal components in which the field can be decomposed, say, the components x and y. The second is an orientable *polarizer*, that is, an optical element capable of transmitting only the wave component that has the vibration plane (of the electric field) along a given direction (the transmission axis). Finally, the third is an intensity meter. Let us place these three elements in succession in front of our plane wave and prepare ourselves to measure the intensity $I(\theta,\phi)$ transmitted by the first two by varying the angle θ between the axis of the polarizer and the x-axis, and the phase ϕ of the compensator.

To ensure that the measures make sense, there are some conditions for the phase ϕ, which must be substantially constant within the frequency band of the wave to be measured. This depends on the device used as a compensator, which is usually a birefringent plate or a Fresnel's rhombus (we will study both in detail later). In particular, with an ordinary birefringent plate, which introduces a time delay between the x and y components, the phase shift varies in proportion to the frequency, thus the wave must have a relatively narrow band. An equivalent statement is that this delay is smaller than the *coherence time* of the wave. If, instead, we use a rephasing by total reflection, as with a Fresnel's rhombus of ordinary optical glass, the introduced phase will be practically constant over the entire visible band.

If we denote the analytic functions corresponding to the x and y components of the incident electric field by $\mathscr{E}_x$ and $\mathscr{E}_y$, the one corresponding to the field transmitted by the compensator and polarizer is given by

$$\mathscr{E}(t;\theta,\phi)=\mathscr{E}_x(t)\cos\theta+\mathscr{E}_y(t)e^{i\phi}\sin\theta\,, \tag{1.6.13}$$

thus, for Eqs. (1.3.37) and (1.5.23), the measured intensity is

$$\begin{aligned}I(\theta,\phi)&=\frac{c\varepsilon}{2n}\langle\mathscr{E}(\theta,\phi)\mathscr{E}^*(\theta,\phi)\rangle\\&=\frac{c\varepsilon}{2n}\Big[\langle\mathscr{E}_x\mathscr{E}_x^*\rangle\cos^2\theta+\langle\mathscr{E}_y\mathscr{E}_y^*\rangle\sin^2\theta\\&\quad+\langle\mathscr{E}_x\mathscr{E}_y^*\rangle e^{-i\phi}\cos\theta\sin\theta+\langle\mathscr{E}_x^*\mathscr{E}_y\rangle e^{i\phi}\cos\theta\sin\theta\Big]\,.\end{aligned} \tag{1.6.14}$$

The terms in parentheses may be arranged to form a matrix:

$$\mathfrak{J} = \begin{pmatrix} \langle \mathscr{E}_x \mathscr{E}_x^* \rangle & \langle \mathscr{E}_x \mathscr{E}_y^* \rangle \\ \langle \mathscr{E}_x^* \mathscr{E}_y \rangle & \langle \mathscr{E}_y \mathscr{E}_y^* \rangle \end{pmatrix}, \tag{1.6.15}$$

which is called a *coherency matrix*. Besides a coefficient, its diagonal terms ($\mathfrak{J}_{xx}$ and $\mathfrak{J}_{yy}$) represent the intensities of the components *x* and *y*, and thus the trace of $\mathfrak{J}$ is proportional to the total intensity of the wave. The non-diagonal elements ($\mathfrak{J}_{xy}$ and $\mathfrak{J}_{yx}$) represent the correlation between these components and they are generally complex; they are also conjugated between them, for which $\mathfrak{J}$ is *Hermitian*. Their modulus, normalized to the geometric mean of $\mathfrak{J}_{xx}$ and $\mathfrak{J}_{yy}$, is a measure of the degree of coherence and their phase is a measure of the actual phase difference between the *x* and *y* components of the field. Finally, applying the Schwarz inequality to the average value of the product of the fields, we have

$$\mathfrak{J}_{xx}\mathfrak{J}_{yy} - \mathfrak{J}_{xy}\mathfrak{J}_{yx} \geq 0, \tag{1.6.16}$$

that is, the determinant of $\mathfrak{J}$ is not negative. It can easily be shown that both the trace and the determinant of the coherency matrix are independent from the choice of the Cartesian axes *x* and *y*. Finally, Eq. (1.6.14) suggests the way to determine the $\mathfrak{J}$ elements with simple measures of intensity, appropriately selecting some combinations of θ and ϕ. Among the various possibilities, I quote here those relating to the determination of the Stokes' parameters, which give a representation of the polarization equivalent to that of the coherency matrix. These new parameters are

$$\begin{aligned} s_0 &= \frac{c\varepsilon}{2n}\left(\mathfrak{J}_{xx} + \mathfrak{J}_{yy}\right) = I(0°,0) + I(90°,0), \\ s_1 &= \frac{c\varepsilon}{2n}\left(\mathfrak{J}_{xx} - \mathfrak{J}_{yy}\right) = I(0°,0) - I(90°,0), \\ s_2 &= \frac{c\varepsilon}{2n}\left(\mathfrak{J}_{xy} + \mathfrak{J}_{yx}\right) = I(45°,0) - I(135°,0), \\ s_3 &= i\frac{c\varepsilon}{2n}\left(\mathfrak{J}_{xy} - \mathfrak{J}_{yx}\right) = I(45°,\pi/2) - I(135°,\pi/2). \end{aligned} \tag{1.6.17}$$

The first is the total wave intensity, while the others are differences in intensity, which are measured by pairs of mutually orthogonal polarizations, in the following order: between linear polarizations along the *x* and *y*-axes, between linear polarizations along axes at 45° from the first and between opposite circular polarizations. In principle, we only need four measures in all: the total intensity, the one that passes through a polarizer aligned along the *x*-axis, the one with the polarizer axis at 45°, and the one that has, in succession, a lamina λ/4 and a polarizer with the axis at 45° from those of the lamina (we will better study this combination later).

For the Stokes parameters, Eq. (1.6.16) becomes

$$s_0{}^2 \geq s_1{}^2 + s_2{}^2 + s_3{}^2 \,. \tag{1.6.18}$$

Let us now look at some special cases. For natural light, the intensity measured does not vary with θ and ϕ for which it must necessarily be

$$s_1 = s_2 = s_3 = 0, \quad \text{that is,} \quad \mathfrak{J}_{xx} = \mathfrak{J}_{yy} \ \text{ and } \mathfrak{J}_{xy} = \mathfrak{J}_{yx} = 0. \tag{1.6.19}$$

In the case of a monochromatic wave, for which

$$\mathscr{E}_x = a_x e^{i\phi_x - i\omega t} \,, \quad \mathscr{E}_y = a_y e^{i\phi_y - i\omega t} \,, \tag{1.6.20}$$

where a_x and a_y are two real amplitudes, while the phase is evidenced by ϕ_x and ϕ_y, the coherency matrix becomes

$$\mathfrak{J} = \begin{pmatrix} a_x{}^2 & a_x a_y e^{i\delta} \\ a_x a_y e^{-i\delta} & a_y{}^2 \end{pmatrix},$$

where $\delta = \phi_x - \phi_y$, from which it is clear that it has a zero determinant:

$$\mathfrak{J}_{xx}\mathfrak{J}_{yy} - \mathfrak{J}_{xy}\mathfrak{J}_{yx} = 0 \,. \tag{1.6.21}$$

It follows that, for a monochromatic wave, we have

$$s_1{}^2 + s_2{}^2 + s_3{}^2 = s_0{}^2 \,. \tag{1.6.22}$$

Therefore, the parameters s_1, s_2 and s_3 can be thought of as the coordinates of a point on a sphere of radius s_0 called *Poincaré's sphere*.

Even a non-monochromatic wave can satisfy Eqs. (1.6.21) and (1.6.22). In this case, the coefficients a and ϕ depend on time, but it is sufficient that the ratio of the amplitudes and the difference of the phases are independent of time.

Let us now look at an important aspect of the coherency matrix and Stokes parameters, which gives meaning to their use. Consider the superposition of *independent* waves, that is, uncorrelated between them, which propagate in the same direction. The elements of the overall coherency matrix are:

$$\mathfrak{J}_{kl} = \sum_n \left\langle \mathscr{E}_k^{(n)} \mathscr{E}_l^{(n)*} \right\rangle + \sum_{n \neq m} \left\langle \mathscr{E}_k^{(n)} \mathscr{E}_l^{(m)*} \right\rangle ,$$

where the summations are extended over all of the waves. If, therefore, the various waves are uncorrelated between them, the terms of the second sum are zero, and therefore the overall coherency matrix is given by the sum of the matrixes of each wave. The same result is also obtained for the Stokes parameters, which can therefore be considered as a vector and are often presented with the four values in columns, in the manner of the Jones vectors. This allows us to state that a partially polarized wave can be decomposed into two waves, one non-polarized and one polarized, mutually independent. Therefore, if $\boldsymbol{s} = (s_0, s_1, s_2, s_3)^{\mathrm{T}}$ is an overall Stokes vector, its decomposition $\boldsymbol{s} = \boldsymbol{s}_{\mathrm{n}} + \boldsymbol{s}_{\mathrm{p}}$ has the result:

$$\begin{aligned} \boldsymbol{s}_{\mathrm{n}} &= \left(s_0 - \sqrt{s_1^2 + s_2^2 + s_3^2},\ 0,\ 0,\ 0\right)^{\mathrm{T}}, \\ \boldsymbol{s}_{\mathrm{p}} &= \left(\sqrt{s_1^2 + s_2^2 + s_3^2},\ s_1,\ s_2,\ s_3\right)^{\mathrm{T}}, \end{aligned} \tag{1.6.23}$$

where $\boldsymbol{s}_{\mathrm{n}}$ and $\boldsymbol{s}_{\mathrm{p}}$ represent the unpolarized and the polarized part, respectively. One can define the degree of polarization p of the original wave as the ratio between the intensity of the polarized part with the total intensity:

$$p = \frac{I_{\mathrm{pol}}}{I_{\mathrm{tot}}} = \frac{\sqrt{s_1^2 + s_2^2 + s_3^2}}{s_0}. \tag{1.6.24}$$

A more extensive discussion of the coherency matrix and the Stokes parameters can be found in Born and Wolf (1980).

1.7 Reflection and refraction on plane interface

After propagation, the phenomena of refraction and reflection are the most important in optics, and their interpretation has always interested researchers. This is difficult material, especially if one wishes to bring it down to the electromagnetic microscopic level.[7] So, generally, we limit ourselves to a macroscopic description based on the boundary conditions on a regular interface between two media. I note only that the refraction, seen as a deviation of the direction of propagation, occurs even when the properties of the medium, like the refractive index, vary with continuity. Instead, the reflection requires a *discontinuity*, namely, a variation of these properties over short distances compared with the wavelength. In fact, the dipoles that form the media, put into oscillation by the wave, in turn radiate secondary waves in all directions. Still, these annihilate each other for destructive interfer-

[7] A microscopic interpretation of these phenomena occurs with the *extinction theorem of Ewald-Oseen* described by Born and Wolf (1980), page 100 et seq. [see also Mansuripur (2002), p. 168 et seq.]. They are still the subject of study and debate. A different approach, which originates with Fresnel, is given, for example, by Sivoukhine (1984), §68 and 69.

ence, except in the expected direction from the macroscopic equations, with the main effect of altering the propagation speed compared to that in vacuum. Similarly to the diffraction produced by the edge of an obstacle, the reflection is caused by a discontinuity that prevents this complete annihilation. If the interface discontinuity were instead distributed in a continuous manner on a thickness of a few wavelengths, the reflected wave would not be perceptible.

1.7.1 Laws of reflection and refraction for homogeneous dielectric media

Consider a harmonic plane wave incident on a plane surface of separation between two different dielectric media (Fig. 1.10). This wave is partly transmitted and partly reflected from this interface. So, let us try a self-consistent solution that consists of three plane waves described by complex equations:

$$\begin{cases} \boldsymbol{E}^{(j)} = \boldsymbol{E}_o^{(j)} e^{i\boldsymbol{k}_j \cdot \boldsymbol{r} - i\omega_j t}, \\ \boldsymbol{H}^{(j)} = \dfrac{1}{\mu\omega_j} \boldsymbol{k}_j \times \boldsymbol{E}^{(j)}, \end{cases} \tag{1.7.1}$$

with j = I, R, T indicating, respectively, the incident, reflected and transmitted fields. The second equation is derived from Eq. (1.4.18.a).

The three waves can be put in relation to each other once they are determined by imposing certain boundary conditions on the surface of discontinuity between the two media. The first of these conditions, also called *kinematics*, relates to the direction of the propagation vectors. Besides a constant amplitude term, the dependence on the space-time of the three waves is given by the exponential of Eq. (1.7.1.a). In order that there can be a relationship between the three waves, independent from time and from the position on the surface, which is supposedly separating two *homogeneous constant* media, it is necessary that the three phases be equal to each other at any time, and at any interface point, for which

$$\omega_I = \omega_R = \omega_T = \omega \tag{1.7.2}$$

and

$$\boldsymbol{k}_I \cdot (\boldsymbol{r} - \boldsymbol{r}_o) = \boldsymbol{k}_R \cdot (\boldsymbol{r} - \boldsymbol{r}_o) = \boldsymbol{k}_T \cdot (\boldsymbol{r} - \boldsymbol{r}_o), \tag{1.7.3}$$

where $\boldsymbol{r}$ and $\boldsymbol{r}_o$ indicate any two points on the interface itself. From this arbitrariness, it follows that the projections of the three vectors $\boldsymbol{k}_j$ on the interface plane are all equal, and therefore $\boldsymbol{k}_I$, $\boldsymbol{k}_R$ and $\boldsymbol{k}_T$ are *coplanar*. The plane on which these vectors lie is called the *plane of incidence*, and it is necessarily perpendicular to the interface plane. In particular, we can choose a Cartesian coordinate frame O*xyz*

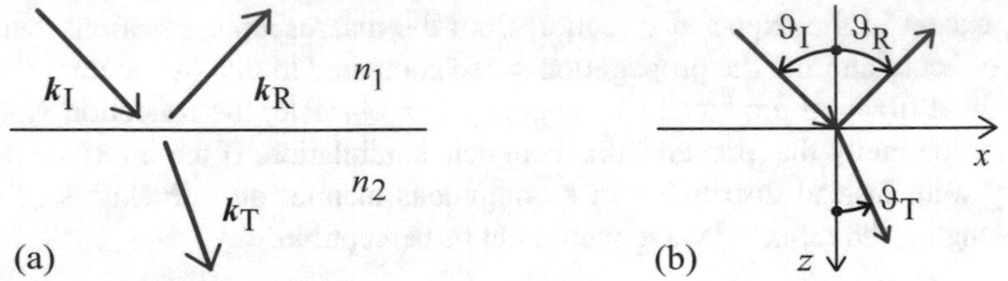

Fig. 1.10 Reflection and refraction

such that, for example, the *xy*-plane will coincide with the interface, and the *zx*-plane with that of incidence. With reference to Fig. 1.10 (b), in which the angles are taken to be positive in the direction indicated by the arrows, Eq. (1.7.3) becomes

$$k_{\mathrm{I}} \sin \vartheta_{\mathrm{I}} = k_{\mathrm{R}} \sin \vartheta_{\mathrm{R}} = k_{\mathrm{T}} \sin \vartheta_{\mathrm{T}}\,, \tag{1.7.4}$$

where the k_j are the various modules of the wave vectors and the ϑ_j are the angles of incidence that these vectors make with the normal to the surface. Both incident and reflected waves travel in the same medium (here, we consider it isotropic), for which $k_{\mathrm{I}} = k_{\mathrm{R}}$. The first of Eqs. (1.7.4) thus leads to the familiar laws of reflection:

$$\vartheta_{\mathrm{I}} = \vartheta_{\mathrm{R}}\,. \tag{1.7.5}$$

Instead, for the transmitted wave, by applying Eqs. (1.4.15), (1.7.2) and (1.7.4), we find the Snell's law:

$$\frac{k_{\mathrm{I}}}{k_{\mathrm{T}}} = \frac{n_{\mathrm{I}}}{n_{\mathrm{T}}} = \frac{\sin \vartheta_{\mathrm{T}}}{\sin \vartheta_{\mathrm{I}}}\,, \tag{1.7.6}$$

where n_{I} and n_{T} are the indices of refraction for the incident and transmitted waves, respectively.

1.7.2 Fresnel's formulas

Now, consider the amplitudes of the waves reflected and transmitted. As usual, we denote the components of the fields that are normal to the plane of incidence with ⊥ and the parallel components to the plane of incidence with ||, for which

$$\boldsymbol{E}_{\mathrm{o}} = \boldsymbol{E}_{\perp} + \boldsymbol{E}_{\|}, \quad \boldsymbol{H}_{\perp} = \frac{1}{\mu\omega} \boldsymbol{k} \times \boldsymbol{E}_{\|}, \quad \boldsymbol{H}_{\|} = \frac{1}{\mu\omega} \boldsymbol{k} \times \boldsymbol{E}_{\perp}. \tag{1.7.7}$$

The choice of the positive directions for the parallel components is shown in

Fig. 1.11
Conventions in choosing positive directions

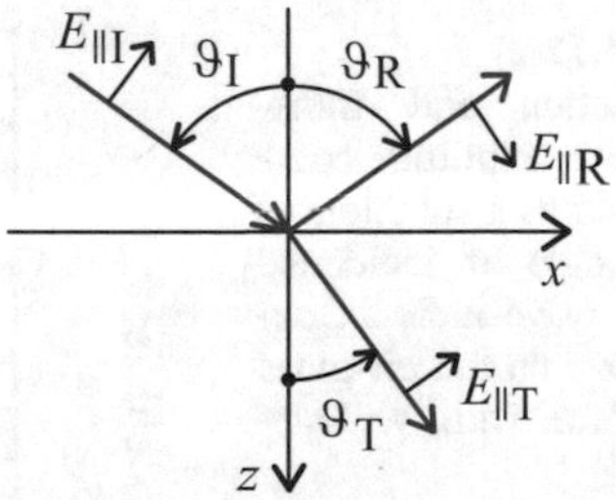

Fig. 1.11, while that for the normal components coincides with the positive direction of the y-axis. This option has the advantage that the positive directions for the amplitudes $E_\perp$ and $E_\|$ of the three waves are the same in the case of normal incidence, but not all authors use this same convention, for example, Born and Wolf (1980) do not. The components x, y, z of field for the three-wave are:

$$\begin{aligned} E_x^{(j)} &= E_\|^{(j)}\cos\vartheta_j; \quad & H_x^{(j)} &= s_j\frac{n_j}{\mu_j c}E_\perp^{(j)}\cos\vartheta_j \\ E_y^{(j)} &= E_\perp^{(j)}; \quad & H_y^{(j)} &= -s_j\frac{n_j}{\mu_j c}E_\|^{(j)} \quad \text{where } s_j = \begin{cases} -1 \text{ per } j\text{=I, T} \\ 1 \text{ per } j\text{=R} \end{cases} \\ E_z^{(j)} &= s_j E_\|^{(j)}\sin\vartheta_j; \quad & H_z^{(j)} &= \frac{n_j}{\mu_j c}E_\perp^{(j)}\sin\vartheta_j \end{aligned} \tag{1.7.8}$$

The boundary conditions discussed in §1.2.2 require that, through the interface, the tangential components of **E** and **H** are continuous:

$$\begin{aligned} E_x^{(\mathrm{I})} + E_x^{(\mathrm{R})} &= E_x^{(\mathrm{T})}, \quad & E_y^{(\mathrm{I})} + E_y^{(\mathrm{R})} &= E_y^{(\mathrm{T})}, \\ H_x^{(\mathrm{I})} + H_x^{(\mathrm{R})} &= H_x^{(\mathrm{T})}, \quad & H_y^{(\mathrm{I})} + H_y^{(\mathrm{R})} &= H_y^{(\mathrm{T})}, \end{aligned} \tag{1.7.9}$$

which, by applying to them Eqs. (1.7.8), constitute 4 linear equations for the four unknowns $E_\|^{(\mathrm{R})}$, $E_\perp^{(\mathrm{R})}$, $E_\|^{(\mathrm{T})}$, $E_\perp^{(\mathrm{T})}$, when $E_\|^{(\mathrm{I})}$ and $E_\perp^{(\mathrm{I})}$ are assigned. The boundary conditions for the components **D** and **B** normal to the interface are automatically fulfilled; this can be verified, remembering that $\mathbf{D} = \varepsilon\mathbf{E}$ and $\mathbf{B} = \mu\mathbf{H}$, by applying Snell's law and that of reflection [Guenther 1990].

Substituting Eqs. (1.7.8) in Eqs. (1.7.9), we have:

$$\cos\vartheta_\mathrm{I}\left(E_\|^{(\mathrm{I})} + E_\|^{(\mathrm{R})}\right) = E_\|^{(\mathrm{T})}\cos\vartheta_\mathrm{T}, \qquad E_\perp^{(\mathrm{I})} + E_\perp^{(\mathrm{R})} = E_\perp^{(\mathrm{T})},$$

$$\frac{n_\mathrm{I}}{\mu_\mathrm{I}}\left(E_\|^{(\mathrm{I})} - E_\|^{(\mathrm{R})}\right) = E_\|^{(\mathrm{T})}\frac{n_\mathrm{T}}{\mu_\mathrm{T}}, \qquad \frac{n_\mathrm{I}}{\mu_\mathrm{I}}\cos\vartheta_\mathrm{I}\left(E_\perp^{(\mathrm{I})} - E_\perp^{(\mathrm{R})}\right) = E_\perp^{(\mathrm{T})}\cos\vartheta_\mathrm{T}\frac{n_\mathrm{T}}{\mu_\mathrm{T}}.$$

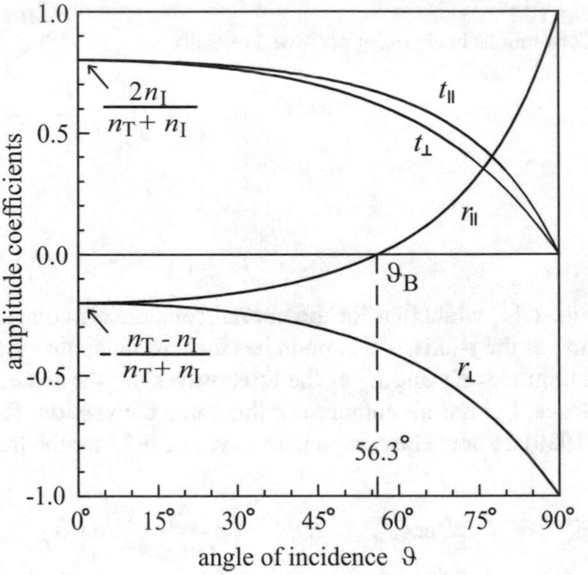

Fig. 1.12(a) Reflection and transmission amplitude coefficients as a function of the angle of incidence for a wave incident *externally* on the air-glass interface, with $n_T/n_I = 1.5$

The two equations at left contain only those components parallel to the incidence plane, while the other two contain only those perpendicular. Therefore, we can decompose the field into two types of wave that are independent of each other. The wave whose electric field oscillates parallel to the plane of incidence is generally indicated as a TM wave (transverse magnetic), or alternately as wave s, while the wave whose electric field oscillates perpendicular to the incidence plane is indicated with TE (transverse electric) or as wave p.[8]

Finally, the reflection and transmission coefficients for the two types of waves are:

TE waves:

$$r_\perp = \frac{E_\perp^{(R)}}{E_\perp^{(I)}} = \frac{\frac{n_I}{\mu_I}\cos\vartheta_I - \frac{n_T}{\mu_T}\cos\vartheta_T}{\frac{n_I}{\mu_I}\cos\vartheta_I + \frac{n_T}{\mu_T}\cos\vartheta_T}, \quad t_\perp = \frac{E_\perp^{(T)}}{E_\perp^{(I)}} = \frac{2\frac{n_I}{\mu_I}\cos\vartheta_I}{\frac{n_I}{\mu_I}\cos\vartheta_I + \frac{n_T}{\mu_T}\cos\vartheta_T}, \tag{7.10}$$

TM waves:

$$r_\parallel = \frac{E_\parallel^{(R)}}{E_\parallel^{(I)}} = -\frac{\frac{n_T}{\mu_T}\cos\vartheta_I - \frac{n_I}{\mu_I}\cos\vartheta_T}{\frac{n_T}{\mu_T}\cos\vartheta_I + \frac{n_I}{\mu_I}\cos\vartheta_T}, \quad t_\parallel = \frac{E_\parallel^{(T)}}{E_\parallel^{(I)}} = \frac{2\frac{n_I}{\mu_I}\cos\vartheta_I}{\frac{n_T}{\mu_T}\cos\vartheta_I + \frac{n_I}{\mu_I}\cos\vartheta_T}. \tag{1.7.11}$$

[8] The ones shown here are the most commonly used terms. The most explicit are TE and TM. Unfortunately, for further confusion, there are others: P = s = TE, S = p = TM.

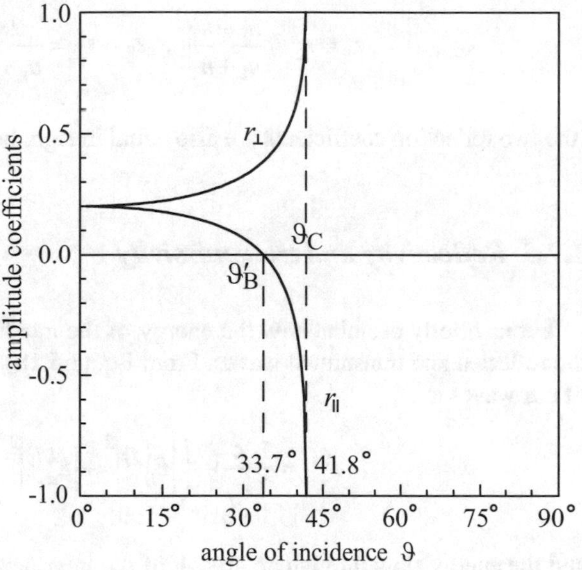

Fig. 1.12(b) Reflection amplitude coefficients r as a function of the angle of incidence for a wave *internally* incident on the air-glass interface, with $n_I/n_T = 1.5$. Beyond the critical angle ϑ_C, the r coefficients are complex and, even if they are of unitary modulus, cannot be represented in this graph

Note that, for the TE wave, one has $r_\perp + 1 = t_\perp$, regardless of the angle of incidence, ensuring the continuity of the tangential components of the field. Contrastingly, this relation is not true for the TM wave. With normal incidence, the distinction between the two types of wave falls and, in this case, we have $r + 1 = t$.

Applying the Snell's law, and limiting ourselves to consideration of the normal transparent dielectric media, for which $\mu_I = \mu_T = \mu_o$, Eqs. (1.7.10) and (1.7.11) can be written in a more compact form:

$$r_\perp = -\frac{\sin(\vartheta_I - \vartheta_T)}{\sin(\vartheta_I + \vartheta_T)}, \quad t_\perp = \frac{2\cos\vartheta_I \sin\vartheta_T}{\sin(\vartheta_I + \vartheta_T)},$$
$$r_\parallel = -\frac{\tan(\vartheta_I - \vartheta_T)}{\tan(\vartheta_I + \vartheta_T)}, \quad t_\parallel = \frac{2\cos\vartheta_I \sin\vartheta_T}{\sin(\vartheta_I + \vartheta_T)\cos(\vartheta_I - \vartheta_T)}. \tag{1.7.12}$$

The signs of these expressions determine the phase of the reflected and transmitted waves, however, they are arbitrary in the sense that they depend on the choice made for the positive directions of the fields (Fig. 1.11), and other authors report different signs. I remark that

with the choice made, the positive directions of the fields of the three waves are the same for $\vartheta \to 0$ (*de gustibus*!).

The tendency of these coefficients to vary the angle of incidence is shown in Fig. 1.12(a) and 1.12(b) in the two cases of external incidence, with $n_I < n_T$, and internal incidence, with $n_I > n_T$. In the case of normal incidence (and with $\mu_I = \mu_T = \mu_o$) from Eqs. (1.7.10) and (1.7.11), we immediately obtain that

$$r_{\parallel} = r_{\perp} = \frac{n_{\mathrm{I}} - n_{\mathrm{T}}}{n_{\mathrm{I}} + n_{\mathrm{T}}}\,, \quad t_{\parallel} = t_{\perp} = \frac{2n_{\mathrm{I}}}{n_{\mathrm{I}} + n_{\mathrm{T}}} \tag{1.7.13}$$

(the two reflection coefficients are also equal in sign, because of the choice made).

1.7.3 Reflectivity and transmissivity

Let us briefly examine how the energy of the incident field is divided between the reflected and transmitted waves. From Eq. (1.5.18), it follows that the intensity of our waves is

$$I^{(j)} = \frac{1}{2}\frac{c}{n_j}\varepsilon_j\left(\left|E_{\parallel}^{(j)}\right|^2 + \left|E_{\perp}^{(j)}\right|^2\right), \tag{1.7.14}$$

and the energy flow through an area A of the interface (i.e., the incident power P) is then

$$P^{(j)} = \frac{1}{2}\frac{c}{n_j}\varepsilon_j\left(\left|E_{\parallel}^{(j)}\right|^2 + \left|E_{\perp}^{(j)}\right|^2\right)A\cos\vartheta_j\,, \tag{1.7.15}$$

where it is to be noted that, together with the field modules, the properties of the medium and the angles of incidence are also involved. Thus, it is found that the *reflectivity R* and *transmissivity T*,[9] which are a different thing from the coefficients r and t obtained earlier for the fields, are (still for $\mu_{\mathrm{I}} = \mu_{\mathrm{T}} = \mu_{\mathrm{o}}$):

TE waves:

$$R_{\perp} = \frac{P_{\perp}^{(\mathrm{R})}}{P_{\perp}^{(\mathrm{I})}} = \frac{\sin^2(\vartheta_{\mathrm{I}} - \vartheta_{\mathrm{T}})}{\sin^2(\vartheta_{\mathrm{I}} + \vartheta_{\mathrm{T}})}, \quad T_{\perp} = \frac{P_{\perp}^{(\mathrm{T})}}{P_{\perp}^{(\mathrm{I})}} = \frac{\sin(2\vartheta_{\mathrm{I}})\sin(2\vartheta_{\mathrm{T}})}{\sin^2(\vartheta_{\mathrm{I}} + \vartheta_{\mathrm{T}})}, \tag{1.7.16}$$

TM waves:

$$R_{\parallel} = \frac{P_{\parallel}^{(\mathrm{R})}}{P_{\parallel}^{(\mathrm{I})}} = \frac{\tan^2(\vartheta_{\mathrm{I}} - \vartheta_{\mathrm{T}})}{\tan^2(\vartheta_{\mathrm{I}} + \vartheta_{\mathrm{T}})}, \quad T_{\parallel} = \frac{P_{\parallel}^{(\mathrm{T})}}{P_{\parallel}^{(\mathrm{I})}} = \frac{\sin(2\vartheta_{\mathrm{I}})\sin(2\vartheta_{\mathrm{T}})}{\sin^2(\vartheta_{\mathrm{I}} + \vartheta_{\mathrm{T}})\cos^2(\vartheta_{\mathrm{I}} - \vartheta_{\mathrm{T}})}. \tag{1.7.17}$$

[9] These terms are usually used to indicate the ratio between refracted or reflected power and incident power for the case of spatially coherent monochromatic waves. In radiometry, the terms *reflectance* and *transmittance* are used. These refer to the ratio between the flow of energy reflected or transmitted with the incident flux for generally incoherent radiation.

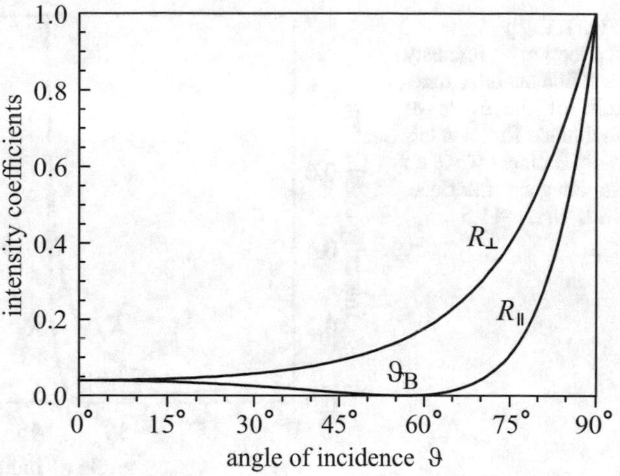

Fig. 1.13(a) Reflection intensity coefficients as a function of the angle of incidence for a wave incident externally on the air-glass interface, with $n_T/n_I = 1.5$

One can easily verify that $R_\perp + T_\perp = 1$, $R_\parallel + T_\parallel = 1$. In particular, in the case of normal incidence, we obtain (still for $\mu_I = \mu_T = \mu_o$)

$$R_\parallel = R_\perp = \frac{(n_I - n_T)^2}{(n_I + n_T)^2}\,, \quad T_\parallel = T_\perp = \frac{4 n_I n_T}{(n_I + n_T)^2}\,. \tag{1.7.18}$$

The trend of the reflectivity when varying the angle of incidence is drawn in Fig. 1.13, for the cases of external and internal incidence. It should be noted that, in this last case, the reflectivity varies very steeply at the critical angle.

1.7.4 Brewster's angle

As seen from Fig. 1.12, the reflection and transmission coefficients of Eqs. (1.7.12) are monotonic functions of the angle of incidence; however, $r_\parallel$ has the remarkable property of changing its sign by passing through zero. In other words, the reflectivity for the TM wave vanishes at a particular angle of incidence ϑ_B, called *Brewster's angle* (Fig. 1.13). Indeed, in the equation for $r_\parallel$, the denominator diverges to infinity when

$$\vartheta_B + \vartheta_T = \pi/2\,. \tag{1.7.19}$$

Applying Snell's law of, we find that

$$n_I \sin\vartheta_B = n_T \sin\left(\frac{\pi}{2} - \vartheta_B\right),$$

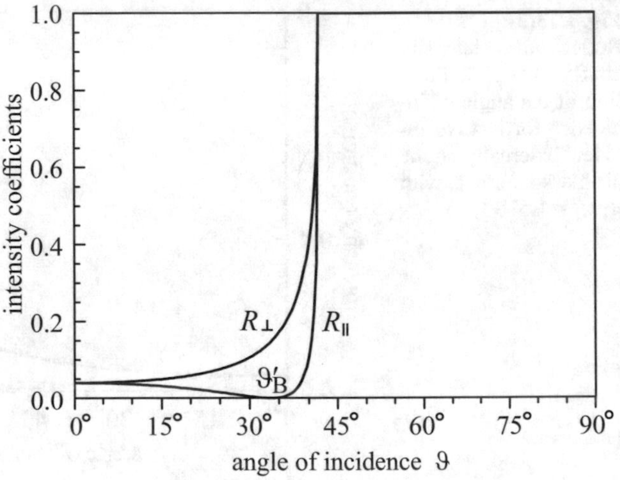

Fig. 1.13(b) Reflection intensity coefficients as a function of the angle of incidence for an internally incident wave on an air-glass interface, with $n_I/n_T = 1.5$

from which

$$\vartheta_B = \arctan\left(\frac{n_T}{n_I}\right). \tag{1.7.20}$$

This expression applies to both internal and external incidence: just swap the *value* of the refractive indices in the argument of the arc tangent. For example, for a normal glass with refractive index $n = 1.5$, the external Brewster's angle (in the air) is 56.3° and the inner one (in glass) is 33.7°. It should then be noted that these two angles are related by Snell's law, so that, if a beam of light passes through a glass plate with parallel faces with incidence at Brewster's angle, the TM component has no reflection on either the first or the second face of the plate. It is also to be noted that, when the ratio of the indices tends to 1, the Brewster's angle tends to 45°. When $\mu_I \neq \mu_T$, the situation is a bit more complicated, and we must use Eqs. (1.7.11), but the trend of the coefficients remains similar.

To this effect, one can give an intuitive interpretation, useful for easily remembering the oscillation direction of the electric field of the wave that is not reflected, and for recognizing the axes of a polarizer, by viewing a reflection through it from some non-metallic surface. Imagine sending a TM wave on a glass slide (Fig. 1.14). The dipoles of the glass are put in oscillation in the same direction in which the electric field of the transmitted wave oscillates, and, in turn, they radiate a wave (remember that **P** is the source of the electromagnetic field, and, at the microscopic level, this wave must be understood in the vacuum). For the discontinuity on the interface, part of the re-emitted energy appears in the form of the reflected wave. Furthermore, the amplitude of the reflected wave depends on the projection of the dipoles' oscillation in the direction perpendicular to that of its propagation. Well, this projection is canceled when $\vartheta_I + \vartheta_T = 90°$, for which the dipoles excited by the TM wave are not able to re-emit the reflected wave. This

Fig. 1.14
Brewster's angle

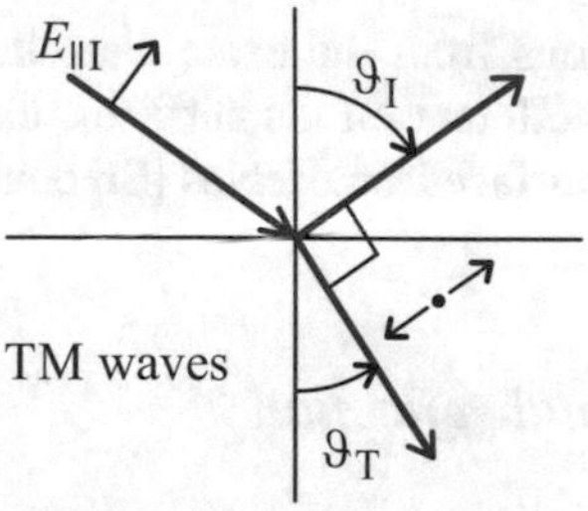

naïve reasoning becomes more cumbersome in the case of internal incidence and can even lead to an incorrect conclusion if the right premises are not taken. I leave these to the imagination of the readers

Apology of the Brewster's angle

The property of Brewster's angle of canceling the reflected TM wave has many applications. This is thanks primarily to the fact that it has a wide spectral band: the index *n* of the transparent common substances, in fact, varies very slowly with the wavelength. Therefore, a window placed at Brewster's angle often has a negligible reflectivity (for the TM wave) for the entire visible band. Moreover, such a window is cheaper and more robust than one made with an anti-reflective coating. For these reasons, windows and surfaces at the Brewster's angle are often used whenever the field is linearly polarized. Inside of a laser, the various surfaces at the Brewster's angle, all aligned with one another, are the ones that fix the polarization of the laser beam: indeed, for the TM wave the losses are negligible, while the TE wave has losses on the order of 15% for each surface with the index of the common optical glasses.

A pile of various parallel plates at the Brewster's angle are good polarizers, useful, in particular, in the infrared, where there was no alternative,[10] but we can use substances with a high refractive index for which the TE wave reflection becomes stronger.

Furthermore, in terms of intensity, Brewster's angle is not too critical, since the reflectivity varies quadratically with the angular deviation from the optimum value. In other words, there is a whole angular range in which the reflectivity for the TM wave remains small. This fact is widely used in scientific applications and makes life easier for experimenters, but is also used in everyday life, for example, with Polaroid sunglasses to mitigate road reflections: the glasses are worn with their axis of transmission vertical, and thus block most of the TE wave reflected from the street, despite its irregularities. Even the sea looks different and more beautiful when viewed with these glasses. It should be noted that the Brewster angle does not exist for metal surfaces: indeed, Polaroid sunglasses do not work with aluminated mirrors.

---

[10] Recently, polarizers for the infrared constituted by wires deposited on a transparent substrate, have been put on the market. They have good efficiency and an excellent extinction ratio.

Deviations from the exact cancellation of the reflection could happen due to layers of pollutant on the surfaces, but also for the microscopic properties of the dielectric surfaces themselves [Sivoukhine 1984, §70].

1.7.5 Total reflection

We return now to examining Snell's law, considering, in particular, the case in which the incident wave propagates in the denser medium, and thus is $n_I > n_T$. It therefore has

$$\frac{\sin\vartheta_T}{\sin\vartheta_I} = \frac{n_I}{n_T} > 1 ,$$

from which we see that, as ϑ_I increases, we always have $\vartheta_I < \vartheta_T$, until ϑ_T reaches the value of $\pi/2$ (90°) at an angle $\vartheta_I = \vartheta_c$, said *critical angle*, given by

$$\sin\vartheta_c = \frac{n_T}{n_I} . \tag{1.7.21}$$

What happens then to the transmitted wave when ϑ_I exceeds ϑ_c? Moreover, what happens to ϑ_T? Well, the solution of Snell's equation for ϑ_T now leads to complex values of the type

$$\vartheta_T = \frac{\pi}{2} - i\,\vartheta'_T \quad (\pm m\pi, \text{ with } m \text{ integer}) .$$

The generalization of the sine and cosine functions is indeed easily obtained using their decomposition in exponential. It thus occurs that

$$\begin{aligned} \sin\vartheta_T &= \sin\left(\frac{\pi}{2} - i\,\vartheta'_T\right) = \cosh\vartheta'_T , \\ \cos\vartheta_T &= \cos\left(\frac{\pi}{2} - i\,\vartheta'_T\right) = i\sinh\vartheta'_T . \end{aligned} \tag{1.7.22}$$

The first expression allows us to calculate the imaginary part of the angle ϑ_T:

$$\vartheta'_T = \cosh^{-1}\left(\frac{n_I}{n_T}\sin\vartheta_I\right) , \tag{1.7.23}$$

although we still do not really need it now. Eqs. (1.7.22) make us remember that the most general solution for a plane wave admits a complex wave vector $\boldsymbol{k}$, for which the wave is inhomogeneous: this consists of the transmitted wave. However, given that it deserves special attention, we postpone the detailed discussion to the next section.

In conclusion, the Fresnel formulas derived above are still valid for $\vartheta_I > \vartheta_c$, with the only warning being that $\cos\vartheta_T$ is now imaginary:

$$\cos\vartheta_T = i\sinh\vartheta'_T = i\sqrt{\left(\frac{n_I}{n_T}\sin\vartheta_I\right)^2 - 1}\,. \tag{1.7.24}$$

Finally, from Eqs. (1.7.10) and (1.7.11), it is seen that, for $\vartheta_I > \vartheta_c$, the reflection coefficients are complex, but with modulus equal to 1, for which *all of the energy* is reflected from the interface. Formally, this happens because, in these equations, the denominators are the complex conjugate of the numerators.

In particular, for $\mu_I = \mu_T = \mu_o$, we have:

TE waves:
$$r_\perp = \frac{\cos\vartheta_I - i\sqrt{\sin^2\vartheta_I - 1/n^2}}{\cos\vartheta_I + i\sqrt{\sin^2\vartheta_I - 1/n^2}}, \tag{1.7.25.a}$$

TM waves:
$$r_\parallel = -\frac{\cos\vartheta_I - i n^2\sqrt{\sin^2\vartheta_I - 1/n^2}}{\cos\vartheta_I + i n^2\sqrt{\sin^2\vartheta_I - 1/n^2}}, \tag{1.7.25.b}$$

where $n = n_I/n_T$. Putting Eqs. (7.1.25) in the form $r = e^{i\delta}$, we can derive the phase shifts $\delta_\perp$ and $\delta_\parallel$ for the two types of wave: just make the ratio between the real part and the imaginary part, for which, having taken that $r = (a - ib)/(a + ib)$, we have

$$\tan\frac{\delta}{2} = -\frac{b}{a},$$

from which:

$$\delta_\perp = -2\arctan\left(\frac{\sqrt{\sin^2\vartheta_I - 1/n^2}}{\cos\vartheta_I}\right), \quad \text{for the TE waves,} \tag{1.7.26.a}$$

$$\delta_\parallel = \pi - 2\arctan\left(n^2\frac{\sqrt{\sin^2\vartheta_I - 1/n^2}}{\cos\vartheta_I}\right), \quad \text{for the TM waves,} \tag{1.7.26.b}$$

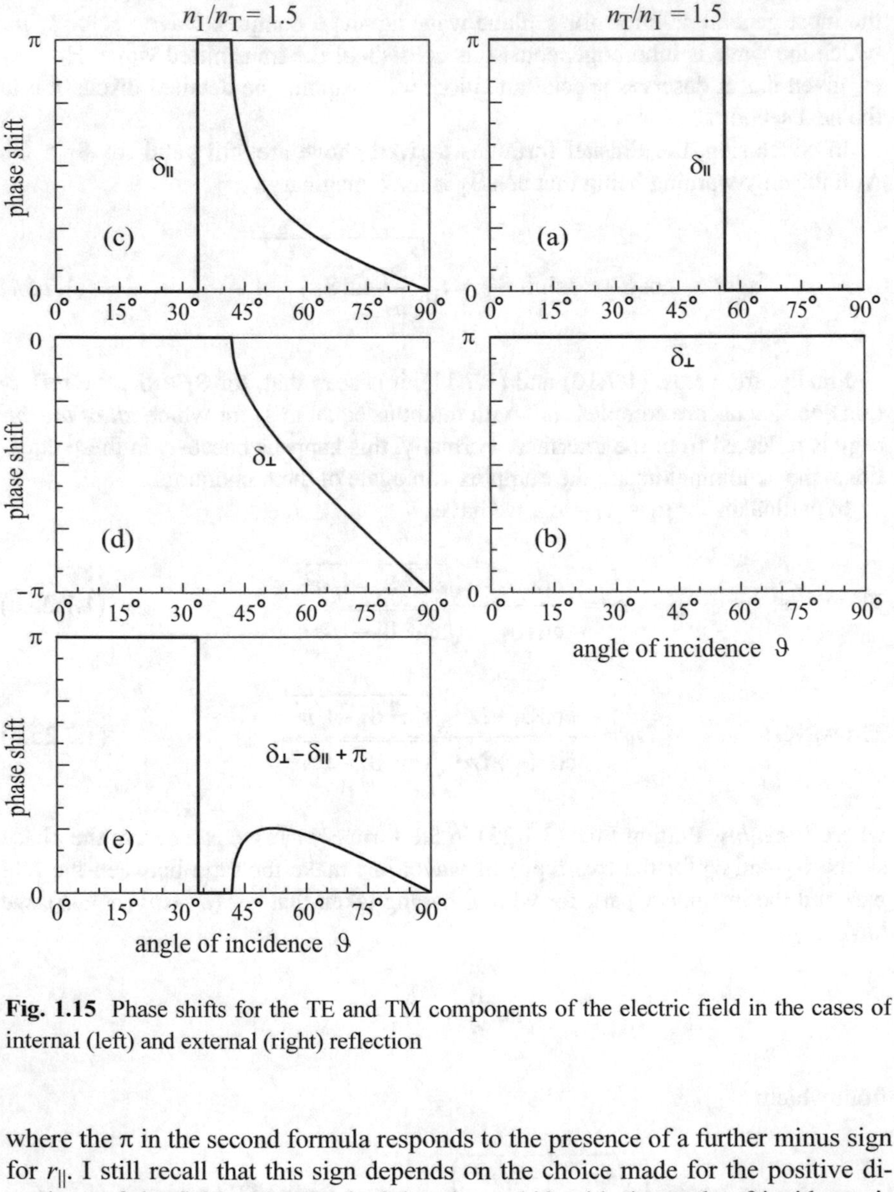

Fig. 1.15 Phase shifts for the TE and TM components of the electric field in the cases of internal (left) and external (right) reflection

where the π in the second formula responds to the presence of a further minus sign for $r_{\|}$. I still recall that this sign depends on the choice made for the positive directions of the fields. The trend of the phase shift with the angle of incidence is shown in Fig. 1.15 for both cases of external and internal reflection. For the TE and TM waves, the phase shifts varies differently with ϑ_I, and therefore, if the incident wave is linearly polarized, the reflected wave will be phase shifted between the two components, i.e., it will be generally elliptically polarized, with ellipticity that depends on the angle of incidence.

The phase shift due to reflection ($\delta = \delta_{\perp} - \delta_{\|}$) can be obtained from the prod-

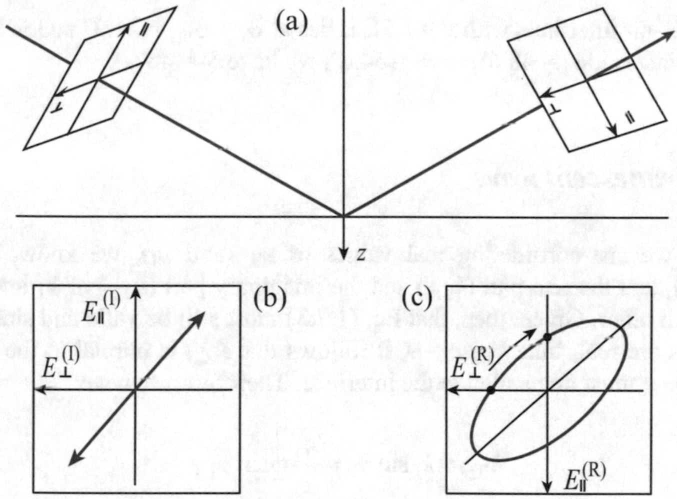

Fig. 1.16 Total reflection for an incident field with the linear polarization axis inclined: in the lower panes, the field is represented as it is viewed looking towards the wave source. The arrows for axis ⊥ and || follow the convention adopted in Fig. 1.11. Here, the reflection field is right elliptical. The opposite happens by reversing the initial inclination

uct $r_{\perp}r_{\parallel}^* = \exp[i(\delta_{\perp} - \delta_{\parallel})]$, from which, proceeding as before, we have

$$\delta = 2\arctan\left(\frac{\cos\vartheta_I\sqrt{\sin^2\vartheta_I - 1/n^2}}{\sin^2\vartheta_I}\right) - \pi\,. \tag{1.7.27}$$

For this type of problem, however, it is better to define a new variable: if we look at each wave, incident or reflected, toward the direction from which it came (Fig. 1.16), the phase shift that is measured is:

$$\delta' = \delta + \pi$$

(the term $-\pi$ disappears); the trend of δ' with ϑ_I is shown in Fig. 1.15(e). In particular, the phase shift δ' vanishes for $\vartheta_I = \vartheta_c$ and for $\vartheta_I = \pi/2$. In addition, we note that, for $\vartheta_I > \vartheta_c$, the argument of the arc tangent of Eq. (7.1.27) is positive, and therefore also δ' is positive: it follows that the TM wave is *advanced* compared to the TE wave, as shown in Fig. 1.16 (c).

The phase shift is maximum for $\cos\vartheta_{m\,I} = \sqrt{(n^2-1)/(n^2+1)}$ and is

$$\delta'_m = 2\arctan\left[\frac{1}{2}\left(n - \frac{1}{n}\right)\right].$$

For a glass-air interface, with $n = 1.51$, it lies at $\vartheta_I = \vartheta_{mI} = 51.3°$ and is $\delta'_m = 45.9$ °. When, instead, $\vartheta_I = 48.6°$, or $\vartheta_I = 54.6°$, we have $\delta' = 45°$.

1.7.6 Evanescent wave

Since we are considering real values of n_T (and n_I), we know, from Eq. (1.4.20.b), that the real part ($\boldsymbol{k}_{T\Re}$) and the imaginary part ($\boldsymbol{k}_{T\Im}$) of $\boldsymbol{k}_T$ are orthogonal to each other. Given, then, that Eq. (1.7.3) must still be valid and since its first two terms are real, thus $\boldsymbol{k}_{T\Im}\cdot\boldsymbol{r} = 0$, it follows that $\boldsymbol{k}_{T\Im}$ is normal to the interface, and that $\boldsymbol{k}_{T\Re}$ must be parallel to the interface. Therefore, we have:

$$k_{T\Re} = k_I \sin\vartheta_I = \frac{n_I}{c}\omega \sin\vartheta_I\,, \qquad (1.7.28)$$

while, from Eqs. (1.4.20.a), we have the imaginary part of $\boldsymbol{k}$:

$$k_{T\Im} = \frac{\omega}{c}\sqrt{n_I{}^2 \sin^2\vartheta_I - n_T{}^2}\,. \qquad (1.7.29)$$

Now, if we apply Snell's law and Eqs. (1.7.22), we finally have

$$\boldsymbol{k}_T = \frac{n_T\,\omega}{c}\left(\mathbf{x}\cosh\vartheta'_T + i\,\mathbf{z}\sinh\vartheta'_T\right). \qquad (1.7.30)$$

By comparison with Eq. (1.4.21) it is seen that ϑ'_T coincides with the parameter ϑ used there, while, in A and A' of Eqs. (1.4.22-23), we can identify the amplitudes $E_\parallel$ and $E_\perp$ respectively, choosing, as versors, $\mathbf{e}_3 = \mathbf{x}$, $\mathbf{e}_2 = \mathbf{z}$, $\mathbf{e}_1 = \mathbf{y}$. For the transmitted wave, Eqs. (1.7.8) coincide with Eqs. (1.4.22-23) as soon as we apply the substitution of Eqs. (1.7.22). So, we find that the amplitude of the transmitted wave decays exponentially moving away from the interface, with the law:

$$\boldsymbol{E}_z = \boldsymbol{E}_o e^{-k_{T\Im} z}\,,$$

where z is the distance from the surface. For this reason, this wave is called an *evanescent wave*. Its extension in space is well represented by the decay distance for the intensity, given by

$$d_e = \frac{1}{2k_{T\Im}} = \frac{\lambda_o}{4\pi}\Big/\sqrt{n_I^2 \sin^2\vartheta_I - n_T^2}\,. \qquad (1.7.31)$$

This distance grows by decreasing ϑ_I and is normally very small (a fraction of

Fig. 1.17
Total reflection in a right angle prism

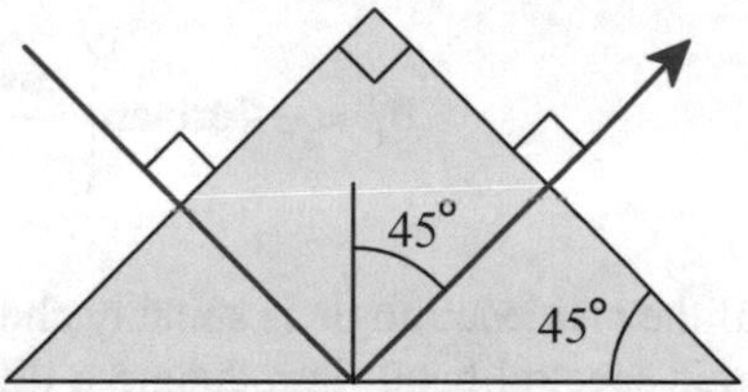

the wavelength λ_o in vacuum), except when ϑ_I approaches the critical angle, for which d_e indeed diverges to infinity. It must be remembered, however, that the solutions found here apply in the strict sense only for a plane wave, infinitely extended. The treatment is more complicated for a light beam of limited size.

1.7.7 Application of total internal reflection

Far from exhausting the applications of this phenomenon here, let us look at some significant examples of it. The first is the right angle prism, used as a mirror at 45° (Fig. 1.17). In fact, with a common optical glass with $n = 1.5$, the critical angle is already 41.8°; thus, with a 45° incidence angle, one has total reflection.

Compared to metallic or dielectric mirrors, to be used at 45°, the right angle prism has many advantages: it is a large spectral band, robust, relatively cheap and may have an antireflective treatment on the two faces of input and output at little expense. Thus, losses can be <1%. However, some care must be placed with respect to the polarization: in fact, if the prism is used with only TE or only TM (linear) polarization, there are no problems. Nevertheless, with an inclined linear polarization, or elliptical in general, it must be remembered that the phase shift that it introduces between the two reflected components TE and TM, that for $n = 1.5$, is, well, 36.9°! For an incoming light beam, linearly polarized in any direction, we can therefore have a strong ellipticity in the outbound beam, which, however, can be compensated apart.

A second interesting application, which exploits the phase shift produced by total reflection, is done with the *Fresnel's rhombus.* It is used to produce circularly polarized light from a linear polarization, by means of two total reflections (Fig. 1.18). In fact, the phase shift will sum constructively in both reflections and is

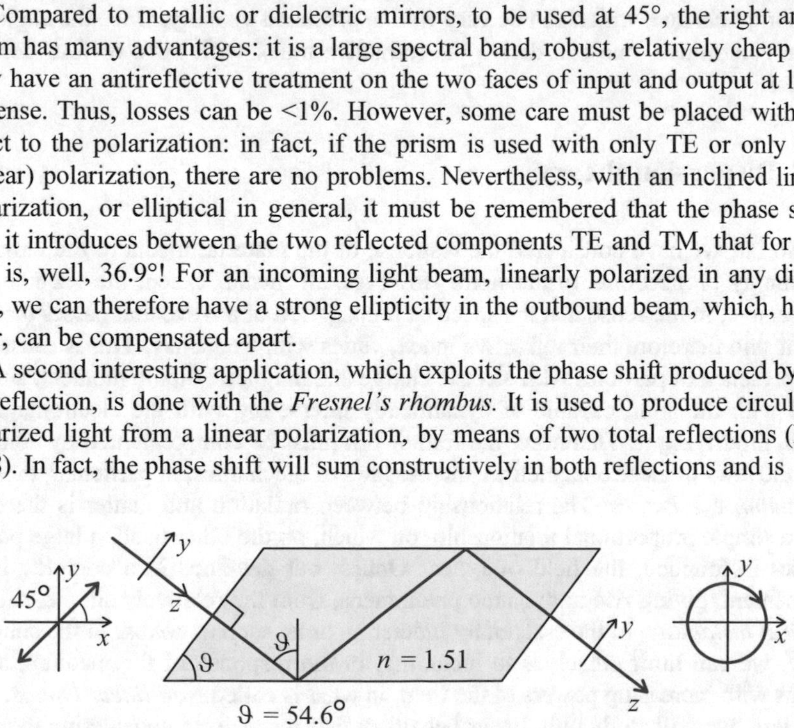

Fig. 1.18 Fresnel's rhombus

$$\delta_F = -\,4\arctan\left(\frac{\cos\vartheta_I\sqrt{\sin^2\vartheta_I - 1/n^2}}{\sin^2\vartheta_I}\right). \tag{1.7.32}$$

If the incidence angle is suitably chosen, this prism constitutes a $\lambda/4$ plate with a wide spectral band, introducing a difference in optical path of a quarter wavelength between the TE and TM components. Contrary to the birefringent plates (which we will study further on), its phase shift varies little with the wavelength, i.e., varies only because of the dependence of n on λ. In practice, with a common glass, the prism constitutes a good $\lambda/4$ plate for the entire visible spectrum. The right circular polarized light is obtained with a linear polarization input oriented at + 45° clockwise from the plane of incidence (first quadrant, looking toward the source): indeed, in this way, the amplitudes of the TE and TM waves are balanced. Vice versa, we have a left circular polarization if the input angle is – 45°.

With respect to the sensitivity to the angle of incidence, we note that, in the example of Fig. 1.18, a variation of $\pm 1°$ in the angle produces a variation in the phase shift of about $\mp 1°$ (one of $\pm 5° \to {}^{-7.6°}_{+1.2°}$).

A third major application is in the measurement of the refractive index, many refractometers are, in fact, based in the critical angle determination. From Fig. 1.14, it is seen that, around the critical angle, the internal reflection coefficient has an abrupt change, which can be very well appreciated visually. Once the angle has been determined, one can derive the refractive index with three or four decimal places.

1.8 Dispersion theory

So far, we have considered the response of the material media to the radiation as simply proportional to the field. However, all media, except the vacuum, are dispersive, in the sense that the speed of propagation of a wave that passes through them, and therefore their refractive index, varies with frequency. This is caused by the presence of particles with electric charge and magnetic dipole moment, associated with the spin, capable of dynamically interacting with the electromagnetic field, modifying it. Therefore, the framework must be complemented by combining the laws of electromagnetism and the laws of mechanics, in particular, those of *Quantum Mechanics*. The relationship between radiation and matter is therefore not a simple proportional relationship, on which, on the other hand, a large part of optics is founded, the field of Linear Optics, but develops in a complex intertwinement, giving rise to dynamic phenomena, from the relatively simple, such as *optical bistability*, to the essentially indeterminable, such as *chaos*. In the simplest case, we can limit ourselves to including, in the response of the medium, some terms with increasing powers of the field, in what is called *Non-linear Optics*.

Here, we will study only linear behaviors that result from considering the matter as being composed of damped harmonic oscillators, forced by the radiation.

Cauchy was the first to try to explain the dispersion: he proposed a development of the index of refraction n in a series of even powers of $1/\lambda_o$, where λ_o is the wavelength in vacuum, interpreting the optical properties of the medium as the mechanical ones of an elastic solid. Subsequently, in 1869, Maxwell suggested, in an examination subject, that atoms and molecules act as oscillators and have natural frequencies. Two years later, Sellmeier independently used this idea by deriving his eponymous equation $n^2 = 1 + A\lambda_o^2/(\lambda_o^2 - \lambda_r^2)$, where λ_r is the wavelength corresponding to a resonance of the medium. The classical theory of absorption and dispersion was then developed by H.A. Lorentz for dielectric media and by P.K.L. Drude for metals. Both treat the charges of the material as a damped oscillator with its own resonance frequency, equal to ω_o. While, in the Lorentz model, the electrons are considered to be tied to the nuclei by a harmonic restoring force, in the Drude model, they are considered as free, with a zero restoring force, which is equivalent to putting their resonance frequency at zero. A detailed review of the classical derivation of the macroscopic equations of Maxwell, from the microscopic equations describing the dynamics of particles and fields, was put forward by Van Kranendonk and Sipe (1977).

1.8.1 Dispersion and extinction in dilute media

A gas generally consists of neutral molecules in which the negative charge of the electrons compensates the positive one of the nuclei. Although these charges are in continuous movement according to the rules of Quantum Mechanics, classically, we can imagine the molecules as consisting of nuclei spaced apart from each other in positions of equilibrium and surrounded by a cloud of electrons. The charge center of this cloud does not necessarily coincide with that of the nuclei and, in this case, the molecule is called *polar* because it has a permanent electric dipole moment. In a non-polar molecule, these charge centers coincide, but an external electric field moves them away from their equilibrium position by inducing a dipole moment. We assume, for simplicity, that this moment is due to a single electron per molecule.[11] When the applied field is much smaller than the local fields, we can assume that the electron is subject to an isotropic harmonic potential described by an equation of classical dynamics,

$$\mathfrak{m}_e\left(\ddot{\boldsymbol{r}} + \Gamma\dot{\boldsymbol{r}} + \omega_o^2\boldsymbol{r} - \gamma\dddot{\boldsymbol{r}}\right) = \mathfrak{e}\mathbf{E}_{loc}\,, \tag{1.8.1}$$

where $\boldsymbol{r}$ is the displacement from the equilibrium point, $\mathfrak{m}_e$ and $\mathfrak{e}$ are respectively the mass and the charge of the electron, and $\mathfrak{m}_e\omega_o^2$ is the coefficient of the restor-

[11] Empirically, it appears that the number of electrons that contribute to the dispersion of a gas is, in good approximation, equal to the valence number of the molecules that constitute it [Sommerfeld 1949].

ing force. Γ is a damping coefficient that is introduced as a frictional force to take into account the energy loss due to collisions of the molecules with each other, or by the interaction with phonons in a solid medium. $\mathbf{E}_{loc}$ is the *local* electric field applied, which does not coincide with the wave field, since the molecules are also subject to the field generated by their closer companions in response to the wave field. Lastly,

$$\gamma = \frac{e^2}{6\pi\varepsilon_0 m_e c^3} \tag{1.8.2}$$

is the constant of radiative damping for the radiation produced by the accelerated charge, which, in quantum terms, corresponds to the *spontaneous emission* of a photon by *Rayleigh scattering*, when the atom or molecule returns to its initial state, or by *Raman scattering*, when the decay happens in a different state. The frequency of the Rayleigh scattering is centered on that of the incident wave, with a spectrum broadened by the Doppler effect. That of the Raman scattering is also shifted for the energy difference between the initial and final states. The emitted radiation is also called *fluorescence* when the incident wave has a frequency close to a resonance. It is particularly intense and, in the presence of collisions that destroy the coherence with the incident wave, its spectrum is centered on the resonance.

In Eq. (1.8.1), it is also assumed that the potential is isotropic, while, in general, the molecules do not have a symmetry such as to render them optically isotropic. However, if their orientations are random, as in a fluid, the medium is isotropic and the approximation made is sufficiently accurate. In these reasonings, the contribution of the nuclei has not been considered, but, in the infrared region, the vibration motion of the nuclei in molecules is dominant, together with that of rotation; however, we can expect that the behavior is similar by replacing e and m_e, respectively, with the nuclear charge and the nuclear mass. Here, the magnetic field is not considered, for simplicity; however, it intervenes both in the term $e\mathbf{v}\times\mathbf{B}$ of the Lorentz force and in the interaction with the magnetic dipoles associated with the spin of the charges. Now, we consider only the interaction of the electric dipole, generally prevailing in the infrared, visible, and ultraviolet regions, but other forms of interaction are allowed, such as those of the electric quadrupole and the magnetic dipole: the latter becomes relevant in the microwave and far-infrared regions. Even these interactions can be treated in terms of a harmonic potential.

When the electric field is that of a monochromatic wave with angular frequency ω and a much larger wavelength than the molecular size, the solution of forced oscillation from Eq. (1.8.1), with $\mathbf{E}_{loc}=\Re e[\boldsymbol{E}_{oloc}\exp(-i\omega t)]$, is $\boldsymbol{r}=\Re e[\boldsymbol{r}_o\exp(-i\omega t)]$, where

$$\boldsymbol{r}_o = \frac{e}{m_e}\frac{1}{\omega_o^2-\omega^2-i\Gamma\omega-i\gamma\omega^3}\boldsymbol{E}_{oloc}\,. \tag{1.8.3}$$

The induced polarization is therefore

$$\boldsymbol{P}_{\mathrm{o}} = N e \boldsymbol{r}_{\mathrm{o}} = N \varepsilon_{\mathrm{o}} \alpha \boldsymbol{E}_{\mathrm{oloc}}, \tag{1.8.4}$$

where N is the *density* of the charges and α is the *polarizability*[12] of the molecules described by the *Lorentz equation*

$$\alpha = \frac{e^2}{\varepsilon_{\mathrm{o}} m_e} \frac{1}{\omega_{\mathrm{o}}{}^2 - \omega^2 - i\Gamma\omega - i\gamma\omega^3}. \tag{1.8.5}$$

The gas structure is such, however, that its atoms or molecules possess countless resonance frequencies whose interpretation is possible only by means of Quantum Mechanics. The quantum theory of susceptibility is well exposed by Boyd (2008), who, in particular, for a rarefied isotropic medium in which the molecules lie in their ground state g, reports that

$$\alpha(\omega) = \sum_u \frac{1}{3\varepsilon_{\mathrm{o}}\hbar} \left|\mu_{ug}\right|^2 \left[\frac{1}{\omega_{ug} - \omega - i\gamma_{ug}} + \frac{1}{\omega_{ug} + \omega + i\gamma_{ug}}\right], \tag{1.8.6}$$

where, for simplicity, the case in which the state g is non-degenerate, with zero angular momentum was taken. The summation is conducted on all of the higher states. $|\mu_{ug}|^2$ is the square modulus of the dipole element between the states u and g, $\hbar\omega_{ug}$ is the energy of the state u with respect to g, $\hbar$ is the Planck's constant, and γ_{ug} is the constant of *dephasing*, i.e., of the coherence decay between these states, either due to spontaneous decay or to other *homogeneous* broadening processes, such as collisions. Here, the inhomogeneous broadening, such as that due to the Doppler effect, is not yet considered. The two terms in square brackets correspond to the two Feynman diagrams for the linear susceptibility in the interaction of the field with the atomic system, between the g and u levels [Delone and Krainov 1988]. For positive frequencies, only the first term can be resonant, because, since g is the ground state, ω_{ug} is positive for all u. The second term is called anti-resonant and is often overlooked. In proximity of a resonance, with $\omega \approx \omega_{ug}$, the polarizability can then be approximated by

$$\alpha(\omega) = \alpha_{\mathrm{o}} + \frac{\left|\mu_{ug}\right|^2}{3\varepsilon_{\mathrm{o}}\hbar} \frac{\omega_{ug} - \omega + i\gamma_{ug}}{\left(\omega_{ug} - \omega\right)^2 + \gamma_{ug}{}^2}, \tag{1.8.7}$$

---

[12] Here, I used Feynman's definition [Feynman 1969], for which $\chi_e = N\alpha$. Hopf and Stegeman (1985) refer to polarizability as that which we here call the quantity $\varepsilon_{\mathrm{o}}\alpha$.

where α_o is a background value due to all other states. The imaginary part of this expression is formally a Lorentzian function, while the real part is a dispersive function. Both are represented in Fig. 1.1 (b), where the susceptibility has been implicitly taken as proportional to the polarizability.

The γ values are typically small compared to the resonance frequencies, for which the spectrum of molecules appears as a series of relatively narrow absorption lines spaced apart from each other. Therefore, adding the two terms in the square bracket of Eq. (1.8.6) and neglecting the term in γ^2, we get

$$\alpha(\omega) = \frac{e^2}{\varepsilon_0 m_e} \sum_u \frac{f_{ug}}{\omega_{ug}{}^2 - \omega^2 - 2i\omega\gamma_{ug}}, \tag{1.8.8}$$

where

$$f_{ug} = \frac{2 m_e \omega_{ug} \left|\mu_{ug}\right|^2}{3\hbar e^2} \tag{1.8.9}$$

is the *oscillator strength* between the levels g and u. For these dimensionless coefficients, it is worth the sum rule

$$\sum_u f_{ug} = 1 \tag{1.8.10}$$

and, g being the fundamental state, for which $\omega_{ug} > 0$, their value is bounded between 0 and 1. The dephasing constant γ_{ug} is typically given[13] by two contributions

$$\gamma_{ug} = \frac{1}{2}\gamma_{\|ug}^{\text{sp.}} + \gamma_{ug}^{\text{col.}}, \tag{1.8.11}$$

where the second term is due to collisions, while

$$\gamma_{\|ug}^{\text{sp.}} = \frac{\omega_{ug}{}^3 \left|\mu_{ug}\right|^2}{3\pi\varepsilon_0 \hbar c^3} = \frac{e^2 \omega_{ug}{}^2}{2 m_e \pi \varepsilon_0 c^3} f_{ug} \tag{1.8.12}$$

is the coefficient A of Einstein for the spontaneous emission. With these substitutions, the denominator of the fractions in Eq. (1.8.8) is formally equal to that of the Lorentz classical equation for the polarizability.

The polarizability, and thus also the susceptibility $\chi_e = N\alpha$, and the permeability

[13] Another cause of population decay is given by *quenching*, in which the energy of the excited state is subtracted for collision.

$$\varepsilon = \varepsilon_0 (1 + N\alpha) \qquad (1.8.13)$$

are now complex variables. In analogy to Eq. (1.4.16), we can define a complex refractive index given by

$$\tilde{n} = \sqrt{\varepsilon\mu / \varepsilon_0\mu_0} = c\sqrt{\varepsilon\mu}\,, \qquad (1.8.14)$$

which is made explicit in the real and imaginary part as

$$\tilde{n} = n + i\kappa\,, \qquad (1.8.15)$$

where n is the real index of refraction and κ is called the extinction index.

For a rarefied gas, still assuming $\mu = \mu_0$, we can take $\boldsymbol{E}_{\text{loc}} = \boldsymbol{E}$ and, for the complex refractive index, we have

$$\tilde{n}^2 - 1 = \chi\,. \qquad (1.8.16)$$

In conditions for which $\chi << 1$, we can make the approximation $\tilde{n}^2 - 1 = (\tilde{n} - 1)(\tilde{n} + 1) \cong 2(\tilde{n} - 1)$ and, from Eq. (1.8.8), we then have

$$\tilde{n}(\omega) = 1 + \frac{Ne^2}{2\varepsilon_0 m_e} \sum_u \frac{f_{ug}}{\omega_{ug}{}^2 - \omega^2 - 2i\omega\gamma_{ug}}\,. \qquad (1.8.17)$$

For the real and imaginary parts of $\tilde{n}$, we have

$$n(\omega) = 1 + \frac{Ne^2}{2\varepsilon_0 m_e} \sum_u \frac{f_{ug}\left(\omega_{ug}{}^2 - \omega^2\right)}{\left(\omega_{ug}{}^2 - \omega^2\right)^2 + 4\omega^2\gamma_{ug}{}^2}\,, \qquad (1.8.18.a)$$

$$\kappa(\omega) = \frac{Ne^2}{\varepsilon_0 m_e} \sum_u \frac{f_{ug}\,\omega\gamma_{ug}}{\left(\omega_{ug}{}^2 - \omega^2\right)^2 + 4\omega^2\gamma_{ug}{}^2}\,. \qquad (1.8.18.b)$$

Since γ is generally small compared to ω, Eqs. (1.8.18) are not inconsistent with Eq. (1.8.7), since, in the vicinity of a resonance with the level u, we have $\omega_{ug}{}^2 - \omega^2 \cong 2\omega(\omega_{ug} - \omega)$. Thus, despite the abundance of squares into these expressions, the trend of κ is that of a sum of Lorentzian, each with half-width at half maximum (FWHM) equal to γ_{ug}. Meanwhile, the trend of n is that of a sum of dispersive curves, with a typical oscillatory behavior, with a minimum at $\omega_n = \omega_{ug} + \gamma_{ug}$ and a maximum at $\omega_x = \omega_{ug} - \gamma_{ug}$. Between these two values, over the resonance, the refractive index decreases with the frequency, and therefore increases with the wavelength. This variation is called *anomalous dispersion*, while, on the tails, outside of these values, the opposite occurs. Therefore, between two successive resonances, the refractive index decreases with wavelength and its variation is called *normal dispersion*.

Recalling that the meaning of $\tilde{n}$ is to determine the propagation of a monochromatic plane wave according to the rule

$$E_\omega(\boldsymbol{r},t) = E_o e^{ikz-i\omega t} = E_o e^{i\frac{\omega\tilde{n}}{c}z-i\omega t} = E_o e^{i\frac{\omega n}{c}z-i\omega t-\frac{\omega\kappa}{c}z},$$

where, for simplicity, the direction of $\boldsymbol{k}$ is taken as being parallel to the z-axis, the Lorentzian of $\kappa(\omega)$ corresponds to absorption lines of the medium. For many molecules, the spectrum consists of a series of lines in the far infrared, due to the degrees of freedom of rotation, a series in the infrared for those of vibration and a series in the visible and ultraviolet for those that are electronic. The big transparency of air and water in the visible region of the spectrum is precisely due to the interval left free between the ro-vibrational transitions in the infrared and those that are electronic in the ultraviolet.

In a gas, we must also consider the *inhomogeneous broadening* for the Doppler effect caused by the motion of translation of the molecules. Whereby, near the resonance with the level u, instead of Eqs. (1.8.18), we have the convolution with the Gaussian distribution of the velocity v_z in the z direction of the wave,

$$\kappa(\omega) \cong \kappa_o + \frac{Ne^2}{\varepsilon_0 m_e}\frac{f_{ug}}{4\omega_{ug}}\frac{1}{\sigma\sqrt{\pi}}\int_{-\infty}^{\infty}\frac{\gamma_{ug}e^{-v_z^2/\sigma^2}}{\left[\omega_{ug}-\omega\left(1-n_o v_z/c\right)\right]^2+\gamma_{ug}^2}dv_z, \quad (1.8.19)$$

where κ_o and n_o are the background values left by other resonances and σ is a parameter for the Doppler broadening. The line profile that is obtained by the convolution (1.8.19) between a Lorentzian and a Gaussian is called a *Voigt*.

In spectral regions far from the resonances, κ is negligible, for which we have

$$n(\lambda_o)^2 - 1 = \sum_j \frac{A_j\lambda_o^2}{\lambda_o^2-\lambda_{oj}^2}, \quad (1.8.20)$$

where the refractive index is expressed in a function of the wavelength in vacuum, $\lambda_o = 2\pi c/\omega$, and λ_{oj} are the wavelengths of the resonances. Eq. (1.8.20) is called *Sellmeier's equation*, and it is particularly effective in describing the trend of the index of refraction of many substances, including those that are solid. This is true for two reasons: the equation is based on realistic physical principles, and its A_j and λ_{oj} coefficients are determined experimentally. A tabulation of these equations is given in the *Handbook of Optics* [Tropf et al 1995] for many substances.

1.8.2 Dispersion and extinction in dense media

It is now necessary to specify the *local field* that intervenes in these equations [Hopf and Stegeman 1985]. Unless we consider isolated molecules in vacuum, it is different from the *average* wave field, due to the neighboring molecules to that considered. This distinction becomes significant when the refractive index is significantly different from 1, either in the vicinity of resonances, or in high-pressure gas, liquids or solids. The calculation of this local field is, in general, difficult, since it involves the microscopic details of the arrangement of the atoms. Its treatment would therefore require use of the microscopic Maxwell's equations for the fields $\boldsymbol{E}$ and $\boldsymbol{B}$ in a vacuum in the presence of charges consisting of electrons and nuclei. However, for isotropic media, we can evaluate it with a reasoning that microscopically treats only a spherical region of space, which includes a single molecule of the medium, while using the macroscopic equations for the medium in the surrounding space. From this approximation, it follows that [Born and Wolf 1980]

$$\boldsymbol{E}_{loc} = \boldsymbol{E} + \frac{\boldsymbol{P}}{3\varepsilon_o}. \tag{1.8.21}$$

For anisotropic media, this relationship can be corrected by adding a further structure factor [Van Kranendonk and Sipe 1977].

Assuming the constitutive equation

$$\boldsymbol{P} = \varepsilon_o N\alpha \boldsymbol{E}_{loc} \tag{1.8.22}$$

and substituting $\boldsymbol{E}_{loc}$ from Eq. (1.8.21), we get

$$\boldsymbol{P} = \varepsilon_o \chi \boldsymbol{E} = \varepsilon_o \frac{3N\alpha}{3 - N\alpha} \boldsymbol{E}. \tag{1.8.23}$$

The susceptibility diverges for values of $\Re e(N\alpha) \to 3$, $\Im m(N\alpha) \to 0$, while, for $N\alpha \to \infty$ in any direction in the complex plane, $\chi \to -3$. With $\mu = \mu_o$, the refractive index is given by

$$\tilde{n} = \sqrt{1+\chi} = \sqrt{1 + \frac{3N\alpha}{3 - N\alpha}}. \tag{1.8.24}$$

Fig. 1.19 depicts, as an example, the case in which the summation in Eq. (1.8.8) is limited to a single resonance ω_r, for which

$$N\alpha(\omega) = \frac{{\omega_p}^2}{{\omega_r}^2 - \omega^2 - 2i\omega\gamma}, \tag{1.8.25}$$

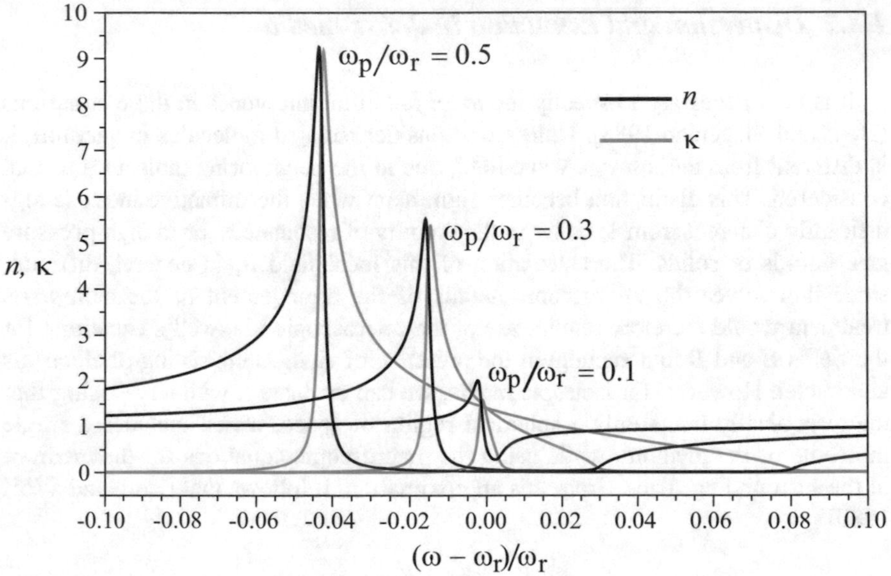

Fig. 1.19 Complex refractive index (real part in black and imaginary part in gray) from Eqs. (1.8.24,25) for some values of ω_p/ω_r, with $\gamma = 0.001\omega_r$

where ω_p is a coefficient called a *plasma frequency* and is proportional to the square root of N. The peak to the left for the two larger values of ω_p/ω_r coincides with $\Re e(N\alpha) - 3 = 0$ in the denominator of Eq. (1.8.24) while in resonance $\kappa \rightarrow \sqrt{2}$. The imaginary part κ of the refractive index describes the extinction rate of the wave according to the law $\exp(-2\pi\kappa z/\lambda_o)$, thus the propagation extends only over distances on the order of the wavelength for values κ of the order of unity. With increasing ω_p, the extinction band of the medium expands, and at the same time, increases the dispersion. The transparency of many substances, despite their dispersion, is relevant, because the value of γ is generally very small compared to the resonance frequency. Therefore, κ decays rapidly outside of the anomalous region, even if this is very extensive, either for the density of the oscillators, or for their inhomogeneous broadening, or for the simultaneous presence of different oscillators.

By solving Eq. (1.8.24) for α, we find that

$$\alpha = \frac{3}{N}\frac{\tilde{n}^2 - 1}{\tilde{n}^2 + 2}. \tag{1.8.26}$$

which is called the *Lorentz-Lorenz equation*.[14] Experimentally, it appears that, far

[14] In addition to H.A. Lorentz, who obtained it from the electromagnetic theory, this law was also found independently a few months earlier by L. Lorenz in Copenhagen, from the theory of elastic media.

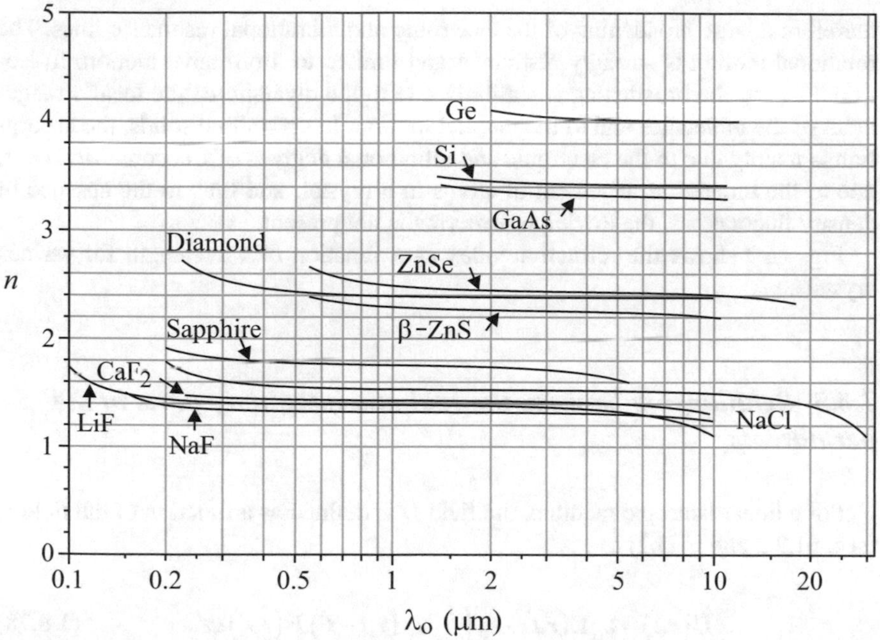

Fig. 1.20 Index of refraction for various crystalline substances. The curves are calculated from the Sellmeier's formulas given in *Handbook of Optics* [Tropf et al 1995]

from the resonance, the polarizability is almost independent of the density, for which this law allows us to extend the estimate of the index of refraction to the high-pressure gas and to liquids with good accuracy, once we know the intrinsic polarizability of the molecule. This law also extends to the determination of the index of refraction of mixtures of several substances, each with density N_1, N_2, N_3, ... and polarizability $\alpha_1, \alpha_2, \alpha_3$, ..., for which, in the transparency regions where κ is negligible,

$$3\frac{n^2-1}{n^2+2} = N_1\alpha_1 + N_2\alpha_2 + N_3\alpha_3 + \cdots. \qquad (1.8.27)$$

It should be specified that the correction for the local field is already incorporated into all of the measured and tabulated values, as the susceptibility and the refractive index, which are used to describe the behavior of the media of optical interest. Therefore, in practice, the induced polarization is obtained directly by multiplying the tabulated susceptibility for the field $\boldsymbol{E}$ of the wave, and we need not worry about the local field.

As regards the absorption lines, the situation in liquids and solids is more complicated than in gases, since each molecule is more subject to the field generated by those nearby. In liquids, the local disposition is different from molecule to molecule and changes rapidly for the thermal motion. This causes scattering, and therefore a large broadening of the electronic and vibrational resonance lines. The rotational motion is strongly restrained and limited to a diffusive motion. In isotropic solids, the broadening is similarly due to the diversity of the local arrangement of the molecules and to the thermal motion. In crystalline solids, the absorption is mainly due to the electronic and vibrational degrees of freedom. Moreover, due to the ordered arrangement of atoms in a crystal, and thus to the absence of density fluctuations, the Rayleigh scattering is not present.

Fig. 1.20 shows the refractive index as a function of wavelength for various crystals.

1.8.3 Relationship between the real and imaginary parts of the permittivity

For a linear isotropic medium, the field **D** is defined as a function of the field **E** (see. §1.2.2 and §1.3.1) as

$$\mathbf{D}(\boldsymbol{r},t)=\varepsilon_0\mathbf{E}(\boldsymbol{r},t)+\varepsilon_0\int_{-\infty}^{t}\widetilde{\chi}_e(\boldsymbol{r},t-t')\,\mathbf{E}(\boldsymbol{r},t')\,dt'. \qquad (1.8.28)$$

This equation represents the most general *spatially local*, *linear* and *causal* relationship that can be written between these two real fields for a *temporally invariant medium*. On the other hand, the relationship between these two fields is *not temporally local*; however, it is limited by the relaxation times γ^{-1} of the oscillators of the medium. The spatial locality is instead a valid approximation, as long as the motion of the oscillators is contained in distances much smaller than the wavelength and the spatial variations of the field amplitude [Jackson 1974]. The motion of the bound oscillators is on the order of the atomic dimensions, or smaller, for which this condition is well verified up to frequencies well beyond those of the visible spectrum. Instead, the motion of oscillating free charges can be macroscopic and is restrained by the collisions. In certain conditions, when the collision frequency is low, for example, at cryogenic temperatures or in superconductors, this approximation becomes invalid.

In the space of temporal frequencies, for Eq. (1.3.28), the permittivity of a *homogeneous* medium is given by

$$\varepsilon(\omega)=\varepsilon_0+\varepsilon_0\chi(\omega)=\varepsilon_0+\varepsilon_0\int_0^{\infty}\widetilde{\chi}_e(t)\,e^{i\omega t}\,dt, \qquad (1.8.29)$$

where the dependence of the susceptibility on the position was removed. From this expression, we can get an important general relationship between the real part and the imaginary part of ε using the theory of holomorphic functions. Let us assume that ω is a complex variable given by

$$\tilde{\omega} = \omega' + i\omega''.$$

As a result of the principle of causality, the integral of Eq. (8.1.29) thus becomes a Laplace transform, thanks to the fact that the integration interval is delimited between 0 and $+\infty$.

Furthermore, since **E** and **D** are real fields, $\hat{\chi}_e(t)$ is also a real function. Therefore, splitting the exponential as

$$e^{i\omega t} = e^{i\omega' t}e^{-\omega'' t} = \cos(\omega' t)e^{-\omega'' t} + i\sin(\omega' t)e^{-\omega'' t},$$

we find at once that, with $\varepsilon = \varepsilon' + i\varepsilon''$,

$$\varepsilon(-\tilde{\omega}) = \varepsilon^*(\tilde{\omega}^*). \tag{1.8.30}$$

In particular, $\varepsilon(\tilde{\omega})$ is real for $\tilde{\omega} = i\omega''$, that is, for $\omega' = 0$.

The convergence of the integral in Eq. (1.8.29), defined as a Laplace transform, requires that $\omega'' > l$, where l is a real value called the *abscissa of convergence*. On the other hand, for physical reasons, it cannot be [Cicogna 1969]

$$\left|\hat{\chi}_e(t)\right| \approx e^{at},$$

with $a > 0$ for large (positive) t; it follows that $l \leq 0$, and therefore $\varepsilon(\tilde{\omega})$ is holomorphic in the upper half-plane $\omega'' > 0$.

Here, we follow the further assumption that $\hat{\chi}_e(t) \in L^2(\mathbb{R})$, i.e., it is a function whose square modulus is integrable over the whole real axis $\mathbb{R}$. In this case, its Fourier transform $\chi(\tilde{\omega} = \omega')$ exists and is equal to the limit of the holomorphic function $\chi(\tilde{\omega})$ for $\omega'' \to 0^+$. On the other hand, the width of the range of t, starting from 0, for which $\hat{\chi}_e(t)$ is significantly different from 0, is on the order of magnitude of the polarization relaxation times. Thus, for the dielectrics, it happens that $\hat{\chi}_e(t) \to 0$ for $t \to \infty$, and thus the assumption is verified.

For conducting media, we must also consider the contribution of the conductivity, which depends, like **P**, on the field **E**. It can be incorporated into **D** and adds a pole of the first order in $\tilde{\omega} = 0$. We will examine this case in the following section.

Let us now consider the function

$$\frac{\varepsilon(\tilde{\omega}) - \varepsilon_0}{\tilde{\omega} - \omega_0},$$

where ω_0 is any real value. This function too is holomorphic in the upper half plane, with one pole at $\tilde{\omega} = \omega_0$ and a possible pole at $\tilde{\omega} = 0$ if the medium is conductive. We can then apply Cauchy's theorem to this function in its holomorphic

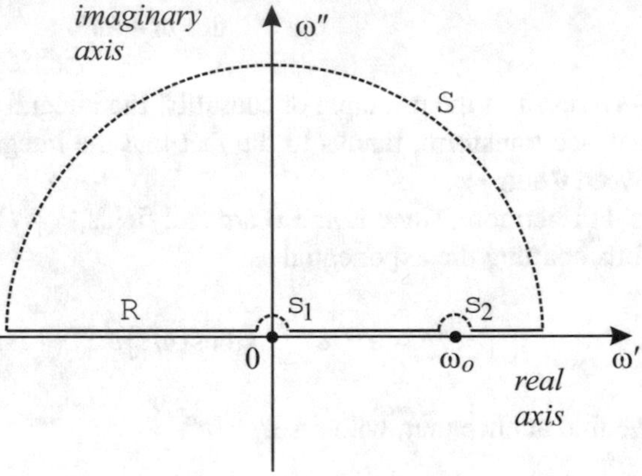

Fig. 1.21
Integration circuit

field, for which

$$\oint_{\mathrm{C}} \frac{\varepsilon(\tilde{\omega})-\varepsilon_0}{\tilde{\omega}-\omega_o} d\tilde{\omega} = 0\,, \tag{1.8.31}$$

where C is a closed path in the upper half-plane, free of singularity. We choose, in particular, the path shown in Fig. 1.21, where the R line is parallel to the real axis at a distance $h \to 0^+$ from this, S is a semicircle centered at 0 of radius $R \to \infty$, and s_1 and s_2 are two semicircles of radius $r \to 0$ centered on the poles, respectively, at 0 and ω_o on the real axis. For $\omega_o > 0$, the integral (8.1.31) can then be split into

$$\left(\int_{-R}^{-r} + \int_{+r}^{\omega_o - r} + \int_{\omega_o + r}^{+R} \right) \frac{\varepsilon(\omega' + ih) - \varepsilon_0}{\omega' + ih - \omega_o} d\omega' + \left(\int_{\mathrm{s}_1} + \int_{\mathrm{s}_2} + \int_{\mathrm{S}} \right) \frac{\varepsilon(\tilde{\omega}) - \varepsilon_0}{\tilde{\omega} - \omega_o} d\tilde{\omega} = 0\,, \tag{1.8.32}$$

and in a similar way if $\omega_o < 0$. If the medium is non-conductive, we can eliminate the integration on s_1 and reduce the number of integrals on R to two. We must now conduct the limit for $R \to \infty$: the trend of $\chi(\omega)$ for large ω can be put in relation with that of $\chi_e(t)$ for small t. Developing, in an asymptotic series, the integral of Eq. (8.1.29) for real and large values of ω, it results that [Jackson 1974]

$$\Re e\left[\varepsilon(\omega) - \varepsilon_o\right] = \mathrm{O}\left(\frac{1}{\omega^2}\right), \quad \Im m\left[\varepsilon(\omega) - \varepsilon_o\right] = \mathrm{O}\left(\frac{1}{\omega^3}\right).$$

Furthermore, $\chi(\tilde{\omega})$ decays to 0 for $\omega'' \to +\infty$ for the presence of the exponential $e^{-\omega'' t}$ in the integral argument. Therefore, the integration on the semicircle S tends to 0 for $R \to \infty$. Then, as discussed above, in the limit for $h \to 0^+$, the integration on R coincides with that taken on the real axis. Finally, within the limit $r \to 0$, we have

$$\lim_{r\to 0}\int_{S_2}\frac{\varepsilon(\tilde{\omega})-\varepsilon_0}{\tilde{\omega}-\omega_o}d\tilde{\omega}=-i\pi\left[\varepsilon(\omega_o)-\varepsilon_0\right]. \tag{1.8.33}$$

In conclusion, for a non-conducting medium, from Eqs. (1.8.32-33), separating the real and imaginary parts, we obtain the dispersion relations

$$\begin{aligned}\varepsilon'(\omega_o)-\varepsilon_o&=\frac{1}{\pi}\mathcal{P}\int_{-\infty}^{+\infty}\frac{\varepsilon''(\omega)}{\omega-\omega_o}d\omega,\\ \varepsilon''(\omega_o)&=-\frac{1}{\pi}\mathcal{P}\int_{-\infty}^{+\infty}\frac{\varepsilon'(\omega)-\varepsilon_0}{\omega-\omega_o}d\omega,\end{aligned} \tag{1.8.34}$$

where $\mathcal{P}$ indicates that the integral should be calculated according to the principal part. These equations are called *Kramers-Kronig relations*, where the real and imaginary parts of the permittivity – ε_0 are, respectively, the direct and inverse *Hilbert transform* one of the other. These relations have a very general validity, in that they are obtained by the hypotheses of causality and linearity and by a few other mathematical hypotheses that are "physically reasonable". In particular, $\tilde{n}(\tilde{\omega})$ also results to be a holomorphic function in the upper half-plane of $\tilde{\omega}$, and therefore the Kramers-Kronig relations also apply between $n-1$ and κ. Since the symmetry rule

$$\tilde{n}(-\omega)=\tilde{n}^*(\omega). \tag{1.8.35}$$

is also valid for $\tilde{n}$, these relations are usually written as

$$\begin{aligned}n(\omega_o)-1&=\frac{2}{\pi}\mathcal{P}\int_{0}^{+\infty}\frac{\omega\kappa(\omega)}{\omega^2-\omega_o^2}d\omega,\\ \kappa(\omega_o)&=-\frac{2\omega_o}{\pi}\mathcal{P}\int_{0}^{+\infty}\frac{n(\omega)-1}{\omega^2-\omega_o^2}d\omega.\end{aligned} \tag{1.8.36}$$

These are useful, in particular, for calculating the trend of n when κ is determined experimentally by means of absorption measurements, or vice versa.

1.8.4 Dispersion in conducting media

Returning to the classical model, in a conductor gas, the charges are free, for which the restoring force is zero. In these conditions, starting again from Eq.

(1.8.5), we can take $\omega_o = 0$, and, also neglecting the term in ω^3, we have the *Drude equation*

$$\varepsilon(\omega) = \varepsilon_0 (1 + N\alpha) = \varepsilon_0 \left(1 - \frac{Ne^2}{\varepsilon_0 m_e} \frac{1}{\omega^2 + i\Gamma\omega} \right), \tag{1.8.37}$$

where Γ is a damping constant due to collisions and N is the density of free charges. The term in ω^3 is omitted, since the Drude model assumes that $\Gamma\omega$ is dominant on $\gamma\omega^3$. For the complex refractive index, we have

$$\tilde{n}^2 = 1 - \frac{\omega_p{}^2}{\omega^2 + i\Gamma\omega}, \tag{1.8.38}$$

where

$$\omega_p = \sqrt{\frac{Ne^2}{\varepsilon_0 m_e}} \tag{1.8.39}$$

is the (angular) *plasma frequency*. For negligible Γ, when $\omega < \omega_p$, the refractive index is imaginary and, in these conditions, a wave incident on the medium is completely reflected. For example [Guenther 1990; Jackson 1974], in the ionosphere, at altitudes above 70 km, the density of the charges is on the order of $10^{11}/m^3$, for which $\omega_p \approx 1.8 \times 10^7$ rad/s. This situation is known to radio amateurs who exploit the reflection of radio waves from the ionosphere at frequencies below 3 MHz for communication on a worldwide scale.

Another approach is to go back to the Maxwell's equations (1.2.6), also considering the contribution of currents, while that of the free charge density can be overlooked, as we will see later. Even for the currents, we can introduce a constitutive equation analogous to (1.3.25) for the polarization, for which

$$\mathbf{J}(\boldsymbol{r},t) = \int_{-\infty}^{t} \tilde{\sigma}(\boldsymbol{r},t-t')\, \mathbf{E}(\boldsymbol{r},t')\, dt', \tag{1.8.40}$$

where $\tilde{\sigma}$ represents the dynamic memory for the currents similar to $\tilde{\chi}$ for polarization. This expression is a temporally non-local generalization of *Ohm's law*, while the approximation of spatial locality is still implied, with the limitations mentioned in the preceding paragraph. From the Fourier transform of Eq. (8.1.40), we find

$$\hat{J}(\omega) = \sigma(\omega)\, \hat{E}(\omega), \tag{1.8.41}$$

where

$$\sigma(\omega) = \int_0^\infty \tilde{\sigma}(t) e^{i\omega t}\, dt \tag{1.8.42}$$

is the complex conductivity of the medium. On the other hand, in the Ampere-Maxwell equation for the curl of the field **H**, the current intervenes together with the temporal derivative of the field **D**. For a monochromatic wave, we have

$$i\boldsymbol{k} \times \boldsymbol{H} = -i\omega t\, \boldsymbol{D} + \sigma \boldsymbol{E}\,.$$

We can then redefine the permittivity as

$$\varepsilon(\omega) = \varepsilon_0 \left[1 + \chi_{ep}(\omega) + i\frac{\sigma(\omega)}{\varepsilon_0 \omega} \right], \tag{1.8.43}$$

where χ_{ep} is the susceptibility of the bound charges. In this way, $\varepsilon(\omega)$ presents a pole at $\omega = 0$, which corresponds to the zero resonance frequency of the Drude model for the conductors. Comparing Eqs. (1.8.37) and (1.8.43), we have

$$\sigma(\omega) = \frac{Ne^2}{m_e} \frac{1}{\Gamma - i\omega}\,. \tag{1.8.44}$$

For copper at room temperature [Jackson 1974], from the static value of σ and the density of the charges, we can evaluate that $\Gamma \approx 3\times10^{13}$ sec^{-1}. Therefore, the conductivity remains predominantly real up to frequencies of $\approx 5\times10^{12}$ Hz, whereas, at higher frequencies, in the infrared and beyond, it has to be considered complex in the way described qualitatively by Eq. (1.8.44).

From this comparison, it is clear that, for frequencies different from 0, the distinction between free and bound charges is essentially artificial, since the contribution of the conduction can be interpreted as a resonance similar to the others. The dispersion properties of the medium can therefore be described either by a complex susceptibility, as a functions of the frequency, or by a complex conductivity and a complex susceptibility, both as function of the frequency.

If, therefore, we also include in $\tilde{\chi}_e(t)$ the contribution of the conductivity, it results that $\tilde{\chi}_e(t) \to \bar{\sigma}/\varepsilon_0$ for $t \to \infty$, where $\bar{\sigma} = \sigma(0)$ is the static conductivity of the medium at $\omega = 0$. This static value adds one pole to the permittivity at $\omega = 0$ and, resuming the reasonings of the previous paragraph, for a conductive medium, the integral on S_1 is

$$\lim_{r\to 0} \int_{S_1} \frac{\varepsilon(\tilde{\omega}) - \varepsilon_0}{\tilde{\omega} - \omega_0} d\tilde{\omega} = -\pi \frac{\bar{\sigma}}{\omega_0}$$

and the corresponding Kramers-Kronig relations become

$$\varepsilon'(\omega_o)-\varepsilon_0=\frac{1}{\pi}\mathcal{P}\int_{-\infty}^{+\infty}\frac{\varepsilon''(\omega)}{\omega-\omega_o}d\omega,$$
$$\varepsilon''(\omega_o)=-\frac{1}{\pi}\mathcal{P}\int_{-\infty}^{+\infty}\frac{\varepsilon'(\omega)-\varepsilon_0}{\omega-\omega_o}d\omega+\frac{\overline{\sigma}}{\omega_o}. \tag{1.8.45}$$

1.8.5 Phase, group and signal velocities

So far, we have considered essentially only monochromatic waves, which are an ideal case and are of constant amplitude by definition. In general, we instead have situations in which the radiation is constituted by a more or less broad spectrum of frequencies. An important case is that of propagation of a "packet" of waves, which is delimited both temporally and spatially, such as that generated by a pulsed laser. The propagation of such a pulse is subject to the dispersion and the dissipation of the medium, whereby the pulse changes its shape over time. Limiting ourselves to the one-dimensional case of a medium that is *linear*, *homogeneous*, and with negligible magnetization, for Eq. (1.4.3), a wave is generally expressed as a function of space and time such that the amplitude E of the field (an electric field component) is constrained by the equation

$$\frac{\partial^2}{\partial z^2}\mathrm{E}(z,t)-\frac{1}{c^2}\frac{\partial^2}{\partial t^2}\mathrm{E}(z,t)=\mu_o\frac{\partial^2}{\partial t^2}\mathrm{P}(z,t), \tag{1.8.46}$$

where

$$\mathrm{P}(z,t)=\varepsilon_0\int_{-\infty}^{t}\tilde{\chi}_e(z,t-t')\,\mathrm{E}(z,t')\,dt'. \tag{1.8.47}$$

Besides, a possible contribution of conductivity can be incorporated into $\tilde{\chi}_e$. The spatial and temporal Fourier transform of Eq. (1.8.46) is

$$k^2\hat{E}(k,\omega)-\frac{1}{c^2}\omega^2\hat{E}(k,\omega)=\frac{1}{c^2}\omega^2\chi(\omega)\hat{E}(k,\omega), \tag{1.8.48}$$

from which

$$[k(\omega)]^2=\frac{1}{c^2}\omega^2[1+\chi(\omega)], \tag{1.8.49}$$

expressing a constraint between the modulus of the wave vector and the angular frequency. Therefore, if we know the spectrum $\hat{E}\left(k(\omega),\omega\right)$ of our wave packet, creating the inverse transform on it, we again find its space-time form given by

$$\mathrm{E}(z,t)=\frac{1}{2\pi}\int_{-\infty}^{\infty}\hat{E}(\omega)e^{ik(\omega)z-i\omega t}d\omega\,, \tag{1.8.50}$$

where, for the constraint mentioned above, the integration remains only for the temporal frequencies. Inverting the relation between k and ω, that is, expressing ω as a function of k, Eq. (1.8.50) can be written in a similar form, integrating in k instead of ω. Suppose now that the wave spectrum is significantly different from 0 only on an interval $2\Delta\omega$ centered at an angular frequency $\overline{\omega}$ such that $\Delta\omega \ll \overline{\omega}$. We can then rewrite Eq. (1.8.50) as

$$\mathrm{E}(z,t)=\frac{e^{i\bar{k}z-i\overline{\omega}t}}{2\pi}\int_{-\Delta\overline{\omega}}^{\Delta\overline{\omega}}\hat{E}(\omega'+\overline{\omega})e^{i\left[k(\omega'+\overline{\omega})-\bar{k}\right]z-i\omega' t}d\omega'+c.c.\,, \tag{1.8.51}$$

where $\bar{k}=k(\overline{\omega})$. On the other hand, if k varies little within the range $2\Delta\omega$, we can develop it in series around $\overline{\omega}$:

$$k(\omega'+\overline{\omega})=\bar{k}+\frac{\partial k}{\partial\omega}\omega'+\frac{1}{2}\frac{\partial^2 k}{\partial\omega^2}\omega'^2+\cdots. \tag{1.8.52}$$

Limiting ourselves to the first order in ω' and rearranging the exponential argument under the sign of integration as

$$i\left(\left.\frac{\partial k_{\Re}}{\partial\omega}\right|_{\overline{\omega}}z-t\right)\omega'-\left.\frac{\partial k_{\Im}}{\partial\omega}\right|_{\overline{\omega}}z\omega'\,,$$

where k has been divided here into its real and imaginary parts. If we can neglect the dissipation in the medium, the integral assumes a value proportional to the amplitude E_o of the field, at the position $x = 0$ and at the time $t' = t - x/v_g$, where

$$v_{\mathrm{g}}=\left.\frac{\partial\omega}{\partial k_{\Re}}\right|_{\bar{k}}=\left(\left.\frac{\partial k_{\Re}}{\partial\omega}\right|_{\overline{\omega}}\right)^{-1} \tag{1.8.53}$$

is called the *group velocity*. Then, Eq. (1.8.51) becomes

$$\mathrm{E}(z,t)=\frac{1}{2}e^{i\bar{k}z-i\overline{\omega}t}E_{\mathrm{o}}\left(z-v_{\mathrm{g}}t\right)+c.c.\,, \tag{1.8.54}$$

where the exponential represents a carrier wave with a *phase velocity* given by

$$\upsilon_\mathrm{p} = \frac{\overline{\omega}}{k_\Re(\overline{\omega})}. \tag{1.8.55}$$

Eq. (8.1.54) then represents the propagation of a quasi-monochromatic plane wave in a (weakly) dispersive medium.[15] The square modulus of E_o is, on the other hand, proportional to the energy density, so that, in *this approximation*, υ_g is the energy transport velocity itself, and therefore coincides with the signal transmission velocity, which, in Relativity, must always be less than or equal to the speed of light. If, as usual, we express that $k_\Re = 2\pi n(\omega)/\lambda = \omega n(\omega)/c$, we then have

$$\upsilon_\mathrm{p} = \frac{c}{n(\overline{\omega})}, \quad \upsilon_\mathrm{g} = \frac{c}{n(\overline{\omega}) + \overline{\omega}\, \partial n(\omega)/\partial\omega|_{\overline{\omega}}}. \tag{1.8.56}$$

The physical sense of *velocity* in these expressions is essentially limited to the regions of normal dispersion, where the dissipation is negligible and $\tilde{n}$ can be considered real. The phase velocity is higher or lower than the speed of light, depending on whether n is less than or greater than unity, while, for the group velocity in the denominator, a further positive term is added to n, because, in the normal region, n grows with ω.

In the anomalous dispersion regions, υ_p and υ_g become complex, with a real part that can exceed the speed of light or even be negative. On the other hand, when this occurs, the υ_g correspondence with the transport speed of the energy or the signals eventually falls off.

It is important to note that the *pulse shape remains unchanged* when the terms higher than the first of the development of k in Eq. (1.8.52) are negligible and the dissipation is absent, which is when, in good approximation

$$k = 2\pi a + b\omega, \quad \text{from which} \quad n = a\lambda_\mathrm{o} + bc\,,$$

where a and b are real constants, and thus the refractive index varies *linearly* with the wavelength along the entire pulse spectrum. In this case, the *group velocity dispersion* (GVD), which is defined as

---

[15] Let us consider here, for simplicity, the case of $\overline{k}_\Im = 0$. When $\upsilon_\mathrm{p} = \upsilon_\mathrm{g}$, the argument of the exponential also becomes a function of $z - \upsilon_\mathrm{g} t$ and Eq. (1.8.54) is reduced to the classical form of a wave packet that propagates at the group velocity without changing its shape. If the two speeds are different, we can rewrite Eq. (1.8.54) in the form

$$\mathrm{E}(z,t) = \frac{1}{2} e^{-i2\pi\nu_\mathrm{ceo} t} e^{i\overline{k}_\Re (z - \upsilon_\mathrm{g} t)} E_\mathrm{o}(z - \upsilon_\mathrm{g} t) + c.c.\,,$$

where $\nu_\mathrm{ceo} = \overline{k}_\Re(\upsilon_\mathrm{p} - \upsilon_\mathrm{g})/(2\pi)$ is the frequency of a phase modulation of the wave packet (ceo = *carrier-envelope offset*).

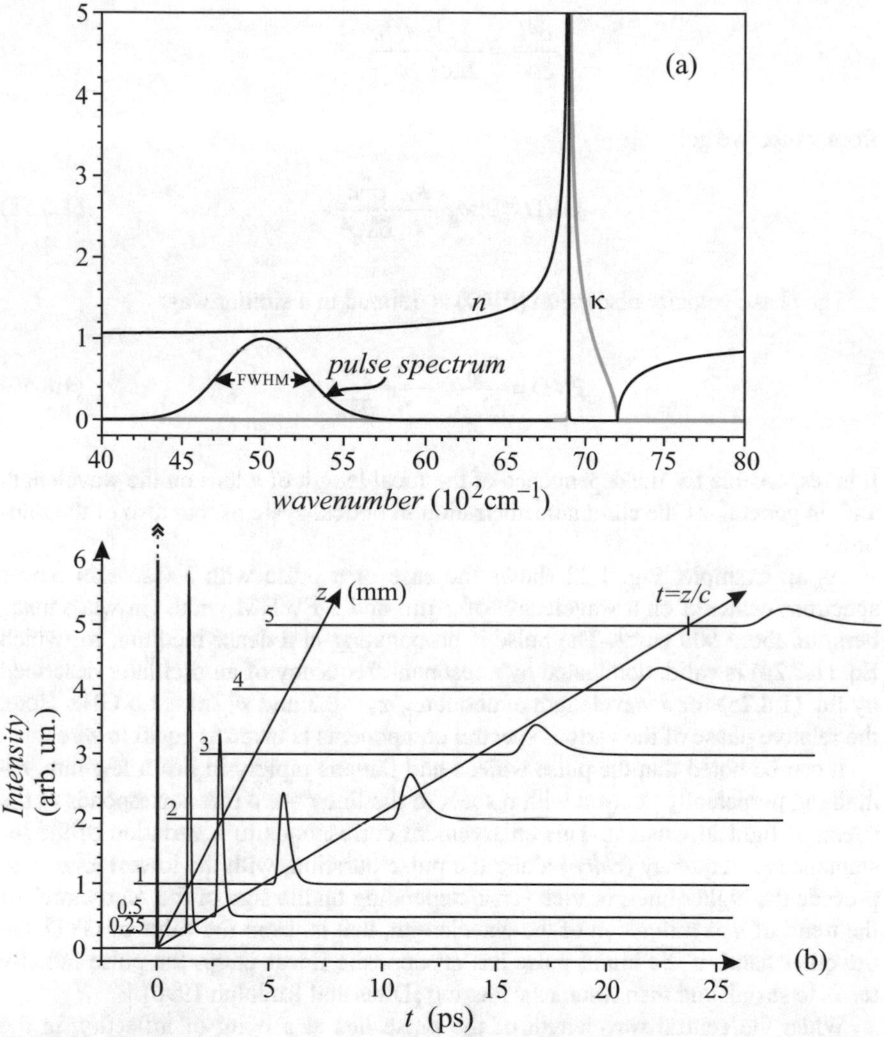

Fig. 1.22 Propagation of a pulse in a dispersive medium. (a) Pulse spectrum, assumed Gaussian; refractive index n and extinction index κ of the medium. The central wavelength of the pulse is taken here as equal to 2 μm. (b) Pulse shape versus time for some distances traveled from the entrance in the medium

$$\mathrm{GVD} \equiv \frac{\partial v_g}{\partial \lambda_o} = \frac{2\pi c v_g{}^2}{\lambda_o{}^2}\frac{\partial^2 k}{\partial \omega^2}, \tag{1.8.57}$$

is zero. In most cases, this does not happen, and it is necessary to consider at least the second term of the development of k. With some passages, it is

$$\frac{\partial^2 k}{\partial \omega^2} = \frac{\lambda_o^3}{2\pi c^2} \frac{\partial^2 n}{\partial \lambda_o^2},$$

from which we get

$$\text{GVD} = 2\pi v_g^2 \frac{\lambda_o}{c} \frac{\partial^2 n}{\partial \lambda_o^2}. \tag{1.8.58}$$

The phase velocity dispersion (PVD) is defined in a similar way:

$$\text{PVD} \equiv \frac{\partial v_p}{\partial \lambda_o} = -\frac{c}{n^2} \frac{\partial n}{\partial \lambda_o}. \tag{1.8.59}$$

It is responsible for the dependence of the focal length of a lens on the wavelength and, in general, of the chromatic aberration in optical systems, but also of the rainbow!

As an example, Fig. 1.22 shows the case of a pulse with a Gaussian power spectrum centered on a wavelength of 2 μm and a FWHM width (in wave numbers) of about 600 cm^{-1}. The pulse is propagating in a dense medium, for which Eq. (1.8.24) is valid, dominated by a resonant frequency of an oscillator described by Eq. (1.8.25) for a wavelength of about $\omega_p/\omega_r = 0.3$ and $\gamma/(2\pi) \cong 1.5$ GHz. Here, the relative phase of the various spectral components is taken as equal to zero.

It can be noted that the pulse widens and flattens rapidly in just a few mm, remaining perpetually delayed with respect to the line $t = z/c$ that corresponds to the speed of light in vacuum. This enlargement corresponds to a variation of the instantaneous frequency (*chirp*) along the pulse duration, with the lowest tones that precede the higher ones, or vice versa, depending on the sign of the "curvature" of the trend of n as a function of the wavelength, that is, from the value of GVD. On the other hand, if the initial pulse has an opposite linear chirp, the pulse initially tends to shrink and then resumes widening [Diels and Rudolph 1996].

When the central wavelength of the pulse lies at a point of inflection in the normal region of the spectrum between two absorption bands, the GVD is zero. This happens, for example, at 1.3 μm for fused silica, which is the very pure material used in optical fibers. This wavelength, however, does not coincide with the minimum of losses, which is at 1.55 μm. However, in the guided propagation, the dispersion also depends on the size of the *core*: with a particular choice of its diameter, the GVD can be canceled at just 1.55 μm.

Fig. 1.23 instead shows the case of a pulse centered on a resonant frequency of the medium, with a spectrum about 10 times wider than the absorption band. The propagation quickly extinguishes the central part of the spectrum and the pulse is deformed rapidly: the oscillation is essentially due to the beat of the two remaining parts of the spectrum or, equivalently, to the local felling of the spectrum.

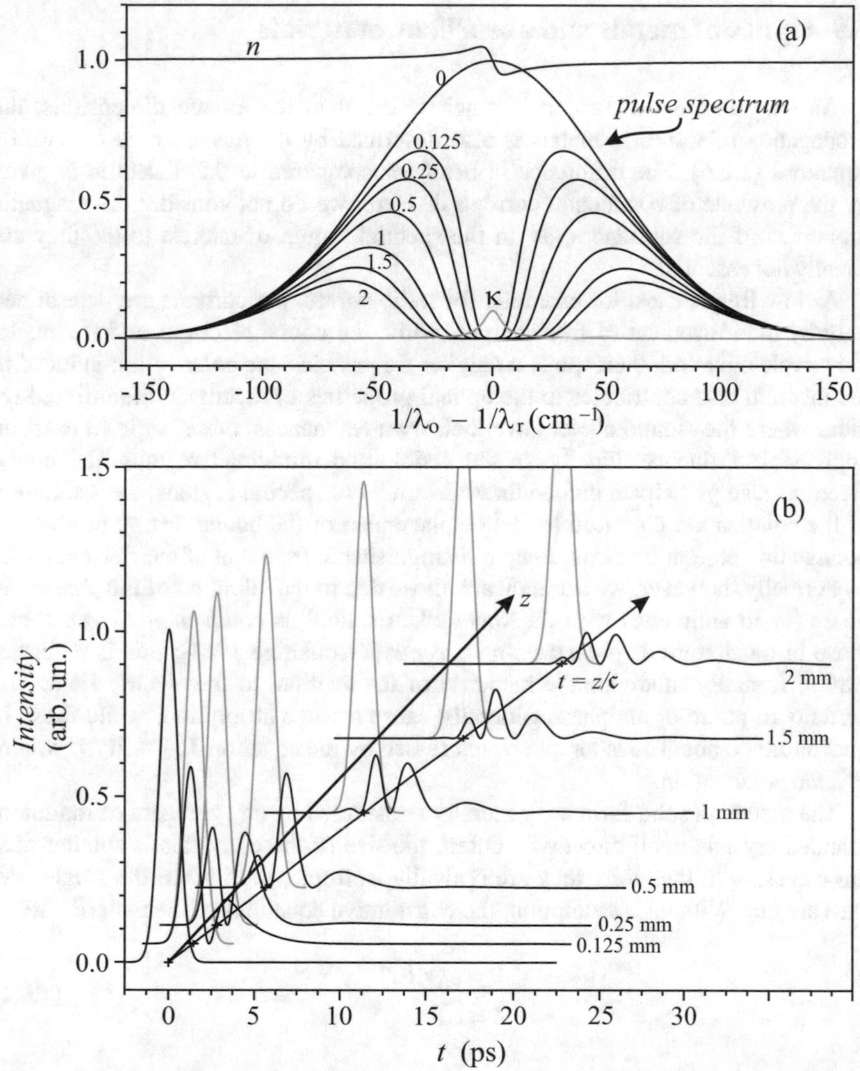

Fig. 1.23 Example of propagation of a pulse whose frequency carrier is centered on a resonant frequency of the medium, with $\lambda_r = 2$ μm. (a) refractive index *n* and extinction index κ, with $\gamma/(2\pi) = 150$ GHz and $\omega_p/\omega_r = 0.02$ of Eq. (1.8.25). The normalized spectrum of the pulse is shown as a function of the wave number. The numbers in the figure indicate the distance in mm traveled from the entrance in the medium. (b) pulse shape for the same penetration distances in the medium. The peaks in gray represent the propagation in vacuum

At the resonance frequency, the group velocity calculated according to Eq. (8.56.b) is negative. However, we can see how, in these conditions, this concept of speed loses physical sense.

1.9 Optics of metals and absorbent materials

As long as the wavelength is much larger than the atomic dimensions, the propagation of waves in metals is also described by the macroscopic Maxwell's equations (1.2.6). The difference in behavior compared to the dielectric is given by the presence of conduction currents **J**. Here, we do not consider the magnetic properties of the substances, as, in the spectral region of interest to us, they are usually not essential.

At low frequencies, for example, for radio waves, the currents are determined only by the movement of free charges, while the bound electrons and the nuclei play a role only with their spins. At higher frequencies, the polarization induced in the medium also contributes to the optical properties of metals: in the infrared region, where the atomic nuclei have their own resonances, these begin to exert an appreciable influence. Finally, in the visible and ultraviolet regions, the bound electrons also participate in the vibrations. In these spectral regions, the oscillation of the polarization produced by the displacement of the bound charge produces a polarization current that is no longer distinguishable from that of the free charges.

Formally, however, we can indicate those due to the vibration of the charges in phase (or in antiphase) with the wave electric field as *conduction currents*, and those in quadrature as *polarization currents* [Sivoukhine 1984]; this is a further way to formally subdivide the response of the medium to the electric field. The currents in phase or antiphase generally cause a wave absorption, while those in quadrature do not. The latter can be expressed by the equation $\mathbf{J}_{\text{pol}} = \partial\mathbf{P}/\partial t$, where **P** is the polarization.

The metals in solid form are generally constituted by an aggregate of randomly oriented crystals in all directions. Often, the size of these crystals is smaller than the wavelength; therefore, they are optically isotropic, even when the single crystals are not. With this assumption, the constitutive equations of the material are

$$\begin{aligned} \mathbf{P} &= \varepsilon_0 \chi' \mathbf{E}, \\ \mathbf{J} &= \sigma \mathbf{E}, \end{aligned} \tag{1.9.1}$$

where, with the previous scheme, we assume that the susceptibility χ' and the *conductivity* σ are real. These expressions, however, give only an approximate and heuristic description of the behavior of metals and, for a good agreement with experiment, especially in high-frequency regions starting from infrared, a quantum treatment is required.

On the other hand, the relation of proportionality assumed in Eq. (1.9.1.b) between current and electric field implies that the motion of the charges is strongly restrained by impacts of these with other particles, namely, that the friction dominates the inertia. The electric field therefore produces work on the charges, which is immediately converted into heat by the Joule effect. Eq. (1.9.1.b) is equivalent to Ohm's law in a conductor, commonly applied in electronics, where the resistance of a wire with length l and section A is given by $R = l/(A\sigma)$. So, it is not surprising that the metals are highly absorbent media, to the point that, as we shall see, the radiation that tries to cross them is extinguished in a small fraction of a wavelength (however, they become more transparent in the ultraviolet). This strong absorption makes them particularly reflective, although not in a uniform way in the visible spectrum, for which some of them have a characteristic color.

Even the dielectric media possess spectral regions of strong absorption, although generally lower than that of metals, for which, in some cases, they exhibit a metallic appearance by reflecting the light. This happens in coincidence with resonant transitions between the quantum energy levels of the material. Classically, we can interpret the behavior of their charges as that of a damped oscillator, in which there is again dissipation of energy of an incident wave into heat, or also in diffuse radiation re-emitted by spontaneous decay.

Together with Eqs. (1.9.1), we also have $\mathbf{D} = \varepsilon'\mathbf{E}$. Moreover, even assuming the relationship $\mathbf{B} = \mu\,\mathbf{H}$, the macroscopic Eqs. (1.2.6) become

$$\begin{cases} \nabla\bullet\mathbf{E} = \rho_m/\varepsilon', & \nabla\bullet\mathbf{H} = 0, \\ \nabla\times\mathbf{E} = -\mu\,\partial\mathbf{H}/\partial t & \nabla\times\mathbf{H} = \varepsilon'\partial\mathbf{E}/\partial t + \sigma\,\mathbf{E}. \end{cases} \tag{1.9.2}$$

On the other hand, in a metal, the charge density ρ_m very quickly relaxes to zero in times on the order of 10^{-18} sec [Born and Wolf, 1980]. In fact, taking the divergence of the fourth equation and substituting into it the first, we find that

$$\varepsilon'\dot{\rho}_m + \sigma\rho_m = 0\,, \tag{1.9.3}$$

which expresses an exponential decay with a time constant equal to ε'/σ. Therefore, we can take $\rho_m = 0$, whence $\nabla\bullet\mathbf{E} = 0$.[16] From Eqs. (1.9.2), we then get the wave equation

$$\nabla^2\mathbf{E} - \mu\varepsilon'\frac{\partial^2}{\partial t^2}\mathbf{E} - \mu\sigma\frac{\partial}{\partial t}\mathbf{E} = 0\,, \tag{1.9.4}$$

where the term in $\partial\mathbf{E}/\partial t$ implies that the wave is attenuated on its way.

For a monochromatic wave of angular frequency ω, in the complex representation, the last two Maxwell's equations become

$$\begin{cases} \nabla\times\boldsymbol{E} = i\omega\mu\,\boldsymbol{H}, \\ \nabla\times\boldsymbol{H} = -i\omega\left(\varepsilon' + i\dfrac{\sigma}{\omega}\right)\boldsymbol{E}, \end{cases} \tag{1.9.5}$$

[16] For the continuity equation, this also means that $\nabla\bullet\mathbf{J} = 0$, but this is still in agreement with Eq. (1.9.2.d): for example, with a Gaussian laser beam, the currents follow close lines instant by instant, such as those of the electric field.

which are formally equivalent to the equations for a dielectric, taking, as permittivity, the complex variable

$$\tilde{\varepsilon} = \varepsilon' + i\varepsilon'' = \varepsilon' + i\,\sigma/\omega\,, \tag{1.9.6}$$

with which we redefine the field $\boldsymbol{D}$ as $\boldsymbol{D} = \tilde{\varepsilon}\boldsymbol{E}$. This equation is an approximate rewriting of Eq. (1.8.43), taking, for σ, the value of conductivity at ω = 0. As long as ε′ and σ can be considered real, comparing Eqs. (1.9.6) and (1.8.14), we have

$$n^2 - \kappa^2 = c^2\mu\varepsilon', \quad n\kappa = \frac{c^2}{2\omega}\mu\sigma\,, \tag{1.9.7}$$

where the real refractive index n and the extinction coefficient κ were introduced by Eq. (1.8.15). The solutions for n and κ are

$$n^2 = \frac{c^2}{2}\left[\sqrt{(\mu\varepsilon')^2 + \frac{\mu^2\sigma^2}{\omega^2}} + \mu\varepsilon'\right], \quad \kappa^2 = \frac{c^2}{2}\left[\sqrt{(\mu\varepsilon')^2 + \frac{\mu^2\sigma^2}{\omega^2}} - \mu\varepsilon'\right], \tag{1.9.8}$$

where the root was taken with the positive sign, having assumed n and κ to be real. The term *absorption* would be improper for κ, because there may be extinction even without absorption: for example, with a plasma, it may happen that $\tilde{\varepsilon}$ is real and negative, for which $\kappa \neq 0$ and $n = 0$ [Sivoukhine 1984].

On the other hand, the wave equation that is obtained from Eq. (1.9.4) is

$$\nabla^2\boldsymbol{E} + \omega^2\mu\tilde{\varepsilon}\boldsymbol{E} = 0\,. \tag{1.9.9}$$

Therefore, we can formally extend to the absorbent materials the same laws of propagation, reflection, and transmission that we have for dielectric media. The essential difference that occurs with these is that the wave vector is now necessarily a complex variable expressed by Eq. (1.4.19). For example, the inhomogeneous plane wave solution to Eq. (1.9.9) is given by

$$\boldsymbol{k}\bullet\boldsymbol{k} = \omega^2\mu\tilde{\varepsilon} \tag{1.9.10}$$

and, solving for the real part $\boldsymbol{k}_{\Re}$ and the imaginary part $\boldsymbol{k}_{\Im}$ of $\boldsymbol{k}$, in place of Eqs. (1.4.20), we have

$$k_{\Re}^2 - k_{\Im}^2 = \frac{\omega^2}{c^2}\left(n^2 - \kappa^2\right), \quad \boldsymbol{k}_{\Re}\bullet\boldsymbol{k}_{\Im} = \frac{\omega^2}{c^2}n\kappa\,. \tag{1.9.11}$$

Therefore, with $\kappa \neq 0$, the direction of $\boldsymbol{k}_{\Re}$ and $\boldsymbol{k}_{\Im}$ are not orthogonal. The electric and magnetic fields of the corresponding inhomogeneous wave are therefore given by

$$\boldsymbol{E} = \boldsymbol{E}_o e^{-\boldsymbol{k}_{\Im}\cdot\boldsymbol{r}} e^{i\boldsymbol{k}_{\Re}\cdot\boldsymbol{r}-i\omega t}, \quad \boldsymbol{H} = \boldsymbol{H}_o e^{-\boldsymbol{k}_{\Im}\cdot\boldsymbol{r}} e^{i\boldsymbol{k}_{\Re}\cdot\boldsymbol{r}-i\omega t}. \tag{1.9.12}$$

As in the case of the evanescent waves, the phase of the field increases along the direction of $\boldsymbol{k}_{\Re}$, while its amplitude decreases proceeding in that of $\boldsymbol{k}_{\Im}$. Assuming $\rho_m = 0$, Eqs. (1.9.2) for the divergence of the fields impose the orthogonality of the electric and magnetic fields with the wave vector, for which Eqs. (1.4.17) still apply. Furthermore, for the first of Eqs. (1.9.5), we have

$$\boldsymbol{H}_o = \frac{1}{\omega\mu}\boldsymbol{k}\times\boldsymbol{E}_o. \tag{1.9.13}$$

Since $\boldsymbol{k}$ is now a complex vector, we note that, for a medium with $\kappa \neq 0$, the magnetic field is not in phase with the electric field. In particular, for Eq. (1.5.8), the wave intensity is now given by

$$\begin{aligned} I &= \frac{1}{2\omega\mu}\left|\Re e\left[\boldsymbol{E}_o\times\left(\boldsymbol{k}^*\times\boldsymbol{E}_o^*\right)\right]\right| e^{-2\boldsymbol{k}_{\Im}\cdot\boldsymbol{r}} \\ &= \frac{1}{2\omega\mu}\left|\Re e\left[\left|\boldsymbol{E}_o\right|^2\boldsymbol{k}^* - \left(\boldsymbol{k}^*\cdot\boldsymbol{E}_o\right)\boldsymbol{E}_o^*\right]\right| e^{-2\boldsymbol{k}_{\Im}\cdot\boldsymbol{r}}, \end{aligned} \tag{1.9.14}$$

where μ was considered real. Only if the vectors $\boldsymbol{k}_{\Re}$ and $\boldsymbol{k}_{\Im}$ are parallel to each other or if $\boldsymbol{E}_o$ is perpendicular to both can we bring the intensity back to the more familiar expression

$$I = \frac{\left|\boldsymbol{E}_o\right|^2}{2\omega\mu}\left|\boldsymbol{k}_{\Re}\right| e^{-2\boldsymbol{k}_{\Im}\cdot\boldsymbol{r}}. \tag{1.9.15}$$

Consider the simplest case in which these two vectors are parallel and oriented along the z-axis, while the field $\boldsymbol{E}$ is oriented along the x-axis. The wave is then represented by the equations

$$E_x = E_{ox} e^{ik_{\Re}z - k_{\Im}z - i\omega t}, \quad H_y = \frac{1}{\mu\omega}\left(k_{\Re} + ik_{\Im}\right)E_x \tag{1.9.16}$$

and, for Eq. (1.5.8), the average Poynting vector is here given by

$$\overline{S}_z = \frac{1}{2}\Re e\left(E_x H_y^*\right) = \left|E_{ox}\right|^2 \frac{k_{\Re}}{2\mu\omega} e^{-2k_{\Im}z} = \left|E_{ox}\right|^2 \frac{n}{2\mu c} e^{-2\frac{\omega}{c}\kappa z}, \tag{1.9.17}$$

where, in the last equality, the parallelism between $\boldsymbol{k}_{\Re}$ and $\boldsymbol{k}_{\Im}$ was exploited. For metals, at least up to the near infrared frequencies, we have $\sigma/\omega >> \varepsilon'$, for which

$$n \approx \kappa \approx c\sqrt{\frac{\mu\sigma}{2\omega}}, \tag{1.9.18}$$

and the penetration distance in a metal is approximately

$$d = \frac{c}{2\omega\kappa} \approx \frac{1}{2}\sqrt{\frac{\lambda_o}{\pi c\mu\sigma}}. \tag{1.9.19}$$

If we assume that σ has the same value of the static conductivity at room temperature, for example, for aluminum, with $\sigma = (1/2.7)\ 10^6$ mho/cm, $\lambda_o = 10\ \mu$m and $\mu = \mu_o$, we have $d \approx 24$ nm.

More generally, if $\boldsymbol{k}_{\Im}$ is still oriented along the z-axis, but $\boldsymbol{k}_{\Re}$ has components $(k_{\Re x}, 0, k_{\Re z})$, the condition of orthogonality between the electric field and the wave vector imposes the constraint

$$k_{\Re x} E_x = -\left(k_{\Re z} + ik_{\Im}\right) E_z, \tag{1.9.20}$$

while E_y is free to take any value. If, in particular, $E_y = 0$, the magnetic field has only the y component different from zero. Similar reasoning applies equally to the magnetic field components.

1.9.1 Reflection and refraction by absorbent materials

For the formal equality of the wave equations, the rules to be applied to get the laws of reflection and refraction with absorbent media, for example, from metals, are the same as those that we have seen for the dielectric. Thus, for a given plane wave, incident from a dielectric medium to an absorbent medium, we still have a transmitted wave and a reflected wave expressible with Eqs. (1.7.1). However, the transmitted wave is now inhomogeneous and has a complex wave vector given by $\boldsymbol{k}_T = \boldsymbol{k}_{T\Re} + i\boldsymbol{k}_{T\Im}$. By imposing the constancy of the phase differences, which derives from the temporal and spatial isotropy on the interface between the two media, we necessarily have that the three waves have the same angular frequency ω and that the projection (here indicated by the symbol $\lrcorner$) of their wave vectors on the plane interface must be the same, so

$$\begin{aligned} \boldsymbol{k}_{T\Re\lrcorner} &= \boldsymbol{k}_{I\lrcorner} = \boldsymbol{k}_{R\lrcorner}, \\ \boldsymbol{k}_{T\Im\lrcorner} &= 0, \end{aligned} \tag{1.9.21}$$

where I, T and R still indicate, respectively, the incident, transmitted, and reflected waves. The imaginary part of the vector $\boldsymbol{k}_{\mathrm{T}}$ must therefore be oriented perpendicular to the interface. The transmitted wave is then characterized by wavefronts perpendicular to the direction of $\boldsymbol{k}_{\mathrm{T}\Re}$ and by surfaces of equal amplitude parallel to the interface. The wave vectors $\boldsymbol{k}_{\mathrm{I}}$, $\boldsymbol{k}_{\mathrm{R}}$, $\boldsymbol{k}_{\mathrm{T}\Re}$ and $\boldsymbol{k}_{\mathrm{T}\Im}$ are still coplanar on the plane of incidence orthogonal to the interface. For the angles of incidence, ϑ_{I}, and reflection, ϑ_{R}, the law of equality (1.7.5) is still valid. For refraction, let us call θ the angle between the direction of $\boldsymbol{k}_{\mathrm{T}\Re}$ and the normal to the interface, and we define a real "index of refraction" n_ϑ with

$$n_\vartheta = n_{\mathrm{I}} \frac{\sin\vartheta_{\mathrm{I}}}{\sin\theta}, \tag{1.9.22}$$

where n_{I} is the refractive index of the medium for the incident and reflected waves. For the moduli of the wave vectors, we have

$$k_{\mathrm{I}} = \frac{\omega}{c} n_{\mathrm{I}}, \quad k_{\mathrm{T}\Re} = \frac{\omega}{c} n_\vartheta, \quad k_{\mathrm{T}\Im} = \frac{\omega}{c} \kappa_\vartheta, \tag{1.9.23}$$

where, in the third equation, we have defined the extinction coefficient κ_ϑ in analogy to the second equation. The wave penetration into the absorbing medium in the direction perpendicular to the interface is then given by

$$d_\vartheta = \frac{c}{2\omega\kappa_\vartheta}. \tag{1.9.24}$$

By substituting Eqs. (1.9.23.b-c) in Eq. (1.9.11), we have

$$n_\vartheta^2 - \kappa_\vartheta^2 = n^2 - \kappa^2 = a, \quad n_\vartheta \kappa_\vartheta \cos\theta = n\kappa = b/2, \tag{1.9.25}$$

where the quantities a and b are called *Ketteler's invariant*, from E. Ketteler (1836-1900), who established these equations, to be compared with Eqs. (1.9.7). For Eq. (1.9.22), the second equation can be rewritten as

$$n_\vartheta^2 \kappa_\vartheta^2 - n_{\mathrm{I}}^2 \kappa_\vartheta^2 \sin^2\vartheta_{\mathrm{I}} = n^2\kappa^2, \tag{1.9.26}$$

and then

$$\begin{aligned} \kappa_\vartheta^2 &= \frac{1}{2}\left[\sqrt{\left(a - n_{\mathrm{I}}^2 \sin^2\vartheta_{\mathrm{I}}\right)^2 + b^2} - a + n_{\mathrm{I}}^2 \sin^2\vartheta_{\mathrm{I}}\right], \\ n_\vartheta^2 &= \frac{1}{2}\left[\sqrt{\left(a - n_{\mathrm{I}}^2 \sin^2\vartheta_{\mathrm{I}}\right)^2 + b^2} + a + n_{\mathrm{I}}^2 \sin^2\vartheta_{\mathrm{I}}\right], \end{aligned} \tag{1.9.27}$$

where the sign in front of the roots was chosen as positive because κ_ϑ and n_ϑ are defined as real entities. By substituting the value obtained for n_ϑ in Eq. (1.9.22), we can calculate the angle θ.

The second step is to calculate the coefficients of reflection from the surface of a medium characterized by the complex index $\tilde{n} = n + i\kappa$. As we noted at the end of the preceding section, we can still decompose the three waves involved in reflection and transmission, according to Eqs. (1.7.7), into TM and TE waves, for which the magnetic and electric fields are respectively orthogonal to the plane of incidence. By choosing a Cartesian system such as that shown in Fig. 1.11, we have, for example, that, for a TM wave, the ratio between the components z and x of the amplitudes of the electric field of the inhomogeneous refracted wave, for Eq. (1.9.20), is given by

$$\tan\vartheta_{\mathrm{T}} = -\frac{E_z}{E_x} = \frac{k_{\Re x}}{k_{\Re z} + ik_{\Im}} = \frac{n_\vartheta \sin\theta}{n_\vartheta \cos\theta + i\kappa_\vartheta},$$

where ϑ_{T} is the complex angle of refraction, which does not coincide with the angle θ introduced with Eq. (1.9.22). In the last step, Eqs. (1.9.23) were applied. By squaring and applying Eqs. (1.9.25), we obtain

$$\frac{\sin^2\vartheta_{\mathrm{T}}}{1-\sin^2\vartheta_{\mathrm{T}}} = \frac{n_{\mathrm{I}}{}^2 \sin^2\vartheta_{\mathrm{I}}}{(n+i\kappa)^2 - n_{\mathrm{I}}{}^2\sin^2\vartheta_{\mathrm{I}}},$$

from which we ultimately find that

$$\sin\vartheta_{\mathrm{T}} = \frac{n_{\mathrm{I}}}{n+i\kappa}\sin\vartheta_{\mathrm{I}}, \quad \cos\vartheta_{\mathrm{T}} = \sqrt{1-\frac{n_{\mathrm{I}}{}^2}{(n+i\kappa)^2}\sin^2\vartheta_{\mathrm{I}}}\,. \tag{1.9.28}$$

Formally, then, the Snell's equation and the Fresnel's equations (1.7.10) and (1.7.11) are still valid under the condition of using, as an index of refraction of the medium for the transmitted wave, the complex index $\tilde{n} = n + i\kappa$ and the complex value of $\cos\vartheta_{\mathrm{T}}$ given by Eq. (1.9.28.b).

For normal incidence, we have

$$r = \frac{n_{\mathrm{I}} - n - i\kappa}{n_{\mathrm{I}} + n + i\kappa}, \quad t = \frac{2n_{\mathrm{I}}}{n_{\mathrm{I}} + n + i\kappa}, \tag{1.9.29}$$

from which the reflectivity and the transmissivity are given by

$$R = \frac{(n_{\mathrm{I}} - n)^2 + \kappa^2}{(n_{\mathrm{I}} + n)^2 + \kappa^2}, \quad T = \frac{4nn_{\mathrm{I}}}{(n_{\mathrm{I}} + n)^2 + \kappa^2}, \quad R + T = 1, \tag{1.9.30}$$

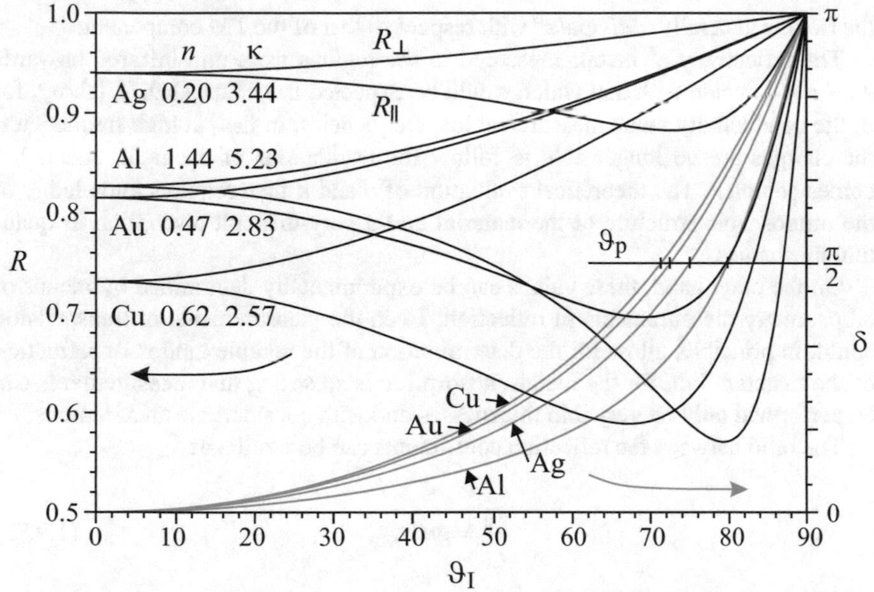

Fig. 1.24 Reflectivity R (in black) and relative phase shift δ (in gray) calculated for known values of n and κ of some metals at $\lambda_o = 589.3$ nm. The horizontal line at π/2 indicates the values of the *principal angle of incidence* ϑ_P, which *does not* coincide with the minimum reflectivity of $R_{||}$

where the transmissivity T refers to the wave immediately after the interface. For metals, the reflectivity can be very high and close to 1, in particular, when n is small and κ is large. For sodium, for example, we have $n \cong 0.044$, $\kappa \cong 2.42$ and $R \cong 0.97$ at $\lambda_o = 589.3$ nm.

For non-normal incidence and $\mu_I = \mu_T$, the reflection coefficients are

$$r_\perp = \frac{n_I \cos\vartheta_I - (n+i\kappa)\cos\vartheta_T}{n_I \cos\vartheta_I + (n+i\kappa)\cos\vartheta_T}, \quad r_{||} = -\frac{(n+i\kappa)\cos\vartheta_I - n_I\cos\vartheta_T}{(n+i\kappa)\cos\vartheta_I + n_I\cos\vartheta_T}, \quad (1.9.31)$$

where $\cos\vartheta_T$ is given by Eq. (1.9.28.b). The reflectivity is given by the square modulus of these coefficients and its trend with the angle of incidence is shown in Fig. 1.24 for the known values of some metals at a wavelength of 589.3 nm. In particular, for gold, at 4.3μm and $\vartheta_I = 45°$, $R_\perp = 0.984$, $R_{||} = 0.969$, and $\delta = 3.05°$. The analytical calculation of these values is very complicated, and we might as well proceed numerically. The curves for $R_{||}$ have a minimum similar to the case of reflection from the dielectrics, at the Brewster's angle. In contrast to these, however, the relative phase shift $\delta = \delta_{||} - \delta_\perp$ between the TM and TE reflections varies continuously, from 0 at normal incidence to π at grazing incidence. Therefore, there is an angle of incidence that takes the name of the *principal angle*, for which the phase shift is π/2. In a fixed position, the oscillation of the TE component of the field is generally *anticipated* with respect to that of the TM component.

The reflectivity of metals measured in the regions from mid-infrared onwards does not coincide with that which would be expected from Eqs. (1.9.7), taking, for σ, the conductivity value measured at low frequencies. In fact, at high frequencies, the charges are no longer able to follow the oscillations of the field, and σ becomes complex. The theoretical evaluation of n and κ then requires knowledge of the microscopic structure of the material and a very difficult discussion in quantum mechanics.

On the other hand, these values can be experimentally determined by means of ellipsometry measurements in reflection. Even the measurements in transmission could, in principle, allow for the determination of the complex index of refraction of the material, but, for the metals, absorption is so strong that measurements can be performed only on very thin thicknesses and with considerable uncertainties.

The ratio between the reflection coefficients can be written as

$$\frac{r_{\parallel}}{r_{\perp}} = \rho e^{i\delta} . \tag{1.9.32}$$

With some passages from Eq. (1.9.31), we get

$$\frac{1-\rho e^{i\delta}}{1+\rho e^{i\delta}} = \frac{\sqrt{(n+i\kappa)^2 - n_I^2 \sin^2 \vartheta_I}}{n_I \tan\vartheta_I \sin\vartheta_I}, \tag{1.9.33}$$

from which it is possible to obtain the complex index of refraction of the material

$$(n+i\kappa)^2 = n_I^2 \sin^2 \vartheta_I \left[\left(\frac{1-\rho e^{i\delta}}{1+\rho e^{i\delta}} \right)^2 \tan^2 \vartheta_I + 1 \right]. \tag{1.9.34}$$

For the principal angle of incidence, we have $\delta = \pi/2$, $e^{i\delta} = i$, and thus

$$\begin{aligned} n^2 - \kappa^2 &= n_I^2 \sin^2 \vartheta_p \left[\cos 4\Psi_p \tan^2 \vartheta_p + 1 \right], \\ 2n\kappa &= -n_I^2 \sin^2 \vartheta_p \tan^2 \vartheta_p \sin 4\Psi_p, \end{aligned} \tag{1.9.35}$$

where the replacement $\rho = \tan\Psi$ has been made.

The ellipsometry consists of sending a monochromatic collimated beam along the surface under study, linearly polarized with an azimuth Ψ_I of 45°, for which the amplitudes of the TE and TM components are equal, and to measure the polarization of the reflected beam. It possesses an elliptical polarization whose axes are rotated with respect to the incidence plane, except for the principal angle of incidence. On this beam is placed a variable birefringence compensator in order to cancel the relative phase shift caused by the reflection, and return a linear polarization of the emerging beam, whose azimuth Ψ is measured by means of a polarizer. The compensator may be replaced with a birefringent plate "lambda/4", with its "fast" axis on the plane of incidence. Looking for the principal value of the angle of incidence ϑ_p, for which the polarization emerging from the plate is linear, the principal azimuth Ψ_p can be measured. Having obtained these values, it is possible to determine the constants of the material through Eqs. (1.9.35).

The modern ellipsometers perform the analysis of surfaces and thin films at various wavelengths and possess actuators with which the angle of incidence is changed with continuity; the reflected radiation is analyzed by means of compensators and polarizers, and even the degree of depolarization is determined. They are interfaced to a computer with a program that determines various parameters, according to a predetermined model, appropriate to the test surface.

Bibliographical references

VV. AA.*Enciclopedia della Scienza e della Tecnica*, A. Mondadori ed., Milano (1963).

Barnett S.M., *Resolution of the Abraham-Minkowski Dilemma*, Phys. Rev. Lett. **104**, 070401 (2010).

Becker R. and Sauter F., *Electromagnetic Fields and Interactions, Vol. I, Electromagnetic Theory and Relativity*, Blaisdell Publishing Company, New York (1964).

Born M. and Wolf E., *Principles of Optics*, Pergamon Press, Paris (1980).

Boyd R.W., *Nonlinear optics*, Academic Press, Orlando (2008).

Cicogna G., *Appunti di Metodi Matematici della Fisica*, Università degli Studi di Pisa (1969).

Delone N.B. and Krainov V.P., *Fundamentals of Nonlinear Optics of Atomic Gases*, John Wiley & Sons, New York (1988).

Diels J-C. and Rudolph W., *Ultrashort Laser Pulse Phenomena*, Academic Press, San Diego (1996).

Feynman R.P., Leighton R.B., Sands M., *The Feynman Lectures on Physics, Vol II*, Addison-Wesley publishing company, Reading, MA (1969).

Garrison J.C. and Chiao R.Y., *Canonical and kinetic forms of the electromagnetic momentum in an ad hoc quantization scheme for a dispersive dielectric*, Phys. Rev. A **70**, 053826 (2004).

Gibson A.F. et al., *A study of Radiation Pressure in a Refractive Medium by the Photon Drag Effect*, Proc. R. Soc. Lond. A **370**, 303-311 (1980).

Gordon J.P., *Radiation Forces and Momenta in Dielectric Media*, Phys. Rev. A **8**, 14 (1973).

Guenther R., *Modern Optics*, John Wiley & Sons, New York (1990).

Hopf F.A. and Stegeman G.I., *Applied Classical Electrodynamics, Vol. I: Linear Optics*, John Wiley & Sons, New York (1985).

Huard S., *Polarization of Light*, John Wiley & Sons, New York, Masson, Paris (1997).

Jackson J.D. , *Classical Electrodynamics*, John Wiley & Sons, New York (1974).

Landau L. and Lifchitz E., (II), *Théorie du Champ*, MIR, Mosca (1966a). (VIII), *Électrodynamique des Milieux Continus,* MIR, Mosca (1966b).

Landsberg G.S., *Ottica*, MIR, Mosca (1979).

Mansuripur M., *Classical Optics and its Applications*, Cambridge Univ. Press, Cambridge (2002).
Nahin P.J., *Oliver Heaviside*, Le Scienze **264**, 74-80 (August 1990).
Novozhilov Yu.V. and Yappa Yu.A., *Electrodynamics*, MIR, Mosca (1981).
Panofsky W.K.H. and Phillips M., *Elettricità e Magnetismo*, Ambrosiana, Milano (1966).
Peierls R., *The Momentum of Light in a Refracting Medium*, Proc. R. Soc. Lond. A **347**, 475-491 (1976).
Sivoukhine D., *Optique*, MIR, Mosca (1984).
Someda C.G., *Onde elettromagnetiche*, UTET, Torino (1986).
Sommerfeld A., *Optics*, Academic Press, New York (1949).
Tropf W.J., Thomas M.E., and Harris T.J., *Properties of crystals and glasses*, in *Handbook of Optics*, Cap. 33, M. Bass et al. ed., McGraw-Hill, Inc., New York (1995).
Van Kranendonk J. and Sipe J.E., *Foundation of the macroscopic electromagnetic theory of dielectric media*, in *Progress in Optics XV*, 245-356, E. Wolf ed., North-Holland Publishing Company, Amsterdam (1977).

Part II
Geometrical Optics

Historical notes: From Empedoklēs to Huygens

The dawning

The phenomenon of reflection has been known since prehistoric times, and mirrors were the first optical devices to be realized: ancient metal mirrors, by now corroded, can still be seen in the archaeological museums. Already in the eighth century BC, the bronze mirrors of Kórinthos and Athína were famous for the delicacy and originality of their processing. Their first written mentions are in the Book of Exodus (38,8), in which it is described that Bezalel used the copper of the mirrors of the women to build the laver for ablutions and its pedestal, for the altar of the Sinaitic Alliance. Moreover, in the *Book of Job* (37,18), the brightness of the sky is compared with that of the molten metal mirrors (polished bronze). Lenses were also known in antiquity: the Athenian Aristophanes (480–388 BC), in his comedy *Nephèlai* (The clouds) (424 BC), alludes to a clear and transparent glass sold in shops and commonly used to initiate fire. The historian Plinius (23–79 AD) describes how the Romans owned burning glasses; various glass spheres have been found in archaeological excavations, including a convex lens in Pompei. Also, Seneca (3 BC. - 65 A.D.) noted that a water-filled glass globe could be used to magnify objects.

The first theoretical developments were done in Greek antiquity. The starting point for the ancient Greek philosophers was humans themselves and their senses. There did not exist the sharp separation between the psyche and external reality that we perceive now. More than the nature of light, they were interested in the process of vision, and light phenomena were interpreted in a comprehensive manner, according to a syncretic conception of nature, in an attempt to establish relationships between two elements identified as fundamental: the eye and the object seen [Bevilacqua and Ianniello 1982].

Light was an integral part of the sense of sight, up to the point of imagining that the light *came out* of the eyes in order to see things, as, with the sense of touch, the hands are used *to touch* things. This, very briefly, was the basic idea, the *metaphysics*, of one of the ancient Greek schools of thought, which dates back to Pythagoras and had, as its supporters, Eukléidēs, Heron, and Ptolemaíos. The eye was considered as a point from which *the rays* are *emitted*, each imagined as a kind of "visual tentacle", which opens in the shape of "cone" or "pyramid", with its base on what was being seen. Yet, this was a very prolific line of thought, since it had, as a guiding principle, the datum of the rectilinear propagation of light: from it, the idea of Perspective and Geometrical Optics were developed. Eukléidēs (~ 330–260 BC), in his book *Optics,* for the first time, structured the optics of vision and perspective in axioms and theorems. There, we find the law of rectilinear propagation of visual rays, the conditions under which things can be visible or not and, implicitly, also the idea of the resolving power of the eye, so that things that are too small or far away may escape observation, while «those things that are seen under more angles, we see more clearly». In *Catoptrics*, which is a book attributed to Eukléidēs,[1] there is a geometrical demonstration of the reflection law, for which «the visual rays are reflected at equal angles», both by flat mirrors and by curved.

Heron of Alexandria (first century A.D.), too, in his *Catoptrics*, dealt with reflection, spherical and cylindrical mirrors, and the various optical tricks that one can do with these. Heron was among the first to notice an analogy between reflection and the mechanical example of a projectile over a surface. If the surface is soft, the projectile stops, and, similarly, visual rays are reflected from shiny surfaces rather than rough ones. Heron believed that shiny surfaces, as opposed to rough ones, are so compact as not to allow the rays to fall within the fissures. Instead, glass and water reflect light only partially, since they consist of both compact parts that reflect the rays that touch them and porous parts that let them pass through. Not satisfied by the geometric affirmations of Eukléidēs, Heron looked for a more profound reason for the rectilinear propagation of rays and the law of reflection. He first brought both of these laws back to the teleological principle of minimum distance, which has demonstrated a great heuristic capacity in the development of Science, and which we now interpret as a variational principle. In common with other philosophers of his school, he believed that the speed of ray propagation is infinite. Heron gives a splendid demonstration of this: «when, after closing the eyes we open them and look at the sky, the visual ray does not need time to reach the sky. In fact we see the stars in the same instant in which we look them, although the distance is, so to say, infinite».

Refraction was studied by Kleomedes (~ 50 AD), and then by Kláudios Ptolemâios (~ 90 – ~168), who, in his book *Optics*, treated the relationship between light and eye, and the visibility of bodies with respect to their shape, color and motion, subsequently explaining why we see only one image despite having two eyes, and ultimately studying reflection and refraction. In addition to proving the law of reflection experimentally, he also sought an explanation with the mechanical example of a ball thrown against a wall, noting that the movement of the ball is opposed in the direction perpendicular to the wall, but not tangentially. Ptolemaíos ultimately measured, with good precision, the angles of refraction versus the angle of incidence between air and water, air and glass, and water and glass: measures that posterity wrongly interpreted as a linear relationship between the two angles. However, the absolute constancy of the second differences in his tables (0.5° for all three cases!), with a step of 10° for the angle of incidence, suggests that he wanted to accommodate the measures with a parabolic law, which, in any case, matches very well with the real data. As for reflection, he compared refraction

[1] Some believe that Catoptrics contains elements from many centuries later, written by subsequent authors.

with a mechanical example, that of the passage of a projectile from one medium to another of different density. Ptolemaíos was a distinguished astronomer, author of the *Almagest*, and discussed atmospheric refraction, underlining that only at the zenith did the position of the stars coincide with that apparent.

Other schools were also active. One was that of the atomists, such as Leúkippos (~ 480– ~ 420 BC), Dēmókritos (~ 460 – ~ 360 BC), Epíkouros (340–270 BC) and Lucretius (~97–55 BC), who anticipated the corpuscular hypothesis. They argued that any sensation takes place by virtue of a contact, but, since our soul does not travel outside of us to touch the external bodies, it is necessary that these come to us so as to touch our soul, and, since the bodies will not actually come to meet us, it is necessary that they send the images that represent them to our soul. So, in their theory, a flow of *sloughs*, or *simulacra* (ειδολα), come off of the bodies while retaining their shape and color, and invest themselves in the eyes, determining vision, in analogy to the bodies that emit sounds and the fragrant substances that emit particles that affect the sense of smell. Democritus believed that the colors were only *appearances*: in his conception, atoms are individually devoid of *qualities* such as color, while compound substances present a color that depends on the disposition in which their elements are arranged; the appearances of color are thus presented through four kinds: white, black, yellow and red. Lucretius, in a beautiful page from his *De rerum natura* (IV, vv. 379-461), in a section dealing with optic illusions, was among the first to notice the difference between reality and the representation that our psyche conjures from it, asserting that, «nec possunt oculi naturam noscere rerum: proinde animi vitium hoc oculis adfingere noli»[2] (IV, 385-386).

The schools of Empedoklēs (~ 490 – ~ 435 BC) and Plátōn (427–347 BC), student of Socrates, both asserted the hypothesis of a double emission, combining the two preceding theories. Plátōn supported the necessity that vision is determined by the meeting of an external agent, the glimmer that comes off of the object and produces color, with the visual fire that is inside of us, emitted by the eyes. In this way, he could explain the inability to see in the dark, and even dreams, as being a result of the internal fire. Aristotélēs (384–322 BC.), in turn, a pupil of Plátōn, instead considered light not as a corporeal substance, but rather as a modification of the medium (the *diaphanous*) induced by the bodies, anticipating the idea of an *aether*, of which the eye perceives the "movement". Unlike Empedoklēs, this modification was conceived by Aristotélēs as a quality that the whole medium acquires simultaneously, and thus he was among the first to support the idea that light is propagated instantaneously. All of these schools coexisted for many centuries and, in each of them, we can find ideas, purified by time, that still survive in modern theories.

As an application of an optical device, must be mentioned (according to the Roman historians) that Archimedes (287-212 BC) put one of Eukléidēs' state-

2 "eyes cannot know the nature of the things: do not then impute to the eyes the error of the spirit".

ments in the *Catoptrics* (the 31st) into pratice, that in which it is possible to light a fire with a concave mirror opposite the Sun. Indeed, he used *burning mirrors* to set fire to the Roman ships that were besieging Siracusa. Still today, we speak of *focal* planes and *to focus*! Solar thermal furnaces operate with the same disposition as Archimedes' mirrors.

There is one aspect of ancient philosophy that needs to be considered for its serious consequences in the development of Science: Plátōn, in particular, insisted thoroughly on the consideration that the senses are imperfect, and therefore misleading; thus, the "apparent world" represented by the psyche is not guaranteed to correspond to the "real world". This line of reasoning, shared by all ancient philosophers, had two important effects. The first was to establish a hierarchy of reliability of the senses. To touch was granted the greater confidence, while sight was considered the most fallacious sense. Any tool placed between the eye and the object seen, "to complicate things", was treated with contempt. The second effect was more subtle: the only reliable tool that remained available for comprehension of the world was the mind; abstract speculation, and, in particular, perfect, mathematic idealization, was given, until the Renaissance, exaggerated status by scholars compared to observation and experimental research [Ronchi 1983].

The reduction of the eye to a single point had constituted a good schematization for the development of perspective with the effective idea of visual rays. However, it was an obstacle for the comprehension of the intimate mechanism of vision. The following quality jump was made by a physician, Claudius Galenus of Pergamon (129–200), who started a process of conceptual separation between physical and physiological processes. He studied the anatomy of the eye and gave body to the Platonic idea of a double flow: a *luminous pneuma*, coming from the brain, flows along the optic nerve and the retina to interact with the light that reaches the eye from the outside. This interaction then produces the images in the *crystalline* (!), which are finally transmitted back to the brain via the retina and the optic nerve.

The Arab period

From the Greek-Roman world, the torch of culture passed to the Arab world, and, in the 9th and 10th centuries a major medical school flourished in Baghdad. One particularly developed sector was the study of the sicknesses of eyes, and the Arab doctors translated the works of Galenus into Arabic and dealt with the structure of the ocular globe and the visual apparatus.

A great scientist who worked to found a theory of vision on physical and physiological bases was Abu Yūsuf Ya'qūb ibn 'Isḥāq aṣ-Ṣabbāḥ al-Kindī (~ 813 – ~ 873), who lived in Basra and in Baghdad. He was known in the West under the name of Al-Kindi and, among his works translated into Latin, two that are particularly worth mentioning are the *De radiis stellatis* and the *De aspectibus*. The first one is essentially an astrological work that deals with the stellar influences on the terrestrial world. Here, Al-Kindi introduced the idea of the "rectilinear propagation" of forces: he believed that the Sun, stars, magnets, fire, and sound all irradiate a power on the objects that surround them that somehow modifies them. Therefore, these forces might explain natural phenomena. Concordantly, in *De aspectibus*, Al-Kindi affirms that vision does not happen by means of simulacra or visual rays, but as a result of the effect of rectilinear rays emitted by each point of the illuminated bodies capable of acting on the eye. This last assumption finally freed light from the Greek/Roman conceptions in which things were considered visible only globally, through a *cone* of rays, or as the entrance into the eye of a coherent, atomistic forms that still had not borne any fruit.

About a century later, Abū ʿAlī al-Ḥasan ibn al-Haytham (~ 965–1039), known as Alhazen, made many observations of a physical and physiological character, including a careful description of the eye's anatomy. He concluded that light is an external agent that penetrates into the eyes and propagates in a straight line, arguing that the movement of light takes a finite, although imperceptible, time. To him, we owe the first description of a darkroom with a small hole that he used to observe solar eclipses. He then rejected the Platonic compromise of a double flow and, finally, he created a synthesis between the idea of light as an external entity and the idea of a ray. Reversing the conceptions of perspective, the light flow now has, for a vertex, the points of the luminous object and, for a base, the pupil of the eye. Nevertheless, Alhazen was considering only the principal ray of this cone, normal to the surface of the eye, and was identifying the seat of the imaging in the crystalline as the correspondence point to point with the object. However, he was still speaking of light as consisting of *forms* coming from the visible objects, which, reaching the eye are then "fixed" to the surface of the crystalline. He knew the principle of the *camera obscura*, and, with this, he probably wanted to avoid the problem of the image inversion that occurs on the retina [Ronchi 1983]. The optic nerve was then identified as the channel leading the light signal to the brain. Among other things, he explained the phenomenon of binocular vision, assuming that the signals of the two eyes are rejoined in a visual nerve. He then described the optic illusions that derive from a false interpretation of a figure by the human mind and its imagination.

Alhazen also studied the refraction by glass spheres and reflection, correcting the mistake of the Greeks concerning the position of the images generated by curved mirrors. In particular, he resumed the mechanical concepts of Heron and Ptolemaíos on reflection, attributing to reflecting bodies the properties of a repulsion or opposition force that is greater for the polished bodies compared to those that are rough. Like Heron, he noticed their similarity with hard or soft bodies and attributed the responsibility of reflection to the compactness, and not to the rigidity. Nevertheless, he declared that some rough bodies diffuse the light, since they are composed of small shiny parts variously oriented, while the pores between the compact pieces of the rough bodies permit a part of the light to penetrate the bodies without then re-emerging outside. Following Ptolemaíos, Alhazen developed considerations of mechanical nature, with the example of an iron or copper sphere that hits an iron mirror: he explained the reflection through the decomposition of the movement in a tangential component and a perpendicular one to the reflecting surface. While the tangential component is not modified, the perpendicular is first canceled by the collision with the mirror and then the opposite movement is acquired by the force of repulsion, for which the reflective body acts as a spring. In this way, he was explaining how light is reflected at equal angles on the plane of incidence, orthogonal to the mirror. More complex is the case of refraction. Alhazen assumed that the speed of light was a property of the medium and that, in particular, this speed was greater in rarefied bodies than in those that are dense: because of their density, all transparent media resist the movement of the light, and the denser they are, the greater the resistance they offer. In refraction, he offered, as examples, a projectile, thrown against a table, and a sword that strikes a piece of wood, noticing that the effect of the greatest breakage is when the motion is orthogonal to the surface. He therefore assumed that, in refraction, the medium offers a greater resistance in the direction parallel to the surface, for which the light is deflected towards the normal direction.

Two Arab philosophers, Abū ʿAlī Ḥosayn ibn Sīnā, known as Avicenna (Bukhara, 980–1037), and Abū l-Walīd Muḥammad ibn ʾAḥmad ibn Rushd, known as Averroes (Cordoba, 1126–1198), began the debate on the distinction between *lumen* and *lux* [Bevilacqua and Ianniello 1982]. Indeed, Avicenna distinguished the *lux*, or "brightness", which is seen in bright objects like fire and Sun, from the *lumen*, or "splendor", which would be the effect of the lux on the medium and on the illuminated bodies. He also emphasized the psychological aspects of the lux and reduced the physical ones of the lumen to material forms or species, stimulating a revival of the atomistic theory of the emission of simulacra from objects, while leaving aside the details of the nature of the simulacra themselves. The relationship between lux and lumen, however, implied a change in the Aristotelian concepts in which light was a state of the medium, whereas it was now conceived as a quality of luminous objects that is perceived through the lumen. The problem then arose of defining the nature of lumen and color in the medium. We have already seen that, for Alhazen, the lumen was conceived as a *form* of lux. Averroes instead distinguished between the spiritual existence and the corporeal existence of the light and the colors: in the soul, they would have a spiritual existence, in transparent bodies, an intermediate existence between the spiritual and the corporeal. Over a long perio of of evolution, the lumen was taking on the meaning of an objective external physical agent of vision, while the subjective, psychologic aspects remained in the lux, such as, for example, the color, of a process that the ancient authors considered to be unitary. While the first term became the field of study of physics, particularly from 1600 onwards, the second was, for a long time, an apanage of philosophers who put the debate over the lux into that of the "metaphysics of light" to which, prior to Averroes, St. Augustinus (354 –430), in particular, had given impulse. Instead, the *sloughs*, which had taken the name of *species*, and the lumen were discussed for the specific purpose of determining whether they were "substances", "accidents" or "qualities".

The Middle Ages

Between 1100 and 1200, many Arab works were translated into Latin. In addition, the fundamental Greek works, which the Arabs had preserved, were retranslated from Arabic and returned to the West, where they rekindled the philosophical debate. The first Universities were born, such as those of Bologna, Paris, and Oxford, which initially served exclusively for the education of the clergy. Among the various thinkers who worked there, we can recall Robert Grosseteste, the Franciscan Roger Bacon, and John Pecham. At Oxford, Grosseteste (1175–1253), who was bishop of Lincoln, acquired, from the Arabic culture, the conception, particularly that of Al-Kindi, of a universe dominated by species that propagate through space and are the cause of all phenomena: in his metaphysical conception, the lux is the fundamental corporeal form and the principle of motion, from which the universe was formed. He thus assigned to Geometrical Optics the role of the main science of nature. Grosseteste also had the merit of writing a book about the rainbow, *De Iride*, which finally put forward the hypothesis that the colorful arc derives from the refraction of solar light in a cloud, just as colors are formed when light passes through a water carafe [Hall and Hall 1979]. Roger Bacon (1214–1294), a disciple of Grosseteste and heir to the neo-Platonic, mathematicising tradition, resumed Alhazen's work, during which he defended the idea of a finite speed of light, against the opinion of Aristotélēs and Al-Kindi. Bacon identified Alhazen's forms with the neo-Platonic species, and unified the Euclidean tradition with that of Aristotélēs and Galenus. In particular, he thought that the species emitted by the visible bodies reached the eye of the observer adapting themselves to the surface of the crystalline with the same spatial order as the points that had generated them. Advocating the need to experiment, he studied the formation of images in a darkroom, refraction, and the rainbow, for which he gave an almost exact value for the angle between the Sun, the center of the rainbow and the observer's eye. He also associated the rainbow's colors with those produced by a prism, discovered the spherical aberration of mirrors, and conducted experiments with parabolic mirrors and plano-convex lenses. John Peckham (~1230–1292), archbishop of Canterbury, wrote a text that formed the basis for the Renaissance's pictorial perspective. A detailed study of the rainbow, with its primary and secondary arches, was also made by a German Dominican, Theodoric of Freiberg (~1250–~1310), in the wake of the Aristotelian tradition. He conducted several experiments in support of the idea that the rainbow is caused by two refractions and one reflection on the surface of spherical droplets, and in support of his explanation, retrieved from Averroes, that colors come from different mixtures of light and dark.

However, the principal mediator of the studies of Alhazen was the Polish mathematician Erazm Ciolek (1230–1275), much better known as Witelo or Vitellione, who lived and studied in Italy. His work was widely studied in the Middle Ages, first in manuscript editions and later published in 1533. Among his contributions were the explanation of the colors of the rainbow as having originated in refraction and reflection in water droplets, and the proposal of a parabolic mirror for focusing sunlight. Like Alhazen and Bacon, he believed that transparent media resist the propagation of light according to their density, and gave an explanation of refraction similar to that of Alhazen. Finally, in 1572, the Alhazen manuscript was translated into Latin and published in a volume in folio, followed by the 10 books by Witelo.

The Renaissance

After Alhazen and Bacon, the principal contributions in the field of Optics did not come from learned men, but rather from the artisans. Through methods that were initially random and confused, but gradually became more systematic and exact, they developed techniques and tools that would serve as one of the principal causes of the scientific revolution. In particular, lenses (or glass "lentecchie" = lentils), rediscovered and probably imported from Islam, spread in the West. Towards the end of the 13th century, eyeglasses were invented to correct presbyopia, but their inventor, most probably an inhabitant of Pisa, remains unknown [Rosen 1966]. With the development of Venetian glass art, eyeglasses were already being produced and traded in 1301. The first painter to represent them was Tommaso da Modena, in a painting from 1352, *Il cardinale Ugo di Provenza*, preserved in the Episcopal Seminary of Treviso. The way that lenses functioned caused a crisis in the theories of late medieval scholars, still struggling with the species (the heirs of simulacra) or the visual rays. In opposition, they maintained that the lenses merely produced optical illusions, on the edge of magic, in a historical period when it was highly suspicious for a science of vision to be so tied to philosophy and theology.

The most distinguished artisan was Leonardo da Vinci (1452–1519), who made a series of observations, mainly of Physiological Optics, related to the study of perspective. Among these, he noted that the pupil dilates in darkness, and discussed how glasses can help our sight. His idea of light was that of species that propagate in a straight line and that emanate from illuminated bodies; he also worked on photometry, describing an apparatus for comparing light sources, based on the shadows produced by a common obstacle. With this concept, Leonardo explicitly associated the functioning of the eye with that of the darkroom: he compared it to the formation of images that are produced on a screen as the result of the light passing through a small hole, and also noted that these images are reversed. Based on his anatomical studies, Leonardo surmised that the rays were focused in the eye twice, first by the cornea and then by the crystalline, to which he assigned the task of straightening the images that were formed on the "sensitiva", that is, on the retina. He drew the crystalline as if it had a spherical shape, because of the deformation that the eyes underwent during his *boiling* treatments, and this prevented him from understanding the real path of the rays.

Leonardo also dealt with lenses and mirrors. In his manuscripts, there is a project concerning a machine for grinding concave mirrors, a discussion of the production of lenses for eyeglasses, a description of how to use a single lens to observe distant objects as enlarged, and, in codex F, folio 25, it seems that there is even a description of a telescope with two lenses of the Galilean type [Argentieri 1939]. However, his notes remained hidden, and were substantially without effect in the development of optics.

In the Italian Renaissance, we must also remember the learned men Giovanni Rucellai (1475–1525), Gerolamo Cardano (1501–1576) and Daniele Matteo Alvise Barbaro (1514–1570). Rucellai was the first to take advantage of a concave mirror for use as a simple microscope, using it to observe bees. Cardano described the "rainbow colors" produced by glass prisms in the light of the sun, and, for the first time, mentioned the application of a lens in a hole in the wall of a darkroom, for the purpose of seeing images «of what happens on the road». However, it was Barbaro who furnished the most accurate description of the images projected onto a white screen by a converging lens, proposing its use in the pictorial art of perspective.

According to the general rule that science proceeds whenever a new tool that demonstrates value and practical utility is introduced, in this case, lenses and subsequently the telescope, a true revolution shocked the academic world. This happened through the work of people of great talent, some of whom dared to challenge even that infamous institution that was the court of the inquisition.

Giovan Battista della Porta (1535–1615) spent much of his life wandering in Europe interviewing scholars and artisans, looking for «maraviglie» (wonders) and natural attractions. In 1589, he published his *Magia Naturalis*, a miscellany of unusual phenomena and beliefs of every kind, devoting an entire chapter to optical devices and lenses, including a description of the application of a lens to the hole of a darkroom and recognizing the analogy with the eye. Among many other things, along the same line as Heron's tricks, in this book, he describes how to use plane, concave, and cylindrical mirrors to produce some amazing effect, and also the combination of flat mirrors. Besides this, he mentions the numerous uses for lenses: as viewing devices, as simple microscopes, to light fires, and the «composition» of a positive lens and a negative lens (!) to improve vision. He decidedly opposed the tradition that regarded sight as a deceptive sense, and the lens as a fallacious device, and, indeed, argued that lenses are useful and even necessary to human life. Lenses had, in fact, already become widespread in correcting both presbyopia and myopia, while the academic scholars limited themselves to ignoring them, being unable to explain how lenses worked with the optical theories of that time. Della Porta instead made this gap evident to a wide audience, as proof of the "magical" vein of the spread of the experimental method. Therefore, these theories had to be reformed, and, although in a confused and substantially incorrect way, he was the first to attempt a theoretical treatment of lenses and mirrors, publishing it in his *De Refractione* in 1593. When Galileo published his astronomical discoveries, Della Porta tried to claim the priority of the invention of the telescope, which was, after all, credited to him by the members of the *Accademia dei*

Lincei, to which he belonged. Already old and sick, he began to write a book on the telescope, but, in the attempt to explain its functioning, he encountered so much difficulty that it disheartened him deeply. The manuscript was lost, and was only recently found in the library of the *Accademia dei Lincei* and published by V. Ronchi.

The contribution of Francesco Maurolico da Messina (1494–1574) was of extraordinary scientific value, but, unfortunately, it was long hidden to his contemporaries. Maurolico was a Franciscan father who must have had significant knowledge of the work of Alhazen, being the son of an exiled doctor from Byzantium. Among other things, he was greatly concerned with finding the scientific works of the past: it is to his collecting that we owe thanks for what has survived of the works of Archimedes, Theodoros of Cyrene, Apollonios of Perga, and others. 50 years ahead of Kepler, he resolved, or at least had frequent success in setting up, many optical and vision problems. In particular, he defined rays as being emitted continuously in all directions from each point of a bright body. He considered them as having been provided with different "densities" and immediately applied this concept to the illumination intensity produced by beams of different densities and inclinations. He studied shadows, diaphragms, refraction, and reflection with flat and spherical mirrors, eventually arriving at an explanation of the formation of images by spherical mirrors, noticing that they must have a smaller diameter than the radius of curvature and that, in general, the rays coming from a single point form caustics. He determined the anatomy of the eye well beyond the knowledge of Galenus and Alhazen. He attributed to the crystalline the task of refracting the rays in analogy with lenses, explained myopia and presbyopia attributing them to a faulty curvature of the crystalline lens compared to the depth of the eye, and ultimately came to consider the retina as the seat of image formation, but upright! This despite having well understood that these images are formed not only from the principal ray, but also for the convergence of all rays collected through the pupil and refracted on the retina, in correspondence with each point of the thing observed. He wore glasses to correct his presbyopia and correctly explained the functioning of them, although only intuitively, together with that of the divergent lenses for correcting the "short-sight" of young people. His manuscripts *Photismi de lumine et umbra ad perspectivam, et radiorum incidentiam facientes* and *Diaphanorum libri* were dated to 1555 and 1554, respectively, but were, unfortunately, only published posthumously in 1611.

The New Science

The Italian Renaissance should be remembered not only for its arts, but also for its culture. The Italian Universities of the 16th century exceeded all others for the level and originality of their teachings, also attracting many personalities from abroad. Padova excelled in medicine and in research on the logic of experimental science. Bologna was famous for astronomy: notably, Mikołaj Kopernik, better known as Nicolaus Copernicus (1473–1543), studied there. The Ptolemaic astronomical system of cycles and epicycles, impregnated by the hierarchical conception of Aristotélēs, was beginning to show its flaws: it was complicated and its predictions could be as much as weeks off from what was learned through observations. The same papacy had stimulated research into a new system so as to rectify the errors of the calendar, upon which the calculation the date of Easter depended. The accurate prediction of astronomic phenomena was also becoming crucial for the determination of the longitude in oceanic navigation and geographic exploration. Among the various attempts, Kopernik's was successful, and his tables were subsequently used for thc reform of the calendar of Pope Gregorio XIII, introduced in October of 1582. While Ptolemaíos was limited to giving the apparent position of the planets as seen from Earth, Kopernik also wanted to determine their distances, and thus obtain their real motion. Driven by a Platonic conception of the harmony of the world, he found that, putting the Sun in the center of the system, the orbits of the planets were simplified and, above all, that their orbital periods formed a regular progression with the distances from the Sun. However, he considered only circular orbits, and therefore the epicycles were not all eliminated (the Sun also had one, concentric with the Earth's orbit; nevertheless, for that time, it was a very good approximation). The next step was accomplished by Kepler, with the discovery that orbits are elliptical. Nevertheless, a persuasive proof that the new astronomical order was real, and not just a geometrical artifice, was still lacking. This task fell to Galileo.

In 1590, unknown artisans, probably opticians, built the first telescope ever recorded, just one year after the publication of *Magia Naturalis*, in which Della Porta made an obscure consideration on the combination of a converging lens with a diverging one. According to a document from 1634, one of these initial specimens was built in Italy and brought to the Netherlands, where, in 1604, it was reproduced by Zacharias Jansen (1588–1632) and distributed commercially by the Dutch artisans [Ronchi 1968; Herzberger 1966]. In particular, reading the archives of Den Haag (The Hague), it turns out that Hans Lippershey (1587–1619) asked for a patent on the telescope on October 2, 1608. However, these instruments, made with spectacle lenses, were affected by strong aberrations, and their enlargement was limited to three: it could not be increased without degrading the images. With a quality so poor and an enlargement so low, these early telescopes were almost useless; therefore, they did not get much of a favorable welcome from the public. However, in 1609, this novelty came to the attention of Galileo Galilei (1564–1642), who, at that time, was a lecturer at the University of Padova. There, he had installed a thriving workshop that produced scientific instruments that were in high demand in Europe, including his "geometric and military compass".

Galileo became so enthusiastic about the telescope that he soon built an exemplar, and then many others. Quite soon, he realized that the "goodness" of the lenses was a crucial factor, and therefore determined only to select the best, quickly achieving remarkable results. After a few weeks, he built a telescope with an initial magnification of six, and then eight. Galileo realized that he had a revolutionary object in his hands, and that he would need to overcome the hostility of the academic world. Very cleverly, he circumvented this situation by referring directly to those who were most powerful at the time. First, he wrote to the Doge of Venezia, Leonardo Donato, claiming to have constructed a useful tool for the military field, and presented his telescope with eightfold magnification to the Senate of Venezia. Although the Doge did not start military production of the telescope, Galileo achieved his purpose: his fame spread, and the Doge confirmed his professorship for life at Padova with an excellent salary *ad personam*. Nevertheless, this sparked a violent reaction from the other academics, who also believed that his increased prestige had been derived from a hoax. Galileo did not stop: in the following months, he built better and better telescopes, with up to twenty times magnification, and used them to observe the Moon and the planets. To achieve such a good result, Galileo concluded that he needed to use large objectives and then to diaphragm them appropriately according to their individual characteristics. Galileo realized that he had found a treasure in his astronomic observations. With the discovery of Jupiter's moons, he dealt a fatal blow to the conceptions on the Universe of that time: those same satellites represented a small Copernican system within the solar system. The phases of Venus showed then that this planet was positioned relative to the Earth, sometimes closer, sometimes farther away than the Sun, and this was enough to delegitimize the Ptolemaic system. The Moon displayed mountains similar to the terrestrial. However, the hostility of academics, the theological implications, and the rupture with the Aristotelian and Thomistic tradition were so serious that the protection of the Doge did not seem to be enough to him. Desiring to return to his native land,[3] he then made contact with the Medici of Firenze and, with their approval, he presented the satellites of Jupiter as *Medicea Siderea* in his *Sidereus Nuncius*. He sent a copy of this publication, and a specimen of his telescopes, to the Grand Duke Cosimo II dei Medici, who was his student in the summer of 1605, and, through the Grand Duke, he even sent them to various European princes, so that their experts could give judgment on the instrument and the content of the book. In April 1610, Galileo traveled to Pisa to present his telescope to the Grand Duke and the academics of that University. In September of that year, he left Padova to settle in Firenze, desiring and ultimately obtaining, in addition to the title of Mathematician, that of *Philosopher*. Meanwhile, the opposition of the professors was spreading to all universities, but the favorable review of Galileo's observations released by Kepler at the end of 1610, and by the fathers of the Collegio Romano in April 1611, marked Galileo's (slow) ascent to victory and the birth of the new optics.

On the other hand, the objections that were put forward were methodologically

---

[3] This choice to go back to Toscana was an unhappy one: the Medici were little more than vassals of papal domination, and the inquisition had control of the territory. On the other hand, Galileo was passionately confident as to the truth of his discoveries and *wanted* to meet with Cardinals so that he might convince them that the New Science was, as well as being true, also useful to the Faith.

relevant. In the face of the *experimental evidence* proclaimed by Galileo, the objection was raised that the problem was one of theoretical character, given that, according to the theories of that time, the telescope could not be considered a valid tool, since it had been built with lenses that "deform" reality. A theory that explained the functioning of the telescope still did not exist, and Galileo avoided replying on the theoretical level, since he had worked only empirically. Moreover, at least in matters of Geometrical Optics, he used the same medieval conceptions of species and visual rays as his opponents. He would not even respond to the solicitations that come to him from his friend Sagredo, who did not believe in the existence nor the usefulness of these rays and who, as self-taught man, became more of an expert in Geometrical Optics than Galileo himself.

Nevertheless, Galilco's contribution was, above all, methodological: he realized that it was necessary to start from mathematics to understand the world and argued that the laws of nature are mathematical, the same all over the Universe, and that we could determine not just the qualities, but the quantities as well. The true father of the modern scientific method, *hypothetical-deductive*, he understood that it was necessary to start from simple problems, by conducting experiments under appropriate conditions to be analyzed with reasonable approximations, so as to minimize friction. His invention of the *thought experiment* also allowed him to establish conclusions without performing the experiment itself.

Consistent with his mechanical and atomistic conception of the world, Galileo believed that light was composed of a fast flow of *quanta*, considered the latest indivisible particles in which matter could be subdivided by heat or other means. In the *Giornata Prima* of the *Dialoghi delle nuove scienze* of 1638, Galileo describes his attempt to measure the speed of light in which two people, with about a mile between them, cover and uncover a lantern, one in response to the other. For the shortness of the route, the perceived delay in the round-trip signal was too small to be estimated, and Galileo concluded that the propagation of light, if it is not instantaneous, must be very fast.

Galileo was also one of the inventors of the compound microscope [Vavilov 1965]. He had quickly noticed that nearby objects could be observed by a telescope simply by lengthening it, and that the magnification would increase. Thus, already in 1610, he was using the telescope as a microscope; however, this instrument was very long and impractical. Later, in 1624, Galileo constructed an entirely new microscope, the "occhialino", with lenses of very small focal length, so that the tube was considerably shortened, and with a fine adjustment of the objective position. With strong business acumen, Galileo not only put the telescope in commerce in reply to all of the requests that were coming to him from each part of Europe, but also the microscope. Some traces of this development can be found in various letters, although he did not publish anything on the microscope. Just as the telescope spurred a new era in astronomy, with the microscope began a new era in the life sciences, and, as early as 1625, Francesco Stelluti had already published a book on bees in which he included the anatomical observations he had made with one of Galileo's microscopes.

Galileo believed that physics could solve the mysteries of nature with thoughts, observations, and experiments that were accessible to everyone. He had personally experienced the bitterness of not being believed and the necessity that others repeat one's observations. He was a true man of Faith who believed that there is only one truth; therefore, there cannot be any contradiction between religion and science. In the beautiful letter he wrote in 1615 to Madame Christine de Lorraine, he expounded upon the scientific program of his New Science,[4] claiming the relationship of complementarity that must exist between Science and Faith and the refinement that Faith can receive from natural science, because Nature is «osservantissima essecutrice de gli ordini di Dio».[5] Responding to the accusation against him that he was refuting the Holy Scriptures, he offered a solution that was theologically exemplary and very modern: the Bible should not be taken literally, but must be properly interpreted in the historical, social and religious context of the time in which it was written.[6] Aiming for the salvation of souls: «alla salute dell'anime», it is natural that, in the Bible, «l'intenzione dello Spirito Santo essere d'insegnarci come si vadia al cielo e non come vadia il cielo»,[7] a phrase that Galileo heard directly from the Cardinal Baronio. Galileo was an overly polemic person and this caused him much trouble, but the real mistake, and it was a very grievous one, was accomplished in the field of theology by his opponents, who, in their extreme greed for power, completely misinterpreted the meaning of Scripture, immensely betraying the Gospel message of Faith and the Christian commandment of Charity.

Along with Galileo, we must also remember some other individuals: his friend, Giovanfrancesco Sagredo (1571–1620), who, while selecting lenses for Galileo, discovered the law of composition of the focal distance for thin lenses; his craftsman Ippolito Francini, also known as *Tordo*, who, in Firenze, was producing the objectives for Galileo and introduced a polishing machine with a vertical axis that is still in use today; and his last student, Evangelista Torricelli (1608-1647), who assisted him in the last three months of his life. After the death of Galileo, Torri-

4 This letter was implicitly addressed to Cardinal Bellarmine, the inquisitor who, in the year 1600, had ordered Giordano Bruno burned at the stake.

5 "very observant executor of God's command"

6 Galileo made use of this intuition of the way in which to interpret the Scripture to support the Copernican theory and to safeguard his freedom of research, but, limited by the constraints of his own time, he could not go further. It was necessary to wait until the Second Vatican Council before he would finally see the firm core of truth contained within this hermeneutic intuition ratified [Card. Martini 1996]. On the other hand, reading the Bible (Joshua 10, 1-43), we understand that, in the episode in which the Sun and Moon are stopped in the sky, the *redactor*, attributing to God the will to *exterminate* "every being that breathes", men, women, children and even animals, who already inhabited the South of Palestine, had only meant to remind the Jews of the benefits that God had allowed them. They had taken this episode literally, although, certainly, this extermination was not brought about, nor even desired, by God. Therefore, the inquisition justified the killing of "heretics", and the Vatican hierarchy justified the Thirty Years' War against the Protestants.

7 "intention of the Holy Spirit is to teach us how to go to heaven, not how heaven works"

celli was appointed as Mathematician to the court of the Grand Duke of Toscana, to continue the studies. He spent the last years of his short life, in addition to his scientific work, perfecting the technique of grinding and polishing lenses, with remarkable results [Molesini 2010].

At the same time as Galileo, between 1608 and 1620, the microscope was invented and built independently in Holland by Jansen himself, who, like Galileo, had derived it from a telescope, and by Lippershey, Mezius and Gans, and finally, in England, by Drebbel in 1621. With the progress of optical processing techniques, from about 1650 onward, optical systems with multiple lenses and precision optical instruments began to be developed. In particular, the use of the microscope spread among physicians and biologists, inspiring a remarkable impulse among their disciplines, especially thanks to the simple microscope, which continued to demonstrate superior performance to that of the compound for a long time. Thus, Antony van Leeuwenhoek (1632–1723), a simple usher of the States General of the Netherlands, was very successful in building simple microscopes with small lenses of short focal length (~ 2 mm), with which he devoted himself to numerous observations.

The turning point, which eventually revolutionized optics in the modern sense, arrived thanks to Johannes Kepler (1571–1630) with the solution of the problem of vision. He was an assistant of Tycho Brahe at the alchemical astrological institute of Praha (Prague). In 1604, he published the book *Ad Vitellionem Paralipomena* ("additions to Witelo"), in which he resumed and surpassed the tradition of Alhazen and Bacon. Kepler analyzed the nature and behavior of light and the processes of formation and localization of images, at both the physical-physiologic and psychic levels. In the first place, as Maurolico had already noted, he declared that, from each point of the object, light rays leave in all directions. Nevertheless, these rays do not have a physical consistency, but are rather like the trajectories of moving objects. Like Alhazen, he used mechanical analogies to explain reflection (with the decomposition of motions) and refraction.

For Kepler, light in itself is colorless but acquires color in reflection with colored bodies or in their crossing; in addition, it travels at infinite speed. He thus measured the refraction in glass, coming to establish a law of proportionality, approximate for small angles, between the angle of incidence and that of refraction, as is still done in Paraxial Optics. However, according to Herzberger (1966), he did not come to formulate the law of refraction, because he let himself be misled by some erroneous tables created by Witelo. Kepler explains vision with the idea of a *double cone* of rays: each object point is the vertex of a cone with the pupil as the base; with the refraction through the transparent tunics of the eye, a second cone is formed with its vertex on the retina. Therefore, the refraction creates an inverted image of the observed scene on the retina, and the optic nerve, with its numerous fibers, carries its signals to the brain. The psyche then has the task of rebuilding and locating in the apparent space, correctly oriented, the ghosts of the objects seen.

Kepler's key step, which enabled him to consider the formation of sharp images on the retina, was that of attributing to the pupil the task of reducing the surfaces of the cornea and the crystalline to spherical caps. In this way, he could explain how to avoid the formation of a caustic, which, at that time, was well-known and had been observed with glass spheres. There was still the problem of depth perception, that is, the distance at which things are located. Kepler solved this problem with the concept of the *distance-measuring triangle*, which had, as its vertex, the observed point and, as its base, the diameter of the pupil: he thought that the psyche was able to perceive the angular opening of this triangle and then its height. On the grounds of these concepts, Kepler made an important distinction between *picturae*, which are images seen on a screen, and *imagines rerum*, which are revealed directly by the eyes and include, in particular, virtual images of objects that are seen through mirrors and lenses. Finally, these could no longer be looked at as fallacious devices that deform reality. The similarity with the work of Maurolico, which was published only after the *Paralipomena*, is remarkable. Nonetheless, Kepler does not cite it, and there is no evidence that he had a manuscript of it, while he does indeed praise Della Porta.

In *Paralipomena*, Kepler did not give much space to the treatment of lenses, but he was later called upon to give an opinion on Galileo's telescope. Initially, he remained neutral, until he came into possession of one of said telescopes, one of those that had been sent to the Elector of Köln (Cologne) through the Gran Duke of Toscana. He examined it and was impressed with it, subsequently repeating the observation of Jupiter's satellites. In September 1610, he published a very favorable report on the instrument. A similar situation also occurred with the fathers of the Collegio Romano, when they too received a good telescope built and sent them by Antonio Santini of Venezia.

Kepler soon resumed his study of optics and, in 1611, he published *Dioptrice*, in which we have the first modern theoretical treatment of lenses and the telescope. In this book, Kepler presents the phenomenon of total reflection, while, for small angles, he assumes a law of proportionality between the angle of incidence and the angle of refraction and determines the focal length of the lenses. Thereby, he develops the laws of optics to the first order for thin lenses and their combinations, coming to propose a new telescope made of two positive lenses, which bears his name.

With Kepler having finally solved the problem of vision, *lux* and *lumen* were identified and their study was separated into distinct sciences, Psychological Optics and Physical Optics. On the physical side, interest turned towards the physical, objective nature of *lumen*, which, over time, improperly came to be called *light* in scientific language, as well as in that of laypersons, despite centuries of debate over distinguishing the two Latin terms. In our studio, we are dealing with the *lumen*, but we must not forget the physiological, psychological, and spiritual implications of the *lux*. After all, our sense of sight gives us the best of itself, but it may not represent the whole reality of the world to us, at least not for now.

The law of refraction and the speed of light

The resumption of experimentation allowed Thomas Harriot (~ 1560–1621) to improve Ptolemâios's tables of the angles of refraction and to identify the law in geometric terms, but he did not publish the results. Around the same time, Willebrord Snel (1591–1626), better known as Snell, professor of Physics at Leiden in the Netherlands, wrote a book on optics containing the law of refraction, which he had found empirically and which the developments of trigonometry had enabled him to express as a ratio of cosecants. However, this discovery was confined to Leiden. René Descartes (1569–1650) must be regarded as one of the discoverers of the law of refraction, too. Indeed, he had suggested the use of plane-hyperbolic lenses to focus light of a collimated beam three years before visiting Leiden in 1630; in any case, he attributed the discovery to himself and published the law of the ratio of the sines in his book *Dioptrique* in 1637. As the refraction law had been determined experimentally, Descartes considered this situation as unsatisfactory and unscientific, because the law had not been inserted in a theoretical context, and so it was not deducible from axioms that are more general. He succeeded in giving it a derivation by using a mechanical analogy and a *causality* principle as the conservation of momentum. Later, Fermat derived the same law from a different conceptual basis and, in particular, according to a principle of *finality* such as the variational principle of minimum time.

Descartes, despite having made important contributions to mathematics, such as the coordinate method for describing the position of a geometrical point, bringing geometry back to algebra, merely described the physical phenomena qualitatively using *analogies*. He built for himself an entire physics and a whole cosmology that was purely *mechanistic*, in the sense that he was considering actions only for contact, conceiving the motion of the planets as being dragged by a vortex of microscopic particles in a state of perennial collision between them. He despised the kinematics of Galileo, since, according to him, Galileo had not understood the *meaning* of gravity and movement. He also rejected Kepler's celestial physics, since he considered it philosophically erroneous. Instead, Descartes required metaphysics to establish and justify science, with the first step being to search for indubitable axioms: the certainty of our own existence and that of God, and since God created both our reason and all things, Descartes derived from it a principle of truth in which things that are clearly and distinctly understood are true; the definition of matter as a pure spatial extension, infinitely divisible; the non-existence of the vacuum; the constancy of the "quantity of motion" of the Universe. According to Descartes, extension and movement are the only primary physical reality, while other factors such as color, taste, and smell are secondary. Beyond these, there extends the region of passion, desire, love, and faith, which remain inaccessible to physics. His was therefore a necessary, deductive, demonstrative, and rational science, according to the Aristotelian ideal, proceeding *deductively* from *a priori* considerations, which constituted, for him, the general principles of all physical explanations, while experiments were relegated to a secondary role of confirming the theoretical conclusions. He replaced the pragmatic reason of Aristotélēs with mathematical reason. However, the world that Descartes derived from its axioms was still a world of paper, and its scientific program, purely *deductive*, was just the opposite of the *inductive* of Francis Bacon, and stood in contrast with those of Galileo, Mersenne, and Pascal. Yet, he had a great influence among scholars and some of his statements were prophetic.

From the metaphysical principle that *movement* is the only existing power in nature, and proceeding through analogies, Descartes derived his conception of the nature of light. In *Dioptrique,* he formulated the hypothesis that light is «un certain mouvement ou une action reçue en une matière très subtile, qui remplit les pores des autres corps»[8] (the aether, even if he did not use this expression), and dealt the final blow to the theory of the *forms* of Alhazen and Witelo and to that of the *species* of Grosseteste and Roger Bacon. Among his principles, he also postulated that light is propagated instantaneously (as many of his predecessors had done, with the few exceptions mentioned above), feeling comforted by the fact that lunar eclipses appear only when the three celestial bodies, the Sun, the Earth and the Moon are well aligned: this is a large-scale variant of Galileo's experiment with lanterns. Despite his dogmatic temperament, he performed experiments with glass spheres filled with water (described in *De Iride* and *Principia Philosophiae*) to study the colors produced by the rainbow, with his primary and secondary arches, finding that these were caused by refraction and that only one refraction was enough, as he verified with a glass prism. He then interpreted the colors, associating them with a rotating motion impressed upon the particles of the subtle matter in striking the refracting surface; this was in contrast with his predecessors, who believed that the colors were due to a different combination of light and dark or a finite number of "primary" colors. Descartes was the first to recognize the individual value of the colors of the spectrum, determining them to correspond to a particular rotation speed. He also conducted experiments with human and animal eyes, observing that a sharper image of an object in front of an eye can be formed on the retina by slightly increasing pressure on the eye. Therefore, eyes are able to accommodate focus. From the refraction law, he determined the shape of the surface of separation between two different transparent media that can give a punctiform image of an assigned object point. This surface is now called the *Cartesian oval*, of which the hyperboloid is a special case for a point object at infinity.

The case of the law of refraction is exemplary of the way in which completely different metaphysical conceptions can lead to the same logical-mathematical expression and simultaneously to different predictions, as in this case for the speed of light. The explanation that Descartes gave of this phenomenon is obtained from a mechanical analogy with the motion of a ball, following a tradition that, from Heron up to Kepler, had continued to consolidate. The initial assumption is that the propagation velocity is different in the two media, with a ratio n assigned: $v_T =$

[8] "certain movement or action received in a very subtle matter, which fills the pores of the other bodies".

$n v_I$. According to Descartes, the component of velocity parallel to the surface of separation between the two media was to be retained, and therefore $v_T \sin\vartheta_T = v_I \sin\vartheta_I$, from which the law of refraction for the incidence angle, ϑ_I, and the transmission angle, ϑ_T, is immediately found. Compared to the ballistic example, Descartes noted that, for a ball, the speed in the densest medium decreases, while the *strength* or the *ease* with which light propagates should instead increase to match the experience: Descartes does not speak of increasing the speed of light, since, for him, it was infinite. He also managed to find a justification for this, resorting to an analogy with a ball that «roule moins aisément sur un tapis, que sur une table toute nue, ainsi l'action de cette matière subtile peut beaucoup plus être empêchée par les parties de l'air, qui, étant comme molles et mal jointes, ne lui font pas beaucoup de résistance, que par celles de l'eau, qui lui en font davantage».[9]

Even before the *Dioptrique* was published, Mersenne sent Descartes' *Discours de la méthode* to Pierre de Fermat (1601–1665), asking for an opinion. As early as September 1637, Fermat replied, deeming the mechanical decomposition of the motion of light, made therein, as arbitrary and even contradictory to Descartes' other hypothesis that the light would propagate instantaneously. In the long dispute between the two, there was also a misunderstanding about the terms used. In fact, Descartes was using the term *determination* of the motion, with which he meant the *vector* concept. Instead, Fermat interpreted it as *direction*, for which the Descartes' demonstrations seemed absurd to him.

The turning point that led Fermat to formulate his own idea on the refraction of light came in 1657, when his friend Marin Cureau de La Chambre sent him a copy of his book entitled *La Lumière*, published that year. Reclaiming Heron's idea, which also turned out to be correct for the rectilinear propagation of light, de La Chambre explained reflection with a metaphysical principle, namely, the teleological principle that nature achieves its purposes in the easiest and cheapest way, which, in the case of light, was interpreted as the shortest path. Nevertheless, de La Chambre was also troubled by two objections: the first was that, in certain situations with concave mirrors, light instead travels the longest path; the second was that this principle did not apply to refraction. Fermat then began to revise this idea and concluded that the principle of economy should be interpreted not as the shortest path, but as the one of least resistance. In other words, accepting that light propagates at a finite speed, this principle is equivalent to seeking the path that requires the least time. In the case of refraction, the calculations would have been a little complicated, and Fermat took his time, promising to make them whenever his friend wished him to.

The application of this variational principle leads, for refraction, to the con-

9 "… rolls less easily on a carpet than on a bare table, so that the action of this subtle matter can be much more easily prevented by the parts of the air, which, being soft and poorly joined, do not give it much resistance, than by those of water, which favor it": we can see an analogy with velocity of sound, which is faster in water than in air.

servation of the components parallel to the interface of the vector $\mathbf{s}/\upsilon$, where $\mathbf{s}$ is the versor of the ray and υ is the velocity of propagation. Therefore, it leads to the relationship $\sin\vartheta_I/\upsilon_I = \sin\vartheta_T/\upsilon_T$.

In modern language, it is the component parallel to the surface of the *wave vector* that is preserved in the refraction and not that of the speed. The law of refraction that results is the same, but the speeds are in the inverse ratio. The first to be astonished by this coincidence was Fermat himself, as he recounted in the report [Fermat 1894] that he wrote in 1664 to an unknown correspondent (M. de ★★★★). Fermat, in fact, had supported the validity of the teleological principle, but had not then proceeded in the calculations, especially as Descartes' law had been verified by experience: «de sorte qu'il me sembloit inutile d'en aller chercher quelque autre por mon principe».[10]

It was only through the «insistances» of his friend de La Chambre that, in 1662, he pursued his analysis, winding up with the exact same proposition as Descartes, when he had hoped to obtain one that would give similar results, but not identical (except for the inverse ratio between speeds). Even M. de ★★★★ had shown him a counter-example to his minimum principle, for which the reflection from a concave surface could have a maximum path; in the same letter, Fermat answered him with the same argument used to reassure de La Chambre, namely, that what matters is that the path is minimal with respect to a plane tangent to the surface! In substance, we know today that the determination of the path of a ray does not necessarily correspond to seeking a minimum, but rather one that is extreme, that was the exact result of the technique he introduced anticipating the differential calculus.

Fermat finally believed that he had found the true demonstration of the law of refraction, leaving the merit of discovery to Descartes. Nevertheless, this position was not accepted by the Cartesians, who instead accused him of conducting noncausal demonstrations and ignored him. Therefore, the idea that the speed of light must increase in the more dense media dominated optical research for two centuries. Among the few to consider Fermat's principle as valid was Huygens: at first, he distrusted it, but then, having redone the calculations, he accepted it without further reservation as a useful research tool.

The debate over the nature of light

The question of whether the lumen was matter or movement increasingly came to the fore: the first hypothesis led to the belief that the lumen consisted of particles, while the second resulted in the wave theory. However, in the beginning, these two extreme conceptions were not well defined and, indeed, the possibility was not ruled out of some "complementarity", as, in their own way, Grimaldi and

10 "So that it seemed unnecessary for me to seek some other with my own principle"

Newton had conceived of it. Francesco Maria Grimaldi (1618–1663), Jesuit father at the Collegio di Bologna, was seriously committed to solving this dilemma and, for the first time, he tested the postulate of the rectilinear propagation of light in an attempt to determine the size of a light ray [Ronchi 1983]. Instead, he discovered a new phenomenon, which was not referable to a treatment of the geometric type, and gave it the name of diffraction. His study was unfortunately only published posthumously in 1665 under the title *Phisico-Mathesis de lumine, coloribus et iride*. It consists of a proem and two books, and its 535 pages contain an accurate description of the author's numerous experiments and the conclusions that he had gradually reached. The style is that of a dialogue, in which the hypothesis that the lumen is a substance is defended against other hypotheses, including the undulatory one.

He immediately starts with experiments on diffraction, using a darkroom with a tiny hole through which sunlight penetrates. In the cone of light that it forms, he inserts various objects and observes their shadow on a white screen, noting that the edge is not composed only of the penumbra that was to be expected from the finished dimensions of the hole. Rather, the shadow appears dilated and, around this, there are three *series lucidae*, that is, three luminous and colored fringes, bluish towards the shade and reddish towards the outside, that faithfully follow the edge of the obstacles. He notes as well that, behind long narrow bodies, immersed in the light cone, under certain conditions, there are iridescent fringes, even in the shade, that vary in number and characteristics (today, we know that these are essentially interference fringes for the diffracted light from the two edges). In another series of experiments, he interposes another diaphragm with a small hole in the solar light cone. Now, the spot of the light beam, produced by the two successive openings on the white screen, is much larger than expected from Geometrical Optics: the central part is white and the edges are colored red and blue. In the two cases, he conducted many tests, varying the position of the white screen and changing the shape, color, substance and arrangement of the obstacles or the diaphragm. As to the fringes, he notes: «earum progression fieri per lineam, quae neque est in directum cum priori linea, quae à foramine recta extenditur ad extremum opaci inserti in cono radioso, neque recta a foramine ad eam partem lucidae basis coni, super qua illae representatur pictae, ac terminate in tabella».[11] Moreover, they remain unchanged whatever «sit corpus illud opacum, quod lucido cono inseritur, sive densum, sive rarum, & sive laeve ac politum, sive asperum ac inaequale, sive denique durum sit, sive molle»[12]. He excludes that the fringes are due to direct sunlight, as they progress in a strange way by varying the position of the screen; he also excludes that they are due to the refracted or reflected light,

[11] Prop. I, n. 17; "[T]heir progression along a line is neither straight as the first line passing through the hole and tangent to the obstacle in the light cone nor on a straight line joining the hole to the point of the screen on which they are observed".

[12] Prop. I, n. 18; "... the opaque body inserted into the cone of light, whether dense, rarefied, smooth and shiny, rough and uneven, tough, soft, might be".

since they do not depend on the substance of the obstacle, neither do they move while rotating the obstacle around its edge. Therefore, they must be produced by a new phenomenon, *diffraction*!

In an attempt to explain this behavior, he compares the light to a fluid that moves very fast, and the way in which water forms waves around a rock that obstructs its path, so he thinks that the fringes should develop in the shade of an obstacle immersed in the cone of light. Grimaldi finally admits that, "at least sometimes", the light also propagates in an "undulatory way" in transparent bodies. Leonardo, referring to the waves on the water, had already written that «il moto dell'impressione fia solamente accompagnato dall'impeto e non dal moto della medesima acqua».[13] Instead, for Grimaldi, the undulatory motion of the luminous fluid accompanies the motion in the direction of propagation with non-periodic "sinuouse crispatae" trajectories, and, for this fluid, he recognizes the need for a finite speed of propagation.

Grimaldi then goes on to examine the properties of the lumen and its interactions with matter in various situations: in diaphanous bodies and in those that are opaque, refraction, reflection, and diffraction, in vision, and, even more interestingly, in the coloring. The general opinion at the time, apart from that of Descartes, was that light was pure and white, while the colors were a property of bodies and were called *permanent*. However, there were also mysterious colors, like those seen in the Rainbow and in soap bubbles, that, since they were not the result of a colored material, were called *apparent*. Supposing the existence of a white lumen, Grimaldi describes numerous experiments with which a colored light is obtained, and he concludes that the color must be a property of the lumen itself. We particularly have him to thank for the realization of spectra obtained for diffraction from finely striated surfaces and by reticular structures like tissues. In his book, he wrote: «Luminis modificatio, vi cuius illud tam permanenter, quam (ut aiunt) apparenter coloratur, seu potius sit sensibile sub ratione coloris, non improbabiliter dici potest esse determinata ipsus undulatio minutissime crispata, et quidem velut tremor diffusionis, cum certa fluitatione subtilissima, qua fit ut illut propria, ac determinata applicatione afficiat sensorium».[14] He then declares that the colors are differentiated by the nature and speed of the undulatory motion, and, in the figure with which he describes this motion, the oscillation is represented by a zigzag line in the *transverse* direction. As an example, he mentions sound, which gives such a variety of tones with as many different vibrations. Among the colors, he defines a small group of them as simple: red, yellow, and blue, and considers the others a combination of these. While, in the first book, he had vigorously de-

[13] "the undulating motion is accompanied only by the impetus, and not by the motion of the same water".

[14] Prop. XLIII: "The modification of the lumens, for which it is colored permanently and apparently (as they say), or, rather, it makes itself sensible with the qualification of color can, not improbably, be said to be a certain undulation, minutely fluctuating, as a tremor of diffusion, with a certain fine oscillation, which, with a given application, stimulates the sensory"

fended the thesis that the lumen was a *substance*, in the second book, he returns to giving scientific legitimacy to the *accidental* hypothesis, but again considered it definitively determined that color results from a modification of the lumen.

Even de La Chambre had gone to battle for the idea that permanent color, apparent colors, and images were unified within the light: rejecting the idea that the colors were produced by a mixture of light with darkness or opacity, he had instead felt that they were a tonality of light, similar to the grave or acute tone of a sound.

In the 1600s, the renewed interest in science led to the formation of scientific societies with the aim of enabling the rapid communication of ideas and discoveries, so as to promote technical development and, ultimately, to obtain funding for research. Credit for this development must also go to Francis Bacon (1561–1626), philosopher of the new science, who strove to free the experiment from the doubtful context to which it had been relegated that it might become the indisputable basis of scientific explanation. Being the democratic intellectual that he was, he understood that only the collective effort of many could penetrate the complexity of nature and urged the formation of scientific institutions. Already, the abbot Marin Mersenne (1588–1648) was a tireless correspondent of all European scientists of his time, among whom he carried out the task of collecting and exchanging letters and manuscripts. The first scientific associations were the *Accademia dei Lincei* in Roma (1603–1630) and the *Accademia del Cimento* in Firenze (1651–1667), which, among other things, repeated Galileo's experiment for the measurement of the speed of light. These Societies, which were creations of private patronage, had a short life, because of the atavistic eternal insipience of the Italic rulers.[15] Better luck was had by the *Royal Society* in London (1662) and the *Académie Royale des Sciences* in Paris (1666), which became the appointed place to make public the work and discoveries of researchers.

The secretary and curator of the experiments of the *Royal Society* was Robert Hooke (1635–1703), who, in 1665, published a book entitled *Micrographia*, in which he reported his microscopic observations, including the discovery of cells. He was mainly interested in finding an explanation for the phenomenon of color. In the course of these observations, he particularly dealt with the coloring of the thin layers that form on steel after it has been heated, the iridescence of soap bubbles and of glass blown into fine leaves, and of the fringes observed on thin sheets of mica or between two glasses pressed together with an interposed liquid of various kinds. He was the first to conduct a systematic investigation of that which is today called *fringes of equal thickness*, or even "Newton rings"! From it, he deduced various circumstances that are still considered valid today, including that the coloring always appears with a very thin lamina of transparent material delimited by media of different refraction from that of the lamina. Moreover, the coloration is uniform if the thickness is uniform. If it is not, the color can take the form

[15] This is Italy: we have smart guys and wonderful land, while the others have good administrators!

of concentric rings of various colors, as in the case of a lamina with spherical surfaces; or of lines, as in the case of two inclined plates that form a wedge in the interposed medium; or, finally, it is irregular if the thickness varies randomly. In the case of the glasses pressed together, he notes that the fringes move if the pressure is varied. He also notes that the thickness of the lamina must be between a maximum and a minimum: if the thickness is excessive, only a uniform and white illumination is seen. However, he admits to never having been able to measure the thickness of the laminae and that the existence of thin foils on substrates such as steel is only his hypothesis, since he has looked for them with his most powerful microscopes without being able to observe them.

Following Descartes, and thus violating one of the Royal Society's rules forbidding dogmatism, Hooke tried to explain the properties of light through the action of a thin material that fills all of the bodies and space. However, reviewing the Cartesian assumptions and highlighting their difficulties, he was able to propose more plausible conjectures with which he advanced the theory of aether. In particular, instead of a circular motion, he assumed a vibratory motion for the particles of the fine matter. Hooke did not explicitly assert that the speed of light must be finite, but he actually favored this idea by claiming that the aether vibrations propagate through space in a straight line with a constant speed in a homogeneous medium, generating spherical waves similar to those that can be seen on the surface of water that has had a stone thrown into it. However, he did not speak on the frequency of vibrations or waves. Rather, he considered these vibrations as pulses, and, in this, he did not wander very far from the mechanistic tradition that had led to Galileo's quanta or Newton's corpuscles.

Thinking about waves, he adhered to the opposite extreme of what we are accustomed to using with monochromatic waves! Nevertheless, since his pulses were delimited temporally but not spatially, Hooke introduced the concept of the wave front, perpendicular to the direction of propagation, and applied it to refraction, assuming the law of sines to be valid, and supposed that the propagation speed was greater in denser bodies, in accordance with the position of Descartes. The construction that he obtained is analogous to that which was later used by Huygens, with the difference that, in the refracted wave, the wavefront is no longer perpendicular to the direction of propagation. In regard to Descartes, he did not accept the theory of colors, finding it inconsistent with his experiences with thin foils. Indeed, according to Descartes, this color should only have been able to be observed at the transition between light and shadow; furthermore, in passage through a lamina with its two subsequent refraction, the second one would have cancelled the rotating motion impressed by the first one upon the aether particles, and therefore would have canceled out any coloring. Instead, Hooke attempted to explain the colors produced by refraction with the oblique wavefronts of his model: if the front is not tilted, it generates the sensation of white light, while, in the case of an oblique front, he reclaimed one of Descartes' mechanical concepts of in order to attribute a different strength at the two ends of the front. Therefore, if the weak part of an impulse of light precedes the strong, one has a blue light, meaning one has a red light in the opposite case. Finally, the other colors produced by refraction must correspond to a "dilution" of red or blue or a combination thereof.

In his interpretation of the colors of the fringes of thin laminae, Hooke attributes equal importance to the reflections from the first and second faces of the lamina: the pulses reflected by both sides would be "combined" with a time delay, depending on its thickness. If the foil is very thin, the two pulses would be superimposed and no color would be apparent to the observer; however, for a greater thickness, the second pulse, being a bit more "weak", when combined with the first would have given rise to a single pulse whose strongest part precedes the weaker, creating the appearance of yellow. Increasing the thickness even more, the two pulses would further separate, creating the appearance of red. Lastly, to explain the periodicity of the fringes, he hypothesizes that the pulse sequence is periodic![16] Thus, when the weak reflected pulse from the second face is located between two strong pulses reflected by the first face, one would have the appearance of purple light, and, continuing to increase the thickness, we would have a weak pulse that precedes and is combined with a strong one, creating the appearance of blue, and so on. In his interpretation of the colors of the fringes, Hooke was therefore just one step away from expressing the principle of interference, as was later acknowledged by Young, who wrote that, if he had not come to this principle alone, Hooke's idea «might have led me earlier to a similar opinion».

Following his long dispute with Newton on the "composition" of white light, Hooke reconsidered his theory of colors, and he was the first to express the superposition principle of waves for light: if each color corresponds to a particular vibratory motion, white light can be *imagined* in a mathematic sense like the combination of thousands of these motions and still be "uniform".[17] However, remaining faithful to his dualistic theory of color, Hooke regarded this hypothesis as unnecessary and refused the uncouth one of Newton.

A few years later, the Jesuit father Ignatius G. Pardies (1636–1673) developed considerations similar to those of Hooke in following Descartes' attempt to explain the properties of light. He still, in fact, used the law of sines, but came to the conclusion that the wavefront remains perpendicular to the propagation direction even after refraction, and that, in agreement with Fermat, the speed of light is smaller in the denser media. With this geometrical construction, Pardies anticipated Huygens, but it still does not seem that he had managed to come up with his own demonstration of the Snell-Descartes law as Huygens had done (his manuscript on refraction has gone lost). In turn, Huygens studied the work of Pardies and Hooke with interest and he was influenced by them.

A new phenomenon, the double refraction from crystals from the spar of Iceland (calcite), was discovered in 1669 by a Danish professor of mathematics, Rasmus Bartholin (1625–1698), who described, through different experiments, the

---

[16] This looks strange, but is exactly what happens with a femtosecond laser comb!

[17] This insight found a mathematical formulation only in 1886, with M. Gouy, who applied the Fourier spectral decomposition theorem to it [Gouy, M. 1886].

physical, chemical, and optical properties of these crystals, obtained by a sailor who had collected them in Iceland. He noticed that they had the form of oblique parallelepipeds and that, when a small object was observed through the two opposite faces of the crystal, the object image appeared to be doubled up. He attributed this behavior to the splitting of a beam incident on the first crystal face into two rays. He described how one of these rays behaved in an ordinary way, following the normal rules of refraction, while the other did not. In particular, he noted that, by rotating the crystal around an axis orthogonal to the two faces through which one looks, the first image of the object remains immobile while the second moves following the rotation. As happened with Bartholin's intentions when he published his observations, this discovery stimulated the imagination of the researchers, and Huygens, in particular, was able to insert the double refraction into his theory of wavelets.

In 1676, the Dane Olaf Römer (1644-1710), who had been a student of Bartholin and had later married his daughter, became a professor of mathematics in France and informed the Academy of Sciences about his demonstration that light propagates with finite speed, giving his estimate of this speed. As part of his research into the determination of longitude at sea, he had observed the eclipses of Jupiter's primary satellite for several years, with the Italian Giovanni Domenico Cassini and other members of the *Académie*, finding that the eclipses are delayed when the Earth recedes from Jupiter and are anticipated in the opposite case. Cassini was the first to suggest, in 1675, that this could be due to the finite speed of light, but shortly afterwards changed his mind [Sabra 1981]. Römer instead estimated that, to explain these irregularities, the light must employ 11 minutes to travel a distance equal to a terrestrial orbit radius, that is, with a speed of approximately 214000 km/sec.[18] But this measure, which was a "side effect" of a research dedicated to developing a method for the determination of longitude, was not immediately accepted by all, given that the irregularities could result from some other effect and that other satellites of Jupiter did not appear to exhibit the same behavior.

In the second half of the 1600s, two new personages entered into the debate on the nature of light: Newton and Huygens. Generally, two opposite theories are attributed to them, the corpuscular one and the undulatory one. However, the deeper difference between them was in regard to the method of scientific investigation, which, for Newton, had to be purely inductive, without assuming any *a priori* hypothesis, according to the presumption already implicit in the method of Francis Bacon, whereby the "truth" could be reached in a "certain" way through the use of *experimentum crucis*. Instead, Huygens deemed that he could use the hypothetical-deductive method of Galileo, closer to the Cartesian method, but purified of the latter's excesses, so the assumptions and their consequences are compared with the data of experience on the basis of *likelihood* or probability. On the other hand,

---

18 The difference with the current determination is largely the result of Römer having not considered Jupiter's movement during the observation period [Bevilacqua and Ianniello 1982].

Newton was not at all exempt from making hypotheses a priori, as he claimed, and indeed, with his corpuscular prejudices, he seriously spoiled his own "inductions".

Christiaan Huygens (1629–1695) was born in Den Haag in the Netherlands into a middle class family. His father Constantijn was a diplomat and a Latinist who was a friend and correspondent of various intellectuals of that time, including Descartes, who sometimes visited his house. Huygens soon showed his penchant for mechanics, drawing, and mathematics. In 1645, he entered the University of Leiden, where he studied mathematics and law. Later on, he acquired a European reputation as a mathematician and astronomer: thanks to the improvements that he introduced in the grinding and polishing of lenses for telescopes, in 1655, he discovered a satellite of Saturn, in 1656, he distinguished the stellar components of the Orion Nebula, and, in 1659, he discovered the true form of Saturn's rings. In 1656, the need for a good device for the measurement of time prompted him to build a pendulum clock with a cycloid isochronous movement. He continued his studies in mechanics, to which he made important contributions, such as the general idea of the energy conservation law, the concepts of moment of inertia and centrifugal force and the laws of collision. Huygens ultimately described these studies in 1673 in his book *Horologium Oscillatorium*. In 1666, he became one of the founding members of the *Académie des Sciences*, and lived in Paris until 1681, when he was forced to leave that city for political reasons and was never able to return.

Together with astronomy and mechanics, Huygens worked on optics and wrote two books on this subject: *Dioptrica* and his famous *Traité de la Lumière*. The first was published posthumously in 1703 and the second was published only in 1691, twelve years after he had written and communicated it to the *Académie des Sciences* in 1679, adding only the section on birefringence to the publication. He treated only specific aspects of optics, not considering diffraction and dealing very little with color.

Dioptrica is a book of Geometrical Optics in which Huygens introduced considerable progress in our understanding of the formation of optical images. It was written over an extended period of time and is divided into three parts. The first part, written in 1653, deals in detail with the refraction produced by dioptres and lenses with finite thickness, the apparent size of objects as seen through an optical system, and, finally, the telescope. The second part, written in 1666, examines of the aberrations. Finally, the third part, written between 1685 and 1692, returns to dealing with the telescope and the microscope. The fundamental concepts of Gaussian Optics are already present there, including the fact that, if, in an optical system, the position of eye and object are exchanged, while the optical components do not move, the object would been seen by the eye as having the same apparent size and the same orientation as before. In modern terms, this is equivalent to saying that the product of three terms, namely, refractive index, angular magnification and lateral magnification, is constant throughout the optical system. In the study of the reciprocal positions of object and image, Huygens found, before Newton, the law of inverse proportionality for distances, when these are measured according to their respective focal planes. In the study of aberrations, he obtained an approximate formula for determining the shape of a lens with minimal spherical aberration.

In the *Traité de la Lumière*, Huygens instead investigated the nature of radiation. He resumed the Descartes' conception that explanations in physics must be given in geometric terms of size, shape, and motion, and that all actions must take place by direct contact; he accepted the idea of the extension of Newton's gravity to the entire solar system, but only as a result of the mediation of a subtle imperceptible matter that fills the entire universe and not as action at a distance. To this subtle matter, which he called aether, he attributed the task of justifying gravity and propagating light signals. Then, moving away from Descartes and resuming the work of Hooke and Pardies, Huygens conceived light as an actual motion of aether particles and not as a tendency to motion. In fact, he did not accept that a particle could have a "tendency" to motion in two opposite directions when two light beams proceed simultaneously in the opposite directions. On the other hand, he did not even accept the corpuscular theory, thinking that, in the intersection of two swarms of corpuscles, these would spread for their mutual collisions. In his conception, aether was the mechanical support for light, as air is for sound. However, while air transmits sound by compressions and rarefaction, aether was conceived with rather unusual properties: it must permeate all transparent bodies and also fill the vacuum recently discovered by Torricelli and investigated by Robert Boyle (1626–1691). This vacuum had shown the ability to transmit light, but not sound. The way in which Huygens conceived the aether was rather complicated [Sabra 1881], but, in his considerations on light, he assumed it to be a fluid consisting of identical particles. These are perfectly elastic and are all in contact with each other, as in the case of a row of balls. Therefore, a pulse imprinted upon the first is transmitted to the last ball of the row without a shift of the intermediate spheres. Under these conditions, the speed of light must be finite, and, simultaneously, Huygens explained why this speed was much greater than that of sound. Contrastingly, Hooke (to whom, among other things, we owe the concept of the modulus of elasticity!) was imagining an aether composed of rigid particles, so the propagation speed that resulted would be infinite. When Römer published his results, Huygens considered them simply as a confirmation of his ideas. He also realized that the evidence brought by Descartes with the lunar eclipses was insignificant for such rapid speeds.

For Huygens, light waves were longitudinal, the ether being a fluid, and nonperiodical: on this last point, he was not wrong, treating the radiation of natural sources. Rather, he was imagining them as being constituted of a random succession of pulses. To explain the rectilinear propagation of the light, like Hooke and Pardies, he was thinking that, from the light source, these pulses would propagate, forming a *main* wavefront. Since each aether particle, excited by the arrival of a pulse, is in contact with all of the surrounding ones, it becomes the source of a *secondary wavelet*. At this point, Huygens introduced his famous principle that wavelets produced at all points of the front mutually reinforce themselves only in the *envelope* of their surfaces (producing a new front), thus giving rise to the propagation of the main front. The word "envelope" still does not render well the sense of what Huygens meant [de Lang 1992]: he actually speaks of a *region*, and not merely a point of tangency, where the *composition* of the wavelets becomes effective. He thus defines, in the modern sense, a physical beam of light as a thin region around the geometric ray, even if he does not indicate its actual thickness. Furthermore, in a purely geometric construction, there is also an envelope surface of wavelets in the retrograde direction! To resolve this difficulty, Huygens appealed to a physical principle, that is, to the mechanical analogy of the balls, so, if these are all equal, the impulse can only propagate forward, without generating a signal dispersion in the other directions. The *obliquity factor*, invoked by Fresnel nearly a century and a half later, had its first draft here.

It is striking that Huygens did not deal with diffraction [Ronchi 1983]. Although the concept of interference and phase began vaguely to emerge, both in the *accesses* of Newton, who had also examined the interference of the waves on the surface of water, and in Huygens' concept of *composition*, the two scholars did not have the mathematical tools to quantitatively evaluate the interference effects and, in particular, they lacked the concept of destructive interference. Indeed, that distinction that we find natural between field and intensity had not yet developed: only the latter was considered in the reasoning. Here, then, Newton considered the wave model to be completely incompatible with the rectilinear propagation of light, because he considered that the waves would expand in all directions beyond an opening, as happens with sound.

Conversely, Huygens believed that, where there is not a well-defined envelope, the light would have a completely negligible intensity, as he himself wrote: «bien que les ondes particulières … se répandent aussi hors de cet espace [outside the light beam], toutefois elles ne concourent point en même instant à composer ensemble une onde».[19] Therefore, he stiffened on the concept of rectilinear propagation, without ever defining whether the envelope area, behind an obstacle, ends in a continuous or discontinuous way. To understand Huygens' words, we must leave aside that shortcut established by the spectral decomposition in harmonic waves with which we usually treat radiation, so as to reason instead in the time domain. He did indeed regard light as being composed of very short pulses: thus, the differences in arrival times play the same role as the phase differences between harmonic waves that interfere [de Lang 1992].

Huygens' greatest success was to be able to explain, in a simple way, the laws of reflection and refraction. The comparison between his model and observation led him to conclude, in accordance with Fermat, that propagation is slower in denser mediums, but now the law of the sines was explained in a causal way and not according to a principle of finality. Furthermore, while Pardies *assumed* that, in refraction, the wavefront is maintained perpendicular to the direction of propa-

---

19 [Huygens 1690, p.23]. "although the particular waves … also spread out of this space, however they do not contribute at the same instant to compose together a wave"

gation, this now appeared to be a *consequence* of the envelope of wavelets. Among other things, with his construction, Huygens could also give a reason for the phenomenon of total internal reflection, since, in this case, there is no envelope of the refracted wavelets. Nevertheless, if the geometric aspects of the model were balancing well, the physical ones were presenting difficulties. Indeed, how can one mechanically explain the phenomenon of partial reflection? In addition, why is the propagation slower in the densest media? Huygens was looking for an explanation at a microscopic level that it is difficult and complicated even with modern theories. For example, he believed that internal reflection was produced by the particles of air outside, but recognized that this explanation could not be given in the case of a tube in which there is vacuum.

In 1677, Huygens achieved another important success with his explanation of double refraction in Iceland spar. He supposed that the ordinary ray propagates with spherical wavelets in the aether present in the crystal, while the extraordinary ray would be transmitted by both the aether particles and those of the crystal. However, the macroscopic crystal structure suggested to him that these particles are arranged in an orderly anisotropic manner, as we can see from the figures reported in the *Traité de la Lumière*. He therefore believed that the extraordinary signal must propagate at different speeds in different directions associated with the crystal axes. Between many possible shapes that could be hypothesized for the extraordinary wavelets, he chose that of an ellipsoid of rotation, which, for his own admission, was the simplest case to consider after the sphere. The application of the principle of the envelope of wavelets to refraction thus leads to two distinct directions of propagation: one that still follows the law of sines and the other that doesn't. Huygens verified the consequences of his idea through experience, finding an agreement that, as he put it, was "merveilleusement" confirmed. During these investigations, he observed an unexpected phenomenon by arranging two crystals separated in succession in the path of a beam of light. When their faces were all parallel, in the first, there was the usual division of the beam into two, but neither of these underwent further division in the second crystal. However, by rotating the two crystals between them along an axis perpendicular to the faces crossed by the light beams, this subdivision reappears, giving place to four bundles of light of variable intensity, only to disappear again every 90°. Huygens thought that this was due to some property *acquired* by the light crossing the first crystal (which is now known as the polarization of the waves). Whereby, when the light meets the second crystal lattice in a certain position, it is still able to move the two different types of matter, which give rise to the two refractions, while, for another disposition, it may move only one.

Huygens wrote «Mais pour dire comment cela se fait, je n'ai rien trouvé jusqu'ici qui me satisfasse».[20] However, he also wrote that «Car bien que je n'en aie pas pû trouver jusqu'icy la cause, je ne veux pas laisser pour cela de l'indiquer,

---

20 [Huygens 1690, p. 91]. "But to say how this is done, I have found nothing so far that satisfies me".

afin de donner occasion à d'autres de la chercher. Il semble qu'il faudroit faire encore d'autres suppositions outre celles que j'ay faites; qui ne laisseront pas pour cela de garder toute leur vrai-semblance, après avoir esté confirmées par tant de preuves».[21] We are the heirs of this wish to discover the mysteries of Nature: courage, imagination, and radically new approaches will be necessary, but, as good sons and daughters, we will not forget the precious inheritance of our fathers.

Suggested readings

These historical notes are derived from the following texts:

Argentieri A., *L'ottica di Leonardo*, in the book *Leonardo da Vinci*, Curated edition of the exhibition on L. da Vinci, 1939, p 405.

Bernal John D., *Storia della Scienza*, of Editori Riuniti, Roma (III ed., 1969).

Bevilacqua F. and Ianniello M.G., *L'ottica dalle origini all'inizio del '700*, Loecher ed., Torino (1982). This is an annotated anthology of various authors, from Plátōn to Newton, with the intention of historically unifying the optics of the various disciplines (physical, physiological and psychological) into which we subdivide it today and to show the link between Physics and Metaphysics, conceptual models, regulative principles and experimental results.

de Lang H., *Christiaan Huygens, originator of wave optics*, in "*Huygens' principle 1690-1990, theory and applications*", Ed. da H. Blok, H.A. Ferweda, H.K. Kuiken, North-Holland Elsevier Science Publishers B.V. (1992).

Drake Stillman, *Galileo*, Ed. Il Mulino, Bologna (1988). Drake was the most eminent historian of Galileo. His book is an accurate chronological biography of Galileo, in which the intellectual itinerary of this great scientist and the events of the people that were close to him are reconstructed.

Fermat Pier de, *Euvres*, P. Tannery and C. Henry ed., Paris (1894), II, pp. 485-89.

Gouy M., *Sur le mouvement lumineux*, J. Phys. Theor. Appl. **5**, 354-362 (1886).

Hall A.R. and Boas Hall M., *Storia della scienza*, Ed. il Mulino, Bologna (1979).

Herzberger Max, *Optics from Euclid to Huygens*, Applied Optics **5**, 1383 (1966). This is a summary of the salient contributions of authors to optics from ancient to modern times, with an interesting bibliography.

Huygens C., *Traité de la Lumiere*, Pierre Vander Aa Ed. (1690).

Lenoble R., *Le origini del pensiero scientifico moderno*, Universale Laterza ed., Bari (1976). This describes, in a vivacious way, the history of the development of scientific thought and of the personages who contributed to it in the century when the philosophy of science and modern science finally began deeply to influence the history of humanity.

Martini Card. Carlo Maria, *L'ira di Dio*, Ed. Euroclub, Longanesi & C., Milano (1996).

Molesini G., *Telescope lens-making in the 17th century: The legacy of Vangelista Torricelli*, Optics & Photonics News (April 2010), p. 26-31.

---

[21] [Huygens 1690, p. 88-89]. "For, although, until now, I have not been able to find its cause, I do not want to refrain from indicating it, in order to give occasion to others to seek it. It seems that still other suppositions would have to be made besides the ones I have made; which, nevertheless, will not fail to preserve all of their probability, after having been confirmed by so many proofs".

Molesini G. and Greco V., *Galileo Galilei: Research and development of the telescope*, in *Trends in Optics, Research, Development and Applications*, ed. by Anna Consortini, ICO, Vol. 3, p. 424-438, Academic Press, London (1996). Based on historical documents, this discusses the realization of the telescopes of Galileo and reports the analysis made on one of Galileo's original objectives to determine its optical quality.

Polvani Giovanni, *Storia delle ricerche sulla natura della luce*, Istituto della Enciclopedia Italiana, Roma (1934). After a brief reference to the ancient authors, it describes the history of optics between 1600 and 1900 and, in particular, the research on aether.

Ronchi Vasco, *Storia della luce, da Euclide a Einstein*, Laterza ed., Bari (1983); this is the last extended edition of *Storia della luce*, by Vasco Ronchi, N. Zanichelli ed., Bologna (1939), which contains many beautiful illustrations that were unfortunately omitted in the new edition. It is a book dense with news and passionate criticism on the concepts, innovative ideas and mistakes of ancient and modern authors, but you will read in one breath. *Scritti di Ottica*, Edizioni il Polifilo, Milano (1968). This is a ponderous annotated anthology of various Italian authors, from Lucretius to Amici. *Storia del Cannocchiale*, Pontificia Accademia delle Scienze, Città del Vaticano (1964).

Rosen Edward, *The Invention of Eyeglasses*, J. of the History of Medicine and Allied Sciences **11**, 13-46; 183-218 (1956).

Sabra A.I., *Theories of Light, from Descartes to Newton*, Cambridge University Press, Cambridge (1981). This is a fundamental book that analyzes, step by step, the work of scientists of the 1600s, who, with their legacy from previous authors, their philosophical ideas and their disputes, contributed to the theories on the nature of light and its properties.

Vavilov S.I., *Galileo in the history of optics*, Soviet Phys. Usp. **7**, 596 (1965). As well as discussing Galileo's activity in Optics, it also relates the story of his precursors.

Wolf Emil, *The life and work of Christiaan Huygens*, in "*Huygens' principle 1690-1990, theory and applications*", Ed. da H. Blok, H.A. Ferweda, H.K. Kuiken, North-Holland Elsevier Science Publishers B.V. (1992).

Chapter 2
Geometrical Optics

> All what we see,
> it is seen in a rectilinear direction.
> Pseudo-Eukleidēs, *Catoptrics*, 2nd postulate

Introduction

Geometrical Optics is one of the oldest of the physical sciences, but still remains the most effective approach for explaining a good part of the most common optical phenomena. It is particularly useful for tracing the propagation of light in inhomogeneous media and for describing or designing optical instruments. The emphasis of this discipline is to find the path of light rays, imagined as geometric lines along which energy flows. It is based on a few simple observations:

a) light propagates in a straight line in homogeneous media and, in particular, it is possible to produce thin *beams of light*, similar to geometrical rays within the physically unattainable limit of an infinite subtlety;
b) the laws of reflection and refraction;
c) different light beams propagate without disturbing each other;
d) "natural" sources are generally uncorrelated between them, for which their light beams overlap without showing interference.

On the other hand, the electromagnetic field associated with visible light is characterized by very small wavelengths, on the order of $10^{-6} \div 10^{-7}$ m. Therefore, the phenomena that violate the first and the last of the above observations can be observed only with accurate experiments. Indeed, the effects of diffraction or interference are almost hidden using natural sources, for which the visibility of the fringes is reduced. The diffraction phenomena appear when there are rapid changes in the amplitude of the field, such as that produced by a sharp obstacle, particularly when some dimension of the optical system, such as the diameter of an aperture, is comparable to the wavelength; or in the neighborhood of a focal point; or over long distances compared to the transverse dimension of a wave, particularly when there is a delimitation imposed upon it. Lastly, point (c) follows from the linearity of the media at the ordinary beams' intensity.

In this chapter, we will explore the consequences of such observations taken as empirical data. However, we will derive the laws of Geometrical Optics by Maxwell's equations within the limit at which the wavelength λ tends to zero. We will also see that, within such a limit, the intensity can be deduced from the transverse dimension of a thin pencil of rays and that the polarization state can be associated with each ray. Therefore, in Geometrical Optics, the rays are associated with the

---

Electronic supplementary material The online version of this chapter (https://doi.org/10.1007/978-3-030-25279-3_2) contains supplementary material, which is available to authorized users.

G. Giusfredi, *Physical Optics*, UNITEXT for Physics,
https://doi.org/10.1007/978-3-030-25279-3_2

transport of energy and, more generally, of information. In particular, we will focus on the special form of information that is contained in optical images.

2.1 Derivation of Geometrical Optics for $\lambda \to 0$

It can reasonably be assumed that the laws of Geometrical Optics can be derived from Maxwell's equations within the limit of the wavelength λ tending to zero [Sommerfeld and Runge 1911]. A more general theory for the solution of the wave equations is given by the asymptotic representation of the field as a Luneburg-Kline series, of which Geometrical Optics is the lowest order of approximation [Luneburg 1948, 1964; Kline and Kay 1965]. A brief summary of this theory is illustrated by Solimeno, Crosignani, and Di Porto (1986). Leaving this discussion to theorists or those who are interested in microwaves or radio waves, we see now a simple quantitative derivation of Geometrical Optics from Maxwell's equations.

Let us consider the case of monochromatic wave that propagates in a *linear* and *isotropic* medium, but not necessarily a homogeneous one. In this case, we can apply the Maxwell's Eqs. (1.2.15) and the wave Eqs. (1.4.6). In particular, we intend to find a solution to these equations with a form that is little more general than that of Eq. (1.4.10):

$$\mathbf{E}(\boldsymbol{r},t) = \Re e\left[\boldsymbol{E}_0(\boldsymbol{r})e^{i\phi(\boldsymbol{r},t)}\right], \quad \mathbf{H}(\boldsymbol{r},t) = \Re e\left[\boldsymbol{H}_o(\boldsymbol{r})e^{i\phi(\boldsymbol{r},t)}\right], \tag{2.1.1}$$

where $\boldsymbol{E}_0$ and $\boldsymbol{H}_o$ are complex functions of the position and

$$\phi(\boldsymbol{r},t) = k_o L(\boldsymbol{r}) - \omega t\,. \tag{2.1.2}$$

The coefficient $k_o = 2\pi/\lambda_o$, where λ_o is the wavelength in vacuum, is introduced to simplify subsequent calculations, and $L(\boldsymbol{r})$ is assumed to be a *real*, *continuous* and *univocal* function of the coordinates. The complications arising from a not univocal L, such as in the presence of reflections or caustics, will be considered later. The locus of points for which ϕ = const. represents a generic moving wavefront, whose dynamics we are going to determine.

Then, if we apply the differential operators to the analytic function $\mathscr{E} = \boldsymbol{E}_0\exp[i\phi(\boldsymbol{r},t)]$ of the real field $\mathbf{E}$, we get

$$\begin{aligned}
&\nabla\cdot\mathscr{E} = \{\nabla\cdot\boldsymbol{E}_0 + ik_o\nabla L\cdot\boldsymbol{E}_0\}e^{i\phi(\boldsymbol{r},t)},\\
&\nabla\times\mathscr{E} = \{\nabla\times\boldsymbol{E}_0 + ik_o\nabla L\times\boldsymbol{E}_0\}e^{i\phi(\boldsymbol{r},t)},\\
&\nabla^2\mathscr{E} = \left\{\nabla^2\boldsymbol{E}_0 - k_o^2(\nabla L)^2\boldsymbol{E}_0 + 2ik_o(\nabla L\cdot\nabla)\boldsymbol{E}_0 + ik_o\left(\nabla^2 L\right)\boldsymbol{E}_0\right\}e^{i\phi(\boldsymbol{r},t)},
\end{aligned} \tag{2.1.3}$$

$$\frac{\partial}{\partial t}\mathscr{E} = -i\omega \boldsymbol{E}_{\mathrm{o}} e^{i\phi(r,t)}, \quad \frac{\partial^2}{\partial t^2}\mathscr{E} = -\omega^2 \boldsymbol{E}_{\mathrm{o}} e^{i\phi(r,t)}. \tag{2.1.4}$$

Therefore, by applying Eqs. (2.1.1-4) to Eqs. (1.2.15), we obtain:

$$\begin{aligned} \nabla\bullet(\varepsilon\mathbf{E}) = 0 \quad &\rightarrow \quad \nabla L\bullet \boldsymbol{E}_{\mathrm{o}} = \frac{i}{k_{\mathrm{o}}}\left(\boldsymbol{E}_{\mathrm{o}}\bullet\frac{\nabla\varepsilon}{\varepsilon} + \nabla\bullet\boldsymbol{E}_{\mathrm{o}}\right), \\ \nabla\times\mathbf{E} = -\mu\frac{\partial\mathbf{H}}{\partial t} \quad &\rightarrow \quad \nabla L\times\boldsymbol{E}_{\mathrm{o}} - \mathrm{c}\mu\boldsymbol{H}_{\mathrm{o}} = \frac{i}{k_{\mathrm{o}}}\nabla\times\boldsymbol{E}_{\mathrm{o}}, \end{aligned} \tag{2.1.5}$$

where, as I will also do later on, I have written only the pair of equations for **E**, since the symmetry of Eqs (1.2.15) allows for obtaining the corresponding equations for **H** with the simple and simultaneous substitution

$$\boldsymbol{E} \leftrightarrow \boldsymbol{H}, \ \varepsilon \leftrightarrow -\mu. \tag{2.1.6}$$

If, then, $\boldsymbol{E}_{\mathrm{o}}$, ε, $\boldsymbol{H}_{\mathrm{o}}$, μ vary little within distances on the order of λ, the terms in $1/k_{\mathrm{o}}$ in Eqs. (2.1.5) can be neglected; therefore, such equations become

$$\begin{aligned} \nabla L\bullet\boldsymbol{E}_{\mathrm{o}} &= 0, & \nabla L\bullet\boldsymbol{H}_{\mathrm{o}} &= 0, \\ \nabla L\times\boldsymbol{E}_{\mathrm{o}} &= \mathrm{c}\mu\boldsymbol{H}_{\mathrm{o}}, & \nabla L\times\boldsymbol{H}_{\mathrm{o}} &= -\mathrm{c}\varepsilon\boldsymbol{E}_{\mathrm{o}}. \end{aligned} \tag{2.1.7}$$

The gradient of $L(\boldsymbol{r})$ therefore has the role that $\boldsymbol{k}/k_{\mathrm{o}}$ has in Eqs. (1.4.13, 17-18), where we have considered plane waves. Eqs. (2.1.7) impose a constraint for L. Comparing the last two equations, we indeed get

$$\mathrm{c}\varepsilon\boldsymbol{E}_{\mathrm{o}} = -\frac{1}{\mathrm{c}\mu}\nabla L\times(\nabla L\times\boldsymbol{E}_{\mathrm{o}}) = \frac{1}{\mathrm{c}\mu}(\nabla L)^2\boldsymbol{E}_{\mathrm{o}} - \frac{1}{\mathrm{c}\mu}(\nabla L\bullet\boldsymbol{E}_{\mathrm{o}})\nabla L.$$

However, for Eq. (2.1.7.a), the last term is null, so we ultimately have

$$(\nabla L)^2 = n^2, \tag{2.1.8}$$

where $n = \mathrm{c}\sqrt{\varepsilon\mu}$ is the index of refraction of the medium. This same relation between L and n can even be derived substituting Eqs. (2.1.1) in the wave equations (1.4.6). For a monochromatic wave with angular frequency ω, Eq. (1.4.6.a) may be transcribed as

$$\nabla^2\mathbf{E} - \frac{n^2}{\mathrm{c}^2}\frac{\partial^2}{\partial t^2}\mathbf{E} = -\frac{\nabla\mu}{\mu}\times(\nabla\times\mathbf{E}) + \nabla\left(\mathbf{E}\bullet\frac{\nabla\mu}{\mu}\right) - 2\nabla\left(\mathbf{E}\bullet\frac{\nabla n}{n}\right), \tag{2.1.9}$$

where the terms on the right have been rearranged. In particular, if the medium is non-magnetic, that is, for $\mu = \mu_0$, we have

$$\nabla^2 \mathbf{E} - \frac{n^2}{c^2}\frac{\partial^2}{\delta t^2}\mathbf{E} = -2\nabla\left(\mathbf{E}\bullet\frac{\nabla n}{n}\right). \tag{2.1.10}$$

From Eqs. (2.1.1) and (2.1.9), with long calculations, by applying Eqs. (2.1.1-2), (2.1.3.c) and (2.1.4.b) to Eq. (2.1.9), we obtain the following expression for the amplitude of the electric field:

$$\begin{aligned} & n^2 \boldsymbol{E}_0 - (\nabla L)^2 \boldsymbol{E}_0 \\ & + \frac{i}{k_o}\left[\left(\nabla^2 L\right)\boldsymbol{E}_0 + 2(\nabla L\bullet\nabla)\boldsymbol{E}_0 + 2\left(\boldsymbol{E}_0\bullet\frac{\nabla n}{n}\right)\nabla L - \left(\frac{\nabla\mu}{\mu}\bullet\nabla L\right)\boldsymbol{E}_o\right] \\ & = \frac{1}{k_o{}^2}\left[-\nabla^2 \boldsymbol{E}_0 - 2\nabla\left(\boldsymbol{E}_0\bullet\frac{\nabla n}{n}\right) + \nabla\left(\boldsymbol{E}_0\bullet\frac{\nabla\mu}{\mu}\right) - \frac{\nabla\mu}{\mu}\times(\nabla\times\boldsymbol{E}_o)\right]. \end{aligned} \tag{2.1.11}$$

Now, consider the member on the right side of this equation. The first term is negligible when the amplitude E_j of any component of $\boldsymbol{E}_0$ varies slowly enough to verify the conditions

$$\left|\lambda\frac{\partial E_j}{\partial x}\right| \ll E_j\,, \quad \left|\lambda\frac{\partial^2 E_j}{\partial x^2}\right| \ll \frac{\partial E_j}{\partial x}$$

for any x-axis direction [Sivoukhine 1984, IV-1, page 43]. The other terms are negligible when ε and μ vary very little within a distance λ. With these approximations, Eq. (2.1.11) can be decomposed into the following two equations:

$$\begin{cases} (\nabla L)^2 = n^2\,, \\ \left(\nabla^2 L\right)\boldsymbol{E}_0 + 2(\nabla L\bullet\nabla)\boldsymbol{E}_0 + 2\left(\boldsymbol{E}_0\bullet\frac{\nabla n}{n}\right)\nabla L - \left(\frac{\nabla\mu}{\mu}\bullet\nabla L\right)\boldsymbol{E}_0 = 0\,. \end{cases} \tag{2.1.12}$$

The first of these equations is identical to Eq. (2.1.8); as we shall see, it is alone enough to determine the trajectory of a ray in a medium of known characteristics, starting with given initial conditions. The second equation, instead, allows to calculate the way in which the field propagates along a predetermined ray. Eqs. (2.1.12) constitute the *equation system* of Geometrical Optics, valid under the conditions mentioned above (and for relatively short propagation distances). Such conditions are not verified, for example, on the frontier of a shadow or in proximity of a "convergence focus" of the wavefront: there, diffraction effects become important. It must also be pointed out that, if Eq. (2.1.11) is integrated for large distances of propagation, the term $\nabla^2 E_o / k_o^2$ may not be negligible. For example, if a plane wave reaches an opening of diameter D that is in any way large, the transmitted field presents evident diffraction effects for distances $\gtrsim D^2/\lambda$. So, even a "collimated" laser beam tends to expand for large distances.

Eq. (2.1.8) is called the *eikonal equation* and L is called e*ikonal*, from the Greek εικων, which means image; it associates the gradient of L with the refractive index n. The surfaces where the eikonal is constant are called *wavefronts*, and therefore a displacement $\delta \boldsymbol{r}$ taken on such front necessarily corresponds to $\delta L = 0$, and thus

$$\delta L = \nabla L \bullet \delta \boldsymbol{r} = 0 \ .$$

Thus, ∇L *is normal to the wavefront* and Eq. (2.1.8) can be written in a vector form:

$$\nabla L = n\mathbf{s} \ , \tag{2.1.13}$$

where $\mathbf{s}$ is the versor normal to the wavefront and, with the choice of sign that we will make in Eq. (2.1.14) below, $\mathbf{s}$ is oriented in the positive direction of propagation. The speed of the wavefront can now be obtained in just a few steps. Eq. (2.1.13) also tells us that

$$\partial L / \partial s = n \ .$$

In addition, we define the function

$$\phi(\boldsymbol{r};t) = k_o L(\boldsymbol{r}) - \omega t = \text{constant.} \tag{2.1.14}$$

It expresses a *family* of wavefronts, ordered according to the parameter t. They depend implicitly upon $n(\boldsymbol{r})$. Differentiating this equation, we get $\omega dt = k_o dL$, whence

$$\omega dt = \frac{\omega}{c}\frac{\partial L}{\partial s} ds = \frac{\omega}{c} n ds \ .$$

Therefore, the normal speed of the wavefront is

$$\frac{ds}{dt} = \frac{c}{n} = \upsilon \ , \tag{2.1.15}$$

which is the same as that of a plane wave. Thus, we can expect that, in a small region of space, a small portion of the wavefront behaves like a portion of a plane. As noted by Born and Wolf (1980), it is essentially from this approximation, generally valid for visible radiation, that the simplicity and effectiveness of Geometrical Optics arises. Indeed, this result allows for building the wavefront F_2 at

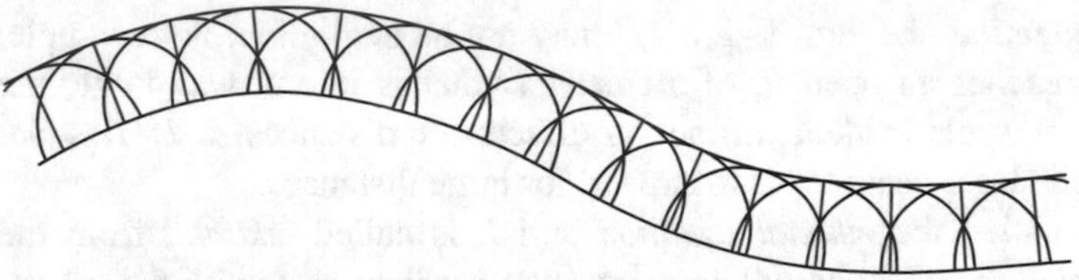

Fig. 2.1 Generation of a new front by the envelope of wavelets

time $t + dt$ if we know the shape of the wavefront F_1 at time t. To do this, it is enough to draw from each point of F_1 a segment of length $v\,dt$ along the normal. Then, joining the ends of all segments, we obtain F_2. This construction is equivalent to the *Huygens' principle*, from 1679, reproduced here below in a simplified way (Fig. 2.1):

Each point of a primary wavefront serves as the source of a spherical secondary wavelet, so that, at a later time, the front of the primary wave is the envelope of these wavelets. Furthermore, the wavelets advance with the same speed and frequency as the primary wavefront at each point in space.

Thus formulated, this principle concerns the *envelope* of the secondary waves, instead of their interference, so it is fully justified only in the approximation of Geometrical Optics. By itself, it does not necessarily imply that optics can be deduced from the propagation of waves or, in other words, that the luminous radiation is made of waves, but only that the propagation of the optical signals is something that is not localized, that is, assigned collectively to a family of rays. However, we have already seen in the historical notes that Huygens' original hypothesis was more refined and contained an idea of *composition* of the wavelets. Newton, who refused all forms of ad hoc hypothesis, did not let himself be persuaded by the approximate argumentation of Huygens and remained convinced that the individualistic optics of the corpuscles was the most acceptable. After one century, around 1819, Fresnel finally modified Huygens' principle replacing the envelope with the interference of the waves discovered by Young. Later, Kirchhoff showed that the Huygens-Fresnel-Young principle is a direct consequence of the wave equations. We will return to this point by studying Diffraction.

2.2 Propagation properties in Geometrical Optics

2.2.1 The light rays and the radiant energy propagation

From the above, it is natural to define rays of light as the orthogonal trajectories to the family of wavefronts defined by Eq. (2.1.14). For consistency with observation (a) of the introduction, it is now necessary to show that the energy actually flows along these trajectories. In other words, it is necessary to verify that the Poynting's vector is also normal point by point to the wavefront. Nevertheless,

Fig. 2.2
Propagation of energy in a ray tube

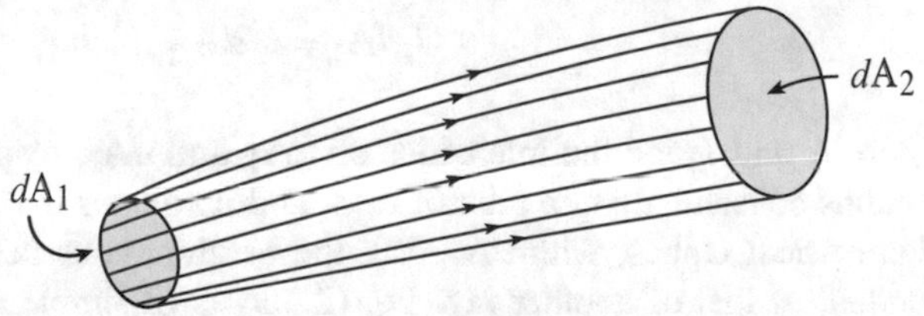

this follows directly from Eqs. (2.1.7) and (2.1.13), from which we see that the amplitudes $\boldsymbol{E}_0$ and $\boldsymbol{H}_o$, as in the case of a plane wave, are parallel to the wavefront and orthogonal to each other. Therefore, with no more than the foresight to replace $\boldsymbol{k}$ with $k_o\nabla L$, the results of §1.5.2 are still applicable and, for Eq. (1.5.13.c), the Poynting's vector averaged in time is

$$\overline{\mathbf{S}} = \frac{c\varepsilon}{2n}\left|\boldsymbol{E}_0\right|^2 \mathbf{s} = \upsilon w \mathbf{s}\,, \qquad (2.2.1)$$

where w is the average density of electromagnetic energy and $\upsilon = c/n$. In other words, even for the case of a non-planar wave in an inhomogeneous *isotropic* medium, the Poynting versor $\mathbf{s}$ is oriented along the normal to the wavefront, and the energy density propagates in the same direction with speed $\upsilon = c/n$. The geometric light rays can therefore be consistently defined as trajectories, point by point orthogonal to the wavefront, along which the energy propagates. It should be noted that this definition of a ray is appropriate only for isotropic media. Indeed, in anisotropic media, the Poynting vector is not generally oriented along the normal to the wavefront, while it remains the pointing vector for the energy flux,[1] as we will see in the last chapter.

The intensity corresponds to the modulus of $\overline{\mathbf{S}}$, thus the intensity that "flows" along a radius, is

$$I = \upsilon w$$

and the law of energy conservation (1.5.19) may be reformulated as follows:

$$\operatorname{div}(I\mathbf{s}) = 0\,. \qquad (2.2.2)$$

To see the implications of this relation, we consider a thin tube made up of all of the rays that start from an element dA_1 of a wavefront defined by $L(\boldsymbol{r}) = a_1$ (=const.), and come in a corresponding element dA_2 of another wavefront, for which $L(\boldsymbol{r}) = a_2$ (Fig. 2.2). Applying the Gauss theorem and Eq. (2.2.2) to the volume thus defined, as was done for Eq. (1.5.20), and observing that $\mathbf{s}$ is parallel to the lateral surface of the tube while it is normal to dA_1 and dA_2, we find that

[1] What a name: the Poynting's vector *is* the pointing (intensity) vector!

$$I_1 dA_1 = I_2 dA_2 , \tag{2.2.3}$$

where I_1 and I_2 are the intensities on dA_1 and dA_2, respectively. Therefore, IdA remains constant along a tube of rays, and it follows that, in the approximation of Geometrical Optics, with $dA \to 0$, the brightness associated with a ray is independent of that of another ray. Eq. (2.2.3) is of simple and frequent application, and it constitutes the law of intensity in Geometrical Optics.

The intensity along a ray can be determined when it is known at one point on the ray. In fact, through Eq. (2.1.13), Eq. (2.2.2) can be developed as follows:

$$0 = \operatorname{div}(I\mathbf{s}) = \operatorname{div}\left(\frac{I}{n}\nabla L\right) = \frac{I}{n}\nabla^2 L + \nabla L \cdot \nabla \frac{I}{n} = \frac{I}{n}\nabla^2 L + n\frac{\partial}{\partial s}\left(\frac{I}{n}\right),$$

where $\partial/\partial s$ is an operation of the derivative to be performed along the ray. Integrating, we have

$$\frac{I}{n} = \frac{I_o}{n_o}\exp\left(-\int_{s_o}^{s}\frac{\nabla^2 L}{n}ds'\right), \tag{2.2.4}$$

where I_o and n_o are, respectively, the intensity and the refractive index at the starting point and the integral is taken along the ray for a path of length $s - s_o$. On the other hand, to calculate the integral, it is necessary to know the eikonal L or, equivalently, it is also necessary to determine the path of all rays near the one considered.

2.2.2 Propagation of polarization

Eq. (2.1.12.b) constitutes the law of propagation of the amplitude vector of the field. It is not easy to solve, but it is also of little use; however, it is important when we want to consider the state of polarization in the propagation in inhomogeneous isotropic media. Through Eq. (2.1.13) and the identity $\mathbf{s}\cdot\nabla = \partial/\partial s$, Eq. (2.1.12.b) can be reformulated as follows:

$$n\frac{\partial \boldsymbol{E}_o}{\partial s} = -\frac{1}{2}\boldsymbol{E}_o\nabla\cdot(n\mathbf{s}) - n\mathbf{s}\left(\boldsymbol{E}_o\cdot\frac{\nabla n}{n}\right) + \frac{1}{2}\boldsymbol{E}_o\left(\frac{\nabla\mu}{\mu}\cdot n\mathbf{s}\right).$$

The first and third terms on the right side can be collected together and, ultimately, it becomes

$$n\frac{\partial \boldsymbol{E}_o}{\partial s} = -\frac{1}{2}\mu\boldsymbol{E}_o\nabla\cdot\left(\frac{n}{\mu}\mathbf{s}\right) - n\mathbf{s}\left(\boldsymbol{E}_o\cdot\frac{\nabla n}{n}\right). \tag{2.2.5}$$

The first term on the right is parallel to $\boldsymbol{E}_0$, thus it determines the variations of the modulus of the field. Instead, the second term, which is null in homogeneous media, causes changes in the orientation of the field. It should thus be noted that Eq. (2.2.5) is consistent with Eq. (2.2.2); in fact, with a scalar multiplication of both sides for $\boldsymbol{E}_0{}^*$, by summing the result with its complex conjugate and recalling that $\boldsymbol{E}_0$ and $\boldsymbol{E}_0{}^*$ are perpendicular to $\mathbf{s}$, we get

$$n\mathbf{s}\cdot\nabla E_0^2+\mu E_0^2\nabla\cdot\left(\frac{n}{\mu}\mathbf{s}\right)=0. \tag{2.2.6}$$

where E_0 is the modulus of $\boldsymbol{E}_0$. Collecting the two terms, it becomes

$$\mu\nabla\cdot\left(E_0^2\frac{n}{\mu}\mathbf{s}\right)=0,$$

where the argument of the divergence is $2I\,\mathbf{s}/c$, and we regain Eq. (2.2.2). Thus, in addition to Eq. (2.2.4) for the intensity, we can write a similar equation for the modulus of the amplitude E_0 [Solimeno et al 1986]:

$$E_0(s)=E_0(s_o)\exp\left(-\frac{1}{2}\int_{s_o}^{s}\frac{\nabla^2 L}{n}ds'\right). \tag{2.2.7}$$

Replacing $\boldsymbol{E}_0$ in Eq. (2.2.5) with $\mathbf{e}E_0$, where $\mathbf{e}$ is the complex versor of $\boldsymbol{E}_0$, we get an equation for the propagation of polarization:

$$n\frac{\partial\mathbf{e}}{\partial s}=-\frac{1}{2E_0}\left[n\mathbf{s}\cdot\nabla E_0^2+\mu E_0^2\nabla\cdot\left(\frac{n}{\mu}\mathbf{s}\right)\right]\mathbf{e}-n\left(\mathbf{e}\cdot\frac{\nabla n}{n}\right)\mathbf{s}.$$

Thanks to Eq. (2.2.6), we ultimately have

$$\frac{\partial\mathbf{e}}{\partial s}=-\left(\mathbf{e}\cdot\frac{\nabla n}{n}\right)\mathbf{s}. \tag{2.2.8}$$

2.3 General properties of rays

2.3. Tracing law of rays in inhomogeneous media

Rays were defined as orthogonal trajectories to a given family of wavefronts determined by the relation (2.1.14), where t is a parameter that orders the succession of the fronts. A generic trajectory can, in turn, be defined parametrically with a function $\boldsymbol{r}(s)$, where s is the path traveled along the same curve, starting from an initial point P_o ($\equiv \boldsymbol{r}_0$). In particular, the unit vector $\mathbf{s}$ tangent to the curve at a point P ($\equiv \boldsymbol{r}$) is given by

$$\mathbf{s} = \frac{d\boldsymbol{r}}{ds} . \tag{2.3.1}$$

In the case in which the curve corresponds to a ray of light, from Eq. (2.1.13), then we have that

$$n\frac{d\boldsymbol{r}}{ds} = \nabla L . \tag{2.3.2}$$

This equation allows us to calculate the trajectory of a ray that passes through a given point P_o, once we know $L(\boldsymbol{r})$.

On the other hand, we can imagine "launching" a ray, with position and initial direction assigned, in a medium of known refractive index $n(\boldsymbol{r})$, and let our ray "decide for itself" where to go, according to changes in the index n. We will now see, with a few passages, which law our ray must obey. For now, we consider the case in which $n(\boldsymbol{r})$ are a continuous function of $\boldsymbol{r}$, but, with some contrivances, we can then extend the discussion to reasonable discontinuity, as a glossy interface between two different media.

In terms of the derivative in the direction of the ray, Eq. (2.1.13) becomes

$$\frac{dL}{ds} = n .$$

Hence, taking the gradient of both sides, we get:

$$\frac{d}{ds}\nabla L = \nabla n ,$$

where, on the left, the order of the derivatives has been exchanged. Furthermore, with Eq. (2.3.2), we obtain

$$\frac{d}{ds}\left(n\frac{d\boldsymbol{r}}{ds}\right) = \nabla n , \tag{2.3.3}$$

which is called the *Euler-Lagrange equation*. Finally, developing on the left, this last equation takes the form

$$\left(\nabla n \cdot \frac{d\boldsymbol{r}}{ds}\right)\frac{d\boldsymbol{r}}{ds} + n\frac{d^2\boldsymbol{r}}{ds^2} = \nabla n , \tag{2.3.4}$$

or even

$$\begin{cases} \dfrac{d\mathbf{s}}{ds} = \dfrac{\nabla n}{n} - \left(\dfrac{\nabla n}{n} \bullet \mathbf{s}\right)\mathbf{s}, \\ \dfrac{d\boldsymbol{r}}{ds} = \mathbf{s}. \end{cases} \tag{2.3.5}$$

Interestingly, if we replace the path s with the parameter

$$\tau = \int^{s} \frac{ds'}{n(s')},$$

Eq. (2.3.3) becomes

$$\frac{d^2\boldsymbol{r}}{d\tau^2} = \nabla\left(\frac{1}{2}n^2\right),$$

representing the motion of a particle of unit mass in a potential $V = -n^2/2$ [Solimeno et al 1986].

Eq. (2.3.3) also has the characteristics of *reciprocity*, in the sense that, going along the ray in the opposite direction, with the transformation $s = -s'$, $ds = -ds'$, $\boldsymbol{r}'(s') = \boldsymbol{r}(s)$, we have that, if $\boldsymbol{r}$ is the solution to Eq. (2.3.3) with respect to the order parameter s, $\boldsymbol{r}'$ is also such with respect to s':

a beam that is reflected back on itself, returns along the same path

In conclusion, these equations allow us to calculate how to continue a ray once it is assigned an initial direction and a point of departure.

2.3.2 Curvature and torsion of the rays

The trend of rays in space is conveniently described by the tangent versor $\mathbf{s}$, by the *normal* or *curvature* versor $\mathbf{c}$ and, finally, by the *binormal* or *torsion* versor $\mathbf{T} = \mathbf{s} \times \mathbf{c}$. These form a set of three orthogonal versors that generally rotates in space by moving along the curve, obeying the *Frénet's equations*:

$$\begin{cases} \dfrac{d}{ds}\mathbf{s} = \dfrac{\mathbf{c}}{\rho} = c, \\ \dfrac{d}{ds}\mathbf{c} = -\dfrac{\mathbf{s}}{\rho} + \dfrac{\mathbf{T}}{\tau}, \\ \dfrac{d}{ds}\mathbf{T} = -\dfrac{\mathbf{c}}{\tau}, \end{cases} \tag{2.3.6}$$

where $\boldsymbol{c}$ is the *curvature vector*, lying in the plane of curvature, while ρ, τ are, respectively, the *curvature radius* and the *torsion* of the curve at the point that is considered. From Eqs. (2.3.5.a) and (2.3.6.a), it is

$$\nabla n = \frac{dn}{ds}\mathbf{s} + \frac{n}{\rho}\mathbf{c} \ , \tag{2.3.7}$$

from which it is seen that the gradient of the refractive index is situated in the plane of curvature ($\mathbf{c}$,$\mathbf{s}$). Next, there are the equations

$$\frac{1}{\rho} = \mathbf{c} \bullet \frac{\nabla n}{n} \ , \quad \frac{1}{\tau} c \bullet \nabla n = \mathbf{T} \bullet \left(\frac{d}{ds} \nabla n \right). \tag{2.3.8}$$

The first is obtained by scalarly multiplying Eq. (2.3.7) for $\mathbf{c}$, while the second is obtained from the derivative of Eq. (2.3.7) for s, by applying Eqs. (2.3.6.a-b) and then multiplying scalarly for $\mathbf{T}$.

Since ρ is always positive, Eq. (2.3.8) implies that, proceeding along $\mathbf{c}$, the refractive index increases. In other words, the beam will always bend toward the region of higher refractive index. In the simplest case, that is, in a homogeneous medium for which $\nabla n = 0$, we have that the light rays are straight lines. Instead, if n is "stratified" in the x direction, we have

$$\frac{d}{ds}\left(n \frac{dy}{ds} \right) = 0, \quad \frac{d}{ds}\left(n \frac{dz}{ds} \right) = 0 ,$$

so that

$$n \frac{dy}{ds} = \text{cost.}, \quad n \frac{dz}{ds} = \text{cost.}.$$

The ray thus remains in the plane dy/dz = const., that is, on a plane containing the x-axis whose normal is oriented in any direction y, z; the coordinate system can then be chosen such that the ray lies in the xy plane. So, we can write

$$n \frac{dy}{ds} = \text{cost.}, \quad \frac{dz}{ds} = 0. \tag{2.3.9}$$

If we denote with ϑ the angle between $\mathbf{s}$ and the x-axis, we have:

$$\frac{dx}{ds} = \cos\vartheta, \quad \frac{dy}{ds} = \sin\vartheta \, .$$

Finally, by the first of Eqs. (2.3.9), it becomes

$$n \sin \vartheta = \text{cost.} \tag{2.3.10}$$

which is nothing but the Snell-Descartes law adapted to a continuous variation of the refractive index.

2.3.2.1 Propagation in media with spherical symmetry

A particularly interesting case is when $n = n(r)$ is a function only of the distance r from a center of symmetry O. For a generic ray defined by the curve $\boldsymbol{r}(s)$, with $r = |\boldsymbol{r}|$, from Eq. (2.3.7), it follows that its curvature plane in $\boldsymbol{r}$ contains the point O, since the gradient of n is parallel to $\boldsymbol{r}$; therefore, varying s, the ray is always on the same plane. Indeed, following Born and Wolf (1980), we have

$$\frac{d}{ds}(\boldsymbol{r} \times n\mathbf{s}) = \frac{d\boldsymbol{r}}{ds} \times n\mathbf{s} + \boldsymbol{r} \times \frac{d}{ds}(n\mathbf{s}) = \boldsymbol{r} \times \nabla n = 0\,,$$

since the first term vanishes for the definition of $\mathbf{s}$ and the second term does the same because the gradient of n is parallel to $\boldsymbol{r}$. Therefore,

$$\boldsymbol{r} \times n\mathbf{s} = a, \quad \text{that is,} \quad nr \sin \vartheta = a\,, \tag{2.3.11}$$

where ϑ is the angle between $\boldsymbol{r}$ and $\mathbf{s}$, while a is a constant. This last expression is called *Bouguer's formula*, and it is formally the same as the conservation of angular momentum of a particle moving in a central field; here, n plays the role of impulse. With some passages, it turns out that, on the plane where it lies, the ray is represented in polar coordinates r, φ by

$$\varphi_2 - \varphi_1 = a \int_{r_1}^{r_2} \frac{dr}{r\sqrt{n^2 r^2 - a^2}}\,. \tag{2.3.12}$$

2.3.2.2 Atmospheric refraction

The case of a medium with spherical symmetry occurs, for example, in the propagation in the atmosphere: let us think of a mountaineer topographer, who, looking at the valley bottom, wants to know how high he is located (Fig. 2.3). In conditions of quiet, on a clear day and without a strong wind, the refractive index is determined by the profile of pressure, temperature, and composition of the atmosphere as a function of height, so the problem is, in principle, quite complex. However, since the effect of refraction is small, we can make some simplifying assumptions. In particular, the temperature profile T is taken as linear with the height h up to the limit altitude h_t of the troposphere, which, in temperate latitudes, is around 12 km in height:

$$T = T_o - ah\,, \tag{2.3.13}$$

where T_o is the absolute temperature at sea level and $a = 6.5$ K/km, as suggested by the *International Standard Atmosphere model*. For $h > h_t$, the temperature can be taken as constant, even though, after 20 km to about 50 km of altitude, it grows into the stratosphere with a slope of about 1 K/km. In the troposphere, the pressure is then given approximately by[2]

$$p(h) = p_o\left(1 - \frac{ah}{T_o}\right)^{\frac{mg_o}{ak_B}}, \tag{2.3.14}$$

where p_o is the pressure at zero altitude, $g_o = 9.80665$ m/s^2 is the acceleration of gravity to the ground, $k_B = 1.38065 \cdot 10^{-23}$ J/K is the Boltzmann's constant and $m = 28.96 \times 1.6605 \cdot 10^{-27}$ Kg is the mean mass of the air molecules. Instead, for $h > h_t$, the pressure can be approximated by

$$p(h) = p_t \exp\left[\frac{mg_o}{k_B T_s}(h - h_t)\right], \quad \text{with } p_t = p(h_t),\ T_s = T_o - ah_t\,. \tag{2.3.15}$$

Finally, the index of refraction is described approximately, for a wavelength of 633 nm and temperatures between 0 and 35 °C, by[3]

$$n = 1 + 7.86\cdot 10^{-4}\, p/(273 + t) - 1.5\cdot 10^{-11}\,\mathrm{RH}\left(t^2 + 160\right), \tag{2.3.16}$$

where t is the temperature in degrees Celsius, p is the pressure in kPa and $0 \le \mathrm{RH} \le 100$ is the relative humidity in percentage.

Making n explicit via these equations in the integral of Eq. (2.3.10), it is therefore possible to obtain the path of a ray in the atmosphere. However, in topographic altimetry, it is of use to do other approximations that do not explicitly require carrying out that integral. In fact, at the topographic altitudes, the refractive index gradient can be assumed to be constant and since the ray deviates little from a straight line, its curvature remains practically constant. Therefore, it is assumed

---

2 For this and the following expression, see, for example, Fermi (1972) and Néda and Volkán (2002).

3 See the web page http://emtoolbox.nist.gov di J.A. Stone and J.H. Zimmerman of NIST. Accurate expressions that also consider the wavelength are given by Ciddor (1996) and by Edlén (1966).

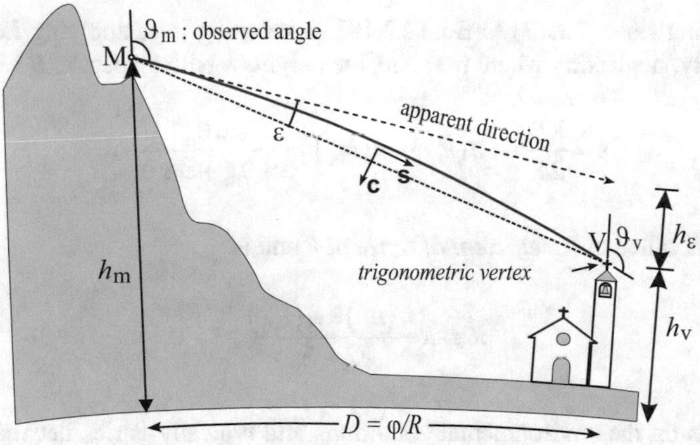

Fig. 2.3 Curvature of a ray into the atmosphere. R is the radius of the geoid in this section, φ is the angle at the center of the Earth between M and V, and D is the distance on the geoid between the feet of these points. The altitude h_m and h_v of M and V are measured relative to the geoid, while h_ε is the difference between the apparent altitude of V and the actual. Finally, ϑ_m and ϑ_v are the angles that the versor of the ray makes with the vertical in M and V

that it coincides with an arc of a circle or that the apparent directions of observation with respect to the vertical, respectively, ϑ_m and $\pi - \vartheta_v$, from the point M on the mountain to the one V in the valley, and vice versa, deviate from the real ones, θ_m and $\pi - \theta_v$, for the same angle ε:

$$\theta_m = \vartheta_m + \varepsilon, \quad \theta_v = \vartheta_v - \varepsilon \,. \tag{2.3.17}$$

With reference to the definitions given in Fig. 2.3, for Bouguer's formula,[4] we then have

$$n(h_m)(R + h_m)\sin\vartheta_m = n(h_v)(R + h_v)\sin\vartheta_v \,, \tag{2.3.18}$$

while, for the law of sines applied to the MVC triangle, where C is the center of the Earth, we have [5]

$$(R + h_m)\sin\theta_m = (R + h_v)\sin\theta_v \,. \tag{2.3.19}$$

---

[4] Pierre Bouguer (1698-1758) was a mathematician, geophysicist, geodesist and French astronomer.

[5] θ_m is the complementary angle to the inner angle at M in the MVC triangle, where C is the center of the Earth: their sines coincide, while the cosines are opposed. Instead, θ_v coincides with the angle inside V.

By applying Eqs. (2.3.17) to Eq. (2.3.18), developing in ε, applying Eq. (2.3.19) and, finally, neglecting where possible, the heights h with respect to R we obtain

$$\varepsilon = \frac{KD}{2R} \cong \left[n(h_{\mathrm{m}}) - n(h_{\mathrm{m}})\right] \frac{\sin\theta_{\mathrm{m}}}{\cos\theta_{\mathrm{m}} + \cos\theta_{\mathrm{v}}}, \tag{2.3.20}$$

where K is called the *coefficient of refraction* and is

$$K \cong R \frac{n(h_{\mathrm{v}}) - n(h_{\mathrm{m}})}{h_{\mathrm{m}} - h_{\mathrm{v}}}. \tag{2.3.21}$$

It depends on the environmental conditions and typically varies between 0.1 and 0.2. Finally, with additional approximations, we have that the difference of height caused by the refraction is

$$h_{\varepsilon} \cong \frac{KD^2}{2R}, \tag{2.3.22}$$

to which must be added the correction for the curvature of the Earth given by $D^2/(2R)$.

In the case of astronomical observation, one must proceed to the effective integration of Eq. (2.3.12), and the deviation is made progressively larger by increasing the zenith angle. On the horizon, it is about 34.5′ of arc, a little higher than the angular diameter of the Sun and Moon. An approximate formula for the deviation was given by Bennett (1982):

$$\varepsilon = \varepsilon' - 0.06\sin(14.7\varepsilon' + 13), \quad \text{where } \varepsilon' = \frac{1}{\tan\left(90 - \vartheta + \dfrac{7.31}{94.4 - \vartheta}\right)}, \tag{2.3.23}$$

where ε is expressed in minutes of arc, the trigonometric functions have, as arguments, sexagesimal degrees, and ϑ is the apparent azimuthal angle expressed in degrees. To this formula must be added the corrections for pressure and temperature. Exercise: why does the Moon appear red in a lunar eclipse?

2.3.3 The optical path

The integral $\int_C n\,dr$ taken along a curve C, where dr is an element of the path on C, is called the *optical length* of the curve C. If the curve coincides with a ray of light that connects two points P_1 and P_2, we have

$$\int_{P_1}^{P_2} n\,ds = \int_{P_1}^{P_2} \frac{\partial L}{\partial s} ds = \int_{P_1}^{P_2} dL = L(P_2) - L(P_1), \tag{2.3.24}$$

where Eq. (2.1.13) is used. Thus, the *optical path* travelled along a light ray connecting two points is just the difference between the corresponding eikonals. Furthermore, given that the energy propagates with speed $\upsilon = \mathrm{c}/n$ along the ray, it also results that

$$\int_{P_1}^{P_2} n\,ds = \int_{P_1}^{P_2} \frac{\mathrm{c}}{\upsilon} ds = \mathrm{c}\int_{t(P_1)}^{t(P_2)} dt = \mathrm{c}\left[t(P_2) - t(P_1)\right], \tag{2.3.25}$$

where dt is the time elapsed by light to travel a distance ds along the ray. We will see in the following sections the important consequences of these simple considerations.

2.3.4 Collective properties of rays

In differential geometry, a set of curves that fill an entire volume where, in general, each of its points is crossed by only one of these curves (exceptions are allowed) is called *congruence*. In addition, if these curves intersect a given surface perpendicularly, the congruence is said to be *normal* [Born and Wolf 1980]. We have seen that, from an assigned eikonal $L(\mathbf{r})$, it is possible to generate a family of rays that have the property of remaining always perpendicular to a wavefront that is associated with $L(\mathbf{r})$ by Eq. (2.1.14), while the front itself is moving in the time. Therefore, these rays constitute a normal congruence.

Rarely, however, does one know the eikonal $L(\mathbf{r})$ a priori. In practical applications, an initial wavefront (at a fixed time) is assigned. Then, the family of rays is built through the Euler-Lagrange equation (2.3.3), launching them from each point of the wavefront and perpendicularly to it. A new surface is constructed from the set of points that are obtained going along each ray for the same optical path. If we know in advance that the eikonal is unique in the crossed volume, we also know from Eq. (2.3.24) that the new surface is actually a wavefront, and is therefore perpendicular to the rays.

However, it remains to be clarified whether this perpendicularity persists even when the uniqueness of $L(\mathbf{r})$ is violated in some way. This uniqueness is not always guaranteed; just think of a wave reflected from a mirror. In general, even starting from a "regular" wavefront, by propagating it with the construction mentioned above, it may split or bend upon itself. Therefore, it may happen that the same point is reached by two different rays of the same family, on which generally different optical paths can be measured from the initial wavefront. An example of such a situation occurs in the caustic generated by a lens.

So, in agreement with Eq. (2.3.24), let us accept that the eikonal corresponds to the path taken on the rays propagated with Eq. (2.3.3) from an initial wavefront, on which we assume, for example, that $L(\boldsymbol{r}) = 0$. Well, we may have some spatial regions in which a plurality of eikonal $L_m(\boldsymbol{r})$ is present. One way to overcome this complication is to consider only the eikonal that evolves continuously in the neighborhood of the path of a given ray and then select only those rays that correspond to that eikonal.

If, then, in an *open volume* V, the eikonal is a *continuous* and *univocal* function of the coordinates (if necessary, by making a selection with the criteria mentioned above), the rays contained therein are the "lines of force" of a *vector field* $n\mathbf{s}$ determined by Eq. (2.1.13). In particular, taking the curl of both members of Eq. (2.1.13), we have

$$\nabla \times (n\mathbf{s}) = 0 . \tag{2.3.26}$$

Finally, applying the Stokes' theorem, we get

$$\oint_{C} n\mathbf{s} \cdot d\boldsymbol{r} = 0, \tag{2.3.27}$$

for which the circulation of $n\mathbf{s}$ along any closed curve C contained in V is null. It is important to notice that Eq. (2.3.26) alone, that is, without imposing conditions for L, is a necessary but not sufficient condition in order that Eq. (2.3.27) be valid for all of the possible *closed* curves of V. If all we have is a field of vectors $n\mathbf{s}$ that obey Eq. (2.3.26), in order that Eq. (2.3.27) be valid for all of the curves in V, it is necessary that the $n\mathbf{s}$ field be defined without singularities at *all* points of V and, lastly, that V is a *simply connected open* domain. For example, the volume between two concentric spheres is simply connected, while the volume of a torus is not. The verification of the topological properties of V allows for explaining the paradoxes that we will encounter later in dealing with Fermat's principle.

The integral of Eq. (2.3.27) is known as the *Lagrange's invariant integral* and implies that the integral

$$\int_{P_1}^{P_2} n\mathbf{s} \cdot d\boldsymbol{r}, \tag{2.3.28}$$

taken between any two points in V, is independent of the integration path. The integration of $n\mathbf{s} \cdot d\boldsymbol{r}$ is therefore clearly distinct from the integration of $n\,d\boldsymbol{r}$, which instead varies depending on the path taken. However, if P_1 and P_2 are two points connected by a ray and the integration is done on it, $\mathbf{s} \cdot d\boldsymbol{r} = ds$ and the integral (2.3.28) coincides with the optical path calculated using Eq. (2.3.24).

Implicitly, we have so far considered that the refractive index was *continuous*, although variable with $\boldsymbol{r}$. However, with a few considerations, the Geometrical Optics hitherto treated can be extended in the very important case of heterogeneous media. This is the subject of the next section.

2.3.5 *The laws of refraction and reflection in Geometrical Optics*

Now, let us look at the case of fundamental importance in Geometrical Optics in which there is a separation surface T between two media of different refractive indexes n_1 and n_2. For the validity of the following derivations, we must assume that the surface is *smooth*, that is, the radii of curvature are >> λ. We then take a family of rays generated from a wavefront and for which we can apply Eq. (2.3.26). Furthermore, to overcome the difficulties, we replace the discontinuity surface T with a transition layer in which the refractive index changes quickly but continuously, by linking the n_1 and n_2 indexes at the two sides of T, as in Fig. 2.4.

In this figure, $\mathbf{n}_{12}$, $\mathbf{t}$, $\mathbf{b}$ indicate a set of three orthogonal versors, where $\mathbf{n}_{12}$ is normal to the surface T, while $\mathbf{t}$ and $\mathbf{b}$ are parallel to T. Consider a rectangle A on the plane ($\mathbf{n}_{12}$, $\mathbf{t}$) placed across T, with sides P_1P_2 and Q_1Q_2 of height δh and parallel to $\mathbf{n}_{12}$, while the sides P_1Q_1 and P_2Q_2 are parallel to T. To this rectangle, we apply Eq. (2.3.27):

$$\oint n\mathbf{s}\cdot d\boldsymbol{r} = 0\,,$$

where the integral is taken with $\boldsymbol{r}$ that goes along the circuit $P_1Q_1Q_2P_2P_1$. By calculating the limit $\delta h \to 0$, the contribution of the sides P_1P_2 and Q_1Q_2 vanishes. Then, since the length of the sides PQ is arbitrary, it is necessary that

$$\mathbf{t}\cdot\left(n_2\mathbf{s}_2 - n_1\mathbf{s}_1\right) = 0\,. \qquad (2.3.29)$$

Using the relationship

$$\mathbf{t} = \mathbf{n}_{12} \times \mathbf{b}\,,$$

Eq. (2.3.29) can be rewritten as

$$\mathbf{b}\cdot\left[\mathbf{n}_{12} \times \left(n_2\mathbf{s}_2 - n_1\mathbf{s}_1\right)\right] = 0\,.$$

Given that the orientation of A, namely, that of $\mathbf{b}$ (and $\mathbf{t}$) around $\mathbf{n}_{12}$, is arbi-

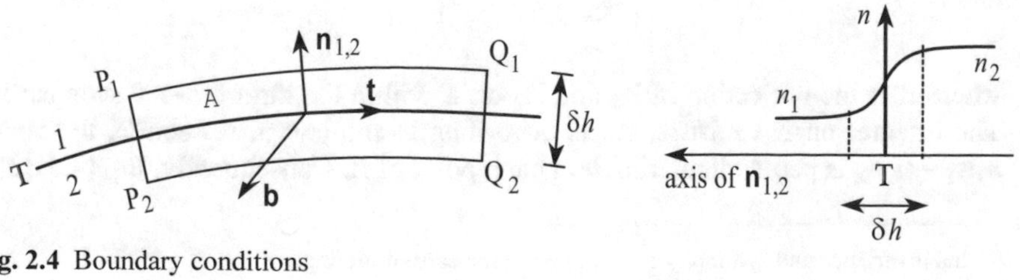

Fig. 2.4 Boundary conditions

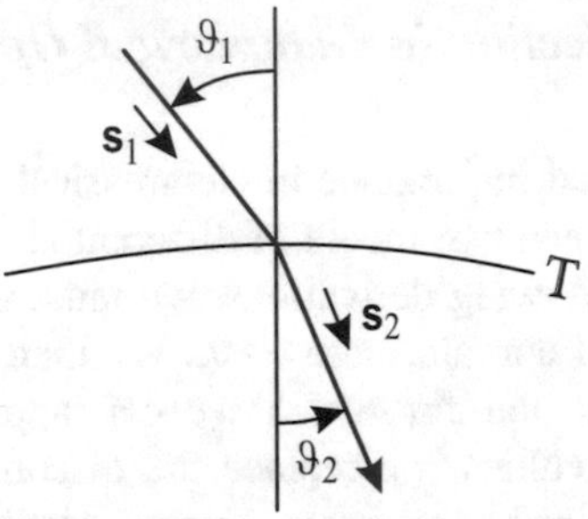

Fig. 2.5 Refraction

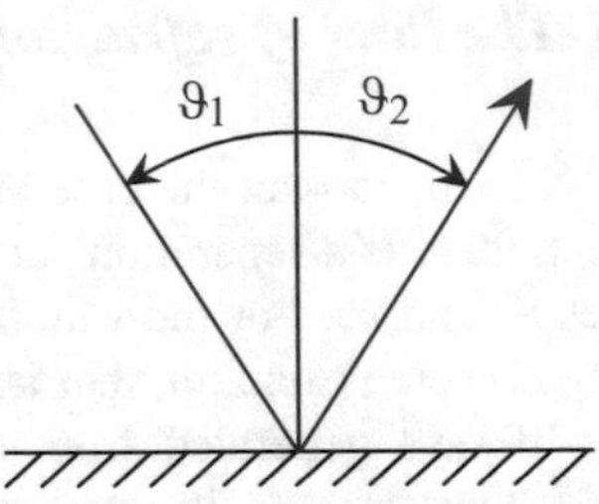

Fig. 2.6 Reflection

trary, it is necessary that the term within the square brackets is zero. Therefore, it follows that $\mathbf{s}_1$, $\mathbf{s}_2$ and $\mathbf{n}_{12}$ must lie on the same plane, and also that

$$n_2 \sin \vartheta_2 = n_1 \sin \vartheta_1 \,, \tag{2.3.30}$$

where ϑ_1 and ϑ_2 are the angles of incidence for a generic ray at the two sides of the surface T, as in Fig. 2.5. The Eq. (2.3.30) is the Snell-Descartes law in its classical form, and all refracting optical instruments are based on it.

From the surface T, we also expect a reflected wave: we can also imagine a reflection mechanism in which n is not discontinuous, like the one used for refraction (just think of mirages). However, for the translational invariance of the surface and its rotational symmetry with respect to the normal direction,[6] the reflected beam must remain on the plane of incidence, and also

$$\vartheta_2 = \vartheta_1 \,, \tag{2.3.31}$$

with the angles defined as in Fig. 2.6.

With the help of the law of refraction, we can now easily see that the Lagrange integral (2.3.28) is invariant even when the paths between points P_1 and P_2 intersect a surface of discontinuity between two media. For the proof, we consider a circuit as given in Fig. 2.7 and split it into two new circuits $\mathrm{C}_1 + \mathrm{K}_1$ and $\mathrm{C}_2 + \mathrm{K}_2$, for each of which Eq. (2.3.27) is valid. By adding the two integration together, we have

$$0 = \int_{\mathrm{C}_1} n_1 \mathbf{s}_1 \cdot d\boldsymbol{r} + \int_{\mathrm{C}_2} n_2 \mathbf{s}_2 \cdot d\boldsymbol{r} + \int_{\mathrm{K}} \left(n_2 \mathbf{s}_2 - n_1 \mathbf{s}_1 \right) \cdot d\boldsymbol{r} \,, \tag{2.3.32}$$

where K is the projection of K_1 and K_2 on T within the limit $\delta h \to 0$ seen earlier. The integral on K vanishes, since, according to the law of refraction, the vector $n_1 \mathbf{s}_1 - n_2 \mathbf{s}_2$ is perpendicular to $d\boldsymbol{r}$ at each point of K. Consequently, Eq. (2.3.32) is

[6] This invariance and symmetry are violated in the case of gratings.

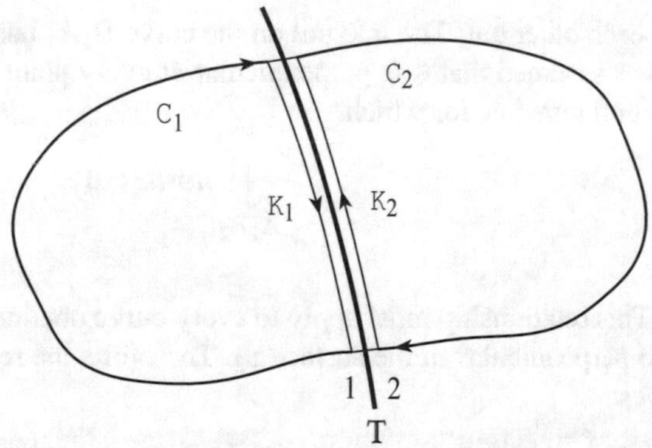

Fig. 2.7
Lagrange integral for a circuit that runs through an interface

reduced to Eq. (2.3.27).

2.3.6 The Malus-Dupin theorem

We can now easily answer a question that we had set for ourselves at the beginning of §2.3 on the relationship between a wavefront and a family of rays. The rays are, in fact, defined as normal trajectories to the wavefronts; on the other hand, from a family of rays that starts from an initial front S_1, it is possible to obtain a new surface S_2 that equals the optical paths.

Is this surface a new wavefront, in particular, even after any refractions or reflection? To answer this, consider an initial wavefront S_1 from which a ray A is launched orthogonally to this front in a dielectric, even heterogeneous, medium (Fig. 2.8). We indicate, with A_1, the starting point of the ray A and, with A_2, a point in the same ray such that the optical path traveled on it is L. Do the same with the points B_1 taken in the neighborhood of A_1 on S_1. The set of points B_2, each taken with the same optical path L, constitutes the surface S_2. Now, we apply Eq. (2.3.27) to the route $A_1A_2B_2B_1A_1$:

$$\int_{\mathrm{A}} n\,ds + \int_{\mathrm{A_2B_2}} n\mathbf{s}\cdot d\boldsymbol{r} - \int_{\mathrm{B}} n\,ds + \int_{\mathrm{B_1A_1}} n\mathbf{s}\cdot d\boldsymbol{r} = 0\,.$$

The integrals on A and B are both equal to L by hypothesis; therefore, they can-

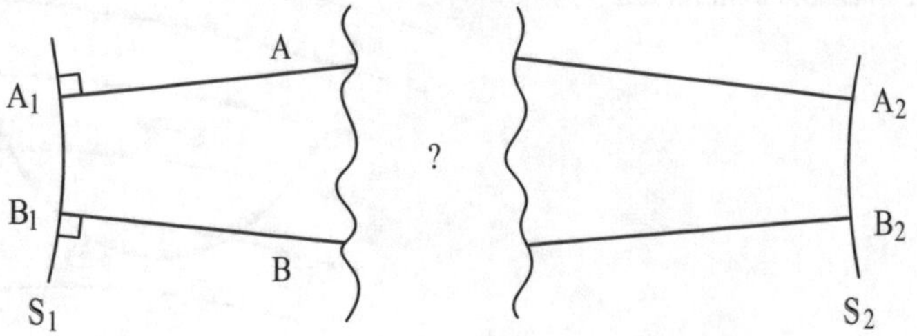

Fig. 2.8
The theorem of Malus-Dupin

cel each other out. The integral on the curve B_1A_1 taken on S_1 also vanishes, since it was assumed that **s** is perpendicular at every point to S_1. Thus, only the second integral remains, for which

$$\int_{A_2B_2} n\mathbf{s}\cdot d\mathbf{r} = 0 \quad .$$

This relationship must apply to every curve obtained on S_2, and thus each ray is also perpendicular to the surface S_2. Therefore, the response to the initial question is *yes*:

S_2 is a wavefront of the family of rays generated by the ray tracing from S_1.

2.3.7 *Fermat's principle*

We can now discuss the most classic principle of Geometrical Optics, which some consider even more fundamental than Maxwell's equations, and therefore take it as a starting point [Pauli 1973].

Pierre de Fermat formulated this principle by relying on the law of refraction, thus perfecting the principle of Heron of Alexandria of the shortest path, in turn derived from the law of reflection by plane mirrors. It says, "the light always chooses to go from one point to another along the path that requires the shortest time". Fermat justified it with the teleological principle according to which Nature always tries to achieve its purposes in the most efficient manner. Perhaps he was right, given that similar principles are the foundation for all of Physics (see the calculus of variations). However, we just need to check the compatibility of this principle with our starting postulates, namely, Maxwell's equations.

Before proceeding to the demonstration (which, however, will remain somehow incomplete here, as we are unable, for brevity's sake, to extend it to all possible sub-cases), I want to premise the modern and perfected version of Fermat's principle:

Fig. 2.9
Fermat's principle for the case of an absolute minimum

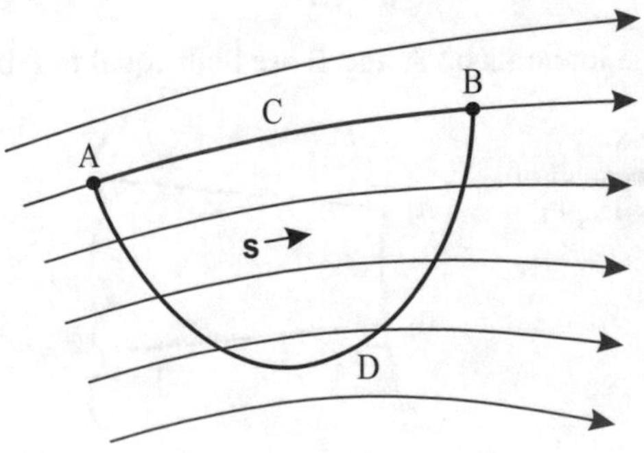

Fig. 2.10
Fermat's principle for the case of a local minimum

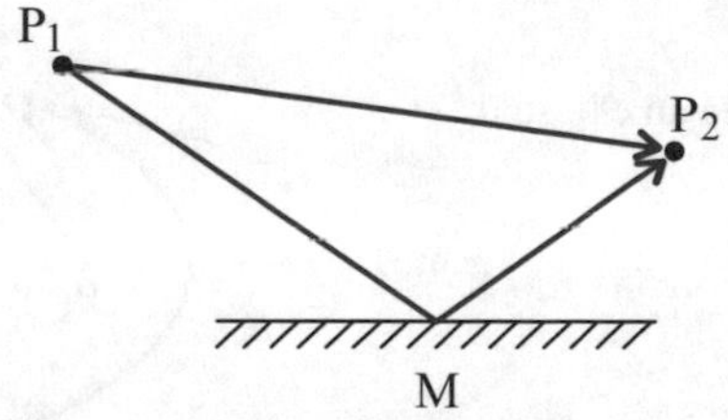

The path of a light ray passing through two assigned points has a *stationary* travel time with respect to variations of that path.

The word *stationary* replaces the word minimum, by extending Fermat's original principle to situations that are more complicated than that observed by him. We will see these cases later. For the moment, we consider the case in which the eikonal L is a *univocal* function of the coordinates, and therefore Eq. (2.3.2) can be applied. Besides, we choose two points A and B that lie on one of the rays (Fig. 2.9) and build a circuit in the following way: a branch (ACB) covers the ray from A to B and another branch joins B with A along a generic curve (BDA). Therefore,

$$\int_{ACB} n\mathbf{s}\cdot d\mathbf{r} = \int_{ADB} n\mathbf{s}\cdot d\mathbf{r}\,.$$

Since, on ACB, $\mathbf{s}$ is parallel to $d\mathbf{r}$, there, it is $\mathbf{s}\cdot d\mathbf{r} = ds$; instead, on ADB $\mathbf{s}\cdot d\mathbf{r} = \cos\left(\widehat{\mathbf{s}d\mathbf{r}}\right)ds \le ds$, and therefore

$$\int_{ACB} n ds \le \int_{ADB} n ds\,. \tag{2.3.33}$$

From here, it follows that the optical path along the ray is the lowest of all. However, this conclusion can be violated if the eikonal is not a unique function of the coordinates. The simplest example is given by the reflection from a plane mirror (Fig. 2.10): in this case, the P_1MP_2 path shown in the figure is a *local* minimum for small variations of the trajectory, with the physical constraint of passing by the mirror's surface, but it is not an absolute minimum. In fact, there are shorter paths that do not respect this constraint. In general, one way to recognize the possible violation of Eq. (2.3.33) is to observe whether a ray of light that joins two points P_1 and P_2 is intersected by another ray that starts from P_1.

Another instructive example is given by the internal reflections in an ellipsoid (Fig. 2.11). If, indeed, the points P_1 and P_2 are the two foci of an ellipsoid E, all optical paths $P_1M_EP_2$ with M_E taken upon E have the same length. Then, if we substitute, for E, a plane mirror P that is tangent to E on M_o, at this point, we still have a reflection of light from P_1 to P_2, but the path $P_1M_oP_2$ has a local *minimum* in length with respect to the other path $P_1M_PP_2$ with M_P on P. If, finally, instead of P, we use a tangent surface S with a curvature greater than that of E, then the

Fig. 2.11
Reflection in an ellipsoid

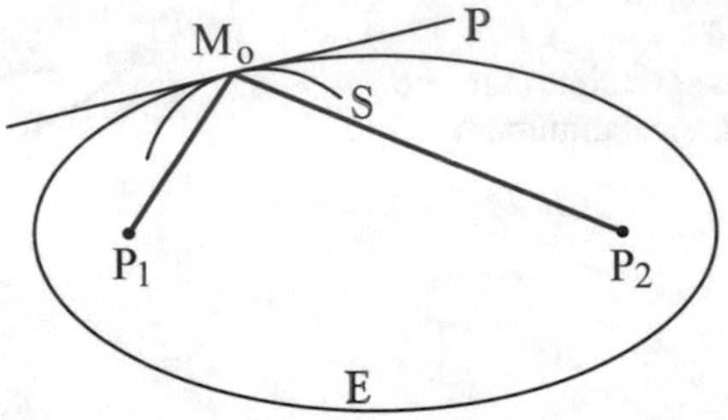

reflection at M_o has a local *maximum* in length. In general, the optical path of a ray corresponds to an *extremum*, also with refractive surfaces.

To demonstrate the applicability of the Fermat principle in its modern version, we must use the calculus of variations. From this calculation, it follows that, in an inhomogeneous and continuous medium, if the integral

$$\int_{P_1}^{P_2} nds$$

is stationary for small variations of the path, the trajectory on which it is calculated is also a solution to the Euler-Lagrange Eqs. (2.3.3-5), thus it is a possible optical ray. Even the inverse relationship is valid: if a curve joining two points P_1 and P_2 is a solution to Eqs. (2.3.3-5), then its optical length is stationary.

Let us now look at a demonstration following Wolfgang Pauli, for a medium with a continuous index of refraction $n = n(\boldsymbol{r})$ [Pauli 1973]. The Fermat's principle is equivalent to the expression

$$\delta L \equiv \delta \int_{P_1}^{P_2} nds = 0, \quad \text{with } P_1 \text{ and } P_2 \text{ fixed,}$$

where the integration represents the optical length of a curve that joins the points P_1 and P_2, while the differentiation δ indicates the difference in optical length with respect to a neighboring curve, also taken between P_1 and P_2. The two curves can both be expressed as a function of a parameter u (which, if it can match the length s held on the first curve, is not for the second curve):

$$\boldsymbol{r} = \boldsymbol{r}(u), \quad \boldsymbol{r}' = \boldsymbol{r}(u) + \delta \boldsymbol{r}(u)\,.$$

On the other hand, ds can be rewritten as $\left|\frac{d\boldsymbol{r}'}{du}\right| du = \sqrt{\frac{d\boldsymbol{r}'}{du} \bullet \frac{d\boldsymbol{r}'}{du}} du$, and the optical length of the second curve becomes

$$L = \int_{u_1}^{u_2} n \left|\frac{d\boldsymbol{r}'}{du}\right| du\,,$$

where u_1 and u_2 represent the values of u for which $\boldsymbol{r}(u_i) \equiv \mathrm{P}_i$. Taking into account that the differentiations δ and d are independent, and therefore we can switch between them, we have

$$\delta L = \int_{u_1}^{u_2} \left\{ \left| \frac{d\boldsymbol{r}}{du} \right| \nabla n \bullet \delta \boldsymbol{r} + n \left| \frac{d\boldsymbol{r}}{du} \right|^{-1} \frac{d\boldsymbol{r}}{du} \bullet \frac{d}{du} \delta \boldsymbol{r} \right\} du \, .$$

Integrating, by parts, the second term, we have

$$\delta L = \int_{u_1}^{u_2} \left\{ \left| \frac{d\boldsymbol{r}}{du} \right| \nabla n - \frac{d}{du} \left(\left| \frac{d\boldsymbol{r}}{du} \right|^{-1} n \frac{d\boldsymbol{r}}{du} \right) \right\} \bullet \delta \boldsymbol{r} du \; + \left| \frac{d\boldsymbol{r}}{du} \right|^{-1} n \frac{d\boldsymbol{r}}{du} \bullet \delta \boldsymbol{r} \Bigg|_{\mathrm{P}_1}^{\mathrm{P}_2} \, .$$

Since, by hypothesis, $\delta \boldsymbol{r} = 0$ on P_1 and P_2, the last term is null. On the other hand, for the arbitrariness of $\delta \boldsymbol{r}$, δL may be canceled if and only if the expression between the curly brackets vanishes in turn.

Now, making sure that u coincides with s for our curve $\boldsymbol{r}(u)$, so that $|d\boldsymbol{r}/du| = 1$, we have:

$$\delta L = 0 \Leftrightarrow \nabla n = \frac{d}{ds} \left(n \frac{d\boldsymbol{r}}{ds} \right) . \tag{2.3.34}$$

We have thus found an equivalence relation between Fermat's principle and the Euler-Lagrange law for continuous media.

In the case of discontinuous media, which is heterogeneous, in the presence of reflection or refraction, Fermat's principle can now be justified by applying the same techniques used in §2.3.4 to remove the discontinuity, or by applying the same laws of reflection and refraction.

I also quote here an intuitive justification of Fermat's principle in terms of interference, although not rigorously, for the sake of brevity. The transit time of the "true" path is, in a first approximation, equal to that of the adjacent paths, since, in this case, there is a zero first derivative with respect to changes of the trajectory. Therefore, the waves that "travel" along the routes adjacent to the true trajectory arrive in phase and reinforce each other (the phase is stationary). If we had chosen a trajectory for which $\int n ds$ is not stationary, then we would have destructive interference among the various waves of the adjacent paths.

2.3.8 Wavefronts and caustics in a homogeneous medium

Let us now return to the study of the propagation of wavefronts for the simplest case of a homogeneous medium with refractive index $n = 1$. Consider, therefore, a given eikonal $L(\boldsymbol{r})$, which is differentiable at least up to the second derivatives in a neighborhood of a point $\boldsymbol{r}_0$, and the corresponding congruence of rays crossing it. This congruence is descriptively called a *pencil* (from Latin *penicillus*, which was a fine brush made of camel hair). Developing L around $\boldsymbol{r}_0$, we find that [Solimeno et al 1986]

$$L(\boldsymbol{r}_0 + \Delta\boldsymbol{r}) = L(\boldsymbol{r}_0) + \Delta\boldsymbol{r} \cdot \nabla L + \frac{1}{2}\Delta r_i \Delta r_j \frac{\partial^2 L}{\partial x_i \partial x_j} + \cdots, \tag{2.3.35}$$

where the summation over repeated indices for the three Cartesian components of $\Delta\boldsymbol{r}$ and $\boldsymbol{r}$ is implied. Thanks to Schwarz's theorem, the tensor

$$\mathfrak{Q} = \frac{\partial^2 L}{\partial x_i \partial x_j} \tag{2.3.36}$$

is symmetrical, as well as real, and therefore it is diagonalizable. We also remember that the elements of $\mathfrak{Q}$ are obtained by the development of the eikonal around the point $\boldsymbol{r}_0$, after which they depend on its position. I will insist on this fact later.

For simplicity, we take $\boldsymbol{r}_0$ to be the origin of coordinates and then we assume that the direction of $\nabla L = n\mathbf{s}$ calculated on $\boldsymbol{r}_0$ is that of the z-axis, which is to say that the ray passing for $\boldsymbol{r}_0$ coincides with the z-axis. This radius is called the *principal ray* of the pencil. With this choice, we have that the elements of $\mathfrak{Q}$ that are a partial derivative in z, are zero. Indeed, given that, in a homogeneous medium, $\mathbf{s}$ is constant along its direction,

$$\left[\frac{\partial}{\partial z}\nabla L\right]_{x=0,y=0,z=0} = \left[\frac{\partial}{\partial z}[\nabla L]_{x=0,y=0}\right]_{z=0} = \mathbf{0}.$$

In other words, the tensor $\mathfrak{Q}$ is the sum of two dyads:

$$\mathfrak{Q} = Q_1\mathbf{t}_1\mathbf{t}_1 + Q_2\mathbf{t}_2\mathbf{t}_2, \tag{2.3.37}$$

where $\mathbf{t}_1$ and $\mathbf{t}_2$ are the versors of the *principal directions*, orthogonal to $\mathbf{s}(\boldsymbol{r}_0)$, wherein the matrix $\mathfrak{Q}$ becomes diagonal, and the coefficients Q_1 and Q_2 are the *curvatures* of the wavefront along these two directions, as we will see shortly. Therefore, we can assume that the x and y axes are respectively oriented along $\mathbf{t}_1$ and $\mathbf{t}_2$ and, in this system, the tensor $\mathfrak{Q}$ is represented by the matrix

$$\mathfrak{Q} = \begin{pmatrix} Q_1 & 0 & 0 \\ 0 & Q_2 & 0 \\ 0 & 0 & 0 \end{pmatrix}. \tag{2.3.38}$$

In the system x, y, z just chosen, let us now consider the intersection of the wavefront, defined by $L(\boldsymbol{r}) = L(\boldsymbol{r}_0)$, with the zx-plane or with the yz-plane. For a displacement $\Delta\boldsymbol{r}$ along these intersecting lines, Eq. (2.3.35) becomes, respectively,

$$z + \frac{1}{2} Q_1 x^2 + \cdots = 0, \qquad z + \frac{1}{2} Q_2 y^2 + \cdots = 0, \tag{2.3.39}$$

and therefore, these lines are locally parabolas with curvature Q_1 and Q_2, respectively. On the other hand, the curve generated by the intersection of the front with a generic plane containing the z-axis has an intermediate curvature between Q_1 and Q_2, one of these values is the maximum curvature and the other is the minimum. Therefore, in accordance with Solimeno et al. (1986), with the definition assigned above to the curvatures $Q_{1,2}$, a wavefront viewed from the positive z-values appears convex if the values of $Q_{1,2}$ are positive.[7] Expressing Eq. (2.3.35) in the x, y, z frame, we have

$$L(\boldsymbol{r}_0 + \Delta\boldsymbol{r}) = L(\boldsymbol{r}_0) + \Delta z + \frac{1}{2} Q_1 (\Delta x)^2 + \frac{1}{2} Q_2 (\Delta y)^2 + \cdots. \tag{2.3.40}$$

where Δx, Δy, Δz are the components of $\Delta\boldsymbol{r}$. Differentiating it in $\Delta z = 0$, we get how the versor $\mathbf{s}$ of the ray varies on the plane xy in the neighborhood of $\boldsymbol{r}_0$,

$$\mathbf{s}(\boldsymbol{r}_0 + \Delta\boldsymbol{r}) = \mathbf{s}(\boldsymbol{r}_0) + Q_1 \Delta x \mathbf{t}_1 + Q_2 \Delta y \mathbf{t}_1 + \cdots. \tag{2.3.41}$$

so, when $Q_1 \neq Q_2$, $\mathbf{s}(\boldsymbol{r}_0 + \Delta\boldsymbol{r})$ is coplanar to $\mathbf{s}(\boldsymbol{r}_0)$ (that is, with the z-axis) if and only if $\Delta\boldsymbol{r}$ lies on the xz-plane or the yz-plane. Since the medium considered here is homogeneous and the rays are straight, all of (and only those) the rays passing through the intersections of the given wavefront with one of these planes intersect the z-axis. Therefore, these rays lie on the xz-plane or on the yz-plane that we have chosen above so that they contain the principal ray, and are respectively oriented parallel to the principal directions $\mathbf{t}_1$ or $\mathbf{t}_2$ in $\boldsymbol{r}_0$. Since the wave fronts of the family always remain perpendicular to the rays, it appears that the two principal directions do not depend on the position $\boldsymbol{r}_0$ along the z-axis. Therefore, the planes generated by the principal ray and one of these directions are called *principal planes*.

Any other plane containing the z-axis is not generally perpendicular point by point to the wavefront along the intersection between such a plane and the wavefront itself. Instead, in the degenerate case $Q_1 = Q_2$, all transverse directions are

[7] Here, unfortunately, the choice of the sign of the curvatures and the radii of curvature of the wavefront is opposite to that usually assigned for the optical surfaces. Nevertheless, I have preferred to keep, for the wavefronts, the convention that is usually taken for the Gaussian beams.

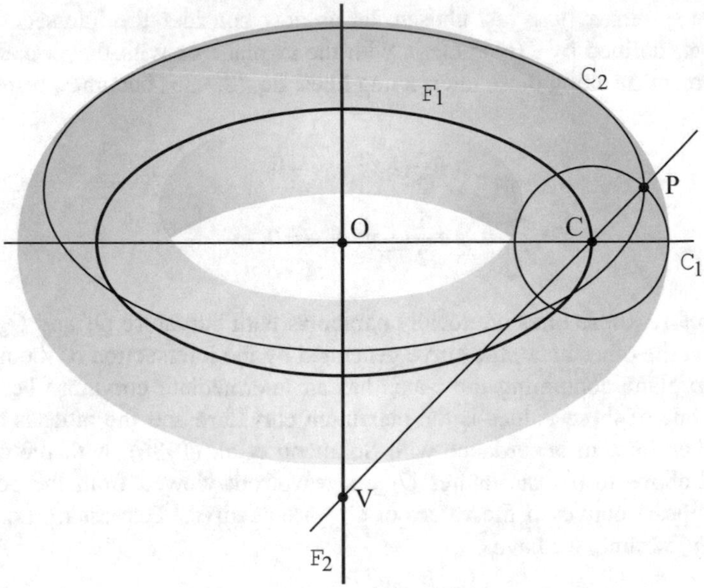

Fig. 2.12
Study of the centers of curvature of a torus. C_1 and C_2 are the principal lines of curvature passing through the point P on the surface of the torus, while F_1 and F_2 are the focal lines. F_2 coincides with the axis of the torus that has its center in O, while F_1 is a circle with its center in O

equivalent.

Having neglected the terms of order higher than the second in the development of L, we can consider the sections of the wavefront represented by Eqs. (2.3.41) as arcs of the circle of radius $R_1 = 1/Q_1$ and $R_2 = 1/Q_2$, with their centers of curvature on the z-axis, that is, on the principal ray of the pencil, at the positions $z_1 = -R_1$ and $z_2 = -R_2$. Therefore, we have that, along the principal ray, the principal radii of curvature vary according to the law

$$R_1(z) = R_1(0) + z, \quad R_2(z) = R_2(0) + z\,. \tag{2.3.42}$$

Having approximately resolved the way in which the pencil rays behave on the xz and yz-planes, we can now ask ourselves how the whole pencil propagates. The Eq. (2.3.40) appears to indicate that the curvatures Q_1 and Q_2 of the wavefront do not depend on the transverse coordinates, but, in general, the situation is more complex, since, as mentioned earlier, the curvatures vary with the choice of the principal ray.

Consider the particular but instructive case of a wavefront that has the shape of a torus (Fig. 2.12). At each point P of its surface, we can identify two tangents and orthogonal directions, $\mathbf{t}_1$ and $\mathbf{t}_2$, for which the curvature matrix is diagonalized. The first lies on the plane containing the axis of the torus and the second on the

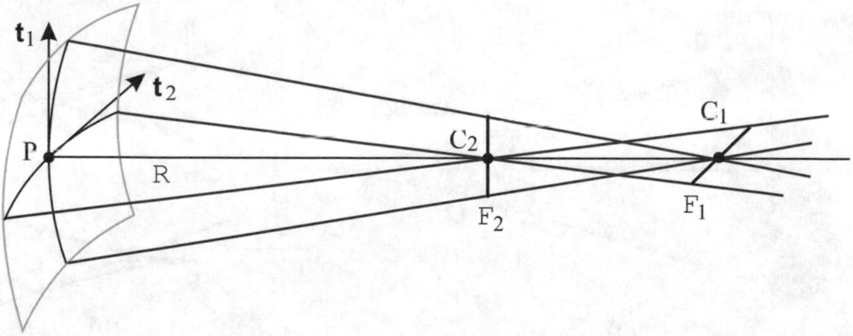

Fig. 2.13 Focal lines of a thin pencil. $\mathbf{t}_1$ and $\mathbf{t}_2$ are the principal directions of curvature of the wavefront and C_1, C_2 their principal centers of curvature along the principal ray R passing for P. The focal lines F_1 and F_2 are perpendicular to R only if the wavefront has reflection symmetry with respect to the planes containing R and parallel to $\mathbf{t}_1$ or $\mathbf{t}_2$

plane perpendicular to that axis. The sections of the torus with such planes (containing P) form two circles, respectively, C_1 and C_2, which intersect in P. All of the straight lines normal to the surface that pass for the points of C_1 intersect themselves in its center C. By varying the position of P, these centers form a focal circular line F_1 around the axis of the torus. Instead, the straight lines normal to the surface that passes for C_2 form the outer surface of a cone, all intersecting on the cone vertex V, which lies on the axis of the torus. The position of V is determined from the intersection of the line that goes through P and C with the axis of the torus. Therefore, at varying P, these vertices form a focal straight line F_2 corresponding to this axis. The curvatures Q_1 and Q_2 of the surface, respectively along the curves C_1 and C_2, are thus constant, even if the curvature $Q_2 = 1/\overline{PV}$ associated with C_2 depends on the choice of P. Nevertheless, the torus represents an ideal case of astigmatism. One surface closely related to a torus is that of a "spindle" generated by the rotation of a circular arc around its chord.

Consider now a wavefront that coincides with a small portion of the surface of a torus and, in particular, we assume that the point P from which the principal ray of the pencil passes is on the plane bisecting the torus orthogonally to its axis (Fig. 2.13). The vertex V coincides, in this case, with the center O of the torus, and thus the curvature Q_2 now varies only quadratically moving P along C_1. The focal lines, both assimilable to straight segments, are now both perpendicular to the principal ray, and the behavior of the pencil is therefore expressible in terms of simple astigmatism. If, however, P is chosen in another position for which the axial point V does not coincide with the center of the torus, the line F_2 is no longer perpendicular to the principal ray.

Along the z-axis, that is, on the principal ray, from Eqs. (2.3.36) and (2.3.42), we have that

$$\nabla^2 L = \mathrm{Tr}\,\mathbb{Q} = \frac{1}{R_1(0)+z} + \frac{1}{R_2(0)+z}.$$

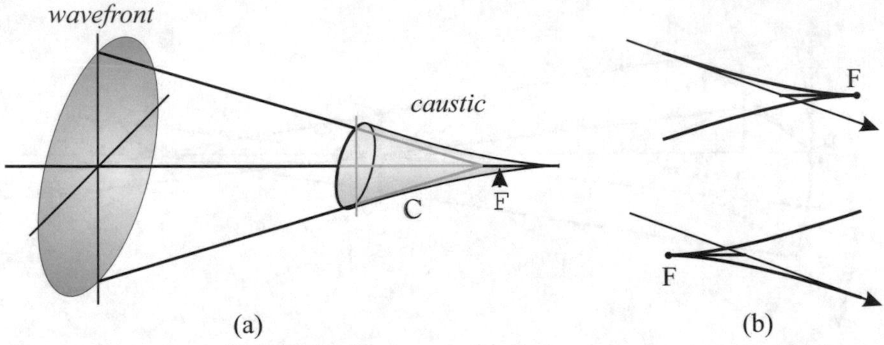

Fig. 2.14 Caustic produced by a wavefront with rotational symmetry. (a) perspective view: C is the surface of the caustic and F is the axial focal line. (b) sections of the caustic in the two cases in which the marginal rays intersect the axis before or after the paraxial focus F

Therefore, from Eq. (2.2.7), with $n = 1$, we obtain a simple equation for the trend of the amplitude E_o of the field [Solimeno et al 1986]:

$$E_o(z) = E_o(0)\sqrt{\frac{R_1(0)}{R_1(0)+z}}\sqrt{\frac{R_2(0)}{R_2(0)+z}}, \tag{2.3.43}$$

in which each root must be taken as a positive real or positive imaginary. In other words, the phase of E_o undergoes a $\pi/2$ jump when z crosses the z coordinate of each of the two centers of curvature, and a jump of π when these coincide. This effect was observed by L. Georges Gouy at the end of the 19th century, and it has been designated as a *phase anomaly*. We will see then, studying Gaussian beams, that this jump is replaced by a continuous variation of the phase.

The product of the principal curvatures $Q_1Q_2 = 1/(R_1R_2)$ is called a *Gaussian curvature* of the surface, and Eq. (2.3.43) indicates that

the intensity at each point of a straight ray is proportional to the Gaussian curvature of the wavefront at that point [Born and Wolf 1980].

In the case of the torus, we have that the focal lines are just lines. Actually, in most cases, this does not occur: the same definition of the center of curvature is derived from the limit for $x \to 0$ or $y \to 0$ in the neighborhood of $\mathbf{r}_o$, and is therefore independent of the transverse coordinates only for wavefronts whose sections along the principal directions are truly circular. For example, in the development of L, a convergent parabolic wavefront differs from a spherical wavefront with the same curvature Q for the term $Q^3r^3/8$ and higher, where $r^2 = x^2 + y^2$. Therefore, the ray that passes through the wavefront at the height r intersects the z-axis at a distance of about $Qr^2/2$ after the center of curvature obtained within the limit for $r \to 0$ (as we will see studying spherical aberration); this difference corresponds to the "arrow" z on the wavefront.

In general, for each point of the wavefront there are two centers of curvature

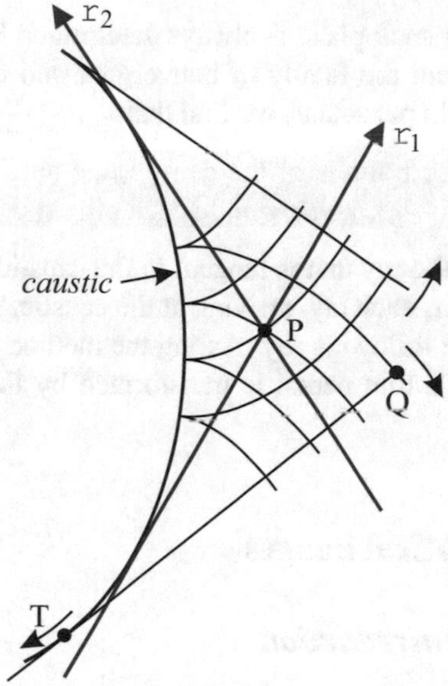

Fig. 2.15
Wavefronts in the vicinity of a caustic

calculated by the above limit, and for the entire front, the set of these centers constitutes two surfaces called *caustics*[8] (which are the "bete noire" of designers). In the case in which the wavefront has rotational symmetry about an axis, one of the two surfaces is reduced to an axial focal line and the other to a (portion) of surface of rotation around the axis. If, then, the principal ray lies on the axis of rotation, such as in the case of a conic of revolution, this surface resembles a cone with a cusp in correspondence with one of the extremes of the axial focal line (Fig. 2.14).

Let us now take a point P near the convex side of a caustic; two tangent rays pass for this point, $\mathtt{r}_1$, which has left the caustic surface, and $\mathtt{r}_2$, which approaches it (Fig. 2.15). Each of them corresponds to a different value of the eikonal and, in P, they intersect two wavefronts, the first (normal to $\mathtt{r}_1$) diverging and the second (normal to $\mathtt{r}_2$) converging on the plane defined by these rays. In the amplitude of the field, there is interference between the two waves; also, it often happens that other rays tangent to other parts of the caustic pass for P, for which the interference pattern is even more complex. Consider the section of the caustic with the plane formed by the rays $\mathtt{r}_1$ and $\mathtt{r}_2$; furthermore, imagine that, on this plane, we have a piece of twine that is wrapped without slipping on the caustic and that we hold tense from one extreme Q. In this way, the twine follows an arc on the caustic to the point of tangency T and then continues straight up to Q. By moving the point Q, it goes along an arc that gradually has its center of rotation at the point of tangency T, and therefore always remains orthogonal to the segments TQ, thus the arc coincides with the section of a wave front.

By varying the length of the twine and the plane on which it lies, but making

---

8 Instead, in inhomogeneous media, the concept of the center of curvature loses meaning and the caustics are only the envelope of the congruence of rays.

sure that such plane is always determined by the congruence of rays, we can then reconstruct the family of converging and diverging wavefronts that originate the caustic. In particular, we find that,

> at each point T of the caustic, the corresponding convergent and divergent wavefronts are tangent to each other,

Indeed, the ray that is tangent to the caustic in T must belong to both wavefronts[9]; moreover, each ray, arriving at the caustic, goes from one wavefront to the other.

In the following, by treating the method of matrices, we shall see how the wave front of a thin pencil is transformed by the refraction or reflection on a smooth surface.

2.4 Optical images

2.4.1 Introduction

Geometrical Optics is primarily the science of image formation and, for this purpose, it has its own jargon: let us now look at a number of important concepts and definitions.

If a light beam originated by a point P_o converges at a point P_i because of reflexions, refractions, or for the deflections undergone by its rays in a non-homogeneous media, the point P_i is called the *optical image* of the point P_o. If the convergence is geometrically perfect, it is said that the point P_i is a *stigmatic image* of P_o. If the convergence is not perfect, but the error is negligible, it is said that the image is *sharp*: in this case, we cannot rigorously speak of a single image point, however, for simplicity of language, we will assume the central point, around which the rays are converging, to be an image of P_o. The point P_i is also called the *focus* of convergence of the rays. The image P_i is also said to be *real* if the light rays actually converge towards the point P_i, while it is said to be *virtual* if the beam of rays (extended rectilinearly) appears to be coming from or going to P_i. The point P_o is, in turn, the image of P_i, given that, reversing the direction of propagation, the path of the rays does not change. The points P_o and P_i are therefore said to be *conjugates* of each other. Finally, the beam of rays that joins (also virtually) P_o and P_i is said to be *homocentric*. The simplest example of conjugation between two points is given by the foci of a rotational ellipsoidal mirror: note that, for definition of ellipse, all of the paths connecting the foci (with a reflection on the ellipsoid) are equal. This property of equality of the optical paths is general for all well-designed optical instruments. One of the points, object or image, can also be at *infinity*, and, in the above example, the ellipsoidal mirror becomes pa-

---

[9] For an analytical treatment of the caustics and their discussion in terms of the field amplitude, see, for example, Solimeno et al (1986).

raboloidal.

The system of mirrors, lenses, etc., employed to deflect the rays is said to be an *optical system*. For a good optical instrument, each point P_o of a three-dimensional region of space, said to be an *object space*, has an image P_i that is, at least, sharp. The set of image points defines the *image space*. Generally, not all of the rays emerging from P_o reach the image space. Some, for example, are excluded by the diaphragms of the instrument. It is customary to say that those rays, which instead reach the image space, are located in the *field of the instrument*.

When P_o describes a curve C in the object space, P_i describes a conjugated curve C′. The two curves are not necessarily geometrically similar, but, if each curve C is geometrically similar to its image C′, we can say that a space is the *perfect* image of the other. In practice, however, perfect images, or even stigmatic ones, are not obtained, except in special cases. The imperfection of the real instruments is not only due to their finished opening, namely, to diffraction, but is also due to the "geometric" limits of such devices. It so happens that the images are affected by *geometrical aberrations* that consist either of incorrect similarities with their corresponding object or their reduced sharpness. In the first case, we speak of *distortion* and *field curvature*, while, in the second case, the aberrations are classified as *spherical aberration*, *chromatic aberration*, *coma*, *astigmatism*, and in their combinations of higher order. When these offenses to the sharpness are negligible, it is said that the instrument is *diffraction limited*

The condition of strict similarity is generally abandoned in the design of optical instruments. However, the instruments that are perfect in the sense defined above are of remarkable conceptual interest: now, I will give a brief reference to some general theorems, but without reporting their demonstration. For a detailed discussion, see Born and Wolf (1980) and Herzberger (1958).

2.4.2 General theorems

An optical system that forms *stigmatic images* of a three-dimensional domain is known as an *absolute instrument*; it was demonstrated that [Maxwell 1858],

> in an absolute instrument, the optical length of each curve in an object space is equal to the optical length of its image.

This theorem, formulated by Maxwell and then perfected by H. Bruns, F. Klein, H. Liebmann and, finally, by C. Carathéodory, also applies to isotropic heterogeneous media. From it, we can draw some interesting conclusions. If, in fact, a given absolute instrument has its own object and image spaces with homogeneous[10] refractive indices n_o and n_i, any figure placed in an object space will be depicted with a figure in the image space in which all distances are in the ratio n_o/n_i

[10] These concepts are generalizable to inhomogeneous media.

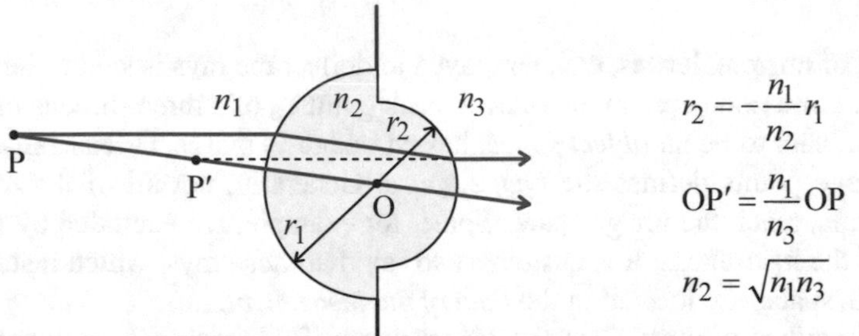

Fig. 2.16 Each real (virtual) point P placed at left (right) of the lens has a stigmatic virtual (real) image P′ placed on the same side

with the corresponding distances in the original figure. Therefore, all of the angles are also reproduced equally, and the figure depicted is similar to the original one. Thus has the following theorem of Caratheodory been established:

the process of reproduction of an absolute instrument between homogeneous spaces is a projective transformation, an inversion, or a combination of the two [Carathéodory 1926].

Therefore, if n_o and n_i are constant in their respective spaces, an absolute instrument is also perfect in the sense specified earlier and has a magnification ratio equal to n_o/n_i. In particular, if $n_i = n_o =$ constant, the magnification is 1, for which

the perfect reproduction between two homogeneous spaces with the same index of refraction is always trivial in the sense that it produces an image that is congruent with the object.

There are few known examples of perfect instruments. A plane mirror (or a combination of plane mirrors) is the simplest example and substantially also the only one that is of practical utility. M. Herzberger reports the case of spherical and concentric refractive surfaces, with a precise relationship between the refractive indices and radii of curvature: the simplest combination is described in Fig. 2.16.

Finally, an example of an absolute instrument that uses an inhomogeneous medium and curvilinear rays is given by Maxwell's "fish eye". It consists in a medium whose index varies with a spherical symmetry according to the radial law

$$n(r) = \frac{n_o}{1+(r/a)^2}, \tag{2.4.1}$$

where r is the distance from a fixed center O, while n_o and a are constants. In this medium, the rays are circular lines and the rays that pass from a given point P form a system of circles that all intersect at a second point Q aligned with OP. The distances of these two points from O respect the relationship

$$r_P r_Q = a^2 .$$

This device has the advantage of reproducing real images, but, in practice, it "lives" only in books about optics. Another interesting optical system is Luneburg's lens, which is characterized by an index of refraction

$$n = n_o\left(2 - \frac{r^2}{R^2}\right)$$

within a sphere of radius R, and $n = n_o$ for $r \geq R$, where r is the distance from the center. A source placed on its surface generates an elliptical-ray system inside of the sphere, which all emerge parallel to each other.

In conclusion, from the above description, we see that, to get non-trivial representations between two homogeneous spaces of equal refractive index (or with a different magnification from n_o/n_i if different), the request for exact astigmatism or close similarity between the object and image must be abandoned.

It is natural now to ask ourselves whether at least some *surfaces* may, in a non-trivial way, be reproduced perfectly, or at least sharply by an optical instrument. This argument has been investigated by several authors,[11] who found that,

> when the object and image spaces are homogenous (apart from special cases in which the entire volume is reproduced stigmatically), no more than two object surfaces can be reproduced stigmatically by an optical system with rotational symmetry.

The conditions for which this is possible are described in Herzberger (1958).

2.4.3 Aplanatic surfaces of a refractive sphere

An interesting case occurs in the reproduction of a spherical surface by a homogeneous sphere with index n immersed in a homogeneous medium with index n'. We consider here the case in which $n > n'$. Take an axis (say, z) that diametrically intersects the sphere of radius R. Assuming the center O of the sphere as the origin of the axes, consider two points P_o and P_i placed on the semi-axis $z > 0$ with coordinates $z_o = (n'/n)R$ and $z_i = (n/n')R$, respectively. Finally, we take a point A on the surface of the sphere with coordinates $z_A \leq z_o$, as shown in Fig. 2.17(a). The two triangles AOP_o and AOP_i are similar! In fact, they have an angle in common and two sides with the same proportion:

---

[11] Among whom are: H. Boegehod and M. Herzberger, T. Smith. See quotes in Born and Wolf (1980), page 149, or Herzberger (1958).

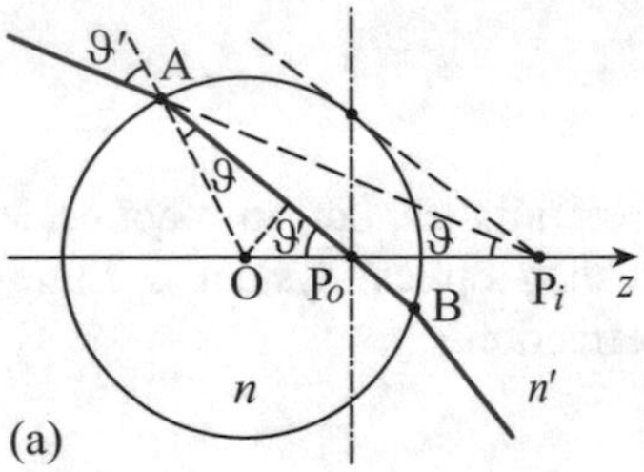

$$\overline{OP_o} = \frac{n'}{n} r$$

$$\overline{OP_i} = \frac{n}{n'} r$$

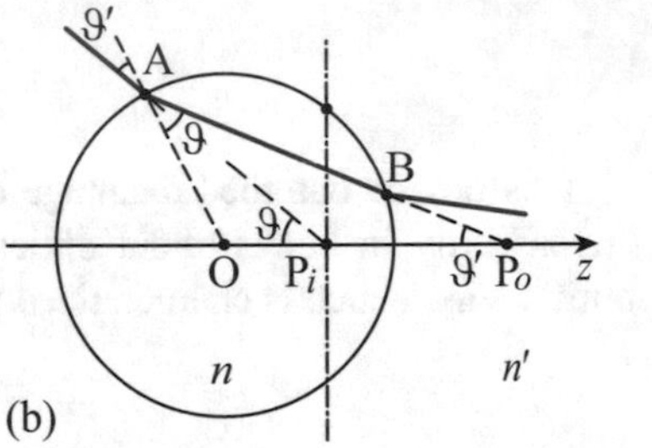

Fig. 2.17 Aplanatic points of a sphere. (a): $n > n'$, (b): $n < n'$

$$\overline{OP_o} : \overline{AO} = \overline{AO} : \overline{OP_i} = n' : n\,.$$

Then, $O\hat{A}P_o = O\hat{P}_i A = \vartheta$, $O\hat{A}P_i = O\hat{P}_o A = \vartheta'$. Finally, drawing, from O, the normal to the segment AP_o, we see that

$$\frac{\sin\vartheta'}{\sin\vartheta} = \frac{\overline{AO}}{\overline{OP_o}} = \frac{n}{n'}\,.$$

Therefore, the sines of angles ϑ and ϑ' are in the ratio prescribed by the Snell-Descartes law. In other words, any ray directed toward the point P_i intersecting the sphere at a point A is deflected towards the point P_o. On the other hand, by symmetry of rotation around the center O, a whole spherical surface of radius z_o has, as virtual stigmatic image, a concentric spherical surface with radius z_i. These surfaces are called aplanatic surfaces of a sphere. In particular, the magnification between the two surfaces is given by the *square* of the ratio between the refractive indices. A similar reasoning can be conducted in the opposite case in which $n < n'$, as shown in Fig. 2.17(b).

It should be noted that, in the case considered above, when $z_A = z_o$, ϑ corresponds to the critical angle (internal) between the two media. The rays coming from P_o that reach the surface of the sphere at points B for which $z_B > z_o$, instead emerge as rays that do not give rise to any stigmatic image.

The properties of the spherical refractive surfaces shown above are used in the construction of immersion objectives of high magnification microscopes.

2.4.4 Aspherical surfaces

Lenses and mirrors with aspherical surfaces of suitable shape are used to collect or concentrate rays of light on much wider angular fields than those generally available with only spherical surfaces, with similar aberration entities. In reflection, these surfaces can have the familial forms (paraboloids, ellipsoid, and hyper-

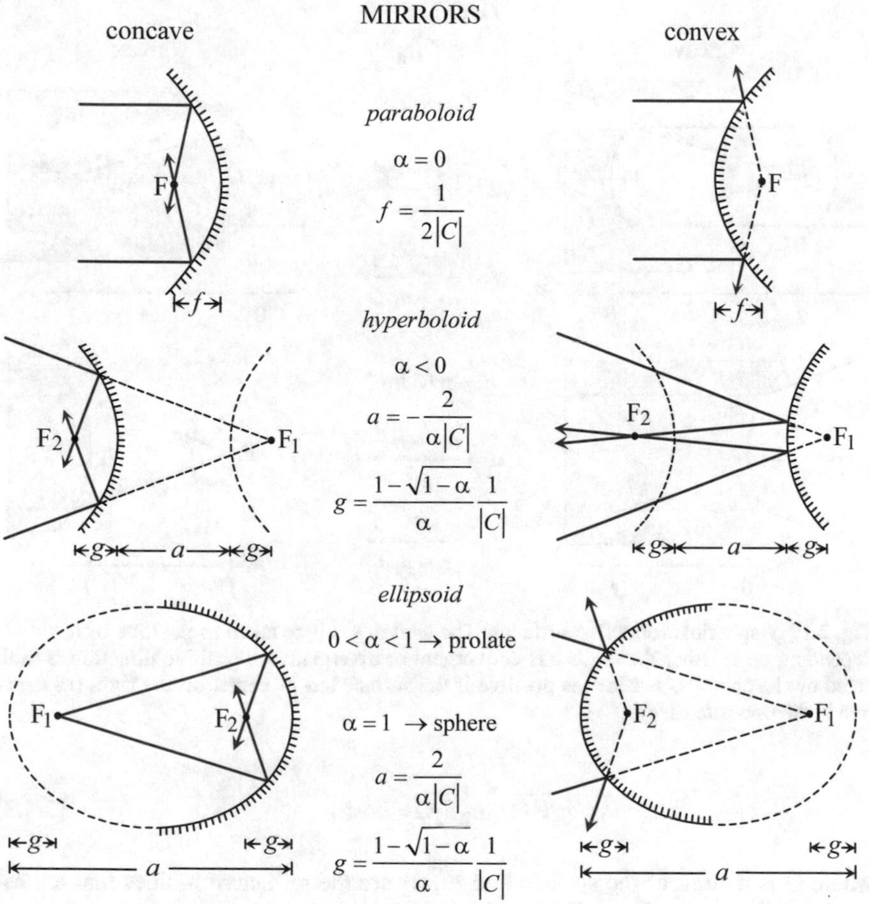

Fig. 2.18 Aspherical specular surfaces with symmetry of revolution around an axis. For simplicity, the quantities f, g, a are all taken positive

boloids) learned in the study of geometry that can be described by the formula

$$z = \frac{1}{2} C \left(x^2 + y^2 + \alpha z^2 \right), \tag{2.4.2}$$

where C is the curvature at the point (0, 0, 0) where the surface intersects the z-axis, while α determines the asphericity. Surfaces of this type, with the exclusion of *oblate* ellipsoids for which $\alpha > 1$, possess two *foci* (one can be at infinity), i.e., two points F_1, F_2 on each of which the surface can produce, by reflection, the stigmatic image of the other (Fig. 2.18). This property can also be obtained in refraction between two homogeneous media. In both cases, one way to determine the surface that produces the stigmatic image of F_1 in F_2 is to apply the constant optical path rule:

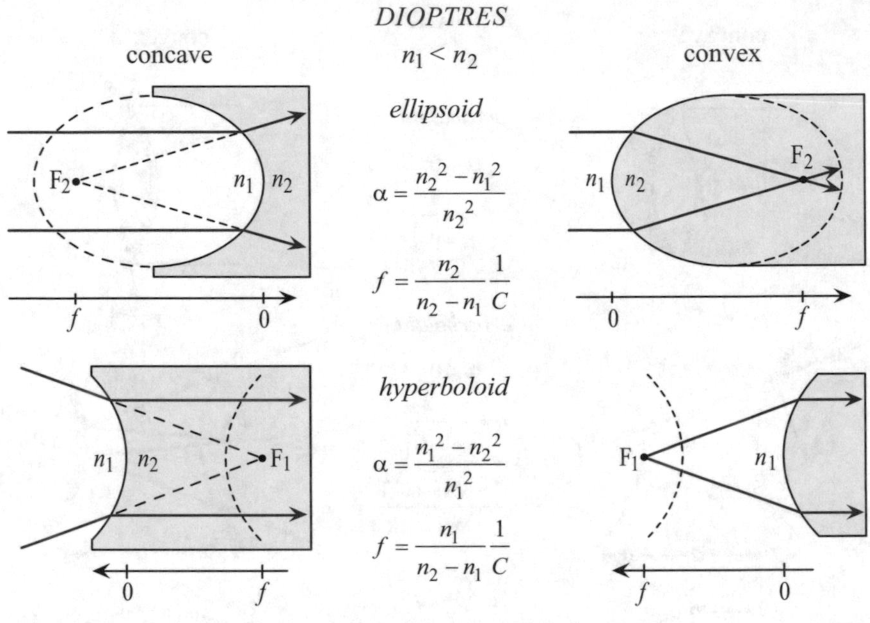

Fig. 2.19 Aspherical refractive surfaces. The sign of f is here taken as positive or negative, depending on whether the *dioptre* is convergent or divergent: the positive direction is indicated by the arrow. C is taken as positive if the surface has its center on the right (as negative in the opposite case)

$$n_1 \overline{F_1O} + n_2 \overline{F_2O} = \text{cost.}, \tag{2.4.3}$$

where O is a point of the surface and n_1, n_2 are the refractive indices that are assigned to the two segments F_1O and F_2O, respectively. In reflection, for which $n_1 = n_2$, the development of Eq. (2.4.3) leads to Eq. (2.4.2). Instead, in refraction, so that $n_1 \neq n_2$, the development of Eq. (2.4.3) leads to a very complicated expression, with terms of 4° grade in the coordinates, that describes a surface called the *oval of Descartes*.

If one of the points is at infinity, the oval of Descartes is reduced to an ellipsoid or a hyperboloid (Fig. 2.19). The refractive surfaces of generic form, with rotational symmetry around an axis, are called *dioptres*.[12] Two of these surfaces can be combined together to form a lens that can produce an image of a stigmatic object point (Fig. 2.20). Aspherical lenses of more complicated shape are also sometimes used to compensate (and thus to correct) the aberrations introduced by the other elements of an optical system. A famous example is the Schmidt telescope, in which a corrector plate is coupled with a spherical mirror, compensating for the spherical aberration in an excellent way.

[12] *Dioptre* is a French word, *diottro* in Italian. From Greek: di- 'through' + optos 'visible'.

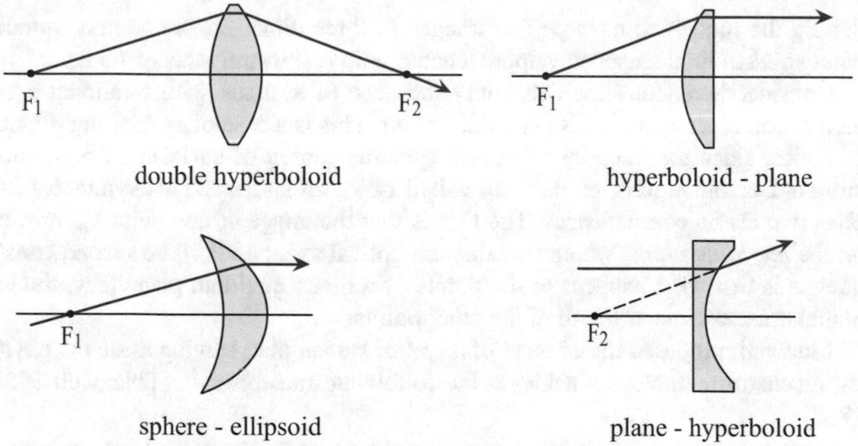

Fig. 2.20 Elementary aspherical lenses

Although the aspherical surfaces may carry out certain tasks in an excellent manner, they are difficult to manufacture with a quality that approaches that of the spherical surfaces, whose production technique is inherently simple and well established.

However, a new generation of diamond cutting machines has developed, controlled by computer, which make it possible to achieve the precision and quality required in optical instruments. In particular, it has become common to use the good objectives of cameras containing aspherical lenses, whose major cost is offset by the reduction of optical elements that are required to correct aberrations. Furthermore, the use of aspheric surfaces has been asserted, even for the ophthalmic lenses, which, being constituted by a single optical element, must correct visual defects while minimizing the introduction of other aberrations.

Single aspherical lenses are now used as collimators of the radiation emitted by laser diodes and for coupling the laser beam in optical fibers. Previously, the use of aspherical lenses or mirrors was limited to cases in which there was no classic alternative solution, as in astronomical instruments, or to systems that collect and collimate the light of lamps, or, finally, when a mass production reduces their single cost. One case of the latter is plastic lenses, produced by molding and used in certain low cost cameras.

2.5 Ideal images

2.5.1 The homography

We saw earlier that a perfect representation of homogeneous spatial domains must necessarily be a projective transformation of ordinary type. However, abandoning the request to have perfect images in three dimensions, one may wonder what an *ideal* limit reasonably approachable with real instruments could be.

Consider here, for simplicity, only the case of systems with rotational symmetry around an *optical axis* (say, the z-axis). This is a case of special importance in optics, since the majority of optical systems consist of surfaces of revolution around a common axis, so they are called *centered systems*. This symmetry implies two useful consequences. The first is that the image of any point P_o must be on the *meridian plane*, which contains the optical axis and P_o. The second consequence is that what happens to the points of a given meridian plane (say, the yz-plane) must also occur for all of the other points.

Maxwell proposed the concept of an *ideal system* that, leaving aside the physical mechanism employed, achieves the following transformation [Maxwell 1856, 1858]:

> Each figure placed on an object plane $x_o y_o$, normal to the z-axis (with $z = z_o$, fixed), is reproduced sharply in a geometrically similar figure on an image plane $x_i y_i$ (with $z = z_i$, fixed). The constant of proportionality is instead left as a function of the nature of the system and the position of the two planes.

Assume then that the object and image spaces (even if virtual) are homogeneous: since we want to achieve the transformation with rays of light, one must have that

> the points of each (straight) ray placed in the object space are reproduced on a straight line in the image space, and then form a new ray.

It happens, and we will soon check this, that these conditions are fulfilled by an axial projective transformation expressed by the equations

$$x_i = \frac{a}{bz_o + c} x_o, \quad y_i = \frac{a}{bz_o + c} y_o, \quad z_i = \frac{dz_o + e}{bz_o + c}, \tag{2.5.1}$$

which is also called *collineation* or *homography* [see the beautiful book by Gino Giotti 1931]. There, a, b, c, d, e are constants characteristic of the system: given that, in the coefficients, only the ratio counts, one could be eliminated; however, this notation will be convenient later.

Eqs. (2.5.1) are not linear in the z coordinates, and the image of a three-dimensional object is deformed compared to the original. It remains remarkable that each straight line is really transformed into a straight line, as we will soon see (but beware of certain singularity!). It is very important to study this Maxwellian ideal of the formation of images, since the actual optical systems can approach it pretty well, at least for a narrow space region around the optic axis. Let us thus examine in detail the properties of Eqs. (2.5.1), which later will be useful in the study of the so-called *paraxial approximation*.

2.5.2 Conventions on signs

Before starting, it should be noted that there are essentially two opposite conventions with regard to the z-axis positive direction for the object space. Faced with this disagreement, the various authors usually take a position in favor of one of these choices, both of which have strengths and weaknesses.

The first is the "classic" convention, in which the distances are taken as positive moving away from refracting or reflecting interfaces. It has the advantage of considering the object space as *specularly symmetric* with respect to the image space, making the focal distances f_o and f_i, respectively assigned to the system for these spaces, agree in sign. This is the convention that is found in the most basic optical manuals [Hecht 1987; Jenkins and White 1957], including the excellent *Optics Guide* by Melles Griot. In fact, the specular symmetry assumed between object and image spaces facilitates the understanding of the behavior of the optical systems. Its flaw is that this symmetry is not true for all of the variables! In fact, the radii of curvature of the optical surfaces, not being reasonably attributable to one or the other of the spaces, are taken as positive if the centers of curvature are in the "right" of the respective surface. By "right" we mean the direction toward which the light travels, a usage born of the fact that we Westerners write from left to right. In addition, when analyzing instruments composed of various elements, we risk getting lost in a Babel of parity transformations.

The second is the "modern" convention used in the design of optical systems and in ray tracing. This convention has the advantage of assigning the same positive direction for all axes Z in the spaces that follow each other in an optical system made up only of refractive surfaces, with rotational symmetry about an axis z. However, this unity of sign is illusory when dealing with complex optical systems, including the non-planar, containing prisms or mirrors. Furthermore, the various authors who use this convention are often reticent to define the variables well and do not agree on everything. A defect of this convention is that, for example, for a *converging* lens, it assigns a *negative* value to the focal distance in the image space [Born and Wolf 1980; Sivoukhine 1984] or to the focal distance in object space [Landsberg 1979; Guenther 1990]. Another debatable point is that almost all of the "modern" authors treat reflection by assigning a *negative* refractive index associated with the image space! Even if a *geometrical* problem is apparently solved, I believe that this choice is an unacceptable forcing of an essential *physical* property of the medium.

Here, I would like to leave the choice to the reader, writing the various expressions in a form that remains valid in both cases. For this purpose, I introduce an *ad hoc* symbol Ξ, which has the value of a + or – depending on the preferred convention, even if the notation is a little complicated and does not solve all of the discrepancies between the various authors. Thus it is:

$\Xi = +1$ for the classical convention of specular symmetry between the object and image spaces;

$\Xi = -1$ for the modern convention in which that of a *central* or *principal ray* (or *chief ray*) is taken as the z-axis, not necessarily coincident with a straight line, refracted and reflected from the surfaces of the optical system, generally neither axial nor planar. As with the positive direction, the one that the light follows in each section along this ray is chosen. For the focal distances, I follow the minority choice of Landsberg.

On the signs of variables, we will see a summary at the end of §2.5. A more detailed discussion of these cases will be given later in the treatment of the method of 4x4 matrices.

2.5.3 Non telescopic homography

2.5.3.1 Focal planes

The plane $z_o = -c/b$, for which there is a divergence in Eqs. (2.5.1), is called the *focal plane* of the object space, and a similar plane for which $z_i = d/b$ is also present for the image space. In the special case that is $b = 0$, both planes are located at infinity; then, the transformation is called *affine* or *telescopic*. This case will be discussed later. The points F_o and F_i, with coordinates $(0, 0, z_o = -c/b)$ and $(0, 0, z_i = d/b)$, respectively, are called *principal foci*.

2.5.3.2 Conjugate points

For non-telescopic systems (with $b \neq 0$), it soon appears evident that Eqs. (2.5.1) can be conveniently simplified by choosing two distinct reference systems for the object and image spaces to be centered in their respective principal foci. We then apply the replacements

$$Z_o = -\Xi\left(z_o + \frac{c}{b}\right), \quad Z_i = z_i - \frac{d}{b}.$$

The two new coordinates for the z-axis are also indicated with capital letter Zs in order to highlight that their axes, in addition to having translated origins, often do not "physically" agree as to direction. In addition, I introduce the *focal lengths*

$$f_o = \Xi\frac{a}{b}, \quad f_i = \frac{cd - be}{ab}. \tag{2.5.2}$$

The symbol Ξ is worth +1 or –1, according to the chosen convention. Its introduction is equivalent to changing the sign of the variables Z_o, f_o, and s_o (which will be introduced later in the text). Substituting the new variables, Eqs. (2.5.1) ultimately become

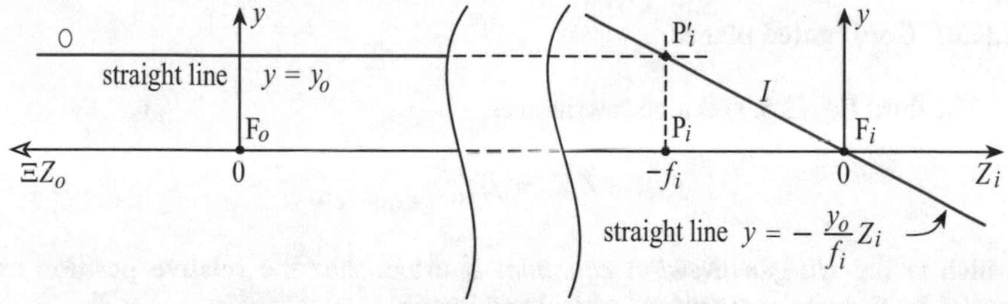

Fig. 2.21 Transformation of a straight line parallel to the optical axis. The two straight lines O and I are each the image of the other, but the point F_i (with $Z_i = 0$) has, as conjugated, the points at $Z_o = \pm\infty$. Similarly, for $Z_o \to \pm 0$, the points at $Z_i \to \pm\infty$. The empty arrow indicates the positive direction for Z_o when $\Xi = +1$, which reverses when $\Xi = -1$

$$x_i = -\frac{f_o}{Z_o} x_o, \quad y_i = -\frac{f_o}{Z_o} y_o, \quad Z_i = \frac{f_i f_o}{Z_o}. \tag{2.5.3}$$

To clarify the way in which the transformation (2.5.3) acts, we consider the points of a straight line object O parallel to the optical axis, with coordinates $x = 0$, $y = y_o$ (Fig. 2.21). With Eqs. (2.5.3), we have:

$$y_i = -\frac{f_o}{Z_o} y_o = -\frac{Z_i}{f_i} y_o .$$

Therefore, these points are transformed into a new straight line I that intersects the optical axis at F_i. The two lines are each the image of the other, but attention must be paid: the points of I in the neighborhood of F_i are conjugate to the points of O in the neighborhood of $Z_o = \pm\infty$. Similarly, the points of O with Z_o in the neighborhood of 0, that is, around the object focal plane, are conjugated with points of I in the neighborhood of $Z_i = \pm\infty$.

In addition, any straight line in the plane yZ expressed by equation $y_o = A + \Xi B Z_o$ is transformed into a straight line. Indeed, its image has coordinates

$$y_i = -\frac{f_o}{Z_o} A - \Xi f_o B = -\frac{A}{f_i} Z_i - \Xi f_o B .$$

In particular, at $Z_i = 0$, the straight line intersects the image focal plane image in $y_i = -\Xi f_o B$, irrespective of the value of A. Finally, for the independence in the transformation between the coordinates x and y, it appears that any straight-line object is conjugated with a straight-line image. In addition, all of this is indicative of an important fact: any object beam of parallel rays among them is transformed (deflected) in a converging beam of rays "focused" into a single point on the focal image plane, and this point coincides with F_i if the incident beam is parallel to the optical axis.

2.5.3.3 Conjugated planes

The third Eq. (2.5.3) can be rewritten as

$$Z_i Z_o = f_i f_o \,, \tag{2.5.4}$$

which is the *Huygens-Newton equation*. It establishes the relative position assumed by two conjugate planes, object and image.

2.5.3.4 Lateral and longitudinal magnification

The *lateral magnification* between figures object and image, each placed on its conjugate plane (for which this magnification is also called the *conjugate ratio*), is easily obtained from Eqs. (2.5.3):

$$m_{\perp} = \frac{x_i}{x_o} = \frac{y_i}{y_o} = -\frac{f_o}{Z_o} = -\frac{Z_i}{f_i} \,. \tag{2.5.5}$$

This magnification varies with the position of the conjugated planes, in particular, it is proportional to Z_i and the divergence at $Z_o = 0$ corresponds to an image in $Z_i = \pm\infty$.

The *longitudinal magnification* is a measure of how a variation of Z_o is reflected in Z_i. Differentiating Eq. (2.5.3.c), it is defined as

$$m_{\|} = \frac{\partial z_i}{\partial z_o} = -\Xi\frac{\partial Z_i}{\partial Z_o} = \Xi\frac{f_i f_o}{Z_o^2} = \Xi\frac{Z_i}{Z_o} \,. \tag{2.5.6}$$

where the sign of $m_{\|}$ is chosen according to the transformation (2.5.1).

2.5.3.5 Principal planes

When $Z_o = -f_o$, it is simultaneously true that $Z_i = -f_i$ and the lateral magnification between the conjugated planes is unitary. These particular planes are called *principal* or *unitary planes* and are placed at a focal distance from the corresponding focal planes. A figure posed on one of them is reproduced identically on the other plane! The intersection of the principal planes with the optical axis gives rise to two points, P_o and P_i, which are called *principal points*. They constitute the origin of the other two axis systems, in which the longitudinal distances are measured starting from the principal planes rather than from the focal planes. This means that, instead of the variables Z_o and Z_i, one uses the variables

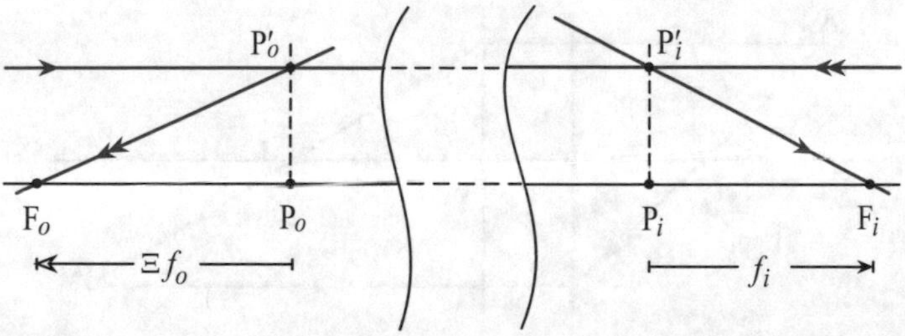

Fig. 2.22 Graphic procedure for fixing the relative positions of foci and principal planes. The arrows associated with the focal distances f_o and f_i indicate the direction in which they are conventionally taken positive when Ξ=+1

$$s_o = Z_o + f_o, \quad s_i = Z_i + f_i\,. \tag{2.5.7}$$

In particular, the Huygens-Newton equation takes the form

$$\frac{f_o}{s_o} + \frac{f_i}{s_i} = 1, \tag{2.5.8}$$

which is called the *Gauss equation*.[13] In addition, the conjugate ratio becomes

$$m_\perp = -\frac{s_i}{s_o}\frac{f_o}{f_i}\,. \tag{2.5.9}$$

Compare this expression with Eq. (2.5.6).

This axis system is probably more familiar than that of Huygens and Newton based on focal planes. However, I can assure you from experience that the Gaussian form in the calculations can lead to convoluted and incomprehensible expressions, while the Huygens-Newton form leads more easily to elegant expressions, rich in physical meaning. The reason for this lies in the fact that the singularities existing on focal planes are hidden when using the principal planes as a reference.

2.5.3.6 Cardinal points and graphical constructions

For an ideal system, the position of the principal points P_o and P_i can also be defined with a graphically equivalent method. Consider, as in Fig. 2.21, an object ray parallel to the optical axis O and extend it in the image space until it intersects its own image straight line I. The point P'_i is the only point of I with the same transverse coordinates (*x*,*y*) of its conjugate P'_o, wherever it is in the object space. A similar construction can also be used to fix the position of the object principal

[13] Born and Wolf (1980) write $f_o/s_o + f_i/s_i = -1$, since they uses opposite signs for f_o and f_i.

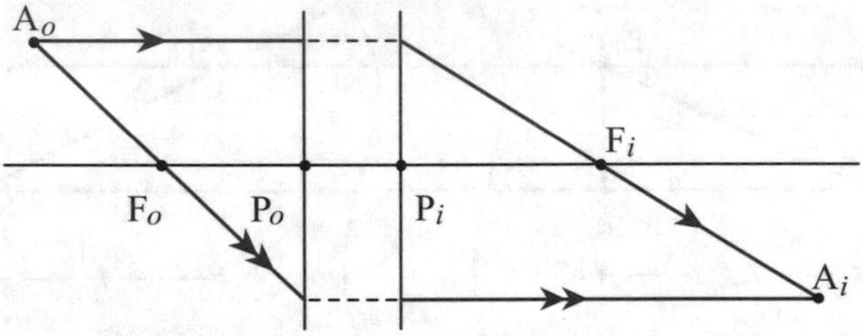

Fig. 2.23 Graphic determination of an image point

plane (Fig. 2.22).

By definition, F_o and P_o are always points in the object space and F_i and P_i are always points of the image space. However, it may happen, for example, that F_i and P_i are physically on the input side of the optical system, the one from which the object rays originate, and they are therefore virtual.

The four points F_o, P_o, F_i and P_i, placed along the optical axis, fully set the properties of an ideal optical system. They can be used to determine the image of any object, by means of the familiar graphic construction used to study the formation of images with lenses. Fig. 2.23 gives us an exhaustive example of it: note that it is enough to consider only rays parallel to the optical axis and their transformed versions, which intersect the axis in F_o and F_i. It should be noted that there is a small inconsistency in the notation used here. With the indices *o* and *i*, the object points *o* are generally distinguished by their conjugate image *i*, but F_o and F_i are an exception, since *they are not* each the conjugate of the other. However, a different way to indicate the focal points may be inappropriate.

On the optical axis, we can usefully define a third pair of points, N_o and N_i, called *nodal points*. They are such that:

Any ray that reaches N_o emerges from N_i parallel to the initial direction.

Fig. 2.24 shows their graphical determination starting from the other 4 points F_o, P_o, F_i and P_i. A ray r_1 is drawn from F_o. It is transformed into a ray parallel to the optical axis and reaches the focal plane image in F_i'. A second ray r_2 is then launched from F_i' so that it is parallel to r_1. In the object space, the radius r_2 must emerge as still parallel to r_1, since r_1 and r_2 intersect the focal plane image at the same point F_i'. The intersections that the two straight lines of r_1 make with

Fig. 2.24 Graphical determination of the nodal points

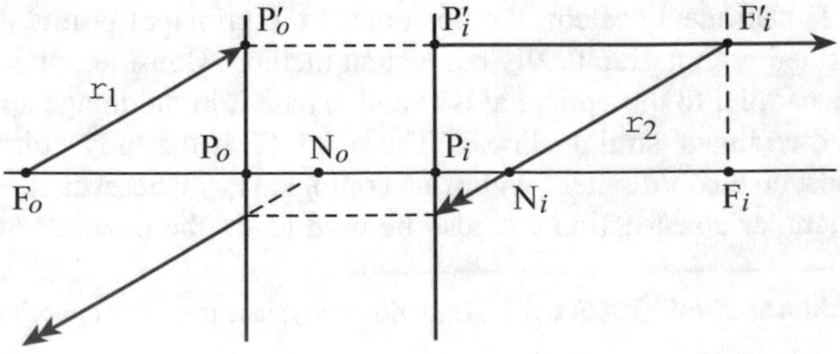

the optical axis are thus the points N_o and N_i that we have searched.

For the similarity of the triangles $F_oP_oP'_o$ and $N_iF_iF'_i$, we have that

$$\overrightarrow{N_iF_i} = \Xi f_o \text{ and similarly } \overrightarrow{N_oF_o} = -f_i, \tag{2.5.10}$$

where, with the sign $\rightharpoonup$, I indicate that the segment here is taken with a defined direction: see the note in Table 2.1. If the two focal distances are equal, the nodal points N of an ideal system coincide with the principal points P.

Even the set of four points F_o, N_o, F_i, N_i is enough to define the properties of an ideal system and, together, the six points F_o, P_o, N_o, F_i, P_i, N_i are called *cardinal points* and can stay on the optical axis in any order, provided that the Eqs. (2.5.10) are respected.

2.5.4 The telescopic case

As we saw earlier, a telescopic transformation is characterized by having both focal planes at infinity. The coefficient b in Eqs. (2.5.1) is zero, and so they are reduced to

$$x_i = \frac{a}{c} x_o, \quad y_i = \frac{a}{c} y_o, \quad z_i = \frac{dz_o + e}{c}. \tag{2.5.11}$$

Even in this case, it is convenient to choose two different coordinate systems for the object and image spaces. As the origin of the two systems, we can choose any pair of *conjugate* points O_o and O_i. According to the convention on the choice of signs for Eqs. (2.5.3), and with reference to the new system of axes, the transformation (2.5.11) takes the form

$$x_i = m_\perp x_o, \quad y_i = m_\perp y_o, \quad Z_i = -\Xi m_\parallel Z_o, \tag{2.5.12}$$

where Z_o, Z_i are the distance from O_o and O_i, respectively, and

$$m_\perp = \frac{a}{c} \text{ and } m_\parallel = \frac{d}{c} \tag{2.5.13}$$

are the lateral and longitudinal magnifications, respectively; they are therefore constant in this case. An important peculiarity of the telescopic systems is that a beam of parallel rays between them, namely, a collimated beam, emerges from the system still collimated; in particular, a ray parallel to the optical axis, at a distance h from it, emerges as shifted at a new distance $m_\perp h$. Even the angular magnification between two conjugate rays is constant; such magnification is defined as the ratio $\tan\vartheta_i/\tan\vartheta_o$, where ϑ_i and ϑ_o are the angles that the two rays make with the

Fig. 2.25
Angular magnification

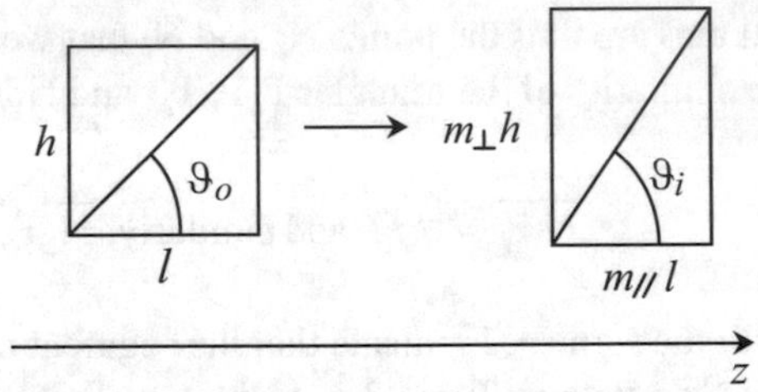

optical z-axis (Fig. 2.25). Therefore, it is

$$m_\angle = \frac{\tan \vartheta_i}{\tan \vartheta_o} = \frac{m_\perp}{m_\parallel}. \tag{2.5.14}$$

You must note that this angular magnification corresponds to the ratio of the angle only within the limit where both of these angles are small.

2.5.5 Combination of two homographies

Consider now the case of two non-telescopic homographies combined in succession (Fig. 2.26); they perform the transformation

$$x_{ig} = -\frac{f_{og}}{Z_{og}} x_{og}, \quad y_{ig} = -\frac{f_{og}}{Z_{og}} y_{og}, \quad Z_{ig} = \frac{f_{ig} f_{og}}{Z_{og}}, \tag{2.5.15}$$

where $g = 1, 2$ is the index that distinguishes the two successive homographies. The image space of the first homography is taken as an object space of the second homography and, indicating the distance between the focal points F_{i1} and F_{o2} corresponding to these spaces with d, we also have

$$x_{o2} = x_{i1}, \quad y_{o2} = y_{i1}, \quad \Xi Z_{o2} = d - Z_{i1}. \tag{2.5.16}$$

Case $d \neq 0$. To ease the subsequent calculations, we immediately introduce ad hoc the two following translations, in the first object space and in the second image space, respectively:

$$Z_o = Z_{o1} - \frac{f_{i1} f_{o1}}{d}, \quad Z_i = Z_{i2} - \Xi \frac{f_{i2} f_{o2}}{d}. \tag{2.5.17}$$

Well, we will immediately demonstrate that these new values represent the distances of object and image from the respective focal planes of the combination, whose focal distances are

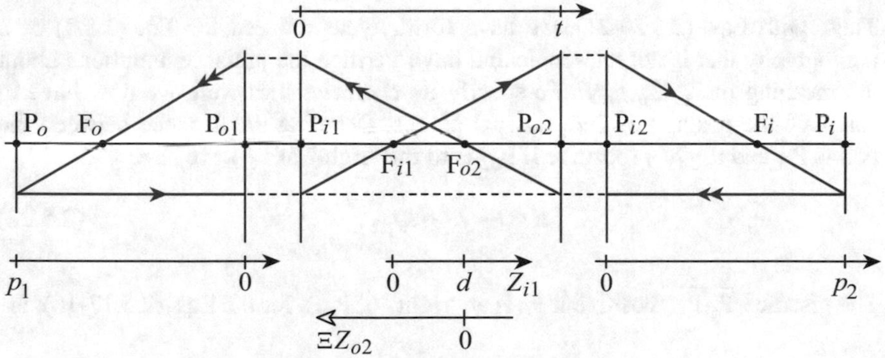

Fig. 2.26 Two converging homographies whose combination is divergent

$$f_o = -\Xi \frac{f_{o1} f_{o2}}{d}, \quad f_i = -\frac{f_{i1} f_{i2}}{d}. \tag{2.5.18}$$

By eliminating the coordinated Z_{i1} in Eq. (2.5.16.c) through Eq. (2.5.15.c) for $g = 1$ and applying Eq. (2.5.17.a), we have:

$$\Xi Z_{o2} = \frac{d}{Z_{o1}}\left(Z_{o1} - \frac{f_{i1} f_{o1}}{d} \right) = \frac{d}{Z_{o1}} Z_o = \frac{d Z_o}{Z_o + f_{i1} f_{o1}/d}. \tag{2.5.19}$$

In addition, with Eqs. (2.5.15-16), (2.5.18) and the second passage of Eq. (2.5.19), $x_i \equiv x_{i2}$ can be expressed as a function of $x_o \equiv x_{o1}$:

$$x_i = \frac{f_{o2}}{Z_{o2}} \frac{f_{o1}}{Z_{o1}} x_o = -\frac{f_o}{Z_o} x_o. \tag{2.5.20}$$

x_o and x_i respectively represent the transverse coordinate in the x direction for the object and image spaces of the overall transformation. There is also a similar equation for the y coordinates. Besides, from Eq. (2.5.15.c) and the last step of Eq. (2.5.19), for the Z, it is:

$$Z_{i2} = \frac{f_{i2} f_{o2}}{Z_{o2}} = \Xi \frac{f_{i2} f_{o2}}{d}\left(1 + \frac{f_{i1} f_{o1}}{d} \frac{1}{Z_o} \right).$$

Thanks to Eqs. (2.5.17. b) and (2.5.18), rearranging the terms, we ultimately have

$$Z_i = \frac{f_i f_o}{Z_o}. \tag{2.5.21}$$

Thus, with Eqs. (2.5.20-21), we have formally reproduced the Eq. (2.5.3) of a homography that is not telescopic and have verified the initial assumptions about the meaning of Z_o, Z_i, f_o, f_i. To specify its characteristics well, we now have to calculate the position of the principal planes. Denoting the distance between the points P_{i1} and P_{o2} by t (positive if P_{o2} is to the "right" of P_{i1}), we have

$$d = t - f_{i1} - \Xi f_{o2}\,. \tag{2.5.22}$$

The distance $\overrightarrow{P_{o1}P_o}$ (positive if P_o is at "right" of P_{o1}), for the Eqs. (2.5.17-18), is

$$\begin{aligned} p_1 &= \overrightarrow{F_{o1}P_o} - \overrightarrow{F_{o1}P_{o1}} = \overrightarrow{F_{o1}F_o} + \overrightarrow{F_oP_o} - \overrightarrow{F_{o1}P_{o1}} = \Xi\left(f_o - f_{o1} - \frac{f_{i1}f_{o1}}{d}\right) \\ &= -\Xi f_{o1}\left(1 + \frac{f_{i1} + \Xi f_{o2}}{d}\right) = -\Xi f_{o1}\left(1 + \frac{t-d}{d}\right) = -\Xi f_{o1}\frac{t}{d} = \frac{f_o}{f_{o2}}t. \end{aligned} \tag{2.5.23.a}$$

Similarly, the distance $\overrightarrow{P_{i2}P_i}$ (with the same convention) is

$$p_2 = f_{i2}\frac{t}{d} = -\frac{f_i}{f_{i1}}t. \tag{2.5.23.b}$$

Case $d = 0$. In this case, we have that $\Xi f_o = f_i = \infty$, so that the equivalent homography is telescopic. Indeed, since, in this case, $\Xi Z_{o2} = - Z_{i1}$, by applying Eq. (2.5.15) and the Huygens-Newton equation, we find that

$$x_i = -\Xi\frac{f_{o2}}{f_{i1}}x_o, \quad y_i = -\Xi\frac{f_{o2}}{f_{i1}}y_o, \quad Z_i = -\Xi\frac{f_{i2}f_{o2}}{f_{i1}f_{o1}}Z_o, \tag{2.5.24}$$

where Z_o and Z_i are now coordinates taken from any pair of conjugate points, between which there is also the couple F_{o1}, F_{i2}. For a telescopic combination of the two homographies, the constants $m_\perp$ and $m_{||}$ of Eqs. (2.5.12) become

$$m_\perp = -\Xi\frac{f_{o2}}{f_{i1}}, \quad m_{||} = \frac{f_{i2}f_{o2}}{f_{i1}f_{o1}} \tag{2.5.25}$$

and the angular magnification is

$$m_\angle = \frac{m_\perp}{m_{||}} = -\Xi\frac{f_{o1}}{f_{i2}}. \tag{2.5.26}$$

The case in which one or both of the homographies to be combined is telescopic is left as an exercise for the readers.

2.5.6 Final remarks

We just verified that two (or more) ideal instruments can be combined together and the result will still be an ideal instrument. So, what has been expounded in §2.5 expresses in general and self-consistent terms, a biunique correspondence between object and image. On the other hand, in the actual processes of the formation of an image, it is not generally possible to ensure that all of the rays emerging from a point object will again converge at a single image point, nor that the images have only the longitudinal deformation prescribed by the ideal case. These discrepancies are referred to as aberrations, and therefore it should be emphasized that the imaging processes are essentially more complicated than in the ideal case that we have just discussed. Yet, it is this ideal that is usually sought in the design of optical systems and its basic concepts are the foundation of practical optics.

Fig. 2.27 graphically summarizes the various terms of the homography. As usual, the object space is drawn on the *left* and the image space on the *right*. However, these terms are somewhat inappropriate. In reality, it would be more appropriate first to determine the path of the central ray, in general, a broken line, and refer the positive direction in the definition of the variables to it: then, with respect to each surface, there is no right and left, but rather there is (respectively) a *before* of refraction/reflection and an *after*.

Therefore, in Table 1, which summarizes the definition of the variable signs, I used the latter terms instead of the traditional ones. For uniformity with the usual conventions, in the case of the reflecting surfaces, the sign of the radius of curvature is given according to the position of the curvature center relative to the initial

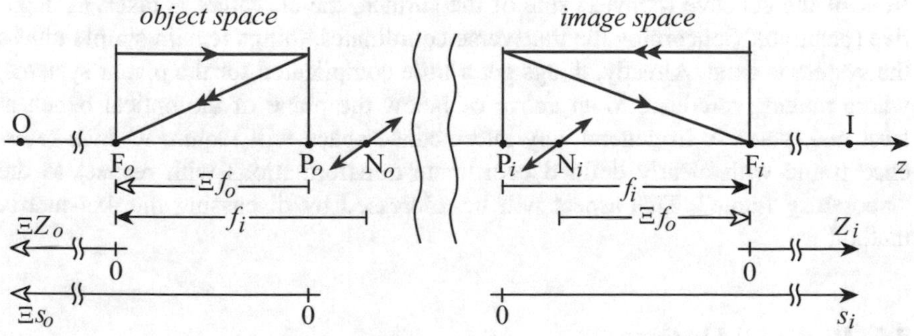

Fig. 2.27 Graphical representation of the various terms.

F_o , F_i : focal points (which are not conjugated to each other)

P_o , P_i : principal points

N_o , N_i: nodal points

Z_o , Z_i : Newtonian coordinates of object and image

s_o , s_i : Gaussian coordinates of object and image

s_o	distance from the object principal plane	Ξ object before P_o
s_i	distance from the image principal plane	+ image after P_i
Z_o	distance from the object focal plane	Ξ object before F_o
Z_i	distance from the image focal plane	+ image after F_i
f_o	object focal distance	Ξ if F_o is before P_o
f_i	image focal distance	+ if F_i is after P_i
F	effective focal distance $(=\Xi f_o = f_i$ se $n_o = n_i = 1)$	+ for a converging lens
y_o, y_i	heights on the axis	+ above the axis
R, $C=1/R$	radius of curvature of the surfaces , curvature	+ if the center is after the surface
$\overrightarrow{AB}$	axial distance between points A and B	+ if B is after A

Note:	Caution: with the symbol $\overrightarrow{AB}$, I do not mean a vector, but only the length of the projection of the segment AB on the optical axis (or on the central ray), with a positive or negative sign, depending on whether the projection of A precedes or follows that of B, respectively

Table 1 Conventions commonly used by opticians

direction of incidence. Therefore, if the incident and reflected rays are located in front of the concave (convex) side of the surface, the curvature is taken as negative (positive). Concerning the transverse coordinates, things remain simple unless the system is axial. Already, things get a little complicated for the planar systems, where making reference to an above or below the plane of the optical bench at least makes sense. In general, any intermediate space will require its own reference frame with clearly defined coordinate transformations with respect to the "laboratory frame". This aspect will be addressed by discussing the 4x4-matrix method.

2.6 Paraxial Optics

2.6.1 Domain of validity

Let us now see how the Maxwellian ideal of image can be approached with actual instruments constructed with rotational symmetry around an optical axis, which also do not have singularities such as cusps or others. These systems are

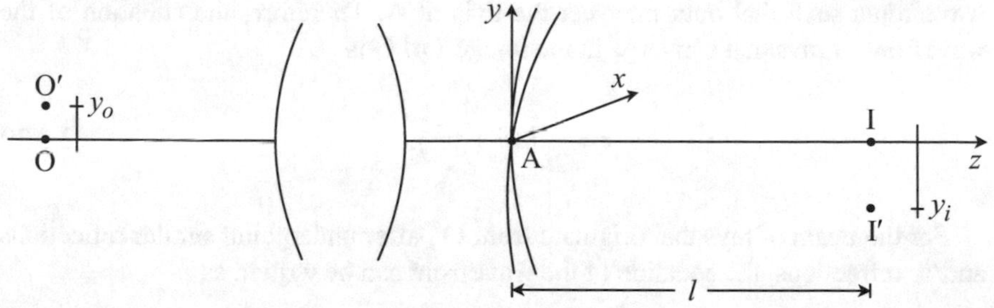

Fig. 2.28 Determination of a wavefront in the image space

said to be *centered*. Thanks to this symmetry, the various optical surfaces of the instrument can be represented as a function of $r^2 = x^2 + y^2$, with r being equal to the distance from the axis. The same reasoning applies to the wavefronts of a point source located on the axis. Therefore, both the equation of the surfaces and that of these wavefronts can be developed in a series of powers in r^2. For example, for a spherical surface of radius R, we have

$$z = \frac{1}{2R}\left(x^2 + y^2\right) + \frac{1}{8R^3}\left(x^2 + y^2\right)^2 + \ldots \,. \tag{2.6.1}$$

Now, the idea comes naturally of limiting the field of the instrument so that we can neglect terms of order higher than the first in r^2 in the development of optical surfaces and wavefronts. This entails limiting this field to those rays that are fairly close to the optical axis, and that are at small angles with it, and therefore it is possible to approximate the sines with the angles in the application of Snell's law. Let us examine the consequences of this approximate treatment, which was mainly introduced by Gauss, and is therefore called *Gaussian Optics*, or Paraxial Optics.

Therefore, in the domain of this approximation, it follows that the wavefronts of an axial point source, after each refraction or reflection, retain the shape

$$z = \frac{C}{2}\left(x^2 + y^2\right) \tag{2.6.2}$$

that cannot be distinguished from a sphere with radius $1/C$. This simple fact ensures that, if the field of the instrument is sufficiently limited, one gets stigmatic images of object points located on the optical axis.

In addition to this result, it remains to be seen whether such a property extends to the points in the neighborhood of the optical axis. Thanks to the symmetry of the system in examination, it is enough for us to consider the case of a point object O′ knocked off its axis, for example, on the yz-plane and with coordinate $y = y_o$. Let O be a point on the axis with the same coordinate z as O′ and I its image. In addition, let us take, as the origin in the z-axis, a point A located at some distance l from I (Fig. 2.28). For the two points O and O′, consider the two corresponding wavefronts such that both intersect the axis at A. Therefore, the equation of the wavefront of rays that converge in the image I of O is

$$z = \frac{1}{2l}\left(x^2 + y^2\right). \tag{2.6.2'}$$

For the beam of rays that originate from O′, after undergoing similar reflections and/or refractions, the equation of the wavefront can be written as

$$z = \frac{1}{2l}\left(x^2 + y^2\right) + F\left(x, y, y_o\right), \tag{2.6.3}$$

where F is a function that expresses the difference between the two fronts and must vanish for $y_o = 0$. Developing this expression only up to the quadratic terms in x, y, y_o, we have:

$$z = \frac{1}{2l}\left(x^2 + y^2\right) + y_o\left(ax + by + cy_o + d\right).$$

Now, c and d must be zero, since, for hypothesis, $z = 0$ for $x = y = 0$, because we have chosen a wavefront that still passes through A. Even the term in x must vanish, given that, for the symmetry present around the yz-plane, in the development, there can only be even terms in x, and therefore $a = 0$ as well. So, finally, with the approximations made, the wavefront generated by O′ still has the shape of a sphere of radius l with the center placed at a distance $y_i = -lby_o$ above I. Indeed, by making a rotation of the coordinates on the yz-plane around the point A by an angle $\alpha \cong -y_i/l$, neglecting the terms of degree higher than the second, we again find Eq. (2.6.2′). Therefore, within the limits of the Gaussian domain, the points around the axis also possess stigmatic images whose transverse coordinates (x, y) are proportional to those of the corresponding object points.

With this, we have already reproduced, with simple analytical considerations, one of the properties of ideal images. This result has been reached by neglecting, in the calculations of the wavefront, the terms of order beyond the second in the transverse coordinates of the objects and the system surfaces. It is therefore natural to associate the aberrations with these terms, while the Gaussian domain extends in the region where they are negligible. A criterion for determining when this happens quantitatively was suggested by Rayleigh in 1880. It derives from the observation that, as long as a wavefront differs from a true sphere for a distance less than one-quarter of a wavelength, the field that arrives at the center of the sphere is reinforced with a contribution in phase from each of the wavefront's elements. We can thus affirm that the terms of higher order in the calculation of the optical paths are negligible if they amount to less than a quarter of a wavelength [Welford 1974].

In the subsequent sections, we will complete the demonstration that the instruments with axial rotational symmetry reproduce, in the Gauss region, the properties of axial projective transformations. We will also refer to these with the names of Gauss homographies.

2.6.2 *Refraction through a spherical surface*

Let us take a concrete look at the properties of one of the simplest optical elements, the *dioptre*, in the paraxial approximation. Consider, then, a single refractive surface S with curvature $1/R$, which separates two media of index n_o and n_i, as shown in Fig. 2.29. Let F_o and F_i be two wavefronts of a beam of light emanating from a point O on the axis and located, respectively, immediately before and after the refraction (in Fig. 2.29, it happens that F_i is virtual, but this does not matter here). A ray that starts from O meets the wavefronts in M_o and M_i and the refracting surface in N. Since the optical paths between the two wavefronts of the same family of rays must be equal and, for the Malus-Dupin theorem, this applies even in the presence of a refraction, we have:

$$n_o \overrightarrow{M_oN} = n_i \overrightarrow{M_iN} \ . \qquad (2.6.4)$$

In the Gaussian region, this equation is correct with no other restrictions, however, within this region, we have

$$\overrightarrow{M_oN} = \frac{1}{2}\frac{h^2}{R} + \Xi\frac{1}{2}\frac{h^2}{s_o} \ ,$$

where h is the height of incidence of the beam and s_o is the distance between the refractive surface and the point O. The signs applied in this expression reflect the convention transcribed in Table 1. A similar equation there is for $\overrightarrow{M_iN}$:

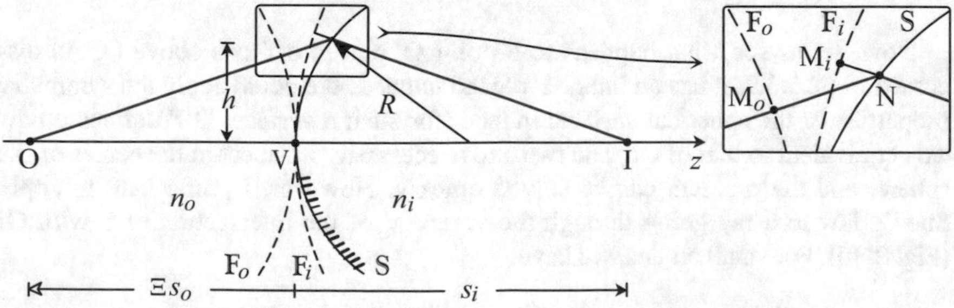

Fig. 2.29 Refraction of a spherical wavefront on a dioptric surface. Here, the radius R of the surface is taken as positive and $n_i > n_o$, for which this is a converging *dioptre*

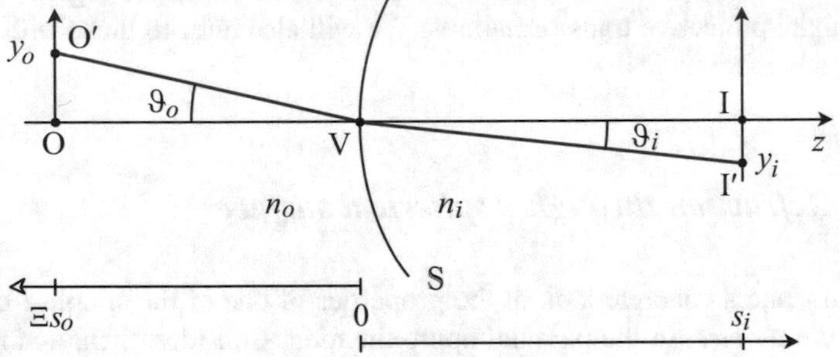

Fig. 2.30 Refraction of a ray at the vertex V of a dioptric surface

$$\overrightarrow{\mathrm{M}_i\mathrm{N}} = \frac{1}{2}\frac{h^2}{R} - \frac{1}{2}\frac{h^2}{s_i}.$$

Substituting these expressions in Eq. (2.6.4) and simplifying, we get:

$$n_o\left(\frac{1}{R} + \Xi\frac{1}{s_o}\right) = n_i\left(\frac{1}{R} - \frac{1}{s_i}\right),$$

that is, rearranging the terms,

$$\Xi\frac{n_o}{s_o} + \frac{n_i}{s_i} = \frac{n_i - n_o}{R}, \tag{2.6.5}$$

where you can recognize the form of the Gauss Eq. (2.5.8) for conjugated distances. From the comparison, it is

$$f_o = \Xi n_o\frac{R}{n_i - n_o}, \quad f_i = n_i\frac{R}{n_i - n_o}. \tag{2.6.6}$$

Now, let us see what happens to a point O′ placed off-axis above O. As discussed in §2.5.3.2, it has an image I′ placed under I. We could apply the symmetry properties of the spherical surface: indeed, for such a surface, O′I′ defines a new axis equivalent to that of OI. The two axes necessarily intersect in the center of the sphere, and the problem can be solved quickly. However, I prefer here to apply Snell's law to a ray going through the vertex V of the intersection of S with OI (Fig. 2.30). For small angles, we have

$$\Xi n_o\frac{y_o}{s_o} = -n_i\frac{y_i}{s_i}. \tag{2.6.7}$$

Therefore, the transverse magnification is

$$m_{\perp} = \frac{y_i}{y_o} = -\Xi \frac{s_i}{s_o}\frac{n_o}{n_i}. \qquad (2.6.8)$$

Thus, we have finally confirmed that a *dioptre*, in the paraxial region, reproduces the properties of the homographies discussed in §2.5. In particular, with respect to such an optical element, the principal planes coincide and $P_o = P_i = V$, In addition, the nodal points are both located in the center of curvature of the dioptric surface. We can then apply all of rules and tricks of ideal images to it. The term

$$\mathcal{P} = \frac{n_i - n_o}{R} \qquad (2.6.9)$$

is called the *power of the surface* and is measured in *diopters* (1 diopter = 1 m^{-1}). If $\mathcal{P} > 0$, and therefore $\Xi f_o, f_i > 0$ as well, it is said that the surface is *positive* or *converging*. Conversely, for $\mathcal{P} < 0$, the surface is called *negative* or *diverging*. In Ophthalmic Optics, the terms $\Xi n_o/s_o$ and n_i/s_i of Eq. (2.6.5) are called *reduced vergence*, and their sum is the power of the dioptric surface; by *vergence* we mean the converging or diverging wavefront curvature, and it too is measured in diopters.

2.6.3 Reflection from a spherical mirror

For the case of a mirror, we can proceed in a similar manner to that which was done for the dioptric surface. Adopting the conventions shown in Fig. 2.31, in

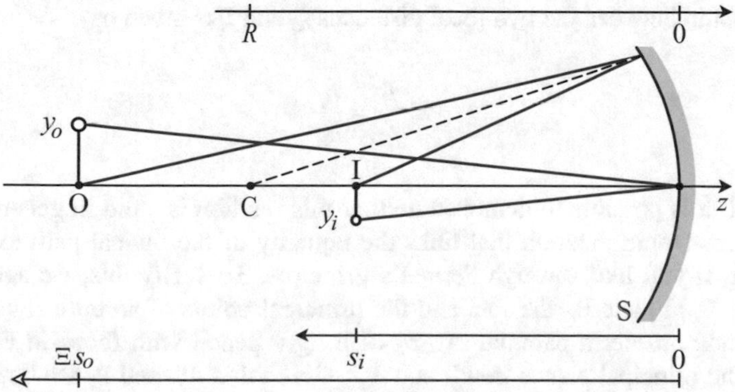

Fig. 2.31 Reflection from a spherical mirror. In this convention, the positive direction for Ξs_o and s_i is the opposite to that for R, thus, in the case of this figure, Ξs_o, $s_i > 0$ and $R < 0$

which the arrow indicates the positive directions, for a mirror of radius R, we have

$$\Xi\frac{1}{s_o}+\frac{1}{s_i}=-\frac{2}{R}, \quad x_i=-\Xi\frac{s_i}{s_o}x_o, \quad y_i=-\Xi\frac{s_i}{s_o}y_o \tag{2.6.10}$$

and, in particular,

$$\Xi f_o=f_i=f=-\frac{R}{2}, \quad \mathcal{P}=-\frac{2}{R}. \tag{2.6.11}$$

Also, in this case, the principal points coincide with the vertex of the mirror, and the nodal points with its center of curvature.

While a parabolic mirror focuses all of the rays parallel to its axis in its focal point, in the same situation, a spherical mirror suffers from *spherical aberration.* However, the parabolic mirror suffers from *coma* and *astigmatism* for distant sources off of its axis, while an optical system consisting of a spherical mirror and a small aperture diaphragm placed on its center of curvature produces a sharp image of such sources on a spherical *focal surface*, concentric with that of the mirror.

2.6.4 Relationship between the two focal lengths

In the two previous sections, we have seen how the two main elementary optical elements reproduce the properties for the formation of the ideal image. Given that the combination of two or more homographies is still a homography (and, shortly, we will see some examples), we can now easily extend the discussion to any axial instrument by combining the transformations that correspond to its optical surfaces. Before proceeding, however, it is appropriate to develop some concepts that will be very useful later. We have seen that, for a single surface, the relationship between the two focal distances f_o and f_i is given by

$$\Xi\frac{f_i}{f_o}=\frac{n_i}{n_o}. \tag{2.6.12}$$

Well, it is possible to demonstrate that this relation is valid in general, through the Malus-Dupin theorem that links the equality of the optical path to the wavefront or, if you like, through Fermat's principle. To verify this, we again denote, with F_o, P_o, F_i and P_i, the foci and the principal points of an optical system (Fig. 2.32) and consider a paraxial ray $\mathtt{r}_2$ of a light pencil with focus in F_o. This ray meets the principal planes in P_o' and P_i'. Given that P_i' and P_i are both points of the *plane* wavefront of the pencil of rays coming from F_o, the equality of optical paths requires that

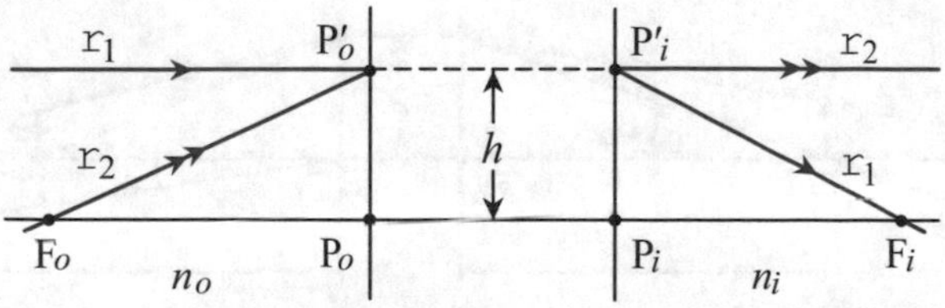

Fig. 2.32 Determination of the relationship between the two focal lengths

$$n_o \mathrm{F}_o \mathrm{P}'_o + [\mathrm{P}'_o \mathrm{P}'_i] = \Xi n_o f_o + [\mathrm{P}_o \mathrm{P}_i],$$

where the square brackets mean the words "optical path length of". Thus, if the distance $\mathrm{P}_o \mathrm{P}'_o$ (= $\mathrm{P}_i \mathrm{P}'_i$) is indicated with h, expanding the distance $\mathrm{F}_o \mathrm{P}'_o$ in powers of h and limiting the expansion up to the term in h^2, we have

$$\Xi \frac{n_o h^2}{2 f_o} = [\mathrm{P}_o \mathrm{P}_i] - [\mathrm{P}'_o \mathrm{P}'_i] .$$

Similarly, taking a ray $\mathtt{r}_1$ that enters the system parallel to the axis and with the same height h, we have

$$\frac{n_i h^2}{2 f_i} = [\mathrm{P}_o \mathrm{P}_i] - [\mathrm{P}'_o \mathrm{P}'_i].$$

From a comparison of these two expressions, we get Eq. (2.6.12), which is then valid in general. It can be re-expressed in the form

$$\mathcal{P} = \Xi \frac{n_o}{f_o} = \frac{n_i}{f_i} . \tag{2.6.13}$$

$\mathcal{P}$ is called the *power* of the instrument and its inverse is said to be the *effective focal distance* (or *equivalent*), which is, in fact, the focal length of a system with the same value $\mathcal{P}$, but immersed "in the air" (that is, with $n_o = n_i = 1$) at both sides.

2.6.5 The lateral, angular and longitudinal magnification, and the Smith-Helmholtz-Lagrange optical invariant

Let us now examine an important relationship involving the *convergence angles* of the rays, that is, the angle that they make with the optical axis. Let us take, as in Fig. 2.33, two conjugated points O′ and I′ and their projection on the axis, O

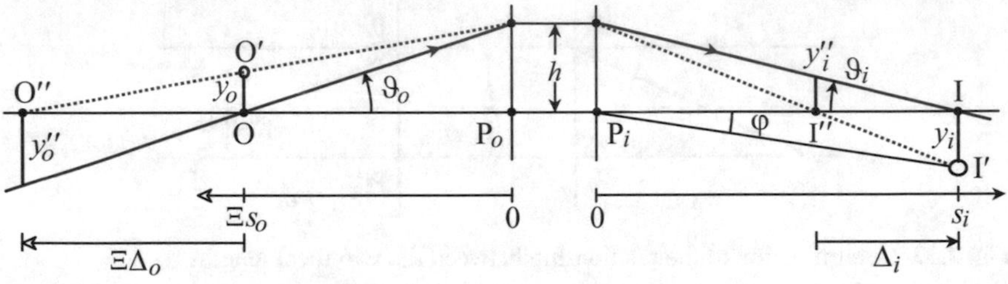

Fig. 2.33 Angular magnification

and I, which are conjugated as well. The *lateral magnification* between the distances OO′ and II′, as we know from Eqs (2.5.9) and (2.6.12), is

$$m_\perp = \frac{y_i}{y_o} = -\frac{s_i}{s_o}\frac{f_o}{f_i} = -\Xi\frac{s_i}{s_o}\frac{n_o}{n_i}\,. \tag{2.6.14}$$

Rearranging the terms, we get

$$\frac{y_i}{s_i}n_i = -\Xi\frac{y_o}{s_o}n_o\,. \tag{2.6.15}$$

If h is the height at which a ray passing through O meets the principal planes (Fig. 2.33), the angles subtended by this ray with the optical axis in the two spaces, in this order of precision, are

$$\vartheta_o = \Xi\frac{h}{s_o}, \quad \vartheta_i = -\frac{h}{s_i}\,.$$

So, multiplying both sides of Eq. (2.6.15) for h, we get

$$\vartheta_i y_i n_i = \vartheta_o y_o n_o = H\,. \tag{2.6.16}$$

Even the planes defined by O″ and I″ in Fig. 2.33 are conjugated, and, from Eq. (2.6.15), by replacing s_o with $-\Delta_o h/y_o''$ and s_i with $\Delta_i h/y_i''$, we find that

$$\frac{y_i'' y_i}{\Delta_i}n_i = \frac{y_o'' y_o}{\Xi\Delta_o}n_o\,.$$

Suppose that our system from Fig. 2.33 is followed by another: now, the output values ϑ_i, y_i, n_i of the first must be assigned to the variables ϑ_o, y_o, n_o of the second, since the image of the first system becomes the object for the second and both the refractive index and the angle of convergence are the same. So, the quantity

$$H = \vartheta y n \tag{2.6.17}$$

is invariant through the two systems and, therefore, it must be invariant, from the object to the image space, through all of the intermediate spaces of each axially symmetrical optical system. This invariant was introduced by Robert Smith in 1738 in his book *Compleat System of Optics*, and it was subsequently taken up by Lagrange and Helmholtz; it is indicated by the name of the *Smith-Helmholtz-Lagrange optical invariant* (one or two names are usually omitted!). Among the applications of this optical invariant, there is the derivation of the relationship between lateral magnification and angular magnification. From Eq. (2.6.17), we immediately find that the *angular magnification*

$$m_{\angle} = \frac{\vartheta_i}{\vartheta_o} = \frac{n_o}{n_i}\frac{y_o}{y_i} = \frac{n_o}{n_i}\frac{1}{m_{\perp}}, \tag{2.6.18}$$

among the angles of convergence, it is inversely proportional to the lateral one, and the constant of proportionality does *not* depend on the position of the object-image conjugate planes.

It may happen that, in one of the spaces of an optical system, the object or one intermediate image is at infinity. As described by Eq. (2.6.16), in this case, H becomes indeterminate there, since $\vartheta \to 0$ and $y \to \infty$. This indetermination can be resolved by again placing

$$\vartheta = -\frac{h}{s} \quad \text{and} \quad y = \varphi s\,,$$

where φ is a field angle, as defined in Fig. 2.33. Therefore, in this space, we can write

$$H = -nh\varphi\,. \tag{2.6.19}$$

It should be noted that, for small angles, this relationship is valid even if the quantities h, s, φ are taken with respect to any plane (normal to the axis), i.e., not only with respect to a principal plane, as in Fig. 2.33.

Lastly, let us look at an important relation between *longitudinal magnification* and *lateral magnification*. From Eq. (2.5.6) and the Gauss Eq. (2.5.8), we have

$$m_{\parallel} = \Xi\frac{Z_i}{Z_o} = \Xi\frac{s_i - f_i}{s_o - f_o} = \Xi\frac{s_i}{s_o}\frac{1-\dfrac{f_i}{s_i}}{1-\dfrac{f_o}{s_o}} = \Xi\frac{s_i}{s_o}\frac{\dfrac{f_o}{s_o}}{\dfrac{f_i}{s_i}} = \Xi\frac{{s_i}^2}{{s_o}^2}\frac{f_o}{f_i}$$

and, remembering Eq. (2.6.12), we ultimately have

$$m_{\parallel} = \frac{s_i^2}{s_o^2}\frac{n_o}{n_i}. \tag{2.6.20}$$

This equation can be compared to Eq. (2.6.14) for the lateral magnification. In particular,

$$m_{\parallel} = m_{\perp}{}^2 \frac{n_i}{n_o}, \tag{2.6.21}$$

and, in the frequent case in which $n_o = n_i$, it follows that the longitudinal magnification is the square of the lateral one. It is this fact that allows us to see and take pictures with a good *depth of field*: meters of thickness (diverging to infinity for more distant objects) are compressed on the retina or on the film!

2.6.6 The lens

This is definitely the most important of all optical devices. The lens is of utmost usefulness to our own lives. Lenses of marvelous perfection can be found in nature and even in ourselves: they are a precious gift from God. In the broadest sense, the lens is a refractive device that is used to reshape wavefronts in a controlled manner. The medium used is not necessarily homogeneous. For example, it is possible to reconfigure a wavefront by passing it through a medium in which the refractive index varies according to a certain law. These lenses are called GRIN, i.e., gradient-index. One example is the *crystalline lens* of the eye. In the more traditional sense, a lens is an optical system consisting of two or more refracting interfaces between homogeneous media, of which at least one is curved. A lens that consists of a single element of transparent material, namely, that only has two refractive surfaces, is called a *simple lens* or *singlet*; instead, if it is made up of multiple elements, it is said to be a *compound lens*. These lenses are typically composed of two or three elements (*doublets* or *triplets*) and are designed to minimize certain aberrations. The lenses are also classified as thin or thick, according to their thickness if it is actually negligible or not.

The lenses are then distinguished by the shape of their surfaces, which is chosen for the particular application for which they are intended. In particular, eyeglass lenses are only nominally simple, since they must serve to correct various visual flaws and frequently come with aspheric or toric surfaces. Here, we will limit ourselves to considering lenses with rotational symmetry around an axis. Such a lens is called *converging* or *positive* when it is thicker in the center than at the edges (and, simultaneously, its refractive index is greater than that of the surrounding medium, as usually happens). Indeed, it delays the beam's wavefront largely at the center, rather than at the edges, and then increases the forward bending (with the sign defined in Table 2.1). The lenses that produce the opposite effect are called *diverging* or *negative*.

Let us now look at the case of a simple lens, constituted by the combination of two *dioptre*, with radius of curvature R_1 and R_2, whose vertices V_1 and V_2 are spaced apart by a distance t. Moreover, we denote, with n_1, n_2, n_3, the indexes of refraction of three media the surfaces of which separate in sequence (Fig. 2.34). From Eqs. (2.6.6), the focal lengths of the first and second surfaces are given, respectively, by

$$f_{o1} = \Xi\frac{n_1}{\mathcal{P}_1}, \quad f_{i1} = \frac{n_2}{\mathcal{P}_1}, \quad \text{with } \mathcal{P}_1 = \frac{n_2 - n_1}{R_1}, \tag{2.6.22.a}$$

$$f_{o2} = \Xi\frac{n_2}{\mathcal{P}_2}, \quad f_{i2} = \frac{n_3}{\mathcal{P}_2}, \quad \text{with } \mathcal{P}_2 = \frac{n_3 - n_2}{R_2}. \tag{2.6.22.b}$$

The focal lengths of the combination are still given by Eqs. (2.5.18), where d is now the distance between the two foci F_{i1} and F_{o2} of Fig. 2.26. For the usual convention on the signs shown in Fig. 2.27, and for Eqs. (2.5.22) and (2.6.22), such a distance is

$$d = t - f_{i1} - \Xi f_{o2} = t - n_2\left(\frac{1}{\mathcal{P}_1} + \frac{1}{\mathcal{P}_2}\right),$$

where, in turn, the distance t between the principal points P_{i1} and P_{o2} is given by the thickness of the lens on its axis. From Eq. (2.5.18), introducing the parameter

$$\mathcal{P} = -\frac{\mathcal{P}_1\mathcal{P}_2 d}{n_2}, \tag{2.6.23}$$

we ultimately find that

$$f_o = \Xi\frac{n_1}{\mathcal{P}}, \quad f_i = \frac{n_3}{\mathcal{P}}, \tag{2.6.24}$$

where, as a function of the thickness t,

$$\mathcal{P} = \mathcal{P}_1 + \mathcal{P}_2 - \frac{t}{n_2}\mathcal{P}_1\mathcal{P}_2 \tag{2.6.25}$$

represents the total power of the two *dioptres* combined together, namely, the power of the lens. It should be noted that, for a thin lens, with $t << R_1, R_2$, the value of $\mathcal{P}$ is simply the sum of the power of the two *dioptres*. We will also encounter

Fig. 2.34
Simple thick lens

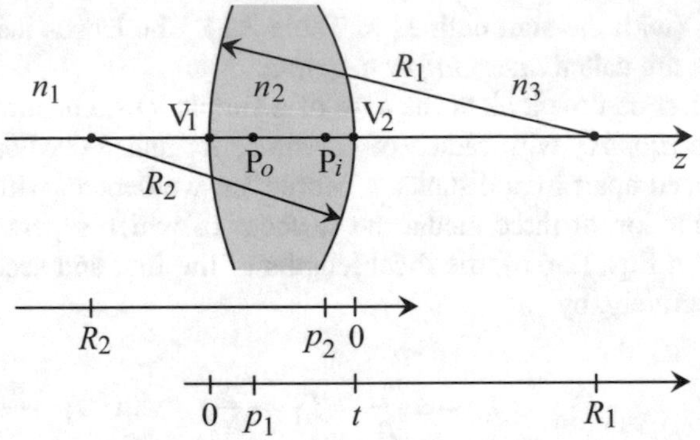

a relationship similar to Eq. (2.6.25) in the combination of two lenses.

To specify fully the Gaussian properties of our lens, we must now determine the position of the principal points P_o and P_i relative to the lens itself. This is easily done with the help of Eqs. (2.5.23). With respect to the vertex V_1 of the lens (I remember that both of the principal points of the first surface are located there), the position of the object principal point P_o of the combination (Fig. 2.34), for Eqs. (2.6.22-23), is

$$p_1 = -\Xi f_{o1}\frac{t}{d} = \frac{n_1}{n_2}\frac{\mathcal{P}_2}{\mathcal{P}}t\,. \tag{2.6.26.a}$$

Similarly, the distance of P_i from V_2 is

$$p_2 = f_{i2}\frac{t}{d} = -\frac{n_3}{n_2}\frac{\mathcal{P}_1}{\mathcal{P}}t\,. \tag{2.6.26.b}$$

In the simplest case of a *lens immersed in the air*, that is, with indices $n_1 = n_3 = 1$ and $n_2 = n$, Eqs. (2.6.24-25) are simplified in the well-known *lens-maker's equation*:

$$f = \left[(n-1)\left(\frac{1}{R_1} - \frac{1}{R_2} + \frac{n-1}{n}\frac{t}{R_1 R_2}\right)\right]^{-1}, \tag{2.6.27}$$

with

$$f = \Xi f_o = f_i = \frac{1}{\mathcal{P}}\,. \tag{2.6.28}$$

In other words, the two focal lengths now coincide with each other. This has the important immediate consequence that the nodal points of the lens coincide with their principal points: $N_o = P_o$, $N_i = P_i$. On the other hand, we can quickly see, from Eq. (2.6.13), that this coincidence happens whenever, in a generic in-

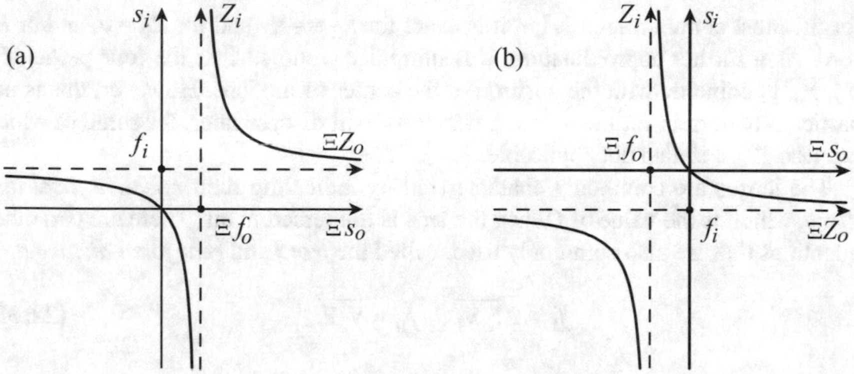

Fig. 2.35 Hyperbolas whose coordinates s_o, s_i satisfy the Gauss equation. (a) converging system, (b) diverging system

strument, the object space and the image space indices are equal. In the case in which the lens is immersed in a generic medium with index $n_\text{m} = n_1 = n_3$, Eq. (2.6.27) can be formally recovered by normalizing the indexes with n_m, i.e., placing $n = n_2/n_\text{m}$.

Together with Eq. (2.6.27), the positions of the principal planes of a lens immersed in air are

$$\overrightarrow{\text{V}_1\text{P}_o} = -\frac{n-1}{n}\frac{f}{R_2}t, \quad \overrightarrow{\text{V}_2\text{P}_i} = -\frac{n-1}{n}\frac{f}{R_1}t, \tag{2.6.29}$$

where I preferred here to make explicit the points involved.

A further important simplification can be made in the case of a thin lens for which $t << |R_1|, |R_2|, |R_1 - R_2|$. This approximate idea of lens is very useful, for its simplicity, in the first phase of the design of optical systems, or when great precision is not necessary. For a thin lens, Eqs. (2.6.27) and (2.6.29) are respectively reduced to

$$f = \left[(n-1)\left(\frac{1}{R_1} - \frac{1}{R_2}\right)\right]^{-1}, \tag{2.6.30}$$

$$\overrightarrow{\text{V}_1\text{P}_o} = -\frac{R_1}{R_2 - R_1}\frac{t}{n}, \quad \overrightarrow{\text{V}_2\text{P}_i} = -\frac{R_2}{R_2 - R_1}\frac{t}{n}, \tag{2.6.31}$$

and the distance between the principal planes is approximately

$$\overrightarrow{\text{P}_o\text{P}_i} = t\left(1 - \frac{1}{n}\right) \tag{2.6.32}$$

for the most common lenses, but it is exact for $R_2 = -R_1$ and for $R_2 = \infty$ and/or $R_1 = \infty$. As a further approximation, it is normally assumed that the four points V_1, V_2, P_o, P_i coincide with the position of the center of the lens. However, this is not particularly correct for the *menisci*, which we will discuss later, for some of which Eq. (2.6.32) it is also not applicable.

The lenses are commonly characterized by indicating their *effective focal distance*, which is the value of f when the lens is immersed in air. There are two other quantities that are also commonly used, called the *front* and *rear focal distance*:

$$f_{\mathrm{f}} = \Xi\overrightarrow{\mathrm{F}_o\mathrm{V}_1}\,, \quad f_{\mathrm{p}} = \overrightarrow{\mathrm{V}_2\mathrm{F}_i}\,, \tag{2.6.33}$$

measured by the vertices V_1 and V_2 of the lens immersed in air and, therefore, *not* to be confused with f_o and f_i. In particular, in Ophthalmic Optics, use is made of the *rear* or *back power*; for Eq. (2.6.26.b), it is explicitly given by

$$\frac{1}{f_{\mathrm{p}}} = \mathcal{P}_{\mathrm{p}} = \frac{\mathcal{P}_1 + \mathcal{P}_2 - (t/n)\mathcal{P}_1\mathcal{P}_2}{1-(t/n)\mathcal{P}_1}\,, \tag{2.6.34}$$

where the denominator constitutes a relatively small change of Eq. (2.6.25). Finally, for a lens in air, the Gauss equation and that of Huygens-Newton become, respectively,

$$\Xi\frac{1}{s_o} + \frac{1}{s_i} = \frac{1}{f}\,, \quad Z_o Z_i = f^2\,. \tag{2.6.35}$$

The graphical solution to these equations is shown in Fig. 2.35.

2.6.7 Variety of simple lenses

Thick lens immersed in air. From Eq. (2.6.27) and (2.6.29), it is seen that there are 4 degrees of freedom, t, n, R_1 and R_2, that allow for varying the characteristics of a lens. Among these, t is generally reduced to the minimum compatible with the diameter and mechanical strength and to improve the optical quality of the lens. Indeed, the aberrations of the lens generally increase with its thickness. The refractive index n has only a limited variability: in the field of visible radiation, the values n of optical glasses vary between about 1.5 (the most common glass) and 2 (for some special glass). Glasses of high index have the advantage of leading to minor geometrical aberrations, but they are also more dispersive and absorbent. For visible radiation, the ideal would be a glass with index 1.9 and an anti-reflective coating with a single layer of magnesium fluoride. Of the two remaining degrees of freedom, one can be indicated in $1/R_1 - 1/R_2$, which, for the case of a

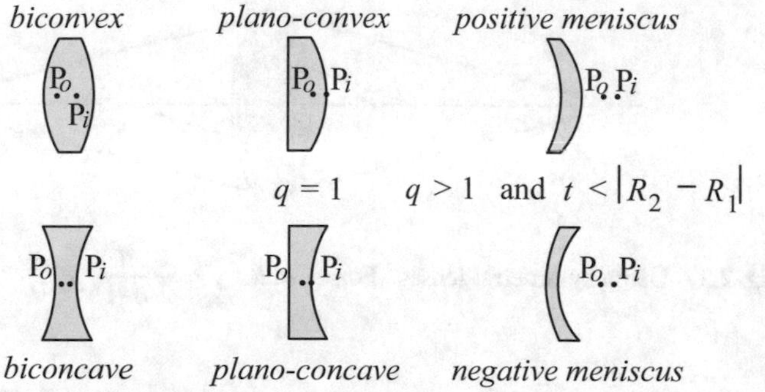

Fig. 2.36 Variety of lenses

thin lens, is proportional to the power of the lens. Thus, by setting the focal length, it still remains a free parameter, which is conventionally indicated by the ratio

$$q = \frac{R_1 + R_2}{R_1 - R_2}, \tag{2.6.36}$$

which is called *Coddington's form factor*.

There is a specific reason that justifies the variety of forms in which lenses are made (Fig. 2.36). Indeed, the quality of images produced by a lens depends strongly on how the lens itself is used and by its shape. The ability to change the shape without changing the Gaussian properties of the lens (this operation is called *bending*) is therefore useful for balancing the aberrations. In other words, for each conjugate ratio assigned to a lens with fixed focal length, for a particular application, we can choose the value of q that allows (for example) for the smallest spherical aberration. A practical criterion for the choice of the best form is to equalize the average angles that the marginal rays, i.e., the rays that intersect the lens near its edge, make with the two surfaces of the lens.

These concepts will be further elaborated in the study of aberrations. For the moment, it is interesting to note how the position of the principal points varies, by changing the shape of the lens. Eqs. (2.6.29), expressed as an explicit function of R_1 and R_2, become

$$\overrightarrow{\mathrm{V_1P}_o} = -\frac{tR_1}{n(R_2 - R_1) + t(n-1)}, \quad \overrightarrow{\mathrm{V_2P}_i} = -\frac{tR_2}{n(R_2 - R_1) + t(n-1)}, \tag{2.6.37}$$

which, for a thin lens, are reduced to Eqs. (2.6.31) given above. The distance between the principal points P_o and P_i is called the *hiatus* or *gap*. Since

$$\overrightarrow{\mathrm{P}_o\mathrm{P}_i} = t - \overrightarrow{\mathrm{V_1P}_o} + \overrightarrow{\mathrm{V_2P}_i},$$

according to Eq. (2.6.37), we get

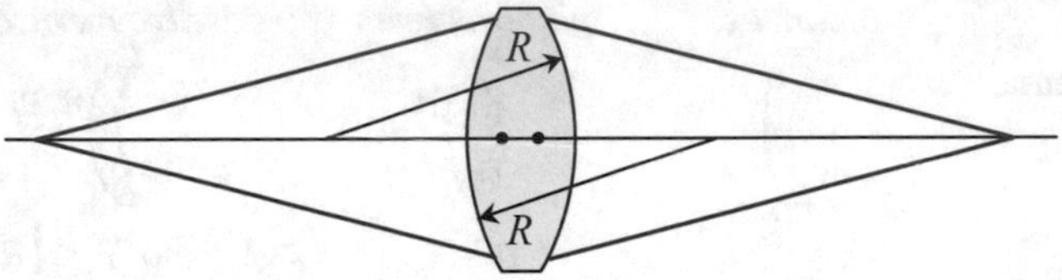

Fig. 2.37 Using symmetric lenses. For $t \ll R$, $f = \frac{1}{2}\frac{R}{n-1}$, $\overrightarrow{P_oP_i} = t\frac{n-1}{n}$

$$\overrightarrow{P_oP_i} = t\left[1 - \frac{R_2 - R_1}{n(R_2 - R_1) + t(n-1)}\right] = t\frac{(n-1)(R_2 - R_1 + t)}{n(R_2 - R_1) + t(n-1)}. \qquad (2.6.38)$$

Symmetric lens. In this case, we have $R_1 = -R_2 = R$ (Fig. 2.37). The position of the principal points can be quickly obtained from Eqs. (2.6.37), but, for a symmetric *thin lens*, with $t \ll R$, one simply has

$$\overrightarrow{V_1P_o} = \overrightarrow{V_2P_i} = \frac{t}{2n}, \quad \overrightarrow{P_oP_i} = \frac{n-1}{n}t. \qquad (2.6.39)$$

Since the refractive index of the most common glasses is ~ 1.5, as a *rule of thumb*, we can remember that:

in a symmetrical lens, the principal planes are located, from the vertices, at about 1/3 of the central thickness of the lens, within the lens itself.

This applies to both converging and diverging lenses.

Lamina with plane and parallel faces. The solution to this case is obtained by bringing $R \to \infty$ in the previous case, and the position of principal planes is still given by Eqs. (2.6.39). The property of these laminae is that each ray that passes through them is only moved parallel to itself. It may seem like a paradox; however, a lamina interposed on a beam of light moves its focal point *forward*, whether it is real or virtual, by an amount equal to the hiatus between the principal planes (Fig. 2.38). As a rule of thumb, in the conditions described above (i.e., for $n \sim 1.5$), we can remember that

Fig. 2.38
Effect of a lamina with plane faces

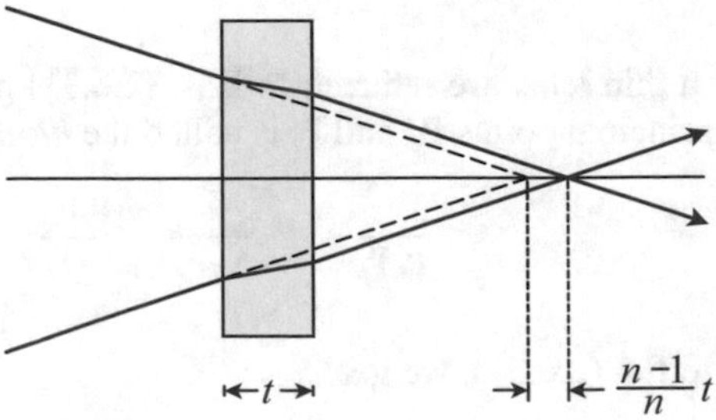

Fig. 2.39
Spherical lens: $f = \dfrac{nR}{2(n-1)}$

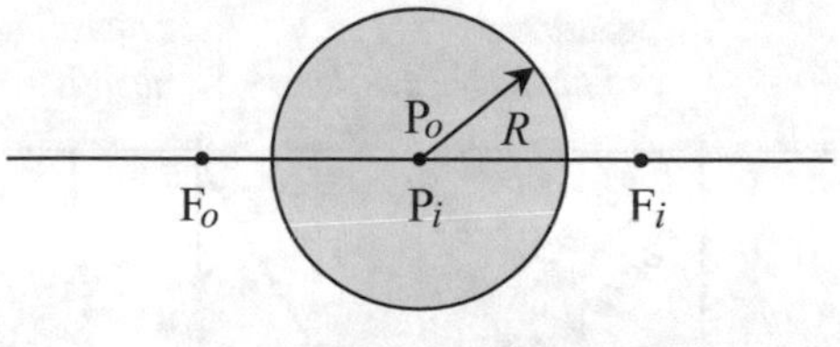

the displacement introduced by a lamina with plane faces is about 1/3 of its thickness.

Please keep this in mind when you wish to focus a laser beam through the window of a cell!

Spherical lens. This is another special case of a thick symmetric lens, so $R_1 = -R_2 = t/2$ (Fig. 2.39). In this case, the points P_o and P_i coincide with the center of the sphere, and this immediately leads one to think that, in the cases considered here, the nodal points coincide with the main ones. The focal distance of a spherical lens is

$$f = \frac{n}{2(n-1)} R. \tag{2.6.40}$$

Plano-convex or plano-concave lens. On the basis of the criterion of "best form" mentioned at the beginning, the symmetrical lenses may be appropriate for conjugated ratios near 1 (Fig. 2.37). Instead, in the case of a conjugate ratio close to zero or ∞, the most appropriate form is near that of a *plano-convex* or *plano-concave* lens, and the curved side should be oriented toward the direction in which the beam is (more) collimated (Fig. 2.40). For these lenses, Eqs. (2.6.29) immediately indicate that one of the principal points coincides with the corresponding vertex on the curved surface, while the other is located inside the lens. The hiatus is

$$\overrightarrow{P_oP_i} = t\frac{n-1}{n}. \tag{2.6.41}$$

Again, for the most common glasses, $\overrightarrow{P_oP_i} \approx (1/3)t$.

Meniscus lenses. These take their name from the Greek μηνισκος, which indicates the Moon shortly before or after the new moon, and were patented by Wol-

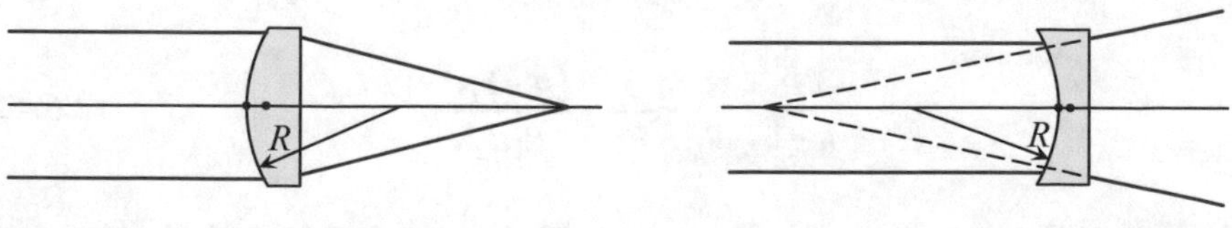

Fig. 2.40 Plano-convex and plano-concave lens. For $t << R$, $f = \dfrac{R}{n-1}$, $\overrightarrow{P_oP_i} = t\dfrac{n-1}{n}$

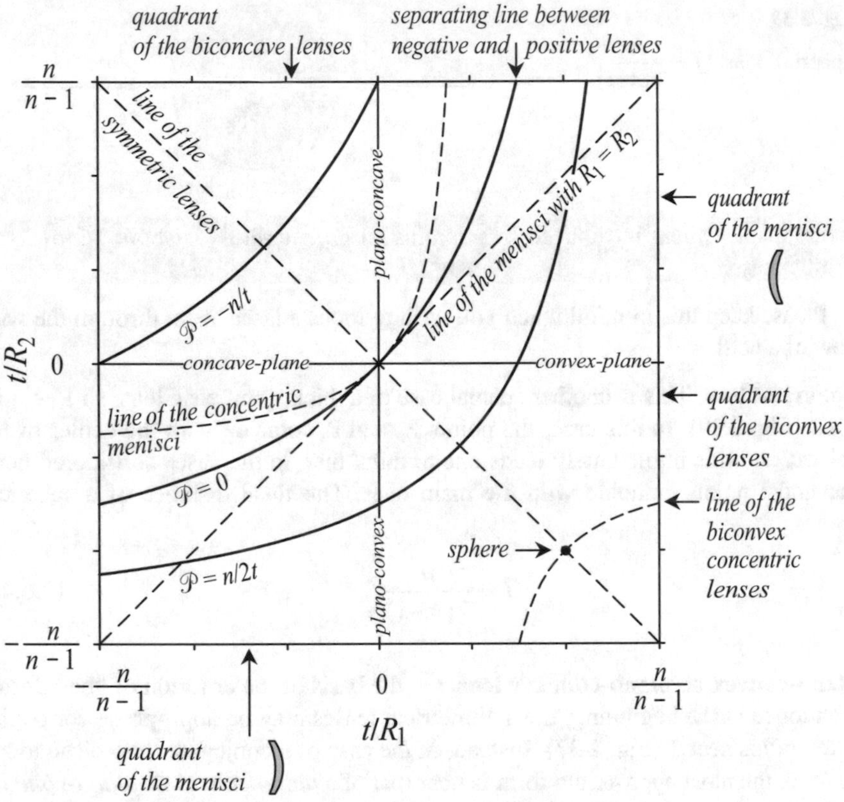

Fig. 2.41 Phase diagram of the lenses (with $n = 1.5$)

laston in 1804 as eyeglass lenses. They allow the eyes to see through them, from the center to the edges, without appreciable distortion. Menisci are generally used, often in combination with other lenses, to collect rays with large angular fields. One or both of the principal points of a meniscus lens are located outside the lens itself, even at great distance. The behavior of a meniscus in relation to these points is quite odd and is described briefly below with the help of Fig. 2.41.

In this figure, the central lines, vertical and horizontal, correspond to the plano-convex and plano-concave lenses, and divide the space of the parameters t/R_1 and t/R_2 into four quadrants, two of menisci, one of a biconcave lens, and one of a biconvex lenses. The solid lines are a hyperbole that solve the equation

$$\frac{t\mathcal{P}}{n-1} = \frac{t}{R_1} - \frac{t}{R_2} + \frac{(n-1)t^2}{R_1 R_2 n} \tag{2.6.42}$$

for three different values of $\mathcal{P}$. The curve with $\mathcal{P} = 0$ coincides with the boundary between converging lenses, at the bottom right, and diverging lenses, in the upper

Fig. 2.42
Conventions in the measurement of distances S_o, S_i from the vertices of a lens

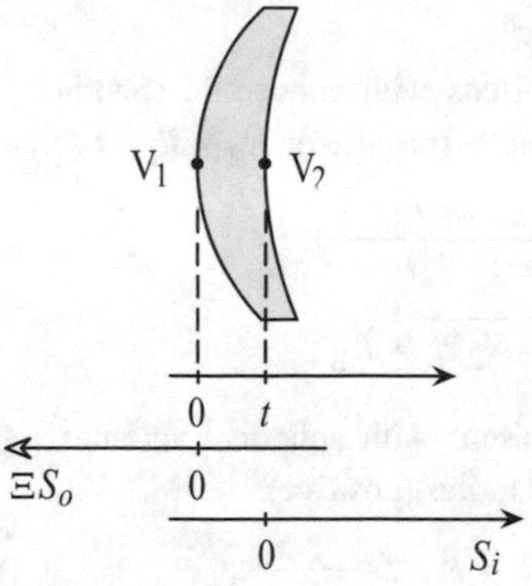

left. On this line, the principal point positions diverge to infinity [Eqs. (2.6.26)], as well as the distance $\overrightarrow{P_oP_i}$.

The menisci of the dotted diagonal line, with $R_1 = R_2$, for Eqs. (2.6.37), have $\overrightarrow{P_oP_i} = t$. The two lines of the concentric menisci and the concentric biconvex lenses solve the equation $R_1 - R_2 = t$. For them, P_o coincides with P_i. In the area between the line of the concentric menisci and the line with $\mathcal{P} = 0$, the point P_o has overtaken the point P_i. The same is true in the lower right corner, below the line of the concentric biconvex lenses. The divergences that we have for $\mathcal{P} \to 0$ in the position of the principal points should not worry us too much. They express only the inadequacy of using them as the origin of the axes for $\mathcal{P}$ in the neighborhood of zero. In fact, if we instead use the vertices V_1 and V_2 of the lens as origins, the distance for the object and the image from these points are (Fig. 2.42)

Fig. 2.43 Aplanatic menisci

(a) positive meniscus:

$$R_2 = R_1\left(1+\frac{1}{n}\right) - t\,,$$

$$f = \frac{R_1}{n-1}\left(n+\frac{R_1}{R_1-t}\right),$$

$$\overrightarrow{V_1P_o} = -t\frac{R_1}{R_1-t}, \quad \overrightarrow{V_1P_i} = -\frac{t}{n}\frac{R_1}{R_1-t}\,;$$

(b) negative meniscus:

$$R_1 = R_2\left(1+\frac{1}{n}\right) + t\,,$$

$$f = -\frac{R_2}{n-1}\left(n+\frac{R_2}{R_2+t}\right),$$

$$\overrightarrow{V_2P_o} = \frac{t}{n}\frac{R_2}{R_2+t}, \quad \overrightarrow{V_2P_i} = t\frac{R_2}{R_2+t}$$

(a)

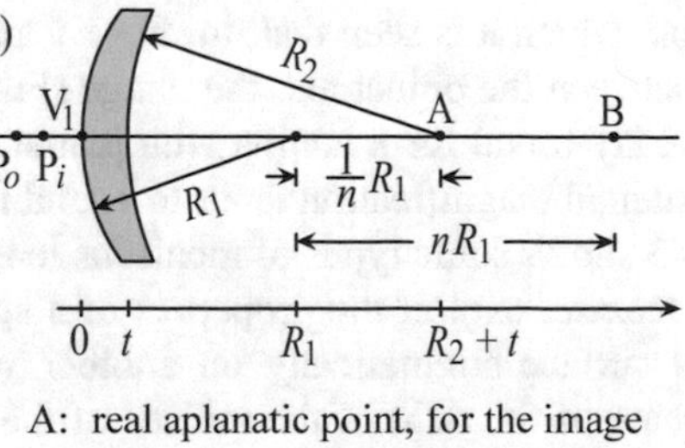

A: real aplanatic point, for the image space.
B: virtual aplanatic point, for the object space.

(b)

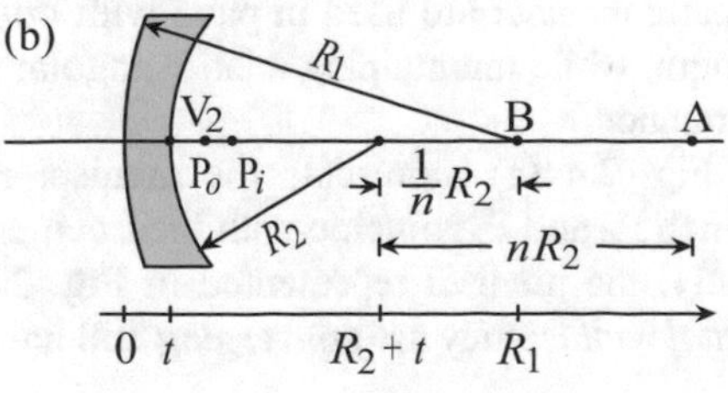

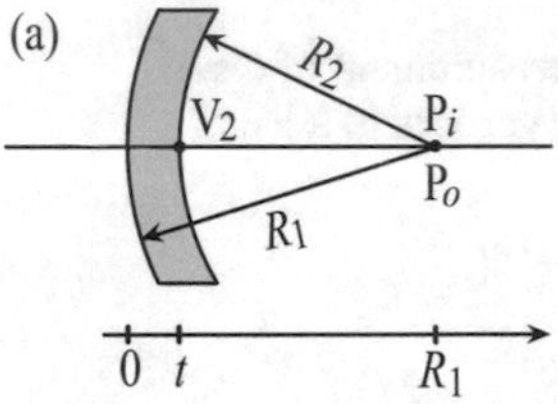

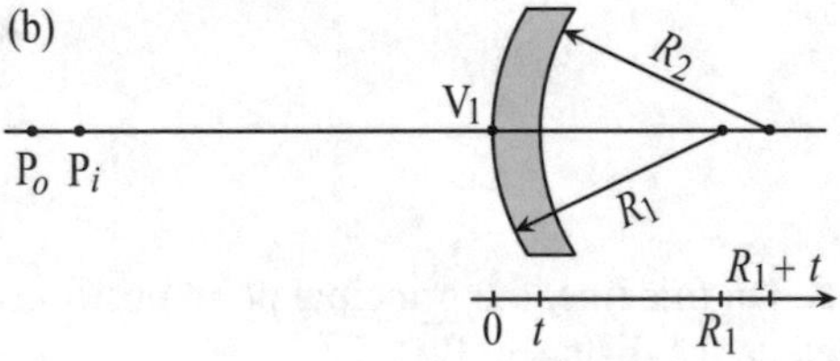

Fig. 2.44
(a) meniscus with concentric spherical surfaces (negative): $R_2 = R_1 - t$,

$$f = -\frac{n}{n-1}\frac{R_1(R_1-t)}{t},$$

$$\overrightarrow{V_2P_o} = \overrightarrow{V_2P_i} = R_2 ;$$

(b) meniscus with spherical surfaces of equal radius (positive):

$$R_2 = R_1 = R\,,\ f = -\frac{n}{(n-1)^2}\frac{R^2}{t},$$

$$\overrightarrow{V_2P_o} = -\frac{R}{n-1},\ \overrightarrow{V_2P_i} = t - \frac{R}{n-1}$$

$$\Xi S_o = \Xi s_o + \overrightarrow{V_1P_o}\ ,\quad S_i = s_i - \overrightarrow{V_1P_i} \tag{2.6.43}$$

and, with a small amount of calculations that we are better off not reporting here, from the Gauss equation, we can derive the relation between these new distances:

$$\Xi S_o + S_i + \frac{t}{n} = \Xi\mathcal{P}S_oS_i + t\frac{n-1}{n}\left(-\frac{\Xi S_o}{R_1} + \frac{S_i}{R_2}\right), \tag{2.6.44}$$

from which it is seen that, for $\mathcal{P} \to 0$ and R_1, $R_2 >> S_o$, S_i, t, the distance $S_i + S_o + t$ between the object and the image (Fig. 2.42) is $t(n-1)/n$, which is equal to that already found for a lamina with planar faces. Under the same conditions, the longitudinal magnification tends to 1, and therefore the lateral one does the same. Fig. 2.43 shows some types of meniscus lens that deserve some attention. The *aplanatic menisci* exploit the properties of a sphere (see §2.4.3) to reproduce one spherical surface stigmatically on another when the conjugated ratio is equal to the square of the ratio of the indices of the external and internal media of the sphere. The meniscus is then a portion of this sphere intersected by a second sphere centered on one of the aplanatic points; so, this point possesses a stigmatic image with an angular field limited only by the size of the lens (Fig. 2.43). Therefore, the aplanatic menisci are used in pairs with doublet or triplet lenses to reduce the focal length, while maintaining a large angular range, without deterioration of their performance.

Fig. 2.44(a) represents the menisci made with two concentric spheres, for which P_o and P_i coincide with their center of curvature; they are *diverging*. I note, lastly, the menisci represented in Fig. 2.44(b), which are made with *spheres of equal radius*; they are *converging* and have a hiatus equal to t.

2.6.8 Optical center of a lens

Let us now look at a particular property of simple lenses. Consider a thick lens consisting of two *spherical dioptres* and immersed in air (Fig. 2.45). From the centers of curvature C_1 and C_2 of its surfaces, we draw two straight parallel lines until they intersect the corresponding surfaces in S_1 and S_2; these lines represent the normals at such points. The segment S_1S_2 intersects the surfaces with equal internal incidence angles ι, since the two normals were taken parallel. Therefore, a ray that coincides with this segment will emerge from the lens with the same external angles of incidence ε on both sides, since the external medium is the same. Hence, altogether, this ray receives a translation from the lenses, but not a deflection. The point of intersection O with the optical axis of the segment S_1S_2 (or its extension the case in which the lens is a meniscus) is called the *optical center* of the lens. This point does not depend on the inclination of the two lines passing through C_1 and C_2. Indeed, the triangles C_1OS_1 and C_2OS_2 are similar to each other, so that the ratio

$$\frac{\overrightarrow{OC_1}}{\overrightarrow{OC_2}} = \frac{R_1}{R_2}, \tag{2.6.45}$$

where R_1 and R_2 are the radii of curvature of the two dioptric surfaces, is constant. This demonstration also applies under non-paraxial conditions. On the other hand, the distance between the centers of curvature is

$$\overrightarrow{C_1C_2} = -R_1 + R_2 + t = -\overrightarrow{OC_1} + \overrightarrow{OC_2},$$

where t is the thickness of the lens. The signs of the variables are taken in accordance with the previous conventions. With some steps, we find that

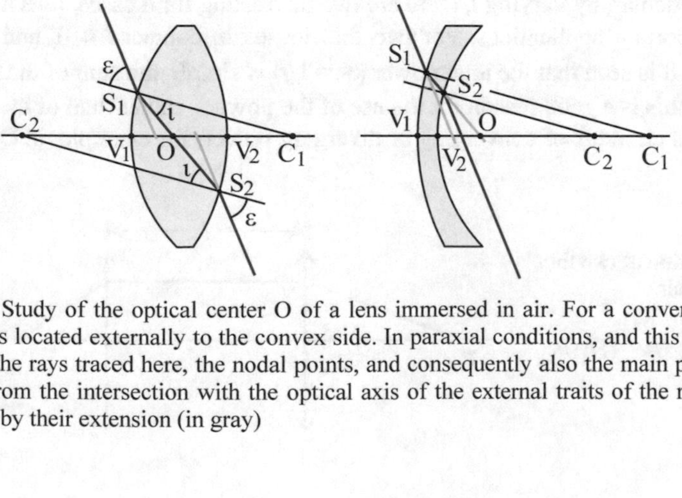

Fig. 2.45 Study of the optical center O of a lens immersed in air. For a converging meniscus, O is located externally to the convex side. In paraxial conditions, and this is not the case with the rays traced here, the nodal points, and consequently also the main points, are obtained from the intersection with the optical axis of the external traits of the ray, either directly or by their extension (in gray)

$$\overrightarrow{V_1O} = \frac{R_1 t}{R_1 - R_2}, \quad \overrightarrow{V_2O} = -\frac{R_2 t}{R_1 - R_2}. \tag{2.6.46}$$

2.6.9 Combination of two thin lenses

Let us now examine the basic but very important case of the combination of two thin lenses. For simplicity, we also assume that they are immersed in air. The solution to this problem is quite similar to that obtained for the combination of two *dioptres*. In fact, if we denote, by f_1 and f_2, the focal lengths of the two lenses and, with l, the distance between them, starting from Eqs. (2.5.18) and in analogy with what was done to derive Eq. (2.6.25), we now have that the focal length of the combination is

$$\frac{1}{f} = \frac{1}{f_1} + \frac{1}{f_2} - \frac{l}{f_1 f_2} \tag{2.6.47}$$

and, from Eqs. (2.5.23), the position of the main planes with respect to the position L_1, L_2 of the two lenses (Fig. 2.46) is

$$\overrightarrow{L_1P_o} = \frac{f}{f_2} l = \frac{f_1}{f_1 + f_2 - l} l, \quad \overrightarrow{L_2P_i} = -\frac{f}{f_1} l = -\frac{f_2}{f_1 + f_2 - l} l. \tag{2.6.48}$$

Eq. (2.6.47) indicates an important thing, namely, that, by varying the distance between the two lenses, we can vary the power of their combination. This finds application, for example, in the telescopic objectives of the cameras, in which the power of the first converging lens, usually of a large opening, is partially compensated by a second diverging lens, thus obtaining a shorter system than is required with a single equal power lens.

In particular, by varying l, there are two interesting limit cases, plus a third case with important applications. For two thin lenses in contact $l \approx 0$, and from Eq. (2.6.47), it is seen that the total power ($\mathcal{P} = 1/f$) is simply the sum of the individual powers; this is a good reason for the use of the powers, rather than of the focal distances, as an index of converging or diverging power, for example, in Ophthalmic Optics.

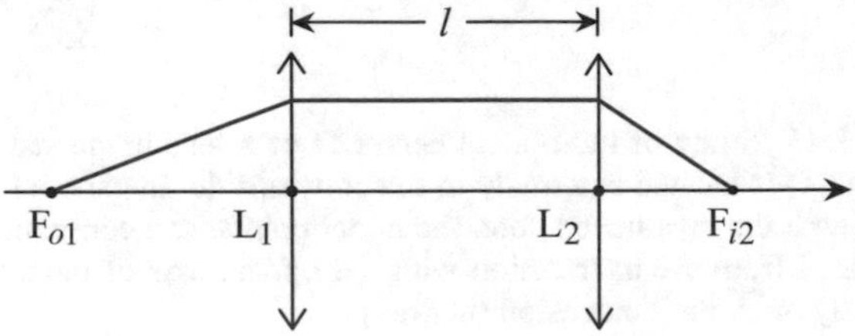

Fig. 2.46
Combination of two thin lenses in air :

$\mathcal{P} = \mathcal{P}_1 + \mathcal{P}_2 - l\,\mathcal{P}_1\mathcal{P}_2$

2.6.9.1 Telescopes

The other interesting situation is for telescopic combinations for which $l = f_1 + f_2$; in this case, the total power is zero and the position of the principal and focal planes, diverging to infinity, loses its practical significance. We have already seen that, for telescopic systems (see §2.5.4-5), the homography takes a simple analytical form, taking, as the origin of coordinates, in the object and image spaces, any two conjugate points, for example, the points F_{o1} and F_{i2} (Fig. 2.46). From Eqs. (2.5.24), we soon get, for two lenses in air,

$$x_i = -\frac{f_2}{f_1}x_o,\quad y_i = -\frac{f_2}{f_1}y_o,\quad Z_i = -\Xi\frac{f_2^2}{f_1^2}Z_o. \tag{2.6.49}$$

I also recall that the telescopic magnification is independent of the position of the conjugate points. In particular, a collimated beam that reaches a telescopic system emerges as still collimated. Thus, if our lenses are *adjusted to infinity*, that is, with $l = f_1 + f_2$, a beam of diameter D, intercepted by the first lens, emerges from the second lens with a diameter $D' = |m_\perp|D = |f_2/f_1|D$, as can immediately be seen from Eqs. (2.5.25.a) and (2.6.49).

Moreover, for Eq. (2.5.26), the angular magnification is

$$m_\angle = -\frac{f_1}{f_2} = \frac{1}{m_\perp}, \tag{2.6.50}$$

and is therefore the inverse of that lateral. This result can also be reached via the optical invariant; it is noteworthy that, for a telescopic system, such as the one discussed here, the principal planes lie at infinity, however, Eq. (2.6.18) is independent of their position, and thus, for each pair of conjugate points, it is applicable without problems. As an example of conjugate points at infinity, let us consider the case of a telescope pointed at a double star; now, what matters is the angular separation at which its two components (say, A and B) are observed and, in particular, the angular magnification of the telescope. For this case, we use the optical invariant with its form in Eq. (2.6.19) and, as reference planes for its application, we take those on which the two lenses of the telescope lie. In particular, we assume that the axis of the telescope is pointed toward the star A. Its rays are therefore parallel to the axis, and, if one of these reaches the first lens with a height h from the axis, it exits the second lens with a height $h' = -(f_2/f_1)h$. Thus, from Eq. (2.6.19), the angular magnification made by the telescope is readily obtained (for a telescope in air):

$$m_\angle = \frac{\phi'}{\phi} = \frac{h}{h'} = -\frac{f_1}{f_2}, \tag{2.6.51}$$

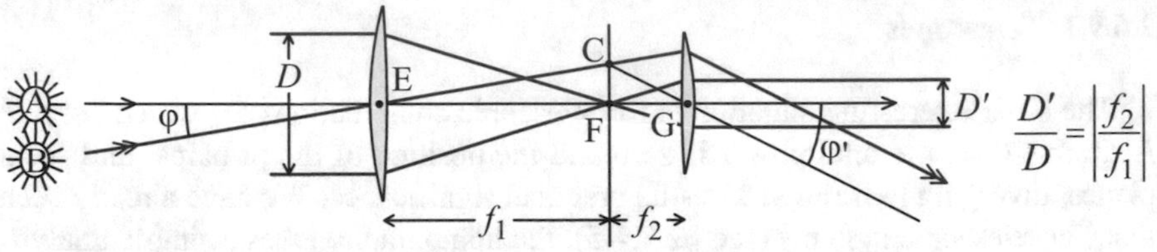

Fig. 2.47 Input and output pupils, D and D', and angular magnification of a telescope. The angular magnification is directly determined by considering the two right-angled triangles EFC and GFC in which the cathetus FC is common and the cathets EF and FC have lengths equal to the focal distances f_1 and f_2 of the two lenses, respectively. The text follows a different reasoning

and this expression coincides with the angular magnification given by Eq. (2.6.50). It is interesting to note here that the angular magnification of the separation between the two stars is precisely the reciprocal of the lateral magnification (the "shrinking") between the diameters of the input and output beams from the telescope (Fig. 2.47).

The relationship between angular and lateral magnification described by Eq. (2.6.18) was obtained above in a somewhat cumbersome way through the optical invariant; however, it is thus demonstrated that it is valid for any telescopic system, with the only variation being to introduce the ratio of the refractive indices of the initial and final spaces if they are not equal.

In visual observation, a telescope is focused by adjusting the distance of the lens so that the image produced by the telescope is positioned at the *remote point of distinct vision* of the observer's eye, which corresponds to the state of relaxation of the muscles of the crystalline lens. For an eye with no visual defects, this point is located at infinity, so the radiation coming from a distant object emerges collimated by the eyepiece and the image focal point of the objective coincides with that object of the eyepiece. Otherwise, for a *myopic* eye, the lenses are approached producing a diverging emerging wave, while, for a *hyperopic* eye, they are moved away and the emerging wave is converging. The configuration of the two lenses of a telescope can also be used to observe even nearby objects as magnified, in a similar way to that which is also created with a compound microscope. In this case, for a healthy eye, the focus corresponds to the position for which the image produced by the objective is located in the object focal plane of the eyepiece, thus the angular magnification is now given by

$$m_\angle = -\frac{s_{i1}}{f_2}, \tag{2.6.52}$$

where s_{i1} replaces f_1 and corresponds to the distance of the image produced by the objective from its image principal plane.

2.6.9.2 Microscopes

Finally, a case similar to that of the telescopes is that of the *compound microscopes*. The difference is only in the positions of the initial object plane and the final image plane, at infinity in telescopes and finite in microscopes, while the intermediate image and object plane still coincide (Fig. 2.48). In a microscope, the objective is constituted by a lens system whose overall focal length is very short. The object plane is located very close to the vertex of the first lens, while the corresponding image plane ends up at a remove from the image focal plane. If we consider the nodal points of the system, we immediately see that a substantial magnification is obtained by leveraging. It is given by the ratio of the distances between the object and image planes with the respective nodal points. The microscope objectives are designed to cancel the aberrations at a specific position of the object plane, and then with a specific magnification m_{ob} that is marked on their barrel [Davidson 2016]. The distance of the image plane from the second focal plane of the objective is usually 160 mm, though systems that are corrected at infinity are currently spreading, requiring an additional lens between the objective and the ocular. The intermediate image is the object for the eyepiece. Unlike the case of telescopes, the eyepiece of a microscope is adjusted to produce the final image at a finite distance, conventionally 25 cm from the eye of the observer. In visual observation, this distance corresponds to the best resolution without tiring the muscle that moves the crystalline lens. Therefore, to an eyepiece with focal f_{oc}, there would be an approximate corresponding magnification of $m_{oc} = 25\text{cm}/f_{oc}$.

The overall magnification of a microscope is then given by $m_{tot} = m_{ob} \times m_{oc}$. This does not mean that the magnification can be increased arbitrarily by sending the final image plane to infinity. Indeed, we must also consider the eye's optical system and the image that is formed on the retina. m_{tot} must rather be understood

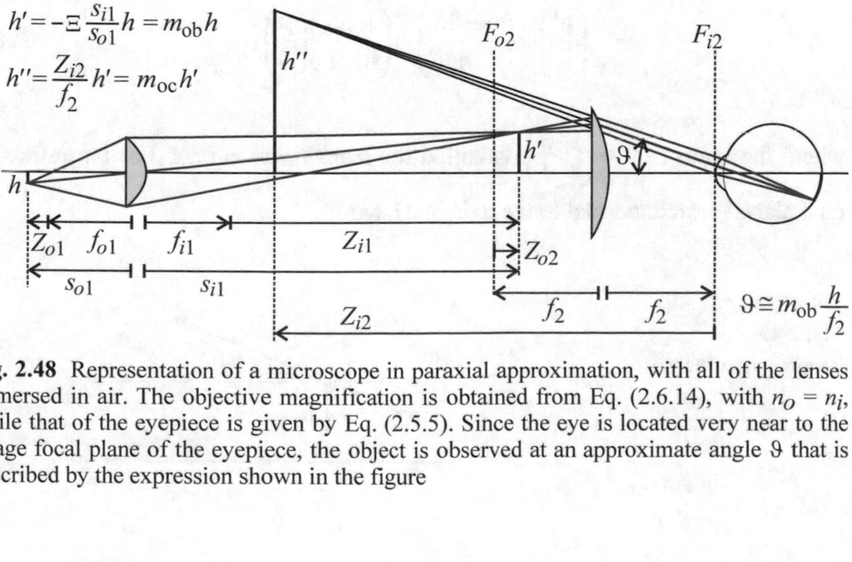

Fig. 2.48 Representation of a microscope in paraxial approximation, with all of the lenses immersed in air. The objective magnification is obtained from Eq. (2.6.14), with $n_o = n_i$, while that of the eyepiece is given by Eq. (2.5.5). Since the eye is located very near to the image focal plane of the eyepiece, the object is observed at an approximate angle ϑ that is described by the expression shown in the figure

as the angular magnification that occurs when comparing the object observed with the naked eye under the best conditions, i.e., conventionally, at a distance of 25 cm, or through the microscope.

2.7 The method of matrices

2.7.1 2x2 matrices

In Paraxial Optics, we can use a simple and elegant method for identifying the behavior of an axially centered system of lenses and/or mirrors. It is based on the propagation of the individual rays and on their deflection at the various surfaces. This method is equivalent to the approximate treatment of Paraxial Optics; in fact, it uses only rays that are paraxial and coplanar with the optical axis. They can be described with a "height" ρ and an angle ϑ with which they intersect a given plane normal to the axis, and these coordinates can be both positive and negative. In addition, $\sin\vartheta$ is still approximated with ϑ. A ray is thus described, for a given z position, by a column vector (Fig. 2.49)

$$\begin{pmatrix} \rho \\ \vartheta \end{pmatrix}. \tag{2.7.1}$$

Its propagation can be described as follows through multiplication by a suitable matrix 2×2. For example, in the case of free propagation to a distance d in a homogeneous medium, we have

$$\begin{pmatrix} \rho' \\ \vartheta' \end{pmatrix} = \begin{pmatrix} \rho + \vartheta d \\ \vartheta \end{pmatrix} = \begin{pmatrix} 1 & d \\ 0 & 1 \end{pmatrix} \begin{pmatrix} \rho \\ \vartheta \end{pmatrix}, \tag{2.7.2}$$

where the matrix $\mathfrak{T} = \begin{pmatrix} 1 & d \\ 0 & 1 \end{pmatrix}$ is called the *translation matrix*. For the refraction on a plane surface, normal to the axis, we have

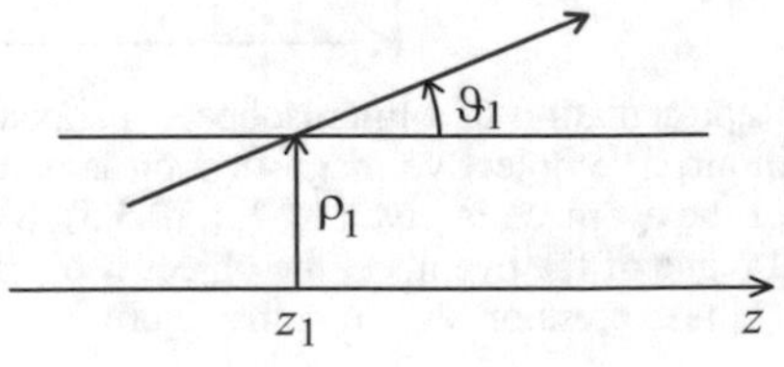

Fig. 2.49
Parameters of a ray

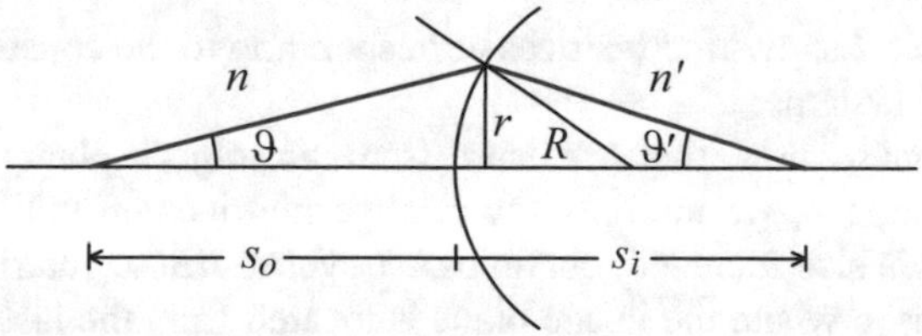

Fig. 2.50
Deflection of a beam through a dioptric surface

$$\begin{pmatrix}\rho'\\ \vartheta'\end{pmatrix}=\begin{pmatrix}\rho\\ \dfrac{n}{n'}\vartheta\end{pmatrix}=\begin{pmatrix}1 & 0\\ 0 & n/n'\end{pmatrix}\begin{pmatrix}\rho\\ \vartheta\end{pmatrix}, \tag{2.7.3}$$

where the vectors are assigned to the same *z*-position (that of the interface), "before" and "after" the refraction. For a spherical *dioptre*, we can use Eq. (2.6.5) with the substitutions $\Xi\rho/s_o = \vartheta$, $\rho/s_i = -\vartheta'$ (Fig. 2.50), obtaining

$$\begin{pmatrix}\rho'\\ \vartheta'\end{pmatrix}=\begin{pmatrix}\rho\\ \dfrac{n-n'}{n'R}\rho+\dfrac{n}{n'}\vartheta\end{pmatrix}=\begin{pmatrix}1 & 0\\ \dfrac{n-n'}{n'R} & \dfrac{n}{n'}\end{pmatrix}\begin{pmatrix}\rho\\ \vartheta\end{pmatrix}. \tag{2.7.4.a}$$

Similarly, for a spherical mirror, from Eqs. (2.6.6) and with the same substitutions, we have

$$\begin{pmatrix}\rho'\\ \vartheta'\end{pmatrix}=\begin{pmatrix}\rho\\ \dfrac{2}{R}\rho+\vartheta\end{pmatrix}=\begin{pmatrix}1 & 0\\ \dfrac{2}{R} & 1\end{pmatrix}\begin{pmatrix}\rho\\ \vartheta\end{pmatrix}, \tag{2.7.4.b}$$

where, in agreement with Eq. (2.7.2), the angles are positive when ρ increases proceeding in the positive direction of the beam propagation, remembering that, in the space, said direction changes at reflection. In accordance with the choice made in Fig. 2.31, *R* is negative for a concave mirror and positive for a convex one.

Thus, in general, as long as the changes induced by the various elements of an optical system can be approximated by linear equations in ρ and ϑ, each element can be described by a matrix 2×2, known as a *transfer matrix*, or *ABCD matrix*. Therefore, for the properties of matrices, in paraxial approximation, an entire optical system can still be described by a 2×2 matrix that is the product, in succession, of the matrices of the individual elements and the various propagations. The resulting *ABCD* matrix expresses the transformation undergone by each ray between any two assigned planes, *not necessarily conjugated*, among which the matrix is calculated, whereby

$$\begin{pmatrix}\rho'\\ \vartheta'\end{pmatrix}=\begin{pmatrix}A & B\\ C & D\end{pmatrix}\begin{pmatrix}\rho\\ \vartheta\end{pmatrix}. \tag{2.7.5}$$

Table 2 shows the matrices corresponding to the constituent elements of simple optical systems.

Suppose now that we start from an object plane: for each element distance/surface, we multiply the relative matrices up until the last surface, whose matrix is also included, obtaining an overall *ABCD* matrix. We can now be interested as to where the image plane is located from the last position reached: we indicate this unknown distance with *d*. The polar coordinates in the image plane of a given initial ray, expressed in the object plane by the vector $(\rho_o, \vartheta_o)^T$, are

$$\begin{pmatrix} \rho_i \\ \vartheta_i \end{pmatrix} = \begin{pmatrix} 1 & d \\ 0 & 1 \end{pmatrix} \begin{pmatrix} A & B \\ C & D \end{pmatrix} \begin{pmatrix} \rho_o \\ \vartheta_o \end{pmatrix} = \begin{pmatrix} A+dC & B+dD \\ C & D \end{pmatrix} \begin{pmatrix} \rho_o \\ \vartheta_o \end{pmatrix}. \tag{2.7.6}$$

Given that, in the image plane, ρ_i must not depend on ϑ_o (all of the rays that start with coordinated ρ_o on the object arrive with coordinate ρ_i on the image), it is necessary that $B + dD = 0$. Therefore, the "missing" distance *d* from the image is

$$d = -\frac{B}{D}. \tag{2.7.7}$$

We will now show that

the determinant of an ABCD matrix is equal to the ratio between the refractive indices n_o and n_i of the initial and final space:

$$\begin{vmatrix} A & B \\ C & D \end{vmatrix} = AD - CB = \frac{n_o}{n_i}. \tag{2.7.8}$$

In fact, once a matrix $\mathfrak{S}$ that describes the behavior of a system between any two planes has been assigned, the overall matrix $\mathfrak{S}_c$ between two conjugate planes is

$$\mathfrak{S}_c = \begin{pmatrix} A_c & B_c \\ C_c & D_c \end{pmatrix} = \begin{pmatrix} 1 & d_i \\ 0 & 1 \end{pmatrix} \begin{pmatrix} A & B \\ C & D \end{pmatrix} \begin{pmatrix} 1 & d_o \\ 0 & 1 \end{pmatrix} = \mathfrak{T}_i \, \mathfrak{S} \, \mathfrak{T}_o, \tag{2.7.9}$$

where d_o and d_i are the missing distances between the conjugated planes and the initial and final planes of the matrix $\mathfrak{S}$, respectively, in the object and image spaces. From algebra, it is known that the determinant of the product of several matrices is equal to the product of the individual determinants; therefore, the determinant of $\mathfrak{S}_c$ is equal to that of $\mathfrak{S}$. On the other hand, as we have seen before, between two conjugate planes, the B_c value is null, for which

$$|\mathfrak{S}| = |\mathfrak{S}_c| = A_c D_c. \tag{2.7.10}$$

If we now consider two rays, characterized by the vectors

<table>
<tr><td>Homogenous medium</td><td>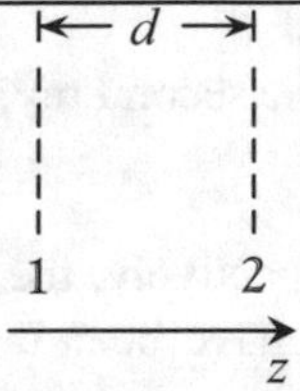</td><td>$$\begin{pmatrix} 1 & d \\ 0 & 1 \end{pmatrix}$$</td><td>Free path between the surfaces 1 and 2 for the distance $d = z_2 - z_1$</td></tr>
<tr><td>Plane surface between dielectric media with indexes n_1 and n_2</td><td>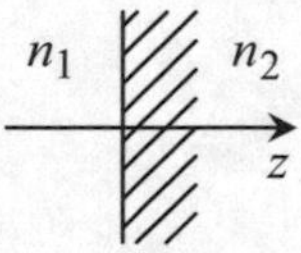</td><td>$$\begin{pmatrix} 1 & 0 \\ 0 & \frac{n_1}{n_2} \end{pmatrix}$$</td><td>Only the crossing of the surface</td></tr>
<tr><td>Plane parallel plate with index n</td><td>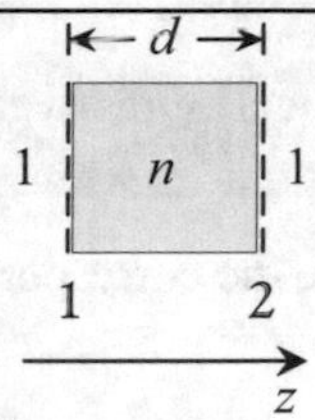</td><td>$$\begin{pmatrix} 1 & \frac{d}{n} \\ 0 & 1 \end{pmatrix}$$</td><td>Path between the two surfaces, 1 and 2, including refractions</td></tr>
<tr><td>Dioptric surface between media with indexes n_1 and n_2, with radius r. If it is convex, as in the figure, $r > 0$; if it is concave, $r < 0$.</td><td>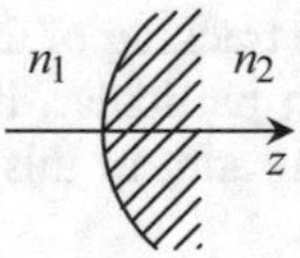</td><td>$$\begin{pmatrix} 1 & 0 \\ \frac{1}{r}\left(\frac{n_1}{n_2} - 1\right) & \frac{n_1}{n_2} \end{pmatrix}$$</td><td>Only the crossing of the surface</td></tr>
<tr><td>Lens with focal length f between two identical media (also applies to a mirror with radius r, taking $f = -r/2$)</td><td>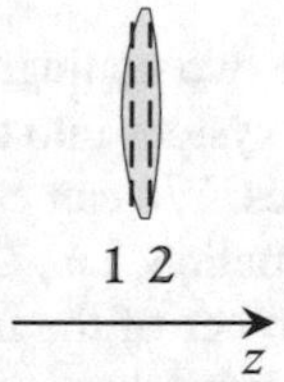</td><td>$$\begin{pmatrix} 1 & 0 \\ -\frac{1}{f} & 1 \end{pmatrix}$$</td><td>Path between the principal planes, 1 and 2, including refractions</td></tr>
<tr><td>Optical fiber or GRIN lens with a radially graded index with parabolic trend</td><td>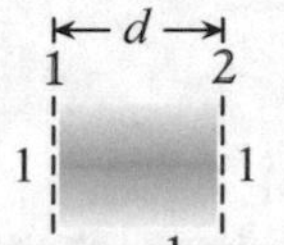$n = n_0 - \frac{1}{2} n_2 r^2$</td><td>$$\phi = d\sqrt{n_2/n_0}$$ $$\begin{pmatrix} \cos\phi & \frac{\sin\phi}{\sqrt{n_2 n_0}} \\ -\sqrt{n_2 n_0}\,\sin\phi & \cos\phi \end{pmatrix}$$</td><td>Path between the two surfaces, 1 and 2, including refractions</td></tr>
</table>

Table 2.2 *ABCD* matrix for some elementary optical systems

$$\begin{pmatrix} \rho_o \\ 0 \end{pmatrix}, \text{ the first, and } \begin{pmatrix} 0 \\ \vartheta_o \end{pmatrix}, \text{ the second ray,}$$

we immediately realize that A_c and D_c are, respectively, the lateral magnification $m_\perp$ and the angular magnification $m_\angle$ as they have been defined in §2.6.5, and, from Eq. (2.6.18), we have

$$|\mathfrak{S}| = m_\perp m_\angle = \frac{n_o}{n_i}, \tag{2.7.11}$$

which is what we wanted to demonstrate. In particular,

the matrices calculated between spaces of equal refractive index have a unitary determinant.

In conclusion, between two conjugate planes, the corresponding matrix is

$$\begin{pmatrix} m_\perp & 0 \\ -1/f_i & m_\angle \end{pmatrix}, \tag{2.7.12}$$

where the value of C_c is easily evinced from the tracking of the first ray mentioned above. The expression (2.7.12) also applies, in particular, between the principal planes of an optical system, for which $m_\perp = 1$, and, in this case, the matrix becomes

$$\begin{pmatrix} 1 & 0 \\ -1/f_i & n_o/n_i \end{pmatrix}. \tag{2.7.13}$$

In addition to generating the matrix between two conjugate planes, Eq. (2.7.9) can also be used to transform the matrix of the system into another pair of planes taken into homogeneous object and image spaces. We can easily note that the coefficient C remains unchanged by this transformation, i.e., $C = C_c$, so its value is always equal to $-1/f_i = -\mathcal{P}/n_i$, where $\mathcal{P}$ is the power of the instrument. In particular, C is zero in telescopic systems. Ultimately remarkable is the matrix between the focal planes, which is

$$\begin{pmatrix} 0 & \Xi f_o \\ -1/f_i & 0 \end{pmatrix}, \tag{2.7.14}$$

whereby the position and inclination of a ray exchange their roles. In fact, the position (inclination) on the image focal plane can only depend on the inclination (position) of the ray on the object focal plane, for which the diagonal elements of the matrix between these two planes are null. For an *ABCD* matrix calculated between any two planes, with $C \neq 0$, this allows us to calculate the distances $d_{\mathrm{F}o}$ and

Table 2.3
Relation between the parameters of the matrix and the position of the focal and principal points, with respect to the points L_o and L_i, which are the intersection of the optical axis with the initial and final planes, respectively, between which the matrix is calculated

$\overrightarrow{F_oL_o} = -\dfrac{D}{C}$	$\overrightarrow{P_oL_o} = -\dfrac{D}{C} + \dfrac{n_o}{n_i}\dfrac{1}{C}$
$\overrightarrow{L_iF_i} = -\dfrac{A}{C}$	$\overrightarrow{L_iP_i} = -\dfrac{A}{C} - \dfrac{1}{C}$

d_{Fi} missing from the object and image focal planes, respectively:

$$d_{Fo} = -\frac{D}{C}, \quad d_{Fi} = -\frac{A}{C}. \tag{2.7.15}$$

Table 2.3 summarizes the relationship between the homography of a non-telescopic optical system and the parameters of its *ABCD* matrix.

Eq. (2.7.5) can be inverted to obtain the vector of the initial radius from the knowledge of the final one, for which

$$\begin{pmatrix} \rho \\ \vartheta \end{pmatrix} = \frac{n_i}{n_o}\begin{pmatrix} D & -B \\ -C & A \end{pmatrix}\begin{pmatrix} \rho' \\ \vartheta' \end{pmatrix}, \tag{2.7.16}$$

where the ratio n_i/n_o is the inverse of the determinant of the original ABCD matrix. The minus sign that appears in front of the B and C coefficients may be surprising. Indeed, the power of the system remains the same, apart from the ratio between the indexes, either proceeding from left or right. On the other hand, in the reverse path, i.e., reversing the propagation of the rays, the ϑ inclinations change sign, while, in Eq. (2.7.16), we kept the forward direction of propagation of both the input and output vectors from the system.

In other words, for a system represented by the *ABCD* matrix of Eq. (2.7.5), the matrix of the inverted system, calculated by multiplying, in reverse order (as compared to the non-inverted system), the matrices of the individual elements (which are also reversed), is given by

$$\frac{n_i}{n_o}\begin{pmatrix} D & B \\ C & A \end{pmatrix}. \tag{2.7.17}$$

2.7.2 Generalized tracing of rays by means of 4x4 matrices in the absence of axial symmetry

So far, we have only taken into account: a) optical systems with axial rotational symmetry, b) rays slightly inclined to the optical axis and c) small angles of incidence on the optical surfaces. However, it is also possible to extend the paraxial treatment to asymmetric optical systems, such as those that contain cylindrical lenses that are oriented anyway and to cases when the angles of incidence are "large", as normally happens with the use of prisms. Such an extension is possible in the case of thin *pencils* of rays, as happens, for example, with a laser beam in an optical ring cavity, even when the systems are not planar. In these cases, instead of the optical axis of the system, we take, as the axis, the *central* (or *principal*) ray of the pencil. Such an axis is a broken line composed of several segments joined together in correspondence with the optical surfaces. Then, condition b) becomes that the rays of the pencil make small angles with such an axis, and condition c) becomes that the *variations* of the angle of incidence are small. For the validity of the paraxial approximation, it is still necessary that the curvature of the surfaces be continuous in the area covered by the pencil of rays on each of them.

An analytical treatment of such a system quickly becomes very complicated, because of the presence of a large number of parameters. We might as well then construct an algorithm that considers one optical element at a time and that can be implemented in a calculation program. The matrix method is well suited for this purpose, and ultimately allows for describing the whole system with 4x4 numeric values, once all of the necessary parameters are assigned. On the other hand, the procedure that we will pursue is essentially a compact way to follow the reasoning of the *generalized ray tracing*, introduced by Allvar Gullstrand in the early 1900s [Gullstrand 1906]. This method facilitates, in addition to the tracking of the central beam, the calculation of the shape of the wavefront in a small neighborhood of the same ray and, in particular, its principal radii of curvature. In this way, we can easily determine the shape of the caustic near the image plane and the properties of a Gaussian beam leaving the optical system, as we shall see later.

Meanwhile, it is necessary to define a vector that exhaustively describes any ray of the pencil that, in an optical system devoid of symmetry, is generally astigmatic [Born and Wolf 1980, §4.6]. Thus, as such a vector was earlier defined by small deviations of position and direction from the optical axis of the system, these deviations must now refer to the central ray, assuming it to be a broken axis. At each of its segments, we can associate a Cartesian reference frame $\Re$ in which the ζ-axis is oriented along the same segment. There are no stringent conditions for the orientation of the ξ and η-axes in the transverse plane to the segment; nevertheless, it is worthwhile to take the ξ–axes as all being parallel to an assigned plane, say, the horizontal one of the bench on which the elements of the optic system are placed. Only if the broken axis completely lies on this horizontal plane are the η-axes parallel to each other and all vertical. A generic ray of the pencil is then describable by a vector of four elements whose transpose is defined as

$$\left(\rho_\xi, \rho_\eta, \vartheta_\xi, \vartheta_\eta\right), \tag{2.7.18}$$

where the variables ρ and ϑ of the previous paragraph are decomposed into their coordinates on the $\zeta\xi$ and $\eta\zeta$–planes of the reference frame chosen for each segment: the pairs ρ_j and ϑ_j (with j=ξ,η) describe the projections of the ray on these planes. The propagation from a transverse plane to another along the broken axis can still be calculated by applying a matrix to the vector, in this case, one with 4x4 elements by necessity, which represents the system between those two planes [Arnaud 1970]. However, it should be kept in mind that, in the propagation of such a vector between one segment and the subsequent one, there are three matrices involved in succession. The first is a rotation matrix on the $\xi\eta$-plane of the frame of the first segment by the angle φ_i between the η-axis and the normal **S** to the incidence plane on the involved surface. The second is the matrix representing the surface, calculated for the angle of incidence of the principal ray. Finally, the third is a rotation matrix in the frame of the second segment, similar to the preceding, but with an inverse rotation. The overall matrix of the system will then be the product of all of these matrices in succession, one surface after the other.

2.7.3 Algorithm for the calculation of the matrix

Now, we can finally find an algorithm to determine the matrix of the system and, in particular, we will use the ray tracing techniques. Concerning these methods, there is a significant amount of literature [Feder 1951, 1963; Spencer and Murty 1962; Collins 1970] and there are many computer programs; the most comprehensive of these is the CODE V. Here, I mean only to summarize succinctly, in a list, the concepts that are used for the calculation of our matrix, in addition to providing a theoretical basis for the ray tracing concepts that I have used in the attached source program in the extra material for this book.

In our analysis, I will consider four different types of reference frame for the coordinates: (i) the *laboratory frame* $\mathfrak{L}$, which is unique, for which lowercase characters will herein be used; (ii) the *central ray frames* $\mathfrak{R}_m$, each associated with its segment *m*, in which the coordinates are expressed by Greek letters and which are, from time to time, thereafter rotated around the direction of propagation so as to agree with the planes of incidence at the surfaces; (iii) the *frames* $\mathfrak{S}_m$ *of each surface m*, in which the variables are expressed in uppercase characters; (iv) finally, the *local frames* ℓ_m centered on the point of intersection of the central ray with the surface *m*, which will be needed only for the explanation of the problem. The index *m* affixed to the versors will indicate that they are specifically associated with the central ray, but, in some variables, such as the ratio between the refractive indices and angles of incidence, it will be omitted for simplicity. Lastly, the various vectors or operators, indicated with bold characters, are to be understood in an "absolute" sense, while their coordinates are expressed according to the chosen frame of reference.

Finally, I admit that, in the literature, there are other ways to refer to the variables and define the reference systems; perhaps the most important is that of O. N. Stavroudis (1972).

a) System definition. First of all, it is necessary to define the optical system by assigning the parameters of all of the M optical surfaces and the refractive indices n_m of the interposed media. This can be done by expressing, in Cartesian coordinates, the vertices $\mathbf{v}$ of the surfaces and the versors $\mathbf{N}_v$ normal to the surfaces on their vertices in the system $\mathfrak{L}$. Each surface can also be defined in its own reference frame $\mathfrak{S}_m$, which simplifies its analytical form, and, in particular, the assignment of the parameters of its curvature. Such a frame can be defined by assigning its Euler angles with respect to the frame $\mathfrak{L}$.

In particular, let us consider only surfaces defined by a *continuous* function, which is *unique* and *continuously differentiable* in its X and Y variables:

$$Z = Z(X,Y), \tag{2.7.19}$$

where X, Y, Z are the coordinates in the frame $\mathfrak{S}_m$ of our surface.

b) Coordinate transformations. We must now define two types of transformation. The first concerns the rotation, around a common origin, between the laboratory frame $\mathfrak{L}$ and the frame $\mathfrak{R}$ associated with each segment of the central ray, whose versor is $\mathbf{R}$. To respond to the constraints described above, once we have assigned the components r_x, r_y, r_z of $\mathbf{R}$ in the frame $\mathfrak{L}$, we define the unitary transformation (when $\mathbf{R}$ does not coincide with the y-axis) as

$$\mathfrak{R}_r = \begin{pmatrix} r_z/r_{zx} & 0 & -r_x/r_{zx} \\ -r_y r_x/r_{zx} & r_{zx} & -r_y r_z/r_{zx} \\ r_x & r_y & r_z \end{pmatrix}, \quad \text{where } r_{zx} = \sqrt{1-r_y^2}, \tag{2.7.20}$$

from the laboratory frame $\mathfrak{L}$, with coordinates x, y, z to the frame $\mathfrak{R}$, with coordinates ξ, η, ζ. The inverse transformation is simply its transpose. One can easily verify that the direction of the versor $\mathbf{R}$ comes to coincide with the ζ-axis and that the ξ-axis lies on the zx-plane, and is thus perpendicular to the y-axis, responding to the aforementioned choice in the previous §2.7.2 for that axis. A further rotation around $\mathbf{R}$ is then necessary for the calculation of the refraction and reflection, as stated at the end of §2.7.2. The case when $\mathbf{R}$ coincides or is next to the y-axis should be treated separately, for example, by rotating the z and y-axes around the x-axis by 90° and redefining the coordinates of $\mathbf{R}$. In this case, the transformation is

$$\mathfrak{R}'_r = \begin{pmatrix} r_y/r_{yx} & -r_x/r_{yx} & 0 \\ r_z r_x/r_{yx} & r_y r_z/r_{yx} & -r_{yx} \\ r_x & r_y & r_z \end{pmatrix}, \quad \text{where } r_{yx} = \sqrt{1-r_z^2}. \tag{2.7.20.bis}$$

A similar rotation can also be used in the transformations between the frames $\mathfrak{L}$

and $\mathbb{S}_m$, taking, in place of **R**, the versor $\mathbf{N}_\mathrm{v}$ normal to the surface at its vertex. To complete the rotation from $\mathbb{L}$ to $\mathbb{S}_m$, it will suffice to perform a further rotation around the *Z*-axis along which $\mathbf{N}_\mathrm{v}$ lies.

In both cases, it will generally be necessary to perform translations, before and after the rotation.

c) Initial conditions. It is necessary to choose the *central ray* and draw it through the optical system. Then, we fix the initial position $\boldsymbol{c}_0$ and direction $\mathbf{R}_0$ of the ray in the frame $\mathbb{L}$.

d) Propagation of the central ray. For each surface *m* of the system, there are a starting point $\boldsymbol{c}_{m-1}$ and the versor $\mathbf{I}_m=\mathbf{R}_{m-1}$ of the incident ray, and we have to calculate the point of intersection $\boldsymbol{c}_m$, and the versor $\mathbf{R}_m$ of the refracted or reflected ray. To simplify calculations, it is appropriate to use the transformation of these initial quantities in the frame $\mathbb{S}_m$ of the surface and calculate the coordinates *X*, *Y*, *Z* of the intersection point in this frame. For their determination, an iterative procedure is generally required, while there are analytical solutions for the classical surfaces, such as those of Eq. (2.4.2). In addition, we must verify that the intersecting point of the ray with the surface is the first among the possible ones. With an inverse transformation towards the frame $\mathbb{L}$, we get the coordinates of the new starting point $\boldsymbol{c}_m$.

e) Incidence versors. Differentiating Eq. (2.7.19), we can determine the versor $\mathbf{N}_m$ normal to the surface in $\boldsymbol{c}_m$:

$$dZ - Z_{;X}dX - Z_{;Y}dY = 0\,, \tag{2.7.21}$$

where the coordinates are those in the frame $\mathbb{S}_m$ and, for brevity, I indicate, in subscripts (after the ";" to distinguish them from the normal indication of coordinates), the partial derivatives of the function *Z*. Eq. (2.7.21) can be interpreted as the scalar product of an infinitesimal vector $(dX, dY, dZ)^\mathrm{T}$ lying on the surface and a normal vector to this last one. The versor normal to the surface at the coordinates *X* and *Y* is then expressed in $\mathbb{S}$ by

$$\mathbf{N} = \frac{1}{\sqrt{r}}\begin{pmatrix} -Z_{;X} \\ -Z_{;Y} \\ 1 \end{pmatrix}_{\mathbb{S}}, \quad \text{where} \quad r = Z_{;X}{}^2 + Z_{;Y}{}^2 + 1\,. \tag{2.7.22}$$

His positive direction is arbitrary: here, it is taken as concordant with the positive direction of *Z*. Therefore, by calculating **N** at the coordinates *X*, *Y* of $\boldsymbol{c}_m$, we get $\mathbf{N}_m$.

There are various ways to calculate the reflection and refraction; one possibility is as follows.[14] The cosine of the angle of incidence is given by

---

[14] One different method is, for example, that indicated by Donald P. Feder [Feder 1968].

$$\cos\vartheta_{\mathrm{I}} = \mathbf{I}_m \cdot \mathbf{N}_m\,. \tag{2.7.23}$$

Its sign depends on the concordance in the direction between these two versors. Furthermore, the projection of $\mathbf{I}_m$ on the tangent plane to the surface in $\boldsymbol{c}_m$ is

$$\boldsymbol{p} = \mathbf{I}_m - \cos\vartheta_I \mathbf{N}_m\,, \tag{2.7.24}$$

whose modulus corresponds to the sine of the angle of incidence:

$$\sin\vartheta_{\mathrm{I}} = \operatorname{sgn}\left(\cos\vartheta_{\mathrm{I}}\right)\sqrt{1-\cos^2\vartheta_{\mathrm{I}}} = \operatorname{sgn}\left(\cos\vartheta_{\mathrm{I}}\right)\left|\boldsymbol{p}\right|, \tag{2.7.25}$$

where the sign function sgn is introduced for convenience, as we will see later. Apart from the degenerate case of normal incidence that must be treated properly, the *incidence plane* I_m is defined, on which the versors $\mathbf{I}_m$ and $\mathbf{N}_m$ lie. Normalizing the vector $\boldsymbol{p}$, we get the versor $\mathbf{P}_m$ that is parallel to the surface and to I_m:

$$\mathbf{P}_m = \frac{\boldsymbol{p}}{\sin\vartheta_{\mathrm{I}}}\,. \tag{2.7.26}$$

Therefore, the incidence versor can be decomposed as

$$\mathbf{I}_m = \sin\vartheta_{\mathrm{I}}\,\mathbf{P}_m + \cos\vartheta_{\mathrm{I}}\,\mathbf{N}_m \tag{2.7.27}$$

and the form of this equation does not depend on the positive direction of $\mathbf{N}_m$. Finally, by the product

$$\mathbf{N}_m \times \mathbf{P}_m = \mathbf{S}_m\,, \quad \text{from which it follows that} \quad \mathbf{N}_m = \mathbf{P}_m \times \mathbf{S}_m\,, \tag{2.7.28}$$

we get the *sagittal versor*, which is perpendicular to I_m. The definition of $\sin\vartheta_{\mathrm{I}}$ in Eq. (2.7.25) is such that the positive direction of $\mathbf{S}_m$ is independent from that of $\mathbf{N}_m$, while that of $\mathbf{P}_m$ depends on it. In this way, only the local frame ℓ is affected by the choice of the sign of $\mathbf{N}$.[15]

f) Versors of refraction and reflection. We now determine the versors $\mathbf{R}'_m$ and $\mathbf{R}''_m$ of the refracted and reflected rays, respectively. In the case of *refraction*, from Snell's law, we obtain the following procedure:

[15] I note here that I have chosen a different convention than that present in the literature, in which $\mathbf{S}$ is instead obtained from the normalized vector product $\mathbf{I} \times \mathbf{N} \,/\, |\mathbf{I} \times \mathbf{N}|$ for which the direction of $\mathbf{S}$ depends on that of $\mathbf{N}$, while $\mathbf{P}$ does not depend on it. This dependence is usually remedied by choosing $\mathbf{N}$ such that its angle with $\mathbf{I}$ is acute.

$$\mu = \frac{n_{m-1}}{n_m}, \tag{2.7.29}$$

$$\begin{cases} \sin\vartheta_{R'} = \mu \sin\vartheta_I, \\ \cos\vartheta_{R'} = \mathbf{R'}_m \cdot \mathbf{N}_m = \mathrm{sign}(\cos\vartheta_I)\sqrt{1-\sin^2\vartheta_{R'}}, \end{cases} \tag{2.7.30}$$

$$\mathbf{R'}_m = \sin\vartheta_{R'}\mathbf{P}_m + \cos\vartheta_{R'}\mathbf{N}_m, \tag{2.7.31}$$

where $\vartheta_{R'}$ is the angle of refraction and the sgn function makes the calculation of $\mathbf{R'}_m$ independent from the positive direction of $\mathbf{N}_m$. In the case of *reflection*, instead of Eqs. (2.7.30), we simply have

$$\sin\vartheta_{R''} = \sin\vartheta_I, \quad \cos\vartheta_{R''} = -\cos\vartheta_I, \tag{2.7.32}$$

formally equivalent to taking $\vartheta_{R''} = \pi - \vartheta_I$, and the versor $\mathbf{R''}_m$ of the reflected beam is obtained as in Eq. (2.7.31):

$$\mathbf{R''}_m = \sin\vartheta_{R''}\mathbf{P}_m + \cos\vartheta_{R''}\mathbf{N}_m. \tag{2.7.33}$$

g) Local reference frame ℓ_m. For the surface m, the set of three mutually orthogonal versors $\mathbf{P}_m$, $\mathbf{S}_m$, $\mathbf{N}_m$ provides a convenient basis for defining the *local reference frame* ℓ_m centered in $\boldsymbol{c}_m$ and the variables of reflection or refraction. Here, a change of the sign of the versor $\mathbf{N}_m$ corresponds to a rotation of π radians around the axis of $\mathbf{S}_m$,[16] as shown in Fig. 2.51(a-b). In particular, the plane defined by the versors $\mathbf{P}_m$ and $\mathbf{S}_m$ has the same role as planes normal to the optical axis and tangent to the optical surfaces on their vertices in systems with axial symmetry. In fact, in the paraxial approximation, an incident ray and the corresponding reflected or refracted ray intersect this plane at the same point, with coordinates ρ_P and ρ_S along the axes defined by $\mathbf{P}_m$ and $\mathbf{S}_m$, respectively. Besides, the direction changes in accordance with a development at the first order of the refraction or reflection laws, in the variation of the angle of incidence from that of the central ray.

h) Tangential versors and tangential coordinates. Around each surface of reflection or refraction, consider two orthogonal planes between them: the *tangential plane* and *the sagittal plane*. The tangential plane coincides with the plane of incidence, which contains the central ray and is generated by the versors $\mathbf{N}_m$ and $\mathbf{P}_m$. Even the sagittal "plane" contains the central ray and is then broken in coincidence with the reflection or refraction. The central ray is the only one that belongs to both.

A further versor to which we must refer and which lies on the tangential plane

[16] Contrary to what is generally assumed in the literature, in which the rotation is taken to be around the axis $\mathbf{P}_m$.

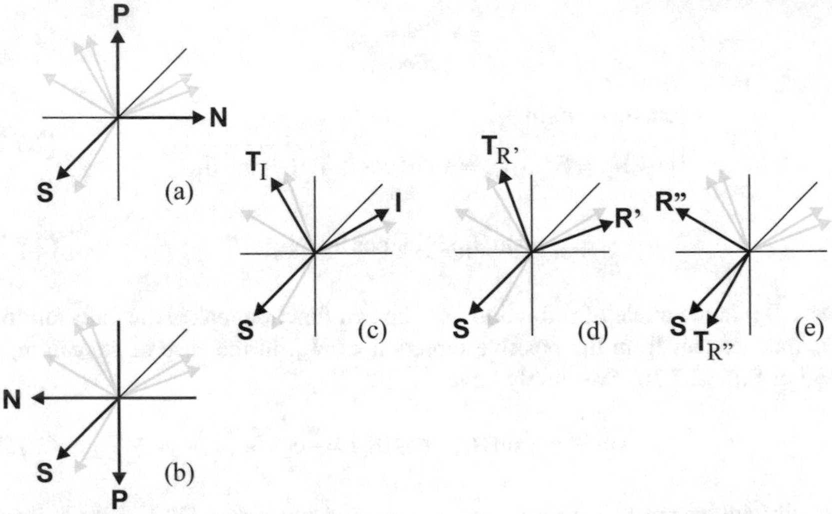

Fig. 2.51 Axonometric representation of the reference frames. The figures show the various frames implied, in which the corresponding versor set is highlighted in black with respect to the others (in gray) (the index *m* is implied here). Versor **S** belongs to all of the sets and is perpendicular to the incidence plane, while all of other versors lie on such a plane. (a): local frame concordant with the incident ray, (b): local frame discordant, (c): frame of the incident ray, (d): frame of the refracted ray, (e): frame of the reflected ray

is the *tangential versor*

$$\mathbf{T}_{m\mathrm{I}} = \mathbf{S}_m \times \mathbf{I}_m. \tag{2.7.34}$$

Fig. 2.51(c) shows the set $\mathbf{T}_{m\mathrm{I}}$, $\mathbf{S}_m$, $\mathbf{I}_m$, which constitutes another reference frame, and, in particular, it is along the direction of $\mathbf{T}_{m\mathrm{I}}$ that we will measure the tangential coordinate ρ_{TI} of the incident rays from the origin c_m. Similarly, the tangential coordinates $\rho_{\mathrm{TR}'}$ and $\rho_{\mathrm{TR}''}$ of the refracted and reflected rays will be respectively measured along the direction of the versors

$$\mathbf{T}_{m\mathrm{R}'} = \mathbf{S}_m \times \mathbf{R}'_m\,, \quad \mathbf{T}_{m\mathrm{R}''} = \mathbf{S}_m \times \mathbf{R}''_m\,. \tag{2.7.35}$$

It is important to note that, in the case of reflection, $\mathbf{T}_{m\mathrm{I}}$ and $\mathbf{T}_{m\mathrm{R}''}$ are oriented specularly between them relative to the plane generated by the versors $\mathbf{S}_m$ and $\mathbf{N}_m$. With these other two tangential versors, we also have the sets $\mathbf{T}_{m\mathrm{R}'}$, $\mathbf{S}_m$, $\mathbf{R}'_m$ and $\mathbf{T}_{m\mathrm{R}''}$, $\mathbf{S}_m$, $\mathbf{R}''_m$, which are the bases of the two reference frames shown in Fig. 2.51(d-e), respectively, for the refracted and reflected rays. For the definition of **S** chosen here, the orientation of these three reference frames does not depend on the direction of the versor $\mathbf{N}_m$.

i) Parameters of a generic ray. Let us now consider a generic incident ray on the surface and the corresponding refracted and reflected rays. In the respective frames, they are represented by the vectors

$$(\rho_{TI},\rho_{SI},\vartheta_{TI},\vartheta_{SI})^T,\ (\rho_{TR'},\rho_{SR'},\vartheta_{TR'},\vartheta_{SR'})^T,\ (\rho_{TR''},\rho_{SR''},\vartheta_{TR''},\vartheta_{SR''})^T$$

in correspondence with the common origin of the systems themselves. The three rays must intersect at the same point on the surface, with coordinates ρ_P and ρ_S of the local frame along the versors **P** and **S**. At the first order of approximation in ρ_P and ρ_S, we can assume that

$$\begin{aligned}&\rho_{TI}=\rho_P\cos\vartheta_I,\quad \rho_{TR'}=\rho_P\cos\vartheta_{R'},\quad \rho_{TR''}=\rho_P\cos\vartheta_{R''},\\&\rho_{SI}=\rho_{SR'}=\rho_{SR''}=\rho_S,\end{aligned} \tag{2.7.36}$$

where the sign of ρ_{TI}, $\rho_{TR'}$ and $\rho_{TR''}$ is independent of the sign of $\mathbf{N}_m$. Regarding the inclination components ϑ of the rays, we can note that all of the tangential components correspond to rotations around the same sagittal axis **S**, and therefore they are equal in both the ray frames and the local one. Instead, the sagittal angular components correspond to rotations around the respective tangential axes **T**. However, they also all correspond to changes of direction of the versors of the rays in the direction of **S**. In conclusion, in the local frame, for the three generic rays, incident, refracted and reflected, we have that the coordinates P, S, N of their versors, developed to the first order in the tangential inclination ϑ_T (from Prosthaphaeresis formulas) and in the sagittal one ϑ_S, are, respectively,

$$\mathbf{I}=\begin{pmatrix}\sin\vartheta_I+\vartheta_{TI}\cos\vartheta_I\\ \vartheta_{SI}\\ \cos\vartheta_I-\vartheta_{TI}\sin\vartheta_I\end{pmatrix}_\ell, \tag{2.7.37.a}$$

$$\mathbf{R}'=\begin{pmatrix}\sin\vartheta_{R'}+\vartheta_{TR'}\cos\vartheta_{R'}\\ \vartheta_{SR'}\\ \cos\vartheta_{R'}-\vartheta_{TR'}\sin\vartheta_{R'}\end{pmatrix}_\ell,\quad \mathbf{R}''=\begin{pmatrix}\sin\vartheta_{R''}+\vartheta_{TR''}\cos\vartheta_{R''}\\ \vartheta_{SR''}\\ \cos\vartheta_{R''}-\vartheta_{TR''}\sin\vartheta_{R''}\end{pmatrix}_\ell. \tag{2.7.37.b-c}$$

This is only a first step in finding a relation between the inclinations of the rays.

j) Matrix of curvature of the surface. For the calculation of the refraction or reflection of a generic ray, it is necessary to know the versor normal on the surface at the point of intersection, and it is therefore necessary to differentiate the versor **N** with respect to the variations of the coordinates ρ_P and ρ_S.

In the frame $\mathbb{S}$, the partial derivatives of $1/\sqrt{r}$ in Eq. (2.7.22) are

$$\frac{\partial}{\partial X}\frac{1}{\sqrt{r}}=-\frac{Z_{;XX}Z_{;X}+Z_{;XY}Z_{;Y}}{r\sqrt{r}},\quad \frac{\partial}{\partial Y}\frac{1}{\sqrt{r}}=-\frac{Z_{;XY}Z_{;X}+Z_{;YY}Z_{;Y}}{r\sqrt{r}},$$

and, differentiating **N** with respect to the variables *X*, *Y*, we obtain the Jacobian matrix

$$\frac{dN_i}{dj}=\frac{1}{\sqrt{r}}\begin{pmatrix} Z_{;XX}\left(N_X{}^2-1\right)+Z_{;XY}N_XN_Y & Z_{;XY}\left(N_X{}^2-1\right)+Z_{;YY}N_XN_Y \\ Z_{;XX}N_XN_Y+Z_{;XY}\left(N_Y{}^2-1\right) & Z_{;XY}N_XN_Y+Z_{;YY}\left(N_Y{}^2-1\right) \\ Z_{;XX}N_XN_Z+Z_{;XY}N_YN_Z & Z_{;XY}N_XN_Z+Z_{;YY}N_YN_Z \end{pmatrix}, \tag{2.7.38}$$

where the indices $i = X, Y, Z$, $j = X, Y$ respectively denote row and column of the matrix. Now, consider the finite differentiation of **N** as a function of the coordinates ρ_P and ρ_S, respectively corresponding to the versors $\mathbf{P}_m$ and $\mathbf{S}_m$ that generate the tangent plane to the surface at $\boldsymbol{c}_m$. It is the sum of two contributions resulting from the projection of the vector $\rho_P\,\mathbf{P}_m + \rho_S\mathbf{S}_m$ on the *X*, *Y* plane of the frame $\mathbb{S}$ on which the surface is defined, for which

$$\Delta N_i=\frac{dN_i}{dj}P_j\rho_P+\frac{dN_i}{dj}S_j\rho_S, \tag{2.7.39}$$

where P_j and S_j are the components, on the frame $\mathbb{S}$, of the versors mentioned above, and the repeated summation on index $j = X, Y$ is also implied. Since **N** is a versor normal on the surface, the vector $\Delta\mathbf{N}$ lies on the tangent plane at the surface, and its components in the directions of versors $\mathbf{P}_m$ and $\mathbf{S}_m$ are obtained from the scalar product of $\Delta\mathbf{N}$ with each of them:

$$\begin{aligned}\Delta N_P&=-C_{PP}\rho_P-C_{SP}\rho_S,\\ \Delta N_S&=-C_{PS}\rho_P-C_{SS}\rho_S,\end{aligned} \tag{2.7.40}$$

where the coefficients *C* are the elements of the *curvature matrix* expressed in the local frame

$$\mathbf{C}=\begin{pmatrix}C_{PP} & C_{SP}\\ C_{PS} & C_{SS}\end{pmatrix}=\begin{pmatrix}-\dfrac{dN_i}{dj}P_jP_i & -\dfrac{dN_i}{dj}S_jP_i\\ -\dfrac{dN_i}{dj}P_jS_i & -\dfrac{dN_i}{dj}S_jS_i\end{pmatrix}. \tag{2.7.41}$$

This matrix is a symmetric one for which $C_{SP} = C_{PS}$, In fact, the difference of the off-diagonal terms is

$$\frac{dN_i}{dj}S_jP_i-\frac{dN_i}{dj}P_jS_i=\frac{dN_Y}{dX}N_Z-\frac{dN_Z}{dX}N_Y-\frac{dN_X}{dY}N_Z+\frac{dN_Z}{dY}N_X, \tag{2.7.42}$$

where Eq. (2.7.28b) is exploited. Finally, by replacing the elements of the matrix (2.7.38), we can verify that this expression is zero. In addition, if the surface is spherical with a radius R, it can be shown that the curvature matrix is diagonal, with C_{PP} and C_{SS} being equal to each other and to $1/R$ except for a sign.

Eqs. (2.7.40) are the finite difference form of the Weingarten's equations [do Carmo 1976]. Given a local reference system centered on the point of intersection $\boldsymbol{c}$ of the central beam with the surface, they allow us to calculate, to the first order of approximation, the variation of the orientation of the versor normal to the surface as a function of the position on the tangent plane to the surface at $\boldsymbol{c}$. In further support of this, I refer to Eq. (2.7.19) and to the discussions in §2.3.8. In the local reference frame, around the point of intersection with the central ray, the surface can be approximated to the second order by

$$z = \frac{1}{2} C_{SS} \rho_S{}^2 + C_{SP} \rho_S \rho_P + \frac{1}{2} C_{PP} \rho_P{}^2 , \tag{2.7.43}$$

where, contrary to §2.3.8 for the curvature of the wavefronts, the plus sign in front of z follows from the choice made in Table 2.1 of §2.5 for the sign of the curvature radius of the optical surfaces. If we differentiate this expression according to the method from Eq. (2.7.21), we get

$$\Delta z - \left(C_{SS} \rho_S + C_{SP} \rho_P \right) \Delta \rho_S - \left(C_{PP} \rho_P + C_{SP} \rho_S \right) \Delta \rho_P = 0 , \tag{2.7.44}$$

for which one finds that the versor normal to the surface varies according to Eqs. (2.7.40). Moreover, by developing the surface locally, for small values of ρ_S and ρ_P, here, the parameter r of Eq. (2.7.22) is about 1. The minus signs in Eqs. (2.7.40) and (2.7.41) in the definition of the coefficients C are then introduced for consistency with that which is generally assumed in Geometrical Optics. Therefore, the curvature is positive if the center of curvature of the surface is located "after" its vertex, that is, in the positive direction of the optical axis of the principal ray, and, in this case, in the positive direction of the **N**-axis of the local frame.

The principal curvatures of the surface are obtained by diagonalizing the curvature matrix, and are given by

$$\begin{cases} C_1 = C_{SS} \cos^2 \Phi + C_{SP} \sin(2\Phi) + C_{PP} \sin^2 \Phi, \\ C_2 = C_{SS} \sin^2 \Phi - C_{SP} \sin(2\Phi) + C_{PP} \cos^2 \Phi, \end{cases}$$
$$\text{where} \quad \Phi = \frac{1}{2} \arctan\left(\frac{2C_{SP}}{C_{SS} - C_{PP}} \right) \tag{2.7.45}$$

is the angle between the principal direction 1 and the sagittal axis **S**, while the inverse relations are

$$
\begin{aligned}
C_{SS} &= C_1 \cos^2 \Phi + C_2 \sin^2 \Phi, \\
C_{PP} &= C_1 \sin^2 \Phi + C_2 \cos^2 \Phi, \\
C_{SP} &= \sin(2\Phi)(C_2 - C_1).
\end{aligned}
\tag{2.7.46}
$$

k) Surface matrix. At this point, we can calculate the paraxial matrix of the surface associated with the central ray. This matrix allows for transforming the sagittal and tangential coordinates of the vector relative to the incident ray in those of the refracted or reflected ray:

$$
\begin{pmatrix} \rho_{TR} \\ \rho_{SR} \\ \vartheta_{TR} \\ \vartheta_{SR} \end{pmatrix} = \mathfrak{M}_S \begin{pmatrix} \rho_{TI} \\ \rho_{SI} \\ \vartheta_{TI} \\ \vartheta_{SI} \end{pmatrix}, \tag{2.7.47}
$$

where R alternately means R′ or R″. So, we take the set $\mathbf{P}_m$, $\mathbf{S}_m$, $\mathbf{N}_m$, as a local axis frame P, S, N with origin in $\boldsymbol{c}_m$. We also need to distinguish between refraction and reflection. In the first case, the law of *refraction* can be rewritten as

$$
\mathbf{R}' = \mu \mathbf{I} + (\mathbf{N} \cdot \mathbf{R}' - \mu \mathbf{N} \cdot \mathbf{I}) \mathbf{N}, \tag{2.7.48}
$$

where μ is the ratio between the refractive indices given by Eq. (2.7.29), $\mathbf{I}$, $\mathbf{R}'$ are the incident and refracted versors of any given ray, and $\mathbf{N}$ is the versor normal at the surface at the point of intersection of the rays. In the *local frame*, in paraxial approximation, $\mathbf{I}$ and $\mathbf{R}'$ are given by Eqs. (2.7.37a-b), while $\mathbf{N}$ can be expressed as

$$
\mathbf{N} = \begin{pmatrix} \Delta N_P \\ \Delta N_S \\ 1 \end{pmatrix}_\ell . \tag{2.7.49}
$$

Then, at the first order, we have

$$
\begin{aligned}
&\mathbf{N} \cdot \mathbf{R}' - \mu \mathbf{N} \cdot \mathbf{I} = \\
&(\Delta N_P \sin \vartheta_{R'} + \cos \vartheta_{R'} - \vartheta_{TR'} \sin \vartheta_{R'}) - \mu (\Delta N_P \sin \vartheta_I + \cos \vartheta_I - \vartheta_{TI} \sin \vartheta_I),
\end{aligned}
$$

and, with a few steps from Eqs. (2.7.37) and (2.7.48-49), we simply get

$$
\begin{aligned}
\vartheta_{TR'} \cos \vartheta_R &= \mu \vartheta_{TI} \cos \vartheta_I + \gamma_o \Delta N_P, \\
\vartheta_{SR'} &= \mu \vartheta_{SI} + \gamma_o \Delta N_S,
\end{aligned}
\tag{2.7.50}
$$

where μ is given by Eq. (2.7.29) and

$$\gamma_o = \cos\vartheta_{R'} - \mu\cos\vartheta_I . \tag{2.7.51}$$

Finally, by virtue of these equations, and from Eqs. (2.7.36) and (2.7.40), which express ΔN as a function of the coordinates ρ, we obtain the matrix associated with the paraxial refracting surface between the ray vectors expressed for the tangential and sagittal coordinates

$$\mathfrak{M}_s = \begin{pmatrix} \dfrac{\cos\vartheta_{R'}}{\cos\vartheta_I} & 0 & 0 & 0 \\ 0 & 1 & 0 & 0 \\ -\dfrac{\gamma_o}{\cos\vartheta_I\cos\vartheta_{R'}}C_{PP} & -\dfrac{\gamma_o}{\cos\vartheta_{R'}}C_{PS} & \dfrac{n_{m-1}\cos\vartheta_I}{n_m\cos\vartheta_{R'}} & 0 \\ -\dfrac{\gamma_o}{\cos\vartheta_I}C_{SP} & -\gamma_o C_{SS} & 0 & \dfrac{n_{m-1}}{n_m} \end{pmatrix} . \tag{2.7.52}$$

We note that the curvature of the surface intervenes only in the 4 elements in the bottom left corner of the matrix, in a manner similar to the shape of the 2x2 matrices of the *dioptre*, to which Eq. (2.7.52) can be traced back in the case of normal incidence.

In the case of reflection, instead of Eq. (2.7.47), we have

$$\mathbf{R}'' = \mathbf{I} + (\mathbf{N}\cdot\mathbf{R}'' - \mathbf{N}\cdot\mathbf{I})\mathbf{N} , \tag{2.7.53}$$

where the versors are given by Eqs. (2.7.37) and (2.7.49). The different direction of the versor $\mathbf{R}''$ will result from the opposite sign between $\cos\vartheta_{R''}$ and $\cos\vartheta_I$ in Eq. (2.7.32.b). We can therefore immediately obtain the result using the expressions of refraction with some simplifications, such as that γ_o is now replaced by $2\cos\vartheta_{R''}$. The matrix associated with the reflective surface is therefore

$$\mathfrak{M}_s = \begin{pmatrix} -1 & 0 & 0 & 0 \\ 0 & 1 & 0 & 0 \\ -\dfrac{2}{\cos\vartheta_I}C_{PP} & -2C_{PS} & -1 & 0 \\ 2C_{SP} & 2\cos\vartheta_I C_{SS} & 0 & 1 \end{pmatrix} . \tag{2.7.54}$$

The sign – for the transformation of the tangential components ρ_T and ϑ_T follows from keeping, even for the reflected beam, the same type of Cartesian frame (say, right), with versors $\mathbf{T}_{mR}$, $\mathbf{S}_m$, $\mathbf{R}''_m$, in the definition of the tangential and sagittal positions and inclinations.

l) Rotation matrices. We now need to apply a rotation to our matrix around the incident ray and one around the reflected or refracted ray in order to refer the transformation to the transverse coordinates ξ and η assigned to each segment of the central ray. These rotations act on the ρ and ϑ variables without mixing them, thus their matrices have a diagonal form as

$$\mathfrak{R}(\varphi)=\begin{pmatrix} \cos\varphi & \sin\varphi & 0 & 0 \\ -\sin\varphi & \cos\varphi & 0 & 0 \\ 0 & 0 & \cos\varphi & \sin\varphi \\ 0 & 0 & -\sin\varphi & \cos\varphi \end{pmatrix}, \tag{2.7.55}$$

where φ is the angle of rotation around the axes ζ of the various segments. The values of $\sin\varphi$ and $\cos\varphi$ can simply be determined by the transformation of the coordinates of the versor **S** from the frame $\mathfrak{S}_m$ to the frame $\mathfrak{L}$ and then in the frames $\mathfrak{R}_{m-1}$ and $\mathfrak{R}_m$ of the incident ray and the reflected or refracted ray, respectively. Indeed, **S** is orthogonal to both the incident ray and to that which is reflected or refracted, and thus lies on the transverse $\xi\eta$-planes of $\mathfrak{R}_{m-1}$ and $\mathfrak{R}_m$, whereby its components on those planes are the exact values that we seek:

$$\cos\varphi_{\mathrm{I}} = S_{\eta\mathrm{I}}, \quad \sin\varphi_{\mathrm{I}} = S_{\xi\mathrm{I}}, \quad \cos\varphi_{\mathrm{R}} = S_{\eta\mathrm{R}}, \quad \sin\varphi_{\mathrm{R}} = S_{\xi\mathrm{R}}\,. \tag{2.7.56}$$

m) Matrices of reflection or refraction. The overall matrix associated with the single surface is then

$$\mathfrak{s} = \mathfrak{R}^{-1}(\varphi_{\mathrm{R}})\,\mathfrak{M}_{\mathrm{s}}\,\mathfrak{R}(\varphi_{\mathrm{I}})\,. \tag{2.7.57}$$

n) Translation matrix. The translation matrix is now

$$\mathfrak{T}=\begin{pmatrix} 1 & 0 & d & 0 \\ 0 & 1 & 0 & d \\ 0 & 0 & 1 & 0 \\ 0 & 0 & 0 & 1 \end{pmatrix}, \tag{2.7.58}$$

where d is the distance traveled along the ray.

o) Overall system matrix. In conclusion, by multiplying the various matrices in succession, we can obtain the matrix that represents the whole system between two assigned planes. It can be imagined as the composition of 4 matrices

$$\mathfrak{S}=\begin{pmatrix} \mathfrak{A} & \mathfrak{B} \\ \mathfrak{C} & \mathfrak{D} \end{pmatrix} \tag{2.7.59}$$

and the relationship between the starting and ending vectors becomes

$$\begin{pmatrix} \boldsymbol{\rho}' \\ \boldsymbol{\vartheta}' \end{pmatrix} = \mathfrak{S} \begin{pmatrix} \boldsymbol{\rho} \\ \boldsymbol{\vartheta} \end{pmatrix}, \tag{2.7.60}$$

where $\boldsymbol{\rho}$ and $\boldsymbol{\vartheta}$ are vectors that express two components in a compact form, respectively, the position and the slope of the ray.

In the case in which the system is planar, is spread on the *zx*-plane, and each surface has a main direction oriented along the *y*-axis, the submatrices $\mathfrak{A}, \mathfrak{B}, \mathfrak{C}, \mathfrak{D}$ are diagonal. In this case, the ξ and η components are separable and the system can be treated with two independent 2x2 matrices. This can happen even if the system is non-planar, if the four submatrices $\mathfrak{S}$ are simultaneously diagonalized with a rotation in the departure $\xi\eta$-plane and one in the arrival plane. However, in general, a non-planar system exhibits non-trivial behavior due to the combined effect of rotations and translations.

2.7.4 Curvature of a quadric surface

As an example, consider a *quadric surface* defined by

$$Z = \frac{1}{2}\left(CX^2 + BY^2 + AZ^2\right),$$

with the principal axes aligned along the Cartesian axes X, Y, Z and the vertex on the origin. Resolved for *Z,* it generally has two values. The solution with $Z = 0$ for $X = Y = 0$ is given by

$$Z = \frac{1}{A} - \frac{1}{A}\sqrt{1 - A\left(CX^2 + BY^2\right)} = \frac{CX^2 + BY^2}{1 + \sqrt{1 - A\left(CX^2 + BY^2\right)}}.$$

For brevity, we take that

$$\zeta = \sqrt{1 - A\left(CX^2 + BY^2\right)}, \quad ACX^2 + ABY^2 + \zeta^2 = 1.$$

Z has real solutions when the argument of the root is real and positive. In this case, ζ is also real and positive. The partial derivatives of *Z* with respect to *X* and *Y* are

$$Z_{;X} = \frac{C}{\zeta}X, \quad Z_{;Y} = \frac{B}{\zeta}Y.$$

Therefore, according to Eq. (2.7.22), the versor normal to the surface is

$$N_X = -\frac{1}{\sqrt{r}}\frac{C}{\zeta}X, \quad N_Y = -\frac{1}{\sqrt{r}}\frac{B}{\zeta}Y, \quad N_Z = \frac{1}{\sqrt{r}},$$

where

$$r = 1 + \frac{C^2}{\zeta^2}X^2 + \frac{B^2}{\zeta^2}Y^2, \quad \zeta\sqrt{r} = \sqrt{\zeta^2 + C^2X^2 + B^2Y^2}.$$

Even for a "simple" ellipsoidal surface, the general equations for the curvature are very complicated so as to be advantageous with respect to a numerical treatment. I will therefore confine myself to expressing them only for the points on the principal YZ-plane. On this plane, we have $X = 0$ and $N_X = 0$, and the Jacobian matrix (2.7.38) simplifies to

$$\frac{dN_i}{dj} = \frac{1}{\sqrt{r}}\begin{pmatrix} -C/\zeta & 0 \\ 0 & -B/(\zeta^3 r) \\ 0 & -B^2Y/(\zeta^4 r) \end{pmatrix}.$$

For the ellipsoid, for points on the surface in the YZ-plane, when this coincides with the plane of incidence, we have $P_X = 0$, $P_Y = -N_Z$, $P_Z = N_Y$, $S_Y = S_Z = 0$, $S_X = 1$ and the curvature matrix is diagonal:

$$\begin{pmatrix} C_{\mathrm{PP}} & C_{\mathrm{SP}} \\ C_{\mathrm{PS}} & C_{\mathrm{SS}} \end{pmatrix} = \begin{pmatrix} \frac{1}{\zeta\sqrt{r}}\frac{1}{\zeta^2 r}B & 0 \\ 0 & \frac{1}{\zeta\sqrt{r}}C \end{pmatrix},$$

it can now be said that

$$\zeta\sqrt{r} = \sqrt{1 + B^2Y^2(1 - A/B)}.$$

Finally, if the ellipsoid is replaced with a conic of revolution around the axis Z defined by Eq. (2.4.2), with curvature at the vertex $C = B$ and asphericity $\alpha = A/C$, its principal curvatures, sagittal C_s and tangential C_t, with which the curvature matrix is diagonalized for a ray on the meridian plane, are given by

$$C_s = \frac{C}{\left[1 + (1-\alpha)C^2(X^2 + Y^2)\right]^{1/2}}, \quad C_t = \frac{C}{\left[1 + (1-\alpha)C^2(X^2 + Y^2)\right]^{3/2}}. \quad (2.7.61)$$

2.7.5 Transformation of the wavefront

Besides considering the rays as taken singly, it is also interesting to discuss the "collective" case of a family of rays belonging to the same eikonal, to which we can attribute a wavefront. In particular, we are interested in determining how this wavefront is transformed in the passage from "immediately before" the refraction or reflection to "immediately after". Even a wavefront is characterized, in the paraxial approximation, by a curvature matrix, and therefore we want to determine how the curvature matrix $\mathfrak{Q}_{I}$ of an incident wave turns into the matrix $\mathfrak{Q}_{R'}$ of the refracted wave or the matrix $\mathfrak{Q}_{R''}$ of the reflected wave. The connection between the tracing of rays and the propagation of wavefront is similar to that of Eq. (2.7.40) for the variations in the normal versor to the surface as a function of the position on it. Such a connection is given by Weingarten's equation, whereby

$$\begin{pmatrix}\vartheta_{TI}\\ \vartheta_{SI}\end{pmatrix} = \mathfrak{Q}_{I}\begin{pmatrix}\rho_{TI}\\ \rho_{SI}\end{pmatrix},\quad \begin{pmatrix}\vartheta_{TR'}\\ \vartheta_{SR'}\end{pmatrix} = \mathfrak{Q}_{R'}\begin{pmatrix}\rho_{TR'}\\ \rho_{SR'}\end{pmatrix},\quad \begin{pmatrix}\vartheta_{TR''}\\ \vartheta_{SR''}\end{pmatrix} = \mathfrak{Q}_{R''}\begin{pmatrix}\rho_{TR''}\\ \rho_{SR''}\end{pmatrix}, \qquad (2.7.62)$$

where, here, the curvature matrices $\mathfrak{Q}$ are unfortunately defined with the opposite sign with respect to the curvature matrices **C** of surfaces. This is now in accordance with the choice made [see also Solimeno et al 1986] in section §2.3.8 dedicated to astigmatic pencils, and especially with the sign typically assigned to the curvature of the wavefront of Gaussian beams, for which the curvature is positive for the divergent wave, *after* the waist position in the propagation of the beam. Applying these equations to the transformation induced by the matrix (2.7.52), we have, for the refraction,

$$\begin{aligned}&(Q_{TTR'}\cos\vartheta_{R'}\rho_{TI} + Q_{TSR'}\cos\vartheta_{I}\rho_{SI})\cos\vartheta_{R'} =\\ &\mu(Q_{TTI}\rho_{TI} + Q_{TSI}\rho_{SI})\cos^2\vartheta_{I} - \gamma_o C_{PP}\rho_{TI} - \gamma_o\cos\vartheta_{I} C_{PS}\rho_{SI},\end{aligned}$$

$$\begin{aligned}&Q_{STR'}\cos\vartheta_{R'}\rho_{TI} + Q_{SSR'}\cos\vartheta_{I}\rho_{SI} =\\ &\mu(Q_{STI}\rho_{TI} + Q_{SSI}\rho_{SI})\cos\vartheta_{I} - \gamma_o C_{SP}\rho_{TI} - \gamma_o C_{SS}\cos\vartheta_{I}\rho_{SI}.\end{aligned}$$

Since the coefficients of the matrices do not depend on the value of the coordinates ρ, we have four equations, two of which are the same, because they relate to the off-diagonal elements of the matrices, which are symmetrical, so we have

$$\begin{aligned}Q_{TTR'}\cos^2\vartheta_{R'} &= \mu Q_{TTI}\cos^2\vartheta_{I} - \gamma_o C_{PP},\\ Q_{SSR'} &= \mu Q_{SSI} - \gamma_o C_{SS},\\ Q_{TSR'}\cos\vartheta_{R'} &= \mu Q_{TSI}\cos\vartheta_{I} - \gamma_o C_{SP}.\end{aligned} \qquad (2.7.63)$$

For reflection, with Eq. (2.7.54), we instead have

$$
\begin{aligned}
Q_{TTR''} &= \quad Q_{TTI} + \frac{2}{\cos\vartheta_I} C_{PP},\\
Q_{SSR''} &= \quad Q_{SSI} + 2\cos\vartheta_I\, C_{SS},\\
Q_{TSR''} &= -Q_{TSI} \quad -2\,C_{SP}.
\end{aligned}
\tag{2.7.64}
$$

These equations were introduced by Stavroudis and then obtained in a different way by Burkhard and Shealy, who refer to torsion as the non-diagonal term of the curvature matrix [Burkhard and Shealy 1981]. They coincide with Coddington's equations when the torsion is zero [Kingslake 1994]. They express a linear relationship between the curvature of the refracted or reflected wave and that of the incident one, modulated by the cosines of the angles of incidence (and refraction) and modified in additive form by the curvature of the surface. The direction of the principal axes and the principal curvatures of the wavefront can be determined from Eqs. (2.7.45) by replacing C with $-Q$.

A suitable choice of the surface can leave the stigmatism of the incident pencil unchanged or correct its astigmatism. An example is given by the conic surfaces of revolution. In the case of a paraboloid mirror ($\alpha = 0$), a collimated pencil with its principal ray parallel to the axis is transformed into one with a spherical wavefront converging on the focus of the paraboloid. However, for a thin and stigmatic pencil, oriented in the same way, also removing the condition of collimation, whereby its center of curvature is at a finite distance, the reflected pencil remains stigmatic. This also occurs in reflection by a *prolate* ellipsoid ($1 \geq \alpha > 0$) or a hyperboloid ($0 > \alpha$) for a pencil with its principal ray passing for one of the two foci. At the point where the main beam strikes the surface, it has two different principal curvatures, sagittal and tangential, but, for that particular angle of incidence, its "effective" curvature is spherical [Greco and Giusfredi 2007], whereby $C_{SS}\cos\vartheta_I = C_{PP}/\cos\vartheta_I$ and $C_{SP} = 0$. It should be noted that, in this case, thanks to Eq. (2.7.45c) with $-Q$ instead of C, the angle Φ of the principal directions remains unchanged. Conversely, an appropriate inclination of a curved surface can compensate for the astigmatism of an incident pencil.

In addition to a single surface, the calculation of the curvature matrix of the emerging wavefront can be extended to an entire optical system represented by the matrix (2.7.59). Indeed, if, for the wavefront at the entry of the system, we have

$$
\begin{pmatrix} \vartheta_\xi \\ \vartheta_\eta \end{pmatrix} = \mathfrak{Q} \begin{pmatrix} \rho_\xi \\ \rho_\eta \end{pmatrix},
\tag{2.7.65}
$$

and, from the relationship between the components of the column vectors relative to the initial and final planes of an optical system represented by the submatrices $\mathfrak{A}, \mathfrak{B}, \mathfrak{C}, \mathfrak{D}$, we get

$$\begin{pmatrix}\rho'_\xi\\ \rho'_\eta\end{pmatrix}=\mathfrak{A}\begin{pmatrix}\rho_\xi\\ \rho_\eta\end{pmatrix}+\mathfrak{B}\begin{pmatrix}\vartheta_\xi\\ \vartheta_\eta\end{pmatrix}=\left(\mathfrak{A}+\mathfrak{B}\mathfrak{Q}\right)\begin{pmatrix}\rho_\xi\\ \rho_\eta\end{pmatrix},$$
$$\mathfrak{Q}'\begin{pmatrix}\rho'_\xi\\ \rho'_\eta\end{pmatrix}=\begin{pmatrix}\vartheta'_\xi\\ \vartheta'_\eta\end{pmatrix}=\mathfrak{C}\begin{pmatrix}\rho_\xi\\ \rho_\eta\end{pmatrix}+\mathfrak{D}\begin{pmatrix}\vartheta_\xi\\ \vartheta_\eta\end{pmatrix}=\left(\mathfrak{C}+\mathfrak{D}\mathfrak{Q}\right)\begin{pmatrix}\rho_\xi\\ \rho_\eta\end{pmatrix}. \tag{2.7.66}$$

By replacing the first into the second equation, remembering that the relation obtained in this way is valid for every pair ρ_ξ, ρ_η and solving for $\mathfrak{Q}'$, we ultimately get

$$\mathfrak{Q}'=\left(\mathfrak{D}\mathfrak{Q}+\mathfrak{C}\right)\left(\mathfrak{B}\mathfrak{Q}+\mathfrak{A}\right)^{-1}, \tag{2.7.67}$$

which we might call the "DCBA law" and which allows for determining the curvature matrix of the wavefront emerging from the optical system. It should be noted that this expression is not simply scalar, as it involves matrices that, moreover, generally do not commute between each other. We can verify that Eqs. (2.7.63-64) for a single refraction or reflection are compatible with Eq. (2.7.67), and the same happens in the case of a rotation or propagation for a homogeneous space whose matrices are given, respectively, by Eqs. (2.7.55) and (2.7.58). We also note that, as in the case of 2x2 matrices, the subsequent application of Eq. (2.7.67) produces an expression with the same shape, where $\mathfrak{A}$, $\mathfrak{B}$, $\mathfrak{C}$, $\mathfrak{D}$ coincide with the submatrices of the product of the 4x4 matrices of the individual elements of the optical system. Indeed, if the above-mentioned sub-matrices result from the product of the matrices ($\mathfrak{A}'$, $\mathfrak{B}'$, $\mathfrak{C}'$, $\mathfrak{D}'$) and ($\mathfrak{A}''$, $\mathfrak{B}''$, $\mathfrak{C}''$, $\mathfrak{D}''$) of a first and a second element of a system, we have

$$\begin{aligned}\mathfrak{Q}''&=\left(\mathfrak{D}''\mathfrak{Q}'+\mathfrak{C}''\right)\left(\mathfrak{B}''\mathfrak{Q}'+\mathfrak{A}''\right)^{-1}\\&=\left[\mathfrak{D}''\left(\mathfrak{D}'\mathfrak{Q}+\mathfrak{C}'\right)\left(\mathfrak{B}'\mathfrak{Q}+\mathfrak{A}'\right)^{-1}+\mathfrak{C}''\right]\left[\mathfrak{B}''\left(\mathfrak{D}'\mathfrak{Q}+\mathfrak{C}'\right)\left(\mathfrak{B}'\mathfrak{Q}+\mathfrak{A}'\right)^{-1}+\mathfrak{A}''\right]^{-1}\\&=\left[\left(\mathfrak{D}''\mathfrak{D}'+\mathfrak{C}''\mathfrak{B}'\right)\mathfrak{Q}+\left(\mathfrak{D}''\mathfrak{C}'+\mathfrak{C}''\mathfrak{A}'\right)\right]\left[\left(\mathfrak{B}''\mathfrak{D}'+\mathfrak{A}''\mathfrak{B}'\right)\mathfrak{Q}+\left(\mathfrak{B}''\mathfrak{C}'+\mathfrak{A}''\mathfrak{A}'\right)\right]^{-1},\end{aligned} \tag{2.7.68}$$

where, in the last step between the two square brackets, the identity given by the product $\left(\mathfrak{B}'\mathfrak{Q}+\mathfrak{A}'\right)\left(\mathfrak{B}'\mathfrak{Q}+\mathfrak{A}'\right)^{-1}$ was inserted.

One case of particular interest is the one in which the incident wave is generated by a source point O taken on the principal ray so that its wavefront is spherical with $\mathfrak{Q}=\left(1/d_0\right)\mathfrak{I}$, where $\mathfrak{I}$ is the identity and d_0 is the distance from O of the first surface considered by the matrix of the system. In this case, at the exit from the system, we have

$$\mathfrak{Q}'=\left(\mathfrak{D}+d_0\mathfrak{C}\right)\left(\mathfrak{B}+d_0\mathfrak{A}\right)^{-1}, \tag{2.7.69}$$

and, from the diagonalization of $\mathfrak{Q}'$ and the reciprocal of the principal curvatures, recalling that, here, a convergent wave is taken with negative curvature, we can get the centers of curvature on the emerging principal ray of the wavefront outgoing from the system. By calculating the system matrix for a grid of outgoing rays from O and collected by the entrance pupil, we can then reconstruct the corresponding caustic near the image plane of O: the caustics of Figs. 63 and 65 were obtained with this procedure.

We will resume this discussion in the study of the propagation of laser beams, which can be considered as narrow pencils, but, as we shall see, due to diffraction, in the propagation, their curvature is also influenced by their transverse dimensions.

Finally, we can use Eq. (2.7.67) to determine the focal points of the optical system expressed by the matrix (see Table 2.3). In fact, for a plane wave incoming or outgoing, for which $\mathfrak{Q}$ or $\mathfrak{Q}'$ are respectively null, we have

$$\mathfrak{Q}' = \mathfrak{C}\mathfrak{A}^{-1}, \quad \text{or} \quad \mathfrak{Q} = -\mathfrak{D}^{-1}\mathfrak{C} \,. \tag{2.7.70}$$

The distances of the focal points from the initial position of the matrix on the principal ray, in object space, and from the final position, in the image space, may be obtained as above by respectively diagonalizing $\mathfrak{Q}$ and $\mathfrak{Q}'$. On the other hand, the focal lengths for the image space of the system are given by $-1/C_{\mathrm{t}}$ and $-1/C_{\mathrm{s}}$, where C are the principal curvatures corresponding to $\mathfrak{C}$ (with t for *tangential* and s for *sagittal*), achievable with appropriate rotations, one in the object space and one in the image space. The focal lengths for the object space are obtained by multiplying those of the image space for n_o/n_i.

2.7.6 Symmetry relations and optical path

In 1970, two important works were published independently, just two months apart, the first by S.A. Collins and the second by J.A. Arnaud, establishing the paraxial properties of non-orthogonal optical systems [Collins 1970, Arnaud 1970]. These properties concern, in particular, the symmetry relationships of the 4x4-matrices describing the system and the calculation of the optical path between two generic points on the initial and final planes.[17]

---

[17] Both Collins and Arnaud define an optical system in which the refractive index has a generic value n_o immediately after the initial plane 1 and n_i immediately before the final plane 2. However, their analysis can be understood by assuming that initial and final spaces have an index of unitary refraction. In other words, two plane interfaces, which coincide with planes 1 and 2, are "added" to the optical system, which imply an index jump to the unitary value. Their ABCD matrices include these elements and have, in fact, a unitary determinant. Here, I have preferred to avoid this addition, so, in this notation, the coefficients A, B, C, D of Arnaud's expressions are replaced, respectively, with A, B/n_o, Cn_i, Dn_i/n_o. For those of Collins, it is

For completeness, I mention here the symmetry relationships for a generic 4x4-matrix. They can be demonstrated by induction by noting that they are valid for each elementary matrix and that, if two 4x4-matrices respect them, these relations will also be preserved for their product. In our notation, the relations of Collins are

$$\begin{aligned} &\mathfrak{D}\mathfrak{C}^{\mathrm{T}} = \mathfrak{C}\mathfrak{D}^{\mathrm{T}}, \quad \mathfrak{B}\mathfrak{A}^{\mathrm{T}} = \mathfrak{A}\mathfrak{B}^{\mathrm{T}}, \\ &\mathfrak{D}^{\mathrm{T}}\mathfrak{B} = \mathfrak{B}^{\mathrm{T}}\mathfrak{D}, \quad \mathfrak{C}^{\mathrm{T}}\mathfrak{A} = \mathfrak{A}^{\mathrm{T}}\mathfrak{C}, \end{aligned} \tag{2.7.71}$$

where the superscript T indicates the transpose.[18] In other words, $\mathfrak{D}\mathfrak{C}^{\mathrm{T}}$ is a symmetric matrix (but this *does not* imply that $\mathfrak{D}^{\mathrm{T}}\mathfrak{C}$ is symmetric too), and the same goes for the others. Furthermore,

$$\begin{aligned} &\mathfrak{D}\mathfrak{A}^{\mathrm{T}} - \mathfrak{C}\mathfrak{B}^{\mathrm{T}} = \frac{n_o}{n_i}\mathfrak{I}, \\ &\mathfrak{D}^{\mathrm{T}}\mathfrak{A} - \mathfrak{B}^{\mathrm{T}}\mathfrak{C} = \frac{n_o}{n_i}\mathfrak{I}, \end{aligned} \tag{2.7.72}$$

where n_o is the refractive index for the initial space and n_i is the one for the final space, while $\mathfrak{I}$ represents the 2x2 identity matrix. They correspond to Eq. (2.7.8) for the determinant of 2x2-matrices.

Collins uses these expressions to determine the relationship between the coordinates and slopes on the initial and final planes of a generic ray that crosses the system. He expresses the two pairs of coordinates and slopes by the $\boldsymbol{X}$ and $\boldsymbol{\theta}$ vectors that, in this present notation, are respectively defined as

$$\boldsymbol{X}^{\mathrm{T}} = \left(\rho_x, \rho_y, \rho'_x, \rho'_y\right), \quad \boldsymbol{\theta}^{\mathrm{T}} = \left(-n_o\vartheta_x, -n_o\vartheta_y, n_i\vartheta'_x, n_i\vartheta'_y\right). \tag{2.7.73}$$

The minus signs in front of ϑ_x and ϑ_y express the sign change in the slope that happens while reversing the direction of propagation in the system. With this exchange of variables, Eq. (2.7.60) becomes

$$\begin{aligned} -\mathfrak{D}\vartheta + \vartheta' &= \mathfrak{C}\rho, \\ -\mathfrak{B}\vartheta \quad &= \mathfrak{A}\rho - \rho', \end{aligned}$$

which, with the definition (2.7.73), can be rewritten as

$$\begin{pmatrix} \mathfrak{D}/n_o & \mathfrak{I}/n_i \\ \mathfrak{B}/n_o & \mathfrak{O} \end{pmatrix} \boldsymbol{\theta} = \begin{pmatrix} \mathfrak{C} & \mathfrak{O} \\ \mathfrak{A} & -\mathfrak{I} \end{pmatrix} \boldsymbol{X}, \tag{2.7.74}$$

---

necessary that we also do the exchanges $A \leftrightarrow D$ and $B \leftrightarrow C$.

[18] I remember that, for each pair of matrices $\mathfrak{a}$, $\mathfrak{b}$ between which it is possible to do the product, it is $(\mathfrak{a}\mathfrak{b})^{\mathrm{T}} = \mathfrak{b}^{\mathrm{T}}\mathfrak{a}^{\mathrm{T}}$. In addition, for each scalar c, it is $c = c^{\mathrm{T}}$.

where $\mathfrak{O}$ represents the zero 2x2-matrix. Therefore,

$$\theta = \mathfrak{L}X, \quad \text{where } \mathfrak{L} = \begin{pmatrix} \mathfrak{D}/n_o & \mathfrak{I}/n_i \\ \mathfrak{B}/n_o & \mathfrak{O} \end{pmatrix}^{-1} \begin{pmatrix} \mathfrak{C} & \mathfrak{O} \\ \mathfrak{A} & -\mathfrak{I} \end{pmatrix}. \tag{2.7.75}$$

After several passages, applying the previous symmetry relations, as long as the determinant of $\mathfrak{B}$ is different from 0, we finally get

$$\mathfrak{L} = \begin{pmatrix} n_o \mathfrak{B}^{-1}\mathfrak{A} & -n_o \mathfrak{B}^{-1} \\ -n_o \mathfrak{B}^{-1\mathrm{T}} & n_i \mathfrak{D}\mathfrak{B}^{-1} \end{pmatrix}. \tag{2.7.76}$$

The 4x4-matrix $\mathfrak{L}$ is symmetric, since, for Eq. (2.7.71.b-c), $\mathfrak{B}^{-1}\mathfrak{A}$ and $\mathfrak{D}\mathfrak{B}^{-1}$ are also symmetric.

Finally, Collins shows that, in the paraxial approximation, the optical path through the system, between a point of coordinates ρ_ξ, ρ_η on the initial plane and a point of coordinate ρ'_ξ, ρ'_η on the final plane, is given by

$$L = L_0 + \frac{1}{2} X^T \mathfrak{L} X, \tag{2.7.77}$$

where L_0 is the optical path of the principal ray, between the same planes on which the matrix is defined. As we will see later, this allows for calculating the Fresnel's diffraction integrals, even through an optical system. Arnaud, in the article mentioned above, extends this analysis to the propagation of laser beams and the study of resonant cavities.

In his demonstration, Collins uses the calculation of variation to find the stationary path between the initial and final points through the optical system. Here, we can follow a simpler demonstration using the results already obtained in §2.3.8. In fact, in paraxial approximation, the matrix of the system already describes each ray at its start and on its arrival. This allows us to reconstruct the final wavefront of the congruence defined by the initial wavefront. Then, between the two wavefronts, the rays belonging to that congruence travel the same optical path.

Consider a wave that starts backwards from the point O_2 at the origin of the arrival plane 2 of the system (Fig. 2.52) and the corresponding wavefront S_1 tangent to the initial plane 1, in its origin O_1. Therefore, the optical path between O_2 and S_1 is equal to that of the principal ray, L_0. In particular, let us consider a ray $\mathtt{r}_1$ that, from O_2, reaches a generic point P on the initial plane. This is possible as long as the planes 1 and 2 are not conjugated. Knowing the system matrix, we can write

$$\mathbf{0} = \mathfrak{A}\rho_1 + \mathfrak{B}\vartheta_1,$$

Fig. 2.52
Study of the optical path

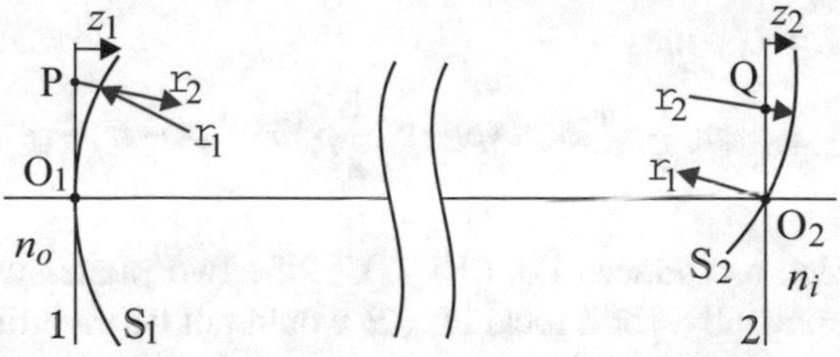

where ϑ_1 and ρ_1 are, respectively, the position and slope of $\mathtt{r}_1$ in P. Resolving for ϑ_1 in function of ρ_1, we have $-\vartheta_1 = \mathfrak{B}^{-1}\mathfrak{A}\rho_1$, where $\mathfrak{B}^{-1}\mathfrak{A}$ is the curvature matrix of the surface S_1, which is therefore defined by

$$z_1 = \frac{1}{2}\rho_1^{\mathrm{T}}\mathfrak{B}^{-1}\mathfrak{A}\rho_1,$$

where the sign of z_1 is taken as positive forwards. The optical path between O_2 and P is therefore approximated by $L_1 = L_o + n_o z_1$. Let us now consider a wave generated in P and the corresponding wavefront S_2 close to the plane 2, such that it intersects it in O_2. In this way, the optical path between P and S_2 is still L_1. Take now a ray $\mathtt{r}_2$ from P to a generic point Q on plane 2, for which we have

$$\rho_2 = \mathfrak{A}\rho_1 + \mathfrak{B}\vartheta_1',$$
$$\vartheta_2 = \mathfrak{C}\rho_1 + \mathfrak{D}\vartheta_1',$$

where ϑ_2 and ρ_2 are, respectively, the position and slope of $\mathtt{r}_2$ in Q, while ϑ'_1 is the slope in P. Resolving ϑ_2 as a function of ρ_2, we have

$$\vartheta_2 = \mathfrak{D}\mathfrak{B}^{-1}\rho_2 + \left(\mathfrak{C} - \mathfrak{D}\mathfrak{B}^{-1}\mathfrak{A}\right)\rho_1 = \mathfrak{D}\mathfrak{B}^{-1}\rho_2 - \frac{n_o}{n_i}\mathfrak{B}^{-1\mathrm{T}}\rho_1,$$

where, in the last step, the symmetry of $\mathfrak{D}\mathfrak{B}^{-1}$ and Eq. (2.7.75.b) were exploited. The term in ρ_2 represents the curvature of the surface S_2, while the one in ρ_1 represents its inclination in O_2. This surface is then defined by

$$z_2 = -\frac{1}{2}\rho_2^{\mathrm{T}}\mathfrak{D}\mathfrak{B}^{-1}\rho_2 + \frac{n_o}{n_i}\rho_2^{\mathrm{T}}\mathfrak{B}^{-1\mathrm{T}}\rho_1,$$

where the sign of z_2 is also positive forwards. The mixed term in ρ_1 and ρ_2, being scalar, has the same value as its transpose and can be subdivided into two terms of equal value. Finally, the optical path between P and Q, also called the *point characteristic*, is approximated by

$$L = L_1 - n_i z_2$$
$$= L_o + n_o \frac{1}{2} \rho_1{}^{T} \mathfrak{B}^{-1} \mathfrak{A} \rho_1 - n_o \frac{1}{2} \rho_1{}^{T} \mathfrak{B}^{-1} \rho_2 - n_o \frac{1}{2} \rho_2{}^{T} \mathfrak{B}^{-1T} \rho_1 + n_i \frac{1}{2} \rho_2{}^{T} \mathfrak{D} \mathfrak{B}^{-1} \rho_2,$$

which reproduces Eq. (2.7.77). If the two planes were conjugated, even only astigmatically, for a focal line, $\mathfrak{B}$ would not be invertible and Eq. (2.7.77) would not be applicable.

Lastly, I recall one of Arnaud's definition, concerning the "scalar product" between two rays $\mathbf{R}_T \equiv (\rho_T, n\vartheta_T)$ and $\mathbf{R}'_T \equiv (\rho'_T, n\vartheta'_T)$ crossing the system

$$(\mathbf{R}, \mathbf{R}') = n\rho^{T}\vartheta' - n\rho'^{T}\vartheta \,, \tag{2.7.78}$$

where n is the refraction index. Applying the transformation rules from plane 1 to plane 2 for the variables n, ρ_T, ϑ_T, ρ'_T, ϑ'_T, we find that this quantity is constant. It constitutes a generalization of the Smith-Helmholtz-Lagrange invariant in paraxial approximation.

2.8 Apertures of an optical system

2.8.1 Aperture diaphragm and the entrance and exit pupils

Inevitably, real-world optical systems use only a part of the field of view available for the obvious limitation of their physical size, such as the diameter of the lens. On the other hand, even when they have arrangements such as that used to widen the field of application of Paraxial Optic, they provide a satisfactory image only if the angular width and angular opening of the bundles of rays used is suitably limited.

Any optical shooting system of a scene, eye, machine, or projection device (excluding holography) ultimately produces images on a surface, retina, photosensitive plate, or screen, while objects are often three-dimensional. A drastic way to obtain sharp images from a three-dimensional scene is to operate a projection through a tiny hole in an opaque wall. The use of a lens greatly increases the collection of light, but this is paid for in terms of the sharpness of objects that are not *in focus*, although, as we have seen, the greatest longitudinal shrinkage compared to the lateral one is of great help in creating photographs. Here, then, we will see that a diaphragm can be inserted into the optic system to graduate the selection of the rays up to the desired sharpness, or *depth of field*.

In any case, willingly or not, it is necessary to take note of the various apertures present in optical systems. By way of example, we consider the case of a system consisting of two lenses and a diaphragm interposed between them (Fig. 2.51). In general, the beam restriction occurs differently for the rays coming out of the dif-

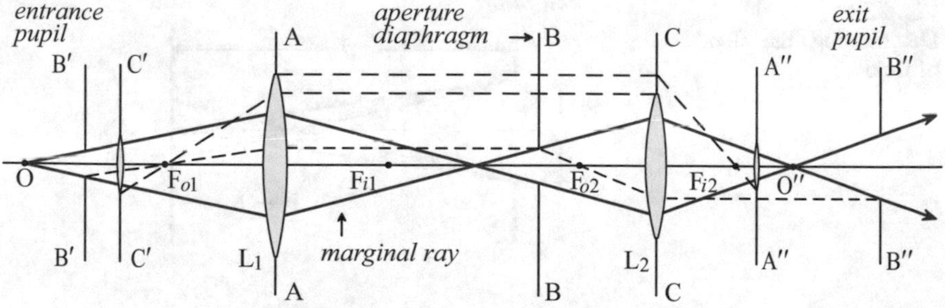

Fig. 2.53 Diaphragms of an optical system

ferent points of the object. The diaphragm that most limits the beam from a given point O on the axis of the system is called the *aperture diaphragm* and the rays grazing its edge are called *marginal*. In the example from Fig. 2.53, the role of aperture diaphragm (the one that most limits the beam of light from the point O) is assumed by the diaphragm BB. A useful way to study the effects of the various apertures is to locate the images (real or virtual) of each element of the instrument in the object space or the image space. Thus, in the figure, the image B′B′ of the diaphragm BB and the image of the lens L_2 are drawn in the object space, while, in the image space, the image B″B″ of the diaphragm and that of the lens L_1 are shown. The real aperture or its image that limits the entry beam more than all of the other openings, as they are seen from the object point O is called the *entrance pupil*. The aperture or its image that limits the outgoing beam from the optical system is called the *exit pupil*. It is obvious that the entry and exit pupils are conjugated to each other by the entire optical system.

In some important cases in which the object to be reproduced is an illuminated aperture, even a slide or transparency, the aperture diaphragm may be the frontier of the light source or the elements (condenser lens, among others) used for lighting and must be taken into account when determining the pupils. In any case, the brightness of the image depends on the area of the pupil, which is proportional to the area of the aperture diaphragm.

2.8.2 Relative aperture, or f-number, and numerical aperture

Anyone who has used an old normal camera (not one that does everything all by itself) may have noted that the diaphragm positions (which limit the useful diameter D of the objective) are indicated by a characteristic series of numbers. They grow step by step by a factor of about $\sqrt{2}$, corresponding to a decrease of a factor of 2 in the collected light. Indeed, every step forward of the diaphragm can be counterbalanced by a doubling of exposure time. If $\hat{I}_o$ is the intensity emitted forward for the unit solid angle from a surface element A_o, the total intensity received on its image A_i, according to Eq. (2.6.14), is

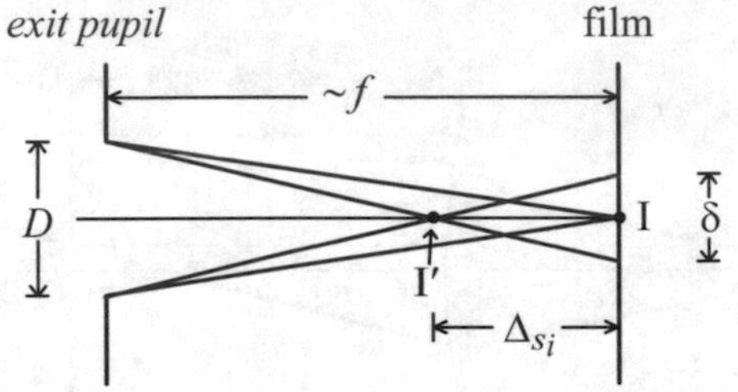

Fig. 2.54 Determining the depth of field

$$I_i \cong \hat{I}_o \frac{\pi}{4}\frac{D^2}{s_o{}^2}\frac{A_o}{A_i} = \hat{I}_o \frac{\pi}{4}\frac{D^2}{s_i{}^2} \cong \hat{I}_o \frac{\pi}{4}\frac{D^2}{f^2},$$

where the final step is valid in normal conditions of the use of a photo camera. The numbers indicated on the diaphragm are thus the dimensionless ratio *f/D*, called the *relative aperture* or *f-number*:

$$f\text{-number} \equiv \frac{f}{D}. \tag{2.8.1}$$

This is used to characterize the lenses, in particular, to indicate their ability to concentrate the light. The symbol *f/#* is a peculiar way to indicate the *f*-number of a lens; for example, a lens with 50 mm focal length and a 25 mm aperture is indicated by the symbol *f*/2, where the 2 stands for its *f*-number. In agreement with the photographic objectives, a lens is called *fast* if it has a small *f*-number, because, with it, the exposure time would be smaller, and *slow* in the opposite case.

The *numerical aperture* (abbreviated to NA) is, by definition, the *sine* of the angle ϑ_m that a marginal ray passing through the focal point makes with the optical axis:

$$\mathrm{NA} = \sin \vartheta_m \approx \frac{1}{2f\text{-number}}, \tag{2.8.2}$$

where the last equality holds *only* in the paraxial limit, for a small angle.

Lastly, a nod to the *depth of field* (Fig. 2.54). With a good objective, the points on the object plane O conjugated to the film are sharply reproduced; instead, the points on another plane O′ produce a circular spot (if the diaphragm is circular) with a diameter δ that grows with the distance Δs_o between the planes O and O′, and proportionally to the diameter *D* of the diaphragm. Now, for simplicity, we assimilate the objective to a thin lens, with the diaphragm coinciding with it. Taking the distances from the focal planes, and remembering the Huygens-Newton Eq. (2.6.35.b), for which $\Delta Z_i = f^2 \Delta(1/Z_o)$, we have

$$\delta \cong \frac{D}{f}\Delta s_i = Df\,\Delta\left(\frac{1}{Z_o}\right) \cong Df\,\Delta\left(\frac{1}{s_o}\right).$$

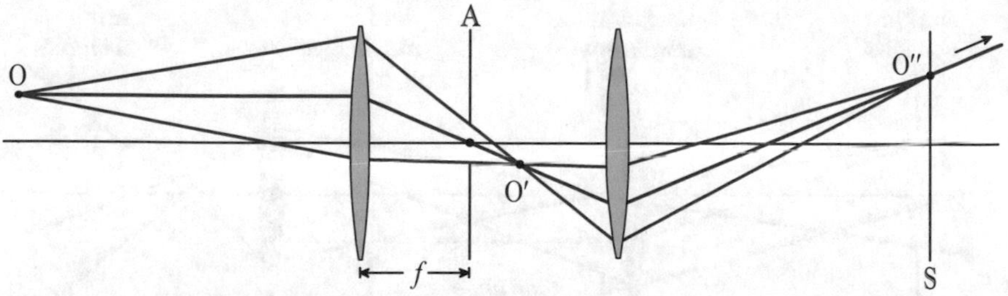

Fig. 2.55 Telecentric system: the aperture diaphragm, A, is located on the second focal plane of the optical elements that precede it, in this case, the first lens. The height of the point O″ does not depend on the longitudinal position of the point O

An alternative way to do this would be to use Eq. (2.6.20) of longitudinal magnification. If we now fix a maximum value for δ, say, 0.02 mm, we can get the amplitude of the intervals Δs_{o1}, Δs_{o2} around s_o, i.e., the depth of field, so the image will still be sufficiently sharp:

$$\Delta s_{o1} \cong \frac{1}{\frac{1}{s_o}+\frac{\delta}{Df}} - s_o, \quad \Delta s_{o2} \cong \frac{1}{\frac{1}{s_o}-\frac{\delta}{Df}} - s_o \,. \tag{2.8.3}$$

With the value of δ considered above, and with $f = 5$ cm, $f/\# = f/4$, $s_o = 5$ m, the overall depth of field is ≈ -0.7 m $\leftrightarrow +1$ m.

2.8.3 Principal rays

The rays that pass through the center of the aperture diaphragm are called *principal rays.* In the paraxial approximation, they also pass through the center of the entrance and exit pupils. These centers can be thought of as the projection centers between object and image, with an angular magnification ratio given by Eq. (2.6.18), where $m_\perp$ is the magnification between the pupils.

2.8.4 Field diaphragm and entrance and exit windows

The position of the entrance pupil with respect to the position of the object has a certain importance. Here, I want to point out only one particular case: when the aperture diaphragm is located in the focal plane of the set of system elements that preceded it, then the entrance pupil is located at infinity. Consequently, all of the principal rays in the object space are parallel to the optical axis. Under these conditions, the system is called *telecentric*. This type of configuration finds applica-

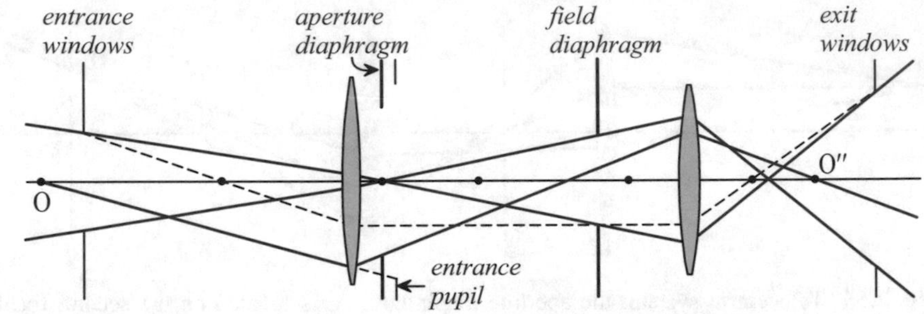

Fig. 2.56 Field diaphragm and entrance and exit windows

tion in measuring microscopes and in optical scanners. In the first case, we take advantage of the properties of these systems whereby, if an extended object goes out of focus, its image blurs without changing its size.

To understand this fact, let us return to the example from Fig. 2.53, however, placing the aperture diaphragm on the focal plane of the first lens and taking, as our source, a bright spot O put off the axis, as shown in Fig. 2.55. If the object point O is moved longitudinally (parallel to the optical axis, and therefore along its principal beam if the system is telecentric), its images O′ and O″ move along the same ray in the respective spaces and the light spot on the screen enlarges, but does not translate along the screen. If we have two points O, the centers of the respective spots on the screen remain at the same distance.

We have understood that the aperture diaphragm corresponds to one of the system apertures, that is, the *smaller* one, as is "seen" from the object plane. The remaining apertures have the effect of limiting the *field of view* of the instrument. In particular, the diaphragm that most limits the angular aperture of the principal rays is called the *field diaphragm*; its image in the object space is called the *entrance window*, and, similarly, in the image space, there is an *exit window* conjugated to it (Fig. 2.56).

2.8.5 Vignetting

The principal rays constitute only a particular set of the multitude that contributes to image formation, and the finite size of the entrance pupil generally shades the edge of the instrument's field of view. In other words, the various points of the object illuminate the image plane with an efficiency that depends on their position and that generally decreases continuously from the axis to the edge. This is due to the shadow and penumbra effects produced by the succession of openings (Fig. 2.57). This variation of illumination produced by the instrument is called *vignetting*. To avoid this effect, the output window is placed on the image plane (and, at the same time, the input window is placed on the object plane). However, this does not always ensure that a third opening will not produce any other vignetting.

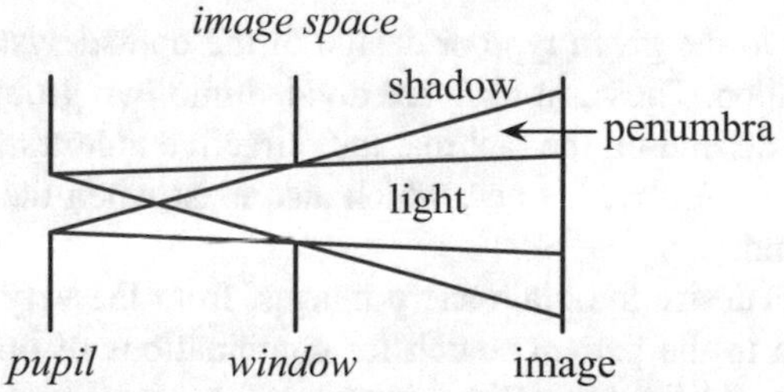

Fig. 2.57
Vignetting

Typically, in visual instruments "adjusted for infinity", the field diaphragm is placed in the object focal plane of the eyepiece.

2.9 Aberrations

So far, we have limited ourselves to considering optical systems in the paraxial approximation for which each point of an object is reproduced in a "sharp" manner, that is, we have considered ideal optical systems in which all of the rays that start from an object point are deflected to a single point image. This is usually only possible by limiting the field of the instrument, thus limiting the collected light energy; or in particular cases, for which Fermat's principle of the equality of the optical paths of all of the rays that form the image is exactly applied. Of these cases, we have seen or recalled some examples of aspheric surfaces, aplanatic lenses and media with variable refractive indexes.

In order to have luminous optical systems, it is often necessary to abandon the paraxial theory. However, it is not always possible to use particular optical components, both for convenience and for economy. Systems with aspherical surfaces are not free from defects for object points that are not in the special positions for which the surfaces are made, for example, points not near the optical axis. Moreover, at least for now, aspherical surfaces or variable index media cannot be produced with good optical quality unless at a high cost, so they are often justified only in special applications: the most remarkable example is the case of astronomical telescopes. Conversely, the production of homogeneous optical glasses with optically polished spherical surfaces has reached a high level of excellence and good economic efficiency.

In fact, it is not possible to produce such a sharp and distortion-free image as the one predicted by the homography, in which all of the rays of an object point are deflected so ideally. The discrepancies between the real and the ideal image generally take the name of *aberrations*. However, it must be noted that defects in real images can have multiple causes. The first defects come from the irregularity of the optical surfaces and the inhomogeneity of the optical materials. There are also *errors of positioning* in the mechanical mounting in the system. These defects are individual, in the sense that they vary from sample to sample. The second type of defect is classified as *aberrations* in the strictest sense, being that they are inherent to the given type or design of the optical system, and do not depend on its realization. These, in turn, are divided into two groups: the *chromatic* aberrations, which depend on the fact that the refractive indexes vary with the wavelength, and the *monochromatic* ones, which also exist when the radiation has a narrow spectral band.

This desire to obtain sharp images, from the very beginnings of Technical Optic, led to the patient search for combinations of optical elements that best compensate for the aberrations introduced by each surface of the system and by the chromatic dispersion of the media.

The study of aberrations can be said to have begun before that of images: in the fifteenth and sixteenth centuries, a study was made of the caustics produced by spherical mirrors with large angular openings, and then of those produced by glass balls, in this case, experimentally, since the law of refraction had not yet been clarified. For the presence of these caustics, the possibility of obtaining geometric images, as we know it today, was not at all clear at that time. Kepler was the first, at the beginning of the seventeenth century, while studying the functioning of the eye, to have the insight of the importance of a diaphragm, the *pupil*, which, when placed in front of a transparent sphere, allowed the formation of sharp images (*Dioptrice*, 1611). The first lenses corrected for chromatic aberrations were designed around 1729 by Chester Moor Hall (1704-1771) and John Dollond (1706-1761) to be used in astronomy. It then fell to the invention of photography, which required large openings and large angular fields, to give a decisive push to the construction of complex objectives for the correction of aberrations. In particular, J. M. Petzval, around 1840, designed and built his homonymous camera lens with which he demonstrated the success of his calculations, extending Gaussian theory with higher order powers, for the inclination of the rays with the axis. Subsequently, in 1856, L. P. von Seidel developed a formal theory of *third degree aberrations*, to which we will soon see a brief nod. Then, upon the development of these analytical methods, E. Abbe contributed by identifying a particular condition, which takes his name, for the reduction of one of the Seidel aberrations, the *coma*.

With the advent of computers, analytical methods remain an important tool for the first stage of optical system design: the analysis provides the general principles for navigating amongst the innumerable possibilities of realizing the systems, while the accumulated experience over time has made optical design a true art. Once the purposes of the system and its constitution have been established, the optimization of the parameters is entrusted to the calculation programs, assigning a merit factor appropriately weighted for the various aberrations, depending on the chosen purposes. Optimization methods generally consist of tracing a set of rays from a given point object through the system and minimizing the deviation from the image point. The result is a merit function that looks like an erratic terrain with peaks and valleys in a multidimensional space, where the program can determine a minimum. However, nothing ensures that this is *the* minimum, if not the experience and/or the repeated tests in other parameter regions.

There are basically two ways to characterize aberrations. The first is to reconstruct the wavefront on the output pupil and measure its deviation from a spherical reference surface. The second is to measure the distance with which the rays lack their paraxial image point. One of the ways to display the aberrations graphically is thus to draw these deviations as a function of one of the ray parameters, in particular, the height of its intersection on the pupil. Another way is to consider a finite number of rays that, for a given object point, intersect the pupil on a regular pattern of points: with these rays, their arrival point is then marked on the image plane. The figure that is obtained (called the *spot diagram*) allows for having an approximate idea of the intensity distribution on the image plane. However, the actual intensity is much more difficult to calculate, and, in general, diffraction phenomena, which are particularly evident with quasi-monochromatic light, should be taken into account. A further method is to draw the caustic, i.e., the surface formed by the curvature centers of the emerging wavefront from the system, for various positions, in axis and off axis of an object point [Shealy 1976]. These centers can be determined through generalized tracking of the rays (see §2.7.2).

The choice to consider the exit pupil as a place on which to determine the wavefront emerging from the instrument may seem arbitrary, but it is not. It represents the aperture diaphragm (or its image), which is decisive in varying the magnitude of the aberrations, both by its size and its position.

2.9.1 The aberration functions for an axial optical system

Consider an optical system with symmetry of rotation. We will devote ourselves only to aberrations in the strictest sense, omitting those arising from position defects or irregularities of the optical elements. For this system, we choose an object point O and we denote its paraxial image with I (Fig. 2.58). We then trace a ray that reaches the image space from O intersecting the paraxial image plane in I′, generally different from I. Near the exit pupil, we also consider a wavefront F of the rays emitted from O and a spherical surface S centered in I (Fig. 2.59). We choose both surfaces passing through the center C of the pupil. Our radius intersects the surfaces F and S at two points that we respectively denote with F and S. The optical path difference Λ = [FS] between these two points represents a measure of ray aberration. It can be written as Λ = [OS] – [OF] = [OS] – [OC], where we used the fact that the points C and F are on the same wavefront, and it is a function of the coordinates of the points O and S. By choosing the optical axis as the z-axis, the aberration function of the wavefront is given by

$$\Lambda = V\left(x_O, y_O, z_O; x_S, y_S, z_S\right) - V\left(x_O, y_O, z_O; 0, 0, z_C\right), \tag{2.9.1}$$

where $O = (x_O, y_O, z_O)$, $S = (x_S, y_S, z_S)$, and $C = (0, 0, z_C)$, while V is the *point characteristic* [Born and Wolf 1980] introduced by Hamilton and is simply the optical path between the two extremes of the ray from which it is determined. Eq. (2.9.1)

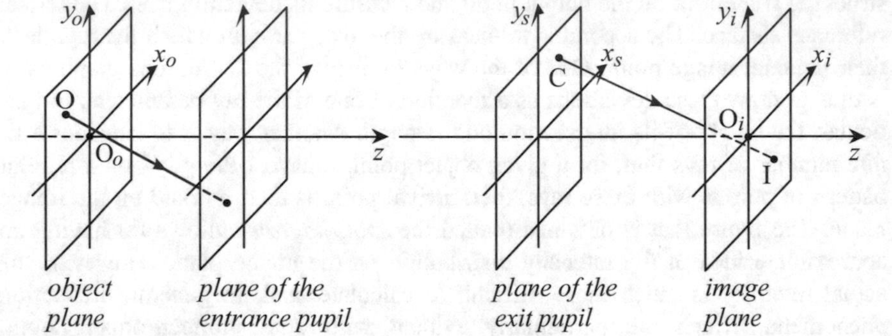

Fig. 2.58 Characteristic planes for determining aberrations

allows us to highlight the dependence of Λ from the coordinates of the points O and S. On the other hand, the coordinates of S are not independent, because they have to respect the condition

$$\left(x_s - x_i\right)^2 + \left(y_s - y_i\right)^2 + \left(z_s - z_i\right)^2 = R^2, \tag{2.9.2}$$

where I = (x_i, y_i, z_i) is the paraxial image point of O and R is the radius of the sphere S.

If we fix the position of the object plane, that of the image plane is also determined. Let us take separate origins on these planes, respectively, O_o and O_i, for the object space and the image space (Fig. 2.59). So, we have $z_o = 0$, $z_i = 0$ and $z_c = -d$, where d is the distance between image and pupil plane. In conclusion, Λ can be described only in terms of the transversal coordinates of O and S[19]:

$$\Lambda = \Lambda\left(x_o, y_o; x_s, y_s\right). \tag{2.9.3}$$

On the other hand, x_o, y_o and x_s, y_s represent the coordinates of two vectors, $\boldsymbol{r}_o$ and $\boldsymbol{r}_s$, respectively. For the rotational symmetry assumed for the system, we have that Λ does not vary if these vectors simultaneously rotate around the optical axis. So, Λ can only depend on

$$r_o^2 = x_o^2 + y_o^2, \quad r_s^2 = x_s^2 + y_s^2, \quad \boldsymbol{r}_o \bullet \boldsymbol{r}_s = r_o r_s \cos\theta = x_o x_s + y_o y_s, \tag{2.9.4}$$

where θ is the azimuthal angle between the two vectors. Therefore, Λ must be a function of even degree in the coordinates. If we develop it in series, we have only

[19] Normally, in texts that deal with the aberration theory in depth, the special Seidel variables are used, which, in particular, replace, the ordinary coordinates taken on the pupils and the object and image planes, suitably rescaling them. For the sake of simplicity, this rescaling is not used here, and therefore the polynomial coefficients of the aberration function do not coincide with those reported in these texts, but must also be rescaled for a comparison.

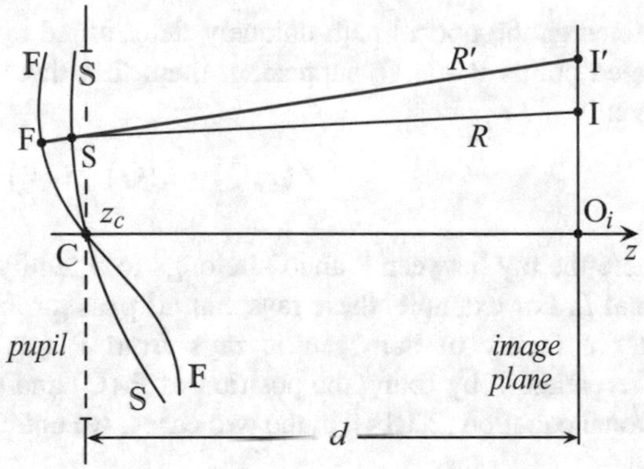

Fig. 2.59 Aberration of the wavefront and aberration of the rays. The FI′ ray does not necessarily lie on the meridian plane of the point I

terms of even degree:

$$\Lambda = \Lambda^{(0)} + \Lambda^{(2)} + \Lambda^{(4)} + \Lambda^{(6)} + \dots \tag{2.9.5}$$

In particular, the development of Λ in polar coordinates up to the fourth degree is:

$$\begin{aligned}\Lambda \cong\ & \Lambda_{000} + \Lambda_{200} r_o{}^2 + \Lambda_{400} r_o{}^4 + \\ & + \Lambda_{020} r_s{}^2 + \Lambda_{111} r_o r_s \cos\theta + \\ & + \Lambda_{040} r_s{}^4 + \Lambda_{131} r_o r_s{}^3 \cos\theta + \Lambda_{222} r_o{}^2 r_s{}^2 \cos^2\theta + \\ & + \Lambda_{220} r_o{}^2 r_s{}^2 + \Lambda_{311} r_o{}^3 r_s \cos\theta,\end{aligned} \tag{2.9.6}$$

where the indexes of the coefficients Λ_{ijk} indicate the power of each of the variables r_o, r_s, $\cos\theta$. The *piston* terms in $\Lambda_{000}, \Lambda_{200}, \Lambda_{400}$ are not geometric aberrations: they are necessarily null if we adhere to the hypothesis that the surfaces F and S intersect the optical axis on the center C of the exit pupil. However, they appear, for greater generality, if the various fronts F, each generated by one of the points on the object plane, are instead determined by taking the same optical path for all of the points, equal to that which, on the optical axis, joins the origin O_o of the object plane to point C at an assigned wavelength. The piston terms imply a variation of the phase with the position of the object point or the wavelength from which their coefficients depend, like all of the other coefficients. They should be taken into account when there is interference between multiple waves.

Of the second-degree terms of the second line, the first is associated with a translation of the image and the second corresponds to a variation of magnification. These terms do not involve true aberrations in the monochrome case and are canceled for the hypothesis that point I is the paraxial image of O, while they are to be considered for chromatic aberrations. The other terms of Eq. (2.9.6) represent the lowest order geometric aberrations of the wavefront on the exit pupil.

From this characterization, we can now determine the deviation of the rays on the image plane from the paraxial image I of O. Since the *point characteristic* is defined as the optical path uniquely determined by a ray traced between two assigned points P and Q, supposing, then, that there is only one of these rays, we have:

$$V(\mathrm{P;Q}) = L(\mathrm{Q}) - L(\mathrm{P}),$$

where the ray between P and Q belongs to a family of rays associated with an eikonal L. For example, these rays can all pass through Q or through P. If we consider a family of homocentric rays from P (or, vice versa from Q) and we differentiate V by fixing the position of P (Q) and changing that of Q (P), for the eikonal equation (2.1.14), in the two cases, we obtain, respectively,

$$\nabla^{(\mathrm{Q})}V = n_\mathrm{Q}\mathbf{s}_\mathrm{Q}, \quad \nabla^{(\mathrm{P})}V = -n_\mathrm{P}\mathbf{s}_\mathrm{P}, \tag{2.9.7}$$

where $\boldsymbol{r}$, n, $\mathbf{s}$ respectively indicate the position, the refractive index and the versor of the ray on the point Q or on the point P, while the suffix in parentheses indicates the point on which we calculate the gradient. Assuming a constant refractive index n_i for the image space, we can now easily compute, from Eq. (2.9.6), the intersection of the ray on the image plane. Indeed, if we associate the point P with an object point O and the point Q with a point S on the reference sphere (hence, not to a point F on the aberrated wavefront), differentiating the function Λ of Eq. (2.9.3), we have

$$\begin{aligned}\frac{\partial \Lambda}{\partial x_s} &= \frac{\partial V}{\partial x_s} + \frac{\partial V}{\partial z_s}\frac{\partial z_s}{\partial x_s},\\ \frac{\partial \Lambda}{\partial y_s} &= \frac{\partial V}{\partial y_s} + \frac{\partial V}{\partial z_s}\frac{\partial z_s}{\partial y_s}.\end{aligned} \tag{2.9.8}$$

From Eq. (2.9.7.a), with $\mathbf{s}_s = (\cos\alpha, \cos\beta, \cos\gamma)$, we then have

$$\begin{aligned}&\frac{\partial V}{\partial x_s} = n_i \cos\alpha = n_i \frac{x_i' - x_s}{R'}, \quad \frac{\partial V}{\partial y_s} = n_i \cos\beta = n_i \frac{y_i' - y_s}{R'},\\ &\frac{\partial V}{\partial z_s} = n_i \cos\gamma = -n_i \frac{z_s}{R'},\end{aligned} \tag{2.9.9}$$

where R' is the distance between S and the point I′ with coordinates $(x'_i, y'_i, 0)$ (Fig. 2.59). Moreover, from Eq. (2.9.2) (with $z_i = 0$), we have

$$\frac{\partial z_s}{\partial x_s} = \frac{x_i - x_s}{z_s}, \quad \frac{\partial z_s}{\partial y_s} = \frac{y_i - y_s}{z_s}, \tag{2.9.10}$$

Inserting Eqs. (2.9.9-10) into Eq. (2.9.8), we finally get

$$x_i' - x_i = \frac{R'}{n_i}\frac{\partial \Lambda}{\partial x_s}, \quad y_i' - y_i = \frac{R'}{n_i}\frac{\partial \Lambda}{\partial y_s}. \tag{2.9.11}$$

Therefore, the lateral aberrations on the paraxial image plane are expressed by a polynomial in which each term is derived by lowering the corresponding term in the development of the aberrations of the wavefront on the exit pupil by one degree. However, Eqs. (2.9.11) contain an implicit dependence on the coordinates x_i, y_i and x'_i, y'_i, as well as x_s, y_s, through the presence of R':

$$R' = d\left[1 + \frac{x_i'^2 + y_i'^2 - 2x_s\left(x_i' - x_i\right) - 2y_s\left(y_i' - y_i\right)}{d^2}\right]^{1/2}. \tag{2.9.12}$$

Given that x_i, y_i and x'_i, y'_i also depend on x_o, y_o and x_s, y_s, even R' can be expressed as a function of these coordinates:

$$R' = R'\left(x_o, y_o; x_s, y_s\right).$$

Usually, the transverse coordinates on the image plane are small in relation to d, and thus one can proceed to a polynomial development of R', which, for the cylindrical symmetry of the optical system, must only contain the products of even degree of the coordinates already described by Eqs. (2.9.4). In particular, the first two terms of this development are

$$R' \approx d - \frac{m_\perp^2}{2d}\left(x_o^2 + y_o^2\right) \cong R, \tag{2.9.13}$$

where $m_\perp$ is the lateral magnification of the system.

In conclusion, the polynomial development of Eqs. (2.9.11) contains only odd degree terms in the transverse coordinates of the object point and the ray position on the exit pupil. Let us now put the point object on the y-axis, so that $x_o = 0$, and we consider the variables r_s and θ as a function of x_s and y_s, whereby

$$x_s = r_s \sin\theta, \quad y_s = r_s \cos\theta. \tag{2.9.14}$$

Therefore, in polar coordinates, the operators $\partial/\partial x_s$ and $\partial/\partial y_s$ become

$$\frac{\partial}{\partial x_s} = \sin\theta \frac{\partial}{\partial r_s} + \frac{1}{r_s}\cos\theta \frac{\partial}{\partial \theta},$$
$$\frac{\partial}{\partial y_s} = \cos\theta \frac{\partial}{\partial r_s} - \frac{1}{r_s}\sin\theta \frac{\partial}{\partial \theta},$$

so that, from Eqs. (2.9.6) and (2.9.11), we have that the lateral aberration expressed up to the third order is

$$\begin{aligned} x'_i - x_i \cong\ & 2\Delta_{020} r_s \sin\theta + \\ & +4\Delta_{040} r_s^3 \sin\theta + \Delta_{131} y_o r_s^2 \sin 2\theta + 2\Delta_{220} y_o^2 r_s \sin\theta, \end{aligned} \tag{2.9.15.a}$$

$$\begin{aligned} y'_i - y_i \cong\ & 2\Delta_{020} r_s \cos\theta + \Delta_{111} y_o + \\ & +4\Delta_{040} r_s^3 \cos\theta + \Delta_{131} y_o r_s^2 (2+\cos 2\theta) + \\ & +2(\Delta_{222} + \Delta_{220}) y_o^2 r_s \cos\theta + \Delta_{311} y_o^3, \end{aligned} \tag{2.9.15.b}$$

where the new coefficients Δ_{ijk} are associated with the previous ones by

$$\Delta_{ijk} \approx \frac{d}{n_i} \Lambda_{ijk}. \tag{2.9.16}$$

This expression comes from having approximated R' with d: if we consider the development of R', we find that it is correct for the coefficients of the first and third order (except Δ_{220} and Δ_{311}), while it is only approximate for the coefficients of the higher order terms. It should also be noted that the aberration functions have been expressed here in terms of the position of the beam on the exit pupil, while, in the tracing of the rays, it is more natural to consider the location on the entrance pupil. However, once the object point for which we want to calculate the aberrations is chosen, the two positions are uniquely linked to each other, in first approximation from a linear relationship whose coefficient is the lateral magnification between the two pupils. In any case, the polynomial development of the aberration functions does not change shape, with only the coefficients varying.

It is also a useful expression for the points of intersection with a different, but still nearby, plane from that image. Just as the deviations of Eqs. (2.9.15) are taken from the paraxial image point I, on this new plane, we can take them from the point of intersection with the principal ray coming from O, which, by definition, crosses the output pupil in its center C and reaches I. Since we are considering a homogeneous image space, where the rays are straight, the deviation varies linearly with the position z of the considered plane, and therefore we simply have

$$\begin{aligned} [x'_i - x_i]_z &\cong \left(1+\frac{z}{d}\right)[x'_i - x_i]_0 - \frac{z}{d} x_s, \\ [y'_i - y_i]_z &\cong \left(1+\frac{z}{d}\right)[y'_i - y_i]_0 - \frac{z}{d} y_s, \end{aligned} \tag{2.9.17}$$

where the deviations are indicated as a function of z (Fig. 2.60). It should be noted that, in these equations, there is a linear dependence on coordinates taken on the

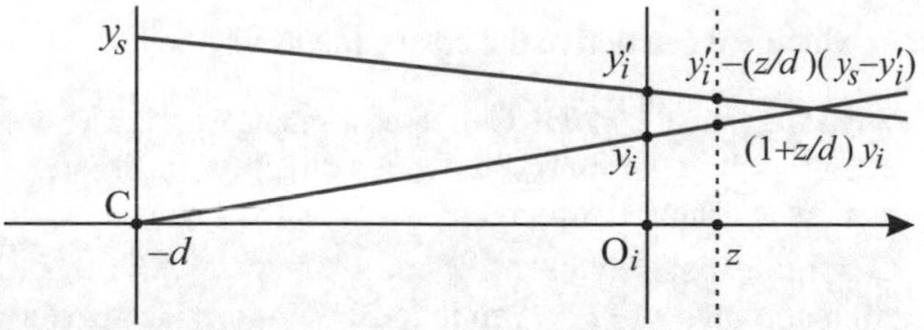

Fig. 2.60 Effect of the image plane translation

plane of the pupil: this justifies what has been stated earlier by asserting that the terms in Λ_{020} and then in Δ_{020} correspond to a translation of the image along the optical axis.

The third order terms correspond to the five monochromatic *Seidel aberrations*: *spherical aberration*, *coma*, *astigmatism*, *field curvature*, and *distortion*, which we will discuss briefly later. Normally, it is not possible to cancel all of these aberrations simultaneously; however, it is useful to consider them one at a time, assuming that we have been able to cancel the others by adjusting the optical system appropriately.

2.9.2 The cosine theorem, Abbe's sine condition and Herschel's condition

Paraxial Optics is based on the assumption that the following quantities are small enough, whereby, in the tracing of the rays, the angles can be approximated by their tangents or their sines. They are: (i) the size of the objects with respect to the longitudinal distances, that is, the *angular field of view* of the instrument, and (ii) the *aperture angle* of the rays admitted by the instrument. Nevertheless, there are cases in which the field of view is limited to small objects near the optical axis, while we want to collect as much light as possible from these objects with a large aperture. To this end, two simple criteria, known as *Abbe's sine condition*, and *Herschel's condition*, are helpful. They are related to the stigmatism of images that can be obtained in these cases. Both conditions can easily be derived from a more general theorem, also valid for inhomogeneous media.

Let us consider the characteristic function V(P,Q) determined between two points P and Q and differentiate it on both extremes. From Eqs. (2.9.7), we can obtain the relationship[20]

$$dV = n_Q \mathbf{s}_Q \cdot d\mathbf{r}_Q - n_P \mathbf{s}_P \cdot d\mathbf{r}_P\,, \tag{2.9.18}$$

[20] This relation is limited at the first order for the displacements of P and Q and does not contradict the Eq. of Collins (2.7.80), which, instead, considers the second order of it, while the first order is subtracted by aligning the z-axis with the main beam and the displacements are on transverse planes.

from which we can derive the *cosine theorem*:

Suppose that the point Q is the stigmatic image of P. We combine these points with an arbitrary ray whose direction in P and Q is expressed by the versors $\mathbf{s}_p$ and $\mathbf{s}_q$, respectively. Let us then consider two more points P′ and Q′ infinitesimally close to P and Q, respectively. In order for Q′ to be the stigmatic image of P′, even in the case of a wide aperture, it is necessary and sufficient that the difference

$$n_Q \mathbf{s}_q \cdot d\boldsymbol{r}_{q'} - n_P \mathbf{s}_p \cdot d\boldsymbol{r}_{p'}, \tag{2.9.19}$$

where $d\boldsymbol{r}_{p'} = \overrightarrow{PP'}$, $d\boldsymbol{r}_{q'} = \overrightarrow{QQ'}$, is independent of the direction of the ray chosen to join P and Q.

To prove that this condition is necessary, suppose that Q′ is the stigmatic image of P′. In this case, neither *V*(P,Q) nor *V*(P′,Q′) depend on the direction of the ray chosen to join the respective pair of points, and therefore their difference *dV*, for which the expression (2.9.19) must be invariant with respect to $\mathbf{s}_p$ and $\mathbf{s}_q$ doesn't either. Finally, to prove that the condition is also sufficient, we join P′ and Q′ with a ray P′Q′ and trace, from P′, a second ray P′Q″ with the same optical length as the first one. Since the theorem assumes that Q is the stigmatic image of P, for reasons of continuity, even the vector $\overrightarrow{QQ''} = d\boldsymbol{r}_{q''}$ must be infinitesimal, and, from Eq. (2.9.18), we have

$$V(P',Q'') - V(P,Q) = n_Q \mathbf{s}_q \cdot d\boldsymbol{r}_{q''} - n_P \mathbf{s}_p \cdot d\boldsymbol{r}_{p'}.$$

On the other hand, we also have:

$$V(P',Q') - V(P,Q) = n_Q \mathbf{s}_q \cdot d\boldsymbol{r}_{q'} - n_P \mathbf{s}_p \cdot d\boldsymbol{r}_{p'}.$$

Since the optical path of the two rays has been chosen as being the same, so that $V(P',Q') = V(P',Q'')$, we have $\mathbf{s}_q \cdot d\boldsymbol{r}_{q''} = \mathbf{s}_q \cdot d\boldsymbol{r}_{q'}$. However, for the arbitrariness of $\mathbf{s}_q$, this is only possible if Q″ coincides with Q′. In other words, Q′ is the stigmatic image of P′. The theorem is thus demonstrated.

So far, we have considered infinitesimal shifts. In order to obtain a stigmatic image of a finite curve, it is necessary and sufficient that the conditions of the theorem are respected for all of its infinitesimal elements. In the derivation of the conditions of Abbe and Herschel, we will consider the particular case of a pair of stigmatic conjugate points taken on the optical axis; moreover, we will use finite displacements, which is equivalent to considering the first order approximation so as to obtain a stigmatic reproduction of a neighborhood of these points.

From the cosine theorem, we can get *Abbe's sine condition* directly

Suppose that a point P and its stigmatic image Q are located on the optical axis of a system, and consider a ray that unites them intersecting the optical axis, respectively, with angles ϑ_p and ϑ_q. Then, let us consider two points P′ and Q′ such that P and Q are, respectively, their projections on the axis, with heights. y_p and y_q. A necessary and sufficient condition for which Q′ is the stigmatic image of P′ is that

$$n_Q y_q \sin\vartheta_q = n_P y_p \sin\vartheta_p \,. \tag{2.9.20}$$

In fact, in the case considered here, the expression (2.9.19) becomes

$$n_Q y_q \sin\vartheta_q - n_P y_p \sin\vartheta_p \,.$$

This difference must be independent of the ray chosen to join P with Q. Since, for a beam coinciding with the axis, this difference is canceled, Eq. (2.9.20) is obtained immediately.

The stigmatic points that meet the sine condition also take on the characteristic of being aplanatic: for comparison, consider the case of aplanatic points of a refractive sphere. In addition, for small angle angles, the sine condition coincides with the expression of the Smith-Helmholtz-Lagrange invariant. Here, I also add a new concept, namely, the *principal surface*, which is a generalization of the principal plane.

In case the object point P is located at infinity, and thus the point Q is on the focal image plane, we can apply the transformation $\sin\vartheta_p \to h/s_p$, where h is the ray height on the principal surface of the object space, while s_p is the distance of the point P from this. Given that we consider small heights y, we can still apply Eq. (2.6.15), and the sine condition becomes

$$\frac{h}{\sin\vartheta_q} = -f_i \,, \tag{2.9.21}$$

where f_i is the focal distance for the image space. Therefore, it follows that the principal surface for the image space is not a plane, but rather a spherical surface with its center in the image focal point. This fact justifies the definition of a numerical aperture, given by the sine of the angle of the marginal rays, and not by the tangent, as one might be tempted to do.

Let us now consider the case of a displacement along the axis of the system. From the cosine theorem, we get *Herschel's condition*:

Suppose that a point P and its stigmatic image Q are located on the optical axis of a system, and consider a ray that unites them intersecting the optical axis with angles ϑ_p and ϑ_q, respectively. Let us then consider two axial points P′ and Q′ close to P and Q, taken from these at distances l_p and l_q, respectively. A necessary and sufficient condition for which Q′ is the stigmatic image of P′ is that

$$n_Q l_q \sin^2(\vartheta_q/2) = n_P l_p \sin^2(\vartheta_p/2). \tag{2.9.22}$$

In fact, for the invariance provided by the cosine theorem, it must be true that

$$n_Q l_q \cos\vartheta_q - n_P l_p \cos\vartheta_p = n_Q l_q - n_P l_p,$$

where the second member corresponds to the case of an axial ray, so the angles are null. From this expression, we can immediately get Eq. (2.9.22). As long as the distances are small, we can use the relation (2.6.21) between the lateral and longitudinal magnification of the paraxial optic, for which

$$\frac{l_q}{l_p} = \frac{n_Q}{n_P}\left(\frac{y_q}{y_p}\right)^2,$$

so that Herschel's condition can also be written in the form

$$n_Q y_q \sin(\vartheta_q/2) = n_P y_p \sin(\vartheta_p/2). \tag{2.9.23}$$

From this expression, we can see that the sine condition and that of Herschel cannot exist simultaneously, except when $\vartheta_P = \pm\vartheta_q$, i.e., for the nodal or antinodal points of the optical system, in which both the lateral and longitudinal enlargements are equal to the ratio of the refractive indices of the object and image spaces. If, on the other hand, we remain within the limits of the paraxial optic, for small angles, the two conditions coincide, with no other restrictions on the positions of P and Q.

2.9.3 Spherical aberration

The terms of the polynomial development of Λ or Δ that do not depend on the distance r_o of the object point from the optical axis constitute the *spherical aberration*. Therefore, for axial object points, it can be studied separately from other aberrations, which are canceled for $r_o = 0$, but it subsists in the same way, even for non-axial object points. The primary spherical aberration, that is, with the lowest order, is expressed by the terms

$$\text{for the wave: } \Lambda_{040} r_s^4, \quad \text{for the ray: } \begin{cases} \Delta^{(3)} x_i = 4\Delta_{040} r_s^3 \sin\theta, \\ \Delta^{(3)} y_i = 4\Delta_{040} r_s^3 \cos\theta. \end{cases} \tag{2.9.24}$$

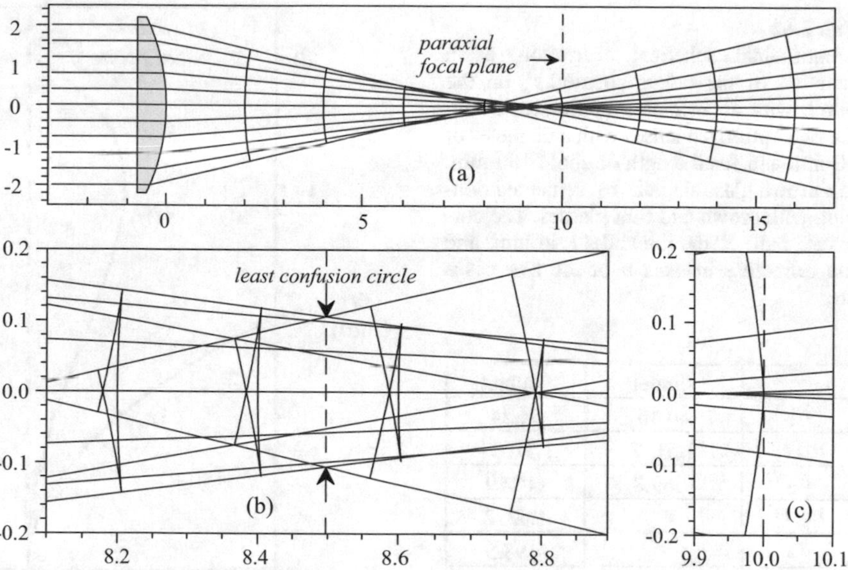

Fig. 2.61 Caustic and wavefronts of a plane-convex lens, with $n = 1.5$, used in the wrong direction for an incoming collimated beam. The vertical and horizontal scales are in the same units of measurement, for instance, cm or mm, with the paraxial focal distance equal to 10 and the lens diameter equal to 4. (b) and (c) are an enlargement of the area around the circle of least confusion and that around the paraxial focal plane, respectively. In this example, the diameter of the circle of least confusion is 0.212, but is reduced to 0.050 if the lens is inverted by placing the convex part facing the collimated beam

An alternative way to characterize this aberration is to express it as a *longitudinal spherical aberration* of the image position, which, at the lowest order, is

$$\left(z'_i - z_i\right)_{ab.sferica} \cong 4d\,\Delta_{040}\,r_s^2\,, \qquad (2.9.25)$$

which can be easily deduced from the second Eq. (2.9.24) for similarity between triangles; for r_s equal to the radius of the pupil, it coincides with the length of the axial focal line of the caustic.

If we divide the pupil into concentric circular zones, we find that the spherical aberration is characterized by the fact that, for each zone, the focal position on the axis depends on their height on the pupil. In particular, the rays of the outer zone, which are called *marginal*, intersect at a *marginal focal point*, which does not coincide with that of the paraxial rays of the innermost zone. As an example, let us focus a collimated light beam from a natural source with a lens and place a screen around the focal plane. Then, we see that, moving away from the lens, the light spot tends to shrink in diameter with an intensity that is larger on the edge, until arriving at an optimum focal position, which corresponds to the *circle of least confusion*, where the light spot becomes smaller. After passing this position, we have a central luminous spot that tends to shrink further, surrounded by a faint halo that

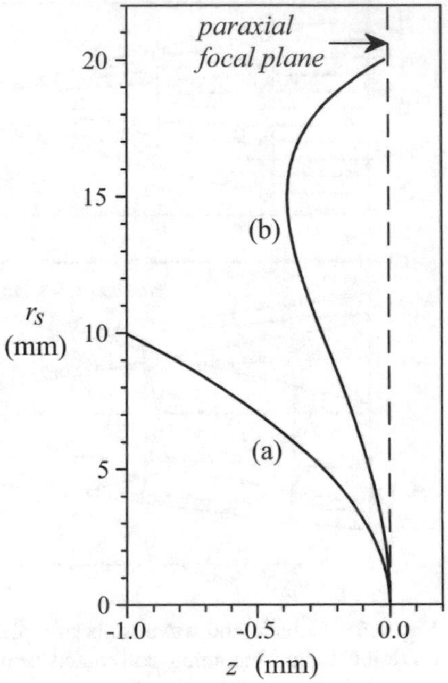

Fig. 2.62
Longitudinal spherical aberration, as a function of the radius position r_s on the pupil, with an incoming collimated beam, for two optimized lenses with a diameter of 40 mm and focal length of about 100 mm. (a): crown glass singlet. (b): cemented doublet, with crown and flint glasses. The curvature radii R, the thickness t, in mm, and the refractive indexes n in the two cases are:

	singlet	doublet
R_1	60.45	54.24
n_1, t_1	1.51, 7	1.51, 11
R_2	-369.8	-44.49
n_2, t_2		1.62, 2
R_3		-198.5

expands rapidly.

Fig. 61 shows the tracing of rays in the case of a plano-convex lens oriented in the direction (the wrong one) that creates the greater spherical aberration. Notice how the wavefront progressively folds in on itself, splitting it into two flaps, which are tangent between them and always remain perpendicular to their generators rays. As the distance from the lens increases, the intersection point of each ray approaches the cusp on the wavefront, which is a point of accumulation for the intersection of the rays (the envelope of the caustic), then moves on the other flap. Finally, on the paraxial focal plane, the wavefront unfolds completely.

If the same lens is turned in the other direction, we get a better focusing of the rays. In fact, spherical aberration depends on the shape of the lens, from its orientation and the conjugate ratio for which it is used. In other words, the lens shape can be optimized for each case so as to minimize this aberration. For an infinite conjugate ratio and common glasses, the plano-convex shape is very close to the best shape. On the other hand, in the laboratory, there is no need to have lenses of all shapes. For any given ratio, just use a pair of plano-convex lenses placed close to each other such that the object focal plane of the first lens is on the object plane and the image focal plane of the second lens is on the image plane. The conjugate ratio for each of the lenses is therefore zero or infinite, and the lenses should be oriented with the curved part in the intermediate space, where the beam of rays originating from each object point is collimated.

In general, simple converging lenses have a *sub-corrected* (negative) spherical aberration, while the opposite is true for divergent lenses, which are generally

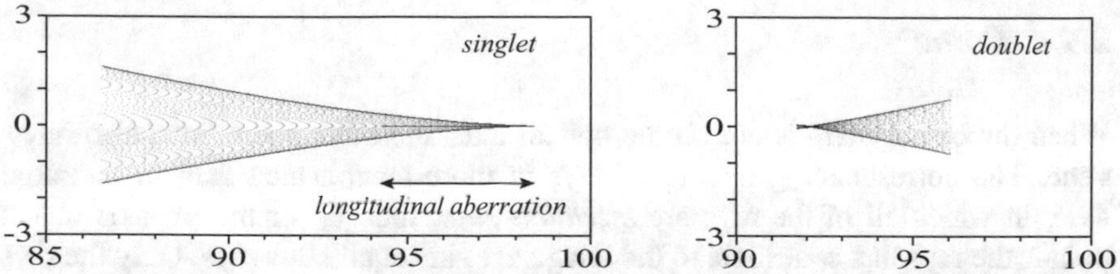

Fig 2.63. Caustics respectively corresponding to the singlet and doublet of the previous figure for a distant axial source. The most marked lines represent the caustic section on a meridian plane, while the points in gray are the projection on this plane of the centers of curvature, calculated with the method of the matrices corresponding to the rays that pass through the points of a squared grid on the input pupil. The abscissa shows the distance in mm from the vertex of the last surface and the ordinate the height from the axis in mm

over-corrected for the same aberration. Combining positive and negative lenses, with opposite spherical aberrations that compensate each other, while keeping the total power to be obtained, we can effectively correct the optical system. This is done, for example, in doublets, in which a positive lens and a negative lens with different refractive indexes are cemented together. In principle, a third-degree spherical aberration could be fully eliminated by adjusting the elements of the optical system; however, for these same elements, there are aberration contribution from the fifth degree up that could not be compensated in the same way. In particular, it may be required that the position of the marginal focus correspond to that of the paraxial one: the position of best focus is an average between those that take place at different r_s and those that generally do not correspond to the paraxial focus (Fig. 2.62). Fig. 63 shows a “dramatization” of a spherical aberration given by the caustics calculated for the lens of the previous figure.

Note that the doublets are generally optimized to be used at full aperture: if narrower beams are used, the result obtained is not the best, and it would be more appropriate to have a doublet of diameter just a little larger than that of the beams. In particular applications, such as the collimation of radiation emitted by laser diodes, a single lens with appropriate aspherical surfaces is used to cancel a spherical aberration.

For a spherical *dioptre* with radius R, we have [Jenkins and White 1957]

$$\frac{n_i}{s_i'(h)} = \frac{n_i - n_o}{R} - \frac{n_o}{\Xi s_o} + \left[\frac{h^2 n_o{}^2 R}{2\Xi f_o n_i}\left(\frac{1}{\Xi s_o} + \frac{1}{R}\right)^2 \left(\frac{1}{R} + \frac{n_i - n_o}{n_o \Xi s_o}\right)\right] + \cdots,$$

where the variables are the same ones already used in §2.6. The focal distance f_o is that of the paraxial limit and s'_i is the actual distance from the vertex of the *dioptre* of the intersection of a ray with the optical axis in the image space. The term in square brackets represents the quadratic correction given by the aberration at the third order (expressed here for the vergence) depending on the height h from the optical axis at which the ray intersects the *dioptre*.

2.9.4 Coma

When the object point is not on the optical axis, the other aberrations also intervene. The corresponding tracing of rays is more complicated than in the axial case, in which all of the rays are *meridian rays*. Indeed, for the off-axis object points, the rays that contribute to the image are almost all *skew rays*. Only the rays contained in the plane determined by the optical axis and the object point are meridians rays. This plane is called the *tangential plane*. Contrastingly, the "broken" plane that, while remaining perpendicular to the tangential plane, contains the principal ray, which changes direction with every refraction or reflection, is called the *sagittal plane*. The rays that lie on this plane are called *sagittal rays*.

Coma and *astigmatism* are caused by a symmetry rupture that results in the principal ray, corresponding to the object point considered, no longer coinciding with the optical axis of the system. The coma owes its name to the aspect of a comet that takes the image of the point object in the absence of other aberrations: substantially, a pointed spot is observed with a bright vertex in correspondence with the paraxial image point and with a tail that fades away from this. For this asymmetry, coma is often considered the worst of aberrations. Analytically, a co-

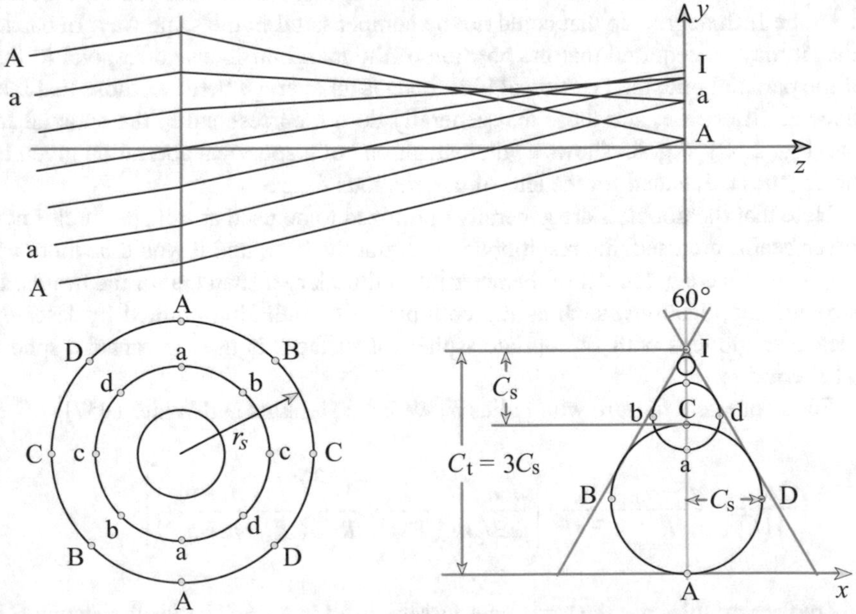

Fig. 2.64 Representation of the coma (for the case of a negative coma). The rays passing through a circle of radius r_s on the pupil arrive at the image plane on a circle whose radius grows with the square of r_s and whose center is transposed relative to the paraxial image point I with the same proportionality. In particular, the two sagittal rays C and the two tangential rays A intersect on the y-axis of the image plane, but each pair does so at a different height

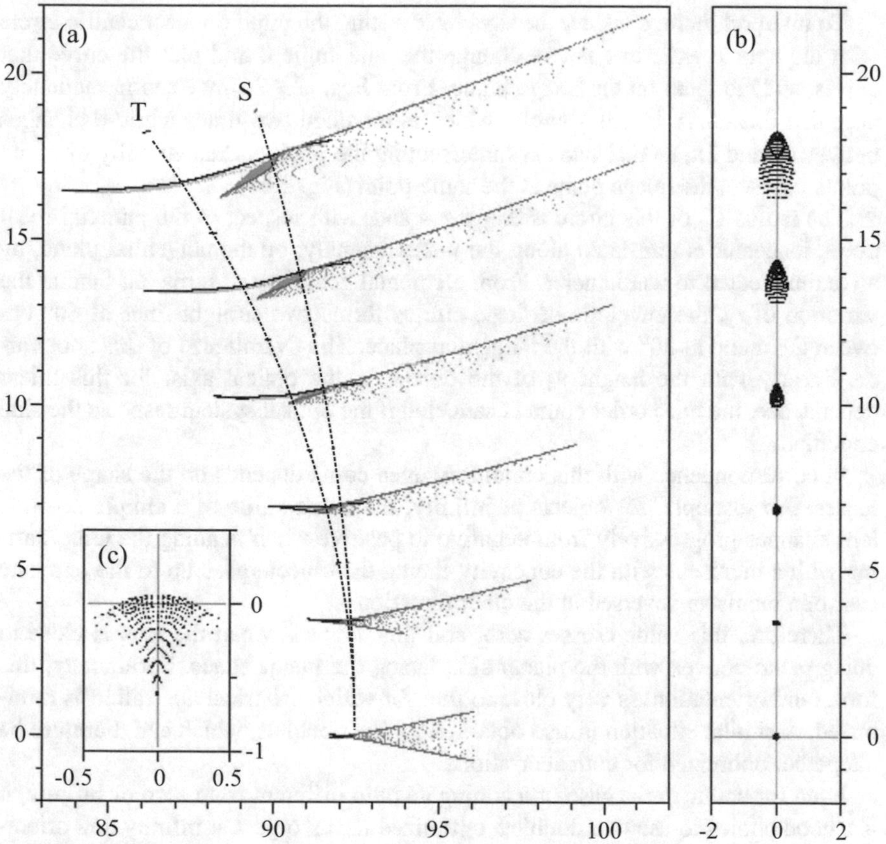

Fig. 2.65 Aberrations of the doublet of Fig. 62. Even these graphics are obtained by the tracing of a set of rays that crosses a regular grid of points on the entrance pupil. (a): caustics corresponding to various angles of inclination ϑ on the plane of the figure of the collimated incident beam, between 0° and 10° with a pitch of 2°. The most marked lines are the section of the caustic surface with the plane of the figure, while the punctuation represents the projection on this plane of the center of curvature, as in Fig. 2.63. The dotted lines T and S also represent the section of the tangential and sagittal focal surfaces. (b): *spot diagrams* on a screen at $z = 92$ mm from the vertex V_3 of the last surface for the same angles ϑ. The origin of the crosshair is placed on the optical axis of the doublet. (c): *spot diagram* for $\vartheta = 10°$. Here, the origin of the crosshair is placed on the central ray of the beam, with a distance from V_3 of 90 mm measured on this ray, halfway between the T and S lines. All scales are in mm

ma of third order is expressed by the terms:

$$\text{for the wave:} \quad \Lambda_{131} r_o r_s^3 \cos\theta,$$

$$\text{for the ray:} \quad \begin{cases} \Delta^{(3)} x_i = \Delta_{131}\, y_o r_s^2 \sin 2\theta, \\ \Delta^{(3)} y_i = \Delta_{131}\, y_o r_s^2 \left(2 + \cos 2\theta\right). \end{cases} \tag{2.9.26}$$

To interpret them, consider the rays intersecting the pupil on a concentric circle with the optical axis, that is, we change the sole angle θ and plot the curve that corresponds to them on the image plane. From Eqs. (2.9.26), we can immediately note that this curve is still a circle, which is travelled two times while θ changes between 0 and 2π, so that two rays intersecting the pupil at diametrically opposite points intersect the image plane at the same point (Fig. 2.64).

The radius C_S of this circle is $\Delta_{131}y_o r_s^2$ and, with respect to the paraxial focal point, its center is translated along the y-axis, namely, on the tangential plane, by an amount equal to its diameter. From elemental geometry, it turns out that, at the variation of r_s, the envelope of these circles forms two straight lines at 60° between them and at 30° with the tangential plane. The overall size of this spot varies linearly with the height y_o of the object on the optical axis: for this linear dependence, the third order coma is canceled if the optical system respects the sine condition.

In correspondence with this condition, even coma depends on the shape of the lenses. For example, for objects at infinity, the coma value of a simple positive lens changes progressively from negative to positive when bending the lens, starting with a meniscus with the concavity facing the object space up to the opposite case of a meniscus reversed in the other direction.

Therefore, this value crosses zero, and this happens when the lens is close to being plano-convex with the planar side facing the image space. Fortunately, this form (and orientation) is very close to that for which spherical aberration is minimized. A similar situation is also obtained for the doublets, which can therefore be altogether optimized for both aberrations.

Even for coma, in the case of a conjugate ratio different from zero or infinite, it is a good choice to use two doublets optimized for an object at infinity and oriented with the convex side in the intermediate space (concave sides and diverging doublets in the case of a negative focal length).

Fig. 2.65 shows the aberrations of the doublet of Fig 2.62 with caustic and spot diagrams.

2.9.5 *Astigmatism and curvature of field*

When the object and its image are relatively far from the optical axis, two other aberrations, *astigmatism* and *field curvature*, become evident. The first produces an effect similar to the homonymous effect that is obtained with a cylindrical lens or a prism. In fact, in relation to the directions indicated by the sagittal and tangential planes, astigmatism manifests itself as a difference of system power in these directions. Instead, the curvature of the field, which does not break the symmetry between the sagittal and tangential planes, causes the rays to intersect on a curved surface instead of the paraxial image plane. Analytically, the terms of the third order of these aberrations are:

for the wave: $\left(\Lambda_{222}\cos^2\theta+\Lambda_{220}\right)r_o{}^2r_s{}^2$, (2.9.27.a)

for the ray:

$$\begin{cases}\Delta^{(3)}x_i=2\Delta_{220}\,y_o{}^2r_s\sin\theta,\\ \Delta^{(3)}y_i=2\left(\Delta_{222}+\Delta_{220}\right)y_o{}^2r_s\cos\theta,\end{cases}=\begin{cases}2\Delta_{220}\,y_o{}^2x_s,\\ 2\left(\Delta_{222}+\Delta_{220}\right)y_o{}^2y_s,\end{cases}\tag{2.9.27.b}$$

where, at the right, they are expressed in Cartesian coordinates for the position on the output pupil. Unlike coma, here, we have a linear dependence on the coordinates on the pupil, and a quadratic dependence on the height of the object, for which these aberrations are especially felt away from the optical axis. Therefore, if we consider a plane different from the paraxial plane, we can cancel the ray dispersion in the x direction or y direction. From Eqs. (2.9.17), we can find two longitudinal positions, z_s and z_t, whereby these two dispersions are respectively canceled:

$$\begin{aligned}&z_s\cong 2d\,\Delta_{220}\,y_o{}^2,\quad\text{for which}\begin{cases}\left(\Delta x_i\right)_s\cong 0,\\ \left(\Delta y_i\right)_s\cong\ 2\Delta_{222}\,y_o{}^2y_s,\end{cases}\\ &z_t\cong 2d\left(\Delta_{222}+\Delta_{220}\right)y_o{}^2\quad\text{for which}\begin{cases}\left(\Delta x_i\right)_t\cong -2\Delta_{222}\,y_o{}^2x_s,\\ \left(\Delta y_i\right)_t\cong 0.\end{cases}\end{aligned}\tag{2.9.28}$$

Then, instead of an image point, we have a *sagittal focal line* for $z=z_s$ and a *tangential focal line* for $z=z_t$. Although the wording could be misleading, the sagittal line lies on the tangential plane and the tangential line lies on the sagittal plane!

If we now consider the quadratic dependence of z_s and z_t by y_o and the axial symmetry of the optical system, we see that, when changing the position of the object point, the tangential and sagittal focal positions form two paraboloidal surfaces. They correspond to a *sagittal focal surface* and a *tangential focal surface*, touching each other on the optical axis. Their curvatures are, respectively,

$$\frac{1}{R_s}=4d\,\Delta_{220},\quad\frac{1}{R_t}=4d\left(\Delta_{222}+\Delta_{220}\right).\tag{2.9.29}$$

For the case of a doublet, these surfaces are shown in Fig. 2.65 by their sections S and T with a meridian plane.

To understand the behavior of astigmatism, it is useful to consider the case of a wheel-shaped object, as in Fig. 2.66. Because of the axial symmetry of the wheel and the optical system, the circle appears in focus on the tangential surface, while the rays are blurred; the opposite happens on the sagittal surface.

When the coefficient Δ_{222} is nulled, the sagittal and tangential surfaces coincide, and therefore the astigmatism vanishes. Then, they lie on what is called

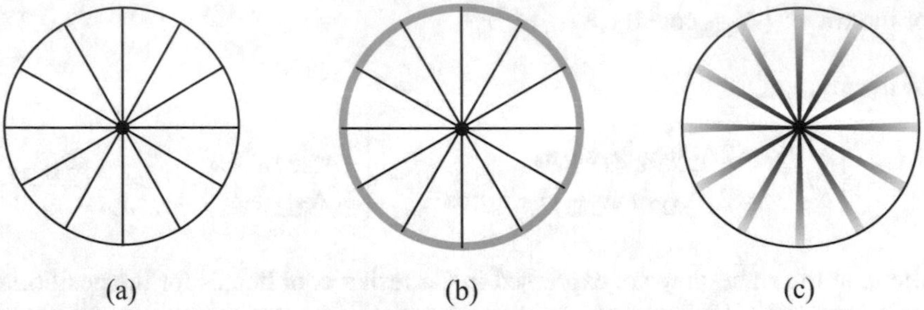

Fig 2.66. Astigmatic images of a wheel. (a): object; (b): image on the sagittal surface; (c): image on the tangential surface

Petzval's surface whose curvature is proportional to the total power of individual system components. Exploiting this property, sometimes a lens called a *field flattener* is placed near the image plane to reduce the field curvature. The Petzval surface does not normally coincide with the surface of the circle of least confusion: in general, the tangential surface is located far from the Petzval surface by a distance three times greater than the sagittal one [Born and Wolf 1980]. Moreover, to the third order of aberration, the two surfaces have the same sign, and thus are on the same side from the Petzval's surface. A doublet corrected for spherical aberration and coma still has a strong astigmatism. This can be reduced with an aperture diaphragm or another suitably arranged lens. With a good optical system consisting of several separate elements, like the Zeiss Sonnar objective, it is possible to compensate for both the astigmatism and the curvature of field; these objectives are therefore called *anastigmatic*. For this purpose, aberrations of superior order are used to fold the tangential surface, making it intersect with the sagittal plane at some distance from the axis, near the image plane. Inside a circle of radius equal to this distance, the image is good, while it degrades quickly outside.

For a thin pencil of rays (think of a laser beam), some results can be obtained beyond the theory of the third order by using the 4x4-matrix method. Thus, for an astigmatic pencil that impinges at an angle ϑ_{I} on a spherical *dioptre* of radius R, with its principal directions of curvature on the tangential and sagittal planes, from Eqs. (2.7.63),[21] we have the Coddington's equations

$$\begin{aligned} \frac{1}{s_{i\mathrm{T}}} &= -\frac{n_o}{n_i}\frac{\cos^2\vartheta_{\mathrm{I}}}{\cos^2\vartheta_{\mathrm{R}}}\frac{1}{\Xi s_{o\mathrm{T}}} + \frac{\cos\vartheta_{\mathrm{R}} - \frac{n_o}{n_i}\cos\vartheta_{\mathrm{I}}}{\cos^2\vartheta_{\mathrm{R}}}\frac{1}{R}, \\ \frac{1}{s_{i\mathrm{S}}} &= -\frac{n_o}{n_i}\frac{1}{\Xi s_{o\mathrm{T}}} + \left(\cos\vartheta_{\mathrm{R}} - \frac{n_o}{n_i}\cos\vartheta_{\mathrm{I}}\right)\frac{1}{R}, \end{aligned} \tag{2.9.30}$$

where $s_{o\mathrm{T}}$ and $s_{o\mathrm{S}}$ are the distances between the pencil centers $\mathrm{O_T}$ and $\mathrm{O_S}$ and the

21 The case in which there is torsion is still solved by Eqs. (2.7.63).

dioptre vertex V, which is taken on the intersection with the central ray of the pencil. ϑ_R is the angle of refraction and $s_{i\mathrm{T}}$ and $s_{i\mathrm{S}}$ are the distances from V of the tangential and sagittal images $\mathrm{O_T}$ and $\mathrm{O_S}$, respectively. All of these distances are then measured along the central ray of the pencil. If this pencil is instead reflected by a spherical mirror of radius R, from Eqs. (2.7.64), we have, with the same meaning for the variables,

$$\begin{aligned}\frac{1}{s_{i\mathrm{T}}} &= -\frac{1}{\Xi s_{o\mathrm{T}}} - \frac{1}{\cos\vartheta_\mathrm{I}}\frac{2}{R},\\ \frac{1}{s_{i\mathrm{S}}} &= -\frac{1}{\Xi s_{o\mathrm{S}}} - \cos\vartheta_\mathrm{I}\frac{2}{R}.\end{aligned} \tag{2.9.31}$$

Finally, if the pencil passes through a thin lens in air on its optical center, we have

$$\begin{aligned}\frac{1}{s_{i\mathrm{T}}} &= -\frac{1}{s_{o\mathrm{T}}} + \frac{1}{\cos\vartheta}\left(\frac{n\cos\vartheta'}{\cos\vartheta} - 1\right)\left(\frac{1}{R_1} - \frac{1}{R_2}\right),\\ \frac{1}{s_{i\mathrm{S}}} &= -\frac{1}{s_{o\mathrm{S}}} + \cos\vartheta\left(\frac{n\cos\vartheta'}{\cos\vartheta} - 1\right)\left(\frac{1}{R_1} - \frac{1}{R_2}\right),\end{aligned} \tag{2.9.32}$$

where ϑ' is the internal angle of incidence in the lens and n is its refractive index. A properly inclined lens, even a small power meniscus, can therefore serve to correct the pencil astigmatism by equalizing the distances $s_{i\mathrm{T}}$ and $s_{i\mathrm{S}}$.

2.9.6 Distortion

If the curvature of field is a longitudinal deformation of the surface on which the image is formed, the *distortion* is its counterpart in directions transverse to the optical axis. The analytical terms to describe the third order are:

$$\text{for the wave: } \Lambda_{311} r_o^3 r_s \cos\theta\,, \qquad \text{for the ray: } \begin{cases}\Delta^{(3)} x_i = 0,\\ \Delta^{(3)} y_i = \Delta_{311} y_o^3.\end{cases} \tag{2.9.33}$$

They express a variation of the magnification when the distance from the optical axis changes. The behavior of the distortion becomes particularly evident when considering a square-shaped object, as in Fig. 2.67. If the magnification decreases with the distance from the axis, the grid image looks like a *barrel*, while, if it increases, the image is deformed to *cushion*. However, distortion by itself does not reduce the sharpness of the image. This type of defect is harmful in photography and in precision applications such as cartography or the production of masks for large-scale integrated circuits, but it can be tolerated in astronomical applications,

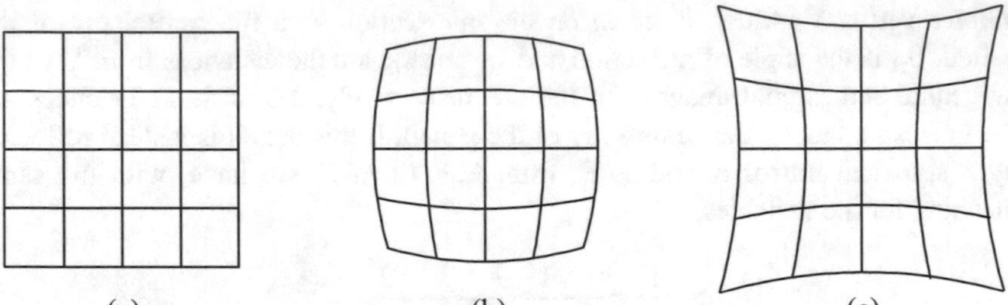

Fig. 2.67 Distorted images of a grid. (a): object; (b): barrel distortion; (c): cushion distortion

in which the most important thing is the resolution of stellar images, while the distortion can be corrected by calculation.

The simplest optical system free of distortion is a simple diaphragm (a hole): in fact, it remains determined by the behavior of the principal rays. In particular, for a symmetrical lens, the distortion is zero: due to the translation of the principal ray, this is strictly true only for a unitary magnification, but that remains true enough for the other working distances. The same can also be said for a generic thin lens, if the aperture diaphragm is put in correspondence with one of its principal planes. If, instead, for a positive lens, the diaphragm is placed before or after these planes, there is, respectively, barrel or cushion distortion. Thus, in general, this aberration can be reduced by using a nearly symmetrical optical system, appropriate for the desired magnification ratio, at the center of which the opening diaphragm is generally positioned. This combination is also useful for reducing astigmatism.

2.9.7 Aberration of a thin lens

For a thin lens of index n in air, with spherical surfaces of radius R_1 and R_2, it is possible to express the Seidel aberrations in a relatively simple manner. Taking the pupils coincident with the principal planes, the coefficients at the 3rd order for the wave are:

spherical aberration:

$$\Lambda_{040} = -\frac{\mathcal{P}^3}{32n(n-1)}\left[\frac{n+2}{n-1}q^2 + 4(n+1)pq + (3n+2)(n-1)p^2 + \frac{n^3}{n-1}\right], \quad (2.9.34)$$

coma:

$$\Lambda_{131} = m_\perp \frac{\mathcal{P}^2}{4ns_i}\left[\frac{n+1}{(n-1)}q + (2n+1)p\right], \quad (2.9.35)$$

astigmatism and curvature of the field:

$$\Lambda_{222} = -\frac{m_\perp{}^2}{2s_i{}^2}\mathcal{P}\,, \quad \Lambda_{220} = -m_\perp{}^2\,\frac{n+1}{4ns_i{}^2}\mathcal{P}\,, \tag{2.9.36,37}$$

distortion:

$$\Lambda_{311} = 0\,, \tag{2.9.38}$$

where $\mathcal{P} = 1/f$ is the power of the lens,

$$q = \frac{R_2 + R_1}{R_2 - R_1}, \quad p = \frac{s_i - \Xi s_o}{s_i + \Xi s_o} = \frac{2}{\Xi s_o \mathcal{P}} - 1 = 1 - \frac{2}{s_i \mathcal{P}} \tag{2.9.39}$$

are the Coddington's form factor and the position factor, respectively, and s_o, s_i are the paraxial object and image positions, respectively.

For an object point on the axis, the deviation on the paraxial image plane is therefore given, from Eqs. (2.9.16), (2.9.24) and (2.9.34), by

$$\Delta^{(3)}r_i = -\frac{s_i r_s{}^3}{8n(n-1)f^3}\left[\frac{n+2}{n-1}q^2 + 4(n+1)pq + (3n+2)(n-1)p^2 + \frac{n^3}{n-1}\right]. \tag{2.9.40}$$

For $n = 1.5$, this deviation can be canceled only within a limited range of values for which $p_2 \geq 21$, $\Xi s_o \leq 0.36f$ for $f > 0$ and $\Xi s_o \geq 0.60f$ for $f < 0$ with R_1, $R_2 < 0$, and for the complementary values that we have while reversing the lens. These values correspond to menisci, including those that are aplanatic, converging with $p = 5$ and $q = -4$ and diverging with $p = -5$ and $q = 4$, for which the coma is also canceled. Differentiating this deviation with respect to q, equating the result to 0 and solving for q, we get

$$q = -\frac{2(n^2 - 1)}{n+2}p\,, \tag{2.9.41}$$

which expresses the relationship between p and q that minimizes the spherical aberration in the neighborhood of p.

The radius of the comatic circle (Fig. 2.64) is

$$C_s = \frac{y_i r_s{}^2}{4nf^2}\left[\frac{n+1}{n-1}q + (2n+1)p\right], \tag{2.9.42}$$

where y_i is the height of the paraxial image point. It is canceled for

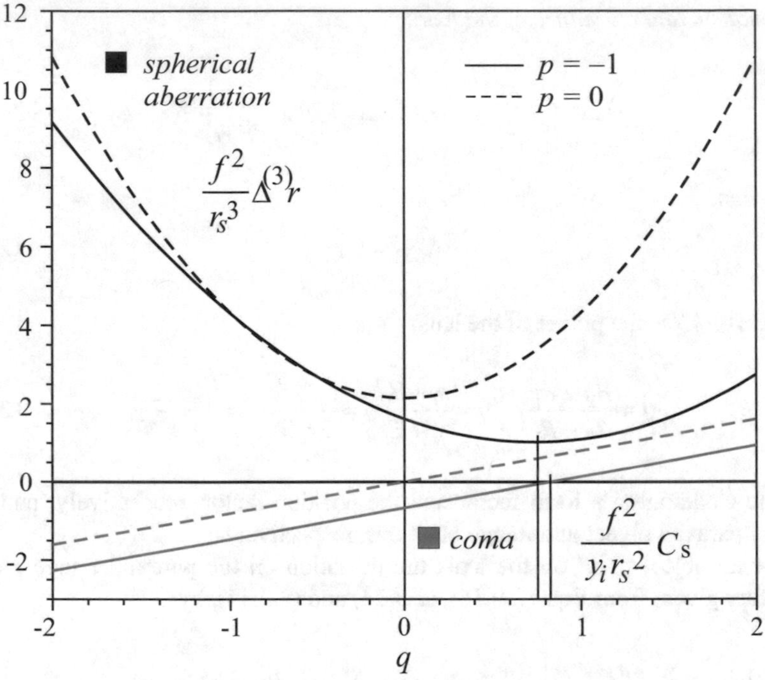

Fig. 2.68 Transverse spherical aberration and radius of the comatic circle of a thin lens with $n = 1.517$ at the variation of the form factor q for two values of the position factor p; (—): $\Xi s_o = -\infty$, (– –): $\Xi s_o = 2f$. The aberration values are normalized

$$q = -\frac{2n^2 - n - 1}{n+1} p \,. \tag{2.9.43}$$

Fig. 68 shows the trend of spherical aberration and coma calculated by means of Eqs. (2.9.40) and (2.9.42).

Finally, the curvatures of the sagittal and tangential surfaces are

$$\frac{1}{R_\mathrm{s}} = -\frac{n+1}{nf}, \quad \frac{1}{R_\mathrm{t}} = -\left(2 + \frac{n+1}{n}\right)\frac{1}{f} \,. \tag{2.9.44}$$

2.9.8 Chromatic aberrations

So far, we have considered the case of non-dispersive media, however, all of the dielectric media, excluding the vacuum, possess a certain degree of dispersion in the refractions that leads to a spectral decomposition of the radiation. This means that, in the optical images, the edges of the figures appear variously col-

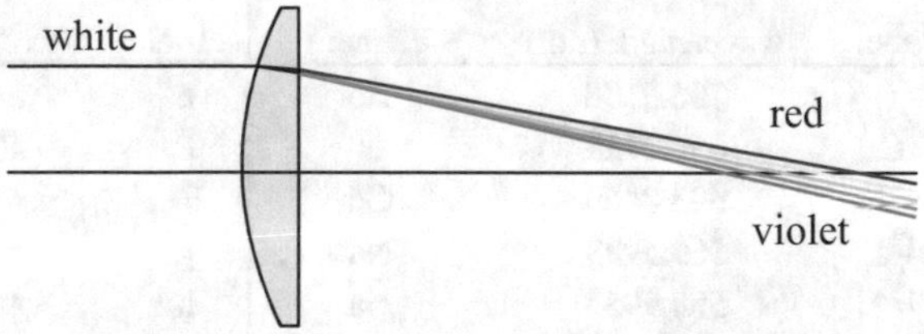

Fig. 2.69 Chromatic dispersion of a simple lens

ored, or even iridescent. This effect can be analyzed by considering the case of monochromatic radiation and assuming a variable refractive index as a function of wavelength for each of the media that make up the instrument. In particular, with a simple converging lens, as shown in Fig. 2.69, and a source point at infinity, the violet component of the light is deflected at a focal point closer to the lens than the red one. Although in different ways, this behavior is common to all passive optical materials, since, in their transparency regions, the refractive index decreases with increasing wavelength (*normal dispersion*). This change accentuates by maintaining the same sign at both ends of the transparency region. To achieve the opposite trend, one should use the material within an absorption band (*anomalous dispersion*), or an *active* medium.

In general, it is found that focal lengths and the principal surfaces of an instrument vary with the wavelength. There are, therefore, two effects of dispersion (Fig. 2.70): the variation in the position of the image plane, which is called *longitudinal* (*axial*) *chromatic aberration*, which corresponds to a loss of sharpness on the image plane at the axis, that is, the *transverse axial chromatic aberration*, and the variation in lateral magnification, that is, the *lateral chromatic aberration*. These aberrations are on the same order of magnitude as those of Seidel, and therefore their correction is equally important. In particular, the lateral aberration depends on the position of the aperture diaphragm and is canceled in the case of a simple lens, when the diaphragm coincides with the lens.

The first person to discover that chromatic aberrations can be reduced by combining optical materials of different dispersing power was Chester Moor Hall, who, in 1733, designed an objective consisting of a crown glass lens and a flint glass lens. By combining these materials, the focal distances and the principal planes can be made equal for two or more wavelengths. For the other wavelengths, however, there remains a residual spectrum. Generally, this coincidence is

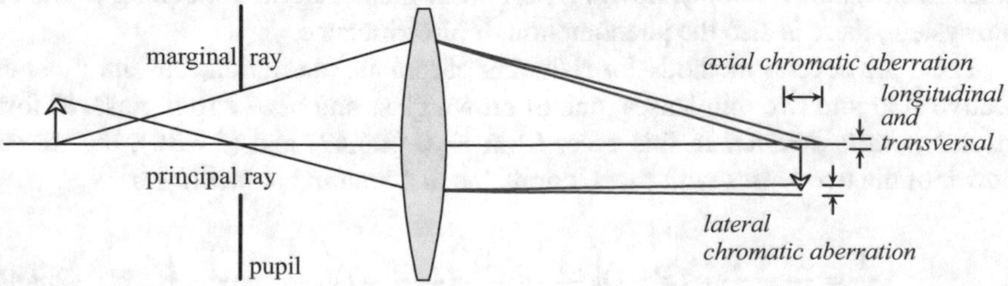

Fig. 2.70 Chromatic aberrations

label	wavelength (nm)	element	label	wavelength (nm)	element
r	706.5188	He	e	546.0740	Hg
C	656.2725	H	F	486.1327	H
C′	643.8469	Cd	F′	479.9914	Cd
D_1	589.5923	Na	g	435.8343	Hg
D_2	588.9953	Na	h	404.6561	Hg
d	587.5618	He			

Table 2.4 Reference wavelengths

obtained for only two wavelengths, and, in this case, the instrument is called *achromatic*, while its spectrum is called secondary. In certain applications, such as in microscopy or astronomy, even this spectrum can be annoying, and therefore one resorts to systems that have a coincidence for three wavelengths. In such instruments, the spectrum is then called *tertiary*. In 1886, at the Zeiss company, Abbe realized microscope objectives in which the position of the image plane has this property, and he introduced the term *apochromatic* [Sivoukhine 1984, §24)]. The choice of wavelengths for which there is the coincidence then depends on the application for which the instrument is being used. For this purpose, some of the wavelengths of the solar spectrum are usually taken as a reference. These were named in a letter from Fraunhofer (Tab. 2.4). The optimization of Commercial Optics is thus achieved with respect to a pair of such radiations, while the *reciprocal* of the *dispersive power* of the dielectric materials is usually indicated by *Abbe's factor*, expressed in terms of the refractive indexes that correspond to three of these wavelengths. In the visible region, the lines that are usually used are the lines F′ (blue), e (green), C′ (red) or the lines F (blue), d (yellow), C (red) and the corresponding Abbe's factors are

$$\mathrm{v_e} = \frac{n_\mathrm{e}-1}{n_{\mathrm{F'}}-n_{\mathrm{C'}}}, \quad \mathrm{v_d} = \frac{n_\mathrm{d}-1}{n_\mathrm{F}-n_\mathrm{C}}, \tag{2.9.45}$$

which, for the optical materials of common use, vary between about 25 and 70. In particular, glasses are called *crowns* when they have a value $\mathrm{v_d}$ greater than 55, or greater than 50 if their index exceeds 1.6, while the others are called *flint*. Higher values for v (and less dispersion) are obtained with crystalline materials such as calcium fluoride (v = 95.4); however, for all of those that do not belong to the cubic system, there is also the phenomenon of birefringence.

There are several methods for reducing chromatic aberration, but the most effective is to join two thin lenses, one of crown glass and one of flint glass, to form an *achromatic doublet*. In this case, from Eqs. (2.6.47) and (2.6.30), the overall power of the two lenses with focal length f' and f'' put in contact ($l = 0$) is

$$\frac{1}{f} = \frac{1}{f'} + \frac{1}{f''} = (n'-1)\left(\frac{1}{R_1'} - \frac{1}{R_2'}\right) + (n''-1)\left(\frac{1}{R_1''} - \frac{1}{R_2''}\right). \tag{2.9.46}$$

Differentiating with respect to the wavelength, we have

$$\Delta\left(\frac{1}{f}\right)_{\mathrm{d}} = \left(\frac{\partial n'}{\partial \lambda_{\mathrm{o}}}\Delta\lambda_{\mathrm{o}} + \cdots\right)_{\mathrm{d}} \frac{1}{(n'_{\mathrm{d}}-1)f'_{\mathrm{d}}} + \left(\frac{\partial n''}{\partial \lambda_{\mathrm{o}}}\Delta\lambda_{\mathrm{o}} + \cdots\right)_{\mathrm{d}} \frac{1}{(n''_{\mathrm{d}}-1)f''_{\mathrm{d}}}, \quad (2.9.47)$$

where line d is taken as the center of differentiation. Considering only the terms to first order in $\Delta\lambda$, the focal length of the lens pair is made achromatic for the lines F and C when

$$\frac{n'_{\mathrm{F}}-n'_{\mathrm{C}}}{n'_{\mathrm{d}}-1}\frac{1}{f'_{\mathrm{d}}} = -\frac{n''_{\mathrm{F}}-n''_{\mathrm{C}}}{n''_{\mathrm{d}}-1}\frac{1}{f''_{\mathrm{d}}}, \quad (2.9.48)$$

where the index n_{d} is taken as an average value between n_{F} and n_{C}. In other words, when

$$\mathrm{v}'_{\mathrm{d}} f'_{\mathrm{d}} = -\mathrm{v}''_{\mathrm{d}} f''_{\mathrm{d}}. \quad (2.9.49)$$

Since, in the normal region, the dispersing powers are always positive, the minus sign in this expression indicates that the focal lengths of the two lenses must be of opposite sign. From the first equality of Eq. (2.9.46), we generally have:

$$f' = f\frac{\mathrm{v}'-\mathrm{v}''}{\mathrm{v}'}, \qquad f'' = -f\frac{\mathrm{v}'-\mathrm{v}''}{\mathrm{v}''}. \quad (2.9.50)$$

From this expression, we see that the focal lengths of the two lenses are proportional to the difference $\Delta\mathrm{v}$ between the Abbe's factors of the glasses used. On the other hand, short focal lenses have the smallest apertures and are subject to greater aberrations than those with long focal length. Therefore, for the realization of achromatic lenses, glasses with $\Delta\mathrm{v} > 20$ are used. The higher order terms in $\Delta\lambda_{\mathrm{o}}$ are then responsible for the secondary spectrum.

Normally, the two lenses are cemented together to form a doublet, so that $R_1' = R_2''$. Once the index n_{d} of the two elements and the overall power are fixed, there is still a degree of freedom to minimize spherical aberration, whereby the doublets in commerce are generally compensated for both of these aberrations.

It should be noted that, with this procedure, we have neglected the thickness of the lenses and, in principle, we only determined the condition for the constancy (to first order) of the focal length of the doublet, neglecting the constancy of position of principal planes. This means that only the transverse chromatic aberration is limited, but, in the case of a cemented doublet, the variation of the principal planes is modest, and one can still speak of *complete achromatism*. In practice, in the case of separate lenses, it is possible to have only a partial achromatism for the focal plane or for the position of the image plane, but not both, unless each lens is already achromatic.

In general, partial achromaticity is also obtained with appropriate combinations of different materials. However, one interesting case is that of the eyepieces of Ramsden and Huygens, which are composed of two lenses made with the same type of glass and separated by a suitable distance for the correction of transverse chromatic aberration. They also have the advantage that their first lens is located near the image plane of the telescope or microscope objective in which they are placed, and thus acts as a field lens, allowing the field of view of the instrument to be increased.

Thus, in the case of two thin lenses separated by a distance l, differentiating Eq. (2.6.47) around the wavelength of line d, we have

$$\Delta\left(\frac{1}{f}\right)_{\mathrm{d}} = \left(\frac{\partial n'}{\partial \lambda_{\mathrm{o}}}\Delta\lambda_{\mathrm{o}} + \cdots\right)_{\mathrm{d}} \frac{1}{(n'_{\mathrm{d}}-1)}\left(\frac{1}{f'_{\mathrm{d}}} - \frac{l}{f'_{\mathrm{d}} f''_{\mathrm{d}}}\right) + \left(\frac{\partial n''}{\partial \lambda_{\mathrm{o}}}\Delta\lambda_{\mathrm{o}} + \cdots\right)_{\mathrm{d}} \frac{1}{(n''_{\mathrm{d}}-1)}\left(\frac{1}{f''_{\mathrm{d}}} - \frac{l}{f'_{\mathrm{d}} f''_{\mathrm{d}}}\right).$$

As in the previous case, keeping only the terms at first order in $\Delta\lambda_{\mathrm{o}}$, the focal length of the lens pair is made achromatic for the lines F and C when

$$l = \frac{\mathrm{v}'_{\mathrm{d}} f'_{\mathrm{d}} + \mathrm{v}''_{\mathrm{d}} f''_{\mathrm{d}}}{\mathrm{v}'_{\mathrm{d}} + \mathrm{v}''_{\mathrm{d}}}. \tag{2.9.51}$$

In the most common case, in which the two lenses are made with the same glass, this condition more simply becomes

$$l = \frac{1}{2}\left(f'_{\mathrm{d}} + f''_{\mathrm{d}}\right) \tag{2.9.52}$$

and the overall power of the lens pair is half of what it would be by joining them.

Since, in a telescope or microscope, the rays coming from the objective are slightly angled relative to the optical axis, it turns out that the angular magnification of the eyepiece depends only on its focal length. This happens either when the eye of the observer is accommodated to infinity, or when it is accommodated at the distinct viewing distance [Sivoukhine 1984, §24)]. This means that the only achromatization of the focal distance of the eyepieces is sufficient for their purpose.

In particular, the Huygens eyepiece consists of two lenses with focal distances and separation in the proportion $f' : l : f'' = 3{:}2{:}1$, which has the property of equally splitting the deviation of the rays between the two lenses, or in proportion 2:3:4, which is preferred in current eyepieces. The Ramsden eyepiece consists of two lenses with the same focal length, but, instead of the proportion 1:1:1, 1:2/3:1 is preferred, even if the chromatic aberration is under-corrected; this can be remedied by replacing the second lens with a properly calculated doublet (Kellner's eyepiece). Indeed, in the first case, the focal planes would coincide exactly with the lenses, so every dust grain deposited therein would be in focus in visual observation.

2.10 Plane mirrors and prisms

Plane mirrors have the remarkable property of reproducing the stigmatic image of the entire object space, and they are therefore very commonly used to deflect light beams without introducing geometrical or chromatic aberrations. In particular, high quality optical planar mirrors are used in two-wave interferometers, of which we will see various examples. Although reflection from a single flat mirror may appear trivial, there is the well-known paradox that, when we look at ourselves in a mirror, it seems that we have exchanged our right with our left, but we do not see our head at the bottom. Actually, the mirror exchanges the front with the back; indeed, we see our face and not the back of our head. In the case of multiple reflection from various mirrors, things become more complicated and not always easy to understand. Here, we will deal briefly with the relationship between the directions of entry and exit of a beam reflected by a series of plane mirrors oriented in any way. This will allow us to calculate the transformation between the corresponding object and image spaces, essentially a rotation and/or reflection, but will not deal with the translation, for which knowledge of the position of each mirror is also required.

In addition to mirrors, a wide variety of prisms has been developed to be used for different purposes in optical instruments and laboratory benches. The prisms are mainly divided into two categories: there are dispersing prisms, used in spectroscopic applications to separate the various spectral components of a beam of light angularly, and reflective prisms, generally with zero power, used to divert the radiation and to rotate or invert the images. They are inserted into optical instruments often just to fold the system in a limited space. There are also rotating, non-dispersive and non-reflective prisms used in optical scans, for example, in cinematography, and anamorphic prisms, which are dispersive, but are used with monochromatic radiation to vary the ratio between the transverse dimensions of a laser beam, in particular, that emitted by a semiconductor laser.

2.10.1 Plane mirrors

The deviation produced by a single plane mirror can be expressed as a linear relationship between the incident and reflected ray versors

$$s_i = \mathbb{S}_{ij} s_j , \tag{2.10.1}$$

where the matrix $\mathfrak{S}$ is given by

$$\mathfrak{S}_{ij} = \delta_{ij} - 2\, p_i\, p_j\,, \tag{2.10.2}$$

where δ_{ij} is 1 for $i = j$ and 0 for $i \neq j$ and p_i are the components of the versor $\mathbf{p}$ perpendicular to the mirror. If the versor $\mathbf{p}$ is reversed, the matrix $\mathfrak{S}$ does not change. It is symmetrical and unitary, with real coefficients, and coincides with its inverse: $\mathfrak{S} = \mathfrak{S}^{-1}$. The three eigenvalues of this matrix are, respectively, −1, 1, and 1, the first of which has $\mathbf{p}$ as an eigenvector: the paradox of the mirror is solved by the fact that just the direction perpendicular to the mirror is *reversed*. The determinant of $\mathfrak{S}$ is equal to −1. It should be noted that, in the case of refraction, there is no linear relationship such as that indicated by Eq. (2.10.1) due to the non-linearity of the law of sines, and not even in the case of a curved reflective surface, in which the deviation also depends on the ray's position.

In the case of a series of reflections from various plane mirrors in succession from 1 to m, the overall matrix is given by the product of the individual matrices:

$$\mathfrak{M} = \mathfrak{S}_m \ldots \mathfrak{S}_2 \mathfrak{S}_1\,. \tag{2.10.3}$$

It is still unitary, and its determinant, which is the product of the individual determinants, is 1 for an even number of reflections and −1 for an odd number. Since, in general, the various matrices $\mathfrak{S}$ do not commute with each other, the matrix $\mathfrak{M}$ does not generally coincide with its inverse, but, as discussed above, it is

$$\mathfrak{M}^{-1} = \mathfrak{S}_1 \mathfrak{S}_2 \ldots \mathfrak{S}_m\,. \tag{2.10.4}$$

Of $\mathfrak{M}$, we can still calculate the eigenvalues Λ and eigenvectors $\mathbf{v}$ for which the following relation holds:

$$\Lambda \mathbf{v} = \mathfrak{M}\, \mathbf{v}\,. \tag{2.10.5}$$

In the three-dimensional space, the solutions to the eigenvalue equation, which has real coefficients, consist necessarily in a real eigenvalue A and a pair of complex eigenvalues B, C conjugated to each other. Therefore, A coincides with one of the values +1 or −1 assumed by the determinant and its eigenvector $\boldsymbol{a}$ corresponds to a beam that, respectively, continues in the original direction or is reflected back. The pair of complex eigenvalues B, C, which correspond to two complex eigenvectors $\boldsymbol{b}$ and $\boldsymbol{c}$, instead indicates a rotation around the axis defined by $\boldsymbol{a}$. Multiplying both sides of Eq. (2.10.5) for $\mathfrak{M}^{-1}$, it is immediately found that the eigenvectors of $\mathfrak{M}^{-1}$ are the same as those of $\mathfrak{M}$, while its eigenvalues are A^{-1}, B^{-1}, C^{-1}. These considerations will be particularly useful in the study of interferometers called *common path interferometers*.

For example, consider a couple of plane mirrors 1 and 2 tilted between them. In this case, $A = +1$, and its eigenvector $\boldsymbol{a}$ coincides with the versor of a ray parallel to both surfaces. A beam perpendicular to $\boldsymbol{a}$ is deflected with opposite rotations for the sequences 1-2 and 2-1 of reflection on the mirrors. If, in particular, the two mirrors are at right angles to each other, the rotation around the axis $\boldsymbol{a}$ is equal to 180°, and the two eigenvalues B, C both assume the value −1. They are therefore degenerate, and any ray that lies on the plane orthogonal to $\boldsymbol{a}$ is reflected back. In the case of three mirrors, however oriented they may be between them and a given sequence of reflections, it is $A = -1$, and therefore there is an eigenvector $\boldsymbol{a}$ for which a ray parallel to it and reflected in that sequence, or, thanks to Eq. (2.10.4), in the opposite sequence, returns back. If the mirrors are arranged, say, vertically, with their normal oriented on the horizontal plane, one of the other eigenvectors lies along the vertical axis. In this case, $B = C = +1$ and the angle of rotation is null.

Finally, if the three mirrors are mutually orthogonal to each other, $A = B = C = -1$. The transformation is thus degenerate and represents an inversion of all three coordinates. Any ray, in any sequence of three reflections, is reflected back.

We will find these effects in some prisms described in the next section.

2.10.2 Reflecting prisms

These prisms use at least one internal reflection, usually for total reflection, otherwise one or more faces are made specular, for example, by means of silvering. The advantage that prisms have in comparison to a mirror combinations is mainly the stability of the angle between the various faces, determined during machining, in addition to the efficiency and simplicity of total reflection, capable of covering a broad spectral band. In general, the entrance and exit faces of the prism

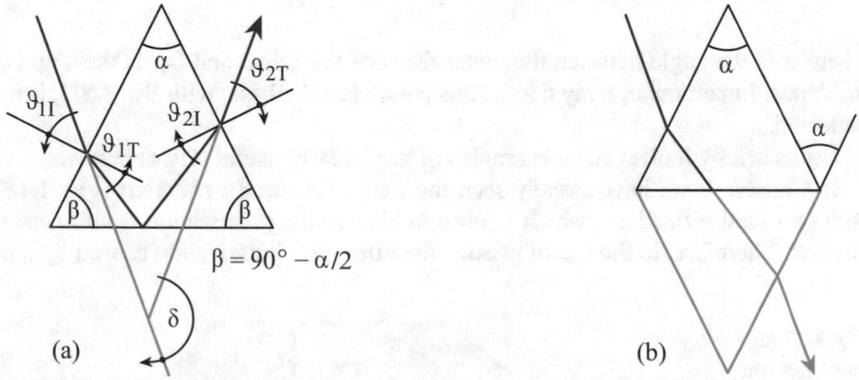

Fig. 2.71 Study of the deviation produced by an isosceles prism. In (b), the prism is unfolded to show the equivalence with a lamina with parallel faces. The arrows indicate the positive directions for the angles ϑ and δ, in accordance with the text

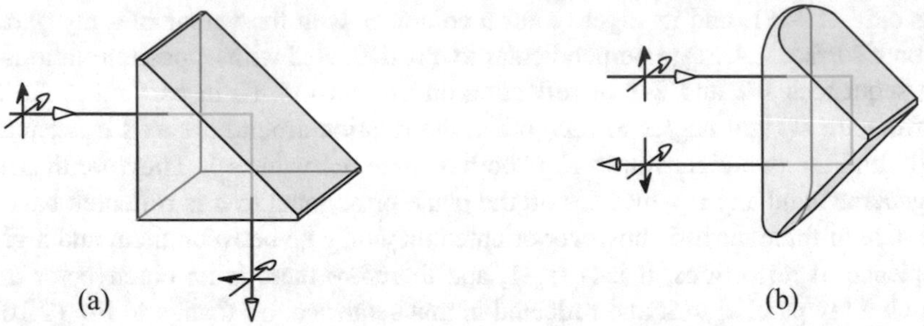

Fig. 2.72 (a): right-angle prism, (b): Porro's prism

are rendered non-reflective with appropriate dielectric layers. In this way, the losses suffered by the beam can be much less than 1%. The disadvantages are mainly due to astigmatism and weight, in the case of large-sized prisms (several cm in size), which are also uneconomical for the cost of the necessary material.

In general, the prisms of this category have their input and output faces arranged so that the (external) angles of incidence on these two surfaces are equal (Fig. 2.71). From the point of view of refraction, the prism thus acts as a parallel face lamina, and the dispersion produces only a translation of the rays corresponding to the different wavelengths, when the incidence is not normal, and not a different deflection.

We can see this by "unfolding" the prism, eliminating any inner reflection and propagating the rays in virtual space beyond each reflecting surface. The most simple example is given in the case of an orthogonal prism with its base in the shape of an isosceles triangle of which the reflection on the unequal face is exploited. If we unfold this prism, we can see that it is equivalent to a parallelepiped; therefore, the emergent ray is always parallel to the incident ray. The actual deflection that a ray undergoes is given by

$$\delta = 2\vartheta_{1I} - \alpha\,, \tag{2.10.6}$$

where α is the angle between the equal faces of the prism and ϑ_{1I} is the angle of incidence. In particular, a ray that beams parallel to the base, with $\vartheta_{1I} = \alpha/2$, is not deflected.

Let us briefly look at some examples of particularly useful prisms below.

In Chapter 1, we have already seen the case of *Fresnel's rhombus* (Fig. 1.18), with two total reflections, which is used to change the polarization from linear to circular. Therefore, in the use of prisms, the effects on polarization caused by total

Fig. 2.73
Dove's prism

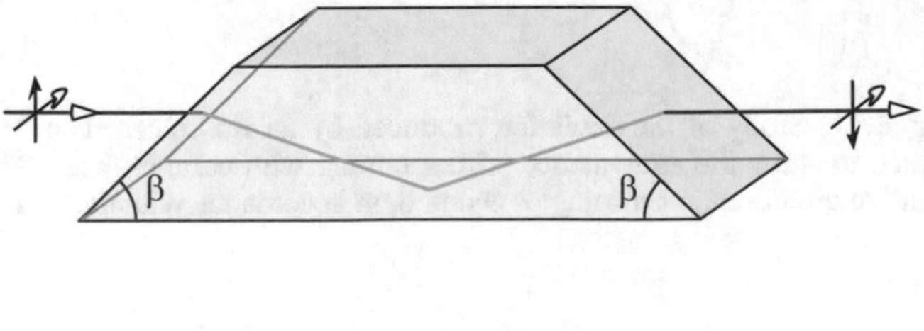

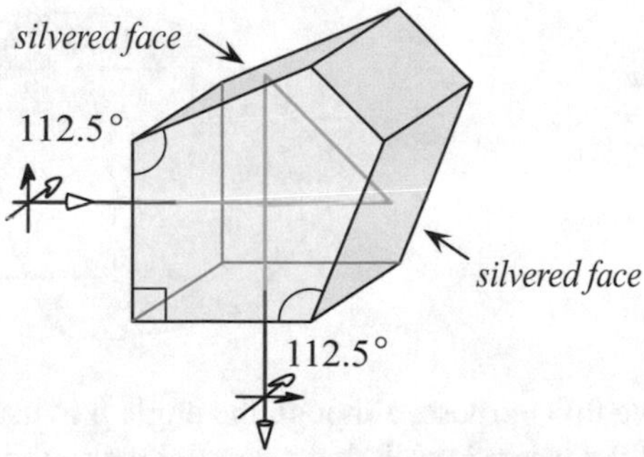

Fig. 2.74
Pentaprism

reflection should be remembered. *Rhomboidal prisms* are also used to translate a beam of light without deviations and keep the orientation of its axes unchanged.

The *right-angle prism* in Fig. 2.72(a) is a particular case of an isosceles prism, which is mainly used for deflecting the radiation by an angle of 90° (see also Fig. 1.17 §1.7.7). Like a mirror, it has the property of producing a mirrored, reversed image of objects that are observed through it.

The *Porro's prism*, depicted in Fig. 2.72(b), has the shape of a right-angle prism, but is employed using its hypotenuse face simultaneously as the input and output face, with a total reflection on each of the cathetus faces. It has the property of reflecting back the incident radiation, with a constant deflection of 180º of the ray projection on the xz-plane perpendicular to the three faces mentioned above, regardless of the prism's rotation about the y-axis, parallel to the edge between the cathetus faces. Since there are two reflections, the image that is observed there (always looking in the direction from which the light comes) does not appear inverted, but rather rotated 180°. In particular, if, on the xz-plane, the z-axis coincides with the incident principal ray, leaving out the translations, the prism operates the transformation $x \to -x$, $y \to y$, $z \to -z$. Porro's prisms are used in pairs in Keplerian type binoculars, with their edges at 90°, in order to straighten the image, while the translation induced by each pair is exploited so as to increase the stereoscopic effect between the two sides of the binoculars. The transformation of the pair is $x \to -x \to -x$, $y \to y \to -y$, $z \to -z \to z$.

Even the *Dove's prism* (Fig. 2.73) comes from an isosceles prism, whose vertex, which is inessential, is cut off to reduce its weight and size. Any isosceles

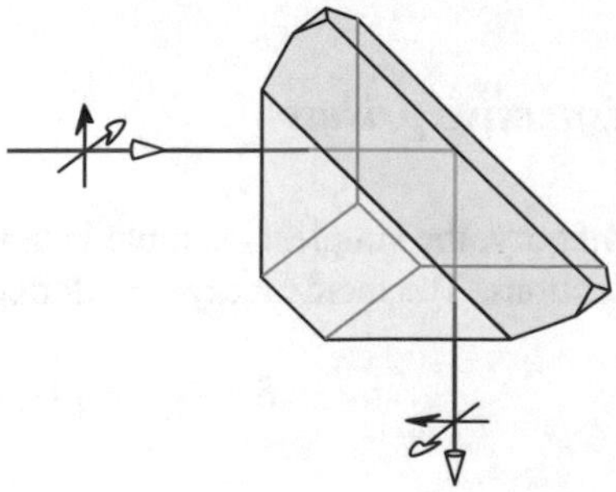

Fig. 2.75
Amici's prism

Fig. 2.76
Reflecting prism
(*corner cube*)

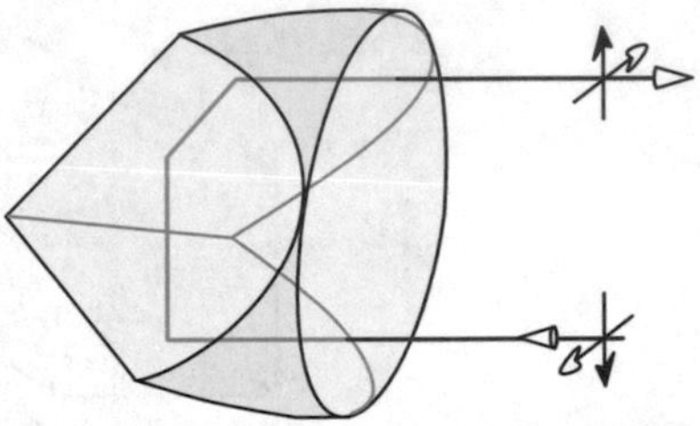

prism can serve this purpose, although the angle β is usually taken as equal to 45º. The cutting of the prism is such that a parallel ray to the reflecting face, incident in the center of the input face, comes out of the center of the output face. It produces an inverted image and it is used in optomechanical devices and periscopes, for its ability to rotate the image at a double speed with respect to the prism's rotation around the longitudinal axis.

The *pentaprism* (Fig. 2.74) is a prism that produces a constant deviation of 90º of incident rays; this occurs by means of two internal reflections, as in the Porro's prism. Thanks to this property, the pentaprism is used as an optical square and as a non-critical alignment element in optical benches. The reflecting faces are at 45° to each other and must be made specular, for example, with silvering, because there is not total reflection for the angle of incidence at which they work. The entrance and exit faces are ultimately at 90° between them

The *Amici's prism* (Fig. 2.75) is similar to a truncated right prism, in which the hypotenuse face is replaced by a *roof*. It consists of two faces at a right angle to each other; it has the effect of dividing the image and exchanging the two halves between them. Therefore, there are two reflections and the image is not inverted, but rather reversed, i.e., rotated 180º. These prisms are expensive, as the angle between the faces of the roof must be accurate within a few seconds of arc to avoid forming two half-images badly connected between them.

The *reflector prism* or *corner cube* (Fig. 2.76) is obtained by cutting the vertex of a cube along a diagonal plane. It has the property of deflecting each ray back parallel in its original direction, after being reflected by each of the three faces of the corner. Some grids of corner cubes installed on the Moon by the astronauts of the Apollo missions and by Soviet probes have allowed for the measurement of the lunar orbit to the precision of cm.

2.10.3 Dispersive prisms

In this category, the simplest example is that of a triangular prism used without internal reflections. The incident rays are deflected by an angle equal to

$$\delta = \vartheta_{I1} + \vartheta_{T2} - \alpha\,, \tag{2.10.7}$$

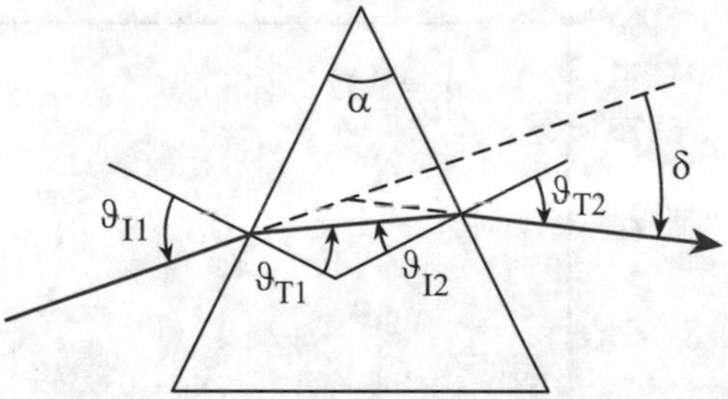

Fig. 2.77
Refraction by a prism

where, with reference to Fig. 2.77, ϑ_{I1} and ϑ_{T2} are, respectively, the angle of incidence on the first face and the transmission angle on the second face, while α is the angle between these faces. In addition, for the inner corners, it is

$$\vartheta_{T1} + \vartheta_{I2} = \alpha\,. \tag{2.10.8}$$

In turn, the angle ϑ_{T2} depends on ϑ_{I1}, α and the refraction index n of the prism. By applying Snell's law and assuming that the prism is immersed in the air, with a refraction index $n_a \approx 1$, it is found that

$$\begin{aligned}\vartheta_{T2} &= \arcsin\left[n \sin\left(\alpha - \vartheta_{T1}\right)\right] \\ &= \arcsin\left[\left(n^2 - \sin^2\vartheta_{I1}\right)^{1/2} \sin\alpha - \sin\vartheta_{I1}\cos\alpha\right].\end{aligned} \tag{2.10.9}$$

The deviation is then

$$\delta = \vartheta_{I1} - \alpha + \arcsin\left[\left(n^2 - \sin^2\vartheta_{I1}\right)^{1/2} \sin\alpha - \sin\vartheta_{I1}\cos\alpha\right]. \tag{2.10.10}$$

Fig. 2.78 shows the typical appearance of the way in which δ depends on ϑ_{I1}. The curve is confined between a maximum deviation, coinciding with the critical angle on the first or second face of the prism, and a minimum deviation. For simple symmetry reasons, the two maxima are equal and the minimum deviation occurs when $\vartheta_{I1} = \vartheta_{T2} = \vartheta_m$, while the internal incidence angles are equal to $\alpha/2$. Indeed, depending on the angle $\gamma = \vartheta_{T1} - \alpha/2$, the ray deviation is given by

$$\delta = \arcsin\left[n \sin\left(\alpha/2 + \gamma\right)\right] + \arcsin\left[n \sin\left(\alpha/2 - \gamma\right)\right] - \alpha\,, \tag{2.10.11}$$

and it turns out that, for $\gamma = 0$, $d\delta/d\gamma = 0$ and $d^2\delta/d\gamma^2 > 0$ for $n > 1$. Therefore, the value $\gamma = 0$ is obtained with the angle of incidence

$$\vartheta_m = \arcsin\left(n \sin\frac{\alpha}{2}\right), \tag{2.10.12}$$

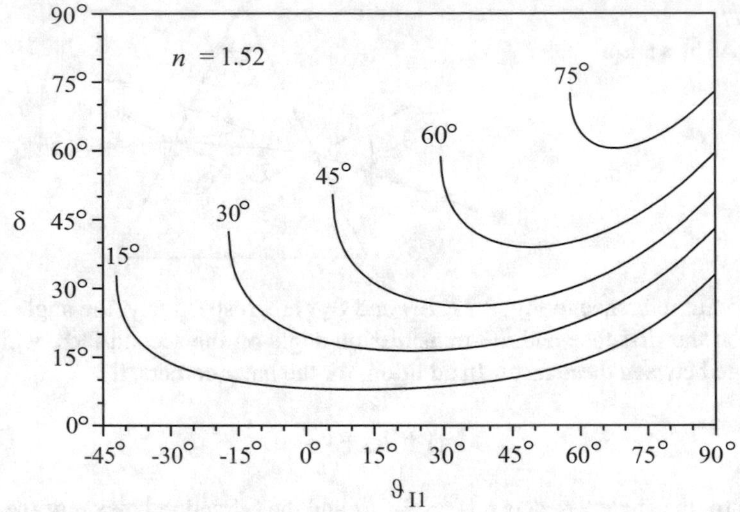

Fig. 2.78 Trend of the deviation δ as a function of the incidence angle ϑ_{I1} for five different values of the angle α between the faces of the prism

which corresponds to the minimum deviation angle

$$\delta_m = 2\arcsin\left(n\sin\frac{\alpha}{2}\right) - \alpha\,. \tag{2.10.13}$$

Not being critical to the rotation of the prism, the measurement of this angle is good for determining its refractive index. Finally, the maximum deviation is

$$\delta_M = \frac{\pi}{2} - \alpha + \arcsin\left[\left(n^2-1\right)^{1/2}\sin\alpha - \cos\alpha\right]. \tag{2.10.14}$$

Associated with the deviation, there is also a variation of the transverse dimensions of the incident beam. In the simplest case of a collimated beam, with the plane of incidence orthogonal to the edge of the prism, there is an anamorphic contraction or expansion of the emerging beam in the transverse direction lying on this plane. The ratio between the widths is given by

$$\frac{l_T}{l_I} = \frac{\cos\vartheta_{T1}}{\cos\vartheta_{I1}}\frac{\cos\vartheta_{T2}}{\cos\vartheta_{I2}} = \left[\frac{1-\frac{1}{n^2}\sin^2\vartheta_{I1}}{1-\sin^2\vartheta_{I1}}\,\frac{1-\sin^2\vartheta_{T2}}{1-\frac{1}{n^2}\sin^2\vartheta_{T2}}\right]^{1/2}, \tag{2.10.15}$$

where ϑ_{T2} can be calculated by Eq. (2.10.9). This ratio varies between 0 and ∞ by changing the incidence angle from its minimum useful value up to 90° (Fig. 2.78).

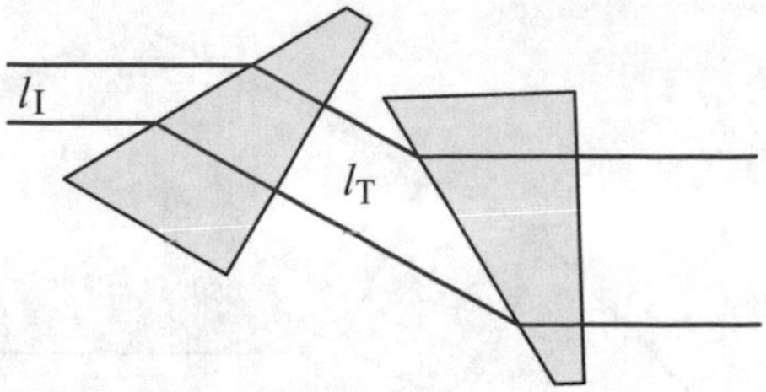

$$\frac{l_T^2}{l_I^2} = 3.1,\ n = 1.762\ \text{(SF11)},\ \alpha \cong 29^\circ$$

Fig. 2.79
Anamorphic prisms for making the beam of a diode laser circular. The input face of each prism is tilted at the Brewster's angle, while the exit face, at almost normal incidence, has an anti-reflection treatment. For a given magnification, the glass type and the angle of the prisms are chosen to approximate these angular conditions, so we have $l_T/l_I = n$

In particular, when the beam is incident normally to the output face, the magnification is

$$\frac{l_T}{l_I} = \frac{\cos\alpha}{\sqrt{1 - n^2 \sin^2\alpha}}. \tag{2.10.16}$$

A prism used for this purpose is called an *anamorphic prism*, and, by rotating it, we may vary a transverse dimension of a monochromatic collimated beam. It is usually preferable to use a combination of multiple prisms on which to divide the variation. A particularly convenient solution is to use two identical prisms arranged with the edges in the opposite direction, so as to cancel the deviation. A pair of anamorphic prisms is generally used to correct the elliptical profile of a diode laser beam (Fig. 2.79).

The dispersive prisms are used in spectrometers or in other apparatuses to decompose the radiation into its components: although their resolving power is generally lower than that of a diffraction grating and their dispersion is non-linear, their spectral brightness is higher and there is no possibility of confusion between overlapping spectral orders. Dispersive prisms are also used in conjunction with diffraction gratings for the sole purpose of separating the various orders from one another, since the direction of dispersion of the prism is arranged perpendicular to that of the grating.

Fig. 80 shows the typical configuration of a prism spectrometer: the dispersion element is placed on a collimated beam produced by an achromatic lens that collects radiation from a slit, while a second lens transforms the angular separation of each monochromatic spectral component into a translation in its focal plane. For each wavelength, on this focal plane, it produces an image of the entrance slit. On this plane, it is possible to put a photographic plate, with which the spectrum is recorded at the same time for all of its wavelengths, or a mobile slit and a detector through which the spectrum is sequentially recorded. In this latter case, the instrument is called a *spectrograph*. Lastly, in the *spectroscopes*, instead of the second lens, a telescope is used to allow for visual observation.

Unlike deviation, dispersion is not symmetrical with respect to the prism's orientation and abruptly rises near the critical angle on the *output* face. However, dis-

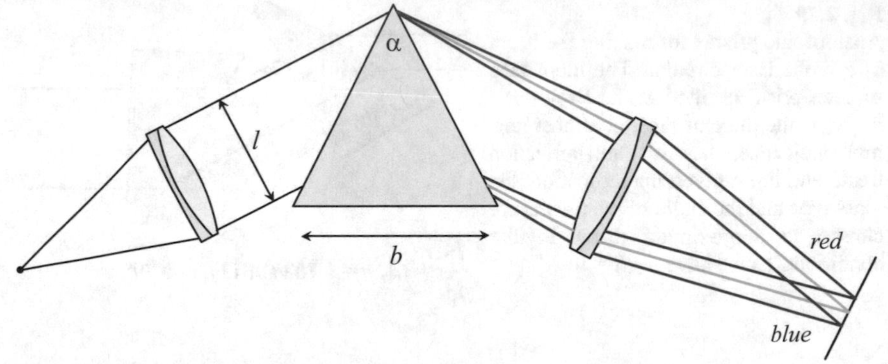

Fig. 2.80 Principle set-up of a prism spectrometer

persing prisms are generally not used in this condition, both for the criticality of alignment and for the loss of resolution associated with diffraction. It is instead preferable to use them near the minimum deviation where the dispersion is more approximately linear and where a monochromatic beam does not undergo anamorphic distortion or astigmatism, in cases in which it is not well collimated. In this position, there is also the minimum loss for reflection of a non-polarized beam. At the minimum deviation angle, for a wavelength λ_c in the center of the spectrum, the prism dispersion is

$$\left.\frac{d\delta}{d\lambda_o}\right|_{\lambda_c,\delta_m} = \left.\frac{dn}{d\lambda_o}\right|_{\lambda_c} \left.\frac{d\delta}{dn}\right|_{\delta_m} . \tag{2.10.17}$$

To complete the calculation, Eq. (2.10.11) is useful: it expresses δ as a function of the angle γ instead of the incidence angle ϑ_{11}. Indeed, at the minimum deviation, both γ and $\partial\delta/\partial\gamma$ are null, and therefore we have

$$\left.\frac{d\delta}{d\lambda_o}\right|_{\lambda_c,\delta_m} = \left.\frac{dn}{d\lambda_o}\right|_{\lambda_c} 2\frac{\sin\frac{\alpha}{2}}{\cos\vartheta_m} = \left.\frac{dn}{d\lambda_o}\right|_{\lambda_c} \frac{b}{l}, \tag{2.10.18}$$

where, in the last step, the ratio of the angular values is expressed by the ratio between the base b of the prism and the width l of the incident beam, supposing that it illuminates the whole entrance face. On the other hand, due to the diffraction, i.e., the limited opening l of the beam, on the focal plane of the second lens, the monochromatic image of the thin input slit consists of a central fringe with the maximum intensity surrounded on both sides by a series of secondary fringes. As we shall see later, the angular separation between the maximum and the first minimum is given by $\Delta\delta = \lambda_o/l$ and, according to Rayleigh's criterion, this separation delimits the spectral resolution between two close spectral lines of equal intensity. The *resolving power* of the prism is therefore given by

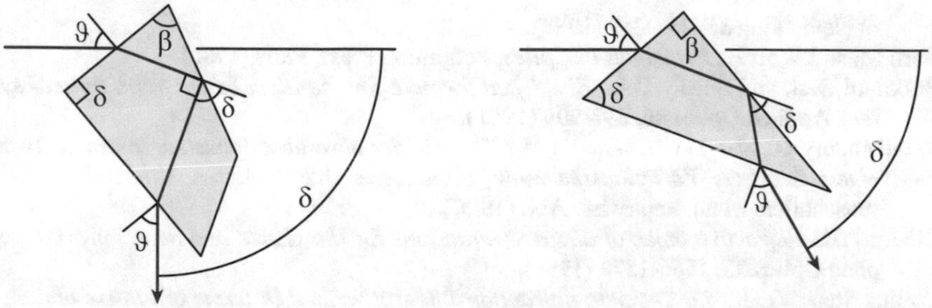

Fig. 2.81 Constant deviation prisms. (a): Pellin-Broca prism, with $\delta = 90°$, $\beta = 75°$. (b): Abbe's prism, with $\delta = 60°$, $\beta = 90°$. The traced ray path is the one with equal angles of incidence ϑ on the input and output faces. For a given value of δ, the angle β adjusts the value of ϑ for which this equality exists; the one used here corresponds to an angle α of 60° for an isosceles prism

$$\mathscr{R} = \frac{\lambda_o}{\Delta\lambda_o} = \frac{\lambda_o \dfrac{dn}{d\lambda_o}\dfrac{b}{l}}{\Delta\delta} = \frac{dn}{d\lambda_o} b \tag{2.10.19}$$

and it grows with its size.

Among the dispersing prisms, one must consider the *Brewster's isosceles prism*, which is cut with an angle δ such that the input and output incidence angles are near the Brewster's angle, and the *Littrow's prism*, which is a right prism, corresponding to a Brewster prism that has been cut in half and whose second face is made specular. These prisms are applied as wavelength selectors inside lasers: for example, in argon lasers, a Brewster prism is used to select one of the possible laser transitions.

Some spectroscopes contain *constant deviation prisms*, like the *Pellin-Broca prism* and the *Abbe's prism*, exploiting a total internal reflection (Fig. 2.81). When these prisms are rotated, the angle of deviation for which they are calculated, 90° and 60°, respectively, retains the minimum deviation angle properties, for which the angle of incidence on the input face and the transmission angle on the output face are equal for a wavelength that varies with the rotation of the prism. Therefore, in order to effect a scan of the radiation spectrum under examination, the detection apparatus is placed at a predetermined angle of view while the prism is rotated by an amount directly calibrated in wavelength.

Bibliographical references

Arnaud J.A., *Nonorthogonal optical waveguides and resonators*, The Bell Systems Technical Journal **49**, 2311-2348 (November 1970).

Bennett G.G., *The Calculation of Astronomical Refraction in Marine Navigation*, Journal

of Navigation **35**, 255-259 (1982).
Born M. and Wolf E., *Principles of Optics*, Pergamon Press, Paris (1980).
Burkhard D.G. and Shealy D.L., *Simplified formula for the illuminance in an optical system*, Applied Optics **20**, 897-909 (1981).
Carathéodory C., *Der Zusammenhang der Theorie der absoluten optischen Instrumente mit einem Satze der Variationsrechnung*, Sitzungsberichte d. Bayer. Akademie d. Wissenschaften. Math.-naturwiss. Abt. (1926).
Ciddor P.E., *Refractive index of air: new equations for the visible and near infrared*, Applied Optics **35**, 1566-1573 (1996).
Collins Stuart A. Jr., *Lens-system diffraction integral written in terms of matrix optics*, J. Opt. Soc. Am. **60**, 1168-1177 (September 1970).
Davidson M.W., *Introduction to Microscope Objectives*, Nikon Instruments Inc. (2016), https://www.microscopyu.com/microscopy-basics/introduction-to-microscope-objectives
do Carmo M., *Differential geometry of curves and surfaces*, Prentice Hall, New York (1976).
Edlén B., *The refractive index of air*, Metrologia **2**, 71-80 (1966).
Feder D.P., *Optical calculations with Automatic Computing Machinery*, J. Opt. Soc. Am. **41**, 630-634 (1951). *Automatic optical design*, Appl. Optics **2**, 1209-1226 (1963). *Differentiation of ray-tracing equations with respect to construction parameters of rotationally symmetric optics*, J. Opt. Soc. Am. **58**, 1494-1495 (1968).
Fermi E., *Termodinamica*, Boringhieri ed. Torino (1972).
Fowles G.R., *Introduction to Modern Optics*, Holt, Rinehart, and Winston, New York (1968).
Gerard A. and Burch J.M., *Introduction to Matrix Method in Optics*, John Wiley & Sons, New York (1968).
Giotti G., *Lezioni di ottica geometrica*, Zanichelli ed., Bologna (1931).
Greco V. and Giusfredi G., *Reflection and refraction of narrow Gaussian beams with general astigmatism at tilted optical surfaces: a derivation oriented toward lens design*, Applied Optics **46**, 513-521 (2007).
Guenther R., *Modern Optics*, John Wiley & Sons, New York (1990).
Gullstrand A., *Die reelle optische Abbildung*, Sv. Vetensk. Handl., **41**, 1-119 (1906).
Hecht E., *Optics, 2nd ed.*, Addison-Wesley, Madrid (1987).
Herzberger M., *Modern Geometrical Optics*, Interscience Publisher Inc., New York (1958).
Iizuka K., *Engineering Optics*, Springer-Verlag, Berlin (1987).
Jenkins F.A. and White H.E., *Fundamental of Optics*, McGraw-Hill, New York (1957).
Kingslake R., *Who discovered Coddington's equations?*, Optics & Photonics News **5**, 20-23 (1994).
Kline M. and Kay I.W., *Electromagnetic theory and geometrical optics*, Wiley Interscience, New York (1965).
Landsberg G.S., *Ottica*, MIR, Mosca (1979).
Luneburg R.K., *Mathematical theory of optics*, Univ. of California Press, Berkeley (1964). *Propagation of electromagnetic waves, mimeographed lecture notes*, New York University (1947-1948).
Maxwell J. Clerk., *On the Elementary Theory of Optical Instruments*, Phil. Mag. **12**, 402-403 (1856). *On the General Laws of Optical Instruments*, Quart. J. Pure Appl. Maths. **2**, 233-246 (1858).
Martin L.C., *Geometrical Optics*, Sir Isaac Pitman & Sons, London (1955).
Möller K.D., *Optics*, University Science Book, Mill Valley (1988).
Néda Z. and Volkán S., *Flatness of the setting sun*, arXiv:physics/0204060 v1, 19 (Apr 2002).

Nussbaum A., *Geometric Optics*, Addison-Wesley, Menlo Park CA (1968).

O'Shea D.C., *Elements of Modern Optical Design*, John Wiley & Sons, New York (1985).

Pauli W., *Pauli lectures on physics*, Vol. 2, *Optics and the theory of electrons*, C.P. Enz ed., The MIT Press, Cambridge (1973).

Shamir J., *Optical Systems and Processes*, SPIE Optical Engineering Press, Bellingam (1999).

Shealy D.L., *Analytical illuminance and caustic surface calculations in geometrical optics*, Applied Optics **15**, 2588-2596 (1976).

Sivoukhine D., *Optique*, MIR, Mosca (1984).

Solimeno S., Crosignani B., Di Porto P., *Guiding, Diffraction, and Confinement of Optical Radiation*, Academic Press, inc., Orlando (1986).

Sommerfeld A., *Optics*, Academic Press, New York (1949).

Sommerfeld A. and Runge J., *Anwendung der Vektorrechnung auf die Grundlagen der geometrischen Optik*, Ann. d. Physik **340**, 277-298 (1911).

Spencer G.H. and Murty M.V.R.K., *General ray-tracing procedure*, J. Opt. Soc. Am. **52**, 672 (1962).

Stavroudis O.N., *The optics of rays, wavefronts, and caustics*, Academic Press, New York (1972).

Welford W.T., *Aberrations of the Symmetrical Optical System*, Academic Press, London, (1974).

Part III
Physical Optics

Historical notes: From Newton to Fresnel

Newton's Optics

In the previous historical notes, we have seen the contributions of those authors who paved the way to the hypothesis of the wave nature of light: Grimaldi, with the discovery and description of diffraction; Descartes, with the hypothesis of the aether and the non-substantiality of light; Hooke, with the study of interference phenomena; Pardies, with the hypothesis of a wavefront perpendicular to the direction of propagation; and Huygens, with the theory of wavelets. This road, though zigzagged with hypotheses that now appear outlandish to us and with various stumbles, led us successfully to optics. Isaac Newton (1642–1726) tried to follow another one, proposing to avoid any particular hypothesis and to find "certain" conclusions uniquely determined by experience instead. This was, in fact, the teaching of Francis Bacon, which Newton attempted to observe scrupulously, and it was also what the members of the Royal Society would have asked him to do.

Newton gained great achievement in the theory of the motion of bodies and, in particular, was able to interpret the motion of the planets mathematically. Starting from the three Kepler's laws, he discovered that the elliptical orbits can be explained by assuming the existence of a central force of attraction between the bodies that varies inversely as the square of the distance. This *force of gravity* was even confirmed quantitatively for him by the motion of the Moon. Newton arrived at this discovery by inventing a new mathematical tool, the *method of fluxions*, essentially equivalent to modern mathematical analysis. In this aspect, therefore, he was the first to continue the work of Galileo in providing mathematical interpretation of natural phenomena, in contrast to the mechanical philosophy of Descartes. However, he was extremely reluctant to publish his findings and made them known only after many years. This fact fueled fierce disputes with his contemporaries on the priority of discoveries [Segrè 1984], as with Hooke for the law of gravity and with Leibniz for mathematical analysis, since, in his opinion, the priority was for the one who had the idea first, even if kept secret! The astronomer Edmund Halley managed to convince him to publish his studies of mechanics, and in 1685, Newton began writing the first book of his famous work, *Philosophiae naturalis principia mathematica*, finally published in 1687, which constituted an enormous progress in the field of Physics.

In optics, however, Newton's story instead shows the failure of the Baconian method of investigation, which was purely inductive: in the first words of his *Treatise of Opticks*, he expressed his intention to avoid any hypothesis, only to have recourse to them many times. In spite of himself, Newton's work in optics was devastating, because of the illusion that he left to his followers of having found some firm conclusions.

Newton had an unhappy childhood: having already lost his father three months before he was born, on December 24, 1642, at the age of two, passed off to his grandmother by his mother, because the latter had a new husband who did not want him to live with them! He returned to his mother's house at the age of 11 years, after his stepfather's death. In 1661, his talent led him to be admitted to the Trinity College, Cambridge, where he immediately began to take charge of luminous phenomena, even before Huygens, at a time when the debate around the nature of light was very alive. Soon, he discovered the renewed atomism of Pierre Gassendi (1592 –1655), who had adapted Lucretius' *De Rerum Natura* to modern culture: Newton preferred it to the natural philosophy of Descartes, and his notebooks from the time show that he was already treating light rays as "globules" of various size and speed. His first works concern the nature and properties of the colors. In particular, he was concerned with the problem, important for that time, of the chromatic aberration of the telescope objectives, which, according to the suggestion of Descartes, was believed to depend on the shape of the lens. In 1665, he examined the possibility of solving the problem through the use of lenses delimited by surfaces of revolution of a conical section, but, at the same time, he studied the causes of color dispersion. He became convinced that the chromatic deviation was not eliminable and designed a reflecting telescope that is exempt of it. In 1668, he built one, subsequently presenting it to the Royal Society in 1671; as a result, he would go on to become a member of the society in January of 1672. However, he was not the first to have such an idea: already in 1652, a reflective telescope had been built by the Jesuit Nicolò Zucchi of Parma, and Mersenne had suggested a Cassegrain-like telescope with two paraboloidal mirrors.

On February 6, 1672, Newton sent an account of his experiments on colors, both with prisms and for reflection and transmission, to Henry Oldenburg, Secretary of the Royal Society. It was the concise fruit of his research from 1661 to 1666. This letter was read (in Newton's absence) at a meeting of members of the society at which Hooke also was present, and was published shortly after in the 80th issue of *Philosophical Transaction* of February 19, 1672, with the title *A new Theory about Light and Colours*.

In it, Newton describes, in fictitious chronological order, the progress of his work exhibited in the Baconian style: an initial experiment, a thorough series of tests to remove legitimate "suspicions" and, finally, a decisive experiment, which he calls *experimentum crucis*, from which he draws his conclusion. However, the account is written rather informally, without much attention to detail and without illustrations that could have clarified its discovery, despite the great skill with which the experiments had been performed. The first of these was to send a beam of sunlight, through a small hole, in a glass prism and to observe the refracted light on a wall. This in itself was a major advance compared to the work of other researchers, who had limited themselves to illuminating the prism without a diaphragm and to observing the refracted radiation on a nearby screen. Being able to make good use of the dispersion of the prism, Newton had noticed, with «very pleasing divertissement», the vivid and intense colors produced on the wall, but what struck him was the geometric aspect of the light spot: «I became surprised to see them in an oblong form; which, according to the receiv'd Laws of Refractions, I expected [the spot shape] would have been circular». Unfortunately, he forgets to state explicitly that the hole was circular and that the prism was orientated for the minimum deviation, so that the transmitted beam would still have appeared circular. This inaccuracy would generate a lot of confusion in the readers (Pardies, in particular, would point this out to him), who found an elongated shape for a generic position of the prism to be completely natural. On the other hand, Newton had reason to be surprised that, despite the fact that this phenomenon was well known even then, no one before him had given importance to the consequences that it could have for the law of refraction.[1] The report then continues with a series of tests to see if this elongation did not depend on a number of causes: the thickness of the glass or a combination of light and shade, according to the Aristotelian tradition; or an irregularity of the prism; or if the elongation was caused by the opening angle of the sunlight; or, lastly, if the transmitted rays moved in curved lines, but «the difference betwixt the length of the image, and the diameter of the hole, through which the light was transmitted, was proportionable to their distance». When all of these *suspicions* had been eliminated, Newton at last described a truly beautiful resolutive experience, i.e., his "experimentum crucis". He had arranged that the light from the prism would fall on a small board, 12 feet away, into which he had drilled a hole through which part of the light could pass. He had then placed another prism behind that board and had observed the transmitted light on the wall. Rotating the first prism, he could make each of the spectrum colors coincide with the hole. In this way, he was obtaining a nearly monochromatic beam of light without changing the angle of incidence on the second prism. Instead, the light spot on the wall was moving, indicating that each color was associated with a different degree of refrangibility. In addition, in the second refraction, the color remained the same as that selected by the hole in the board. His conclusion was that «the true cause of the length of that image was detected to be no other, than that light is not similar or homogenial, but consists of difform rays, some of which are more refrangible than others; so that without any difference in their incidence on the same medium, some shall be more refracted than others; and therefore that, according to their particular degrees of refrangibility, they were transmitted through the prism to divers parts of the opposite wall». A little later, he reiterated that «[white] light is a confused aggregate of rays indued with all sorts of colours».

In his famous *Treatise of Opticks* of 1704, Newton then wrote that if the prism was illuminated through a thin slit of half a millimeter instead of a circular hole, the colors appeared with higher purity. This can be considered the first example of

[1] Newton, and even members of the Royal Society, were unaware of the book by Marcus Marci de Kronland *Thaumantias Liber, de arcu coelesti, deque colorum apparentium natura, ortu et causis*, published in Prague in 1648. He had already done the same experiments with prisms by discovering the bi-univocal correspondence between the colors of the rainbow and the refrangibility, and that «a colored ray that undergoes a further refraction does not change the kind of color » [Ronchi 1983].

solar radiation spectroscopy, although, for lack of collimation and the poor quality of the prism, he did not observe any of the solar absorption lines.

In the second part of the letter, Newton describes his theory of colors that distinguishes them into *primary*, or *simple*, and *compounded*. The primary ones are those that are seen in the solar spectrum decomposed by the prism, which, according to various tests done, both with prisms and by reflection or transmission, are not further decomposed or altered. Those compounded are formed by combining two or more primary colors. In particular, by composing two close primary colors in the spectrum, the intermediate primary color is obtained, but it can, indeed, be modified by reflection or transmission, or resolved into its components by refraction. Newton assigns a particular status to the color white for which «there is no one sort of rays which alone can exhibit this [whiteness]. 'Tis ever compounded; and to its composition, are requisite all the aforesaid primary colours, mix'd in a due proportion». With this, Newton then explains the appearance of complementary colors in reflection and transmission through materials such as thin gold plates or colored glasses. He also successfully interprets, in terms of selective absorption, an experiment by Hooke, who expected the appearance of intermediate colors in the light that passes through two wedge-shaped boxes, one filled with a blue liquid and the other with a red liquid and juxtaposed in opposite directions.

The answers that Hooke, Pardies, and Huygens gave to Newton's theory are similar in two aspects. The first was that they did not appreciate the value of Newton's discovery over the law of refraction and rather considered his article as describing only color formation and chromatic aberration (especially Huygens), as well as because of the inaccurate way in which the article had been written. The second aspect was that they, in particular, Hooke and Huygens, perceived a substantial, namely, corpuscular, interpretation of the nature of the lumen and considered Newton's conclusion merely as a plausible but unnecessary *hypothesis*. Huygens then, cartesianly, asked him to find a mechanical explanation for the diversity of colors. Newton was greatly resentful of this reduction of the value of his theory and tried to convince them that his experimentum crucis was not vitiated by any corpuscular hypothesis, but that it nevertheless showed the heterogeneity of white light in a unique way. For the express purpose of avoiding any association with the corpuscular hypothesis, Newton called "rays" the individual components of light, but, for the properties he insistently assigned to them, he left his interlocutors no choice but to interpret them as synonyms for corpuscles. Certainly, in hindsight, the experiment did not prove the conclusion that Newton assigned to it. At that time, the idea existed that refraction was a "dirtying" of white light and, in fact, the first prism of the experiment split the white light into its spectral components; however, by selecting one of these components, the second prism did not generate other colors. Therefore, the colors were not manufactured by the prism, but must already be present *individually* in the white light, as is the case for a powder composed of many different colored particles. This reasoning had a "hypnotic" effect [Sabra 1981] that has also confused some modern historians. In fact, in the case of color generation, it is not necessary for the second prism to have the same effect on a colored light beam as the first has on white light, or that a property acquired by light in the first prism cannot remain unaltered with the second. On the other hand, Newton was an atomist who believed that matter was composed of solid particles, endowed with various permanent properties from the very beginning of creation, and, extending this conception to the lumen, he was able to find a simple, though incorrect, interpretation of white light and colors. Newton considered a purely undulatory explanation for the nature of light impossible, because, according to him, it «seems impossible, namely, that the waves or vibrations of any fluid can be propagated in straight lines», without a continual spreading and bypassing obstacles, as with sound. Nevertheless, in an attempt to reconcile his theory with the undulatory hypothesis, Newton proposed to Hooke that he consider white light as waves with different «bignesses of vibrations», in analogy to sound, assigning a wavelength to each color of the spectrum: at the violent extreme, the shortest, and, at the red one, the longest! However, neither Hooke nor Huygens accepted this idea, preferring a dualistic hypothesis of colors.

Newton described the compendium of his optical studies in a series of three books entitled *Opticks, or a Treatise of the Reflexions, Inflexions and Colours of Light*[2] that was not published until 1704, a year after the death of Hooke, with whom Newton had been in serious contention over the nature of colors. The same author, in a warning to the readers, claims: «to avoid being engaged in disputes about these matters, I have hitherto delayed the printing, and should still have delayed it, had not the importunity of friends prevailed upon me». He goes on to point out that, when he speaks of colored rays, he does so, to speak, «grosly», for use by the people, not «philosophically and properly» [Newton, 1730, p. 108]. In them, there is only a certain *disposition* to stimulate the sensation of color, and he recognizes the ability of the human mind to perceive colors even in the absence of lumens. However, as a physicist he speaks of the necessity to deal only with the external physical agent of light, not with the psychological and physiological causes. On the other hand, for the mere fact of his having found a correlation between color and physically measurable entities such as different refrangibility, Newton had already introduced that sort of talk into the scientific field, which would soon make it possible for people to forget the centuries of discussion on the distinction between *lumen* and *lux*. So, today, we use the terms 'light' and 'color' improperly for both meanings, and when we talk in a learned way, as physicists, we say *radiation* and *spectral distribution*.

Newton found, in mechanics, the opportunity to explain the law of refraction, as well as a justification in the diversity in the mass of luminous corpuscles for their different refrangibility. In his *Principia,* indeed, he considered the case of a force of attraction of bodies on the material particles in motion. If this force had a short radius of action, it would be effective only near the surface of separation between two media and would be directed perpendicularly to the surface itself,

---

[2] This treatise was then translated into Latin and published in 1706 with the title *Optice: sive de reflexionibus, refractionibus, inflexionibus et coloribus lucis*.

while, inside the body, the attraction exerted by its material particles in all directions would be overall zero. With a few calculations, he simply derived Snell's law of refraction. Extending this mechanical model to the lumen, Newton found himself admitting that his light rays propagate faster in denser bodies. However, in the case of the lumen, Newton speaks only of an action on the rays of unknown nature, except that it is directed perpendicular to the interface between two media. Being able to explain refraction without specifying the nature of this action, he then deceived himself that he had also given a certain demonstration for Snell's law, independent from the nature of the lumen. However, he had implicitly assumed that the rays still had to obey his second law of dynamics. At the same time, attributing their different refrangibility only to a particular *disposition* of light rays (such as the mass of corpuscles), he thought that there was a strict proportionality between the refractive and dispersion indexes, and therefore it would never be possible to construct an achromatic lens. Through a beautiful and difficult experiment, with two prisms oriented at 90° to each other, Newton was able to measure the ratio between these two indices, but he did so only for a single type of glass, without bothering to try others.

While the first book of *Optiks* is devoted to Geometrical Optics, refraction, and colors, Newton addressed, in the second and third book, respectively, the Interference and Diffraction and described there an accurate series of experiments that extended the knowledge already acquired by Hooke and Grimaldi. Nevertheless, despite the success obtained on refraction and colors, Newton was now facing problems that put his corpuscular conception of light into serious crisis. He tried to remedy the contradictions that emerged from the experiments, admitting that the rays of light were also associated with a periodic behavior. He imagined that there was a mutual interaction between the rays and the medium in which they propagate, and that this medium could also be the aether. This would be put into vibration by the rays, for example, in the act of refraction, and this wave would propagate faster than the rays modulating their behavior. From that, he drew an explanation of the interference that is obtained between two very close glass surfaces. Hooke's experiments on thin foils were the starting point from which Newton began the studies in this field. He recognized, on several occasions, the priority of Hooke's many discoveries, including that he had been the first to discover the rings between two convex glass surfaces in contact. Nevertheless, in *Optiks*, after Hooke's death, he failed even to mention his name, writing only summarily, on the first page of the second book, that *others* had observed the presence of various colors depending on the thickness of the plates [Newton, 1730, p. 168]. Unlike Hooke, he was able to determine the distance between the two surfaces and the mathematical relationship between this distance and the fringes. For this purpose, he used two lenses by putting the flat face of the first into contact with the spherical face of the second, with a great radius of curvature, from which he deduced the trend of the thickness of the interstitial air between the two. In addition, using the monochromatic light from a prism, he discovered that the observable rings become much more so. Given that their diameter grows along with the square root of their order number from the center, he deduced that the dark rings are formed when the thickness becomes a multiple of a given value. He also found that the diameter of each ring progressively increases with the color from violet to red. However, the explanation that he has given is rather singular. Meanwhile, he completely neglected the reflection from the first face: Newton, with his mechanical model of light, had been able to explain the total reflection well, but found no justification for the partial one. In the production of the rings, he considered the sole contribution of the first surface to be the refraction, and therefore «Every Ray of Light in its passage through any refracting surface is put into a certain transient constitution or state which in the progress of the ray returns, at equal intervals and disposes the ray at every return to be easily transmitted through the next refracting surface, and between the returns to be easily reflected by it». He then attributed to this constitution the ability to modulate the light transmitted and reflected from the second surface, and called the states in which the rays exist in a given position *vices* or *fits* of easy reflection or easy refraction. He gave the value of a proven *fact* with certainty to the existence of these fits, whereas, by way of assumption, he felt that their physical cause was that, when a ray (i.e., a corpuscle) shocks the first surface, it causes a vibration in the interposed medium. These vibrations, propagating faster than the rays in order to stay ahead of them, would, reaching the second surface, facilitate or impede the motion by adjusting the amount of transmitted light. What to say, then, about the partial reflection? At this point, Newton imagined that the fits that accompany the rays are also produced in the act of being emitted from the sources, so that, by coming upon a non-thin transparent medium, some of the light is reflected and the rest transmitted. But why, then, is it not shared half and half, and, indeed, this ratio vary with the angle of incidence? On these points, Newton did not dwell at all.

Newton knew Grimaldi's work on diffraction and cited it in his book; he described the experiments and performed them with great care, noting, in particular, the non-straightness of the fringes as they change with the obstacle's distance. However, he did not accept the idea of diffraction itself, as it was incompatible with his corpuscular theory, and ignored the experiments with which Grimaldi had been able to demonstrate the peculiar nature of this phenomenon. He indeed took precisely the opposite path by trying to bring the fringes to effects of refraction and reflection. In the first case, he used a term coined by Hooke, *inflexion*, so that the rays would be deflected by the obstacles by some attractive force, and he experimented with sharp edges, while, in the second case, he used objects with smooth edges, such as needles, to highlight reflection. He found that, with the slits, or, respectively with the wires, the spot of light, or the shadow, that is produced on a screen is wider than that provided by Geometrical Optics and that, in this spot, there are also fringes. For example, placing a screen at a certain distance from a slit, one observes a dark fringe in the center of the spot. Newton interpreted it simply as the division of the light beam into two parts, respectively attracted by the two sharp edges of the slit itself, but he remained silent on the fact that, by changing the distance, the dark fringe alternates with a clear one. Moreover, in the illustrations from the book, Newton failed to represent the fringes outside of the spot, which he had not framed in his theory although he had already observed and described them in a previous memory.

The third book of *Optiks* ends with a sense of dissatisfaction and surrender from Newton in the face of so many difficulties. Therefore, he bequeathed to posterity the unresolved issues with a number of *Queries*, extended over 60 pages, summarizing his experiences and his theories on matter and light and revealing a series of tracks to investigate. Like Grimaldi, he came to admit some "complementary behavior" wave-particle of the light. Indeed, while denying, in the 28th *Querie*, the possibility that light propagates by waves in a fluid medium, in the 3rd, he admits: «are not the rays of light in passing by the edges and sides of bodies, bent several times backwards and forwards, with a motion like that of an eel?». Another *Querie* worthy of attention is the 25th, in which he deals with double refraction. In a few lines, Newton dismissed Huygens' construction as fallacious, quoting precisely the phrase with which the author had claimed not to be able to explain a strange phenomenon, namely, the change of intensity in the rays refracted in succession by two birefringent crystals with the variations of the angle between them. Newton realized that this fact could be due to the structure of the crystals and to some asymmetry of the light rays, to which he assigned a special shape with polarities similar to those of small magnets, having planes of orthogonal symmetry passing through their trajectory. This fruitful idea introduced the concept of polarization that we know now, but Newton did not succeed in framing it in some reasonable mechanical process.

For his successes in mechanics and in the explanation of colors, Newton had gained great prestige, and his corpuscular theory prevailed throughout the 1700s on the wave theory, but the tracks he had traced in his *Queries* were sterile and generally impeded the development of optics in that century. The wave-particle duality has since found space, but only as part of a Quantum Mechanics that is radically different from the Newtonian one and still remains a mystery.

The progress of the 18th century

However, the 1700s was a century of intense theoretical and technical preparation: the infinitesimal methods, the mechanics, the observations and determinations, the crystallography, the glass processing, the optical systems and, in general, the physical instruments were developed and perfected. It was also the century of the discovery of many electrical phenomena, which culminated in the creation of the Volta's pile, and from these phenomena would spring the first great unification of the laws of nature.

The invention (or *discovery*?) by Newton and Leibniz of infinitesimal calculus made available an exceptional tool for solving physical problems. The mathematicians of continental Europe, using the Leibniz notation, wich was much more manageable, had the greatest success. It is worth mentioning the members of the Bernoulli family, who, from the 17th to 18th centuries were represented by 11 important mathematicians; Leonhard Euler (1707–1783), to whom we owe the formulas of trigonometry and the important equation $e^{i\varphi} = \cos\varphi + i\sin\varphi$; Jean le Rond d'Alembert (1717-1783); Joseph Louis Lagrange (1736–1813), who rewrote the Newtonian mechanics in his *Mécanique analytique* (1788), which constitutes the basis of modern Physics; and Jean Baptiste Joseph Fourier (1768–1830).

In Astronomy, John Hadley (1682–1744) presented to the Royal Society a Newtonian reflector telescope capable of overcoming the long refractor telescopes.[3] The main mirror, with a 16 cm opening and 1.6 m focal, was parabolic and was made in *speculum*, an alloy made of 68% copper and 32% tin. Further refinements in the polishing of mirrors in speculum were made by William Herschel (1738–1822), who built several reflectors, the first with a 16 cm opening and 2 m focal, with which he discovered Uranus in 1781, one with 30 cm and a focal length of 6.1 m and one with 46 cm and the same length. Lastly, in 1789, he realized a 122 cm aperture reflector with 12.2 m focal length. The era of increasingly large reflectors had begun. The mirror in speculum weighed 960 kg and, a few years later, was replaced with one with twice the thickness, but unfortunately had to be re-polished every two years for a defect in the alloy composition.

At that time, however, there was a turning point that brought refractors back to the forefront. Chester Moor Hall (1704–1771), a London lawyer and amateur astronomer, after studying the chromatic aberration of the lens for a few years, found that, by combining a crown glass lens and a flint glass lens, one could get an achromatic objective and, in 1733, commissioned one of them from the optician George Bass. Later, Bass and other opticians began to realize these objectives. The Swiss mathematician Leonhard Euler was the only one to resume wave theory, accepting Newton's suggestion of considering white light as consisting of a polychrome set of elementary waves. He had observed the lack of an appreciable chromatic aberration for the eye and, around 1754, developed the theory of achromatic objectives through the combination of two glass lenses with different ratio of refraction and dispersion. A skilled craftsman, John Dollond (1706–1761), "stolen" Hall's idea, or, as he claimed, got there with an independent study, obtaining a patent in 1758. He began to produce good quality doublets with typical openings of 6-7 cm and focal lengths of 1.5 m. However, the lack of large meltings of good flint glass did not allow for the realization of large objectives for astronomical use at that time.

Pierre Bouguer (1698–1758), in 1729, published a work, *Traité d'optique sur la gradation de la lumière*, in which he laid the foundations for photometry. In particular, he formulated the law of the exponential attenuation of light intensity in an absorbent medium. Even the mathematician and physicist Johan Heinrich

---

[3] These telescopes were also tens of meters long, with a simple lens as the objective, backed by a large reticular structure. The large focal length of the lens compared to their diameter made spherical aberration and chromatic aberration negligible.

Lambert (1728–1777) took notice of this topic and published the book *Photometria*, in which he distinguishes between «claritas visa» (luminance) and «illuminatio» (illuminance), which depends on the illuminated surface. A source that has the same luminance from any direction one may look at it is said to be *Lambertian*, while his *cosine law* refers to the illuminance.

Finally, it is worth mentioning Gaspard Monge (1746–1818), member of the *Académie Royale des Sciences* and then founder of the *Ecole Polytechnique*, who was wanted by Napoleon. In the wake of the French troops presence in Egypt during the Napoleonic campaign, he described the mirages that plagued the soldiers with cruel deceptions and explained their cause.

The emission theory and the speed of light

Newton had only at first attributed the dispersion produced in a prism to the possible speed diversity of his rays. He then attenuated this opinion, attributing the dispersion to an effect of size, shape, or force of the light particles, but not explicitly to the speed. In fact, the satellites of Jupiter, when they emerged from the shadow of the planet, did not show any color variation. The red radiation was the least deflected by a prism, and was therefore considered to be the fastest in Newton's attractive hypothesis. The satellites should then appear initially red for an appreciable time. On the other hand, in Newton's theory of attraction, the smaller refrangibility of red radiation could simply result from a minor attraction for that type of ray or a greater mass. These rays then had to have a very complicated structure so as also to explain the color and polarization, as well as a speed not well defined.

Newton remained vague and possibilist on many aspects, but his followers took the corpuscular theory much more to the letter of their master. In particular, Alexis Claude Clairaut (1713–1765) developed a ballistic theory of light in which it behaved like a stream of bullets. There was no place for frequency and not even color, and, following Newton's analogy expressed in the *Principia*, the refraction was explained by the attractiveness of dense objects on the corpuscles. Clairaut's theory became known as *emission theory*. It was also turning out to be natural to attribute an effect on light to gravity and the astronomer John Michell (1724–1793), in a letter to Cavendish from 1784, supposed that the light particles would be attracted like all of the other bodies, coming to determine the deflection of radiation by a star, just half of what Einstein would later predict with general relativity. Michell also speculated that the velocity of the corpuscles emitted by a star would be slowed down by the gravitational attraction of the source, until they were brought back if the mass of the star had exceeded 500 solar masses. The same idea was also expressed by Pierre Simon de Laplace (1749–1827), who calculated a lower limit, 250 times the mass of the Sun, beyond which the stars would be «black bodies». They then predicted the existence of *black holes*, observable only by their effects on surrounding bodies.

The stars would then emit the light with different speeds according to their mass. Robert Blair (1748–1828) was the first to argue that the sources would produce corpuscles at different speeds due to the same velocity of the source, as compared to the observer, according to the Galilean principle of relativity. Blair proposed to reveal the difference in the speed of the light emitted by the stars by determining the deviation of their light by a system of 12 achromatic prisms, but failed to carry out the experiment. His article from 1786 was only read and deposited at the Royal Society and was never published [Eisenstaedt 2005].

An important consequence of the improvement in observations was the discovery of stellar aberration,[4] accomplished around 1727 by the English astronomer James Bradley (1693–1762), while trying to determine the distance of stars from parallax measures. He interpreted this fact as a result of the composition of Earth's speed with that of light, from which he obtained both the confirmation of Earth's orbital motion and a new determination of the speed of light, approximately 250000 Km/sec, in agreement with the latest applications of Römer's method.

Among the scholars of that time, we should mention the Jesuit Ruggero Giuseppe Boscovich (1711–1787) of Ragusa in Dalmacija, who was director of the astronomical observatory of Brera. Boscovich had a concept that matter is composed of *physical points* with a sphere of action alternately attractive and repulsive as a function of distance, which he also tried to apply to light, criticizing the inconsistencies of the corpuscular theory and rejecting the principle of rectilinear propagation. He was the first to propose, in 1766, an experiment to discriminate between the corpuscular and the wave theory, based on the different relationship that occurs between the speed of light in vacuum and in dense media in the interpretation of the law of sines. Bradley's discovery suggested to him the idea of pinpointing a star with a telescope filled with water. He expected an angular variation of about 5″, more or less according to whether the speed in water was greater or smaller. This experiment was not performed and, in any case, would not have been able to give a useful result; however, it was the progenitor of research on aether, which, in turn led, to the theory of relativity. The experiment proposed by Boscovich raised objections because, among other reasons, he had neglected the effect of refraction at the surface of entry into the water. It was clear to many, including Arago, that, at least in the corpuscular theory, a telescope filled with water would not show any behavior different from that of an ordinary telescope, as a result of the Earth's motion, both in observing stellar aberration and in the observation of terrestrial objects.

But if, as was then believed, the light emitted by stars had different speeds, then the angle of aberration would vary, but this did not appear. On his return from Af-

---

[4] This effect is not to be confused with the stellar parallax, for which the nearest stars appear in different positions with respect to those that are distant, as the Earth's position changes in its orbit around the Sun. These shifts are less than 1 second in arc. Instead, stellar aberration is the same for all of the stars observed in the same direction. In particular, at the poles of the ecliptic, the stars have an apparent circular motion with a diameter of about 41 seconds of arc.

rica in 1809, Jean Dominique François Arago (1786–1853), «fort jeune encore», at 23 years old, became interested in this problem at Laplace's suggestion. Arago soon realized that, even assuming a variation of speed of 1/20 of the speed of light, the aberration would only be varied by about an 1″ arc, too little to be observable. The study of the aberration was therefore not appropriate for resolving the matter. Instead, the effect of the speed would be appreciable in the deviation produced by a prism with a relatively large angle at the vertex. But even the chromatic dispersion would be great. Arago then resorted to the use of achromatic prisms made up of a couple of crown and flint glasses, in a very simplified version of the Blair scheme. According to the theory developed by Clairaut, the deviation produced in the refractive index was due to an attractive potential of the diaphanous medium on the corpuscles of light, and the law of sines was ensured by the conservation of momentum in the direction parallel to the surface. If we assume that the attractiveness potential is a constant of the material dependent only on the properties of the corpuscles that give rise to color, it was possible to compensate for the chromatic dispersion with the "Dollond method" and retain the effect of a dispersion produced by different velocities of the corpuscles [Møller Pedersen 2000]. The achromatic prisms used by Arago had a total angle of 24° and produced a deviation of about 10°, with an orientation that produced the most well-defined image. They were mounted in front of the objective of a telescope with which Arago measured the passing of several stars on the local meridian in the evening and morning of several days. In another configuration, Arago mounted two prisms, bringing the deviation to about 22°; they covered only half of the objective of a *cercle répétiteur*, allowing for a more direct measure of the deviation. The diurnal fluctuation that Arago expected from the deviations due only to the Earth's motion was 6″ with the first device and 14″ with the second, values that had to change between evening and morning observations. The change in the deviation between a star and the other for their relative motion would have to be much greater. However, the measurements only had fluctuations of a few seconds of an arc that did not follow any regular behavior. In other words, Laplace's theory was disproved. Not knowing how to interpret the result, Arago and Laplace agreed on a very unlikely explanation, i.e., that the stars emit corpuscles with a broad spectrum of speed, and that, in visual observation, the retina was sensitive only to a narrow range of speed. Arago read a memory on his experiment at the *Institut de France* on December 10, 1810, but published it only 42 years later(!) in 1853, when he found it among his papers at the end of his life [Arago 1853].

Young and the principle of interference

Optics, after a century of paralysis, esperienced a golden age: in the early 1800s, there was a flourishing of new discoveries and ideas. This was due to men like Young, Malus, Brewster, Arago, Fraunhofer, Fresnel, and others.

The first to overcome the stalemate caused by the corpuscular theory was the English physician Thomas Young (1773–1829). He was born in Milverton to a wealthy Quaker family and was undoubtedly a genius: at only two years old, he was already reading English and, subsequently, in his infancy and youth, learned, in succession, Latin, Greek, French, Italian, and German, ultimately coming to have a thorough knowledge of 14 languages, including various Eastern Asiatic languages. Displaying a remarkable grasp of culture, throughout his life, he became interested in many subjects from physics, chemistry, physiology, naval architecture, theory of the tides, egyptology and hieroglyphics, and also wrote many entries for the *Encyclopaedia Britannica*.

Following the suggestion of an uncle that he undertake to join medical profession, at age 19, Young went to London and then to Edinburgh to study medicine. In 1793, he took office at St. Bartholomew Hospital, where he studied the accommodation mechanism of vision. His research had excellent results and, at only 21 years old, Young was appointed a *fellow* of the Royal Society. In particular, he realized that the eye was able to adapt its sight from a *far point* to a *near point* and invented a simple tool, the *optometer*, to determine their position for each person. With this tool, he discovered that some had different focusing positions along the horizontal and vertical directions and coined the term *astigmatism*. Finally, through various experiments, he concluded that the adaptation was due to variations in shape of the crystalline lens and not of the cornea or the eyeball.

Young is also the true father of color theory: to him, we owe the *trichromatic theory*, namely, the principle according to which each color can be obtained by mixing in appropriate doses of three fundamental colors. In fact, he suggested that the retina possesses three different types of «particles», each capable of vibrating in response to one of three different «undulations» corresponding to *red*, *yellow* and *blue* and, to a lesser degree, to other frequencies: «for instance, the undulations of green light being nearly in the ratio of 6½, will affect equally the particles in unison with yellow and blue, and produce the same effect as a light composed of those two species; and each sensitive filament of the nerve may consist of three portions, one for each principal colour»[5] [Weiss 2005]. Next, he corrected the main colors in *red*, *green* and *violet*.

Young also became interested in acoustic and optics. He also studied the phenomena of elasticity, the modulus of which was eventually assigned his name. He then intuited, in 1807, that radiant heat, which had recently been investigated by W. Herschel, W. Wollaston and others, had to be differentiated from light only for longer wavelengths.

He was familiar with the work of Newton, but felt that the corpuscular theory was inadequate to explain optical phenomena, including why an intense light has

---

[5] Today, we know that the retina represents an extension of the brain and already itself produces a sophisticated signal processing revealed by its three types of *cones* corresponding to the three main colors, red, green and violet, and *rods*, sensitive only to intensity. In particular, color is coded as a two-dimensional signal given by the difference in intensity between yellow and blue and between red and green.

the same propagation speed as a weak one. He was also not satisfied with Newton's *vices* to explain Hooke's rings. Young noted that this phenomenon could find an adequate explanation in a wave theory: he was familiar with waves, as he knew the laws of sound and wave propagation; in particular, he had discussed the production of the human voice as a thesis. He also knew the first rudiments of interference that Newton had used in a hydrodynamic case to explain the abnormal tides on the coasts of the Gulf of Tonkin, in the «Port of Batsha», near present-day Haiphong. Such tides are caused by the presence of large islands in front of the Gulf acting as a perforated screen on the motion of the oceanic water waves.

In an article entitled *On the theory of light and colors*, Phil. Trans., p. 12, from 1802, Young introduced a series of hypotheses on the wave nature of light, comparing them with those of Newton and pointing out that the wave hypothesis could easily explain a large number of seemingly heterogeneous optical phenomena. Therefore, it deserved to be accepted as a coherent and universal law.

These hypotheses are that: (I) «a luminiferous ether pervades the universe, rare and elastic in a high degree»; (II) «undulations are excited in this ether whenever a body becomes luminous»; (III) «the sensation of different colours depends on the different frequency of vibrations excited by light in the retina»; (IV): «all material bodies have an attraction for the ethereal medium, by means of which it is accumulated within their substance, and for a small distance around them, in a state of greater density, but not of greater elasticity».

Young notes that the first three hypotheses had been suggested by Newton as well, while the fourth is in contrast to the more complicated Newtonian system. In modern terms, aether is space itself, and its "density increase" derives from the electromagnetic polarization of the media.

From the properties of the elastic media, it then follows that (I): «all impulse are propagated in a homogeneous elastic medium with an equable velocity»; (II): «an undulation conceived to originate from the vibration of a single particle, must expand through a homogeneous medium in a spherical form, but with different quantities of motion in different parts». This final definition results from the difficulty of justifying the essentially straightforward propagation of light. Young admits: «it cannot be expected that we should be able to account perfectly for so complicated a series of phenomena as those of elastic fluids. The theory of Huygens, indeed, explains the circumstance in a manner tolerably satisfactory». Such a clarification is also present in proposition (III): «a portion of a spherical undulation, admitted through an aperture into a quiescent medium, will proceed to be further propagated rectilinearly in concentric superficies, terminated laterally by weak and irregular portions of newly diverging undulations». There are four other propositions to explain: (IV) partial reflection; (V) the law of sines, understood in the opposite direction to that of Newton; (VI) total reflection; and (VII) dispersion, attributed to a resonance of the medium, capable of altering the propagation speed for which «the more frequent luminous undulations will be more retarded than the less frequent; and consequently, that blue light will be more refrangible than red». Finally, he expresses proposition (VIII), in which he introduces interference: «when two undulation, from different origins, coincide either perfectly or very nearly in direction, their joint effect is a combination of the motion belonging to each».

In the following, he mentions various phenomena on the colors produced by gratings, thin foils and inflection. Significant is the description of the light behavior that passes through a diffraction grating formed by a micrometric microscope slide, drawn with 500 lines per inch (20 lines/mm). He notes that red light can be observed in 4 distinct diffraction directions whose sines of their respective angles vary in the ratio 1:2:3:4, discovering the law of sines for gratings, with which Young is able to determine the wavelength corresponding to the various colors.

Here, Young also formulates the cosine law, which determines the delay between the external and the internal reflection for a plate with plane parallel faces, with which he can explain the trend of colors that are observed in thin foils as the angle of inclination changes. Among other things, Young uses the measurements made by Newton on the rings to determine, for the first time, the wavelength corresponding to the various colors of the visible spectrum with remarkable accuracy. The value he assigns to *yellow* (translated in nanometers), 576 nm, approaches the value attributed to the average eye according to the modern definitions within 1 nm.

Finally, he closes the article with the proposition (IX): «Radiant light consists in undulations of the luminiferous ether».

Young, however, was not the first to study gratings. One night in 1785, the American Francis Hopkinson, a signer of the *Declaration of independence*, observed a distant street lamp through a thin silk fabric. He noticed that the fabric produced many images of the lamp and communicated this to a friend of his, the astronomer David Rittenhouse, who recognized a diffraction effect. Rittenhouse then built a half-inch wide grating made with a thin wire pulled between the grooves of two fine pitch screws. He also noted that using wires of different diameters could vary the relative diffracted intensity in different orders. Knowing the pitch of screws, he managed to give the first approximate determination of the wavelength of light, and he published his investigation with the title *An optical problem proposed by F. Hopkinson and solved* [Rittenhouse 1786].

Unlike Newton, Young had the happy intuition to consider the reflection on both surfaces of a diaphanous lamina. In addition, applying the concept of periodicity of the wave motion intended as a local vibration of the aether, he thought that these two reflections should add up algebraically in generating the fringes: where the vibrations are reinforced, there is light; where the vibrations cancel each other out, there is darkness, according to the difference of the paths. Young described these considerations in an article published in the *Philosophical Transactions of the Royal Society* of 1802, p. 387, entitled *An Account of some cases of Production of Colours, not hitherto described*, in which he wrote: «The law is, that whenever two portions of the same light arrive at the eye by different routes, either exactly or very nearly in the same direction, the light becomes most intense when the difference of the routes is any multiple of a certain length, and least intense in the intermediate state of the interfering portions; and this length is different for light of different colours».

With this law, he could explain the behavior of the colors of the rings and that of the thin foils with the variation of the inclination, and noticed that, between the outer and inner reflection, there is a delay of half an undulation. In fact, in soap bubbles, the thinnest area is dark (right before exploding), as well as at the point of contact between a convex lens and a flat lamina of the same glass at the center of the rings. He thus designed an experiment in which he interpose, between a flint glass prism and a crown glass lens, a sassafras oil that has an intermediate refractive index, finding, as expected, that the order of rings is reversed, with a white spot in the center instead of a dark one.

In 1804, Young published another work in *Philosophical Transactions*, entitled *Experiments and calculations relative to physical Optics*, in which he analyzed various experiences of interference and diffraction, making wide use of those of Newton and adding only a few new ones (he did not like to experiment himself).

As a source for his experiments, he used the sunlight that leaked from a small hole in the shutter of a window on which he superimposed a card pierced with a fine needle, thereby simulating a point source. With a mirror, he horizontally directed the light cone to a table on which he laid out various screens. One of them was a thin strip of cardboard about a thirtieth of an inch wide. He was particularly interested in the parallel and thin fringes that were seen on a screen within the shadow of the strip. They changed in number by varying the distance of the screen, but they always left the center of the shadow illuminated. Here, he described a new experience: he noticed that these fringes disappeared when the light lapping only one edge of the strip was intercepted by another card, a sign indicating that, to produce these fringes, the contribution of light on both sides of the strip is required. This fact was even more evident in his most important experience, which has become historical, in which Young intercepted the cone of light diffracted from the first hole through a screen with two neighboring holes, similar to what Grimaldi had done with the Sun. However, this historical experiment is not reported in the three articles by Young quoted herein; he described it only briefly, and it is hidden in his ponderous treatise of 1807 in which he collected a cycle of lectures on a large number of topics.[6] It actually only became known thanks to Fresnel, who quoted it in a letter dated May 24, 1816, that he wrote to Young himself [Peacock 1855, p. 377], and in his memory of July 15, 1816 [Senarmont et al 1866, p. 150].

From diffraction experiments, Young found confirmation that «homogeneous» light, that is, monochromatic light, «at certain equal distances in the direction of its motion, is possessed of opposite qualities, capable of neutralizing or destroying each other, and of extinguishing the light, where they happen to be united; that

[6] The most famous of all optical experiment is described by its executor only in a few lines, and no details, between pp. 464 and 465 of his encyclopedic Lectures [Young 1807]!

these qualities succeed each other alternately in successive concentric superficies, at distances which are constant for the same light, passing through the same medium». Thus, he could also well explain the fringes and colors of thin foils and the colors of striated surfaces. Newton had observed that, in replacing the air with water between the two lenses in contact, the rings narrow: Young inferred that «light moves more slowly in a denser, than in a rarer medium»[!] and that «refraction is not the effect of an attractive force directed to a denser medium». He also noted, «there must be some strong resemblance between the nature of sound and that of light». After all, diffraction, as Grimaldi and Boscovich had noticed, implies that light also turns around obstacles.

Young added that diffraction can have effects of practical importance in microscopy, for which you need to be cautious in the appearance of minute bodies, which, due to the diffraction, may have a central spot surrounded by a dark ring «and impress us with an idea of a complication of structure which does not exist»: he was really far ahead of his time!

In this way, Young had demonstrated the validity of his principle of interference. Starting from the observation that a diffractive border appears bright, he also tried to apply this principle to an explanation of the diffraction fringes. Thus, in the case, for example, of a half-plane screen, he interpreted these fringes as being due to interference between the part of the wave that propagates unperturbed and the cylindrical wave generated by a sort of reflection on the edge of the screen, from which it originated. Although there is no reflection in the ordinary sense of the term, Young's intuition is not erroneous, as was demonstrated long afterwards by G.A. Maggi (1888) and then by A. Rubinowicz (1917, 1924, 1957), who transformed the integral formula of Kirchhoff's diffraction into a line integral extended to the opening edge. The Maggi-Rubinowicz theory was then further developed by Miyamoto and Wolf (1962).[7]

However, the times, and especially the mathematical techniques then available, were not yet ripe to appreciate the validity of Young's principle and permit the resumption of the wave theory. Moreover, the interferential phenomena that Young had treated were intimately related to diffraction, and were therefore difficult to quantify. Even notable scientists such as Herschel and Laplace wrote about Hooke's rings and double refraction without mentioning Young's works, claiming the corpuscular theory instead. Disappointed by the opposition that his theory of interference received from the English academics, and, in particular, by the vulgarity that he was publicly subjected to from a boorish lord whose name does not deserve to be mentioned, Young devoted himself to Egyptology studies, including dealing with the deciphering of the Rosetta Stone, by identifying a reading key. Finally, starting in 1818, he was secretary of the *Bureau des Longitudes*, dedicating himself to the *Nautical Almanac*.

Young had a frequent correspondence with William Hyde Wollaston (1766–1828), who had many interests and was concerned with medicine, chemistry,

[7] A discussion of these theories is given in Born and Wolf (1980), p. 449.

physics, and mineralogy. In particular, Wollaston discovered palladium and rhodium. He was the first to observe, in 1802, some black rows in the solar spectrum. In addition, he studied lens aberrations and, in 1812, he found a configuration that would improve the image projected in a darkroom. This was achieved with a convergent meniscus diaphragmed at a suitable distance, minimizing the coma and astigmatism. To Wollaston, we also owe the idea of using a meniscus ophthalmic lens. In mineralogy, he invented the reflecting goniometer with which one can measure the angle between the faces of the crystals, and he also postulated the atomic models of their structure. Finally, we remember him for the invention of the Wollaston prism, which is used to separate two linear polarization components.

In France, Etienne Louis Malus (1775–1812), a young lieutenant colonel of the imperial engineer corps, was induced to try a series of experiments with the spat of Iceland to compete for the prize that, in 1808, the Académie Royale des Sciences had announced for a contest on the mathematical theory of double refraction. It was then known that light passing through one of these crystals acquires special properties that can be revealed with a second crystal. He established what is now called Malus's law, namely, that the intensities of the two components in which a beam of polarized light is divided are in proportion to the square of the cosine of the angle between the original direction of linear polarization and that of the respective components. In addition, looking at the reflected sunlight from a window through one of these crystals, Malus found that the double refraction was not always present, but that it appeared and disappeared at the turn of the crystal, as in the experiment conducted with two crystals. At first, he considered it an atmospheric phenomenon, but then observed the same effect with the light of a candle reflected by a water surface around a particular angle of incidence from the vertical. Malus concluded that these properties could also be acquired through simple reflection, and, in January of 1810, he presented his discovery to the Académie in a memory entitled *Théorie de la double réfraction de la lumière dans les substances cristallisées*, which won the prize.

Malus therein quotes the theories of Huygens and Newton admitting that «l'hypothèse d'Huyghens est sujète à de grandes difficultés; elle semble même incompatible avec les phénomènes chimiques que produit la lumière: celle de l'émission est plus vraisemblable et s'accorde mieux avec nos connaissances physiques. J'adopte donc dans cet ouvrage l'opinion de Newton, non comme une vérité incontestable, mais comme un moyen de fixer les idées et d'interpréter les opérations de l'analyse. C'est une simple hypothèse qui n'a d'ailleurs aucune influence sur les résultats définitifs du calcul».[8] As noted by Ronchi (1983), this is the humiliating condition to which Newton's theory had then been reduced.

---

8 «The hypothesis of Huyghens is subject to great difficulties; it even seems incompatible with the chemical phenomena produced by light: that of emission is more likely and is more in harmony with our physical knowledge. I adopt, therefore, Newton's opinion, not as an incontestable truth, but as a means of fixing ideas, and of interpreting the operations of analysis. It is a mere hypothesis which has no influence on the final results of the calculation»

The mention of the chemical effects of light that Malus brought back into his memory was not, however, a small objection! It had been known since ancient times that nitrate and silver chloride underwent remarkable modifications with sunlight: the Arabic alchemist Geber (Jābir ibn Hayyān, c.721–c.815) had noted, in the eighth century AD, the blackening of the nitrate, and Giorgio Fabricius discovered the effect on chloride in 1556. In 1801, Johann Wilhelm Ritter (1776 –1810) revealed the existence of ultraviolet, which was called *black light*, projecting the solar spectrum on a plate covered with silver nitrate. Photography was about to be invented, but it would be up to Einstein to discover, much later, that emission and absorption take place by quanta.

In his memory, Malus also draws attention to partial reflection and to the fact that «la cause, de cette réflexion partielle a jusqu'ici échappé aux recherches des physiciens».[9] He points out that, for double refraction and partial reflection, «qu'il existe néanmoins une étroite liaison entre ces deux genres de phénomènes, et que sous certains rapports on peut les regarder comme dépendant d'une même cause, puisqu'ils présentent des effets identiques».[10] As an example, he shows that «si on fait tomber un faisceau de lumière sur la surface d'une eau stagnante et sous l'angle de 52° 45′ avec la verticale [the Brewster's angle], la lumière réfléchie a tous les caractères d'un des faisceaux produits par la double réfraction d'un cristal».[11] According to Malus, the cause of this behavior had to reside in the interaction between the luminous corpuscles with the surface of the diaphanous body. Referring to the 26th *Querie* of Newton, he assumes that the particles have asymmetries in their form, meaning that they have *poles*, in analogy to magnetic bodies. In general, the corpuscles would be randomly oriented in natural light, while, in the phenomena of reflection or double refraction, the surface would have the task of orienting the poles, generating a light that he calls *polarized*.

This *bilateralism* of light rays was also invoked by Arago when he discovered the colors exhibited by certain crystals when one illuminates them with polarized light, colors that had already been noticed by Brewster. We will come back to this later.

The Royal Society awarded Malus with the *Rumford Medal*, and it was up to Young, as Foreign Secretary, to notify him of this award in a letter dated March 22, 1811. In this letter, Young wrote to Malus, «vos expériences démontrent l'insuffisance d'une théorie (celle des interférences) que j'avais adoptée, mais elles n'en prouvent pas la fausseté»[12] [Barral 1855, Tome III p. 146]. Such insuf-

[9] «The cause of this partial reflection has hitherto escaped the researches of physicists».

[10] «That there exists nevertheless a close connection between these two kinds of phenomena, and that in certain respects they may be regarded as dependent upon the same cause, since they exhibit identical effects».

[11] «If a beam of light is sent on the surface of stagnant water with the angle of 52° 45′ to the vertical, the reflected light has all the characteristics of one of the beams produced by the double refraction of a crystal».

[12] «Your experiments demonstrate the insufficiency of a theory (that of interference) that I have adopted, but they do not prove its falsity».

ficiency resided in Young's initial idea, then accepted by all supporters of wave theory, that light was a longitudinal wave.

Malus also contributed to the discovery of an important theorem of Geometrical Optics: the *theorem of Malus-Dupin*. By reasoning in terms of the rays that depart from a point source, he found that these form a normal congruence, i.e., they are all perpendicular to a family of surfaces, even after a refraction.

After the discovery of polarization produced by reflection, made by Malus in 1808, and that by refraction, made by Malus and Biot in 1811, there were still others. David Brewster (1781–1868) was born in Jedburgh, Scotland, and, at 12 years old, his family sent him to the University of Edinburgh to become a priest. However, he had a strong inclination to natural sciences and, in 1799, he began studying diffraction. He subsequently authored many articles for major scientific journals. We owe to him the law that establishes the connection between the refractive index and *Brewster's angle* at which the reflection produces total polarization [Brewster 1815; Lakhtakia 1989]. He also made a systematic study of birefringence, studying over 150 different types of crystal. Particularly important was his discovery of biaxial crystals, identifying them in those crystal classes without an axis of rotational symmetry. In addition, he discovered the birefringence induced by pressure and uneven heating [Brewster 1816]. He also dealt with the laws of metal reflection and experimented with the absorption of light. His findings were readily acknowledged and, in 1815, he became a member of the Royal Society. In 1816, he was also awarded by the Institut de France (the Académie). His method of study was essentially empirical and his laws were the result of many experiments. He was also the inventor of the kaleidoscope, from which, despite its great success, he did not obtain any profit, because, although he had patented it, millions of pirated copies flooded the market. He also contributed to the enhancement of the stereoscope, invented by Charles Wheatstone, who used two mirrors to combine two dissimilar binocular images. Brewster's contribution was to use two prisms instead of mirrors, making the device much more compact. Finally, he worked, pre-Fresnel, on the refinement of the beacons of lighthouses with lenses made up of different circular segments, although Fresnel was the first to see his project realized. Brewster was also an astronomer, and wrote a book in which he argued the existence of multiple inhabited worlds in our Universe, analyzing, in particular, the religious aspects of this idea and proving that there is no contradiction with the Scriptures.

In addition to the many opponents of Young's theory, there were also some who appreciated it: Arago, who knew his work, and Fresnel. It was one of Fresnel's discoveries, under the patronage of Arago, that awakened Young's interest in optics. Arago and Fresnel found that, when the polarization on the two interfering paths is orthogonal, the fringes of interference disappear, yet they did not understand the reason why. In 1816, Arago communicated the results of these experiments to Young, who acknowledged their great importance. Young replied to Arago a few months later, with a letter dated January 12, 1817. In this letter [Peacock 1855, p. 380], Young suggests that if one accepts the idea that light waves also possess a transverse movement, then it is possible to give reason for the polarization phenomena and explain why, with two light beams polarized orthogonally to each other, the interference fringes disappear. Nevertheless, he also believed that this transverse movement is very small and was uncertain as to whether it could produce sensitive effects. Arago probably showed the letter to Fresnel, to whom a similar idea had already been suggested by Ampere in 1816, and Fresnel had already reflected on it without finding the opportunity to develop it at that time.

In a subsequent article written in January 1823 for the *Enciclopædia Britannica* [Young 1823], Young recognized Fresnel's paternity of the *physical condition* that attributed a transverse and non-longitudinal movement to aether, noting that it caused difficulty for the very same concept of aether, which should have been «not only highly elastic, but absolutely solid!!!». This seemed so strange that even Arago, unlike Fresnel, never wanted to accept the transversality of light waves.

Fresnel

Whilst Young and other contemporaries had a wide variety of interests, Fresnel devoted himself essentially only to optics, and, with great tenacity, he revolutionized its foundations. In his memoirs, Fresnel proceeded systematically, exploring every detail with great attention, looking for confirmation of the *wave theory* of light, as opposed to the *emission theory* of Newton and his followers. He was a perfectionist and, whatever he faced, he took it to its definitive treatment [Segrè 1984]. He was able to obtain from each new, albeit minute, effect a profound physical explanation that surprises in its ability to maintain its validity even today, although new terms are used. Fresnel is, for me, the scientist who most contributed to optics: his work brought about the fundamental breakthroughs that gave us modern Physical Optics, and all of the sciences and techniques of our day have greatly benefited from it.[13] A description of his work would require rewriting an entire text of Physical Optics: here, I can only mention a few salient facts to indicate the evolution of his thought. In addition to Fresnel, we must also be very grateful to his great friend and supporter Jean Dominique François Arago (1786 –1853), who collaborated with him and knew how to direct him towards a true revolution.

Augustin Jean Fresnel was born on May 10, 1788, in Broglie, Normandy, shortly before the French Revolution began. His father was an architect and his mother belonged to the distinguished Merimée family, amongst the members of which Augustin's uncle Léonor was a well-known painter and his cousin Prosper a

---

[13] It is truly disheartening that textbooks insist on highlighting the history of criminals who have dominated the world with wars, abuse and devastation, neglecting that of the main benefactors of humanity: saints, scientists and artists.

literate author of the novel *Carmen*. As a boy, he distinguished himself for his manual technical skills and, at 16 years old (in 1804), he entered the École Polytechnique, one year after François Arago and one year before Augustin Cauchy, and then continued his studies at the École des Ponts et Chaussées. Not yet twenty, through the strength of his engineering degree, he got a job in the homonymous *Corps des Ponts et Chaussées* directing the construction of roads in Vendée, dans la Drome, and then, in 1812, in Nyons (through which the imperial road from Spain had to pass to arrive in Torino). There, in complete isolation from academic circles, he began to work on optics as a vocation. On May 15, 1814, he wrote to his brother Léonor (namesake of his uncle), asking him for a copy of Hauy's book on Physics and some memory on the polarization of light: «J'ai vu dans le Moniteur, il y a quelques mois, que Biot avait lu à l'Institut un mémoire fort intéressant sur la polarisation de la lumière. J'ai beau me casser la tête, je ne divine pas ce que c'est»[14]. He himself would then finally go on to explain the polarization phenomena recently discovered by his contemporaries.

Fresnel had had a pro-monarchical education and the return of Napoleon from Elba appeared to him as an «attaque contre la civilisation», as Arago wrote [Barral 1854, Tome I, p. 116]. Despite his fragile health, he joined the Royal Army, but returned home in Nyons with his health close to failing. With the change in power, Fresnel was dismissed from his office and was confined under the control of the Napoleonic police in the Mathieu estate at Caën, where his mother resided. However, he was allowed to visit Paris, where he managed to meet Arago, who understood his value, supported him and became his brotherly friend. When Fresnel requested that he refer him to works on light and diffraction, Arago, in a note of July 12, 1815, advised him to read Grimaldi, Newton, Jordan, ... and Young. But Fresnel told him that those texts were not available in Mathieu and that, since he did not know English, he would need help from his brother Fulgence to read Young. Of Newton, whose work he knew very well, he had to have the French translation by Marat of 1787, which, in fact, he cited in his memory of July 15, 1816. With the definitive fall of Napoleon, in July of 1815, Fresnel was reinstated in the Corps of Ponts et Chaussées and recalled in active service at Rennes only at the end of November of that year.

The aether and stellar aberration

Fresnel's interest in optics began with an astronomical question.[15] Bradley had

---

[14] «I saw in the 'Moniteur', a few months ago, that Biot has read at the Institute a very interesting paper on the polarization of light. I'm breaking my head; I don't divine what it is» [Senarmont et al 1866, Vol 2, N° LIX7, p. 819].

[15] Letters of Augustin Fresnel to his brother Léonor from Nyons. July 5 and 6, 1814 [Senarmont et al 1866, Vol. II, n° LIX^8 and LIX^9 p. 820 and 824; Magalhães 2006].

discovered stellar aberration and, from this, he had been able to determine the speed of light on the basis of the corpuscular theory, according to the estimate made by Römer: the light appears to come from a direction moved forward with respect to Earth's motion, as well as the way in which raindrops are seen from a moving carriage.

Since light velocity also depends on the medium crossed, Boscovich had suggested using a water-filled telescope to discriminate between wave and corpuscular theories. In Fresnel's time, this experiment had not yet been executed, however, between 1810 and 1812, Arago had conducted the experiments that we mentioned above, observing the deflection of the light of a star by an achromatic prism placed in front of a telescope, without finding any noticeable effect.

Stellar aberration had found justification in the theory of emission. According to this theory, even a water-filled telescope would measure the same angular deviation of an ordinary one, and therefore Boscovich's experiment would have been useless. Laplace had launched a challenge by writing in his book *Exposition du système du monde* that double refraction and stellar aberration were evidence in favor of the theory of emission and that «Ces phénomènes sont inexplicables dans l'hypothèse des ondulations d'un fluide éthéré» [Laplace 1813, p. 327], even though his theory had been rejected by Arago's experiment with prisms, an experiment that he set aside by not publishing it. Haüy had expressed the same opinion in his *Traité élémentaire de Physique*.

From his isolation in the province, Fresnel wished to emerge with an important contribution to science, and took up the challenge. Emission theory appeared senseless to him: in his years of study at the *École Polytechnique,* it was taught that stellar aberration was a phenomenon resulting from the impact direction of the corpuscles "on the observer's retina" and it was understood that different colors correspond to different speeds.

Was it also possible that the wave theory was in accordance with the observations? On the other hand, only the theory of light as a wave that propagates in the aether seemed capable of explaining the constancy of its speed, regardless of the motion of the sources and of the color, as resulted from astronomical observations.

After 1810, even Arago had moved away from Newton's followers, becoming increasingly interested in wave theory. When Fresnel was sent into internal exile to Mathieu, all of his attention became concentrated on the phenomenon of aberration and, in Paris, he found the right person in Arago. Arago informed him of his experiments with prisms, as well as that of Boscovich and directed him to the study of diffraction.

Fresnel reported the undulatory solution of the aberration in a letter of 1817 to Arago.[16] He considered the case of a plane wave coming from a distant star in the direction perpendicular to the motion of the Earth. In observing of the star with a

---

16 *Lettre d'Augustin Fresnel à François Arago sur l'influence du mouvement terrestre dans quelques phénomènes d'optique*, Annales de chimie et de physique, t. IX, p. 57 (September 1818) [Senarmont et al 1866, Vol. II, al n° XLIX, p. 627].

telescope oriented towards it, the wavefront captured by the lens is made convergent on the focal plane of the objective. Since the telescope moves transversely, the wave does not converge on the same axis, but is shifted in the opposite direction of the motion. To observe the star in axis, the telescope must then be tilted slightly forward.

Fresnel also solved the enigma of the experiment with the prisms, deducing that the aether was to be *partially dragged* from the medium in motion, in proportion to the increase in the density of the aether in the transparent medium. That is, in modern terms, as $1 - 1/n^2$, where n is the refractive index, as we will see in §7.9.5. Fresnel rightly concluded that even the experiment proposed by Boscovich would not have shown a difference between a telescope filled with water and one without. We will continue this story of the aether concept later.

Diffraction and Interference

With his isolation in Mathieu, Fresnel had finally found the opportunity to devote himself entirely to the diffraction experiments that he described in a memoir sent, on October 15, 1815, to Delambre, Secretary of the Académie, entitled *La diffraction de la lumière.*[17] As Ronchi noted, the title already says it all: the original name of the phenomenon is restored, as opposed to the term used by Newton. The first part of this memory is dedicated to highlighting how «la théorie Newtonienne conduit à plusieurs hypothèses improbables», contesting these hypotheses one by one with a passion that only a young man can muster. With a little bit of imagination, he also used a new argument. He argued that, if light were composed of molecules, they would be of the same nature as those hypothesized for heat, and would be rejected by them. Since, then, the atmosphere contains «tant de milliard» of caloric molecules, how is it possible that the speed of a light molecule is not altered?

Instead, for the wave theory, he saw only one serious objection, the one that comes from the comparison of light and sound, which bypasses the obstacles. But he wrote that nothing proves that one can compare the vibration of air, a heavy fluid, with those of the «calorique», a thin fluid. The march of light is much faster than that of sound, so it should be deflected to a far lesser degree by an obstacle. And, he added, «Cette objection, la seule à laquelle il me paraisse difficile de répondre complètement, m'a conduit à m'occuper des ombres portées».[18]

He then looked at the diffraction fringes, as Grimaldi had done, but he wanted

[17] A. Fresnel, *La Diffraction de la lumière, où l'on examine particulièrement le phénomène des franges colorées que présentent les ombres des corps éclairés par un point lumineux*. This first memory was deposited at the Académie on October 23, 1815 [Senarmont et al 1866, Vol 1, N° II, p. 9].

[18] «This objection, the only one to which it seems difficult to answer completely, led me to investigate the shadows».

to find out how they are generated immediately after the edge of the obstacles and how they change by moving away from these. With very few resources available (as an engineer, he had to have his instruments, and the village smith had built some supports for him), but with considerable skill and patience, he made measures with unprecedented precision. The metric system had recently been introduced by the Revolution, and he reported distances expressed in meters with 4 decimal digits up to 8 m! As a source, it was not enough to drill a small hole in a shutter from which a beam of sunlight would leak out, because he needed one that was much smaller. Arago suggested that he use a lens of very short focal immediately after the hole to keep the image of the Sun firm enough during measurements, but the lenses at his disposal had focals that were too long, and thus the image remained too large. Then, he resorted to the expediency of depositing a drop of *honey* into a small hole drilled in a copper plate, obtaining a source of micrometric width that was essentially immobile, even though the cone of light (having a large angular aperture) was moving. The available light was therefore very small, and he used a magnifying glass to produce the image of the fringes on a piece of satin glass observed in transparency. By moving the lens and the screen, he could measure the fringes on planes at different distances from the edge of the diffracting objects. Soon, he realized that, leaving the little light in place to shine directly into his eyes, the fringes appeared much brighter and more defined, after which watched them without the screen. As Grimaldi had already noticed, the fringes did not change shape if the obstacle's edge was sharp or not, as with the wire or the back of a razor, or if the edge was polished or blackened. However, they all seemed to come from the edge itself. He then reported a series of measures in which, as an obstacle, he used an iron wire that he knew was exactly 1 mm in diameter. With this, he determined the separation between the fringes of the first order at the two sides of the shadow, with the precision of a tenth of a mm, with the wire placed at various distances from the source, up to 8 m of distance, and the screen from the wire, from about ten cm to 5 m. As Grimaldi had already observed, the fringes have a non-rectilinear trend. Fresnel was able to establish that such a trend is very near to a hyperbole, with one focus coincident with the source and the other with the edge of the wire. The difference with the theory was very small, but he wanted to be sure of this fact. He then built a simple "micrometer", using a fine silk thread tensed to form a V on a 218 mm long frame. The separation between the two arms of the V ranged from 0 to 5 mm, and could be determined with the precision of 25 μm by placing a graduated rule with 1 mm pitch on the long edge of the frame. Aligning the separation between the wires with the fringes, he confirmed the trend as being in substantial agreement with his theoretical model. He then dealt with the inner fringes in the shadows and found that they follow the same rule as the outer ones, although the foci of the hyperbola are, instead, the two edges of the wire: they disappear if a little card is placed at one edge. Therefore, they are caused by the *interference* between the waves coming from the two edges. Fresnel did not yet know that Young had already reported the same experience in his article of 1804, but he had independently demonstrated the principle of interference.

Fresnel also examined reflection and refraction, pointing to a correct and non-trivial physical justification for why radiation is reflected and refracted according to precise angles without being widespread: it is necessary that the surface roughness is small, but not infinitely small. He also pointed out that the theory of emission was rather deficient in this aspect. Besides this, he also first established the foundation for what would later be called the *Huygens-Fresnel principle*: diffracted radiation complies with the law of sines, as only in that direction are the waves refracted from the various points of the surface in agreement, while, in the other directions «le vibrations des rayons réfractes se contrarient». However, the principle was not expressed in a precise way, although there was finally indication of the path for understanding the rectilinear propagation of radiation, which nobody up to then had been able to indicate, not Huygens, not Young, and Newton least of all, who had explicitly excluded the wave propagation because of his *own lack* of understanding of the phenomenon. Fresnel concluded that the speed of light should be lower in the denser media, as already indicated by Fermat and supported by Huygens, and this fact will thus be considered crucial for deciding once and for all who was right.

On October 15, 1815, Fresnel sent a paper to the secretary of the Institute, as the Académie was now called, a *Complément* to his first memory, inpired by the objections that had been put forward. Meanwhile, for the first time, he introduced the concept that is now called *spatial coherence*. In fact, because of their smallness, the hole or the focal spot produced by a small lens can be treated as a point source, since the light diffracted from the hole or focalized by the lens can be approximated to a spherical wave. He then returned to discussing reflection, considering the case of streaked surfaces, by explicitly writing the law of reflection by gratings, in which the ordinary reflection law, corresponding to the zero order, is generalized for subsequent orders. This same law had already been found by Young, who had only mentioned it in a few lines, and would also be rediscovered by Fraunhofer. To verify it, Fresnel conducted an accurate measurement, but used only two wires. Apparently, he did not realize that the formula of gratings is actually valid if the wires are very numerous, while, using only two wires, the effects are more complicated.

He then dealt again with Hooke's rings and, with particularly accurate measures, explained their behavior when the light is obliquely incident in detail. For this purpose, he approximated the two surfaces as parallel, introducing the *cosine law* for the path difference, noting that this approximation falls for strong inclinations. Again, Young had preceded him, dismissing the law in a few lines and not applying it to the rings. Like Young, Fresnel also identified the important fact that, at the point of contact, there is a dark spot in the reflected light, and he asked himself the reason why. He argued that the reflection from the second face must be out of phase by a half-wave compared to that on the first face; indeed, the cancellation of the reflection occurs even when the separation between the faces is $\lambda/2$. Here, he noticed that something like this also happens in diffraction, in which in the position of the fringes, he found a phase shift of a semi-undulation between the direct rays and those that are "inflected". This unexpected event would eventually lead him to change the interpretation of diffraction radically. Finally, he remarked that, in the transmission, the rings are complementary to those in reflection.

Arago informed him that many of the observations in these memories had already been made by Young, a fact of which Fresnel was unaware, not having seen his work, such as the study of the fringes of the rings and the idea that the diffraction fringes would result from interference with the waves "inflected" by the edge of the obstacles. At the same time, even Arago did not know that the non-rectilinear trend of the fringes had already been observed by Grimaldi, Newton and Young. However, Fresnel's memoirs were much more detailed and quantitative, and there was enough new material to justify their publication in the *Mémoires de l'Académie des Sciences*. Verdet notes that what no one before Fresnel had introduced was the application of the theory of interference to reflection and refraction, which would then lead him to the "true" theory of diffraction and that of the gratings.

After being recalled to service in Rennes, in 1816, at the request of Poinsot and Ampere, Fresnel was allowed to spend a few months working in Arago's laboratory, where he finally had access to the best equipment for his experiments. Here, together with Arago, he repeated the measures with greater precision and completed a second memory that was published in the *Annales de chimie et de physique*, t. I, p. 239 (March 1816). This memory is a rewrite of the previous one, in which he described the diffraction experiments, again demonstrated the laws of reflection and refraction on the basis of the emerging principle of Huygens-Fresnel and, quoting Young, explicitly avoided transcribing the laws of gratings and of cosine.

In particular, he described here the new measures of the fringes produced by the obstacles by filtering the light with a red glass, so as to determine their position up to the 4th order, and confirmed that the outer fringes are hyperbolas whose foci are the source point and the diffracting edge. Therefore, their curvature is concave when viewed from the side of the shadow, opposite to that predicted by Newton. After the hint given in the *Complément* of October 15, 1815, he pointed out that, since the shadow extends beyond the tangent drawn from the source to the diffracting edge, invading the illuminated area, we must assume that «la réflexion apporte un retard d'une demi-vibration dans le progrès des ondes lumineuses»,[19] finding that this hypothesis agreed very well with the observations. Discussing the hyperbolic trend of the fringes, he also pointed out that «Il ne faudrait pas conclure de ces observations que la lumière a un mouvement curviligne».[20] In other words, the energy transport is not located in the fringes.

However, Fresnel was not satisfied with the strange phase change supposed for the light diffracted from the edge and, in addition, the reflection on this was not

[19] «The reflection brings a delay of half a vibration in the progress of the light waves».

[20] «It should not be concluded from these observations that light has a curvilinear motion».

sufficiently intense: the explanation of diffraction was still weighed down by the Newtonian burden. A few months later, on July 15, 1816, he presented to the Académie a *Supplément* to the second memory that was a radical turning point in wave theory. Here, Fresnel reconsidered the whole situation.

Citing Young, however, he believed it useful to present the theory of the rings again, as it «est peu connue», and he had found some new explanation for their understanding. He then resumed the interference produced by the laminae, deriving the cosine law in modern style, and applied it to Newton's measurements on the rings. Then, he assumed, as a principle proven by the experience, that the sum of reflected and transmitted intensities does not change by moving between the rings; thanks to this, he was able to explain an effect discovered by Arago, in which the reflection and transmission rings that are observed with incidence at Brewster's angle are polarized in the same way. But even more impressive is the fact that he here laid the foundation for the explanation of the intensity of the light reflected by a dielectric, which would eventually lead him to write *Fresnel's laws*. These laws are now deduced from electromagnetism, and the approach that Fresnel used is equivalent to that of microscopic theory: reflection does not occur on the surface (this is a simplification of the macroscopic theory, which uses the boundary conditions), but rather it is caused by the abrupt discontinuity of the medium with respect to the wavelength. «En effet, puisque la lumière traverse librement le verre, elle doit en frapper toutes le molécules, qui deviennent alors autant de centres d'ondulations. Comment se fait-il cependant qu'elle ne soit réfléchie qu'à la surface, ou, plus exactement, dans le voisinage de cette surface? C'est ce qu'il s'agit d'expliquer».[21] To simplify, he considered the case of perpendicular incidence on a flat glass sheet that was divided into $\lambda/4$ layers parallel to the surface. If all of the molecules can reflect the light, each of those found in one of the layers is in complete discordance with those found in the preceding layer, and the next one at a distance of $\lambda/4$ from it along the normal. Therefore, the light reflected by an *inner* layer will be deleted from the half of the light reflected by each of the two layers that comprise it. The reflection that is observed is therefore due to the remaining half of the reflected light from the first layer for which deletion is not complete. Here, then, he explained *why* there is a change of sign between the internal and external reflection coefficients discovered in the study of the rings. Compared to the surface, it is as if the reflection happens in the middle of the first layer, and therefore the path taken by the externally reflected light is $\lambda/4$ longer, while that for the internally reflected one is $\lambda/4$ shorter. This physical interpretation of the phenomenon escapes in the macroscopic theory. Fresnel later also extended this explanation to the case in which the surface separates two dense media with different refractive indexes, assigning to each substance a reflective power that depends on that index. In this way, he explained Young's experience of the

---

21 «Indeed, since light passes freely through the glass, it must strike all the molecules, which then become so many centers of undulations. How is it, however, that it is reflected only on the surface, or, more exactly, near this surface? That's what it comes to explaining».

clear central spot, also repeated by Arago, with a flint glass prism and a crown glass lens with interposed sassafras oil.

He then cited Young's experiences of interference, the one with the paper strip and that of the two small holes. However, «pour éloigner toute idée de l'action des bords du corp, de l'écran ou des petits trous, dans la formation et la disparition des franges intérieurs, j'ai cherché à en produire de semblables au moyen du croisement des rayons réfléchis par deux miroirs».[22] Conducting his famous *experiment of the mirrors* requires quite a bit of delicacy, since the angle between the mirrors must be very close to 180 °, without offsets, so that the fringes are produced by the radiation reflected from the center of these and are sufficiently wide. Fresnel succeeded in his intent, and ultimately liberated the principle of interference from the effects of diffraction at the edge of the obstacles. Then, he launched a challenge: «J'engage les physiciens, qui douteraient encore de l'influence mutuelle des rayons lumineux, à répéter cette expérience».[23] The emission hypothesis must be abandoned, «car on ne peut pas espérer de trouver la vérité dans un autre système que celui de la nature».[24]

Fresnel also quoted an Arago experiment in which, by placing a glass sheet on one of the two paths, the fringes disappear. Arago had already observed this effect for the inner fringes in the shadow of a thread. Fresnel attributed this effect to the *delay* induced by the lamina, which moves the zero order fringe out of the common field of the two mirrors. Indeed, Fresnel had suggested that Arago use a thin mica or glass foil: in that case, the fringes would reappear as they were. This is one of the first demonstrations of the effects of what we now call *temporal coherence*. It is noteworthy that Arago, in a letter to Young on July 13, 1816 [Senarmont et al 1866, Vol. 2, n° LVI^2, p. 741], proposed applying this effect to the measurement of small changes in the index of refraction of glass and especially gases.

Fresnel thought, at this point, that he should amend the explanation he had given earlier on diffraction. In fact, several new observations showed discrepancies with the position and width of the fringes deduced from the idea of reflection or inflection at the edge of the obstacles. For example, the central fringe that is observed with a very narrow diaphragm (in Fraunhofer's diffraction) is twice as large as the others. Then, with respect to its calculations, the distance between the outer fringes of the first order at the two sides of a wire is an interval (that is, the period of the fringes) greater, while, for a diaphragm, the distance between the internal ones behind the shadow is one interval smaller. Therefore, the hypothesis that the center of the reflected or inflected wave from the edge coincides with this

---

22 «To remove any idea of the action of the edges of the body, screen, or small holes, in the formation and disappearance of the inner fringes, I have sought to produce similar ones through the intersection of the rays reflected by two mirrors».

23 «I urge physicists who still doubt the mutual influence of the light rays to repeat this experience».

24 «For one cannot hope to find the truth in another system than that of nature».

now appeared to him to be inexact, and Fresnel had no choice but to abandon it.[25] All of this instead left him thinking that the diffracted light also comes from an appreciable distance from the edge.

Then, he resumed Huygens' principle associating it with that of interference and applying it to the diffraction, with a reasoning quite similar to that used for reflection. He had taken into consideration the spherical wavefront emitted by the source at the height of the diffracting edge and divided it into zones, assigning to each one a width resulting from a path difference equal to a half-wave between the source and the point of observation.[26] Only the first zone, which he called the *arc éclairant*, can make a contribution to the diffraction, since the others cancel each other out by destructive interference and erase up to half of the amplitude of the wavelets emitted from the first zone. Among all of the various paths of light, he then identified an *effective ray* that passes halfway through the first zone counted from the edge. The center of the diffracted wave does not coincide with the edge, and the phase shift to be added is $\lambda/4$, not yet enough, but in the right direction, obtaining a better agreement with the observations, both for external and internal fringes. At this point, he was able, *quantitatively for the first time*, to give an explanation of the problem of shadows, whose difficulty was invoked by supporters of the theory of emission as proof against the waves. He explained that the light gradually decreases in the shadows, as, departing from the geometric edge of this, the zones, especially the first one, narrow down. And the decay is rapid because of the smallness of the wavelength.

To confirm his new interpretation of diffraction, he cut a screen out of a copper foil, shaped so as to reproduce the two Young's experiments on interference simultaneously. At the top, the screen was reduced to an opaque strip and, in the lower part, there were two slits whose inner edges were the extension of the strip. He then observed the upper fringes produced internally in the shadow of the strip and compared them with the lower ones produced by the slits. He then discovered that, at a relatively small distance, the lower fringes were sharper than those above, while, if the diffraction were solely due to the edges, it would have to be more confused by the presence of several orders of overlapping fringes. Also, at a distance such that the width of the slits is equal to that of the arc éclairant, the intensity of the central fringe is double that produced from the strip. Finally, such intensities are the same when, moving still further away, the slits cover just half of the arc éclairant; then, the intensity of the lower fringes decays more rapidly than that of the upper fringes, since the arc éclairant becomes increasingly wider and exceeds the width of the slits. All of this confirmed his zone model.

However, the discrepancy in the position of the outer fringes was only halved by the effective-ray model. Fresnel did not yet have an explanation for this and

---

[25] In reality this idea, which was the same as Young, is not entirely wrong: in the calculation of the diffraction integrals, the direct wave from the point source corresponds to a *first order stationary point*, while the one diffracted from the edge corresponds to a *second order stationary point*.

[26] The concept of Fresnel zones is analyzed in this book's chapter on diffraction.

begged the Académie «de juger avec indulgence mes essais dans une théorie aussi difficile».

In the summer of 1816, Arago went to London with Gay-Lussac to talk with Young and presented Fresnel's works to him [Barral 1854, Tome I, p. 292-293]. On his return, he informed Fresnel that Young had stated that he had already described the same phenomena in his *Natural Philosophy* treaty, reiterating the priority over the same discoveries despite Arago's passionate defense. Fresnel was initially greatly discouraged to learn this news, but, towards the end of the year, he worked with Arago to design other experiments (especially on polarization).[27] However, he had come back to Rennes, working in conditions that were not suited to his health, often on the road, under the rain in the cold French winter. The needs of his career separated Fresnel from his scientific activity for about a year, and it was not until the autumn of 1817 that he was authorized to be on leave in Paris, thanks to the efforts of M. Becquey, Director General of *Ponts et Chaussées*.[28] Finally, in the spring of 1818, he was transferred so that he could direct the construction of the *canal de l'Orcq*, and then, in May 1819, he went to the Cadastre of the streets of Paris. Finally, from June 21, 1819, he was assigned to the *Commission des phares*.

On the other hand, thanks to the news that he had been given by Arago, Young's interest in optics was awakened, and, starting in 1816 he began a frequent correspondence with both Arago and, above all, with Fresnel. The relationship between these true gentlemen, lovers of truth above all things, was always one of great esteem, generosity and respect.

The "Mémoire couronné"

In 1817, the supporters of the emission theory, especially Biot and Poisson, alarmed by the successes of Fresnel and Arago, proclaimed a contest, with a prize, under the aegis of the *Académie* whose theme was

« 1° Déterminer par des expériences précises tout les effets de la diffraction des rayons lumineux directs et réfléchis, lorsqu'ils passent séparément ou simultanément près des extrémités d'un ou de plusieurs corps d'une étendue, soit limitée, soit indéfinie, en ayant égard aux intervalles de ces corps, ainsi qu'à la distance du foyer lumineux d'où les rayons émanent;

2° Conclure de ces expériences, par des inductions mathématiques, les mouvements des rayons dans leur passage près des corps.» [29]

---

[27] Letters of September 25, 1816, and October 14, 1816, by A. Fresnel to his brother Léonor [Senarmont et al 1866, Vol. 2, N° LIX[18], p. 836 and p. 837].

[28] Verdet, *Introduction* [Senarmont et al 1866, Vol. 1, p. IX].

[29] «1. Determine by precise experiments all the effects of the diffraction of direct and reflected light rays, when they pass separately or simultaneously near the extremities of one or more bodies of an extent, either limited or indefinite, having regard to the intervals of these bodies, as

As was rightly observed by Ronchi, this was a theme in the Newtonian style. Its purpose was not to challenge Fresnel, but rather to encourage some young scholars to find a Newtonian solution to the phenomenon of diffraction. Arago and Ampère pressed Fresnel to participate in the competition. Fresnel was now ready to formulate his definitive version of diffraction theory, and he presented a monumental memory with the epigraph «Natura simplex et fecunda». It takes up well over 136 pages of his *Œuvre Complètes*. There was another unknown physicist in the competition, who was not even taken into consideration. The commissioners were Biot, Arago, Laplace, Gay-Lussac, and Poisson; Arago was entrusted to draft the final report. Fresnel won the prize, and the second part of his famous *Mémoire Couronné sur la diffraction de la Lumière* was published in the *Annales de chimie et physique*, Tome XI, (July-August 1819).[30]

In the first part of the memory, Fresnel definitively incinerated Newton's optics, burying it with the epigraph «les phénomènes de la diffraction sont inexplicables dans le système de l'émission». In the second part, he made a remarkable leap into modernity as compared to his previous memories. Fresnel recognized the wave motion as a collective behavior extended in both space and time, as opposed to the individual notion of the Newtonian rays. Meanwhile, he quantitatively introduced the interference of waves as an integration on a continuous set of wavelets coming from a common wavefront. For simplicity of calculation, he immediately stated that, in his theory, only monochromatic waves with a given frequency were considered. Among other things, he introduced, for the first time, the convention of indicating the wavelength with λ. As a physical model, he assumed that light is constituted by an oscillating movement of *molécules* of aether or matter and associated what we call field with the *speed* of such molécules. For small oscillations, the terms beyond the linear one in the displacement can be neglected in the forces, so they are harmonic. Not having the theory of complex numbers, he considered each wave as consisting of two quadrature components, with a phase difference of a quarter of a period. His treatment is equivalent to our complex representation of the fields, $E = A_1 \sin\omega t + iA_2 \cos\omega t$. As in the composition of two orthogonal vectors on a plane, the amplitude and the overall phase are obtainable by the amplitudes of these two orthogonal components. He then showed that the sum of two or more harmonic oscillations of the same frequency, however out of phase with each other, is always expressible as the sum of two oscillations in quadrature. He eventually applied this property to Huygens' principle , «le vibration d'une onde lumineuse dans chacun de ses points peuvent être regardées comme la somme des mouvements élémentaires qu'y envieraient au même instant, en agissant isolément, toutes les parties de cette onde considérée dans une quelconque de ses positions antérieures».[31]

---

well as to the distance of the luminous focus from which the rays emanate; 2. Conclude from these experiments, by mathematical inductions, the movements of the rays in their passage near the bodies».

[30] The memory was published in its entirety in 1826 by order of the Académie.

[31] «vibration of a light wave in each of its points can be viewed as the sum of elementary

It remained to establish another hypothesis. Meanwhile, Fresnel assumed that the vibratory movement is oriented in the normal direction to the overall wave surface and propagated only forward.[32] On the other hand, the initial velocity of an isolated molecule, projected in any direction, varies like the cosine of the angle of this direction with the wave normal. Fresnel admitted that the search for the law (now known as the *obliquity factor*) according to which the amplitude of the secondary waves varies with such an angle presents great difficulties. However, it cannot have discontinuities; in addition, the secondary wavelets are canceled significantly between them in directions even slightly inclined with respect to the normal. Therefore, precise knowledge of this law is not necessary.

As a surface on which to integrate the individual elementary contributions, he considered the wavefront positioned at the diffracting screen. In particular, he examined the case in which diffraction is produced by screens with straight and parallel edges. This reduced the two-dimensional integration to one-dimensional. Once a source point C and an observation point P have been fixed, the pathways increase quadratically, moving on the wavefront from the point where it intersects the line CP. Therefore, the quadrature oscillation components in P are given by two *Fresnel's integrals* that do not have an analytical solution in terms of elementary functions. Fresnel ingeniously found an approximate solution in terms of sines and cosines, which could be estimated numerically.

Here, he introduced the important concept that distinguishes intensity and amplitude: «C'est ce que j'appellerai l'*intensité de la lumière*, pour me conformer à l'acception la plus ordinaire de ce mot, réservant l'expression *intensité des vibrations* pour désigner le degré de vitesse des molécules éthérées dans leurs oscillations»[33] (the text in italics is in the original). The intensity at P is then given by the sum of the squares of the two components in quadrature, and thus the sum of the squares of the two integrals. In this way, Fresnel found not only an excellent agreement with the maximum and minimum position of the fringes, but also quantitatively the trend of their intensity.

The wavelength appears as a free parameter of the theory: Fresnel had used a particular red glass as a filter and used the position of the diffraction fringes to determine it, obtaining a value of 638 nm. This value was completely independent of experimental conditions such as distances. In addition, his theory also accurately predicted the pace of the fringes observed with the same filter in the double mirror experiments and the biprism.

In this memory, Fresnel considered, in detail, various cases of diffraction, in-

---

movements that would be the envy at the same time, acting alone, by all parts of this wave seen in any of his previous positions».

[32] In a note following the first writing of the memory, Fresnel corrected this hypothesis to account for polarization. In that note, he stated that the vibratory movements are instead parallel to the wave's surface and that this did not rob his theory of diffraction of any validity.

[33] «This is what I will call the intensity of light, in conformity with the most ordinary meaning of this word, reserving the expression intensity of vibrations to designate the degree of velocity of ethereal molecules in their oscillations».

cluding that of Fraunhofer, by placing a lens, in particular, a cylindrical lens, on the diffractive opening, but merely considered only the one-dimensional case, hinting in a note that, with differently shaped openings, the integration must be performed in two dimensions. In particular, he suggested using his zone method for circular apertures. Poisson noticed this lack and was able to calculate that, in the case of a circular obstacle, a light spot would appear in the center of the shadow, the *Poisson's spot*, with an intensity equal to that which it would be in the absence of the obstacle. This unexpected deduction seemed completely absurd and could have invalidated Fresnel's theory. Arago then experimented with this deduction using an obstacle 2 mm in diameter: the Poisson's spot was actually present! Fresnel, on the other hand, experienced the case of a circular aperture, observing an alternation of light and shadow on the axis of aperture, varying the observation distance from this, in perfect agreement with the theory. Among other things, he also used "white" light, observing a variation in color similar to that found in Hooke's rings.

This memory represented the final victory of the wave theory; however, it made no breach among the supporters of the emission theory. They always remained attached to their beliefs, although they failed to produce a single valid argument. Fresnel's memory, however, won the well-deserved prize and the title of *Mémoire couronné*.

The studies on polarization

Among the various phenomena related to polarization discovered in those years, there was also *optical activity*. In 1811, François Arago found that light, first polarized, then transmitted by a quartz crystal along its optical axis, and finally analyzed with a calcite crystal used as a polarizer, appeared as variously colored. This effect was noted independently in 1812 by Jean Baptiste Biot (1774–1862). He showed that color generation occurs because polarization rotates in the direction of propagation, and this rotation depends on the *length of fit*, a term taken from Newton's interpretation of interference, which Young had shown to be the wavelength. This effect is similar to the color dispersion generated by refraction in a prism and is known as rotating optical dispersion. Biot designated the media that show this effect as *active* and found that it is also present in liquids and gases. The historical record of his studies includes an incident in which the Church where Biot was working burned down after he had installed a long tube filled with vapors of turpentine to study the rotation of the polarization. Just before the boiler producing the vapor exploded, Biot was able to observe the chromatic dispersion through the tube, confirming that optical activity was an intrinsic property of the gas molecules [Applequist 1987].

Another phenomenon, also discovered by Arago in 1811, was that the light, previously polarized, transmitted by a birefringent thin lamina and analyzed with a crystal of calcite, was divided into two rays with complementary colors. Arago believed that the lamina changed the polarization state of the various components of the white light in different ways. Nevertheless, it was still Biot who studied the details of this change, finding that, by varying the thickness of the lamina, there is a periodic return of two different polarizations, separated by intermediate states in which the light appears as a superposition of natural light and polarized light. Biot believed that he had discovered a periodic oscillatory movement of the polarization axes. To explain all of these effects within the corpuscular theory, Biot formulated a set of assumptions that were not so simply related to one another: light molecules are polyhedra, possessing, besides a longitudinal orientation from which easy reflection and transmission fits are generated, a structure provided with symmetry of reflection with respect to a plane containing the propagation direction; the direction normal to this plane determines the polarization axis around which the molecules rotate endlessly, with a frequency dependent on the color, so that their repulsive and attractive extremities present themselves, in turn, in front of the refractive media they encounter; the refraction alters this speed of rotation and tends to bring the axis of rotation perpendicular to the incidence plane; birefringent media induce an oscillation of the axis between two symmetrical positions compared to the principal section of the crystal, etc. All of these hypotheses of a mechanical nature were, however, difficult to justify, but Biot never wanted to abdicate them in favor of the much simpler theory that Fresnel would develop a few years later. His merit was to have conducted numerous experiments on polarization, and also to have determined, at least empirically, *Biot's law* describing the trend of fringes in biaxial crystals, generalizing, by analogy, the corresponding *law of Laplace* for uniaxial crystals. His work, together with that of Brewster, was the basis from which Fresnel was able to move forward.

The transverse nature of waves

However, the fundamental discovery about the nature of polarization was made in 1816 by Fresnel and Arago, when they found, through numerous experiments, that two light beams orthogonally polarized did not give rise to interference fringes.[34] Initially, Fresnel tried to observe the fringes produced by the two images of a bright spot split by a prism of calcite. He took the precaution of placing a suitable glass sheet on one of the two rays to compensate for the difference in optical path in the crystal, or, for the same purpose, using a glass lamina with one ray reflected from the first face and another from the second face. But the fringes never ap-

[34] Fresnel left several reports of these experiments, which have been combined into two distinct *Mémoire sur l'influence de la lumière polarisée dans l'actions que les rayons lumineux exercent les uns sur les autres*, XV (A) and XV (B) of Senarmont, Verdet, Fresnel, Vol. 1. The first is dated August 30, 1816, while the second is dated October 6, 1816, and was deposited the following day at the Institut.

peared. Yet, in another attempt, Fresnel used a thin birefringent lamina of calcium sulphate, through which he watched the light of a candle. The two ordinary and extraordinary beams were now not separated, but the optical path difference produced by the lamina would have to produce a coloration for those colors whose emerging waves of the two polarizations are in phase. Nevertheless, the light of the candle kept appearing white. He concluded, therefore, that «des rayons polarisés en sens contraires n'exercent pas l'un sur l'autre la même influence que les rayons non modifiés ou polarisés dans le même sens».[35]

Arago then suggested that they perform a more direct experiment using a two-slit interferometer such Young had used. Since this test could not be done by polarizing light with a crystal of calcite, Arago also suggested using two piles of thin mica sheets, with the same thickness. By tilting them appropriately at Brewster's angle, it was possible to obtain an excellent degree of polarization for the transmitted light; in addition, orienting them at 90° to each other and arranging them in front of each one of the slits, it was then possible orthogonally to polarize the light that crossed them. Even in this test, the interference fringes were not perceptible, while they reappeared if the two stacks of sheets were oriented in the same way.

In an even simpler experiment, Fresnel took a thin lamina of calcium sulfate cut it into two pieces, posing the first in front of one of the two slits and the other, rotated by 90°, in front of the other. In Young's experiment, using white light, the visible fringes are the white one of order zero surrounded by some colored fringes. In Fresnel's experiment, the lamina introduces a different delay between ordinary and extraordinary rays, and the same polarization that corresponds to the ordinary ray transmitted from a slit coincides with that of the extraordinary ray from the second slit. Therefore, the two pieces of the lamina generate a difference in the path. Thus, the polarization component that is ordinary from the first slit, and extraordinary from the second, gives rise to a system (a package) of fringes shifted with respect to the center of the interference pattern, while the other component produces fringes shifted in the opposite direction. In the center, a fringe package corresponding to the interference between the two orthogonal components should appear. Instead, there is only one uniform white lighting. Among other things, Fresnel indicated the measurement of the separation between the zero order fringes, moved in opposite directions, as a method for obtaining the birefringence of other crystals, and Arago also proposed cutting these crystals in different directions so as to verify Huygens' law, which had been demonstrated only for the calcite. When the two plates of calcium sulphate are rotated at an angle other than 90° between them, even the central fringes reappear. They concluded that the fringes disappear completely only when the two interfering polarizations are orthogonal to each other. Observing the two fringe systems with a calcite prism, Fresnel verified that they were actually polarized orthogonally when the axes of the two plates were at 90° between them.

---

[35] «Polarized rays in opposite directions do not exert one on the other the same influence as the rays unmodified or polarized in the same direction».

By itself, this did not yet imply the transversality of the waves, as the incident radiation used so far had been natural, consisting of an incoherent mixture of two polarization states, which, therefore, could not give rise to interference patterns. But Fresnel conducted other experiments in which *the radiation was previously polarized linearly*. In one of these, he used a whole lamina of calcium sulphate: in this way, the fringes exchanged position compared to the case of the two laminae. Those of equal polarizations now appeared in the center of the interference pattern, while those between opposing polarizations appeared as translated on both sides from the center. Still, in this case, interference fringes were not observed between the crossed polarizations, although there was now coherence between them. And, again, a new fact was in evidence: looking at the fringes with a calcite prism «en ramenant une portion de chaque faisceau au même plan de polarisation»,[36] even the two lateral fringe systems reappeared. Through the prism for each system, two complementary images were observed, «c'est à dire que les bandes obscures de l'une répondent aux bandes brillantes de l'autre, de manière que leur superposition ne présente qu'une lumière blanche continue».[37] Fresnel here attributed this fact to a phase shift corresponding to a semi-undulation between the ordinary ray and the extraordinary one, independent of the optical path and similar to that which he was assuming for the diffraction. He even rearranged the experiment that used the mirror method, using two plates of glass in place of them, towards which the radiation would arrive almost at the Brewster's angle, whereby the reflected radiation was completely polarized. Again, intercepting the reflected waves with a whole lamina and observing the fringes with a prism of calcite, the lateral fringes reappeared and had their maximum brightness when the polarization was at 45° from the axis of the lamina.

Fresnel pondered the physical motive for what hampers the formation of the fringes with orthogonal polarization, writing that he had not been able to arrive at an explanation. Ampère had indicated to him a possible working hypothesis: if, in a fluid, the progressive movement of the molecules were to be modified by another transverse movement perpendicular to the first, the composition of two motions would not give rise to observable effects. Fresnel had in mind another obscure hypothesis, in which the same wave surface is traveled by a transverse oscillation too dense to give perceptible fringes. Verdet notes that the idea of a transverse wave system seemed like a mechanical absurdity to all scholars, especially to Arago, who never wanted to accept it.[38] With so much opposition, Fresnel left these arguments for a better time. On the other hand, at that time, the challenge was to persuade others to accept his diffraction theory, and there was no good reason to add further conflict to the mix.

In the subsequent part of the memory, Fresnel resumes the experiments already

---

[36] «by bringing a portion of each beam to the same plane of polarization».

[37] «that is, the dark bands of one correspond to the bright bands of the other, so that their superposition presents only a continuous white light».

[38] Verdet, *Introduction* [Senarmont et al 1866, Vol. 1, p. LV].

conducted by Arago and Biot so as to study the fringes that are observed by placing a birefringent lamina between two prisms of calcite used as polarizers.[39] The first polarizer is recognized by everyone as being necessary for the appearance of colored fringes: Fresnel explains this by assuming that natural light is composed of polarized oscillations in all directions. In other words, a polarizer does not create polarization: it in some way divides the two components of the motion already present in natural light. He acknowledged that Young had already dealt with this case by contesting the interpretation that Biot gave of his own experiments and noting the analogy of these fringes with those of thin foils. In fact, the birefringent lamina used did not have the conditions to separate the ordinary ray and the extraordinary one spatially, so it merely introduced a delay between the two that gave rise to colored fringes. The only difference was that, with birefringence, the fringes were observed with much greater thicknesses of the laminae. Young had correctly attributed these fringes to a phenomenon of interference. However, as Verdet notes,[40] he had not gone beyond this statement of principle, failing to explain various details of the behavior of the fringes, such as the complementary colors in the two figures divided by the second calcite crystal. Even Fresnel resumed Biot's work using the Malus's law to determine the intensity of the ordinary and extraordinary rays, by limiting himself to studying only the case of normal incidence on the plate, since Young had already somehow considered the general case of oblique incidence. Malus's law, however, applies to *intensity*, and therefore it is not able to provide the complete information required for the *amplitude* of oscillations. Fresnel, explained the complementarity of colors according to the principle of energy conservation, but he was forced to add another $\lambda/2$ difference between the ordinary and extraordinary paths for one of the two images produced by the second calcite prism.

When Young later became acquainted with these experiments by Arago and Fresnel, he immediately assigned great importance to them. He also realized that the disappearance of the fringes could be due to transverse wave motion.[41] According to Young, this motion, however, was more *imaginary* than real, and he remained very cautious in assigning it a physical sense. In his article *Chromatics* of 1817 [Young 1817], which is a difficult text to interpret, he noticed that, even with longitudinal waves, there is a residual transverse motion due to the different directions in which the waves propagate, both in interference and in diffraction by obstacles, and also in the expansion of a spherical wave. He wrote that these motions could be too weak to be perceived, but may have utility as a *mathematical representation* of polarization. They may also have a physical sense, more plausible than the complicated machinery invented in the theory of corpuscles. But he added that there are other similarities between the polarization and the transverse

---

39 The discussion of this case is far from trivial, as we will see in the chapter dedicated to anisotropy.

40 Verdet, *Introduction* [Senarmont et al 1866, Vol. 1, p. XLIX].

41 Letter from Young to Arago of January 12, 1817 [Peacock 1855, p. 380].

motion that force us to consider them. In particular, Young noted, «If we assume as a mathematical postulate, in the undulatory theory, without attempting to demonstrate its physical foundation, that a transverse motion may be propagated in a direct line, we may derive from this assumption a tolerable illustration of the subdivision of polarized light by reflection in an oblique plane». Here, Young found the solution: if a polarizing event decomposes the wave motion into two orthogonally polarized components, the projection of the motion in the two directions takes place in proportion to the cosines of the angles that they make with the original direction. Since, then, the intensity is proportional to the square of the speed, we finally have an explanation of Malus's law.

Unlike Young, at one point, Fresnel accepted as a real fact, without reserve, the complete transversality of light waves, but there are no documents proving when this happened with any certainty. This conversion was induced not only by the interference experiments conducted with Arago, but also by other phenomena observed with birefringent plates, in particular, from that arbitrary addition of a $\lambda/2$ difference that we touched upon earlier. In a note, unfortunately undated,[42] subsequent to his work with Arago, Fresnel finally stated that the effects of polarization can be explained simply by admitting that the vibrations corresponding to two orthogonal polarizations are also orthogonal to each other, and regretted having initially refused this idea. In the same note, he discursively solved the dilemma of the additional delay with the projections of the Cartesian method of vectors, between the initial polarization, the axis system of the lamina, and, finally, that of calcite. Thus, the two vectors, which represent the amplitudes of oscillation of ordinary and extraordinary waves in the lamina, are reassembled in the same direction along a calcite axis, and with opposite directions along the other axis. The strong birefringence of calcite then allows for the separation of two polarized images corresponding to the amplitude of oscillation of these axes.

Reflection

In the years following his studies of diffraction, and, in particular, from 1821 to 1824, Fresnel wrote many notes and memories on reflection and propagation in anisotropic media, which, in large part, overlap and show the progress of his ideas. Most of these were only deposited at the Académie and published many years after his death, some in the *Annales*, thanks to the attention of Arago and Biot, and finally in the collection *Œuvre Complètes d'Augustin Fresnel*, edited by H. de Senarmont, É. Verdet, and L. Fresnel.[43] As regards reflection, the final synthesis

[42] Fresnel, *Note sur la théorie des couleurs que la polarisation développe dans les lames minces cristallisées*. [Senarmont et al 1866, Vol. 1, Fragment at n° XIX (A) p. 523].

[43] As Biot then publicly declared, Fresnel was a tireless inventor: a new memory became the instrument for new research, and it was natural for him to retain his possession for a long time, merely publishing only short extracts.

of this work is contained in his *Mémoire sur la loi des modifications que la réflection imprime a la lumière polarisées* of 1823, which was published posthumously[44] after being rediscovered by Biot among the papers of Fourier, and, for birefringence, there is his *Second mémoire sur la double réfraction*[45] of 1824, which became the fundamental text of study in this matter.

Already in 1819, Fresnel had tried to establish the intensity of reflection through two principles[46]: the conservation of living forces, i.e., energy, and a principle of conservation of motion, that is, a boundary condition at the interface between the two media, at that time supposing that the vibration motion was longitudinal, but without success. When he finally admitted the transversality of waves, he was able to calculate the coefficient of reflection for both internal and external reflection for polarized waves parallel to or perpendicular to the incidence plane.

In the third note published in the Annales of 1821 [Fresnel 1821], Fresnel formulated the laws of reflection from isotropic dielectric media and described their verification by the azimuth measurement of the polarization direction of the reflected light when the incident radiation was polarized at 45° from the plane of incidence. Later, in his memory of 1823, Fresnel gave a detailed account of his theory of reflection. As in the case of propagation in anisotropic media, this theory is formally similar to that which is now assumed with electromagnetism. He assumed the continuity of the speed component of the molecules parallel to the interface as a boundary condition: this choice alone already explained the different behavior between the two TE and TM waves, as they are called today, while the direction of propagation was given by the law of reflection and refraction. On the other hand, the conservation of energy required that the sum of the living forces of the reflected wave and the refracted wave be equal to that of the incident wave. But these energies, in an oscillation period, are proportional to the product of the square of the speed of the molecules by their mass corresponding to a propagation distance equal to the wavelength in the medium. In turn, the mass of the waves is given by the product of the density of the medium for the volume involved, which is given by the product of the cross-sectional area of the beams for the wavelength in the two media. We can now find, in the oscillation speeds, the amplitude of the electric field, and likewise the density in the permeability of the medium. Well, with these hypotheses, Fresnel was able to obtain the correct formulas for reflection. In particular, he could use them to explain the effect of total polarization at Brewster's angle. In addition, these expressions also remained valid for internal reflection, in particular, even when the angle of incidence exceeds the critical angle of total reflection. In this case, however, a new fact emerges: Fresnel's formulas provide a complex unit value to the reflection coefficients, which is different

---

[44] Fresnel, Annales de chimie et de physique, XLVI, 225 (1831), read at the Académie on January 7, 1823 [Senarmont et al 1866, Vol. 1, at n° XXX, p. 767].

[45] Fresnel, tome VII du Recueil de l'Académie des Sciences, p. 45 [Senarmont et al 1866, Vol. 2, n° XLVII, p. 479].

[46] See the *Appendice* at p. 649 reported in the end of n° XXII, Vol I di Senarmont et al (1866).

for TE and TM waves. Fresnel noticed this unexpected peculiarity, and experimental verification confirmed it. In particular, he found that the phase shift between the two waves varies with the angle of incidence. Besides, depending on the number of internal reflections between the parallel faces of a prism, there are angles of incidence such that, when the initial wave is polarized at 45° from the plane of incidence, the final wave has a particular polarization state. It is simply constituted by two rectilinearly polarized waves with the same intensity, but out of phase with each other by a quarter of an undulation with a positive or negative sign. In his memory, Fresnel describes a series of glass prisms, which are now called Fresnel prisms, for which this happens. He himself notes that, with these prisms, the dependence on the wavelength is due to the dispersion alone.

Fresnel then defined this particular state as «polarisation circulaire», and distinguished it from that of «polarisation rectiligne», which was known at that time. He found that these two states have quite similar properties. In particular, there were two distinct circular polarizations that could be transformed into linear ones, and vice versa, through total reflections in his prisms. Besides, the light in one of these states, apparently depolarized, allows observing complementary colors through birefringent plates followed by an analyzer. But not only this. In studying the activity of optical quartz, Fresnel used a combination of three prisms that formed a right parallelepiped [Fresnel 1822]. It was formed by a central isosceles quartz prism glued to other two right prisms obtained from a crystal with opposing rotating properties. Moreover, all of the prisms were cut so that their optical axis was parallel to the base of the isosceles prism. In this way, the light propagated in the parallelepiped nearly exactly along the optical axis, entering and exiting with almost normal incidence while being subjected to refraction on the inclined inner faces. With natural light, this combination acted like a prism of calcite, decomposing now, albeit slightly, the two components of circular polarization. Then, on the inner inclined faces there is refraction and Fresnel concluded that, along the optical axis, the two opposite circular components have a different propagation speed. So, Fresnel had explained the optical activity of quartz, as well as that observed by Biot in liquids and gases, and he supposed that this difference was caused by a helical arrangement of the molecules of such media.

Propagation in anisotropic media

In the three notes published in 1821, *Sur le calcul des teintes que la polarisation développe dans le lames cristallisées* [Fresnel 1821], Fresnel faced three arguments, the third one of which we mentioned above. In the first, he described, this time analytically, the consequences of the hypothesis of the transversality of waves, proving that all of the results of the experiments with the laminae were well explained. Among these were the complementarity of colors, the fact that the contrast is canceled when the axes of the lamina and the calcite prism are parallel or at 90° between them, and is maximum when the lamina axes are at 45° from those of the calcite and from initial polarization. As Young had suggested, even Malus's law found a natural explanation, but, above all, Fresnel gave an accurate explanation of what Biot called *polarisation mobile*. In the dispute with Biot, there were two conflicting interpretations. Biot argued that, in thin laminae, there was no separation between ordinary and extraordinary rays. Inside, the molecules of light oscillate between the initial orientation and a symmetrical one with respect to the optical axis, while the act of establishing the outgoing polarization is only on the output face of the lamina. Fresnel instead argued that the vibrations are already decomposed on the first face of the lamina in ordinary and extraordinary rays, distinguished by a different propagation velocity. At the increasing of the lamina's thickness, recombining the two movements, there is actually a periodic variation of the outgoing polarization. When the difference in optical path between the two waves is equal to a whole number of wavelengths λ, the outbound polarization reproduces the initial one. However, when this difference is equal to an odd number of $\lambda/2$, the wave emerges with a polarization oriented at the opposite angle to the initial one with respect to the optical axis of the lamina. But, for the intermediate thicknesses, the transmitted wave, analyzed with a calcite prism, appears partially polarized. Specifically, when the initial polarization is at 45° from the optical axis and the thickness of the lamina is such that it causes an optical path difference equal to an odd number of $\lambda/4$, the emerging wave appears completely depolarized. Curiously, Fresnel briefly described here, in a note, what he then called *circular polarization*: «une conséquence remarquable de la composition des oscillations dans ce dernier cas, c'est que, dans le système d'ondes résultants, les molécules éthérées, au lieu d'osciller, tournent chacune autour de leurs positions d'équilibre avec une vitesse uniforme»[47] [Fresnel 1821, p. 640]. Later, he also wrote that natural light consists of the rapid succession of polarized waves in all directions and that the polarization act does not consist in creating these transverse movements, but rather in decomposing them in two orthogonal directions by separating the two components.

The second argument discussed in the notes of 1821, and then widely reproduced in the memory of 1824, had the title *Considérations mécaniques sur la polarisation de la lumière*. Here, Fresnel tried to justify the transversality of the vibrations and the behavior of the light in the anisotropic media with mechanical arguments, also setting down an account of how he began to be convinced of the transversality of light waves. When, in September 1816, he was busy writing his memory on the coloration observed with the crystalline lamina, he thought that «le ondes lumineuses agissaient les unes sur le autres comme des forces perpendiculaires aux rayons qui seraient dirigées dans leurs plans de polarisation, puisqu'elles ne s'affaiblissent ni ne se fortifient mutuellement quand ces plans

---

[47] «A remarkable consequence of the composition of the oscillations in the latter case is that in the resultant system of waves the ethereal molecules, instead of oscillating, each revolve around their positions of equilibrium with a uniform speed».

sont rectangulaires»,[48] thus also explaining the sign change mentioned above [Senarmont et al (1866), p. 629]. Even Ampère, to whom he had communicated the results of the experiments, came to the same conclusion. But, together, they wondered what happens to the *longitudinal* part of the oscillations. In their failure to obtain an answer, and their embarrassment over that fact, they put aside this idea. However, after a few months, presumably between the end of 1816 and the beginning of 1817, Fresnel became convinced that the absence of interference effects with orthogonal polarizations necessarily implied that the longitudinal part of the oscillations had to be absent, and therefore the waves were *completely* transversal. Arago remained neutral because he was a member of the Institute, where there were strong opponents of the theory of oscillations, and wave transversality was really too difficult to defend.[49]

Fresnel always acknowledged Young's priority in publishing the idea of transversality, but he again defended the independence of his discovery. In an undated fragment, he wrote, «Je ne dis point ceci pour réclamer une part à l'honneur de cette découverte: il appartient tout à M. Young. Mais j'ai rappelé que la même idée m'était venue sur-le-champ lorsque je cherchai la cause de la coloration des lames cristallisées, pour faire voir combien la théorie des ondulations rendait cette découverte facile».[50] [Senarmont et al 1866, Vol. I, fragment al n° XIX (F), p. 551]. Nevertheless, in the notes on the Annales, Fresnel wrote that a letter from Young, dated April 29, 1818, presented by Arago, convinced him even more of the absence of longitudinal oscillations. From the properties of the two-axis crystals discovered by Brewster, Young would have concluded that the aether vibrations could resemble the transversal ones of a tight rope. Fresnel gave him the merit of being the first to enunciate such a possibility for an elastic fluid! All of this is strange, as, several times, Young had excluded the possibility of transverse waves in a fluid, coming to the conclusion that the aether should have been more than just elastic but also solid. Unfortunately, that letter has disappeared.

As Verdet notes, Fresnel was the first to formulate a theory of waves in elastic media, in the linear approximation of small harmonic motions to which one could apply the principle of superposition. Fresnel then opened a new way of investigation for science, which was then developed by Cauchy, Green, Poisson, and Lamé. At the same time, the mechanical model that Fresnel employed to justify the transversality of light waves had serious flaws. Despite the great prophetic acumen of his intuitions and the obvious success of his theory in accurately explaining all known optical phenomena, his mechanical interpretation of wave transversality

[48] «The light waves acted on each other as forces perpendicular to the rays which would be directed in their polarization planes, since they do not weaken or strengthen each other when these planes are rectangular».

[49] See Verdet's note, on p. 635, Vol. I, di Senarmont et al (1866).

[50] «I do not say this in order to claim a share in the honor of this discovery: everything belongs to Mr. Young. But I have recalled that the same idea occurred to me at once when I sought the cause of the coloration of the crystallized plates, to show how the theory of undulations made this discovery easy».

was neither acceptable to nor accepted by anyone. Transversality was, in fact, necessary as an experimental datum, which found its explanation only with electromagnetism. If we accept it as a postulate, then the mechanical response of the media suggested by Fresnel, and, in particular, also those that are anisotropic, still remains a valid model that is widely used today, in the moment in which we identify this response in the movement of electric charges of the media.

In addition to wave transversality, Fresnel deduced another hypothesis from the behavior of the crystals: that *the propagation speed depends only on the direction of oscillation*.[51] Indeed, in uniaxial crystals, the independence of the propagation speed of ordinary rays from their direction can be explained by admitting that the corresponding oscillations are orthogonal to the optical axis. Instead, the extraordinary rays must have oscillations on a plane containing the optical axis, since their speed depends on the inclination of the beam, and then of the oscillations, with respect to such an axis. Fresnel also deduced that oscillations corresponding to a given linear polarization were orthogonal to the polarization plane defined by Malus.

In his *Premier mémoire sur la double réfraction* [Senarmont et al 1866, Vol. II, n° XXXVIII, p. 261], Fresnel described an experiment with biaxial prisms of topaz. By propagating the light along one of the principal axes, it is divided into two rays whose polarization coincides with the direction of one or the other of the other two axes. By measuring the deviation made by the prism, he still found confirmation that the speed of propagation of each of these rays is determined by the direction of oscillation.

Fresnel derived the theory of propagation in anisotropic media with essentially the same tools that are still used today. That is, he considered *i*) plane waves, *ii*) with a direction of oscillation so as to remain constant and parallel to the wavefront, and *iii*) with a propagation speed dependent only on this direction. But he also wanted to find an experimentally deduced mechanical justification for these conclusions and, as Verdet writes, perhaps unwittingly led his mechanical interpretation towards the results that he knew in advance.

Fresnel attributed to the aether a peculiar behavior that distinguished it from both fluid media and solid media. At the same time, he pointed to it as a fluid, in the slightly vague sense in which the aether was considered in the studies of electric and magnetic phenomena. In 1823, this misunderstanding generated a contrast with Poisson,[52] who instead attributed to the term "fluid" the more concrete meaning of the property that characterizes gases and liquids. According to Fresnel, aether also permeates the solid media, such as crystals, and the molecules of these participate in the motion of those of the aether.

In his monumental memory of 1824, no less than 118 pages long, Fresnel describes his theory of elastic anisotropic media. In the meantime, he considered the

[51] Today, we can identify this direction as that of field $\boldsymbol{D}$ and the velocity is that of phase.

[52] See the *Controverse avec Poisson sur la théorie de la lumière*, al n° XXXIV, p. 183, Vol. II, by Senarmont et al (1866).

reaction force to which a molecule is subject when it is moved in any direction. Decomposing the displacement along the three directions of a Cartesian system then brought back such a force to the nine coefficients of what we now call a tensor. He implicitly excluded the torsion forces, showing that this tensor is symmetric, and therefore diagonalizable, in the sense that, in the medium, there are three mutually orthogonal directions for which the force is directed in the same direction of displacement. In these directions, he identifies the principal axes of the medium: if, along these axes, d_x, d_y, d_z are the versor components of the displacement and A^2, B^2, C^2 are proportional to the three coefficients of the diagonalized tensor, the resultant force has components proportional to $A^2\mathsf{d}_x$, $B^2\mathsf{d}_y$, $C^2\mathsf{d}_z$.

Then, taking as inspiration the example of the vibrating rope, he states that «la vitesse de propagation soit proportionnelle à la racine carrée de l'élasticité mise in jeu» by the vibration. In addition, for a plane wave, he assimilated a homogeneous medium to a rope whose vibrations are parallel to the wavefront. He then assumed that the same law was also valid for that medium, and therefore he found it natural that «l'élasticité mise in jeu par le vibration lumineuses dépend seulement de leur direction et non de celle des ondes».[53]

Therefore, if the waves consist of vibrations parallel to the wavefront, the displacement of one of these molecules is countered by a recall force that, in an anisotropic elastic medium, is generally proportional to the amplitude of the displacement, but it is also oriented in a different direction. According to Fresnel, only the parallel part of this force to the wavefront contributes to propagation of the motion to the other molecules and determination of the wave propagation speed. The justification that Fresnel found for this behavior is that the medium is *uncompressible* in the longitudinal direction and that the molecules are instead confined in layers parallel to the wavefront, where they are relatively free to slide («glisser») collectively, with a recall force generated by the slippage between layers. Despite the substantial inconsistency of this hypothesis with his elastic medium model, Fresnel then assumed that «l'élasticité mise in jeu» corresponds to the *projection* on the plane of the wavefront of the force defined above, obtaining that the speed of propagation $\boldsymbol{v}$ is given by $\boldsymbol{v}^2 = A^2\mathsf{d}_x{}^2 + B^2\mathsf{d}_y{}^2 + C^2\mathsf{d}_z{}^2$.

This equation describes what Fresnel calls *surface d'élasticité*, whose *rayons vecteurs*, that is, the rays joining the center with the points of the surface, are long in proportion to the phase velocity corresponding to a vibration in the direction indicated by such rays.[54] Fresnel associated this with the ellipsoidal wave surface employed by Huygens to explain the propagation of extraordinary beams, but, in the *Extrait* of his first memory that he read to the Institute on November 26, 1821, he recognized that this surface is not an ellipsoid, but a fourth-degree surface with

[53] «The elasticity put in play by light vibration depends only on their direction and not on that of the wave ».

[54] In Chapter 7 of this book, we will see that, taking, for d, the versor of the field $\boldsymbol{D}$, and with the transposition $A^{-2} = \mu\varepsilon_1$, $B^{-2} = \mu\varepsilon_2$, $C^{-2} = \mu\varepsilon_2$, this surface has properties similar to those of the *normal ellipsoid* defined by the equation $c^2/\,n^2 = A^2\mathsf{d}_x{}^2 + B^2\mathsf{d}_y{}^2 + C^2\mathsf{d}_z{}^2$.

semi-axes equal to A, B, C.

On the other hand, the projection of the force on the wavefront also has, in general, a different direction from that of the displacement: this would then cause a continuous variation of the vibration's motion. But, on the plane of the wavefront, one can still find two directions in which to decompose the vibration, such that, for them, the strength remains parallel to the displacement. These two directions are orthogonal, and each corresponds to a state of linear polarization with its own propagation speed. Here is how Fresnel could explain the decomposition of a wave in two, and only two, rays and the phenomenon of birefringence: to determine the two polarization directions and their speed, it is enough then to *i*) dissect the *surface d'élasticité* with a plane passing through the center and parallel to the wave front, *ii*) find the symmetry axes of the section, and *iii*) know that the length of these semi-axes is equal to the (phase) propagation speed. Knowing these speeds, Fresnel suggested that one could even calculate how the two rays refract from the crystal. But not only this. Fresnel found that, by tilting the plane of the section, there are two, and only two, directions, orthogonal to that plane, in which the semi-axes of the section are the same, and these directions are the exact optical axes. His theory had been confirmed in the experiment with topaz prisms, in which the measurement of the speed of propagation of the two beams along the principal axes had allowed him to calculate the angular separation between the optical axes, in good agreement with a direct determination of their direction. In addition to the *surface d'élasticité*, he also defined a surface with similar properties, the *ellipsoïde*, which is now called the *ray ellipsoid*, whose sections are instead orthogonal to the direction of the rays.

With these geometric tools, Fresnel was then able to express all of the essential aspects of propagation in anisotropic media. From the *surface d'élasticité* he derived the equation of the phase velocity surface, while, from the *ellipsoïde* he found that of the group velocity, demonstrating the distinction between these two speeds and that between optical axes and radial axes (to be exact, what Fresnel called optical axes are now called radial axes). He was particularly interested in the surface with *deux nappes* of group velocity, since the planes tangent to it are parallel to the wavefront. In this way, he generalized Huygens's construction of birefringence, replacing the sphere and ellipsoid with this surface. He also theoretically deduced the laws of Laplace and Biot, with which one interpret the fringes observed with birefringent plates. However, Fresnel did not notice the phenomenon of conical refraction resulting from his own theory and which constituted an unexpected test of its validity. It was then theoretically foreseen in 1832 by Hamilton [Hamilton 1837] and was experimentally demonstrated in 1833 by Lloyd [Lloyd 1837].

Epilogue

The first memory on double refraction and its supplements, presented to the

Académie, earned Fresnel the support of Laplace. In fact, at the meeting of August 19, 1822, after reading a report signed by Arago, Fourier, and Ampere, Laplace took the floor, praised these studies by proclaiming their exceptional importance, and declared them superior to those that had been presented to them over a long period of time.[55] This opened the way for Frcsnel's entrance into the ranks of the members of the Académie, which occurred on May 12, 1823. In 1825, the Royal Society also elected him as a foreign member, and the Académie requited in 1827, choosing Young as one of his eight foreign associates. Unfortunately, in the last years of life, bad health prevented Fresnel from continuing his studies with the same vigor. His sense of duty pushed him to dedicate his residual forces to the design of headlights for lighthouses, for which he developed a rotating system of step-down lenses, known as Fresnel's lcnses. This work was of great importance for navigation at sea and was remarkably successful. Fresnel had the satisfaction of seeing the first of these lighthouses built in 1823 in Cordouan, at the mouth of the Gironde.

Fresnel also needed to tap into his own resources to finance his research, and therefore he searched for additional employment. The only job he was able to get, however, was as temporary examiner of the students of the École Polytechnique, a very strenuous and poorly paid position. In 1824, after a violent attack of hemoptysis (coughing with blood loss from the lungs), he had to give up this activity.

In 1827, his illness worsened considerably, and, at the beginning of June, he retired to Ville d'Avray, in the suburbs of Paris. The story of Arago's last visit to Fresnel on his deathbed is very touching [Barral 1854, Tome I, p. 184]. The Royal Society had recently bestowed the Rumford medal upon Fresnel, the highest scientific honor of that time, and Young had instructed Arago to deliver it to him[56]: «Ses forces, alors presque épuisées, lui permirent à peine de jeter un coup d'œil sur ce signe, si rarement accordé, de l'estime de l'illustre Société. Toutes ses pensées s'étaient tournées vers sa fin prochaine, tout l'y ramenait: "Je vous remercie, me dit-il d'une voix éteinte, d'avoir accepté cette mission; je devine combien elle a dû vous coûter, car vous avez ressenti, n'est-ce pas, que la plus belle couronne est peu de chose, quand il faut la déposer sur la tombe d'un ami?"».[57] Eight days later, on July 14, 1827, Augustin Fresnel died in the arms of his mother.[58]

---

55 In the 1824 edition of the *Exposition du systeme du monde*, Laplace eliminated the whole chapter with the segment on the inability of wave theory to explain phenomena.

56 Letters of Young to Arago on March 29, 1827, and of Young to Fresnel on June 18, 1827, respectively, at n° LVI[20] and n° LVI[21], p. 778, Vol. II, in Senarmont et al (1866). See also p. 408-409 of Peacock (1855).

57 «His forces, then almost exhausted, scarcely permitted him to cast a glance upon this sign, so rarely granted, of the esteem of the illustrious Society. All his thoughts were turned to his next end, everything brought there: "I thank you," he said to me in an extinguished voice, "for having accepted this mission; I guess how much it must have cost you, for you have felt, is not it, that the most beautiful crown is a small thing when it is necessary to place it on the tomb of a friend?"».

58 The mortal remains of Augustin Jean Fresnel are interred in the modest grave of his family in

Suggested reading

The historical news in these notes are derived from the following texts:

Abati S., Borchi E., de Cola A., *Storia dell'ottica per immagini*, Fabiano ed., S. Stefano Belbo (1997).

Applequist J., *Optical Activity: Biot's Bequest*, American Scientist, **75**, 59-67 (1987). Reprinted in Lakhtakia.

Arago François, *Mémoire sur la vitesse de la lumière*, Comp. Rend. **36** (2), 38-49 (1853). Barral, Œvre Complètes de François Arago, Tome IV, Notices Scientifiques, *Vitesse de la lumière*, p. 548 - 568.

Barral J.A., *Œvre Complètes de François Arago, Tome I et III, Notices biographiques*, Gide et Baudry ed., Paris (Tome I 1854, Tome III 1855). This contains the biographies of various scientists, including Fresnel, written by Arago. Among these, I recommend reading Arago's autobiography, in Tome I, *Histoire de ma jeunesse*, a real novel on the thrilling vicissitudes suffered by the author in making and bringing back his geodetic measurements that were then used to define the sample meter. This book may help you get some idea of the real life of that time. *Tome IV, Notices scientifiques*, T. Morgand ed., Paris (1865). Accessible on http:// books.google.com.

Bernal J.D., *Storia della Scienza*. Editori Riuniti, Roma (III ed., 1969).

Born Max and Wolf Emil, *Principles of Optics*, Pergamon Press, Paris, 1980.

Brewster David, *On the laws which regulate the polarisation of light by reflexion from transparent bodies*, Phil. Trans. Royal Soc. of London **105**, 125-159 (1815). - *On the communication of the structure of doubly refracting crystals to glass, muriate of soda, fluor spar, and other substances, by mechanical compression and dilatation*, Phil. Trans. of the Royal Soc. of London **106**, 156-178 (1816). *More Worlds than one: the Creed of the Philosopher and the Hope of the Christian*, R. Carter & Brothers, New York, (1854). Accessible on http://books.google.com.

Eisenstaedt J., *Light and relativity, a previously unknown eighteenth-century manuscript by Robert Blair* (1748–1828), Annals of Science **62**, 347-376 (2005). *From Newton to Einstein: a forgotten relativistic optics of moving bodies*, Am. J. Phys. **75**, 741-746 (2007).

Ferreri W., *Il libro dei telescopi*, Il Castello ed., Milano (1998).

Fresnel Augustin Jean, *Sur le calcul des teintes que la polarisation développe dans le lames cristallisées*, Annales de chimie et de physique, t. XVII, p. 102, 167 et 312 (1821). Reprinted on Senarmont et al (1866), Vol. I, n° XXII, p. 609. *Extrait d'un mémoire sur la double réfraction particulière qu présente le cristal de roche dans la direction de son axe*, Bullettin de la Société philomatique, pour 1822, p. 191, reprinted on Senarmont et al (1866), Vol. I, al n° XXVII, p. 719.

Ganci S., *An experiment on the physical reality of edge diffracted waves*, Am. J. Phys., **57**, 370-373 (1989). *Le teorie della diffrazione di T. Young e di A. Fresnel*. Giornale di Fisica, **33**, 199-206 (1992).

Guicciardini N., *Newton, un filosofo della natura e il sistema del mondo*, editorial series *I grandi della scienza*, Anno I, n° 2, Le Scienze s.p.a. ed., Milano (1998).

Hall A.R. and Boas Hall M., *Storia della scienza*. Ed. il Mulino, Bologna (1979).

Hamilton William Rowan, *Third supplement to an essay on the theory of systems of rays*,

---

the very sad cemetery of Père Lachaise, in Paris, within division 14, near the angle between the rue C. Perier and the homonymous roundabout. The tomb is completely abandoned and the epigraph on the tombstone is now almost unreadable. The Fresnel name is engraved on the Eiffel Tower, along with that of the other great scientists who were contemporaries.

Trans. of the Royal Irish Acad., vol. **17**, part 1, 1-144 (1837).

Lakhtakia A., *Would Brewster recognize today's Brewster angle?*, OSA Optics News, 14-18 (June 1989).

Laplace Pierre Simon, *Exposition du systeme du monde*, Quatrieme edition, Mme Ve Courcier ed., Paris (1813).

Loewen E.G. and Popov E., *Diffraction gratings and applications*. Ed. Marcel Dekker, Inc., New York (1997).

Lloyd Humphrey, *On the phenomena presented by light in its passage along the axes of bi-axial crystals*, Trans. Royal Irish Acad. **17**, 145–157 (1837).

Magalhães G., *Remark on a new autograph letter from Augustin Fresnel: Light aberration and wave theory*, Science in Context **19**, 295-307 (Cambridge University Press, 2006).

Maggi G.A., *Sulla propagazione libera e perturbata delle onde luminose in un mezzo isotropo*, Annali di Matematica **16**, 21-48 (1888).

Miyamoto K. and Wolf E., *Generalization of the Maggi-Rubinowicz theory of the boundary diffraction wave*, J. Opt. Soc. Am., **52**, Part I - 615-622, Part II - 626-636 (1962).

Mollon J.D., *The origins of the concept of interference*, Phil. Trans. R. Soc. Lond. A **360**, 807-819 (2002).

Møller Pedersen K., *Water filled telescopes and the pre-history of Fresnel's ether Dragging*, Arch. Hist. Esact Sci. **54**, 499-564 (2000).

Newton Isaac, *Optiks: or a treatise of the reflection, refraction, inflection and colours of light*, William Innys ed., London (1730)

Peacock G., *Miscellaneous works of the late Thomas Young*, John Murray ed., London (1855). Accessibile su http://books.google.com

Polvani Giovanni, *Storia delle ricerche sulla natura della luce*. Istituto della Enciclopedia Italiana, Roma (1934).

Ronchi Vasco, *Storia della luce, da Euclide a Einstein*, N. Zanichelli ed., Bologna (1939), Biblioteca Universale Laterza, Bari (1983).

Rittenhouse D., *An optical problem proposed by F. Hopkinson and solved*, J. Am. Phil. Soc. **201**, 202-206 (1786).

Rubinowicz A., *Die Beugungswelle in der Kirchhoffschen Theorie der Beugungserscheinungen*, Ann. d. Physik, **53**, 257-278 (1917). *Zur Kirchhoffschen Beugungstheorie*, Ann. d. Physik, **73**, 339-364 (1924). *Thomas Young and the theory of diffraction*, Nature **180**, 160-162 (1957).

Sabra A.I., *Theories of Light, from Descartes to Newton*. Cambridge University Press, Cambridge (1981).

Segrè E., *From falling bodies to radio waves*, W.H. Freeman and Company ed. New York (1984).

Senarmont Henri, de, Verdet Émile et Fresnel Léonor, *Œuvre Complètes d'Augustin Fresnel*, Imprimerie Imperial, Paris (1866). Volumes 1 and 2 are accessible on http://books.google.com

Weiss R.J., *Breve storia della luce*, Ed. Dedalo, Bari (2005).

Young Thomas, *Chromatics*, Supplement to the Encyclopædia Britannica, Art. 5 (1817), reprinted in Peacock (1855), p. 332-336. *Theoretical investigation intended to illustrate the Phenomena of Polarization: being an addition made by Dr, Young to M. Arago's 'Treatise on the polarization of light'*, supplement to the Enciclopædia Britannica (1823), reprinted in Peacock (1855), p. 412-417. *A course of lectures on natural philosophy and the mechanical arts*, J. Johnson ed. (1807); accessible on http://books.google.com.

Chapter 3
Interference

> The law is, that "wherever two portion of the same light arrive at the eye by different routes, either exactly or very nearly in the same direction, the light becomes most intense when the difference of the routes is any multiple of a certain length, and least intense in the intermediate state of the interfering portions; and this length is different for light of different color".
>
> Thomas Young [Young 1802, p.387]

Introduction

The most impressive evidence of the wave nature of light is given by the bright and dark bands that, under appropriate conditions, are formed in the overlapping zone of two or more beams of light, and which become observable, for example, by interposing a diffuser screen in such a zone. These bands are called *interference fringes*, and the phenomenon itself is known as *interference*.

Historically, the first documented observations of interference fringes were made independently by Robert Boyle and Robert Hooke, who noted the colored fringes produced by thin films. Hooke, in particular, made a systematic study of them and observed the ones that are improperly called "Newton's rings". Newton himself realized the periodic nature of Hooke's rings, but he gave an explanation of compromise between wave ideas and his corpuscular theory of light that was completely wrong [Landsberg 1979, p. 123]. Finally, Thomas Young, in 1802, correctly interpreted the phenomenon as interference between the reflections from the interfaces, coming to the measurement of the wavelength. To further demonstrate the wave nature of light, Young performed his famous experiment, published in 1807, in which he observed the interference produced by the light coming from two slits [Young 1807]. However, his *principle of interference* was not accepted until Augustin Jean Fresnel removed the various objections to the wave theory, demonstrating the validity of his diffraction theory.

In the following sections, the theory of interference is developed from the equations of electromagnetism in the case of a linear and isotropic medium. In this context, interference finds its justification in the sum of the fields of the various waves and in the principle of superposition. We will first analyze the cases of two-wave interference, the conditions for their observation and the interferometers, with some examples of their use, such as in the examination of the optical quality of the surfaces, in the refractive index measurements, and in stellar interferometry. Next, we will study the most significant cases of interference with many waves: the Fabry-Perot interferometer, as a prototype of resonant cavity, and the dielectric multilayers with their applications.

Electronic supplementary material The online version of this chapter (https://doi.org/10.1007/978-3-030-25279-3_3) contains supplementary material, which is available to authorized users.

G. Giusfredi, *Physical Optics*, UNITEXT for Physics,
https://doi.org/10.1007/978-3-030-25279-3_3

3.1 Generalities on interference

3.1.1 The principle of linear superposition

The phenomenon of interference arises from the fact that Maxwell's equations for linear media are linear in the amplitude of the fields. Consequently, the sum of the solutions of these equations is still a solution, namely, the principle of superposition is justified, for which

electromagnetic field generated by the set of multiple sources is equal to the sum of the various fields $\mathbf{E}_1$, $\mathbf{E}_2$, ..., $\mathbf{E}_n$ individually generated from each source:

$$\mathbf{E} = \mathbf{E}_1 + \mathbf{E}_2 + \; ... \; \mathbf{E}_n \; . \tag{3.1.1}$$

In general, in the presence of matter, this principle is only roughly true, due to non-linear processes present in the mutual interaction between field and matter. In optics, these non-linearities can become particularly evident for high radiation intensity and/or in non-linear special media. Even vacuum is non-linear, because photons may interact through vacuum polarization, but such an effect is very small. Thus, in the study of that which is called Non-linear Optics, other terms of a power series with crossed products of the amplitudes of the fields are added to the sum (3.1.1), when expressing the matter contribution as a function of the fields.

An essential fact for the generation of fringes is that the terms of the summation (3.1.1) are given by an amount that can also elide between them (here, in particular, they are vector quantities). In other words, the operation required to treat the superposition of multiple beams of light *is not* the sum of the individual intensities, which are instead *positive* scalar quantities. If we think that these intensities are nothing more than the flow of energy per unit area, it may seem that the principle of superposition is in violation of the conservation of energy (and that's a principle we trust!). Actually, the superposition principle does not violate the conservation of energy, but does limit itself to producing, in interference, a different spatial distribution of the energy flow or, through the interaction between the sources, a modification of the emitted power.

Thus, limiting ourselves to considering only two beams of light described by the fields $\mathbf{E}_1(\boldsymbol{r},t)$, $\mathbf{H}_1(\boldsymbol{r},t)$ and $\mathbf{E}_2(\boldsymbol{r},t)$, $\mathbf{H}_2(\boldsymbol{r},t)$, we know from, the equations of §1.5, that the density of energy flow, averaged over time, is

$$\overline{\mathbf{S}} = \overline{\mathbf{E} \times \mathbf{H}} = \overline{\mathbf{S}_1} + \overline{\mathbf{S}_2} + \overline{\mathbf{E}_1 \times \mathbf{H}_2} + \overline{\mathbf{E}_2 \times \mathbf{H}_1} \, , \tag{3.1.2}$$

where $\overline{\mathbf{S}_1}$ and $\overline{\mathbf{S}_2}$ are the flow densities individually associated with the two beams, while the other two terms contribute to $\overline{\mathbf{S}}$ with a spatial flow modulation and formally represent the interference itself. In addition, the intensity (energy flow per unit of surface) that is observed on a screen is given by

$$I = \overline{\mathbf{S}} \cdot \mathbf{n}, \tag{3.1.3}$$

where $\mathbf{n}$ is the versor normal to the screen surface. Therefore,

interference fringes consist of a spatial modulation of intensity, which does not correspond to the sum of the intensity generated by individual sources, but ratyher overlaps with this.

On the other hand, as we shall see later, the depth of this modulation, that is, the *visibility* of the fringes, is related to the degree of spatial and temporal *coherence* (i.e., correlation) of the sources. So, this phenomenon is not commonly observed in nature, with the exclusion of interference caused by reflection on thin films. This fact is what concealed from our fathers and mothers, until modern times, the *additional* undulatory nature of the light. When coherence is lacking, the interference terms are canceled, and Eq. (3.1.2) is reduced to the sum of flow densities. The particular conditions mentioned at the beginning, in which the fringes of interference become visible, are those in which a correlation is established between the beams, and this can also be obtained in various ways from natural sources, as will briefly examine shortly.

3.1.2 Interference between monochromatic plane waves

Let us now begin by analyzing the simple case of the interference generated by the sum of two monochromatic plane waves having the same frequency and expressed by

$$\mathbf{E}_j = \Re e\left[\boldsymbol{E}_j e^{i\left(\boldsymbol{k}\cdot\boldsymbol{r} - \omega t + \phi_j\right)} \right], \quad \mathbf{H}_j = \Re e\left[\boldsymbol{H}_j e^{i\left(\boldsymbol{k}\cdot\boldsymbol{r} - \omega t + \phi_j\right)} \right], \tag{3.1.4}$$

with $j = 1, 2$. The phases ϕ_j are introduced for convenience in subsequent considerations. Applying Eq. (1.5.11) and proceeding as in the case of Eqs. (1.5.13), from Eqs. (3.1.2) and (3.1.4), we get the total energy flow density:

$$\begin{aligned} \overline{\mathbf{S}} = {} & \frac{\varepsilon c}{2n}\left\{ \left|\boldsymbol{E}_1\right|^2 + \Re e\left[\boldsymbol{E}_1 \cdot \boldsymbol{E}_2^* e^{i(\Delta\boldsymbol{k}\cdot\boldsymbol{r} + \Delta\phi)} \right] \right\} \mathbf{s}_1 \\ & + \frac{\varepsilon c}{2n}\left\{ \left|\boldsymbol{E}_2\right|^2 + \Re e\left[\boldsymbol{E}_2 \cdot \boldsymbol{E}_1^* e^{-i(\Delta\boldsymbol{k}\cdot\boldsymbol{r} + \Delta\phi)} \right] \right\} \mathbf{s}_2, \end{aligned} \tag{3.1.5}$$

where $\mathbf{s}_1$ and $\mathbf{s}_2$ are the versors of $\boldsymbol{k}_1$ and $\boldsymbol{k}_2$, respectively, and $\Delta\boldsymbol{k} = \boldsymbol{k}_1 - \boldsymbol{k}_2$, $\Delta\phi = \phi_1 - \phi_2$. For example, from Eq. (3.1.3) the intensity present on a plane screen normal to the bisector between $\mathbf{s}_1$ and $\mathbf{s}_2$ is

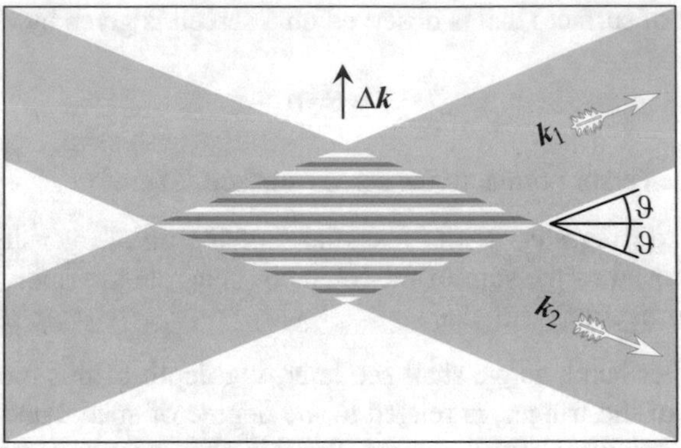

Fig. 3.1 Interference between two plane waves

$$I=\left\{I_1+I_2+2\sqrt{I_1 I_2}\,\Re_e\left[\mathbf{e}_1\cdot\mathbf{e}_2^* e^{i(\Delta\boldsymbol{k}\cdot\boldsymbol{r}+\Delta\phi)}\right]\right\}\cos\vartheta\,, \tag{3.1.6}$$

where 2ϑ is the angle between $\boldsymbol{k}_1$ and $\boldsymbol{k}_2$, I_1 and I_2 are the intensity of the two waves, and, finally, $\mathbf{e}_1$ and $\mathbf{e}_2$ are the (complex) versors of the electric fields. Eq. (3.1.5) is a little complicated, because of the vector nature of the flow itself. There is a similar, but simpler, equation for the density of energy; indeed, for our two plane waves, from Eq. (3.1.1), we have (Fig. 3.1):

$$w=\frac{\varepsilon}{2}\left\{\left|\boldsymbol{E}_1\right|^2+\left|\boldsymbol{E}_2\right|^2+2\Re_e\left[\boldsymbol{E}_1\cdot\boldsymbol{E}_2^* e^{i(\Delta\boldsymbol{k}\cdot\boldsymbol{r}+\Delta\phi)}\right]\right\}. \tag{3.1.7}$$

This equation is useful when the means used to reveal the fringes is insensitive to the direction of propagation. The third term in Eqs. (3.1.6) and (3.1.7) represents a sinusoidal spatial modulation of intensity or energy, oriented along the direction of $\Delta\boldsymbol{k}$, which is in the plane defined by $\boldsymbol{k}_1$ and $\boldsymbol{k}_2$ perpendicular to their bisector. It should be noted that the amplitude of the modulation is such that the sum of the three terms in Eqs. (3.1.6) and (3.1.7) is always ≥ 0. Around the maxima, it is said that there is *constructive* interference, while it is called *destructive* around the minima. The modulation period is

$$p=\frac{2\pi}{|\Delta\boldsymbol{k}|}=\frac{\lambda_o}{n}\frac{1}{2\sin\vartheta}. \tag{3.1.8}$$

For example, for $\lambda_o = 633$ nm, $n = 1$ and $2\vartheta = 1°$, the period is 36.3 μm.

The position of the fringes depends linearly on the phase difference between the two waves, which we have just highlighted here with $\Delta\phi$; if $\Delta\phi$ is constant, it is said that the two waves are mutually coherent and, in particular, the fringes are, in this case, stationary. However, if, for some reason, $\Delta\phi$ fluctuates, the fringes also fluctuate and may lose their visibility.

From the same Eqs. (3.1.6) and (3.1.7), we see that the fringes disappear even when the fields $\mathbf{E}_1$ and $\mathbf{E}_2$ are orthogonal to each other, for example, in the case of two orthogonal linear or circular polarizations. But be careful: here, $\mathbf{E}_1$ and $\mathbf{E}_2$ are three-dimensional, and their orthogonality can also come from having $\boldsymbol{k}_1$ and $\boldsymbol{k}_2$ perpendicular to each other, when both lie on the plane formed by these wave-vectors. This fact was discovered by Arago and Fresnel in 1816 and was explained by Young in 1817, assuming that the vibrations of the light waves are transversal, that is, they are perpendicular to the direction of propagation. Fresnel independently arrived at the same conclusion and justified it with further observations, coming to formulate quantitative laws (that bear his name) of the reflection and refraction of light, and even the theory of propagation of light in biaxial crystals (1821).

With two coherent and slightly divergent plane waves polarized in the same way, the intensity observed on a flat screen, normal to their average direction, is expressed by

$$I = I_1 + I_2 + 2\sqrt{I_1 I_2}\cos\left(2\pi\frac{h}{p}\right), \tag{3.1.9}$$

where p is given by Eq. (3.1.8) and h is the height along the direction of $\Delta\boldsymbol{k}$, with its origin on a maximum. If the two waves have the same intensity I_o, the minima have null intensity, while the maxima have intensity $4I_o$, with, say, 100% modulation. If instead one of the two waves had an intensity smaller than the other, there would be a sort of amplification of its presence: for example, if its intensity was one-hundredth that of the other, modulation would still be about 20%.

3.1.3 Interference produced by two point sources

Many classical interference experiments make use of configurations in which one can identify two sources that are (in first approximation) punctiform and coherent with each other. Therefore, before going into the details of these experiments, let us now look at the interference produced by the spherical waves generated by two punctiform sources S_1 and S_2 separated by a distance d. For simplicity, we limit ourselves, in this case, to considering only the oscillating part of the interference term, paying no attention to the problems related to polarization, which is how to consider the case of scalar spherical waves of the type:

$$U = \frac{A}{r}e^{i(kr-\omega t+\varphi)}, \tag{3.1.10}$$

where A is a real constant and r is the distance from the source. The interference term contained in the square module of the sum of the amplitudes of two such

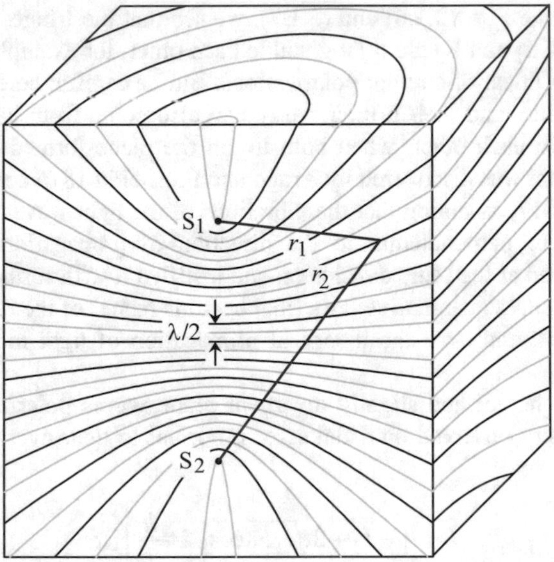

Fig. 3.2 Interference between two spherical waves. The dark lines correspond to $k(r_1 - r_2) = 2m\pi$, while the gray ones correspond to $k(r_1 - r_2) = (2m + 1)\pi$

waves is

$$2\frac{A_1 A_2}{r_1 r_2}\cos\left[k\left(r_1 - r_2\right) + \Delta\varphi\right], \tag{3.1.11}$$

where r_1 and r_2 are the distances from the two sources and $\Delta\phi$ is the phase difference. Unlike in the case of interference between two plane waves, here, the amplitude of modulation varies in space, however, for distances $r_1 \sim r_2 >> d$, this dependence can be neglected in the ratio between the modulation and the background. The energy density is thus mainly dependent on the phase difference between the two waves at the point considered. The surfaces of equal phase difference are given by the relation

$$k\left(r_1 - r_2\right) + \Delta\phi = \text{cost.}$$

and are therefore hyperboloids of rotation around the axis S_1-S_2 (Fig. 3.2), whose foci are still S_1 and S_2. If the fringes are observed on a screen perpendicular to the axis of the sources, they appear as rings, while, if they are viewed on a plane parallel to the same axis, they still appear as hyperbolas. Particularly, in terms of the optical path, there are maxima and minima of brightness when, respectively,

$$n\left(r_1 - r_2\right) + \frac{\lambda_o}{2\pi}\Delta\phi = \begin{cases} m\lambda_o & \text{bright fringe,} \\ \left(m + \dfrac{1}{2}\right)\lambda_o & \text{dark fringe,} \end{cases} \tag{3.1.12}$$

Fig. 3.3
Calculation of the path difference from two sources

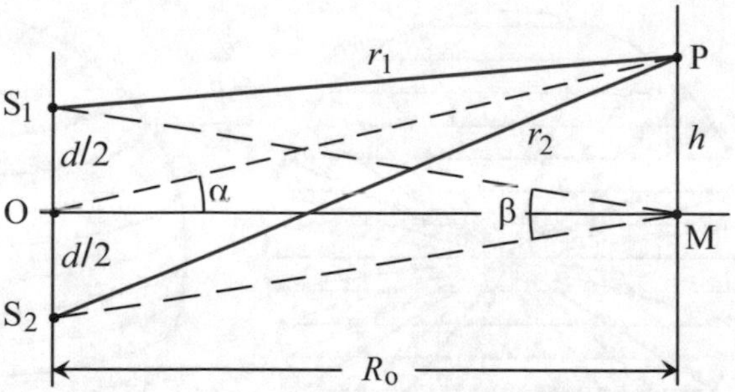

where λ_o is the wavelength in the vacuum and n is the refractive index of the crossed medium. Here, m is an integer, positive or negative, which is called the *interference order*. It should be noted that, for λ_o assigned, the number of possible values of m, that is, the number of fringes, is limited. As can easily be seen in Fig. 3.2, we have

$$m_{\max} - m_{\min} \le 2\frac{dn}{\lambda_o}, \tag{3.1.13}$$

where, again, d is the distance between the sources. In the case of a screen as shown in Fig. 3.3, parallel to the source axis and at a distance R_o from them, it is easy to calculate the difference in path $\Delta r = r_1 - r_2$ for each point P of the screen at a distance h from the median line OM:

$$r_1^2 = R_o^2 + (h - d/2)^2, \quad r_2^2 = R_o^2 + (h + d/2)^2;$$

hence, $r_1^2 - r_2^2 = 2hd$, from which we still obtain

$$\Delta r = \frac{2hd}{r_1 + r_2}. \tag{3.1.14}$$

For distances $R_o >> d$ and $R_o >> h$, it is approximately

$$\Delta r = \frac{hd}{R_o}. \tag{3.1.15}$$

The difference in path length, and thus the phase difference, is therefore, in the first approximation, linear with h. Concluding, around the central point M of the interference pattern, the fringes are almost straight, parallel and equidistant, with spatial period

$$p = \frac{\lambda_o R_o}{nd} = \frac{\lambda_o}{n\beta}, \tag{3.1.16}$$

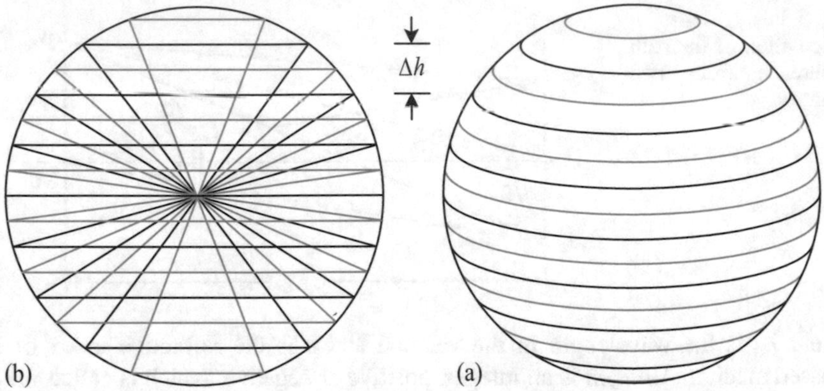

Fig. 3.4 (a): Schematic representation of the fringes produced by two point sources on the *sphere of observation*. (b): section of the sphere and the asymptotic, conical surfaces, of the hyperboloids. The black and gray lines respectively represent the maxima and the minima of interference

where p is the distance between two successive maxima and β is the angular separation with which the two sources are seen from M. For example, if $d = 1$ mm, $R_o = 1$ m, $\lambda_o = 630$ nm, $n = 1$, it is $\Delta h = 0.63$ mm. For large R_o, the angular separation $\Delta\alpha$ between two subsequent maxima is

$$\Delta\alpha = \frac{\lambda_o}{nd}. \tag{3.1.17}$$

In the previous example, it is equal to 0.63 mrad = 0.036°. Around the point M, the waves from the two sources can be considered roughly planar and the intensity observed on the screen is still given by Eq. (3.1.9).

A concept that can sometimes be useful is that of the sphere of observation (Fig. 3.4), valid for observation distances that are large compared to the distance between the sources. In this case, the hyperboloids of Fig. 3.2 merge with their asymptotic conical surfaces, with which they can be approximated. Their intersection with a sphere of radius R, centered on the sources, is therefore a set of parallel circles whose separation in the direction defined by the axis of the two sources is constant, as can be seen from Eq. (3.1.14), taking $r_1 = r_2 = R$, for which

$$p = \frac{\lambda_o R}{nd}. \tag{3.1.18}$$

On the sphere of observation, the fringes are denser in the vicinity of the equatorial plane, for which the order of interference is zero (Fig. 3.3), while rarefying proceeding toward the poles, where the order of interference is maximal.

When we have $d >> \lambda_o/n = \lambda$, the interference term of Eq. (3.1.9) oscillates rapidly in space. This means that the flow through a closed surface containing the sources is roughly equal to the sum of the flows emitted individually. In fact, the integration of the interferential term leads to a mutual cancellation between maxima and minima. However, this deletion becomes less accurate for $d \to 0$. When, in particular, we have $d < \lambda/2$ (and $\Delta\phi \approx 0$), the interference is only constructive, so the total flow is larger than the sum of the individual flows! (And destructive if $\Delta\phi \approx \pi$.) On the other hand, the principle of energy conservation does not necessarily imply that the flows remain the same. In fact, if the sources are close, they interact with each other, and the increase (or decrease) in the flow occurs at the expense (or saving) of the energy reserves available to the sources. The case $d < \lambda$ is hardly feasible in the optical field, but, in the range of radio waves, this phenomenon is used to produce a directional radiation and also to increase the radiated power. On the other hand, for short wavelengths, there is something similar in the case of superradiance and with lasers. In any case, in the calculation of radiated power, the transverse vector nature of electromagnetic waves is important.

3.2 Two-wave interference

3.2.1 Classification of interference methods

In order to observe the interference fringes, the main problem to be solved is to make the interfering waves mutually coherent. If this can be more or less automatic with laser sources, it is not so with natural sources; they are typically composed of a large number of excited atoms, each capable of radiating a "wave train" for times on the order of 10^{-8} s, and the various emissions are uncorrelated between them. Two distinct sources could therefore maintain their relative phase only for similarly short times, and the same happens for the position of the fringes. The arrangement normally employed to observe stable fringes is to use a single source from which two (or more) distinct light beams are obtained.

The methods derived from it can be classified into two categories. In one case, the methods called *wavefront division* use a point (or linear) source: two distinct parts of the wavefront generated by the source, propagating in different directions, are diverted so as to overlap. Here, the smallness of the source is associated with the request that, for each point of arrival in the interference figure, all of the points of the source contribute with path differences (between the two different roads) close to each other (within $\lambda/2$).

The methods of the second case, called *amplitude division*, divide a single beam of light into two or more beams, by partial reflections. Here, there is no need for point sources, since there is a one-to-one correspondence between the reflected wave and the transmitted one (which are then appropriately recombined). These methods can thus be used with extended sources, with the advantage that the intensity of the interference patterns can be much larger than that of the division wavefront systems.

Although the amplitude division methods solve the problem of *spatial coherence* of the sources, they can still be sensitive to their *temporal coherence*. In both types of interferometers, at the microscopic level, with natural sources, it can be said that the interference fringes are caused by interference of each wave train (photon) with itself on the different paths.

Let us now look at a short review of interferometer, and we will then study the limitations associated with the degree of spatial and temporal coherence of the sources in a little more detail. In the following, for simplicity, we will use the symbol $\lambda=\lambda_o/n$ and, for propagation in air, we will take $n = 1$. We will also distinguish the optical path difference with ΔL, and the distance difference with Δr.

3.2.2 Interference by division of the wavefront

3.2.2.1 Young's experiment

The historically most famous experiment to demonstrate the wave nature of light is that of Young; it uses the method, a bit drastic, of sampling light from a source with two small holes (or two slits). This same approach had previously been used by Francesco Maria Grimaldi, who used the sunlight to illuminate a pair of small holes. To increase the degree of mutual coherence of the light that arrives at the two openings, Young had previously filtered sunlight with a third hole S, as shown in Fig. 3.5. For both the first opening and the other two openings, diffraction plays an important role in diffracting and then recombining the light that passes through the openings S_1 and S_2. The precise treatment of this experiment is not therefore very simple, and it was also because of this difficulty that Young could not convince many of his contemporary colleagues. On the other hand, the priority for this experiment is not attributed to Grimaldi, because, apparently, he

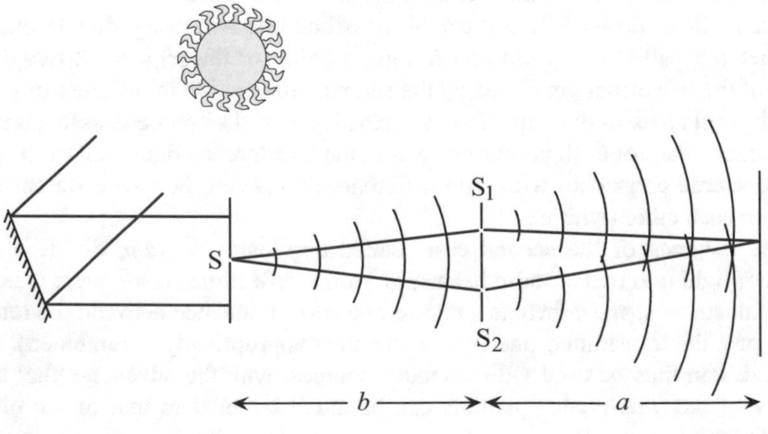

Fig. 3.5 Young's interference experiment

does not mention the presence of any sunlight filtering to illuminate the two holes, and it is also said that the effects he observed were due to diffraction alone. In fact, the Sun is a relatively large source (0.5°) and, as we will see below, the resulting convolution leads to the erasing of the fringes, unless the distance between the holes is less than ≈ 25 μm.

In any case, the two openings S_1 and S_2 can be assimilated to two point sources and the analysis of the interference figure of Young's experiment is, in a first approximation, equal to what we have already seen in §3.1.3. Since (at least in the middle part of the interference figure) the fringes are almost straight, the brilliance of the figure can be greatly increased by using parallel slits, instead of pinholes. Within certain conditions, which would take too long to describe here, the brightness can be increased by focusing the light on the first slit. We will return to this experiment in the study of diffraction.

3.2.2.2 Fresnel's double mirror and biprism

In his interference demonstrations, in 1816, Fresnel removed most of the problems related to diffraction, again using a point source (an opening), by creating two slightly separate images of it by means of two mirrors or a "biprism", as shown in Figs. 3.6 and 7. In the case of the double mirror, the source S has two virtual images S_1 and S_2 and all three points are at the same distance b from the point O of intersection of the mirrors, inclined to each other by a small angle θ. Therefore, $d \equiv \overline{S_1S_2} \cong 2b\theta$. The angle β under which the sources are seen from point M in the center of the interference pattern is

$$\beta \cong \frac{d}{a+b} \cong \frac{2b\theta}{a+b}, \tag{3.2.1}$$

where $a = \overline{\text{OM}}$. Finally, for Eq. (3.1.16), the separation between two successive maxima is

$$\Delta h = \frac{\lambda}{\beta} = \lambda \frac{a+b}{2b\theta}. \tag{3.2.2}$$

The width of the overlapping area of the beams is $\overline{\text{AB}} = 2a\theta$. As a result, the number of fringes that one can see is

$$N = \frac{2a\theta}{\Delta h} = \frac{4ab}{\lambda(a+b)}\theta^2. \tag{3.2.3}$$

For θ = 1°, λ = 633 nm, a = 1 m, b = 10 cm, it is Δh = 0.2 mm, N = 175.

Fresnel's biprism uses refraction to split and superimpose the beams (Fig. 3.7).

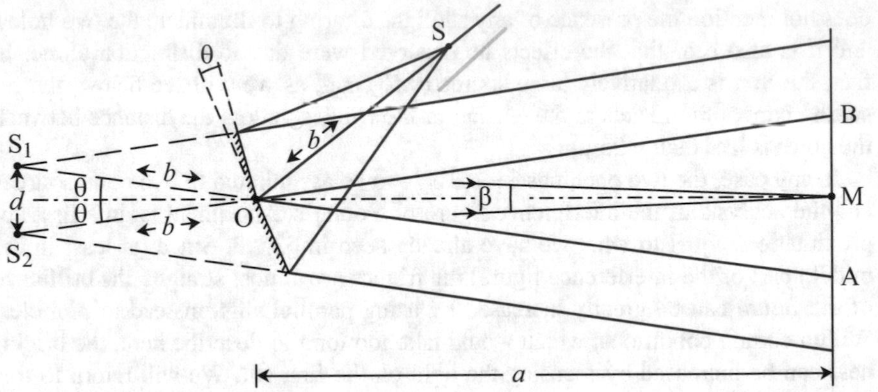

Fig. 3.6 Fresnel's double mirror

In this case, the distance d between the images S_1 and S_2 is $d \cong 2b(n-1)\theta$, where 2θ is now a small angle between the faces of the prism and $b = \overline{SO}$ is the distance between the source S and the vertex O of the biprism. The angular distance with which S_1 and S_2 are viewed from the point M on the screen is

$$\beta \cong \frac{2b(n-1)\theta}{a+b}, \tag{3.2.4}$$

where, again, $a = \overline{OM}$. Therefore, the separation between two maxima is

$$\Delta h = \lambda \frac{a+b}{2b(n-1)\theta}. \tag{3.2.5}$$

For $n = 1.5$, and with the same values as in the previous case for the other variables, we have $\Delta h = 0.2$ mm.

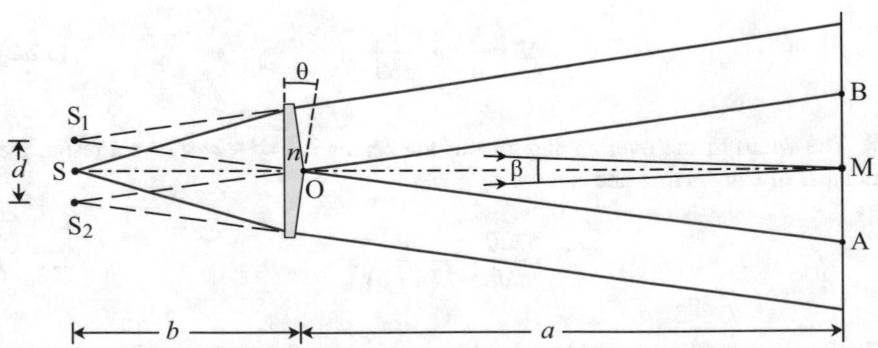

Fig. 3.7 Fresnel's Biprism

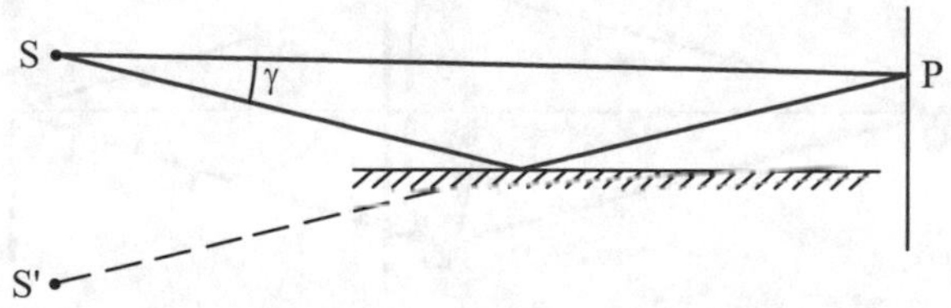

Fig. 3.8 Lloyd's Mirror

The biprism is much easier to construct and handle than the double mirror system, although it has some distortion. In either case, however, the fringes are disturbed by the diffraction on the edge O.

3.2.2.3 Lloyd's mirror

In this, which is the simplest classical configuration, a point source is placed close to the plane of a mirror (Fig. 3.8). Interference is therefore produced by the source and its image. In calculating the intensity at a point P on the screen, it is now necessary also to consider the phase shift produced by reflection.

3.2.3 Tautochrone properties of Optical Systems

The interference of light examined in this chapter considers the emphasis of this phenomenon through special experiments; however, interference permeates the entirety of optics, including Geometrical Optics. If we remember what we have already studied with Fermat's principle and with the production of images by an ideal optical system, we already know that all of the rays that start from a point object O arrive at the corresponding image point I having gone down the same optical path. This can be thought of as a consequence of the Malus-Dupin principle, so that the rays connecting a divergent wavefront ("just" emitted by O) with a converging wavefront (near I) all have the same optical length. In other words, they all employ the same time to go from O to I. The possibility of obtaining a maximum intensity in I, representing the image of the source O, is conditioned by a mutual reinforcement of the single wave portions that reach I without phase difference, since they have followed tautochronous paths. Instead, the paths that lead from O at any other point in space are not optically equal, and, for them, the mutual interference implies a weakening of the light. Thus, an image obtained by an optical system is itself a phenomenon of interference.

Consider, for example, the case of an optical system into which a screen with two small openings is inserted, and a source O, punctiform, or at least far, with a

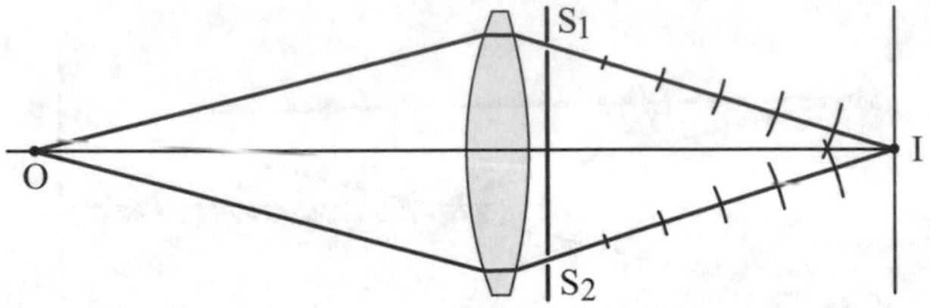

Fig. 3.9 Using a lens in Young's experiment

sufficiently small angular aperture (Fig. 3.9). This configuration is similar to that of Young's experiment, but it allows for superimposing the diffraction patterns produced by the two openings on the image plane of the source, even for a considerable distance between them. The interference figure observed on the image plane of the source therefore has fringes that are still separated by a distance given by Eq. (3.1.16), although, in this case, they will be very dense. If the system is stigmatic, for the equality of the optical paths, in the point image I of the source, there is a bright zero-order fringe, also observable therefore in white light (in the absence of chromatic aberration, or at least when the two openings are equidistant from the axis).

If, on the contrary, the system is not stigmatic, this fringe will be translated by an amount that depends on the extent of the aberrations of the system. Michelson proposed using this fact to perform a quantitative examination of optical systems: In particular, if an aperture is held fixed, for example, on the optical axis, while changing the position of the other aperture, one can determine the difference from the sphericity of the wavefront produced by the system [Michelson 1918].

Later, we will see examples in which a lens is used in interference experiments (even our eyes, which we use to "see" interference figures, contain a lens), but we will have to take into account this tautochrone property between conjugate points.

3.2.4 Importance of the size of the light source

In the previous sections, we presumed the use of point sources. Clearly, if the source dimensions were much smaller than a wavelength (of the radiation considered), a sharp interference pattern would always be obtained, since the variation in the path difference Δr would be negligible in passing from one point to another of the source. However, in practice, the sizes of the sources are much greater, and we must examine how their extension affects the fringes.

For simplicity, we consider here the case in which the emissions between the different points of a source are completely uncorrelated between them. The intensity at each point of the interference pattern is therefore the sum of the intensities relative to the single points of the source, that is, the sum of many interference figures. These figures, albeit similar, are shifted relative to each other so that the

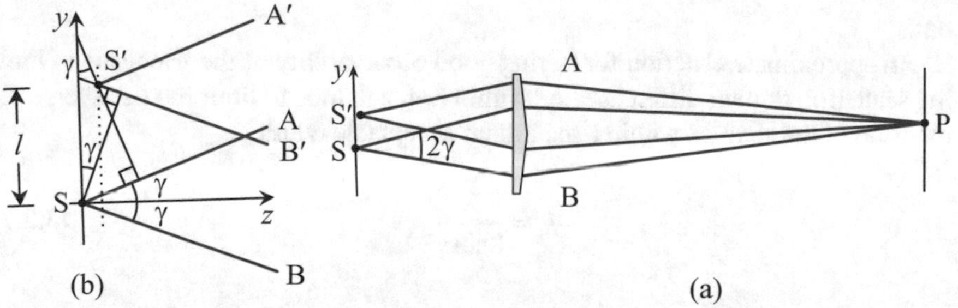

Fig. 3.10 Determination of the path differences for two sources S and S′. (a): path of the A and B rays for the Fresnel biprism; (b): detail around the sources. The z-axis is taken on the ASB internal bisector of the rays, and the y-axis on the external bisector

resulting figure is more or less faded and tends to disappear when we increase the size of the source. The precise dependence with which this occurs depends on the particular configuration of the interferometer; however, at least at the first order, a general rule can be given. To be explicit, by way of example, consider the case of the Fresnel biprism and, in particular, the effect that two sources S and S′ produce on a point P assigned on the figure of interference (Fig. 3.10). From each point (S and S′), two rays start (A and B for the point S and A′, B′ for S′), which, with different paths, reach the point P interfering with each other.

Let S be the origin of a local coordinate system with the z-axis on the A$\hat{\text{S}}$B internal bisector and the y-axis on the external bisector. Now, consider the case in which the source S′ is translated from S of a small distance l and has coordinates (x, y, z); it does not necessarily lie on the y-axis or on the yz-plane. If l is much smaller than the path of the rays up to point P (that is, $l << b$ in the case of Fig. 3.5), the rays remain two by two nearly parallel, but, while one is shortened, say, from A to A′, the other stretches from B to B′, going from S to S′. Denoting, with 2γ, the angle in S between the A and B rays, which is called the *opening angle of the interference*, from Fig. 3.10(b), it is immediately apparent that, at the first order in x, y, z,

$$[\text{A}'] - [\text{A}] = -ny\sin\gamma - nz\cos\gamma \quad \text{and} \quad [\text{B}'] - [\text{B}] = ny\sin\gamma - nz\cos\gamma\,,$$

where the square brackets indicate the optical path of the rays and n is the refractive index at the source. So, the optical path differences $\Delta L = [\text{B}] - [\text{A}]$ (between the rays A and B) and, similarly, $\Delta L'$ (between A′ and B′), differ from each other according to the relation

$$\Delta L' - \Delta L = 2ny\sin\gamma\,. \tag{3.2.6}$$

In particular, when $2y\sin\gamma = \lambda/2$, if, in P, there is a maximum in the interference figure produced by S, there is instead a minimum for that corresponding to S′. In this case, the two systems of fringes are out of phase and cancel each other out.

An approximate criterion for having good observability of the fringes is to limit the scattering of path differences Δr within $\lambda/4$, and thus to limit the transverse dimension of the sources within a maximum acceptable width

$$l_a = \frac{\lambda}{8\sin\gamma} . \tag{3.2.7}$$

The smaller the opening angle γ, the larger the source can be. In the case of a distant source, such as a street lamp, the Sun, or a star, a similar relationship can be obtained in terms of the angular width Λ of the source. For example, in Young's experiment, it is

$$2y\sin\gamma \cong \Lambda b\frac{d}{b} = \Lambda d ,$$

where b is the distance of the source from the two openings and d is their separation. With the same criteria as before, the maximum acceptable angular width is

$$\Lambda_a = \frac{\lambda}{4d} . \tag{3.2.8}$$

So far, we have considered a particular direction (y in Fig. 3.10) in the extension of the sources, parallel to the plane containing the main source S and its two derived sources S_1 and S_2. As we have seen in §3.1.3, fringes extend along the direction (say, x) perpendicular to this plane, so if the source S is moved in this direction, the fringes move parallel to themselves. Then, in place of a point source, one can use a straight line (a thin slit) oriented along the x-direction without altering the sharpness of the fringes, at least for a length where their curvature gives negligible shifts. Indeed, in this case, the difference in optical paths remains practically constant for the various points of the source.

Let us now quantify the effects of a small but non-zero width rectangular source with the shorter side of width l (Fig. 3.11). Let us take an infinitesimal strip of width dy centered in the center S of the source, for which the point M in the figure of interference corresponds to a maximum intensity equal to $4I_o dy/l$, and then, with $\Delta r = 0$. Here, I_o corresponds to the intensity in M, in the absence of interference if one of the two paths is obstructed. The path difference in M that we have for another strip, centered in S′, is then

$$\Delta r' = 2y\sin\gamma ,$$

where y is the distance from S taken in the y-direction and γ is the opening angle of interference, which we here consider to be constant. In the neighborhood of M, the intensity associated with the strip corresponding to S′ is therefore

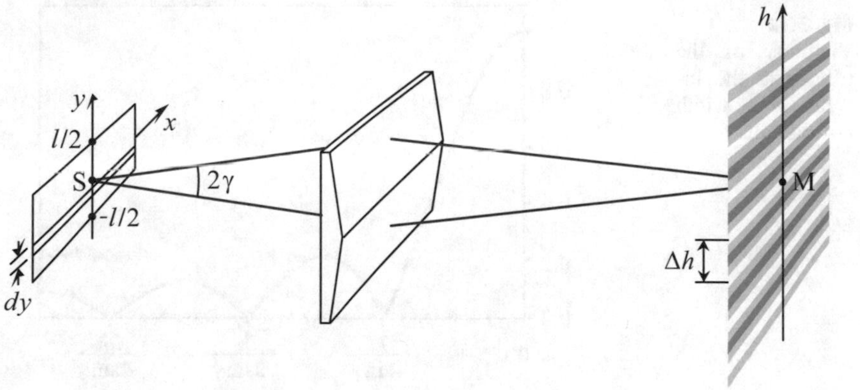

Fig. 3.11 Determination of intensity modulation for a rectangular source

$$dI = \frac{2I_o}{l}dy\left[1+\cos\left(k\Delta r'+\phi\right)\right] = \frac{2I_o}{l}dy\left[1+\cos\left(\frac{2\pi}{\lambda}2y\sin\gamma+\phi\right)\right],$$

where $\phi = 2\pi h/p$, h is the height from M (in the direction perpendicular to the fringes) and p is the pace between two successive maxima in the interference pattern. If we then integrate over the width of the source, assuming that each point emits incoherently with the others, the total intensity at a given point around M is

$$I = \frac{2I_o}{l}\int_{-l/2}^{l/2} dy\left[1+\cos\left(2gy+\phi\right)\right] = 2I_o\left[1+\frac{\sin(gl)}{gl}\cos\phi\right], \quad (3.2.9)$$

with $g = (2\pi/\lambda)\sin\gamma$. Eq. (3.2.9) tells us that, around M, the intensity oscillates with the variation of ϕ and then with the distance h from M. The sinc function determines the amplitude of this modulation: if the slit is narrow, on the maxima, the intensity quadruples when passing from one open path to two, while it only doubles on the zeros of the sinc. In general, the intensity increases from a minimum $I_{\min}$ to a maximum $I_{\max}$ given by

$$I_{\min} = 2I_o\left[1-\frac{\sin(gl)}{gl}\right], \quad I_{\max} = 2I_o\left[1+\frac{\sin(gl)}{gl}\right]. \quad (3.2.10)$$

Following Michelson, we take as an index of the fringe contrast in M their *visibility* $\mathcal{V}$, defined as

$$\mathcal{V} = \frac{I_{\max} - I_{\min}}{I_{\max} + I_{\min}}, \quad (3.2.11)$$

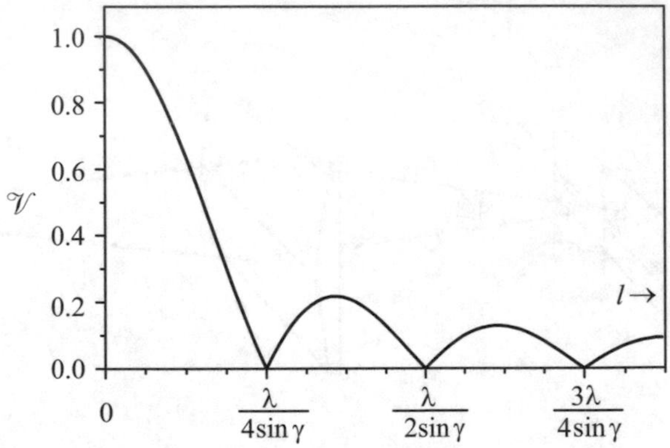

Fig. 3.12 Visibility of the fringes at the increase in the width l of the source

which, in our case of a rectangular source, is

$$\mathcal{V} = \frac{4I_o\left|\frac{\sin(gl)}{gl}\right|}{4I_o} = \left|\frac{\sin(gl)}{gl}\right| = \left|\frac{\sin\left(\frac{2\pi}{\lambda}l\sin\gamma\right)}{\frac{2\pi}{\lambda}l\sin\gamma}\right|, \quad (3.2.12)$$

which has the trend shown in Fig. 3.12 with the variation of l. The zeros correspond to slit widths, so that the difference in travel varies from λ, 2λ, 3λ, ..., from side to side.

These considerations apply in the hypothesis that γ does not vary appreciably with the y-coordinate, and this is what happens with the biprism and the Fresnel double mirror, and in Young's experiment. With Lloyd's mirror, the situation is different, because a shift of the source S perpendicular to the mirror leads to a shift of its image in the opposite direction. Thus, the pace of the fringes in the interference pattern varies with the position of S, i.e., it decreases as S rises from the mirror plane while, at the same time, the opening angle γ of the interference grows. With a slit of defined width, the sharpness of the fringes decreases, therefore, moving away from the mirror plane.

Finally, in order to quantify γ in the case of the biprism and the double mirror, it is enough to do a construction similar to that made for calculating the angle β with which the secondary sources S_1 and S_2 are viewed from the midpoint M on the screen (§3.2.2). Thus, with the same meaning of the symbols from Figs. 3.6 and 7, for the double mirror, we have

$$\gamma \cong \frac{a}{a+b}\theta, \quad (3.2.13)$$

and, for the biprism,

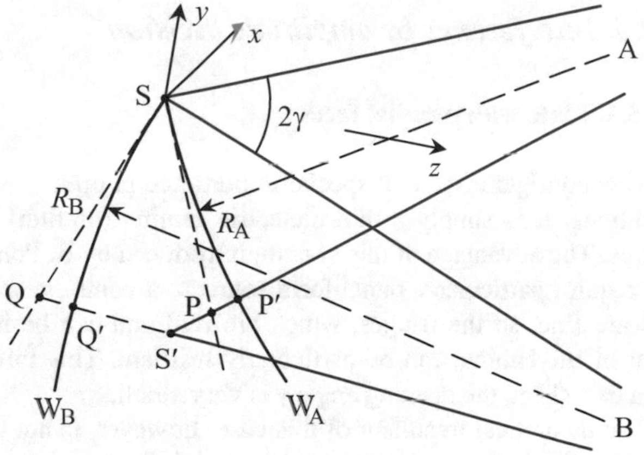

Fig. 3.13 Analysis at the 2nd order of the path differences $\Delta r = l_B - l_A$

$$\gamma \cong \frac{a}{a+b}(n-1)\theta\,. \tag{3.2.14}$$

In the following, we will need a more accurate expression than Eq. (3.2.6), that is, an approximate expression at the 2nd order for the path differences as a function of the position on the source. To evaluate this dependence, we take here a Cartesian reference system with origin at the (central) point of the source S and such that the emerging rays from S towards a given point P of observation, along the two paths A and B, lie in the *yz*-plane. We also take as the *z*-axis the bisector between these two rays (Fig. 3.13). The path differences toward the point P can be calculated assuming that, from the same point P, a spherical wave starts, which is then divided and overlapped by the optical system. In S, therefore, two approximately spherical waves will arrive, with radii of curvature R_A and R_B. Let W_A and W_B be the two wavefronts, which intersect in S, for the waves of the two paths. For a point source S′, the path differences must therefore be calculated with respect to W_A and W_B. Referring to what was done for Fig. 3.10, in which the emerging rays were parallel, the result of Eq. (3.2.6) should now be corrected considering the separations $\overline{QQ'}$ and $\overline{PP'}$ taken on the rays that start from S′ toward P, between the wavefronts W_A and W_B and their respective tangent planes in S. In conclusion, instead of Eq. (3.2.6), we have [Born and Wolf 1980]

$$\Delta r' - \Delta r \cong 2y\sin\gamma + \frac{1}{2}\left(\frac{1}{R_A} - \frac{1}{R_B}\right)\left(x^2 + y^2\right), \tag{3.2.15}$$

which will be useful in studying the visibility of fringes in amplitude division systems, in which the angle γ can be very small.

3.2.5 Interference by amplitude division

3.2.5.1 Plate with parallel faces

This configuration is of special importance in optics, and is found in many applications. It is simply a thin dielectric lamina (or film) illuminated by a small source. The advantage of this system, introduced by R. Pohl in 1884, is that it does not require particularly punctiform sources: a common spectral lamp and a mica leaf are fine, so the fringes, which are real and can be intercepted anywhere in front of the lamina, can be particularly brilliant. This follows from the fact that, for a thin sheet, the opening angle γ is very small.

The theoretical treatment of this case, however, is not trivial. Here, we start to consider the (indeed frequent) case in which the interfaces of the foil have a weak reflectivity: in this way, we can neglect multiple reflections between the two faces. For the moment, we refer to a point source. The sources of coherent waves are therefore its images given by the reflection on the first and second faces of the lamina, of thickness d and index $n_2 = n$, which, for simplicity, we here suppose to be immersed in a medium of index $n_1 = n_3 = 1$. However, refractions at the first face make the second image not stigmatic. In particular, its apparent position (Fig. 3.14) depends on the external incidence angle ϑ according to a somewhat complicated relation (of which I leave the long demonstration for exercise):

$$S_1 = \begin{cases} x = 0, \\ y = -h, \end{cases} \qquad S_2 = \begin{cases} x = \dfrac{(n^2 - 1)\, 2d \sin^3 \vartheta}{(n^2 - \sin^2 \vartheta)^{3/2}}, \\[2ex] y = \dfrac{x}{\tan \vartheta} - \dfrac{2d \cos \vartheta}{(n^2 - \sin^2 \vartheta)^{1/2}} - h. \end{cases} \tag{3.2.16}$$

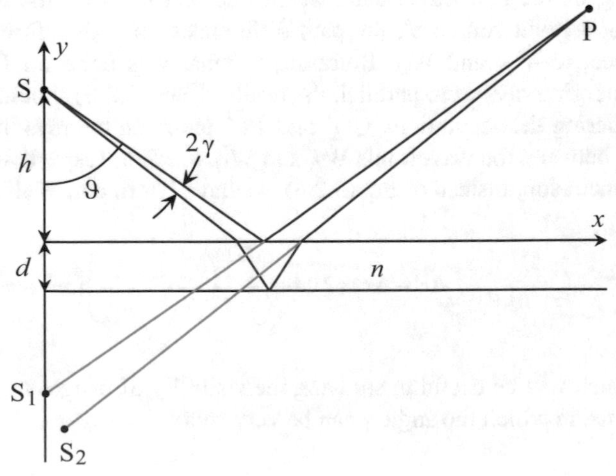

Fig. 3.14 Interference from a dielectric plate with index n_2. S_1 and S_2 are the apparent images of S as they are seen from point P

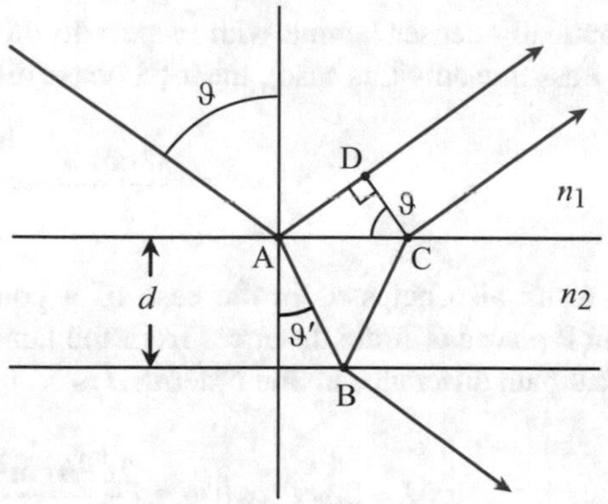

Fig. 3.15
Parallel reflected rays, for the same incident ray

Only for small angles can one say that the second image is backward with respect to the first of a distance *2d/n*. Even the phase relationship between the two images is a complicated function of ϑ. Therefore, in the resolution of these problems, it is preferable to have recourse to a plane waves treatment, which is otherwise adequate for sources and/or places of observation points at a great distance from the lamina, as compared to its thickness *d*. The optical path difference between the two waves reflected by the lamina is the same as that between the two reflected rays from a ray normal to the incident wavefront. This difference, with reference to Fig. 3.15, is

$$n_2\left(\overline{\mathrm{AB}}+\overline{\mathrm{BC}}\right)-n_1\overline{\mathrm{AD}},$$

that is,

$$\Delta L=\frac{2n_2d}{\cos\vartheta'}-2n_1d\tan\vartheta'\sin\vartheta=\frac{2dn_2}{\cos\vartheta'}\left(1-\sin^2\vartheta'\right),$$

where, in the second step, Snell's law has been applied and ϑ' is the internal angle of incidence. Therefore, we have the *cosine law*

$$\Delta L=2n_2d\cos\vartheta', \tag{3.2.17}$$

where you well have to notice that the difference in optical path length *decreases* as ϑ' increases, contrary to what one might think at first. The corresponding phase difference is $4\pi n_2 d\cos\vartheta'/\lambda$; note also that, at the same angle of internal incidence, it does not depend on the refractive index outside of the lamina. To this phase, it is necessary to add that which is derived from the reflection itself: in the case of a dielectric lamina in air, or, in any case, with a lesser (or greater) index of both surrounding media, the Fresnel formulas of Cap. 1 show that the relative phase shift of the second reflection with respect to the first one is $-\pi$ (or π), where the choice of the sign is suggested by the microscopic consideration made by Fresnel. For a monochromatic wave, this choice has no effect on the fringe intensity. Therefore, for an optically denser lamina with respect to the surrounding media (in the following, I assume only this case), the total phase difference is

$$\Delta\phi = 4\pi \frac{n_2 d \cos\vartheta'}{\lambda} - \pi\,. \tag{3.2.18}$$

This result also helps us in the case of a point source S and a point of observation P placed at finite distances from the lamina. With reference to Fig. 3.16, the optical path difference at 2nd order in d is

$$\Delta L = 2n_2 d \cos\vartheta' - n_1 \frac{2d^2 h \tan^2\vartheta'}{l^2} \cos\vartheta\,, \tag{3.2.19}$$

with $\tan\vartheta \approx b/h$ and $\sin\vartheta' \approx n_1 g/(n_2 l)$. Fig. 3.16 also shows that the opening angle of interference is

$$2\gamma \cong \frac{q}{l} = \frac{2d \tan\vartheta' \cos\vartheta}{l} \cong \frac{2dh}{h^2+g^2} \tan\vartheta'\,. \tag{3.2.20}$$

If the source is punctiform, the fringes fill the whole space, and therefore, in this case, they are called *non-localized*. However, in the real case of an extended source, their visibility decreases as the thickness of the lamina increases, due to the increase in the opening angle γ, for assigned points S and P. The fringes do not disappear anywhere: in the next few sections we will discuss the *localized* fringes *that* persist even with considerably extended sources, and which may be made evident with a simple trick.

3.2.5.2 Fringes of equal inclination by a plate with parallel faces

With a parallel face lamina, an effective way of reducing the opening angle γ is to consider viewing points P far away from the lamina. In the limit of an infinite distance, we return to the case of Fig. 3.15, in which a single ray from the source gives rise to two parallel rays whose reciprocal phase difference depends only on the angle of incidence ϑ, according to Eq. (3.2.18). So, the maximum of interference, or the minimum, will be arranged along a direction corresponding to the inclination ϑ, whereby

$$\left(4\frac{n_2 d \cos\vartheta'}{\lambda_o} - 1\right)\pi = 2m\pi\,, \tag{3.2.21}$$

with $m = 1, 2, 3, \ldots$ for the maxima,

$m = \frac{1}{2}, \frac{3}{2}, \frac{5}{2}, \ldots$ for the minima,

Fig. 3.16
Graphic study of optical path difference, with $h = h_1 + h_2$ and:

$$\Delta L = [\mathrm{SABCP}] - [\mathrm{SEP}],$$

$$[\mathrm{SEP}] = n_1 l,$$

$$[\mathrm{SABCP}] = [\mathrm{SAQ}] + 2dn_2 \cos\vartheta'.$$

$$l = \overline{\mathrm{S_1P}} = \sqrt{h^2 + g^2}$$

Consider the right triangle $\mathrm{S_1QP}$:

$$\overline{\mathrm{PQ}} = \overline{\mathrm{DC}} = q = 2d \tan\vartheta' \cos\vartheta,$$

$$[\mathrm{SAQ}] = n_1 a = n_1\sqrt{l^2 - q^2},$$

from which:

$$\Delta L = $$

$$= 2n_2 d \cos\vartheta' + n_1\sqrt{l^2 - q^2} - n_1 l$$

$$\cong 2n_2 d \cos\vartheta' - \frac{n_1 q^2}{2l}$$

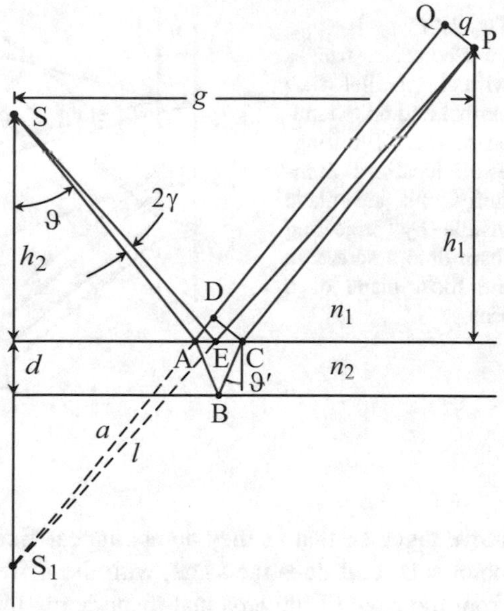

where, as usual, m indicates the order of interference and ϑ, ϑ' respect Snell's law.

In order to be able to observe these fringes, in addition to placing a screen at a great distance from the lamina (relative to its thickness), you can look towards the lamina itself by accommodating the eye to infinity. In a similar way, you can collect the reflected light through a positive lens and observe the interference pattern that forms on a screen placed on the second focal plane of the lens itself (Fig. 3.17). Instead, the source must be such as to illuminate the lamina with a relatively large angular field. The two parallel rays of each pair reflected by the lamina are, in fact, rejoined in the focal plane of the lens, in a position that depends solely on their inclination. Different inclinations, with different phase differences, "illuminate" different screen positions. It is immediately apparent that this interference pattern is maintained even if the source is unlimitedly extended. Often, it is the opening of the lens, or the eye's pupil, that limits the part of the source that is actually used for each point on the screen. In fact, the rays originating from different source points, albeit incoherent with each other, undergo the same law of phase difference dependent on the inclination alone, and thus contribute to making the figure of interference more brilliant. The fringes of this figure are called *fringes of equal inclination*, simply because they outline the points of the screen itself for which there is the same inclination in reflection from the lamina.

When the illumination and observation take place with their axis approximately perpendicular to the lamina, on the focal plane of the lens (or the retina of the observer), the fringes take the form of rings, called *Haidinger's fringes*, after the man who described them in 1855. These rings (Fig. 3.18) are centered on the axis of the lens. With an extended source, if the lens or observer moves, the fringes

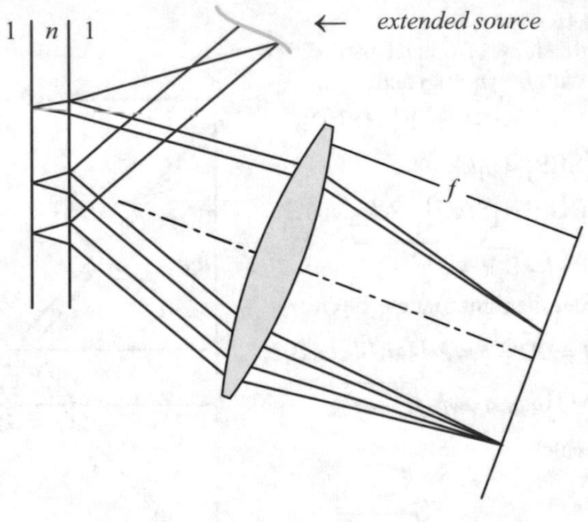

Fig. 3.17 Interference fringes with a parallel face lamina and an extended source. The fringes are localized at infinity, but are made visible by projecting them onto a screen in the focal plane of a lens

move together, that is, they do not appear fixed with respect to the lamina. With a point source, it does the same, with the difference of being progressively cut off from the edge of the lens that diaphragms the angular field of the source. In any case, the number of visible fringes depends on the angular field allowed by the optical system, depending on the size and distances between the source, the lamina, and the lens. The interference order in the center of the figure (for $\vartheta = \vartheta' = 0$) is the highest, and is

$$m_o = \frac{2n_2 d}{\lambda_o} - \frac{1}{2}; \tag{3.2.22}$$

m_o is not necessarily an integer, and we can write

$$m_o = m_1 + e, \tag{3.2.23}$$

where m_1 is the integer order of the innermost luminous fringe and e (<1) is called the fractional order in the center. For the various fringes, numbered with p (integer > 0) from the first, we have

$$\frac{2n_2 d}{\lambda_o}\cos\vartheta'_p - \frac{1}{2} = m_p = m_1 - p + 1 = m_o - p + 1 - e,$$

from which, for Eq. (3.2.21), we have

$$\frac{2n_2 d}{\lambda_o}\left(1 - \cos\vartheta'_p\right) = p - 1 + e. \tag{3.2.24}$$

As long as ϑ' remains small, we can write

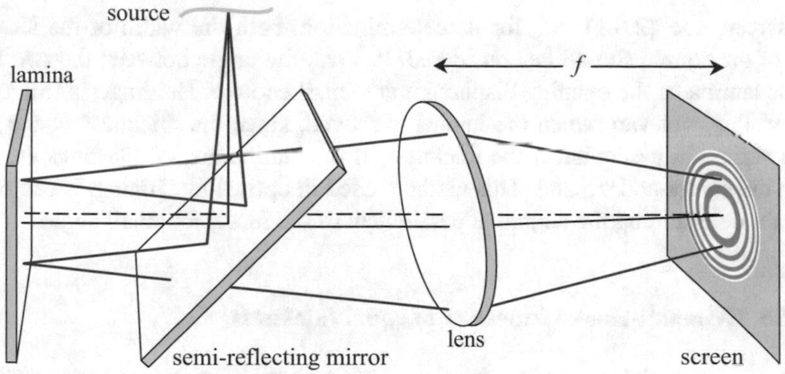

Fig. 3.18 Haidinger's fringes

$$1-\cos\vartheta'_p \cong \frac{1}{2}\vartheta'^2_p \cong \frac{1}{2}\left(\frac{n_1}{n_2}\right)^2 \vartheta_p^2 \,,$$

where ϑ_p is the angle of external incidence coinciding with the p-th fringe, hence

$$\vartheta_p \cong \frac{1}{n_1}\sqrt{\frac{n_2\lambda_o}{d}(p-1+e)} \,. \tag{3.2.25}$$

Unless the lamina is very thin, the fringes have very high interference orders, they are dense and they are not visible in white light. In fact, if we consider, for example, a glass plate with $d = 1$ mm, $n_1 = 1$ and $n_2 = 1.5$, the order of interference in the center is about 5000 for $\lambda_o = 600$ nm and a very small variation of the wavelength is enough to switch from one order to the next. In the case considered here, $\Delta\lambda \cong 2n_2d/m_o{}^2 = 0.12$ nm and, to be able to observe fringes, the spectral band of radiation used should already be lower than 50 GHz. Again, with this example, the angular separation between the first two fringes is about 1.7° and there are about 34 fringes within an incidence angle of 10°.

The case dealt with in this section is, on the other hand, an extreme case; often, one has to make do with plates that have non-perfectly parallel faces, so the treatment becomes considerably complicated. At first approximation (for a very small angle between the faces), the shift between the two reflections is still given by Eq. (3.2.18), but the thickness d now becomes an important *variable* to consider. In the configuration of Figs. 3.16 and 17, the visibility of the fringes is now compromised by the fact that reflections from different points of the lamina, with different phase differences, will contribute to the field at a given point of the screen. Thus, Haidinger's fringes disappear when the phase difference varies more than $\pi/2$ within the area of the lamina actually used by the optical system. The width of this zone depends on the aperture of the system itself (taking as the object a point on the screen, see §2.8.1) and, for its determination, both the width of the lens and that of the source should be considered. If, then, the angle between the two faces of the lamina or the opening diaphragm are small enough, Haidinger's fringes are still visible, but vary when the lamina is moved, since the thickness of the zone used varies. In particular, if the thickness of the lamina grows, the rings are born from the center and expand. This method, used in optical machining, is one of the best for determining the degree of parallelism of the faces, for small angles.

3.2.5.3 Grimaldi-Hooke fringes or of equal thickness

However, the fringes of interference with which we are more familiar are those that are observed *near* thin sheets or between adjacent laminas, with relatively large light sources, and, in any case, such that the *cosine* of the angle of incidence ϑ_i varies little. In this case, $n_2 d$ is the dominant parameter rather than ϑ_i. These fringes are called fringes of equal thickness. In white light, the iridescences observed on soap bubbles, between two overlapping slides, in an oil film on water, and also on oxidized metal surfaces, are all the result of optical thickness variations. The interference bands like this are similar to the contour lines of topographic maps. In particular, when n is constant, they are useful in determining the quality of the surface of optical elements. For example, a surface to be examined can be brought into contact with a reference sample surface that is "optically flat". Between the two surfaces, there is a thin layer of air whose thickness generally varies. The interference pattern that is formed by the reflections from the two surfaces marks the variations in thickness of the air layer, and thus the shape of the surface under examination. Therefore, if the two surfaces are flat and slightly inclined to each other, the fringes that are formed are straight and regularly spaced.

The general treatment of these fringes is very difficult, and we will have to settle for various approximations. The case that I will consider here is that of a wedge-shaped lamina with a refraction index n, immersed in air, and whose faces are inclined at a very small angle α. Let us start by determining the angle φ between the two reflected rays from a ray lying on a plane orthogonal to both surfaces and with an incidence angle ϑ. With reference to Fig. 3.19 for Snell's law applied to refractions in A and C, we have, respectively,

$$n \sin \vartheta' = \sin \vartheta,$$
$$n \sin(\vartheta' + 2\alpha) = \sin(\vartheta + \varphi).$$

Subtracting member to member and using, for the sine difference, the prosthaphaeresis formula, we have

$$\sin\frac{\varphi}{2} = \frac{n \sin\alpha \cos(\vartheta' + \alpha)}{\cos(\vartheta + \varphi/2)}. \tag{3.2.26}$$

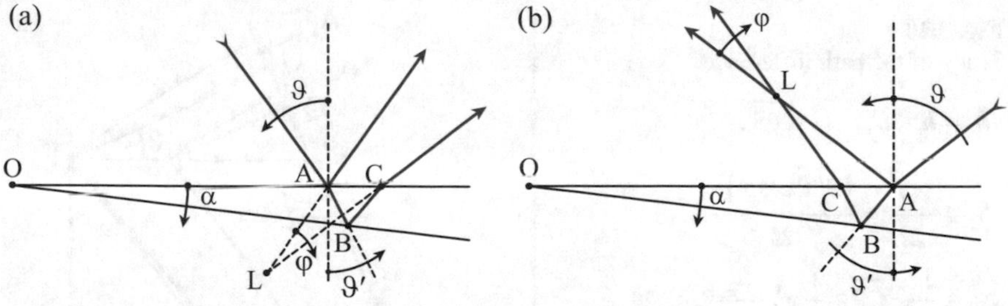

Fig. 3.19 Reflections from a wedge-shaped lamina: (a) case with $\vartheta, \vartheta', \alpha, \varphi > 0$, (b) case with $\vartheta, \vartheta' < 0, \alpha, \varphi > 0$

For angles ϑ not too close to $\pi/2$, for small α, at 1st order in α, we have

$$\varphi \cong 2n\alpha \frac{\cos\vartheta'}{\cos\vartheta} . \tag{3.2.27}$$

This is therefore approximately the angle between the two reflected plane waves of a plane wave coming at the incidence angle ϑ.

We intend now to calculate the difference in optical path length of two rays from a point source S reaching an observation point P, as in Fig. 3.20, taking into account the terms up to the second order of parameters α and d, where the latter is the thickness of the lamina at point A (taken perpendicular to OB). Meanwhile, we calculate the length of segments AB, BC, and AC of Figs. 3.19(a) and 20, using the law of sines from trigonometry. Let O be the point of intersection between the faces of the lamina in the incidence plane of the rays. For the OAB triangle, it is

$$\overline{AB} = \frac{\overline{OA}\sin\alpha}{\cos(\vartheta'+\alpha)} = \frac{d}{\cos(\vartheta'+\alpha)} \cong \frac{d}{\cos\vartheta'}(1+\alpha\tan\vartheta') . \tag{3.2.28}$$

For the ABC triangle, it is also

$$\overline{BC} = \overline{AB}\frac{\cos\vartheta'}{\cos(\vartheta'+2\alpha)} \cong \frac{d}{\cos\vartheta'}(1+3\alpha\tan\vartheta'), \tag{3.2.29}$$

$$\overline{AC} = \overline{AB}\frac{\sin(2\vartheta'+2\alpha)}{\cos(\vartheta'+2\alpha)} = \frac{2d\sin(\vartheta'+\alpha)}{\cos(\vartheta'+2\alpha)} \cong 2d\tan\vartheta' + 2d\alpha\left(1+2\tan^2\vartheta'\right), \tag{3.2.30}$$

$$\overline{AB}+\overline{BC} \cong \frac{2d}{\cos\vartheta'}(1+2\alpha\tan\vartheta') . \tag{3.2.31}$$

Compared to the formulas in the previous section, for $\overline{AC}$ and $\overline{AB}+\overline{BC}$, the terms in α represent a 1st order correction for inclination between the faces. Fig.

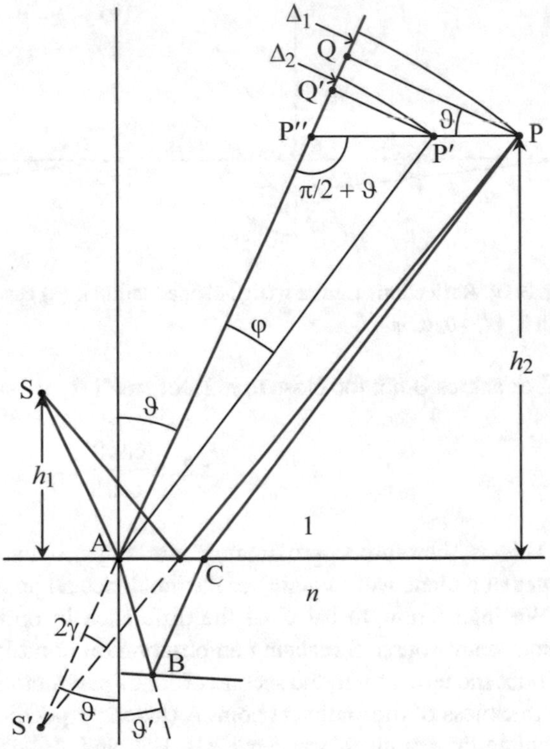

Fig. 3.20
Study of the path difference.

$h = h_1 + h_2, \quad l = \overline{PS'}.$

$$\Delta_1 \cong \frac{\overline{PQ}^2}{2l} = \frac{\left(\overline{P''P}\cos\vartheta\right)^2}{2l}.$$

$$\Delta_2 \cong \frac{1}{2}\varphi^2\overline{CP} \cong \frac{1}{2}\varphi^2\frac{h_2}{\cos\vartheta}.$$

$$\overline{Q'Q} = \overline{AC}\sin\vartheta.$$

$$\overline{P''P} = \overline{P''P'} + \overline{AC}$$

$$= \overline{AP'}\frac{\sin\varphi}{\cos\vartheta} + \overline{AC}$$

$$\cong \varphi\frac{h_2}{\cos^2\vartheta} + 2d\tan\vartheta'.$$

The path difference is therefore

$$[\mathrm{ABC}] + \overline{SA} + \overline{CP} - l =$$

$$= [\mathrm{ABC}] + \Delta_2 - \Delta_1 - \overline{Q'Q}.$$

20 is drawn as follows. S′ is the image of S produced by the first surface. From A, a segment AP′ is plotted parallel to CP, while P′P is parallel to AC. The broken line S′AP′ then has the same length of the refracted ray path in air $\overline{SA} + \overline{CP}$. Instead, the segment S′P corresponds to the length l of the reflected ray from the first surface. Lastly, the segments AQ′ and S′Q are the projections of AP′ and S′P, respectively, in the direction of the reflected ray in A. Δ_1 and Δ_2 represent the differences in length between the hypotenuse and their projections. They are

$$\Delta_1 \cong \varphi^2\frac{{h_2}^2}{2l\cos^2\vartheta} + 2d^2\frac{\tan^2\vartheta'\cos^2\vartheta}{l} + 2d\varphi\frac{h_2}{l}\tan\vartheta', \quad \Delta_2 \cong \frac{1}{2}\varphi^2\frac{h_2}{\cos\vartheta}.$$

Finally, the optical path difference between the two rays from S to P is

$$\Delta L = \overline{SA} + n\left(\overline{AB} + \overline{BC}\right) + \overline{CP} - l = n\left(\overline{AB} + \overline{BC}\right) - \overline{AC}\sin\vartheta - \Delta_1 + \Delta_2$$
$$\cong 2nd\cos\vartheta' + 2d\alpha\sin\vartheta\left(1 - 2\frac{h_2}{h}\right) + 2n^2\alpha^2\frac{h_1h_2}{h}\frac{\cos^2\vartheta'}{\cos^3\vartheta} - 2d^2\frac{\tan^2\vartheta'\cos^3\vartheta}{h}. \tag{3.2.32}$$

The first and last term correspond to those of Eq. (3.2.19). For a distant source ($h >> |h_2|$), it is

$$\Delta L \cong 2nd\cos\vartheta' + 2d\alpha\sin\vartheta + 2n^2\alpha^2 h_2 \frac{\cos^2\vartheta'}{\cos^3\vartheta}. \tag{3.2.33}$$

For almost normal incidence and small h_2, it is therefore

$$\Delta L \cong 2nd. \tag{3.2.34}$$

The fringes that are observed by normal incidence near the lamina are called *Fizeau's fringes*. Similar to Eq. (3.2.21), the maxima occur for

$$\left(4\frac{nd}{\lambda_o} - 1\right)\pi = 2m\pi, \quad m = 0, 1, 2, 3, ..., \tag{3.2.35}$$

that is, for

$$nd_m = \frac{\lambda_o}{2}\left(m + \frac{1}{2}\right). \tag{3.2.36}$$

In particular, for $d = 0$, $m = -1/2$, which corresponds to the "black fringe". The equal thickness fringes can then be understood as contour lines for the optical thickness, with *equidistance* equal to $\lambda_o/2$. In terms of position on the lamina,

$$x_m = \frac{\lambda_o}{2n\alpha}\left(m + \frac{1}{2}\right), \tag{3.2.37}$$

where x_m is the distance from the vertex O, as in Fig. 3.19. The fringes thus appear as regular straight lines with a pace

$$\Delta x = \frac{\lambda_o}{2n\alpha}. \tag{3.2.38}$$

Depending on their separation, the angle α between the faces of the lamina can therefore be evaluated.

In white light, similar to that which we have already seen for Haidinger's fringes, these fringes are visible only if the lamina is very thin. In particular, the fringe corresponding to $m = 0$ appears partially decomposed in the colors of the spectrum and, with increasing thickness, for subsequent orders, these colors tend to separate and eventually confuse, erasing the visibility of the fringes. A good example of this iridescence is observed in soap bubbles, or between two adjacent pieces of glass. As mentioned earlier, when the faces of the lamina are not flat, but are more

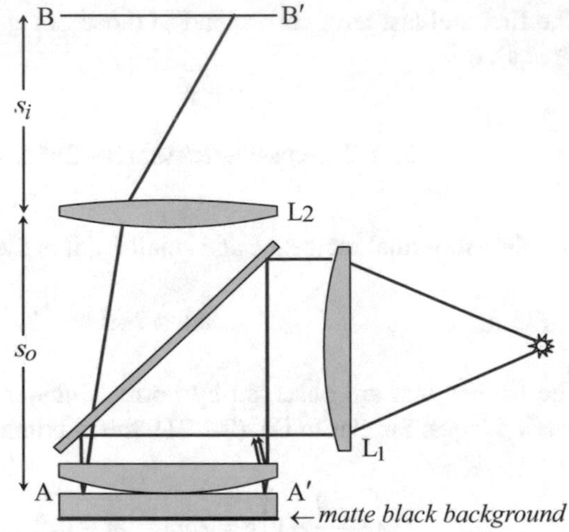

Fig. 3.21 Newton's device for observing Hooke's rings. The lens L_1 produces a collimated beam that illuminates the surfaces in contact. The lens L_2 produces an image of the plane AA′ on the screen BB′

or less irregular, the fringes run along lines of equal optical thickness, and are therefore useful in mapping the surfaces. One case of historical interest is that of Hooke's rings, which are obtained, for example, by putting a spherical surface in conctact with a plane one (Fig. 3.21), as in the case of Newton's device. Taking the distance from the point of contact for r and calling d the thickness of the interspace, we have

$$r^2 = 2Rd\,,$$

where R is the radius of the spherical surface. For the rotational symmetry of the system, the fringes are annular with the maxima at the radii

$$r_m = \sqrt{2Rd_m} = \sqrt{R\lambda\left(m - 1/2\right)}, \quad \text{with } m = 1,\ 2,\ 3,\ \ldots, \tag{3.2.39}$$

where Eq. (3.2.36) is used with $n = 1$. Similarly, from Eq. (3.2.39), the minima are still taking instead the values $m = 1/2, 3/2, 5/2, \ldots$.

At the point of contact, there is a minimum when the interspace has a lower or greater index of both dielectrics that delimit it. Contrary to Haidinger's fringe, the order m grows away from the center. If the two surfaces are moved away, the fringes contract and disappear into the center. In white light, for Eq. (3.2.39), the first bright fringe is quite extended and the various colors of the spectrum appear to be almost separated. The various orders have a complex set of colors by which Newton was impressed [Landsberg 1979, I, p. 125].

The device in Fig. 3.21 is that typically used in optic workshops to check the surfaces under work for comparison with reference surfaces. Sodium or mercury spectral lamps are normally used as sources. With laser sources, very similar devices are used, with the advantage of being able to position the optical elements even at great distances between them.

3.2.6 Localization of interference fringes

In the previous sections, we were interested in two types of fringe: those that can be observed in a large volume of space, which are then called *non-localized*, and those that are observed on particular surfaces, such as the fringes of Haidinger and Fizeau, which are called *localized*. Let us now clarify the relationship that exists between them.

We have already seen that the interference between two waves obtained from a punctiform monochromatic source produces net fringes at every point in the space that they illuminate. These non-localized fringes can be observed continuously on a screen moved widely in space that intercepts their light. As long as the source is point-like, this happens for both wavefront and amplitude division interferometers. However, the condition of a punctiform monochromatic source, which is necessary to observe the non-localized fringes, is achievable only in an approximate way. In §5.5, we will analyze the problem of spatial and temporal coherence in detail; here, we continue to consider only the case of spatially incoherent sources. In general, it is observed that, as we have already seen in §3.2.4, the visibility of the fringes decreases as the extension of the source grows, but this happens at a speed that depends on the position of observation. Sometimes, in some particular area, visibility can remain good, while fringes disappear elsewhere. Hence, the fringes that persist are localized and correspond to those observation points such that the path differences from the various points of the source typically vary by less than $\lambda/4$.

As we can see, from Eq. (3.2.15), the various directions in which the source can be extended are not equivalent. Thus, in a wavefront division interferometers, such as Fresnel's biprism or double mirror, a linear source parallel to the edge may be used, continuing to have non-localized fringes. In fact, for each observation point P, the source appears parallel to the edge itself without a transverse extension orthogonal to the xz-plane defined in Fig. 3.11, in which the direction of the x-axis is that of the edge, while the z-axis is dependent on P. However, if the linear source is rotated, the visibility generally disappears or narrows around the observation points for which the source does not extend appreciably in a transversal direction to this plane [Born and Wolf 1980, p. 294].

However, the most relevant cases of localization are those, still quite common, of interferometers with amplitude division, since, in these cases, there are observation points P where the opening angle of interference γ is canceled. In fact, the value of γ generally depends on P, so if, in Eq. (3.2.15), the dependence on R_A and R_B is negligible, as the source extension grows, the fringes are located around those points P for which $\gamma = 0$. The localization region is thus formed by those points (L in Fig. 3.19) that lie at the intersection between the two rays derived from a *single* incident ray from the source. Around these points, γ is negligible, so, according to Eq. (3.2.15), we have

$$\Delta r' - \Delta r = \frac{1}{2}\left(\frac{1}{R_{\mathrm{A}}} - \frac{1}{R_{\mathrm{B}}}\right)\left(x^2 + y^2\right), \tag{3.2.40}$$

where I recall that x and y are coordinates centered on a reference point S on the source, taken in a direction perpendicular to the ray that, from S, reaches the observation point P considered. R_{A} and R_{B} are the curvature radii in S of a reverse wave (on the two paths A and B) starting from P. Then, in general, even for these points L, the fringes disappear when the transverse dimensions of the source are such that $\Delta r' - \Delta r$ exceeds a distance on the order of $\lambda/4$. If, in particular, the wavefronts W_{A} and W_{B} (Fig. 3.13) are spherical, R_{A} and R_{B} are independent of x and y, and the reduction in visibility is independent of the direction in which the source extends transversely. For a circular source, if we consider a $\lambda/4$ variation from the center to the periphery, the visibility remains good as long as the source diameter remains below

$$l_{\mathrm{a}} = 2\sqrt{\frac{\lambda}{2}\frac{R_A R_B}{|R_A - R_B|}}. \tag{3.2.41}$$

In evaluating the visibility of fringes, however, one must consider what the actual extension of the source used by the optical system is. The fringes can therefore be visible even if the source is very wide, provided that the angular field with which they are observed is sufficiently limited.

In the case of a lamina with plane parallel faces, the localization region is at infinity, for which R_{A} and R_{B} are both infinite. Thus, for Eq. (3.2.40), the path difference is constant for all values of x and y, i.e., it does not depend on the position on the source. Haidinger's fringes are therefore a particular case in which the source can be infinitely extended. However, as soon as there is an inclination α between the faces of the lamina, the localization region approaches (even quickly) to the lamina itself, for which, in order to observe the fringes, it is necessary that the source (or the part of it actually used) be sufficiently small. The handling of this case becomes complicated, because the points where there is localization vary with the position of the point source considered.

We see here only the case of a wedge-shaped lamina with plane faces, and a very distant source for which the lamina can be considered invested by a plane wave with incidence angle ϑ. We also assume that the plane of incidence coincides with the principal plane of the lamina, that is, it is perpendicular to the intersection between its two surfaces. Referring again to Fig. 3.19, the distance $\overline{\mathrm{AL}}$, which fixes the position of the localization point for a given incident ray, can still be obtained by means of the sinus theorem on the triangle ALC,

$$\overline{\mathrm{AL}} = \overline{\mathrm{AC}}\frac{\cos(\vartheta + \varphi)}{\sin\varphi} \cong \frac{2d\sin(\vartheta' + \alpha)\cos(\vartheta + \varphi)}{\cos(\vartheta' + 2\alpha)\sin\varphi}.$$

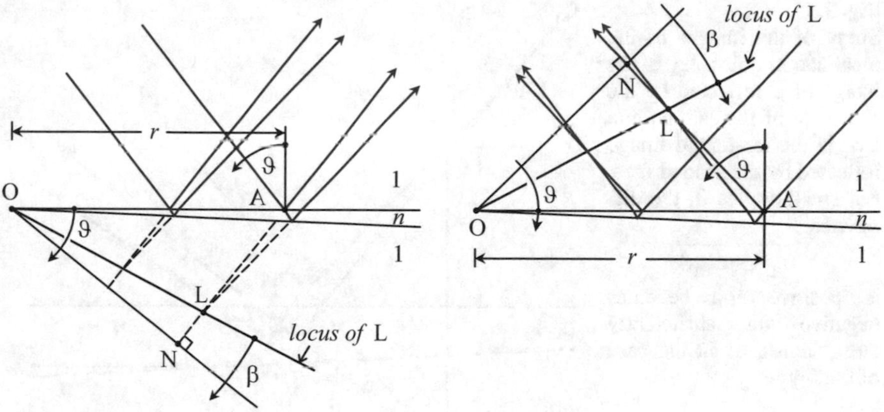

Fig. 3.22 Study of the localization of the fringes for a plane wave

Now, for φ, instead of applying Eq. (3.2.27), we use Eq. (3.2.26), approximating sin(φ/2) with (sinφ)/2. Therefore,

$$\overline{\mathrm{AL}} \cong \frac{d}{\sin\alpha}\frac{\sin(\vartheta'+\alpha)\cos(\vartheta+\varphi)\cos(\vartheta+\varphi/2)}{n\cos(\vartheta'+2\alpha)\cos(\vartheta'+\alpha)}.$$

If, for φ we now use Eq. (3.2.27), developing the second fraction at the 1st order in α, with a few steps, we have

$$\overline{\mathrm{AL}} \cong \frac{d}{\alpha}\frac{\sin\vartheta'\cos^2\vartheta}{n\cos^2\vartheta'} + \frac{d}{n}\frac{\cos^2\vartheta}{\cos\vartheta'}\left(1-3\tan^2\vartheta+3\tan^2\vartheta'\right), \qquad (3.2.42)$$

which, for small angles ϑ, we can approximate with

$$\overline{\mathrm{AL}} \cong \frac{d}{n^2}\frac{\sin\vartheta}{\alpha} + \frac{d}{n}\left[1-3\sin^2\vartheta\left(1-\frac{1}{n^2}\right)\right]. \qquad (3.2.43)$$

On the other hand, $\rho = d/\alpha$ is the distance of A from point O where the two faces intersect (Fig. 3.22), so, for an incident plane wave, $\overline{\mathrm{AL}}$ is proportional to ρ and the locus of points L is a plane passing through the point O, whose inclination β with the wavefront reflected by the first face is obtainable by

$$\tan\beta = \frac{\overline{\mathrm{AN}}-\overline{\mathrm{AL}}}{\overline{\mathrm{ON}}} \cong \left(1-\frac{\cos^2\vartheta}{n^2\cos^2\vartheta'}\right)\tan\vartheta + \frac{\alpha\cos\vartheta}{n\cos\vartheta'}\left(1-3\tan^2\vartheta+3\tan^2\vartheta'\right), \qquad (3.2.44)$$

where N is the projection of O on the reflected ray in A.

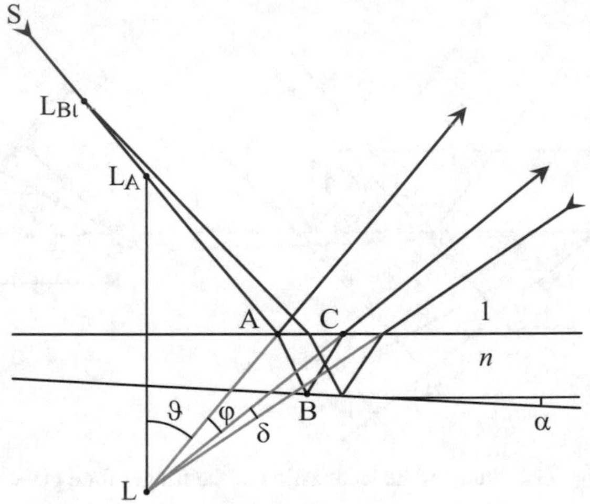

Fig. 3.23
Study of the images of the localization point. L_A is the image of L reflected by the first face of the lamina and L_{Bt} is the tangential image reflected by the second face. For small angles ϑ, the difference

$$\overline{L_A A} - \overline{L_{Bt} A}$$

is positive and becomes negative for sufficiently large angles, as in the case of the figure

The localization surface varies with the angle of incidence ϑ, and, in particular, for $\vartheta = 0$, in the example considered here of a lamina with index n surrounded by air, it is located on an intermediate plane between the faces of the lamina itself. Note, however, from Eq. (3.2.43), that the localization plane moves away in proportion to $\approx \vartheta/\alpha$.

The optical path difference for the points L can now easily be obtained from Eqs. (3.2.33) and (3.2.42), taking $h_2 = -\overline{AL}\cos\vartheta$, so the terms in $d\alpha$ elide each other. Neglecting those of the superior order, we simply have

$$\Delta L(\mathrm{L}) = 2nd\cos\vartheta',$$

where d is the thickness in A taken perpendicular to the 2nd face.

In the present moment, it lacks determination as to the degree to which the source can be extended without appreciably reducing the visibility of the fringes. To this end, in order to apply Eq. (3.2.40), it is necessary to determine, on the source, which are the radii of curvature R_A and which are the radii of curvature R_B on the two paths of an inverse wave emitted by a source point corresponding to a given point L (despite it being virtual). The wave reflected from the first face is spherical with a curvature center located in the image L_A of L reflected by the same front face. The refracted wave is not exactly spherical, for the astigmatism introduced by the refractions. If we consider a narrow light pencil centered on the direction ϑ of the source, tracing only the rays lying on the principal plane of the lamina, that orthogonal to both faces, we can still locate an image L_B of L, from which the wave diverges, and the corresponding radius of curvature on this plane.

Let us therefore consider the localization point L associated with a ray SA (Fig. 3.23). The tangential image L_{Bt} of L can be determined by considering an inverse ray directed towards L and inclined at a small angle δ with respect to the refracted ray on the principal plane of the wedge, or simply by using Eq. (2.7.63.a) of the wavefront transformation from a *dioptre*. With a rather complicated calculation, within the limit of $\alpha \to 0$, recalling that $\overline{\mathrm{AL}} \propto 1/\alpha$ and that d varies with the position, eliminating the residual terms in α, the tangential separation $\overline{\mathrm{L_A L_{Bt}}} = R_A - R_{Bt}$ is[1]

$$\Delta R_t = \frac{2d}{n\cos\vartheta'}\left[\frac{\cos^2\vartheta}{\cos^2\vartheta'} - 3\sin^2\vartheta + 2\tan^2\vartheta'\cos^2\vartheta\right]. \qquad (3.2.45.a)$$

Instead, for Eq. (2.7.63.b), the sagittal separation is

$$\Delta R_s = \frac{2d}{n\cos\vartheta'}\cos^2\vartheta. \qquad (3.2.45.b)$$

When the outer and inner media have the same refractive index, i.e., for $n = 1$, these expressions reduce to $2d\cos\vartheta$. On the other hand, in Fizeau's configuration, for small ϑ, $\Delta R \cong 2d/n$.

Moreover, we have seen that the fringes remain visible as long as the source has a diameter less than $2\sqrt{\lambda R_A R_B/(2\Delta R)}$. For a distant enough source, for which $R_A \approx R_B >> \Delta R$, that which matters for the visibility of fringes is the angular opening of the source, seen from $\mathrm{L_A}$. For small ϑ, fringes are therefore visible if, for this opening, it is

$$\Lambda_a < \frac{1}{R_A}\sqrt{\frac{R_A R_B}{2\Delta R}} \cong \sqrt{\frac{\lambda n}{d}}. \qquad (3.2.46)$$

For $d = 0.1$ mm, $\lambda = 500$ nm and $n = 1.5$, the Fizeau fringes are visible only if the source, or the part of it actually used, has an angular aperture Λ_a less than ≈ 3.5°. In other words, while the Fizeau fringes are commonly observed on a few micron thin sheets, for macroscopic thicknesses, even with the monochromatic light of a spectral lamp, they are observable only with a high degree of collimation of the light beams.

3.2.7 *Two-wave interferometers*

Interference is suitable for high precision measurements, and there are many interferometers that have been designed for various purposes, in fundamental and applied physics, as in spectroscopy, in the measurement of distances and speeds, in measures of wavelengths, refractive index, quality of optical components

[1] From a numerical verification, with a non-approximate trigonometric calculation of ray tracing, this expression is much more accurate than that given by Born and Wolf (1980), which does not match the limit for $\alpha \to 0$.

(Malacara 2007), etc. We will now look at some of the *two-wave* interferometers and the use that can be made of them.

3.2.7.1 Michelson interferometer

In 1880, Michelson developed what became, along with its variants, the most famous and versatile of amplitude division interferometers. He used it for the experiment that proved the constancy of the speed of light [Michelson and Morley 1887; Michelson 1927] and, in 1892, for the first interferometric measurement of the standard sample meter with respect to a cadmium line [Michelson and Benoit 1895].

It essentially consists of a plane semi-reflective mirror D (*beam splitter*), which serves both to decompose and to recombine the radiation of a source S, and two other plane mirrors M_1 and M_2, which reflect back, respectively, the reflected and transmitted waves coming from D (Fig. 3.24). Generally, but not necessarily, D is placed at 45° with respect to the incident radiation, and therefore mirrors M_1 and M_2 form, with D, two arms at 90° between them. The beam splitter D is usually constituted by a lamina of glass with plane faces, one of which has an anti-reflective coating and the other of which is made semi-reflecting with a thin layer of metal or with suitable dielectric layers. Since the substrate of the beam splitter has a certain dispersion, on one of the two arms, that on the side of the semi-reflective mirror, a lamina of the same material and thickness is placed parallel to D; this precaution is essential when using the interferometer with non-monochromatic light. Alternatively, we can use a cube beam splitter (non-polarizing). Generally, at least one of the two mirrors can slide back and forth so as to vary the optical path of one of the two arms relative to the other.

The interference between the two waves that come back from the mirrors can be observed in two directions, one of which is back toward the source. However, the best contrast for the fringes is ensured in the other direction, where each of the two interfering waves has undergone both a reflection and a transmission through D. In the first case, a wave is doubly reflected, the other being doubly transmitted, and therefore a divisor with equal transmissivity and reflectivity is needed.

In Fig. 3.24, the image M'_2 of the mirror M_2 reflected by D is represented, as would be seen by an observer in O. The interference patterns that can be observed with this tool are therefore the same produced by two flat and facing reflective surfaces, separated by air, with the difference that, here, we do not have multiple reflections between them. Therefore, for their study, that which we determined in the previous paragraphs is valid, with the simplification of taking the refractive index of the lamina as being equal to 1 and the same sign for both reflections. In short, with an extended source, if M'_2 is parallel to M_1, Haidinger's circular fringes, localized at infinity, are observed. If, on the other hand, these two surfaces are inclined to each other and close together, Fizeau's fringes are observed along lines of equal separation and localized near M_1 and M'_2. If, finally, the surfaces are in-

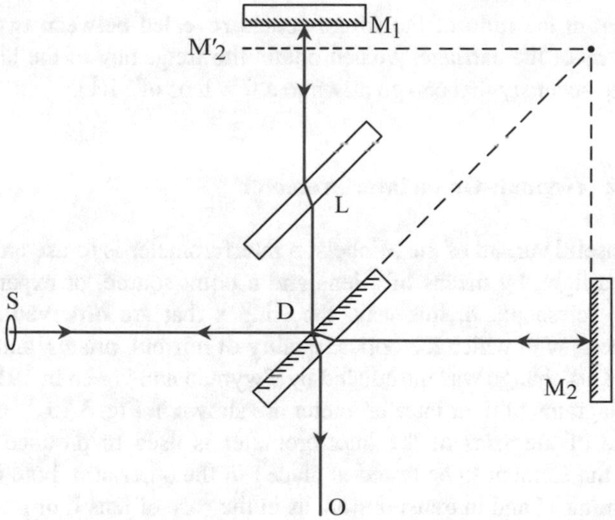

Fig. 3.24 Michelson's interferometer

clined and distant, for the various points in the field of view, there is a diversity of incidence angles on the mirrors that is no longer negligible. Then, the fringes become curved with the convex portion facing the edge of the air wedge between M_1 and M'_2.

Translating one of the mirrors, we can vary the order of interference, and, when the optical path in the two arms is the same, it is said that the two mirrors are in optical contact. However, in determining the zero-order fringe, we must also take into account the difference of phase $\Delta\phi_R$ between internal reflection and external reflection in the divisor. This difference depends on the nature of the reflective treatment, whether it is metallic or dielectric, and the thickness and arrangement of. The zero-order fringe can be recognized by the others using a white light source: it appears white, black, or, in any case, "achromatic", according to $\Delta\phi_R$, while the other fringes appear variously colored according to a characteristic sequence [Landsberg 1979]. This trick was used by Michelson for the equality of the optical paths in his determination of the standard sample meter.

Thanks to the possibility of varying the length of one of the interferometer arms with respect to the other, interference may be created between waves that are emitted from the source at different times. In this way, from the trend of the visibility of the fringes with the difference in path, we can deduce the spectral properties of the source, such as its *coherence time*.

A special variant of Michelson's interferometer is what is called a *lambda-meter*, in which one or both of the mirrors are replaced by a corner cube and mounted on a movable carriage so as to continuously vary the relative length of the arms. Typically, a frequency-stabilized 632.8 nm He-Ne laser beam, the frequency value of which is well known, is sent to the interferometer, and the fringes are detected with a photodiode during the motion of the carriage. A second laser beam is sent in parallel to the first, and its fringes are revealed by another photodiode. From the ratio of the fringe counts revealed between two (almost) extreme positions of the carriage, we can obtain the frequency of the laser to be measured with an accuracy that can go down to a few tens of MHz.

3.2.7.2 Twyman-Green interferometer

A useful variant of the Michelson interferometer is to use an input beam of collimated light, by means of a lens and a point source, or expanding a laser beam with a telescope. In this way, the fringes that are observed are those of equal thickness, with which the optical quality of mirrors, prisms, and lenses can be verified. This change was introduced by Twyman and Green in 1916, and the application diagrams of their interferometer are shown in Fig. 3.25.

One of the arms of the interferometer is used to produce a reference wave, while the element to be tested is placed in the other arm, both in reflection, as for the lamina M and in transmission, as in the case of lens L or prism P. The collimator lenses C are highly corrected and the mirrors M_R are also of high quality. The mirror M''_R is spherical and is appropriately placed to recover a plane wave on the divisor D. This same configuration is also used to control a spherical surface and, in this case, the roles of the lens and the mirror are inverted.

Since the fringes of interference are located near the surface to be controlled, the second collimator lens is used to produce an image of this surface on the detector R, for example, a CCD matrix. In this way, the fringes can be associated with the defects in machining or the homogeneity of the elements under test. In the case of measures in transmission, there may be ambiguity in the position of defects for the double crossing of the wave, so it is good that the mirrors are close to the object to be verified.

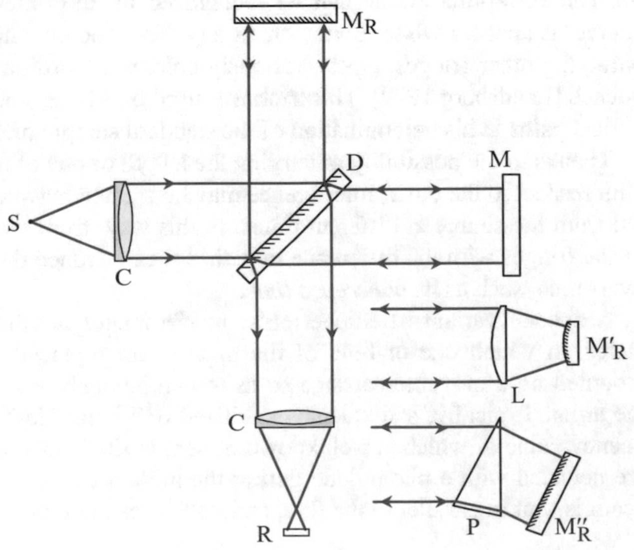

Fig. 3.25
Twyman - Green interferometer

3.2.7.3 Fizeau interferometer

However, the interferometer that is most often used for optical quality measurements is the Fizeau interferometer (Fig. 3.26). It is used in a similar way to the Twyman-Green interferometer, but the divider D works at normal incidence and directly generates the reference wave, so one of the folding mirrors is eliminated and the test element P is placed in the only remaining arm. Since the beams are reflected towards the source, one must still use another divider d to observe the interference pattern, which is recorded, as in the previous case, by the detector R. The divider is necessarily made up of a relatively thin sheet, since it is used on non-collimated waves. To clean the beam from the light diffused by the dust or reflected from other surfaces, an opening A is suitably placed in the focal plane of the collimating lens C.

This type of interferometer is preferred over the others because it allows for maximum measurement accuracy. In fact, both the reference wave and the reflected wave from the optic to be tested travel together towards the detecting device of the fringes. Therefore, the distortions induced by the optical elements that the two waves meet in their path are the same, and thus subtract in the formation of the fringes on the detector. The subtraction is not strictly complete if the optics under test and the reference lamina are inclined to each other: in this case, the two waves are not exactly superimposed, and this occurs even if the tested optic simply presents aberrations.

In the case, for example, of checking the planarity of a lamina, the two surfaces, the reference one and the one being tested, face each other. In this way, there are multiple reflections between the two surfaces and one can also no longer observe the zero-order fringe. However, this configuration is preferred, since it is sufficient to have only one reference surface of excellent optical quality, and the homogeneity of the divider material is no longer so stringent. Modern interferometers employ the radiation of a single-mode laser, whose coherence length extends for tens of meters, and therefore there is no need to place the optics in the test particularly close to the reference lamina and the fringes are no longer localized. However, there are other good reasons to keep such a distance small, such as to limit the effect of turbulence in the air and thermal drifts.

The wavefront that returns from a test surface replicates its form by doubling its deviations from the plane. Likewise, in the case of a lens measured in transmission, the return wavefront doubles its deviations from the reference sphere with

Fig. 3.26 Fizeau's interferometer

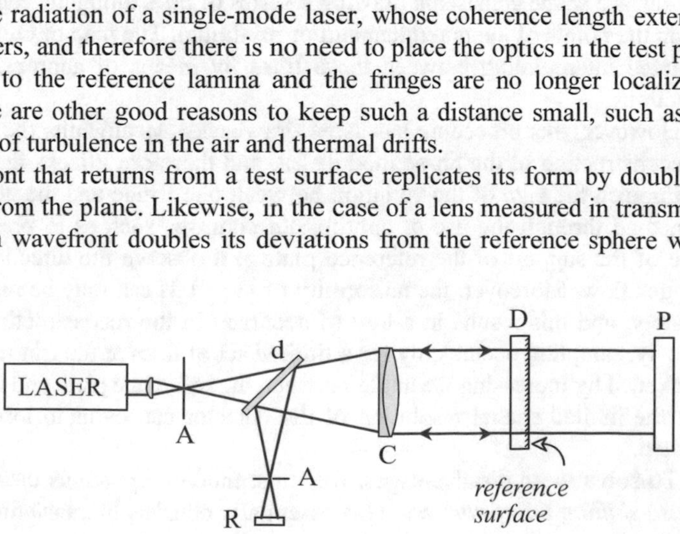

respect to the forward wave propagated by the lens. Therefore, interferograms are a coding of the optical path variation OPD (Optical Path Difference) between the various points in the transverse beam profile: a change in depth Δz on the test surface produces one OPD = $2n_a\Delta z$; operating in air, a refraction index $n_a = 1$ is usually assumed. This corresponds to a phase delay

$$\delta = \frac{2\pi}{\lambda_o}\mathrm{OPD}\,, \tag{3.2.47}$$

where λ_o is the wavelength in vacuum of the laser employed. By neglecting multiple reflections, the wave intensity reaching the detector at the transverse coordinate point (x,y) is given by

$$I(x,y) = I_0\left[1 + M\cos\delta(x,y)\right], \tag{3.2.48}$$

where I_0 is the sum of the intensities that independently return from the divider and the optic under test and M is the modulation coefficient of the fringes; these coefficients are ideally constant within the measuring range of the interferometer. The fringes that are formed are therefore like contour lines with a pace equal to 2π in the phase, which corresponds to a $|\mathrm{OPD}| = \lambda_o$ and a height variation $|\Delta z| = \lambda_o/(2n_a)$.

The first estimate of the quality of a surface can be simply obtained by drawing a straight line that joins the extreme points of a central fringe within the measuring range and by determining the maximum fringe deviation from this line in relation to the pace of the fringes. For a more accurate analysis, the reconstruction of the phase map $\delta(x,y)$ is used instead, or, equivalently, that of OPD(x,y), through the processing of the interferogram with a computer. One of the methods used to give the surface shape consists in drawing a series of lines along the center of the fringes on the points of the maximum and/or minimum. The map of OPD is then reconstructed interpolating between these lines by means of appropriate polynomial functions.

However, this procedure has some drawbacks. Meanwhile, the fringes provide a reconstruction of the phase module 2π, and therefore it lacks the information to distinguish the sign of the variation between one fringe and the next. This can be remedied through the use of appropriate controls, such as to press slightly on a side of the support of the reference plate and observe the direction in which the fringes flow. Moreover, the uniformity of I_0 and M can only be respected approximately, and this results in a loss of accuracy in the reconstruction of the phase. Finally, sampling occurs only for a limited set of lines at the fringes, which can be thickened by increasing the angle between the reference plate and the test element, but the limited spatial resolution of the detector causes us to lose the advantage gained.

To solve these disadvantages, the most modern apparatus uses the method of *phase-shifting interferometry*. This essentially consists in translating the reference plate longitudinally in order to vary the phase in a well-controlled manner. The fringe image is detected by a CCD sensor and, by neglecting the effect of multiple reflections, the intensity that reaches its pixels is given by

$$I\left(x_i, y_j, \phi\right) = I_0\left(x_i, y_j\right)\left\{1 + M\cos\left[\delta(x, y) + \phi\right]\right\}, \tag{3.2.49}$$

where (x_i, y_j) are the coordinates of the pixel and ϕ represents the phase corresponding to the translation applied to the reference plate. There are several measurement methods that differ in the way that the phase is varied and in the calculation algorithm employed in signal processing. One of these is the following (Wyant, 1975), in which the phase ϕ is incremented linearly over time. The signal is then integrated by the CCD detector on $\Delta\phi = \pi/2$ intervals with constant time step and is immediately acquired at the end of each interval. The individual readings are then given by

$$S_n\left(x_i, y_j\right) = \int_{(n-1)\frac{\pi}{2}}^{n\frac{\pi}{2}} I\left(x_i, y_j, \phi\right) d\phi = I_0\left(x_i, y_j\right)\left\{\frac{\pi}{2} + M\left[s_n \sin\delta + s_{n+1}\cos\delta\right]\right\},$$

where $s_n = -, -, +, +, -, -, +, +, \ldots$. The following relations are found:

$$S_4 - S_1 = S_3 - S_2 = 2I_0 M \sin\delta,$$
$$S_1 - S_2 = S_4 - S_3 = 2I_0 M \cos\delta,$$

and so we have

$$\frac{-S_1 - S_2 + S_3 + S_4}{S_1 - S_2 - S_3 + S_4} = \frac{4I_0 M \sin\delta}{4I_0 M \cos\delta} = \tan\delta. \tag{3.2.50}$$

Also, knowing the signs of sinδ and cosδ separately, from this equation, we can derive the value of $\delta' = \delta + \pi/2$, module 2π, corresponding to each pixel without ambiguity of sign. By appropriately unwrapping the phase in the zones delimited by the extreme values 0 and 2π of δ', a continuous map of the phase itself is then obtained, followed by the map of the OPD. Knowing the laser wavelength used and the refractive index of the interposed medium, the measurement result can finally be expressed in metric units as an altimetric map of the test surface with respect to the reference one.

In this reconstruction, it is not necessary that the intensity of the laser beam, the depth of modulation of the fringes and the pixel sensitivities be very uniform. In addition, only signal differences are involved in the calculation, and the background, if present, is subtracted. It should be noted, however, that this method cannot be applied when multiple reflections become important, for example, if the test element is a mirror. In this case, a semi-transparent film or a plane reflective plate is interposed between the mirror and the reference plate.

3.2.7.4 Jamin and Mach-Zehnder interferometers

These interferometers use two beam dividers, the first to generate two coherent waves, which travel along separate paths, and the second to recombine them. In this way, the test element is analyzed in transmission with a single pass along one of the two paths, which are of the same length. These interferometers allow for the observation of the zero-order fringe and, for their great sensitivity, they are used, above all, for measurements of the gas refractive index.

The Jamin interferometer is constituted of two plates with parallel faces of the same thickness, obtained from a single plate optically machined (Fig. 3.27). Each lamina has its rear face rendered reflective and the other is semi-reflective. In this way, the first plate reflects two parallel rays to the other, separated by a few cm. The second lamina recombines the two rays and is slightly inclined with respect to the first so as to form an equidistant fringe set. In the figure, only the rays that are reflected just once for each lamina are shown. Of these, only two are used in the interferometer, namely, the Sadhfl and Sacbek rays, while the others are diaphragmed out. Given that each lamina has parallel faces, these two rays emerge from the interferometer parallel to each other, and therefore, if S is an extended source, the fringes are localized to infinity. Usually, a lens L is used to observe fringes on a screen, or, in the case of visual observation, a telescope is used. From Eq. (3.2.17), it is obtained that the optical path difference between the rays k and l is given by

$$\Delta L = 2nd\left(\cos\vartheta'_A - \cos\vartheta'_B\right). \tag{3.2.51}$$

Since the index *n* and the thickness *d* of the laminae are constant, these are fringes of equal inclination, but now the phase shift depends on the difference between the cosines of the internal incidence angles into the plate. So that, if the plates are parallel, the phase shift is zero.

Since all of the optical surfaces are plane, the apparent sources of the rays k and l emerging from the second plate of the interferometer are separated in proportion to the rotation angle, and are reciprocally aligned in the direction perpen-

Fig. 3.27 Jamin's interferometer

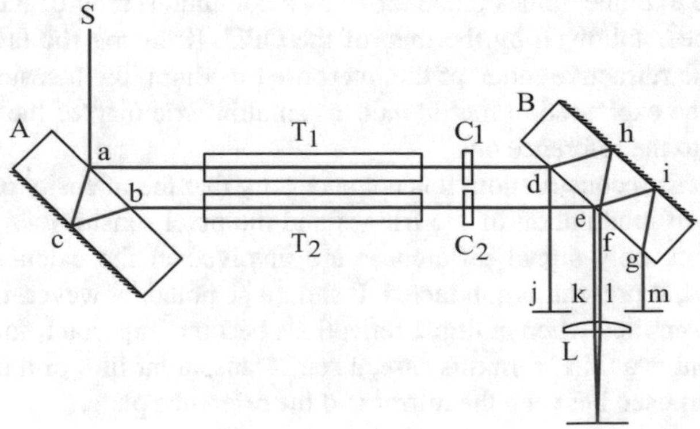

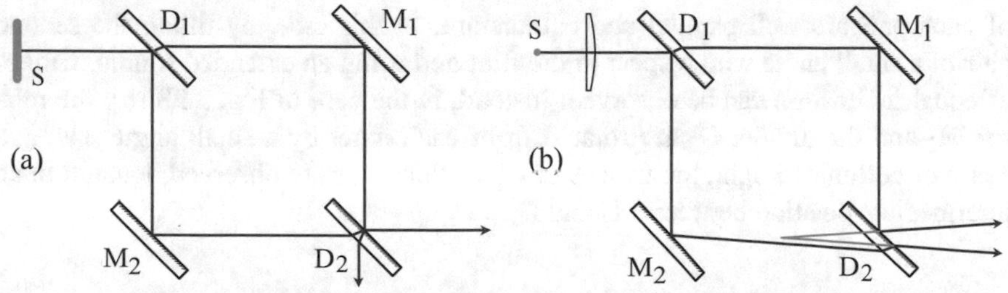

Fig. 3.28 Mach-Zehnder interferometer in two different arrangements of the laminae

dicular to both the axis of rotation between the two plates and the bisector of the normal to them. These three directions thus form a set of three Cartesian axes.

Therefore, the trend of the fringes is that which is deduced from Fig. 3.2 or Fig. 3.4, and thus depends on the direction of observation with respect to the three axes. In the configuration of Fig. 3.27, assuming that it represents the horizontal plane of the instrument, if the second plate is tilted vertically, the fringes are horizontal and of low order. Instead, if the lamina is tilted horizontally, the fringes are vertical, but, as the angle of incidence on the lamina is about 45°, they become of high order [Born and Wolf 1980, p. 310].

A typical application of Jamin's interferometer is to place two identical vacuum tubes T on the two parallel paths (Fig. 3.27). On each path is also placed a compensating lamina C: adjusting the inclination of one of these laminae, operating first in white light and then in monochromatic light, the zero-order fringe is determined, and therefore the optical path difference is canceled. Finally, by slowly introducing a gas into one of the two tubes, the fringes flowing in the field of the interferometer are counted in monochromatic light. Alternatively, one returns to compensate for the optical paths by rotating the other compensator lamina. From the measurement of the rotation angle of this lamina, also knowing the length of the vacuum tubes, one finally obtains the gas refractive index at the given pressure and temperature.

A variant of Jamin's interferometer was developed by Sirks and Pringsheim for the measurement of the refractive index of small objects. In their interferometers, the plates are wedge-shaped, with a small angle between the faces, and are illuminated by a collimated light beam. In this way, by adjusting the inclination between the two plates, the fringe location region may be set in the center of the instrument, in correspondence with the object to be measured [Born and Wolf 1980]

The Mach-Zehnder interferometer is similar to that of Jamin, but each plate is replaced by a beam splitter and a mirror. Despite the greater complexity of the system, in this way, one can create two very separate paths between them, so small variations in the refractive index of large objects can be observed, such as those due to variations in air density in wind tunnels. One modern use also involves putting some optical devices in one of the two arm, as a phase modulator.

Fig. 3.28 shows two different ways of using this interferometer. In (a), the Jamin's configuration is reproduced by providing that the divider D and the mirror S of each pair are well parallel and equidistant. In this case, by tilting the second pair of a small angle with respect to the first and using an extended source, fringes of equal inclination can be observed. Instead, in the case of Fig. 3.28 (b), the mirror M_2 and the divider D_2 are rotated from each other by a small angle. Using a beam of collimated light, the fringes of equal thickness are observed, located in an intermediate position between M_2 and D_2.

3.2.7.5 Common-path interferometers

An optical system consisting of a beam splitter and two or more plane mirrors can be arranged in such a way that the two interfering waves travel the same path, or nearly so, propagating in opposite directions. They can be imagined as the modification of an optical ring cavity in which one of the mirrors is replaced with a beam splitter orthogonal to the mirror and to the plane of incidence. By appropriately aligning the optics, the two paths can overlap completely, producing a uniform field in the interference of the two waves that, at the end of the path, re-emerge from the divider. Tilting a component of the optical system a bit can produce fringes visible even in white light, as the two paths hold almost the same optical length. As in all other cases of two-wave interference, the fringe formation can be more comprehensively determined by the positions of the two images of the source seen through the succession of reflections on the mirrors for the two opposite paths.

This type of configuration is very similar to that employed in sub-Doppler spectroscopy, in which a pump beam and a test beam run through a gas in opposite directions. In this case, one of the mirrors is semi-transparent so as to separate these beams and the interference between the waves retrieved by the divider can be useful for verifying, at least in part, the inclination between the two counter-propagating waves.

The relationship between the direction $\boldsymbol{s}$ of the incoming ray and that of the two emerging rays can be calculated, in general, by the matrices considered in §2.10.1. In particular, Eqs. (2.10.3) and (2.10.4) express the matrices $\mathfrak{M}$ and $\mathfrak{M}^{-1}$ corresponding to the two paths in which the beam splitter and the mirrors are traversed with two opposite sequences of reflections. Thus, the difference in direction between the two rays emerging from the beam splitter in the same side of the input ray can be expressed as

$$\Delta \boldsymbol{s} = \left(\mathfrak{M} - \mathfrak{M}^{-1}\right) \boldsymbol{s} . \tag{3.2.52}$$

Moreover, the complete overlap of the two paths is only possible if the matrices $\mathfrak{M}$ and $\mathfrak{M}^{-1}$ have at least one eigenvalue equal to -1. On the other hand, these paths are superimposed if and only if they are both closed on the input point at the divider. Experimentally, this can be achieved by sending a collimated beam through the divider and aligning the mirrors so as to bring the beam transmitted

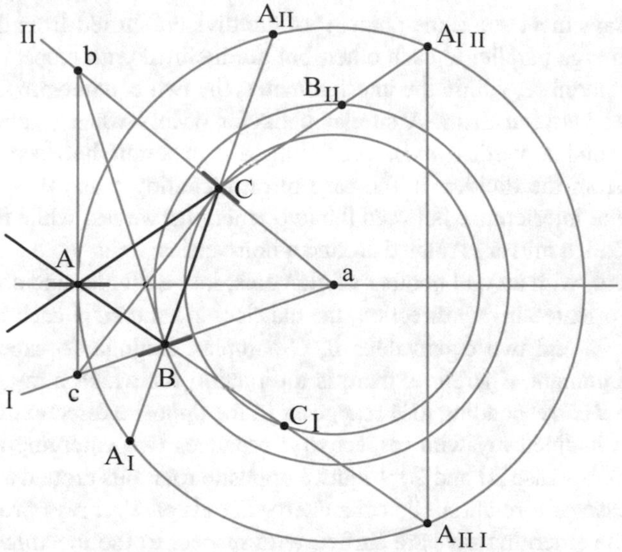

Fig. 3.29 Representation of a common path interferometer with a beam splitter, A, and two mirrors, B, C, orthogonal to the plane of the drawing. The scalene triangle ABC describes the internal paths in the interferometer when these are closed in A and coinciding, while the triangle abc is constructed with two sides tangent to the mirrors B and C, respectively, and a side perpendicular to the divisor A. For a theorem of geometry, the angle bisectors of the triangle ABC coincide with the normals aA, bB, cC of the triangle abc. The points A_I, A_{II}, B_{II}, C_I, $A_{I\,II}$, $A_{II\,I}$ are the images of the points A, B, C through the mirrors I and II

coincident with the input position on the divider. Next, the divider is aligned so that the reflected beam also returns to the same position.

To exemplify the behavior of these interferometers, we consider the case of a ray impinging on a device consisting of a divider and only two plane mirrors. The complete overlap of the two counter-propagating rays produced by the divider happens when they define a triangle lying on a plane of incidence orthogonal to all three optical elements (Fig. 3.29). Let us assume that this plane is horizontal and that the divider and the mirrors are vertically arranged. From this condition, it is concluded that the intersection of the mirror planes must lie on the vertical plane defined by the divider. This is true if the divider substrate has a zero thickness. In practice, this does not happen, but the complete overlap can always be obtained by appropriately translating the mirrors, and even turning them if the substrate is wedge-shaped. Hereinafter, for simplicity, we will *omit* the presence of refractions in the divider substrate.

In the case of Fig. 3.29, we have eigenvalues $A = -1$, $B = C = +1$ for both matrices, so $\mathfrak{M} = \mathfrak{M}^{-1}$ and $\Delta \boldsymbol{s} = 0$ for any direction of $\boldsymbol{s}$. If the mirrors are aligned to form the closed path ABCA, for a given incoming ray parallel to the eigenvector $\boldsymbol{a}$ corresponding to the eigenvalue A, all rays parallel to $\boldsymbol{a}$ form a closed path. In this case, the interference field remains uniform for any incoming spherical wave. Indeed, for an incoming ray whose direction does not coincide with $\boldsymbol{a}$, the two emerging rays meet at a same point A′ of the divisor, shifted from the input point A. They emerge parallel to each other, but are inclined with respect to the incoming beam: therefore, within the interferometer, the two counter-propagating beams are inclined between them. A similar behavior occurs when rotating one of the mirrors around a vertical axis, producing a translation between the two rays emerging from the divider. In the case of an incoming plane wave, there are no fringes in the interference between the two emerging waves, while they do appear when, instead, a mirror is rotated around a non-vertical axis.

In general, with an odd number of elements, i.e., a divider and an even number of mirrors oriented in any direction, the matrices $\mathfrak{M}$ and $\mathfrak{M}^{-1}$ each have an eigenvalue $A = -1$ and two eigenvalues B, C complex conjugates, except for special cases of alignment. Therefore, there is a direction for which a ray parallel to the eigenvalue $\boldsymbol{a}$ corresponding to A reappears in the opposite direction for both paths. Instead, an inclined ray with respect to $\boldsymbol{a}$ produces two emerging rays tilted between them, because $\mathfrak{M}$ and $\mathfrak{M}^{-1}$ induce opposite rotations around $\boldsymbol{a}$.

In the case of a rotation other than zero, not all parallel rays produce a closed path, but the emerging rays are shifted with respect to the incoming ray. Since a succession of reflections from plane mirrors does not produce any magnification, we find that there is a single ray O that emerges without translation for both paths, which are then closed and superimposed. All of this is possible if there are at least four mirrors, arranged to form a non-planar closed path inside the interferometer. In this case, for an incident spherical wave whose center S lies on ray O, there is a uniform interference field between the two emerging waves, while the fringes appear if one of the optical elements is tilted in any direction or the source S is brought out of the ray O.

If the interferometer consists of a divider and an odd number of mirrors, it has an eigenvalue $A = +1$, so the emerging rays can only go back if the alignment of the mirrors is such as to produce two eigenvalues $B = C = -1$. In this case, all of the rays that lie on the plane orthogonal to the eigenvector $\boldsymbol{a}$ are reflected backwards and the two paths inside the interferometer are parallel to each other.

3.2.7.6 Shearing interferometers

There are various interferometric techniques that do not use a reference wave substantially free of defects, such as that used in the Fizeau and Twyman-Green interferometers. In particular, the Mach-Zehnder interferometer allows for the use of a single wave in which the two beams produced by the division in amplitude can be made to interfere, overlapping them only partially, by means of a lateral translation (*shear*), keeping them parallel to each other or not. Bates introduced this technique using a Mach-Zehnder interferometer to determine the optical quality of a converging wavefront [Bates 1947]. He was arranging the interferometer so that, for each of the two paths, the wave converged at the same point F on the semi-reflective surface of the last divider, D_2 in Fig. 3.28. The first divider D_1 and the M_2 mirror were then rotated solidly around the axis of the incident beam so as to separate the focal points F_1 and F_2 on D_2, perpendicularly to the plane of the figure, obtaining a series of fringes parallel to this plane. D_2 was finally rotated around the axis F_1-F_2, and the fronts of the emerging waves were then translated laterally. From the shape of the fringes thus obtained, the aberrations of the incident wave could be deduced with respect to an ideal spherical wave. Two compensating plates were also needed to balance the displacements due to the thickness of the dividers in their rotation. The fringes that were created were of low order and could be observed even in white light.

Subsequently, Drew simplified this rather complicated scheme [Drew 1951], but the most drastic simplification was introduced in 1964 by Murty, who proposed the use of a simple lamina with plane and parallel or inclined faces [Murty 1964]. In this case, the interference produced by the two waves reflected from the faces of the lamina is examined, for which, in addition to a lateral displacement, there is also a non-zero path difference. The fringes are thus of relatively high order, however, they remain visible with nearly monochromatic radiation, like that from a laser, which had been created two years earlier.

This Murty interferometer is easy to use and, in normal laboratory work, is suitable for accurate measurement of laser beam collimation and of optical component features such as the curvature radius of both converging or diverging mirrors and their aberrations. In this section, we will deal with the laser beams measures of divergence, following the analysis of Riley and Gusinow (1977).

Let us consider, therefore, the diagram of Fig. 3.30, in which a laser beam arrives on a wedge-shaped lamina, with the plane of incidence on the first face parallel to the edge of the wedge itself. This is not the most general case, but it is the configuration that is used in practice. Let ϑ be the incidence angle of the ray, α the angle between the wedge faces, t the wedge thickness at the incidence point and n its refractive index. Then, we take as our reference system a set of three axes with origin at the point of incidence A. The z-axis is chosen in the direction of the reflected beam from the first face, with the x-axis lying on the plane of incidence. The coordinates of the exit point C of the refracted ray are then

$$\begin{aligned} x_C &= 2t\cos\vartheta\tan\vartheta' = t\frac{\sin 2\vartheta}{\sqrt{n^2-\sin^2\vartheta}} = s, \\ y_C &= t\tan 2\alpha \qquad \cong 2t\alpha, \\ z_C &= 2t\sin\vartheta\tan\vartheta' = 2t\frac{\sin^2\vartheta}{\sqrt{n^2-\sin^2\vartheta}}, \end{aligned} \tag{3.2.53}$$

where s is the *shear distance* and, in the second equation, the approximation is valid for small α. Neglecting terms of higher order in α, it follows that, between the two reflected beams, the optical path difference on the xy-plane, with $z = 0$, is

$$D \cong 2t\sqrt{n^2-\sin^2\vartheta}\,, \tag{3.2.54}$$

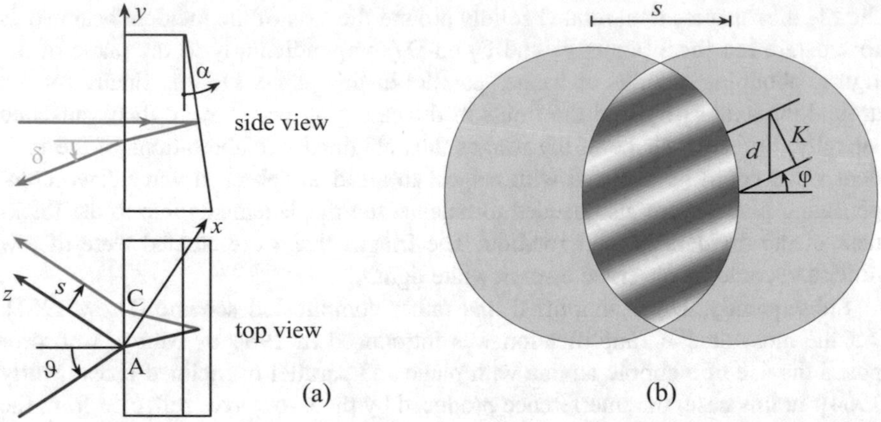

Fig. 3.30 Schematic representation of a Murty interferometer and interference pattern obtained with a spherical wavefront. The rays and angles that do not lie on the plane of the drawing are shown in gray. The inclination between the faces of the lamina is exaggerated so as to highlight the path of the reflected ray from the second surface

while the versor of the ray emerging from C is approximately given by (0,δ,1), where

$$\delta \cong 2\alpha\sqrt{n^2 - \sin^2 \vartheta}\,. \tag{3.2.55}$$

The two reflected rays are therefore shifted between them (in the direction of the *x*-axis) by a distance *s* that depends on the thickness of the wedge and with mutual inclination δ that depends on α. Both variables depend on ϑ and, in particular, *s* has a maximum that, for a glass with index 1.515, corresponds to an angle of incidence of about 49°.

Let us consider a second system of Cartesian axes with the z'-axis coincident with the principal ray of the refracted beam; we take as the origin point $(s, a, 0)$ on the plane $z = 0$ of the previous system, where

$$a = y_C - \alpha z_C \cong 2t\alpha\left(1 - 2\sin^2 \vartheta\right). \tag{3.2.56}$$

The relationship between the two axis systems is defined by the transformation

$$x' = x - s,$$
$$\begin{pmatrix} y' \\ z' \end{pmatrix} = \begin{pmatrix} \cos\delta & \sin\delta \\ -\sin\delta & \cos\delta \end{pmatrix} \begin{pmatrix} y - a \\ z \end{pmatrix} \cong \begin{pmatrix} 1 & \delta \\ -\delta & 1 \end{pmatrix} \begin{pmatrix} y - a \\ z \end{pmatrix}. \tag{3.2.57}$$

In the respective reference systems, the fields of the two waves reflected by the lamina are given by

$$E_1 = E_o(x,y,z)e^{-ikz-i\Phi(x,y,z)},$$
$$E_2 = E_o(x',y',z'+D)e^{-ik(z'+D)-i\Phi(x',y',z'+D)}, \tag{3.2.58}$$

where Φ represents a phase variation in space with respect to a plane wave. The second beam also presents, in general, a certain amount of astigmatism introduced by the refractions, but as long as D is very small compared to the radius of curvature of the wavefront on the lamina, this effect is small, and we have chosen to neglect it here. However, the consequences of this aberration can be minimized by adjusting the inclination of the lamina for the value ϑ that maximizes the distance s [Hegedus et al 1993].

Normally, the two reflected waves are almost of the same intensity, and, if the variations in field amplitude are small within transverse distances comparable to s, the interference figure visible on a screen in z has a modulation of intensity given by

$$|E_1 + E_2|^2 \approx 2|E_o|^2(1+\cos\Theta), \tag{3.2.59}$$

where the argument of the modulation is given by

$$\Theta \approx k\delta y - kD + \Phi(x,y,z) - \Phi(x-s, y-a+\delta z, z-\delta y+D), \tag{3.2.60}$$

where Eqs. (3.2.58) have been used, neglecting the terms of order higher than the first in α.

Let us consider, in particular, the case in which the phase Φ represents a parabolic wavefront with symmetry of rotation around the z-axis, that is,

$$\Phi(x,y,z) = \frac{k}{R(z)}(x^2+y^2), \tag{3.2.61}$$

where R is the local curvature radius of the wavefront. Then, making a series of reasonable approximations that neglect the displacements associated with the angle α between the wedge faces and the difference of optical path D, for which

$$a << \delta z << s < w(z), \quad \delta y << R(z), \quad D << R(z), \tag{3.2.62}$$

where w is the half width of the beam, at the first order in δ, we have

$$\Theta \approx k\delta y - kD - \frac{k}{R(z)}(s^2 - 2xs - 2ya + 2y\delta z)$$
$$\approx \frac{ks}{R(z)}x + k\delta\left[1 - \frac{2z}{R(z)}\right]y + \Phi_o, \tag{3.2.63}$$

where Φ_0 contains terms that do not depend on x or y. Equating this expression to multiples of π, we get the spacing between the fringes and their inclination φ with respect to the x-axis:

$$d = \frac{\lambda}{\delta\left[1 - z/R(z)\right]}, \quad \tan\varphi = -\frac{sd}{\lambda R(z)}, \quad \sin\varphi = -\frac{sK}{\lambda R(z)}, \tag{3.2.64}$$

where d is the separation of the fringes in the direction y and K is that normal to the same fringes. From the measurement of one of these separations and of the angle φ, and by the knowledge of the translation s, we can then obtain the radius of curvature of the wavefront at the position z where the interference is observed:

$$R(z) = -\frac{sK}{\lambda \sin\varphi}. \tag{3.2.65}$$

In the practical use of this interferometer, we must have a lamina with plane faces that are optically well worked and with a suitable thickness with respect to the beam size. The lamina is centered on a rotator, with its axis normal to the faces and horizontal, after which this mounting is put on another graduated rotator with the axis of rotation vertical. Then, this set is placed on a collimated beam with an incidence angle close to zero and the lamina is rotated until the fringes produced are well horizontal: this calibration can be done once and for all. Lastly, the interferometer is placed on the beam to be analyzed, with an incidence angle determined by the second rotator, and the fringes are measured on a screen. If some optical system is used on the reflected beams to enlarge the fringes, account must be taken of both the magnification produced and the fact that the curvature radius that is measured is that corresponding to the conjugated plane to the plane of the screen.

Finally, from Eqs. (3.2.64), we see that the inclination of the fringes is proportional to s, which, in turn, depends on the angle of incidence. Therefore, when the radius of curvature of the wavefront is relatively small, the fringes become curved and lose their parallelism. However, as mentioned above, this distortion is reduced using the angle of incidence that produces the maximum shear.

3.2.7.7 Stellar interferometry

An interesting application of the dependence of visibility of the fringes with the size of the sources is represented by the measurement of the diameter or of the separation of celestial bodies. It was initially suggested by Fizeau, and then realized by Michelson with his stellar interferometer mounted on the reflecting telescope with 2.5 m aperture of the Mount Wilson Observatory [Born and Wolf 1980]. The shape of this device was similar to that of Young's interferometer in Fig. 3.9. The radiation of the object to be measured was collected by two planar mirrors that could be moved transversally to a distance of about 3 m from the axis of the telescope and two other planar mirrors, opposite the first, that deflected the light towards the reflector through two slits. In this way, the angular resolution of the telescope could be varied while the fringes on the image plane remained fixed, with a pace of about 20 microns. Observing the visibility of the fringes around the zero order, in 1921, Michelson was able to measure the angular dimension of some very bright giant stars like Betelgeuse [Michelson and Pease 1921].

Stellar interferometers based on this principle have also been contructed in recent times with two or more telescopes separated by distances of over 100 m. The most powerful are the *Very Large Telescope* consisting of 4 telescopes with 8.2 m aperture and a maximum separation of 200m at the Observatory of Cerro Paranal in Chile and the *Large Binocular Telescope* on Mount Graham in Arizona consisting of two 8.4 m mirrors separated by 6 m [Conti 2001]. An alternative method is to correlate the *intensity* fluctuations of the signals obtained with high-speed photodetectors located in distant telescopes aimed at a star to be measured [Hanbury Brown and Twiss 1956].

In the microwave region, it is instead possible to detect the radiation field itself electronically. In this context, Michelson's stellar interferometer has inspired the development of the *aperture synthesis* technique, which allows for a high resolution by means of radio telescopes arranged to form a linear or matrix array or in two cross directions. High resolutions are then obtained by combining *a posteriori* signals of various radio telescopes on an intercontinental basis (*Very Large Baseline Interferometry*) equipped with atomic clocks, in particular, hydrogen maser, which provide a local oscillator signal [Lipson et al 1995]. The *Square Kilometer Array* is currently being realized, constituted by a central cluster of antennas extended on a diameter of 5 Km and from thousands of orientable antennas arranged in a matrix of 150 Km in diameter and in spirals up to 3000 Km of distance or more.

3.3 Multiple-wave interference

3.3.1 Fabry-Perot interferometers

In 1897, two French physicists from the University of Marseille, Alfred Perot and Charles Fabry, described a new interference device that soon became revolutionary in the field of spectroscopy and is now a fundamental component of lasers. It made use of the interference produced by the multiple reflections that exist between two perfectly aligned semi-silvered plane mirrors (Fig. 3.31). In this way, one can see Haidinger's fringes (§3.2.5.2), but (with almost monochromatic light) these rings are now much thinner. Fabry and Perot soon realized that this subtlety was well suited to precise thickness measurements, in wavelength measurements, and for resolving neighboring spectral lines with great accuracy. They therefore designed and built sophisticated interferometers with which they devoted them-

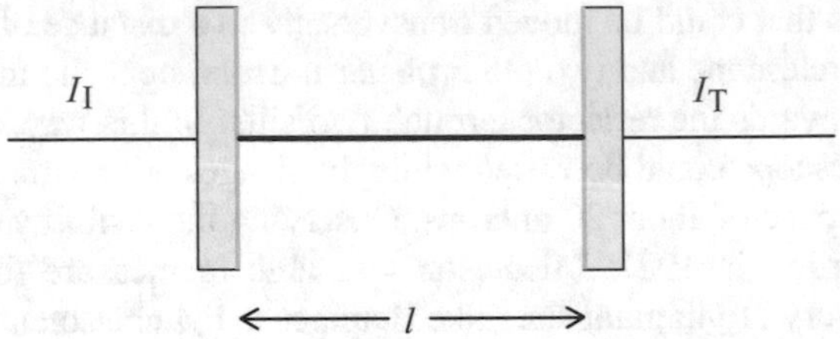

Fig. 3.31 Fabry-Perot interferometer

selves to many spectroscopic measures. In their honor, the devices of this type are called *Fabry-Perot interferometers* (FP). Before them, George Airy had analytically solved the problem of multiple reflections between two parallel glass laminae, obtaining an expression that he published in 1831 in his *Mathematical Tracts*, and which, in fact, describes the behavior of these interferometers. After the observation described by Haidinger in 1855 of the rings bearing his name, M.R. Boulouch, in 1893, applied this technique using the interference between two glass sheets covered with a thin layer of silver and observed, for the first time, the D-lines of sodium, which were resolved spectrally, forming a double set of rings. However, he did not follow up on his work.

In principle, a Fabry-Perot interferometer, in which one can continuously vary the separation between the mirrors, allows for the absolute measurement of the wavelength of the atomic spectral lines by simply counting the blossoming of the fringes while a mirror moves away to a well-defined distance. Another important feature is that the resolving power ($\nu/\Delta\nu$) of this tool can be simply varied by changing the spacing between the mirrors, in contrast, for example, with diffraction gratings, which have a fixed resolution and which are also much more expensive.

The first mirrors were created by silver plating or aluminizing glass substrates, so there was a fair amount of absorption limiting the resolution power of the interferometer. In recent times, from the year 1950 onwards, the development of deposition of dielectric multilayers has allowed for the realization of mirrors virtually devoid of absorption and with reflectivity very close to 100%. Normally, the two mirrors have the same reflectivity and the system is used in transmission.

Together with the proper FPs, in which the mirrors are separate elements, there are also etalons consisting of a single transparent plate with well-paralleled faces onto which a reflective treatment is deposited. With etalons, one loses the ability to tune the interferometer (though, if they are sufficiently thin, the tuning can be effected by tilting them), but they are much simpler and more stable. Also in the same class are interferential filters, in which the central dielectric layer has an optical thickness of $\lambda_o/2$, where λ_o is the wavelength in vacuum of the radiation transmitted by the filter.

The Fabry-Perot interferometers can therefore be used as tunable interference filters, with very high-resolution power ($\nu/\Delta\nu \approx 10^5 \div 10^7$) and, at least in theory (with equal mirrors), they have a transmission equal to 100% for a plane wave on the resonance peak. They can accept large-diameter light beams, though small in divergence: this is a limit in the use of classical sources, but is not so with lasers. However, at comparable resolution power, they are much brighter than diffraction grating monochromators.

The Fabry-Perot with plane mirrors was the progenitor of the *open optical resonator*, which usually has curved (concave) mirrors to limit diffraction losses. The resonant cavities employed in the lasers are of this type. In recent times, supercavities with enormous resolution power and stable resonance frequency within Hz have also been realized!

The advent of lasers revolutionized the spectroscopic techniques that are now largely based on absorption measurements and the Fabry-Perot plane mirrors are now of rare use. On the other hand, the excellent collimating properties of coherent sources are well suited to the use of curved mirror cavities, and optical resonators are now a fundamental tool in spectroscopic laboratories. Whether large or small, they are in evidence to one degree or another pretty much everywhere: they are used to monitor the spectrum of lasers, as filters inserted into the laser cavities themselves and as frequency references; in combination with laser sources, they are then used to study the form of spectral lines, to resolve multiplets' hyperfine structure, etc. A wide discussion of the Fabry-Perot interferometers and their application is reported by Vaughan (1989), with many historical references. In this section, I only consider the Fabry-Perot theory with plane mirrors and their properties. Although some of their limitations appear to have been overcome with curved mirror cavities, many other aspects remain common, and their study is propaedeutic.

3.3.2 Transmission function (Airy function) of a Fabry-Perot with plane and parallel mirrors

Usually, there are two ways to calculate the transmission of a FP: the first considers the subsequent reflections between the two mirrors and could be generalized for cases of a non-constant wave over time, while the second exploits the boundary conditions given by the mirrors themselves, but only applies to "continuous" waves, that is, stationary over time. To introduce a bit of generality, instead of considering the most trivial case, let us consider the case that the mirrors can be different and with an absorption not necessarily zero, and lastly that the interposed medium can introduce additional absorption. This entails little effort more and the simplifications can also be done in the end. For simplicity, we take instead equal to unity the refractive index on both sides of the reflecting surfaces.

3.3.2.1 Interference with multiple plane waves

For the explanation of this method, which follows the analysis by Airy, in Fig.

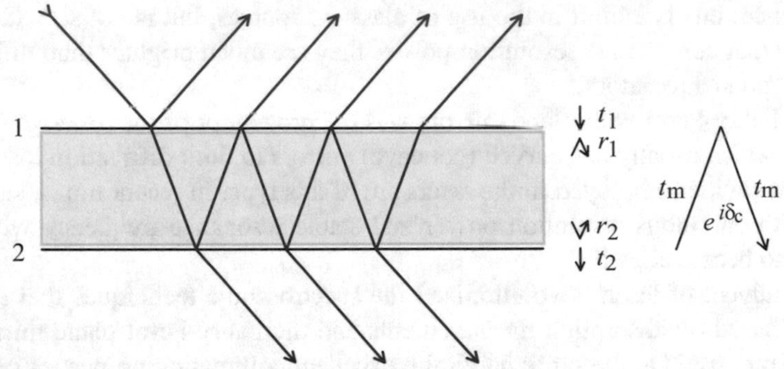

Fig. 3.32 Diagram of the multiple reflections between the mirrors

3.32, the subsequent reflected waves are schematized with rays so as to distinguish them. The mentioned coefficients are those that apply to the field:

t_1, t_2 : mirror transmission coefficients,
r_1, r_2 : reflection coefficients,
t_m : transmission coefficient of the medium in a single pass,
$\delta_c/2$: phase difference in the path between two reflections.

Each of the transmitted beams has in common the factor $t_1 t_2 t_m \exp(i\delta_c/2)$, and the total field of the transmitted wave is ultimately the sum of the fields of the single output waves:

$$E_T = E_I\, t_1 t_2 t_m e^{i\delta_c/2}\left(1 + r_1 r_2 t_m^2\, e^{i\delta_c} + \ldots\right) = E_I\, t_1 t_2 t_m e^{i\delta_c/2} \sum_{j=0}^{\infty}\left(r_1 r_2 t_m^2\, e^{i\delta_c}\right)^j$$

$$= E_I \frac{t_1 t_2 t_m e^{i\delta_c/2}}{1 - r_1 r_2 t_m^2\, e^{i\delta_c}}. \tag{3.3.1}$$

3.3.2.2 Self-consistency of the field with boundary conditions

Schematically, the problem can be described by Fig. 3.33, where E_+ and E_- respectively represent the amplitudes for waves traveling forward and backward, while the other terms are taken as in the previous case. On mirror 1, the transmitted forward field will sum with the reflected field that returns from mirror 2:

$$E_+(1) = t_1 E_I + r_1 E_-(1). \tag{3.3.2}$$

Instead, on mirror 2, for the transmitted wave and the reflected wave, we have

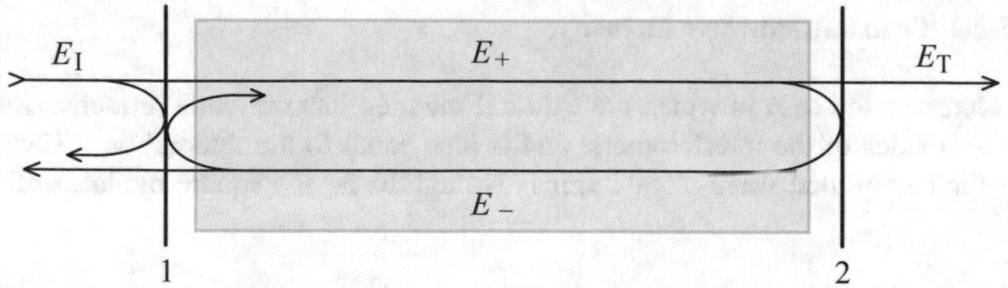

Fig. 3.33 Boundary conditions on the mirrors and self-consistency of the field

$$\begin{aligned} E_T(2) &= t_2 E_+(2), \\ E_-(2) &= r_2 E_+(2). \end{aligned} \tag{3.3.3}$$

For the propagation between the mirrors, we also have

$$\begin{aligned} E_+(2) &= t_m e^{i\delta_c/2} E_+(1), \\ E_-(1) &= t_m e^{i\delta_c/2} E_-(2). \end{aligned} \tag{3.3.4}$$

From all of these equations, it follows that

$$E_+(2) = t_1 t_m e^{i\delta_c/2} E_I + r_1 r_2 t_m^2 \, e^{i\delta_c} E_+(2),$$

and we finally again obtain

$$E_T = E_I \frac{t_1 t_2 t_m e^{i\delta_c/2}}{1 - r_1 r_2 t_m^2 \, e^{i\delta_c}}. \tag{3.3.1'}$$

With a few more steps, we find the reflected wave field:

$$E_R = E_I \left(r_1' + \frac{t_1' t_1 r_2 t_m^2 \, e^{i\delta_c}}{1 - r_1 r_2 t_m^2 \, e^{i\delta_c}} \right) = E_I r_1' \left(1 + \frac{t_1' t_1}{r_1' r_1} \frac{r_1 r_2 t_m^2 \, e^{i\delta_c}}{1 - r_1 r_2 t_m^2 \, e^{i\delta_c}} \right), \tag{3.3.5}$$

which can also be obtained through the summing method of the previous section. Here, r'_1 and t'_1 are the reflection and transmission coefficients, respectively, of the first mirror seen from the outer side of the cavity, since, in general, they are not equal to r_1 and t_1. We will deal with the relations between these coefficients later when we treat multiple layer mirrors: in particular, the ratio $t'_1 t_1 / r'_1 r_1$ turns out to be real and negative for a mirror without absorption, like those obtained with a succession of transparent dielectric layers.

3.3.2.3 Transmitted wave intensity

Consider the case in which the external medium has the same refractive index on both sides of the interferometer and is also equal to the internal one. Then, to get the transmitted wave intensity, it is enough to do the square modulus of Eq. (3.3.1):

$$I_{\mathrm{T}} = I_{\mathrm{I}} \left| \frac{t_1 t_2 t_{\mathrm{m}}}{1 - r_1 r_2 t_{\mathrm{m}}^2\, e^{i\delta_{\mathrm{c}}}} \right|^2 .$$

So far, no hypothesis has been made on the r and t coefficients of mirrors, which, in general, are complex; however, in the calculation of the transmitted intensity, we can introduce the real parameters R and T_{m}, by joining in δ the phase shifts δ_{r} and δ_{m}, respectively due to the reflections and the transmission of the internal medium, whereby

$$\delta = \delta_{\mathrm{c}} + \delta_{\mathrm{r}} + \delta_{\mathrm{m}} . \tag{3.3.6}$$

Then, let

$$T_1 = |t_1|^2 , \quad T_2 = |t_2|^2 \tag{3.3.7}$$

be the transmissivity of the mirrors,

$$R = |r_1 r_2| \tag{3.3.8}$$

the geometric mean of the reflectivity of the mirrors, and

$$T_{\mathrm{m}} = |t_{\mathrm{m}}|^2 \tag{3.3.9}$$

the transmissivity of the medium interposed between the mirrors, for a single pass. If the refractive indexes of the various media are different, Eqs. (3.3.7) should be modified, adapting them with the same considerations used to derive the equations of §1.7.2-3. However, if the first and last media are equal, we can write

$$I_{\mathrm{T}} = I_{\mathrm{I}} \frac{T_1 T_2 T_{\mathrm{m}}}{\left|1 - R T_{\mathrm{m}} e^{i\delta}\right|^2} = I_{\mathrm{I}} \frac{T_1 T_2 T_{\mathrm{m}}}{1 - 2RT_{\mathrm{m}} \cos\delta + R^2 T_{\mathrm{m}}^2}$$

$$= I_{\mathrm{I}} \frac{T_1 T_2 T_{\mathrm{m}}}{\left(1 - RT_{\mathrm{m}}\right)^2 + 4RT_{\mathrm{m}} \sin^2(\delta/2)} ,$$

from which we finally get

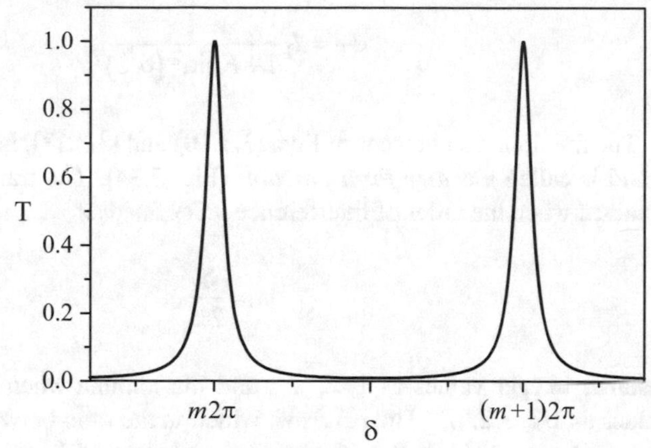

Fig. 3.34
Airy function for a contrast parameter $F = 100$. The corresponding finesse $\mathcal{F}$ is about 16

$$I_{\mathrm{T}} = I_{\mathrm{I}} \frac{1}{1 + F\sin^2(\delta/2)} \frac{T_1 T_2 T_{\mathrm{m}}}{(1 - RT_{\mathrm{m}})^2}, \tag{3.3.10}$$

where the contrast parameter F is defined by the formula:

$$F = \frac{4RT_{\mathrm{m}}}{(1 - RT_{\mathrm{m}})^2}. \tag{3.3.11}$$

If the internal medium is completely transparent and the mirrors are equal, hence $T_1 = T_2 = T$, and

$$R + T + A = 1, \tag{3.3.12}$$

where A is the absorption of mirrors, we simply have

$$I_{\mathrm{T}} = I_{\mathrm{I}} \frac{1}{1 + F\sin^2(\delta/2)} \frac{T^2}{(1 - R)^2} = I_{\mathrm{I}} \frac{1}{1 + F\sin^2(\delta/2)} \left(1 - \frac{A}{1 - R}\right)^2, \tag{3.3.13}$$

where now

$$F = \frac{4R}{(1 - R)^2}. \tag{3.3.14}$$

For good dielectric film mirrors, the term A is largely replaced by losses A_{s} due to diffusion caused by defects in the dielectric layers. When even this term is negligible, Eq. (3.3.13) is reduced to

$$I_T = I_I \frac{1}{1 + F\sin^2(\delta/2)}. \tag{3.3.15}$$

The fraction, also present in Eqs. (3.3.10) and (3.3.13), is a periodic function in δ and is called the *Airy form function* (Fig. 3.34). The transmission maxima are obtained when the order of interference m, defined as

$$m = \frac{\delta}{2\pi}, \tag{3.3.16}$$

assumes integer values as 1, 2, ... , and the minima when assuming half-integer values as 1/2, 3/2, ... The *contrast*, which is the ratio between the maximum and the minimum, is $F + 1$. It is important to note that, if the mirrors are equal and the losses by absorption or diffusion are null, the transmission is unitary on the peaks. Otherwise, there is an attenuation of the transmitted light, but, for the same F, the transmission curve obtained by varying δ is only scaled to the new value of the transmission on the peak.

3.3.2.4 Reflected wave intensity

Similarly, we can calculate the intensity of the reflected wave, so

$$I_R = I_I\left|r_1'\right|^2\left|1+\frac{t_1t_1'}{r_1r_1'}\frac{RT_m e^{i\delta}}{1-RT_m e^{i\delta}}\right|^2 = I_I\left|r_1'\right|^2\frac{\left|1-\left(1-\frac{t_1t_1'}{r_1r_1'}\right)RT_m e^{i\delta}\right|^2}{\left|1-RT_m e^{i\delta}\right|^2}, \tag{3.3.17}$$

where it should be noted that δ incorporates the phase of $r_1r_2t_m{}^2$, as in the case of transmission. The reflected intensity is essentially complementary to that of the transmitted intensity and presents a background value equal to $I_I|r'_1|^2$ furrowed by a succession of periodic dips. Both the numerator and denominator of the fraction have minima with period 2π in δ. While the transmission peaks are symmetrical[2] in δ, the dips are generally asymmetrical: in fact, because of the term $-t_1t'_1/(r_1r'_1) = a+ib = s\exp(i\beta)$, where s is its modulus and β its phase, if β is not zero, there is a shift of the minima of the numerator and a consequent asymmetry of the dips in the reflected intensity. However, there is a physical limit to the extent of this phase shift that is not made explicit in Eq. (3.3.17).

In particular, Eq. (3.3.17) can be rewritten as

---

2 This symmetry is present for the ideal case of a plane wave and unlimited plane mirrors. In practice, a laser beam is spatially delimited and a good alignment is also required: for a tilted beam, the peaks are asymmetrical.

$$I_R = |r_1'|^2 \frac{\left[1-(1+a)RT_m\right]^2 + 4(1+a)RT_m \sin^2(\delta/2) + b^2R^2T_m{}^2 + 2bRT_m \sin\delta}{(1-RT_m)^2 + 4RT_m \sin^2(\delta/2)}.$$

In addition, it has a shift $\Delta\delta$ of the reflection minima with respect to the transmission maxima. Given that the transmission of the mirrors of a Fabry-Perot is normally small, we can assume small δ, a, and b, so that the sines can be replaced by their arguments. Therefore, differentiating with respect to δ, then neglecting the terms of higher order than the first in δ, a and b, and, finally, solving for δ, we find the translation between the reflection minima and the transmissions maxima

$$\Delta\delta \cong -\beta \frac{1-RT_m}{1+RT_m}.$$

3.3.2.5 Finesse

As we can immediately understand, from Eqs. (3.3.14) and (3.3.15), the transmission peaks are becoming increasingly narrow with increasing parameter F, hence with the reflectivity of the mirrors. It is interesting now to calculate the width ξ of the peaks at half height (FWHM, *full width half maximum*) as a function of F. Let us therefore consider one of the peaks for which $\delta = m2\pi$ with integer m. Around our peak, the transmission is 1/2 of the maximum for values of δ equal to

$$\delta_\pm = 2m\pi \pm \frac{\xi}{2},$$

so

$$\frac{1}{1+F\sin^2(\xi/4)} = \frac{1}{2}. \tag{3.3.18}$$

If F is sufficiently large, this equation is verified for values ξ such that the sine can be confused with its argument, and therefore, solving for ξ, we get

$$\xi = \frac{4}{\sqrt{F}}. \tag{3.3.19}$$

Anyway, the parameter that most characterizes the quality of a Fabry-Perot is the ratio, which is called the *finesse*, of the separation between two successive peaks and the FWHM width of a peak. Since the distance between two peaks is 2π in the phase δ, the finesse is given by

$$\mathscr{F} = \frac{\pi}{2}\sqrt{F}\,. \tag{3.3.20}$$

In terms of the finesse, the Airy function can therefore be reformulated as

$$\frac{I_{\mathrm{T}}}{I_{\mathrm{I}}} = \frac{T_{\max}}{1+\dfrac{4}{\pi^2}\mathscr{F}^2\sin^2(\delta/2)}, \tag{3.3.21}$$

where $T_{\max}$ represents the peak transmission.

If, in Eq. (3.3.20), we replace F with Eq. (3.3.14), we get what is called the *finesse for reflection*:

$$\mathscr{F}_R = \frac{\pi\sqrt{R}}{1-R}\,. \tag{3.3.22}$$

This distinction is necessary because various factors, such as the imperfections of mirror flatness, diffraction, the way in which the interferometer is used, and more, contribute to the actual finesse, as we shall see later.

3.3.2.6 Optical path and phase difference between two successive reflections

As we already know from §3.3.1, the difference in optical paths between two successive rays in the series of reflections between two planar and parallel mirrors is given by

$$2L = 2nl\cos\vartheta'\,, \tag{3.3.23}$$

where ϑ' indicates here the incidence angle within the interferometer, n is the refraction index present there and l is the separation between the mirrors. The phase shift due to this difference in path length is then

$$\delta_{\mathrm{c}} = 2\pi\frac{2L}{\lambda_{\mathrm{o}}} = 4\pi\frac{nl}{\lambda_{\mathrm{o}}}\cos\vartheta'\,. \tag{3.3.24}$$

Note that $\delta_{\mathrm{c}}/2 = lk'\cos\vartheta'$ is just the phase difference that one has traveling from one mirror to the other along the direction normal to the mirrors, that is, the projection in this direction of the wave vector k' multiplied by the distance l.

In general, the two reflections give rise to a further phase shift δ_{r}, which also depends on the angle of incidence and the wavelength. The overall phase is therefore

$$\delta = \delta_{\mathrm{c}} + \delta_{\mathrm{r}} + \delta_{\mathrm{m}}\,, \tag{3.3.25}$$

however, it is normally δ_m, $\delta_r << \delta_c$ (but *not* also δ_m, $\delta_r << \pi$) and, in practical applications, δ_m, δ_r can be considered constant or even negligible. In the latter case, recalling Eq. (3.3.16), we can also write

$$2L = m\lambda_o = 2nl\cos\vartheta' . \qquad (3.3.26)$$

3.3.3 *Applications of Fabry-Perot interferometers*

3.3.3.1 The system of fringes in the transmission

As we have already mentioned, the fringes that can be seen in transmission are basically the fringes of Haidinger, in which the bright rings are becoming thinner with increasing finesse of the interferometer. With an extended source, these fringes are localized at infinity, but, in practice, they are normally made visible on the focal plane of a lens placed immediately after the interferometer. Usually, indeed, a Fabry-Perot is used in the manner described by Fig. 3.35, in which the source is also located at the focal plane of a lens placed in front of the interferometer, with the aim of collimating its light. The rings are thus arranged according to the law

$$2m\pi = 4\pi\frac{nl}{\lambda_o}\cos\vartheta'_m + \delta_r ;$$

calling m_o the interference order in the center, so that

$$2m_o\pi = 4\pi\frac{nl}{\lambda_o} + \delta_r ,$$

and subtracting these equations between them, it is found that the phase difference with the center is

$$2(m_o - m)\pi = 4\pi\frac{nl}{\lambda_o}(1-\cos\vartheta'_m) , \qquad (3.3.27)$$

from which, for small ϑ',

$$\frac{\lambda_o}{nl}(m_o - m) \cong \vartheta_m'^2 . \qquad (3.3.28)$$

If we now number the rings as we did for the Haidinger fringes, the radius of the rings in the focal plane of the lens is

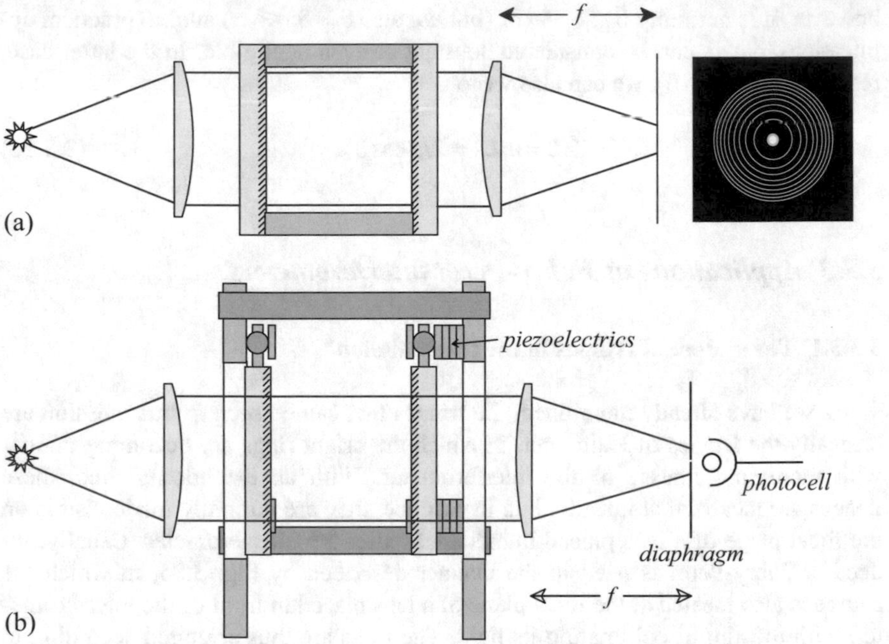

Fig. 3.35 Arrangement used with a Fabry-Perot in the two configurations: (a) etalon and (b) scanning interferometer

$$r_p \cong f \frac{n}{n_e} \sqrt{\frac{\lambda_o}{nl}(p-1+e)}\,, \tag{3.3.29}$$

where e is the *fractional order in the center*, f is the focal distance of the lens in the external medium, and n_e is the refractive index outside of the interferometer. Under ideal conditions, as well as for the purpose of becoming denser, the rings are progressively thinner moving away from the center. For a *null* fractional order, the central fringe is a spot of light with the maximum in the center. As ϑ' increases, the phase difference with the center varies approximately as

$$\delta_\vartheta = 2\pi \frac{nl}{\lambda_o} \vartheta'^2\,, \tag{3.3.30}$$

whereby, by substituting this expression in place of $\xi/2$ in Eq. (3.3.19) and then solving for ϑ', we find that the width at half maximum (FWHM) of the spot is

$$a = 2f \frac{n}{n_e} \sqrt{\frac{1}{\mathscr{F}} \frac{\lambda_o}{2nl}}\,, \tag{3.3.31}$$

where $\mathscr{F}$ is the finesse defined by Eq. (3.3.20).

The number of visible rings is limited by the angular field of the incident radiation. For a point source, the interference fringes are not localized, however, in general, they take on a more complicated aspect than Haidinger's fringes. Indeed, for non-zero angles of incidence, from a single incident ray, we obtain a series of rays that are, however, shifted between them. Therefore, for example, if one uses a divergent laser beam, to observe sharp fringes, it is necessary to use a lens located downstream of the interferometer that recombines the various rays in its focal plane. Moving away from this plane, the rings appear to multiply and their interpretation is complicated.

One would think that even the lens in front of the interferometer is essential; however, this is not completely true. It is used to collimate the radiation of the source, thereby the radiation emitted by each source point is converted into a plane wave on which the above theory is based. In addition, the lens limits the angular field of radiation impinging on the interferometer, to the benefit of the brilliance of the fringes that remain present. In constructing the bright fringes of the interference figure, a plane Fabry-Perot has, in fact, the ability to select rays with particular directions. If the source is not in focus or the lens is absent, the various rays emitted by a single source point are directed to different points on the focal plane of the second lens. But, again, for the multiple interference of the reflection series on the mirrors of each incident ray, the brightness at each point on this plane depends on the angle of incidence, resulting in a series of luminous rings. Different source points give rise to the same figure, and (if the source is spatially incoherent) the overall intensity is the incoherent sum of the contributions of each of the source points.

Lastly, two words on the fringe system in reflection. These are complementary to those in transmission, with thin dark rings in a bright field, again observed in the focal plane of a lens. However, the trend of the intensity of these rings is critically dependent on the presence of a phase shift in the reflection on the mirrors, in the presence of which they become asymmetrical. Their visibility also becomes negligible when the losses on the mirrors are comparable to their transmission [Vaughan 1989, §3.2.2].

3.3.3.2 Free spectral range and overlap of orders

As we have just seen in the previous section, δ is a function of several factors: l, n, λ (or the frequency ν), ϑ', and, by varying one, we can do a scan of the Airy function. For example, if we measure the intensity of a monochromatic laser beam transmitted by the FP with a photodiode and put this signal on the Y-axis of an oscilloscope, while, on the X-axis, we send a signal corresponding to the parameter that is varied, on the screen, we observe traces similar to the one in Fig. 3.36. When the interference order, defined by Eq. (3.3.16), assumes two successive integer values, the distance between two corresponding peaks that one has in terms of the parameter that varies (although one at a time) is:

– for the phase δ, it is, naturally,

$$(\Delta\delta)_{SR} = 2\pi \, ; \tag{3.3.32}$$

– for the optical path $L = nl\cos\vartheta'$, with λ_o, ν constant, it is:

$$(\Delta L)_{SR} = L_{m+1} - L_m = \frac{\lambda_o}{2} = \frac{c}{2\nu} \, ; \tag{3.3.33}$$

– for the wavelength, with L constant, it is:

$$(\Delta\lambda_o)_{SR} = \lambda_m - \lambda_{m+1} \cong \frac{\lambda_o{}^2}{2L} \, ; \tag{3.3.34}$$

– for the frequency $\nu = c/\lambda_o$, with L constant, it is:

$$(\Delta\nu)_{SR} = \nu_{m+1} - \nu_m \simeq \frac{c}{2L} \, . \tag{3.3.35}$$

In particular, from Eqs. (3.3.16), (3.3.24) and (3.3.25), it is found that the cavity resonance frequency comb, at the variation of m, with assigned values of the other parameters, is explicitly

$$\nu_m = \frac{2m\pi - \delta_r - \delta_m}{4\pi nl\cos\vartheta'} c \, . \tag{3.3.36}$$

It is to be noted that, in Eq. (3.3.35), $(\Delta\nu)_{SR}$ is independent of ν. However, this is, in general, only an approximation assuming that the sum of phase shifts due to the mirrors and the internal medium, $\delta_r + \delta_m$, does not depend on ν. This can be achieved by reciprocally compensating the dispersion of the mirrors and the medium.

The SR index in Eqs. (3.3.32-35) means that these differences correspond to the *free spectral range*. To understand its meaning, let we look at an example.

Suppose we want to measure the hyperfine structure of a certain atomic spectral line emitted by a source whose light beam is well collimated on the interferometer. Then, we scan the optical path L of the interferometer while measuring the light transmitted with a detector, as in the configuration of Fig. 3.35(b), with $\vartheta' = 0$. If $g(\nu)$ is a function that represents the spectral distribution of the incident radiation, the signal that is observed is the convolution of g with the Airy function:

$$S(L) = \int_0^\infty g(\nu) \frac{1}{1 + F\sin^2(\pi\nu 2L/c)} d\nu \, . \tag{3.3.37}$$

Since the Airy function is periodic, this integral can be rephrased as a summation of interference order m taken as an integer:

$$S(L)=\sum_{m=1}^{\infty}\int_{-c/4L}^{c/4L} g\left(\Delta\nu+m\frac{c}{2L}\right)\frac{1}{1+F\sin^2\left(\pi\Delta\nu\frac{2L}{c}\right)}d\Delta\nu \qquad (3.3.38)$$

[where, to be strict, with this expression, we assume $g(\nu)=0$ for $\nu < c/4L$].

Generally, only small scans of the length L are used around a mean value L_o, and, in this case, the width of the Airy function peaks on the axis of the parameter $\Delta\nu$ changes little within the range of L values that are considered. If the finesse is sufficiently high, the Airy function assumes negligible values around the integration extremes of Eq. (3.3.38), which can then be replaced with $\pm\, c/(4L_o)$. So, the integral can be written as

$$\frac{\pi c}{2\mathcal{F}2L_o}g'_{L_o}(\nu')=\int_{-c/4L_o}^{c/4L_o} g(\Delta\nu+\nu')\frac{1}{1+F\sin^2\left(\pi\Delta\nu\frac{2L_o}{c}\right)}d\Delta\nu\,, \qquad (3.3.39)$$

where the initial coefficient is introduced for a suitable renormalization, within the limit of a large finesse $\mathcal{F}$.

The new function $g'_{L_o}(\nu)$ represents the convolution of $g(\nu)$ with a single peak of the Airy function, while the index L_o reminds us that this function is calculated for a predetermined value around which to modulate the optical path between the mirrors. Eq. (3.3.38) can now be rewritten as

$$S(L)=\frac{\pi c}{2\mathcal{F}2L_o}\sum_{m=1}^{\infty}g'_{L_o}\left(m\frac{c}{2L}\right). \qquad (3.3.40)$$

If the finesse of the interferometer is now sufficiently high so that the Airy function (at the variation of ν) exhibits very thin peaks with respect to the trend of $g(\nu)$, we can approximate $g'_{Lo}(\nu)$ with $g(\nu)$ itself. On the other hand, for small variations ΔL around L_o, it is

$$\frac{mc}{2L}\cong\frac{mc}{2L_o}-\frac{mc}{2L_o^{\,2}}\Delta L=\nu_m-\Delta\nu'\,, \qquad (3.3.41)$$

and hence the observed signal is given by the superposition of spectral orders, each of which can be attributed to a scale in ν, oriented in the opposite direction from the one on L (Fig. 3.36). So, the order m is associated with a scale in which the frequency $\nu_m = mc/(2L_o)$ corresponds to the position L_o in the length scale,

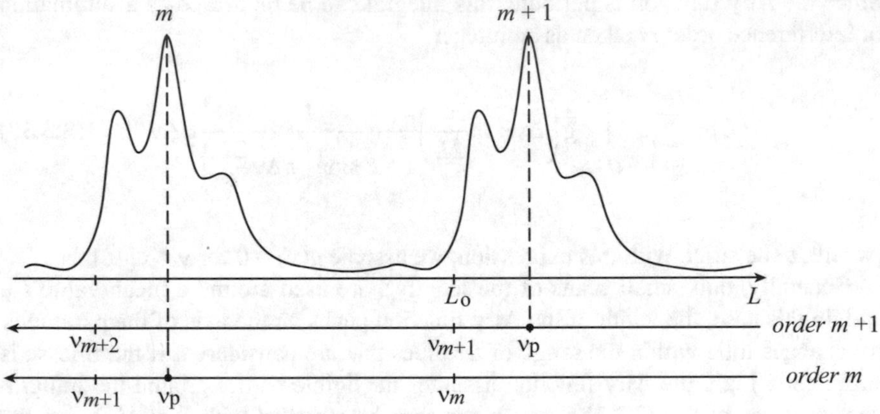

Fig. 3.36 Superposition of spectral orders

with a scale renormalization equal to

$$\frac{\Delta\nu}{\Delta L} = -\frac{m}{L_o}\frac{c}{2L_o}\,. \tag{3.3.42}$$

The various scales in ν are shifted from each other by an amount

$$\nu_{m+1} - \nu_m = \frac{c}{2L_o} = (\Delta\nu)_{SR}\,. \tag{3.3.43}$$

If, therefore, the spectrum of radiation to be studied is contained within a free spectral range, the pattern of the transmitted signal will show several repeated copies of it in succession. If, on the other hand, the spectrum is wider, the various copies overlap, and it may not be easy to reconstruct the original spectrum. These ambiguities can still be resolved by recording the spectra for two or more different values of L_o.

3.3.3.3 Resolving power of a Fabry-Perot

We spoke earlier of the ability of a Fabry-Perot interferometer to resolve the spectral lines. Consider the case of a spectrum consisting of two neighboring lines with frequency ν_1 and ν_2, the separation of which is small compared to the free spectral range, and with the same intensity. As we have seen before, the intensity transmitted by varying the optical path length L is given by the convolution with the Airy function, which, in the case of two nearly monochromatic components narrower than the Airy peaks, from Eq. (3.3.35), can be expressed as the sum

Fig. 3.37
Taylor's criterion for resolving two nearby spectral lines

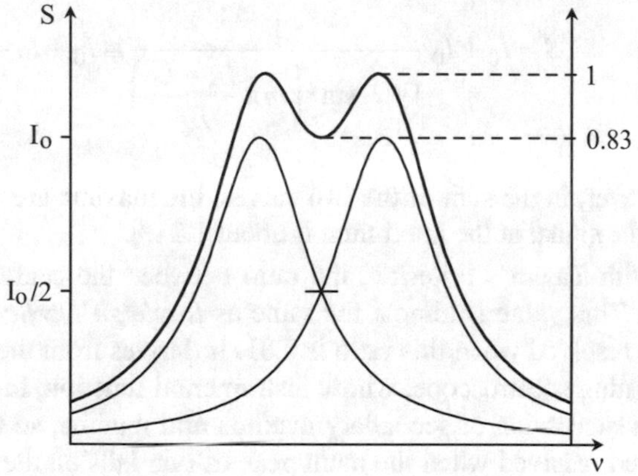

$$S(L) = I_0 \frac{1}{1 + F\sin^2(\pi\nu_1 2L/c)} + I_0 \frac{1}{1 + F\sin^2(\pi\nu_2 2L/c)}. \tag{3.3.44}$$

If we now limit ourselves to considering a given interference order m, the two peaks lie at the values $L_1 = mc/(2\nu_1)$ and $L_2 = mc/(2\nu_2)$. Therefore,

$$S(L) = I_0 \frac{1}{1 + F\sin^2\left(m\pi\frac{L - L_1}{L_1}\right)} + I_0 \frac{1}{1 + F\sin^2\left(m\pi\frac{L - L_2}{L_2}\right)}.$$

Now, it is said that the two frequencies ν_1 and ν_2 are resolved in the scan of L if a depression between the corresponding peaks in the transmitted signal is visible. A useful convention for quantifying this concept is that known as *Taylor's criterion* (Fig. 3.37). It states that two spectral lines (of equal intensity) begin to be resolved when their separation is such that their individual curves intersect at half height. In this case, with L halfway between the two lines, whereby $L - L_1 = L_2 - L = (L_2 - L_1)/2$, with $L_1, L_2 \approx L_o$ we can write

$$S = I_0 \frac{2}{1 + F\sin^2\left(m\pi\frac{L_2 - L_1}{2L_o}\right)} = I_0,$$

from which, for big F, we find that

$$\frac{|L_2 - L_1|}{L_o} = \frac{1}{\sqrt{F}}\frac{2}{m\pi} = \frac{1}{m\mathcal{F}}, \tag{3.3.45}$$

where $\mathcal{F}$ is the finesse defined by Eq. (3.3.20). For this condition of separation, in correspondence to one of the two peaks, the signal is instead

$$S = I_0 + I_0 \frac{1}{1 + F\sin^2\left(m\pi\dfrac{L_2 - L_1}{L_o}\right)} \cong I_0 + I_0 \frac{1}{1+4} = 1.2 I_0 .$$

However, in the sum of the two curves, the maxima are a little closer to each other and the signal at the maximum is about $1.21 I_0$.

With Taylor's criterion, the ratio between the center-to-peak signals is about 0.83. This value is almost the same as *Rayleigh's criterion*, which considers two lines resolved when this ratio is 0.81. It derives from the characteristics of a prism or grating spectroscope, whose instrumental function, in addition to a central peak, has a succession of secondary maxima and minima, so that the two lines are considered resolved when the main peak of one falls on the first minimum of the other.

Finally, in terms of frequency, the minimum separation for considering two lines resolved is

$$\delta\nu = |\nu_1 - \nu_2| = \left|\frac{mc}{2L_1} - \frac{mc}{2L_2}\right| \cong \frac{mc}{2L_o}\frac{|L_2 - L_1|}{L_o} = \frac{(\Delta\nu)_{SR}}{\mathcal{F}} . \tag{3.3.46}$$

The *resolving power* is defined as

$$\mathcal{R}_{\mathcal{P}} = \frac{\nu}{|\delta\nu|} = \frac{\lambda}{|\delta\lambda|} . \tag{3.3.47}$$

Then, finally, for a Fabry-Perot, it results that

$$\mathcal{R}_{\mathcal{P}} = m\mathcal{F} , \tag{3.3.48}$$

and therefore it is determined by the order of interference and the finesse. If Rayleigh's criterion is used, the resolving power becomes $0.97 m\mathcal{F}$.

The resolving power can, in principle, therefore be increased at will by growing the order of interference, and this can be achieved by increasing the separation between the mirrors. However, at the same time, the free spectral range decreases, and this may render problematic the interpretation of the transmitted signal. In practice, one has to look for a suitable compromise for each application, and, in general, it is good that the free spectral range remains considerably larger than the spectral width of the lines to be studied.

3.3.3.4 Recording of spectra

In the two configurations of Fig. 3.35, the first method of recording spectra, using an *etalon*, consists in photographing the fringe system: this can be done with a photographic plate of which the blackening is then measured, or, more modernly, using a photodiode matrix, whose signal is analyzed with a computer. However, this method is not very practical, as the fringes are circular instead of straight, as in the case of a grating spectroscope, and also the dispersion is not linear, as we have already seen with Eq. (3.3.29); consequently, the analysis to be done is rather complicated. On the other hand, this first method has the advantage of recording the entire spectrum simultaneously: it may therefore be necessary when the source emits light for a very short time interval.

When, instead, the source emits a stationary or slowly variable spectrum, it is more convenient to use a scanning interferometer where the optical path changes, while a circular diaphragm selects a small area at the center of the interference pattern, as shown in Fig. 3.35(b). The transmitted radiation is then detected by a photomultiplier or a photodiode: if the spectrum to be studied is centered on a wavelength λ_o, a variation of optical path length L equal to $\lambda_o/2$ is sufficient to scan an entire free spectral range. To obtain a linear recording of the spectrum, the distance l between mirrors and the refractive index n of the interposed medium can be varied linearly. In both cases, the diaphragm must be exactly centered in the center of the interference pattern and with a diameter small enough so as not to excessively reduce the resolution of the interferometer.

In the first case, one of the mirrors is moved by a piezoelectric, i.e., a ceramic material whose length can be varied linearly by some μm by applying a voltage of tens or hundreds of volts to its electrodes. Three piezoelectric devices are often used to allow for fine adjustment of parallelism between the mirrors. With this piezoelectric scan method, the separation between the mirrors is modulated by applying to the piezoelectrics a periodic sinusoidal or triangular signal. The maximum frequency of the modulation is limited by the resonant frequencies of the mechanical mounting, which are generally of some KHz, but can reach dozens of KHz. In general, unless one is able to take advantage of it, it is best to keep the modulation at a frequency far below these resonances, which can cause strong distortions in the scanning of l. The advantage of the piezoelectric method is, however, to facilitate the recording of the entire spectrum in a very short time. Its main disadvantage is due to the hysteresis of piezoelectric material, so that there is always a slight deviation from the proportionality between displacement and applied voltage.

For high-precision measurements, it is preferable to use the pressure scan in which the distance between the mirrors is kept constant but the interferometer is enclosed in a container in which the pressure of a gas varies, which then varies its refractive index n. This type of scan is much slower than the piezoelectric one, but the change of n with the gas pressure can be known and tabulated with great precision by means of a Michelson or Jamin interferometer. From Eq. (3.3.26), we have

$$m\lambda_o = 2nl\,.$$

Then, for an assigned distance l and an order m, differentiating, we get

$$m\delta\lambda_o = 2l\delta n,$$

so by dividing member to member, we have

$$\frac{\delta\lambda_o}{\lambda_o} = \frac{\delta n}{n} \cong \delta n.$$

On the other hand, in terms of λ_o, a free spectral range is given by Eq. (3.3.33), and hence this same interval is observed with a variation of n equal to

$$\Delta n = \frac{\lambda_o}{2nl} \cong \frac{\lambda_o}{2l}. \tag{3.3.49}$$

In the case of air, it is $n - 1 \approx 3 \cdot 10^{-4}$ for an atmosphere, and if, for example, λ_o = 600 nm and l = 10 cm, a whole spectral order can be observed by varying the pressure by about 10 mbar.

3.3.3.5 Limitations to the finesse of a Fabry-Perot

We have seen above that the resolution power of a Fabry-Perot is proportional to the finesse $\mathcal{F}$, which, in the general sense, is defined by the ratio between the distance between two peaks and the width of a peak in the transmission curve. This parameter then indicates the quality of the interferometer. However, many factors contribute to limiting the value of $\mathcal{F}$; they are [Strumia 1973-1974]:

(a) the reflection of mirrors and absorption or other losses in the internal medium, from which we have already determined the function of Airy as an instrumental function, with a corresponding finesse that is obtained by combining Eqs. (3.3.11) and (3.3.20),
(b) an imperfect planarity of the mirrors,
(c) a bad parallelism between the mirrors,
(d) the inhomogeneity of the refractive index in the internal medium,
(e) the finite size of the aperture with which the transmitted light is recorded, the accuracy in the centering of the aperture, the optical quality of the lens that forms the fringes and an incorrect focusing,
(f) the diffraction due to the finished diameter of the mirrors.

To each of these factors, in principle, one can assign its own instrumental function f_i and its relative finesse $\mathcal{F}_i$ given by the ratio between the free spectral range and their FWHM characteristic width. The overall instrumental function $f(\nu)$ is no longer the simple function of Airy, but is rather obtained from the convolution between these functions. The width of $f(\nu)$ is therefore greater than that of the single f_i functions, but less than the sum of their widths. The value of the overall finesse

$\mathcal{F}$ depends on the shape of the f_i; if, for example, we assimilate them to Gaussians, we would have

$$1/\mathcal{F}^2 = \sum_i 1/\mathcal{F}_i^2 ,$$

while, if we assimilate them to Lorentzians, we would have a similar expression, but without the squares.

Normally, the factors in points (a) and (b) are predominant in determining the final value of the finesse. With R given by Eq. (3.3.8) and T_{m} by Eq. (3.3.9), from Eqs. (3.3.11) and (3.3.20), the finesse corresponding to (a) is

$$\mathcal{F}_{\mathrm{a}} = \frac{\pi\sqrt{RT_{\mathrm{m}}}}{1-RT_{\mathrm{m}}}, \qquad (3.3.50)$$

which, for a medium without losses, is reduced to the finesse by reflection of Eq. (3.3.22). This finesse can be increased by enhancing the reflection of the mirrors; however, when 1 – R becomes comparable to the losses, the Fabry-Perot transmission begins to fall rapidly; for example, in the case of Eqs. (3.3.13) the transmission on the peak is reduced to 1/4 when $1 - R = 2A$. This problem was severe when metal coatings were used, but it is now heavily scaled down with modern dielectric mirrors.

At point (b), the imperfect planarity of the mirrors is a factor that seriously limits the quality of a Fabry-Perot compared to the one obtainable for reflection finesse. Flatness defects lead to a more or less wide distribution on the path between the mirrors. The form of this distribution varies from case to case, while, normally, the quality of a lamina is quantified by the manufacturers only by the deviation from flatness in terms of a fraction of the wavelength indicating the maximum peak-to-valley difference, λ/q_{pv}, or the average RMS value on the surface, λ/q_{RMS}, and, in this case the acronym RMS must be clearly specified. If the value of λ is not indicated, it is usually meant by a transition of mercury, at 546.1 nm. There are currently available, at a non-negligible cost, coupled laminae that together are better than $\lambda/200$ peak-to-valley on surfaces of 5 cm or more in diameter. At this level, the reflective treatment must also be extremely well done, uniform in thickness and as devoid as possible of surface tensions, which may considerably deform the surface; even the mounting of the mirrors must be done so as not to distort them.[3] However, higher optical qualities can be achieved if one is satisfied with using a smaller portion of the lamina.

In terms of finesse, the width of the distribution of the paths can be approximate with

$$\mathcal{F}_{\mathrm{b}} = \frac{q_{\mathrm{pv}}}{2}, \quad \text{or} \quad \mathcal{F}_{\mathrm{b}} = \frac{q_{\mathrm{RMS}}}{4}, \qquad (3.3.51)$$

---

[3] See, for example, the three-point support system adopted by Burleigh.

where q indicates the overall quality of the *pair* of mirrors. The problem that usually arises is: what reasonable limit should be given to the finesse for reflection to be required for a given pair of laminae? With $\mathcal{F}_a > \mathcal{F}_b$, the overall instrumental function is heavily influenced by surface defects, with virtually no gain in resolution, but rather by reducing the transmitted intensity. In practice, the reasonable limit value is between $\mathcal{F}_b$ and $\mathcal{F}_b/2$.

Points (c) and (d) are similar to the previous one. If α is the angle between the two mirrors and D their diameter (for circular mirrors), we have

$$\mathcal{F}_c = \frac{\sqrt{3}}{4}\frac{\lambda}{\alpha D}. \tag{3.3.52}$$

Even the inhomogeneity of the medium between the mirrors can be expressed as a fraction of a wavelength, after which we proceed as for point (b).

In the case of an extended source, the opening in point (e) delimits the angular field used for recording. For a well-centered and focused aperture with radius b, the accepted maximum internal angle is approximately equal to

$$\vartheta'_m \cong \frac{n_e}{n}\frac{b}{f}, \tag{3.3.53}$$

where n and n_e are the internal and external refractive indexes at the Fabry-Perot and f is the focal length in the external medium of the lens that forms the fringes. From Eqs. (3.3.30) and (3.3.53), we also notice that the distribution of the phase differences delimited by the aperture has the shape of a rectangle, since, for each increment of the opening *area*, there is a proportional increase in δ_ϑ. From these equations, the opening finesse can therefore be assumed to be the inverse of the spectral fraction allowed by the opening:

$$\mathcal{F}_e = \frac{2\pi}{\delta_{\vartheta m}} = \frac{\lambda_o}{nl}\left(\frac{nf}{n_e b}\right)^2, \tag{3.3.54}$$

where l is still the physical separation between the mirrors; in particular, if $2b$ is chosen as equal to the diameter a of the central spot given by Eq. (3.3.31), it is ~ $2\mathcal{F}_o$, where $\mathcal{F}_o$ is the finesse associated with the other processes. The opening finesse, however, degrades rapidly if the aperture is not well centered or focused, so it needs to be positioned with great care. Moreover, the recorded signal has its maximum when the fractional order at the center of the fringe system is not zero, but rather half of $\delta_{\vartheta m}/(2\pi)$.

For the analysis of a well-collimated laser beam, the discourse becomes different, because the angular field is delimited by the divergence of the beam itself, which can be very small.

Lastly, we consider the effect resulting from diffraction. Since mirrors have a finite extension, at each reflection, part of the radiation is lost by diffraction. To this it is added the diffractive contribution to the width of the fringes in the lens focus. However, these effects are normally completely negligible.

Some of these considerations, in particular, those relating to irregularities in reflective surfaces, are not valid in the case of so-called supercavities, which have concave mirrors with very small transmissions and very low losses, so the diffraction becomes the most significant effect. In fact, they are used with coherent laser radiation with which only the fundamental mode of the cavity is usually injected; moreover, the concavity of the mirrors means that the incoming radiation remains confined even if the mirrors have slight form imperfections. For confocal cavities, the transverse modes are degenerate with the longitudinal ones; however, they are generally used with the radiation that runs through a ring-shaped or V path, whose mode is similar to the fundamental one. The finesses that are obtained in a supercavity coincide almost completely with those of reflection and can exceed one million.

3.4 Dielectric film with multiple layers

3.4.1 Reflection and transmission

As we have learned with the Fresnel formulas, a polished separation surface between different dielectric materials has the property of reflecting the radiation without producing substantial absorption losses, as instead happens for metallic surfaces. The magnitude of this reflection increases with the diversity of the refractive indices at the two sides of the surface. This reflection may be harmful when one wants to transmit the maximum radiant energy or, vice versa, insufficient when one wants to use the surface as a mirror. However, if one or more thin films of different dielectric materials is interposed between the two media, there are more surfaces parallel to each other: in this way, appropriately choosing the material of the films and their thickness, the interference between the various reflections can produce both a cancellation of the overall reflection or its accentuation. This has created a modern technology that finds applications in many fields, from anti-reflective lenses and other optical components to the production of mirrors, interference filters, and more.

There are various methods for dealing with this problem: here, I describe one of them that uses the boundary conditions on each interface sequentially. The calculation that we will undertake refers to the case of *plane and parallel* interfaces between dielectric media, and the field considered is constituted by monochromatic plane waves, similar to what we did in Chapter 1 for the determination of Fresnel's formulas.

More generally, a layered medium can also have a continuous trend of the refractive index and of the electric permittivity and magnetic permeability [Born and

Wolf 1980]. Historically, F. Abelès conducted the first studies of the propagation of electromagnetic waves in these media [Abelès 1950], and all subsequent analysis can be traced back to them.

For an assigned incident plane wave on a discontinuous layered medium, the angles of incidence ϑ_j within each of the layers can be readily determined by the relation:

$$n_j \sin\vartheta_j = n_0 \sin\vartheta_0 , \tag{3.4.1}$$

where n_j is the refractive index of the j-th layer and n_0, ϑ_0 are the initial refractive index and incidence angle. From these angles, we can then determine the reflection and transmission coefficients of the individual interfaces via the Fresnel's formulas (1.7.10) and (1.7.11).

With reference to Fig. 3.38, for each of the TE and TM waves, let us call E^+_{j-1}, E'^+_j and E^-_{j-1}, E'^-_j the amplitude of the fields across the j-th interface (just before, without the prime, and just after, with the prime), where, with indexes "+" and "–", we denote, respectively, the wave that propagates forward and the wave that propagates backward. In addition, we assume that these waves are the result of a single wave incident on the multilayer. Therefore, the transmitted (reflected) wave generated by E^+_{j-1} will sum coherently *in the same direction* of propagation with the reflected (transmitted) wave generated by E'^-_j:

$$\begin{aligned} E'^+_j &= r_{j,-} E'^-_j + t_{j,+} E^+_{j-1} , \\ E^-_{j-1} &= t_{j,-} E'^-_j + r_{j,+} E^+_{j-1} , \end{aligned} \tag{3.4.2}$$

where $t_{j,+}$, $t_{j,-}$, $r_{j,+}$, $r_{j,-}$ are the transmission (t) and reflection (r) coefficients for the forward and backward waves on the j-th interface given by the Fresnel's Eqs. (1.7.10-12): they *are different* for TE and TM waves, but the ⊥ and || indexes are omitted to avoid encumbering the notation.

Resolving for E^+_{j-1} and E^-_{j-1}, we have

$$\begin{aligned} E^+_{j-1} &= \frac{1}{t_{j,+}} E'^+_j - \frac{r_{j,-}}{t_{j,+}} E'^-_j , \\ E^-_{j-1} &= \frac{1}{t_{j,+}} \left(t_{j,+} t_{j,-} - r_{j,+} r_{j,-} \right) E'^-_j + \frac{r_{j,+}}{t_{j,+}} E'^+_j , \end{aligned} \tag{3.4.3}$$

On the other hand, from the Fresnel's equations, we can immediately see that for both TE and TM waves, we have

$$r_{j,+} = -r_{j,-} , \tag{3.4.4}$$

and also that

Fig. 3.38
Fields on the j-th interface between the (j-1)-th medium and the j-th medium

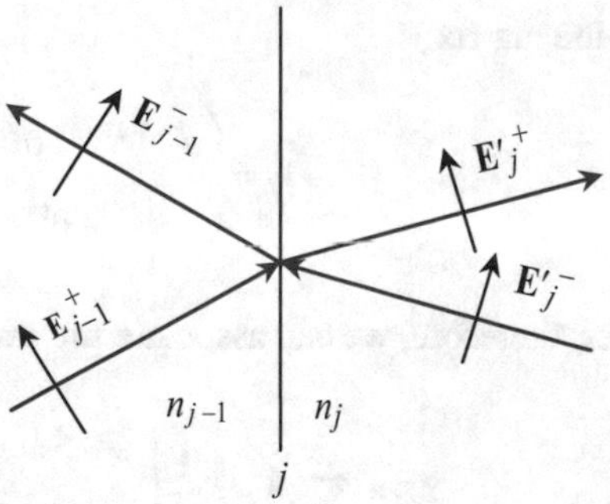

$$t_{j,+}t_{j,-} - r_{j,+}r_{j,-} = 1. \tag{3.4.5}$$

Therefore, the relationship between the fields at the j-th interface can be put into the vectorial form

$$\begin{pmatrix} E_{j-1}^+ \\ E_{j-1}^- \end{pmatrix} = \mathfrak{T}_j \begin{pmatrix} E'^+_j \\ E'^-_j \end{pmatrix}, \tag{3.4.6}$$

where $\mathfrak{T}$ is the 2x2 matrix:

$$\mathfrak{T}_j = \frac{1}{t_j}\begin{pmatrix} 1 & r_j \\ r_j & 1 \end{pmatrix}, \tag{3.4.7}$$

where, for brevity, it is further understood that t_j and r_j refer here to the forward wave (+).

Let us now consider the phase difference that occurs in going through the j-th layer between the interfaces j and $j+1$. As we learned with Eq. (3.3.24), this phase shift is given by

$$\phi_j = \frac{\delta_j}{2} = \frac{2\pi n_j h_j}{\lambda_o} \cos\vartheta_j, \tag{3.4.8}$$

where n_j, h_i, ϑ_j are, respectively, the refractive index, the thickness and the angle of incidence for the j-th layer. The fields at the beginning and end of a layer are then linked to each other by the relationship

$$\begin{pmatrix} E'^+_j \\ E'^-_j \end{pmatrix} = \mathfrak{H}_j \begin{pmatrix} E_j^+ \\ E_j^- \end{pmatrix}, \tag{3.4.9}$$

where $\mathfrak{H}_j$ is the matrix

$$\mathfrak{H}_j = \begin{pmatrix} e^{-i\phi_j} & 0 \\ 0 & e^{i\phi_j} \end{pmatrix}. \tag{3.4.10}$$

To each layer, therefore, we can associate the *transfer matrix*:

$$\mathfrak{L}_j = \mathfrak{T}_j \mathfrak{H}_j = \frac{1}{t_j}\begin{pmatrix} e^{-i\phi_j} & r_j e^{i\phi_j} \\ r_j e^{-i\phi_j} & e^{i\phi_j} \end{pmatrix}. \tag{3.4.11}$$

In conclusion, in the presence of m layers, taking, for the final medium $\mathfrak{L}_{m+1}$ with $\phi_{m+1} = 0$, and by introducing the index $f = m + 1$ for brevity, the overall transfer matrix is given by

$$\mathfrak{L} = \begin{pmatrix} a & b \\ c & d \end{pmatrix} = \mathfrak{L}_1 \mathfrak{L}_2 \cdots \mathfrak{L}_f \tag{3.4.12}$$

and the initial and final fields are related by

$$\begin{pmatrix} E_0^+ \\ E_0^- \end{pmatrix} = \mathfrak{L} \begin{pmatrix} E_f^+ \\ E_f^- \end{pmatrix}. \tag{3.4.13}$$

On the other hand, with an input wave in the sole forward direction, we have $E_f^- = 0$ and, in the opposite case of a backward wave, it is $E_0^+ = 0$. So, the amplitude coefficients for the overall transmission and reflection of the various interfaces are given, respectively, by

$$t_+ = \frac{E_f^+}{E_0^+} = \frac{1}{a}, \quad r_+ = \frac{E_0^-}{E_0^+} = \frac{c}{a}; \quad t_- = \frac{E_0^-}{E_f^-} = \frac{ad - bc}{a}, \quad r_- = \frac{E_f^+}{E_f^-} = -\frac{b}{a}, \tag{3.4.14}$$

where a and c are the elements of the first column of $\mathfrak{L}$; it should be noted that the coefficients are not equal for the forward and backward waves. For the intensity, we have finally that the transmissivity and reflectivity are respectively given by

$$\begin{aligned} T_+ &= \frac{n_f}{n_0}\frac{\mu_0}{\mu_f}\frac{\cos\vartheta_f}{\cos\vartheta_0}\left|\frac{1}{a}\right|^2, \quad R_+ = \left|\frac{c}{a}\right|^2; \\ T_- &= \frac{n_0}{n_f}\frac{\mu_f}{\mu_0}\frac{\cos\vartheta_0}{\cos\vartheta_f}\left|\frac{ad - bc}{a}\right|^2, \quad R_- = \left|\frac{b}{a}\right|^2, \end{aligned} \tag{3.4.15}$$

where n_0, μ_0, ϑ_0 and n_f, μ_f, ϑ_f are, respectively, the refractive index, the magnetic permeability, and the angle of incidence in the initial and the final medium.

An alternative way to deal with this problem is to directly resort to boundary conditions for the fields on the interfaces, especially those relating to the tangential components of fields **E** and **H**. As we have already done with Eq. (1.7.9), here indicating these components with a bar, for each of the TE and TM waves across the j-th interface, it is simply:

$$\underline{E}_j = \underline{E}'_j, \quad \underline{H}_j = \underline{H}'_j, \tag{3.4.16}$$

where, on both sides of the interface, these new variables are now the sum of the tangential components for the waves in the two directions of travel:

$$\underline{E}_j = \underline{E}^+_{j-1} + \underline{E}^-_{j-1} = \underline{E}'^+_j + \underline{E}'^-_j, \quad \underline{H}_j = \underline{H}^+_{j-1} + \underline{H}^-_{j-1} = \underline{H}'^+_j + \underline{H}'^-_j, \tag{3.4.17}$$

where we have to note that, generally, $\underline{E}^+_{j-1} \neq \underline{E}'^+_j$, etc. On the other hand, for Eqs. (1.7.8), in the layer j, the amplitudes of the magnetic field of the wave + and – are related to the amplitudes of the electric field of the same wave, hence

$$\underline{H}_j = \underline{H}'^+_j + \underline{H}'^-_j = \eta_j \left(\underline{E}'^+_j - \underline{E}'^-_j \right), \tag{3.4.18}$$

where η_j is the *effective admittance* of the j-th layer defined by

$$\eta_j = \begin{cases} -\dfrac{n_j}{\mu_j c} \cos \vartheta_j & \text{for the TE wave,} \\ +\dfrac{n_j}{\mu_j c} \dfrac{1}{\cos \vartheta_j} & \text{for the TM wave.} \end{cases} \tag{3.4.19}$$

The difference between the TE and TM waves derives simply from the fact that the electric and magnetic fields are perpendicular to each other and that the vectors in the order ***E***, ***H***, ***k*** form a right-handed set for both TE and TM waves (with the conventions chosen in this text), while the minus sign in front of $\underline{E}'^-_j$ in Eq. (3.4.18) derives from the change of travel direction.

The variables $\underline{E}_j$ and $\underline{H}_j$ are therefore a linear combination of $\underline{E}'^+_j$ and $\underline{E}'^-_j$ (and vice versa) and, in vector form, we can write

$$\begin{pmatrix} \underline{E}_j \\ \underline{H}_j \end{pmatrix} = \mathfrak{M}_j \begin{pmatrix} \underline{E}'^+_j \\ \underline{E}'^-_j \end{pmatrix}, \quad \begin{pmatrix} \underline{E}'^+_j \\ \underline{E}'^-_j \end{pmatrix} = \mathfrak{M}_j^{-1} \begin{pmatrix} \underline{E}_j \\ \underline{H}_j \end{pmatrix}, \tag{3.4.20}$$

where

$$\mathfrak{N}_j = \begin{pmatrix} 1 & 1 \\ \eta_j & -\eta_j \end{pmatrix}, \quad \mathfrak{N}_j^{-1} = \frac{1}{2}\begin{pmatrix} 1 & 1/\eta_j \\ 1 & -1/\eta_j \end{pmatrix}. \tag{3.4.21}$$

In turn, the tangential components $\underline{E}_j^+$, $\underline{E}_j^-$ are related to the amplitudes E_j^+, E_j^- by

$$\begin{pmatrix} \underline{E}_j^+ \\ \underline{E}_j^- \end{pmatrix} = c_j \begin{pmatrix} E_j^+ \\ E_j^- \end{pmatrix}, \tag{3.4.22}$$

where

$$c_j = \begin{cases} 1 & \text{for the TE wave,} \\ \cos\vartheta_j & \text{for the TM wave.} \end{cases} \tag{3.4.23}$$

Also, for the pair of variables $\underline{E}$, $\underline{H}$, we can write a recursive relationship between one interface and the next at the extremes of a single layer. Thanks to Eqs. (3.4.9) and (3.4.10), we have

$$\begin{pmatrix} \underline{E}_j \\ \underline{H}_j \end{pmatrix} = \frac{1}{2}\begin{pmatrix} 1 & 1 \\ \eta_j & -\eta_j \end{pmatrix}\begin{pmatrix} e^{-i\phi_j} & 0 \\ 0 & e^{i\phi_j} \end{pmatrix}\begin{pmatrix} 1 & 1/\eta_j \\ 1 & -1/\eta_j \end{pmatrix}\begin{pmatrix} \underline{E}_{j+1} \\ \underline{H}_{j+1} \end{pmatrix}, \tag{3.4.24}$$

and therefore

$$\begin{pmatrix} \underline{E}_j \\ \underline{H}_j \end{pmatrix} = \mathfrak{M}_j \begin{pmatrix} \underline{E}_{j+1} \\ \underline{H}_{j+1} \end{pmatrix}, \tag{3.4.25}$$

where $\mathfrak{M}$ is the matrix

$$\mathfrak{M}_j = \begin{pmatrix} \cos\phi_j & -i\dfrac{1}{\eta_j}\sin\phi_j \\ -i\eta_j\sin\phi_j & \cos\phi_j \end{pmatrix}, \tag{3.4.26a}$$

which is called the *characteristic matrix* of the j-th layer, or even the *interference matrix*.

If the layer j is an ideal dielectric, without absorption, the $\mathfrak{M}_j$ matrix has a unitary determinant and also has real diagonal elements, while the non-diagonal elements are imaginary. This also applies when the index of the initial medium is greater than that of the layer j and the initial angle of incidence is sufficiently large so that, for Eq. (3.4.1), $\sin\vartheta_j$ exceeds the unit. The wave that permeates the layer j

is then an evanescent wave, with pure imaginary $\cos\vartheta_j$. In this case, for Eqs. (3.4.8) and (3.4.19), $\phi_j = i\phi''_j$ and $\eta_j = i\eta''_j$ are imaginary and Eq. (3.4.26a) becomes

$$\mathfrak{M}_j = \begin{pmatrix} \cosh\phi''_j & -i\dfrac{1}{\eta''_j}\sinh\phi''_j \\ i\eta''_j \sinh\phi''_j & \cosh\phi''_j \end{pmatrix}. \tag{3.4.26b}$$

Similar to Eq. (3.4.13), the interference matrix for a set of m layers is given by the product of the matrices of the individual layers:

$$\mathfrak{M} = \begin{pmatrix} A & B \\ C & D \end{pmatrix} = \mathfrak{M}_1\mathfrak{M}_2\cdots\mathfrak{M}_m. \tag{3.4.27}$$

If, therefore, the multilayer is made from ideal dielectrics, without losses, even the overall matrix $\mathfrak{M}$ is unitary, and its elements A and D are real, while B and C are imaginary.

The matrix of interference, even if it is of less immediate understanding, has the advantage of being expressed directly as a function of the refractive indices and the thickness of the various layers and allows for a faster computation of the overall behavior of the set of layers. For the amplitudes E^+ and E^- of the fields at the ends of the stack of layers, we ultimately have

$$\begin{pmatrix} E_0^+ \\ E_0^- \end{pmatrix} = c_0{}^{-1}\mathfrak{N}_0{}^{-1}\mathfrak{M}\mathfrak{N}_f c_f \begin{pmatrix} E_f^+ \\ E_f^- \end{pmatrix}. \tag{3.4.28}$$

It is therefore apparent that

$$\mathfrak{L} = c_0{}^{-1}\mathfrak{N}_0{}^{-1}\mathfrak{M}\mathfrak{N}_f c_f, \tag{3.4.29}$$

where the elements a, b, c, d of $\mathfrak{L}$, expressed in terms of the elements A, B, C, D of $\mathfrak{M}$, are given by

$$a = \frac{1}{2\eta_0}\frac{c_f}{c_0}\left(A\eta_0 + B\eta_0\eta_f + C + D\eta_f\right),\quad b = \frac{1}{2\eta_0}\frac{c_f}{c_0}\left(A\eta_0 - B\eta_0\eta_f + C - D\eta_f\right),$$

$$c = \frac{1}{2\eta_0}\frac{c_f}{c_0}\left(A\eta_0 + B\eta_0\eta_f - C - D\eta_f\right),\quad d = \frac{1}{2\eta_0}\frac{c_f}{c_0}\left(A\eta_0 - B\eta_0\eta_f - C + D\eta_f\right), \tag{3.4.30}$$

and the reflection and transmission coefficients can still be calculated with Eqs. (3.4.14) and (3.4.15).

3.4.2 Propagation in periodic structures

Of particular importance are the periodic structures, whose analytical study dates back to the historical works of F. Abelès mentioned at the beginning [Abelès 1950] and subsequently to those of A. Yariv and P. Yeh [Yeh et al 1977; Yariv and Yeh 1977]. They have properties of great interest in optics and are used, in addition to the making of dielectric mirrors, in various other applications, such as to obtain *quasi-phase-matching* in nonlinear optics, Bragg effect reflection in a *distributed feedback* laser (DFB) and a *distributed Bragg reflection* laser (DBR). They are now the subject of study in *photonics*, which uses microscopic, two-dimensional, or three-dimensional periodic structures to manipulate radiation and guide it or trap it like electronic circuits. For a general theory of propagation in such structures, see Yariv and Yeh (1984).

The simplest example of such structures is that of a periodic sequence of alternate layers of two thin films with different indexes. Consider here a medium constituted by N cells, each corresponding to a period of modulation, whose characteristic matrix is

$$\mathfrak{M} = \begin{pmatrix} A & B \\ C & D \end{pmatrix} = \begin{pmatrix} \cos\phi_2 & -i\dfrac{1}{\eta_2}\sin\phi_2 \\ -i\eta_2\sin\phi_2 & \cos\phi_2 \end{pmatrix} \begin{pmatrix} \cos\phi_1 & -i\dfrac{1}{\eta_1}\sin\phi_1 \\ -i\eta_1\sin\phi_1 & \cos\phi_1 \end{pmatrix} =$$

$$\begin{pmatrix} \cos\phi_2\cos\phi_1 - \dfrac{\eta_1}{\eta_2}\sin\phi_2\sin\phi_1 & -i\dfrac{1}{\eta_1}\sin\phi_1\cos\phi_2 - i\dfrac{1}{\eta_2}\sin\phi_2\cos\phi_1 \\ -i\eta_2\sin\phi_2\cos\phi_1 - i\eta_1\sin\phi_1\cos\phi_2 & \cos\phi_2\cos\phi_1 - \dfrac{\eta_2}{\eta_1}\sin\phi_2\sin\phi_1 \end{pmatrix}, \tag{3.4.31}$$

where the development of the coefficients A, B, C, D applies in the case of a period consisting of two layers where the indices 1 and 2 refer to the two materials and thicknesses that make up them.

If the two media are free of losses, their characteristics matrices have unitary determinant; therefore, the matrix (3.4.31) has also unit determinant. This allows us to use an important result, obtained by Abelès, which states that the overall matrix of the N cells is given by

$$\mathfrak{M}^N = \begin{pmatrix} A & B \\ C & D \end{pmatrix}^N = \begin{pmatrix} sA - \dfrac{U_{N-2}(x)}{U_{N-1}(x)} & sB \\ sC & sD - \dfrac{U_{N-2}(x)}{U_{N-1}(x)} \end{pmatrix} s^{-N}U_{N-1}(x), \tag{3.4.32.a}$$

where s is generally a complex number of unit modulus and takes the values

$$\begin{cases} s=1 & \text{for } A+D=0, \\ s=\dfrac{|A+D|}{A+D} & \text{otherwise.} \end{cases} \qquad (3.4.32.b)$$

This parameter extends the validity of the equation, even in cases in which $A + D$ is complex. Moreover, U_N are the functions

$$\begin{aligned} U_N(x) &= \frac{\sin[(N+1)\alpha]}{\sin\alpha}, \quad \text{where} \quad \cos\alpha = x, \quad \text{for } x<1; \\ U_N(x) &= \frac{\sinh[(N+1)\alpha]}{\sinh\alpha}, \quad \text{where} \quad \cosh\alpha = x, \quad \text{for } x>1; \\ U_N(1) &= N+1, \end{aligned} \qquad (3.4.32.c)$$

where $x = |A+D|/2$. The U_N are the *Chebyshev's polynomials of second species*, defined as orthogonal with weight function $w(x) = (1-x^2)^{1/2}$ in the interval (−1,1).

The only condition for the validity of Eq. (3.4.32.a) is that the matrix $\mathfrak{M}$ is *unimodal*, namely, that its determinant is equal to one, as stated earlier.[4] The demonstration of this important identity can be derived from the equation

$$\{\mathfrak{R}\mathfrak{M}\mathfrak{R}^{-1}\}^N = \mathfrak{R}\mathfrak{M}^N\mathfrak{R}^{-1},$$

where $\mathfrak{R}$ is a unitary transformation. Let us take the case that A+D is real and positive or null. All other cases can be traced back to this by dividing the matrix coefficients with the parameter s defined above.

For x<1, from the equation for eigenvalues and eigenvectors for the matrix $\mathfrak{M}$ (3.4.31), the eigenvalues turn out to be

$$e^{\pm i\alpha} = (A+D)/2 \pm \sqrt{[(A+D)/2]^2 - 1}.$$

One is the reciprocal of the other, because $\mathfrak{M}$ is unitary. The corresponding normalized eigenvectors are

$$\begin{pmatrix} a_\pm \\ b_\pm \end{pmatrix} = \frac{1}{d_\pm}\begin{pmatrix} B \\ e^{\pm i\alpha} - A \end{pmatrix}, \quad \text{with } d_\pm = \sqrt{B^2 + \left(e^{\pm i\alpha} - A\right)^2}.$$

---

[4] Eq. (3.4.32) can also be applied to the transfer matrix corresponding to a modulation period of media whose refractive index varies continuously. Indeed, this matrix is also unitary because, within a period, the material properties return to the initial values.

Choosing, for $\mathfrak{R}$, the transformation that diagonalizes the matrix $\mathfrak{M}$, we have

$$\mathfrak{R}\mathfrak{M}\mathfrak{R}^{-1} = \begin{pmatrix} e^{i\alpha} & 0 \\ 0 & e^{-i\alpha} \end{pmatrix} .$$

In particular, it turns out that [Yeh et al 1977; Yariv and Yeh 1977]

$$\mathfrak{R}^{-1} = \frac{1}{p}\begin{pmatrix} a_+ & a_- \\ b_+ & b_- \end{pmatrix}, \quad \mathfrak{R} = \frac{1}{p}\begin{pmatrix} b_- & -a_- \\ -b_+ & a_+ \end{pmatrix}, \quad \text{con } p = \sqrt{a_+ b_- - a_- b_+} .$$

Therefore,

$$\mathfrak{M}^N = \mathfrak{R}^{-1}\begin{pmatrix} e^{iN\alpha} & 0 \\ 0 & e^{-iN\alpha} \end{pmatrix}\mathfrak{R} .$$

For $x>1$, just replace $i\alpha$ with α in the two previous equations. In both cases, we ultimately get Eq. (3.4.32).

3.4.3 Reciprocity and time reversal: the Stokes relations

We have seen that, for a single interface between isotropic and transparent dielectric media, according to Fresnel's equations, we obtain the relations (3.4.4) and (3.4.5). In particular, from the first of these equations, the reflection coefficient is opposed between the waves that impact on one side or the other on the surface. Well, such expressions generally exist for all mirrors made of non-absorbing linear materials where the power of the incident wave equals the sum of that of the reflected and refracted waves. This is the consequence of the symmetry by time inversion of Maxwell equations in the absence of absorption mechanisms such as those due to non-null values of the current density. Consider, then, a monochromatic wave A, which propagates in a linear isotropic and transparent medium, whose electric field is expressed by the relation

$$\mathbf{E}(\boldsymbol{r},t) = \frac{1}{2}\boldsymbol{E}_o(\boldsymbol{r})e^{i\boldsymbol{k}\cdot\boldsymbol{r}-i\omega t} + \frac{1}{2}\boldsymbol{E}_o^*(\boldsymbol{r})e^{-i\boldsymbol{k}\cdot\boldsymbol{r}+i\omega t} = \Re e\left[\boldsymbol{E}_o(\boldsymbol{r})e^{i\boldsymbol{k}\cdot\boldsymbol{r}-i\omega t}\right] ,$$

where $\boldsymbol{E}_o$ represents the amplitude of the field, which is slowly variable in the position $\boldsymbol{r}$. Therefore, if we effect the replacement $t \to -t$, Maxwell's equations also accept, as a solution, a wave B whose field is

$$\mathbf{E}_-(\boldsymbol{r},t) = \frac{1}{2}\boldsymbol{E}_o(\boldsymbol{r})e^{-i\boldsymbol{k}_-\cdot\boldsymbol{r}+i\omega t} + \frac{1}{2}\boldsymbol{E}_o^*(\boldsymbol{r})e^{i\boldsymbol{k}_-\cdot\boldsymbol{r}-i\omega t} = \Re e\left[\boldsymbol{E}_o^*(\boldsymbol{r})e^{i\boldsymbol{k}_-\cdot\boldsymbol{r}-i\omega t}\right] ,$$

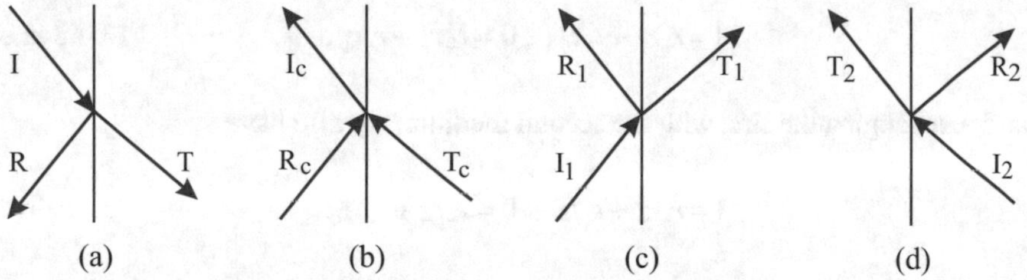

Fig. 3.39 Study of the relations between reflection and refraction coefficients

where its wave vector $\boldsymbol{k}_- = -\boldsymbol{k}$ also replaces that of wave A with the opposite sign. Wave B retraces the appearance of the first exactly backward in the space, and is called the *conjugated wave*; it can be obtained by reflecting wave A with a *phase conjugation mirror*.

As a next step, we consider the set of three waves, namely, incident I, reflected R and transmitted T, which is obtained with a semitransparent mirror, as illustrated in Fig. 3.39(a). We assume this mirror to consist of a thin plane and parallel stratification of dielectric media, and to be free of losses. By inverting the time and direction of the three waves, we get the diagram from Fig. 3.39(b), in which the corresponding conjugated waves R_c and T_c, the first to the reflected wave, and the second to the transmitted wave, interfere so as to generate the conjugated wave I_c and, instead, cancel the corresponding wave in the second medium. For the components $\|$ or $\perp$ of the amplitudes of the fields of conjugated waves R_c and T_c, we have

$$E_{\mathrm{Rc}} = r_+^* E_o^*, \quad E_{\mathrm{Tc}} = t_+^* E_o^*,$$

where E_o is the amplitude of the wave I and r_+, t_+ are, respectively, the coefficients of reflection and transmission in correspondence with the case of TE or TM considered.

In the *absence of losses*, for the conservation of the energy transported by these waves, the wave I_c is the conjugate of wave I, so that

$$E_{\mathrm{Ic}} = E_o^* .$$

The diagram in Fig. 3.39(b) can be understood as the sum of the fields of two sets of three waves I_1, R_1, T_1 and I_2, R_2, T_2 represented in Fig. 3.39(c) and (d). So, for the four directions, we have the relationships

$$\begin{aligned} E_o^* &= r_+ E_{\mathrm{I1}} + t_- E_{\mathrm{I2}}, \qquad & 0 &= t_+ E_{\mathrm{I1}} + r_- E_{\mathrm{I2}}, \\ r_+^* E_o^* &= E_{\mathrm{I1}}, & t_+^* E_o^* &= E_{\mathrm{I2}}, \end{aligned} \tag{3.4.33}$$

from which we finally obtain the Stokes' relations

$$1 = r_+ r_\mp^* + t_- t_\mp^*, \quad 0 = t_+ r_\mp^* + r_- t_\mp^*, \tag{3.4.34a,b}$$

and, exchanging the first with the second medium, we also have

$$1 = r_- r_\pm^* + t_+ t_\pm^*, \quad 0 = t_- r_\pm^* + r_+ t_\pm^*. \tag{3.4.34b,c}$$

From the first and third equations, it turns out that the product $t_- t_\mp^*$ is real, and, from the other two equations, we have that

$$r_- = -r_\mp^* \frac{t_+}{t_\mp^*} = -r_\mp^* \frac{t_-}{t_\pm^*}. \tag{3.4.35}$$

On the other hand, if the media on both sides of the mirror are the *same*, in the absence of losses, we also have $|E_\mathrm{I}|^2 = |E_\mathrm{R}|^2 + |E_\mathrm{T}|^2$, and therefore

$$1 = r_+ r_\mp^* + t_+ t_\mp^*, \quad 1 = r_- r_\pm^* + t_- t_\pm^*, \tag{3.4.36}$$

which, combined with the previous equations, imposes that

$$t_+ = t_-. \tag{3.4.37}$$

From Eqs. (3.4.14-15) and (3.4.30), we can verify the correspondence with the Stokes' relations. In particular, if, in the first and last medium, the wave is ordinary, with η_0, η_f, c_0, c_f real, we have

$$r_+ r_\mp^* + t_- t_\mp^* = \frac{cc* + ad - bc}{aa*} = 1, \quad t_+ r_\mp^* + r_- t_\mp^* = \frac{c* - b}{aa*} = 0,$$

thanks to the fact that, under the above conditions and in the absence of losses, from Eqs. (3.4.30), it results that $b = c^*$ and $d = a^*$. Furthermore,

$$\frac{t_+ t_-}{r_+ r_-} = -\frac{ad - bc}{bc} = \frac{-4\eta_0 \eta_f}{A^2 \eta_0{}^2 + D^2 \eta_f{}^2 - B^2 \eta_0{}^2 \eta_f{}^2 - C^2 - 2\eta_0 \eta_f}, \tag{3.4.38}$$

where it was exploited that, in the absence of losses, the interference matrix is unitary. Finally, this ratio is real and negative, just as bc, ad and $\eta_0 \eta_f$ are real and positive. Recalling the discussion of reflection from a Fabry-Perot cavity with dielectric plane mirrors, the phase shift β of §3.3.2.4 is therefore null. For this not to be the case, it is therefore necessary that there be losses equivalent to an absorption in the media that constitute the entrance mirror of the cavity.

3.4.4 Applications

Do not let the "simplicity" of previous expressions deceive you! It is one thing to numerically calculate the various reflection and transmission coefficients once the thicknesses and refractive indexes of the various layers are known (a simple small iterative program on the computer suffices). Well, it is another thing to design a dielectric multilayer that has the desired characteristics: somewhat like in cryptography or in a book of crime fiction, this is a problem whose difficulty is largely asymmetrical in the two directions, and the analytical setting of a device of even a few layers leads to expressions of very discouraging complexity. The way to do this is both an art and also an industrial secret. There are complex analytical and/or numerical "synthesis" methods [Knittl 1976, Heavens 1955, Delano and Pegis 1969], but we can only look at some well-known solutions here.

The materials used in the manufacture of thin films are briefly mentioned in Table 3.1. They are deposited on the substrate through evaporation under vacuum by placing them in a small pan heated electrically or, more modernly, by bombardment with an electrons beam of a few keV from an "electron gun". In this way, only the material to be deposited is heated, without contamination of other substances, and refractory materials such as aluminum oxide, which would otherwise be impossible to deposit, can also be evaporated. The substrate is also heated to a few hundred degrees Celsius for greater compactness and adherence between

Material		Refraction index (a ~0.55 μm)	Spectral range, μm
cryolite	Na_3AlF_6	1.35	0.15 - 14
magnesium fluoride	MgF_2	1.38	0.11 - 4
silicon dioxide	SiO_2	1.45	0.2 - 9
thorium fluoride	ThF_4	1.52	0.2 - 15
sodium chloride	NaCl	1.54	0.18 - >15
aluminum oxide	Al_2O_3	1.63	0.15 - 6
cerium fluoride	CeF_3	1.63	0.3 - >5
magnesium oxide	MgO	1.7	0.2 - 8
silicon monoxide	SiO	1.9	0.7 - 9
zirconium dioxide	ZrO_2	2.1	0.34 - 12
cerium dioxide	CeO_2	2.2	0.4 - 5
zinc sulphide	ZnS	2.3	0.4 - 14
titanium dioxide	TiO_2	2.3 - 2.9	0.4 - 3
zinc selenide	ZnSe	2.5	0.55 - 15
silicon	Si	3.4 (a 3 μm)	0.9 - 8
germanium	Ge	4.4 (a 2 μm)	1.4 - >20
lead telluride	PbTe	5.6 (a 5 μm)	3.9 - >20

Table 3.1 Typical materials used for thin films

the films and the substrate itself.

Thin films are always either a bit more porous or less compact than the corresponding bulk material for which the film's refractive indexes are typically lower than those of their bulk and tend to grow with the substrate temperature during deposition. For some, the refractive index can also be varied by allowing small amounts of gas into the chamber, like the silicon oxide that, depending on the oxygen pressure, can produce film's index variable between 1.45 and 1.9. Not all films are compatible with each other, due to the mechanical stress that results from their different thermal expansion coefficients. For some, such as magnesium fluoride and cerium dioxide, which are resistant and durable materials, the infrared spectral range is limited by the fact that, for relatively high thicknesses of the film, they split due to the strong stress. Instead, sodium chloride has a very small stress and is used in interferential filters up to 20 μm. As normally happens, the refractive indices of the thin films also exhibit a dependence on the wavelength, and thus an accurate design must consider this. In this respect, it should be noted that the curves shown in the following figures refer to ideal materials, with constant indices, and therefore may differ considerably from what is practically obtained. We will now look at some applications of this technology (for non-magnetic media, as is the norm).

3.4.4.1 Antireflection coatings

Suppose we have a single film with index n deposited on a substrate with index n_S; the coefficient of reflection can be derived from Eqs. (3.4.14.b), (3.4.26) and (3.4.30). For normal incidence, and with $n_0 = 1$, we have:

$$r = \frac{n(1-n_s)\cos\phi - i(n_s - n^2)\sin\phi}{n(1+n_s)\cos\phi - i(n_s + n^2)\sin\phi}, \qquad (3.4.39)$$

where ϕ is the phase shift of the film. If, in particular, the *optical* thickness of the film is $\lambda_o/4$ (where λ_o is the wavelength in vacuum), then $\phi = \pi/2$. The reflectivity for a $\lambda_o/4$ film is then

$$R = |r|^2 = \frac{(n_s - n^2)^2}{(n_s + n^2)^2} \qquad (3.4.40)$$

and, in particular, the reflectivity is null if

$$n = \sqrt{n_s}\ . \qquad (3.4.41)$$

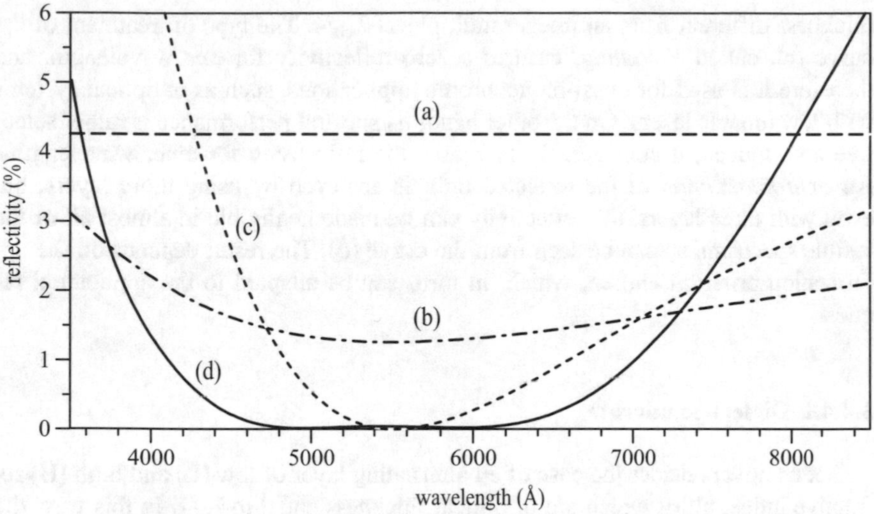

Fig. 3.40 Spectrum of reflection of antireflection coatings on a glass with normal incidence, with initial and final indices of 1/.../1.52. (a): uncoated glass; (b): reflectivity with a single layer with index 1.38 (MgF) and optical thickness $\lambda_{co}/4$, with $\lambda_{co} = 5500$ Å; (c): reflectivity with a double layer *V-coating*, with the sequence of indices 1/1.38, 2.2/1.52 and with optical thicknesses of, respectively, /1.29, 0.23/ $\times\lambda_{co}/4$; (d): reflectivity with a three-layers coating with indices 1/1.38, 2.1, 1.7/1.52 and optical thicknesses of, respectively, /0.93, 1.73, 0.77/ $\times\lambda_{co}/4$

This reduction of reflectivity with a $\lambda_o/4$ optical thickness layer and an intermediate refraction index between the initial and final media has a simple interpretation: a) the incident wave passes from a less (more) dense to a more (less) dense medium for both interfaces, and hence the corresponding coefficients of reflection have the same sign; b) on the other hand, the wave reflected by the second interface runs in a longer optical path (a round trip) equal to $\lambda_o/2$, for which it overall interferes *destructively* with the reflected wave from the first interface.

The most common substrates are glass with a refractive index around 1.52 (the Schott BK7), so if there was a 1.23 index material, a single dielectric layer could have (at a given wavelength) a complete cancellation of the reflection between air and glass. However, among the low-index materials, the only useful one is magnesium fluoride (with $n = 1.38$), which is rather hard and resistant to water and other solvents, while cryolite (with $n = 1.35$) is too soft. Although it does not exactly meet the condition (3.4.41), the reflectivity of such a glass treated with a magnesium fluoride layer with *physical* thickness $\lambda_{co}/(4 \cdot 1.38)$ is reduced to little more than 1% on a large spectral band around λ_{co}, i.e., just over a quarter of that of untreated glass, as it can be seen from Fig. 3.40. Another advantage of single-layer treatments is that it does not give rise to a fast growing reflection for wavelength values outside of the antireflection band around λ_{co}.

Best results are obtained with two or more dielectric layers, and the curves (c) and (d) of Fig. 3.40 report the case of "non-tuned" layers, that is, with optical thickness different from an integer multiple of $\lambda_{co}/4$. The type of treatment of the curve (c), called *V-coating*, ensures a zero-reflectivity for *one* wavelength, and therefore it is used for quasi-monochrome applications, such as in optical systems with low tunable lasers. On the other hand, its spectral performance is rather selective and, indeed, it considerably increases the reflectivity for other wavelengths. An *achromatization* of the reflected light is achieved by using more layers, but even with three layers, the reflectivity can be made negligible in almost all of the visible spectrum, as can be seen from the curve (d). The result depends on the optimization criterion chosen, which, in turn, can be adapted to the application request.

3.4.4.2 Dielectric mirrors

Let us now consider the case of an alternating layer of low (L) and high (H) refractive index, all of which are of optical thickness equal to $\lambda_o/4$. In this way, the sign of the coefficients of reflection alternates between + and – between the various interfaces: this time, the interference between all of the waves reflected back is *constructive*, since the path forth and back in each of the layers leads to a further factor –1 for the amplitude of the fields. With this principle, high-reflectivity dielectric mirrors can be constructed with very low losses, much better than metal mirrors. Let us now look at an analytical treatment. For a couple of layers of optical thickness $\lambda_o/4$ (whereby $\phi = \pi/2$) with refractive indexes n_H and n_L, the interference matrix for normal incidence is:

$$\begin{pmatrix} 0 & -i\dfrac{\mu_0 c}{n_H} \\ -i\dfrac{n_H}{\mu_0 c} & 0 \end{pmatrix}\begin{pmatrix} 0 & -i\dfrac{\mu_0 c}{n_L} \\ -i\dfrac{n_L}{\mu_0 c} & 0 \end{pmatrix} = \begin{pmatrix} -\dfrac{n_L}{n_H} & 0 \\ 0 & -\dfrac{n_H}{n_L} \end{pmatrix}. \tag{3.4.42}$$

So, if we have N pairs of HL layers with an initial and a final medium of indexes, respectively, 1 and n_S, for Eqs. (3.4.30) and (3.4.15.c), the reflectivity for a radiation with wavelength λ_{co}, that is, on the peak of the reflection curve, is:

$$R = \left| \frac{\left(-\dfrac{n_L}{n_H}\right)^N - n_S\left(-\dfrac{n_H}{n_L}\right)^N}{\left(-\dfrac{n_L}{n_H}\right)^N + n_S\left(-\dfrac{n_H}{n_L}\right)^N} \right|^2, \tag{3.4.43}$$

and, with the growth of m, reflection tends to 1. For example, for only 8 alternate layers (1/HLHLHLHL/1.52, $N = 4$) of magnesium fluoride and zinc sulphide de-

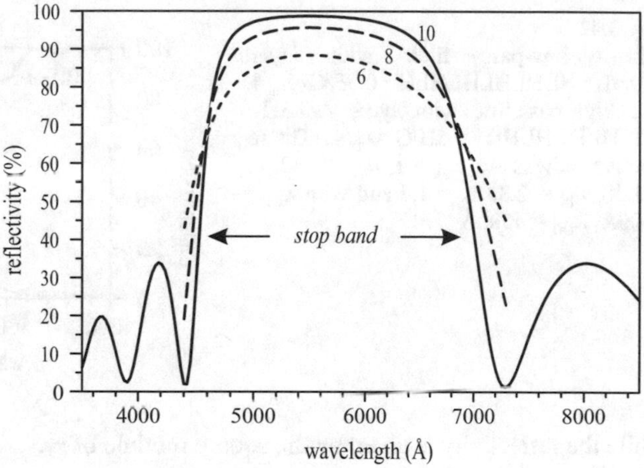

Fig. 3.41 Reflectivity of some dielectric layer mirrors configured as A/HL···HL/G×λ_c/4, with n_H = 2.3, n_L = 1.38, n_A = 1, n_G = 1.52 and λ_{co} = 5500 Å. The numbers inside indicate the total number of layers

posited on a BK7 substrate, the reflectivity is already 0.957. This calculation is done for a given wavelength, but, around this one, there is a whole area of high reflectivity (called a *stop band*), as can be seen from Fig. 3.41, that is obtained numerically. The analytical solution of reflection can be derived from Eqs. (3.4.32). Limiting ourselves to the case of normal incidence, assuming that the phase shift ϕ is the same for the two layers L and H and taking that $\mu_j = \mu_o$, then, neglecting the different index of refraction dependence on the wavelength, for Eq. (3.4.31), the characteristic matrix of a couple of layers is

$$\mathfrak{M} = \begin{pmatrix} A & B \\ C & D \end{pmatrix} = \begin{pmatrix} \cos^2\phi - \frac{n_1}{n_2}\sin^2\phi & -i\sin\phi\cos\phi\left(\frac{1}{\eta_1}+\frac{1}{\eta_2}\right) \\ -i\sin\phi\cos\phi\left(\eta_1+\eta_2\right) & \cos^2\phi - \frac{n_2}{n_1}\sin^2\phi \end{pmatrix}, \quad (3.4.44)$$

where the indices 1 and 2 refer to the first and second medium in succession in the unit cell. The angle α of Eq. (3.4.32.c) is derived from

$$x = \left|\cos^2\phi - \frac{1}{2}\left(\frac{n_1}{n_2}+\frac{n_2}{n_1}\right)\sin^2\phi\right|, \quad \begin{cases} \cos\alpha = x & \text{for } x<1 \\ \cosh\alpha = x & \text{for } x>1 \end{cases} \quad (3.4.45)$$

and the reflection coefficient of the mirror given by Eqs. (3.4.14.b), (3.4.30) and (3.4.32) is

$$r_+ = \frac{c}{a} = \frac{s\left(A\eta_0 - D\eta_f + B\eta_0\eta_f - C\right) - \frac{U_{N-2}(x)}{U_{N-1}(x)}\left(\eta_0 - \eta_f\right)}{s\left(A\eta_0 + D\eta_f + B\eta_0\eta_f + C\right) - \frac{U_{N-2}(x)}{U_{N-1}(x)}\left(\eta_0 + \eta_f\right)}, \quad (3.4.46)$$

Fig. 3.42
(a): low-pass filter with layers A/0.5L · HLHLHLHLHLH · 0.5F/G×$\lambda_a/4$, (b): high-pass filter with layers A/0.5H · LHLHLHLHLHL · 0.5H/G ×$\lambda_b/4$. The refractive indices are $n_A = 1$, $n_G = 1.52$, $n_L = 1.38$, $n_H = 2.3$, $n_F = 1.7$ and with $\lambda_{ao} = 6500$Å, $\lambda_{bo} = 4500$Å

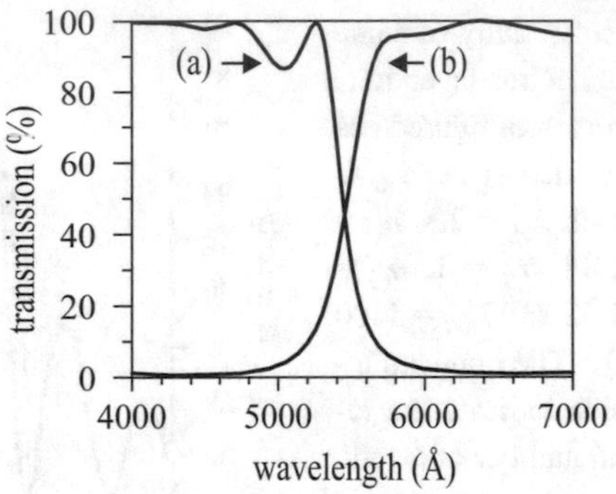

while the reflectivity is given by the square module of r_+.

For $x > 1$, the field decays exponentially within the series of layers and the conditions for which $x = 1$ indicate the size of this "band gap", i.e., the *stop band* that corresponds to the maximum of reflection. Taking that $\phi = (2m+1)/2 + \phi_{sb}$, the half width ϕ_{sb} of the stop band is obtained from

$$\sin\phi_{sb} = \frac{n_H - n_L}{n_H + n_L}. \tag{3.4.47}$$

It is substantially independent of the number of the layers, while it grows with the ratio n_H/n_L. Outside of the stop band, there are oscillations: these can almost be canceled on one of the two sides, while, on the other, they are accentuated, simply by adding a $\lambda/8$ layer to both extremes of the $\lambda/4$ stack. In this way, *low-pass* and *high-pass* filters are realized, such as those in Fig. 3.42.

3.4.4.3 Interference filters

Combining two stacks of HL alternate layers with an intermediate layer of optical thickness $\lambda_o/2$, a resonant cavity of the Fabry-Perot type is built, which presents a thin transmission peak within the stop band. This property is used to construct subtle *interference filters* useful, for example, in selecting single atomic spectral lines from the background radiation. However, at the sides of the stop band, the system returns to transmitting with a trend that is more or less rough, for which a "complete" interference filter is commonly constituted by two substrates: the first one is transparent and the dielectric layers are deposited on it. The second substrate is instead colored to absorb the shortest wavelengths and a second stack of dielectric/metal mixed layers is deposited on it to reflect the radiation of longer wavelengths. The two substrates are then cemented together on the coated side, so that they remain protected (see the Melles Griot Optics Guide).

The filter is then used by sending the radiation toward the "clear" reflecting

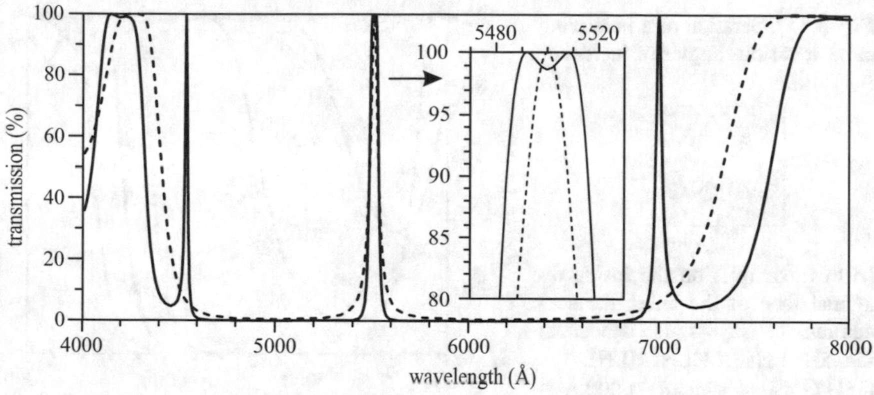

Fig. 3.43 Interference filters with *one* (– – –) and *two* (—) *cavities*.
The configuration of the two filters here is G/HLHLHLH · LL · HLHLHLH/G for the first (with 15 layers) and G/LHLHLHL · HH · LHLHLHLHLHLHL · HH · LHLHLHL/G for the second (with 29 layers), with n_L=1.35 (cryolite) and n_H=2.3 (zinc sulphide), n_G=1.52

side: in this way, one avoids burning it, as might happen if it entered from the absorbing side. This type of filter can also be used as a dichroic mirror to separate two adjacent radiations: one is transmitted and the other reflected.

An important property of interference filters is that they can be tuned, within certain limits, by tilting them: indeed, for Eq. (3.4.8), as the angle of incidence ϑ increases, the phase shift ϕ *decreases*, and therefore the transmission peak moves, for example, from the *green* towards the *blue*!

An interference filter with *one* Fabry-Perot cavity, however, has an uncomfortable drawback: the transmission peak is quite sharp (with a Lorentzian shape) and has extended tails, and, due to angular dependence, the incident beam should be very well collimated and aligned with the filter. With an extended source, like a spectral lamp, much of the radiation would be lost. For this reason, filters usually have two or more cavities $\lambda_o/2$, with which the transmission peak becomes more squared. Among the many possible combinations, Fig. 3.43 shows the comparison between incomplete filters, one having one cavity and one having two cavities. These are theoretical curves with 100% transmission at the peak, however, due to imperfections in the dielectric layers, and their absorption, along with that caused by high and low pass filters, the interference filters that can be achieved have a maximum transmission that may vary from 40% to 70%.

3.4.4.4 Polarizing cube

As the incidence angle increases, the behavior of TE and TM waves differs, and this can be exploited to build cheap polarizers with large surface area. One example is the polarizing cube (Fig. 3.44). It is cut along a diagonal plane on which a high pass filter is basically deposited, however, the angle of incidence between

Fig. 3.44 Operation of a polarizing cube at various angles of incidence ϑ.

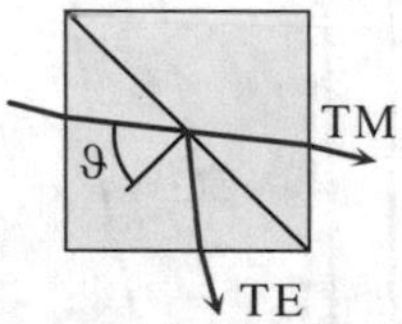

In this example, on the inside diagonal face of the cube, the succession of layers is deposited: G/0.5H · LHLHLHLHLHLHL · 0.5H/G × $\lambda_o/4$ with $\lambda_o = 6000$ Å a $\vartheta=0$, $n_G = 1.52$, $n_H = 2.3$, $n_L = 1.35$

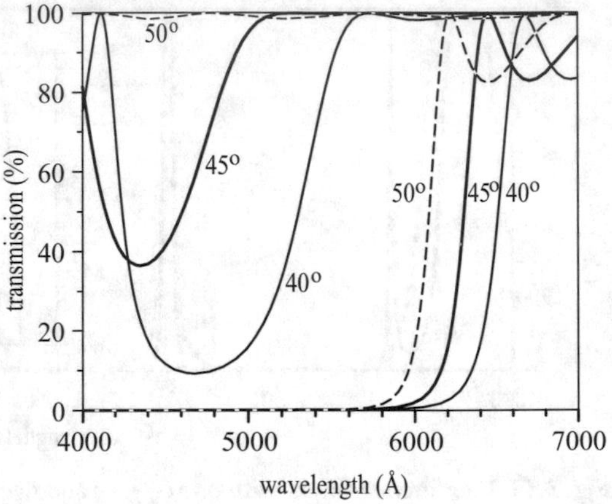

the layers H and L is close to Brewster's angle, so the reflectivity of the various interfaces is small for the TM wave, which is then transmitted, while the TE wave is reflected.

3.4.5 Phase dispersion and amplitude filtering on a wave not monochromatic

We are used to imagining a mirror as a single reflective surface, but, in reality, for a dielectric mirror, the reflection is given by the contribution of many surfaces separated by a small amount of space between them, albeit very little. Therefore, a light pulse is reflected at slightly different times from these surfaces, and then, in general, reflection stretches it over time. In the previous figures, we have limited ourselves to considering the *square module* of the field amplitude, that is, only their intensity. On the other hand, the information on the optical path and the corresponding time elapsed is contained in the phase of the fields themselves and, in the case that interests us here, from the phase ϕ of the amplitude coefficients of reflection ($r = 1/a$) and transmission ($t = c/a$) of Eqs. (3.4.14), which, as usual, for example, for r, is obtained from the factorization $r = Ae^{i\phi}$ with real A and ϕ. An example of these phases is given in Fig. 3.45 for the case of the 10-layer mirror of Fig. 3.41. The deformation of a pulse as a consequence of a reflection can be calculated using the properties of the Fourier transform in a manner similar to what we have already seen in the study of dispersion in Chapter 1. Thus, for a pulse defined by a field $\mathrm{E}(t)$, its transformation into the frequency space is

$$\hat{E}(\omega) = \int_{-\infty}^{\infty} \mathrm{E}(t) e^{i\omega t} dt \,. \tag{3.4.48}$$

When a pulse is reflected by a dispersive surface (or transmitted through a disper-

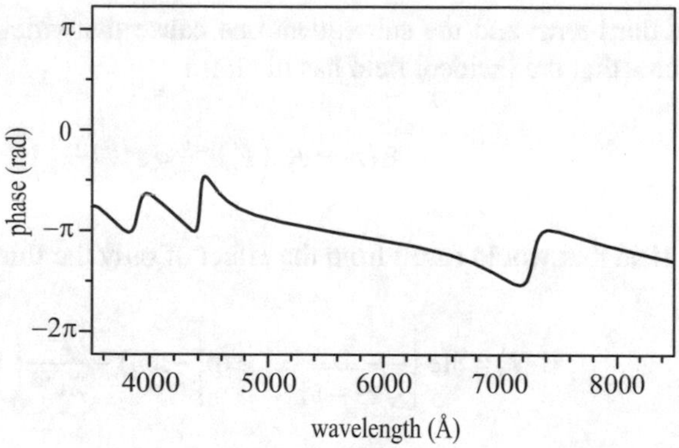

Fig. 3.45 Trend of the phase of the field reflected by a mirror with 10 layers such as that in Fig. 3.31, as a function of wavelength

sive medium), the different spectral components are modulated by different amplitudes $A(\omega)$ and different phases $\phi(\omega)$ by the multiplication for $r(\omega) = A(\omega)e^{i\phi(\omega)}$]. The reflected field can then be calculated with the inverse Fourier transform:

$$\mathrm{E}'(t) = \frac{1}{2\pi}\int_{-\infty}^{\infty} \hat{E}(\omega)A(\omega)e^{i\phi(\omega)}e^{-i\omega t}d\omega\,, \tag{3.4.49}$$

where $\phi(-\omega) = -\phi(\omega)$ and $A(-\omega) = A(\omega) > 0$. $\mathrm{E}'(t)$ is still (necessarily) a real function with $\mathrm{E}' = \frac{1}{2}\mathcal{E}' + \frac{1}{2}\mathcal{E}'^*$, where $\mathcal{E}'(t)$ is its analytic function [see Eqs. (1.3.16-17)].

The elongation in time of a pulse can therefore result from two different aspects of the behavior of an optical component:
– the *amplitude filtering,* which is there when its reflection or transmission characteristics reduce the spectral bandwidth of the incident field: the smaller the resulting bandwidth, the greater the pulse length;
– the *dispersion,* which changes the relative phase between the various components at different wavelengths.

Let us very briefly consider here this second aspect alone. The phase $\phi(\omega)$ can be developed into a power series around a central frequency ω_c:

$$\phi(\omega) = \phi_o + \left.\frac{d\phi}{d\omega}\right|_{\omega_c}(\omega - \omega_o) + \frac{1}{2}\left.\frac{d^2\phi}{d\omega^2}\right|_{\omega_c}(\omega - \omega_o)^2 + ... \tag{3.4.50}$$

The first term yields a constant coefficient that can be taken out of the integral and only involves a phase shift in the field oscillation. Even the second term does not change the shape of the pulse; it simply delays it by a time

$$t_d = \left.\frac{d\phi}{d\omega}\right|_{\omega_c}\,. \tag{3.4.51}$$

The third term and the subsequent one cause the widening of the impulse. If we assume that the incident field has the form

$$\mathrm{E}(t) = \Re e\left(E_o e^{-i\omega_c t} e^{-t^2/2\tau^2}\right), \tag{3.4.52}$$

the field that would result from the effect of only the third term would be

$$\mathrm{E}'(t) = \Re e\left\{\frac{\tau}{\sqrt{\tau^2 - i\delta}} E_o \exp\left[-i\omega_c t - \frac{t^2}{2\tau'^2}\left(1 + i\frac{\delta}{\tau^2}\right)\right]\right\}, \tag{3.4.53}$$

where

$$\delta = \left.\frac{d^2\phi}{d\omega^2}\right|_{\omega_c}, \quad \tau' = \tau\sqrt{1 + \frac{\tau_c^4}{\tau^4}}, \quad \tau_c = \sqrt{|\delta|}\,. \tag{3.4.54}$$

The constant τ_c therefore characterizes the critical time width (limited to the third term in the development of ϕ) of our optical component. In the center of the *stop band* of a mirror, there is an inflection point for the phase ϕ and δ vanishes there. It must be said that, for a dielectric mirror, these enlargement effects are felt only for "femtosecond" pulses, which have a large spectral band.

3.4.6 Chirped mirrors

In the generation of pulses with a duration of a few femtoseconds, it is necessary that the laser cavity be overall devoid of dispersion. In fact, the cavity must remain resonant, and therefore with the same optical length for all of the modes that make up the pulse train spectrum that is generated, a spectrum that extends over many tens of THz. On the other hand, both the active medium and the internal optics introduce a variation of the optical path with the wavelength λ_o. This variation is indicated by the *group delay dispersion* (GDD), expressed in fs^2, and is generally defined as

$$\mathrm{GDD} = \left.\frac{d^2\phi}{d\omega^2}\right|_{\omega_c} = \left.\frac{\lambda^3}{2\pi c^2}\frac{d^2\mathrm{OPL}}{d\lambda^2}\right|_{\lambda_c}, \tag{3.4.55}$$

where ϕ and OPL are, respectively, the phase and the optical path traveled, while ω_c and λ_c are the central reference values of the spectral band.

Typically, the path to the laser cavity is several μm shorter for waves with greater λ_o than those with smaller λ_o. The corresponding GDD is positive and must be compensated in some way. Initially, pairs of prisms were oriented so as to give a negative GDD [Fork et al 1984]. Although these devices may give larger values of GDD (thousands of fs^2), they also present a dispersion of higher order, which limits the generation of ultrashort pulses below 10 fs. To overcome this limit, the *chirped mirrors* were invented, in which the reflection progressively takes place in the thickness of the multilayer, reflecting the various spectral components in a differentiated way [Szipöcs et al 1994]. A *chirped mirror* is formed by several tens of layers that are calculated to compensate the internal dispersion of the laser cavity very precisely. To this end, various types of dispersing mirrors have been introduced. The most complex is the *double chirped mirror*, in which, by proceeding from the surface, a series of layers is first encounter with the property of being an effective *anti-reflective* treatment! This avoids reflection, even partial, due to the index jump between air and mirror. A series of H and L layers then follows, varying in pace and duty cycle. The pace is essentially $\lambda_o/2$, but it increases by moving inwardly to reflect the spectral components with smaller λ_o first and then those with greater λ_o. Also, the relative thickness between the layers H and L is progressively varied to adjust the coupling between the incident and reflected waves. It can also follow a series of layers for which the chirping concerns only the pace, and, finally, there is a series of $\lambda_o/4$ layers at a constant pace to reflect even the end of the spectrum with greater λ_o efficiently. The theory of these structures is described in Matuschek et al (1999). The initial series of antireflection layers avoids the appearance of showy oscillations of GDD, but, at the same time, it is difficult to achieve it with a wide spectral band. Recently, the use of chirped mirrors has been reported in which the anti-reflection structure is absent. By using them in pairs, with a different angle of incidence, the oscillations of the GDD of one are well compensated by opposite oscillations of the other mirror [Pervak 2011].

The GDD is approximately given by the ratio $\Delta\tau/\Delta\omega$, where $\Delta\omega$ is the spectral reflectivity band of the mirror and $\Delta\tau$ is the delay difference for pulses centered on the extremes of this band.

Bibliographical references

Abelès F., *Recherche sur la propagation des ondes electromagnétiques sinusoidales dans les milieux stratifiés. Applications aux couches minces*, Ann. de Physique **5**, 596-640 and 706-782 (1950).

Bates W.J., *A wavefront shearing interferometer*, Proc. Phys. Soc. **59**, 940 (1947).

Born M. and Wolf E., *Principles of Optics*, Pergamon Press, Paris (1980).

Hanbury Brown R. and Twiss R.Q., *Correlation between photons in two beams of light*, Nature **177**, 27-29 (1956).

Conti A., *Gli occhiali sull'Universo*, Le Scienze **393**, 61-67 (may 2001).

Delano E. and Pegis R.J., *Methods of synthesis for dielectric multilayer Filters*, in Progress in Optics, Vol. VII, p.67 (1969).

Ditchburn R.W., *Light*, Vol. I and II, Academic Press Inc., London (1976).
Drew R.L., *A Simplified Shearing Interferometer*, Proc. Phys. Soc. **64**, 1005 (1951).
Fork R.L., Martinez O.E., and Gordon J.P., *Negative dispersion using pairs of prisms*, Opt. Lett. **9**, 150-152 (1984).
Fowles G.R., *Introduction to Modern Optics*, Holt, Rinehart, and Winston, New York (1968).
Gerard A. and Burch J.M., *Introduction to Matrix Method in Optics*, John Wiley & Sons, New York (1968).
Guenther R., *Modern Optics*, John Wiley & Sons, New York (1990).
Heavens O.S., *Optical Properties of Thin Solid Films*, Academic Press Inc., New York (1955).
Hecht E., *Optics, 2nd ed.*, Addison-Wesley, Madrid (1987).
Hegedus Z.S., Zelenka Z., and Gardner G., *Interference patterns generated by a plane-parallel plate*, Appl. Optics **32**, 2285-2288 (1993).
Knittl Z., *Optics of Thin Films*, John Wiley & Sons, New York (1976).
Landsberg G.S., *Ottica*, MIR, Mosca (1979).
Lipson S.G., Lipson H., and Tannhauser D.S., *Optical Physics*, III ed., Cambridge University Press, Cambridge (1995).
Malacara D. , *Optical Shop Testing*, III ed., edited by Malacara D., John Wiley & Sons, Inc., Hoboken, New Jersey (2007).
Matuschek N., Kärtner F.X., and Keller U., *Analytical design of double-chirped mirrors with custom-tailored dispersion characteristics*, IEEE J. of Quantum Electronics **35**, 129-137 (1999).
Michelson A.A., *On the Correction of Optical Surfaces*, Astrophys. J. **47**, 283-288 (1918). *Studies in Optics* (1927), reprinted by University of Chicago Press (1962).
Michelson A.A. and Benoit J.R., *Détermination expérimentale de la valeur du mètre en longueurs d'ondes lumineuses*, Trav. et Mem. Int. Bur. Poids et Mes. **11**, 1 (1895).
Michelson A.A. and Morley E.W., *On the relative motion of the earth and the luminiferous Æther*, Phil. Mag. **24**, 449-463 (1887).
Michelson A.A. and Pease F.G., *Measurement of the diameter of alpha Orionis with the interferometer*, Astrophys. J. **53**, 249–259 (1921).
Möller K.D., *Optics*, University Science Book, Mill Valley (1988).
Murty M.V.R.K., *The Use of a Single Plane Parallel Plate as a Lateral Shearing Interferometer with a Visible Gas Laser Source*, Appl. Opt. **3**, 531-534 (1964).
Pervak V., *Recent development and new ideas in the field of dispersive multilayer optics*, Applied Optics **50**, C55-61 (2011).
Pulker H.K., *Coatings on Glass*, Elsevier, Amsterdam (1984).
Riley M.E. and Gusinow M.A., *Laser beam divergence utilizing a lateral shearing interferometer*, Appl. Optics **16**, 2753-2756 (1977).
Sivoukhine D., *Optique*, MIR, Mosca (1894).
Sommerfeld A., *Optics*, Academic Press, New York (1949).
Strumia F., *Appunti di conduzione elettrica nei gas, Cap. IV - I filtri*. Università degli Studi di Pisa, (1973-1974).
Szipöcs R. and Ferencz K., Spielmann C. and Krausz F., *Chirped multilayer coatings for broadband dispersion control in femtosecond lasers*, Optics Letters **19**, 201-203 (1994).
Vaughan J.M., *The Fabry-Perot Interferometer*, A. Hilger, Bristol (1989).
Wyant J.C., *Use of an ac Heterodyne Lateral Shear Interferometer with Real-Time Wavefront Correction Systems*, Appl. Opt. **14**, 2622 (1975).
Yariv A. and Yeh P., *Electromagnetic propagation in periodic stratified media. II. Birefringence, phase matching, and x-ray lasers*, J. Opt. Soc. Am. **67**, 438 (1977). *Optical*

waves in crystals, John Wiley (1984).

Yeh P., Yariv A., and Hong C.S., *Electromagnetic propagation in periodic stratified media. I. General theory*, J. Opt. Soc. Am. **67**, 423-438 (1977).

Young Thomas, *An account of some case of the production of colours not hitherto described*, Phil. Trans. R. Soc. Lond. **92**, p. 387-397 (1802).

Young Thomas, *A course of lectures on natural philosophy and the mechanical arts*, Volume 1, p. 464, J. Johnson ed. (1807).

Chapter 4
Diffraction

> Lumen propagatur seu diffunditur non solum Directe, Refracte, ac Reflexe, sed ctiam alio quodam quarto modo, DIFFRACTE.
> Propositio I, *De Lumine*, P. Francesco Maria Grimaldi

Introduction

Diffraction, whose name was introduced by Grimaldi in 1665, when he first discovered it and described its effects, has been conveniently defined by Sommerfeld (1949), paraphrasing the Grimaldi's expression, as «any deviation of the light rays from rectilinear paths which cannot be interpreted as reflection or refraction». For example, if an opaque object is placed between a point source and a screen, the shadow thrown by the object does not have an edge as sharp as the one predicted by Geometrical Optics. In fact, careful observation of the shadow edge reveals that a bit of light goes into the shaded area, while darkened fringes appear in the illuminated area. On the other hand, there is also a similarity between the diffraction produced by a diffractive body and the refraction or reflection from a surface: both effects are due to a sudden *discontinuity* of the medium and, indeed, the diffraction fringes can be attenuated by the *apodization* of the obstacle edges, which consists of a gradual variation of their opacity. The phenomenon of the diffraction fringes is quite complex, and there have historically been deep disputes about their origins. Today, we can say that there is both an electromagnetic contribution from the edge of diffractive objects, and a "geometric" contribution due to their form; in this chapter, we will only deal with cases in which this second contribution is relevant and the other can be neglected.

Many diffraction studies are devoted to the effects of objects placed in the path of radiation. The term *diffraction*, however, is also used to indicate a particular method of calculation of the wave field by means of surface or line integrals, as an alternative to methods of propagation through the integration of the wave equations in three-dimensional space.

After that of Grimaldi, the first important contribution to the diffraction theory was given by Fresnel, who perfected the Huygens model by introducing Young's interference principle in place of the envelope of wavelets. Fresnel was thus able to calculate diffraction figures with remarkable precision. Even Young presented a theory of diffraction, based on the visual observation that an aperture edge appears illuminated. Therefore, he suggested that the diffraction fringes were produced by the interference between the radiation transmitted directly from the aperture and a wave coming from the edge, which is now called a *boundary-diffracted wave*. An equivalent to Fresnel's theory was then demonstrated by Maggi and Rubinowicz

---

Electronic supplementary material The online version of this chapter (https://doi.org/10.1007/978-3-030-25279-3_4) contains supplementary material, which is available to authorized users.

G. Giusfredi, *Physical Optics*, UNITEXT for Physics,
https://doi.org/10.1007/978-3-030-25279-3_4

[Maggi 1888; Rubinowicz 1924, 1957] and subsequently perfected by Miyamoto and Wolf, who have transformed the surface diffraction integrals into a line integral along the opening edge [Miyamoto and Wolf (1962)]. Young's idea was also developed by Keller in what he called the *geometrical theory of diffraction*, based on the premise that the propagation of the radiation is a local phenomenon resulting from the relative smallness of the wavelength with respect to the diffracting objects [Keller 1957, 1962]. This, in turn, connects the geometrical theory to the *asymptotic theory of diffraction*, in which the wave vector module tends to infinity.

The Huygens-Fresnel model was formalized mathematically by Kirchhoff on the basis of the wave nature of light, and on assumptions, i.e., approximations, that are *physically* reasonable about the boundary conditions on the diffracting objects. Then, Poincaré, in 1892, and Sommerfeld, in 1894, found that two of Kirchhoff's assumptions were *mathematically* inconsistent with each other, despite the success of the theory.

Indeed, the quantitative evaluation of diffraction is of particular difficulty. In order to find an exact solution, we should also know, in addition to the contribution of the sources, the contribution of the diffracting objects in response to the incident wave, because, in principle, the physical nature of such objects is important. Rigorous results in terms of Maxwell's equations were obtained only in some ideal situations, such as in the case of geometrically simple apertures formed in perfect specular conductors. The use of "blacks" objects is of no help to a theory that wants to be rigorous, given that, if a substance is strongly absorbent, its reflection coefficient is close to 1[!] [Landau and Lifchitz (II) 1966]. So, the concept of a black body constitutes a bad approximation when it is associated with thicknesses $< \lambda$ in which the wave should be extinguished; in other words, the "black color" is obtained when the substance allows the wave to penetrate inside, absorbing it on a distance $>> \lambda$ and/or its surface is fairly irregular and porous such that it will trap the reflected light anyway. On the other hand, Kottler pointed out that, in the visible region of the spectrum, it is possible to realize a surface with absorption very close to 100%, while it is not possible to have a reflectivity as close to 100% for a metal surface, and proposed his own theory. The first rigorous solution to a diffraction problem was given by Sommerfeld in 1896, who treated the two-dimensional case of a plane wave incident on a screen with the shape of a half-plane, infinitely thin and perfectly conducting [Born and Wolf 1980]. Cases of this type may be interesting in the spectral region from the radio waves to the far infrared, but are obviously of little practical utility in the visible region.

However, Rayleigh and Sommerfeld modified Kirchhoff's theory (approximate but reasonable) by means of the functions theory of Green, eliminating the need to make both of those two assumptions mentioned above. In this way, they resolved the inconsistency present in the Kirchhoff model by introducing what is called the *Rayleigh-Sommerfeld theory of diffraction*. It should be noted, however, that these theories contain an important simplification: the light is treated as a scalar phenomenon, neglecting the fact that the various components of the electric and magnetic fields are coupled to one another in Maxwell's equations and cannot be treated independently. However, the experiments conducted in the microwave region have shown that the scalar theory gives accurate results, provided that:

1) the apertures and the diffracting objects are large in comparison to the wavelength considered,
2) the diffracted field is not observed too close to such openings or objects.

The first of these conditions allows for treating the contribution of the diffracting objects as a simple interception of the incident wave, no matter their physical nature. In particular, the electromagnetic contribution of the edge of objects is thus neglected, as it is seen from the observation point.

In conclusion, the difficulties encountered in the research and the application of a rigorous theory mean that we normally have to settle for approximate solutions to the problem of diffraction, which, in many cases of practical interest, are fairly accurate and, at the same time, provide a powerful and handy tool for investigation. In the following discussion, we will only consider the case of a monochromatic wave, that is, of a single spectral component with angular frequency ω.

4.1 Scalar theory of diffraction

4.1.1 General premises

The typical diffraction topology considers two spatial regions I and II, separated by a surface $S = S_1 + S_2$, as shown in Fig. 4.1. Region I contains the sources of radiation, while region II is the space where the diffraction is to be determined. The surface S_2 is generally taken at "infinity", while the S_1 surface delimits the sources and the screen, which consists of opaque areas and openings. This screen interacts with the field generated in region I, absorbing a bit of energy, reflecting a bit of it, and letting some of the wave, altered by the interaction, goes into region II. In the following, we will examine this simplest of situation. There may also be complex situations in which there are reflections and refractions for which the field can be identified by combining various methods, such as ray tracing, interference, etc., and those of diffraction.

The diffraction theory relies, in particular, on a general relationship known as Green's theorem or Green's second identity:

Let $U(\mathrm{P})$ and $G(\mathrm{P})$ be two complex functions of position P, and S be a closed surface that contains a volume V. If U, G and their first and second partial derivatives are unique and continuous in V and on S, then:

$$\iiint_{\mathrm{V}} \left(G\nabla^2 U - U\nabla^2 G\right) d\mathrm{V} = \oiint_{\mathrm{S}} \left(G\frac{\partial U}{\partial n} - U\frac{\partial G}{\partial n}\right) d\mathrm{S}\,, \tag{4.1.1}$$

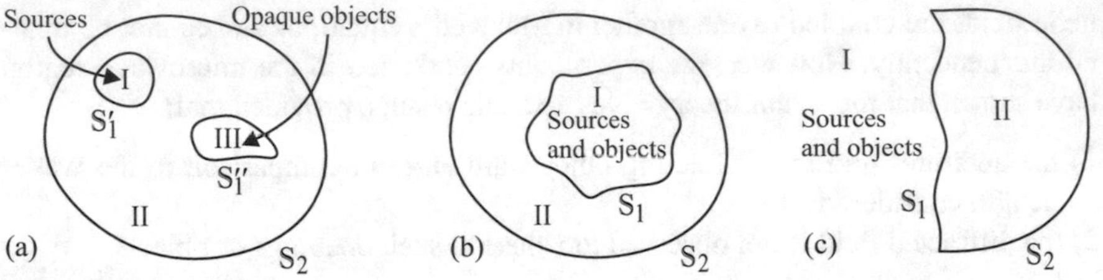

Fig. 4.1 Diffraction topology. Region II is where one wants to determine the field. It surrounds regions with sources and opaque objects, such as in (a) and (b), or not, as in (c). In Figure (a), the region of the sources (I) is distinct from that of the diffracting objects (III); this situation lends itself to study of the case in which there are positions with both direct and diffracted lighting and other positions, in the geometric shadow produced by objects, where the field therein is attributed only to the diffracted wave. The condition of figure (c) is the one most often considered, and it lends itself to dividing the space into two infinite semi-spaces separated by a plane, near which, in region I, there are the apertures

where $\partial/\partial n = \mathbf{n} \cdot \boldsymbol{\nabla}$ denotes the partial derivative in the direction normal to S (whose versor is $\mathbf{n}$) oriented towards the outside (externally to V).

Therefore, let U be a component of the unknown field to be determined in region II that obeys certain conditions at the boundary (that, in particular, will be taken on S_1) and, finally, that obeys the Helmholtz's scalar equation

$$\nabla^2 U + k^2 U = 0\,, \tag{4.1.2}$$

where the zero at the second member indicates the absence of sources that is assumed in region II. As a second function, consider a generic solution G to a similar equation, but having, at the second member, a source term expressed by a three-dimensional Dirac's function, $\delta(\boldsymbol{r} - \boldsymbol{r}_0)$, centered on a "point of observation" $P_o(\boldsymbol{r}_0)$ taken inside the region II:

$$\nabla^2 G + k^2 G = -\delta(\boldsymbol{r} - \boldsymbol{r}_0)\,. \tag{4.1.3}$$

In both equations, the operator ∇^2 acts on the variable $\boldsymbol{r}$. This $G(\boldsymbol{r}, \boldsymbol{r}_0)$ is called *Green's function* and is a *generic* solution to Eq. (4.1.3), in the sense that, without representing any physical entity, it is not required that it satisfy the same boundary conditions assigned to U and it may also be defined in a wider region than region II.

The presence of a Dirac's delta in Eq. (4.1.3) indicates that G has a singularity at the point $\boldsymbol{r}_0$; however, for the properties that are assigned to the delta function, Eq. (4.1.1) can still be used and, in this way, we can concisely get to the result. In short, multiplying Eq. (4.1.2) for G and Eq. (4.1.3) for U, subtracting the two equations member to member, and, finally, integrating on the volume V of region II, we get

$$U(\boldsymbol{r}_0)=\iiint_V \left(G\nabla^2 U-U\nabla^2 G\right)dV=\oiint_{S_1+S_2}\left(G\frac{\partial U}{\partial n}-U\frac{\partial G}{\partial n}\right)dS. \tag{4.1.4}$$

In essence, the delta has acted as a seed that has enabled us to extract the value of U at the point of observation $\boldsymbol{r}_0$ as a function of the field on the surface $S = S_1 + S_2$: the knowledge of the field on this surface is enough then, in principle, to determine the field at all points V. However, it should be noted that, for the applicability of Eq. (4.1.4) the diffracting objects, in addition to sources, must all be outside of the volume enclosed by S, since they constitute a lack of homogeneity of the medium. In other words, the surface S is in the diffraction zone, where the field is not yet known, unless the problem was already solved by other means! So, Eq. (4.1.4), if interpreted strictly, cannot lead to any practical result. However, it suggests that, in the absence of an exact knowledge, if it is possible to identify a surface S where the field can be evaluated with good accuracy, we can also get an acceptable approximate solution to the field inside S. An example of this procedure was introduced by Kirchhoff and will be described in the next section.

4.1.2 The Kirchhoff formulation of the diffraction

Kirchhoff chose, as the Green's function,

$$G(\boldsymbol{R})=\frac{e^{ikR}}{4\pi R},\quad \text{with } \boldsymbol{R}=\boldsymbol{r}-\boldsymbol{r}_0,\quad R=|\boldsymbol{R}|. \tag{4.1.5}$$

This represents a spherical wave that expands over time and is centered at the observation point $\boldsymbol{r}_0$. The coefficient $1/(4\pi)$ is a necessary normalization term, which can be deduced by writing Eq. (4.1.3) in spherical coordinates and then integrating on the volume. With this choice of the Green's function, Eq. (4.1.4) becomes

$$U_K(\boldsymbol{r}_0)=\frac{1}{4\pi}\oiint_{S_1+S_2}\left[\frac{e^{ikR}}{R}\frac{\partial U}{\partial n}-U\frac{\partial}{\partial n}\left(\frac{e^{ikR}}{R}\right)\right]dS, \tag{4.1.6}$$

which is known as the Helmholtz-Kirchhoff integral theorem. The two surfaces together form a closed surface $S = S_1 + S_2$. Although S can, in general, have any shape, with the help of this equation, we now face the case of the diffraction produced by an aperture realized in an infinite plane and opaque screen. As surface S_2 of integration, we choose a sphere centered at the point of observation P_o and with radius R_S that we will tend to infinity (Fig. 4.2). As surface S_1, we choose a plane portion delimited by the intersection with S_2 and placed at an arbitrarily small distance from the screen. In turn, S_1 can be split into the sum of two surfaces $S_1 = A$

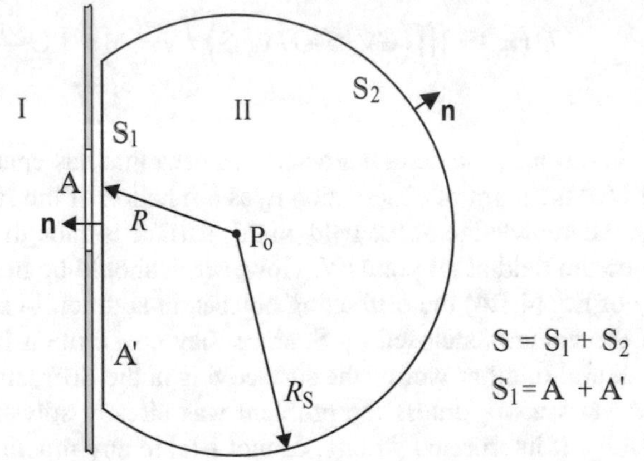

Fig. 4.2
Geometry of diffraction with an infinite plane screen

+ A′, the first coinciding with the opening on the screen. As R_S grows, the fields on S_2 decrease as $1/R_S$. However, the surface S_2 grows as $R_S{}^2$, so it is not obvious whether the contribution of S_2 to the integral can be neglected. Examining the problem in detail, it can be seen that the integral on S_2 is

$$\iint_{S_2} \frac{e^{ikR_S}}{4\pi R_S}\left[\frac{\partial U}{\partial n} - U\left(ik - \frac{1}{R_S}\right)\right] dS = \iint_{\Omega} \frac{e^{ikR_S}}{4\pi}\left(\frac{\partial U}{\partial n} - ikU + U\frac{1}{R_S}\right) R_S d\Omega ,$$

where Ω is the solid angle with which S_2 is seen from P_o and $d\Omega = dS/R_S{}^2$. Therefore, the whole integral on S_2 vanishes when R_S tends to infinity if

$$\lim_{R_S\to\infty} R_S\left(\frac{\partial U}{\partial n} - ikU + U\frac{1}{R_S}\right) = 0 \tag{4.1.7}$$

uniformly for all angles. This request is known as *Sommerfeld's radiation condition* and is satisfied if U fades at least as fast as a *diverging* spherical wave. This is actually true in our case because we are assuming that the field energy flows in region II through the aperture A, so, in spherical coordinates ϑ, φ, R_S, we have

$$U \to f(\vartheta, \phi)\frac{e^{ikR_S}}{R_S}, \quad \frac{1}{U}\frac{\partial U}{\partial R_S} \to ik - \frac{1}{R_S}$$

when R_S tends to infinity (on S_2 it is $\partial/\partial n = \partial/\partial R_S$). For further details, see, e.g., Sivoukhine (1984), Vol. I, p. 296.

The term "diverging wave" has not been used casually. Indeed, intuition tells us that a wave that moves away from a point P (without reflections) is no longer able to act on P. In fact, the information on the direction of propagation is contained in the reciprocal sign between $\partial U/\partial n$ and $-ikU$ in Eq. (4.1.7), and, in this regard, I recall the convention here being used as an abbreviated complex notation of the fields, which implies the further term $\exp(-i\omega t)$. If, in Eq. (4.1.7), the wave direction was reversed, it would have the substitution $-ikU \rightarrow +ikU$ and the limit for $R_S \rightarrow \infty$ would no longer be zero. These considerations throw some light on the meaning of Eq. (4.1.6): the Green's function "collects" those contributions of field U that are traveling *towards* P_o.

Having been able to eliminate the integral on S_2, there remains that on S_1, and the Kirchhoff's integral formula becomes

$$U_K(\boldsymbol{r}_o) = \frac{1}{4\pi}\iint_{S_1} \frac{e^{ikR}}{R}\left[\frac{\partial U}{\partial n} - ik\left(1+\frac{i}{kR}\right)U\cos(\mathbf{n},\boldsymbol{R})\right]dS. \qquad (4.1.8)$$

Both vectors $\mathbf{n}$ and $\boldsymbol{R}$ (see Fig. 4.2) point toward the region I, so it is $\cos(\mathbf{n},\boldsymbol{R}) > 0$. Since the screen considered is opaque except for the opening A, it is intuitively reasonable to assume that the greatest contribution comes from the points located in A. Accordingly, Kirchhoff adopted the following two assumptions:

1) on the surface A, U and $\partial U/\partial n$ are the same as they would be if the screen were not there,

2) on that part of S_1 lying in the geometrical shadow of the screen, both U and $\partial U/\partial n$ are zero.

The second of these assumptions allows us to reduce the integration only on the surface of the opening, on which the field would be unperturbed by the screen. These assumptions simplify the procedure considerably, but none of them is completely correct. The presence of the screen, in fact, necessarily disturbs the field on A, since, particularly on the edge, the field must satisfy certain boundary conditions, which are not required in the absence of the screen. Moreover, the shadow behind the screen is never perfect and the field extends behind it for several wavelengths. However, if the size of the opening is $\gg \lambda$, these edge effects can be neglected and the two assumptions lead to a result in good agreement with experiments.

There is, however, a *mathematical* inconsistency in the second assumption (which has been mentioned above). It can indeed be shown that, if U and $\partial U/\partial n$ (where U is a solution to the Helmholtz's equation) are both null on any finite surface, then $U = 0$ everywhere! Thus, the second assumption is not mathematically acceptable if, as affirmed by the first assumption, a field is radiated through the opening A. It can also be observed that the field recalculated on surface S_1 through Eq. (4.1.8) does not correspond to the assumptions made for U and $\partial U/\partial n$. In the next section, we will see how, at least in principle, we can remedy these problems.

4.1.3 The Rayleigh-Sommerfeld formulation

Suppose that the Green's function G of the Kirchhoff's theory is modified in such a way that, while remaining a solution to Helmholtz's equation, G or $\partial G/\partial n$, one of the two, vanishes over the entire surface S_1. The case in which G vanishes is called the *Dirichlet boundary condition*, while, if $\partial G/\partial n$ vanishes, we have the *Neumann boundary condition*. If, for example, U is known or roughly known on S_1, the Dirichlet-Green function $G_D(\boldsymbol{r},\boldsymbol{r}_0)$ is used such that

$$G_D(\boldsymbol{r},\boldsymbol{r}_0)=0 \quad \text{for } \boldsymbol{r} \text{ on } S_1 . \tag{4.1.9}$$

In this way, the diffraction integral becomes

$$U_I(\boldsymbol{r}_0)=-\iint_{S_1} U(\boldsymbol{r})\frac{\partial G_D(\boldsymbol{r},\boldsymbol{r}_0)}{\partial n}dS, \tag{4.1.10}$$

and a reasonable and fair approximation consists in taking $U(\boldsymbol{r}) = 0$ on the opaque portion of the screen, while, on the opening, it is assumed equal to the incident wave field therein. No hypothesis is now needed for $\partial U/\partial n$. Similarly, if $\partial U/\partial n$ is known with sufficient approximation, the Neumann-Green function $G_N(\boldsymbol{r},\boldsymbol{r}_0)$ is used, for which

$$\frac{\partial G_N(\boldsymbol{r},\boldsymbol{r}_0)}{\partial n}=0 \quad \text{for } \boldsymbol{r} \text{ on } S_1, \tag{4.1.11}$$

and the corresponding integral becomes

$$U_{II}(\boldsymbol{r}_0)=\iint_{S_1}\frac{\partial U(\boldsymbol{r})}{\partial n}G_N(\boldsymbol{r},\boldsymbol{r}_0)dS. \tag{4.1.12}$$

In the special case that S_1 is a plane surface, the two Green functions are given by

$$G_{D,N}(\boldsymbol{r},\boldsymbol{r}_0)=\frac{1}{4\pi}\left(\frac{e^{ikR}}{R}\mp\frac{e^{ikR'}}{R'}\right), \quad \begin{cases} D\leftrightarrow - \\ N\leftrightarrow + \end{cases}, \tag{4.1.13}$$

where $\boldsymbol{R} = \boldsymbol{r} - \boldsymbol{r}_0$, $\boldsymbol{R}' = \boldsymbol{r} - \boldsymbol{r}'_0$, where $\boldsymbol{r}'_0$ is the mirror image of $\boldsymbol{r}_0$ with respect to the surface S_1. In particular, on S_1, for the R and R' modules of these vectors, one has $R = R'$. It is interesting to note that G_D and G_N are composed of waves that expand from those points $\boldsymbol{r}_0$ and $\boldsymbol{r}'_0$, in antiphase for G_D and in phase for G_N. The important thing is that the wave centered in $\boldsymbol{r}'_0$ does not contribute to the left side in Eqs. (4.1.10) and (4.1.12), as $\boldsymbol{r}'_0$ is outside of region II: to realize this fact, it is enough to repeat the calculations that led to Eq. (4.1.4) by adding to Eq. (4.1.3) a Dirac's delta function centered on $\boldsymbol{r}'_{\mathrm{o}}$.

Finally, replacing, for G_{D}, Eq. (4.1.13) in Eq. (4.1.10), we have what is called the Rayleigh-Sommerfeld *first* diffraction integral:

$$U_{\mathrm{I}}(\boldsymbol{r}_{\mathrm{o}}) = -\frac{ik}{2\pi}\iint\limits_{\mathrm{S}_1}\frac{e^{ikR}}{R}\left(1+\frac{i}{kR}\right)\frac{Z_{\mathrm{o}}}{R}U(\boldsymbol{r})\,d\mathrm{S}\,, \tag{4.1.14}$$

where, since S_1 is plane, the substitution $\cos(\mathbf{n},\boldsymbol{R}) = Z_{\mathrm{o}}/R$ has been made, where, in turn, Z_{o} is the distance of the observation point from the plane of the aperture. When R is $>> \lambda$, Eq. (4.1.14) is normally written as

$$U_{\mathrm{I}}(\boldsymbol{r}_{\mathrm{o}}) = -\frac{ik}{2\pi}\iint\limits_{\mathrm{S}_1}\frac{Z_{\mathrm{o}}}{R}\frac{e^{ikR}}{R}U(\boldsymbol{r})\,d\mathrm{S}\,. \tag{4.1.14'}$$

An equivalent expression can be written for Eq. (4.1.12):

$$U_{\mathrm{II}}(\boldsymbol{r}_{\mathrm{o}}) = \frac{1}{2\pi}\iint\limits_{\mathrm{S}_1}\frac{e^{ikR}}{R}\frac{\partial U(\boldsymbol{r})}{\partial n}d\mathrm{S}\,, \tag{4.1.15}$$

which is indicated as the Rayleigh-Sommerfeld *second* diffraction integral. Both were introduced for the first time by Rayleigh. The comparison between Eqs. (4.1.8), (4.1.14), and (4.1.15) shows that the Kirchhoff diffraction integral calculated on a plane surface S_1 is just the average of these expressions:

$$U_{\mathrm{K}}(\boldsymbol{r}_{\mathrm{o}}) = \frac{1}{2}U_{\mathrm{I}}(\boldsymbol{r}_{\mathrm{o}}) + \frac{1}{2}U_{\mathrm{II}}(\boldsymbol{r}_{\mathrm{o}})\,. \tag{4.1.16}$$

If the screen is not flat but curved, there are still two G_{D} and G_{N} functions that meet the above conditions. However, even in the case of a simple spherical surface, the analytical representation of these functions is so complex that it is impractical; therefore, the use of Eqs. (4.1.10) and (4.1.12) is limited only to the case of a plane screen, while the Kirchhoff's integral can extend over a regular surface of any shape.

4.1.4 Diffraction by plane screens: comparison between the theories of Kirchhoff and Rayleigh-Sommerfeld

The at least formal difference between the three equations for U_{K}, U_{I}, U_{II} can be best understood considering the particular case of a wave U incident on a

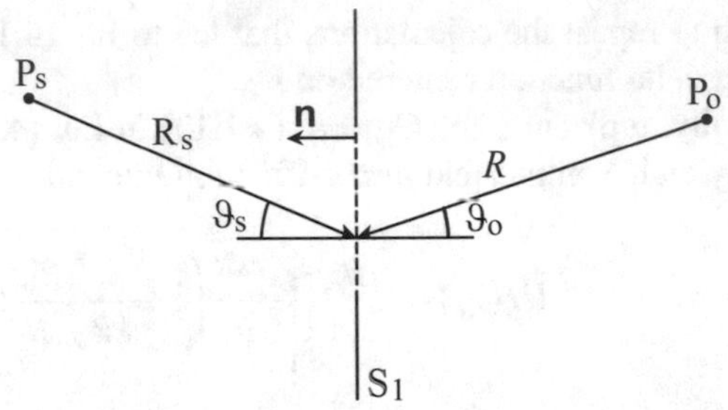

Fig. 4.3 Diffraction with a point source

screen, generated by a point source P_s arranged as in Fig. 4.3. The field S_1 is expressed by

$$U(\boldsymbol{r}) = \frac{e^{ikR_s}}{R_s}, \quad \boldsymbol{R}_s = \boldsymbol{r} - \boldsymbol{r}_s \tag{4.1.17}$$

and all three Eqs. (4.1.8), (4.1.14), (4.1.15), neglecting the terms i/kR, can be collected in the expression

$$U(\boldsymbol{r}_0) = \frac{k}{2\pi i} \iint\limits_A \frac{e^{ikR_s}}{R_s} \frac{e^{ikR}}{R} O(\vartheta_s, \vartheta_o) dA, \tag{4.1.18}$$

where the *obliquity factor* $O(\vartheta_s, \vartheta_o)$ is the only different term. In the three cases, we have

$$O(\vartheta_s, \vartheta_o) = \begin{cases} \cos\vartheta_o & \text{Dirichlet: } U \text{ approximed on } S_1, \\ \cos\vartheta_s & \text{Neumann: } \partial U/\partial n \text{ approximed on } S_1, \\ (\cos\vartheta_o + \cos\vartheta_s)/2 & \text{Kirchoff's approximation,} \end{cases} \tag{4.1.19}$$

where the angles are defined in Fig. 4.3. It can be noted that U_K is the average of the other two solutions U_I and U_{II}, and this is valid in general, even if the wave that affects the opening is not spherical.

One can ask which of the three solutions is the most valid. Certainly, if U or $\partial U/\partial n$ were exactly known, one would opt for U_I or U_{II}, respectively. However, the actual exact value on the screen is practically never known, and therefore approximations in all three cases are normally required. The mathematical inconsistency of Kirchhoff's solution and the incorrect correspondence of the result it gives on S_1 with the starting values may suggest an intrinsic superiority of the other two solutions. However, this is not a correct conclusion, as the field supposed on A (with Kirchhoff's first assumption, which is common to the three cases) does not exactly match that actually present, because of the perturbation induced by the screen. Finding which solution is best is a topic of great debate, and you can consult Stamnes (1986), who makes an accurate comparison between them and others. In practice, the response can only be given in each case by comparison with the experiment.

Significant differences between the three diffraction integrals can only essentially be observed with relatively small openings, wide up to a few tens of wavelengths, and that is why experimental tests were performed mainly in the microwave region. In particular, the situation has been explored in which a plane wave is incident on a circular aperture A, on which U_{I} returns the uniform amplitude of the field assumed in its integration, while U_{II} returns an amplitude with strong oscillations, with a zero minimum in the center of the aperture. Kirchhoff's integral, for Eq. (4.1.16), reconstructs an amplitude that is just the average of those of the other two integrals. Experiments carried out with conductor diaphragms [Ehrlich et al 1955] show similar oscillations of the field on the opening, with a value at the center that accords better with U_{K}. The agreement with Kirchhoff's integral is expected to be even better with "black" diaphragms, as we will discuss later.

The fact that U_{II} exhibits strong oscillations is not in contradiction with the assumption of a uniform value for $\partial U/\partial n$, and, indeed, $\partial U_{\mathrm{II}}/\partial n$ returns the uniform starting value on A. The behavior of U_{II} can be understood through the development in plane waves of the field transmitted from the diaphragm: it is found that, due to the diaphragm, the spectrum of wave vectors $\boldsymbol{k}$ opens up, making the presence of evanescent waves close to the opening itself relevant as well. These are, in particular, the waves with $k_{\mathrm{n}} \approx 0$ that give the greatest contribution to the oscillations of U_{II} on A.

Likewise, the fact that U_{K} does not return a uniform amplitude on A is not a proof of its mathematical inferiority, as it can be split into two contributions I and II, each of which returns the uniform value of its integrand. The consistency of Kirchhoff's theory was first demonstrated by Marchand and Wolf, who showed that the U_{K} field on the aperture can be written as the sum of a direct lighting wave (*geometrical-optics wave*) and a *boundary-diffracted wave* (BDW) from the opening edge, which can be expressed as a line integral along the opening edge [Marchand and Wolf 1966]. In particular, for an opening A in a plane screen normal to the *z*-axis, it is

$$
\begin{aligned}
U_{\mathrm{I}}(x,y,0) &= \varepsilon(x,y)U(x,y,0),\\
U_{\mathrm{K}}(x,y,0) &= \varepsilon(x,y)U(x,y,0)+U_{\mathrm{BDW}},\\
U_{\mathrm{II}}(x,y,0) &= \varepsilon(x,y)U(x,y,0)+2U_{\mathrm{BDW}},
\end{aligned}
$$

where $\varepsilon(x,y)$ is 1 if $(x,y,0) \in$ A, and 0 otherwise, and where

$$
U_{\mathrm{BDW}}(x,y,0)=\frac{1}{2}\varepsilon(x,y)-\frac{1}{4\pi}\iint_{\mathrm{A}}\left[\frac{\partial U(x',y',z)}{\partial z}\right]_{z=0}\frac{e^{ikR}}{R}dx'dy' .
$$

Therefore, in the same condition, U_{I} is composed only of the direct wave, and thus it appears physically less plausible [Stamnes 1986].

In many cases of practical interest, in which the aperture sizes are >> λ (as often happens for the optical region of the spectrum), the diffracted power is confined to a narrow range of angles around the original direction of the wave (on A). This fact is mostly a consequence of the term $\exp(ikr)$, rather than the cosine terms. Indeed, as we will see later, if $U(r)$ is approximately uniform in amplitude and phase on the aperture, when the direction of observation is very inclined with respect to the axis of the opening, $\exp(ikr)$ oscillates rapidly with the position on A. So that, if the dimensions of A are >>λ, the integration on A for relatively large inclinations produces a smaller result than that which is obtained for small inclinations.

This also comforts us in the use of scalar theory for another reason: indeed, from Eqs. (4.1.8), (4.1.14) and (4.1.15), in contrast with the transverse nature of electromagnetic waves, it is found that the diffracted field retains the direction it has on the aperture. Consider, for example, a plane wave that comes perpendicularly on A. It has the electric and magnetic fields parallel to A: the calculated field is still parallel to A for each direction of observation!

Since, fortunately, we can often limit ourselves to considering small angular divergences, the mistake made may be negligible. Besides, if the distance of the sources and observation points from the screen is large with respect to the aperture size, the obliquity factors can be considered as constant terms (with respect to the variable $\boldsymbol{r}$ taken on A) and brought out of the integral. If, then, in particular, we do not move too far away from the axis of the opening, all obliquity factors can be approximated to 1 and the distinction between the various models of Kirchhoff and Rayleigh - Sommerfeld disappears.

4.1.5 The Huygens-Young-Fresnel principle

This principle was initially introduced by men of great genius from some simple physical considerations, rather than through the somewhat abstract formality of Kirchhoff's theory and that of the following authors. The first consideration is that the effect of some local perturbation of the field propagates at a given speed, which is constant, in particular, if the medium is homogeneous, isotropic, and linear: in such conditions, a point disturbance propagates as a spherical wave in the surrounding medium. The second consideration is that the set of perturbations gives rise to effects (the overall wave) that are the linear overlap of those generated individually by their perturbations, in accordance with Young's interference principle. Therefore, making a conceptual leap, each element of a surface struck by a wave (not a material surface, but imagined in the medium) can be regarded as the source of *secondary wavelets* whose interference constitutes the transmitted wave. However, the relationship of amplitude and phase between the incident wave and the wavelets is still to be determined.

Let us analyze, for example, Eq. (4.1.14′). In it, we can recognize the Huygens-

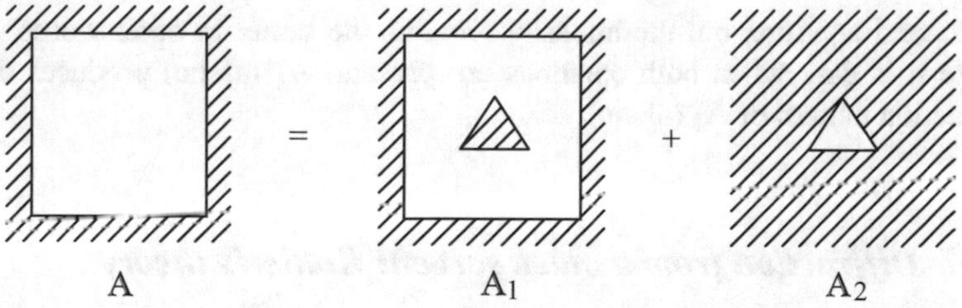

Fig. 4.4 Complementary apertures

Young-Fresnel principle: the first term in the integration describes a spherical wavelet whose amplitude is given by the field present at its center and is further modulated by an obliquity term (the cosine). The integral itself represents the interference of all of these wavelets. Out of the integral, there is the imaginary multiplying coefficient $-i/\lambda$: this coefficient compensates for the phase advance that comes from the integration of the wavelets (in place of their envelope). Its presence can be justified by applying Eq. (4.1.14′) to a plane wave and integrating on an unlimited plane surface perpendicular to the direction of propagation.

Siegman (1986) explains the factor $-i$ with the *Gouy's phase* associated with any wave that passes through a focus or comes from a point source: the initial phase shift $-\pi/2$ between the Huygens wavelets and the field actually on their center compensates for the Gouy phase that each wavelet acquires in its expansion.

4.1.6 Babinet's principle

This is another historically important *principle* asserting that (in the *far field*) complementary diffracting objects produce similar diffraction figures. Consider an aperture A and its partition into two openings A_1 and A_2 (for which $A = A_1 + A_2$), as shown in Fig. 4.4. From the diffraction integrals, it is deduced that the field U_P observed at a point P and transmitted by A is the sum of the contributions from A_1 and A_2, namely:

$$U_P = U_{1,P} + U_{2,P}, \tag{4.1.20}$$

where $U_{j,P}$ is the field that is observed in P when only the aperture A_j is present (j = 1, 2). Babinet's principle is particularly useful when the size of A_2 is small compared to that of A, so that, at a long distance, the diffraction figure of A_2 extends to a far greater degree than that of A. Then, in places where the overall field U_P is negligible, it is

$$U_{1,P} \cong -U_{2,P}\,. \tag{4.1.21}$$

At these points, then, the field phase changes by 180° when replacing the first with the second aperture, but the intensity remains the same. In other words, in areas where it is dark when both openings are present, A_1 (alone) produces the same diffraction pattern of A_2 (alone).

4.1.7 Diffraction from a black screen: Kottler's theory

In his articles from 1923, Kottler pointed out that, despite the fact that the theory of diffraction had been developed for perfectly conducting and reflecting screens, in practice, optical systems such as the objectives of cameras were made with lenses housed in specially blackened mountings, as were their diaphragms, in order to eliminate diffuse or reflected radiation from the walls [Kottler 1923]. The absorption coefficient of this blackening was also much closer to unity in regard to what can be obtained in practice for the coefficient of reflection of the metal surfaces, at least in the field of visible radiation. Therefore, according to Kottler, a theoretical discussion of the diffraction valid for most practical applications had been neglected reducing Kirchhoff's formulation of to a «first approximation to a more exact formulation, although it miraculously gives results which agree well with the experiment». Kottler therefore gave a justification for Kirchhoff's assumptions and the validity of his diffraction integral for the relevant case of "black" diaphragms. To remove physical and mathematical objections to such assumptions, Kottler «has attempted to solve this contradiction … by changing one interpretation and adding a restriction» [Kottler 1966]. The new interpretation was that, instead of considering the assumptions as the imposition of a *boundary value* to the field U and its derivative $\partial U/\partial n$ on the surface of integration, these should instead be considered as a *discontinuity* of U and $\partial U/\partial n$ at the two sides of the surface, thus transforming the assumptions into a problem of *saltus*. The restriction is that the diffracting body is *black*. In his long defense, Kottler introduces the idea that a black body can be assimilated to an *open door* between two Riemann spaces, one physical and one non-physical, in which the wave equations are still valid. The radiation that strikes it simply disappears from the physical space to enter the other space without the need for absorption. In turn, the door, corresponding to the illuminated part of the black body, is an open surface that belongs to both spaces and appears to have a null thickness in each of them. The new interpretation of Kirchhoff's assumptions is then to consider a discontinuity between the two sides of the door in the physical space. So, if U_s is a representative component of the field of the geometrical wave that illuminates the black body, the contour conditions are replaced by the discontinuity

$$U_a - U_b = U_s, \quad \frac{\partial U_a}{\partial n} - \frac{\partial U_b}{\partial n} = \frac{\partial U_s}{\partial n}, \tag{4.1.22}$$

where U_a and U_b represent the limit value of the field, respectively, at the two sides a and b of the door, and **n** is a versor normal to its surface. Analogous reasoning is also done for the surface in the geometric shadow of the black body, for which he assumes that

$$U_{\mathrm{a}} - U_{\mathrm{b}} = 0, \quad \frac{\partial U_{\mathrm{a}}}{\partial n} - \frac{\partial U_{\mathrm{b}}}{\partial n} = 0 . \tag{4.1.23}$$

In other words, there is no discontinuity between the two sides (external and internal to the body) of the surface in shadow; neither is there an imposition that the field there presents nor its derivatives are null.

Trying to translate Kottler's idea, we can thus imagine, with reference to Fig. 4.1(a), that the surface S'_1 encloses a single point source with coordinates $\boldsymbol{r}_{\mathrm{s}}$, while the surface S''_1 closely encloses the black body and is divided into two parts, one illuminated by the source, L, and the other in the shade, O. Each of these is an open door in a non-physical space where the incident radiation is lost forever. Since Green's theorem does not apply to spaces connected by doors (*branch cut*), it is necessary to isolate each of the two doors with a closed surface surrounding it and which can be brought indefinitely close to the null thickness surface that constitutes the door. Within the limit of an infinitesimal distance, the application of Green's theorem to the case of Fig. 4.1(a) leads to the expression

$$\begin{aligned} U(\boldsymbol{r}_{\mathrm{o}}) = U_{\mathrm{s}}(\boldsymbol{r}_{\mathrm{o}}) &+ \oiint_{\mathrm{L}} \left[\left(\frac{\partial U_{\mathrm{a}}}{\partial n} - \frac{\partial U_{\mathrm{b}}}{\partial n} \right) G - (U_{\mathrm{a}} - U_{\mathrm{b}}) \frac{\partial G}{\partial n} \right] d\mathrm{L} \\ &+ \oiint_{\mathrm{O}} \left[\left(\frac{\partial U_{\mathrm{a}}}{\partial n} - \frac{\partial U_{\mathrm{b}}}{\partial n} \right) G - (U_{\mathrm{a}} - U_{\mathrm{b}}) \frac{\partial G}{\partial n} \right] d\mathrm{O}, \end{aligned} \tag{4.1.24}$$

where $U_{\mathrm{s}}(\boldsymbol{r})$ represents the spherical wave centered in $\boldsymbol{r}_{\mathrm{s}}$ generated by the source and G is still given by Eq. (4.1.5). In the two integrals, the contributions to the sides a (external to the body) and b (inside of the body) of each door are summed, taking the versor $\boldsymbol{n} = \boldsymbol{n}_{\mathrm{a}} = -\boldsymbol{n}_{\mathrm{b}}$ oriented in the direction a→b. With Kottler's assumptions (4.1.22) and (4.1.23), this expression is reduced to

$$U(\boldsymbol{r}_{\mathrm{o}}) = U_{\mathrm{s}}(\boldsymbol{r}_{\mathrm{o}}) + \oiint_{\mathrm{L}} \left[G \frac{\partial U_{\mathrm{s}}}{\partial n} - U_{\mathrm{s}} \frac{\partial G}{\partial n} \right] d\mathrm{L}, \tag{4.1.25}$$

which coincides with that of Kirchhoff. This expression also applies to the geometrical shadow zone behind the black body. Indeed, the integral on L can be split into two contributions, as in the case discussed above, that is, in a geometrical wave and a wave diffracted by the edge of the illuminated zone: here, the first identically erases the direct lighting contribution from the source, U_{s}, and only the second remains to irradiate the shadow zone. We can even use this result for the complementary case of an opening in a screen. For this purpose, it is sufficient to exchange the role of the two spaces and simply consider the aperture as an open door A on another space in which the sources are located. Surrounding the door with a closed surface, within the limit coinciding with the sides a and b of A, we get

$$U(\boldsymbol{r}_0) = \oiint_{A} \left[G \frac{\partial U_s}{\partial n} - U_s \frac{\partial G}{\partial n} \right] dA\,, \tag{4.1.26}$$

where the versor $\boldsymbol{n}$ is here oriented back towards the source space. In this way, there is no hypothesis to create for the field in the shadow of the screen, because, in this space, the screen is simply not there, and the surface S_2 (of Fig. 4.2) to be extended at infinity can now be a sphere that also includes the door. The mathematical inconsistencies of Kirchhoff's assumptions are therefore eliminated in their new interpretation.

Kottler argues that, with his interpretation, there is no need to assume that the black body is thin, yet he does not mention the possibility of further diffraction of the diffracted waves. The problem of multiple diffraction at the edge of small openings on black screens was addressed by Marchand and Wolf through a different approach based on the BDW method [Marchand and Wolf 1969].

4.1.8 Outline on the calculation methods of the diffraction integrals

As the diffraction integrals may appear simple, they generally present formidable difficulties when, as usually happens, the integrand exhibits strong oscillations that must be taken into account. The result is essentially what remains from a large mutual cancellation of the contributions of the various elements of the integration surface. An ordinary numerical calculation therefore requires a sufficiently dense sampling of the integrand to represent these oscillations with care, and the required distance between two adjacent sampling points can also be on the order of the wavelength. If one also wants to determine the diffraction pattern produced on a given observation plane, generally, one needs to calculate the integral for a grid of points that is equally dense. On the other hand, analytical solutions of these integrals exist only in a few classic cases, of which we will see various examples in the next sections, in which further approximations are made. To calculate diffraction for the generality of applications, mathematical techniques of considerable complexity have been developed, and there is a wide range of specialized literature, out of which, in particular, I refer the readers to Stamnes (1986) and Solimeno et al (1986) for a broad treatment.

Among the various numerical methods, Stamnes et al (1983) developed a computing program that, taking into account the local conditions with which the phase and amplitude vary in the integration domain, combines various algorithms in which these quantities are locally approximated linearly or parabolically along one direction, while using the Gauss-Legendre integration formula for the other direction.

It is worth mentioning the techniques that, as stated above, exploit the very thing that makes it difficult to calculate the integrals. These are the asymptotic methods, whose mathematical bases are described in the classical text by Erdélyi. There, the author describes, among other things, the asymptotic expansion of functions defined by integrals containing a "big" parameter [Erdélyi 1956]. These methods are applicable to cases in which the apertures are large compared to the wavelength, whereby, with the wave vector $k \to \infty$, the diffraction integral approaches an asymptotic expansion that typically consists of a series of powers in $1/k$. To obtain the qualitative behavior of the diffraction integral, it is usually sufficient to consider only the first term of the expansion, while, for a quantitative assessment, it may be necessary also to consider some of the subsequent terms without a particular increase in the computation time. The most well known and important among the asymptotic methods is that which is said to be *of the stationary phase*, which, in the midst of so much mathematical complexity, at least has the merit of being intuitively understandable with physical considerations.

For three-dimensional waves and an assigned observation point (here implicit), a typical integral of diffraction has the form

$$U = \iint_{\mathrm{A}} g(x, y) e^{ik\, f(x,y)} dx\, dy\,, \tag{4.1.27}$$

where the real function kf represents the phase of the field on the aperture A, and the function g represents its amplitude (as they are seen at the observation point). These functions are assumed continuous and continuously derivable. For sufficiently large values of k, the exponential term oscillates so rapidly with respect to g that the integration on a small finite surface is negligible, except in the neighborhood of points where the phase kf is stationary, or along the edge of the aperture where the cancellation between adjacent zones in antiphase ceases to be effective. The first step in the resolution of the integral is therefore to identify, in the integration domain, the critical points where this occurs. At this stage, the problem is topological, since their position depends on that of the observation point, so, while this moves, they can appear or cancel each other out. Therefore, there is a variety of situations that must be handled with different mathematical artifice. The simplest case is that of an isolated critical point, said to be of *first species*, for which the first derivatives of f with respect to the variables x and y are null. Further cases are those in which a point of first species is on the edge of the opening. Critical points of the *second species* are those along the edge of the aperture where the derivative of f is canceled only in the direction tangent to the edge, and the critical points of the *third species* coincide with the corners of the contour, where there is a discontinuity in the direction tangent to the edge. This is why some parts of the aperture edge appear illuminated in visual observation and others do not.

In particular, for a critical point of the first kind, both functions f and g are developed around it in a Taylor series, and the exponential term is also expanded in a suitable form. The overall series obtained with these developments is then integrated term by term, by taking the limits of integration at infinity, and this is possible if the critical point is sufficiently isolated from others and far from the edge. Finally, collecting the results of the various integral, with the same power in $1/k$, one obtains the asymptotic series sought. Similar development procedures, along with coordinate transformations, are also used in the other cases. The asymptotic development of the first species points can be considered the contribution of the geometric wave passing through the aperture, while the other is that of the wave diffracted from the edge. It turns out that the first term of asymptotic developments for the first, second and third species is proportional to k^{-1}, $k^{-3/2}$ and k^{-2}. We also need separately to consider the situations in which two or more critical points of various species are placed close together. The collision between the points of first species corresponds to the caustic of the geometrical wave, while, if points of second and third species are involved, there are caustics with the diffracted wave. These cases are treated using what are called *uniform asymptotic approximations* [Stamnes 1986].

In cases in which the diffraction integral is reduced to a line integral of the type

$$U = \int_C g(x) e^{ikf(x)} dx \,, \tag{4.1.28}$$

as an alternative to the stationary phase method, the method of *steepest descent* can be used, in which the function of a real variable to be integrated is extended to the complex plane and is rendered holomorphic by excluding the neighborhoods of singular points and inserting a branch cut. Taking advantage of the properties of holomorphic functions, integration can be carried on the complex plane on a path for which the real part of f is constant, while its imaginary part has the highest slope. Integration is then completed with the contribution of the singular points and the branch cuts. With this method, one can get a better asymptotic approximation [Solimeno et al 1986.].

In the next chapter, we will instead discuss the methods based on the plane waves spectrum of the field on an aperture, with their implementation in programs that take advantage of the *Fast Fourier Transform*.

4.1.9 Diffraction with quasi-monochromatic radiation

So far, we have considered the case of an ideal monochromatic illumination. What happens, instead, when the radiation employed has a finite bandwidth $\Delta\nu$ around a carrier frequency ν_0? In particular, let us consider the case in which the incident wave has an amplitude given by

$$u(\boldsymbol{r},t) = \int_0^\infty U(\boldsymbol{r},\nu) e^{-i2\pi\nu t} d\nu \,, \tag{4.1.29}$$

where U is the Fourier's transform in the space of the frequencies of the real field u, of which $\boldsymbol{u}$ is the analytic function. This transformation between t and ν will be better defined in the next chapter. Following Goodman (1996), we then apply the integration on the frequencies to the first diffraction integral of Rayleigh-Sommerfeld.[1] For $R \geq Z_o >> \lambda$, we have

$$\boldsymbol{u}_{\mathrm{I}}(\boldsymbol{r}_0, t) = -i \int_0^{\infty} \frac{\nu}{\upsilon} \iint_{S_1} \frac{Z_o}{R^2} e^{i2\pi\upsilon\nu R - i2\pi\nu t}\, U(\boldsymbol{r}, \nu)\, dS\, d\nu\,, \tag{4.1.30}$$

where $\upsilon = c/n$ is the speed of light in the considered medium, $\lambda = \upsilon/\nu$, $R = |\boldsymbol{r}_0 - \boldsymbol{r}|$ and Z_o is the distance of the observation point from the plane of the aperture. On the other hand, differentiating Eq. (4.1.29) over time, we have

$$\frac{\partial}{\partial t} \boldsymbol{u}(\boldsymbol{r}, t) = \int_0^{\infty} (-i2\pi\nu) U(\boldsymbol{r}, \nu) e^{-i2\pi\nu t} d\nu\,. \tag{4.1.31}$$

Therefore, exchanging the integration order into Eq. (4.1.30) and applying Eq. (4.1.31), we get

$$\begin{aligned} \boldsymbol{u}_{\mathrm{I}}(\boldsymbol{r}_0, t) &= \frac{1}{2\pi\upsilon} \iint_{S_1} \frac{Z_o}{R^2} dS \int_0^{\infty} (-i2\pi\nu) e^{-i2\pi\nu(t - R/\upsilon)}\, U(\boldsymbol{r}, \nu) d\nu \\ &= \frac{1}{2\pi\upsilon} \iint_{S_1} \frac{Z_o}{R^2} \frac{\partial}{\partial t} \boldsymbol{u}(\boldsymbol{r}, t - R/\upsilon) dS\,, \end{aligned} \tag{4.1.32}$$

where, here, the function $\boldsymbol{u}$ in the integral is taken at the *delayed time* $t - |\boldsymbol{r}_0 - \boldsymbol{r}|/\upsilon$ so as to take into account the propagation time between $\boldsymbol{r}$ and $\boldsymbol{r}_0$.

In particular conditions, we can find a similar expression to the diffraction integrals of the monochromatic case. Let us restart from Eq. (4.1.30): if U has a limited extension in an interval $\Delta\nu$ around ν_o in frequency space and if the aperture has a limited extension for which R varies between a minimum R_a and a maximum R_b around an average value $\bar{R}$ $(\boldsymbol{r}_0)$, the argument of the exponential can be rewritten as

$$i2\pi\left[\frac{1}{\nu}(\nu' + \bar{\nu})(R' + \bar{R}) - \nu t\right] = i2\pi\left[\frac{1}{\upsilon}\nu' R' - \nu\left(t - \frac{\bar{R}}{\upsilon}\right) + \frac{1}{\upsilon}\bar{\nu} R - \frac{1}{\upsilon}\bar{\nu}\bar{R}\right],$$

where $R' = R - \bar{R}$ and $\nu' = \nu - \bar{\nu}$. By inverting the integration order and approximating the term of amplitude ν/υ with $\bar{\nu}/\upsilon$, Eq. (4.1.30) becomes

[1] Born and Wolf (1980) instead starts from Kirchhoff's diffraction integral. In this way, the integration surface can have any shape. Unfortunately, the expressions that are obtained are much more complicated.

$$u_{\mathrm{I}}(\boldsymbol{r}_{\mathrm{o}},t) =$$
$$-\frac{i\overline{\nu}}{\upsilon} e^{-i2\pi\frac{n}{c}\overline{\nu}\overline{R}} \iint_{S_1} \frac{Z_{\mathrm{o}}}{R^2} e^{i\frac{2\pi}{\upsilon}\overline{\nu}R} \int_0^{\infty} U(\boldsymbol{r},\nu) e^{i\frac{2\pi}{\upsilon}\nu' R' - i2\pi\nu\left(t - \frac{\overline{R}}{\upsilon}\right)} dS\, d\nu. \tag{4.1.33}$$

If, finally, the spectral band of u is sufficiently limited, so that

$$\Delta\nu << \frac{\upsilon}{R_{\mathrm{b}} - R_{\mathrm{a}}}, \tag{4.1.34}$$

the term in $\nu' R'$ can be neglected, and, with Eq. (4.1.29), the diffraction integral becomes

$$u_{\mathrm{I}}(\boldsymbol{r}_{\mathrm{o}},t) = -\frac{i\overline{\nu}}{\upsilon} e^{-i\frac{2\pi}{\upsilon}\overline{\nu}\overline{R}} \iint_{S_1} \frac{Z_{\mathrm{o}}}{R^2} e^{i\frac{2\pi}{\upsilon}\overline{\nu}R} \boldsymbol{u}\left(\boldsymbol{r}, t - \frac{\overline{R}}{\upsilon}\right) dS. \tag{4.1.35}$$

The term $t - \overline{R}/\upsilon$ is typical of the retarded potentials and $\overline{R}/\upsilon$ represents an average time delay, while the argument of the exponential out of the integral is the average phase shift of the wave that arrives in $\boldsymbol{r}_{\mathrm{o}}$. This phase shift is compensated with that due to the time delay. Indeed, factorizing $\boldsymbol{u}$ around the carrier frequency $\overline{\nu}$ as (please note the change of font)

$$\boldsymbol{u}(\boldsymbol{r},t) = u(\boldsymbol{r},t) e^{-i2\pi\overline{\nu}t}, \tag{4.1.36}$$

we can rewrite Eq. (4.1.34) as

$$u(\boldsymbol{r}_{\mathrm{o}},t) = -\frac{i\overline{\nu}}{\upsilon} \iint_{S_1} \frac{Z_{\mathrm{o}}}{R^2} e^{i\frac{2\pi}{\upsilon}\overline{\nu}R} u\left(\boldsymbol{r}, t - \frac{\overline{R}}{\upsilon}\right) dS. \tag{4.1.37}$$

In the monochromatic case, $u(\boldsymbol{r},t)$ is a constant amplitude over time, and thus we again find the diffraction integral (4.1.14′).

Therefore, if the spectral band is sufficiently limited, the diffraction integral has substantially the same shape as that of a monochromatic wave. When, instead, the condition (4.1.34) is not respected, there is a reduction in the visibility of the diffraction fringes, with an increasing deletion effect as the distance of the observation point from the axis of the aperture increases, due to the increase in the difference between R_{a} and R_{b}. On the other hand, in evaluating R_{a} and R_{b}, it is enough to consider only the points on the surface S_1 that contribute significantly to the field at the observation point, i.e., only the stationary phase areas of S_1 seen from this point. We will return to these considerations in §5.5 which is dedicated to coherence.

4.2 Fresnel and Fraunhofer diffraction

4.2.1 Evolution of the diffraction pattern with the distance from the aperture

Let us imagine having an opaque screen with a small opening A lit by a distant source: in this way, the incident wave is approximately plane on A. With the help of a sheet of paper, held parallel to the screen, you can observe a spot of light that draws the diffraction shape. This type of situation is shown in Fig. 4.5 in which the images were obtained with a numerical simulation that will be described later in treating the Fourier diffraction. If the sheet is near the screen, the shape of the opening can be recognized in the spot of light, although some fringe is visible around the edges (they appear as a swing in the intensity). Moving the sheet away, the spot becomes more and more complex and the fringes become more pronounced. For large distances, the diffraction pattern becomes very wide but continues to have little or no resemblance with the shape of the aperture. Moving the sheet further away, the spot of light continues to widen, but its shape tends to stabilize. This limit case is called *Fraunhofer's diffraction* or *far field*.

The regime at intermediate distances is called *Fresnel's diffraction*, while that near the aperture, which can still reasonably be treated with Geometrical Optics, is called the *near field*. The border between these types of diffraction is blurred anyway, and we will come back to it shortly.

In the Fraunhofer's region, the diffraction pattern corresponds to the angular spectrum of the field diaphragmed by the aperture. It should be noted, however, that, in the case of a triangular opening such as that in Fig. 4.5, this limit is reached slowly, while the similarity would be reached faster for a rectangular opening or, generally, for apertures delimited by pairs of parallel sides between them.

In the preceding sections the results of the scalar theory have been presented in their most general form, however, in calculating the diffraction figures, some further approximations are normally introduced. They are of a different degree depending on the distance from the openings, and thus are still called by the names of Fresnel and Fraunhofer. The advantage that is obtained on Eqs. (4.1.8), (4.1.14) and (4.1.15) (which are, however, relatively simple), is the possibility of performing calculations using the Fast Fourier Transform.

4.2.2 Initial approximations

Then, let A be an aperture practiced on a plane opaque screen and S_0 a plane of observation parallel to A and placed at a distance Z_0 from it. Following the considerations made at the end of §4.1.4, we limit our attention to when:

Fig. 4.5 Simulation of the diffraction pattern produced by a triangular aperture on a monochromatic plane wave. The height of the triangle is 1.5 mm and the wavelength is 1 μm. The figures are taken at distances from the opening indicated at the top, and their width is 4mm × 4mm for the first eight and 8mm × 8mm for the other two. The figure here to the right represents, instead, the angular spectrum of the field on the opening and is about 14.6 mrad wide. In the figures, the grayscale represents the module of the field amplitude, rather than the intensity, so as to highlight the weaker details that would otherwise not be appreciable. Moreover, the maxima are normalized to the same value

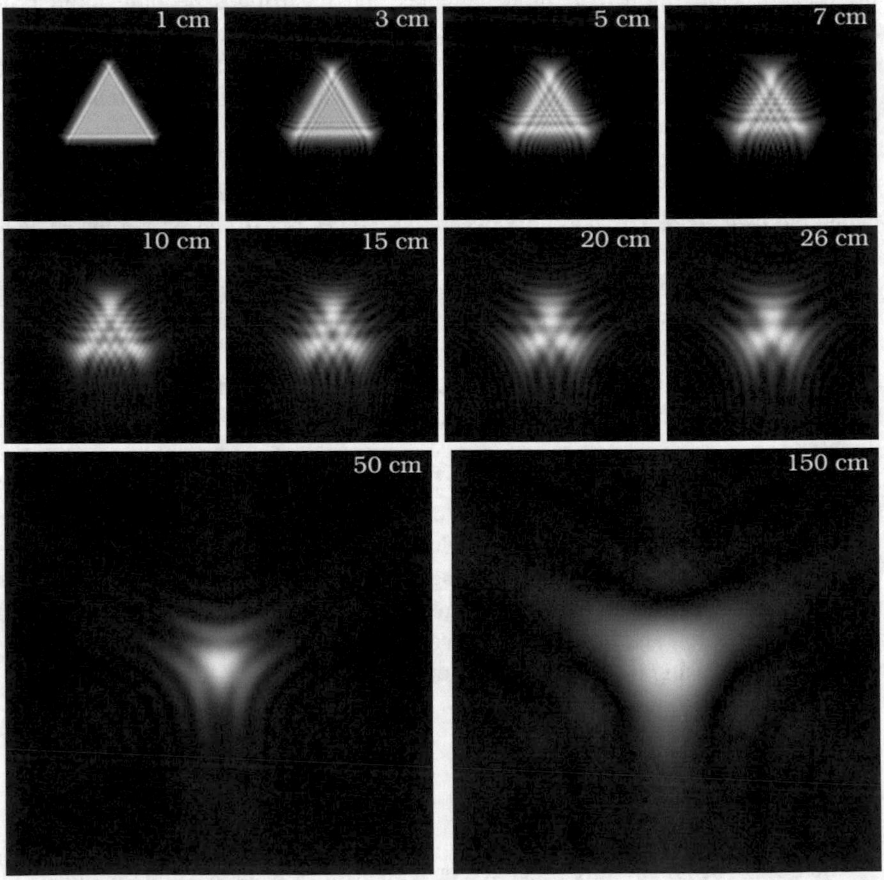

i - the sources and the observation points are approximately aligned with an axis passing through the aperture A. Without too much loss of generality, we can assume that this axis is normal to A and S_o and consider it as the *z*-axis of a Cartesian reference system *x*, *y*, *z* centered in A;

ii - the coordinates *x*, *y* of the observation points are small compared to Z_o;

iii - the size of A is small compared to Z_o, but also $\gg\lambda$;

iv - the radii of curvature of the incident wavefront on A are large compared to the size of A, i.e., the sources are distant from A.

Therefore, within the limits of these assumptions, we can approximate the obliquity factors to 1, and the diffraction integral (4.1.14) becomes

$$U_o(x_o, y_o) = -\frac{i}{\lambda}\iint_A \frac{e^{ikR}}{R} U(x, y)\, dx\, dy\,, \tag{4.2.1}$$

where

$$R = |\mathbf{r}_0 - \mathbf{r}| = \left[(x_o - x)^2 + (y_o - y)^2 + (z_o - z)^2\right]^{1/2}. \tag{4.2.2}$$

For brevity, I have omitted the *z* coordinates in U_o and *U*: Eq. (4.2.1) represents the transfer operation of the field from the plane of the aperture to the observation plane. Even Eqs. (4.1.8) and (4.1.15) can be approximated to Eq. (4.2.1); it is sufficient to note that, thanks to the conditions (*i*) and (*iv*), on the aperture, we have $\partial U/\partial n = -\,\partial U/\partial z = -\,ikU$. As an additional approximation, the term *R* in the denominator can be replaced with

$$Z_o = |z_o - z| \tag{4.2.3}$$

and brought out of the integral.

4.2.3 Fresnel's approximation

The simple replacement $R \to Z_o$ cannot be done in the exponent, since the resulting error would be multiplied by *k*, obtaining a very large number ($kR \gg 1$). To simplify the integrand further, however, we can develop *R* with a binomial series of the root in Eq. (4.2.2):

$$\sqrt{1+b} = 1 + \frac{1}{2}b - \frac{1}{8}b^2 + \cdots \quad \text{valid for } |b| < 1.$$

Hence, if $(x_o - x)^2 + (y_o - y)^2 < R_o^2$,

$$R = Z_o\sqrt{1+\frac{(x_o-x)^2+(y_o-y)^2}{{Z_o}^2}} =$$

$$= Z_o + \frac{(x_o-x)^2+(y_o-y)^2}{2Z_o} - \frac{\left[(x_o-x)^2+(y_o-y)^2\right]^2}{8{Z_o}^3} + \cdots .$$

When Z_o is large enough, for which the terms after that in $1/Z_o$ can be neglected, it is said to be a case of Fresnel's diffraction. Under these conditions, we assume the relation

$$R = Z_o + \frac{1}{2Z_o}\left[(x_o-x)^2+(y_o-y)^2\right], \tag{4.2.4}$$

which is called *Fresnel's approximation.* On the other hand, if the error made in neglecting the term in $1/{Z_o}^3$ were already excessive, it would be better to deal with the problem in another way. Applying Eq. (4.2.4) to Eq. (4.2.1), one obtains

$$U_o(x_o, y_o) = -\frac{ie^{ikZ_o}}{\lambda Z_o}\iint_A e^{i\frac{k}{2Z_o}\left[(x_o-x)^2+(y_o-y)^2\right]} U(x,y)\,dxdy\,. \tag{4.2.5}$$

As a condition of sufficient accuracy for this formula, we may require that the error made in R stopping at the term in $1/Z_o$ is $<< \lambda/4$. If, for example, $Z_o = 10$ cm and $\lambda = 500$ nm, the angle subtended by $\boldsymbol{R}$ should be $<< 3.2°$ ($<< 9.4°$ if the terms in $1/{Z_o}^3$ were also included). Fortunately, this condition is not really necessary. In order that the Fresnel's approximation remain valid, it is only required that the terms of higher order in the development of R do not appreciably change the value of the diffraction integral. For distances Z_o sufficiently small to violate the condition that the error in R is $<< \lambda/4$, for example, when the term in $1/{Z_o}^3$ is $\lambda/4$, that is

$$\frac{\rho^4}{8{Z_o}^3} = \frac{\lambda}{4}, \quad \text{with} \quad \rho^2 = (x_o-x)^2+(y_o-y)^2,$$

the term in $1/Z_o$ is

$$\frac{\rho^2}{2Z_o} = \frac{{Z_o}^2}{\rho^2}\frac{\rho^4}{2{Z_o}^3} = \frac{{Z_o}^2}{\rho^2}\lambda,$$

and is therefore generally very large compared to λ (in the above mentioned example, it is 316 λ). Thus, when the point (x, y) explores the integration domain, the function $R' = \rho^2/(2\,Z_o)$ + [phase of $U(x, y)$]$/k$ generally changes by many wavelengths, and then the exponential under the integration sign oscillates many times by changing the sign of the real and the imaginary part of the integrand. Consequently, the contributions from the various elements of surface of the aperture generally cancel each other out (*destructive interference*).

However, the situation is different for a surface element that is centered on a point (called a *critical point* or pole) where, for a given point of observation, R' is stationary. Here, the integrand varies much more slowly, and we can expect that it give a significant contribution here. Therefore, the integral value is substantially determined by the behavior of R' in the neighborhood of the points where R' is stationary. This is the principle of the *stationary phase method*, and for a wider and rigorous discussion of it, see Appendix III of Born and Wolf (1980). At the stationary phase points, the contribution of the highest order terms in $1/Z_o$ is often entirely negligible. So, in conclusion, the validity of Eq. (4.2.5) generally extends to a far greater degree than expected from the condition that the error on R is $\ll \lambda/4$. It should be noted that when the aperture becomes microscopic, the diffracted light extends over a large solid angle, so the approximations described above become problematic and the same scalar theory becomes insufficient.

Accepting the validity of Fresnel's approximation, we can continue in the study of Eq. (4.2.5). It can be interpreted as the convolution of $U(x, y)$ with the function

$$h(x,y) = -i\frac{e^{ikZ_o}}{\lambda Z_o} e^{i\frac{k}{2Z_o}\left[x^2+y^2\right]}. \tag{4.2.6}$$

Eq. (4.2.5) can also be transcribed in an equivalent form by developing Eq. (4.2.4):

$$R = Z_o + \frac{{x_o}^2 + {y_o}^2}{2Z_o} - \frac{xx_o + yy_o}{Z_o} + \frac{x^2 + y^2}{2Z_o}, \tag{4.2.7}$$

so that the diffraction integral becomes

$$\begin{aligned} U_o(x_o, y_o) = &-\frac{ie^{ikZ_o}}{\lambda Z_o} e^{i\frac{k}{2Z_o}\left({x_o}^2+{y_o}^2\right)} \times \\ &\times \iint\limits_A e^{-i\frac{2\pi}{\lambda Z_o}(xx_o+yy_o)} e^{i\frac{k}{2Z_o}\left(x^2+y^2\right)} U(x,y)\,dx\,dy. \end{aligned} \tag{4.2.8}$$

Then, apart from the phase and amplitude terms outside of the integral sign, the field $U(x_o, y_o)$ can be derived from the Fourier transform of

$$\exp\left[i\frac{k}{2Z_o}\left(x^2+y^2\right)\right]U(x,y),$$

evaluated at the *spatial frequencies*

$$f_x = \frac{x_o}{\lambda Z_o} \quad f_y = \frac{y_o}{\lambda Z_o}. \tag{4.2.9}$$

An alternative to the approximation of R given by Eq. (4.2.4) is to calculate the diffracted field not on a plane surface located at a distance Z_o from the aperture, but on a hemispherical surface of radius S with its center on the coordinates (x=0, y=0, z=0) in the center of the opening, whereby

$$x_o^2 + y_o^2 + z_o^2 = S^2$$

$$R = \sqrt{(x - x_o)^2 + (y - y_o)^2 + S^2 - x_o^2 - y_o^2} = \sqrt{x^2 + y^2 - 2(xx_o + yy_o) + S^2},$$

and Eq. (4.2.4) is replaced with

$$R \cong S + \frac{x^2 + y^2}{2S} - \frac{xx_o + yy_o}{S}, \tag{4.2.10}$$

while Eq. (4.2.8) is replaced with

$$U_o(x_o, y_o) = -\frac{ie^{ikS}}{\lambda S} \iint_A e^{-i\frac{2\pi}{\lambda S}(xx_o + yy_o)} e^{i\frac{k}{2S}(x^2 + y^2)} U(x, y)\, dx\, dy. \tag{4.2.11}$$

This expression is more precise and allows for calculating the diffracted field, even with large angular openings, without introducing complications in the calculation of the diffraction integral, which can still be expressed in terms of a Fourier transform, evaluated at the frequencies

$$f_x = \frac{x_o}{\lambda S}, \quad f_y = \frac{y_o}{\lambda S}. \tag{4.2.12}$$

However, the diffraction profile is obtained on a spherical surface!

4.2.4 Self-consistency of Fresnel's diffraction

We have seen before that Fresnel's approximation is valid for distances Z_o much larger than the size of both the aperture and the transverse coordinates of the observation point. However, it also happens that this approximation is applied with no concern in other cases. As partial justification for this use, we can invoke a self-consistency property of the integral (4.2.5). Suppose that we divide the distance into two parts, for which $Z_o = Z_1 + Z_2$, and duplicate the integration in two operations to be done in succession:

$$U_o(x_o, y_o) = -\frac{e^{ikZ_o}}{\lambda^2 Z_1 Z_2} \int_{-\infty}^{\infty}\int_{-\infty}^{\infty} dx' dy' e^{i\frac{k}{2Z_2}\left[(x_o - x')^2 + (y_o - y')^2\right]} \times \iint_A e^{i\frac{k}{2Z_1}\left[(x'-x)^2 + (y'-y)^2\right]} U(x,y)\,dx\,dy, \tag{4.2.13}$$

where an infinite integration is added in an intermediate plane x', y'. By collecting the arguments of the exponentials under the sign of integration, with some steps, we get

$$\begin{aligned} & i\frac{k}{2}\frac{1}{Z_1 Z_2}\left[Z_1(x_o - x')^2 + Z_1(y_o - y')^2 + Z_2(x' - x)^2 + Z_2(y' - y)^2\right] \\ & = i\frac{k}{2}\frac{1}{Z_o}\left[(x_o - x)^2 + (y_o - y)^2\right] \\ & + i\frac{k}{2}\frac{Z_o}{Z_1 Z_2}\left[\left(\frac{Z_1}{Z_o}x_o + \frac{Z_2}{Z_o}x - x'\right)^2 + \left(\frac{Z_1}{Z_o}y_o + \frac{Z_2}{Z_o}y - y'\right)^2\right]. \end{aligned} \tag{4.2.14}$$

With an appropriate translation of the variables x' and y', Eq. (4.2.13) becomes

$$U_o(x_o, y_o) = -\frac{e^{ikZ_o}}{\lambda^2 Z_1 Z_2} \iint_A e^{i\frac{k}{2Z_o}\left[(x_o - x)^2 + (y_o - y)^2\right]} U(x,y)\,dx\,dy \times \int_{-\infty}^{\infty}\int_{-\infty}^{\infty} dx'' dy'' e^{i\frac{2Z_o}{\lambda Z_1 Z_2}\frac{\pi}{2}\left(x''^2 + y''^2\right)}. \tag{4.2.15}$$

The second double integral can be factorized in two independent terms in x'' and y''. Each of them can, in turn, be expressed by Fresnel integrals and is equal to

$$\int_{-\infty}^{\infty} dx'' e^{i\frac{2Z_o}{\lambda Z_1 Z_2}\frac{\pi}{2}x''^2} = 2\sqrt{\frac{\lambda Z_1 Z_2}{2Z_o}} \int_0^{\infty} dt\left[\cos\left(\frac{\pi}{2}t^2\right) + i\sin\left(\frac{\pi}{2}t^2\right)\right] = \frac{1+i}{\sqrt{2}}\sqrt{\frac{\lambda Z_1 Z_2}{Z_o}}.$$

By replacing this value in Eq. (4.2.15), we again find Eq. (4.2.5) of the Fresnel diffraction. In conclusion, the subdivision of Z_o into Z_1 and Z_2 is arbitrary, and each of the diffraction integrals resulting from this decomposition may not respect the validity conditions of Fresnel's approximation. However, if Z_o is large enough, the overall result is still valid.

On the other hand, as will be seen later, the Huygens-Fresnel diffraction integral is mathematically equivalent to the paraxial equation used to derive the propagation of laser beams. Therefore, as Siegman (1986) writes, it is the paraxial nature of the wave taken into account that makes Fresnel's approximation valid: «the paraxial or Fresnel approximation is a physical property of the optical beam, not a mathematical property of the Huygens-Fresnel formulation».

4.2.5 Fraunhofer's approximation

When Z_o is very large with respect to the aperture size, for which

$$\frac{\left(x^2+y^2\right)_{max}}{2Z_o} << \frac{\lambda}{4}, \tag{4.2.16}$$

Eq. (4.2.7) can be further simplified by obtaining

$$R = Z_o + \frac{{x_o}^2+{y_o}^2}{2Z_o} - \frac{xx_o+yy_o}{Z_o}, \tag{4.2.17}$$

which is called *Fraunhofer's approximation*. Eq. (4.2.8) is reduced to

$$U_o\left(x_o, y_o\right) = -\frac{ie^{ikZ_o}}{\lambda Z_o} e^{i\frac{k}{2Z_o}\left({x_o}^2+{y_o}^2\right)} \iint_A e^{-i\frac{2\pi}{\lambda Z_o}\left(xx_o+yy_o\right)} U(x,y)\,dx\,dy\,, \tag{4.2.18}$$

which allows for obtaining the field U in the Fraunhofer's region.

Apart from the terms out of the integral, Eq. (4.2.18) is simply the Fourier transform of the field on the opening, and is evaluated (as before) at the frequencies $f_x = x_o/(\lambda Z_o)$, $f_y = y_o/(\lambda Z_o)$. The qualitative behavior of the diffraction described in §4.2.1 is thus understood. For large distances, the diffraction shape has an aspect fixed by the Fourier transform, but its size grows with Z_o. Indeed, for unchanged spatial frequencies f_x, f_y, the coordinates x_o, y_o vary as $Z_o\lambda f_x$, $Z_o\lambda f_y$, respectively.

Instead, the variability of the diffraction pattern in the intermediate Fresnel's region corresponds to the decrease, with increasing Z_o, of the contribution of

$$e^{i\frac{x^2+y^2}{2Z_o}}.$$

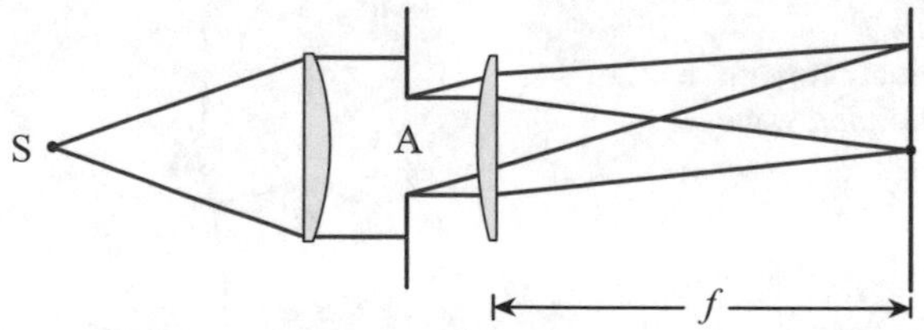

Fig. 4.6 Method for observing Fraunhofer's diffraction figures

At optical wavelengths, the conditions required for the validity of Eq. (4.2.17) may be very severe. If D is the width of the aperture (e.g., the diameter, if A is circular), Eq. (4.2.16) can be rewritten as

$$Z_o >> \frac{D^2}{2\lambda}, \tag{4.2.19}$$

and many authors consider a conventional threshold value $Z_F = D^2/\lambda$. For example, with $D = 5$ mm and $\lambda = 500$ nm, $Z_F = 50$ m. However, Fraunhofer's diffraction is of great practical importance, thanks to its simplicity. Moreover, diffraction patterns of this type can be obtained without resorting to very large distances. For example, consider the diffraction pattern produced by a plane wave incident on A and observed at a great distance from A. The same figure (or, better yet, the same intensity profile), apart from a scale factor, can be obtained with a convergent wave, rather than plane, on an observation plane located at a relatively small distance, that is, on the convergence point of the incident wave. The same can be done by keeping the incident wave plane, by laying a converging lens downstream of the opening with its focus on the observation plane.

Both in the case of the converging wave and that of the lens, the effect is to eliminate, compensating it, the term in $x^2 + y^2$ in the integrand. In the case of the converging wave, this is quite obvious, while the effect of a lens will be studied in detail in a section of Chapter 5. Here, I anticipate only that the normal experimental arrangement for observing Fraunhofer's diffraction is of the type shown in Fig. 4.6. The first lens produces a wave (at least approximately) plane on the aperture A. The second lens converts the direction of each of the plane waves of which we can think of the diffracted wave being composed, at a point in the focal plane. The two lenses must be large enough to not, in turn, be a diffracting diaphragm.

4.2.6 Debye's approximation

In the case of a convergent wave focused with large numerical aperture, one generally uses what is called *Debye's approximation*, as an alternative to Fresnel's approximation, which, instead, is valid for apertures smaller than the observation distance, and therefore with small numerical apertures. In Debye's approximation, the field in the focal region is treated as a superposition of plane waves whose

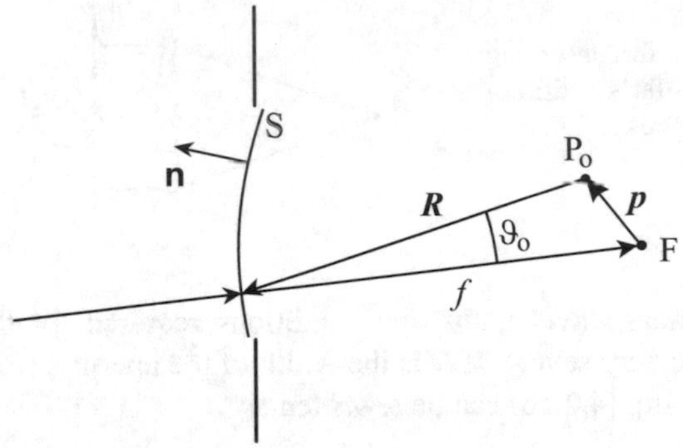

Fig. 4.7 Diffraction with a converging wave

wave vectors fall within the geometric cone, having, as a vertex, the focal point and, as a basis, the opening. In other words, this approximation consists of a truncation in the angular spectrum of the diffracted field.

Let us consider, therefore, the case in which an approximately spherical converging wave illuminates an opening and has its center at a point F. As the integration surface for the calculation of the diffraction, we take a spherical surface of radius f with a center in F delimited by the edge of the aperture itself (Fig. 4.7). On this surface, the incident wave field can be expressed as

$$U = P(\boldsymbol{r})\frac{e^{-ikf}}{f}, \tag{4.2.20}$$

where the complex function P takes into account both the field amplitude variations and a small deviation of the wavefront from the sphere, as the position $\boldsymbol{r}$ on the integration surface changes. We also assume that the aperture size and the distance f are much larger than the wavelength, and we are limited to the observation points P_o whose distance from F is small compared to f. The vectors $\mathbf{n}$, $\boldsymbol{R}$, $\boldsymbol{f}$, $\boldsymbol{p}$ are thus defined in Fig. 4.7.

By applying these conditions to Kirchhoff's diffraction integral, assuming the point F as the origin for the positions and ignoring the term in $\partial P/\partial n$ in the derivative of the field, we get

$$U_K(p) = \frac{-ik}{4\pi}\iint_S P(\boldsymbol{r})\frac{e^{ik(R-f)}}{Rf}\left[1+\cos(\mathbf{n},\boldsymbol{R})\right]dS. \tag{4.2.21}$$

If the distance p between F and P_o is also small compared to f, the cosine can be approximated by 1, while the term $R-f$ can be approximated as

$$R - f = \sqrt{f^2 + p^2 + 2\boldsymbol{p}\cdot\boldsymbol{f}} - f \cong \boldsymbol{p}\cdot\mathbf{s},$$

where $\mathbf{s}(\boldsymbol{r})$ is the versor of $\boldsymbol{f}$. Finally, taking $dS/(Rf) \cong dS/f^2 = d\Omega$, the integral

(4.2.21) becomes an integral on the solid angle Ω subtended by the aperture, whereby

$$U_K(\boldsymbol{p}) = \frac{-i}{\lambda} \iint_{\Omega} P(\boldsymbol{r}) e^{i\boldsymbol{k}\cdot\boldsymbol{p}} d\Omega \,, \qquad (4.2.22)$$

where $\boldsymbol{r}$ and $\boldsymbol{k} = k\mathbf{s}$ are now functions of the angular coordinates. This is called Debye's diffraction integral, and the approximations used for its derivation are called *Debye approximations* .

The expression (4.2.22) can be considered as the superposition of plane waves whose amplitude is proportional to the field at individual points of the spherical surface of integration. In other words, each plane wave corresponds to a single geometric ray passing through the aperture, with the phase and intensity associated with it. On the other hand, it neglects the contribution of plane waves diffracted outside of the geometrical cone, which can only be neglected when the *Fresnel's number* is >> 1. This number is defined as the number of Fresnel zones $N_F = a^2/(4\lambda Z_o)$ corresponding to a circular aperture of diameter a, illuminated by a plane wave and observed at the longitudinal distance Z_o (see §4.4). Here, Z_o coincides with f.

4.2.7 Diffraction through an optical system described by an 𝔄𝔅ℭ𝔇 *matrix*

In his major article from 1970, Stuart Collins introduced a diffraction integral that is the generalization of Fresnel's integral through an optical system, by linking the diffraction theory with Geometric Optics [Collins 1970]. Indeed, in Fresnel's integral, the amplitude of the field at each point on the output plane is the sum of the elementary contributions from the points of the input plane, weighed by a phase and an amplitude that depend on the geometric optical path between the two points. The same result was published independently shortly afterwards by J.A. Arnaud in a hard-to-read work [Arnaud 1970]. In Collins' integral, the kernel of the operation is still a function of the optical path, which I express here in terms of the 𝔄𝔅ℭ𝔇 matrix describing the system, already defined in Eq. (2.7.52), between an initial plane 1 and a final plane 2. Its result is limited to paraxial approximation, but it also applies to asymmetric optical systems, in which the main beam follows a non-axial path, with the optics placed in space even in a non-planar manner. Optical surfaces can also be astigmatic, ellipsoidal or cylindrical, and rotated around the optical axis. For the validity of the integral, it is, however, necessary that, between the initial and final planes, there are no diffracting elements, such as apertures that intercept the energy flux of the waves that pass through the system. Therefore, the present optics must be large enough. This integral is particularly well suited to the study of optical resonant cavities, and we will use it for the propagation of laser beams.

The kernel of the integral, to be applied to the amplitude of the field, is therefore expressed as

$$K(x_1, y_1; x_2, y_2) = a(x_1, y_1; x_2, y_2)\, e^{ik_o L(x_1, y_1; x_2, y_2)}, \tag{4.2.23}$$

where, again, $k_o = 2\pi/\lambda_o$ is the wave vector module in the vacuum, while L is the optical path and a is a complex amplitude. Both depend on the positions of the points considered on the initial and final planes. After determining the value of L through the system (see §2.7.6), Collins applies the same approximations used to derive the Fresnel diffraction integral, essentially assuming that a is constant with respect to the positions, although depending on the parameters of the optical system. This is equivalent to replacing the amplitude term $1/R$ with $1/Z_o$ and then having it moved out of the integral (4.2.1). Therefore, the amplitude of the electric field at the observation point is

$$E_2(x_2, y_2) = a \iint_{-\infty}^{\infty} e^{ik_o L(x_1, y_1; x_2, y_2)} E_1(x_1, y_1)\, dx_1\, dy_1 \ . \tag{4.2.24}$$

Finally, for the determination of a, Collins applies the principle of energy conservation. If the initial and final media are ordinary dielectrics, for Eq. (1.5.14), we can assume that[2]

$$\iint_{-\infty}^{\infty} n_2 \left|E_2(x_2, y_2)\right|^2 dx_2 dy_2 = \iint_{-\infty}^{\infty} n_1 \left|E_1(x_1, y_1)\right|^2 dx_1\, dy_1 \ , \tag{4.2.25}$$

where n_1 and n_2 are, respectively, the refractive indexes in the initial and final spaces. Eq. (4.2.25) differs from Collins' original one because he implicitly considers a unitary index for both of these spaces (see footnote in §2.7.6).

Replacing Eq. (4.2.24) in (4.2.25) and applying Eqs. (2.7.73, 76 and 77) for the optical path, Collins finds the module of a, except for a complex multiplicative constant of unit module. For comparison with the Fresnel integral for the case in which the optical system is the simple propagation in a homogeneous space, it is found that this constant is $-i$. Finally, we have

$$a = -i\sqrt{\frac{n_1}{n_2}}\, \frac{n_1 k_o}{2\pi\sqrt{|\mathfrak{B}|}},$$

where $|\mathfrak{B}|$ denotes the *determinant* of $\mathfrak{B}$ and the presence of the refractive indices

2 The assumption of such preservation means that the various transmitting optical surfaces have good anti-reflection treatment, and that the mirrors are highly efficient, otherwise the discrepancies may be significant.

generalizes Collins' result for the case of initial and final indexes other than 1. Collins's diffraction integral is therefore

$$E_2(x_2,y_2) = -i\sqrt{\frac{n_1}{n_2}}\frac{n_1 k_o}{2\pi\sqrt{|\mathbf{B}|}}e^{ik_o L_o} \times \iint_{-\infty}^{\infty} e^{ik_o L_1(x_1,y_1;x_2,y_2)} E_1(x_1,y_1)\,dx_1\,dy_1\,, \tag{4.2.26}$$

where L is decomposed in the sum of two terms: $L = L_o + L_1$, of which

$$L_1 = \frac{1}{2}\left(\rho_1{}^{\mathrm{T}};\rho_2{}^{\mathrm{T}}\right)\begin{pmatrix} n_1\mathbf{B}^{-1}\mathbf{A} & -n_1\mathbf{B}^{-1} \\ -n_1\mathbf{B}^{-1\mathrm{T}} & n_2\mathbf{D}\mathbf{B}^{-1}\end{pmatrix}\begin{pmatrix}\rho_1\\ \rho_2\end{pmatrix},\quad \rho_1 = \begin{pmatrix}x_1\\ y_1\end{pmatrix},\quad \rho_2 = \begin{pmatrix}x_2\\ y_2\end{pmatrix}, \tag{4.2.27}$$

while L_o is the optical path along the principal ray (see §2.7.6). I remember here that L_1 is defined only if the determinant of $\mathbf{B}$ is not zero, that is, when the planes 1 and 2 are not conjugated with each other. If this is the case, the field U_2 should be calculated in another way. Moreover, in this integration, the effects of any finite apertures are not considered; however, as Collins points out, this restriction can be removed with due caution by calculating the field up to the opening, multiplying it by the opening function and again applying the integral to the rest of the system. The same procedure is preferable when the plane 2 is next to be conjugate with the plane 1.

Lastly, if the system matrix is diagonalizable along two directions, in other words, if the system can be described by two independent 2x2 matrices $A_xB_xC_xD_x$ and $A_yB_yC_yD_y$, Eq. (4.2.26) takes the following form:

$$E_2(x_2,y_2) = -i\sqrt{\frac{n_1}{n_2}}\frac{n_1 k_o}{2\pi\sqrt{B_x B_y}}e^{ik_o L_o}\iint_{-\infty}^{\infty} E_1(x_1,y_1)\,dx_1\,dy_1 \times e^{i\frac{k_o}{2}\left[n_2\left(\frac{D_x}{B_x}x_2{}^2+\frac{D_y}{B_y}y_2{}^2\right)-2n_1\left(\frac{x_2x_1}{B_x}+\frac{y_2y_1}{B_y}\right)+n_1\left(\frac{A_x}{B_x}x_1{}^2+\frac{A_y}{B_y}y_1{}^2\right)\right]}. \tag{4.2.28}$$

4.3 Examples of Fraunhofer's diffraction

In the following, we will consider, for simplicity, only the case of a plane wave incident on an aperture A, that is, a wave with constant module and phase on A and with intensity equal to I_A. Here, I introduce some abbreviations common to all of the examples. Eq. (4.2.18) of the Fraunhofer diffraction can be rewritten as

Fig. 4.8
Definition of variables for a rectangular aperture

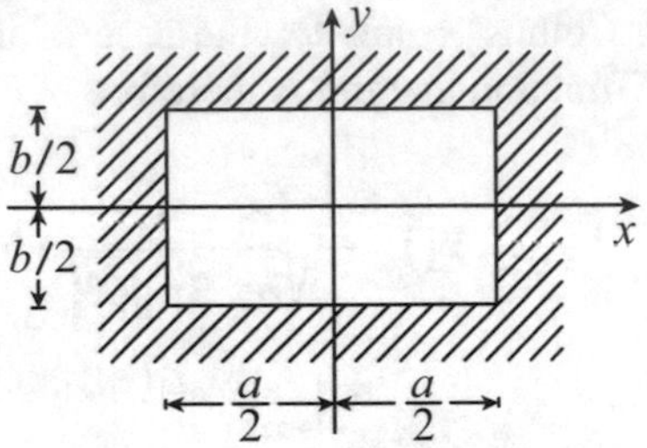

$$U_o(x_o, y_o) = C \iint_A e^{-i2\pi(f_x x + f_y y)} dx\, dy, \tag{4.3.1}$$

where the spatial frequencies f_x and f_y are given by Eq. (4.2.9), and

$$C = -U_A \frac{i}{\lambda Z_o} e^{ikZ_o} e^{i\frac{k}{2Z_o}\left(x_o{}^2 + y_o{}^2\right)}, \tag{4.3.2}$$

where U_A is the amplitude of the field on the aperture.

4.3.1 Rectangular aperture

For a rectangular aperture with the size and position of the axes defined as in Fig. 4.8, Eq. (4.3.1) becomes

Fig. 4.9
Intensity profile along an axis in the Fraunhofer's diffraction from a rectangular aperture

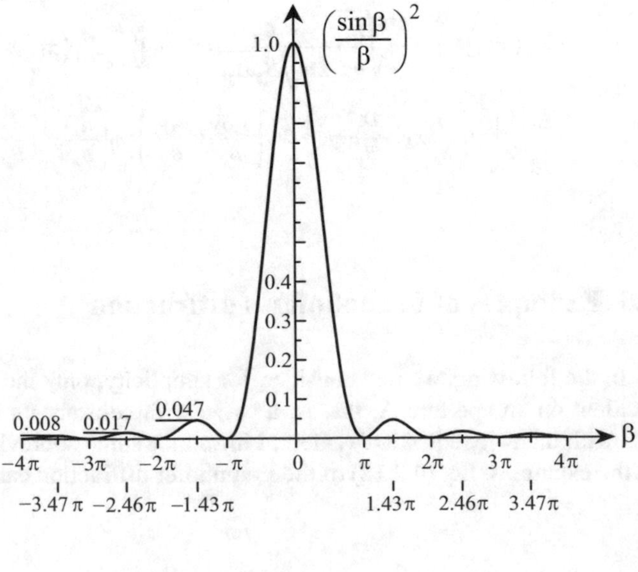

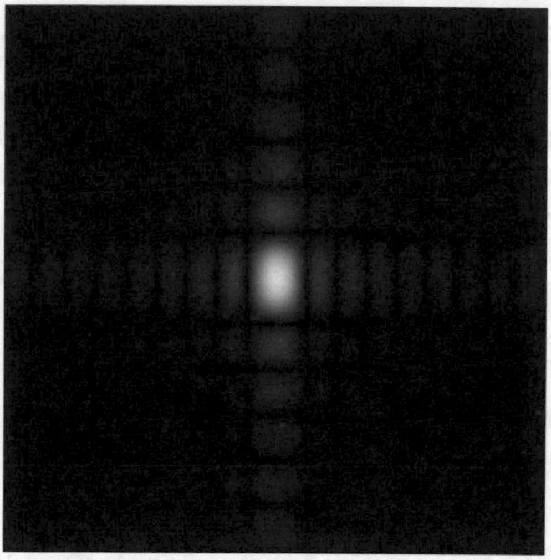

Fig. 4.10
Fraunhofer diffraction pattern produced by a rectangular aperture. This image is obtained from a numerical simulation in the same conditions as those used in Fig. 4.5. It represents the angular spectrum generated by an aperture with sides of 1 mm × 1.5 mm, illuminated by a plane wave with $\lambda = 1\,\mu\text{m}$. The figure extends for about 12 mrad for each side

$$U_o(x_o, y_o) = C \int_{-a/2}^{a/2} \int_{-b/2}^{b/2} e^{-i2\pi\left(f_x x + f_y y\right)} dxdy = abC \,\mathrm{sinc}\left(\pi f_x a\right) \mathrm{sinc}\left(\pi f_y b\right). \quad (4.3.3)$$

The luminous intensity of the diffraction pattern is

$$I(x_o, y_o) = I_A \frac{a^2 b^2}{\lambda^2 Z_o^2} \mathrm{sinc}^2\left(\pi f_x a\right) \mathrm{sinc}^2\left(\pi f_y b\right), \quad (4.3.4)$$

where the dependence on each of the x and y coordinates is expressed by the function $\mathrm{sinc}^2(\beta)$ represented in Fig. 4.9. The typical overall appearance of the diffrac-

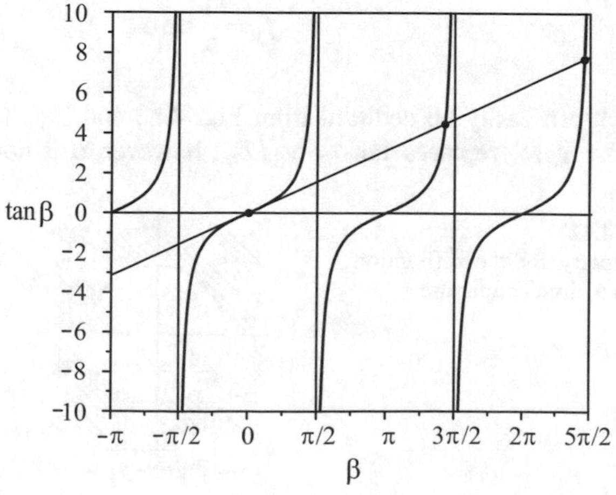

Fig. 4.11
Graphical determination of the maximum position

tion pattern is shown in Fig. 4.10: in particular, it is seen that it is furrowed by a regular grid of null intensity lines defined by $\pi f_x a = \pm\pi, \pm 2\pi, \cdots$ and by $\pi f_y b = \pm\pi, \pm 2\pi, \cdots$. Moreover, its extension and the pace of the grid in x and y varies *inversely proportional* to a and b, respectively. In terms of observation direction, the angular separation between the center and the first minimum (for small angles) is given by

$$\alpha_o = \frac{\lambda}{a}, \quad \beta_o = \frac{\lambda}{b}. \tag{4.3.5}$$

The position β_m of the intensity maxima can be obtained by deriving the function $\text{sinc}^2(\beta)$, from which we have

$$\tan\beta_m = \beta_m, \tag{4.3.6}$$

which can be resolved graphically as in Fig. 4.11. The central peak contains about 81.5% of the power that passes through the aperture and the eight side peaks that surround it contain about 8.7% of this power altogether.

4.3.2 Circular aperture

To calculate the diffraction pattern from a circular aperture of radius a, we can exploit the symmetry of the problem by choosing a direction of observation such that $x_o = 0$ and $y_o = Z_o \tan\alpha$, where α is the angle of observation. For all points (x_o, y_o) with the same viewing angle, the result will be the same. Therefore, we can take $f_x = 0$, so that, thanks to the circular symmetry, we can assume, for f_y, a somewhat more precise value than we get with Fresnel's approximation, that is,

$$f_y = \frac{1}{\lambda}\sin\alpha, \tag{4.3.7}$$

which can easily be deduced from Fig. 4.12 and Eqs. (4.2.11-12). In particular, $\sin\alpha = y_o/S$ replaces $\tan\alpha = y_o/Z_o$, however, α is normally small and the dif-

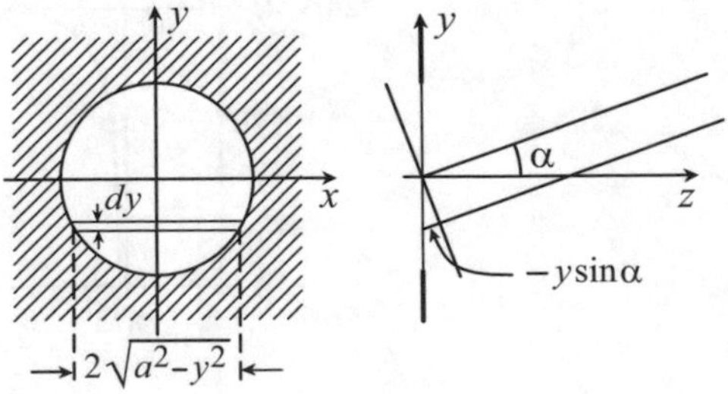

Fig. 4.12 Geometry for the diffraction from a circular aperture

ference is often negligible. The field that is observed at the angle α is therefore:

$$U(\alpha) = C\int_{-a}^{a} e^{-i\rho\frac{y}{a}} 2\sqrt{a^2 - y^2}\,dy\,,$$

with

$$\rho = 2\pi\frac{\sin\alpha}{\lambda}a\,. \tag{4.3.8}$$

Now, exploiting one of the known integral representations of the Bessel function J_1, whereby

$$J_1(\rho) = \frac{2\rho}{\pi}\int_{-1}^{1} e^{-i\rho u}\sqrt{1-u^2}\,du\,, \tag{4.3.9}$$

and taking $y = au$, we finally find that

$$U(\alpha) = C\pi a^2\,\frac{2J_1(\rho)}{\rho}\,, \tag{4.3.10}$$

where ρ (function of α) is given by Eq. (4.3.8).

The luminous intensity of the diffraction pattern is now given by:

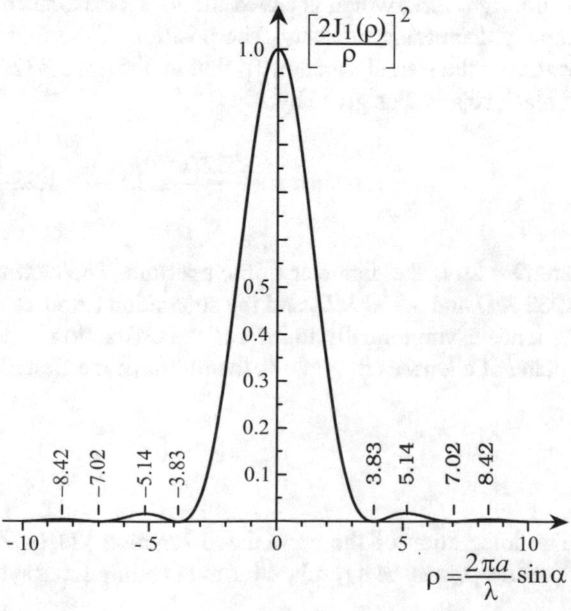

Fig. 4.13 Intensity profile along a radial axis of the Fraunhofer's diffraction pattern from a circular aperture. With the peak normalized to unity, the height of the first secondary maximum is only 0.0175, and only 0.0042 for the second

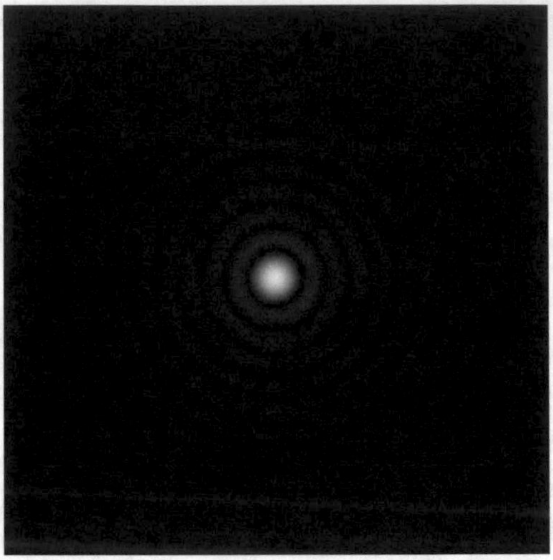

Fig. 4.14
Airy's rings with a circular aperture. Even this image is obtained by a numerical simulation under the same conditions used for Fig. 4.5. It represents the angular spectrum produced by a circular aperture of radius 1 mm, illuminated by a plane wave with $\lambda = 1\,\mu\text{m}$. The figure extends for about 12 mrad for each side

$$I(\alpha) = I_{\mathrm{A}} \left(\frac{\pi a^2}{\lambda R_o} \right)^2 \left(\frac{2\,\mathrm{J}_1(\rho)}{\rho} \right)^2, \tag{4.3.11}$$

which is called Airy's formula. The intensity distribution is characterized by the function $[2\mathrm{J}_1(\rho)/\rho]^2$, which is shown in Fig. 4.13: notice its resemblance to function $\mathrm{sinc}^2(\beta)$.

The diffraction shape is now circularly symmetrical (Fig. 4.14) and consists of a central light disk, which is called the *Airy's disk*, surrounded by concentric rings of rapidly decreasing intensity. The position of the first black ring is given by the first zero of the Bessel function J_1, that is, for $\rho \cong 3.832$. The angular radius of this first black ring is thus given by

$$\alpha \cong \sin\alpha \cong \frac{3.832}{2\pi}\frac{\lambda}{a} \cong 0.61\frac{\lambda}{a} = 1.22\frac{\lambda}{D}, \tag{4.3.12}$$

where $D = 2a$ is the diameter of the aperture. The next two black rings occur at $\alpha \cong 2.232\,\lambda/D$ and $3.238\,\lambda/D$, and the separation ($\Delta \sin\alpha$) between two neighboring rings tends asymptotically to λ/D. If the diffraction pattern is observed on the focal plane of a lens, as in Fig. 4.6, the radius of the first black ring is given by

$$r_{\mathrm{A}} = \alpha f = 0.61\frac{\lambda f}{a}. \tag{4.3.13}$$

The integration of the normalized function $[2\mathrm{J}_1(\rho)/\rho]^2/\pi$ on a circular area with radius ρ_o has, as a result, a formula credited to Rayleigh:

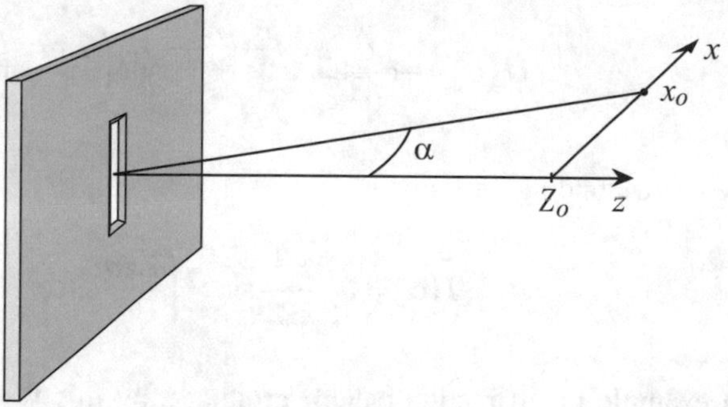

Fig. 4.15
Rectilinear slit

$$L(\rho_o) = 1 - J_0^2(\rho_o) - J_1^2(\rho_o), \tag{4.3.14}$$

which represents the fraction of energy that flows within the radius ρ_0 and where J_0 is the Bessel function of the first kind of order 0. Thus, the energy flowing through the Airy's disk is about 83.8% of the total flow, while that of the first ring is about 7.2% and that of the second ring is about 2.8% of the total flow.

4.3.3 Single slit

In the case of a rectangular opening having sides *a* and *b* that are very different from each other, such as $b >> a$ (Fig. 4.15), it may be possible that, for a given distance Z_o of observation, Fraunhofer's condition is only respected in one direction, say, *x* parallel to the shorter side *a*. Instead, in the direction *y*, parallel to the longer side *b*, the fringes may be poorly developed and the light that arrives on the observation plane can still appear as the geometric projection of the opening. Under these conditions, the problem can be treated in a one-dimensional way. With a few calculations, which I do not show here for brevity, for a plane wave perpendicularly incident to the aperture A, the field observed at the angle α from the normal to A is, with good approximation, given by

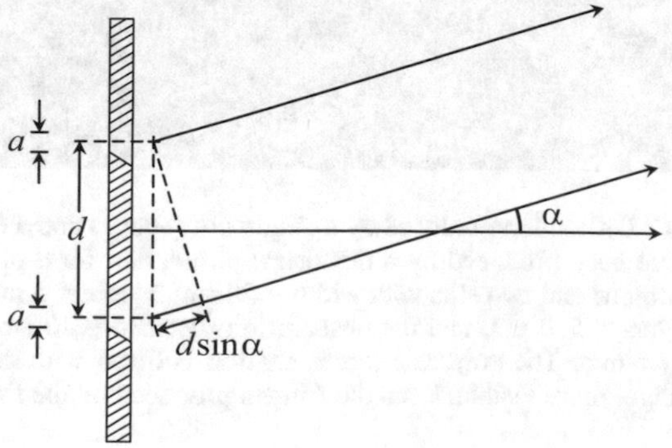

Fig. 4.16
Double slit

$$U(\alpha) = -i\frac{U_A\, a}{\sqrt{\lambda Z_o}} e^{i\frac{\pi}{4}+i\frac{kZ_o}{\cos\alpha}} \operatorname{sinc}\left(\frac{2\pi a}{\lambda}\sin\alpha\right). \qquad (4.3.15)$$

The corresponding intensity is

$$I(\alpha) = I_A \frac{a^2}{\lambda Z_o} \operatorname{sinc}^2\left(\frac{2\pi a}{\lambda}\sin\alpha\right). \qquad (4.3.16)$$

An example of diffraction pattern produced by this type of aperture is given in Fig. 4.17(a).

4.3.4 *Double slit*

This historically important case was proposed by Young for the demonstration

Fig. 4.17 Diffraction produced by a single slit (above) and a double slit (below). These figures have been produced by a numerical simulation for a plane wave with $\lambda = 1$ μm, incident on one and two slits with width = 50 μm, height = 5 mm; the separation between the slit centers = 500 μm, and the observation distance = 10 cm. The size of the image is 20 mm x 7.5 mm. The grayscale representation is linear with the field module so as to make side fringes more visible. Even the fringes produced by the finite height of the slits are visible

of its principle of interference. Let us consider two equal and parallel slits, whose mutual distance, taken from the center of each, is d (Fig. 4.16). Up to a constant phase factor, the field observed at the angle α is now

$$U(\alpha)=-i\frac{U_A\, a}{\sqrt{\lambda Z_o}}e^{i\frac{kZ_o}{\cos\alpha}}\operatorname{sinc}\left(\frac{2\pi a}{\lambda}\sin\alpha\right)\left(e^{ik\frac{d}{2}\sin\alpha}+e^{-ik\frac{d}{2}\sin\alpha}\right),$$

which, for small angles α, can be approximated by

$$U(\alpha)=-i\frac{U_A\, a}{\sqrt{\lambda Z_o}}e^{ikZ_o}2\operatorname{sinc}\left(\frac{2\pi a}{\lambda}\alpha\right)\cos\left(\frac{kd\alpha}{2}\right). \tag{4.3.17}$$

The intensity profile is now:

$$I(\alpha)=4\frac{I_A\, a^2}{\lambda Z_o}\operatorname{sinc}^2\left(\frac{2\pi a}{\lambda}\alpha\right)\cos^2\left(\frac{\pi d}{\lambda}\alpha\right). \tag{4.3.18}$$

Compared to the previous case, the diffraction pattern is now also modulated by a square cosine term. An example is shown in Fig. 4.17(b).

4.4 Examples of Fresnel's diffraction

4.4.1 Fresnel's zones

As we have seen, Fresnel's diffraction occurs when the sources or observation plane, or both, are close enough to the aperture so that the wavefront curvature becomes important. Because we are not working with plane waves here, the Fresnel diffraction is a bit more mathematically difficult to treat than that of Fraunhofer. However, it is most frequently observed, since it only needs a source, a diffracting aperture and an observation screen. Despite the mathematical complexity of the problem, Fresnel has introduced an ingenious method with which we can easily interpret various diffraction phenomena.

Let us consider a plane aperture A illuminated by a point source S and a point of observation P such that the plane of A is perpendicular to the line connecting S to P (Fig. 4.18). Let us denote, by O, the intersection point of the SP line with the plane of the aperture and, by Q, a generic point on that plane. The distances between these points are R, R_S, Z_o, Z_S, and r as indicated in Fig. 4.18. Consider, then, the path $\overline{\mathrm{SQP}} = R_s + R$ and imagine tracing a series of circles on the plane, centered in O, to indicate, for each circle m = 1, 2, 3, ..., the locus of points such that

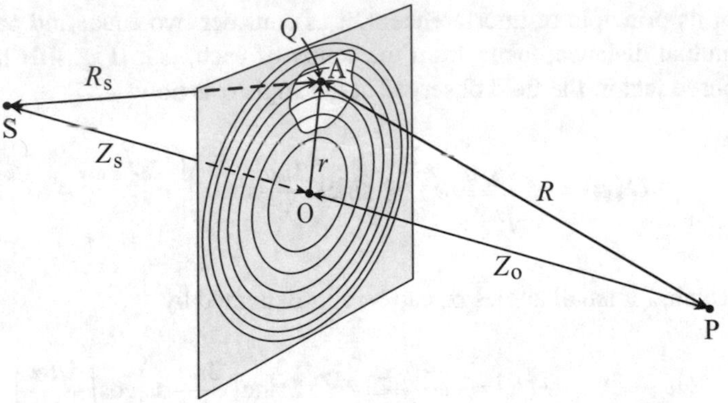

Fig. 4.18 Determination of Fresnel's zones

$$R + R_s = \text{costant} = Z_o + Z_s + m\frac{\lambda}{2} . \tag{4.4.1}$$

In other words, the path $\overline{SQP}$ differs by λ/2 from one circle to the next. The various circles delimit a series of concentric annular regions on the plane, called Fresnel's zones, which are homogeneous in the sense that, taking two points Q and Q′ within one of them, the difference in path is $\left| \overline{SQP} - \overline{SQ'P} \right| < \lambda/2$. With some calculations, it is found that, for $m \ll 2Z_o/\lambda$ and $m \ll 2Z_s/\lambda$, the radius r_m of the circle that externally delimits the m-th zone is

$$r_m = \sqrt{m\lambda \frac{Z_o Z_s}{Z_o + Z_s}} . \tag{4.4.2}$$

Typically, r can be very small; for example, if $Z_o = Z_s = 1$ m and λ = 500 nm, one finds that $r_1 = 0.5$ mm and $r_{100} = 5$ mm. In particular, with a source at infinity, the number of Fresnel's zones comprised in a given radius r is

$$N_F = \frac{r^2}{\lambda Z_o} . \tag{4.4.3}$$

This quantity is called *Fresnel's number*.

The path corresponding to the average radius $\bar{r} = (r_{m-1} + r_m)/2$ of the zone m is

$$\left\langle \overline{SQP} \right\rangle_m = \bar{R} + \bar{R}_s = \sqrt{\bar{r}^2 + Z_s^2} + \sqrt{\bar{r}^2 + Z_o^2} . \tag{4.4.4}$$

For Z_o, $Z_s \gg \lambda$, the width $\Delta r = r_m - r_{m-1}$ of each Fresnel's zone can be obtained by differentiating Eq. (4.4.4) and then imposing that $\left(\partial \overline{SQP}/\partial \bar{r}\right)\Delta r$ is λ/2. One has it so that

$$\Delta r \cong \frac{\lambda}{r_{m-1}+r_m}\frac{\overline{R}\,\overline{R}_\mathrm{S}}{\overline{R}+\overline{R}_\mathrm{S}} \quad \text{for } m >> 1, \tag{4.4.5}$$

where $\overline{R}$ and $\overline{R}_\mathrm{s}$ also implicitly depend on m. For $m \to \infty$, Eq. (4.4.5) is more accurate than that which can be obtained from Eq. (4.4.2), although it remains congruent with this for small m. The area of each zone is therefore

$$A_m = \Delta r \cdot \pi\left(r_{m-1}+r_m\right) = \pi\lambda\frac{\overline{R}\,\overline{R}_\mathrm{S}}{\overline{R}+\overline{R}_\mathrm{S}}, \tag{4.4.6}$$

so that each of them contributes to the integral of diffraction, expressed by Eq. (4.1.18) with

$$\begin{aligned} U_m &\cong -i\frac{k}{2\pi}\pi\lambda\frac{\overline{R}\,\overline{R}_\mathrm{S}}{\overline{R}+\overline{R}_\mathrm{S}}\left\langle\frac{e^{ik(R+R_s)}}{R\,R_\mathrm{S}}O\left(\vartheta_\mathrm{S},\vartheta_\mathrm{o}\right)\right\rangle_m p_m \\ &\cong -i\pi\frac{O_m\left(\vartheta_\mathrm{S},\vartheta_\mathrm{o}\right)}{\overline{R}+\overline{R}_\mathrm{S}}\left\langle e^{ik(R+R_\mathrm{S})}\right\rangle_m p_m, \end{aligned} \tag{4.4.7}$$

where the average $<>_m$ is here understood over the interval Δr, corresponding to $\Delta\left(\overline{R}+\overline{R}_\mathrm{S}\right) = \lambda/2$. The factor p_m is the fraction of the zone that coincides with the aperture A. Finally, by averaging the exponential, one get

$$U_m \cong -2\frac{O_m\left(\vartheta_\mathrm{S},\vartheta_\mathrm{o}\right)}{\overline{R}+\overline{R}_\mathrm{S}}(-1)^m p_m e^{ik(Z_\mathrm{o}+Z_\mathrm{S})}, \tag{4.4.8}$$

so the U_m constitutes an alternating succession tending to zero for $m \to \infty$, thanks to the obliquity factor O_m and the term $1/(\overline{R}+\overline{R}_\mathrm{S})$. Therefore, the total field module at point P is

$$|U_\mathrm{P}| = |U_1| - |U_2| + |U_3| - \cdots. \tag{4.4.9}$$

Suppose now that $p_m = 1$ for all m, i.e., that there is no diffracting screen. Now, we group the terms of Eq. (4.4.9) as follows:

$$|U_\mathrm{P}| = \frac{1}{2}|U_1| + \left(\frac{1}{2}|U_1| - |U_2| + \frac{1}{2}|U_3|\right) + \left(\frac{1}{2}|U_3| - |U_4| + \frac{1}{2}|U_5|\right) + \cdots. \tag{4.4.10}$$

As a result, the total field in P is approximately half of the contribution of the first Fresnel zone! Indeed, since, in this case, $|U_m|$ decreases slowly with m, each parenthesis contains terms that erase each other in good approximation. Thus,

$$|U_{\mathrm{P}}| \cong \frac{1}{2}|U_1|. \tag{4.4.11}$$

On the other hand, in the absence of a screen, the field in P can be directly calculated, and, assuming an obliquity factor O_1 equal to 1, the amplitude of the field is exactly equal to half of U_1 calculated with Eq. (4.4.8).

So far, we have considered, for simplicity, the "axial" case of a plane screen perpendicular to the line between the source and the point of observation. In particular, moving away the point P from the axial position, the Fresnel's zones move proportionally in the same direction, remaining centered on the intersection between the straight line SP and the surface containing the aperture (Fig. 4.19). Therefore, the Fresnel's zone method is quite general, and can also be applied to non-planar screens and arbitrary inclinations. Similarly, it is not necessary for the source to be point-like, just enough to know the shape of the wavefront on the aperture: in general, the Fresnel's zones will no longer be annular, but will depend on the shape of the wavefront.

4.4.2 Circular aperture

As an example of an application of the Fresnel zones, we now consider a circular aperture centered in O and with variable radius r_{A}. If only the first Fresnel's zone is uncovered, the amplitude of the field in P is just U_1, i.e., double the amplitude that it would be for $r \to \infty$. So, the intensity in P is four times what there would be in the absence of the screen.

By increasing $\mathrm{r_A}$ up to the point so as to uncover the second Fresnel's zone, the intensity in P decreases to almost zero for the mutually and almost complete cancellation between U_1 and U_2. In addition, around P, a bright ring is forming, which can be interpreted as follows. From the points (off axis) of this ring (Fig. 4.19), the first Fresnel's zone is fully visible, while the second and third zones are only partially visible and their contribution, at least partially, is mutually compensated.

Continuing to increase r_{A}, the intensity in P oscillates with the minima and maxima, respectively, at an even or odd number of uncovered zones. An example of this behavior is shown in Fig. 4.20.

A quantitative evaluation of this phenomenon can be given quite simply when P and S are on the axis of the opening. For brevity, we use the variables below:

$$R' = 2\frac{Z_{\mathrm{o}} Z_{\mathrm{s}}}{Z_{\mathrm{o}} + Z_{\mathrm{s}}} \quad \text{and} \quad Z_{\mathrm{T}} = Z_{\mathrm{o}} + Z_{\mathrm{s}}. \tag{4.4.12}$$

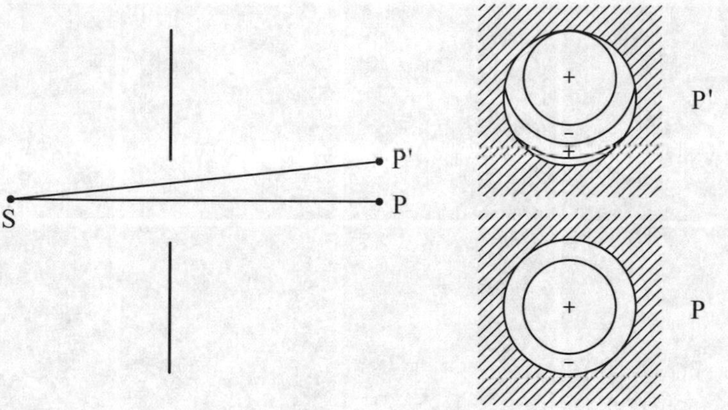

Fig. 4.19 Circular aperture seen from two different positions P and P′

The field $U(x,y)$ on the opening can be approximated with

$$U(x,y) = A\frac{e^{ikZ_s}}{Z_s}e^{i\frac{k}{2Z_s}\left(x^2+y^2\right)},$$

where A is a constant that fixes its amplitude. Eq. (4.2.8) becomes:

$$U(\mathrm{P}) = -iA\frac{e^{ikZ_T}}{\lambda Z_o Z_{os}}\int_0^{r_A}\int_0^{2\pi} e^{i\frac{k}{R'}r^2}\, r\,dr\,d\phi, \tag{4.4.13}$$

which yields

$$U(\mathrm{P}) = A\frac{e^{ikZ_T}}{Z_T}\left(1-e^{i\frac{r_A{}^2 k}{R'}}\right). \tag{4.4.14}$$

The intensity measured on the axis of the aperture is therefore:

$$I(\mathrm{P}) = I_A\frac{4Z_s{}^2}{Z_T{}^2}\sin^2\left(\pi\frac{r_A{}^2}{\lambda R'}\right), \tag{4.4.15}$$

where I_A is the intensity on the aperture. Fig. 4.21 shows the relative intensity I/I_A for the case $Z_s = \infty$. By varying Z_o, this quantity has an infinite number of maxima and minima with their accumulation point at $Z_o = 0$. These oscillations will instead stop for $Z_o \to \infty$, i.e., in the Fraunhofer's region.

Eq. (4.4.15) also gives an intensity that oscillates without converging for $r_A \to \infty$, as well as for $Z_o \to 0$. This is an artifact resulting from the approximations in-

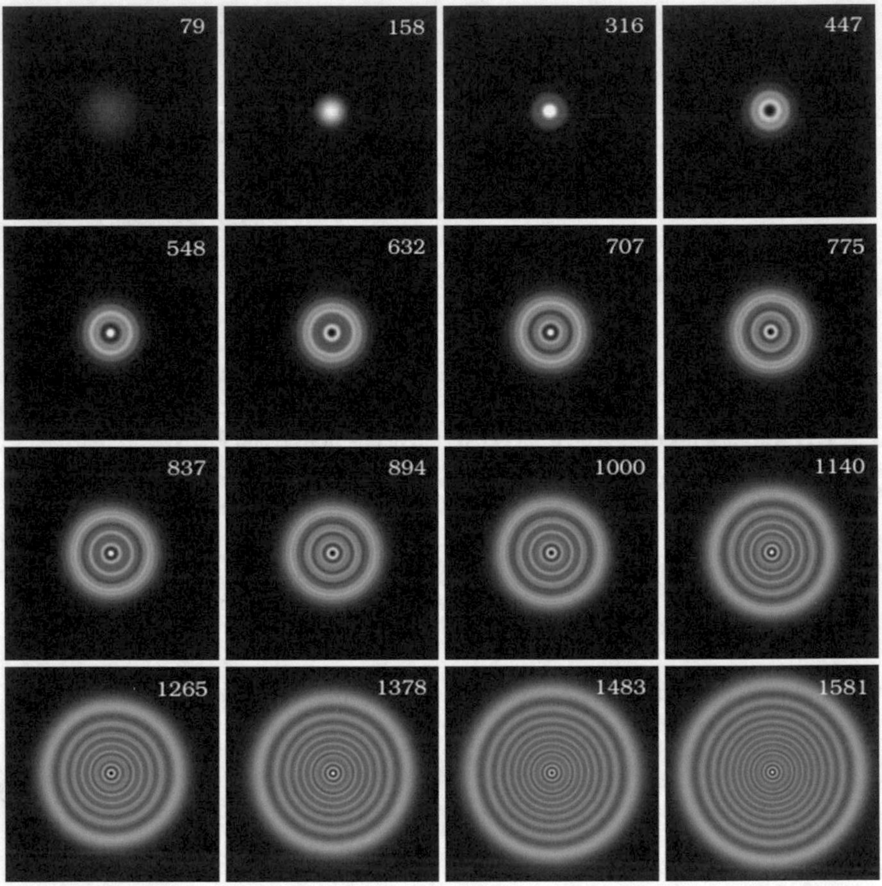

Fig. 4.20 Diffraction figures for circular apertures of increasing radius. Even these figures have been produced by a numerical simulation for an incident plane wave with $\lambda = 1$ μm and an observation screen placed at 10 cm from the openings. The number indicates the aperture radius in microns. Viewed from the screen, these radii respectively correspond to apertures of equal size, the first two at 1/4 and 1/2 of the first Fresnel's zone, the subsequent ones to the 1, 2, 3, 4, 5, 6, 7, 8, 10, 13, 16, 19, 22, 25 Fresnel's zones. The width of each figure is 3.2 mm. The grayscale representation is linear with intensity, whose value on the aperture corresponds to approximately 45% of the scale. In the figures with an odd number of Fresnel zones, the central peak has an intensity that exceeds the upper limit of the scale by almost double and is saturated. In the first two figures, the intensity was increased by 16 and 4 times, respectively

serted into the Fresnel diffraction integrals (4.2.5) and (4.2.8) relative to the obliquity factor and the distances R and R_S, and it does not represent the real physical behavior of the field.

However, for apertures sufficiently small compared to R', Eq. (4.4.15) is sufficiently accurate.

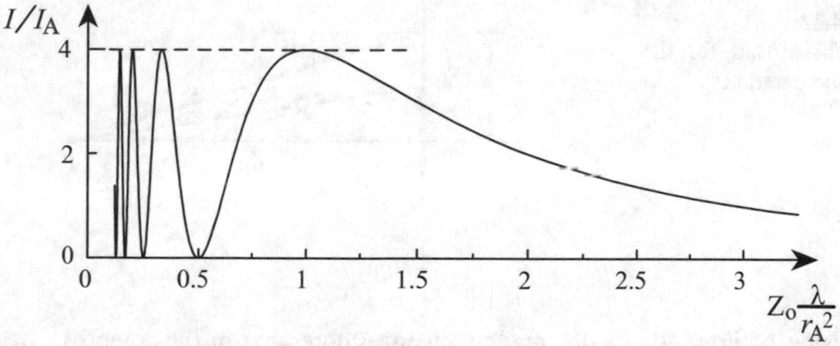

Fig. 4.21 Trend of the intensity on the axis of a circular aperture, illuminated by a plane wave, by varying the distance Z_o of the screen from the aperture

On the other hand, the case we are considering here is a singularity: the lack of convergence originates from the equidistance of the paths that, from S to P, skim the edge of the opening.

In the case of a plane wave impinging normally on a circular aperture, there is also an analytical solution from the Rayleigh-Sommerfeld first diffraction integral for intensity on the *axis* of the aperture. This "exact" solution is [Osterberg and Smith 1961; Sheppard and Hrynevych 1992]:

$$I(z) = I_A\left[1+\frac{z^2}{z^2+r_A{}^2}-\frac{2z}{\sqrt{z^2+r_A{}^2}}\cos\left(\frac{kr_A{}^2}{z+\sqrt{z^2+r_A{}^2}}\right)\right].$$

where z is the distance from the opening and r_A is its radius. From this expression, it follows that, for $z \to 0$, the amplitude of the intensity oscillation on the axis decays to zero, while, for $z >> r_A$, it can be approximated by Eq. (4.4.15) calculated within the limit of a source at infinity.

4.4.3 Poisson's paradox

An astonishing diffraction effect can be observed with circular obstacles, and the story of its discovery is quite interesting. When, in 1818, Fresnel presented his theory of diffraction to a competition organized by the French Academy, the jury included, among other eminent scientists of the time, S.D. Poisson, who supported the emission (corpuscular) theory. From Fresnel's theory, Poisson deduced a remarkable and seemingly absurd conclusion: he showed that, in the middle of the shadow projected by an opaque disk, one would have to observe a rather brilliant spot of light. At this objection, Fresnel and Arago, who was another member of the jury, readily responded with an experimental test: the spot of light was actually present! Fresnel's work won the prize and the title of *Mémoire Couronné*. This

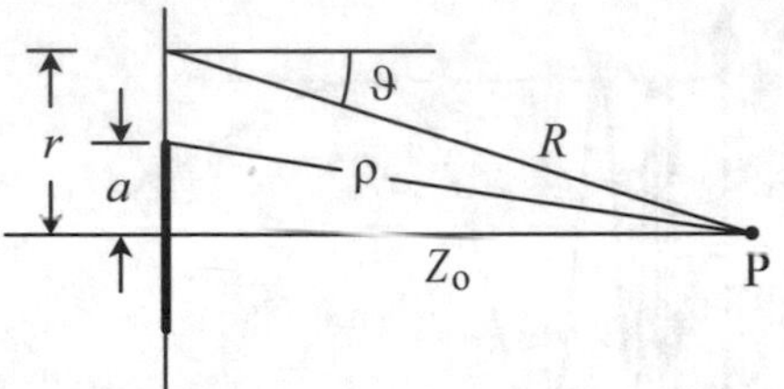

Fig. 4.22
Variables used for the Poisson paradox

prevision of Fresnel's theory made a strong impression in the scientific world at the time and was one of the deciding factors in favor of the wave theory of light. The Poisson-Arago spot, as it is now called, had been observed many years before (1723) by Giacomo Filippo Maraldi (1665-1729), but his work had been forgotten.

We can reformulate Poisson's reasoning as follows. Once a source S and an observation point P are fixed on the axis of a circular obstacle, we can trace, as before, a series of Fresnel zones whose edges meet the relationship

$$\overline{SQP} = \overline{SQ_A P} + m\frac{\lambda}{2}, \quad m = 1, 2, \cdots, \tag{4.4.16}$$

where Q is a point on the outer edge of the zone m considered and Q_A is a point on the edge of the obstacle. We can then reapply the reasoning that led to Eq (4.4.11): the field in P is half of the contribution of the Fresnel's zone immediately adjacent to the edge of the obstacle. The intensity on the axis, for distances of S and P from the obstacle large compared to the size of this, is almost that measured in the absence of the obstacle! As a quantitative example, we consider the case in which S is far from the obstacle for which the incident field can be treated as a plane wave. For the circular symmetry of this case, the intensity in P can be calculated easily and more precisely starting from Eq. (4.1.14), rather than using the subsequent approximations. With the various parameters defined as in Fig. 4.22, we have

$$R^2 = Z_o{}^2 + r^2, \quad r\,dr = R\,dR, \quad \cos\vartheta = \frac{Z_o}{R}, \quad \rho = \sqrt{Z_o{}^2 + a^2},$$

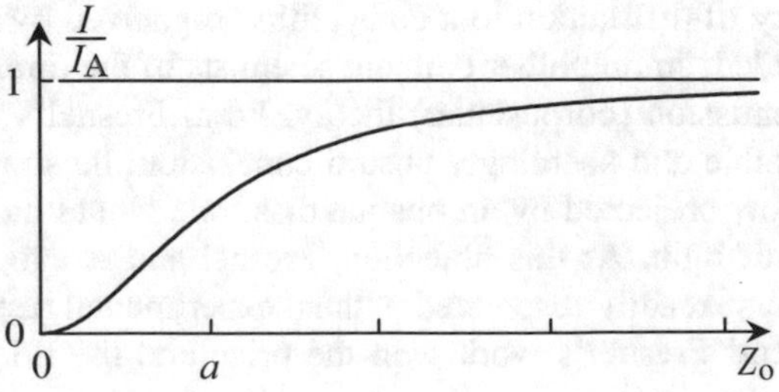

Fig. 4.23
Intensity in the center of the Poisson-Arago spot versus the distance from the obstacle, for a plane wave incident with intensity I_A. The obstacle is circular with radius a

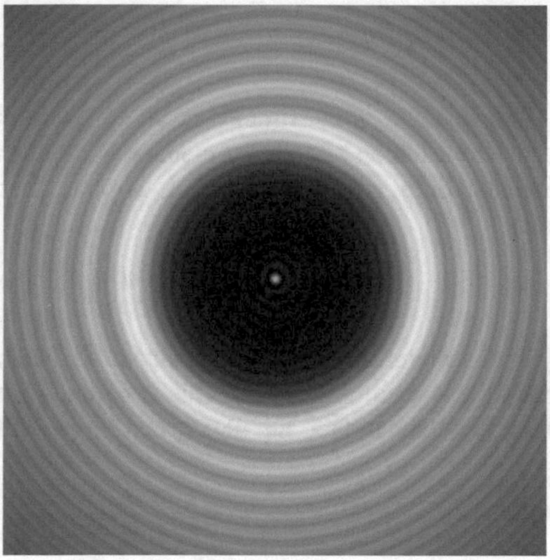

Fig. 4.24
Diffraction figure produced by a circular obstacle. Numerical simulation for a 1 mm radius obstacle on the waist plane of a Gaussian incident beam with $w = 5$ mm and $\lambda = 1$ μm. The figure is 5 mm wide and is produced on a screen placed at 20 cm from the obstacle. Grayscale representation is linear with intensity

so that

$$U(\mathrm{P}) = -\frac{i}{\lambda} U_{\mathrm{A}} \int_{\rho}^{\infty} \int_{0}^{2\pi} \frac{e^{ikR}}{R} \frac{Z_{\mathrm{o}}}{R} R\, dR\, d\phi = -i \frac{2\pi}{\lambda} Z_{\mathrm{o}} U_{\mathrm{A}} \int_{\rho}^{\infty} \frac{e^{ikR}}{R} dR\,,$$

where U_{A} is the field amplitude on the plane of the obstacle. The remaining integral can be expressed through the sine and cosine integral functions $\mathrm{Si}(x), \mathrm{Ci}(x)$ [Abramowitz and Stegun 1972]:

$$U(P) = i \frac{2\pi}{\lambda} Z_{\mathrm{o}} U_{\mathrm{A}} \left[\mathrm{Ci}(k\rho) + i \mathrm{Si}(k\rho) - i \frac{\pi}{2} \right] \cong -i \frac{Z_{\mathrm{o}} U_{\mathrm{A}}}{\rho} e^{-ik\rho}\,,$$

where the last equality is valid for $k\rho >> 1$. Therefore, the intensity observed in P is (Fig. 4.23)

$$I(\mathrm{P}) = I_{\mathrm{A}} \frac{Z_{\mathrm{o}}^2}{Z_{\mathrm{o}}^2 + a^2}\,, \tag{4.4.17}$$

where I_{A} is the intensity of the wave on the plane of the obstacle. For $Z_{\mathrm{o}} \to \infty$, this equation gives $I(\mathrm{P}) \to I_{\mathrm{o}}$: this is due to having taken a plane wave as the incident field.

In order for the Poisson-Arago spot to be visible, the disk irregularities should be less than the width of its first Fresnel zone. In addition, approaching the obstacle, the spot becomes smaller and smaller down to the size of the wavelength when the distance drops below a. A simulation of the diffraction produced by a small sphere is shown in Fig. 4.24.

4.4.4 Zone plate

If an opening is constructed so as to obstruct, with respect to a point P, the Fresnel zones in an alternating manner, such as those that are even, then all of the remaining ones give a contribution in phase between them, and Eq. (4.4.9) becomes

$$|U_P| = |U_1| + |U_3| + |U_5| + \cdots .$$

$|U_P|$, and therefore also the intensity in P, becomes much larger than in the absence of this kind of aperture. This is called a *zone plate*, and it acts in a manner similar to a lens; in fact, Eq. (4.4.2) can be rewritten as

$$\frac{1}{Z_o} + \frac{1}{Z_s} = \frac{m\lambda}{r_m^{\ 2}}, \tag{4.4.18}$$

which has the same shape as the Gauss's Eq. (2.6.35.a), where the variables s_o and s_i have been replaced with Z_s and Z_o, respectively. Thus, a zone plate has a focal length characterized by

$$f = \frac{r_m^{\ 2}}{m\lambda}, \tag{4.4.19}$$

for which the radiation emitted by a source placed on the axis of the lamina at a distance Z_s is refocused at a distance Z_o calculable from Eq. (4.4.18). This "lens", however, has a strong chromatic aberration from the dependence of f on λ. In addition, unlike normal lenses, the lamina has a multitude of focal distances: $\pm f$, $\pm f/3$, $\pm f/5$, ... so that, in addition to the main focus, the light is also focused to a lesser extent, on other points for which the transparent rings of the lamina correspond to an odd number of adjacent Fresnel's zones. Along with the converging diffracted waves, there are also diverging waves that correspond to the negative focal lengths.

4.4.5 Rectangular aperture

Let us suppose that a rectangular aperture, with the sides of amplitude a and b, assigned as in Fig. 4.8, is illuminated with normal incidence by a plane wave. From Eq. (4.2.5), one get

$$U_o(x_o, y_o) = C \int_{-a/2}^{a/2} \int_{-b/2}^{b/2} e^{i\frac{k}{2Z_o}\left[(x_o - x)^2 + (y_o - y)^2\right]} dx\, dy, \tag{4.4.20}$$

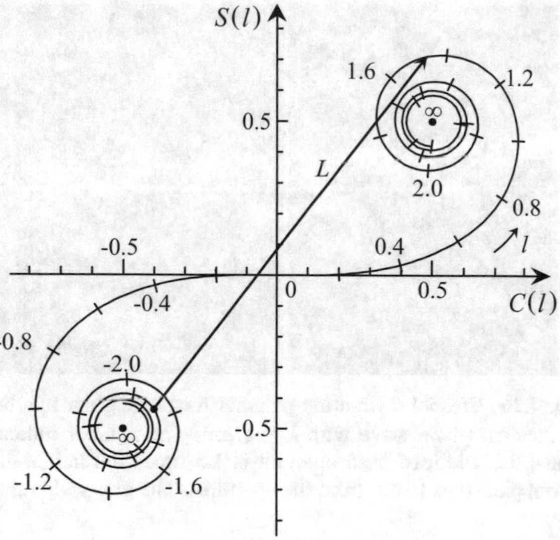

Fig. 4.25
Cornu's spiral: this curve is plotted taking, as coordinates, the Fresnel integrals by varying the parameter l, which, in turn, measures the path taken on the same curve from the origin. L is the length of a vector that joins two points of the curve

where C is now given by

$$C = -i\frac{e^{ikZ_o}}{\lambda Z_o}U_A\,,$$

where, in addition, U_A is the amplitude of the field on the aperture. Eq. (4.4.20) can be simplified by translating and scaling the variables x and y:

$$\xi = \sqrt{\frac{k}{\pi Z_o}}\left(x - x_o\right),\quad \eta = \sqrt{\frac{k}{\pi Z_o}}\left(y - y_o\right) \tag{4.4.21}$$

with a corresponding change in the integration limits:

$$\begin{aligned}\xi_1 &= -\sqrt{\frac{k}{\pi Z_o}}\left(\frac{a}{2} + x_o\right),\quad \eta_1 = -\sqrt{\frac{k}{\pi Z_o}}\left(\frac{b}{2} + y_o\right),\\ \xi_2 &= \sqrt{\frac{k}{\pi Z_o}}\left(\frac{a}{2} - x_o\right),\quad \eta_2 = \sqrt{\frac{k}{\pi Z_o}}\left(\frac{b}{2} - y_o\right),\end{aligned} \tag{4.4.22}$$

from which one gets

$$U\left(x_o, y_o\right) = C\frac{\pi Z_o}{k}\int_{\xi_1}^{\xi_2} e^{i\frac{\pi}{2}\xi^2}\,d\xi \int_{\eta_1}^{\eta_2} e^{i\frac{\pi}{2}\eta^2}\,d\eta$$

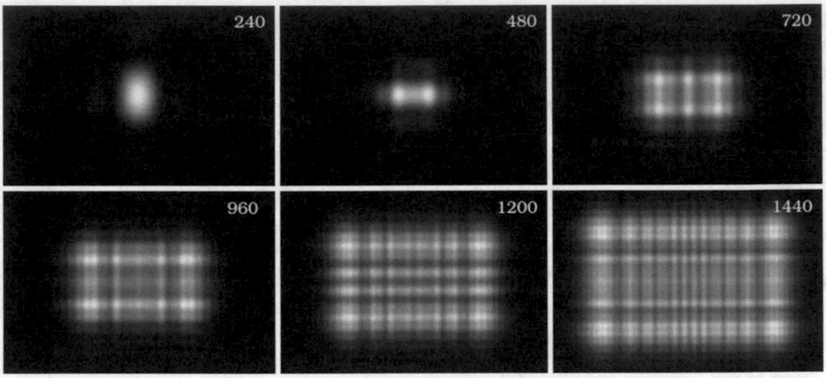

Fig. 4.26 Fresnel diffraction patterns for rectangular apertures. Numerical simulation for an incident plane wave with $\lambda = 1$ μm. The number indicates the aperture height in microns; the width of each aperture is 1.5 times the height. The figures are produced on a screen placed at 10 cm from the apertures. The grayscale representation is linear with intensity

from which

$$U(x_o, y_o) = -i\frac{e^{ikZ_o}}{2}U_A\left\{\left[C(\xi_2)-C(\xi_1)\right]+i\left[S(\xi_2)-S(\xi_1)\right]\right\} \\ \times\left\{\left[C(\eta_2)-C(\eta_1)\right]+i\left[S(\eta_2)-S(\eta_1)\right]\right\}, \tag{4.4.23}$$

where the functions C and S are Fresnel's integral (see appendix B at the end of this book). The corresponding distribution of intensity is

$$I(x_o, y_o) = \frac{I_A}{4}\left\{\left[C(\xi_2)-C(\xi_1)\right]^2+\left[S(\xi_2)-S(\xi_1)\right]^2\right\} \\ \times\left\{\left[C(\eta_2)-C(\eta_1)\right]^2+\left[S(\eta_2)-S(\eta_1)\right]^2\right\} \tag{4.4.24}$$

and, as usual, I_A is the intensity on the opening. The behavior of Fresnel integrals can be illustrated with a geometric representation introduced by A. Cornu. $C(l)$ and $S(l)$ are taken as the Cartesian coordinates of a point of a parametric curve: as the parameter l changes from $-\infty$ to $+\infty$, a double spiral is formed, called Cornu's spiral (Fig. 4.25). Since $C(0) = S(0) = 0$, the curve goes through the origin and, since

$$C(l) = -C(-l), \quad S(l) = -S(-l),$$

it is anti-symmetric with respect to both axes. Parameter l has the interesting property of being equal to the length of the curve measured from the origin. Indeed, if

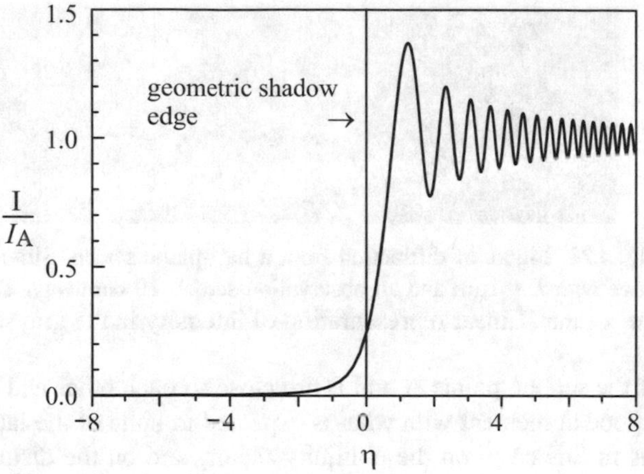

Fig. 4.27 Diffraction fringes produced by a rectilinear edge

ds is a curve arc element, we have

$$(ds)^2 = (dC)^2 + (dS)^2 = \left[\cos^2\left(\frac{\pi}{2}l^2\right) + \sin^2\left(\frac{\pi}{2}l^2\right)\right](dl)^2 = (dl)^2.$$

The intensity can thus be evaluated as the product

$$I(x_o, y_o) = \frac{I_A}{4} L^2(\xi_1, \xi_2) L^2(\eta_1, \eta_2), \tag{4.4.25}$$

where the function $L(\xi_1, \xi_2)$ is the length of the vector $\boldsymbol{L}$ joining two points that are found, respectively, going along the Cornu's spiral for the two distances ξ_1 and ξ_2 from the origin.

Let us consider, for example, the case of scanning the coordinate x_o, keeping $y_o = 0$. For $x_o = 0$, $\boldsymbol{L}$ has its two extremes at two diametrically opposite points on the spiral. As x_o grows, ξ_1 and ξ_2 change linearly with x_o, but, while one of the two points, A, wraps away from the origin, the other, B, unrolls approaching the origin. As a result, the observed intensity is oscillating with increasing amplitude. Finally, when x_o reaches the geometrical shadow edge, i.e., for $x_o = a/2$, point B comes to the origin. From that point on, growing x_o again, both points wrap themselves in the same direction away from the origin. When Z_o is small, the point A is already very internal to the spiral for which the intensity decays monotonically. When, instead, Z_o is large, i.e., close to or immersed in the Fraunhofer region, the intensity decays oscillating.

In particular, for very small values of Z_o compared to the size of the aperture, and for points of observation in the geometric projection area of the aperture, the two points A and B are both very close to one and the other of the two opposing centers (−½,−½) and (½,½), respectively, around which the spiral wraps, whereby $L \approx \sqrt{2}$, so that $I \approx I_A$. Instead, if the observation point is in the geometric shadow

Fig. 4.28 Figure of diffraction from a half-plane screen. Simulation for an incident plane wave with λ = 1μm and an observation screen 10 cm away. The figure has dimensions 4 mm x 1 mm. Linear representation of intensity in the grayscale

of the screen, points A and B are close to each other and $I \approx 0$. Therefore, there is a good agreement with what is expected in spite of the large approximations made (as in this case) on the obliquity factor, and on the distances. Some examples of Fresnel's diffraction on rectangular apertures are given in Fig. 4.26.

4.4.6 Slit and straight edge

A Fresnel diffraction from a long slit can be treated as in the case of a rectangular opening; taking $\xi_1 = -\infty$ and $\xi_2 = +\infty$ in Eq. (4.4.23), we get

$$U(x_o, y_o) = \frac{e^{ikZ_o}}{1+i} U_A \left\{ \left[C(\eta_2) - C(\eta_1) \right] + i \left[S(\eta_2) - S(\eta_1) \right] \right\}. \quad (4.4.26)$$

The rectilinear edge is, in turn, a limit case of the slit, taking $\xi_1 = -\infty$; so, assuming $a = 0$, we have

$$U(y_o) = \frac{e^{ikZ_o}}{1+i} U_A \left\{ \left[\frac{1}{2} - C\left(\sqrt{\frac{2}{\lambda R_o}}\, y_o \right) \right] + i \left[\frac{1}{2} - S\left(\sqrt{\frac{2}{\lambda R_o}}\, y_o \right) \right] \right\}. \quad (4.4.27)$$

A drawing of how the intensity varies as a function of $\eta = \sqrt{2/(\lambda Z_o)}\, y_o$ is given in Fig. 4.27 and an image of this type of diffraction is shown in Fig. 4.28.

Bibliographical references

Arnaud J.A., *Nonorthogonal optical waveguides and resonators*, The Bell Systems Technical Journal **49**, 2311-2348 (November 1970).

Abramowitz M. and Stegun I.A., *Handbook of Mathematical Functions*, Dover Publications Inc., New York (1972).

Born M. and Wolf E., *Principles of Optics*, Pergamon Press, Paris (1980).

Collins S.A. Jr., *Lens-system diffraction integral written in terms of Matrix optics*, J. Opt. Soc. Am. **60**, 1168-1177 (September 1970).

Ditchburn R.W., *Light*, Vol. I and II, Academic Press Inc., London (1976).

Ehrlich M.J., Silver S. and Held G., *Studies of the Diffraction of Electromagnetic Waves by Circular Apertures and Complementary Obstacles: The Near-Zone Field*, J. Appl. Phys. **26**, 336-345 (1955).

Erdélyi A., *Asymptotic expansion*, Dover Publications, inc., New York (1956).

Fowles G.R., *Introduction to Modern Optics*, Holt, Rinehart, and Winston, New York (1968).

Guenther R., *Modern Optics*, John Wiley & Sons, New York (1990).

Goodman J.W., *Introduction to Fourier Optics*, II ed., McGraw-Hill, San Francisco (1996).

Hecht E., *Optics, 2nd ed.*, Addison-Wesley, Madrid (1987).

Jenkins F.A. and White H.E., Fundamental of Optics, McGraw-Hill, New York (1957).

Keller J.B., *Diffraction by an Aperture*, J. Appl. Phys. **28**, 426-444 (1957). *Geometrical Theory of Diffraction*, J. Opt. Soc. Am. **52**, 116-130 (1962).

Kottler F., *Zur Theorie der Beugung an schwarzen Schirmen*, Ann. Physik **375**, 405-456 (1923). *Elektromagnetische Theorie der Beugung an schwarzen Schirmen*, Ann. Physik **376**, 457-508 (1923). *Diffraction at a black screen, part I: Kirchhoff theory*, Progress in Optics **IV**, 281-314 (1966), reprinted in Scalar Wave Diffraction, K. E. Oughstun ed., SPIE Milestone Series Vol. MS51, p. 108 (1992). *Diffraction at a black screen, part II: electromagnetic theory*, Progress in Optics **VI**, 331-377, Elsevier Science Publ. (1967).

Landau L. and Lifchitz E., (II), *Théorie du Champ*, MIR, Mosca (1966).

Landsberg G.S., *Ottica*, MIR, Mosca (1979).

Lipson S.G., Lipson H. and Tannhauser D.S., *Optical Physics*, III ed., Cambridge University Press, Cambridge (1995).

Maggi G.A., *Sulla propagazione libera e perturbata delle onde luminose in un mezzo isotropo*, Annali di Matematica **16**, 21-48 (1888).

Mansuripur M., *Classical Optics and its Applications*, Cambridge Univ. Press, Cambridge (2002).

Marchand E.W and Wolf E., *Consistent Formulation of Kirchhoff's Diffraction Theory*, J. Opt. Soc. Am. **56**, 1712-1722 (1966). *Diffraction at Small Apertures in Black Screens*, J. Opt. Soc. Am. **59**, 79-90 (1969).

Miyamoto K. and Wolf E., *Generalization of the Maggi-Rubinowicz theory of the boundary diffraction wave*, J. Opt. Soc. Am. **52**, *Part I*: 615-625, *Part II*: 626-637 (1962).

Möller K.D., *Optics*, University Science Book, Mill Valley (1988).

Osterberg H. and Smith L.W., *Closed solutions of Rayleigh's diffraction integral for axial points*, J. Opt. Soc. Am. **51**, 1050-1054 (1961).

Rubinowicz A., *Die Beugungswelle in der Kirchhoffschen Theorie der Beugungserscheinungen*, Ann. Physik **358**, 257-278 (1917). *Zur Kirchhoffschen Beugungstheorie*, Ann. Physik **378**, 339-364 (1924). - *Thomas Young and the theory of diffraction*, Nature **180**, 160-162 (1957).

Sheppard C.J.R. and Hrynevych M., *Diffraction by a circular aperture: a generalization of Fresnel diffraction theory*, J. Opt. Soc. Am. A **9**, 274-281 (1992).

Siegman A. E., *Lasers*, University Science Books, Mill Valley, California (1986).

Sivoukhine D., *Optique*, MIR, Mosca (1984).

Solimeno S., Crosignani B., Di Porto P., *Guiding, Diffraction, and Confinement of Optical Radiation*, Academic Press, inc., Orlando (1986).

Sommerfeld A., *Lectures on Theoretical Physics: Optics*, Academic Press, New York (1949).

Stamnes J.J., *Waves in Focal Regions*, A. Hilger, Bristol (1986).
Stamnes J.J., Spjekavik B. and Pedersen H.M., *Evaluation of diffraction integrals using local phase and amplitude approximations*, Opt. Acta **30**, 227-222 (1983).

Part IV
Fourier's Optics

Historical notes: From Fourier to Fizeau

The mathematicians

Jean Baptiste Joseph Fourier (1768–1830) did not deal directly with optics, but without the mathematical concepts introduced by him to explain the propagation of heat, it would not be possible to write the theory of waves and what is now called Fourier's Optics. As is well known, heat has the peculiarity of evolving over time in a privileged direction, not attributable to the mechanical behavior of simple physical systems, which are, in principle, reversible. From here arose the difficulty for scholars such as Pierre Simon Laplace (1746–1827) to give it a justification. In 1807, and then, in more extended form, in 1811, Fourier presented to the Institut de France a long memory in which he was able to tackle and solve the problem by means of a differential equation at the partial derivatives for the evolution of heat in a body and a particular method for integrating it. The basis of the Fourier method was assuming that "any" function could be represented by a series of harmonic functions. In the case of an infinite length wire, Fourier replaced the series with an integral. However, the assumption that the development was extensible to "any" function had difficulties that were resolved only after many decades. His memory was rewarded, as it was acknowledged to contain the right differential equations, but was criticized by Lagrange for insufficient mathematical rigor and was not "crowned" by the publication. However, Fourier reformulated his theory, enriching the study of many situations, and, in 1822, he published the book *Théorie analytique de la chaleur*, in which heat is considered as an indestructible fluid. The influence of this text was remarkable, since, in order to represent the continuous flow of heat, he developed the mathematical tool of the partial differential equations, which are the mathematical basis for Maxwell's equations and waves, and the development in series of orthogonal functions.

As a director of the Gottinga Observatory, the great mathematician Carl Friedrich Gauss (1777–1855) also dealt with optics, noting that the aberrations in centered optical systems become negligible with slightly sloping rays with respect to the axis of the system; hence, the paraxial approximation takes his name. Nevertheless, Gauss was also the father of differential geometry, which plays an essential role in the whole optical design.

One disciple of Fresnel's work was another great mathematician, Augustin Louis Cauchy (1789–1857), to whom we owe many important concepts and theorems of mathematical analysis. He also worked on Physical Optics, succeeding in reformulating Fresnel's wave theory on more solid mathematical bases. To treat propagation and reflection from absorbent media, Cauchy generalized the refractive index in complex form. From this also followed the generalization of Fresnel's formulas for reflection, with important applications in microscopy. For transparent media, he formulated a law bearing his name for the dispersion of refractive indexes as a development in power series of $1/\lambda^2$. For anisotropic media, he clarified the role of the quadric surfaces introduced by Fresnel and studied the mechanical bases for the transverse waves. However, as Born and Wolf (1980) note, ordinary mechanical elasticity remained irreconcilable with the transverse nature of waves, which could only be justified by Maxwell's electromagnetic theory. In particular, it lacked a justification for which, at the interface between two media, the transmitted wave is a transverse wave. The aether should therefore be a very special solid medium, perfectly elastic and incompressible.

New glasses and optical instruments

Joseph Fraunhofer (1787–1826) was born near Monaco into a very poor family. His father had a glassworks where Joseph also worked, but he became an orphan at just 11 years old. Then, he was taken in as an apprentice of a manufacturer and glass polisher with the obligation to work six years without pay! In 1801, the house where the master was lodging him collapsed, burying him. Fraunhofer was saved thanks to the solicitude of Maximillian IV Joseph Wittelsbach, Elector of Bavaria, who presented him to his counselor, Joseph von Utzschneider. They took him under their protection and directed him in his studies. Fraunhofer was finally able to have books to study, in secret from his master, showing both a real passion for optics and a great enterprising spirit, admirable in a young boy.

Meanwhile, Utzschneider, in company with G. Reichenbach, began producing optical and mathematical instruments. Only their ability to produce lenses was lacking, as they did not have good crown and flint glasses. After being informed about their manufacture, Utzschneider built a furnace in a former convent in Benedictbauern, near Monaco, and appointed a famous optician, Pierre Louis Guinand (1748–1824), for its management. Guinand was a Swiss artisan in Brenèts, near Neuchâtel, who became passionate about optics and had begun to produce flint glass around 1783. In 1798, he had introduced a glass mixing technique during fusion and, in 1805, with a subsequent improvement, he was the first to produce excellent flint glasses with great homogeneity and no defects. The mixing guaranteed a uniform distribution of the various glass components, which had very different densities, and eliminated air bubbles. In Benedictbauern, Guinand managed to produce regular castings of large flint glass and crown glass disks, and, between 1806 and 1807, the company began producing its first lenses.

At that time, Fraunhofer had finally managed to release himself from his master, but he was in difficulty. Hesitant, he went to Utzschneider, who welcomed him very amicably. Recognizing his value, Utzschneider employed him at his company, and eventually took him on as a partner. Fraunhofer had already studied aberrations of mirrors for telescopes and had designed polishing machines to produce aspherical surfaces. At the company, he was initially engaged in the designing and polishing methods of large lenses to be used in telescopes. In 1811, after learning from Guinand the best art in glass manufacturing, he was also put in charge of the direction of its production. Dealing with lenses, he soon recognized the importance of the quality of glass in the performance of a lens and, with fine scientific planning, studied how to improve the quality of the glass further, managing to produce good flint without streaks and then even crowns. He also developed new processing for lenses and mirrors. This enabled him to produce the finest lenses for telescopes of the time, including that of the refractor at Dorpat (present-day Tartu) with 24 cm in diameter and a 4.3 m focal length, made in 1824 for W. Struve. In addition to the optics, Fraunhofer also directed the production of refined mechanical parts that complemented the instruments. The Institute of Optics, the name of Utzschneider and Fraunhofer's company, came to have more than 50 employees and represented the beginning of German supremacy in precision mechanics and optical instruments. Many German optics scholars, such as Joseph Max Petzval, Karl August Steinheil and Ernst Abbe, and optics firms such as Zeiss and Leitz, descended from Fraunhofer [Segré 1984].

Due to uncertainty about the value of dispersion, even after obtaining good homogeneous glasses, the realized objectives were not as achromatic as Fraunhofer had expected from his calculations. The dispersion of his glasses turned out to be different from the English one, and Fraunhofer contrived how to measure it at various wavelengths. On the other hand, it seemed difficult to quantify it, since the spectrum has no defined references and the transition between one color and another happens too gradually. It was in studying the dispersion of the glass (with a prism spectrograph) that Fraunhofer rediscovered the absorption lines in the solar spectrum (Wollaston had already observed some of them in 1802). Fraunhofer had found that, in the spectrum of light produced by a flame, there is always a thin yellow-orange line (the D-lines of sodium) that could be used as a reference. He wanted to see if the same line existed even in the solar spectrum. Instead, he found a forest of dark lines that appeared independent of the material used for the prism. Fraunhofer studied them systematically and compiled a catalog of 590 lines, which finally provided him with a reference to describe the chromatic dispersion. This important discovery was published in the Memories of the Academy of Monaco [Fraunhofer 1817], and Fraunhofer's classification is still used today. He also discovered that the spectra of Sirius and other bright stars are different from one another and that of the Sun, thus founding stellar spectroscopy.

Speculating on the nature of these lines, Fraunhofer was led to the idea that they derived from a phenomenon of interference. Thus, he undertook a systematic study of the diffraction (in far field), independently and in the same years as Fresnel's studies. In particular, he used a heliostat to illuminate a slit, whose width was adjustable, and measured the diffraction angles by observing the diffracted light with a theodolite [Fraunhofer 1822]. He did the same with a wire grating. He knew the work of Young, and therefore also the law of the gratings. Then, he found a way to measure the wavelength of these black lines of the solar spectrum accurately. Since a very small grating period was necessary, Fraunhofer built the first *grating engine* with which the first diffraction gratings on a glass substrate covered with a gold foil was realized. He studied the tracing process, discovering the need for extraordinary accuracy (1% of the grating period). In the measurements, he used collimated light on the grating and made angular measurements of the diffracted beams by means of a theodolite. He also used a cross-dispersion system, with a prism placed on the entrance slide. In this way, he avoided the overlap between the various diffraction orders, managing to determine the wavelength of the black lines of the spectrum with remarkable precision.

Unfortunately, in October 1825, Fraunhofer was also hit by a lung disease from which he did not recover. Hoping to come to Italy to heal, he died on June 7, 1826, only 39 years old.

Giovanni Battista Amici (1786–1863) was also an important manufacturer of optical instruments and was gifted with great genius. He was born in Modena, and graduated in Bologna in 1808 as an engineer and architect. After graduation, he devoted himself to the teaching of mathematics, but, in 1825, he was exempted from teaching to devote himself exclusively to research and to his instruments, which he built in his workshop in Modena. In 1831, he was called to Firenze by the Granduca Leopoldo II to direct the Museum of Natural Physics and History and the Specola Observatory in Boboli, in addition to being given the title of Professor of Astronomy at the University of Pisa. At the end of that year, he moved to Firenze with his family and two workers.

Amici built many excellent telescopes, of great aperture for those times, both reflectors and refractors, for various astronomical observatories. At that time, the two types of telescope were equivalent in quality. The mirrors were furnished out of polished metal, the *speculum*, an alloy of copper that was white, but fragile and more difficult to polish, and required frequent reworking to maintain their gloss. Contrastingly, the achromatic objectives were very heavy, and it was difficult to obtain a sufficient homogeneity of the glass.

In addition to telescopes, Amici provided Herschel and other European astronomers with various accessory tools, such as micrometers for measuring the size of planets and satellites and the separation of double stars. However, Amici's greatest achievements were the realization of excellent achromatic microscopes, for which many microscopists came expressly to Firenze to buy them. Amici traveled several times through France and England and knew practically all of the scientists of his time. He carried his microscopes with him to compare them with those of other manufacturers, and his were always the best. His first microscope was in reflection, similar to a Newtonian telescope used for the opposite purpose, with an objective consisting of an elliptical mirror with about five cm of focal length, which faced a flat mirror at 45 degrees roughly halfway between the mirror and the first focal point. Subsequently, Amici investigated achromatic objectives consisting of three juxtaposed doublets made with crown and flint glasses. He realized that, with this arrangement, the lens close to the object was too big, preventing very small focal distances and large apertures at the same time. Then, he had the resolutive idea of replacing the first doublet with a simple hemispherical lens, and corrected the aberrations with an appropriate design of the other two doublets.

In 1847, he invented immersion microscopes. The first were by immersion in water, with which he got an opening angle up to 160°. They were made of lenses with six types of glass of different refraction and dispersion. He also built oil immersion objectives around this time.

To get special glasses, Amici turned to several suppliers, such as Fraunhofer, and even Faraday. In 1844, during his second trip to Paris and London, Amici bought, from Henri Guinand, son of Pierre, two flint glass disks that he used to build the *Amici II*, an objective for the museum's observatory, with a 238 mm aperture and 3.18 m focal distance. Guinand's sons had moved to France at Choisy-le-Roy, near Paris, where, associating themselves with Georges Bontemp, they continued the art of their father

With the powerful microscopes that he built, Amici also devoted his time to biology and to the study of plant cells, investigating the diseases of plants and, in particular, their insemination, with important results. Amici was ultimately one of the main architects of the optical tradition in Toscana.

Research on the speed of light and on the aether

After the death of Fresnel, Arago wanted to devise an experiment to discriminate between the emission theory and the wave theory. He had known that Wheatstone had been able to measure the speed of electrical signals in a wire using a rotating mirror at 800 revolutions per second, with which he was directing the light of a spark produced by the electric signal at the end of the wire. Encouraged by this news, in 1838, Arago proposed to apply the same rotating mirror system so as to try to determine, once and for all, whether the speed of light was greater or lesser in dense media, i.e., between a path in the air and one in the water. Since water absorption is appreciable after just one meter, the walkable distance was reduced to about ten meters. The travel times were therefore very short and needed rotation speeds of over 1000 revolutions per second, comparable to those of a modern turbo-molecular pump! Arago considered various configurations, including the one with subsequent reflections between rotating mirrors, and, with the help of one of his friends, L. Breguet, made various devices. However, none of these could reach the required rotation speeds.

The weakening of his sight eventually forced Arago to abandon the enterprise, and he passed it on, around 1850, to two young self-taught researchers, Armand Hyppolite Louis Fizeau (1819–1896) and Lèon Foucault (1819–1868). From the early 1840s, they had generally worked together, beginning with a study of the chemistry of daguerreotype production. Fizeau never had an academic position and financed all of his research himself, while Foucault obtained a position at the Observatory of Paris, thanks to Arago's support.

Fizeau had determined, in 1849, the speed of light in air using a toothed wheel shutter system and an 8633 m path, getting 315000 km/s, in excess of the true value due to an incorrect estimate of the rotation velocity of the wheel. A year later, the measurement was repeated by Foucault, who used a rotating mirror by Breguet who had managed to perfect it. He obtained the value of 298000 ± 500 km/s, which can be considered the first accurate determination of the speed of light.

The experiment proposed by Arago was accepted by Fizeau and Foucault, first in collaboration and then in competition, because of a bitter disagreement between them on how to proceed. The first to conclude was Foucault, who used a rotating mirror to achieve approximately 800 revolutions per second on which the light from a source, collimated by a lens, was reflected. For a small angular interval, the reflected light reached a distant mirror S that reflected it back toward the rotating prism. The image of the source so reflected twice by this mirror appeared significantly shifted due to the time spent between going and returning. Placing a second mirror S′, making the same reflection game of S, and interposing a tube filled with water on one of the two paths, he could compare the propagation speeds in air and water. A similar experiment was also done by Fizeau in collaboration with Breguet. Finally, it was proved that the velocity in water was less than that in air, definitively refuting the corpuscular theory. The result of these experiments was made public at the Académie meeting on May 6, 1850, and Arago had the satisfaction of seeing the wave theory confirmed before he died in 1853.

Fizeau had begun as early as 1847 to reason and experiment on the *aether drag* by matter [Frercks 2005]. Fizeau faced three hypotheses: the first that there was no drag, the second the partial drag planned by Fresnel, and the third the total drag proposed by Stokes. The detection method identified by Fizeau was to create two interfering light beams, each passing through a pipe in which air or water flowed. The flow had the same direction of propagation of light for one beam and the opposite direction for the other. At the beginning of 1850, Fizeau studied this effect in collaboration with Foucault, exploiting a fast flow of air, but turbulence prevented a clear result. The disagreement that arose between them caused the separation of the two scientists, and Fizeau continued the research alone. In the spring of 1851, he performed a new experiment with an airflow using an interferometer that he had made more symmetrical. This time, he observed that there was no appreciable shift of fringes, and therefore excluded the hypothesis of the total drag. Finally, in the autumn of 1851, Fizeau concluded his experiment with a flow of water using a common path interferometer, in which the light from a source crossed a half-silvered mirror and was then divided into two beams, each entering in one of the two tubes with counter-propagating streams. For both beams, the light that had passed through a tube was reflected back, entering the other tube by means of a lens and a mirror arranged as a "cat's eye". One of the two returning beams therefore had passed through a co-propagating flow twice, while the other had done the same path in the opposite direction, facing a counter-propagating stream twice. These two beams were then observed after being reflected by the same half-silvered mirror, interfering with each other. This time it was possible to observe a shift of fringes due to the motion of the water, and such displacement was compatible within the error only with Fresnel's partial dragging hypothesis.

After a long examination, Fizeau finally decided to publish the result of his research on the aether [Fizeau 1851, 1859].

The photography

During the nineteenth century, the advent of industrial production turned up new social classes, eager to reach the status of the more wealthy. It became increasingly important to fix images that represented that status, or that described the increasingly complex objects of production, or even faithfully illustrated the events spread by the media at that time, that is, by the press [De Paz 1989]. Until then, the only way to produce such images was through the agency of painters and designers, and fidelity to the subject was generally their main goal. To respect this fidelity, designers, architects, and painters have long used a dark room, sometimes a portable tent, where the image of the outside scene, projected by a lens onto a translucent paper screen, was traced by hand with a stylus. Subsequently, this design was used as a basis for painting or lithographic printing. However, few could afford this very slow and costly method.

However, in that century's change towards the Industrial Revolution, chemistry also began to play a successful role, promising new solutions. Thomas Wedgwood (1771–1805) and Humphry Davy (1778–1829) succeeded in reproducing images over layers of silver nitrate between 1790 and 1802. However, these images were not fixed and, after a short time, they were lost because the light diffused from the environment produced a uniform blackening. Since scientists do not usually have great practical sense, Davy abandoned these intents. Instead, an old passionate lithographer, Joseph Nicéphore Niepce (1765–1833), used Judea's bitumen to produce stable images in relief and applied this method to zincography; in 1824, he succeeded in obtaining images by means of the darkroom, capturing poses over a period of 8 hours in full sun. He used a plate of silver-plated copper coated with a bitumen layer. Exposed to light, the bitumen polymerizes, and the non-illuminated parts are removed in a bath of petroleum and lavender essence. The light is given by the remaining bitumen, while the dark portions are reflected by the silver plate. After learning this technique in 1826, Louis Jacques Mandé Daguerre (1787–1851), who was a painter of theatrical scenery, entered into partnership with Niepce in 1829. Daguerre used silver-coated copper plates exposed to iodine vapor in darkness that formed a thin layer of purple iodide silver on the slab. Exposed to light in a darkroom, this layer could form images with exposures of 5 hours, but Daguerre was able to reduce the exposure times drastically down to 4 or 5 minutes with the discovery of the *latent image*, which was developed by treating the plate with mercury vapor. With two subsequent washings in a kitchen salt solution, the unexposed silver iodide was removed, fixing the image. In 1839, after a proposal by Arago, the French government acquired the manufacturing rights to the *daguerreotype*, and Arago himself made the first public explanation of it in front of the gathered members of the Academies of Sciences and Fine Arts. The daguerreotype was an image with great definition, and was greatly successful, but it had the defect of being a unique, unreproducible piece. In 1819, John Herschel discovered that silver halides dissolve in a sodium thiosulphide bath. This substance was then used in place of salt to fix the images.

Important progress had also been made independently by the English physicist William Henry Fox Talbot (1800–1877). In 1834, he was able to fix images on a sheet of paper brushed with a silver nitrate solution. However, the quality of these images was not good, and he discontinued the work until he heard the news of Arago's report. He then resumed his investigations, and, in 1841, he filed a patent diffusing his process, which he called *photography*, throughout France with a letter addressed to the *Académie des sciences*. It consisted of two steps. In the first, Talbot would expose a sheet of translucent paper, impregnated with silver iodide, in the image plane of the objective of a darkroom. On this sheet, a latent image would form, which he then fixed in a bath of sodium thiosulfate, obtaining a negative image, a *calotype*. Subsequently, Talbot would draw a positive copy for contact, developed through a similar process.

Further improvements to the Talbot method were made by Blanquart-Evrard. He would immerse the sheet to be sensitized in a silver nitride solution, so as to impregnate it deeply, and then he would reveal the latent image with Gallic acid, obtaining a remarkable reduction of the exposure times. The great advantage of photography over the daguerreotype was the fact that, with photography, it was possible to obtain a large number of copies. However, the quality of the images obtained was lower due to the texture of the paper. Claude Felix Abel Niepce de Saint-Victor (1805–1870), cousin of Niepce, proposed using a slab of glass covered with egg albumin and then dried, with potassium iodide added, for the negative. Before exposure, the plate was sensitized with a vinegar solution of silver nitrate, and was subsequently developed with gallic acid. Using these plates, in 1857, he discovered, for the first time, the radioactivity of uranium salts, distinguishing it from the phenomena of fluorescence and phosphorescence and attributing it to an invisible radiation.

However, albumin required much longer exposure times and was replaced with collodion by Frederick Scott Archer (1813–1857). Thus, the photograph became very sharp, definitively supplanting the daguerreotype. In a short time, photography had become a common practice, and, since then, it has been increasingly perfected. In parallel to the progress of chemistry, there was a strong push to improve the quality of the lenses and begin the design of sophisticated objectives to produce strong numerical apertures so as to reduce exposure times.

The studies on the color and the physiology of the eye

Among his fundamental contributions to science, between 1850 and 1870, Maxwell also devoted himself to the study of color vision, resuming Young's studies with new experiments [Peruzzi 1998]. During his first years of university study in Edinburgh, there were several scholars who were dealing with colors: the painter David Ramsey Hay, the physical naturalist James Forbes, the chemist George Wilson, the physician William Swan, and the physicist David Brewster. The interest of these scholars in colors had been awoken by Young's theory of the three receptors, which was not yet fully accepted, as Young had inserted it into the wave optics theory. There was still confusion between the actual physicality of light and that which was highlighted by spectral analysis, for example, with a prism, and the chromatic, perceptive properties of light. This, in turn, caused difficulty in regard to color classification and the understanding of color combinations through mixing. Helmholtz was the first to conclude that it is necessary to distinguish between the additive process of lights and the subtractive one by absorption from pigments, starting from the observation that the mixing of azure and yellow light beams does not produce green. However, in his article of 1852 on compound colors, he also considered 5 primary colors as being necessary to reproduce the colors of the spectrum, and it was only after an article by Herman G. Grassmann (1809–1877) from 1853 that he accepted the theory of the three receptors. Grassmann had pointed out that a color could be described by only three parameters: «the tint, the intensity of the colour, and the intensity of the intermixed white» [Grassmann 1853]. In turn, Forbes had resumed Young's studies and classified the colors as the inner points at an equilateral triangle whose vertices represented the three basic colors.

Guided by Forbes, of whom he was a student, Maxwell performed various mixing experiments, at first using disks in rapid rotation on which were placed colored sheets arranged in circular sectors. The amplitude of these sectors could be varied by changing the angle at which these sheets, in the form of radially cut disks, overlapped. In this way, Maxwell and Forbes could give a *quantitative measure* of the color as a combination of three colors chosen as fundamental. Subsequently, in 1854, on the central part of the disk, Maxwell overlapped another disk with smaller diameter, with one or two other sheets cut radially to form sectors according to the same principle, such as a white sheet and a black one. This configuration allowed for performing a comparison of the color and the brightness perceived by an observer between the outer and inner parts. In this way, Maxwell could compare the results obtained with different observers, who had to indicate when the color or brightness of the two areas coincided [Maxwell 1855]. As a further refinement of the method, Maxwell also built some "color boxes", in one of which a beam of white light was first decomposed spectrally by a prism, and, with a system of three movable slits, variable in width, he would select three distinct regions of the visible spectrum. The three beams of light that passed through them were then recombined with another prism, thus obtaining a mix of colors that was directly associated with the visible spectrum. With a more complicated, but also more compact, instrument, using, instead, two prisms in succession crossed twice, Maxwell could thus prove that all colors can be described by only three parameters, spectral color, saturation and brightness, with the first two being represented in a triangle inside of which the white color is placed. Each color is then represented by the angle, and the saturation by the distance, at which the point that represents it is located with respect to that of the white. Maxwell had therefore succeeded in establishing the trichromy on quantitative bases, thus establishing a bridge between the mind and the external physical reality through the physiology of the eye [Maxwell 1871].

Maxwell was the first to produce a color photo, with the help of Thomas Sutton, inventor of the *reflex camera*, who photographed a staple made with a colorful Scottish fabric with separate exposures through colored filters, one red, one blue, and one green. The result was imperfect, although apparently effective, as the film was insensitive to red but sensitive to ultraviolet, which the red fabric reflected and to which the red filter was transparent.

We must also remember Hermann Ludwig Ferdinand von Helmholtz (1821 –1894), who was an important German physician and physicist to whom we also owe the formulation of the principle of conservation of energy and the concept of free energy in thermodynamics. As a physician, he attended to the physiology of the eye and, in 1850, invented the *ophthalmoscope*, with which he studied the retina and its nervous system; he also invented the *ophthalmometer*, with which he studied the crystalline curvature and accommodation problems. In 1856, he published a book on these topics, *Handbuch der Physiologischen Optik*, which was, over various editions, the reference text for physiological optics for many decades.

Suggested readings

The historical data in these notes are derived from the following texts:

Various authors, Enciclopedia della scienza e della tecnica, Mondadori ed., Milano (1966), in the *fotografia* entry.

Bottazzini U., *I "politecnici" francesi, e La "moderna analisi"; E. Bellone, L'esposizione del sistema del mondo*. In Storia della Scienza, Vol. 3, directed by P. Rossi, Gruppo ed. l'Espresso, Istituto Geografico De Agostini s.p.a. (2006).

Born M. and Wolf E., *Principles of Optics*, Pergamon Press, Paris (1980).

Brewster D., *Memoir of the life of M. le Chevalier Fraunhofer*, The Edinburgh Journal of Science, Vol VII, 1-11 (1827). Accessible at http://books.google.com.

De Paz A., *L'occhio della modernità. Pittura e fotografia dalle origini alle avanguardie storiche*, Cooperativa Libraria Universitaria Ed., Bologna (1989).

Fizeau H., *Sur les hypothèses relatives à l'éther lumineux, et sur une expérience qui paraît démontrer que le mouvement des corps change la vitesse avec laquelle la lumière se propage dans leur intérieur*, Compte rendus **33**, 349-355 (1851), Annales de Chimie et de Physique **57**, 385-404 (1859).

Fraunhofer J., *Bestimmung des Brechungs- und Farbenzer-streuungs-Vermögens verschiedener Glasarten, in bezug auf die Vervollkommnung achromatischer Fernröhre*, Denkschriften der Königl. Akademie der Wissen-schaften zu München für die Jahre 1814 und 1815, **5**, 193-226 (1817). *Neue modification des Lichtes durch gegenseitige Einwirkung und Beugung der Strahlen, und Gesetze derselben*, Denkschriften der Königl. Akademie der Wissenschaften zu München für 1821 und 1822, **8**, 1-76

(1824).
Frercks J., *Fizeau's research program on ether drag: a long quest for a publishable experiment*, Physics in Perspective **7**, 35-65 (2005).
Grassmann H.G., *Theory of Compound Colors*, Philosophical Magazine **4**, 254-264 (1854). Translated from Annalen der Physik und Chemie, **19**, 53-60 (1853).
Maxwell J. Clerk, *Experiments on colours, as perceived by the eye*, Trans. Roy. Soc. Edimburgh **21** part II, 275-299 (1855). *On colour vision*, Nature **4**, 13-16 (1871).
Peruzzi G., *Maxwell, dai campi elettromagnetici ai costituenti ultimi della materia*, series "I grandi della scienza", anno I, n. 5, November 1998, Le Scienze S.p.A. ed. (1998).
Proverbio E., *The production of achromatic objectives in the first half of the nineteenth century: the contribution of Giovanni Battista Amici*, Mem. Soc. Astronomica Italiana **61**, 829-875 (1990), NASA Astrophysics Data System.
Segrè E., *From falling bodies to radio waves*, W.H. Freeman and Company ed. New York (1984).

Chapter 5
Fourier's Optics

Introduction

We have seen that, in Fresnel's approximation the propagation of a wave between two parallel planes can be expressed in terms of a Fourier's transform. Here, we examine an alternative technique, relying on the fact that the field present on the first plane can be represented by its spectrum in plane waves, for which, in a homogeneous space, one can determine their propagation in a simple way. By recombining these waves, one can therefore easily rebuild the field on the second plane with an inverse transform. This fact has two important applications, the first concerns the mathematical and numerical techniques that can be used to calculate the diffracted field, and the second is, in a certain sense, opposite to the first and concerns the processing of signals by optical means. In particular, we will briefly discuss some topics that make use of the Fourier's transform, including sampling theorems and the numerical techniques for the calculation of diffraction, the formation of images and analysis of the quality of optical systems, the theory of coherence and some of its applications, spatial filtering and finally diffraction gratings.

5.1 Mathematical preliminaries

5.1.1 Some special, frequently used functions

In optics one often refers to the concept of a point source, when, for example, its dimensions are very small compared to the resolving power of the instrument and then the source is assimilated to a point with a finite emissive power, but also of infinite intensity. Other times, then, an extended source is assimilated into a continuous collection of point sources, with an overall finite intensity. Mathematically, the concept of a point source is rendered by Dirac's δ function, which is not an ordinary function, but rather belongs to the category of generalized functions, endowed with particular discontinuity, which have meaning only within an integral. A solid discussion of these functions requires the theory of distributions, but, here, we limit ourselves to understanding them as the limit of a sequence of ordinary functions with "good behavior". Some functions of this type and others that are particularly useful in optics [Goodman 1996] are as follows.

Electronic supplementary material The online version of this chapter (https://doi.org/10.1007/978-3-030-25279-3_5) contains supplementary material, which is available to authorized users.

G. Giusfredi, *Physical Optics*, UNITEXT for Physics,
https://doi.org/10.1007/978-3-030-25279-3_5

Dirac's generalized δ function. This can be defined as the limit of a sequence of ordinary even functions f_n such that

$$\int_{-\infty}^{\infty} f_n(x)\,dx = 1 \quad \text{per } \forall n,$$

whose *width* decreases *uniformly* for $n \to \infty$, for which

$$\lim_{n\to\infty} \int_{-\infty}^{\infty} f_n(x - x_o)\,g(x)\,dx = g(x_o), \tag{5.1.1}$$

where $g(x)$ is limited and continuous in x_o. Eq. (5.1.1) constitutes the property of *sifting* of the Dirac's delta. Among the possible equivalent limits, two of them are

$$\delta(x) = \lim_{n\to\infty} n \exp\left(-n^2 \pi x^2\right), \tag{5.1.2}$$

$$\delta(x) = \lim_{n\to\infty} \frac{\sin(n\pi x)}{\pi x} = \lim_{n\to\infty} n\,\mathrm{sinc}(n\pi x). \tag{5.1.3}$$

The *two-dimensional Dirac's δ function* (one of the possible definitions) is

$$\delta(x, y) = \lim_{N\to\infty} N \exp\left[-N^2 \pi \left(x^2 + y^2\right)\right]. \tag{5.1.4}$$

Their integration in x, or in x and y in the two-dimensional case, is unitary. One important property is the following:

$$\delta(ax) = \frac{1}{|a|}\delta(x), \quad \delta(ax, by) = \frac{1}{|ab|}\delta(x, y). \tag{5.1.5}$$

Since the function can be defined as the limit of a sequence of differentiable functions, it is also natural to define its derivative $\delta^{(n)}$ as the limit of the sequence of the derivatives of these functions. Integrating by parts, taking into account the limits of integration, we have

$$\int_{-\infty}^{\infty} \delta^{(n)}(x - x_o)\,g(x) = (-1)^n\, g^{(n)}(x_o), \tag{5.1.6}$$

where g is a function that is (at least) n-times derivable. In turn, the function δ can be understood as the derivative of the *step* function described below.

Generalized comb function

$$\operatorname{comb}(x)=\sum_{n=-\infty}^{\infty}\delta(x-n). \tag{5.1.7}$$

Generalized step and sign functions

$$\operatorname{step}(x)\equiv\begin{cases}0 & \text{for } x<0,\\ 1/2 & \text{for } x=0,\\ 1 & \text{for } x>0,\end{cases}\quad \operatorname{sgn}(x)\equiv\begin{cases}-1 & \text{for } x<0,\\ 0 & \text{for } x=0,\\ 1 & \text{for } x>0,\end{cases} \tag{5.1.8}$$

so that we have sgn(x) = 2step(x) − 1. The step function can be understood as the limit of $h(kx)$ for $k\to\infty$, where

$$h(x)=\begin{cases}\exp(-1/x) & \text{for } x>0,\\ 0 & \text{for } x\le 0\end{cases}$$

is an infinitely derivable function.

Generalized function 1/*x*

$$\mathcal{P}\left(\frac{1}{x}\right)=\lim_{\varepsilon\to 0}\frac{x}{x^2+\varepsilon^2}, \tag{5.1.9}$$

where $\mathcal{P}$ indicates the principal value according to Cauchy and is sometimes omitted. While the ordinary function $1/x$ is not defined for $x=0$, the generalized function is 0 for $x=0$.

Triangle function

$$\operatorname{tri}(x)=\begin{cases}1-|x| & \text{for } |x|\le 1,\\ 0 & \text{otherwise.}\end{cases} \tag{5.1.10}$$

Rectangle function

$$\operatorname{rect}(x)=\begin{cases}1 & \text{for } |x|<1/2,\\ 1/2 & \text{for } |x|=1/2,\\ 0 & \text{otherwise.}\end{cases} \tag{5.1.11}$$

Sinc function (classical definition)

$$\operatorname{sinc}(x)=\frac{\sin x}{x}. \tag{5.1.12}$$

Circle function

$$\mathrm{circ}\left(\sqrt{x^2+y^2}\right)=\begin{cases}1 & \text{for } \sqrt{x^2+y^2}<1,\\ 1/2 & \text{for } \sqrt{x^2+y^2}=1\\ 0 & \text{otherwise.}\end{cases} \tag{5.1.13}$$

5.1.2 Fourier's transform

The Fourier's transform (one-dimensional) is defined as the operation on a complex function $g(x)$ with real x, such that

$$\mathcal{F}:\quad \hat{G}(w)=\int_{-\infty}^{\infty} g(x)e^{-iwx}dx\,, \tag{5.1.14.a}$$

where $G(w)$ is also a complex function of a real parameter w. The inverse operation is

$$\mathcal{F}^{-1}:\quad g(x)=\frac{1}{2\pi}\int_{-\infty}^{\infty}\hat{G}(w)e^{+iwx}dw\,. \tag{5.1.14.b}$$

In order for these transformations to be performed, $g(x)$ and $\hat{G}(w)$ must satisfy certain regularity conditions, only some of which we will now briefly examine, deferring the discussion to mathematical texts.[1] If the complex function $g(x)$ of the real variable x is absolutely integrable, i.e., the integral of its module between $-\infty$ and $+\infty$ is finite, then g belongs to L^1, its transform $\hat{G}(w)$ exists and it is continuous and limited. However, this does not guarantee that the inverse transform of $\hat{G}$ exists; indeed, $\hat{G}$ does not necessarily belong to L^1. One example is that of the Dirac's δ: it belongs to L^1 but its transformation is worth 1, which is said to belong to L^∞. To dispose of the inverse transform, it is necessary to exclude certain functions that "behave badly". These functions are said to be of *unlimited variation*, i.e., they diverge infinitely or oscillate infinitely fast, such as $\cos(1/x)$. The functions with *limited variation* are instead such that their graph has a finite length in each finite interval of their variable x: for the L^1 functions of this type, with, at most, a finite number of discontinuities, there is also the inverse transform. However, the function that is obtained by transforming and anti-transforming may not coincide *exactly* with the original function. Indeed, if g has a discontinuity in x_o, we have

$$\mathcal{F}^{-1}\{\mathcal{F}g\}(x_o)=\frac{1}{2}\lim_{\varepsilon\to 0}\left[g(x_o-\varepsilon)+g(x_o+\varepsilon)\right]. \tag{5.1.15}$$

[1] See also Barrett and Myers (2004), which is a wonderful, monumental text of image science, with extensive mathematical discussion.

One class of particularly important functions in optics is the set of functions L^2, whose square module is integrable, according to Lebesgue, with finite result. For these functions, the norm is defined as[2]

$$\|g\|_{L_2} = \int_{-\infty}^{\infty} |g(x)|^2 \, dx < \infty . \tag{5.1.16}$$

In the two-dimensional case, when g describes the amplitude of the field, the norm corresponds exactly to the power of the radiation crossing an infinite surface: to have a physical sense, this power has to be finite, as is precisely determined by the inequality. Therefore, the L^2 functions are also called *deterministic* [VanderLugt 1992]. Among Lebesgue spaces, L^2 is the only one that constitutes a Hilbert space, because, in addition to the norm, a scalar product between two functions g and h is also defined:

$$(g,h) \equiv \int_{-\infty}^{\infty} g^*(x) h(x) \, dx . \tag{5.1.17}$$

In addition, the norm and the scalar product having been defined, the important *Cauchy-Schwarz inequality* is also valid in L^2, and it expresses itself here as

$$\int_{-\infty}^{\infty} |g(x)|^2 \, dx \int_{-\infty}^{\infty} |h(x)|^2 \, dx \geq \left| \int_{-\infty}^{\infty} g^*(x)\, h(x) \, dx \right|^2 . \tag{5.1.18}$$

Plancherel's theorem states that, if $g \in L^2$, the function

$$\hat{G}_L(w) = \int_{-L}^{L} g(x) e^{-iwx} dx$$

for $L \to \infty$ converges on average, according to the L^2 norm, to a function $\hat{G} \in L^2$, and the same is true for the inverse transform. In practice, the functions of L^2 are transformed into functions of L^2.

The definitions of Eqs. (5.1.14) are the most common ones among mathematicians; however, in optics, it is usually preferable to use an equivalent form of the Fourier's transform, say, $\overline{\mathscr{F}}$ and $\overline{\mathscr{F}}^{-1}$, where, instead of the "pulsation" w, the "frequency" $f = w/(2\pi)$ is used:

$$\overline{\mathscr{F}}: \; G(f) = \int_{-\infty}^{\infty} g(x)\, e^{-2\pi i x f} dx, \quad \overline{\mathscr{F}}^{-1}: \; g(x) = \int_{-\infty}^{\infty} G(f)\, e^{+2\pi i x f} df . \tag{5.1.19}$$

---

2 The one shown here is the limit on the entire real line for the space $L_2(a,b)$ of the square-integrable functions in the finite interval (a,b).

This has the advantage of eliminating the $1/(2\pi)$ term from the inverse transformation. The two-dimensional transformations are quite similar:

$$\begin{aligned}\overline{\mathcal{F}}:&\quad G\left(f_x,f_y\right)=\int_{-\infty}^{\infty}\int_{-\infty}^{\infty}g(x,y)e^{-2\pi i\left(f_x x+f_y y\right)}dx\,dy,\\ \overline{\mathcal{F}}^{-1}:&\quad g(x,y)=\int_{-\infty}^{\infty}\int_{-\infty}^{\infty}G\left(f_x,f_y\right)e^{+2\pi i\left(f_x x+f_y y\right)}df_x\,df_y.\end{aligned}\tag{5.1.20}$$

where x, y are the *real space* coordinates and f_x, f_y are the coordinates in the *frequency space*. These transforms satisfy several interesting theorems. Let us look at a short list. Consider two functions $g(x,y)$ and $h(x,y)$ for which their transformations exist:

$$G\left(f_x,f_y\right)=\overline{\mathcal{F}}\{g(x,y)\},\quad H\left(f_x,f_y\right)=\overline{\mathcal{F}}\{h(x,y)\},\tag{5.1.21}$$

in the transformation from the coordinates x, y to coordinates f_x, f_y, and vice versa, given by Eq. (5.1.20). For these functions, we have:

Fourier's integral theorem
At each point of continuity of g, we have

$$\overline{\mathcal{F}}^{-1}\overline{\mathcal{F}}\{g(x,y)\}=g(x,y);\tag{5.1.22}$$

Linearity theorem

$$\overline{\mathcal{F}}\{\alpha g+\beta h\}=\alpha G+\beta H,\tag{5.1.23}$$

with $\alpha,\beta\in\mathbb{C}$ and constant;

Similarity theorem

$$\overline{\mathcal{F}}\{g(ax,by)\}=\frac{1}{|ab|}G\left(\frac{f_x}{a},\frac{f_y}{b}\right),\tag{5.1.24}$$

with $a,b\in\mathbb{R}$ and constant;

Shift theorem

$$\overline{\mathcal{F}}\{g(x-a,y-b)\}=G\left(f_x,f_y\right)\exp\left[-2\pi i\left(af_x+bf_y\right)\right],\tag{5.1.25.a}$$

$$\overline{\mathcal{F}}\left\{g(x,y)e^{+2\pi i(ax+by)}\right\}=G\left(f_x-a,f_y-b\right),\tag{5.1.25.b}$$

with $a,b\in\mathbb{R}$ and constant;

Symmetry theorem

$$\overline{\mathcal{F}}\left\{G\left(f_x,f_y\right)\right\}=g\left(-x,-y\right); \qquad (5.1.26)$$

Complex conjugation theorem

$$\overline{\mathcal{F}}\left\{g*\left(x,y\right)\right\}=G^*\left(-f_x,-f_y\right), \qquad (5.1.27.a)$$

$$\overline{\mathcal{F}}^{-1}\left\{G^*\left(f_x,f_y\right)\right\}=g*\left(-x,-y\right); \qquad (5.1.27.b)$$

Derivative theorem (one-dimensional case)

$$\overline{\mathcal{F}}\left\{\left(\frac{d}{dx}\right)^q g\left(x\right)\right\}=\left(2\pi i f_x\right)^q G\left(f_x\right), \qquad (5.1.28.a)$$

$$\overline{\mathcal{F}}\left\{\left(\frac{d}{df_x}\right)^q G\left(f_x\right)\right\}=\left(-2\pi i x\right)^q g\left(x\right); \qquad (5.1.28.b)$$

Parseval's theorem (for $g \in L^2$)

$$\int_{-\infty}^{\infty}\int_{-\infty}^{\infty}\left|g\left(x,y\right)\right|^2 dx\,dy=\int_{-\infty}^{\infty}\int_{-\infty}^{\infty}\left|G\left(f_x,f_y\right)\right|^2 df_x\,df_y\,; \qquad (5.1.29.a)$$

The Parseval's theorem actually expresses the unitary properties of the Fourier's transform in L^2, and it applies in generalized form for the scalar product of two functions g and $h \in L^2$ and their transforms:

$$\int_{-\infty}^{\infty}\int_{-\infty}^{\infty}h^*\left(x,y\right)g\left(x,y\right)dx\,dy=\int_{-\infty}^{\infty}\int_{-\infty}^{\infty}H^*\left(f_x,f_y\right)G\left(f_x,f_y\right)df_x\,df_y\,; \qquad (5.1.29.b)$$

Convolution theorem

$$\overline{\mathcal{F}}\left\{\int_{-\infty}^{\infty}\int_{-\infty}^{\infty}g\left(\xi,\eta\right)h\left(x-\xi,y-\eta\right)d\xi\,d\eta\right\}=G\left(f_x,f_y\right)H\left(f_x,f_y\right) \qquad (5.1.30)$$

and, similarly,

$$\mathcal{F}^{-1}\left\{\int_{-\infty}^{\infty}\int_{-\infty}^{\infty} G(\xi,\eta)H\left(f_x-\xi,f_y-\eta\right)d\xi d\eta\right\}=g(x,y)h(x,y); \quad (5.1.31)$$

Autocorrelation theorem

$$\mathcal{F}\left\{\int_{-\infty}^{\infty}\int_{-\infty}^{\infty} g(\xi,\eta)g^*(\xi-x,\eta-y)d\xi d\eta\right\}=\left|G\left(f_x,f_y\right)\right|^2 \quad (5.1.32)$$

and, similarly,

$$\mathcal{F}\left\{\left|g(x,y)\right|^2\right\}=\int_{-\infty}^{\infty}\int_{-\infty}^{\infty} G(\xi,\eta)G^*\left(\xi-f_x,\eta-f_y\right)d\xi d\eta. \quad (5.1.33)$$

Lastly, a note about convolution and correlation. The convolution is indicated by the symbol * and is *commutative*:

$$[g*h](x)=[h*g](x)=\int_{-\infty}^{\infty} g(\xi)h(x-\xi)d\xi=\int_{-\infty}^{\infty} h(\xi)g(x-\xi)d\xi. \quad (5.1.34)$$

Instead, the correlation of *g* and *h* functions *is not* commutative and may be indicated by the symbol ★, but is usually explicitly indicated by an integral

$$[g\star h](x)=\int_{-\infty}^{\infty} g(\xi)h(x+\xi)d\xi=\int_{-\infty}^{\infty} g(-x+\xi)h(\xi)d\xi=[g_r*h](x), \quad (5.1.35)$$

with $g_r(x)=g(-x)$. In both cases, the notation $[g\cdots h](x)$ is preferable to $g(x)\cdots h(x)$, where $\cdots$ stands for ★ or $*$, especially when the variable x is replaced by a rescaling such as ax and, even worse, if the rescaling is different for g and h. Indeed, in Eqs. (5.1.34-35), x represents the translation to be applied between the two functions and the integration variable is commensurable with x. In the case of a scale change, it is ambiguous as to whether one has to integrate in ξ or in $a\xi$. The best choice would be to use auxiliary functions where the scaling is embedded in the function itself. Here, I will use, for brevity, the notation $a{\cdot}x$ to indicate that the coefficient a is an integral part of the function that acts on x. In any case, when $[g\cdots h](ax)$ is written, it should be understood that the variable of integration has the same dimensions as ax and therefore, if, in Eqs. (5.1.34-35), x is replaced by ax, ξ is also replaced by $a\xi$. In the study of the expressions, it is useful to make a dimensional control, remembering that, in the convolution and correlation, the result has the dimensions of the product of those of the two functions and, as mentioned, of that of their argument.

A similar discourse should also be made for the Fourier's transform. If the variable x is replaced by ax, it is to be understood that f_x is replaced by f_x/a, and that the dimensions of the transform are those of the original function multiplied by those of ax, given that dx is also replaced by adx. If you do not want to use auxiliary functions that incorporate the scaling, you should always explicitly indicate the coordinates of the space domain and that of the frequencies between which the transformation is performed.

5.1.3 Hankel's transform

As we shall see in the next section, the Hankel's transform intervenes in the calculation of a two-dimensional Fourier's transform when using polar coordinates. In its general form, the Hankel's transform of order n is

$$\mathcal{H}_n: \quad \hat{g}_n(w) = \int_0^\infty r g(r) \, \mathrm{J}_n(wr) \, dr, \tag{5.1.36}$$

where J_n is the Bessel's function of order n. The inverse transform is

$$\mathcal{H}_n^{-1}: \quad g(r) = \int_0^\infty w \hat{g}_n(w) \, \mathrm{J}_n(wr) \, dw, \tag{5.1.37}$$

and this can be demonstrated thanks to the properties of the Fourier's transform. $\mathcal{H}_n$ and $\mathcal{H}_n^{-1}$ are completely symmetrical so, if $\hat{g}_n(w)$ is the transform of $g(r)$, $g(w)$ is the transform of $\hat{g}_n(r)$.

5.1.4 Fourier's transform in polar coordinates

When the optical system and the field have a circular symmetry, it may be useful to use a polar coordinate system for calculation of the Fourier's transform. In particular, we can reformulate Eq. (5.1.20.a) with the following replacements:

$$\begin{aligned} x &= r\cos\phi, \quad f_x = f_r \cos\varphi, \\ y &= r\sin\phi, \quad f_y = f_r \sin\varphi; \end{aligned} \tag{5.1.38}$$

in addition, we denote g and G in the new coordinates as

$$g(x,y) = g_\mathrm{p}(r,\phi), \qquad G(f_x, f_y) = G_\mathrm{p}(f_r, \varphi). \tag{5.1.39}$$

Thus, Eq. (5.1.20.a) becomes

$$G_{\mathrm{p}}(f_r,\varphi)=\int_0^\infty\int_0^{2\pi} g_{\mathrm{p}}(r,\phi)e^{-2\pi i f_r r\cos(\phi-\varphi)}r\,dr\,d\phi \quad . \tag{5.1.40}$$

The function $g_{\mathrm{p}}(r,\phi)$ can be developed in a Fourier's series as follows:

$$g_{\mathrm{p}}(r,\phi)=\sum_{-\infty}^{\infty}{}_n\, g_n(r)e^{in\phi}, \tag{5.1.41}$$

where, in turn,

$$g_n(r)=\frac{1}{2\pi}\int_0^{2\pi} g_{\mathrm{p}}(r,\phi)e^{-in\phi}d\phi . \tag{5.1.42}$$

Replacing Eq. (5.1.41) in Eq. (5.1.40), we have:

$$G_{\mathrm{p}}(f_r,\varphi)=\sum_{-\infty}^{\infty}{}_n\, e^{in\varphi}\int_0^\infty rg_n(r)dr\int_0^{2\pi} e^{-2\pi i f_r r\cos(\phi-\varphi)+in(\phi-\varphi)}d\phi . \tag{5.1.43}$$

The argument of the second integral is periodic with period 2π, for which both the variable of integration and the extremes of integration can be independently shifted. In particular, Eq. (5.1.43) can be simplified by means of the Bessel's functions J_n described in §B.3 of appendix B. Thus, the transformation in polar coordinates is finally obtained:

$$\mathscr{F}_{\mathrm{p}}:\;\; G_{\mathrm{p}}(f_r,\varphi)=2\pi\sum_{-\infty}^{\infty}{}_n\, i^{-n}e^{in\varphi}\int_0^\infty rg_n(r)\mathrm{J}_n(2\pi f_r r)dr\;, \tag{5.1.44}$$

where the functions g_n are given by Eq. (5.1.42). It can be immediately noticed that even $G_{\mathrm{p}}(f_r,\phi)$ can be expressed in a series similar to that of $g_{\mathrm{p}}(r,\varphi)$:

$$G_{\mathrm{p}}(f_r,\varphi)=\sum_{-\infty}^{\infty}{}_n\, G_n(f_r)e^{in\varphi}, \tag{5.1.45}$$

where $G_n(f_r)$ is $i^{-n}\,2\pi$ times the Hankel's transform of order n of $g_n(r)$:

$$G_n(f_r)=i^{-n}2\pi\int_0^\infty rg_n(r)\mathrm{J}_n(2\pi f_r r)dr . \tag{5.1.46}$$

$g(x,y)$	$G(f_x,f_y)=\overline{\mathscr{F}}\{g(x,y)\}$
$\frac{1}{ab}\exp\left[-\pi\left(\frac{x^2}{a^2}+\frac{y^2}{b^2}\right)\right]$	$\exp\left[-\pi\left(a^2 f_x{}^2+b^2 f_y{}^2\right)\right]$
$\frac{1}{ab}\mathrm{rect}\left(\frac{x}{a}\right)\mathrm{rect}\left(\frac{y}{b}\right)$	$\mathrm{sinc}(\pi a f_x)\,\mathrm{sinc}(\pi b f_y)$
$\frac{1}{ab}\mathrm{tri}\left(\frac{x}{a}\right)\mathrm{tri}\left(\frac{y}{b}\right)$	$\mathrm{sinc}^2(\pi a f_x)\,\mathrm{sinc}^2(\pi b f_y)$
$\frac{1}{ab}\delta\left(\frac{x}{a},\frac{y}{b}\right)$	1
$\frac{1}{ab}\mathrm{comb}\left(\frac{x}{a}\right)\mathrm{comb}\left(\frac{y}{b}\right)$	$\mathrm{comb}(a f_x)\,\mathrm{comb}(b f_y)$
$g_{\mathrm{p}}(r)$	$G_{\mathrm{p}}(f_r)=\mathscr{B}\{g_{\mathrm{p}}(f_r)\}$
$\frac{1}{a^2}\mathrm{circ}\left(\frac{r}{a}\right)$	$\frac{\mathrm{J}_1(2\pi a f_r)}{a f_r}$
$\exp\left(i\pi\beta r^2\right)$	$\frac{i}{\beta}\exp\left(-\frac{i\pi f_r{}^2}{\beta}\right)$
$g(x)$	$G(f_x)=\overline{\mathscr{F}}\{g(x)\}$
$\mathcal{P}\left(\frac{1}{x}\right)$	$-i\pi\,\mathrm{sgn}(f_x)$
$\mathrm{step}(x)$	$\frac{1}{2}\delta(f_x)+\mathcal{P}\left(\frac{1}{2\pi i f_x}\right)$
$\lvert x\rvert^{-1/2}$	$\lvert f_x\rvert^{-1/2}$
$\exp\left(i\pi\beta x^2\right)$	$\sqrt{\frac{i}{\beta}}\exp\left(-\frac{i\pi f_x{}^2}{\beta}\right)$

Table 5.1 Transforms for some (separable) functions in Cartesian coordinates or in circular coordinates [see also Barrett and Myers (2004), In particular, page 134, on the discussion of the quadratic phase factor in the last line]. The constants a, b and β are real and positive

The inverse Fourier's transform in polar coordinates can be similarly obtained from Eq. (5.1.20.b):

$$\mathscr{F}_{\mathrm{p}}^{-1}: \quad g_{\mathrm{p}}(r,\phi) = 2\pi \sum_{-\infty}^{\infty} {}_n\, i^n e^{in\phi} \int_0^{\infty} f_r G_n(f_r) \mathrm{J}_n(2\pi f_r r)\, d\rho \; . \qquad (5.1.47)$$

The two equations for $\mathscr{F}_{\mathrm{p}}$ and $\mathscr{F}_{\mathrm{p}}^{-1}$ are therefore similar, except for a substitution of i^n for i^{-n}.

A particularly interesting case is when $g_{\mathrm{p}}(r,\varphi)$ has circular symmetry for which $g_{\mathrm{p}}(r,\varphi) = g_{\mathrm{p}}(r)$, so that Eqs. (5.1.44) and (5.1.47) are reduced to expressing (apart from factors 2π from the standard definition) the zero order Hankel's transform between $g_{\mathrm{p}}(r)$ and $G_{\mathrm{p}}(f_r)$:

$$\begin{aligned} \mathscr{B}: &\quad G_{\mathrm{p}}(f_r) = 2\pi \int_0^{\infty} r g_{\mathrm{p}}(r) \mathrm{J}_0(2\pi f_r r)\, dr, \\ \mathscr{B}^{-1}: &\quad g_{\mathrm{p}}(r) = 2\pi \int_0^{\infty} f_r G_{\mathrm{p}}(f_r) \mathrm{J}_0(2\pi f_r r)\, d\rho. \end{aligned} \qquad (5.1.48)$$

They are generally indicated with the name of Fourier-Bessel transforms. Some functions with their transforms are given in Table 5.1.

5.1.4 Uncertainty relation

In the Fraunhofer's diffraction, we have seen that there is an inverse relationship between the relative width of the openings, d/λ, and the angular divergence of the diffracted wave. In other words, the product between these two quantities does not depend on the size of the scale. This fact is a direct consequence of the similarity theorem (5.1.24). The same behavior will also be found with Gaussian laser beams: the more thin they are, the more they diverge. Physically, this effect arises from the application to the radiation of *Heisenberg's uncertainty principle*, for which the product of the uncertainties in the measurement of two conjugate variables, such as position and momentum, cannot fall below a minimum value. On the other hand, in Quantum Mechanics, a physical system can be described by a wave function, and therefore the uncertainty principle has an interpretation based on the mathematical Fourier's transform [Barrett and Myers 2004, p. 217].

Let us limit ourselves here to the one-dimensional case. To find the general relation between the "width" of a function $g(x)$ in L^2, in the space of positions, and that of its transform $G(f)$, in the frequency space, we must first define what is meant by *width*. For this purpose, using the concepts of statistical analysis, we can take $|g(x)|^2$ and $|G(x)|^2$ as weight functions and define a centroid as

$$\overline{x}=\frac{\int_{-\infty}^{\infty}x\left|g(x)\right|^2dx}{\int_{-\infty}^{\infty}\left|g(x)\right|^2dx},\quad \overline{f}=\frac{\int_{-\infty}^{\infty}f\left|G(f)\right|^2df}{\int_{-\infty}^{\infty}\left|G(f)\right|^2df}\tag{5.1.49}$$

and a *variance* as

$$\sigma_x{}^2=\frac{\int_{-\infty}^{\infty}(x-\overline{x})^2\left|g(x)\right|^2dx}{\int_{-\infty}^{\infty}\left|g(x)\right|^2dx},\quad \sigma_f{}^2=\frac{\int_{-\infty}^{\infty}(f-\overline{f})^2\left|G(f)\right|^2df}{\int_{-\infty}^{\infty}\left|G(f)\right|^2df}.\tag{5.1.50}$$

The respective widths are therefore the square root of these variances.

In these two expressions, with a change of variable, we can take both zero $\overline{x}$ and $\overline{f}$ without loss of generality.[3] Therefore, using the derivative theorem (5.1.28.a) and the Parseval's theorem, we have

$$\sigma_f{}^2=\frac{\int_{-\infty}^{\infty}f^2\left|G(f)\right|^2df}{\int_{-\infty}^{\infty}\left|G(f)\right|^2df}=\frac{1}{4\pi^2}\frac{\int_{-\infty}^{\infty}\left|\mathcal{F}\{g'(x)\}\right|^2df}{\int_{-\infty}^{\infty}\left|g(x)\right|^2dx}=\frac{1}{4\pi^2}\frac{\int_{-\infty}^{\infty}\left|g'(x)\right|^2dx}{\int_{-\infty}^{\infty}\left|g(x)\right|^2dx},$$

where $g'(x)$ is the first derivative of $g(x)$. In addition, for Cauchy-Schwarz inequality (5.1.18), we have

$$\sigma_x\sigma_f\geq\frac{1}{2\pi}\frac{\left|\int_{-\infty}^{\infty}g^*(x)\,xg'(x)\,dx\right|}{\int_{-\infty}^{\infty}\left|g(x)\right|^2dx}.\tag{5.1.51}$$

Thanks to the relationship

$$\frac{d}{dx}\left[xg(x)\right]=xg'(x)+g(x),\tag{5.1.52}$$

we can replace $xg'(x)$ in Eq. (5.1.51). Integrating by parts and recalling that $g(x)\to 0$ for $x\to\pm\infty$, the integral in the numerator of this inequality becomes

$$\int_{-\infty}^{\infty}g^*(x)\,xg'(x)\,dx=-\int_{-\infty}^{\infty}g(x)\,xg'^*(x)\,dx-\int_{-\infty}^{\infty}\left|g(x)\right|^2dx,\tag{5.1.53}$$

from which

[3] For the shift theorem, a phase term appears in the new function g and its transform G, however, it disappears in the square module in the integrals of Eqs. (5.1.48).

$$2\Re e\int_{-\infty}^{\infty} g^*(x)\, x g'(x)\, dx = -\int_{-\infty}^{\infty} \left|g(x)\right|^2 dx\,. \tag{5.1.54}$$

Since, then, the module of a complex number is greater than the absolute value of its real part, we finally find the desired inequality

$$\sigma_x \sigma_f \geq \frac{1}{4\pi}, \tag{5.1.55}$$

which expresses the minimum value that can have the product of the widths in the space of positions and frequencies. The sign ≥ becomes an equality when $g'(x)$ is proportional to $x g(x)$, that is, when $g(x)$ is a Gaussian function.

5.2 Sampling theorems

5.2.1 Fourier's series

In the theory of diffraction that we have examined so far, the field can be approximated by continuous functions of the coordinates and, in particular, we have seen that diffraction integrals are essentially Fourier's integrals. Their analytical solution is absent in most cases; on the other hand, any numeric procedure requires us to be content to perform the calculation on and with a finite number of coordinate values. Thus, the Fourier's transform goes from being continuous to being *discrete*, since sampling needs to be performed both in the variables of the function to be transformed and in those of its transform.

Typically, these samplings are done at a constant pace and, in this case, we will see that the discrete transformation we need is closely related to the *Fourier's series* that are defined for periodic functions. For this purpose, we see how these series can be traced back to a particular case of Fourier's integral. For brevity, we confine ourselves here to the one-dimensional case [Oran Brigham 1974].

A periodic function p expressed as a Fourier's series is given by the expression

$$p(x) = \sum_{n=-\infty}^{\infty} c_n e^{i2\pi n f_o x}\,, \tag{5.2.1}$$

where f_o is the fundamental frequency, which corresponds to the period $X_o = 1/f_o$ in the coordinate x of the function p. In this expression, it was preferable to use a complex exponential notation instead of the most familiar summation on sines and cosines; the coefficients c_n are generally complex and are given by the integrals

$$c_n = \frac{1}{X_o}\int_{-X_o/2}^{X_o/2} p(x)\, e^{-i2\pi n x/X_o} dx\,. \tag{5.2.2}$$

We now develop the function p in terms of its Fourier's transform. We can imagine that p is the convolution of a function $q(x)$, which is defined as different from zero only in the interval $-X_o/2 < x < X_o/2$, where q is equal to p, with the periodic function $(1/X_o)\,\text{comb}(x/X_o)$:

$$p(x) = \frac{1}{X_o} q(x) * \text{comb}\left(\frac{1}{X_o}\cdot x\right), \tag{5.2.3}$$

where the asterisk indicates the convolution integral on the variable x that must be executed between the two terms. The transform of q is

$$Q(f) = \int_{-\infty}^{\infty} q(x)\, e^{-i2\pi x f} dx = \int_{-X_o/2}^{X_o/2} p(x)\, e^{-i2\pi x f} dx\,. \tag{5.2.4}$$

In particular, for $f = n f_o = n/X_o$, from Eq. (5.2.2), we have

$$Q\left(\frac{n}{X_o}\right) = X_o c_n\,. \tag{5.2.5}$$

On the other hand, for the theorems of similarity and convolution, the transform of p is given by

$$\begin{aligned} P(f) &= Q(f)\,\text{comb}(X_o f) = Q(f) \sum_{n=-\infty}^{\infty} \delta(X_o f - n) \\ &= \sum_{n=-\infty}^{\infty} Q(n f_o)\,\delta(X_o f - n) = \sum_{n=-\infty}^{\infty} c_n\, \delta(f - n f_o), \end{aligned} \tag{5.2.6}$$

from which we see that the spectrum of p is constituted by a succession of pulses with a constant pace f_o whose "area" (integrating in f) is given by the coefficients c_n of the development in Fourier's series of p. This fact implies something that one should always keep in mind so as to avoid surprises:

> every time that we sample a spectrum in the frequency domain, with a constant pace f_o, the sampled values correspond to a periodic function in the domain of the positions, with pace $X_o = 1/f_o$.

For the properties of symmetry of Fourier's transform, the opposite also applies:

> the spectrum of a function sampled with a constant pace in the domain of the positions is periodic.

5.2.2 Sampling theorem in Cartesian coordinates

While the actual physical variables, such as the electromagnetic field, are defined at all points in space and are, in principle, represented by a function of the coordinates, in the analysis of experimental data and in numerical simulation of theoretical models, we have to limit ourselves to a finite number of points that constitute a sample of those functions. If the sampling is sufficiently dense, the original function can be faithfully reconstructed by interpolating between samples. This intuitive consideration is based on a mathematical theorem by E.T. Whittaker, whereby the functions with limited spectral band can be reconstructed *exactly* if the interval between samples does not exceed a certain limit [Whittaker 1915]. This theorem was used by Shannon in his studies on the information theory [Shannon 1949]. For his demonstration, we will follow Goodman (1996), who considers a two-dimensional extension to the case discussed by Shannon. In Fig. 5.1, the steps of this procedure are represented.

Suppose, therefore, that a function g is sampled on a rectangular grid of points, and we define a sampling function

$$g_S(x,y) = \mathrm{comb}\left(\frac{x}{\Delta_x}\right)\mathrm{comb}\left(\frac{y}{\Delta_y}\right)g(x,y), \tag{5.2.7}$$

which consists of a lattice of functions $\delta(x/\Delta_x - n,\, y/\Delta_y - m) \times g(n\Delta_x, m\Delta_y)$ regularly spaced with pitch Δ_x and Δ_y in the directions of the axes x and y, respectively. Here, we have chosen to match the "volume" (and, in the one-dimensional case, the "areas") of the pulses, namely, the integration of g_S on each δ, with the value of g on the corresponding lattice point multiplied by $\Delta_x\Delta_y$. For the transformation properties of the comb function (see Table 5.1), the spectrum G_S of g_S is the convolution on the frequencies f_x and f_y

$$G_S\left(f_x, f_y\right) = \Delta_x\Delta_y\left[\mathrm{comb}\left(\Delta_x \cdot f_x\right)\mathrm{comb}\left(\Delta_y \cdot f_y\right)\right] * G\left(f_x, f_y\right). \tag{5.2.8}$$

Therefore, the spectrum of the sampled function is equal to the repetition of the spectrum of g around each point $(n/\Delta_x,\, m/\Delta_y)$ in the frequency space:

$$G_S\left(f_x, f_y\right) = \sum_{n=-\infty}^{\infty}\sum_{m=-\infty}^{\infty} G\left(f_x - \frac{n}{\Delta_x}, f_y - \frac{m}{\Delta_y}\right). \tag{5.2.9}$$

If the function g has a limited band, its spectrum G is null outside of a finite region S in the frequency space. If, therefore, the sampling g is sufficiently dense, the various replicas of G that constitute G_S are separated from each other without overlapping.

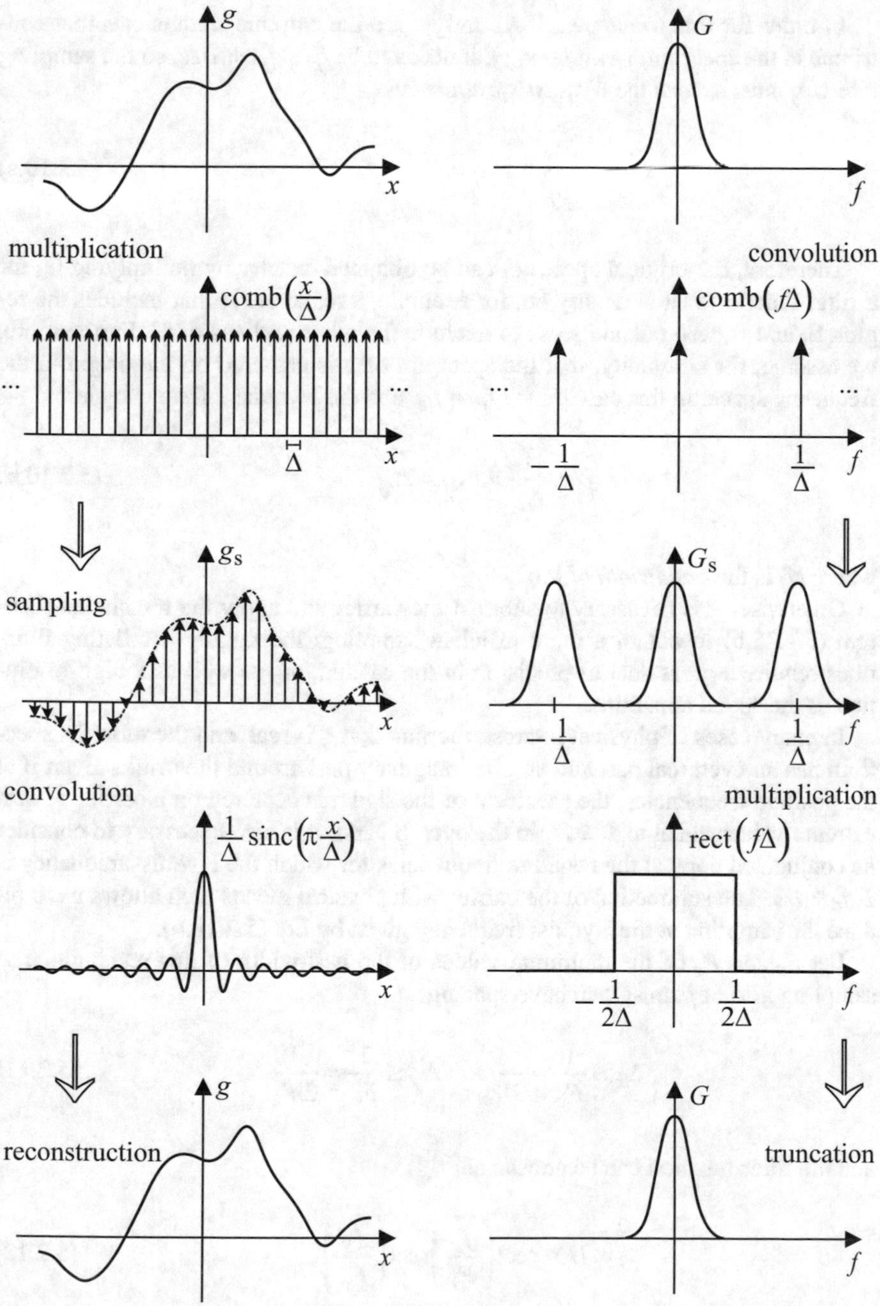

Fig. 5.1 Qualitative graphical representation of the sampling theorem. The figures on the left describe functions in the space of positions, while those to the right indicate their transforms in frequency space

In order for this to happen, if f_{j-} and f_{j+} are the extreme frequencies that contribute to the spectrum (with $j = x, y$), it needs to be $f_{j+} < f_{j-} + 1/\Delta_j$, so the sampling rate $1/\Delta_j$ must *exceed* the *Nyquist's frequency*

$$\frac{1}{\Delta_j} > F_{Nj} = f_{j+} - f_{j-} \,. \tag{5.2.10.a}$$

Therefore, the original spectrum can be obtained exactly by multiplying G_s for a filter function that is unitary on, for example, a rectangle R that includes the region S, and is zero outside so as to exclude the other replicas of G. Furthermore, we assume, for simplicity, that the spectrum of g is centered on the origin in the frequency space. In this case, $f_{j-} = -f_{j+} = f_{jb}$ and the Nyquist's frequency is

$$\frac{1}{\Delta_j} > F_{Nj} = 2W_j \,, \tag{5.2.10.b}$$

where W_j is the *bandwidth* of g.

Otherwise, it is necessary to subtract the carrier and apply the translation theorem (5.1.25.b) to obtain a more efficient sampling: the rapidly oscillating functions require a great deal of precision in the calculation, as well as a high resolution of the actual apparatus.

In many cases of physical interest, the function g is real, and therefore its spectrum has an even real part and an odd imaginary part around the origin. Even if at the positive frequencies, the spectrum of the signal is centered on a carrier f_p with extreme values equal to $f_p \pm W$, in the overall band, it is also necessary to consider its conjugated copy at the negative frequencies for which the Nyquist frequency is $2(f_p + W)$. The subtraction of the carrier with physical means then allows us to reduce the sampling at the Nyquist frequency given by Eq. (5.2.10.b).

Let F_x and F_y be the minimum values of the half-width of this rectangle. The sampling grid of g must then have spacing

$$\Delta_x \le \frac{1}{F_x} = \frac{1}{2W_x}, \quad \Delta_y \le \frac{1}{F_y} = \frac{1}{2W_y}, \tag{5.2.11}$$

and the filter function can be chosen as

$$H = \text{rect}\left(\frac{f_x}{F_x}\right)\text{rect}\left(\frac{f_y}{F_y}\right), \tag{5.2.12}$$

whereby

$$G_s\left(f_x, f_y\right)H\left(f_x, f_y\right) \equiv G\left(f_x, f_y\right),$$

whose transformation is

$$\left[\operatorname{comb}\left(\frac{1}{\Delta_x}\cdot x\right)\operatorname{comb}\left(\frac{1}{\Delta_y}\cdot y\right)g(x,y)\right]*\left[F_xF_y\operatorname{sinc}(\pi F_x\cdot x)\operatorname{sinc}(\pi F_y\cdot y)\right].$$
$$=g(x,y).$$

By developing the convolution integral, we obtain

$$g(x,y)=F_xF_y\Delta_x\Delta_y \times\sum_{n=-\infty}^{\infty}\sum_{m=-\infty}^{\infty}g(n\Delta_x,m\Delta_y)\operatorname{sinc}\left[\pi F_x(x-n\Delta_x)\right]\operatorname{sinc}\left[\pi F_y(y-m\Delta_y)\right], \tag{5.2.13}$$

which, in the case of taking the intervals Δ_x and Δ_y as being equal to $1/F_x$ and $1/F_y$ respectively, becomes

$$g(x,y)=\sum_{n=-\infty}^{\infty}\sum_{m=-\infty}^{\infty}g\left(\frac{n}{F_x},\frac{m}{F_y}\right)\operatorname{sinc}\left[\pi F_x\left(x-\frac{n}{2F_x}\right)\right]\operatorname{sinc}\left[\pi F_y\left(y-\frac{m}{2F_y}\right)\right]. \tag{5.2.14}$$

This is the result of the *Whittaker-Shannon sampling theorem*. Notice that, for each pair (x, y), this operation involves *all* g samples. It is therefore not a simple interpolation between neighboring points, which would not be able to eliminate the beat between the components of the various replicas of the g spectrum. If you wanted to proceed for interpolation, it would be necessary to sample at a frequency at least twice that of Nyquist.

5.2.3 Sampling theorem in polar coordinates

The usefulness of the Fourier-Bessel transformation results from the elimination of one of the integrations. However, to have a practical effect in numerical calculation, it is necessary to find a sufficiently precise and fast technique. It should be borne in mind that J_0 is an oscillating function, with a non-constant period, and there is a need for caution in choosing the integration step.

There is a theorem that, similarly to the Cartesian coordinate case, resolves our problem [Papoulis 1968]. It is based on the properties of the same J_0 function. The function $J_0(x)$ has an infinite number of zeros corresponding to the values x_n for which $J_0(x_n) = 0$. These x_n are called the *zeros of* J_0. Redefining the function argument x with $x = \alpha r$ and assigning to α the values

$$\alpha_n = \frac{x_n}{a}, \quad \text{with} \quad n = 1, 2, 3, \ldots, \tag{5.2.15}$$

where a is a predetermined value, from J_0, one can build an infinite set of functions of r, that is, the functions $J_0(\alpha_n \cdot r)$. It turns out that this set of functions constitutes a basis of orthogonal functions for the generic functions $f(r)$, defined within a range $(0, a)$, that satisfy some regularity conditions (absolute integrability and limited variations [Papoulis 1968]). In particular, it is

$$\int_0^a r J_0(\alpha_n r) J_0(\alpha_m r)\, dr = \begin{cases} \frac{a^2}{2} J_1(\alpha_n a) & \text{for} \quad n = m, \\ 0 & \text{for} \quad n \neq m. \end{cases} \tag{5.2.16}$$

With this basis, the function $f(r)$ has a development

$$f(r) = \sum_{n=1}^{\infty} C_n J_0(\alpha_n r), \quad \text{with} \quad 0 < r < a, \tag{5.2.17}$$

where the coefficients C_n can be calculated by multiplying both sides of Eq. (5.2.17) for $r\, J_0(\alpha_n r)$ and integrating between 0 and a. Applying Eq. (5.2.16) and Eq. (5.1.36) for the zero-order Hankel's transform case, we have

$$C_n = \frac{2}{a^2} \frac{\hat{f}_o(\alpha_n)}{J_1^2(\alpha_n a)} = \frac{2}{a^2} \frac{\hat{f}_o(\alpha_n)}{J_{1n}^2}, \tag{5.2.18}$$

where J_{1n} are the values of J_1 corresponding to the zeros x_n of J_0 quoted above, and $\hat{f}_o(w)$ is the Hankel's transform of the function

$$f_o(r) = \begin{cases} f(r) & \text{for} \quad 0 < r \le a, \\ 0 & \text{for} \quad r > a. \end{cases} \tag{5.2.19}$$

Thanks to the symmetry properties of the Hankel's transform, a similar development can also be conducted for the transform $\hat{f}(w)$ of $f(r)$. Let us consider, in particular, the case of a function $f(r)$ that has a zero-order Hankel's transform $\hat{f}(w)$ with a *limited bandwidth*:

$$\hat{f}(w) = 0 \quad \text{for} \quad w > b, \tag{5.2.20}$$

where b now means the limit of an interval $(0, b)$ in the space of frequencies w. With the same procedure as before, we find that the coefficients $\hat{C}_n$ rewritten for the development of $\hat{f}(w)$ are

$$\hat{C}_n = \frac{2}{b^2} \frac{f(r_n)}{J_{1n}^{\ 2}}, \tag{5.2.21}$$

where $r_n = x_n/b$ are the points where $f(r)$ is sampled. Therefore, we obtain that $\hat{f}(w)$ can be derived from this sampling by the formula:

$$\hat{f}(w) = \frac{2}{b^2} \sum_{n=1}^{\infty} \frac{f(r_n)}{J_{1n}^{\ 2}} J_0(r_n w) P_b(w), \tag{5.2.22}$$

where $P_b(w)$ is an impulse function, that is, $P_b(w) = 1$ for $0 < w \le b$ and $P_b(w) = 0$ for $w > b$. With this limited spectral band hypothesis, the whole function $f(r)$ can be derived from its sampling: doing the transform of both sides of Eq. (5.2.22), we have

$$f(r) = \frac{2}{b^2} \sum_{n=1}^{\infty} \frac{f(r_n)}{J_{1n}^{\ 2}} \int_0^{\infty} w J_0(r_n w) P_b(w) J_0(rw) dw .$$

With a few calculations [Papoulis 1968], and still exploiting the fact that $J_0(r_n b) = 0$, we find that

$$f(r) = \frac{2}{b} \sum_{n=1}^{\infty} \frac{r_n f(r_n)}{J_{1n}} \frac{J_0(br)}{r_n^{\ 2} - r^2}. \tag{5.2.23}$$

Note that, for $r = r_n$, it is $J_0(rb) = 0$.

We can finally reformulate Eq. (5.2.22) for the calculation of the Fourier-Bessel transform: from the comparison of Eqs. (5.1.36) and (5.1.48.a), we find that, for a given function $g(r)$,

$$G_p(f_r) = 2\pi \hat{g}_0(2\pi f_r),$$

where $G_p(f_r)$ and $\hat{g}_0(w)$ are the Fourier-Bessel and Hankel's transforms of zero order, respectively, of the function $g(r)$ with $w = 2\pi f_r$. Therefore, the condition (5.2.20) becomes

$$G_p(f_r) = 0 \quad \text{for} \quad f_r > c, \quad \text{with} \quad c = b/(2\pi).$$

Finally, for $f_r \le c$, we have

$$G_p(f_r) = \frac{1}{\pi c^2} \sum_{n=1}^{\infty} \frac{g(r_n)}{J_{1n}^{\ 2}} J_0(2\pi r_n f_r), \tag{5.2.24}$$

where I ultimately remember that r_n are the points where the function g is sampled, chosen so that $J_0(2\pi r_n c) = 0$.

5.2.4 The Discrete Fourier's Transform

Let us finally consider a pair of continuous functions g and G that are each others' Fourier's transform. By exploiting the sampling theorem, under the appropriate conditions, in place of g and G, we can obtain a couple of discrete functions, which are a satisfactory approximation of the original ones. We will now see how to convert the Fourier's transform into a numeric procedure applicable to a digital computer. The operation that is obtained is called the *discrete Fourier's transform*. For simplicity, we will limit ourselves here to the case of the one-dimensional transformation in Cartesian coordinates [Oran Brigham 1974]. The steps of this procedure are shown in Fig. 5.2.

The first step is to sample the function $g(x)$ with constant pace Δ_x. For this purpose, we use a non-normalized function $\text{comb}(x/\Delta_x)$

$$g_s(x) = g(x)\,\text{comb}\left(\frac{x}{\Delta_x}\right) = \sum_{n=-\infty}^{\infty} g(n\Delta_x)\,\delta\left(\frac{x}{\Delta_x} - n\right). \quad (5.2.25)$$

With this (arbitrary) choice of the sampling function, the area below each pulse of g_s is equal to Δ_x times the corresponding value of g. Its transform is the periodic repetition of the spectrum G of g

$$G_s(f) = \Delta_x G(f) * \text{comb}(\Delta_x \cdot f) = \sum_{n=-\infty}^{\infty} G\left(f - \frac{n}{\Delta_x}\right). \quad (5.2.26)$$

If Δ_x is not taken to be sufficiently small, the copies of G overlap, and this gives rise to the phenomenon of *aliasing*: something that appears to be contributing to the frequency f can instead be, under a false name, a copy of that at a frequency $f - n/\Delta_x$ with n as a positive or negative integer.

The remedy is to sample according to the rules outlined in §5.2.2 above. However, if g does not have a limited bandwidth, but its spectrum does at least decline away from its center, we should try to thicken the sampling until aliasing is bearable or filter g before sampling it, something that must always be done in real equipment to reduce the effects of noise. It is clear that, in these cases, we will only have an approximate representation of our function or signal.

The second step is to make a truncation in the sampling; indeed, we can handle only a finite number N of samples. To do this, we multiply the sample function by a rectangle function

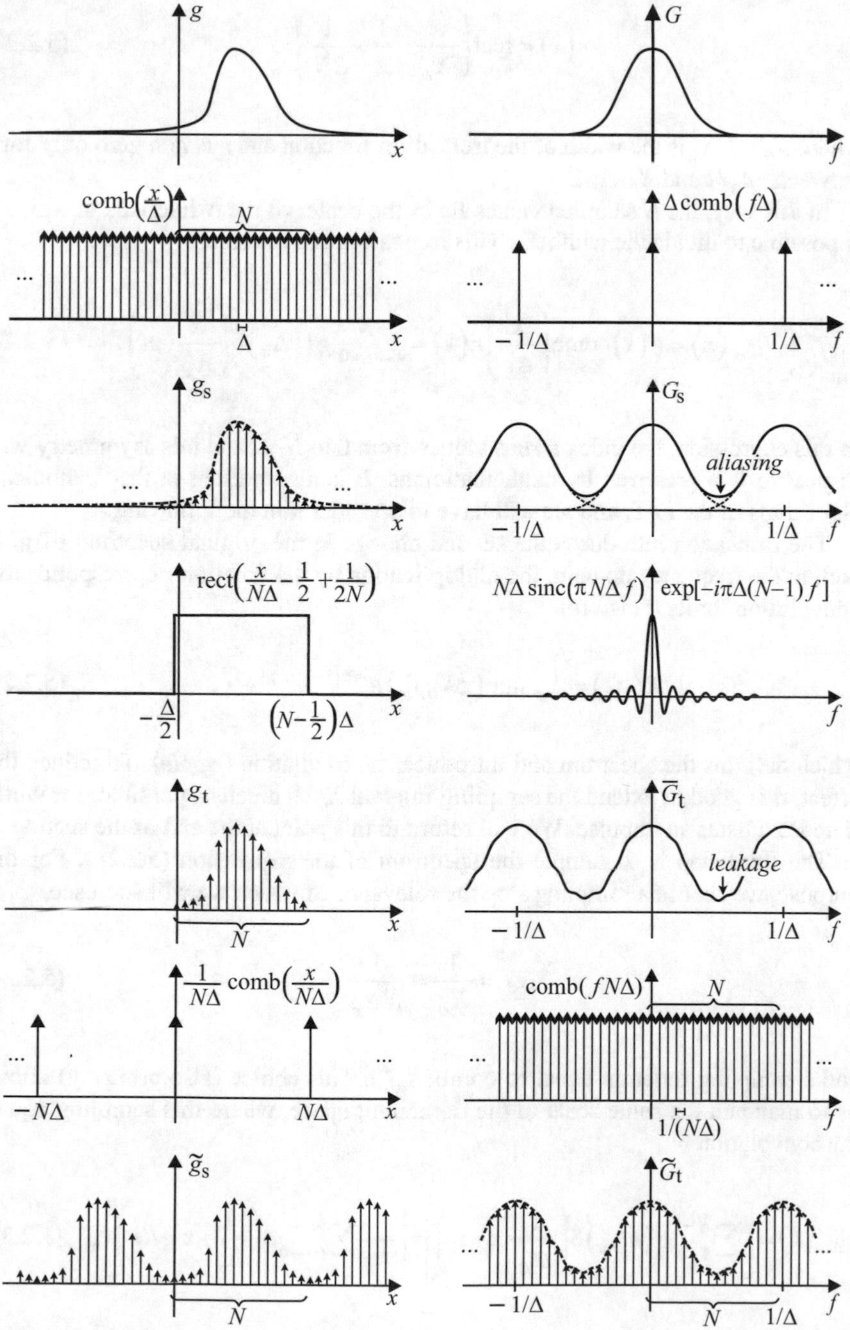

Fig. 5.2 Qualitative graphic derivation of the discrete Fourier's transform

$$r(x) = \text{rect}\left(\frac{x}{X_o} - \frac{1}{2} + \frac{1}{2N}\right), \tag{5.2.27}$$

where $X_o = N\Delta_x$ is the width of the truncation function and r is non-zero only for x between $-\Delta_x/2$ and $X_o - \Delta_x/2$.

In this way, the N sampled values lie in the center of the N intervals in which it is possible to divide the width X_o. This truncation has the result

$$g_t(x) = g(x)\text{comb}\left(\frac{x}{\Delta_x}\right)r(x) = \sum_{n=0}^{N-1} g(n\Delta_x)\delta\left(\frac{x}{\Delta_x} - n\right). \tag{5.2.28}$$

In this expression, the index n runs values from 0 to $N-1$, and this asymmetry with respect to 0 is preferred by mathematicians. It is also present in the "canonical" algorithms of the FFT, and we will have to consider it in the following.

The truncation introduces the second change in the original spectrum of g. In fact, in the frequency domain, the multiplication by the function r corresponds to a convolution for its transform

$$R(f_x) = X_o \,\text{sinc}(\pi X_o f_x) e^{-i\pi(1-1/N)X_o f_x}, \tag{5.2.29}$$

which deforms the spectrum and introduces an oscillation (*ripple*). To reduce this effect, it is good to extend the sampling interval X_o as much as possible, for which R approximates an impulse. We will return to this point at the end of the section.

The third step is to sample the spectrum of the expression (5.2.28). For this purpose, we choose a sampling step the relevance of which we will soon see,

$$\Delta_f = \frac{1}{X_o} = \frac{1}{N\Delta_x}, \tag{5.2.30}$$

and a sampling function equal to $\text{comb}(X_o f)$. This choice (also arbitrary) allows us to maintain the same scale in the domain of space, where this sampling equals the convolution

$$\tilde{g}_s(x) = \left[\sum_{n=0}^{N-1} g(n\Delta_x)\delta\left(\frac{1}{\Delta_x}\cdot x - n\right)\right] * \left[\frac{1}{\Delta_x}\sum_{l=-\infty}^{\infty} \delta\left(\frac{1}{\Delta_x}\cdot x - lN\right)\right], \tag{5.2.31}$$

where $\tilde{g}_s$ indicates an approximation of the sample function g_s and where the delta property (5.1.5) was used. With a few steps, we find that

$$\tilde{g}_s(x) = \sum_{l=-\infty}^{\infty} \sum_{n=0}^{N-1} g(n\Delta_x)\,\delta\!\left(\frac{x}{\Delta_x} - n - lN\right). \tag{5.2.32}$$

We can note that this expression is periodic with period $X_o = N\Delta_x$ and, in particular, that it consists of a regular succession of pulses with pitch Δ_x; it would not have had this property with a different choice of Δ_f. Besides, the order $l = 0$ of $\tilde{g}_s$ coincides with g_s, just as we wanted.

Let us now resume the results of the Fourier's series discussion: since $\tilde{g}_s$ is periodic we can apply Eq. (5.2.6) to it, so that its transform is given by

$$\tilde{G}_s(f) = X_o \sum_{m=-\infty}^{\infty} \tilde{G}_m \delta(X_o f - m), \tag{5.2.33}$$

where, for Eq. (5.2.2), the individual pulses of $\tilde{G}_s$ have area

$$\tilde{G}_m = \frac{1}{X_o} \int_{-\Delta_x/2}^{X_o-\Delta_x/2} \tilde{g}_s(x) e^{-i2\pi m x/X_o} dx\,. \tag{5.2.34}$$

Replacing Eq. (5.2.32) in (5.2.34), and considering that the integration is only over a period, we get

$$\tilde{G}_m = \frac{1}{N} \sum_{n=0}^{N-1} g(n\Delta_x)\, e^{-i2\pi nm/N}\,. \tag{5.2.35}$$

Even this expression is periodic in m with period N: just replace m with m'+N to check it, since $\exp(i2\pi n) = 1$ for the n integer.

Therefore, both $\tilde{g}_s$ and its Fourier's transform G_s are periodic functions of pulse trains. Similarly, we denote, by

$$\tilde{g}_n = \Delta_x g(n\Delta_x), \tag{5.2.36}$$

the pulse area of $\tilde{g}_s$, as can be deduced from Eq. (5.2.32) as well. We can then establish the relationship between these areas as

$$\tilde{G}_m = \frac{1}{X_o} \sum_{n=0}^{N-1} \tilde{g}_n\, e^{-i2\pi nm/N}, \quad \text{with } m = 0, ..., N\text{–}1. \tag{5.2.37}$$

The inverse relationship is

$$\tilde{g}_n = \Delta_x \sum_{m=0}^{N-1} \tilde{G}_m\, e^{i2\pi nm/N}, \quad \text{with } n = 0, ..., N\text{–}1, \tag{5.2.38}$$

as can be seen with some simple steps.

These relationships could be considered to be the discrete transformation and its inverse that we are looking for. However, it still lacks a relationship between these coefficients and the Fourier's transform of the original function g. This relationship is made quite complicated by the truncation.

To understand its effects, consider the case in which g is a simple sinusoid with frequency f. Its spectrum is then made up of two Dirac's deltas at frequencies $\pm f$. The convolution with R produces two phenomena: the two pulses widen and a series of satellite spikes invades the entire spectrum. This effect (which is closely related to the diffraction produced by an opening) is called *leakage*.

One way to counter it is to choose a truncation interval that is a whole number of periods of the sinusoid. In fact, in this case, the sampling of the spectrum made with the rules discussed above produces only two non-null values at the frequencies $\pm f$, whereas all of the other samples coincide with the zeros of both sinc of the spectrum of the truncated function. The sampled spectrum that is obtained is thus equal, less a constant, to what would occur without truncation.

If, however, the truncation interval is different from a whole multiple of the oscillation period, the spectrum we find does correspond to a periodic function, but it is the repetition of the sampled portion, and the lack of coincidence between the extremes of the sampled curve produces a discontinuity that feeds the spectrum of spurious components and aggravates the aliasing.

In this case, there may be a strong deviation from the spectrum of g. This situation occurs in the general case of a non-periodic and spatially non-delimited function. One remedy is to sample a relatively long interval such that we are able to connect the ends with the least possible discontinuity. Another remedy is to attribute, at point 0 or point N, the average of the values sampled at points 0 and N. This trick is also useful in the discontinuity points of piecewise continuous functions, such as square waves, whose spectrum is not delimited.

The sampling of a signal can thus have rather negative aspects. However, there are two cases that can be treated easily, as we shall see in the next two sections.

5.2.4.1 Spatially bounded functions with limited bandwidth

Let us consider the case in which the function g is zero outside of the truncation interval: this operation has no effect here, and since, within the integration interval of Eq. (5.2.34), it is $\tilde{g}_s = g_s$, as mentioned above, according to Eq. (5.2.34), it results that $G_s\,[m/(N\Delta_x)] = X_o\tilde{G}_m$. If g also then has a limited spectrum and the sampling is sufficiently dense to make the aliasing negligible, from Eq. (5.2.26), we have

$$G\left(\frac{m}{N\Delta_x}\right) \cong X_o\tilde{G}_m\,, \quad \text{that is,} \quad \tilde{G}_m \cong \Delta_f G\left(\frac{m}{N\Delta_x}\right), \quad \text{for } m = 0,\cdots,N-1\,. \tag{5.2.39}$$

In the case we are dealing with here, for Eq. (5.2.35), the *discrete Fourier's transform* is then expressed by

$$G\left(\frac{m}{N\Delta_x}\right)=\Delta_x\sum_{n=0}^{N-1} g\left(n\Delta_x\right)e^{-i2\pi nm/N}, \tag{5.2.40}$$

and its *inverse* can be obtained from Eq. (5.2.38):

$$g\left(n\Delta_x\right)=\frac{1}{N\Delta_x}\sum_{m=0}^{N-1} G\left(\frac{m}{N\Delta_x}\right)e^{i2\pi nm/N}. \tag{5.2.41}$$

The symmetry between these two relationships can be seen simply by remembering that $1/(N\Delta_x) = \Delta_f$. Therefore, the division for N that appears before the sum of Eq. (5.2.41) is not surprising.

5.2.4.2 Periodic functions with limited bandwidth

Let us now consider the case in which the function g is periodic with a period X_o. Its Fourier's transform consists of pulses at frequencies $f_m = m/X_o$ whose area, for Eqs. (5.2.6) and (5.2.2), is

$$G_m=\frac{1}{X_o}\int_{-\Delta_x/2}^{X_o-\Delta_x/2} g(x)e^{-i2\pi mx/X_o}dx\ . \tag{5.2.42}$$

Sampling the spectrum at frequencies f_m, and by exploiting the periodicity of g, we can bring Eq. (5.2.32) to (5.2.25): we still have $\tilde{g}_s = g_s$, but, this time, on the whole x-axis. Therefore, the effect of truncation is also canceled in this case. In particular, it is now $\tilde{G}_s = G_s$. On the other hand, Eq. (5.2.26) states that, if the spectrum of g has a limited bandwidth and the sampling in the space is sufficiently dense to render aliasing negligible, the zero order of G_s coincides with G. Therefore, for Eq. (5.2.35), we also have

$$G_m=\tilde{G}_m=\frac{1}{N}\sum_{n=0}^{N-1} g\left(n\Delta_x\right)e^{-i2\pi nm/N}\ . \tag{5.2.43}$$

This result can also be obtained by transforming the integral of Eq. (5.2.42) in a summation, with the replacements $x \to n\Delta_x$, $dx \to \Delta_x$, and $X_o \to N\Delta_x$. However, in this way, the proof would have made the bounds imposed by the aliasing less visible. The corresponding inverse transform is

$$g(n\Delta_x) = \sum_{m=0}^{N-1} G_m \, e^{i2\pi nm/N} \,. \tag{5.2.44}$$

5.2.5 The Fast Fourier Transform

Mathematically, the canonical discrete Fourier's transform is expressed in the dimensionless form

$$G(m) = \sum_{n=0}^{N-1} g(n) e^{-i2\pi nm/N} \tag{5.2.45}$$

and its inverse by

$$g(n) = \frac{1}{N} \sum_{m=0}^{N-1} G(m) \, e^{i2\pi nm/N} \,, \tag{5.2.46}$$

to which, for example, Eqs. (5.2.40-41) can be traced back, respectively, by choosing a scale of x such that $\Delta_x = 1$ and, consequently, for Eq. (5.2.30), a scale for f such that $\Delta_f = 1/N$, simplifying the notation for the argument of G. The calculation of these expressions involves N output values, each of which requires a sum of N input values, each multiplied by a complex number. Therefore, the direct application of the transformation formula requires $N(N-1)$ sums and N^2 complex multiplications.

In the two-dimensional case, as for the calculation of the diffraction, the situation is aggravated even more for which a grid of $N \times N$ values would require $N^2(N-1)^2$ sums and N^4 complex multiplications. For $N = 1000$, we would have an exorbitant number of 1,000 billion sums and multiplications!

Fortunately, J.W. Cooley and J.W. Tukey, as well as other independent authors, introduced a mathematical algorithm called the *Fast Fourier Transform* (FFT) to calculate the discrete Fourier's transform [Cooley and Tukey 1965]. It eliminates the redundant operations of the direct method and reduces the number of necessary operations to approximately $2N\log_2 N$. We have no need here to enter into the complex details of the derivation of this algorithm, of which, in addition, there are various "canonical" versions [Oran Brigham 1974]. However, the values that we can attribute to N are not arbitrary: in order to obtain maximum efficiency, it is necessary that N be highly factorizable. The most common algorithm is to base 2 so that N must be a power of 2; there are also base-4, 8, and 16 versions that are more efficient and in which N must be a power of these bases. What matters is that they solve the calculation of Eqs. (5.2.45-46) efficiently. The base 2 algorithm can be found in all mathematical software packages[4], such as the *Numerical Recipes*; what is striking is that, transcribed, for example, in Fortran, it only requires one

[4] In any case, it is always good to check the exact correspondence with the canonical Eqs. (5.2.45-46).

page of instructions, far less than what it takes to explain it.

Frequently, both the signal and the corresponding spectrum are centered on zero in their respective scales. In such cases, to apply the FFT, it is also necessary to perform a translation of the samples before and after the transformation. Indeed, the discrete transformation assumes that, both in the domain of space and that of the frequencies, the functions $g(n)$ and $G(m)$ are periodic with period N. Consider the mathematically more elegant case of the C language: the FFT algorithm places the zero value of position and frequency in coincidence with the indices n and m = 0, respectively. In addition, the values m = 1, ..., N/2–1 correspond to the positive frequencies, while, for the periodicity of the spectrum, the negative ones are assigned to the remaining values, with $m = N/2, ..., N–1$. The same happens for the positions. Therefore, before executing the FFT, the values $g(n)$ at the negative positions n = –N/2, ..., –1 must be translated into $g(n+N)$. Wishing, then, to recompose the frequency scale, it is necessary to translate the values $G(m)$ in $G(m–N)$ for m = N/2, ..., N –1. In the case of programming in Fortran, 1 must be added to all indexes n and m.

The FFT algorithm is only one-dimensional, but it is easily extensible to the two-dimensional case. For this purpose, it is necessary to perform the transformation (including the translations) on each row, and then on each column. In the case of dealing with the diffraction of a spatially delimited field, it may be useful to embed the samples in a larger matrix so as to have a frame of null values (better if it is extrapolated without discontinuity) around the original samples. In this way, the spectrum that is obtained will be defined with a denser sampling, with smaller Δ_f.

5.3 Applications of the Fourier's transform to the diffraction

As we have seen in the chapter dedicated to diffraction in the Fresnel approximation, the propagation of an electromagnetic wave in a homogeneous medium can be determined by means of a Fourier's transform of the product of the field, which is defined on the plane crossed by the radiation, by a phase term that varies radially, according to Eq. (4.2.8). This fact is exploited to calculate the propagation numerically: the input field is sampled on a regular grid of points and the output field is returned on a similar grid with the same number of points.

Another approach consists in calculating the angular spectrum of the field on the aperture and, from this, reconstructing the diffracted field: in this case, it is necessary to calculate two Fourier's transforms to get the field at a given distance z from the opening. As we will see, this method is effective in the case of near field diffraction, where the Fresnel approximation falls, but becomes unfeasible in Fraunhofer's diffraction.

Various conditions are required for the success of these applications. The first one can be easily overlooked or may seem overly obvious, but it is not: it is necessary that the integrals, which are to be computed, really have the shape of a Fourier's transform. For example, Kirchhoff's diffraction integral does not have this property. In the binomial development of the root, we have a typical term of the transform, which is the mixed bilinear product of the coordinates of the points on the aperture and those on the observation plane, but there are also other mixed terms of higher power. Instead, in a true Fourier's transform, the function to be transformed cannot contain any dependence on the variables conjugated to those on which it is integrated, unless they may be factorized and removed from the integral. Therefore, the diffraction integral can be traced back to a Fourier's transform only in the Fresnel's and Fraunhofer's approximations, in which the additional mixed terms are neglected. This fact should be kept in mind when, for example, one wants to determine the resolving power of an optical instrument (see §5.4) with a large numerical aperture.

Another thing to remember is that the Fourier's transform method applies to entities that sum up scalarly, as is the case with acoustic longitudinal waves in a gas, in which the wave amplitude is given by the difference in pressure compared to that in the absence of sound. However, we will apply this transform for a *single* Cartesian component of the amplitude of the field of an electromagnetic wave, within the limit of the paraxial approximation.

In the case of Fresnel diffraction, two other conditions are also required. The first is that *the field on the starting plane is spatially delimited*; in fact, numerically, we can sample only a finite portion of space. The second condition is that *the field on the arrival plane is also spatially delimited* so as to satisfy Eq. (5.2.11); indeed, the spacing between the samples must necessarily be finite. This will allow for applying the Whittaker-Shannon sampling theorem.

Sometimes, instead of considering a field that decays to zero outside of a finite region of space, a field is considered that extends periodically throughout the space along both transversal coordinates. In this case, its spectrum consists of a two-dimensional grid of Dirac's delta. For reasons of transverse symmetry, although the field profile changes, the same periodicity must be maintained at any propagation distance. In this case, Fresnel's diffraction cannot be applied, where the periodicity of the function to be transformed is erased by the phase term that varies with the square of the distance from the axis. With some contrivance, we can instead apply the angular spectrum method to this case. It should also be noted that, contrary to Fraunhofer's diffraction, the diffraction of a periodic field does not tend to its spectrum as the distance increases. It is also true that a truly periodic field, and therefore an infinitely extended one, is not physically feasible.

5.3.1 Application of the FFT to the Fresnel diffraction

Let us now consider the case that on the departure plane the field u (transmitted from a possible aperture) is zero or at least sufficiently negligible outside a rectangle with sides a_x and a_y centered on the origin, as not to give appreciable overall contributions to the diffraction integral from there. To proceed with the analysis we introduce the spatial frequencies

$$f_x = \frac{x_o}{\lambda Z_o}, \quad f_y = \frac{y_o}{\lambda Z_o}. \tag{5.3.1}$$

The diffraction integral (4.2.8) becomes

$$u_o\left(f_x \lambda Z_o, f_y \lambda Z_o\right) = -\frac{i}{\lambda Z_o} e^{ikZ_o} e^{i\pi\lambda Z_o\left(f_x{}^2 + f_y{}^2\right)} \times$$
$$\times \int_{-\frac{a_x}{2}}^{\frac{a_x}{2}} \int_{-\frac{a_y}{2}}^{\frac{a_y}{2}} \left[e^{i\pi \frac{1}{\lambda Z_o}\left(x^2+y^2\right)} u(x,y) \right] e^{-i2\pi\left(f_x x + f_y y\right)} dxdy. \tag{5.3.2}$$

Let us go back to the analysis of the previous section. Observing the structure of Eq. (5.3.2), we see that the function

$$V\left(f_x, f_y\right) = i\lambda Z_o e^{-ikZ_o} e^{-i\pi\lambda Z_o\left(f_x{}^2 + f_y{}^2\right)} u_o\left(f_x \lambda Z_o, f_y \lambda Z_o\right) \tag{5.3.3}$$

corresponds to the Fourier's transform of the function

$$\upsilon(x,y) = e^{i\pi \frac{1}{\lambda Z_o}\left(x^2+y^2\right)} u(x,y). \tag{5.3.4}$$

In addition to the scale transformation of variables, u_o differs from V only for the term λZ_o and a phase factor. Therefore, if $u_o(x_o, y_o)$ is sufficiently limited within a rectangle with sides b_x, b_y *centered around the origin*, the integration can be reduced to a summation on a grid of $N \times M$ points with coordinates (x_n, y_m), where the corresponding spacing Δ_x and Δ_y in the x and y coordinates meet the conditions

$$\Delta_x \le \frac{\lambda Z_o}{b_x}, \quad \Delta_y \le \frac{\lambda Z_o}{b_y}, \tag{5.3.5}$$

arising from Eqs. (5.2.11) and (5.3.1). In particular, we take that

$$x_n = n\Delta_x, \quad -\frac{N}{2} \le n \le \frac{N}{2} - 1, \quad n \in \mathbb{Z}, \quad \text{with } N \text{ even}; \tag{5.3.6.a}$$

$$y_m = m\Delta_x, \quad -\frac{M}{2} \le m \le \frac{M}{2} - 1, \quad m \in \mathbb{Z}, \quad \text{with } M \text{ even}. \tag{5.3.6.b}$$

On the other hand, the sampling grid must cover the area $a_x \times a_y$ in which $u(x,y)$ is confined. So, it is necessary that

$$N \geq \frac{a_x}{\Delta_x} \geq \frac{a_x b_x}{\lambda Z_o}, \quad M \geq \frac{a_y}{\Delta_y} \geq \frac{a_y b_y}{\lambda Z_o}. \tag{5.3.7}$$

These conditions impose a minimum limit on the number of grid samples, which we will discuss at the end of this section. In the following, we denote the side lengths of the sampled area with

$$X_a = N\Delta_x \geq a_x, \quad Y_a = M\Delta_y \geq a_y. \tag{5.3.8}$$

In turn, υ in Eq. (5.3.4) represents the spectrum of V and differs from u only for a phase factor; it is band-limited, given that we have assumed u to be zero outside of a delimited region in the transverse coordinates. Therefore, u_o can also be sampled without loss of information.

On the other hand, the FFT algorithm produces a linear transformation between sets with the same number of elements; hence, the calculation of u_o obtained through this algorithm will still be executed on a grid of N points $f_{x,n'}$ by M points $f_{y,m'}$, defined by

$$f_{x,n'} = n'\frac{F_x}{N}, \quad -\frac{N}{2} \leq n' \leq \frac{N}{2} - 1, \quad n' \in \mathbb{Z}, \quad \text{for } N \text{ even}; \tag{5.3.9.a}$$

$$f_{y,m'} = m'\frac{F_y}{M}, \quad -\frac{M}{2} \leq m' < \frac{M}{2} - 1, \quad m' \in \mathbb{Z}, \quad \text{for } M \text{ even}, \tag{5.3.9.b}$$

where

$$F_x = \frac{1}{\Delta_x}, \quad F_y = \frac{1}{\Delta_y} \tag{5.3.10}$$

are the maximum frequencies that derive from the sampling of the intervals X_a and Y_a, respectively. Therefore, the values assumed by $f_{x,n'}$ and $f_{y,m'}$ are

$$f_{x,n'} = \frac{n'}{N}\frac{1}{\Delta_x}, \quad f_{y,m'} = \frac{m'}{M}\frac{1}{\Delta_y}. \tag{5.3.11}$$

For the choice made on Δ_x and Δ_y with the conditions (5.3.5), we have

$$X_b = \lambda Z_o F_x = \frac{\lambda Z_o}{\Delta_x} \geq b_x, \quad Y_b = \lambda Z_o F_y = \frac{\lambda Z_o}{\Delta_y} \geq b_y, \tag{5.3.12}$$

with which it occurs that the sampling at the arrival plane covers the region where the field is significantly different from zero.

With Eqs. (5.3.11), Eq. (5.3.2) becomes

$$u_o\left(\lambda Z_o \frac{n'}{N\Delta_x}, \lambda Z_o \frac{m'}{N\Delta_y}\right) = -ie^{ikZ_o} e^{i\pi\lambda Z_o\left[\frac{n'^2}{N^2\Delta_x^2}+\frac{m'^2}{M^2\Delta_y^2}\right]} \frac{\Delta_x\Delta_y}{\lambda Z_o}$$
$$\times \sum_{n=-\frac{N}{2}}^{\frac{N}{2}-1} \sum_{m=-\frac{M}{2}}^{\frac{M}{2}-1} \left[e^{\frac{i\pi}{\lambda Z_o}\left(n^2\Delta_x^2+m^2\Delta_y^2\right)} u\left(n\Delta_x, m\Delta_y\right)\right] e^{-i2\pi\left(\frac{nn'}{N}+\frac{mm'}{M}\right)}. \tag{5.3.13}$$

In the far field regime (Fraunhofer's diffraction), $b_x \sim \lambda Z_o F_x$ and $b_y \sim \lambda Z_o F_y$. Therefore, the necessary number $N\times M$ of grid samples given by the inequalities (5.3.7) is independent of Z_o. Instead, in the near field regime, it is usually $b \sim a$, and, in such case, $N\times M$ increases with decreasing Z_o.

Between the size of the initial and final grid, a magnification intervenes given by

$$m_{\perp x} = N\frac{\lambda Z_o}{X_a^2}, \quad m_{\perp y} = M\frac{\lambda Z_o}{Y_a^2}. \tag{5.3.14}$$

If, in particular, we require a unitary magnification, which can, for example, occur in the study of resonant cavities, it is necessary that the initial plane be sampled on a rectangle of sides $X_c = \max(X_a, X_b)$ and $Y_c = \max(Y_a, Y_b)$, whereby

$$N = \frac{X_c^2}{\lambda Z_o}, \quad M = \frac{Y_c^2}{\lambda Z_o}. \tag{5.3.15}$$

To conclude, Eq. (5.3.13) can ultimately be resolved with the following steps: *i*) multiplying each value of field u for the corresponding curvature term, *ii*) translating the samples of the matrix thus obtained, *iii*) applying the two-dimensional FFT, *iv*) translating the matrix back, *v*) multiplying each value for the corresponding term of curvature and amplitude.

It should be noted that the fractions that appear in Eqs. (5.3.7) represent a generalization of the concept of Fresnel's number (multiplied by 4) defined by Eq. (4.4.3). This value consists of the number of Fresnel zones, for the observation distance Z_o, in which the aperture has to be divided so as to obtain uniform contributions in the diffraction integral with a plane incident wave. Here, N and M indicate the number of samples required, also taking into account the field structure on the input window, and therefore its spectral band.

5.3.2 Angular spectrum of the field on an opening

We will now see a different approach to diffraction theory, which I would call "Fourier's diffraction". While the Fresnel-Kirchhoff diffraction uses a development of the field in spherical *wavelets* that are individually generated by each point on the surface of an opening, here, we instead consider the development in plane waves that are "overall generated" by the field on the whole opening. Both methods can be considered as limiting cases of a field decomposition in propagation modes.

Let us take the case of a monochromatic wave that crosses an *xy*-plane immediately placed behind an opening in the positive direction of the *z*-axis. As in the case of Kirchhoff, let us suppose that we know the complex field $u(x,y,0)$ on this plane and want to determine the field $u(x,y,z)$ in the half-space $z > 0$. Further, suppose that the space in which the waves propagate is homogeneous and isotropic. Through the *xy*-plane, the function U has a two-dimensional Fourier's transform given by

$$A\left(f_x, f_y; z=0\right) = \int_{-\infty}^{\infty}\int_{-\infty}^{\infty} u(x,y,0)\, e^{-2\pi i\left(f_x x + f_y y\right)} dx\, dy\,, \tag{5.3.16}$$

and, reciprocally, we have

$$u(x,y,0) = \int_{-\infty}^{\infty}\int_{-\infty}^{\infty} A\left(f_x, f_y; 0\right) e^{2\pi i\left(f_x x + f_y y\right)} df_x\, df_y\,. \tag{5.3.17}$$

On the other hand, a monochromatic plane wave of unit amplitude that propagates with direction cosines (α,β,γ) is simply expressed by

$$o(x,y,z) = e^{ik(\alpha x + \beta y + \gamma z)}\,,$$

where $k = 2\pi/\lambda$, with λ equal to the wavelength in the considered medium, and

$$\gamma = \sqrt{1-\alpha^2-\beta^2}\,.$$

Therefore, the field $u(x,y,0)$ can be considered as the superposition in $z = 0$ of the fields of a plane wave spectrum whose direction cosines are

$$\alpha = \lambda f_x,\quad \beta = \lambda f_y,\quad \gamma = \sqrt{1-\left(\lambda f_x\right)^2 - \left(\lambda f_y\right)^2}\,, \tag{5.3.18}$$

and Eq. (5.3.17) can be rewritten as

$$u(x,y,0)=\int_{-\infty}^{\infty}\int_{-\infty}^{\infty}A\left(\frac{\alpha}{\lambda},\frac{\beta}{\lambda};0\right)e^{ik(\alpha x+\beta y)}d\frac{\alpha}{\lambda}d\frac{\beta}{\lambda}, \tag{5.3.19}$$

where $A(\alpha/\lambda,\beta/\lambda;0)$ is called the *angular spectrum* of u, because it represents the spectral amplitude of the plane waves that compose u. The calculation of A can be performed, in general, on every plane parallel to the xy-plane, at a distance z from it:

$$A\left(\frac{\alpha}{\lambda},\frac{\beta}{\lambda};z\right)=\int_{-\infty}^{\infty}\int_{-\infty}^{\infty}u(x,y,z)e^{-ik(\alpha x+\beta y)}dx\,dy\,. \tag{5.3.20}$$

5.3.3 Reconstruction of the diffracted field

In principle, according to our knowledge of the angular spectrum, we can calculate u for every distance z:

$$u(x,y,z)=\int_{-\infty}^{\infty}\int_{-\infty}^{\infty}A\left(\frac{\alpha}{\lambda},\frac{\beta}{\lambda};z\right)e^{ik(\alpha x+\beta y)}d\frac{\alpha}{\lambda}d\frac{\beta}{\lambda}. \tag{5.3.21}$$

On the other hand, for a given planar component of the spectrum, the phase difference between two positions at a distance z between them (while keeping the values of x and y unchanged) is simply $(2\pi/\lambda)\gamma z$, whereby

$$A\left(\frac{\alpha}{\lambda},\frac{\beta}{\lambda};z\right)=A\left(\frac{\alpha}{\lambda},\frac{\beta}{\lambda};0\right)e^{ikz\sqrt{1-\alpha^2-\beta^2}}, \tag{5.3.22}$$

as we can also check by applying the Helmholtz's equation $\nabla^2 u + k^2 u = 0$ to Eq. (5.3.20) [Goodman 1966].

The plane-wave components are ordinary for α and β values such that

$$\alpha^2+\beta^2<1\,. \tag{5.3.23}$$

However, if the field that affects the opening is rapidly modulated in space, such as can happen with diffraction gratings, the spectrum of u can be extended to values of α and β for which

$$\alpha^2+\beta^2>1\,. \tag{5.3.24}$$

In this case, the corresponding plane wave component is inhomogeneous, that is, an evanescent wave. However, in this type of situation, as well as in the case in which $\alpha^2 + \beta^2 = 1$, a vector treatment of diffraction is needed.

From the knowledge of the angular spectrum A for $z = 0$, the field u in the half-space $z > 0$ can be calculated as

$$\begin{aligned} u(x,y,z) &= \int_{-\infty}^{\infty}\int_{-\infty}^{\infty} A\left(\frac{\alpha}{\lambda},\frac{\beta}{\lambda};0\right) e^{ik\left(\alpha x+\beta y+\sqrt{1-\alpha^2-\beta^2}\,z\right)} d\frac{\alpha}{\lambda}\, d\frac{\beta}{\lambda} \\ &= \int_{-\infty}^{\infty}\int_{-\infty}^{\infty} A\left(f_x, f_y;0\right) e^{ikz\sqrt{1-(\lambda f_x)^2-(\lambda f_y)^2}}\, e^{i2\pi\left(f_x x+f_y y\right)} df_x\, df_y . \end{aligned} \tag{5.3.25}$$

If z is large enough, the contribution of evanescent waves can be neglected and integration can be delimited within a circular area defined by (5.3.23). The field for coordinates $z >> \lambda$ can therefore be calculated by two Fourier's transforms. With the first, we obtain the angular spectrum of $u(x,y,0)$, which is then multiplied by the *transfer function*

$$T_z\left(f_x, f_y\right) = \begin{cases} e^{ikz\sqrt{1-(\lambda f_x)^2-(\lambda f_y)^2}} & \text{for } f_x^2 + f_y^2 < (1/\lambda)^2 , \\ 0 & \text{otherwise} \end{cases} \tag{5.3.26}$$

and, finally, we apply an inverse transform to obtain $u(x,y,z)$.

In particular, we consider the case of a spherical wave expressed by e^{ikr}/r, where r is the distance from the origin of coordinates. Using its symmetry and applying Eq. (5.1.48) to this function on the plane $z = 0$, its Fourier-Bessel transform is [Abramowitz and Stegun 1972, Eq. 11.4.39]

$$2\pi\int_0^{\infty} e^{i\frac{2\pi}{\lambda}r}\, \mathrm{J}_0(2\pi\rho r)\, dr = \frac{1}{\sqrt{\rho^2 - 1/\lambda^2}} ,$$

where ρ is the radial coordinate of the spatial frequency defined by Eqs. (5.1.38). The result is real if $\rho > 1/\lambda$, and imaginary for $\rho > 1/\lambda$. Returning in Cartesian coordinates, the plane wave spectrum of our spherical wave is given by

$$A\left(f_x, f_y;0\right) = \frac{i\lambda}{\sqrt{1-(\lambda f_x)^2-(\lambda f_y)^2}} = \frac{i\lambda}{\gamma}$$

and its development in plane waves is given by [Weyl 1919; Stamnes 1986]

$$\frac{e^{i\frac{2\pi}{\lambda}r}}{r} = \int_{-\infty}^{\infty}\int_{-\infty}^{\infty} i\frac{\lambda}{\gamma} e^{i2\pi\left(f_x x+f_y y+\frac{\gamma}{\lambda}z\right)} df_x\, df_y .$$

The method of the plane wave spectrum does not solve the problems arising from the edge effects that we have already discussed for Kirchhoff's diffraction, since, even in this case, field knowledge is required on a plane beyond the opening (it can be obtained only through the application of *rigorous* vector methods). However, it allows for ignoring various approximations introduced for the calculation of the Fresnel's and Fraunhofer's diffraction. Moreover, in this way, we can bring back from the window that which we had excluded in departing, that is, the ability to rebuild the field vectorially, albeit approximately.

Suppose, in fact, that we know both the amplitudes E_x and E_y of the field on the plane $z = 0$. From the transformation (5.3.16), we can then derive two spectral amplitudes A_{Ex} and A_{Ey} for each plane wave of the field spectrum. Since, for each of them, one knows the wave vector $\boldsymbol{k}$, the component A_{Ez} can be derived from the Maxwell's equation for the field divergence, i.e., from $\boldsymbol{k}\cdot\boldsymbol{A}_E = 0$, for which

$$A_{Ez}\left(f_x, f_y;0\right) = -\frac{A_{Ex}\left(f_x, f_y;0\right)\lambda f_x + A_{Ey}\left(f_x, f_y;0\right)\lambda f_y}{\sqrt{1-(\lambda f_x)^2-\left(\lambda f_y\right)^2}}. \qquad (5.3.27)$$

In addition, the spectral amplitudes for the magnetic field $\boldsymbol{B}$ ($\boldsymbol{H}$) can be obtained from the Faraday's equation $\boldsymbol{k}\times\boldsymbol{A}_E = \omega\boldsymbol{A}_B$ $(= \omega\boldsymbol{A}_H/\mu)$. Taking the inverse transform, given by Eq. (5.3.25), of the three components of $\boldsymbol{A}_E$ and of $\boldsymbol{A}_B$, it is thus possible to obtain the amplitudes of the electric and magnetic fields of the wave in their three Cartesian components. Finally, we can derive the Poynting vector $\boldsymbol{S}_o = \boldsymbol{E}_0\times\boldsymbol{H}_o$, that is the actual depositary of the diffracted wave energy flow.

5.3.4 Application of the FFT to a spatially delimited field with limited bandwidth

Summing up, let us assume that our field is well spatially limited and with limited bandwidth and try to solve the diffraction calculation with the FFT. The first thing to do is to calculate the discrete angular spectrum, that is,

$$A\left(\frac{j}{N\Delta_x}, \frac{l}{M\Delta_y}\right) = \Delta_x\Delta_y\sum_{n=-\frac{N}{2}}^{\frac{N}{2}-1}\ \sum_{m=-\frac{M}{2}}^{\frac{M}{2}-1} u_{\mathrm{A}}\left(n\Delta_x, m\Delta_y\right)e^{-2\pi i\left(\frac{jn}{N}+\frac{lm}{M}\right)}, \qquad (5.3.28)$$

where $j = -N/2, ..., N/2-1$, $l = -M/2, ..., M/2-1$ and u_{A} is the field on the opening. The applicability conditions of this formula are those already discussed for the sampling theorem. Thus, for example, in the case of a plane wave diffracted by an aperture of diameter d, it should be taken that $1/\Delta_x$, $1/\Delta_y > p/d$, where p is the number of rings that we want to rebuild. To this end, it should be noted that, wishing to use the angular spectrum to reconstruct the field at any distance from the opening, it is necessary that the energy associated with the set of all of the excluded rings be negligible to render the aliasing harmless. In addition, if sampling of the opening edge in the Cartesian grid is insufficiently dense, the result is showily altered by "floral" fringes that can be very beautiful, but not true. If, then, the field in the opening is spatially modulated, we have to verify that we effectively include its spectrum.

In particular, in optical instruments, we usually have to consider the field diffracted by an aperture of radius r_A whose wavefront is approximately spherical with curvature C. The fastest oscillations in the field are near the opening edge with a spatial period of $\sim\lambda(1+C^2 r_A^2)^{1/2}/(|C| r_A)$. For an incident plane wave, Mansuripur suggests a sampling Δ_x, $\Delta_y \leq r_A/10$ and, in the case of a curvature different from zero, [Mansuripur 1989],

$$\Delta_x, \Delta_y < \min\left\{\frac{r_A}{10}, \lambda\frac{\left(1+C^2 r_A{}^2\right)}{2|C| r_A}\right\}. \tag{5.3.29}$$

Eq. (5.3.28) can be calculated by the FFT, taking into due consideration the necessary translations and variations of scale. This operation can be duplicated for the x and y components of the field, and, finally, the z component of the spectrum can also be derived from Eq. (5.3.27).

Lastly, the field at a distance z from the opening is given by

$$u\left(n\Delta_x, m\Delta_y, z\right) = \frac{1}{NM\Delta_x\Delta_y} e^{ikz} \times$$

$$\sum_{j=-\frac{N}{2}}^{\frac{N}{2}-1} \sum_{l=-\frac{M}{2}}^{\frac{M}{2}-1} A\left(\frac{j}{N\Delta_x}, \frac{l}{M\Delta_y}\right) e^{ikz\left[-1+\sqrt{1-\left(\frac{j\lambda}{N\Delta_x}\right)^2-\left(\frac{l\lambda}{M\Delta_y}\right)^2}\right]} e^{i2\pi\left(\frac{jn}{N}+\frac{lm}{M}\right)}. \tag{5.3.30}$$

Unfortunately, the term with the square bracket grows indefinitely with increasing z, and therefore its exponential becomes rapidly oscillating. To avoid aliasing problems, it is then necessary to increase, in proportion to z, the number $N \times M$ of spectrum samples. On the other hand, the diffracted wave expands, and it is therefore necessary to extend the sampled area $N\Delta_x \times M\Delta_y$ around the opening, or reconstruct a denser spectrum by means of Eq. (5.2.14).

In the previous example of a circular opening, it is necessary to take that [Mansuripur 1989]

$$N\Delta_x, M\Delta_y > \max\left\{10 r_A, \frac{2z\sigma_{\max}}{\left(1-\sigma_{\max}^2\right)^{1/2}}\right\}, \tag{5.3.31}$$

where

$$\sigma_{max} = \frac{|C| r_A}{\left(1 + C^2 r_A{}^2\right)^{1/2}} + \frac{10\lambda}{2 r_A} \tag{5.3.32}$$

is the angular component that delimits the spectrum.[5]

5.3.5 *Extending the range of application of the angular spectrum method*

One method to extend the applicability of FFT was introduced by M. Mansuripur. This concerns two cases: the first one extends the validity of the calculation in the Fraunhofer region, allowing us to limit the sampled area on the opening plane. The second concerns a very important case in optics, the one in which the field on the opening is constituted by an approximately spherical wave. This is indeed the typical form of the field on the output pupil of a good optical instrument.

First case: extension of the method in the Fraunhofer region. We follow, in part, Mansuripur restarting from Eq. (5.3.25) and rewriting it as

$$u(x,y,z) = e^{ikz} \int_{-\infty}^{\infty} \int_{-\infty}^{\infty} A'\left(f_x, f_y; z\right) e^{-i\pi\lambda z\kappa\left(f_x{}^2 + f_y{}^2\right)} e^{+i2\pi\left(f_x x + f_y y\right)} df_x \, df_y, \tag{5.3.33}$$

where

$$A'\left(f_x, f_y; z\right) = A\left(f_x, f_y; 0\right) e^{ikz\left\{\sqrt{1-(\lambda f_x)^2-(\lambda f_y)^2} - 1 + \frac{\kappa}{2}\left[(\lambda f_x)^2 + (\lambda f_y)^2\right]\right\}} \tag{5.3.34}$$

is a modified angular spectrum in which the carrier exp(ikz) has been subtracted and a parabolic curvature term is added. The phase $kz\delta$ in the exponential, with

$$\delta = \sqrt{1 - \alpha^2 - \beta^2} - 1 + \frac{\kappa}{2}\left(\alpha^2 + \beta^2\right), \tag{5.3.35}$$

where α and β are still given by Eqs. (5.3.18), represents the difference between the spherical and parabolic wavefront that approaches it. The coefficient κ is a free

---

[5] Compared to Mansuripur's expression [Mansuripur 1989], I added a factor 10 to the second term. I recall that $1.22\lambda/(2r_A)$ is the angular radius of the first black ring in Fraunhofer's diffraction of a circular aperture with radius r_A and it must be abundant with respect to this value.

parameter to minimize the maximum slope of δ (later indicated with $\dot{\delta}_{max}$) with the change of α and β. κ is on the order of unity and its optimum value will be discussed at the end of this section. This optimization can also be effected taking into account the curvature contribution that is present in the spectrum *A*.

Eq. (5.3.33) has the form of the inverse transform of the product of two functions, and therefore can be converted into the convolution of their inverse transforms ($f_x, f_y \to x, y$), respectively,

$$a'(x,y;z) = \mathcal{F}^{-1}\left\{A'\left(f_x, f_y; z\right)\right\} \text{ and}$$
$$\frac{-i}{\lambda z\kappa} e^{\frac{i\pi}{\lambda z\kappa}\left(x^2+y^2\right)} = \mathcal{F}^{-1}\left\{e^{-i\pi\lambda z\kappa\left(f_x{}^2+f_y{}^2\right)}\right\}, \tag{5.3.36}$$

that is,

$$u(x,y,z) = e^{ikz}\int_{-\infty}^{\infty}\int_{-\infty}^{\infty} a'(x',y';z)\frac{-i}{\lambda z\kappa} e^{\frac{i\pi}{\lambda z\kappa}\left[(x-x')^2+(y-y')^2\right]}dx'\,dy'. \tag{5.3.37}$$

By developing the exponential, we have

$$u(x,y,z) = e^{ikz+\frac{i\pi}{\lambda z\kappa}\left(x^2+y^2\right)}$$
$$\times\int_{-\infty}^{\infty}\int_{-\infty}^{\infty} a'(x',y';z)\frac{-i}{\lambda z\kappa} e^{\frac{i\pi}{\lambda z\kappa}\left(x'^2+y'^2\right)} e^{-\frac{i2\pi}{\lambda z\kappa}\left(xx'+yy'\right)}dx'\,dy'.$$

We now replace the variables x', y' with $X' = x'/(\lambda z\kappa)$, $Y' = y'/(\lambda z\kappa)$. Therefore,

$$u(x,y,z) = -i\lambda z\kappa\, e^{ikz+\frac{i\pi}{\lambda z\kappa}\left(x^2+y^2\right)}\int_{-\infty}^{\infty}\int_{-\infty}^{\infty} a'(\lambda z\kappa X', \lambda z\kappa X'; z)$$
$$\times e^{i\pi\lambda z\kappa\left(X'^2+Y'^2\right)} e^{-i2\pi\left(xX'+yY'\right)}dX'\,dY', \tag{5.3.38}$$

where the integral now has the form of a Fourier's transform from the space of the "positions" X', Y' to that of the "frequencies" x, y:

$$u(x,y,z) =$$
$$-i\lambda z\kappa\, e^{ikz+\frac{i\pi}{\lambda z\kappa}\left(x^2+y^2\right)}\underset{X',Y'\to x,y}{\mathcal{F}}\left\{a'(\lambda z\kappa X', \lambda z\kappa X'; z)\, e^{i\pi\lambda z\kappa\left(X'^2+Y'^2\right)}\right\}. \tag{5.3.39}$$

In conclusion, to use this method, one must:

i) set the intervals and the sampling extension of the field at the initial plane. In particular, the sampling should extend over a surface with sides X_a, Y_a, whereby the sampling intervals in the space of the frequencies f_x and f_y are equal to $1/X_a$, $1/Y_a$. To choose the appropriate values of X_a and Y_a, it is necessary to establish the maximum frequencies $\lambda\sigma_{max}$ of the initial field. In the case of an aperture with a spherical wavefront, σ_{max} is given by Eq. (5.3.32). Then, we calculate the values of κ and $\dot{\delta}_{max}$ with the procedure described at the end of the section. The extension to be sampled is finally given by

$$X_a, Y_a > \max\left\{10r_A, 2z\dot{\delta}_{max}\right\}. \tag{5.3.40}$$

The sampling intervals depend on the behavior of the initial field and, for example, for a spherical wavefront, they must respect the condition (5.3.29). To this must be added the condition

$$\Delta_x, \Delta_y < \frac{\lambda z\kappa}{2\left(r_A + z\dot{\delta}_{max}\right)}, \tag{5.3.41}$$

so that the sampling in the frequency space is sufficiently extended to perform the last transform. Note that, for large z, this condition becomes independent of z.

ii) calculate, with the first FFT, the angular spectrum A as in the previous case;

iii) calculate the function A' from Eq. (5.3.34), and then calculate a' with an inverse FFT;

iv) lastly, calculate the field u from Eq. (5.3.39) with another FFT, from the scaled coordinates X', Y' to the coordinates x, y. The extension of sampling in X', Y' is given by $X_a/(\lambda z\kappa)$, $Y_a/(\lambda z\kappa)$, respectively. Then, the final sampling in x, y extends on a rectangle of sides $N\lambda z\kappa/X_a$ and $M\lambda z\kappa/Y_a$, respectively. This allows one to calculate the field on larger distances than the ordinary method, as long as the inequality (3.5.40) is respected.

Second case: spherical wave carrier of the field on the opening. Suppose that the field on the initial plane (taken with $z = 0$) has a spherical wave carrier with center on a point C with coordinates (x_c, y_c, z_c), and therefore we factor it as

$$u(x,y,z=0) = u_o(x,y)\,e^{-ikz_c\sqrt{1+\frac{(x-x_c)^2+(y-y_c)^2}{z_c^2}}}, \tag{5.3.42}$$

where the amplitudes u and u_o represent a field component on the plane (x,y). In other words, they indicate the *projection* of the wave field on this plane[6]. The am-

[6] This "precaution" does not apply in the case of a longitudinal acoustic wave.

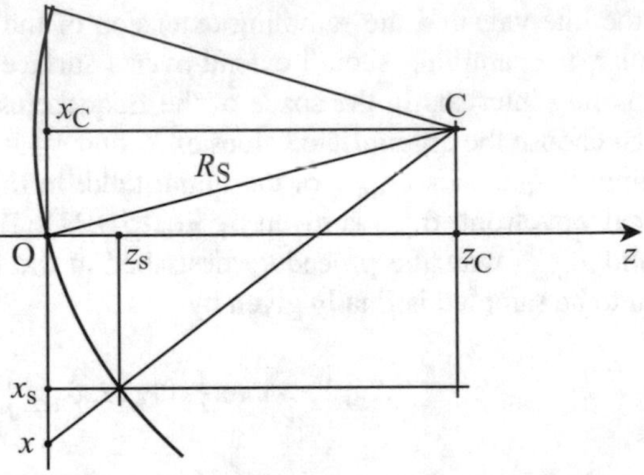

Fig. 5.3 Geometric study of a wave with a spherical carrier

plitude u_o may also include aberration terms relative to a spherical wave and the transmission effects of an aperture at $z = 0$. This method is useful for calculating the field near the center of curvature of the spherical wave carrier.

In the case in which the field is not assigned on the plane of the aperture, but rather on the sphere centered at C and passing through the origin with a radius

$$R_S = \sqrt{x_c{}^2 + y_c{}^2 + z_c{}^2}\,, \tag{5.3.43}$$

so that it is known that

$$u_s(x_s, y_s) = u(x_s, y_s, z = z_s), \tag{5.3.44}$$

where

$$z_s = z_c - \sqrt{z_c{}^2 - x_s{}^2 + 2x_s x_c - y_s{}^2 + 2y_s y_c}\,, \tag{5.3.45}$$

the amplitude u_o can be approximately obtained from the amplitude on the sphere through the "geometric" projection

$$u_o(x, y) = T(x, y)\frac{R_s}{R}e^{ik(R-R_s)}u_s\left(\frac{R_s}{R}(x - x_c) + x_c, \frac{R_s}{R}(y - y_c) + y_c\right), \tag{5.3.46}$$

where it is to be noted that the distance R is still a function of x and y

$$R = \sqrt{(x - x_c)^2 + (y - y_c)^2 + z_c{}^2}\,, \tag{5.3.47}$$

while the coefficient R_s/R scales the amplitude in function of the distance R from the center C and the function T represents the transmission of an opening on the plane $z = 0$.

Since the spherical carrier can give rapid oscillations to the amplitude for large numerical apertures, we perform a further factorization similar to that of the previous case by introducing a new amplitude

$$u_{\mathrm{p}}(x,y) = u_o(x,y)\,e^{-ikz_{\mathrm{c}}}\,e^{-ikz_{\mathrm{c}}\left[\sqrt{1+\frac{(x-x_{\mathrm{c}})^2+(y-y_{\mathrm{c}})^2}{z_{\mathrm{c}}^2}}-1-\frac{\kappa}{2}\frac{(x-x_{\mathrm{c}})^2+(y-y_{\mathrm{c}})^2}{z_{\mathrm{c}}^2}\right]}, \quad (5.3.48)$$

where, again, κ is a parameter that can be optimized for a minimum number of samples to be taken on the initial plane and is on the order of the unit. The expression (5.3.35) is replaced with

$$\delta = \sqrt{1+\alpha^2+\beta^2} - 1 - \frac{\kappa}{2}\left(\alpha^2+\beta^2\right), \quad (5.3.49)$$

where now $\alpha = (x-x_{\mathrm{c}})/z_{\mathrm{c}}$, $\beta = (y-y_{\mathrm{c}})/z_{\mathrm{c}}$. Eq. (5.3.41) then becomes

$$u(x,y,z=0) = u_{\mathrm{p}}(x,y)\,e^{-i\pi\frac{\kappa}{\lambda z_{\mathrm{c}}}\left[(x-x_{\mathrm{c}})^2+(y-y_{\mathrm{c}})^2\right]}. \quad (5.3.50)$$

The spectrum A of u is then the convolution of the transforms

$$U_p\left(f_x,f_y\right) = \overline{\mathcal{F}}\left\{u_p(x,y)\right\} \quad (5.3.51)$$

and

$$\overline{\mathcal{F}}\left(e^{-i\pi\frac{\kappa}{\lambda z_{\mathrm{c}}}\left[(x-x_{\mathrm{c}})^2+(y-y_{\mathrm{c}})^2\right]}\right) = -i\frac{\lambda z_c}{\kappa}\,e^{i\pi\frac{\lambda z_c}{\kappa}\left(f_x^2+f_y^2\right)}\,e^{-i2\pi\left(x_c f_x+y_c f_y\right)},$$

whereby

$$A\left(f_x,f_y,z=0\right)$$
$$= -i\frac{\lambda z_c}{\kappa}\iint U_p(\xi,\eta)\,e^{-i2\pi\left[x_c\left(f_x-\xi\right)+y_c\left(f_y-\eta\right)\right]+i\pi\frac{\lambda z_c}{\kappa}\left[\left(f_x-\xi\right)^2+\left(f_y-\eta\right)^2\right]}d\xi\,d\eta.$$

By developing the argument of the exponential, the integral can be traced back to a direct Fourier's transform, whereby

$$A\left(f_x,f_y,z=0\right)=\\-i\frac{\lambda z_c}{\kappa}e^{-i2\pi\left(x_c f_x+y_c f_y\right)}e^{i\pi\frac{\lambda z_c}{\kappa}\left(f_x{}^2+f_y{}^2\right)}H\left(f_x\frac{\lambda z_c}{\kappa},f_y\frac{\lambda z_c}{\kappa}\right),\tag{5.3.52}$$

where

$$\begin{aligned}H(X,Y)&=\underset{f_x,f_y\to X,Y}{\mathcal{F}}\left\{h\left(f_x,f_y\right)\right\}\\h\left(f_x,f_y\right)&=U_p\left(f_x,f_y\right)e^{i2\pi\left(x_c f_x+y_c f_y\right)+i\pi\frac{\lambda z_c}{\kappa}\left(f_x{}^2+f_y{}^2\right)}.\end{aligned}\tag{5.3.53}$$

Finally, from Eq. (5.3.25), we can get the field at the distance z by using the inverse Fourier's transform of

$$A\left(f_x,f_y,z\right)=-i\frac{\lambda z_c}{\kappa}H\left(f_x\frac{\lambda z_c}{\kappa},f_y\frac{\lambda z_c}{\kappa}\right)e^{i\frac{2\pi}{\lambda}(z-z_c)\sqrt{1-(\lambda f_x)^2-(\lambda f_y)^2}}\\\times e^{i\frac{2\pi}{\lambda}z_c\left\{\frac{1}{2\kappa}\left[(\lambda f_x)^2+(\lambda f_y)^2\right]+\sqrt{1-(\lambda f_x)^2-(\lambda f_y)^2}\right\}}e^{-i2\pi\left(x_c f_x+y_c f_y\right)},\tag{5.3.54}$$

for which

$$u(x,y,z)=\mathcal{F}^{-1}\left[A\left(f_x,f_y,z\right)\right].\tag{5.3.55}$$

By summing up, this method also requires the application of three transforms with the following steps:

i) knowing the amplitude u_o of the field on the plane of the opening, to which the spherical carrier of the effective amplitude u has been extracted with Eq. (5.3.42), to calculate u_p on a grid of points whose spacing satisfies the inequality

$$\Delta<\min\left\{\frac{r_A}{10},\frac{\lambda}{2\dot{\delta}_{\max}}\right\}.\tag{5.3.56}$$

The values of κ and $\dot{\delta}_{\max}$ will be discussed shortly. The extension of the function h in the plane f_x, f_y follows that of U_p and is given by

$$\frac{\sigma_{\max}}{\lambda}=\frac{\dot{\delta}_{\max}}{\lambda}+\frac{10}{2r_A}\tag{5.3.57}$$

(where I still took a more conservative condition than Mansuripur). On the other hand, for the exponential term, h has oscillations whose maximum frequency is

$$v = r_c + \frac{z_c}{\kappa}\sigma_{max}, \quad \text{con } r_c = \sqrt{x_c{}^2 + y_c{}^2}\,, \tag{5.3.58}$$

and therefore the extension of the sampling in x, y must be greater than twice that of this value

$$X_a, Y_a > \max\left\{10r_A, 2r_c + \frac{z_c}{\kappa}\left(\dot{\delta}_{max} + \frac{10\lambda}{2r_A}\right)\right\}; \tag{5.3.59}$$

ii) calculating the transform of u_p;

iii) calculating the function h and its transform from f_x, f_y to X, Y from Eqs. (5.3.53). Although this transform is direct, the extension of the H sampling in these new variables is again X_a, Y_a. Therefore, the sampling of spectrum A of the field in its variables f_x, f_y extends on $X_a\kappa/(\lambda z_c)$, $Y_a\kappa/(\lambda z_c)$, respectively;

iv) Finally, calculating the spectrum A at distance z with Eq. (5.3.54) and its inverse transform. Then, the final sampling of the field u in x, y extends on a rectangle of sides $N\lambda z_c/(\kappa X_a)$ and $M\lambda z_c/(\kappa Y_a)$, respectively.

Calculation procedure of κ and $\dot{\delta}_{max}$. This is not the same as making a fit between a sphere and a paraboloid, but rather minimizes the absolute maximum value of the slope of δ. For details, see the work by Mansuripur (1989). In short, it is necessary to establish the corresponding integration interval in the variables α and β of Eqs. (5.3.35) and (3.5.49) for δ, depending on the case considered. In practice, for the polar symmetry of δ, it is sufficient to consider a single radial variable ρ with $0 \le \rho_1 \le \rho \le \rho_2 \le 1$, where ρ_1 and ρ_2 are the limits of integration. Differentiating δ with respect to ρ, one gets a new function $\dot{\delta}$. With a further derivation in ρ, we have determined the position ρ_m of the maximum $\dot{\delta}_m$ of $|\dot{\delta}|$, which may or may not fall into the integration interval. In addition, the values $\dot{\delta}_1 = |\dot{\delta}(\rho_1)|$, $\dot{\delta}_2 = |\dot{\delta}(\rho_2)|$ are also determined. Each of these quantities depends on the parameter κ. One should therefore look for the value of κ so that the maximum between $\dot{\delta}_1$, $\dot{\delta}_2$, and even $\dot{\delta}_m$ when ρ_m is between ρ_1 and ρ_2, is the minimum possible.

5.3.6 Talbot's effect

When a wave is transmitted by a set of equal apertures distributed to form a periodic figure, or, more generally, when the wave is imprinted with a periodic modulation in amplitude and/or phase, either for transmission or for reflection, this same modulation is periodically reproduced along the direction of propagation of the wave. This effect was described in 1836 by H. F. Talbot and bears his name [Talbot 1936].

Consider the case in which on the plane $z = 0$, the field is periodic along two

Cartesian directions x and y (orthogonal to the z-axis) with periods X_o and Y_o, respectively. If its spectral band is limited, it will consist of a finite number of pulses at frequencies $f_n = n/X_o$ and $f_m\ m/Y_o$ with n, m integer. Then, Eq. (5.3.25) can be reduced to the sum of a finite number of plane waves

$$U(x,y,z) = \sum_{n=-N}^{N} \sum_{m=-M}^{M} A_{n,m}\, e^{ikz\sqrt{1-(\lambda n/X_o)^2-(\lambda m/Y_o)^2}}\, e^{i2\pi(xn/X_o+ym/Y_o)}, \quad (5.3.60)$$

where $A_{n.m}$ are the amplitudes associated with these waves and where N and M are integers large enough to include all spectrum frequencies. From this expression, we can observe that, for each position z, the field retains its periodicity in the transverse directions, but the amplitudes of each wave are multiplied by a phase term that depends on z, and therefore the field profile varies with z as in a kind of kaleidoscope.

If, however, N/X_o, $M/Y_o << \lambda$, that is, if the spectrum is sufficiently limited so that the paraxial approximation is valid, Eq. (5.3.60) can be simplified in

$$U(x,y,z) = e^{ikz} \sum_{n=-N}^{N} \sum_{m=-M}^{M} A_{n,m}\, e^{-i\pi n^2 \frac{z\lambda}{X_o^2} - i\pi m^2 \frac{z\lambda}{Y_o^2}}\, e^{i2\pi(xn/X_o+ym/Y_o)}. \quad (5.3.61)$$

Consider, now, the case in which the periods X_o and Y_o are in a rational relationship,

$$P = \nu X_o = \mu Y_o,$$

where ν, μ are coprime integers, so that P is the least common multiple of the two periods. Then, at the distance z_q such that

$$z_q = q\frac{P^2}{\lambda}, \text{ with } q \text{ integer}, \quad (5.3.62)$$

we have that the first exponential in Eq. (5.3.61) becomes

$$T_{n,m} = e^{-i\pi q\left(n^2\nu^2+m^2\mu^2\right)}. \quad (5.3.63)$$

Therefore, for $q = 2, 4, 6, ...,$ i.e., even, we have that $T_{n,m} = 1$ uniformly for all of the plane waves of the spectrum; in other words, at the corresponding distances z_q, the field U is identical to the initial one. This means that the field has a periodic behavior along the z-axis with period $2P^2/\lambda$.

Other q values also have interesting situations. For simplicity, let us consider the one-dimensional case with $\nu = 0$ and $\mu = 1$ where modulation is along the y-

axis. So, if q is an odd integer, e.g., when $z = Y_o^2/\lambda$, we have that $T_m = (-1)^m$. Further, suppose that our periodic function $f(y)$ is composed of the repetition of an isolated figure $f_o(y)$ on a null background. Its spectrum F can be considered to be the product of the comb function for the spectrum F_o of the single figure. Lastly, we assume that F_o is wide and regular with respect to the delta sampling. At z distances for which q is an odd integer, the spectrum can then be decomposed into a sum of two combs, one for m even and one for m odd, each multiplied by the spectrum F_o. The two combs therefore have period $2/Y_o$; that corresponding to odd m is shifted by $1/Y_o$ and its sign is inverted by the multiplication with T_m. In the y-space each of the two combs is transformed into a series of deltas with period $Y_o/2$ and the multiplication by the F_o spectrum becomes a convolution of the delta with f_o, giving rise to the repetition of figures with the same width and shape as the original one, but with a halved period. However, in our case, the delta in y corresponding to the odd m comb has a minus sign, due to T, and a phase $2\pi iy/Y_o$, due to the translation, which becomes a further alternating sign between a delta and the next. Finally, the field at distance z considered here is the sum of these two sequences of figures. Therefore, for the hypothesis made on F_o, there is a substantial cancellation in the sum of the figures that are located in the original position, and what appears is a figure similar to the initial one, but shifted by a half period.

In the case, then, in which q is half-integer, for example, when $z = Y_o^2/(2\lambda)$, the spectrum F can still be decomposed into 2 sets of delta with period $2/Y_o$ and

$$T_m = e^{-i\pi m^2/2}\text{, whereby } T_{2j}=1,\ T_{2j+1}=-i\text{, with } j \text{ integer.} \tag{5.3.64}$$

Again, in this case, the field on the observation plane can be considered as the sum of two periodic repetitions of f_o. Now, the sum does not involve a deletion, but the various figures alternate with amplitude ½$(1-i)\,f_o$ and ½$(1+i)\,f_o$ with a halved period compared to the figure on the initial plane.

5.4 Analysis of optical systems by means of the theory of linear systems

An optical system can be considered as a tool that transforms an input signal into an output signal. The peculiarity of optical systems is that these signals are functions of spatial coordinates in three-dimensional spaces. In addition, they can usually be traced back to the field amplitudes of the waves on an object surface and an image surface, respectively. For them, the principle of superposition is valid (in the absence of non-linear optical phenomena), whereby, at each point of the image surface, the field is the linear composition of that at each point of the object surface according to a characteristic transfer function of the instrument. What has been said here is true in cases of *coherent* illumination; however, we shall see that, even in the case of *incoherent* illumination, the principle of linear superposition can apply for the intensity, rather than the amplitude.

Therefore, the image-forming properties of an optical system can be analyzed according to the theory of linear systems for which the following property is required. Let us suppose that the function $g_A(x_A,y_A)$ represents the input signal and $g_B(x_B,y_B)$ the output one by a system defined by an operator T, so that

$$g_B(x_B,y_B) = T\{g_A(x_A,y_A)\}. \tag{5.4.1}$$

The system is said to be linear if, for any function f and g and any constants a and b, we have

$$T\{af(x_A,y_A)+bg(x_A,y_A)\} = aT\{f(x_A,y_A)\}+bT\{g(x_A,y_A)\}. \tag{5.4.2}$$

This property allows for treating the system in terms of a response to certain "elementary" functions that form a basis in which the input signals can be decomposed. In particular, we are already accustomed to decomposing an input image into its object points, and this results in representing the generic function g_A with the convolution integral [Goodman 1996]

$$g_A(x_A,y_A) = \int_{-\infty}^{\infty}\int_{-\infty}^{\infty} g_A(\xi,\eta)\delta(x_A-\xi,y_A-\eta)\,d\xi d\eta; \tag{5.4.3}$$

we can therefore consider $g_A(\xi,\eta)$ as a weight assigned to the elemental function $\delta(x_A-\xi, y_A-\eta)$. If we now apply the transformation T to $g_A(x_A,y_A)$, thanks to the linearity of T expressed by Eq. (5.4.2), this operator can be brought within the integral to act on the individual δ, and we get

$$g_B(x_B,y_B) = \int_{-\infty}^{\infty}\int_{-\infty}^{\infty} g_A(\xi,\eta)T\{\delta(x_A-\xi,y_A-\eta)\}\,d\xi d\eta. \tag{5.4.4}$$

Finally, we introduce the function $h(x_B, y_B; \xi, \eta)$ representing the system response at point (x_B, y_B) in the image space to an impulse signal centered on the object point (ξ, η). This function is called the impulse response of the system; it is

$$h(x_B,y_B;\xi,\eta) = T\{\delta(x_A-\xi,y_A-\eta)\}, \tag{5.4.5}$$

for which, by exchanging the variables $\xi, \eta \leftrightarrow x_A, y_A$, Eq. (5.4.4) becomes

$$g_B(x_B,y_B) = \int_{-\infty}^{\infty}\int_{-\infty}^{\infty} g_A(x_A,y_A)h(x_B,y_B;x_A,y_A)\,dx_A dy_A. \tag{5.4.6}$$

In the particular case in which the impulsive response h is *spatially invariant*, whereby

$$h(x_B, y_B; x_A, y_A) = h(x_B - m_\perp x_A, y_B - m_\perp y_A), \tag{5.4.7}$$

where $m_\perp$ is the magnification constant between the coordinates on the objects and image planes, Eq. (5.4.6) simply becomes the convolution of the input signal, whose scale is enlarged by a factor $m_\perp$, with the impulse response function of the system:

$$g_B(x_B, y_B) = \int_{-\infty}^{\infty}\int_{-\infty}^{\infty} \tilde{g}_A(\tilde{x}_A, \tilde{y}_A) h(x_B - \tilde{x}_A, y_B - \tilde{y}_A) d\tilde{x}_A d\tilde{y}_A, \tag{5.4.8}$$

where I introduced the transformations

$$\tilde{x}_A = m_\perp x_A, \quad \tilde{y}_A = m_\perp y_A, \tag{5.4.9}$$

and the function

$$\tilde{g}_A(\tilde{x}_A, \tilde{y}_A) = \frac{1}{m_\perp^2} g_A\left(\frac{\tilde{x}_A}{m_\perp}, \frac{\tilde{y}_A}{m_\perp}\right). \tag{5.4.10}$$

These transformations are useful in the study of image formation and will be used in the following.

An optical system with a spatially invariant impulse response is called *isoplanatic*. However, due to aberrations, the impulse response generally varies with the position of the object point considered, so optical systems are rarely isoplanatic along their entire field of view; this can be remedied by dividing the field into approximately isoplanatic regions on which to estimate the performance of the system.

Eq. (5.4.8) allows for analyzing an optical isoplanatic system in spectral terms; indeed, in the space of spatial frequencies, it simply becomes the product

$$G_B(f_x, f_y) = \mathrm{TF}(f_x, f_y) G'_A(f_x, f_y), \tag{5.4.11}$$

where G'_A represents the transform of $\tilde{g}_A(\tilde{x}_A, \tilde{y}_A)$ and the function TF, called the *transfer function*, is the transform of h. This transfer function, generally complex, represents the effect of filtering of the spatial frequencies that the system makes on input signals, similarly to what happens, for example, in an electronic amplifier on the temporal frequencies of the amplified signal.

In the next sections, several functions will be introduced to characterize the image quality produced by optical systems in two limit cases of illumination, coherent and incoherent. For coherent illumination, we will study the *amplitude point spread function* (APSF), closely linked to the impulse response, and its transformation, the *coherent transfer function* (CTF). In the case of incoherent illumination, we have the functions called the *point spread function* (PSF) and its transform, the *optical transfer function* (OTF), whose module is called the *modulation transfer function* (MTF), and its phase, the *phase transfer function* (PTF). When the system is not isoplanatic, the PSF is replaced by the *point response function* (PRF).

The evaluation of the performance of optical instruments in terms of spatial frequency analysis and linear systems theory is relatively recent. The first studies in this direction were made by Ernst Abbe (1840-1905) and by Lord Rayleigh (1842-1919), who recognized that the finite opening of an instrument constituted a cut of the higher spatial frequencies generated by an object figure. Subsequently, during the 1930s, the use of sinusoidal test patterns for the analysis of optical systems became common, and P.M. Duffieux was the first to publish a book, in 1946, on the use of Fourier's methods in optics [Duffieux 1946]. In that same period, Otto H. Schade applied the methods of linear theory and communication theory to the analysis and improvement of the objectives of television cameras [Schade 1948]. In the 1950s, H.H. Hopkins developed the physical concepts of the OTF, generating an exhaustive basis for optical design and evaluation [Hopkins 1956].

One of the qualities often required in an optical system is its resolution, which is generally understood as the ability to discriminate two nearby sources. Many different criteria have been proposed for this purpose, the best known among them being *Rayleigh's criterion*. In addition, many methods have been proposed to increase the resolution of a system beyond the classical limit, including through a posteriori analysis of detected images. A review of the work on resolution is given by den Dekker and van den Bos (1997).

On the other hand, an object figure consists of information, and the optical system that produces an image can be considered as a means of transmitting information: the greater the amount of information transmitted, the more detailed the image being reproduced. These concepts have led to the development of a new branch of optics called *Optical Signal* (or *Information*) *Processing*, which exploits the parallel nature (in the spatial coordinates) of the signal transmission. Important contributions to this discipline were given by Norbert Wiener [Wiener 1953], P. Elias [Elias 1953], Peter Fellgett [Fellgett 1953], Giuliano Toraldo di Francia [Toraldo di Francia 1955] and E.L. O'Neill [O'Neill 1956].

The result of these studies was the development of spatial optical filtering techniques, image recognition, image restoration, holography, aperture synthesis, optical processing methods of non-optical signals, etc., on which extensive literature exists. Here, in addition to Goodman (1996), I refer readers to the book by Anthony VanderLugt: *Optical Signal Processing* [VanderLugt 1992], and that by Francis T.S. Yu: *Entropy and Information Optics* [Yu 2000]. Lastly, an extensive discussion of transfer functions is given by C.S. Williams and O.A. Becklund: *Introduction to the Optical Transfer Function* [Williams and Becklund 1989].

5.4.1 Degrees of freedom of optical signals

There are several ways to characterize a signal. The first is to give their form in space or in time; those that are physically realizable have a limited extension in space and time, though they are often idealized with unlimited extensions, such as plane waves.

An important parameter is therefore the time duration T_0 or the extension X_0, Y_0, … of the signal in the space. Another way is to describe its spectrum, which, in turn, can be characterized by a limited bandwidth, W_t in the temporal frequencies, and W_x, W_y, … in the spatial ones, introduced in §5.2.2. This limitation may also be due to the finite resolution of the detector.

In the preceding sections, we have appreciated these concepts in connection with the sampling theorem, for which signals with limited spectral band can be reconstructed exactly by sampling the signal with a frequency higher than the Nyquist's frequency defined by Eqs. (5.2.10). If, then, the signal is also delimited in the time (or space) domain, it can be determined by a finite number N of samples, in other words, it can be sampled without information loss. To be rigorous, delimited signals in both domains of time (or space) and frequencies are not possible, but, in practice, this assumption can be obtained with reasonable accuracy. Then, from the sampling theorem, we know the minimum number of samples required to represent the signal accurately. It is given, for each coordinate, by

$$N_t = 2T_o W_t, \quad N_x = 2X_o W_x, \quad N_y = 2Y_o W_y, \quad \cdots. \tag{5.4.12}$$

For an image on the xy-plane, the total number of samples is then

$$N = 4X_o W_x Y_o W_y. \tag{5.4.13}$$

The products between the space extension and the corresponding bandwidth, $X_o W_x$, $Y_o W_y$, which are called the *length bandwidth product, height bandwidth product*, are a standard measure of the complexity of a signal, and they play an important role in optimizing optical systems.

In addition, each sample can be characterized by a variety of other parameters. For example, to define an image, it is necessary to give its intensity, which, in turn, can be quantized in a finite number a of discrete levels that constitute the *grayscale*. This scale corresponds to the dynamic range of the signal itself or the detection or reproduction system. The smallest level that can be resolved is that which corresponds to the noise level present in the signal or due to the system; possibly, the minimum level is taken with a signal-to-noise ratio of 3 or 6 dB. The highest level is instead defined by the maximum one allowed by the system. Another possibility is to match the highest level to the maximum value of the signal. This corresponds to normalizing the signal by setting aside an amplitude coefficient. Many cameras, and even eyes, have incorporated an automatic signal level control system that performs this operation.

In addition to the intensity, we can then assign *c* levels to color, *s* levels to saturation, *p* levels to polarization; hence, $g = acsp$ values are required to characterize each sample.

Finally, considering all of these samples, we have that the number of *degrees of freedom* of a signal is given by *gN*.

5.4.2 Behavior of a lens in the paraxial approximation

In the study of the behavior of an optical system within the theory of linear systems, we take, as a model, the case of a thin lens immersed in air. The results we will obtain here are in paraxial approximation and monochromatic radiation. The spatially incoherent quasi-monochromatic illumination case will be discussed in §5.4.7 and thereafter.

In the stigmatic case in which a plane wave is transformed into a spherical wave, we could represent (the effect of) a lens (on the wave amplitude) at the output pupil plane with the function

$$e^{i\frac{2\pi}{\lambda_o}n_a\left(f-f\sqrt{1+r^2/f^2}\right)}p(r),$$

where λ_o is the vacuum wavelength; n_a is the refraction index of air; f is the distance of the convergence point from the plane of the pupil, positive if the wave is convergent and negative if it is divergent; r is the distance from the optical axis on the plane of the pupil; finally, $p(r)$ represents the pupil, that is, the transmission function associated with the lens aperture.

A ray of light passing through a thin lens, entering with coordinates (x,y), emerges roughly with the same coordinates (therefore, we can collapse the two principal planes on a single surface S); however, it is subjected to a phase delay proportional to the traversed thickness, which, in paraxial approximation, is given by

$$\phi(x,y) = \frac{2\pi}{\lambda_o}n\Delta_o - \frac{\pi n_a}{\lambda_o f}\left(x^2+y^2\right), \tag{5.4.14}$$

where n is the refraction index of the lens, Δ_o is the thickness on its axis, which is at point $(0,0)$, while f is the focal distance. Since the term in Δ_o is constant, i.e., it does not depend on the transverse coordinates, in what follows, it will be neglected for simplicity.

Let us now examine the configuration shown in Fig. 5.4 in detail. When calculating the way in which the field changes from the plane S_A to the plane S_B, three

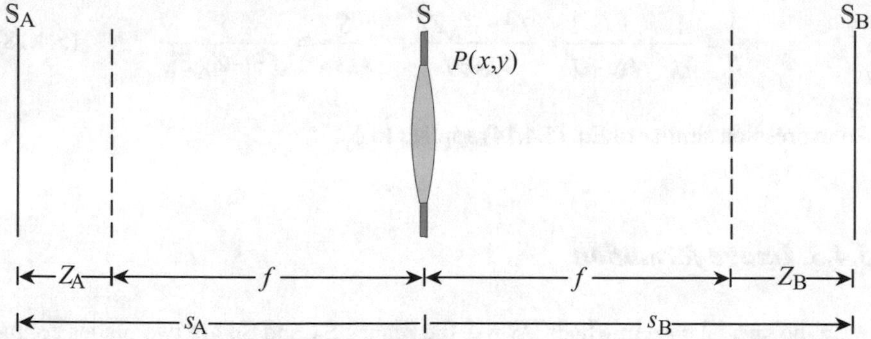

Fig. 5.4 Distances for the calculation of the diffraction between the planes S_A and S_B

operations must be performed. The first is the propagation from S_A to S, the plane of the lens. The second is the multiplication by the dephasing term produced by the lens and the amplitude term of its pupil:

$$p(x,y)e^{-i\frac{\pi}{\lambda f}\left(x^2+y^2\right)}, \tag{5.4.15}$$

where $\lambda = \lambda_o/n_a$ and p is equal to 1 in the area of the lens and 0 outside. The third operation is the propagation from S to S_B. Overall, using the Fresnel diffraction integral (4.2.5), we have:

$$u_B(x_B,y_B) = -\frac{e^{i\frac{2\pi}{\lambda}(s_A+s_B)}}{\lambda^2 s_A s_B}\iint_{S_1}\iint_{S} p(x,y)u_A(x_A,y_A)e^{i(\phi_x+\phi_y)}dx_A\,dy_A\,dx\,dy\,, \tag{5.4.16}$$

where ϕ_x and ϕ_y are functions of the various coordinates. For ϕ_x, we have

$$\phi_x = \frac{\pi}{\lambda}\left[\frac{(x-x_A)^2}{s_A}+\frac{(x-x_B)^2}{s_B}-\frac{x^2}{f}\right].$$

By rearranging the terms, we have

$$\phi_x = \frac{\pi}{\lambda}\left[\frac{x_A{}^2}{s_A}+\frac{x_B{}^2}{s_B}+\frac{x^2}{S}-2x\left(\frac{x_A}{s_A}+\frac{x_B}{s_B}\right)\right], \tag{5.4.17}$$

where, for brevity, we introduced the parameter S, whereby

$$\frac{1}{S}=\frac{1}{s_A}+\frac{1}{s_B}-\frac{1}{f}=\frac{f^2-Z_A Z_B}{s_A s_B f};\quad \frac{S}{s_A s_B}=\frac{f}{f^2-Z_A Z_B}. \tag{5.4.18}$$

An expression similar to Eq. (5.4.14) applies to ϕ_y.

5.4.3 *Image formation*

In the special case in which $1/S = 0$, the planes S_A and S_B are two planes conjugated by the lens, so that $s_A = s_o$, $s_B = s_i$. Eq. (5.4.16) becomes

$$u_B(x_B, y_B) = -e^{i\frac{2\pi}{\lambda}(s_o+s_i)} C_B(x_B, y_B) \times \iint_{S_o} \text{APSF}(x_B, y_B; \tilde{x}_A, \tilde{y}_A)\tilde{C}_A(\tilde{x}_A, \tilde{y}_A)\tilde{u}_o(\tilde{x}_A, \tilde{y}_A)\, d\tilde{x}_A\, d\tilde{y}_A, \tag{5.4.19}$$

where the transformations (5.4.9) and (5.4.10) were respectively used for the coordinates of the object point and for the amplitude of the field on the object plane. In addition, for brevity, I introduced the terms of field curvature

$$\tilde{C}_A = e^{i\frac{\pi}{s_o\lambda m_\perp^2}\left(\tilde{x}_A^2+\tilde{y}_A^2\right)},\quad C_B = e^{i\frac{\pi}{s_i\lambda}\left(x_B^2+y_B^2\right)}. \tag{5.4.20}$$

APSF is the *amplitude point spread function*[7] of the lens obtained by integrating only the terms in x and y:

$$\text{APSF}(x_B, y_B; \tilde{x}_A, \tilde{y}_A) = \frac{1}{\lambda^2 s_o s_i}\iint_S p(x, y)\, e^{-i2\pi\left(x f_x + y f_y\right)} dx\, dy = \frac{P\left(f_x, f_y\right)}{\lambda^2 s_o s_i}, \tag{5.4.21}$$

where, now,

$$f_x = \frac{1}{\lambda}\left(\frac{\tilde{x}_A}{m_\perp s_o}+\frac{x_B}{s_i}\right),\quad f_y = \frac{1}{\lambda}\left(\frac{\tilde{y}_A}{m_\perp s_o}+\frac{y_B}{s_i}\right) \tag{5.4.22}$$

---

7 Unlike other texts, I preferred explicitly to distinguish here the impulsive response from the APSF. In fact, the terms of the curvature of field, in addition to the phasor of propagation, are often implicitly incorporated in this response or in the field amplitudes. Here, instead, I consider these terms as separate.

and P is the Fourier's transform ($\mathcal{F}$) of p. In other words, for each object point $\boldsymbol{r}_\mathrm{A} = (x_\mathrm{A}, y_\mathrm{A})$, the *point spread function*, which is the image of $\boldsymbol{r}_\mathrm{A}$, is given by the Fraunhofer's diffraction pattern of the pupil, centered (with $f_x = f_y = 0$) at point $\boldsymbol{r}_\mathrm{B} = (x_\mathrm{B}, y_\mathrm{B}) = m_\perp \boldsymbol{r}_\mathrm{A}$.

With the change in variables (5.4.9), remembering that the lateral magnification is given by $m_\perp = -s_i/s_o$, and using Eqs. (5.4.22), Eq. (5.4.19) becomes

$$u_\mathrm{B}(x_\mathrm{B}, y_\mathrm{B}) = -\frac{e^{i\frac{2\pi}{\lambda}(s_o+s_i)}}{\lambda^2 s_o\, s_i} C_\mathrm{B}(x_\mathrm{B}, y_\mathrm{B})$$

$$\times \int_{-\infty}^{\infty}\int_{-\infty}^{\infty} P\left(\frac{x_\mathrm{B}-\tilde{x}_\mathrm{A}}{\lambda s_i}, \frac{y_\mathrm{B}-\tilde{y}_\mathrm{A}}{\lambda s_i}\right) \tilde{C}_\mathrm{A}(\tilde{x}_\mathrm{A}, \tilde{y}_\mathrm{A})\, \tilde{u}_\mathrm{A}(\tilde{x}_\mathrm{A}, \tilde{y}_\mathrm{A})\, d\tilde{x}_\mathrm{A}\, d\tilde{y}_\mathrm{A}. \tag{5.4.23}$$

The terms $\tilde{C}_\mathrm{A}$ and C_B constitute a non-trivial complication. C_B disappears in the intensity calculation. Instead, $\tilde{C}_\mathrm{A}$ is not trivially eliminable. However, if the pupil is large, the transform P of p has a narrow central peak and, by varying $\tilde{x}_\mathrm{A}, \tilde{y}_\mathrm{A}$ within that peak, the term of field curvature varies little and can be approximated as[8]

$$\tilde{C}_\mathrm{A} \cong e^{\frac{i\pi}{\lambda s_o m_\perp^2}\left(x_\mathrm{B}^2 + y_\mathrm{B}^2\right)} \tag{5.4.24}$$

and be brought out of the integral. In Eq. (5.4.23), we can recognize the convolution product between the field u_1 (multiplied by $\tilde{C}_\mathrm{A}$) and the diffraction pattern resulting from the lens pupil. With the approximation (5.4.24), Eq. (5.4.23) can be rewritten as:

$$u_\mathrm{B}(x, y) = -\frac{1}{\lambda^2 s_o\, s_i} e^{i\frac{2\pi}{\lambda_o}L(x,y)} \left[P\left(\frac{x}{\lambda s_i}, \frac{y}{\lambda s_i}\right) * \tilde{u}_\mathrm{A}(x, y) \right], \tag{5.4.25}$$

where convolution is performed on the variables x, y and

---

[8] The condition for which this approximation is valid is obtained from the comparison on the image plane between the width of the PSF peak, which, for a circular pupil with radius a, is contained in a radius equal to $1.22\lambda|s_i|/(2a)$, and the distance Δr within which the phase $-\pi\left(\tilde{x}_\mathrm{A}^2 + \tilde{y}_\mathrm{A}^2\right)/\left(\lambda s_o m_\perp^2\right)$ has a variation equal to $\pi/2$. Therefore, it is necessary that $a >> 1.22(\lambda|s_o|/2)^{1/2}$ for an object point in axis, and $a >> 2.44 r_\mathrm{A}$ for points outside of the axis at a distance r_A.

$$L(x,y) = n_a(s_o + s_i) + \frac{n_a}{2s_i}\left(1 + \frac{1}{m_\perp}\right)(x^2 + y^2) + [P_oP_i]. \tag{5.4.26}$$

where I have also here reintegrated $[P_oP_i] = \Delta_o(n - n_a)$, which is the optical path between the principal object and image planes.

5.4.4 As a lens does the Fourier's transform

Let us now go back to examine Eqs. (5.4.16-17). If $1/S$ is different from zero, i.e., when the planes S_A and S_B are not conjugated, we can rewrite Eq. (5.4.17) in the form

$$\begin{aligned}
\phi_x &= \frac{x_A^2}{s_A} + \frac{x_B^2}{s_B} + \frac{1}{S}\left[x - S\left(\frac{x_A}{s_A} + \frac{x_B}{s_B}\right)\right]^2 - S\left(\frac{x_A}{s_A} + \frac{x_B}{s_B}\right)^2 = \\
&= x_A^2 \frac{S}{s_A}\left(\frac{1}{S} - \frac{1}{s_A}\right) + x_B^2 \frac{S}{s_B}\left(\frac{1}{S} - \frac{1}{s_B}\right) - 2S\frac{x_A x_B}{s_A s_B} + \frac{1}{S}\left[x - S\left(\frac{x_A}{s_A} + \frac{x_B}{s_B}\right)\right]^2 \\
&= \frac{Z_B x_A^2 + Z_A x_B^2 - 2f x_A x_B}{f^2 - Z_A Z_B} + \frac{1}{S}\left[x - S\left(\frac{x_A}{s_A} + \frac{x_B}{s_B}\right)\right]^2 .
\end{aligned} \tag{5.4.27}$$

The impulse response of the lens for the amplitude is now

$$h(x_A, y_A; x_B, y_B) = \frac{1}{\lambda^2 s_A\, s_B} \int_{-\infty}^{\infty}\int_{-\infty}^{\infty} p(x,y)\, e^{-i\frac{\pi}{\lambda S}\left\{\left[x - S\left(\frac{x_A}{s_A} + \frac{x_B}{s_B}\right)\right]^2 + \left[y - S\left(\frac{y_A}{s_A} + \frac{y_B}{s_B}\right)\right]^2\right\}} dx\, dy. \tag{5.4.28}$$

Let us consider, for simplicity, the case in which the pupil is infinitely extended, that is, $p = 1$ everywhere. In this limit case, with a suitable change of variables, Eq. (5.4.28) becomes

$$h = \frac{1}{\lambda^2 s_A\, s_B} \int_{-\infty}^{\infty}\int_{-\infty}^{\infty} e^{-i\frac{\pi}{\lambda S}(x^2 + y^2)} dx\, dy = \frac{i}{\lambda}\frac{S}{s_A\, s_B} = \frac{i}{\lambda}\frac{f}{f^2 - Z_A Z_B}, \tag{5.4.29}$$

where the impulse response h is now reduced to a constant. Therefore, Eq. (5.4.16) becomes

$$u_B(x_B, y_B) = -\frac{i}{\lambda}\frac{f}{D} e^{i\frac{2\pi}{\lambda}(s_A+s_B)} e^{i\frac{\pi}{\lambda}\frac{Z_A}{D}\left(x_B^2+y_B^2\right)}$$
$$\times \int_{-\infty}^{\infty}\int_{-\infty}^{\infty} u_A(x_A, y_A) e^{i\frac{\pi}{\lambda}\frac{Z_B}{D}\left(x_A^2+y_A^2\right)} e^{-i2\frac{\pi}{\lambda}\frac{f}{D}(x_A x_B + y_A y_B)} dx_A\, dy_A, \tag{5.4.30}$$

with $D = f^2 - Z_A Z_B$,

which is very similar to Eq. (4.2.8) for the Fresnel diffraction. If we consider the particular case $Z_B = 0$, i.e., when S_B coincides with the second focal plane of the lens, then D is reduced to f^2 and the term in $x_A^2 + y_A^2$ is canceled. Therefore, apart from a phase and amplitude term, the field in S_B is the Fourier's transform of the field in S_A evaluated at the spatial frequencies

$$f_x = \frac{x_B}{\lambda f}, \quad f_y = \frac{y_B}{\lambda f}. \tag{5.4.31}$$

Finally, if, additionally, $Z_A = 0$, i.e., S_A coincides with the first focal plane, even the remaining field curvature term disappears, and we have

$$u_B(x_B, y_B) = -\frac{i}{\lambda f} e^{i\frac{4\pi}{\lambda}f} \int_{-\infty}^{\infty}\int_{-\infty}^{\infty} u_A(x_A, y_A) e^{-2i\pi\frac{1}{\lambda f}(x_A x_B + y_A y_B)} dx_A dy_A$$
$$= -\frac{i}{\lambda f} e^{i\frac{4\pi}{\lambda}f} U_A\left(\frac{x_B}{\lambda f}, \frac{y_B}{\lambda f}\right), \tag{5.4.32}$$

where

$$U_A\left(f_x, f_y\right) = \mathcal{F}\left\{u_A(x_A, y_A)\right\}. \tag{5.4.33}$$

Therefore, the amplitude of the field on the image focal plane corresponds to the transform of the field amplitude on the object focal plane.

5.4.5 *Behavior in three dimensions*

Finally, we consider the case in which the plane where the image is revealed is out of focus. Thus, let us write Eq. (5.4.18) in the form

$$\frac{1}{S} = \frac{1}{s_o + z_A} + \frac{1}{s_i + z_B} - \frac{1}{f}, \tag{5.4.34}$$

where s_o and s_i represent the positions of two mutually conjugate planes, while z_A and z_B are the deviations from these positions. If $z_A << s_o$ and $z_B << s_i$, we have that

$$\frac{1}{S} \approx \frac{1}{{s_i}^2}\left(m_\perp^2 z_A - z_B\right), \tag{5.4.35}$$

where $m_\perp$ is the lateral magnification produced by the lens between the above conjugated planes. This approximation allows for introducing a spatial invariance for the axial direction as well. Indeed, by applying Eq. (5.4.35) to Eq. (5.4.16), approximating the terms s_1 and s_2 in the denominator, respectively, with s_o and s_i, and applying the substitutions (5.4.9-10), we obtain

$$\begin{aligned} u_B(x_B, y_B) \cong & -\frac{1}{\lambda^2 s_o s_i} e^{i\frac{2\pi}{\lambda}(s_o+s_i+z_A+z_B)} C_B(x_B, y_B) \\ & \times \iint_{S_A} \tilde{u}_A(\tilde{x}_A, \tilde{y}_A) \tilde{C}_A(\tilde{x}_A, \tilde{y}_A) \mathrm{APSF}(x_B - \tilde{x}_A, y_B - \tilde{y}_A, z_B - \tilde{z}_A) d\tilde{x}_A\, d\tilde{y}_A, \end{aligned} \tag{5.4.36}$$

where $\tilde{C}_A$ and C_B are still given by Eqs. (5.4.20); moreover,

$$\tilde{z}_A = {m_\perp}^2 z_A \tag{5.4.37}$$

and

$$\mathrm{APSF}(x', y', z') = \iint_S p(x, y)\, e^{-i\frac{\pi}{\lambda s_i}\left[2xx' + 2yy' + \frac{1}{s_i} z'\left(x^2 + y^2\right)\right]} dx\, dy \tag{5.4.38}$$

represents the three-dimensional *point spread function* for the amplitude. An extensive discussion of these three-dimensional properties is given by Min Gu [Gu 2000].

5.4.6 Image formation with a generic optical system

In the case of an optical system composed of various elements, we can still resort, within certain limits, to the results from Geometrical Optics. If the system is well designed, we can define an input pupil and an output pupil. Both are images of the aperture of the system, which is sufficiently small to render the diffraction produced by the other diaphragms present negligible. At the same time, it is sufficiently large to allow the calculation of the propagation between the two pupils with only the tracking of rays. In this way, we can limit the diffractive calculations to propagation only from object to input pupil and from output pupil to a generic surface, on which we wish to analyze the field pattern. In other words, the diffraction effects of the diaphragm can be attributed to the entrance pupil or to the exit one

The first interpretation was proposed by Ernst Abbe in 1873: in his theory, the diffracted waves from an extended object are decomposed into a spectrum of plane waves, which, only in part and only in the lowest orders, are intercepted by the entrance pupil. Therefore, this pupil performs a filtering that excludes the spatial high frequency components. In particular, if the object is constituted by a grating, the optical system intercepts only a few diffraction orders, typically the lower ones, of the diffracted wave. This effect is therefore responsible for the limited resolution in the reproduction of the object.

From his theory, Abbe also obtained a justification for the *sine condition*. Suppose that the object has a periodic structure with period q_o along the y direction, for example, it is constituted by a series of lines parallel to the x-axis and uniformly spaced. When it is illuminated by a coherent plane wave, it produces a diffracted field constituted by a series of plane waves, called diffracted orders, whose direction on the zy-plane is given by

$$\sin\vartheta_j = j\lambda / q_o\,,$$

where $j = 0, \pm1, \pm2, \pm3, \ldots$, is the index of the diffracted order. In paraxial approximation, each of these plane waves is converted by the optical system (in the non-telescopic case) in a spherical wave with its origin on a line parallel to the y-axis on the image focal plane. Here, however, for each diffracted order, I consider only the local behavior of the wave around one of its rays passing through the object point O on the optical axis and reaching the axial image point I (Fig. 5.5). This ray forms angles ϑ_j and ϑ'_j with the optical axis, respectively, in the object and image spaces. If we consider a pair of orders with j equal in absolute value and opposite in sign, their interference figure on the image plane has a pitch equal to

$$\frac{q_i}{j} = \frac{\lambda}{\sin\vartheta'_j}\,.$$

The pitch of the fringes decreases with the increase of j, and therefore we can say that

> the finest details observed in an image are given by the highest diffraction orders transmitted by the system.

On the other hand, for the correct reproduction of the object, q_i must be independent of j. Therefore, it is necessary that

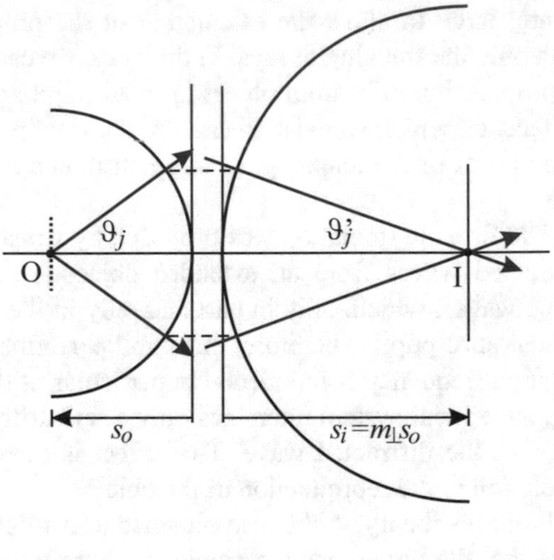

Fig. 5.5 Graphic representation of Abbe's sine condition. The principal object and image planes are replaced with spherical surfaces with center in O and center in I, respectively

$$\frac{q_i}{q_o} = \frac{\sin\vartheta_j}{\sin\vartheta'_j} = m_\perp ,$$

where $m_\perp$ is the transverse magnification and where the second equality expresses the sine condition, which is required to avoid aberrations, particularly when the system has a large angular opening and the paraxial approximation is invalid.

The second way to understand the system's functioning, that of attributing diffraction to the output pupil, was presented by Rayleigh in 1896, and that is what we will consider in the following. In this case, we still take the object as the overlapping of point sources. For each of them, we can construct the PSF with two steps; the first is the reconstruction of the wavefront and the amplitude of the field on the output pupil by means of ray tracing from the source point; the second is the calculation of the diffraction integral of the reconstructed field, from the plane of the output pupil up to the image plane (or another plane).

Consider a monochromatic wave that crosses the object plane of an optical system. It can be interpreted as the composition of wavelets with their center at each point object (x_o, y_o) on that plane, and the overall field amplitude can be decomposed into an integral of δ functions thanks to their sifting property

$$u(x_o, y_o) = \iint\limits_{S_o} u(x'_o, y'_o)\,\delta(x_o - x'_o, y_o - y'_o)\,dx'_o dy'_o \,.$$

With limited aberrations, for each wavelet, the wave that emerges from the system can be likened to a portion of a spherical wave whose carrier converges in the paraxial image point of coordinates $\tilde{x}_o = m_\perp x_o$, $\tilde{y}_o = m_\perp y_o$. In paraxial approximation, we can express the amplitude of the wave generated by each point on the output pupil plane as

$$u_p(x,y;\tilde{x}_o,\tilde{y}_o)=e^{i\frac{2\pi}{\lambda_o}L_o(\tilde{x}_o,\tilde{y}_o)}\tilde{u}_o(\tilde{x}_o,\tilde{y}_o)p(x,y;\tilde{x}_o,\tilde{y}_o)$$
$$\times\frac{i}{\lambda d_i}e^{-i\frac{\pi}{\lambda d_i}\left[\left(x^2-2x\tilde{x}_o\right)+\left(y^2-2y\tilde{y}_o\right)\right]},\tag{5.4.39}$$

where the second part of the expression represents the spherical wave carrier, and where $i/(\lambda d_i)$ represents its amplitude in paraxial approximation,[9] $\lambda=\lambda_o/n_i$ is the wavelength, n_i is the refractive index in the image space, and d_i is the distance between the output pupil and the image plane. Lastly, $\exp(i2\pi L_o/\lambda_o)\,\tilde{u}_o p$ represents the amplitude of the reconstructed field on the plane of the output pupil, to which the paraxial carrier was subtracted. In particular, $2\pi L_o/\lambda_o$ corresponds to the phase shift due to the optical path between the object point and the center of the exit pupil, and therefore already includes the term C_1 of field curvature that we encountered in dealing with the case of a lens. Here, p is assumed null outside of the pupil, but admits amplitude modulation inside and includes aberrations, including the gap between the paraxial curvature and the ideal sphere; with these additions, the function p is called the *generalized pupil*.

Again applying the diffraction integral (4.2.5) for the distance d_i and integrating on the object plane, we get the amplitude of the field at the image plane

$$u_i(x_i,y_i)=e^{i\frac{2\pi}{\lambda}d_i+\frac{i\pi}{\lambda d_i}\left(x_i^2+y_i^2\right)}$$
$$\times\iint_{\tilde{S}_o}e^{i\frac{2\pi}{\lambda_o}L_o(x_i,y_i)}\tilde{u}_o(\tilde{x}_o,\tilde{y}_o)\,\mathrm{APSF}(x_i,y_i;\tilde{x}_o,\tilde{y}_o)\,d\tilde{x}_o d\tilde{y}_o,\tag{5.4.40}$$

where $\tilde{S}_o$ essentially represents the object surface, scaled for the transverse magnification, and the APSF is given by

$$\mathrm{APSF}(x_i,y_i;\tilde{x}_o,\tilde{y}_o)=$$
$$\frac{1}{\lambda^2 d_i^2}\iint_{A}p(x,y;\tilde{x}_o,\tilde{y}_o)e^{-i\frac{2\pi}{\lambda d_i}\left[x(x_i-\tilde{x}_o)+y(y_i-\tilde{y}_o)\right]}dxdy,\tag{5.4.41}$$

where A is the surface of the pupil and the APSF has the inverse size of an area.

If the system is isoplanatic, that is, if p is not dependent on the coordinates of the object point, we can define a new point spread function

[9] The coefficient $+i/(\lambda d_i)$ is introduced in accordance with the notation of Born and Wolf (1980), §9.5.1, unlike Goodman (1996), Chap. 6.

$\tilde{u}_o(x_i,y_i)$ object amplitude	$\overline{\mathscr{F}} \rightarrow$ $\leftarrow \overline{\mathscr{F}}^{-1}$	$\tilde{U}_o(f_x,f_y)$ object amplitude spectrum
$*$		$\times$
APSF(x_i,y_i) *point spread function* for the amplitude	$\overline{\mathscr{F}} \rightarrow$ $\leftarrow \overline{\mathscr{F}}^{-1}$	$\mathrm{CTF}(f_x,f_y)$ *coherent transfer function*
$=$		$=$
$\hat{u}_i(x_i,y_i)$ image amplitude	$\overline{\mathscr{F}} \rightarrow$ $\leftarrow \overline{\mathscr{F}}^{-1}$	$\hat{U}_i(f_x,f_y)$ image amplitude spectrum

Tab. 5.2 Formation of images with coherent radiation. Here, the functions $\tilde{u}_o$ and $\hat{u}_i$ are expressed in the same spatial coordinates x_i, y_i, which are those of the image space

$$\mathrm{APSF}(x_i',y_i') = \frac{1}{\lambda^2 d_i^2}\iint_A p(x,y)\, e^{-i\frac{2\pi}{\lambda d_i}(xx_i'+yy_i')}\, dxdy = \frac{1}{\lambda^2 d_i^2} P\left(\frac{x_i'}{\lambda d_i},\frac{y_i'}{\lambda d_i}\right), \tag{5.4.42}$$

where $x_i' = x_i - \tilde{x}_o$ and $y_i' = y_i - \tilde{y}_o$ are the difference between the coordinates of the observation point from the ideal image point. In other words, the APSF is proportional to the transform P of the pupil, including any amplitude and phase modulations. If, in particular, the aberrations are negligible and the pupil is circular, so that we can write $p(r) = \mathrm{circ}(r/a)$, where a is its radius, we have

$$\mathrm{APSF}(r') = \frac{a}{\lambda d_i r'} \mathrm{J}_1\left(2\pi\frac{ar'}{\lambda d_i}\right). \tag{5.4.43}$$

Furthermore, applying the same approximation to L_o as was made to $\tilde{C}_A$ in Eq. (5.4.24), we can rewrite Eq. (5.4.40) as

$$u_i(x_i,y_i) = e^{i\frac{2\pi}{\lambda_o}L(x_i,y_i)} \iint_{S_o} \mathrm{APSF}(x_i-\tilde{x}_o, y_i-\tilde{y}_o)\,\tilde{u}_o(\tilde{x}_o,\tilde{y}_o)\, d\tilde{x}_o d\tilde{y}_o, \tag{5.4.44}$$

where

$$L(x_i,y_i) = L_o(x_i,y_i) + \frac{2\pi}{\lambda_o} n_i d_i + \frac{\pi n_i}{\lambda_o d_i}\left(x_i^2+y_i^2\right). \tag{5.4.45}$$

Eq. (5.4.44) represents a convolution product, so, by performing the Fourier transform, we get

$$\widehat{U}_i\left(f_x,f_y\right)=\mathrm{CTF}\left(f_x,f_y\right)\widetilde{U}_o\left(f_x,f_y\right), \tag{5.4.46}$$

where $\widehat{U}_i$ is the transform of $\widehat{u}_i=\exp(-ikL)u_i$. The function CTF is called the *coherent transfer function*, and it is

$$\mathrm{CTF}\left(f_x,f_y\right)=\overline{\mathscr{F}}\{\mathrm{APSF}(x_i',y_i')\}=\frac{1}{\lambda^2 d_i^{\,2}}\overline{\mathscr{F}}\left\{P\left(\frac{x_i'}{\lambda d_i},\frac{y_i'}{\lambda d_i}\right)\right\}. \tag{5.4.47}$$

Applying the theorems of similarity and symmetry, we finally get the important equation

$$\mathrm{CTF}\left(f_x,f_y\right)=p\left(-\lambda d_i f_x,-\lambda d_i f_y\right). \tag{5.4.48}$$

With Eqs. (5.4.46) and (5.4.48), we have traced the image-forming process to that of a spatial frequency filter, in which the filter function is given "simply" by the 180°-*rotated* (both axes inverted) generalized pupil of the optical system. We can recognize in this fact a justification of Debye's diffraction integral, which we have studied in §4.2.6. For an ideal system, *diffraction-limited*, the pupil function is null outside of the system exit pupil and is constant in its interior. In the case of a circular pupil with radius r, the filtering is reduced to a truncation of the spatial frequencies at the cutoff frequency

$$f_r=\frac{r}{\lambda d_i}. \tag{5.4.49}$$

Although, within this frequency, the spectrum remains constant, if there are aberrations, there is, nevertheless, an alteration of the phase between the various spectral components that reduces the sharpness of the reproduction. However, it should be remembered that spectral analysis of image quality makes sense only when the system is isoplanatic. Therefore, in the presence of aberrations, with some exceptions, a transfer function could not be introduced. However, when the analysis is limited to small portions around a given object point, we can invoke an approximation of local invariance of the system operation.

A summary of the functions that are used in the analysis of coherent imaging is given in Table 5.2.

5.4.7 Images with incoherent illumination

So far, we have considered the case of monochromatic illumination. More commonly, images are formed by incoherent natural radiation, and the physical entity that constitutes them is expressed by the intensity rather than by the amplitude of the field. In fact, with non-monochromatic and non-correlated sources, the interference phenomena are canceled, and the diffraction ones are attenuated in a kind of diffusion. Therefore, in this situation, we expect a linear behavior in which the overall intensity at an image point is given by the sum of the intensities of the individual sources. However, unlike the amplitude of the field, the intensity is always defined as positive. A serious discussion requires the study of the spatial and temporal coherence of the sources and of the illumination conditions of the object from which the image is produced [Born and Wolf 1980]. Here, I will only mention some commonly accepted simplifications. In §5.5, we will address the case of partial coherence.

If the source radiation is strictly monochromatic, with angular frequency ω, and objects (and sources) are immobile and invariable, we necessarily have a complete spatial and temporal coherence of radiation that interacts with the objects. In this case, we can imagine that the space is pervaded by a stationary amplitude wave emitted from the sources and is diffracted, diffused, transmitted and reflected by the material present therein. Therefore, the field oscillates in time in unison at every point with frequency ω, and has a phase and amplitude that varies spatially, even in a very fast and complex way, but remains constant over time. For example, illuminating a scene with a monochromatic laser source in the presence of diffusion from secondary sources, the scene appears granularly illuminated by *speckles*, to the point where objects in the scene are difficult to perceive in detail.

In the case of *quasi-monochromatic* illumination, we can express the field as the product of a carrier for an amplitude slowly varying over time. The temporal coherence then has an extension limited to an interval Δt of about $1/(2\pi\Delta\nu)$, where $\Delta\nu$ is the illumination bandwidth [Born and Wolf 1980, §10.7.3]. The spatial coherence can still be greatly extended: just think, for example, of a spectral lamp whose radiation is filtered through a hole a few wavelengths in size. Conversely, the individual emitters of the lamp, although they emit radiation with approximately the same wavelength, have random emission phases, and are therefore uncorrelated between them. Therefore, this extended source is spatially incoherent. Illumination must now be treated with statistical methods, so, instead of the randomly fluctuating field, it is necessary to consider the intensity. In essence, it is necessary to take into account the integration time of the detector and the source can be considered spatially incoherent only if this time is higher than $1/\Delta\nu$. Consider the following example. One way of obtaining an intense, extended, and spatially incoherent quasi-monochromatic source is to use an expanded monochromatic laser beam that impinges on a frosted glass set in rotation. If the rotation is slow, the scene will be illuminated by speckles in motion, and only if the rotation is fast will the scene be lit evenly, since the speckles' movement can no longer be followed by the detector, such as an eye or a camera with a given opening time of its shutter.

In §4.1.9, we saw a condition required for quasi-monochromatic radiation to apply diffraction integrals. It places a constraint between the bandwidth and the maximum difference in path length between the point of observation and those on the surface of integration, namely, that the difference must be less than the coherence length of the radiation. This condition is more restrictive than the one required for Geometrical Optics, in which an extended band may produce chromatic aberration. However, in the formation of images, instruments are used that make the paths of rays from object points to their image points (nearly) equal, so this new condition is not actually particularly restrictive.

If, lastly, the radiation has a band that exceeds the quasi-monochromatic limit, it is said to be *polychromatic*. In this case, we can divide its spectrum into a finite number of quasi-monochromatic bands, each of which is then treated separately.

Finally, we mathematically analyze the formation of images with *spatially incoherent illumination* by a generic optical system. Consider the case of an object that diffuses quasi-monochromatic radiation so that the amplitude of the field on its surface is expressed as a function of time, as well as of coordinates. Suppose, in addition, that the differences of the path between the surface object and image plane are small enough so as to still be able to apply the diffraction integral as we discussed in §4.1.9; this condition is generally verified for pairs of conjugate points by the system. Besides, we neglect the time delay and phase shift due to the propagation, such as those present in Eq. (4.1.32); indeed, these terms have no effect on the intensity obtained with incoherent illumination. Lastly, we assume that the system is (at least locally) isoplanatic. With a small change, Eq. (5.4.44) becomes

$$\widehat{u}_i(x_i,y_i,t)=\iint_{\tilde{S}_o}\mathrm{APSF}(x_i-\tilde{x}_o,y_i-\tilde{y}_o)\tilde{u}(\tilde{x}_o,\tilde{y}_o,t)\,d\tilde{x}_o d\tilde{y}_o\,, \qquad (5.4.50)$$

where $\widehat{u}_i$ is without the phase term mentioned above and the APSF is still computed by Eq. (5.4.42), but with λ replaced by the mean wavelength of the radiation in its spectral band. The intensity of the image is then given by

$$\begin{aligned} i_i(x_i,y_i)=a&\iint_{\tilde{S}_o} d\tilde{x}_o d\tilde{y}_o \\ &\times\iint_{\tilde{S}_o}\mathrm{APSF}(x_i-\tilde{x}_o,y_i-\tilde{y}_o)\mathrm{APSF}^*(x_i-\tilde{x}'_o,y_i-\tilde{y}'_o) \\ &\times\left\langle\tilde{u}(\tilde{x}_o,\tilde{y}_o,t)\tilde{u}^*(\tilde{x}'_o,\tilde{y}'_o,t)\right\rangle d\tilde{x}'_o d\tilde{y}'_o, \end{aligned} \qquad (5.4.51)$$

where the triangular brackets indicate the time average, a is the coefficient to be applied for the transition from the field amplitude u to the intensity i, and

$$i_i(x_i,y_i)=a\left\langle\widehat{u}_i(x_i,y_i,t)\widehat{u}_i^*(x_i,y_i,t)\right\rangle=a\left\langle u_i(x_i,y_i,t)u_i^*(x_i,y_i,t)\right\rangle.$$

We also assume that the field on the surface object is spatially incoherent and, in particular, that

$$a\left\langle \tilde{u}(\tilde{x}_o,\tilde{y}_o,t)u^*(\tilde{x}'_o,\tilde{y}'_o,t)\right\rangle = \tilde{i}_o(\tilde{x}_o,\tilde{y}_o)A_c\delta(\tilde{x}_o-\tilde{x}'_o,\tilde{y}_o-\tilde{y}'_o), \quad (5.4.52)$$

where A_c is a normalization constant that has the dimensions of an area and corresponds to the area of spatial coherence of the field. Having introduced the δ, this representation is not exact, since one must expect a minimum spatial coherence extended over at least a wavelength; however, for practical purposes, it is acceptable. If it were not, we would have to resort to a much more complicated treatment, appropriate to the case of partial coherence, which is needed with objects that are secondary sources, and this is the most common case. From an important theorem, established by P.H. van Cittert and F. Zernike, which we will study later (see §5.5.3.4), one finds that, with a primary circular source, the spatial coherence area of a secondary source has a diameter of 0.32 λ_m/ϑ. Here, λ_m is the average wavelength of the radiation and ϑ is the angular half-aperture with which the primary source is seen from the secondary one [van Cittert 1934; Zernike 1938; Born and Wolf 1980, §10.4.2]. For example, a scene illuminated by the Sun, with $\lambda_m \sim 0.55$ μm, has a coherence diameter on the order of 19 μm, and the partial coherence area extends up to a diameter about 4 times greater.

Replacing Eq. (5.4.52) in Eq. (5.4.51), we get

$$i_i(x_i,y_i) = A_c\iint_{S_o} \mathrm{PSF}(x_i-\tilde{x}_o,y_i-\tilde{y}_o)\tilde{i}_o(\tilde{x}_o,\tilde{y}_o)\,d\tilde{x}_o d\tilde{y}_o, \quad (5.4.53)$$

where

$$\mathrm{PSF}(x_i-\tilde{x}_o,y_i-\tilde{y}_o) = \left|\mathrm{APSF}(x_i-\tilde{x}_o,y_i-\tilde{y}_o)\right|^2 \quad (5.4.54)$$

is the *point spread function* for the intensity with incoherent illumination and has the dimensions of the inverse square of an area.

5.4.8 Resolution

The problem of evaluating the performance of an optical system and image quality is one of the most felt in optics, but, among the many varieties of measurements proposed, none of these is quite satisfactory. An important concept is that of *fidelity* of reproduction, for which various methods of measurement have been proposed, such as the root mean square deviation point by point between the real object and image. In itself, this should be an obvious criterion. However, it is affected by errors that can be considered important or not, such as a magnifying error, a small rotation or distortion, a variation in intensity or contrast, or a color change. Therefore, the concept of fidelity also ends up depending on the purpose for which the system is used.

Besides, one often resorts to the concept of *resolution*, that is, the ability to discriminate, by means of the image produced by the system, the real objects that determine it and their shape. In addition, it is required that the evaluation be "objective", determined by the characteristics of the system itself, and then given *a priori*, calculated, and therefore, regardless of the images actually detected, always affected by noise, which is left to subsequent considerations.

The PSF defined above is suitable for this purpose: in fact, it is determined by the aberrations of the system and, in the case of optimal systems limited by diffraction alone, by its opening. Many resolution criteria were formulated on this basis, particularly for the case of two incoherent point sources of equal intensity. All are affected by arbitrariness and are subject to various criticisms. Two of these are the ones that at least have the merit of being simple to formulate and are able to give a first estimate of the performance of an optical system. The best known and most used is the *Rayleigh's criterion*, which states that [Rayleigh, 1879, 1880]

> two point sources are just resolved on the image plane if the central maximum intensity of the diffraction pattern produced by a source coincides with the first zero of the diffraction figure produced by the other source.

In the case of an ideal system in which the PSF is given by Airy's figure, the resolution is expressed by the radius of the first dark ring around the central peak. With more complicated diffraction figures, for example, in the presence of aberrations, this criterion is generalized, requiring that, between the two maxima, there is a depression such that the sum of the intensities produced by two equal sources is 81% of that on the maxima.

The *Sparrow criterion* is less restrictive and considers that [Sparrow 1916]

> two point sources of equal intensity are just resolved if the secondary derivative of the intensity halfway along the connection between the position of their centers is canceled.

In other words, the sources are still resolved when the intensity decrease between the two maxima is canceled.

Both of these criteria may be useful for a comparison between optical systems, but do not give an absolute measure of the resolution limit. For example, using a reasonable model, such as the sum of two Airy functions, you can make a *fit* of the observed image and get the distance between the two sources. The limit to this distance measurement is given by the noise present in the image, so the measurement error can be much lower or much higher than the estimated resolution with these criteria.

Therefore, if the instrument pupil is circular with diameter D, for example, a telescope, angular resolution is 1.22 λ/D with the Rayleigh criterion and 0.95 λ/D with that of Sparrow.

On the other hand, Rayleigh and Sparrow's criteria can work well for evaluating the performance of a telescope in solving the structure of a binary star, but are essentially inadequate to evaluate the quality of reproduction of extended objects. For this situation, we rely on the spectral analysis summarized by the *Optical Transfer Function*

5.4.9 The Optical Transfer Function

Even Eq. (5.4.53) has the form of a convolution, and, if we calculate its Fourier's transform, we get the product

$$I_i\left(f_x, f_y\right) = A_c\, \mathrm{O}\left(f_x, f_y\right) \tilde{I}_o\left(f_x, f_y\right), \tag{5.4.55}$$

where the function O is

$$\mathrm{O}\left(f_x, f_y\right) = \overline{\mathcal{F}}\left\{ \mathrm{APSF}\left(x_i', y_i'\right) \mathrm{APSF}^*\left(x_i', y_i'\right) \right\}. \tag{5.4.56}$$

This represents the spatial frequency filtering operated by the optical system in the reproduction of the object for the case of incoherent illumination. Applying the autocorrelation theorem (5.1.33) with the substitution $\xi = -x/(\lambda d_i)$, $\eta = -y/(\lambda d_i)$, and Eqs. (5.4.47-48) with the replacement $f_x = x_p/(\lambda d_i)$, $f_y = y_p/(\lambda d_i)$, where x_p and y_p are the transverse coordinates of the translation taken on the pupil's plane, we have

$$\mathrm{O}\left(\frac{x_p}{\lambda d_i}, \frac{y_p}{\lambda d_i}\right) = \frac{1}{\lambda^2 d_i^2} \int_{-\infty}^{\infty} \int_{-\infty}^{\infty} p(x, y)\, p^*\left(x + x_p, y + y_p\right) dx\, dy\,. \tag{5.4.57}$$

The function O corresponds to the *autocorrelation* of the generalized pupil (Fig.

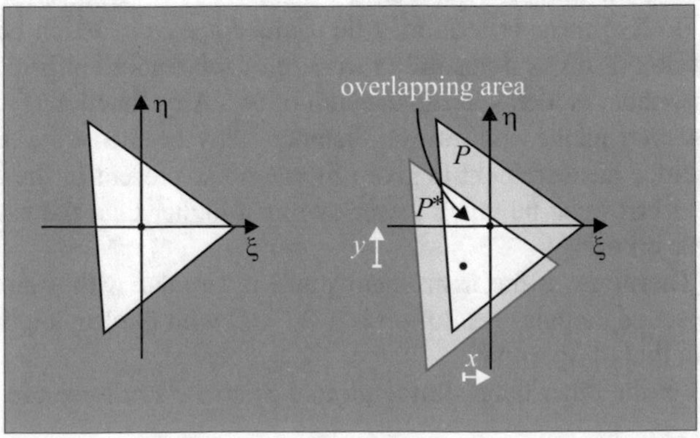

Fig. 5.6 Geometric interpretation of the OTF by the autocorrelation of the pupil, given by the integral (on the overlap area) of the product of the pupil function with its copy shifted and conjugated

$\tilde{i}_o(x_i,y_i)$ object intensity	$\overline{\mathcal{F}}\rightarrow$ $\leftarrow\overline{\mathcal{F}}^{-1}$	$\tilde{I}_o(f_x,f_y)$ spectrum of the object intensity
$*$		$\times$
$\mathrm{PSF}(x_i,y_i)$ *point spread function* for the intensity	$\overline{\mathcal{F}}\rightarrow$ $\leftarrow\overline{\mathcal{F}}^{-1}$	$\mathrm{O}(f_x,f_y)$ *optical transfer function* (not normalized)
$=$		$=$
$i_i(x_i,y_i)$ image intensity	$\overline{\mathcal{F}}\rightarrow$ $\leftarrow\overline{\mathcal{F}}^{-1}$	$I_i(f_x,f_y)$ spectrum of the image intensity
$\mathrm{APSF}(x_i,y_i)$	$\overline{\mathcal{F}}\rightarrow$ $\leftarrow\overline{\mathcal{F}}^{-1}$	$p(-\lambda d_i f_x,-\lambda d_i f_y)$
$\times$		$*$
$\mathrm{APSF}^*(x_i,y_i)$	$\overline{\mathcal{F}}\rightarrow$ $\leftarrow\overline{\mathcal{F}}^{-1}$	$p^*(\lambda d_i f_x,\lambda d_i f_y)$
$=$		$=$
$\mathrm{PSF}(x_i,y_i)$	$\overline{\mathcal{F}}\rightarrow$ $\leftarrow\overline{\mathcal{F}}^{-1}$	$\mathrm{O}(f_x,f_y)$

Tab. 5.3 Formation of images with incoherent radiation. Even here, the functions $\tilde{i}_o$ and i_i are expressed in the same spatial coordinates x_i, y_i, which are those of the image space. Note that the convolution for the calculation of O is between the *180°-rotated* (both axes *inverted*) pupil and the *conjugated* pupil (see the symmetry and conjugation theorems and Eq. (5.1.35), and is thus equal to the autocorrelation of the pupil itself. Here, convolution is understood by integrating on the scale of f_x and f_y

5.6). It is normalized to its value at frequencies $f_x = 0$, $f_y = 0$, obtaining what is called the *Optical Transfer Function* (OTF), which is given by

$$\mathrm{OTF}\left(\frac{x_p}{\lambda d_i},\frac{y_p}{\lambda d_i}\right)=\frac{\int_{-\infty}^{\infty}\int_{-\infty}^{\infty} p(x,y)\, p^*\left(x+x_p,y+y_p\right)dx\,dy}{\int_{-\infty}^{\infty}\int_{-\infty}^{\infty}\left|p(x,y)\right|^2 dx\,dy}. \tag{5.4.58}$$

It has the following properties

$$\begin{aligned}&\mathrm{OTF}(0,0)=1,\\&\mathrm{OTF}\left(-f_x,-f_y\right)=\mathrm{OTF}^*\left(f_x,f_y\right),\\&\left|\mathrm{OTF}\left(f_x,f_y\right)\right|\le\left|\mathrm{OTF}(0,0)\right|.\end{aligned} \tag{5.4.59}$$

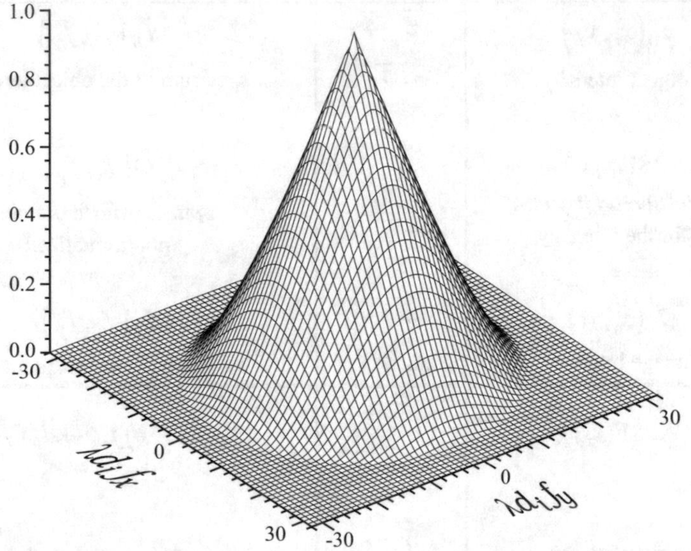

Fig. 5.7 Axonometric representation of the OTF of a circular pupil. This figure is obtained by the autocorrelation of a circle with a diameter equal to 26 steps of the grid

The first one trivially follows the normalization done. The second is a direct consequence of the fact that, for Eq. (5.4.56), the OTF is proportional to the transformation of a real function, and hence is Hermitian. Most important is the third property that can be demonstrated by applying Schwarz's inequality to Eq. (5.4.58).

In the case in which the pupil function is symmetrical for *inversion*, for which

$$p(\xi,\eta) = p(-\xi,-\eta),$$

the corresponding OTF is real. This symmetry exists for circular or rectangular openings, but not for triangular openings.

A summary of the relations between the functions that are used in the analysis of incoherent imaging is given in Table 5.3.

Frequently, the OTF is expressed in terms of module and phase

$$\mathrm{OTF}\left(f_x, f_y\right) = \mathrm{MTF}\left(f_x, f_y\right) e^{i\,\mathrm{PTF}\left(f_x, f_y\right)}, \tag{5.4.60}$$

where MTF and PTF are real functions; these are called the *Modulation Transfer Function* and the *Phase Transfer Function*, respectively.

Let us now consider the case in which there are no aberrations and the pupil function is constant and equal to 1 inside of the output pupil and 0 outside. The OTF simplifies in a simple real geometric function of the pupil's shape, and two

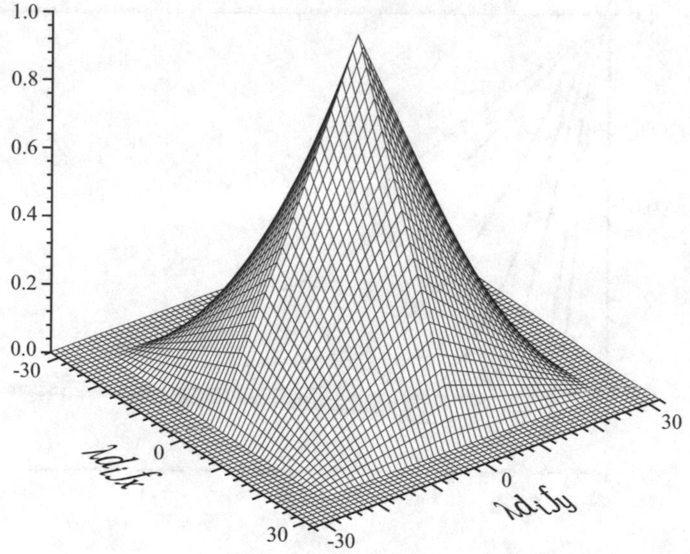

Fig. 5.8 Axonometric representation of the OTF of a square pupil with sides parallel to the x and y axes. This figure is obtained from the autocorrelation of a square with sides equal to 26 steps of the grid

examples are shown in Figs. 5.7 and 5.8 for circular and square pupil cases. The shape of these OTFs is similar to a cone or a pyramid, respectively, whose vertex corresponds to null spatial frequencies. What is most impressive is that there is a linear (at least initially) decay of the system's frequency response to the growing (in modulus) of the spatial frequency. The extent to which the OTF is non-zero is here limited to a circle or a square.

In the common case of a circular pupil of radius r, the radius of the base of the cone of the OTF, i.e., its cut-off frequency, is equal to

$$f_r = \frac{2r}{\lambda d_i}\,. \tag{5.4.61}$$

which is twice that of the CTF for the same pupil. One might therefore think that the incoherent illumination gives a better resolution than the coherent one. Apart from the fact that the CTF is constant in its band, while the OTF decreases, so that there is a sort of compensation, the two transfer functions are not comparable. Indeed, the CTF refers to the *amplitude* of the field, while the OTF refers to the *intensity*, which are the quantities to which the linear superposition principle can be applied, in a mutually exclusive way, in the two cases of illumination,

This behavior, pointed around the origin of the frequencies, is also there for pupils of more complicated form, made with ring masks or composed of separate apertures. In spite of all of that, it is due to the wave nature of the radiation, even in the context of incoherent illumination, where the wavelength does not appear

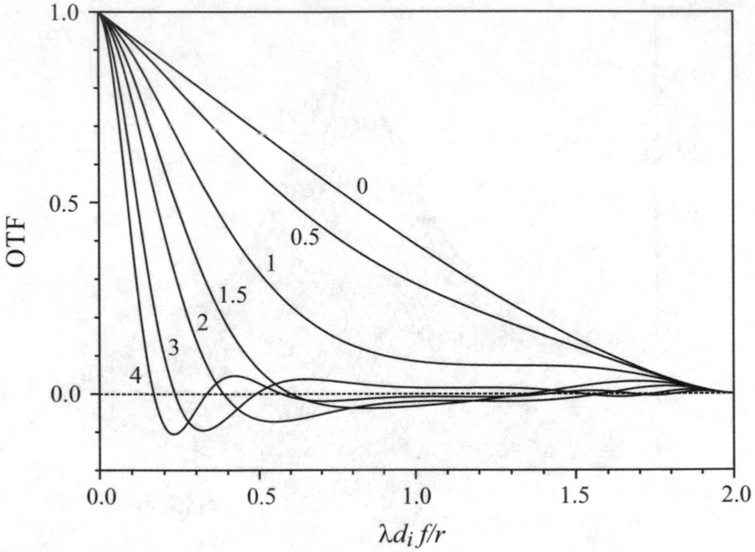

Fig. 5.9 Radial profiles of the OTF with a circular aperture of radius r for the case of *defocusing* as a function of the normalized spatial frequency. The numbers associated with each curve represent the deviation from the ideal sphere at the edge of the pupil, expressed in units of wavelength λ. The curve marked with 0 corresponds to the ideal case

explicitly, if not in scale relationship with the frequency space. On the other hand, the mathematical complexity with which the OTF is calculated is likely to obscure its physical origin. Consider, for example, a circular pupil and a spatial frequency value near the cutting frequency. This case corresponds to calculating the overlapping area between two equal circles whose centers are separated by little less than one diameter. It is evident that this overlap exists due to two subtle areas of the pupil at two diametrically opposite sides. In other words, the contribution to the spatial frequency comes from the radiation coming from both of those areas. If even one alone was obscured, that contribution would be null: we have already considered this situation by studying Young's interference experiment.

Let us imagine, therefore, that our pupil is made up of two small circular openings separated in each direction in the x direction, at positions $\pm l$. In this case, the OTF is given by three thin cones; one is centered at the origin of frequencies, while the other two are symmetrically arranged around the frequencies $f_x = \pm l/\lambda d_i$, $f_y = 0$, as can be seen from Eqs. (5.3.1) and (4.2.18).

In the image of a source point, these two peaks then produce a modulation of intensity, namely, a series of fringes approximately rectilinear and oriented along the y-axis, whose spatial separation is given by $\lambda d_i/(2l)$. In other words, these fringes are due to the interference between the radiation (originated by a given source point) that crosses the two openings. When the radiation is not monochromatic, the central zero-order fringe remains visible, while the others disappear progressively with increasing spectral band. Instead, the central peak of the OTF

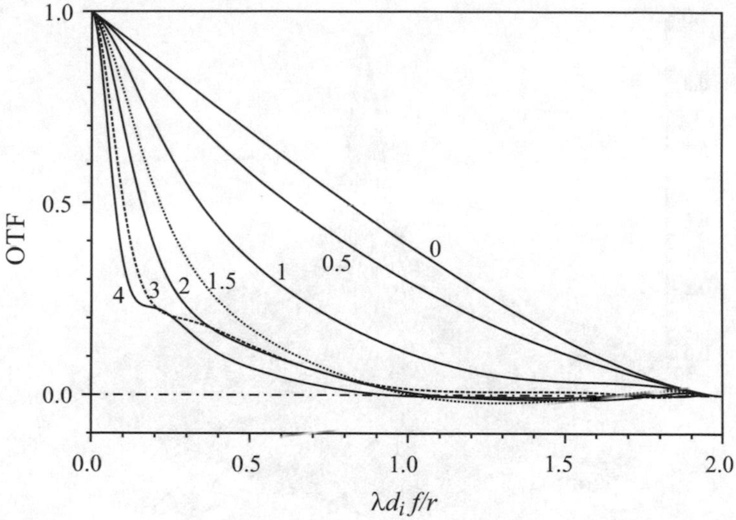

Fig. 5.10 Radial profiles of the OTF with a circular aperture of radius r for the case of *spherical aberration* as a function of the normalized spatial frequency. The numbers associated with each curve represent the deviation from the ideal sphere at the edge of the pupil, expressed in units of wavelength λ. This OTF is evaluated on the paraxial image plane and can partially be improved with a bit of defocusing by placing the screen on the plane where there the *circle of minimal confusion* is

around the origin of the frequencies gives rise to a uniform background intensity, which ensures a non-negative value everywhere for the intensity of the image, as occurs in the interference between two plane waves (v. §3.1.2).

In the presence of aberrations, to the extent that the system can be considered isoplanatic, the OTF is generally a complex function where its real part can also assume negative values. Specifically, the phase (modulo 2π) can also assume the value of π for some values of spatial frequencies. In this case, an object composed of a sinusoidal modulation at those frequencies is reproduced "in negative", with the maximum in place of the minimum intensity, and vice versa, and it has what is called *contrast reversal*. This phenomenon is well represented by the case in which the screen plane is out of focus. If the aperture is also symmetrical by inversion, the OTF is a real function; some of its profiles in this situation are shown in Fig. 5.9 for various distances from the image plane. It can be seen how the spectral band is rapidly degraded as the distance from the image plane increases. It is precisely this effect that allows us to find the optimal focus of a camera (if our sight is good!).

Instead, Fig. 5.10 shows the contour of the OTF in the case of spherical aberration for some values of the deviation with the ideal sphere at the edge of the pupil. Even in this case, the pupil is assumed to be circular and, since the aberration function is still symmetric with respect to the center of the pupil, the OTF is real. The two-dimensional aspect of the OTF is represented in Fig. 5.11 for a deviation at the edge equal to one wavelength.

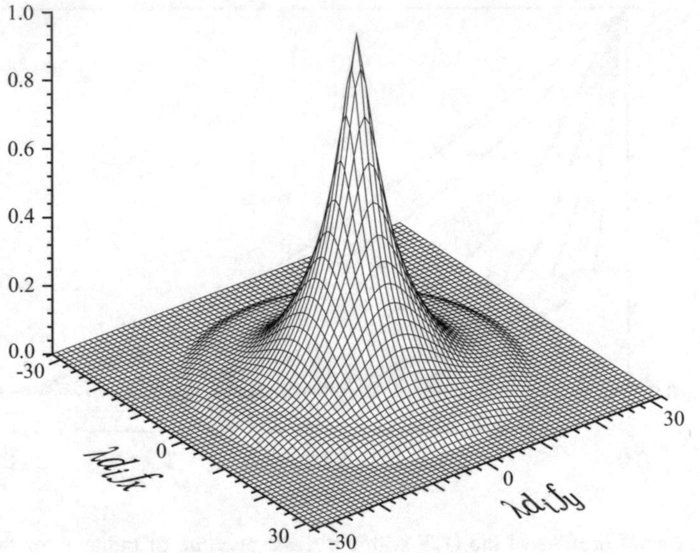

Fig. 5.11 Representation of the OTF in the presence of *spherical aberration* for a circular pupil, under conditions similar to those of Fig. 5.6. Here, the deviation at the edge of the pupil is taken as equal to one wavelength

5.4.10 Apodization

So far, we have seen the effect of openings with a sharp edge. For example, for a circular aperture, the central peak of the Fraunhofer diffraction pattern is surrounded by a series of rings of decreasing intensity that, in certain situations, is inconvenient. In the Fresnel's diffraction, it is clear that the fringes derive from a sudden interruption of the series of zones with alternate phase. To mitigate these fringes, the sharp edge of the apertures is replaced with a transmission that gradually decays. The openings of this type are called *apodized*, from the Greek αποδοσ (without feet), and *apodization* literally means "to take off the feet"; here, it indicates the elimination of the fringes. In an optical instrument, this is at the expense of the power transmitted by the aperture and its spatial resolution. For example, in a telescope, the figure of the Airy diffraction that constitutes the image of a star is replaced with a spot that is a little wider, but substantially devoid of rings, favoring the identification of nearby weak stars. For the apodization, various transmission functions have been proposed that depend on the shape of the opening. In the one-dimensional case, some of them are:

$$\textit{Barlett}: \quad t(x) = \mathrm{rect}(x/L)\left(1 - 2|x|/L\right),$$

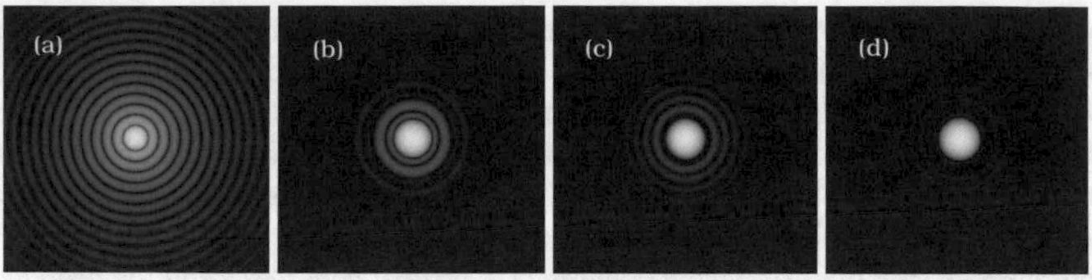

Fig. 5.12 PSF of circular openings with various apodizations. (a) not apodized, (b) Barlett, (c) truncated Gaussian with A=5, (d) Hamming. The peak value is normalized and the grayscale is logarithmic and covers a range of 50 dB

$$\textit{truncated Gaussian}: \quad t(x) = \mathrm{rect}(x/L)\, e^{-2Ax^2/L^2},$$

$$\textit{Hamming}: \quad t(x) = \mathrm{rect}(x/L)\left[0.53836 + 0.46164\cos(2\pi x/L)\right],$$

where L is the total aperture width. The Barlett is a triangular function that, in the face of a noticeable loss of resolution, does not produce a sufficient reduction in the lateral lobes. The truncated Gaussian works better, with a reduction in side lobes that depends on the parameter A. The Hamming produces a good reduction of the lobes for the same loss of resolution as the other two. Even more effective is the Kaiser's function, and many others have been proposed (see *www.wikipedia.org* for the entry *Window function*). Fig. 5.12 shows the effect of some radially applied apodization functions to a circular opening.

The practical realization of openings or mirrors with these transmission functions is, however, difficult. In astronomy, apodization is sparking new interest in the detection of extrasolar planets, and its realization has been studied by means of masks with binary openings of special shape or with pairs of special aspheric mirrors [Kasdin et al 2005; Traub and Vanderbei 2003]. In terms of Geometrical Optics, these mirrors remap a bundle of parallel rays by keeping them as such, but concentrating them at the center of the aperture and diluting them at the edge. To observe these planets, the side lobes must be below -100dB with respect to the central peak.

5.5 Coherence

We have repeatedly encountered the concept of *coherence* in the illumination conditions from an extended source to determine the visibility of the fringes of interference and diffraction and in the analysis of the quality of optical systems. However, the only two cases handled so far were the extreme conditions of *coherent* and *incoherent* illumination, in which we have applied, respectively, the *linear* superposition principles of the *amplitudes* and that of *intensity*. These cases do not represent the full story, as such conditions are only mathematical idealizations that become inadequate in various situations, such as in microscopy, especially that of phase contrast and dark-field illumination; in "coherent" optical systems such as spatial filtering, optical processing, and holography; in spectroscopy and interferometry, and also in photography. Here, we will define coherence in a more general sense as *correlation* between the amplitudes of the field, and we will analyze the intermediate situations of *partial coherence*, *temporal coherence* and the *propagation of coherence*.

The first experiments on coherence were carried out by E. Verdet, who, in 1865 measured the spatial length of coherence of an extended source. In 1869, he showed that fringes of interference could be observed with two pinholes illuminated directly by the Sun, when their distance was already 100 μm, and the contrast increased by approaching them. Michelson studied the connection between spatial and spectral distribution of the sources with the visibility of the fringes of interference, which he applied, in particular, to the determination of the diameter of the stars. These experiments showed that even incoherent primary sources could give rise to coherent fields on small regions of space as a function of their angular amplitude: Michelson found that the star Betelgeuse has, on Earth, a coherence region about 3m wide. In addition, the fringe contrast increases when the spectral band of the source is reduced with a filter. The non-linearity of behavior, the impossibility of directly measuring the field at optical frequencies and the dependence on many parameters make the partial coherence particularly difficult to deal with, and thus new mathematical techniques are required.

The first correlation measures were performed by M. von Laue and by M. Berek. The classical theory of coherence was then developed by P.H. van Cittert, by F. Zernike, and by H.H. Hopkins [van Cittert 1934, 1939; Zernike 1938; Hopkins 1953]. It was later generalized by E. Wolf, who introduced the concept of mutual coherence [Wolf 1955]. Still, in astronomy, R. Hanbury Brown and Richard Q. Twiss invented a new interferometric technique based on the field intensity fluctuations revealed in two different positions [Hanbury Brown and Twiss 1957, 1958]. Though Hanbury Brown and Twiss had anticipated this effect, which is now known as *photon bunching*, derived from only Classical Optics, they also demonstrated its consistency with Quantum Optics. After the advent of lasers, the modern concept of coherence was developed by Roy J. Glauber [Glauber 1963, 1965] and by Leonard Mandel. It is linked to the statistical properties of radiation, in which even the quantum mechanism of photon detection should be considered. In quantum mechanics, the electromagnetic field is also quantized by obtaining peculiar non-classical properties, interpreted by photon statistic. The experimental observation of these properties is one of the main tests in favor of Quantum Theory. Important contributions to photon statistics, especially with laser sources, were made by F.T. Arecchi [Arecchi 1965; Arecchi et al 1972]. An account of classical and quantum coherence is reported by Mandel and Wolf (1965), while, for photon statistics and laser fluctuations, see *Laser Handbook, Part A*, Arecchi, Schultz, Du Bois, ed. (1972).

5.5.1 Random processes

A generic *random process* is described by a collection of functions of one or more variables, indexed by a value ζ representing the various independent realizations of the same process, for example, $g(\boldsymbol{r},t;\zeta)$, where $\boldsymbol{r}$ is the position in space and t is the temporal coordinate [Barrett and Myers 2004].

Experimentally, it is normal to consider the value of g in a discrete number N of points, for example, the values $t_1, t_2, ..., t_N$, for a function of a single variable t. Then, $g(t_1), g(t_2), ..., g(t_N)$ are a set of random variables characterized by a hierarchy of joint probability distributions (*probability density function*, PDF) [Arecchi 1977]

$$\Pr(g_1), \Pr(g_1,g_2), \ldots, \Pr(g_1,g_2,\ldots,g_N), \tag{5.5.1}$$

where, for example, $\Pr(g_1,g_2)$ is the joint probability of obtaining the values g_1 in t_1 and g_2 in t_2, in the same realization. The statistical properties of a generic random process are also described by statistical autocorrelation functions of increasing order

$$\langle g_1 g_2 \cdots g_N \rangle \equiv \int_\infty g_1 g_2 \cdots g_N \Pr(g_1,g_2,\ldots,g_N)\, dg_1 dg_2 \ldots dg_N, \tag{5.5.2}$$

where the triangular brackets indicate the *expectation value* obtained from the *ensemble average* on the various realizations and the integral extends between $-\infty$ and $+\infty$ on all variables g_m. The first of these expectations is simply the statistical average of the variable at time t,

$$\langle g(t) \rangle \equiv \int_\infty g(t) \Pr(g(t))\, dg. \tag{5.5.3}$$

The second expectation value is the statistical autocorrelation of the function g between two values t_1 and t_2,

$$R(t_1,t_2) = \langle g_1 g_2 \rangle \equiv \int_\infty g_1 g_2 \Pr(g_1,g_2)\, dg_1 dg_2. \tag{5.5.4}$$

When g is a complex function, in the product of the various values g_m, one also takes combinations in which some terms are the corresponding complex conjugate. For example, for an electromagnetic wave, particular importance is held by the statistical autocorrelation function

$$R(t_1,t_2) \equiv \langle E_1^* E_2 \rangle, \tag{5.5.5}$$

which, for $t_1 = t_2$, corresponds to the intensity

$$I \equiv \frac{nc}{2}\varepsilon_0 \left\langle E_1^* E_1 \right\rangle. \tag{5.5.6}$$

A process is said to be (temporarily) *stationary in the strict sense* when all autocorrelation functions, and therefore all of its statistical properties, are independent of the origin of time. Usually, we limit ourselves to *stationary* processes *in the broad sense,* for which it is only required that the expectation $\langle g(t)\rangle$ is not dependent on t and that $\langle g(t_1)g(t_2)\rangle$ depends only on the difference $\tau = t_1 - t_2$.

A stationary process is then said to be *ergodic* if the average on the set of realizations can be replaced with an average on the time of a single realization. The degree of ergodicity follows that of stationarity, given by the order of autocorrelation up to which there is independence from the origin. In particular, the first two expectations of a stationary, at least in the broad sense, and ergodic process, are given by

$$\langle g\rangle = \bar{g}(\zeta_o) = \lim_{T\to\infty}\frac{1}{T}\int_{-T/2}^{T/2} g(t;\zeta_o)\,dt, \tag{5.5.7.a}$$

$$R(\tau) = \bar{R}(\tau,\zeta_o) = \lim_{T\to\infty}\frac{1}{T}\int_{-T/2}^{T/2} g(t;\zeta_o)\,g(t+\tau;\zeta_o)\,dt, \tag{5.5.7.b}$$

where, in each expression, the assumption of ergodicity is indicated by the first equality, for which the ensemble average is equal to the average over time (represented by the bar) performed on a single realization ζ_0 of the process. Thus, the statistical autocorrelation function $R(\tau)$ is expressed here by the integral of autocorrelation of a single realization divided by the integration time.

The processes that are encountered in optics are often approximate as stationary, because the humanly appreciable time intervals are much larger than the period of field oscillations, though each process has a beginning and an end. Ergodicity is not trivially demonstrable. However, these processes are generally assumed to be ergodic: this allows us to assign the intensity of a wave averaging the signal on the response time of the detector. Spatial stationarity is more rarely present and considered: a uniformly gray image is of little interest. However, background noise (such as graininess) in the image can often be treated as a spatially stationary and ergodic process.

5.5.2 The power spectrum of random processes

The signals obtained physically are always subject, by varying degrees, to a noise that alters their useful information content and typically changes from measure to measure. Therefore, it is necessary to know the statistical characteristics of this noise so that it can be subtracted or attenuated. A useful indication comes from its spectrum, but it has to be treated with particular methods to which we will now see a nod for the one-dimensional case.

The Fourier's transform of a signal is often indicated by the name of *amplitude spectrum* and its square module is called the *power spectrum*. Indeed, the Parseval's theorem assures us that the total energy of the signal obtained by integrating the square module of the field on the coordinates is equal to the integral on the frequencies of the square module of the Fourier's transform of the field. The autocorrelation theorem then expresses the important concept that the power spectrum is the Fourier's transform of the autocorrelation of the signal itself. For a one-dimensional signal extended over time, we have

$$\mathcal{F}\left\{\int_{-\infty}^{\infty} g(\xi)\, g^*(\xi - t)\, d\xi\right\} = |G(\nu)|^2 . \tag{5.5.8}$$

All of this is possible if the signal is deterministic, for which these integrals are finite. However, in the case of non-deterministic signals, such as noise without spatial or temporal limits, we can resort to the concept of media. Thus, for a *stationary* random signal, Norbert Wiener defined an autocorrelation function given by [Wiener 1930; Barrett and Myers 2004; VanderLugt 1992].

$$s_g(t) = \lim_{T\to\infty} \frac{1}{2T} \int_{-T}^{T} g(t')\, g^*(t' - t)\, dt' . \tag{5.5.9}$$

Since s_g is generally a function in L_1, we can calculate its Fourier's transform

$$S_W(\nu) = \int_{-\infty}^{\infty} s_g(t)\, e^{-2\pi i \nu t} dt , \tag{5.5.10}$$

which is called the *spectral power density*.

The stationary condition is not strictly applicable to physical signals, as they have a beginning and an end, and this is more apparent for extended signals in space, where, for example, the edges of a figure are always present. However, even in these cases, the concept of stationarity is a useful approximation. When one has a single realization, one can think of dividing it into separate portions that are translated to the same origin and apply the ensemble averages to these. In practice, it is possible to dispose only of finite portions of a signal. Then, instead of Wiener's definition, one can use the previous one of A. Schuster that proposed specifying the spectrum of $g(t)$ by the *periodogram* defined by [Schuster 1906]

$$S_p = \lim_{T\to\infty} \frac{1}{T} |G_T(\nu)|^2 , \tag{5.5.11}$$

where

$$G_T(\nu)=\int_{-T/2}^{T/2}g(t)e^{-2\pi i\nu t}dt \tag{5.5.12}$$

is the Fourier's transform of the function $g(t)$ truncated in an interval of width T. For each function g for which S_W is finite, S_W and S_p coincide.

With increasing T, it increases the frequency resolution, called the *resolution bandwidth*, with which the spectrum is obtained, but the periodogram does not converge uniformly for a simple spectral function either, but rather it increases the speed of its oscillations. For this reason, electronic spectrum analyzers have an additional filter with a width indicated by the term *video bandwidth*. A further way of improving the convergence is to mediate on many distinct realizations of the function $g(t)$.

A stationary random process in the broad sense $g(t)$ that goes through a spatially invariant linear system produces a new random process $g'(t)$ given by

$$g'(t)=\int_{-\infty}^{\infty}g(t')h(t-t')dt', \tag{5.5.13}$$

where the integration can be extended between $-\infty$ and $+\infty$, given that the impulse response h of the system is zero outside of a finite interval. It can be demonstrated [VanderLugt 1992] that the spectral power density at the system output is simply given by

$$S_{\text{out}}(\nu)=\left|H(\nu)\right|^2 S_{\text{in}}(\nu), \tag{5.5.14}$$

where S_{in} is the input density and H is the transform of h.

5.5.3 Theory of partial coherence

Consider a polychromatic (real) scalar[10] field u($\boldsymbol{r}$,t), which can be represented by its (complex) analytic function,

$$\boldsymbol{u}(\boldsymbol{r},t)=\int_0^{\infty}U(\boldsymbol{r},\nu)e^{-2\pi i\nu t}d\nu, \tag{5.5.15}$$

where $U(\boldsymbol{r},\nu)$ is the *temporal* Fourier's transform of the real field u($\boldsymbol{r}$, t). If the field is quasi-monochromatic, its analytical function can be factored as

---

[10] The vector theory of partial coherence for the electromagnetic field (including polarization) is reported by Tervo, Setälä, and Friberg (2004).

$$\boldsymbol{u}(\boldsymbol{r},t)=u(\boldsymbol{r},t)e^{-2\pi i\bar{\nu}t}, \tag{5.5.16}$$

where $\bar{\nu}$ is the carrier frequency and u is a complex amplitude (taken from u) that varies *randomly* in space and time (please note the change of font). This decomposition can be done without loss of generality, even if the field spectrum is broadband. In this case, a suitable value can be given to $\bar{\nu}$, for example, the center frequency of the spectrum. Given that the phase of u varies randomly, its mean value is zero

$$\langle \boldsymbol{u}(\boldsymbol{r},t)\rangle = 0. \tag{5.5.17}$$

The triangular brackets denote an ensemble average, but we assume here that the emission is ergodic, so we can also interpret them as an average over time. If, however, the wave is monochromatic, the temporal average is not zero, and the case in which the wave is the sum of the monochromatic components with a phase relationship established between them should be treated with caution.

Consider two points P_1 and P_2 in space-time with coordinates $(\boldsymbol{r}_1,t_1)$ and $(\boldsymbol{r}_2,t_2)$. If the process is temporarily stationary, its complex space-time autocorrelation function $R_{\boldsymbol{u}}$ depends only on the difference $\tau = t_1 - t_2$ and defines the *mutual coherence function* as

$$\Gamma(\boldsymbol{r}_1,\boldsymbol{r}_2,\tau) \equiv R_{\boldsymbol{u}}(\boldsymbol{r}_1,t+\tau;\boldsymbol{r}_2,t) = \langle \boldsymbol{u}(\boldsymbol{r}_1,t+\tau)\boldsymbol{u}^*(\boldsymbol{r}_2,t)\rangle . \tag{5.5.18}$$

Applying the factorization (5.5.16), a second version of the mutual coherence is also defined,[11] which Goodman (1996) calls the *mutual intensity*:

$$J(\boldsymbol{r}_1,\boldsymbol{r}_2,\tau) \equiv R_u(\boldsymbol{r}_1,t+\tau;\boldsymbol{r}_2,t) = \langle u(\boldsymbol{r}_1,t+\tau)u^*(\boldsymbol{r}_2,t)\rangle . \tag{5.5.19}$$

Between these two definitions, there is the relationship

$$\Gamma(\boldsymbol{r}_1,\boldsymbol{r}_2,\tau) = J(\boldsymbol{r}_1,\boldsymbol{r}_2,\tau)e^{-i2\pi\tau}. \tag{5.5.20}$$

For a quasi-monochromatic wave, with respect to Γ, the function J has the advantage of varying more slowly with τ, just as u represents a slowly variable amplitude modulating the carrier.

Further, here and below, we assume that the process is ergodic, so the ensemble average is equal to the time average. In addition to these functions, a normalized version called the *complex degree of mutual coherence* is also defined [Born and

[11] For $\tau = 0$, the two functions coincide: Born and Wolf (1980) define J only for this value of τ. In contrast, Barrett and Myers (2004) define as the mutual coherence function, which they call Γ, what Goodman (1996) calls J, such as here.

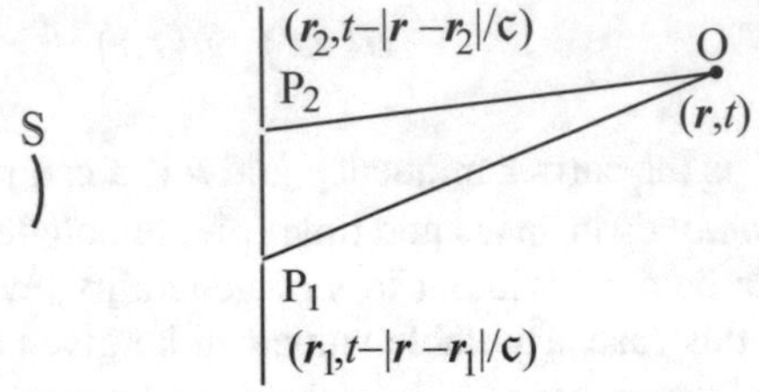

Fig. 5.13
Diagram of the evaluation of the mutual coherence function

Wolf 1980]

$$\gamma(\boldsymbol{r}_1,\boldsymbol{r}_2,\tau)\equiv\frac{J(\boldsymbol{r}_1,\boldsymbol{r}_2,\tau)}{\sqrt{\left\langle\left|u(\boldsymbol{r}_1,t)\right|^2\right\rangle\left\langle\left|u(\boldsymbol{r}_2,t)\right|^2\right\rangle}}, \tag{5.5.21}$$

which, as we shall see now, corresponds to the visibility of the fringes of interference between two waves coming from two different origins. The denominator shows, except for a constant factor, the geometric mean of the intensity produced by the two sources separately.

Let us restart from Young's experiment as a way to select the field of a wave in two assigned positions P_1 and P_2 (Fig. 5.13), considering two identical holes in an opaque screen illuminated by an ordinary thermal source (optionally provided with spatial coherence): this allows us to replace the ensemble averages with temporal averages. With laser sources, this is not possible, and such a case must be treated separately. Following Barrett and Myers (2004), the irradiance that is observed, except for constant factors,[12] on a screen is given by the temporal mean

$$\overline{I}(\boldsymbol{r})=\left\langle\left|u_1(\boldsymbol{r},t)+u_2(\boldsymbol{r},t)\right|^2\right\rangle_t, \tag{5.5.22}$$

where $u_1(\boldsymbol{r},t)$ is the analytic function for the field amplitude, at time t at the observation point O, defined by the vector $\boldsymbol{r}$, for the wave coming from the hole in P_1, and, similarly, for $u_2(\boldsymbol{r},t)$, from the hole in P_2. By developing the square module, we have

$$\overline{I}(\boldsymbol{r})=\overline{I}_1(\boldsymbol{r})+\overline{I}_2(\boldsymbol{r})+2\Re e\left\langle u_1(\boldsymbol{r},t)u_2^*(\boldsymbol{r},t)\right\rangle_t, \tag{5.5.23}$$

where

$$\overline{I}_1(\boldsymbol{r})=\left\langle\left|u_1(\boldsymbol{r},t)\right|^2\right\rangle_t,\quad \overline{I}_2(\boldsymbol{r})=\left\langle\left|u_1(\boldsymbol{r},t)\right|^2\right\rangle_t. \tag{5.5.24}$$

12 Since this is a scalar treatment, $\overline{I}$ does not exactly correspond to the Poynting vector module averaged over time. Rather, being defined as the square module of the amplitude of the field, it is proportional to the density of energy.

On the other hand, $\boldsymbol{u}_1$ is proportional to the field at the hole in P_1 at the time $t-(\boldsymbol{r}-\boldsymbol{r}_1)/c$, and similarly for $\boldsymbol{u}_2$. For the normalized interference term, we can then write

$$\frac{\left\langle u_1(\boldsymbol{r},t)u_2^*(\boldsymbol{r},t)\right\rangle_t}{\sqrt{\bar{I}_1(\boldsymbol{r})\bar{I}_2(\boldsymbol{r})}}=\frac{\left\langle u\left(\boldsymbol{r}_1,t-|\boldsymbol{r}-\boldsymbol{r}_1|/c\right)u*\left(\boldsymbol{r}_2,t-|\boldsymbol{r}-\boldsymbol{r}_2|/c\right)\right\rangle_t}{\sqrt{\bar{I}(\boldsymbol{r}_1)\bar{I}(\boldsymbol{r}_2)}}. \quad (5.5.25)$$

Factoring $\boldsymbol{u}$ with Eq. (5.5.16) and defining the time difference

$$\tau=\frac{1}{c}|\boldsymbol{r}-\boldsymbol{r}_1|-\frac{1}{c}|\boldsymbol{r}-\boldsymbol{r}_2|, \quad (5.5.26)$$

with Eqs. (5.5.18-21), and with the assumption of stationarity, the time dependence is reduced to the sole dependence on τ, and we have

$$\frac{\left\langle u_1(\boldsymbol{r},t)u_2^*(\boldsymbol{r},t)\right\rangle_t}{\sqrt{\bar{I}_1(\boldsymbol{r})\bar{I}_2(\boldsymbol{r})}}=\gamma\left(\boldsymbol{r}_1,\boldsymbol{r}_2,\tau\right)e^{-2\pi i\bar{\nu}\tau}. \quad (5.5.27)$$

Therefore, the irradiance in $\boldsymbol{r}$ is

$$\bar{I}(\boldsymbol{r})=\bar{I}_1(\boldsymbol{r})+\bar{I}_2(\boldsymbol{r})+2\left|\gamma(\boldsymbol{r}_1,\boldsymbol{r}_2,\tau)\right|\sqrt{\bar{I}_1(\boldsymbol{r})\bar{I}_2(\boldsymbol{r})}\cos\left[2\pi\bar{\nu}\tau-\Phi_\gamma(\boldsymbol{r}_1,\boldsymbol{r}_2,\tau)\right], \quad (5.5.28)$$

where γ is decomposed into module and phase

$$\gamma(\boldsymbol{r}_1,\boldsymbol{r}_2,\tau)=\left|\gamma(\boldsymbol{r}_1,\boldsymbol{r}_2,\tau)\right|e^{i\Phi_\gamma(\boldsymbol{r}_1,\boldsymbol{r}_2,\tau)}. \quad (5.5.29)$$

If we now apply Michelson's expression (3.2.11) for the visibility of the fringes, we get

$$\mathcal{V}=\frac{I_{max}-I_{min}}{I_{max}+I_{min}}=2\left|\gamma(\boldsymbol{r}_1,\boldsymbol{r}_2,\tau)\right|\frac{\sqrt{\bar{I}_1(\boldsymbol{r})\bar{I}_1(\boldsymbol{r})}}{\bar{I}_1(\boldsymbol{r})+\bar{I}_1(\boldsymbol{r})}. \quad (5.5.30)$$

If, in particular, $\bar{I}_1=\bar{I}_2$, we have $\mathcal{V}=|\gamma|$, and therefore the degree of mutual coherence coincides with the visibility of the fringes, which is experimentally measurable. On the other hand, if $\bar{I}_1\neq\bar{I}_2$, just measure these two values to get $|\gamma|$ from $\mathcal{V}$. In any case, the coherence between the fields in $\boldsymbol{r}_1$ and $\boldsymbol{r}_2$ is total if $|\gamma|=1$, is zero if $|\gamma|=0$, and is *partial* for $0<|\gamma|<1$. Instead, *in principle,* the phase Φ_γ of γ can be obtained from the position of the maximum and the minimum of the fringes of interference, comparing them in position with those that would be obtained by

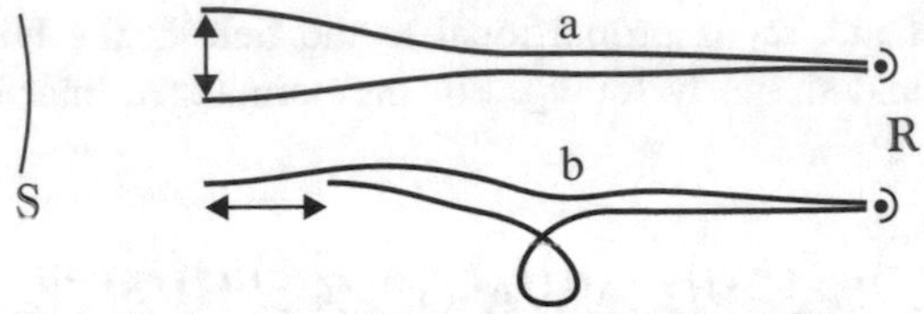

Fig. 5.14 Conceptual determination of coherence by means of a stethoscope with two optical fibers and a detector R. (a) lateral coherence and (b) temporal coherence of the radiation emitted by a source S

illuminating (with zero phase difference) the two holes with a monochromatic wave with frequency $\overline{\nu}$.

In optical signal processing, Eq. (5.5.28) confirms that, with incoherent illumination, the optical system is linear with intensity, while, with coherent illumination, it is linear with amplitudes. Instead, the case of illumination with partial coherence is generally to be avoided, since the system does not respond linearly either in intensity or amplitude [VanderLugt 1992]!

The function γ is 1 for $\boldsymbol{r}_1 = \boldsymbol{r}_2$ and $\tau = 0$ and, ordinarily, it is significantly different from zero only in a finite neighborhood of τ from 0 and in a finite neighborhood of $\boldsymbol{r}_1 - \boldsymbol{r}_2$ from 0, with a gradual decay away from the origin. We can characterize it in two complementary ways (Fig. 5.14), as an aid to understanding the principles of coherence. The complex degree of coherence is a function of seven parameters, of which six spatial are the coordinates of $\boldsymbol{r}_1$ and $\boldsymbol{r}_2$ of the two holes, and one temporal, τ. On the other hand, τ corresponds to the travel time difference toward the observation point from those two positions, and therefore depends on the geometry of the system. In Young's interferometer, the value of τ is generally very small compared to the propagation time from the holes to the observation point. However, with a different device, such as Michelson's interferometer, τ can be made very large, keeping the distance between P_1 and P_2 small, where the field is under examination.

5.5.3.1 Temporal coherence

By approximating γ with a regular peaked function, its half-width τ_c at half-height versus the change of τ (only) is called the *coherence time*, and the corresponding optical path difference $c\tau_c$ is called the *coherence length*. These values give us an indication of the spectral bandwidth $\Delta\nu$ of the source. Indeed, for Eqs. (5.5.9-10), the spectral power density is just the Fourier's transform of the autocorrelation function. Therefore, due to the uncertainty relation (5.1.55), we know that

$$\Delta\nu \geq \frac{1}{4\pi\tau_c} . \tag{5.5.31}$$

Usually, we can assume the approximate relationship $\Delta\nu \approx 1/\tau_c$. When τ_c is large compared to each time difference in a given situation, the radiation is said to be *quasi-monochromatic*, since, in that situation, the radiation behaves as if it was monochromatic with frequency $\bar{\nu}$.

To assess the coherence length of laser sources, and thus their spectral band, kilometric optical fibers are also used as delay lines for one of the two paths. With frequency-stabilized laser sources with spectral bands within the range of kHz and below, this solution becomes impractical due to the magnitude of the coherence length, and then the analysis of the beating with other similar or better sources is used.

5.5.3.2 Coherence and spatial correlation

Let us now consider the complementary case in which the observation is made at $\tau = 0$ by moving the two holes. In this case, the function γ is usually written in abbreviated form as $\gamma(\boldsymbol{r}_1,\boldsymbol{r}_2) = \gamma(\boldsymbol{r}_1,\boldsymbol{r}_2,\tau=0)$, and the same for the function Γ. The half-width (or the square root of the variance) L_c of γ indicates the degree of *spatial coherence* of the incident wave on the holes, which, with ordinary sources, coincides with the *correlation length*.

For natural sources, γ is generally a strictly peaked function around $\boldsymbol{r}_1 - \boldsymbol{r}_2 = 0$, and it is possible to approximate it with a Dirac's delta. For example, when $\boldsymbol{r}_1$ and $\boldsymbol{r}_2$ lie on the surface of such a source, this approximation is written as

$$\gamma\left(\boldsymbol{r}_1,\boldsymbol{r}_2\right) \approx A_c\delta\left(\boldsymbol{r}_1-\boldsymbol{r}_2\right), \tag{5.5.32}$$

where the vectors $\boldsymbol{r}$ are taken as two-dimensional and A_c is a constant called the *coherence area*. This expression has meaning only within an integral on the surface of the source, where the other factors present vary slowly in comparison to $\gamma(\boldsymbol{r}_1,\boldsymbol{r}_2)$. When this substitution is valid, it is said that the source is *spatially incoherent*. We have already taken advantage of this approximation in the preceding sections. The sources that most closely approach this situation are the *Lambertian sources*, including the thermal ones called *black bodies*. Adriaan Walther showed that, for a Lambertian quasi-monochromatic source, the complex degree of mutual coherence must have the form [Walther 1968]

$$\gamma\left(\boldsymbol{r}_1,\boldsymbol{r}_2\right) = \frac{\sin\left(\bar{k}\left|\boldsymbol{r}_1-\boldsymbol{r}_2\right|\right)}{\bar{k}\left|\boldsymbol{r}_1-\boldsymbol{r}_2\right|}, \tag{5.5.33}$$

where $k = 2\pi/\lambda$ is the modulus of the wave vector corresponding to the carrier [Barrett and Myers 2004]. Unfortunately, for this function, the variance defined as in Eq. (5.1.50) diverges. On the other hand, the function γ of an extended Lambertian source is often employed inside of an integral on the surface of the source, so the oscillations of the sinc are essentially deleted, while the contribution of the central peak remains. Therefore, we can assume, for γ, a width (from peak to first zero) of λ/2. If thc optical system has a large PSF with respect to this width, the approximation of γ with a delta may be adequate, but it is not entirely justified with optical systems that can resolve details on the order of the wavelength.

The situation can radically change with laser radiation. In fact, suppose the radiation diffused by a rough object reaches the screen of the two holes. On this screen, one will be able to observe *speckles*, which are characterized by rapid fluctuations in the intensity and phase of the wave field as the position changes. If laser, diffuser, and screen are stuck to each other, these speckles are immobile. Beyond the two holes, we now consider another screen to observe the interference fringes of the radiation that passes through them. By moving the holes, the fringes will change their position randomly and their visibility will fluctuate in a remarkable way, but without decay. Our measure will then determine a high degree of *spatial coherence*, even though the spatial *correlation length* of the field on the first screen may be particularly small [Shamir 1999].

If the diffusing object is placed in rapid motion, the speckles move, and therefore the value of γ obtained with the temporal average of Eq. (5.5.25) can give the correct value of the spatial correlation of the field. On the other hand, the field on the first screen is rapidly modulated in amplitude and phase by the movement of the object and is no longer monochromatic.

5.5.3.3 Wave equations for the mutual coherence

Following Born and Wolf (1980), let us now consider a statistically stationary wave in the vacuum and let $u(\boldsymbol{r}_1,t_1)$, $u(\boldsymbol{r}_2,t_2)$ be the wave amplitudes in two different space-time positions. With the assumption of ergodicity, the mutual coherence function is given by

$$\Gamma(\boldsymbol{r}_1,\boldsymbol{r}_2,t_1,t_2)=\left\langle u(\boldsymbol{r}_1,t_1+t)u*(\boldsymbol{r}_2,t_2+t)\right\rangle_t$$
$$=\lim_{T\to\infty}\frac{1}{2T}\int_{-T}^{T}u(\boldsymbol{r}_1,t_1+t)u*(\boldsymbol{r}_2,t_2+t)\,dt\,.$$

Applying, to Γ, the Laplacian operator

$$\nabla_1^{\,2}=\frac{\partial^2}{\partial x_1^{\,2}}+\frac{\partial^2}{\partial y_1^{\,2}}+\frac{\partial^2}{\partial z_1^{\,2}},$$

where x_1, y_1, z_1 are the coordinates of $\boldsymbol{r}_1$, and exchanging the order of the operators, we have

$$\nabla_1^2\Gamma(\boldsymbol{r}_1,\boldsymbol{r}_2,t_1,t_2) = \lim_{T\to\infty}\frac{1}{2T}\int_{-T}^{T}\left[\nabla_1^2 u(\boldsymbol{r}_1,t_1+t)\right]u^*(\boldsymbol{r}_2,t_2+t)\,dt\,.$$

As we know from §1.4.2, the wave equation (1.4.7) is also applicable to the analytic function u of the field. Therefore, we can replace, in the integral, ∇_1^2 with $(1/v^2)(\partial^2/\partial t_1^2)$, where v is the velocity in the medium. By again inverting the order of operations, we can write

$$\nabla_1^2\Gamma(\boldsymbol{r}_1,\boldsymbol{r}_2,t_1,t_2) = \frac{1}{v^2}\frac{\partial^2}{\partial t_1^2}\Gamma(\boldsymbol{r}_1,\boldsymbol{r}_2,t_1,t_2)\,. \tag{5.5.34.a}$$

Repeating the procedure for point 2, we also have

$$\nabla_2^2\Gamma(\boldsymbol{r}_1,\boldsymbol{r}_2,t_1,t_2) = \frac{1}{v^2}\frac{\partial^2}{\partial t_2^2}\Gamma(\boldsymbol{r}_1,\boldsymbol{r}_2,t_1,t_2)\,. \tag{5.5.34.b}$$

Finally, having assumed the stationarity for which Γ depends only on the difference between the times $\tau = t_1 - t_2$, we have $\partial^2\Gamma/\partial t_1^2 = \partial^2\Gamma/\partial t_2^2 = \partial^2\Gamma/\partial\tau^2$. In conclusion, the function of mutual coherence obeys two wave equations as the space-time position of the two points considered is varied.

5.5.3.4 Van Cittert-Zernike theorem

As a consequence of Eqs. (5.5.34), we can then also apply the mathematical methods of the theory of diffraction to the calculation of the mutual coherence, when this is known on a surface. Consider the particular case of a primary incoherent extended source on a given plane surface S taken as perpendicular to the z-axis (Fig. 5.15). Applying the first Rayleigh-Sommerfeld diffraction integral for a non-monochromatic wave (4.1.31) and taking into account that we have to satisfy the two wave equations (5.5.34), we then have[13]

$$\Gamma(\boldsymbol{r}_1,\boldsymbol{r}_2',t_1,t_2) = \frac{1}{2\pi v}\iint_S \frac{z_1}{R_1^2}\frac{\partial}{\partial t_1}\Gamma(\boldsymbol{r}_1',\boldsymbol{r}_2',t_1 - R_1/v,t_2)\,d\boldsymbol{r}_1'$$

and, in succession,

---

[13] The choice of this integral here is only for mathematical convenience. Born and Wolf (1980) report a similar procedure, but using the Kirchhoff diffraction integral, applicable to surfaces of any form. Unfortunately, the final expression is much more complicated, including many terms, which are instead reduced here to only the second derivative over time.

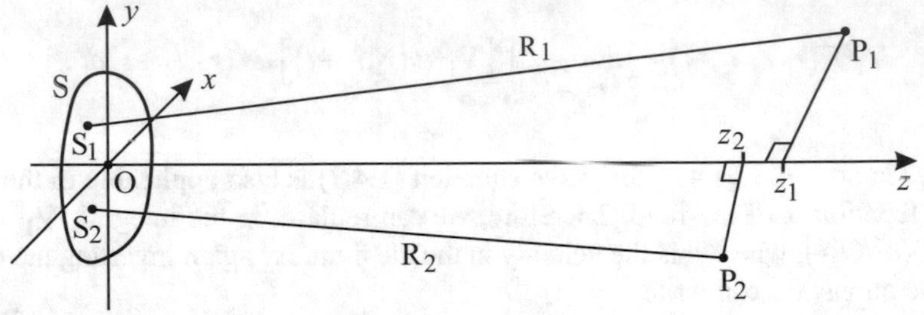

Fig. 5.15 Propagation of the mutual coherence

$$\Gamma\left(\boldsymbol{r}_1,\boldsymbol{r}_2,t_1,t_2\right)=\frac{1}{2\pi\upsilon}\iint\limits_{S}\frac{z_2}{R_2{}^2}\frac{\partial}{\partial t_2}\Gamma\left(\boldsymbol{r}_1,\boldsymbol{r}_2',t_1,t_2-R_2/\upsilon\right)d\boldsymbol{r}_2',$$

where, with respect to the origin O, $\boldsymbol{r}_1$ and $\boldsymbol{r}_2$ indicate *two* observation points, P_1 and P_2, in front of the surface S; z_1 and z_2 are, respectively, the distance of P_1 and P_2 from this surface (in the *z* direction), while $\boldsymbol{r}_1'$ and $\boldsymbol{r}_2'$ indicates two points, S_1 and S_2, on S; lastly, $R_1 = |\boldsymbol{r}_1 - \boldsymbol{r}_1'|$ and $R_2 = |\boldsymbol{r}_2 - \boldsymbol{r}_2'|$. Finally,

$$\Gamma\left(\boldsymbol{r}_1,\boldsymbol{r}_2,t_1,t_2\right)=\frac{1}{4\pi^2\upsilon^2}\iint\limits_{S}\frac{z_2}{R_2{}^2}d\boldsymbol{r}_2'\iint\limits_{S}\frac{z_1}{R_1{}^2}\frac{\partial^2}{\partial t_2\partial t_1}\Gamma\left(\boldsymbol{r}_1',\boldsymbol{r}_2',t_1-R_1/\upsilon,t_2-R_2/\upsilon\right)d\boldsymbol{r}_1'. \quad (5.5.35)$$

We can consider the source radiation quasi-monochromatic when the condition (4.1.34) is respected:

$$\Delta\nu \ll \left|\frac{\upsilon}{R_a - R_b}\right|, \quad (5.5.36)$$

where R_a and R_b are the minimum and maximum distances $|\boldsymbol{r}_1 - \boldsymbol{r}_1'|$ (or $|\boldsymbol{r}_2 - \boldsymbol{r}_2'|$) from the points on the surface S belonging to the stationary phase areas (mentioned in §4.1.8), and thus significantly contribute to the field at the observation point P_1 (or P_2). In this case, we can apply the approximations adopted for the integral (4.1.37), where the factorization (5.5.16) is used, obtaining

$$J\left(\boldsymbol{r}_1,\boldsymbol{r}_2,t_1,t_2\right)=\frac{\overline{\nu}^2}{\upsilon^2}\iint\limits_{S}\iint\limits_{S}\frac{Z_1}{R_1{}^2}\frac{Z_2}{R_2{}^2}e^{i2\pi\frac{\overline{\nu}}{\upsilon}(R_1-R_2)}J\left(\boldsymbol{r}_1',\boldsymbol{r}_2',t_1-\overline{R}_1/\upsilon,t_2-\overline{R}_2/\upsilon\right)d\boldsymbol{r}_2'd\boldsymbol{r}_1', \quad (5.5.37)$$

where $\overline{R}_1$ and $\overline{R}_2$ represent the average value of R_1 and R_2 at the variation of the points S_1 and S_2 on the integration surface, respectively.

Finally, if the source is spatially incoherent, with its degree of coherence approximated by Eq. (5.5.32), we can reduce Eq. (5.5.37) to a single surface integral. In paraxial approximation, taking that $\tau = 0$ and assuming that $\left|\overline{R}_1 - \overline{R}_2\right|/\upsilon << \tau_c$, we have

$$\Gamma(\boldsymbol{r}_1,\boldsymbol{r}_2,0) = J(\boldsymbol{r}_1,\boldsymbol{r}_2,0) = \frac{\overline{\nu}^2}{\upsilon^2}\frac{A_c}{Z_o^2}\iint_S e^{i2\pi\frac{\overline{\nu}}{\upsilon}(R_1-R_2)}\overline{I}(\boldsymbol{r},0)\,d\boldsymbol{r}\,, \tag{5.5.38}$$

where $R_1 = |\boldsymbol{r}_1 - \boldsymbol{r}|$, $R_2 = |\boldsymbol{r}_2 - \boldsymbol{r}|$, $\boldsymbol{r}$ is now the position of a point S on the surface S on which the points S_1 and S_2 converge, and Z_o is the geometric mean between z_1 and z_2. This is the *Van Cittert-Zernike theorem*, which can be interpreted as the solution to a diffraction problem. Let us take the point P_2 as fixed and the point P_1 as variable: imagine that the source is an opening in a screen illuminated by a spherical wave that converges in P_2. In the opening, there is then a transparency that modulates the wave *amplitude* in proportion to $I(\boldsymbol{r},0)$. The mutual coherence function, as the position of P_1 changes, is therefore proportional to the amplitude of the wave diffracted by that aperture. As a result, we have that, if we can also consider a primary source as completely incoherent, a secondary source may instead be partially coherent, especially when the primary source appears at the secondary one under a small solid angle. On the other hand, if the secondary source is much wider than its spatial coherence width, the objects are still incoherently illuminated by this.

As a further simplification, we consider the case $z_1 = z_2 = Z_o$. In paraxial approximation, as for Fresnel's diffraction integrals, for the argument in the exponential of Eq. (5.5.38), we have [Born and Wolf 1980]

$$R_1 - R_2 \cong \frac{x_1^2 + y_1^2 - x_2^2 - y_2^2}{2Z_o} - \frac{(x_1 - x_2)x + (y_1 - y_2)y}{Z_o}\,, \tag{5.5.39}$$

where x_1, y_1, x_2, y_2, and x, y are, respectively, the transverse coordinates of $\boldsymbol{r}_1$, $\boldsymbol{r}_2$, and $\boldsymbol{r}$. Eq.(5.5.38) becomes

$$\Gamma(\boldsymbol{r}_1,\boldsymbol{r}_2,0) = \frac{\overline{\nu}^2}{\upsilon^2}\frac{A_c}{Z_o^2}e^{i\psi}\iint_S e^{-i2\pi(f_x x + f_y y)}\overline{I}(\boldsymbol{r},0)\,d\boldsymbol{r}\,, \tag{5.5.40}$$

where

$$\psi = 2\pi\frac{\overline{\nu}}{\upsilon}\frac{x_1^2 + y_1^2 - x_2^2 - y_2^2}{2Z_o}, \quad f_x = \frac{\overline{\nu}}{\upsilon}\frac{x_1 - x_2}{Z_o}, \quad f_y = \frac{\overline{\nu}}{\upsilon}\frac{y_1 - y_2}{Z_o}. \tag{5.5.41}$$

In the particular case of a uniform circular source, with radius a and its center in O, the *complex degree of coherence* on the observation plane (which is normalized to 1 in $P_1 = P_2$) is (see Table 5.1)

$$\gamma_{12} = \frac{2\,J_1(2\pi a f_r)}{2\pi a f_r} e^{i\psi}, \tag{5.5.42}$$

where $f_r = \sqrt{f_x^2 + f_y^2}$. Therefore, with $\overline{\nu}/\upsilon = 1/\overline{\lambda}$ and $a/Z_o = \alpha$, we have

$$\gamma_{12} = \frac{J_1(2\pi\alpha r/\overline{\lambda})}{\pi\alpha r/\overline{\lambda}} e^{i\psi}, \tag{5.5.43}$$

where r is the distance between P_1 and P_2. Taking, as the radius of the coherence area, that corresponding to the first zero of the Bessel function J_1, we have

$$r_0 \cong 0.61\frac{\overline{\lambda}}{\alpha}. \tag{5.5.44}$$

In experiments in which it is necessary to have a high degree of coherence between the points P_1 and P_2, Born and Wolf (1980) take, as the limit distance, the value to which it corresponds to $\gamma_{12} = 0.88$, that is, by solving Eq. (5.5.43) for r,

$$r_c \cong 0.16\frac{\overline{\lambda}}{\alpha}. \tag{5.5.45}$$

With this value, we have justified that which was anticipated in §5.4.7 for sunlight.

5.5.4 *Hanbury Brown and Twiss' interferometer*

An alternative to Michelson's stellar interferometer, which requires high phase stability on the two optical paths, can be found in of the interferometer developed by Robert Hanbury Brown and Richard Q. Twiss [Hanbury Brown and Twiss 1954]. It only requires the measure of the intensity by two detectors that may be very distant. In their experiment, they observed the Sirius star with two small telescopes that were about 6 meters apart. By correlating the intensity fluctuations recorded in the two signals, it is possible to determine the size of the star source. The correlation function used is

$$\left\langle E(\mathbf{r}_1,t)E^*(\mathbf{r}_1,t)E(\mathbf{r}_2,t-\tau)E^*(\mathbf{r}_2,t-\tau)\right\rangle,$$

where $\boldsymbol{r}_1$ and $\boldsymbol{r}_2$ are the positions of the two detectors. The corresponding complex degree of fourth-order coherence is

$$\gamma^{(4)}(\boldsymbol{r}_1,\boldsymbol{r}_2,\tau)=\frac{\langle I(\boldsymbol{r}_1,t)I(\boldsymbol{r}_2,t-\tau)\rangle}{\langle I(\boldsymbol{r}_1,t)\rangle\langle I(\boldsymbol{r}_2,t)\rangle}.$$

For a stationary thermal source, we have that the distribution of the intensities $P(I)$ is Gaussian around the mean value I_o. In this case, it has been shown that [Wolf 1957]

$$\gamma^{(4)}(\boldsymbol{r}_1,\boldsymbol{r}_2,\tau)=1+\frac{1}{2}|\gamma(\boldsymbol{r}_1,\boldsymbol{r}_2,\tau)|^2.$$

We also have that the correlation of the intensity fluctuations defined by

$$\delta I=I-\langle I\rangle$$

is given by

$$\begin{aligned}\langle\delta I(\boldsymbol{r}_1,t)\delta I(\boldsymbol{r}_2,t-\tau)\rangle&=\langle[I(\boldsymbol{r}_1,t)-\langle I(\boldsymbol{r}_1,t)\rangle][I(\boldsymbol{r}_2,t-\tau)-\langle I(\boldsymbol{r}_2,t)\rangle]\rangle\\&=\langle I(\boldsymbol{r}_1,t)I(\boldsymbol{r}_2,t-\tau)\rangle-\langle I(\boldsymbol{r}_1,t)\rangle\langle I(\boldsymbol{r}_2,t)\rangle,\end{aligned}$$

that is,

$$\frac{\langle\delta I(\boldsymbol{r}_1,t)\delta I(\boldsymbol{r}_2,t-\tau)\rangle}{\langle I(\boldsymbol{r}_1,t)\rangle\langle I(\boldsymbol{r}_2,t)\rangle}=\gamma^{(4)}(\boldsymbol{r}_1,\boldsymbol{r}_2,\tau)-1=\frac{1}{2}|\gamma(\boldsymbol{r}_1,\boldsymbol{r}_2,\tau)|^2.$$

Therefore, by measuring the correlation of the intensity fluctuations, it is possible to determine the modulus of the complex degree of coherence of the radiation, measured on the Earth, emitted from the star, and then proceed to determine the diameter of the star as in the case of the stellar interferometer.

5.5.5 *A classic experiment*

In the experiment with the Fresnel's biprism (see §3.2.4), we found that the visibility of the fringes is reduced down to zero and is then partially recovered with a periodic trend as the source size increases. The same "revival" of visibility is also observed in Young's experiment as the distance between the two openings (centered in P_1 and P_2 in Fig. 5.13) increases, when the illumination produced on them

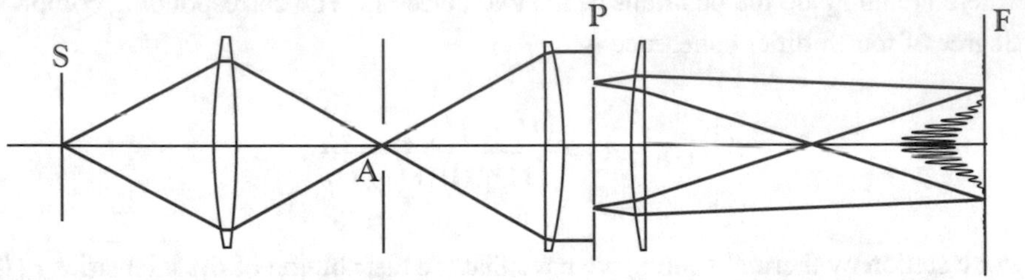

Fig. 5.16 Study of mutual coherence by observation of interference fringes

by the source S is partially coherent. This has been demonstrated experimentally by Thompson and Wolf (1957). In their apparatus (Fig. 5.16), a lens produces an image of a quasi-monochromatic extended source S on a circular aperture A. This aperture is smaller than the source image, but is larger than the Airy disk produced by the lens aperture for a single point of the source. In this way, the opening A corresponds to a still spatially incoherent secondary source. A second lens produces a collimated beam, which is then refocused by a third lens. Among them is placed a screen P with two circular holes, whose diffracted radiation is observed on the focal plane F of the third lens on which the Airy figure is produced corresponding to the Fraunhofer diffraction from each of the two holes.

The overlapping of the two figures creates fringes of interference whose pitch is inversely proportional to the distance between the two holes. Increasing their distance, the fringes become denser, but their visibility follows the oscillating trend mentioned above due to the variation of the mutual coherence with which the pinholes are illuminated. In fact, for the van Cittert-Zernike theorem, the mutual coherence function on the plane of the screen P also has the shape of an Airy figure, the opening A being circular. The relative displacement of the two holes radially explores this figure and the annulment of visibility corresponds to the zeros of the Bessel function J_1, while the reappearance corresponds to the rings. If the source is quasi-monochromatic, the visibility of the fringes on the plane F is not dependent on the position. If, instead, the source is polychromatic, the difference in path length also affects the visibility, which decreases away from the center of the figure due to the limited temporal coherence, and thus to the limited longitudinal coherence of the radiation [Reynolds et al 1989].

5.5.6 Propagation of coherence in an optical system

The results of the preceding sections can be generalized for an optical system that produces images. The case of non-monochromatic wave in a system that is (at least locally) isoplanatic may be treated by the analytic function of the field on the image plane calculated by integrating Eq. (5.4.44) on the frequencies ν

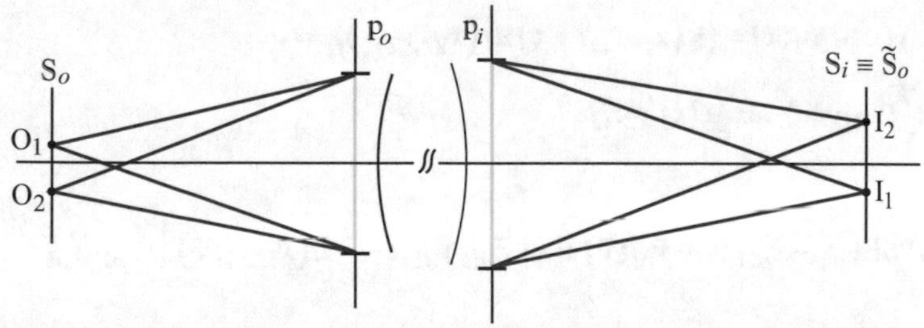

Fig. 5.17 Study of the propagation of the mutual coherence. S_o and S_i are, respectively, the object and image planes, while p_o and p_i are the corresponding pupils

$$u(x,y,t)=\int_0^\infty d\nu\, e^{-i2\pi\nu\left[t-\frac{1}{c}L(x,y;\nu)\right]}$$
$$\times\iint_{\tilde{S}_o}\text{APSF}(x-\tilde{x}_o,y-\tilde{y}_o;\nu)\tilde{U}_o(\tilde{x}_o,\tilde{y}_o;\nu)\,d\tilde{x}_o d\tilde{y}_o, \tag{5.5.46}$$

where $\tilde{U}_o$ is the spectrum of the field on the object plane, rescaled to the coordinates $\tilde{x}_o$ and $\tilde{y}_o$ according to Eqs. (5.4.9-10), and c is the speed of light in the vacuum. Here, L plays the role of $\bar{R}$ in Eq. (4.1.37).

The dependence on frequency ν appears in various positions; however, if the spectral bandwidth is small enough to overlook even the chromatic aberrations of the system, we can ignore this dependence in L and in the APSF. We can then reverse the integration order and integrate with respect to ν using Eq. (5.5.15), obtaining

$$u(x,y,t)=\iint_{\tilde{S}_o}\text{APSF}(x-\tilde{x}_o,y-\tilde{y}_o;\bar{\nu})\,\tilde{u}_o\left(\tilde{x}_o,\tilde{y}_o,t-\frac{1}{c}L(x,y;\bar{\nu})\right)d\tilde{x}_o d\tilde{y}_o\,. \tag{5.5.47}$$

With the factorization (5.5.16), we then have

$$u(x,y,t)=e^{i2\pi\frac{\bar{\nu}}{c}L(x,y;\bar{\nu})}$$
$$\iint_{\tilde{S}_o}\text{APSF}(x-\tilde{x}_o,y-\tilde{y}_o;\bar{\nu})\,\tilde{u}_o\left(\tilde{x}_o,\tilde{y}_o,t-\frac{1}{c}L(x,y;\bar{\nu})\right)d\tilde{x}_o d\tilde{y}_o. \tag{5.5.48}$$

Let us now consider two points I_1 and I_2 on the same image plane and their object points O_1 and O_2 (Fig. 5.17). The function J of mutual coherence evaluated on the image plane on points I_1 and I_2 is then given by

$$J(x_1,y_1;x_2,y_2;\tau)=\langle u(x_1,y_1,t+\tau)\,u^*(x_2,y_2,t)\rangle_t =$$

$$\Bigg\langle e^{i2\pi\frac{\overline{\nu}}{c}L(x_1,y_1;\overline{\nu})-i2\pi\frac{\overline{\nu}}{c}L(x_2,y_2;\overline{\nu})}$$

$$\times\iint_{\tilde{S}_o}\mathrm{APSF}(x_1-\tilde{x}_{o1},y_1-\tilde{y}_{o1};\overline{\nu})\,\tilde{u}_{o1}\left(\tilde{x}_{o1},\tilde{y}_{o1},t-\frac{1}{c}L(x_1,y_1;\overline{\nu})\right)d\tilde{x}_{o1}d\tilde{y}_{o1}$$

$$\times\iint_{\tilde{S}_o}\mathrm{APSF}^*(x_2-\tilde{x}_{o2},y_2-\tilde{y}_{o2};\overline{\nu})\,\tilde{u}^*_{o2}\left(\tilde{x}_{o2},\tilde{y}_{o2},t-\frac{1}{c}L(x_2,y_2;\overline{\nu})\right)d\tilde{x}_{o2}d\tilde{y}_{o2}\Bigg\rangle_t. \tag{5.5.49}$$

Taking the average over time inside the integrals, we finally get

$$\begin{aligned}&J(x_1,y_1;x_2,y_2;\tau)=e^{i2\pi\frac{\overline{\nu}}{c}\Delta L_{12}}\\&\times\iint_{\tilde{S}_o}\iint_{\tilde{S}_o}\mathrm{APSF}(x_1-\tilde{x}_{1o},y_1-\tilde{y}_{1o};\overline{\nu})\,\mathrm{APSF}^*(x_2-\tilde{x}_{2o},y_2-\tilde{y}_{2o};\overline{\nu})\\&\times J(x_{1o},y_{1o};x_{2o},y_{2o};\tau-\Delta L_{12}/c)\,d\tilde{r}_{1o}d\tilde{r}_{2o},\end{aligned} \tag{5.5.50}$$

where

$$\Delta L_{12}=L(x_1,y_1;\overline{\nu})-L(x_2,y_2;\overline{\nu}). \tag{5.5.51}$$

Usually, in a good optical system, it is required that, among all of the geometric paths that connect an object point to its image, the path changes are small or at least no bigger than the typical wavelength for which the system is built. In other words, the APSF is a very narrow function and the contributions to the integral on $\tilde{S}_o$ essentially derive only from a small neighborhood of I_1 and I_2. Moreover, the optical path difference ΔL_{12} between the two object points and the respective images is generally small compared to the path length. Therefore, if the two object points are close enough to each other, with quasi-monochromatic radiation, we can have $\Delta L_{12}/c \ll \tau_c$. In that case, we can ignore the term in ΔL_{12}, in the argument of J in the integral.

If, then, the illumination on the object plane is totally incoherent, for which we can apply the approximation (5.5.32), for $\tau = 0$, we have

$$\begin{aligned}&J(x_1,y_1;x_2,y_2;0)=A_c e^{i2\pi\frac{\overline{\nu}}{c}\Delta L_{12}}\\&\times\iint_{\tilde{S}_o}\mathrm{APSF}(x_1-\tilde{x}_o,y_1-\tilde{y}_o;\overline{\nu})\,\mathrm{APSF}^*(x_2-\tilde{x}_o,y_2-\tilde{y}_o;\overline{\nu})\,I(\tilde{x}_o,\tilde{y}_o)\,d\tilde{r}_o,\end{aligned} \tag{5.5.52}$$

and, if the two observation points coincide, we again find Eq. (5.4.53).

If the intensity on the object plane is constant, we can interpret the integral of Eq. (5.5.50) as an autocorrelation, this time of the APSF itself. Therefore, if the system is isoplanatic, J depends only on the differences $x_{12} = x_1 - x_2$, $y_{12} = y_1 - y_2$, and, if the APSFs are limited to a small extension in the integration surface, changing the integration variable into $\boldsymbol{r}'_o = \boldsymbol{r}_o - \boldsymbol{r}_1$, we can rewrite J as

$$J(x_{12}, y_{12}) = A_c e^{i2\pi\frac{\overline{\nu}}{c}\Delta L_{12}} \overline{I}_o \times \iint_{\infty} \mathrm{APSF}(-x'_o, -y'_o; \overline{\nu})\, \mathrm{APSF}^*(x_{12} - x'_o, y_{12} - y'_o; \overline{\nu})\, dr'_o. \tag{5.5.53}$$

For Eqs. (5.4.47-48) and the autocorrelation theorem (5.1.33), its Fourier's transform is then

$$\mathscr{F}\{J(x_{12}, y_{12})\} \propto \left| p(-\lambda d_i f_x, -\lambda d_i f_y) \right|^2, \tag{5.5.54}$$

and therefore it does not depend on the aberrations of the system implied in the phase modulation of the generalized pupil. In other words, the coherence area is not increased by system aberrations, although its PSF may be more extensive than that of a system limited by diffraction alone; think of the simple case of a focus error. On the other hand, for the validity of the relation (5.5.54), it is necessary that the source on the object plane be large compared to its coherence area. We also assumed that the system could be considered isoplanatic in a neighborhood of the image point larger than the PSF extension. If these conditions are not valid, then aberrations can make the illumination generated on the image plane partially coherent.

Then, if Eq. (5.5.54) is valid and the pupil is circular with radius a and it is illuminated uniformly, we can assume that

$$\left| p(-\lambda d_i f_x, -\lambda d_i f_y) \right|^2 \propto \mathrm{circ}\left(\frac{\lambda d_i f_r}{a} \right). \tag{5.5.55}$$

By anti-transforming (see Table 5.1), the mutual intensity on the image plane is given by

$$J(r_{12}) \propto \pi \left(\frac{a}{\lambda d_i} \right)^2 \frac{2\, \mathrm{J}_1\left(2\pi \frac{a r_{12}}{\lambda d_i} \right)}{2\pi \frac{a r_{12}}{\lambda d_i}}, \tag{5.5.56}$$

where r_{12} is the distance between the observation points I_1 and I_2, and where J_1 is the Bessel function of the same name.

5.5.7 Partial coherence and microscopy

Although, generally, with incoherent illumination, the correlation length L_c on the surface of the illuminated objects is very small, the situation changes when these objects are observed with a microscope, especially when it is capable of resolving distances closer to or lower than L_c. In this situation, the coherence of the illumination makes the optical system non-linear with respect to intensity: consequently, in the enlarged image, we have the presence of "fringes" that introduce false details, the transformation of a phase modulation in amplitude modulation and the apparent displacement of the edges of objects, especially in high contrast conditions. What is worse is that these effects are not trivially predictable, simply because of their non-linearity, and they vary from object to object. The most prudent action to take is therefore to work in conditions to avoid them, which bear the burden of a substantial loss of ability to distinguish and interpret the observed objects. In other words, it is necessary that the resolution of the microscope not be such as to be able to investigate the objects within the coherence area of their illumination.

In a microscope, the radiation outflow from the observed surface dilutes on a much larger surface at the image plane. Therefore, this loss of brightness must be compensated by a very concentrated illumination, which can come from behind, through a *condenser*, or from the front, through the same objective (*epi-illumination*). There are essentially two ways to concentrate the source radiation on the object plane (Fig. 5.18). The first, called *critical illumination*, is to use a slightly extended source and to produce a shrunk image of it on the object plane. However, the non-uniformity of intensity of the source is reproduced on such a plane. The second method, called *Köhler's illumination*, is to produce an enlarged image of the source on the first focal plane of the condenser, which focuses the radiation at its second focal plane, where the object plane of the microscope is located [Köhler 1899]. In this way, the non-uniformity of the source does not give rise to a spatial non-uniformity of illumination, but they affect its angular spectrum. In modern microscopes, the lamp is contained in a body separate from the microscope, and the radiation is conducted to the condenser through a bundle of optical fibers that are mixed to produce as much uniformity as possible.

The surprising result of the previous section justifies another road for the Zernike's conclusion, resumed by Born and Wolf (1980), when he states that the mutual intensity on the image plane *is independent of the aberrations of the system* and that *the condenser aberrations have no influence on the resolution power of a microscope*. This is contrary to the occasional assertion that the condenser must be well corrected to incoherently illuminate the object plane. This applies to cases of both critical and Köhler's illumination [Born and Wolf 1980]. In the example of the focusing error alone, with critical illumination, the condenser produces a longitudinally shifted image of the source, but, for the van Cittert-Zernike theorem, it suffices that this source is large compared to its coherence area and that it looks angularly large from the plane of the microscope so as to generate incoherent il-

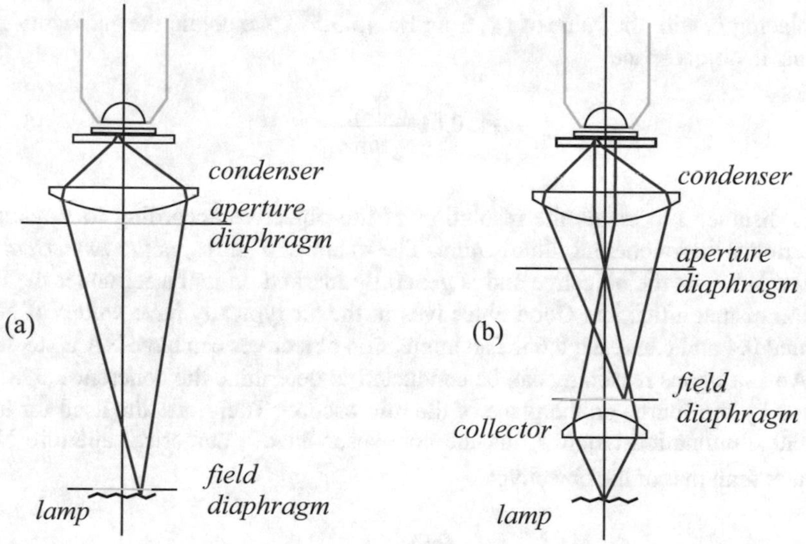

Fig. 5.18 (a) critical illumination; (b) Köhler's illumination

lumination. On the other hand, in order to obtain a high resolution, microscopes work with very large numerical apertures for which paraxial approximations and the same scalar theory on which the conclusion of the previous section is based are no longer valid.

However, we can estimate the resolution of a microscope as follows. Take, for example, the diagram in Fig. 2.48 and suppose that the principal image surface of the objective coincides with its exit pupil and that the lighting on it is uniform. In the region between the objective and the eyepiece, the propagation can be considered paraxial; then, on the intermediate image plane, the Airy's figure, generated by the diffraction from the circular edge of the pupil, has a radius given by Eq. (4.3.13)

$$r_{\mathrm{A}i} = 0.61 \frac{\lambda_o s_{i1}}{n_i a}, \tag{5.5.57}$$

where a is the radius of the pupil and s_{i1} is the distance of the intermediate image plane from the principal plane of the objective; λ_o is the average wavelength in the vacuum and n_i is the refractive index of the intermediate space. Although the propagation is not paraxial in the object space of the objective, we can invoke the sine condition, for which, between the heights h_o and h_i of an object and its image, the refractive indexes n_o and n_i and the angles α_o and α_i of a marginal ray, passing through the base of the object, the following relation holds:

$$h_o n_o \sin\alpha_o = h_i n_i \sin\alpha_i \cong h_i n_i \frac{a}{s_i}.$$

Replacing h_i with the value of r_{Ai} from Eq. (5.5.57), we obtain the analogous Airy radius in object space

$$r_{Ai} \cong 0.61 \frac{\lambda_0}{n_o \sin \alpha_o} . \tag{5.5.58}$$

This distance represents the resolution of the objective according to Rayleigh's criterion with incoherent illumination. The quantity $n_o \sin \alpha_o$ is the *numerical aperture* NA_{ob} of the objective and is generally marked on its barrel under the indication of magnification. Good objectives in the air typically have values of NA_{ob} around 0.4 and can reach 0.65. The immersion objectives can have NA up to 1.4.

An analogous reasoning can be conducted to determine the coherence area produced by the source on the plane of the microscope. Therefore, the need for incoherent illumination requires that the condenser have a numerical aperture NA_{co} greater than that of the objective:

$$m = \frac{NA_{co}}{NA_{ob}} > 1 .$$

In particular, H.H. Hopkins and P.M. Baraham showed that the best resolution (according to Rayleigh) is obtained with a ratio $m = 1.5$ [Hopkins and Baraham 1950]. This value is obtained considering the partial coherence that is on the object plane and the uniform illumination that is on the condenser. The corresponding resolution is still expressed by Eq. (5.5.58) by replacing the numeric term 0.61 with a slightly lower value [Born and Wolf 1980].

On the other hand, sometimes the microscopists narrow the aperture of the condenser to produce a partially coherent illumination. In fact, in many biological objects, the contrast is very poor and the information is mainly contained in a phase modulation. Thus, by reducing the numerical aperture and by bringing the object slightly out of focus, it is possible to transform the phase modulations into amplitude modulation by making it possible to perceive the objects under examination. In fact, it is sometimes preferable to have an imperfect image rather than a perfect but invisible one.

Various techniques have been proposed since the beginnings of microscopy to increase image contrast [Goldstein 1999]. For example, to avoid losing resolution, one uses oblique illumination, or rather an annular *dark field illumination*, so that only diffracted radiation from objects is collected by the objective. One alternative is to place a circular stop on the center of the second focal plane of the objective in order to block the direct radiation (*central dark field*). However, both systems alter the contribution of spatial frequencies, and the resulting image is not faithful. In any case, the diaphragm produces an alteration of the PSF attributable to the condenser, with non-trivial consequences on the illumination coherence.

The decisive solution for making the transparent biological objects visible, was proposed by Fritz Zernike through the introduction of a retarding lamina on the image focal plane of the objective. This technique is called *phase contrast microscopy*. Typically, the delay is applied to an annular zone on that plane, which acts on direct radiation, also produced in an annular shape with a diaphragm opening at the condenser. This is a special case of spatial filtering that we will see shortly, and has the advantage over previous methods of producing a linear conversion from phase modulation to amplitude modulation, at least when the observed objects are optically thin enough.

A further measure to increase the contrast is to limit the aperture of the field by illuminating only the portion of interest on the object plane (this requires that the condenser be well corrected). In this way, we avoid the problem of the radiation coming from other areas producing a background, because of the diffusion from various mechanical elements, as well as the optical of the objective.

In recent times, combining different techniques, such as non-linear fluorescence, spatial modulated laser illumination, computer image processing, etc., has brought optical microscopy into the nanometer range. See, for example, the site https://en.wikipedia.org/wiki/Super-resolution_microscopy.

5.6 Spatial filtering

The peculiar property of an optical system to produce the Fourier's transform of an input signal allows for filtering and processing its spectral content. This fact has many applications. The simplest is to filter out unwanted signal components, for example, to clean a laser beam of the non-uniformities and fringes introduced by defects in the optics, such as dust or scratches, parasitic reflections, or diaphragms. In some situations, a selective filtering is needed to eliminate a disturbance without excessively deteriorating the signal content. In other cases, as with phase contrast microscopy, it is necessary to accentuate the variations of a weakly imprinted signal on a uniform background. Further applications consist of elaborating signal information content such as character recognition or automatic detection of particular objects.

In itself, the filtering *reduces* the information contained in the signal. Sometimes, this is just the simplification we want to get: a "velvety" youthful face (or a beautiful TEM_{00} mode) is generally more "attractive" than a "wrinkled" one (marked by diffraction rings), which, on the other hand, contains more information about a "lived life" (of dust and scratches on the optics). However, in many cases, one wants to extract particular details from a scene that is complex or affected by noise or aberrations. The filtering must then be well calibrated to avoid false details or false attributions, that is, any information of its own that it adds.

In general, the term *spatial filtering* means the multiplication of the spectrum of a signal with a given transmission function. It can be realized in the most varied forms. The simplest case is that of the diaphragms, or masks with appropriate shape, which are called *binary filters*, since their transmission assumes only two values, 0 and 1. The filtering can also be performed with multiple discrete levels

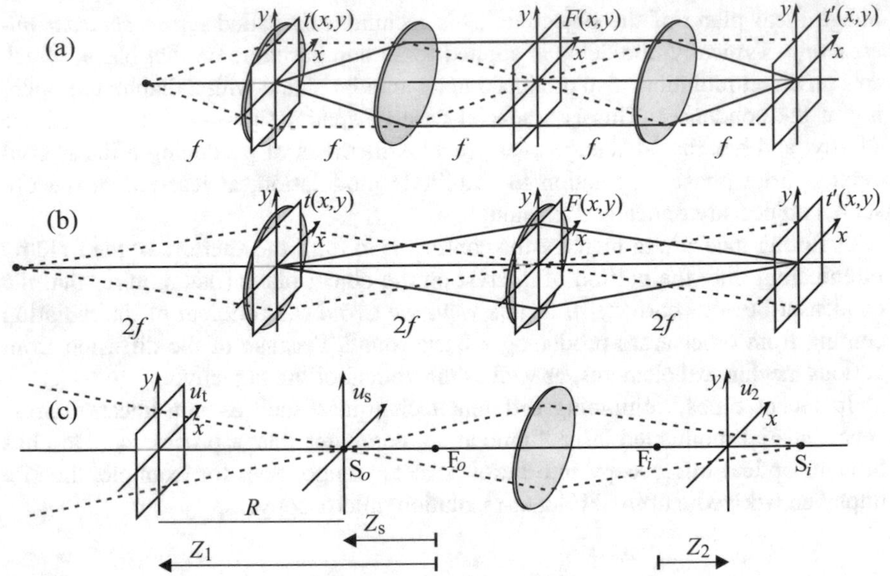

Fig. 5.19 Schemes of optical spatial filtering. (a): "4*f*" filtering system; the first lens produces a collimated beam on the transparency *t* and the other two lenses perform two transforms in succession: on the intermediate focal plane, a space filter *F* is placed and, finally, on the image focal plane of the last lens, one gets the filtered signal *t*′. (b): filtering system with convergent illumination. (c): study of the Fourier's transform properties with a convergent wave

or in continuous form, for example, with transparencies, also obtained photographically, which modulate the intensity transmitted, and, in this case, one speaks of *intensity filters*. Another way is to vary the relative phase of the spectrum components, for example, by depositing thin films with given shapes and thickness on transparent optical laminae, which are therefore *phase filters*. Finally, the most complex case is that of *intensity* and *phase* filters that modulate both quantities.

An important difference should be noted with the temporal filtering of electrical circuits. In fact, for a (minimum-phase) electronic filter, the module and phase of its transfer function are not independent, but rather are linked by the Hilbert transform. Instead, the same quantities for a spatial filter can be assigned independently. For example, a circular opening behaves like a spatial low-pass filter with intensity modulation and no phase change. If the input signal is a square symmetric pulse, the result is a pulse with beveled edges, but still symmetrical. Instead, a low pass electronic filter transforms a rectangular pulse into an asymmetric signal. This is due to the directionality of the time and the principle of causality, which does not apply for the space.

The spatial Fourier's transform can be obtained in more varied forms, with respect to that which has been discussed in §5.4.4. Indeed, in many applications, the signal to be processed is recorded on a thin lamina, a transparency such as a photographic film, a spatial intensity and/or phase modulator, uniformly illuminated by a plane or spherical monochromatic wave with radius of curvature R and wavelength λ. The conceptually simplest case is that shown in Fig. 5.19(a), which is called a "4f" filtering system; this method has the disadvantage of producing vignetting (mainly) in the first transformations, unless one uses a large enough lens to collect diffracted radiation, even from the marginal areas of the transparency. One system that uses one lens less and that does not suffer from vignetting is that of Fig. 5-19(b) [Goodman 1996]. In this case, the illumination on the transparency is convergent.

Let us consider, for example, the case in which the signal is imprinted on a transparency placed on the plane 1 with a transmission function for the field equal to $t(x_1,y_1)$ and that the transparency is lit by a wave with radius of curvature R (for a converging wave, R is negative) as in Fig. 19(c). The transmitted amplitude is given by

$$u_t(x_1,y_1)=t(x_1,y_1)\,u_o\,e^{i\frac{\pi}{\lambda R}\left(x_1^2+y_1^2\right)}. \tag{5.6.1}$$

Applying this signal to the Fresnel diffraction integral (4.2.8), for propagation in a homogeneous medium up to the distance $d=-R$ from the transparency, on plane S, where the illuminating wave converges, the amplitude is

$$u_s\left(\lambda d f_x,\lambda d f_y\right)= -\frac{i}{\lambda d}e^{i\frac{2\pi}{\lambda}d}\,e^{i\pi\lambda d\left(f_x^2+f_y^2\right)}\,u_o\int_{-\infty}^{\infty}\int_{-\infty}^{\infty}t(x_1,y_1)\,e^{-i2\pi\left(x_1 f_x+y_1 f_y\right)}\,dx_1\,dy_1. \tag{5.6.2}$$

Therefore, the amplitude diffraction figure $u_s(x_s,y_s)$ that is formed on plane S corresponds, apart from a field curvature term, to the transform of $t(x_1,y_1)$ valued at the spatial frequencies $f_x=x_s/\lambda d$, $f_y=y_s/\lambda d$.

Now, we add the effects of a lens with focal f, as shown in Fig. 19(c). If, in Eq. (5.4.30), which determines the field after the lens on a generic plane screen 2, we replace the field u_1 on the object plane with u_t, we get

$$u_2(x_2,y_2)=-\frac{i}{\lambda}\frac{f}{H}e^{ik(s_1+s_2)}e^{i\frac{\pi}{\lambda}\frac{Z_1}{H}\left(x_2^2+y_2^2\right)} \times u_o\int_{-\infty}^{\infty}\int_{-\infty}^{\infty}t(x_1,y_1)\,e^{i\frac{\pi}{\lambda}\left(\frac{Z_2}{H}+\frac{1}{R}\right)\left(x_1^2+y_1^2\right)}\,e^{-i2\frac{\pi}{\lambda}\frac{f}{H}\left(x_1x_2+y_1y_2\right)}\,dx_1\,dy_1, \tag{5.6.3}$$

with $H=f^2-Z_1Z_2$,

where Z_1 is the distance of the plane of transparency from the object focal plane of the lens and Z_2 is the distance of the screen 2 from the image focal plane.

With some passages, it is found that the term $Z_2/H+1/R$ is canceled when

$$f^2-(Z_1-R)Z_2=0\,,$$

i.e., when the plane 2 on which the field is calculated coincides with the image plane of the wave source point that illuminates the transparency. Therefore, aside from a term of curvature of field, u_2 is here given by the Fourier's transform of u_ι evaluated at the frequencies

$$f_x=\frac{f}{\lambda\left(f^2-Z_1Z_2\right)}x_2,\ f_y=\frac{f}{\lambda\left(f^2-Z_1Z_2\right)}y_2 \tag{5.6.4}$$

and, in these conditions, Eq. (5.6.3) becomes

$$u_2\left(\frac{\lambda H}{f}f_x,\frac{\lambda H}{f}f_y\right)=-\frac{i}{\lambda}\frac{f}{H}e^{ik(s_1+s_2)}e^{i\pi\frac{Z_1H}{f^2}\left(f_x{}^2+f_y{}^2\right)}$$
$$\times u_o\int\limits_{-\infty}^{\infty}\int\limits_{-\infty}^{\infty}t(x_1,y_1)\,e^{-i2\pi\left(x_1f_x+y_1f_y\right)}dx_1\,dy_1. \tag{5.6.5}$$

If $Z_1=0$, then the term of curvature out of the integral is also reduced to unity. However, with respect to the case represented by Eqs. (5.4.31), we can now change the transformation scale by varying Z_1 and keeping the source position fixed with respect to the lens, so that $Z_1-R=Z_s$ is constant.

We can understand that this property also applies to more complex optical systems, by inserting the transparency on the path of radiation between a source point and its image plane. On this plane, aside from a term of field curvature, we have the Fourier's transform of the signal recorded on the transparency. It is therefore not necessary to limit ourselves to optical systems in which the *plane of the transform* lies on a focal plane. In general, it is located on the plane where the spherical carrier wave finds its center of convergence, which, it should be remembered, corresponds to the origin of the spectrum, with $f_x=f_y=0$, on which the spatial filter is to be centered.

In the diagram from Fig. 5.19(b), we have, therefore, that the spectrum of the transparency is formed on the image plane of the source produced by the first lens, and the spatial filter is placed therein. Behind this plane, we place the second lens, which produces the filtered image of the transparency on the final plane.

5.6.1 Binary filters

The first studies on spectrum manipulation of an image were made by Ernst Abbe in 1873, and then by A.B. Porter in 1906, with the aim of verifying Abbe's theory of image formation with a microscope [Abbe 1873; Porter 1906]. Their es-

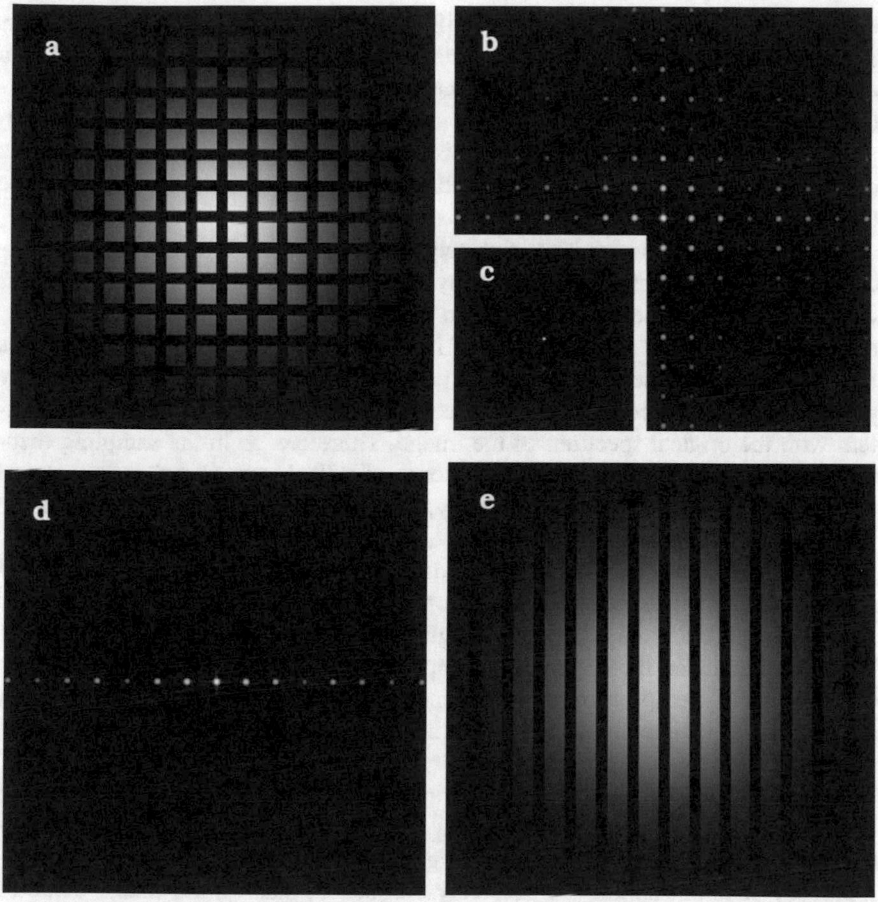

Fig. 5.20 Numerical simulation of spatial filtering produced by an optical system with two lenses in configuration 4*f*. Diffraction is calculated according to Kirchhoff, using the FFT. The incoming Gaussian beam has its waist on the focal plane of the first lens (coherent lighting). For $\lambda = 1$ μm, the waist is 2.5 mm, the lens focal distance is 25 cm and the figures have 8 mm side. (a) linear scale representation of the field intensity profile on the object focal plane of the first lens; (b) logarithmic scale representation of the signal present on the plane of the transform: the grayscale is extended to 5 decades (50 dB); (c) central portion of the spectrum in linear scale; (d) filtered spectrum; (e) signal on the image focal plane of the second lens

sence is to illuminate a thin periodic grid of wires with a coherent and collimated beam and reproduce the image with a lens.

On the focal plane of the lens (or in the center of convergence of the carrier wave), the spectrum of the grid is formed, consisting of a series of isolated, regularly spaced components with a diffusion degree and a structure that depends on the lens pupil and the extension of the grid or the illumination. By placing various types of *diaphragm* on this plane, one can get different spatial filtering effects.

The effectiveness of this system is illustrated by the example shown in Fig. 5.20, in which a grid formed by horizontal and vertical lines is superimposed on a Gaussian beam. Here, the "4*f*" configuration with two lenses was used. By intercepting, in the plane of the transform, the signal spectrum with a rectangular diaphragm that only passes the components aligned with the horizontal axis in the center of the spectrum, we have the vertical lines removed from the beam profile at the image plane of the system.

Binary filters are not the best in principle, but they are often effective, and, because of their simplicity, they find many applications. Filtering with diaphragms can be successfully used, for example, to eliminate the raster of television images, and the half-tone screening that is used in photo printing to represent continuous tone image. In fact, the screening has a periodic structure and the spectrum of the printed photo can be considered as the convolution of a regular Dirac's delta pattern with the original spectrum of the image. Therefore, as in the sampling theorem, this spectrum can be retrieved by means of a diaphragm allowing only one of the diffused light spots to pass (typically, the one of the zero order, which is the brightest).

Binary filters can also be used to extract periodic signals in the presence of noise or to eliminate unwanted periodic components whose frequency is contained within the spectral band of the input signal (*notches filters*). One particular case involves eliminating the continuous component by placing an opaque diaphragm in the center of the spectrum of such dimensions as to intercept the wave carrier, as it is focused by the optical system in the absence of diffracting transparency on which the input signal is modulated.

Thus, with a faded illumination of fringes on a dark background, the lines of discontinuity or the variations of intensity present in the input signal are highlighted in the filtered image[14]. By increasing the diameter of the obstacle, the low-frequency components are progressively eliminated, making the image contours sharper with the penalty of a drastic decrease in the intensity of the filtered signal.

5.6.2 Schlieren filter

With the word *schlieren*, which means *stria*, the unevenness of the refractive

---

[14] Following Kirchhoff's theory, a line of discontinuity in the intensity of the signal is reproduced with a dark line between two series of bright fringes parallel to the line itself. Indeed, a filter function given by $1 - \mathrm{rect}(f_x/a, f_y/a)$ has, as its transform, $\delta(x,y) - a^2\mathrm{sinc}(a\pi x, a\pi y)$. Consider a figure representing a step in the amplitude. The convolution of this step with the transform of the filter produces the difference between the original step and its copy "blunt" by the convolution with the sinc. Far away from the discontinuity, the resulting amplitude is zero, while, near it, the amplitude presents oscillation around zero. Then, the intensity has two bright fringes on the sides of a dark one at the step and other weaker parallel fringes, due to the oscillating shape of the sinc. This situation may change if we consider the electromagnetic edge contribution, which is neglected in theory.

index in transparent materials is indicated. They are not visible with ordinary optical methods, but are a serious defect if they are present in glass for optics. The schlieren are also commonly present in air produced by wind, flame, heat or moving objects. These striae were described for the first time by Robert Hooke in 1665, after he observed the effect on the light by hot air rising from the flame of a candle.

The *schlieren method* is mainly attributed to August Toepler, who designed an optical system for detecting the striae present in the glass to be used for making lenses. It essentially consists in the use of a point source for illuminating, in transmission, the object to be observed and a subsequent converging lens. At the image plane of the source, which is the plane of the transform, a binary filter is placed that simply consists of a blade; it intercepts the signal so that its edge passes to the center of the spectrum by intercepting, in part, the continuous component. In this way, about half of the spectrum is transmitted in a given transverse direction. The schlieren method, in general, allows for observing phase objects and defects in the form of optics in a simple way using only the spatial coherence of the source, and therefore works even in "white" light. In particular, *schlieren* photography, introduced by Ernst Mach in 1864, is now largely used in aeronautical engineering to visualize the flow of air around objects.

The mathematical treatment of this system is not trivial. For simplicity, we consider here a one-dimensional case in which the signal spectrum is given by a function $U(f_x)$. The presence of the blade leads to this spectrum being multiplied by a step function that can be written as

$$t(f_x) = \frac{1}{2}(1 + \operatorname{sgn} f_x), \tag{5.6.6}$$

whose transform is given by

$$T(x) = \frac{1}{2}\delta(x) - \frac{i}{2\pi x}. \tag{5.6.7}$$

On the image plane, the signal u_i is thus represented by the convolution

$$u_i(x) = T(x) * u(x) = \int_{-\infty}^{\infty} u(x')T(x - x')dx' = \frac{1}{2}u(x) - \frac{i}{2\pi}\int_{-\infty}^{\infty} \frac{u(x')}{x - x'}dx', \tag{5.6.8}$$

where u is the transform of U and corresponds to the field on the image plane without the blade. The integrand diverges towards $\pm\infty$ for $x \to \mp 0$, so it is not trivially treatable. The integral can be rewritten as

$$-\int_0^{\infty} \frac{u(x + x') - u(x - x')}{x'} d(x'), \tag{5.6.9}$$

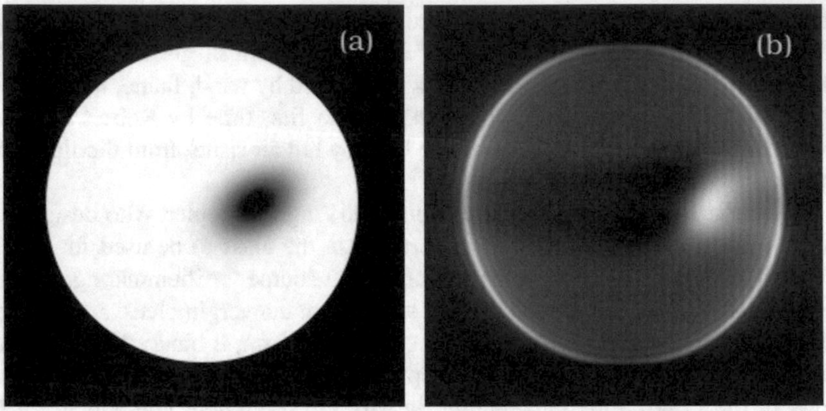

Fig. 5.21 Simulation of schlieren filtering. The blade is arranged vertically and the thin vertical fringes are due to aliasing. (a) gray scale representation between 0 (white) and π (black) of a *phase* object within a circular aperture. (b) image of the aperture filtered by the blade. The phase variation is transformed into a variation of intensity. This figure is particularly unfaithful and ugly due to the excessive amplitude of the phase modulation

and can be considered to be the average of the derivatives of u in x calculated at the finite difference of pairs of values u taken symmetrically from x. The presence of the imaginary unit as a factor of the integral implies that u_i receives contributions from both u and iu. Suppose that the object imprints a pure phase modulation to the transmitted field, so that

$$u(x) = C\exp\left[i\phi(x)\right], \tag{5.6.10}$$

where C is a complex constant and ϕ is a real function with $|\phi| \ll \pi$. It turns out that the intensity of the image can be approximated as [Goodman 1996]

$$I(x) = \frac{|C|^2}{4}\left[1 - \frac{2}{\pi}\int_{-\infty}^{\infty}\frac{\varphi(x')}{x-x'}\,dx'\right]. \tag{5.6.11}$$

Therefore, a phase modulation is converted into an amplitude modulation, making it visible with a normal detector. An example of this method is shown in Fig. 5.21, in which a large phase defect (up to π) is induced on the converging spherical wave just after the first lens of the diagram of Fig. 5.19(b).

5.6.3 Intensity and phase filters

The intensity filters are, in turn, transparencies on which the filtering is represented by a continuous transmission function between 0 and 1 and are usually created photographically.

For example, with a suitable attenuation of the continuous component with a semi-transparent obstacle, you can increase the contrast of the image. Similarly, filters can also be used to equalize the OTF of an optical system at the expense of its brightness.

These filters are most frequently used in combination with phase filters. If, in particular, the phase filter is binary, with induced phase shift of 0 or π, the overall filter is said to be *real-valued* and its transmission is expressed by a real function with values between -1 and 1. With this type of filter, one can correct some aberrations for which the OTF takes negative values [Maréchal and Croce 1953; Goodman 1996]. Another example is one in which the filter performs a mathematical differentiation along, for example, the x direction. The amplitude transmission function for this filter is

$$H\left(f_x, f_y\right) = \begin{cases} af_x & \text{for } \; 0 \le af_x \le 1, \\ af_x e^{i\pi} & \text{for } -1 \le af_x < 0, \end{cases} \tag{5.6.12}$$

where a is a constant to keep the filter passive. On the image plane, this filter intensifies the regions where the signal varies along the x direction. However, even in this case, one has a strong attenuation of the resultant image brightness.

One particularly important filter, but also one that is not trivially feasible, is indicated by the name of *matched filter*, in which both intensity and phase are modulated continuously and constitute the optimal filter for separating well-defined signals by a stationary additive noise [Van Vlek and Middleton 1946; Zadeh and Ragazzini 1952]. Indeed, the linear filter that maximizes the signal-to-noise ratio in the presence of stochastic additive noise has a transmission expressed by [VanderLugt 1992]

$$H\left(f_x, f_y\right) = a\frac{G^*\left(f_x, f_y\right)}{S_\mathrm{n}\left(f_x, f_y\right)}, \tag{5.6.13}$$

where S_n is the spectral power density of the noise, G is the transform of the signal being searched and a is a normalization constant.

Let us assume that the signal consists of a small figure; when this figure is present, centered at the position x_o, y_o, in the scene observed by the system, its spectrum is formed on the plane of the transform, and the spectrum amplitude, transmitted by the filter, is proportional to

$$G\left(f_x, f_y\right)e^{2\pi i\left(x_o f_x + y_o f_y\right)} H\left(f_x, f_y\right) = \frac{\left|G\left(f_x, f_y\right)\right|^2}{S_\mathrm{n}\left(f_x, f_y\right)} e^{2\pi i\left(x_o f_x + y_o f_y\right)}, \tag{5.6.14}$$

where the exponential represents a plane wave carrier oriented in function of the position of the figure in the scene. Since the indicated fraction is real, the filtered wavefront coincides with that of the plane wave of the carrier corresponding to the object; therefore, on the image plane of the system, one does not observe the same figure, but rather a bright spot in correspondence with the position where the figure is centered, indicating its presence.

5.6.4 Phase contrast microscopes

Often, in microscopy, it happens that the biological objects to be observed have little or no absorption. Indeed, when light passes through such objects, the predominant effect is that of a spatial phase modulation, mainly due to refractive index variations, and is not observable with sensors that respond instead to the intensity. This is often remedied by using special dyes that bind specifically to the structures of the objects themselves. A further increase in contrast is obtained with fluorescent dyes that emit visible radiation when they are illuminated with ultraviolet radiation. An alternative that is good for observing living biological entities in vitro, without altering their functions, as with dyes, is to transform phase modulation into amplitude modulation. As we have seen, the schlieren method produces such a conversion, but it has no simple interpretation. A better solution was found by Fritz Zernike, who, in 1935, proposed acting on the spectrum of spatial frequencies by changing the phase in a neighborhood of the component at null frequency [Zernike 1935]. This invention earned him the Nobel Prize in 1953. Suppose, indeed, that we illuminate the object to be observed with coherent radiation. We can express the transmitted amplitude as

$$A(x_o, y_o) = A_o e^{i\phi(x_o, y_o)}, \tag{5.6.15}$$

where ϕ represents phase modulation. If ϕ has small variations $\Delta\phi$ compared to 2π radians around an average value ϕ_o, we can approximate the amplitude transmitted with

$$A(x_o, y_o) \cong A_o e^{i\phi_o}\left[1 + i\Delta\phi(x_o, y_o)\right], \tag{5.6.16}$$

where the terms in powers of $\Delta\phi$ above 1 are neglected. In the square brackets, the unit represents the illumination carrier wave, supposedly uniform, while the other term is the modulation effect of the object. The two terms are in phase quadrature, so, $\Delta\phi$ being supposed small, even in this approximation, there is no appreciable modulation of intensity. On the plane of the Fourier's transform of the microscope, we have a light spot corresponding to the image of the source, surrounded by the spatial spectrum of $i\Delta\phi$, which, by definition, is null at zero frequency. For a point source, the central spot would correspond to a Dirac's delta; yet both spectral contributions of Eq. (5.6.16) are convoluted with the impulse response of the objective. Therefore, the size of the central spot is at least that of the central peak of the figure of Fraunhofer diffraction of the objective on the plane of the transform. Zernike then proposed putting a binary phase filter on this plane, consisting of a transparent substrate; at its center, a dielectric disk with a diameter equal to that of the central spot is deposited such as to induce on the carrier a phase shift δ of π/2 or 3π/2 radians with respect to that on the diffracted light. On the image plane of the microscope, we then have an intensity that is given by

$$\begin{aligned} &I_i(x_i,y_i) \cong I_{io}\left|+i+i\Delta\phi(x_o,y_o)\right|^2 \cong I_{io}\left[1+2\Delta\phi(x_o,y_o)\right] \qquad \text{for } \delta=\pi/2 \\ &I_i(x_i,y_i) \cong I_{io}\left|-i+i\Delta\phi(x_o,y_o)\right|^2 \cong I_{io}\left[1-2\Delta\phi(x_o,y_o)\right] \qquad \text{for } \delta-3\pi/2, \\ &\text{with} \quad x_o = x_i/m_\perp, \quad y_o = y_i/m_\perp, \end{aligned} \tag{5.6.17}$$

where $m_\perp$ is the transverse magnification of the microscope. The images obtained for these two cases are one negative of the other and, for small $\Delta\phi$, the resultant intensity modulation is linear to that of phase. To increase the contrast, the deposited disk may also have an absorption so as to attenuate the carrier.

The actual phase contrast microscopes have special objectives and special condensing lenses that are used to illuminate the object. The light coming from the lamp is intercepted by an annular diaphragm located at the focal plane of the condenser lens; therefore, the source takes the form of a thin ring and its annular image is formed on the plane of the transform, at the second focal plane of the objective. On this plane, a phase filter is centered on which, instead of a disc, a ring is deposited coinciding with the image of the source. The special lenses allow for the insertion of phase filters with different levels of absorption. These filters now have a more complex form, and they are called *apodized*; they have two additional absorbing rings located inside and outside with the phase ring, with a reduced absorption compared to this. In this way, a spurious diffraction effect, which consists of a bright halo around objects or in a reversed contrast, is significantly reduced.

5.6.5 VanderLugt's filter

The maximum performance of a filter is obtained when its transmission may vary in a continuous manner with complex values. In principle, this can be accomplished by superimposing an intensity filter and a phase filter. However, a compound filter of this type is not simple to realize. VanderLugt proposed instead to encode the phase modulation in the same intensity filter [VanderLugt 1964], in fact, using a holographic method or, put another way, of heterodyne.

The scheme of principle for creating this coding is given in Fig. 5.22. The radi-

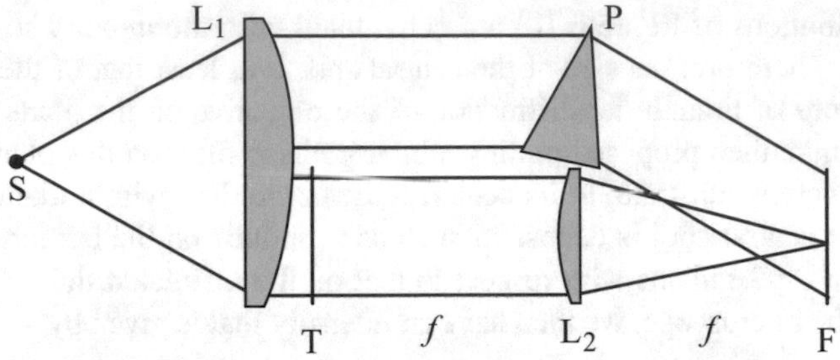

Fig. 5.22 Registering a VanderLugt Filter. T is a transparency on which it is imprinted a sample figure to be recognized and F is a film on which the interference of its spectrum with a plane wave deflected by the prism P is photographed

ation of a point source S is collimated by lens L_1. A part of the wave passes through the transparency T whose amplitude transmission is proportional to the desired impulse response h obtaining the signal to be encoded. The lens L_2 produces the Fourier's transform H of h on its second focal plane, where a photographic film F is placed. A reference wave, constituted by a second part of that collimated by L_1, is deflected by the prism P on this same plane. For Eq. (5.4.32), on the film, we have, therefore, the interference of two waves with amplitude

$$u_s \frac{1}{\lambda f} H\left(\frac{x}{\lambda f}, \frac{y}{\lambda f}\right) \quad \text{and} \quad u_r e^{-i2\pi\eta y}, \quad \text{with } \eta = \frac{\sin\vartheta}{\lambda}, \tag{5.6.18}$$

where, for simplicity, following Goodman (1996), I removed the phase factors resulting from the propagation and the transform produced by L_2. In fact, the phase of the reference wave can be varied as desired by moving the prism: this produces a translation of the fringes of interference, which, however, is irrelevant to the result. The corresponding intensity is calculated by Eq. (3.1.2) applied to this particular case, obtaining

$$\begin{aligned} I = {} & I_s \frac{1}{(\lambda f)^2} \left| H\left(\frac{x}{\lambda f}, \frac{y}{\lambda f}\right) \right|^2 + I_r \\ & + I_{rs} \frac{1}{\lambda f} H\left(\frac{x}{\lambda f}, \frac{y}{\lambda f}\right) e^{i2\pi\eta y} + I_{rs} \frac{1}{\lambda f} H^*\left(\frac{x}{\lambda f}, \frac{y}{\lambda f}\right) e^{-i2\pi\eta y}, \end{aligned} \tag{5.6.19}$$

where I_s, I_r, I_{rs} are constant real coefficients that incorporate the (non-trivial) geometric conditions, such as the direction with respect to the film, and the polarization of the two interfering waves. Since common films capture a negative image, one can take another photographic step on another film to get a positive image, but this is not essential to produce the VanderLugt filter. In any case, the films are exposed and developed to produce transparency with a real coefficient of transmission of the *amplitude* (rather than the intensity, as is usually the case) *linear* with the intensity of the exposure

$$t = I_s \frac{1}{(\lambda f)^2}|H|^2 + I_r + I_{rs}\frac{1}{\lambda f}He^{i2\pi\eta y} + I_{rs}\frac{1}{\lambda f}H^* e^{-i2\pi\eta y}. \quad (5.6.20)$$

Denoting, with ϕ, the phase of H, we can write this expression as

$$t = I_s \frac{1}{(\lambda f)^2}|H|^2 + I_r + 2I_{rs}\frac{1}{\lambda f}|H|\left[\cos(\varphi + 2\pi\eta y)\right], \quad (5.6.21)$$

from which we see how the phase H is encoded in the position of the fringes on a solely absorbing filter that functions as a diffraction grating.

The interferometer in Fig. 5.22 works with wavefront division and requires a large lens L_1. However, amplitude division systems have been proposed that use a beam splitter and require smaller aperture optics.

The VanderLugt filter is used like the other filters in one of the devices described above, but it is necessary to arrange the system so that, on the Fourier's plane, the desired signal transform coincides in size, orientation, and centering with that recorded on the filter. Let g be the signal to be analyzed and G its transform. The amplitude transmitted by the filter is then

$$G' = I_s \frac{1}{(\lambda f)^3}G|H|^2 + I_r \frac{1}{\lambda f}G + I_{rs}\frac{1}{(\lambda f)^2}GHe^{i2\pi\eta y} + I_{rs}\frac{1}{(\lambda f)^2}GH^* e^{-i2\pi\eta y}. \quad (5.6.22)$$

The first and second terms are without utility and, at the image plane of the system, they give rise to two spots centered in the origin. Instead, the last two terms are the interesting ones; for the diffraction produced by the fringes recorded on the filter, they produce two distinct spots on the image plane, diametrically opposite out of the axis. The term GH corresponds to the *convolution* of g and h, centered on the transverse coordinates $(0, -\eta\lambda f)$, while the term in GH^* corresponds to their *correlation* centered on $(0, \eta\lambda f)$ on the image plane. In order that these terms be separated from the others, it is necessary that the spatial frequency η of the reference wave in the filter recording be large enough. As in the case of the matched filters, the correlation spot indicates the presence of the object in the scene.

5.7 Diffraction gratings

The diffraction gratings are, in a broad sense, periodic or quasi-periodic spatial structures in their geometrical shape and in their electromagnetic properties. They are usually two-dimensional in their macroscopic appearance. But they can also be three-dimensional, as in the case of acousto-optic modulators, for which the lattice is created by a traveling acoustic wave of "volume" with a pitch on the order of 10 μm, or in the case of crystals with a lattice pitch on the order of Ångströms and capable of diffracting X-rays

In a strict sense, the diffraction gratings are planar or concave surfaces corrugated periodically. They are used both in reflection and in transmission, and their peculiarity and importance consists in spatially resolving the spectrum of an incident wave, in a manner similar to the chromatic dispersion produced by a prism, but much more effectively.[15]

The first to produce these devices systematically was Fraunhofer, who built the first ruling machine (*ruling engine*) in the second decade of the 1800s and discovered that it was necessary to have a very high regularity of the pitch. He mentioned its performance without giving a description of it. With his discovery of absorbing lines in the spectrum of sunlight, he started one of the most important disciplines in Physics: Spectroscopy. Subsequently, from 1850 until his death in 1881, F. Norbert produced gratings etched on glass. L.M. Rutherfurd was the first to build reflective gratings by engraving them on metal. Considerable progress was achieved by Henry Augustus Rowland, who, in 1882, presented gratings that were much larger and more accurate than those available at that time. In particular, he invented and produced concave gratings that, by eliminating the need to use lenses for collimating and focusing, enabled a simplification of spectrometers. Additional contributions were given by Albert Michelson, who, in 1900, decided to build a ruling machine, believing that he could complete the work in five years. Instead, he dedicated the last three decades of his life to this work: a good example of how things often require much more time than you expect. His second machine, begun in 1910, was sold to the Baush and Lomb Co. in 1947; it is still working, with various changes made by Wiley, and an interferometric control was finally added in 1990. This idea was only applied to a ruling machine for the first time in 1945, by George Harrison of M.I.T.. Michelson had proposed it 30 years earlier, but he lacked sources, electronic detectors and adequate instruments of counter-reaction.

The gratings with their derived optical devices, such as spectrographs, monochromators, and spectrophotometers, are the principal instrument of Spectroscopy, which led to the scientific revolution of the transition from Classical Physics to Quantum Physics, with countless applications in all fields of science and with a historically significant impact. Their main use is still in Spectroscopy, to analyze the spectral content of radiation and identify absorption and emission lines of substances to be analyzed for medical, chemical, environmental, and productive use, as well as and many others. The use of spectroscopes is very important in astron-

[15] An excellent text entirely dedicated to gratings is that of Loewen and Popov (1997), which describes their optical properties, based on accurate simulations, and their manufacture. It also has a large bibliography and historical introduction.

omy, as radiation is the main carrier of information on the nature of the cosmos. The gratings are also used to select the emission frequency of laser diodes, and are often integrated directly into the structure of the diodes themselves. They are then used as bandpass filters, compressors and expanders of laser pulses and in various other optical applications.

The gratings are used with a pitch of their periodicity that is greater, but often only slightly more so, than the typical half-wavelength of the spectrum of the radiation used. Therefore, the typical simplifications of diffraction theory, which consist in neglecting the electromagnetic effects of the edge of the openings and in the paraxial approximation, cannot be applied to them. On the contrary, the electromagnetic properties of the specific material of which the grating is made are of crucial importance. In fact, to determine their behavior, it is necessary to start from Maxwell's equations, and only in recent times have complex theories and sufficiently reliable calculation programs been developed. However, the periodicity of a lattice allows for determining, in a simple way, certain geometric aspects of their response to incident wave, and, in particular, to assign to the spectrum an accurate wavelength scale. The gratings have thus found numerous applications in spite of the rather uneven response of the amplitude diffracted along the same scale.

5.7.1 Reflection and refraction law of the gratings

Although they are called diffraction gratings, their law of reflection and refraction is based on interference. Indeed, let us consider a plane surface periodically corrugated by identical grooves parallel between them with pace *d*. Consider, then, a plane wave that arrives on the grating with an angle of incidence ϑ_I and with its plane of incidence orthogonal to the grooves. Compared to the case of a plane interface, the gratings have lost their isotropy in the direction orthogonal to the grooves, remaining identical to themselves only when they are shifted by an integer number of steps. We can then subdivide the surface into strips of width *d*, each corresponding to a groove, and consider the amplitude of the re-emitted radiation in various directions, treating it as the diffraction from a slit. Finally, we add the amplitude contribution of the various strips, knowing that, for each of these, there is a phase shift with respect to the previous one. In the case of *reflection,* it is (Fig. 5.23)

$$\phi = (\sin\vartheta_I - \sin\vartheta_R)\, n_I \frac{2\pi d}{\lambda_o},$$

where ϑ_R is the direction of observation, n_I is the refractive index of the medium of the incident wave and λ_o is the wavelength in the vacuum. To obtain constructive interference, it is then necessary that $\phi = -m2\pi$, where m is an integer called the *order of diffraction*. From this constraint, we get the *equation of gratings*

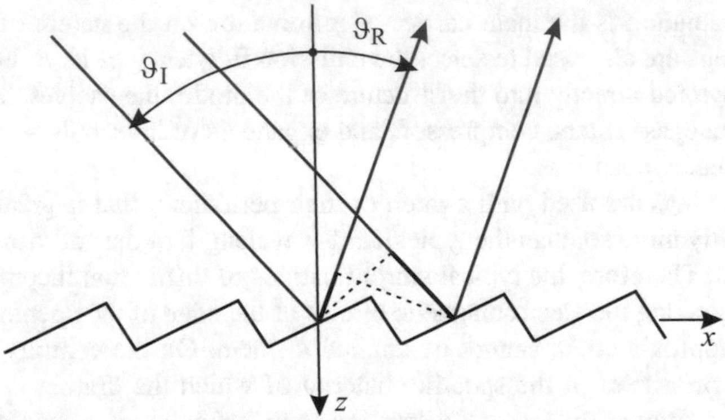

Fig. 5.23 Diagram to determine the phase shift between two diffracted rays in reflection from adjacent grooves

$$\sin \vartheta_{R,\,m} = \sin \vartheta_I + m\frac{\lambda_o}{n_I d}, \tag{5.7.1.a}$$

where $\vartheta_{R,m}$ are the directions along which the incident wave is diffracted.

For transmission, we instead have

$$n_T \sin \vartheta_{T,\,m} = n_I \sin \vartheta_I + m\frac{\lambda_o}{d}, \tag{5.7.1.b}$$

where n_T is the refractive index of the medium of the transmitted wave. If the grating is partially reflecting and transmitting, one has the diffracted waves of both cases. For m=0, there is the ordinary reflection or transmission, which has no dispersion. The sign of m is here chosen in such a way the positive orders are diffracted forward with respect to ordinary reflection and transmission and the negative ones back. The angles $\vartheta_{R,m}$ depend on the pace d of the grating and on the wavelength, as well as on ϑ_I, but not on the profile and the depth of the groove, on which the refracted intensity instead depends.

Let us suppose that we flatten the grooves to return an isotropic surface. It then happens that, in the diffraction integral, for one of the $\vartheta_{R,m}$ viewing directions, going across a strip by its width, the integrand remains constant in module, but its phase varies linearly, aside from a constant, from 0 to $m2\pi$. Therefore, the overall contribution of any strip is null in itself, except for m=0.

The above arguments can also be derived by making the Fourier's transform of the field of the reflected wave, taken on a plane tangent to the lattice, knowing only that its periodicity is linked to the reticular pitch and the progressive shift is due to the inclination of the incident wave. One further way is to consider the conditions for the phase on this same plane, similarly to what we saw for reflection and refraction from plane interfaces. Let $\boldsymbol{k}_I$, $\boldsymbol{k}_R$ and $\boldsymbol{k}_T$ be the wave vectors for the in-

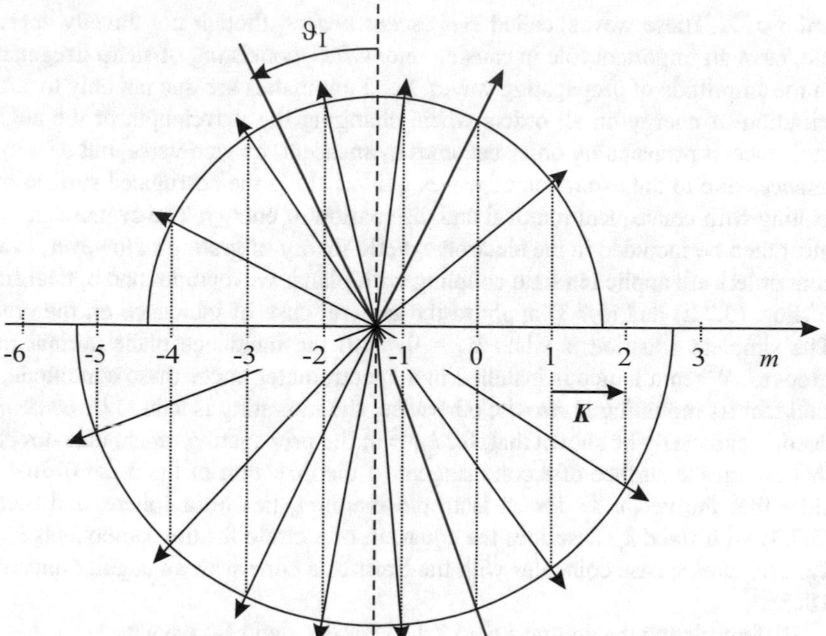

Fig. 5.24 Classical mounting. Graphic determination of diffracted orders. For generality, we here indicate even those in refraction, which are absent in the reflection gratings, when they are opaque in transmission

cident, reflected and refracted wave; choosing, as Cartesian axes, *z* along the normal direction to the grating, *y* that is parallel to the grooves and *x* as the perpendicular one, with such a direction that the components of $\boldsymbol{k}_I$ are positive, we have that

$$\begin{cases} k_{xR,m} = k_{xT,m} = k_{xI} + mK, \\ k_{yR} = k_{yT} = k_{yI}, \end{cases} \tag{5.7.2}$$

where $K=2\pi/d$ is the module of the *grating vector* $\boldsymbol{K}$ that lies on the plane of the grating, orthogonal to the grooves (Fig. 5.24). The *z* components of these vectors are, on the other hand,

$$k_{zj,m} = s_j \sqrt{\frac{4\pi^2 n_j{}^2}{\lambda_o{}^2} - k_{xj,m}{}^2 - k_{yj}{}^2}, \tag{5.7.3}$$

where *n* is the refractive index of the considered medium, with j = R, T and $s_R = -1$, $s_T = 1$. In order to speak of a diffracted order, known as a *propagating order*, *m* must be such that k_{zj} is real. Otherwise, the corresponding wave is an evanescent wave, which can be detected only within distances from the grating on the order of λ. These waves, called *evanescent orders*, though not directly appreciable, have an important role in causing *anomalies*, consisting of sharp irregularities in the amplitude of propagating waves. Such anomalies are due not only to a redistribution of energy on all orders when, changing the wavelength or the angle of incidence, a propagating order becomes evanescent, or vice versa, but also to resonances due to the excitation of waves guided along the corrugated surface of the grating with consequent removal and dissipation of energy. The evanescent waves must then be included in the electromagnetic theory of gratings. However, evanescent orders are applied in field coupling with planar waveguides and optical fibers.

Eqs. (5.7.2) and (5.7.3) apply to the general case of incidence on the grating. The simplest situation is when $k_{yI} = 0$, with the incidence plane normal to the grooves. When a lattice is installed in a spectrometer under these conditions, it is said that its mounting is *classic*; otherwise, the mounting is said to be *conical*. Indeed, it can easily be shown that, for $k_{yI} \neq 0$, the propagating orders take directions that are on the surface of a cone tangent to the direction of the order 0. Just consider that the vector $\boldsymbol{k}_j$, for an isotropic medium, lies on a sphere, and that Eq. (5.7.3) with fixed k_{yj} describes the equation of a circle for the components k_{xj} and k_{zj}. The classic case coincides with the limit of a cone with an angular opening of 180°.

Differentiating the grating Eq. (5.7.1.a) for $\vartheta_{R,m}$ and λ_o, we get

$$\frac{\partial \vartheta_{R,m}}{\partial \lambda_o} = \frac{m}{n_I d \cos \vartheta_{R,m}} = \frac{1}{\lambda_o} \frac{\sin \vartheta_{R,m} - \sin \vartheta_I}{\cos \vartheta_{R,m}}. \tag{5.7.4}$$

This means that, for a given wavelength, the angular dispersion is only a function of the incidence and diffraction angles.

The angle between the incident and the diffracted ray is called the *angular deviation* (A.D.) (Fig. 5.25). When a grating is used with null deviation for which we observe the diffracted radiation in the opposite direction to the incident one, that is for $\vartheta_{R,m} = -\vartheta_I$, it is said to be mounted in a *Littrow* configuration or autocollimation. In this case, we have

$$\frac{\partial \vartheta_{R,m}}{\partial \lambda_o} = \frac{2}{\lambda_o} \tan \vartheta_{R,m}, \tag{5.7.5}$$

which increases indefinitely, increasing ϑ_I up to the grazing incidence. It should be noted that the dispersion depends only on the angular conditions of each order and that m appears only as an index. Therefore, once ϑ_I has been determined, it is necessary to choose whether to use a grating with a fine pitch and a low order m, typically −1, or a grating with a coarse pitch and a big order in absolute value.

A typical defect of the gratings is that the angular dispersion of the various orders overlaps; it could also result in errors of attribution of the wavelength. This becomes particularly relevant with relatively coarse gratings used in high order.

As in the case of interferometers, the concept of free spectral range is used that can be deduced from the equation of the gratings. In fact, the same diffraction angle, and hence the same value $m\lambda$, for two successive orders can be obtained at two wavelengths λ_1 and λ_2 (still in the vacuum) such that

$$m\lambda_2 = (m+1)\lambda_1,$$

from which we get

$$FSR_{\lambda_o} = \frac{\lambda_o}{m}. \tag{5.7.6}$$

In terms of wave number $\eta = 1/\lambda_o$, the free interval is

$$FSR_\eta = \frac{1}{\lambda_1} - \frac{1}{\lambda_2} = \frac{\lambda_2 - \lambda_1}{\lambda_1 \lambda_2} \approx \frac{\eta}{m}, \tag{5.7.7}$$

where the final approximation applies to large orders for which $\lambda_1\lambda_2 \approx \lambda_o^2$.

Finally, for Eq. (5.7.1.a), the wavelengths corresponding to the passages of the order m from propagating to evanescent, and vice versa, which cause anomalies on other orders, verify the condition

$$\left| \sin\vartheta_I + m\frac{\lambda_o}{n_I d} \right| = 1 \quad \text{for } m = \pm 1, \pm 2, \cdots. \tag{5.7.8}$$

5.7.2 *Diffraction efficiency*

While the calculation of the diffraction directions of a lattice represents a "simple" generalization of reflection and refraction laws, the energy that flows in the various diffracted orders, and thus the diffraction efficiency, follows a complicated trend that cannot be attributed to simple models. In the past, it was not even reproducible from one grating to another, although they were produced the same way. Many parameters need to be considered to solve Maxwell's equations at the surface of the grating and determine the energy flow that, for relatively deep corrugations, also has vortexes in the grooves. In particular, in order to obtain a good agreement with the actual behavior, it is not possible to prescind from the material constituting the surface of the lattice, and the calculation of efficiency is not simply a multiplication of the one obtained for some ideal material, an infinite conductivity metal, for the reflection response of a polished material.

One major aspect is the disparity of efficiency between the two TE and TM polarizations. For example, for the order −1, the efficiency trend for TE waves is es-

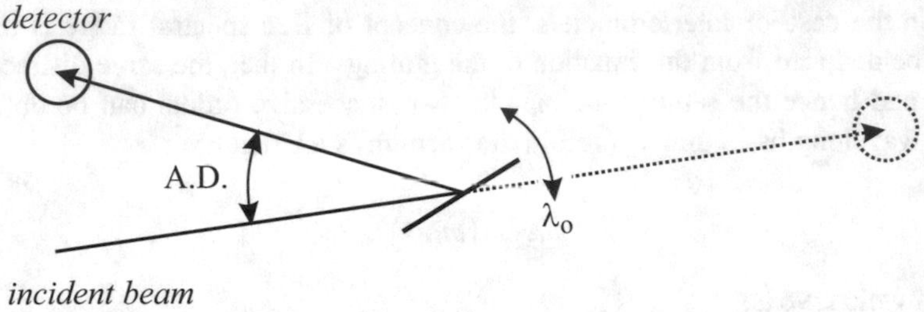

Fig. 5.25 Simplified scheme for measuring the efficiency of a grating. The angular deviation A.D. is held fixed to a given value, while the grating is rotated. The dotted circle indicates the position of the detector to calibrate the source

sentially given by an asymmetric bell curve that is relatively narrow in the spectrum of wavelengths, while that for the TM waves is generally more squared and extended in the spectrum; however, it suffers most of the anomalies that become denser at short wavelengths. Many useful examples are presented by Loewen and Popov (1997), under various conditions of use.

In practice, it is important to check the data sheets of commercial gratings to determine whether their absolute efficiency is adequate to the application you want to give them.

Experimentally, efficiency can be measured in various ways. One is shown in Fig. 5.25, by means of a monochromatic source obtained, for example, by another grating, and a detector. The angular deviation (A.D.) between the source and the detector is held fixed, while the reticle rotates at the variation of the wavelength of the source, maintaining a specific order of diffraction directed to the detector. The measurement is repeated without the grating, placing the detector so as to look directly at the source. From the ratio between the two measures, the efficiency is obtained. This type of calibration is the most useful, since, in many applications, the angular deviation is fixed.

Some classical arguments, such as the diffraction theory and the boundary conditions for the field, at least allow for a qualitative understanding of the general properties of gratings. Consider the case of a grating constituted by a series of M grooves of arbitrary shape equally spaced with a pitch d. The lighting conditions are also such that the grating can be brought back to a one-dimensional structure along a path orthogonal to the grooves, with the uniform illumination of a plane wave and classic mounting. The aperture of the grating can then be considered as the *convolution* of a finite comb of M Dirac's delta, in which we embody the phase resulting from the position of the grooves and the incidence angle of the incident wave, and a function $g(x)$ representing the field present in a *single* groove, in reflection or in transmission. Therefore, let us calculate the diffracted plane wave spectrum corresponding to the field emerging on a plane A close to the grating surface, with the periodicity described above. Using the convolution theorem, we have that it is given by the *product* of two functions: one is the transform $G(f_x)$ of $g(x)$ and the other is the transform H of the delta series. It constitutes the generalization of Eq. (4.3.17) of the diffracted field from a double slit. For the calculation of H, it is easier to follow an "ordinary" derivation:

$$H(f_x) = \int e^{-i2\pi x f_x} \sum_{q=0}^{M-1} \delta(x - qd) e^{ik_{Ix}x} dx = \frac{1 - e^{iM(-2\pi f_x + k_{Ix})d}}{1 - e^{i(-2\pi f_x + k_{Ix})d}}, \quad (5.7.9)$$

where $k_{Ix} = 2\pi(n_I / \lambda_o) \sin\vartheta_I$. The square module of H is the *interference function*

$$|H(f_x)|^2 = \frac{\sin^2\left(M\dfrac{\pi p d}{\lambda_o}\right)}{\sin^2\left(\dfrac{\pi p d}{\lambda_o}\right)}, \quad (5.7.10)$$

where

$$p = f_x \lambda_o - n_I \sin\vartheta_I . \quad (5.7.11)$$

Expressing f_x angularly according to Eqs. (5.3.18), we have

$$f_x = \frac{n_d \sin\vartheta_d}{\lambda_o}, \quad (5.7.12)$$

$$p = n_d \sin\vartheta_d - n_I \sin\vartheta_I , \quad (5.7.13)$$

where ϑ_d indicates the diffraction angle for a grating in both transmission and reflection and n_d is the corresponding refractive index. Finally, the spectral intensity is given by

$$I\left(\frac{n_d \sin\vartheta_d}{\lambda_o}\right) = \left|H\left(\frac{n_d \sin\vartheta_d}{\lambda_o}\right)\right|^2 \left|G\left(\frac{n_d \sin\vartheta_d}{\lambda_o}\right)\right|^2, \quad (5.7.14)$$

where the square module of G is the *intensity function of a groove.*

The typical aspect of these functions is shown in Fig. 5.26. Both the denominator and the numerator of Eq. (5.7.10) simultaneously vanish, for the values of p for which

$$\frac{pd}{\lambda_o} = m \quad \Rightarrow \quad n_d \sin\vartheta_d - n_I \sin\vartheta_I = \frac{m\lambda_o}{d}, \quad (5.7.15)$$

where m is an integer, again finding the law of gratings. Locally, around each of these values of p, the square module of H behaves like a sinc^2 with a principal peak of height M^2 surrounded by secondary peaks. This is due to the hypothesis of a precise number M of equally illuminated grooves. It should be noted that, of all of the main peaks of $|H|^2$, only some correspond to propagating orders.

Between two main maxima, there are M–1 zeros at the values

$$M\frac{pd}{\lambda_o}=mM+q \quad \text{with} \quad q=1,2,\cdots,M-1, \tag{5.7.16}$$

where the numerator of Eq. (5.7.10) vanishes. Amid these zeros, there are M–2 secondary maxima whose height I_q, for M large, tends to [Sivoukhine 1984]

$$I_q=\frac{4}{(2q+1)^2\pi^2}I_{\text{pr}} \quad \text{per} \quad q<<M, \tag{5.7.17}$$

where I_{pr} is that of the closest principal maximum. For $q = 1$, we have $I_1 = 0.045\ I_{\text{pr}}$, which is not really negligible. However, the number of illuminated grooves is generally very large, up to hundreds of thousands, for which the secondary peaks go to constitute a continuous background in which the modulation is filtered off by various factors that contribute to limiting the resolution, for example, the finite size of the entry slit or the smooth intensity decreasing of a Gaussian beam.

So far, we have only considered the "geometric" aspects of the problem. The real difficulty arises from the evaluation of the function g of the field on a single groove. A very rough example is to consider the pattern as consisting of slits with aperture $s < d$ and apply the ordinary approximations of diffraction, those of Kirchhoff. The field on the slit is therefore proportional to

$$g(x)=\begin{cases} e^{ik_{\text{I}x}x} & \text{for } |x|<s/2, \\ 0 & \text{otherwise,} \end{cases} \tag{5.7.18}$$

where, again, $k_{\text{I}x} = 2\pi(n_\text{I} / \lambda_o) \sin\vartheta_\text{I}$. The Fourier's transform of g is

$$G(f_x)=s\,\text{sinc}\left[\pi s\left(f_x-\frac{n_\text{I}}{\lambda_o}\sin\vartheta_\text{I}\right)\right]. \tag{5.7.19}$$

An example of this case is given in Fig. 5.26, aside from a peak normalization, with $s = 0.2\ d$. Eq. (5.7.14) becomes

$$I\left(\frac{n_d\sin\vartheta_d}{\lambda_o}\right)=s^2\frac{\sin^2\left(M\dfrac{\pi pd}{\lambda_o}\right)}{\sin^2\left(\dfrac{\pi pd}{\lambda_o}\right)}\text{sinc}^2\left(\frac{\pi ps}{\lambda_o}\right), \tag{5.7.20}$$

with p again given by Eq. (5.7.13). The function G modulates the height of the maximum of the function H, and sometimes deletes some of them. For example,

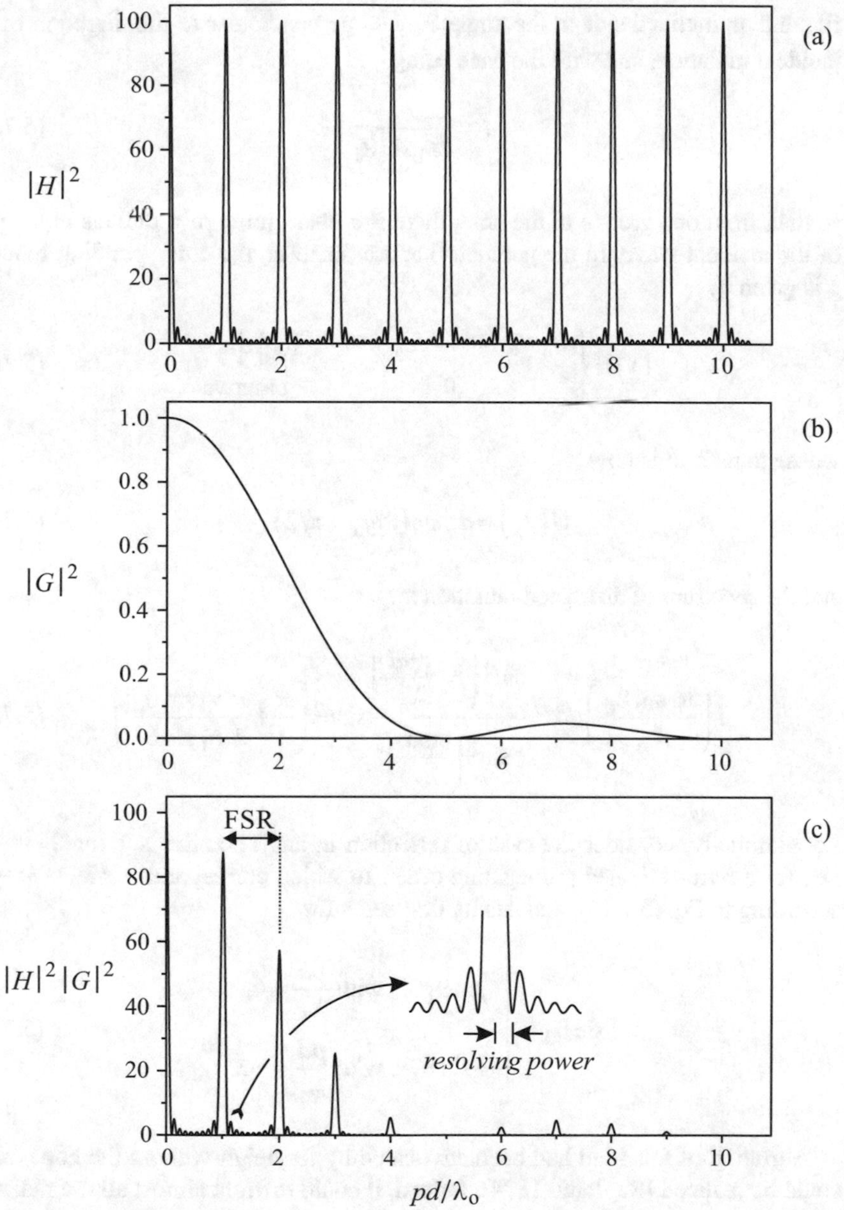

Fig. 5.26 (a) Interference function calculated for $M = 10$ at the variation of pd/λ_o. (b) Square module of the spectrum for a single groove, here represented by the function $\mathrm{sinc}^2(\pi ps/\lambda_o)$ with $s = 0.2\,d$. (c) Resulting intensity

for $s = d/2$, even orders are zeroed, as they coincide with the zeros of G.

A particularly interesting case is given by a groove with a right triangular profile and an inclined side at the angle ϑ_I, i.e., perpendicular to the direction of the incident radiation, and with the pace being

$$d = \frac{\lambda_o}{2n_I \sin \vartheta_I}, \tag{5.7.21}$$

so that, from one groove to the next, there is a phase jump of π radians in the path of the incident wave. In the geometric approximation, the corresponding function g is given by

$$g(x) = \begin{cases} e^{ik_{Ix}x} e^{-i2\pi x/d} = e^{-i\pi x/d} & \text{for } |x| < d/2, \\ 0 & \text{otherwise,} \end{cases} \tag{5.7.22}$$

whose transform is now

$$G(f_x) = d\,\mathrm{sinc}(\pi d f_x + \pi/2), \tag{5.7.23}$$

and the spectrum of diffracted radiation is

$$I\left(\frac{n_d \sin \vartheta_d}{\lambda_o}\right) = d^2 \frac{\sin^2\left(M\dfrac{\pi p d}{\lambda_o}\right)}{\sin^2\left(\dfrac{\pi p d}{\lambda_o}\right)} \mathrm{sinc}^2\left(\frac{\pi}{2} + \frac{\pi}{2}\frac{n_d \sin \vartheta_d}{n_I \sin \vartheta_I}\right). \tag{5.7.24}$$

For simplicity, consider the case of reflection in air. The sinc is 1 for $\vartheta_d = -\vartheta_I$, i.e., for a retro-reflected propagating order, to which corresponds $pd/\lambda_o = m = -1$ according to Eq. (5.7.15), and has its first zeros for

$$\sin \vartheta_d = \begin{cases} \sin \vartheta_I & \text{with } \dfrac{pd}{\lambda_o} = 0, \\ -3\sin \vartheta_I & \text{with } \dfrac{pd}{\lambda_o} = -2. \end{cases} \tag{5.7.25}$$

A grating of this kind had been advocated by Rayleigh without the hope that it could be realized [Rayleigh 1874]. Indeed, it could diffract almost all the radiation in the counter-propagating order $m = -1$, erasing, in particular, the order 0, corresponding to the ordinary reflection, and also the order −2. The doubt is legitimate, because the above calculations are based on a rough approximation that neglects the electromagnetic effects of the grating. However, nature has been benign and, at least for the TM polarization, this behavior is achieved with good efficiency. The gratings that are able to concentrate the incident radiation in a single diffracted order are called *blazed*, and it is not really necessary that they have a triangular

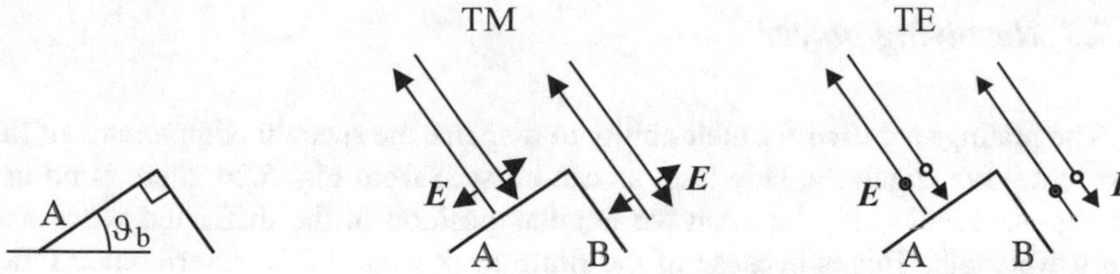

Fig. 5.27 Representation of the wave vectors of the electric field for the incident and diffracted waves along the faces A and B of a grating with triangular cross-section and vertex angle of 90°. ϑ_b is the blazing angle of the grating

cross-section, although this is the most commonly realized. When a lattice allows for only two diffracted orders, with $m = 0$ and -1, it is possible to extinguish the zero order by adjusting the depth h of the grooves, as in the case of sinusoidal gratings in reflection with a h/d modulation of 40% and pitch of 1800 lines/mm [Loewen and Popov 1997]. As you can see from Eq. (5.7.24), compared to Fig. 5.26, a blazed grating has its intensity function of the grooves centered on the desired diffraction order, rather than on the zero order.

Blazed patterns that support more propagating orders are made with a right triangular section, and there is a valid electromagnetic argument that justifies their efficiency for TM waves. Let us consider the case of such a reflective grating whose surface is made of a perfectly conductive metal (Fig. 5.27). In the case of TM waves, the boundary conditions on face B are satisfied, as the stationary wave produced by the two counter-propagating directions has its own electric field orthogonal to the surface. Instead, for the TE waves, the electric field of the stationary wave lapping the "parasite" surface B is parallel to it, and therefore the boundary conditions cannot be satisfied for this polarization. On the "useful" surface A, the boundary conditions are satisfied in both cases, assuming that the amplitudes of the incident and reflected waves are opposite in sign and equal in module.

Regarding the dielectric gratings in transmission, the situation is less favorable, because the boundary conditions cannot be met on the parasite face. One remedy would be to cover it with a very thin conductive metal layer on which the same arguments made above for TM waves can be applied, but it is a road that is not practicable [Loewen and Popov 1997].

One very important concept is the one derived from the principle of reciprocity. As we have seen by studying the reflection from the dielectric surfaces (v. §1.7.2), and, in general, with interfaces and materials *devoid of absorption*, due to the principle of conservation of energy and the time reversal symmetry of Maxwell's equations, the reflection coefficient, aside from a phase shift, is the same by virtue of exchanging the incident and reflected waves. The same arguments can also be applied to the grating, with the precaution to consider the relationship between the powers of the waves, since the incident and reflection beams generally have different widths. Thus, *in the absence of absorption*, reflectivity and transmissivity, in other words, the coupling efficiency, *remain the same by exchanging reflected and diffracted waves*.

5.7.3 Resolving power

The gratings are used for their ability to disperse the spectral components of the incident wave angularly. However, as can be seen from Fig. 5.26, there is no unambiguous relationship between the angular position in the diffracted wave and the wavelength. This is because of the finite peak width of the interference function with which the radiation spectrum is convoluted in the angular dispersion process. The ability of a grating to separate neighboring spectral lines is called resolution or *resolving power*. It is conventionally estimated using the Rayleigh criterion, based on the distance between the position of a major peak and that of the first zero adjacent to the maximum, as shown in the insert of Fig. 5.26 (c). This distance, $\Delta\lambda_o$, is equal to the fraction $FSR_{\lambda o}/M$, whereby, from Eq. (5.7.6),

$$\mathcal{R} = \frac{\lambda_o}{\Delta\lambda_o} = |m| M = \frac{Md}{\lambda_o} p = \frac{W}{\lambda_o} |n_d \sin\vartheta_d - n_I \sin\vartheta_I| , \qquad (5.7.26)$$

where $W = Md$ is the width of the grating surface illuminated by the incident wave, as, on the other hand, we expect, since the width of Fraunhofer's diffraction peaks is inversely proportional to the width of the openings. So, it is not important how small the pitch of the gratins is, if not for its FSR; rather, what counts is its enlightened width. Incidence and diffraction angles also occur in the calculation, and the maximum resolution is when the diffracted wave, for the propagation order considered, proceeds in the opposite direction to the incident wave. In air, for grazing incidence, the angular factor is, at most, 2, and in these conditions, we have

$$\mathcal{R}_{max} = \frac{2W}{\lambda_o} . \qquad (5.7.27)$$

The resolution one gets can be considerable; for example, at W=20 cm and λ_o = 500 nm, we have $\mathcal{R}_{max}$ = 800000. By comparison, we can consider the resolution obtained with a SF11 flint glass prism with $dn/d\lambda_o \approx 0.16\ \mu m^{-1}$ for visible radiation. From Eq. (2.10.19), with a base of 20 cm, the prism resolution is 32000.

Several factors conspire to reduce the achievable resolution. The regularity of the grating pitch must be maintained within $\lambda/100$ over the entire surface, which, in turn, must deviate from a plane or sphere for less than $\lambda/4$. This also applies to the quality of the incident wavefront. Periodic variations of the grating pitch caused the appearance of "ghost" lines in the spectrum, which characterized the first grating built in the past.

The resolution can be increased by using double-grating spectrometers in cascade, but what is especially improved in this way is the extinction ratio between the secondary peaks and the main ones.

5.7.4 Types of grating

The variety of gratings is so wide that it is not easy to classify them. A sensible way is to subdivide them according to their constitutive parameters and to the technique used to realize them, accepting them almost individually for what they are. A more ambiguous way would be to distinguish them by their use.

Slit gratings or *Ronchi rulings*, are made by depositing thin metal strips on a glass substrate, or are produced photo-lithographically. They work by spatially modulating the reflected or transmitted intensity with a relatively small spatial frequency on the order of 100 lines/mm or less. They were introduced by Vasco Ronchi as a device for assessing the quality of optically machined surfaces and have various other applications, for example, the calibration of microscope magnification. As we have seen before, symmetric gratings of this type, with an opening equal to half the pitch, have only odd orders in addition to the zero order. In particular, in transmission, 11% of power is deflected on each of the orders 1 and –1, and much less on the higher orders.

To observe diffraction effects, it is not essential that the pitch be very dense. Even a net can serve this purpose: try to observe a distant light with good binoculars through the mosquito net of your window. For visible radiation, the scalar theory can be sufficiently accurate, but, for the infrared and microwave radiation, the difference in the behavior between TE and TM waves is such that the metal strips serve as polarizers.

The *phase gratings* do not have a corrugation of the surface, but a modulation of the refractive index of a more or less thin layer of dielectric material. Examples of this type are the holograms, in which the recording is made in the volume of the photosensitive material, and the acousto-optical devices, in which the density of the material is modulated by an acoustic traveling wave. They are also used in integrated optics and fiber optics.

The most common category is that of *corrugated gratings*. Modern gratings of this type have up to several thousand lines per mm with profiles of various shapes, sinusoidal, trapezoidal, triangular - and various depths. The preferred form in order to concentrate the diffracted radiation in one order, for which the grating is said to be blazed, is triangular, and the best replicas of triangular section gratings are those of odd generation, where the convex edge is well defined, while the hollowed one is quite rounded. The triangular gratings are characterized by the inclination ϑ_b of their face A (Fig. 5.27).

The gratings with a right triangular cross-section of the thinnest pace, used in the first diffraction orders, are called *echelettes*, French for "small stairs". Larger pitch gratings, used at dozens or hundreds of diffraction orders, are called *echelles*. They work at high angles of incidence and diffraction, are less sensitive to polarization, and cover a wider spectral range with good efficiency. Finally, macroscopic gratings used with thousands of orders, built by Michelson joining dozens of glass sheets, now largely disused, are called *echelons*. Rectangular section gratings are called *lamellar* and are useful as beam dividers.

The *holographic gratings* owe their name to the fact that they are produced with an interferential technique and not with ruling machines. Their most common profile is sinusoidal, which can only be made asymmetrical and at least partially blazed through complicated techniques. Their essential advantage is their ability to correct geometric aberrations of spheric and even aspheric concave gratings

The *concave gratings* were introduced by Rowland in order to simplify spectroscopic instruments. Indeed, they can also perform the function of producing an image of the input slit, for each wavelength, on a surface on which the spectrum is generated. This avoids the use of concave collimation mirrors. Their use is now largely limited to ultraviolet, for which there are no efficient mirroring materials.

Other gratings are those used on planar waveguides, especially in the semiconductor laser as distributed mirrors to retro-reflect radiation within the active medium, For this purpose, the laser wave-guide has a corrugated surface, and there are two types of them. One is called the *distributed Bragg reflector* (DBR) and has two corrugations out of the active zone forming the mirrors of the laser cavity. The other one is called *distributed feedback* (DFB) and has a single corrugation extended over the entire active zone. This allows for a monochromatic laser emission that is more easily controllable.

5.7.5 Production

The gratings are still generally produced by ruling machines equipped with a carriage, moved by a special micrometric screw, on which it is fixed the plate to be engraved, and by an arm, to which a diamond tip is attached, which moves in an orthogonal direction so as to draw the grooves. A bending spring, on which the arm is pinned, allows for lowering and raising the tip at the beginning and end of each row. Currently, the movement of the carriage is controlled interferometrically, to eliminate the fluctuations of the pace caused by irregularities of the micrometer screw. The best material on which to draw a lattice is an aluminum film on a low-thermal expansion glass substrate polished with good optical quality. The diamond tip does not engrave it by cutting, but rather by plastic deformation.

The accuracy needed to position the tip is on the order of nanometers and the tracing of a large size grating (some ten cm per side) takes several weeks. A grating produced in this way has a very high cost, so these gratings, called *masters* are then used to produce various replicas through contact. These latter are obtained by depositing, on the master, a thin layer of a material suitable for facilitating the separation from the replica, and subsequently evaporating on it, under vacuum, a metallic film, usually still aluminum, onto which another substrate is then firmly bonded with a low viscosity resin that is slowly left to cure. Finally, the replica is cautiously separated from the master. Each replica, in turn, constitutes a *submaster*, and this operation is repeated for several generations by exponentially producing up to tens of thousands of good replicas from a single original. A typical defect in these gratings is the presence of scattering due to noise in the positioning of the diamond tip and ghost images due to a periodic error of the screw pitch. These fluctuations have been drastically reduced through the use of interferential control methods and are no longer a serious problem, but they have not been completely eliminated.

With the advent of laser and special photoresists, *holographic gratings* obtained by interference are now also produced. This technique allows for obtaining gratings in which the separation between the lines and their curvature varies according to a predetermined rule. This is of particular importance for the production of concave gratings, sometimes with a toroidal shape. The interference is generated here by the radiation of two suitable point sources, automatically producing a grating in which geometric aberrations are largely eliminated. The holographic gratings are, in principle, easier to achieve and are devoid of the typical errors of those obtained mechanically. They produce a spectrum far more free of the spurious effects of the ruled gratings, despite the graininess that might result from the processes of exposure and development of the photoresist. However, in order to achieve the same resolution as the others, it is necessary that high-diameter, *diffraction-free* optics be placed in the path of the interfering laser beams. In addition, the spurious radiation diffused by the optical or mechanical elements present in the field of vision of the grating to be recorded, must be carefully screened, in particular, that diffused from the edge of the optics or even from their surface. Significantly, while it is possible to produce excellent gratings with a sinusoidal relief, it is not possible to produce blazed gratings, except with expensive methods and results that are of poor efficiency compared to those of the ruled gratings. Finally, they are less uniform in the modulation depth due to the inherent Gaussian profile of the laser beams.

Lastly, for special applications, special gratings are produced using lithographic techniques similar to those used for electronic devices, using masks, ultraviolet light sources, electron beams, and ions.

5.7.6 Tools of analysis and spectral tuning

The gratings find their natural location in various optical instruments, both for spectral analysis and as a frequency selection element in laser systems. Among the spectral analysis tools, the most important are *spectrographs* and *monochromators*. They are essentially distinguished by the way in which the spectrum is recorded, and they are sometimes incorporated into a single instrument in which the wavelength is selected by the rotation of a special mirror. The *spectrographs* produce an extended spectrum that was typically recorded on a photographic plate. With the advent of CCD sensors, they are back in vogue, in particular, when using echelles and removing the overlap of the orders by means of the dispersion of a prism or a secondary grating in a direction orthogonal to that of the main grating.

In this way, on the surface of the CCD, the spectrum is divided into several strips that allow for the recording of a wide spectral band without having to rotate the grating, and thus without moving parts.

The *monochromators* instead use an exit slit to select a particular wavelength and a detector, such as a phototube, to measure the power. The registration of the spectrum is achieved by rotating the grating, for which the wavelength transmitted by the slit varies in a controlled manner. Some instruments are built with two or three monochromators in series. In this way, the background resulting from secondary peaks can be considerably reduced. They are useful for Raman spectroscopy that requires high resolutions and the ability to extinguish the nearby strong laser line used for excitation.

The *spectrophotometers* are instruments that use a grating in the same way as the monochromators. However, they use one or more broadband sources, such as lamps, and produce two output beams of collimated monochromatic radiation. On one of the two beams, a sample substance is placed to be analyzed, generally a liquid, but it can also be gaseous or solid, with the foresight that does not deflect the beam. The detection system determines the ratio between the power of the beam transmitted by the sample and that of the other beam. This determines the sample absorption spectrum and can identify its composition. With suitable modifications, one can make measurements in reflection. A particular version of the spectrophotometer is the colorimeter, which typically works in reflection and determines the color coordinates of a sample.

There are many different designs for the construction of a spectrograph or a monochromator; some use plane gratings and others concave gratings. The related problems are the deletion of geometric aberrations, in particular, the coma and the flatness of the surface on which the spectrum is formed.

Among the systems utilizing flat gratings, the most accomplished is the one called the Czerny-Turner design, despite Abney having proposed the same configuration 50 years earlier [Czerny and Turner 1930; de Wiveleslie Abney 1880]. It uses two concave spherical mirrors, the first to collimate the radiation from the entrance slit to the grating and the second to focus the diffracted radiation on the image plane where the exit slit or an image detector, CCD or photographic plate is located. The grating is mounted so that it can rotate around an axis parallel to the grooves and lying in the center of its surface. The movement is carefully controlled to determine the wavelength transmitted from the exit slit. This scheme has the advantage of erasing the coma with an appropriate choice of the angles shown in Fig. 5.28 and the curvature radius of the mirrors [Shafer et al 1964]. This deletion there is for a single wavelength, but, in practice, this is not a problem. In addition, the spherical aberration and astigmatism, which are instead added in the reflection from the two mirrors, are fairly limited. The advantage of this system is the relatively low cost of the three main elements. The spherical mirrors could be replaced with off-axis paraboloids, but, although the astigmatism and the spherical aberration would be canceled, the coma produced by the two mirrors is added instead, unless a different, non-planar scheme is used.

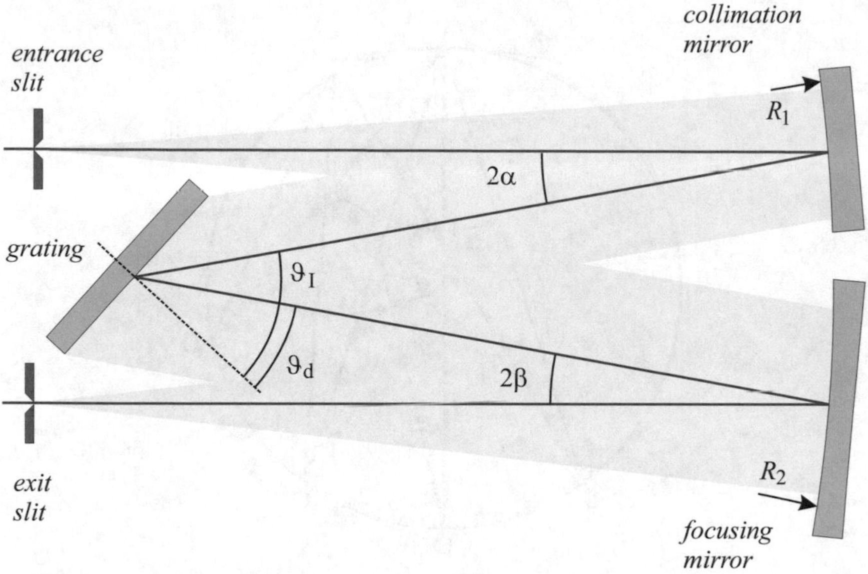

Fig. 5.28 Optical diagram of a Czerny-Turner monochromator

A variant of this type of spectrometer is the Ebert-Fastie design. It uses a single spherical mirror, and is therefore more stable. This solution has been used in various space missions with interplanetary spacecraft [Fastie 1991]. In particular, Fastie noted that the resolution of a spectrometer improves if the slits are suitably curved and, in the case of the Ebert design, with the center of curvature on the mirror axis.

Concave grating spectrometers were the most used after their invention by Rowland in 1882. They have the advantage that, with a single optical element, the concave grating, a good horizontal focusing of the image of the input slit can be obtained, taking care to position both the slit and the detectors along the Rowland circle; it has a radius of half that of the grating and is tangent to it. The proof of this property is, for example, shown by Born and Wolf (1980) .

Consider a ray that, from the slit S placed on the Rowland circle R, reaches the vertex V of the grating G (Fig. 5.29). This radius forms an angle α with the normal VC to the grating, where C is its center of curvature. This ray is reflected with the same angle α towards the point O. Now, consider a ray coming from S to another point W on the grating. We can roughly assume that W is on the circle R, since the opening of a concave grating is normally much smaller than its radius of curvature. Then, for the geometric theorem of the angles subtended by an arc of circumference, the angle $S\hat{V}C = \alpha$ is (almost) equal to the angle $S\hat{W}C$. The same can be said for the angles $O\hat{V}C = \alpha$ and $O\hat{W}C$, so the ray reflected in W also reaches O. Finally, we consider a ray that, from V, instead of being reflected, is diffracted at the angle $\alpha + \beta$ towards D. For the same wavelength, there is an equal diffraction angle in W. Therefore, since, as before, $C\hat{V}D = \alpha + \beta$ and $C\hat{W}D$

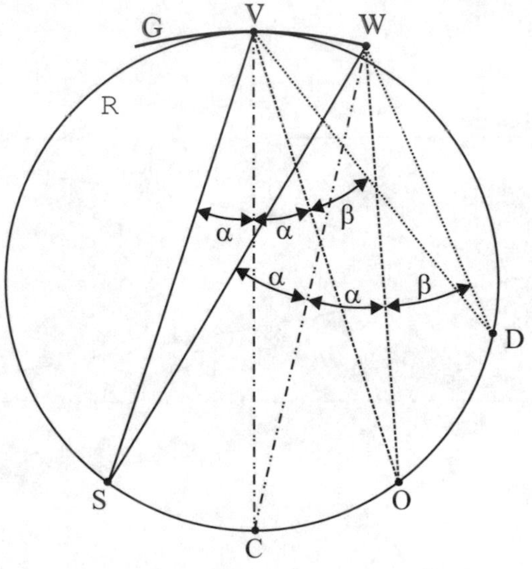

Fig. 5.29 Studying the properties of the Rowland circle

are approximately equal, even the ray diffracted in W comes to D. In conclusion, for a given order and wavelength, all of the radiation emerging from S is diffracted from the grating to a point in the Rowland circle. This focusing occurs only in directions parallel to the plane of the circle.

A discussion of the aberrations of this system is reported by Loewen and Popov (1997). Generally, this kind of spectrometer suffers from aberrations significantly greater than those obtained with plane gratings, resulting in a loss of resolution and photometric efficiency. An improvement is achieved with toroidal surfaces, with which one can reduce astigmatism, but at higher expense. Their use is now restricted to short wavelengths, below 110 nm, where there are no reflective materials with good efficiency and where, therefore, the use of mirrors constitutes a penalty, or to spectrographs with low resolution. A new impetus for the use of concave gratings has come from the development of holographic techniques described earlier to produce the corrugation of the surface. In this way, the aberrations can be greatly reduced and the spectrum can be obtained on a flat surface, more suitable for the use of CCD detectors. However, the efficiency of a holographic grating is generally lower than that produced mechanically, as it is not possible to produce a good blazing of the grooves.

Small plane gratings are applied as tuning elements of semiconductor lasers. There are two ways of using them. One is the *Littrow's mounting* (Fig. 5.30a), wherein the radiation output from the laser is retro-reflected by the grating with diffraction angle zero for the desired wavelength. The grating constitutes one of the mirrors of the laser cavity and the tuning is performed by rotating it. To increase the spectral selection, anamorphic prisms can be used to widen the beam

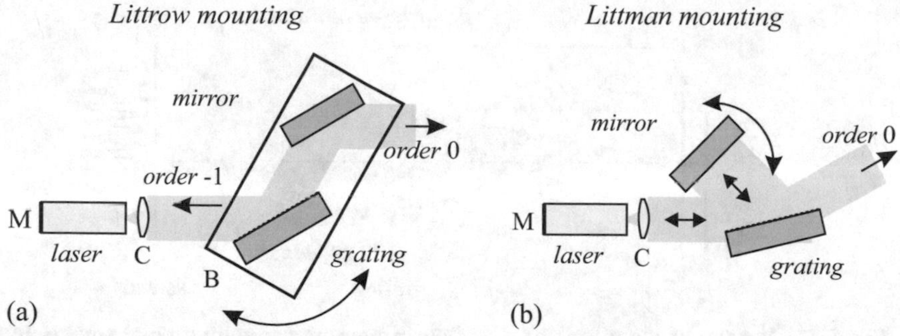

Fig. 5.30 Extended cavity laser systems with grating tuning. *M* is the rear laser mirror, *C* is the collimator, *B* is the base on which the lattice and mirror are joined

and increase the number of illuminated grooves of the reticle. The pattern is also used as a laser *output coupler*, using the zero-order diffracted beam. This bundle is further reflected by an ordinary mirror that is joint with the grating. In this way, when the grating is rotated for the tuning, there is no deviation of the output beam from the laser system.

An alternative is the Littman's mounting (Fig. 5.30b), in which the laser radiation reaches the grating at grazing incidence ($\vartheta_I \approx 89°$); the diffracted order is then reflected back to the reticle by a mirror. Tuning is achieved by rotating the mirror, which does not affect the output beam, obtained, as before, by reflection in the zero order. The signal returning toward the laser is doubly diffracted and the selection is good, as the surface illuminated on the grating at grazing incidence is much larger than the size of the laser beam. However, under these conditions, a grating has relatively low efficiency, 10-20% for single passage, and therefore the feedback on the laser is much weaker.

5.7.7 *Ronchi's ruling*

Let us now consider the case of a slit grating in transmission, previously mentioned, deposited on a thin substrate and illuminated by a point source S_o (Fig. 5.31). Observing this source through the grating, we can see, in addition to the original source corresponding to the diffraction zero order, a set of replicas corresponding to the other orders, arranged on a line parallel to the surface of the grating and perpendicular to the slits. We can indeed represent the various plane waves of the plane wave spectrum of the source with rays whose direction is given by the corresponding wave vectors and whose source is the source itself. From Eqs. (5.7.2), for each order *m* with a fixed wavelength, we have that the emerging rays all stigmatically appear to have originated from the same point S_m. The distance *d* between two successive points, S_m and S_{m+1}, is

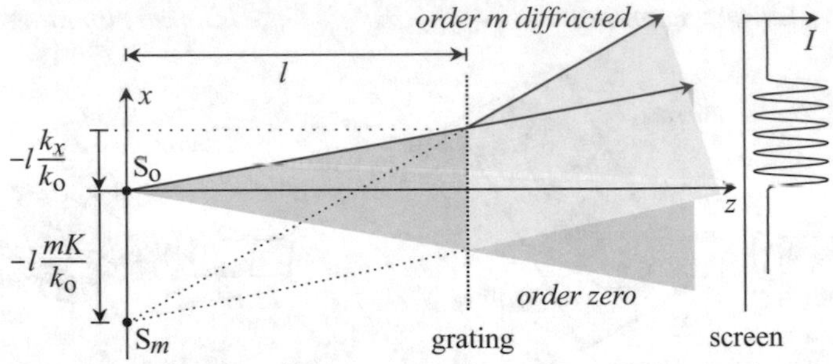

Fig. 5.31 Diagram of the effects of a grating in transmission with a point source. In the overlap zone of two orders, we have interference fringes observable in the intensity I on a screen

$$d = l\frac{K}{k_o} = l\lambda_o\frac{K}{2\pi},$$

where l is the distance between the source S_o and the grating, K is the vector module of the grating and λ_o is the wavelength considered. Since all emerging waves are spherical, in the overlapping zone between two orders, such as zero order and order 1, there are fringes that are (almost) straight, like those discussed in §3.1.3, the spacing of which is proportional to the ratio λ_o/d. However, being that, in this case, d is proportional to λ_o, such fringes are achromatic! The separation between the fringes is also inversely proportional to the distance l for which these widen up to give a uniform field when $l = 0$.

Thanks to these properties, especially the stigmatic property of the diffracted waves, in 1922, Vasco Ronchi proposed the use of low frequency gratings (10 to 200 lines per mm) to analyze, in a very simple way, the quality of an optical system, in particular, the objectives. Indeed, the visible fringes between two diffracted orders are straight and parallel to the lines of the grating only if the wave that illuminates it is spherical.

Laying the grating behind the objective on which a homocentric wave is coming, the fringes that appear on a screen assume shapes that are typical of the kind of aberration that the objective introduces to the beam, which also vary by moving the grating with respect to the focal position of the converging wave. For example, spherical aberration typically produces fringes that vary from being bent as a spindle, wider in the center of the field and thicker towards the edges to the opposite situation, while astigmatism, in which one tilts the lens, still gives the straight fringes, which, however, are inclined with respect to the direction of the lines of the grating. In a suggestive way, Ronchi writes that, in this case, the fringes perform a "somersault" by moving the reticle from before to after the focal position [Ronchi 1963].

Bibliographical references

Abbe E.K., *Beitrage zür theorie des mikroskops und der mikroskopischen wahrnehmung*, Archiv. Microskopische Anat. **9**, 413-468 (1873).

Abramowitz M. and Stegun I.A., *Handbook of Mathematical Functions*, Dover Publications Inc., New York (1972).

Arecchi F.T., *Measurement of the Statistical Distribution of Gaussian and Laser Sources*, Phys. Rev. Lett. **15**, 912–916 (1965). *An introduction to quantum optics*, in "Interaction of Radiation with Condensed Matter", Vol. I, International Atomic Energy Agency, Vienna (1977).

Arecchi F.T., Courtens E., Gilmore R., Thomas H., *Atomic Coherent States in Quantum Optics*, Phys. Rev. A **6**, 2211–2237 (1972).

Arecchi F.T., Schultz E.O., Du Bois, ed., *Laser Handbook, Part A*, North Holland (1972).

Barrett H.H. and Myers K.J., *Foundations of Image Science*, John Wiley & Sons Inc. ed., New York (2004).

Born M. and Wolf E., *Principles of Optics*, Pergamon Press, Paris (1980).

Cicogna G., *Appunti di Metodi Matematici della Fisica*, Università degli Studi di Pisa (1969).

Cooley J.W. and Tukey J.W., *An algorithm for the machine computation of Fourier series*, Maths. of Comput. **19**, 297-301 (1965).

Czerny M. and Turner F., *Über den astigmatismus by spiegelspectrometern*, Z, Physik **61**, 792-797 (1930).

den Dekker A.J. and van den Bos A., *Resolution: a survey*, J. Opt. Soc. Am. **14**, 547-557 (1997).

de Wiveleslie Abney W., *On the photographic method of mapping the long wavelength end of the spectrum*, Phil. Trans. Royal Soc. **171**, II, 653-667 (1880).

Duffieux P.M., *L'Intégrale de Fourier et ses applications à l'optique*, Faculté des Sciences, Besançon, 1946. Reprinted by Masson et Cie, Paris (1970) and by Wiley, New York (1983).

Elias P., *Optics and communication theory*, J. Opt. Soc. Am. **43**, 229-232 (1953).

Fastie W., *Ebert spectrometer reflections*, Physics Today **44**, 37-43 (1991).

Fellgett P., *Concerning photographic grain, signal-to-noise ratio, and information*, J. Opt. Soc. Am. **43**, 271-281 (1953).

Glauber R.J., *Coherent and incoherent states of the radiation field*, Phys. Rev. **131**, 2766-2788 (1963). In "Quantum Optics and Electronics", Les Houches Summer School of Theoretical Physics, University of Grenoble, C. DeWitt, A. Blandin and C. Cohen-Tannoudji, Eds., Gordon and Breach, New York (1965).

Goldstein D.J., *Understanding the light microscope*, Academic Press, London (1999).

Goodman J.W., *Introduction to Fourier Optics*, II ed., McGraw-Hill, San Francisco (1996).

Gu M., *Advanced Optical Imaging Theory*, Springer-Verlag, Berlin (2000).

Hanbury Brown R. and Twiss R.Q., *A new type of interferometer for use in radio astronomy*, Phil. Mag. **45**, 663-682 (1954). *Interferometry of the Intensity Fluctuations in Light. I. Basic Theory: The Correlation between Photons in Coherent Beams of Radiation*, Proc. R. Soc. **A242**, 300-324 (1957). *Interferometry of the Intensity Fluctuations in Light II. An Experimental Test of the Theory for Partially Coherent Light*, Proc. R. Soc. **A243**, 291-319 (1958).

Hopkins H.H., *On the Diffraction Theory of Optical Images*, Proc. R. Soc. **A217**, 408-432 (1953). *The frequency response of optical systems*, Proc. Phys. Soc. (London) Ser. B **69**, 562-576 (1956).

Hopkins H.H. and Baraham P.M., *The Influence of the Condenser on Microscopic Resolution*, Proc. Phys. Soc. B **63**, 737-744 (1950).

Kasdin N.J., Vanderbei R.J., Littman M.G., and Spergel D.N., *Optimal one-dimensional apodizations and shaped pupils for planet finding coronagraphy*, Appl. Optics **44**, 1117-1128 (2005).

Köhler A., *Ein neues Beleuchtungsverfahren für mikrophotographische Zwecke*, Zeits. f. wiss. Mikroskopie **10**, 433-440 (1893); *Beleuchtungsapparat für gleichmässige Beleuchtung mikroskopischer Objecte mit beliebigem, einfarbigem Licht*, Zeits. f. wiss. Mikroskopie **16**, 1-29 (1899).

Loewen E.G. and Popov E., *Diffraction Gratings and Applications*, Marcel Dekker, Inc., New York (1997).

Mandel L. and Wolf E., *Coherence properties of optical fields*, Rev. Mod. Phys. **37**, 231-287 (1965).

Mansuripur M., *Certain computational aspects of vector diffraction problems*, J. Opt. Soc. Am. A **6**, 786-805 (1989). *Classical Optics and its Applications*, Cambridge Univ. Press, Cambridge (2002).

Maréchal A. and Croce P., *Un filtre de frequencies spatiales pour l'amelioration du contraste des images optiques*, C.R. Acad. Sci. **127**, 607-609 (1953).

O'Neill E.L., *Spatial filtering in optics*, IRE Trans. Inf. Theory **IT-2**, 56-65 (1956).

Oran Brigham E., *The fast Fourier Transform*, Prentice-Hall, Inc., Englewood Cliffs, N.J. (1974).

Papoulis A., *System and Transforms with Applications in Optics*, McGraw-Hill, New York (1968).

Porter A.B., *On the diffraction theory of microscope vision*, Phil. Mag. **11**, 154-166 (1906).

Rayleigh (Lord), *On the manufacture and theory of diffraction gratings*, Phil. Mag. Series 4, **47**, 193-205 (1874). *Investigation in optics, with special reference to the spectroscope*, Philos. Mag. **8**, 261-274, 403-411, 477-486 (1879), Philos. Mag. **9**, 40-55 (1880).

Reynolds G.O., DeVelis J.B., Parrent G.B. Jr, Thompson B.J., *The new Physical Optics Notebook: Tutorial in Fourier Optics*, SPIE Optical Engineering Press, Bellingam, and American Institute of Physics, New York (1989).

Ronchi V., at the entry *aberrazione*, Enciclopedia della Scienza e della Tecnica, Mondadori ed., Milano (1963).

Schuster A., *The periodogram and its optical analogy*, Proc. Roy. Soc. **77**, 136-140 (1906).

Shade O.H., *Electro-optical characteristics of television systems*, R. C. A. Rev. **9**, 5-37, 245, 490, 653 (1948).

Shafer A., Megil L., and Droppelman L., *Optimization of Czerny-Turner spectrometers*, J. Opt. Soc. Am. **54**, 879-888 (1964).

Shamir J., *Optical Systems and Processes*, SPIE Optical Engineering Press, Bellingam (1999).

Shannon C.E., *Communication in presence of Noise*, Proc. IRE **37**, 10-21 (1949).

Sivoukhine D., *Optique*, MIR, Mosca (1984).

Sparrow C.M., *On spectroscopic resolving power*, Astrophys. J. **44**, 76-86 (1916).

Stamnes J.J., *Waves in Focal Regions*, A. Hilger, Bristol (1986).

Talbot H. F., *Facts relating to optical science. No. IV*, Philos. Mag. **9**, 401-407 (1836).

Tervo J., Setälä T., Friberg A.T., *Theory of partially coherent electromagnetic fields in space-frequency domain*, J. Opt. Soc. Am. A 21, 2205-2215 (2004).

Thompson B.J. and Wolf E., *Two-beam interference with partially coherent light*, J. Opt. Soc. Am. **47**, 895-902 (1957).

Toraldo di Francia G., *The capacity of optical channels in the presence of noise*, Opt. Acta **2**, 5-8 (1955).

Traub W.A. and Vanderbei R.J., *Two-mirror apodization for high-contrast imaging*, The Astrophysical J. **599**, 695-701 (2003).

van Cittert P.H., *Die Wahrscheinliche Schwingungsverteilung in Einer von Einer Lichtquelle Direkt Oder Mittels Einer Linse Beleuchteten Ebene*, Physica **1**, 201-210 (1934). *Kohaerenz-probleme*, Physica **6**, 1129-1138 (1939).

VanderLugt A., *Signal detection by complex spatial filtering*, IEEE Trans. Inf. Theory, **IT-10**, 139-145 (1964). *Optical Signal Processing*, John Wiley & Sons Inc. ed., New York (1992).

Van Vlek J.H. and Middleton D., *A theoretical comparison of the visual, aural, and meter reception of pulsed signals in the presence of noise*, J. Appl. Phys. **17**, 940-971 (1946).

Walther A., *Radiometry and coherence*, J. Opt. Soc. Am. **58**, 1256-1259 (1968).

Wiener N., *Generalized harmonic analysis*, Acta Math. **55**, 117-258 (1930). *Optics and the theory of stochastic processes*, J. Opt. Soc. Am. **43**, 225-228 (1953).

Williams C.S., Becklund O.A., *Introduction to the Optical Transfer Function*, John Wiley & Sons ed., New York (1989).

Wilson R.G., *Fourier Series and Optical Transform Techniques in Contemporary Optics. An introduction*, John Wiley & Sons ed., New York (1995).

Weyl H., *Ausbrietung elektromagnetischer Wellen über einem ebenen Leiter*, Ann.Phys. Lpz. **60**, 481-500 (1919).

Whittaker E.T., *On the functions which are represented by the expansions of the interpolation theory*, Proc. Roy. Soc. Edinburgh, Sec. A **35**, 181-194 (1915).

Wolf E., *A Macroscopic Theory of Interference and Diffraction of Light from Finite Sources. II. Fields with a Spectral Range of Arbitrary Width*, Proc. Roy. Soc. **A230**, 246-265 (1955). *Intensity fluctuations in stationary optical fields*, Phil. Mag. **2**, 351-354 (1957).

Zadeh L.A. and Ragazzini J.R., *Optimal filters for the detection of signal in noise*, Proc. IRE **40**, 1223-1231 (1952).

Yu Francis T.S., *Entropy and Information Optics*, Marcel Dekker Inc. ed., New York (2000).

Zernike F., *Das phasenkontrastverfahren bei der mikroskopischen beobachtung*, Z. Thec. Phys. **16**, 454-457 (1935). *The concept of degree of coherence and its application to optical problems*, Physica **5**, 785-795 (1938).

Part V
Propagation

Historical notes: From Kirchhoff to Einstein

Relativity

In the development of the theory of electromagnetism, Maxwell did not care to examine how the equations of the fields are transformed from one reference system to another. On the other hand, the equations admitted as solutions were also waves consisting in the propagation of a perturbation in a medium pervading all space, with a speed that was implicit in the same equations [Bergia 1998]. This medium, with mechanical characteristics that were quite strange, was called aether, and the equations were meant to refer to a system in which the aether was at rest. It had long been known that Earth is not immobile in the center of the Universe. In at least one year, its speed υ compared to this privileged aether reference would have to change by tens of km/sec, with consequent dependence on the propagation speed of radiation from the direction of its motion. There were, therefore, various hypotheses between the two extremes: one, by Stokes, maintained that the Earth dragged the aether along with it, the other that the Earth left it undisturbed. Hertz was the first to seek a generalization of Maxwell's equations, based on Stokes' hypothesis, but this was contradicted by the work of Fizeau, who had instead shown a partial dragging by media in motion. Thus, it seemed reasonable to imagine that a substantial «wind of aether» crossed the labs.

Maxwell, in 1878, had found that the effects provided by the change of the reference system in the measurement of travel times, back and forth on a fixed distance, had to be on the order of $(\upsilon/c)^2$, so they might be too small to reveal. Albert A. Michelson (1852–1931) recognized that an interferometric method, by using what is now called a "Michelson interferometer", could be sufficiently accurate to reveal these effects. After an initial experiment in Germany in 1881 with an uncertain outcome, he returned to the United States and, in 1887, performed a second experiment with Edward W. Morley (1838–1923). This time, the interferometer was constructed on a square concrete platform floating on mercury so that it could be rotated. A source was placed on a corner of the platform, creating a beam of collimated light that a semi-reflective mirror divided into two other beams. These were reflected by various mirrors and would travel several times along the two diagonals of the platform. In the recombination of the two beams, the fringes were observed with a small telescope. Even with this new apparatus, there was no appreciable effect attributable to the wind of aether, though the sensitivity was more than enough by a considerable degree.

To explain the negative result of the experiment, Hendrik Antoon Lorentz (1853–1928), who had followed Michelson's work with interest, and, independently, George Francis Fitzgerald (1851 - 1901) suggested that all moving material objects undergo a contraction in the direction of motion of a factor such as to make the travel times equal in the two arms of the interferometer. After all, all substance are governed by electromagnetic forces, which may be affected by the motion as the same radiation, varying the interatomic distances. Lorentz, in particular, considered electrons as charge carriers, and hence as mediators between matter and aether. Pieter Zeeman's (1865–11943) discovery, in 1896, of the effect of a magnetic field on spectral lines was explained by Lorentz assuming that electrons oscillate around equilibrium positions emitting radiation, so a magnetic field, altering their motion, would change the frequency of the radiation. In particular, Lorentz and Max Abraham (1875–1922) supposed that the mass of electrons could be explained, in whole or in part, by the radiation that is generated when they are accelerated. Lorentz was therefore committed to the task of reconciling Maxwell's equations with the situation in which matter is in motion. On the other hand, these equations were confirmed by experiments in the terrestrial reference, and he thus realized that it was necessary to introduce a coordinate transformation that would leave the equations unchanged between different inertial references. This could be done if, besides a contraction in the direction of the relative motion, say, along the x-axis, we also added a transformation of the temporal coordinate t, introducing a "local time" $t' = t - \upsilon x/c^2$, as he called it, without, however, giving it a physical meaning [Bergia 1998]. In this regard, there was the intervention of J. Henri Poincaré (1854–1912), who considered the various ad hoc assumptions unsatisfactory to account for the failure to detect the motion of the Earth through the aether. In response to Poincaré, in 1904, Lorentz published an article in which he finally presented the definitive form of what are now called *Lorentz's transformations*, which make the Maxwell equations invariant between inertial systems, of which the Galileo transformations for the mechanics are only an approximation.

The failure to reveal the Earth's motion convinced Poincaré that this was, in fact, derived from a general «principe de la relativité». In that same year, after reading the work of Lorentz, he stated at a conference that physical laws were to be the same in every inertial reference «de sorte que nous n'avons, et ne pouvons avoir aucun moyen de discerner si nous sommes, oui ou non, emportés dans un pareil mouvement». At that time, he had already given a justification for Lorentz's "local time" with a reasoning on the synchronization of clocks that was later taken up by Einstein; on the other hand, he still believed in the existence of a real time, «temps vrai», as he called it, and considered the speed of light anchored to the aether. But he was also the first to understand that the Lorentz transformations were a group. In 1905, he wrote an article, subsequently published in 1906, introducing the concept of space-time, in which the fourth time coordinate was ict, demonstrating that the Lorentz transformations in this space were rotations that left unchanged the quantity $c^2(\Delta t)^2 - (\Delta x)^2 - (\Delta y)^2 - (\Delta z)^2$ [Bergia 1998].

However, the decisive step was taken by Albert Einstein (1879–1955), as Polvani (1934) notes: «The solution of the problems of optical systems in motion by Einstein was sought and found not with particular hypotheses on ether ... or with particular hypotheses on the bodies ... , but with the revision of the concepts of time and space». In the last of his articles from his "annus mirabilis", 1905, with the title *On the Electrodynamics of Moving Bodies* [Einstein 1905a], Einstein resumed the principle of relativity. It states that all physical laws must have the same form in the inertial systems, and added the principle of *constancy of the speed of light* in empty space, independent of the state of motion of the source and of the inertial system chosen. This also involves the fact that this speed remains the same in all directions, as resulted from the Michelson-Morley experiment) [Einstein 1905b]. In this work, Einstein immediately states that these two principles appear irreconcilable, but that they are not such when a thorough analysis of the concepts of space and time is conducted. Operationally, space and time are measurable, the first by means of a rule, that is, a rigid body, and the second by identical clocks; however, clocks placed in different locations must be synchronized, and this can be done by sending a synchronizing light pulse, back and forth between the two places, with identical travel times in both directions. Then, in a given reference system, two events in different places appear simultaneous when the clocks there mark the same position of the hands. Einstein imagines two other clocks in solidarity with a different inertial system, such that each passes in the same space-time position of these same events, and are synchronized in such a passage with the previous clocks. However, he notes that an observer at rest in the second reference, when applying the same synchronization procedure, assuming the principle of constancy of the speed of light to be equally valid for him, would find that these two clocks are not synchronous between them. Einstein explains that the contradiction arises from the tacit classical assumption that «a moving rigid body is geometrically replaceable by *the same* body, when it is *at rest* in a particular location» and from the absolute meaning attributed to simultaneity: these were the causes of the difficulties encountered up to that point.

On the basis of these considerations, therefore, Einstein built a new kinematics, in which the coordinate transformations are those of Lorentz for the space-time and the two principles of relativity and constancy of the speed of light are mutually compatible. He also founded the electromagnetic theory of bodies in motion by discovering the transformation relations of the fields between two inertial systems.

The most striking consequences of restricted relativity are that: it is not generally possible to give a unique chronological order to events occurring in different places; the speed of light in the vacuum is a limit speed; lastly, the concept of aether is superfluous. Opponents challenged Michelson-Morley's experiment and other subsequent astronomical experiments and observations with two objections: the fact that the source was solidly fixed with the interferometer and that, in the media present, the original wave was replaced by another for the Ewald-Olsen extinction theorem. However, the Theory of Relativity is currently confirmed by other experiments with great accuracy [Jackson 1974; Malykin 2010].

Black body radiation

With the experiments of Fizeau and Foucault, the wave theory of light radiation was finally confirmed, and Maxwell's electromagnetic theory give it a solid mathematical foundation in which the electromagnetic field and the matter were described by continuous functions of space and time. Meanwhile, in the nineteenth century, the foundations for a further revolution, that of the Quantum Mechanics of the next century, were laid. Matter was displaying its discrete properties, beginning with its atomic structure, glimpsed at the beginning of the century by the law of multiple proportions between the masses of compounds in chemical reactions, discovered by John Dalton. It appeared to be composed of atomic elements with peculiar chemical properties, and was therefore "quantized" in atoms that, for each element, are all identical, but that are different from element to element. These atoms could also be isolated and purified by electrochemical methods, establishing a relationship with electrical current and charge concepts. Chemists were thus recovering the atomistic concepts of the ancient Greek philosophers Epíkouros, Dēmókritos, and the Latin poet Lucretius in his *De rerum natura*, in a new dress: chemical bonds had to have their origin in electromagnetic forces.

In the nineteenth century, Thermodynamics was also emerging, thanks to the work of Nicolas Léonard Sadi Carnot (1796–1832), who, in 1824, introduced the second principle in a memory on the efficiency of steam engines, and by James Prescott Joule (1818–1889, who enunciated the principle of equivalence between heat and work (first principle). Using atomistic concepts, considerable progress was then made by William Thomson (Lord Kelvin, 1824–1907), who defined the absolute temperature scale; by Rudolf Clausius (1822–1888), who introduced the term *entropy* in 1865; by James Maxwell, who determined the velocity distribution of molecules in a gas; and, especially, by Ludwig Boltzmann (1844–1906). Discretizing the possible energies, Boltzmann was able to connect the entropy S with the "probability" W, that is, the number of dynamic states corresponding to a thermodynamic state, in the famous equation $S = k_B \ln W$, where k_B is the constant that takes his name. Boltzmann also showed that the statistical distribution that maximizes the value of W is that of thermal equilibrium. In particular, for a finite set of N particles, with total energy assigned, this distribution is given by $n_i = (N/Z)\exp(-qE_i)$ on the various energy states with energy E_i, where Z and q are constants that are used to comply with the given constraints of population and energy. In particular, $q = 1/(k_B T)$, where T is the absolute temperature. In the case of a gas, Boltzmann himself found the Maxwellian distribution of velocities on a more general basis.

After the discovery of spectral lines in sunlight and stars by Wollaston, and especially by Fraunhofer, Spectroscopy was born, with which the various chemical elements contained in the substances could be identified. The initiators of this new discipline were Gustav Robert Kirchhoff (1824–1887) and Robert Wilhelm Bunsen (1811–1899) in their collaboration in the years 1850 to 1875, first at the University of Breslau and later at the University of Heidelberg. Kirchhoff should also be remembered for his youthful contributions to the theory of electrical circuits and the development of diffraction theory.

Bunsen was the chemist who developed, with the help of the mechanical Peter Desaga, a special gas burner, which is now called a *Bunsen burner*, capable of producing a very hot flame and little light, mixing air into the gas before it reaches the flame. Bunsen and Kirchhoff set up a prism spectroscope with which they could analyze substances by observing the spectrum of light that they produced when they were subjected to the flame. In this way, they discovered new elements, including cesium, rubidium, thallium, and indium. Comparing the spectrum of solar radiation with that obtained in the laboratory for sodium, they discovered that some lines emitted by this element coincided with some absorption lines in the solar spectrum. Then, they detected the presence on the Sun of various elements that are also present on Earth, opening the possibility of studying the composition of stars by analyzing their spectrum.

In addition to having great importance in astrophysics, this discovery opened up a new line of investigation into the nature of radiation. Indeed, on the basis of thermodynamic considerations, in 1860, Kirchhoff concluded that, if the emission and absorption of a body are of a thermal nature, the ratio between the emissive power (the irradiated energy per unit of time, per unit of surface, within a wavelength range $d\lambda$) and the absorptive power (ratio between absorbed energy and incident energy for the same wavelength λ) of a substance should be a universal function of the temperature and wavelength, completely independent of the properties of any material body [Polvani 1934]. In particular, if the absorbing power is 1, i.e., when the absorption is total, this function corresponds to the emitting power of the black body. Since there are no blacks materials in this sense, Kirchhoff himself proposed using a small opening on a cavity inside of an obscure opaque material at a uniform temperature as a black body. He also coined the term *black-body radiation* for the radiation present therein in thermal equilibrium with the walls of the cavity, emphasizing the need to experimentally determine its spectral distribution.

This observation opened up a new line of research that, in a few decades, disrupted the whole framework of Classical Physics with experimental results that disagreed with expectations. Maxwell had already found, from electromagnetical theory, that a wave normally incident on an absorbent body would produce a pressure numerically equal to its energy per unit volume and suggested its experimental verification. The Italian Adolfo Bartoli (1851–1896), in 1876, came to the same conclusion, using thermodynamic arguments instead [Day and Van Orstrand 1904]. Three years later, in 1876, analyzing the experimental data of various authors, Joseph Stefan (1835–1893) proposed that the radiant intensity emitted by a black body should be proportional to the fourth power of the temperature. Boltzmann, who had been a student of Stefan, was impressed by Bartoli's work and, resuming his study, he managed to give a theoretical demonstration of Stefan's semi-empirical law. Boltzmann then put together two arguments that were the state of the art of two different disciplines. From Electromagnetism, he derived the relationship $p(T) = (1/3)\rho(T)$, where T is the absolute temperature and p is the pressure exerted by an isotropic radiation, propagating in all directions, with energy density equal to ρ. From general thermodynamic considerations, he obtained the differential equation $Tdp - pdT = \rho dT$. Replacing the first in the second equation and integrating, he found the law[1]

$$\rho = \sigma T^4, \tag{1}$$

where σ was a constant to be determined experimentally. Boltzmann also reported a second demonstration by modifying Bartoli's scheme and imagining a cylinder in which a piston confines the radiation generated by a black body at the base of the cylinder. A clearer demonstration based on a Carnot cycle is reported by Sivoukhine (1984, II p. 299). However, while Stefan considered this law applicable to all bodies, so that the experimental comparison was disappointing, Boltzmann deduced it for the sole black body, for which the agreement with the experiments was good.

Still, it remained to determine the spectral shape of the black body radiation, not only theoretically but also experimentally. On this second front, the difficulty resided in the weak intensity of the radiation to be measured, in calibrating the instrumental functions of the prisms with respect to the wavelength scale, or of the gratings that have an uneven amplitude response, and detector response, such as photographic plates. Even the temperature measurement was not trivial. After the introduction of Kelvin's absolute scale, the most precise determinations were made with hydrogen-expanding thermometers up to a temperature of about 600°C; beyond that, nitrogen-expanding thermometers were used up to 1150°C; even further beyond, one was trusting in a platinum/platinum-rhodium thermocouple up to 1600 °C, calibrated with gas thermometers to the greatest degree possible. On the theoretical level, the first to propose a function for the black body spectrum was the Russian Vladimir Alexandrovich Michelson (1860–1927), who noted that the absolute continuity of the spectrum necessarily had to derive from the complete irregularity of oscillations of the emitting atoms. On the other hand, he suggested that each atom was the seat of periodic vibrations, and that the distribution of these principal modes of vibration were of the Maxwellian type. He therefore proposed that the spectrum be factored into various contributions, including an oscillator distribution, a function of the temperature and one of the oscillation period, with parameters to be determined experimentally [Bellone 2006; Day and Orstrand 1904]. This fact opened up the possibility for a comparison with the experimental data of S.P. Langley, in particular, with respect to the presence of a maximum in the spectrum, but the agreement with the Michelson model was illusory.

In the observations, however, it emerged that the wavelength λ_{max}, corresponding to the maximum amplitude of the spectrum, underwent a shift towards shorter wavelengths with increasing temperature T. This allowed H.F. Weber, in 1888, to establish the relation $\lambda_{max} T$ = constant. The next step in order to derive an accurate analytical function was taken by Wilhelm Wien (1864–1928). In 1893, on the

---

[1] The original symbols used by Boltzmann are f for p and ψ for ρ. It is possible to derive this equation from the consideration that entropy, for a system characterized by an absolute temperature T and a volume V, is an exact differential, so $(\partial U/\partial V)_T = T(\partial p/\partial T)_V - p$, where U is the energy of the system and p is the pressure [Fermi 1972, p. 72], taking that $U = \rho V$.

basis of thermodynamic considerations, he succeeded in giving a demonstration of the displacement law. Indeed, Boltzmann had assumed, but did not prove, that, in an adiabatic process of compression of the radiation, where the energy density increases as a result of the work done in the compression, its spectral composition is the same as that obtained with a temperature rise. Wien proved this equivalence by resorting to the Doppler effect in the reflection on the surface in motion that produces the compression. For example, in a thermally isolated spherical volume that is slowly compressing, the relative variation $\Delta\nu/\nu$ of the frequency of each spectral component of the radiation contained therein is equal to the relative variation $-\Delta r/r$ of the radius r of the sphere [Sivoukhine 1984, II]. In other words, in the adiabatic compression, we have that νr = constant, and, since $r \propto V^{1/3}$, where V is the volume of the vessel, we also have that $\nu^3 V$ = constant. From Boltzmann's reasoning, one also gets the adiabatic invariant $\rho V^{4/3}$ = constant, where ρ is the energy density of the radiation. Therefore, combining the two invariants, one gets ν^4/ρ = constant. Finally, according to the Stefan-Boltzmann equation, the displacement law ν/T = constant follows, or, equivalently, λT = constant. Thus, the displacement law of the maximum is only the particular case of a scale transformation that involves each single component of the spectrum of black body radiation as the temperature varies. From the Stefan-Boltzmann law, in 1893, Wien concluded that the spectral distribution $\rho_\nu(\nu, T)$ of energy density, for which $\rho = \int \rho_\nu d\nu$, must have the form $\rho_\nu(\nu, T) = \nu^3 f(\nu/T)$ [Wien 1894].[2] Two years later, Wien published a new memory [Wien 1896], in which he proposed an explicit form of the spectral distribution of black body radiation in the space of wavelengths: $\rho_\lambda = c_1\lambda^{-5} \times \exp[-c_2/(\lambda T)]$, where c_1, c_2 are two constants, which, in the frequency space, is equivalent to

$$\rho_\nu = \alpha\nu^3 \exp(-\beta\nu/T), \qquad (2)$$

where α and β are two additional constants. Although Rayleigh considered it little more than a conjecture, this law was able to reproduce the experimental data then available and had a significant impact on subsequent developments.

The year 1900 marked a major breakthrough, as O. Lummer and E. Pringsheim, and later H. Rubens and F. Kulbaum, demonstrated that, in the infrared region, the Wien distribution gave values that were not compatible with the measures. Indeed, for large wavelengths, the measured intensity trend was proportional to the temperature, unlike the Wien function. However, not even the new distributions proposed by Max Thiesen that year, with $\rho_\lambda = c_1(\lambda T)^{1/2}\lambda^{-5} \exp[-c_2/(\lambda T)]$, and by Rayleigh (of which we will discuss later), with $\rho_\lambda = c_1 T\lambda^{-4} \exp[-c_2/(\lambda T)]$, to fit the new data for λ large, while respecting the Stefan-Boltzmann and displacement laws, gave values in accordance with the measures

[2] Indeed, $\int_0^\infty \nu^3 f(\nu/T) d\nu = T^4 \int_0^\infty x^3 f(x) dx$ and, since the integral in x is a constant value, Stefan-Boltzmann's law is respected. The corresponding distribution in the wavelength space, on the other hand, has the shape $\rho_\lambda(\lambda, T) = \lambda^{-5}\varphi(\lambda T)$.

[Day and Van Orstrand 1904].

The scientist who found the right distribution was Max Karl Ernst Ludwig Planck (1858–1947), but the story of this discovery is very complex and the demonstration that is given now is very different from Planck's original one [Klein 1961]. At the beginning of his activity, Planck specialized in Thermodynamics and Physical Chemistry, also dealing with the theory of electromagnetic field and the Mechanics of the continuum. He devoted himself mainly to clarifying the meaning of the second law of thermodynamics and to exploring its consequences. However, he was skeptical of the lines drawn by Maxwell and Boltzmann regarding molecular motions, believing that a thermodynamics free of these mechanical concepts would be more appropriate to explain the phenomena and that a continuous physics could explain the irreversibility traced by the continuous increasing of entropy [Bellone 2006].

Planck was attracted by the problem of the black body, because of the universality of the spectral distribution law required by the Kirchhoff theorem. He then began to study the problem of field interaction with a harmonic oscillator, hoping to find a solution to the problem of irreversibility. The choice of this type of resonator was dictated by its simplicity and the fact that black body radiation was independent of the nature of the cavity walls. In particular, he initially considered a field of flat waves interacting with a resonator that reverberated spherical waves, changing the character of radiation in an apparently irreversible manner, and sought the thermodynamic equilibrium conditions between the field energy density and the average energy of the resonator. The form of the equations that he was obtaining allowed him to glimpse an evolution over time so as to avoid the disturbing (for him) conjecture of probabilistic molecular disorder of Boltzmann's work, and, in 1897, he began a series of five publications on irreversible radiation processes.

Boltzmann himself, however, pointed out to him that Maxwell's equations were invariant for temporal inversion, and therefore one could not deduce their irreversibility in the interaction between field and resonator. In other words, the irreversibility had to be implicit in the initial conditions. Planck accepted the criticism, and then considered the case of a "natural radiation" that would ensure irreversibility similarly to the molecular chaos of Boltzmann's theory. In this way, he came to understand an important relation between the spectral density of the field $\rho(\nu,T)$ and the average energy $u_\nu(T)$ of a harmonic oscillator with frequency ν. By equating the emission and absorption rate of the oscillator immersed in the natural radiation field, he found that, at equilibrium,

$$\rho_\nu(T) = (8\pi\nu^2/c^3)\, u_\nu(T), \tag{3}$$

where c is the speed of light.

At this point, Planck had only to determine the average energy of the oscillators to find the spectral density of the field as well. But, instead of using the equipartition theorem of statistical mechanics, he followed a thermodynamic approach, looking for a relationship between entropy S and mean energy u of the oscillator. In the last of his five articles published in the year 1900, he proposed the relationship

$$S = -\frac{u}{a\nu} \ln\left(\frac{u}{be\nu}\right), \tag{4}$$

where e is the Nepero number and $u = u_\nu(T)$. Planck did not give an explanation for the choice of this function, but it is presumed to have been driven by the fact that he got Wien's exact distribution from it. In addition, he was able to demonstrate that, with this definition, the total entropy grew over time. In particular, he was impressed by the fact that $\partial^2 S/\partial^2 u$ was simply proportional to minus u^{-1}.[3] The result that Planck found was therefore [Planck 1900]

$$\rho_\nu(T) = (8\pi b/c^3)\ \nu^3 \exp(-a\nu/T). \tag{5}$$

Planck attributed to a and b the meaning of *universal constants*, just like the speed of light. The constant b was already what he later indicated with the symbol h, and what would then be called *Planck's constant*. From the comparison with the experimental data, he obtained the values $a = 0.4818\cdot10^{-10}$ sec·K, $b = 6.885\cdot10^{-27}$ erg·sec.

Planck was pleased with this result, but the new measures made by Lummer and Pringsheim, and, then, more accurately, in October 1900, by Rubens and Kulbaum, forced him quickly to re-analyze the whole question.[4] He found, therefore, that, in order to satisfy the constraint of an increasing entropy, it was sufficient that $\partial^2 S/\partial^2 u$ was any negative function of u. If, in particular, this second derivative were proportional to $-1/u^2$, the spectral distribution would be proportional to T. So, Planck found that the function $-1/[u(\gamma + u)]$, where γ depends on the frequency, was bringing together Wien's distribution at small wavelengths and the trend proportional to T at long wavelengths, obtaining what is now known as *Planck's distribution*: $\rho_\nu(T) = A\nu^3/[\exp(B\nu/T) - 1]$, here expressed in the frequency space, where A and B are constants.

Shortly after, on October 19, 1900, Planck attended a meeting of the German Society of Physics, where he presented this conclusion with the title *Ueber eine Verbesserung der Wien'schen Spectral-gleichung* (*On an improvement of the Wien spectral equation*). Planck himself has said that, the next morning, he received a visit from Rubens, who told him that he had checked an excellent agreement with his formula at each point of the measures that night. In subsequent measurements, the more accurate they were, the better agreement there was with

[3] From the displacement law and Eq. (3), it follows that $u = \nu f(\nu/T)$. From this, it also follows that $T = \nu F(u/\nu)$. Since $T^{-1} = \partial S/\partial u$, integrating with respect to u, it is found that S must be a function of u/ν. Deriving Eq. (4) with respect to u, and solving for u, we have that $u = b\nu \exp(-a\nu/T)$. Finally, with Eq. (3), we again find the Wien distribution (2) [Klein 1961].

[4] An account of the developments of the black body source is given by Cottington (1986).

Planck's distribution, contrary to those proposed by other authors.

Planck realized the semi-empirical significance of his demonstration, and would then seek a deeper physical explanation of his law. He thus set aside his reservations about Boltzmann's theory and considered the case of a collection of N identical harmonic oscillators with frequency ν, among which would be shared an energy $U_N = Nu$, where u is the average energy of an oscillator. Resuming the Boltzmann equation, the entropy of this collection of oscillators would then be given by $S_N = k_B \ln W$ + constant, where the "probability" W was proportional to the number R of "complexions" with which the energy was partitioned between the oscillators. On the other hand, in order to establish this number, it was necessary, as Boltzmann had done, to divide the energy into an integer number P of equal elements ε, that is, $U_N = P\varepsilon$. The complexions were therefore $R = (N+P-1)!/[P!(N-1)!]$. With some passages, using the Stirling's approximation for large numbers N and P, aside from a constant, he obtained the mean entropy per oscillator

$$S = k_B\left\{(1+u/\varepsilon)\ln(1+u/\varepsilon) - (u/\varepsilon)\ln(u/\varepsilon)\right\}. \tag{6}$$

Since, for the law of displacement, entropy then had to be function of u/ν, it resulted that $\varepsilon = h\nu$, where h was a new constant. In this way, he again found his distribution law (5), where $b = h$ and $a = h/k$. From the comparison with the experimental data, Planck obtained the values $h = 6.55 \cdot 10^{-27}$ erg·sec and $k = 1.346 \cdot 10^{-16}$ erg/K. It was Planck himself who called k "Boltzmann's constant", noting that he had not given it a particular meaning. Subsequently, by the value of k, Planck also obtained the Avogadro number and the charge of the electron, which remained the best determinations of these constants for a long time [Klein 1961]. Planck's distribution law was therefore

$$\rho_\nu(T) = \frac{8\pi h\nu^3}{c^3}\frac{1}{\exp\left[h\nu/(k_B T)\right]-1}, \tag{7}$$

Planck's demonstration was not, however, free of criticism. Indeed, he had not brought the limit for $\varepsilon \to 0$. Neither did he follow Boltzmann's scheme, since he would have had to maximize the total entropy for the various frequencies. If he did, he would have come to the same conclusion, but this fact was only understood many years later.

The contrast between Planck's law and classical physics began to emerge with a short note published by John William Strutt (Lord Rayleigh, 1842–1919) in June 1900 in the journal Philosophical Magazine [Lord Rayleigh 1900]. Strutt proposed a new distribution law of black body radiation using a completely different reasoning to that of Planck. He had, in fact, considered the resonating modes of the electromagnetic field in a cavity, establishing their number within a given frequency range. Assuming that the radiation was in thermal equilibrium at a temperature T, and then using the equipartition theorem of classical statistical dynamics, he had attributed to each of these modes the same average energy, which is proportional to T. It turned out that spectral density had to be proportional to $\nu^2 T$. On the other hand, in order to avoid an evident divergence for increasing frequencies, he suggested that the «complete expression» should be proportional to $\nu^2 T \exp(-a\nu/T)$. Rayleigh recognized that, at the highest frequencies, «although for some reason not yet explained», the Maxwell-Boltzmann law of equipartition failed, however, «it seems possible that it may apply to the graver modes», for which the radiation density was proportional to T, as the new experimental measures were beginning to highlight.

This contrast was made explicit by a series of articles by Rayleigh and by James Hopwood Jeans (1877–1946) that appeared in Nature in 1905. Jeans, in particular, was trying to preserve the general value of the equipartition theorem and criticized Planck's derivation on the two aspects mentioned above: his break with Boltzmann's procedure and the fact that he had not brought the limit for $h \to 0$.

The photon

Even the young Albert Einstein began to deal with Thermodynamics, with a series of articles published between 1902 and 1904 [Bergia, 1998]. In particular, he was concerned with the energy fluctuations of a thermal equilibrium system, maintained at a constant temperature by means of a thermostat. Since the system is not isolated, its energy can fluctuate, and Einstein intended to get the Avogadro number from the extent of these fluctuations. Josiah Willard Gibbs (1839–1903) had estimated that these fluctuations would be too small to be revealed in ordinary systems. During his "annus mirabilis", Einstein became convinced instead that the thermal system of the radiation contained in a cavity at a constant temperature could have detectable fluctuations. In one of his articles of 1905, entitled *On a heuristic point of view concerning the generation and transformation of light* [Einstein 1905a], he dealt with the theory of the black body and with the works of Planck and Wien. Then, he resumed the Planck's equation (3) that links the spectral density ρ to the average energy of an oscillator,[5] but, according to the energy equipartition theorem, this average energy had to be equal to $k_B T$. In this way, he again found the so-called Rayleigh-Jeans formula that Rayleigh himself had found untenable:

$$\rho_\nu(T) = (8\pi\nu^2/c^3)\, k_B T. \tag{8}$$

Einstein was the first to note that Classical Physics clearly led to a result in open conflict with the experience. In fact, Eq. (8) implies that the spectral energy

5 Following Planck, he explicitly considered the case of totally incoherent radiation, generated by "*separate* resonators".

density should diverge to infinity as the frequency increases, according to what Paul Ehrenfest then called *ultraviolet catastrophe*. On the other hand, Einstein also noted that, for small values of ν/T, Eq. (8) converges to the Planck equation, which correctly describes the experimental data, indicating the correctness of its derivation. Therefore, Classical Physics fails in the case of large values of ν/T. At this other limit, there is another formula that works well, that of Wien (2). As we have already seen, Wien had obtained this formula on the basis of thermodynamic considerations in which the radiation behaves in a manner similar to a gas exerting pressure on cavity walls. In his article, Einstein then introduced the heuristic hypothesis (i.e., suitable to facilitate the discovery of new results) that the radiation was composed of grains or *quanta of energy*, distributed in space, such as a gas, that «move without dividing and are absorbed or generated only as a whole». For the case of a monochromatic radiation of frequency ν, he assumed that the quantum has energy $h\nu$. In support of this hypothesis, he considered several examples, including the observation made by Philip Lenard in 1902 that the energies of electrons emitted by the photoelectric effect were independent of the intensity of the incident radiation. Einstein interpreted this fact by foreseeing that the maximum energy of the electrons emitted is given by $E_{max} = h\nu - P$, where the threshold value P is the work that each electron must perform to leave the material on which the radiation is incident. This law was verified with precision only 11 years later, in 1914, by Robert Andrews Millikan (1868–1953), producing a very clean metal surface under vacuum that acted as a photocathode. This law earned Einstein the Nobel Prize in 1921.

As Bergia (1998) notes, in 1905, Einstein did not yet have a dualistic view of the nature of radiation. He recognizes the validity of wave theory, but writes «optical experiments observe only time-averaged values, rather than instantaneous values. Hence, ..., the use of continuous spatial functions to describe light may lead to contradictions with experiments, especially when applied to the generation and transformation of light». However, some years later, in 1909, Einstein resumed the fluctuation issue, finding that, according to Planck's law, the mean square value of the fluctuations of the energy of black body radiation had to be given by a relationship consisting of *two* terms, both present. The first, proportional to $h\nu$, is of corpuscular nature, and the second, proportional to $c^3\rho/(8\pi\nu^2)$, is of undulatory nature (due to interference between «light rays» uncorrelated between them) [Einstein 1909].[6] In his conclusion, Einstein imagines that the «the manifestation of light's electromagnetic waves is constrained at singularity points, like the manifestation of electrostatic fields in the theory of the electron», and, in the presence of many of these singularities densely arranged with respect to the

---

6 In modern language, the variance, that is, the root mean square value of the difference in the number of photons n from its mean value $\langle n \rangle$, is given by [Arecchi 1970]: $\langle \Delta n^2 \rangle = \langle n \rangle + \langle n \rangle^2$. In other words, in photon statistics, Planck's law corresponds to a probability $p(n) = \langle n \rangle^n/(1+\langle n \rangle)^{n+1}$ to detect n photons within a time interval Δt smaller than the inverse of the spectral band used to filter the radiation. This probability is known as "geometric distribution", and $\langle n \rangle$ is the average number of photons revealed in repeated measurements of duration Δt.

extension of their field, one might get an oscillating field similar to that of the electromagnetic theory. Einstein finally deduced that «the two structural properties of radiation according to Planck's formula (oscillation structure and quantum structure) should not be considered incompatible with one another».

In 1911, the experiments of Ernest Rutherford (1871–1937) showed the existence in atoms of an atomic nucleus of positive charge surrounded by electrons, similar to a planetary system. On the other hand, from Classical Physics, it turned out that the orbits of these electrons would be unstable by emission of radiation with a continuous spectrum. However, the experiments with gas discharges instead showed that emissions were done with a discrete set of frequencies. To explain these observations, in 1913, Niels Henrik David Bohr (1885–1962) introduced an atomic orbital model with a quantization rule for which the magnitude of the angular momentum is limited to an integer multiple of $\hbar$, restricting the possible orbits of the electrons to a discrete set of energies. In this way, he was able to explain quantitatively the spectral lines of the hydrogen atom and ions with only one electron as jumps between these energy levels.

In 1916, and later in a 1917 reprint, Einstein returned to discussing Planck's law with a new approach. In his previous works, he had considered it only as an expression that was justified the experimental measurements, but it was contradictory that, in its derivation, it made use of classical arguments that the same law refuted. Then, he tried to give a new demonstration, including, for the first time, together with the energy, the concept that each quantum of light also brings with it a momentum equal to $h\nu/c$. In this article, Einstein considered the case of a cavity containing gas and electromagnetic radiation in thermal equilibrium with the walls of the cavity. Resuming the Bohr model, Einstein stated that the exchanges of energy between radiation and gas would take place through three types of process: *one* of absorption of a quantum of radiation by a molecule from a lower energy state n to a higher one m, and *two* of emission, of which one would be *spontaneous* and one *stimulated*, from the upper to the lower state. The spontaneous process had to have a probability of decay $dW_{\mathrm{sp}} = A_m^n dt$ in the time interval dt, independent of the presence of radiation, while stimulated absorption and emission probabilities were respectively given by $dW_{\mathrm{a}} = \rho B_n^m dt$ and $dW_{\mathrm{st}} = \rho B_m^n dt$, and therefore were proportional to the radiation energy density ρ. At thermal equilibrium, according to the law of the Boltzmann's distribution, also considering the degeneracy of states n and m, Einstein again obtained Planck's law and also the relationship $A_m^n / B_m^n \propto \nu^3$. Less known is that, in this same article, in addition to energy, Einstein also attributed, through complex reasoning, an exchange of momentum $h\nu/c$ between radiation and molecules: each absorbed quantum transfers its impulse to the molecule in its own direction, while, in stimulated emission, each molecule emits a quantum by giving it impulse in the same direction as the quanta that stimulate it. Instead, for the spontaneous emission, the impulse yielded to the quantum can be directed in any direction, and therefore its isotropy is obtained only statistically. In this way, he found that the Maxwellian velocity distribution of the molecules was not altered by the thermal equilibrium radiation with the cavity.[7]

The young Indian physicist Satyendra Nath Bose (1894–1974), being interested in the Planck law in regard to his teaching of quantum physics, found that all derivations of this law still had recourse to classical topics. They were the relationship between the radiation density and the average energy of the oscillators found by Planck, or the displacement law of Wien and the correspondence principle of Bohr in the classical limit, to which Einstein was forced to resort in his 1916 work for a quantitative comparison. Thus, these derivations appeared unjustified to him from a logical point of view. He succeeded in finding a fully quantum derivation of Planck's law, assuming the quanta of light to be particles whose coordinates and moments were represented in a phase space in six dimensions. Quantizing this space in cells of volume h^3 and calculating the number of these cells within the range of frequency $d\nu$, he could calculate the thermodynamic probability of a state defined macroscopically. Unlike Planck, he fully carried out the calculation of maximizing entropy, finally getting Planck's law. On the other hand, as Bergia notes, in his derivation, Bose introduced three new elements of fundamental importance. The first is what would subsequently come to be called the *photon spin* by doubling the number of cells to account for two radiation polarization states (which form a base). The second element was that the number of quanta is not conserved in order to take account of absorption and emission processes: this was translated into the choice of constraints used in maximizing entropy. The third element was the must shocking, in that Bose had unknowingly introduced a new statistic. In fact, in the combinational calculation of probability, he had considered quanta of equal energy as indistinguishable from each other, in sharp contrast to the classical concept of a particle. Bose submitted his work to a British magazine that, however, rejected it. He then called on Einstein, whom he deeply admired, confidently asking him his opinion and possibly for help in the publication. Einstein appreciated the value of what he had accomplished, translating Bose's original text, *Planck's Law and the Hypothesis of Light Quanta*, into German, and with an accompanying letter, presenting it to the magazine Z. Physik, through which he obtained publication on Bose's behalf [Bose 1924].

Later, Einstein extended Bose's treatment to material particles, but it was Paul Ehrenfest who pointed out that they had introduced a new statistic, now called the *Bose-Einstein statistic*, in which one has a loss of statistical independence between identical particles, now known as *bosons*, with integer spin. To this quantum statistic was then added that of *Fermi-Dirac*.

The light quanta were later called photons in 1926 by Gilbert Newton Lewis, in analogy with protons and electrons, and this name has since remained in common use. However, Lewis's work was based on a faulty premise, as it assumed the conservation of their number.

---

[7] This work can be considered as the basis for interpreting the operation of lasers and masers, however, in 1961, the inventor of the Maser, Charles Townes, was inspired by the development of coherent microwave oscillators used in radar for military purposes [Bergia 1998].

Einstein's indication of 1909 on the dual nature of radiation was revived only in 1923 by Louis Victor Pierre Raymond de Broglie (1892–1987), who reinterpreted Bohr's quantization rule as a standing wave condition for electrons, actually extending the wave-particle duality to material particles, revolutionizing all of the Classic Physics in Quantum Mechanics.

Besides the Millikan experiment, other experiments appeared to confirm the heuristic hypothesis of light quanta. Already in 1909, Joseph John Thomson (1856-1940) had suggested to Geoffrey Ingram Taylor (1886–1975) that he should check whether there were any energy units that seemed to exist in ionization experiments and with X-rays, discovered by Wilhelm C. Röntgen (1845–1923) in 1895. In this case, they expected that diffraction phenomena would be altered with a sufficiently low intensity. Taylor then conducted an experiment in which he photographed the shadow of a needle produced by a slit illuminated by a flame strongly attenuated with blackened glasses. The exposure lasted for three months and produced the same fringes visible without attenuation. Thomson and Taylor were unaware of the work of Planck and Einstein, and concluded with an upper limit for these units of energy that was much larger than $h\nu$. Still, the experiment that created a much greater sensation was that carried out in 1923 by Arthur Holly Compton (1892–1962), who produced a frequency shift of X-rays toward lower energies in interaction with electrons, with an angle of diffusion characteristic of a collision between material particles.

During the twentieth century, the concept of the photon gradually changed, and it finds its explanation today in Quantum Electro-Dynamics (QED) which originated from Paul Dirac (1902–1984), with contributions from various authors such as Heisenberg, Pauli, Fermi, and others [Kidd et al 1988]. More recently, in 1963, R.J. Glauber introduced the bases for photon statistics within this theory [Glauber 2006]. Many situations can still be explained by semi-classical approximation in which the propagation is described by the theory of the electromagnetic field, while the interaction occurs for quanta, but it is only the QED that can explain all known phenomena. It has modern application in non-classical entanglement phenomena, quantum computation, and quantum encryption [Zeilinger et al 2005]. The initial concept of the photon expressed by Einstein is now obsolete, as well as that of wave-particle duality when it is taken literally [Kidd et al 1988]. Photons are thus a way of representing the quantum states of the field. But all of this is outside of the scope of this book.

Let me just add my own comment. Even recent history shows that the development of knowledge is a collective process, like a tree out of which new branches are born, with intuitions that are immediately strange. Some appear promising, but then cease their growth, others are disputed, but then they become the main trunk toward heaven, and many give fruit that then nourishes everyone's life. Those who stand in the rearguard, such as I myself, can only at least hope to contribute to that undergrowth of technique and knowledge that fosters the stimulus in new minds of new experiments and new ideas. But the only One who drives the game is God through people of goodwill who also unknowingly dare to reach Him.

Suggested reading

The information in these notes is derived from the following texts:

Arecchi F.T., *Einstein without Relativity*, Annali dell'Istituto e museo di Storia della Scienza di Firenze, year IV, issue 2, 49-70 (1970).

Bartoli A., *Sopra i movimenti prodotti dalla luce e dal calore*, Le Monnier, Firenze (1876).

Bellone E., *La teoria dinamica del calore*, p. 253-269, *L'irreversibilità*, p. 345-360, in Storia della scienza, Vol. 4; *Il corpo nero*, p. 291-310, in Storia della Scienza, Vol. 5; *Albert Einstein*, p.259-311, in Storia della Scienza, Vol. 6. Series directed by P. Rossi, Gruppo Editoriale L'Espresso (2006).

Bergia S., *Einstein: quanti e relatività, una svolta nella fisica teorica*, in "I grandi della scienza", year I, n. 6, Le Scienze S.p.A. (1998).

Boltzmann L., *Ableitung des Stefan'schen Gesetzes, betreffend die Abhängigkeit der Wärmestrahlung von der Temperatur aus der electromagnetischen Lichttheorie*, Annalen der Physik **258**, 291-294 (1884).

Bose S.N., *Plancks Gesetz und Lichtquantenhypothese*, Zeitschrift für Physik **26**, 178-181 (1924).

Day A.L. and Van Orstrand C.E., *The black body and the measurement of extreme temperatures*, The Astrophysical J. **19**, 1-40 (1904).

Cottington I.E., *Platinum and the standard of light*, Platinum Metals Rev. **30**, 84-95 (1986).

Einstein A., *Über einen die Erzeugung und Verwandlung des Lichtes betreffenden heuristischen Gesichtspunkt*, Annalen der Physik **17**, 132–148 (1905a). *Zur Elektrodynamik bewegter Körper*, Annalen der Physik **322**, 891–921 (1905b). *The development of our views on the composition and essence of radiation*, Physikalische Zeitschrift **10**, 817-826 (1909). *Zur Quantentheorie der Strahlung*, Physikalische Zeitschrift, **18**, 121-128 (1917).

Fermi E., *Termodinamica*, Ed. Boringhieri, Torino (1972).

Glauber R.J., *Nobel lecture: One hundred years of light quanta*, Rev. Mod. Phys. **78**, 1267-1278 (2006).

Jackson J.D., *Classical Electrodynamics*, John Wiley & Sons, New York (1974).

Kidd R., Ardini J., Anton A., *Evolution of the modern photon*, Am. J. Phys. 57, 27-35 (1989).

Klein M.J., *Max Planck and the beginnings of the quantum theory*, Archive for History of Exact Sciences **1**, 459-479 (1961).

Lord Rayleigh, *Remarks upon the law of complete radiation*, Philosophical Magazine **49**, 593-540 (1900).

Malikin G.B., *Sagnac effect and Ritz ballistic hypothesis (review)*, Optics and Spectroscopy **109**, 951-965 (2010).

Pike R., *Laser, photon statistics, photon-correlation spectroscopy and subsequent applications*, J. European Opt. Soc., Rap. Publ. **5**, 100475 1-16 (2010).

Planck M., *Ueber irreversible stralungsvorgänge*, Annalen der Physik **306**, 69-122 (1900).

Polvani G., *Storia delle ricerche sulla natura della luce*. Istituto della Enciclopedia Italiana, Roma (1934).

Taylor G.I., *Interference fringes with feeble light*, Proc. Camb. Phil. Soc. Math. Phys. Sci. 15, 114–115 (1909).

Sivoukhine D., *Optique*, MIR, Mosca (1984).

Wien W., *Temperatur und Entropie der Strahlung*, Annalen der Physik **288**, 132-165 (1894). *Ueber die Energievertheilung im Emissionsspectrum eines schwarzen Körpers*, Annalen der Physik **294**, 662-669 (1896).

Zeilinger A., Weihs G., Jennewein T. and Aspelmeyer M., *Happy centenary, photon*, Nature **433**, 230-238 (2005).

Chapter 6
Propagation of laser beams in linear media

The Gaussian, Bessel, and Bessel-Gauss approaches

Introduction

If the medium is homogeneous, linear, isotropic, and non-dispersive, any electromagnetic wave can, in principle, be decomposed into plane waves, which are a simple and effective basis for complex wave equation solutions (1.4.12). But, in real cases, we have to deal with electromagnetic beams of limited size, and the decomposition in plane waves is not always the most appropriate. Consider, for example, the case of a monochromatic, and therefore continuous (CW, that is, *continuous wave*, single-frequency) laser beam; at every point in the space, its electric field oscillates sinusoidally over time. The spatial trend in this field is not equally simple: at first sight, a collimated laser beam is a genuinely good representation of a pencil of parallel geometric rays between them, but, with more careful observation, we find that the beam tends to expand due to the diffraction.

As we have already seen, the solution to diffraction problems is difficult to obtain in general. However, if we limit ourselves to considering waves whose spectrum of wave vectors from the development in plane waves remains contained within a narrow cone centered around a direction of predominant propagation, let's say, the z-axis, the wave equations can be simplified in a less drastic way than for Geometrical Optics, while maintaining (in many cases) an excellent representation of the diffraction (when, along the path of the beam, there are no diffracting objects).

The equation that is obtained is called the *paraxial wave equation*, and its characteristic solutions constitute a basis for replacing the plane waves. The advantage of this treatment consists in the fact that, in many practical cases, it is sufficient to consider a finite number of such solutions, instead of the infinite or very large number that would be required for a plane wave development.

One class of characteristic solutions of the paraxial equation is that of Gaussian beams, which constitute a series of *transverse modes* and are particularly suitable for treating laser beams. Their field decays as $\exp(-r^2/w^2)$, where r is the distance from the beam axis and w is a parameter that depends on the position along the axis.

A different approach to the problem of diffraction derives from the observation that, for a linear medium, the propagation between two planes orthogonal to the z-axis can be seen as the response (output) of a linear system to an excitation input. In particular, as we have seen above, the "classical" treatment of diffraction con-

---

Electronic supplementary material The online version of this chapter (https://doi.org/10.1007/978-3-030-25279-3_6) contains supplementary material, which is available to authorized users.

G. Giusfredi, *Physical Optics*, UNITEXT for Physics,
https://doi.org/10.1007/978-3-030-25279-3_6

sists in calculating the field on the output plane by the convolution of the field profile on the input plane with the impulse response of the system.

In this chapter, we will instead follow the treatment of H. Kogelnik and T. Li, giving more emphasis to the characteristic solutions of the paraxial equation as the basis for generating the response of a system to an incoming excitement [Kogelnik and Li 1966a 1966b]. The advantage of this discussion is that the paraxial equation can be more easily generalized to deal with media with a refractive index that varies radially, as in certain optical fibers, or with non-linear media.

We will also analyze in detail the propagation of Gaussian beams, even in non-axial optical systems, and of the so-called *diffraction free* beams such as Bessel's waves and Bessel-Gauss beams. Lastly, we will study the properties of the resonant cavities, with their resonant frequencies, and a type of multipass cavity.

6.1 The paraxial wave equation

Let us finally look at a derivation of our paraxial equation. For a monochromatic wave, Eqs. (1.4.12) take the form

$$\nabla^2 u + k^2 u = 0\,, \tag{6.1.1}$$

where u represents the spatial trend of any component of the field and k is the wave vector module. If, as mentioned above, the wave propagates predominantly along the z-axis, u can be conveniently factorized in the form:

$$u = \psi(x, y, z)\, e^{+ikz}\,, \tag{6.1.2}$$

where ψ represents the amplitude of the field and its dependence on x, y, z is much less rapid than that of the factor $\exp(ikz)$. Replacing Eq. (6.1.2) in (6.1.1), we have

$$\nabla_\perp^2 \psi + \frac{\partial^2}{\partial z^2}\psi + 2ik\frac{\partial}{\partial z}\psi = 0\,,$$

where $\nabla_\perp^2$ concisely indicates the operator $\partial^2/\partial x^2 + \partial^2/\partial y^2$. If we assume that ψ varies as slowly in z, whereby $|\partial\psi/\partial z| << k\psi$, we can neglect $\partial^2\psi/\partial z^2$ in comparison to $k\partial\psi/\partial z$. So, we finally get the *paraxial wave equation*

$$\nabla_\perp^2 \psi + 2ik\frac{\partial}{\partial z}\psi = 0\,, \tag{6.1.3}$$

for the amplitude of the field in a linear, homogeneous and isotropic medium. A particular solution to this equation is given by what is called the *paraxial spheri-*

cal wave [Siegman 1986]:

$$\psi(\boldsymbol{r},\boldsymbol{r}_0)=\frac{1}{z-z_o}e^{ik\frac{(x-x_o)^2+(y-y_o)^2}{2(z-z_o)}}, \qquad (6.1.4)$$

where $\boldsymbol{r}_0$ represents the center from which this wave expands. This solution constitutes the kernel of the Huygens-Fresnel diffraction integral (4.2.5). In other words, $\psi\,\exp[ik(z\text{-}z_o)]$ replaces the expression for the Huygens's wavelets in Eq. (4.2.1). Therefore, once the field at $z = z_o$ is known, the same diffraction integral of Huygens-Fresnel constitutes the general solution to the paraxial scalar wave equation and is mathematically equivalent to it. J.A. Arnaud has demonstrated that this equivalence persists even in the paraxial propagation calculated by Collins' diffraction integral (4.2.26) through an optical system defined by a generic matrix $\mathfrak{ABCD}$ [Arnaud 1970].

An important aspect that unites the two formulations is that they both deal with the case of monochromatic waves: time does not appear explicitly in the paraxial equation and diffraction integrals. In other words, they do not describe the propagation of optical signals commonly understood as a field evolution over time, as well as in space. The solutions that are also found with the paraxial wave equation are instead meant as "instant photos", in which the only extension is space. Time evolution can only be reconstructed by collecting solutions for the entire spectrum of the temporal frequencies of the optical signals involved.

Eq. (6.1.3) can also be expressed in the form

$$\frac{\partial}{\partial z}\psi=\frac{i}{2k}\nabla_{\perp}^2\psi \qquad (6.1.5)$$

that lends itself to be integrated numerically so as to determine the spatial evolution of the field amplitude, once it is known on an initial plane $z = z_o$. Siegman (1986) notes that the presence of the factor i implies an additional phase that varies along the z-axis in proportion to the second derivative of the amplitude and which increases the wave phase velocity with respect to that of the carrier. As the field is increasingly transversely confined, or has a complicated transverse structure, this effect is accentuated and becomes particularly large around a focal point. In other words, Eq. (6.1.5) also implicitly expresses the evolution of Gouy's phase in continuous form.

6.2 Propagation of the fundamental Gaussian mode

Let us now see how a Gaussian paraxial wave propagates. It is easy to verify that

$$\psi = \exp\left[i\left(P + \frac{k}{2q}r^2\right)\right] \tag{6.2.1}$$

is *a* solution to Eq. (6.1.3). Here, $r^2 = x^2 + y^2$. The parameter $P(z)$ represents a complex phase difference that is associated with the propagation of the beam, while $q(z)$ is another complex parameter describing its form. Inserting Eq. (6.2.1) into Eq. (6.1.3) and comparing the terms with the same power in r, the following relationships are obtained:

$$\frac{dq}{dz} = 1, \quad \frac{dP}{dz} = \frac{i}{q}. \tag{6.2.2}$$

The integration of Eq. (6.2.2.a) between two coordinate points z_1 and z_2 gives us

$$q(z_2) = q(z_1) + z_2 - z_1. \tag{6.2.3}$$

We now introduce, for convenience, two real parameters R and w, whose meaning will soon be clear, linked to q by

$$\frac{1}{q} = \frac{1}{R} + i\frac{\lambda}{\pi w^2}, \tag{6.2.4}$$

where λ is the wavelength in the considered medium, with a refractive index n. Therefore, $k = 2\pi/\lambda = 2\pi n/\lambda_o$. Eq. (6.2.1), for the field amplitude, becomes

$$\psi = \exp\left(iP + i\frac{\pi}{\lambda R}r^2 - \frac{r^2}{w^2}\right). \tag{6.2.5}$$

Then, for an assigned z value,

$R(z)$ is the *curvature radius of the wavefront* measured on the beam axis. In other words, R is the radius of the osculating sphere to the wavefront

$$kz + \frac{1}{2}k\frac{r^2}{R} = \text{cost.}$$

$w(z)$ is the *spot size radius*, i.e., w is the radius for which the amplitude of the field is $1/e$ times that on the axis. A circle of radius w contains about 86.5% of the total beam power (see §6.4.2).

Observing Eq. (6.2.3), it is seen that the real part of q is canceled for a particular position on the z-axis; this position is indicated as the beam *waist*, and we shall soon see the reason for this name. At the waist plane, the parameter q is

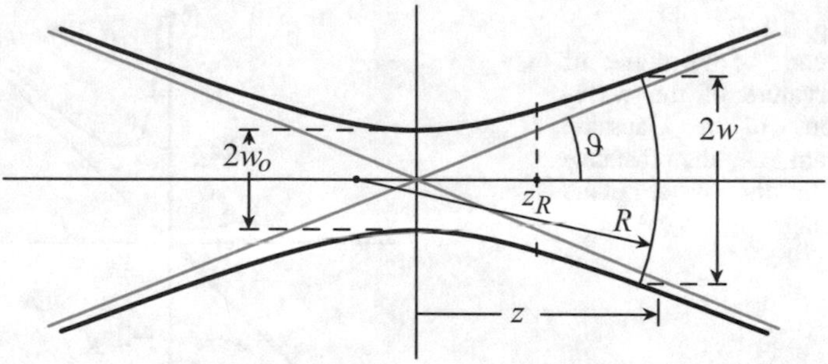

Fig. 6.1 Course of a Gaussian beam with its characteristic parameters

$$q_o = -i\frac{\pi w_o^2}{\lambda} = -i\frac{\pi n w_o^2}{\lambda_o}, \tag{6.2.6}$$

and, necessarily, from Eq. (6.2.4), the wavefront there is plane. Thus, choosing just the waist point as the origin of the coordinates, from Eq. (6.2.3), we have that

$$q(z) = q_o + z = -i\frac{\pi w_o^2}{\lambda} + z = -iz_R + z, \tag{6.2.7}$$

where

$$z_R = iq_o = \frac{k}{2}w_o^2 = \frac{\pi w_o^2}{\lambda} = \frac{\pi n w_o^2}{\lambda_o} \tag{6.2.8}$$

is said to be the Rayleigh's distance. The parameter q then behaves like the radius of curvature of the wavefront in Geometrical Optics as the distance from its curvature center varies; therefore, the term $-iz_R$ represents an imaginary correction of this concept to account for the wave nature of radiation in its propagation. A similar reasoning can be made for $1/q$, which we can indicate as the *complex curvature* of the wavefront, in which the imaginary term $i\lambda/(\pi w^2)$ is intended to take into account the finite and then localized extension of the field. It is important to note that (in our complex representation) the imaginary part of q is negative, so z_R is a positive number and the imaginary part of $1/q$ remains positive for all z values; otherwise, we will have a non-physical solution to the wave equations, with an amplitude of the field that diverges to infinity when the distance from the beam axis increases.

While q varies *linearly* with the position z, the same does not happen for the size w and the curvature radius R of the beam, since the translation (6.2.7) mixes the real and imaginary parts of the complex curvature $1/q$ between them. Therefore, q fits well with the calculation of propagation, while $1/q$ is better suited to the transformations induced by localized optical elements, such as dioptres, as will be discussed in §6.6.

When dealing with Gaussian beams, the origin for the beam axis (our z-axis) is

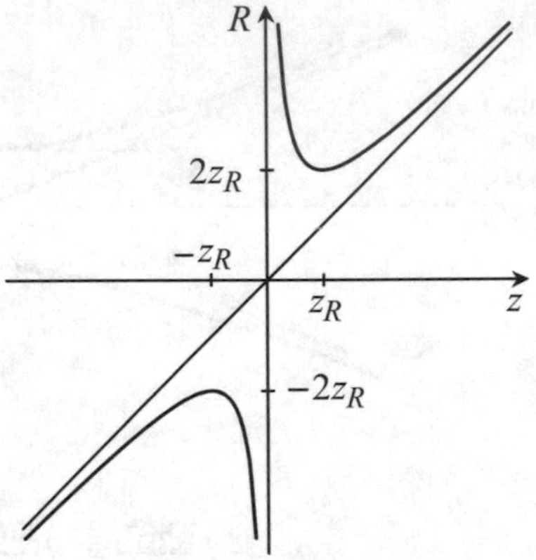

Fig. 6.2
Trend of the radius of curvature of the wavefront of a Gaussian beam as the distance from the waist plane changes

frequently and conveniently taken on the waist point. Combining Eqs. (6.2.4) and (6.2.7) and separating the real and imaginary parts, we obtain how w and R vary with z:

$$w^2(z) = w_o^2\left[1+\left(\frac{z}{z_R}\right)^2\right], \quad R(z) = z\left[1+\left(\frac{z_R}{z}\right)^2\right]. \tag{6.2.9}$$

We note, in particular, that $w(z)$ is minimal in $z = 0$ and, from this fact, we derive that the name of "waist radius" is assigned to w_o (Fig. 6.1); furthermore, $w(z)$ has the form of a hyperbola with asymptotes inclined relatively to the axis of the beam of an angle

$$\vartheta = \frac{\lambda}{\pi w_o} = \frac{w_o}{z_R} \tag{6.2.10}$$

and 2ϑ is the total angle of divergence of the Gaussian beam. It is interesting to note that, once λ is assigned, the three characteristic beam quantities, w_o, z_R, ϑ, *can all derive from one another*. R, instead, has an extreme in $\pm z_R$ (where it is $\pm 2z_R$) and it diverges in $z = 0$ and $z = \pm\infty$ (Fig. 6.2).

A useful relationship is obtained with Eq. (6.2.8) dividing Eq. (6.2.9.a) for Eq. (6.2.9.b):

$$\frac{z}{z_R} = \frac{\lambda z}{\pi w_o^2} = \frac{\pi w^2}{\lambda R}, \tag{6.2.11}$$

which can be used to derive w_o and z in terms of w and R:

$$w_o^2 = w^2\left[1+\left(\frac{\pi w^2}{\lambda R}\right)^2\right]^{-1}, \quad z = R\left[1+\left(\frac{\lambda R}{\pi w^2}\right)^2\right]^{-1}. \tag{6.2.12}$$

Finally, with Eq. (6.2.7), the differential Eq. (6.2.2.b) can be made explicit for the complex phase difference

$$\frac{dP}{dz}=\frac{i}{q}=\frac{i}{z-iz_{\mathrm{R}}},$$

whose integration leads to

$$-iP(z)=\ln\left(1+i\frac{z}{z_{\mathrm{R}}}\right)=\ln\sqrt{1+\left(\frac{z}{z_{\mathrm{R}}}\right)^2}+i\arctan\left(\frac{z}{z_{\mathrm{R}}}\right). \tag{6.2.13}$$

The real part of P reprcsents the phase difference ϕ between the Gaussian beam and a plane wave, while the imaginary part produces an amplitude factor w_o/w that gives the expected decrease in intensity for the beam expansion, so that the spatial trend of the field of the fundamental Gaussian mode can be written in the form

$$u(r,z)=u_o\frac{w_o}{w}e^{i(kz-\phi)+i\frac{kr^2}{2q}}=u_o\frac{w_o}{w}e^{i(kz-\phi)-r^2\left(\frac{1}{w^2}-\frac{ik}{2R}\right)}, \tag{6.2.14}$$

where

$$\phi=\arctan\left(\frac{z}{z_{\mathrm{R}}}\right)=\arctan\left(\frac{\lambda z}{\pi w_o^{\ 2}}\right)=\arctan\left(\frac{\pi w^2}{\lambda R}\right) \tag{6.2.15}$$

is the Gouy's phase gained by the beam from the waist position and where w and R are given by Eqs. (6.2.9). Unlike Geometrical Optics, in which there is a discontinuous jump of π around the focal point of a spherical wave, here the Gouy's phase varies with continuity from $-\pi/2$ to $\pi/2$ crossing the waist position. The term in r^2 in the exponent of Eq. (6.2.14) can be made explicitly in function of z as follows:

$$-r^2\left(\frac{1}{w^2}-i\frac{k}{2R}\right)=-\frac{r^2}{w^2}\left(1-i\frac{\pi w^2}{\lambda R}\right)=-\frac{r^2}{w_o^{\ 2}}\frac{1}{1+iz/z_{\mathrm{R}}}, \tag{6.2.16}$$

where, in the last step, Eqs. (6.2.9.a) and (6.2.11) have been exploited. Directly using the first equality of Eq. (6.2.13), the amplitude of the field of our monochromatic Gaussian wave is explicitly given by

$$u(r,z)=u_o\frac{q_o}{q_o+z}e^{ikz+i\frac{kr^2}{2(q_o+z)}}=u_o\frac{1}{1+iz/z_{\mathrm{R}}}e^{ikz-\frac{r^2}{w_o^2(1+iz/z_{\mathrm{R}})}}. \tag{6.2.17}$$

The integration in the x, y transverse coordinates of the square module of u is

$$\int_{-\infty}^{\infty}\int_{-\infty}^{\infty}\left|u(r,z)\right|^2 dxdy = \left|\frac{u_o q_o}{q}\right|^2 \int_{-\infty}^{\infty}\int_{-\infty}^{\infty} e^{i\frac{k}{2}\left(\frac{1}{q}-\frac{1}{q^*}\right)\left(x^2+y^2\right)} dxdy = \left|u_o\right|^2 \frac{\pi}{2} w_o{}^2 . \quad (6.2.18)$$

from which it follows that the power carried by the beam is independent of z. In other words, even the paraxial wave equation, which is an approximation of electromagnetic wave equations, conserves energy.

6.3 Modes of higher order

Along with Eq. (6.2.1), there exist other solutions to Eq. (6.1.3) that form a complete and orthogonal basis of functions that are called modes of propagation. Thus, any arbitrary distribution of monochromatic light, at a fixed z, can be expanded in terms of these modes. Of course, the choice of the base is arbitrary. Below, I describe the two most used bases and anticipate that these Gaussian modes propagate while retaining the shape of their profile.

6.3.1 Modes in Cartesian coordinates

When we have a system with a rectangular geometry, we can search a base of solutions to Eq. (6.1.3) by factorizing the amplitude as follows:

$$\psi = \psi_x \psi_y , \quad (6.3.1)$$

with

$$\psi_j = G_j F_j e^{-i\phi_j + i\frac{k}{2q_j}x_j{}^2} , \quad (6.3.2)$$

where $j = x, y$; $x_j = x, y$. Besides, $G_j = G_j(x_j,z)$, $F_j = F_j(z)$, $\phi_j = \phi_j(z)$ are real functions of the coordinates, while $q_j = q_j(z)$ is a complex function. For each component j, the wave equation for the amplitudes (6.1.3) becomes

$$\frac{\partial^2 \psi_j}{\partial x_j{}^2} = -2ik\frac{\partial \psi_j}{\partial z} . \quad (6.3.3)$$

Replacing Eq. (6.3.2) in Eq. (6.3.3), and by requiring that the terms in $x_j{}^2$ vanish among them, we still find that q_j can be expressed as

$$q_j = -iz_{\mathrm{R}\,j} + z - z_{oj}\,. \tag{6.3.4}$$

We also again take that

$$\frac{1}{q_j} = \frac{1}{R_j} + i\frac{\lambda}{\pi w_j{}^2}, \tag{6.3.5}$$

so, from Eq. (6.3.4), it is found that R_j and w_j obey equations similar to Eqs. (6.2.9):

$$w_j{}^2(z) = w_{\mathrm{o}j}{}^2\left[1+\left(\frac{z_j}{z_{\mathrm{R}j}}\right)^2\right], \quad R_j(z) = z_j\left[1+\left(\frac{z_{\mathrm{R}j}}{z_j}\right)^2\right], \tag{6.3.6}$$

with $z_j = z - z_{\mathrm{o}j}, \quad z_{\mathrm{R}j} = \dfrac{\pi w_{\mathrm{o}j}{}^2}{\lambda}$,

where $z_{\mathrm{o}j}$ and $w_{\mathrm{o}j}$ are, respectively, the waist positions and the corresponding waist radius. In particular, the waist planes for the *x*-axis and the *y*-axis do not necessarily coincide, i.e., they may have different coordinates $z_{\mathrm{o}x}$ and $z_{\mathrm{o}y}$. Note also that $R_j(z)$ and $w_j(z)$ do not depend on the form of G, F and ϕ, and therefore these parameters are the same for all modes of the base that we are looking for. Assuming, then, that

$$F_j(z) = \sqrt{\frac{z_{\mathrm{R}j}}{|q_j(z)|}} = \sqrt{\frac{w_{\mathrm{o}j}}{w_j(z)}}, \tag{6.3.7}$$

$$G_j(x_j, z) = g\left(\frac{\sqrt{kz_{\mathrm{R}j}}}{|q_j(z)|}x_j\right) = g\left(\frac{\sqrt{2}}{w_j(z)}x_j\right), \tag{6.3.8}$$

$$\phi_j(z) = \left(m+\frac{1}{2}\right)\arctan\left(\frac{z-z_{\mathrm{o}j}}{z_{\mathrm{R}j}}\right), \quad \text{with } m = 0,\ 1,\ 2,\ \ldots \tag{6.3.9}$$

we find, for g, the differential equation

$$g_m''(\eta) - 2\eta g_m'(\eta) + 2mg_m(\eta) = 0\,, \tag{6.3.10}$$

where the apexes $'$ and $''$ indicate the first and second derivatives and $\eta = \sqrt{2}\,x/w$. It has, for solution, a Hermite polynomial of order m. These polynomials are de-

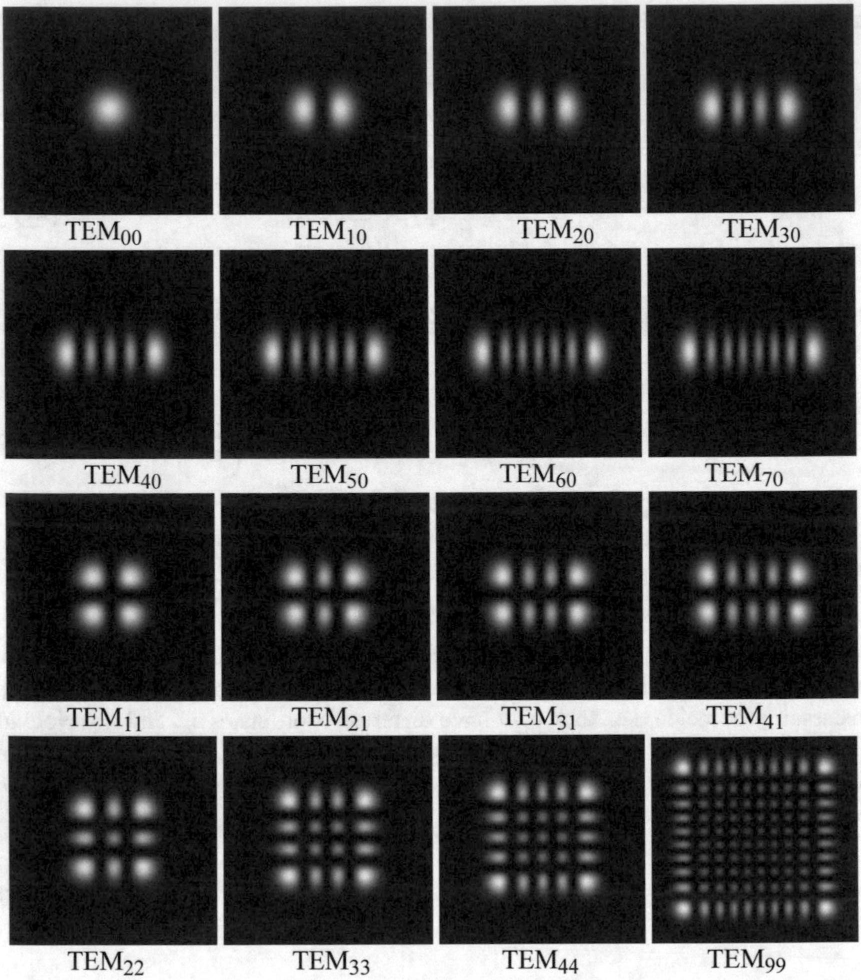

Fig. 6.3 Simulated examples of Gauss-Hermite modes; linear representation of the intensity in the grayscale. For comparison, see the measured modes reported by Kogelnik and Li (1966a)

noted by H_m and form a base of orthogonal functions that verify the conditions[1]

$$\int_{-\infty}^{\infty} e^{-\eta^2} H_m(\eta) H_n(\eta) d\eta = 0 \quad \text{for } n \neq m; \; n, m = 0, 1, 2, \ldots,$$

$$\int_{-\infty}^{\infty} e^{-\eta^2} H_m{}^2(\eta) d\eta = \sqrt{\pi}\, 2^m m! \quad \text{for } m = 0, 1, 2, \ldots .$$

Finally, for the two amplitudes with which ψ is factored, we have:

[1] A detailed discussion of orthogonal polynomials, of which Hermite polynomials are a particular case, is given by Abramowitz and Stegun (1972).

$$\psi_{mj} = \psi_{oj}\sqrt{\frac{w_{oj}}{w_j}}H_m\left(\sqrt{2}\frac{x_j}{w_j}\right)e^{-i\left(m+\frac{1}{2}\right)\arctan\left(\frac{z_j}{z_{Rj}}\right)-x_j^2\left(\frac{1}{w_j^2}-\frac{ik}{2R_j}\right)}. \quad (6.3.11)$$

Therefore, the functions ψ_{mj} form the basis sought. Their square module integrated along the x_j axis is given by

$$\int_{-\infty}^{\infty}\left|\psi_{jm}\left(x_j,z\right)\right|^2 dx_j = \left|\psi_{oj}\right|^2\frac{w_{oj}}{w_j}\int_{-\infty}^{\infty} e^{-2\frac{x_j^2}{w_j^2}} H_m\left(\sqrt{2}\frac{x_j}{w_j}\right)dx_j$$
$$= \left|\psi_{oj}\right|^2\sqrt{\frac{\pi}{2}}w_{oj}2^m m!, \quad (6.3.12)$$

which is still independent of the position along the z-axis.

Finally, for a non-astigmatic beam, the overall amplitude is

$$u_{nm}\left(x,y,z\right) = \psi_{nm}e^{ikz}$$
$$= \psi_o\frac{w_o}{w}H_n\left(\sqrt{2}\frac{x}{w}\right)H_m\left(\sqrt{2}\frac{y}{w}\right)e^{ikz-i(n+m+1)\arctan\left(\frac{z}{z_R}\right)-r^2\left(\frac{1}{w^2}-\frac{ik}{2R}\right)}, \quad (6.3.13)$$

where z is the distance from the waist position. The modes of this shape are called *Gauss-Hermite modes* and are indicated by the symbol TEM_{nm}; some of these modes are shown in Fig. 6.3. In particular, for $n = m = 0$, we again get Eq. (6.2.14). The first 6 Hermite polynomials are

$$H_0(x)=1, \qquad H_1(x)=2x,$$
$$H_2(x)=4x^2-2, \qquad H_3(x)=8x^3-12x, \quad (6.3.14)$$
$$H_4(x)=16x^4-48x^2+12, \quad H_5(x)=32x^5-160x^3+120x.$$

6.3.2 Modes in cylindrical coordinates

After a few calculations, for a system having a cylindrical symmetry with coordinates (r, φ, z), a base of normalized circular modes is found with the form

$$u_p^{(l)}\left(r,\varphi,z\right) = \psi_p^{(l)}\left(r,\varphi,z\right)e^{ikz} =$$
$$= \sqrt{\frac{2p!}{\pi\left(p+|l|\right)!}}\frac{1}{w}\left(\sqrt{2}\frac{r}{w}\right)^{|l|}L_p^{(l)}\left(2\frac{r^2}{w^2}\right)e^{i\left(kz-\phi_{p,|l|}+l\varphi\right)-r^2\left(\frac{1}{w^2}-\frac{ik}{2R}\right)}, \quad (6.3.15)$$

where $w(z)$ and $R(z)$ are still given by Eqs. (6.2.9) and

$$\phi_{p,|l|}(z) = (2p + |l| + 1)\arctan\left(\frac{z}{z_R}\right). \tag{6.3.16}$$

The normalization is such that the integral of $|u|^2$ on the transverse coordinates is 1. Furthermore, $L_p^{(|l|)}$ is the generalized Laguerre polynomial of radial order p and angular order l, with p, $|l| = 0, 1, 2, \ldots$. The sign of l determines the helicity of the wavefront. Some low order Laguerre polynomials are:

$$\begin{aligned}
&L_0^{(|l|)}(x) = 1,\\
&L_1^{(|l|)}(x) = |l| + 1 - x,\\
&L_2^{(|l|)}(x) = \frac{1}{2}(|l|+1)(|l|+2) - (|l|+2)x + \frac{1}{2}x^2,\\
&L_3^{(|l|)}(x) = \frac{1}{6}(|l|+1)(|l|+2)(|l|+3) - \frac{1}{2}(|l|+2)(|l|+3)x + \frac{1}{2}(|l|+3)x^2 - \frac{1}{6}x^3.
\end{aligned} \tag{6.3.17}$$

The modes of this form are called *Gauss-Laguerre modes* and are indicated by the symbol TEM_{pl}. The parameter p indicates the number of rings around the central one, while the value of l indicates the phase change in units of 2π around the beam axis. The combination of two modes of the same amplitude with opposite values of l gives rise to an azimuthal amplitude modulation, whereby $2l$ is the number of "petals" that can be observed for each ring in the intensity profile.

It should be noted that, in cases of both the modes of Gauss-Hermite and those of Gauss-Laguerre, the Gouy's phase is an integer multiple of that of the fundamental mode. In other words, it increases with the increasing of the mode structure. This fact is very important in determining the resonance frequencies of a laser cavity.

Also very important is the fact that the wavefront of these modes with $l \neq 0$ has a helical shape. Therefore, they carry an orbital angular momentum, equal to $l\hbar$ per photon, on which a vast literature is being developed [Allen et al 1992]. A recent review on this subject and its applications is reported by Piccirillo et al. (2013).

6.3.3 Polarization of the modes

In the previous derivation, a generic component of the field, however oriented, is considered. The modes may thus have arbitrary polarization, uniform across the

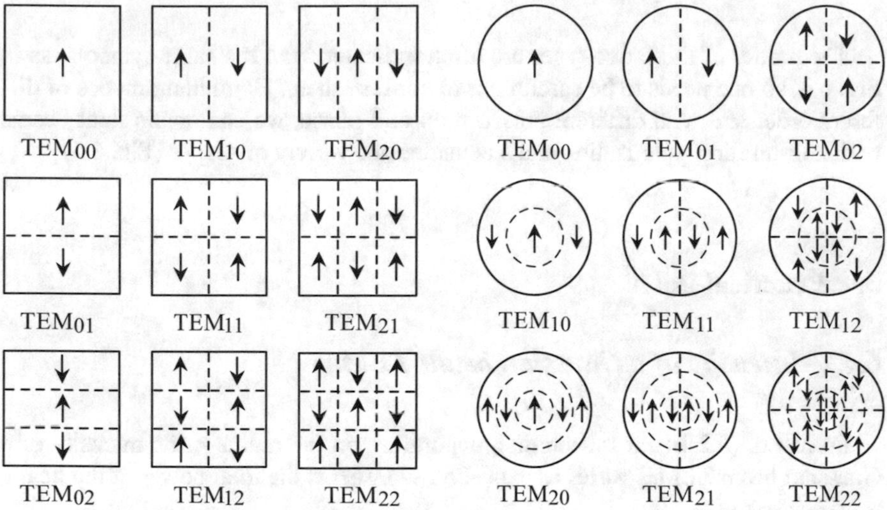

Fig. 6.4 Linear polarization modes of Gauss-Hermite and Gauss-Laguerre, which are obtained, respectively, in resonators with square and circular mirrors. The arrows indicate the direction and the sign of the electric field at a definite instant, that is, arrows in the opposite direction indicate a phase shift of π in the field oscillation. The polarization of the field is uniform over the whole beam profile, but, in general, its direction does not correspond to one of the axes of the mode and can also be circular or elliptical

beam, with axes that do not necessarily coincide with the axes of the mode (Fig. 6.4). In other words, the polarization of a Gaussian beam can be changed, for example, with a birefringent lamina, without its intensity profile being changed.

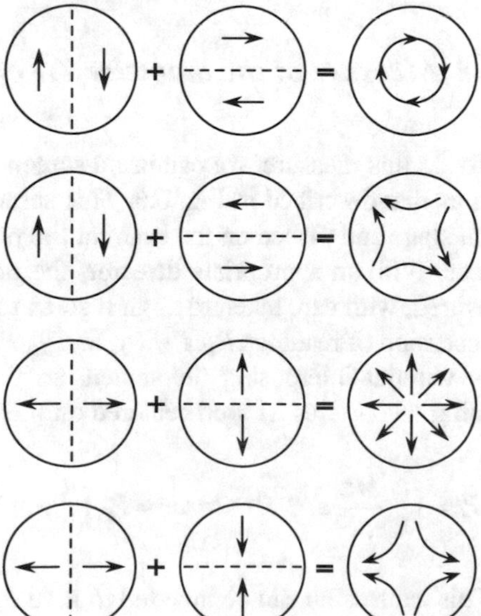

Fig. 6.5 Synthesis of various non-uniform polarizations by the sum of two linearly polarized TEM_{01} modes with equal intensity. The relative phase between the two modes is equal to zero, and the polarization is still locally linear, but changes direction. With a phase difference equal to π/2, there would be configurations in which the polarization goes continuously from linear to circular while the position changes

The modes of these two types are often indicated with the same symbols, as in Fig. 6.4, so one needs to be careful not to confuse them. Combining modes of different order and with different polarization and phase, we can obtain light beams with non-uniform polarization and a considerable variety of shapes (Fig. 6.5).

6.4 Practical notes

6.4.1 Intensity of a Gaussian beam TEM_{00}

From Eq. (6.2.14), it is seen that, depending on the radius r, the intensity of a Gaussian beam TEM_{00} varies as $\exp(-2r^2/w^2)$. If P is the total power of the beam, it turns out that

$$I(r) = \frac{2P}{\pi w^2} e^{-2r^2/w^2} , \tag{6.4.1}$$

where the coefficient $2/(\pi w^2)$ corresponds to the normalization for a two-dimensional Gaussian. The peak intensity is

$$I_{\mathrm{p}} = \frac{2P}{\pi w^2} . \tag{6.4.2}$$

6.4.2 Measure of the diameter $2w$ of a TEM_{00} mode

To do this measure, we can use a screen with a succession of calibrated holes, such as that described in Fig. 6.6. This screen must be mounted on an X-Y translation stage and placed on the beam at the point where you want to measure its diameter. With an appropriate detector, the power transmitted by each hole is then measured, with care taken to align it so as to maximize each reading. The result is a succession of readings $P_1, P_2, P_3, ..., P_n, P_\infty$, where, by the latter, I mean the one done without interposing the screen, so that the power transmitted by the i-th opening with radius R_i, well centered on the beam, is

$$P_i = \int_0^{R_i} \frac{2P_\infty}{\pi w^2} e^{-2r^2/w^2} 2\pi r dr = P_\infty \int_0^{R_i} e^{-2r^2/w^2} d\frac{2r^2}{w^2} = P_\infty \left(1 - e^{-2R_i^2/w^2}\right).$$

This expression can be inverted to give R_i/w as function of P_i/P_∞ . To evaluate

Fig. 6.6
Screen with circular holes for measuring w

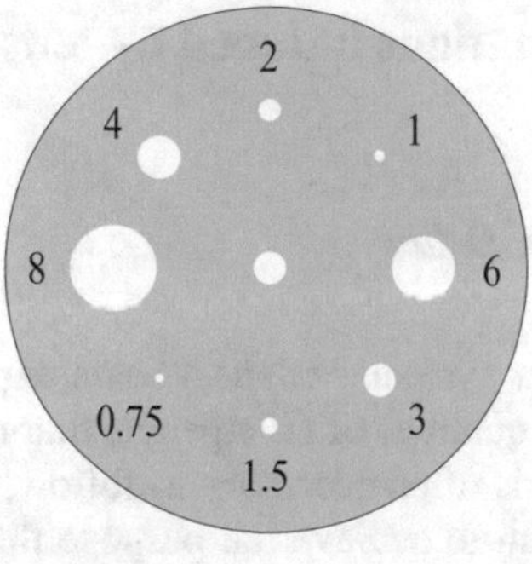

the $2w$ diameter, we proceed as follows:
a) write a table in which each row is:

x	y
R_i	$\sqrt{-\frac{1}{2}\ln\left(1-\frac{P_i}{P_\infty}\right)}=\frac{R_i}{w}$

b) then, make a graphical fit, with a straight line, of the x, y values obtained (Fig. 6.7). The reciprocal of the slope of the line, x/y, is the value w sought. Specifically, the hole that passes 86.5% of the total beam power has a radius approximately equal to w.

Next, as regards the radius of curvature of the wavefront and the collimation of the beam, a simple and effective measure is achieved with the shearing interferometer discussed in §3.2.7.6.

The above method is useful for relatively large beams. In the alternate case of measuring the waist position and the beam radius w_0 of a focused beam, it is still preferable to use a sharp blade, mounted on an xz or yz micrometric translation stage, and to perform a series of measurements at various positions z so as to reconstruct the Gaussian beam profile in the direction x or y, perform a fit to get w in these positions and then determine the waist position and the waist radius w_0. The detector should have a surface sufficiently wide to collect all of the diffracted radiation from the blade.

Fig. 6.7
Graphic construction to determine w

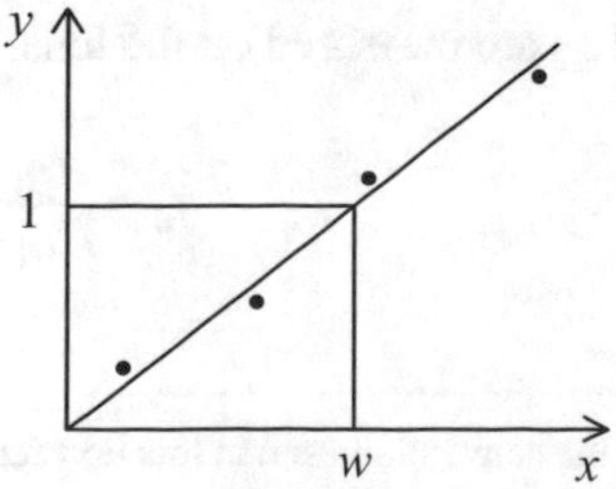

6.5 Transformations induced by lenses and by axial optical systems

6.5.1 The ABCD law

A lens or a lens system, such as a beam expander, can be used to give the laser beam the desired qualities of divergence, diameter, wavefront curvature, and waist position. Thus, it is often necessary to follow, and also understand, the beam path through the system to achieve the purpose that is intended, given that, in certain situations, there is a considerable discrepancy with the conclusions that could be deduced from the Geometrical Optics discussed herein. In this section, we will study an extension of the paraxial optical methods for manipulating Gaussian beams, starting with the behavior of a lens; along with that which Geometrical Optics teach us, we will see how this is enough to understand the operation of any ideal instrument.

An ideal lens does not change the transverse distribution of the field, i.e., each mode remains a mode of the same order when it passes through the lens. However, a lens changes the parameters $R(z)$ and $w(z)$, but, if they are the same for all modes, the lens behaves uniformly for all of the profiles obtainable by the combination of these modes. An ideal thin lens, with focal length f, transforms a spherical wave with a curvature radius R_1 immediately before the lens, into a spherical wave with radius R_2 immediately after the lens, with the rule deduced from the Gauss Eq. (2.6.35.a) of Geometrical Optics:

$$\frac{1}{R_2} = \frac{1}{R_1} - \frac{1}{f}, \tag{6.5.1}$$

where the curvature radii are again positive if the wavefront appears convex when it is "seen" from $z \to +\infty$.[2] Since, for a thin lens, the size of the beam immediately before and after the lens is the same, for Eq. (6.2.4), the parameters q for the incoming and outgoing beams are related by

$$\frac{1}{q_2} = \frac{1}{q_1} - \frac{1}{f}, \tag{6.5.2}$$

where q_1 and q_2 are measured on the lens. This expression can be reformulated as

$$q_2 = \frac{Aq_1 + B}{Cq_1 + D}, \tag{6.5.3}$$

[2] According to the convention used in this text for wavefronts.

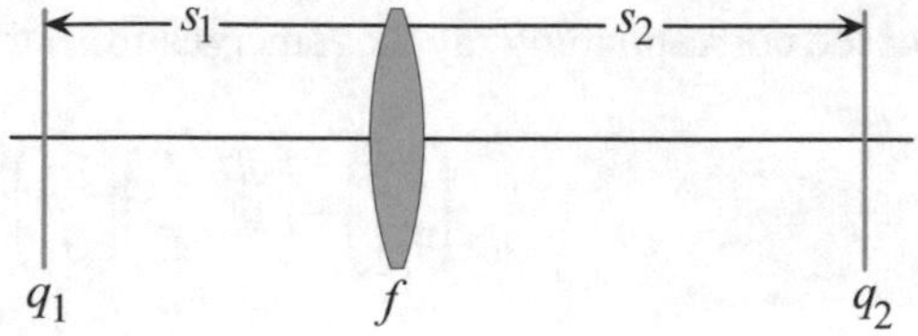

Fig. 6.8
Symbols conventions

where the coefficients $A = 1, B = 0, C = -1/f, D = 1$, are those corresponding to the matrix of a lens between its two principal planes. The same relation also formally applies in the case of a translation s on the z-axis. In this case, the coefficients are $A = 1, B = s, C = 0, D = 1$.

When q_1 and q_2 are finally measured at distances s_1 and s_2 from the lens, as shown in Fig. 6.8, by applying Eq. (6.2.3) to Eq. (6.5.2), we find that

$$q_2 = \frac{(1 - s_2/f)q_1 + (s_1 + s_2 - s_1 s_2/f)}{-(1/f)q_1 + (1 - s_1/f)}, \tag{6.5.4}$$

which can again be written in the form of Eq. (6.5.3). Indeed, as can be easily verified, the coefficients A, B, C, D are precisely the same as in the matrix obtained in paraxial Geometrical Optics

$$\begin{pmatrix} A & B \\ C & D \end{pmatrix} = \begin{pmatrix} 1 & s_2 \\ 0 & 1 \end{pmatrix} \begin{pmatrix} 1 & 0 \\ -1/f & 1 \end{pmatrix} \begin{pmatrix} 1 & s_1 \\ 0 & 1 \end{pmatrix} = \begin{pmatrix} 1 - s_2/f & s_1 + s_2 - s_1 s_2/f \\ -1/f & 1 - s_2/f \end{pmatrix} \tag{6.5.5}$$

for the same choice of distances s_1, s_2 and f, as shown in Fig. 6.8. Note that s_1 and s_2 are not necessarily conjugated distances.

An equivalent demonstration can also be done for a dioptre or for a mirror; importantly, one can easily prove that the successive application of two transformations as in Eq. (6.5.3) gives rise to an expression of the same type, where the coefficients A, B, C, D are those of the matrix resulting from the product of the matrices of each transformation. Therefore, Eq. (6.5.3) generally expresses the transformation of the Gaussian beam parameters q, even for more complicated optical systems. So, it is enough to know the transfer matrix for the paraxial rays of such a system to derive the way in which to transform a Gaussian beam that passes through it.

Eq. (6.5.3) is called *ABCD law* and, in spite of everything, is a simple consequence of a reinterpretation of the column vectors that describe the rays in Geometrical Optics. Indeed, let us consider an initial plane 1 and a final plane 2, among which the ABCD matrix of an optical system and a wave emitted by a point object O placed on the optical axis are calculated. Of this wave, we follow the path of a ray that, starting from O, reaches the initial plane with height ρ_1 and inclination ϑ_1. The curvature $1/R_1$ of the wavefront that reaches this plane is associated with these values with the paraxial relation $1/R_1 = \vartheta_1/\rho_1$. The same can be said for the corresponding ray that emerges from the final plane, for which

$$\begin{pmatrix} \rho_2 \\ \vartheta_2 \end{pmatrix} = \begin{pmatrix} A & B \\ C & D \end{pmatrix} \begin{pmatrix} \rho_1 \\ \vartheta_1 \end{pmatrix} . \tag{6.5.6}$$

The curvature of the wave emerging from the second plane is then

$$\frac{1}{R_2} = \frac{C\rho_1 + D\vartheta_1}{A\rho_1 + B\vartheta_1} = \frac{C + D/R_1}{A + B/R_1} , \tag{6.5.7}$$

which with a simple transformation, takes the form of Eq. (6.5.3). Its application to Gaussian beams is then a natural consequence of extending the concept of curvature as it appears in Eq. (6.2.4).

A good discussion of the matrix method is given by Kogelnik (1965). In §6.5, we will return to this argument by discussing an extension of the paraxial limit in the case of tilted beams with respect to the optical axis, and for the generalization of the ABCD law for the case of a generic non-axial optical system.

6.5.2 *Effects of a thin lens*

Let us now return to the case of a thin lens. An interesting way to rewrite Eq. (6.5.4) is obtained by replacing the variables s_1 and s_2 with the variables Z_1 and Z_2 given by

$$Z_1 = s_1 - f, \quad Z_2 = s_2 - f . \tag{6.5.8}$$

With some steps, Eq. (6.5.4) becomes

$$(q_2 - Z_2)(q_1 + Z_1) = -f^2 , \tag{6.5.9}$$

which closely resembles the Huygens-Newton Eq. (2.5.4). Let us try to understand its significance by considering the particular case in which $Z_1 = Z_o$ and $Z_2 = Z_i$ indicate the waist position for the incoming beam and the outgoing beam, respectively. In this case, as we have already seen with Eq. (6.2.6), the parameters q_1 and q_2 of departure and arrival are both purely imaginary:

$$q_1 = -i\frac{\pi}{\lambda} w_1{}^2 = -i\, z_{R1}, \quad q_2 = -i\frac{\pi}{\lambda} w_2{}^2 = -i\, z_{R2} . \tag{6.5.10}$$

Thus, we see that, between the two waist planes, conceived as object and image, q_1 and q_2 represent an imaginary correction in the Huygens-Newton equation to take account of the finite, non-punctiform dimension of the beam in those planes; moreover, unlike the classic case, the curvature radius of the wavefront present there is infinite, as we know from Eq. (6.2.4). When, instead, Eq. (6.5.9) refers to any Z positions, the real part of the parameters q, plus the values of Z, only reconstructs the position of the waist planes, so the real part of the terms in the two brackets of this equation represents the distance of the waist from the focal plane. Inserting Eq. (6.5.10) into Eq. (6.5.9) and developing the products, for the position of the Z_o and Z_i waist planes, we obtain the relation:

$$f^2 - z_{R1}z_{R2} - Z_o Z_i - i\left(z_{R2}Z_o - z_{R1}Z_i\right) = 0\,.$$

Therefore, our problem has a solution when both the real and imaginary parts are canceled, that is, for

$$\begin{cases} \dfrac{z_{R2}}{z_{R1}} = \dfrac{Z_i}{Z_o}, \\ Z_o Z_i = f^2 - z_{R1}z_{R2}. \end{cases} \tag{6.5.11}$$

Now, wishing to express z_{R2} and Z_i as a function of z_{R1} and Z_o, we find, with a few steps (multiply the first equation for $z_{R1}{}^2$, replace in the second, etc.), that,

$$\begin{cases} z_{R2} = \alpha^2 z_{R1}, \\ Z_i = \alpha^2 Z_o, \end{cases} \tag{6.5.12}$$

with

$$\alpha^2 = \frac{f^2}{Z_o{}^2 + z_{R1}{}^2}\,. \tag{6.5.13}$$

These equations are represented in Fig. 6.9 (a) and (b), respectively, for z_{R2} and Z_i, at the variation of $s_o = f + Z_o$, where s_o is the distance of the waist plane from the lens. Fig. 6.9(a) shows the Lorentzian form of $\alpha^2(Z_o)$ when z_{R1} and f are fixed, while $s_i = f + Z_i$, as a function of s_o, looks similar to a dispersion curve, as shown in Fig. 6.9(b). The maximum of this curve is at $Z_o = z_{R1}$, to which corresponds $Z_i = f^2/z_{R1}$. The slope of the curve in $s_o = f$, that is, for $Z_o = 0$, is $(f/z_{R1})^2$; in particular, for this point, $s_i = f$, that is, if the incident beam has its own waist on the focal plane of the object space, the transmitted beam has the waist on the focal plane of the image space. This result clearly deviates from the one from Geometrical Optics in the relationship between object and image distances, which is represented in Fig. 6.9(b) by the thin curve, which has a divergence at $\pm\infty$ for that value of s_o.

Recalling the relationship (6.2.8) between Rayleigh's distances and the waist radius, from Eq. (6.5.12.a), it is seen that α is the magnification parameter for the transverse dimension of the beam between the waist planes conjugated by the lens:

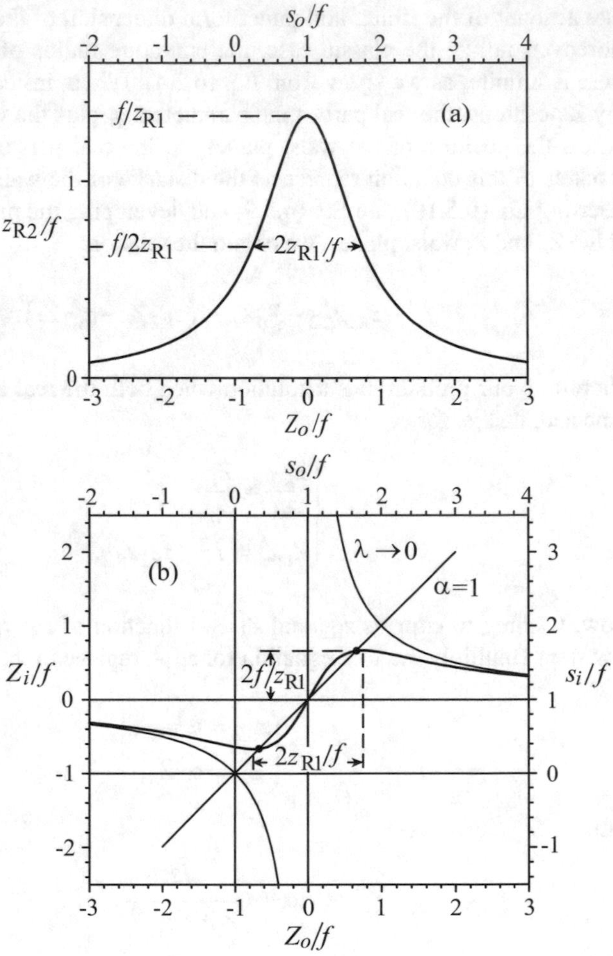

Fig. 6.9 Transformations of the waist of a Gaussian beam induced by a thin lens for an assigned value of $z_{R1}/f = 0.75$. (a): Rayleigh's distance as a function of the waist position. (b): waist position in the image space as a function of that in the object space.

With the variation of z_{R1}/f, there is a family of curves that, when this parameter is decreased, tend to the limit of the Geometrical Optics for $\lambda \to 0$, represented in (b) by the two hyperbola, while, in (a), the curve becomes increasingly high but also more narrow. The intersection with the diagonal line indicates the positions for which the magnification is $\alpha = 1$

$$w_2 = \alpha w_1 = \frac{|f|}{\sqrt{Z_o{}^2 + \left(\frac{\pi}{\lambda} w_1{}^2\right)^2}} w_1 . \tag{6.5.14}$$

It should be noted that the magnification is α for the transverse size and α^2 for the longitudinal displacements, for Eq. (6.5.12.b), similar to what we saw in Geometrical Optics. Also, since waists are defined as positive quantities, α is also defined as positive. This expression applies to both converging and diverging lenses; however, $Z_o = s_o - f$ depends on the sign of f.

In a practical problem, there may be a need for a fixed magnification α, so we have to arrange it by varying Z_o, or f, or both. If Z_o is also fixed, according to Eq. (6.5.13), we can calculate f. However, if we only have a lens with a given focal length (the one still remaining in the lens tray), we may not be able to succeed in our intent: just look at Fig. 6.9(a) to understand that α has a higher limit. Resolving Eq. (6.5.13) for Z_o, we have

$$Z_o = \pm\sqrt{\frac{f^2}{\alpha^2} - z_{R1}{}^2} = \pm\frac{1}{\alpha}\sqrt{f^2 - z_{R1}z_{R2}}\,. \tag{6.5.15}$$

To have one, indeed two(!) real solutions, it must be $f > \alpha z_{R1}$.

As an example we consider the case in which $\alpha^2 = 1$, corresponding to the diagonal line in Fig. 6.9(b), whereby $Z_o = Z_i$ and $z_{R1} = z_{R2} = z_R$. In this case, the total distance between the two waist planes is

$$d = 2f + 2Z_i = 2f \pm 2\sqrt{f^2 - z_R{}^2}\,.$$

If $f >> z_R$, then the two solutions for d are those known from Geometrical Optics, i.e., $d = 0$ and $d = 4f$, that is, the waist planes are either on the lens or both at a distance $2f$ from the lens. If, instead, $f = z_R$, it follows that $d = 2f$ and the beam waists are located at the respective focal planes of the lens.

It should be noted that, when the waist of the beam is located on the object focal plane, the magnification $|\alpha|$ is equal to the ratio $|f/z_{R1}|$ and, in these conditions, Eq. (6.5.14) becomes

$$w_2 = \frac{|f|\lambda}{\pi w_1}\,. \tag{6.5.16}$$

As an example, we consider the case of a laser beam, with $\lambda = 500$ nm, whose waist has a diameter of $2w_{o1} = 1$ mm, is placed at a focal distance $f = 30$ cm before a lens. In such conditions, it is $z_{R1} \cong 157$ cm, thus $\alpha = 0.191$, for which the lens focuses the beam in an area with diameter $2w_{o2} = 0.191$ mm on the second focal plane of the lens.

If, instead, the incident beam has its own waist on the lens, so that $Z_o = -f$, it is

$$Z_i = -f\,\alpha^2, \quad \alpha^2 = \frac{1}{1 + z_{R1}{}^2/f^2} \tag{6.5.17}$$

and, in the conditions of the previous example, we now have $\alpha = 0.188$, $2w_{o2} =$ 0.188 mm, $Z_i = -1.06$ cm.

6.5.3 Collimation of Gaussian beams

We have seen that, due to diffraction, a light beam cannot have a zero divergence except for an infinite extension of the diameter of the beams. We can therefore ask how we can give a definition of a collimated beam, once it has a fixed parameter, such as its diameter on a certain departure plane. There are two possible answers to this question, and this ambiguity can sometimes cause some confusion. The first definition is that the divergence *is the minimum possible*, and the second is that *the distance until the next waist of the beam is the maximum possible*.

Let us consider Eqs. (6.2.12): with a fixed value of w, our control parameter is R. For Eq. (6.2.10), the minimum divergence is when w_o is maximum, i.e., when $R = \infty$, and thus $w_o = w$. In other words, the minimum divergence occurs when the plane of the waist coincides with our departure plane, for which it is

$$\vartheta = \frac{\lambda}{\pi w}. \tag{6.5.18}$$

The maximum distance z, for Eq. (6.2.12.b), occurs when $(1/R) + R\lambda/(\pi w^2)$ is minimal, i.e., for $R = \pi w^2/\lambda$. In such a condition, $w_o = w/\sqrt{2}$, thus

$$\vartheta = \sqrt{2}\frac{\lambda}{\pi w}. \tag{6.5.19}$$

The difference from the previous case is an additional factor $\sqrt{2}$. The correspondent maximum distance is

$$z_{max} = \frac{\pi w^2}{2\lambda}. \tag{6.5.20}$$

A case of particular practical importance is when a first lens focuses a beam on a small spot. We can now ask where we should put a second lens, with assigned focal length $f > 0$, to meet either of the collimation conditions. In this problem, the initial data is the diameter $2w_{o1}$ in the plane of the waist, while the parameter to act on is the distance of the lens from it.

We can immediately see, from Eqs. (6.5.12.a) and (6.5.13), and also from Fig. 6.9(a), that the minimum divergence condition is for $s_o = f$, and therefore $s_i = f$ as well. In this case, we have that the diameter in the plane of the second waist is

$$2w_{o2} = 2\frac{\lambda}{\pi}\frac{f}{w_{o1}}, \tag{6.5.21}$$

which corresponds to a divergence

$$\vartheta = \frac{w_{o1}}{f}. \tag{6.5.22}$$

Instead, for the condition of maximum distance of the new beam waist *from the lens*, from Eqs. (6.5.12.b) and (6.5.13), we find that Z_o must be equal to z_{R1}, as can be seen from Fig. 6.9(b). Therefore,

$$s_o = f + Z_o = f + \frac{\pi}{\lambda} w_{o1}{}^2, \tag{6.5.23}$$

that is, the lens should be placed a little farther along! The maximum distance from the lens at which we can place the new waist in the image space is then

$$s_i = f + \frac{\lambda}{2\pi} \frac{f^2}{w_{o1}{}^2}. \tag{6.5.24}$$

Under these conditions, the new beam waist has a diameter of

$$2w_{o2} = \sqrt{2} \frac{\lambda}{\pi} \frac{f}{w_{o1}}, \tag{6.5.25}$$

which corresponds to a divergence

$$\vartheta = \sqrt{2} \frac{w_{o1}}{f}. \tag{6.5.26}$$

We can immediately notice that, to get a low divergence, it is good that $f >> w_{o1}$, so that if the initial beam is already well collimated, in order to further reduce the divergence, we must first focus it with a short focal length lens and then collimate it again with a long focal length lens, or, in other words, use a "reversed" telescope. The first lens may also be divergent, and, in that case, the telescope is Galilean.

6.5.4 *Effects of a dioptre*

Sometimes, it may be necessary to focus a Gaussian beam inside of a dielectric material; in this case, it is useful to know how a dioptre transforms the beam parameter q. The ABCD matrix including the translation is

$$\begin{pmatrix} 1 & s_2 \\ 0 & 1 \end{pmatrix} \begin{pmatrix} 1 & 0 \\ -\dfrac{K}{n_2} & \dfrac{n_1}{n_2} \end{pmatrix} \begin{pmatrix} 1 & s_1 \\ 0 & 1 \end{pmatrix} = \begin{pmatrix} 1 - \dfrac{s_2 K}{n_2} & s_1 + s_2 \left(\dfrac{n_1}{n_2} - \dfrac{s_1 K}{n_2} \right) \\ -\dfrac{K}{n_2} & \dfrac{n_1}{n_2} - \dfrac{s_1 K}{n_2} \end{pmatrix}, \tag{6.5.27}$$

where n_1 and n_2 are, respectively, the refractive indices in the initial and the final spaces, $K=(n_2-n_1)/R$ is the power of the dioptre of radius R, and ultimately, s_1 and s_2 are the distances from the vertex of the dioptre, with the same convention as in Fig. 6.8. The coefficients of this matrix can then be applied to Eq. (6.5.3) to determine the parameters q_1 and q_2 in the assigned positions.

In the event that the dioptre is plane, it is $K=0$, and we simply have

$$q_2 = q_1\frac{n_2}{n_1} + s_1\frac{n_2}{n_1} + s_2\,, \tag{6.5.28}$$

from which we see that, if s_1 and s_2 are taken on the beam waists, with

$$q_{o1} = -i\frac{n_1\pi}{\lambda_o}w_{o1}^{\,2}\,,\quad q_{o2} = -i\frac{n_2\pi}{\lambda_o}w_{o2}^{\,2}\,, \tag{6.5.29}$$

where the dependence of λ by the refractive index of the two media has been made explicit, then, from the imaginary and real parts of Eq. (6.5.28), we have

$$w_{o2} = w_{o1},\quad s_2 = -s_1\frac{n_2}{n_1}\,. \tag{6.5.30}$$

6.5.5 Determination of the waist position by means of the ABCD law

As an example of an application of ABCD matrices, let us suppose that we want to know the waist position and the waist size of a beam emerging from a given optical system once the waist position and the waist size of the incoming beam is assigned. In this case, one calculates the matrix, with coefficient A, B, C, D, between the plane of the first waist and any plane in the image space, such as that of the last surface of the optical system. It remains to determine the missing distance to reach the plane of the second waist. The overall matrix between the waist planes is then

$$\begin{pmatrix} A_c & B_c \\ C_c & D_c \end{pmatrix} = \begin{pmatrix} 1 & d \\ 0 & 1 \end{pmatrix}\begin{pmatrix} A & B \\ C & D \end{pmatrix} = \begin{pmatrix} A+dC & B+dD \\ C & D \end{pmatrix}. \tag{6.5.31}$$

For Eq. (6.5.10), the parameters q corresponding to the two waists are both purely imaginary and the expression that determines $q_2 = -iz_{R2}$ as a function of $q_1 = -iz_{R1}$ can be rewritten as

$$q_2 = \frac{A_c q_1 + B_c}{C_c q_1 + D_c} = \frac{A_c C_c z_{\mathrm{R1}}^2 + B_c D_c - i \frac{n_1}{n_2} z_{\mathrm{R1}}}{C_c^2 z_{\mathrm{R1}}^2 + D_c^2}, \tag{6.5.32}$$

where, in the second passage, the denominator was made real and, in the numerator, the fact was exploited that, for Eq. (2.7.8), the matrix determinant is equal to the ratio between the refractive indexes between the initial and final spaces. Since q_2 is imaginary, it must be that

$$A_c C_c z_{\mathrm{R1}}^2 + B_c D_c = 0,$$

from which we derive the missing distance to get the new waist

$$d = -\frac{AC z_{\mathrm{R1}}^2 + BD}{C^2 z_{\mathrm{R1}}^2 + D^2}. \tag{6.5.33}$$

Finally, the Rayleigh distance of the emerging beam is given by

$$z_{\mathrm{R2}} = \frac{n_1}{n_2} \frac{z_{\mathrm{R1}}}{C^2 z_{\mathrm{R1}}^2 + D^2}, \tag{6.5.34}$$

where I remember that C and D also depend on n_1 and n_2.

6.5.6 Amplitude and phase of a Gaussian beam transmitted by an optical system with axial symmetry

So far, we have been concerned with how the q parameter of a Gaussian beam is transformed. In some situations, such as interferometry, in the study of resonant cavities, and in holography, it is, however, also necessary to know the variation of the phase induced by the crossing of an optical system. To this end, it is necessary to use the diffraction theory, that is, the other approach to propagation mentioned in the introduction, the one of determining the response of a linear optical system to an input signal. Here, we will only apply the integral of S. A. Collins (see §4.2.7) for determining the field transmitted by a paraxial optical system, characterized by an assigned ABCD matrix [Collins 1970]. In his article, Collins analyzes a system with reduced symmetry, comprising ellipsoidal or cylindrical astigmatic optical surfaces, rotated at any angle around the optical axis; however, here, we will only use a simplified version of his integral (4.2.26) for the case of a system with symmetry of rotation. If, therefore, we consider the field present on two planes normal to the optical axis z, between which our optical system is inserted, the field on the final plane 2 as a function of that on the initial plane 1 is given by

$$E_2(x_2,y_2) = -\frac{i k_o n_1}{2\pi B}\sqrt{\frac{n_1}{n_2}} e^{ik_o L_o} \iint dx_1 dy_1\, E_1(x_1,y_1)$$
$$\times \exp\left\{\frac{i k_o n_1}{2B}\left[A\left(x_1^2+y_1^2\right) - 2\left(x_1 x_2 + y_1 y_2\right) + D\frac{n_2}{n_1}\left(x_2^2+y_2^2\right)\right]\right\}, \tag{6.5.35}$$

where k_o is the wave vector in the vacuum, L_o is the optical path on the axis of the system and n_1, n_2 are the refraction indices of the initial and final media, respectively; moreover, A, B, D are the corresponding elements of the system matrix between the planes 1 and 2. The element C does not appear explicitly, but, in the resolution of Eq. (6.5.35), Eq. (2.7.8) has been exploited for the determinant of the matrix. The matrix notation used by Collins is different from that of this text, and the original expression of the integral has been appropriately converted into Eq. (6.5.35).

Consider, then, the application of Collins' integral to the case of the propagation of a tilted off-axis laser beam into an axial optical system. This generalization with respect to the case of an axial beam will soon be useful and, to this end, we will follow the track of an article by M. Santarsiero [Santarsiero 1996]. First, we determine how the system transforms the axis of the Gaussian beam. In Geometrical Optics, we have seen that such a ray is defined by a vector consisting of the position relative to the axis and the angle of inclination. Here, we will also consider skew rays, but we will deal with their projections on the planes (x,z) and (y,z), and the vectors

$$\begin{pmatrix}\rho_x\\ \vartheta_x\end{pmatrix}, \quad \begin{pmatrix}\rho_y\\ \vartheta_y\end{pmatrix}, \tag{6.5.36}$$

of which we already know the transformation properties by an optical system characterized by an assigned ABCD matrix:

$$\begin{pmatrix}\rho_{2x}\\ \vartheta_{2x}\end{pmatrix} = \begin{pmatrix}A & B\\ C & D\end{pmatrix}\begin{pmatrix}\rho_{1x}\\ \vartheta_{1x}\end{pmatrix}, \quad \begin{pmatrix}\rho_{2y}\\ \vartheta_{2y}\end{pmatrix} = \begin{pmatrix}A & B\\ C & D\end{pmatrix}\begin{pmatrix}\rho_{1y}\\ \vartheta_{1y}\end{pmatrix}. \tag{6.5.37}$$

On the starting plane, the beam field can be written as

$$E_1(x_1,y_1,z_1) = \frac{E_{01} e^{i n_1 k_{oz} z_1}}{1+i z_1/z'_{R1}} e^{i n_1 k_o\left(\vartheta_{1x} x_1 + \vartheta_{1y} y_1\right) - \frac{\left(x_1-\rho_{1x}\right)^2 + \left(y_1-\rho_{1y}\right)^2}{w_{o1}^2\left(1+i z_1/z'_{R1}\right)}}, \tag{6.5.38}$$

where the transverse components of the wave vector $\boldsymbol{k} = n_1 \boldsymbol{k}_o$ have been approximated with $k\vartheta_{1x}$ and $k\vartheta_{1y}$, z_1 is the distance on the z-axis from the waist plane of the incoming beam and E_{o1} is the amplitude of the field on the waist. Also, the ellipticity of the beam section has been neglected. Moreover, the projection of the Rayleigh distance on the z-axis is

$$z'_{\mathrm{R}1} \cong z_{\mathrm{R}1}\sqrt{1-\vartheta_{1x}{}^2-\vartheta_{1y}{}^2}\,.$$

However, in the following, we will take $z'_{\mathrm{R}} = z_{\mathrm{R}}$ and, for brevity, we will use the coefficients

$$a = \frac{1}{w_{o1}{}^2\left(1+iz_1/z_{\mathrm{R}1}\right)} = -i\frac{z_{\mathrm{R}1}}{w_{o1}{}^2\left(z_1 - iz_{\mathrm{R}1}\right)} = -i\frac{n_1 k_o}{2q_1};\quad b = \frac{k_o}{2B}\,. \tag{6.5.39}$$

The integration that is obtained by inserting Eq. (6.5.38) into Eq. (6.5.35) can be factored into two separate integrations in x and y. For example, for x, we have

$$e^{in_2bDx_2{}^2}\int_{-\infty}^{\infty} dx_1 e^{ik_o n_1\vartheta_{1x}x_1 - a\left(x_1-\rho_{1x}\right)^2 + in_1b\left[Ax_1{}^2-2x_1x_2\right]}, \tag{6.5.40}$$

where, for the moment, we keep aside the amplitude and phase term in front of the exponential of Eq. (6.5.38) and the term preceding the double integration into Eq. (6.5.35). With the replacements

$$\xi = x_1 - \rho_{1x},\ \text{and the relationship}\ \rho_{2x} = A\rho_{1x} + B\vartheta_{1x}\,,$$

developing and re-aggregating the arguments of exponentials, we get

$$e^{ib\left[n_2Dx_2{}^2+n_1A\rho_{1x}{}^2-2n_1\left(x_2-B\vartheta_{1x}\right)\rho_{1x}\right]}\int_{-\infty}^{\infty} d\xi\, e^{-\left(a-in_1bA\right)\xi^2-2in_1b\left(x_2-\rho_{2x}\right)\xi}\,.$$

Resolving the integral, we have

$$\sqrt{\frac{\pi}{a-in_1bA}}\, e^{ib\left[n_2Dx_2{}^2-2n_1x_2\rho_{1x}+n_1A\rho_{1x}{}^2+2n_1B\vartheta_{1x}\rho_{1x}\right]-\frac{n_1{}^2b^2\left(x_2-\rho_{2x}\right)^2}{a-in_1bA}}\,.$$

In particular,

$$a - in_1bA = -in_1\frac{k_o}{2B}\left(A+\frac{B}{q_1}\right). \tag{6.5.41}$$

With several steps, repeatedly using Eq. (2.7.8) of the determinant of the ABCD matrix and the transformations (6.5.37), the argument of the exponential can be rewritten as

$$in_2k_o\vartheta_{2x}x_2 - i\frac{n_2k_o}{2}\left[C\rho_{1x}\rho_{2x} + B\vartheta_{1x}\vartheta_{2x}\right] + i\frac{n_2k_o}{2q_2}\left(x_2-\rho_{2x}\right)^2, \qquad (6.5.42)$$

where the ABCD law has also been exploited

$$q_2 = \frac{Aq_1+B}{Cq_1+D}, \quad \text{where } q_1 = z_1 - iz_{R1}, \quad q_2 = z_2 - iz_{R2}. \qquad (6.5.43)$$

The same set of expressions can also be written with respect to the y coordinate, for which the resulting field on the final plane is

$$\begin{aligned} E_2\left(x_2,y_2,z_2\right) &= \frac{-iz_{R1}\,E_{o1}\,e^{in_1k_{oz}z_1}}{\left(Aq_1+B\right)}\sqrt{\frac{n_1}{n_2}} \\ &\times e^{ik_oL_o - i\frac{n_2k_o}{2}\left[C\left(\rho_{1x}\rho_{2x}+\rho_{1y}\rho_{2y}\right)+B\left(\vartheta_{1x}\vartheta_{2x}+\vartheta_{1y}\vartheta_{2y}\right)\right]} \\ &\times e^{in_2k_o\left(\vartheta_{2x}x_2+\vartheta_{2y}y_2\right)-\frac{\left(x_2-\rho_{2x}\right)^2+\left(y_2-\rho_{2y}\right)^2}{w_{o2}^2\left(1+iz_2/z_{R2}\right)}}. \end{aligned} \qquad (6.5.44)$$

In conclusion, the beam emerging from the optical system is still a Gaussian beam that, thanks to Eq. (6.5.43), has the same size and radius of curvature as an axial beam; it also propagates along a path that, within the paraxial limit, simply follows the path of a geometric ray determined from Eqs. (6.5.37). In addition, the expression for the field contains a complex factor that ensures the conservation of total energy and establishes the phase shift introduced by the system.

6.6 Propagation in astigmatic paraxial optical systems

6.6.1 Astigmatic beams

In many applications, it happens that the optical system does not have an axial symmetry, and therefore a laser beam that is propagated therein assumes an astigmatic shape. Consider, for example, a beam, propagating initially with axial symmetry along the z-axis of a Cartesian reference x, y, z, that impinges on a cylindrical lens, whose generatrix is oriented along the x-axis. The wavefront curvature is altered only in the y direction. The emerging beam must then be treated with two distinct parameters q, one for the x-axis, and one for the y-axis, where the expression for the field can be factored. In particular, the waist positions will be in different positions on the z-axis. Similar situations happen for non-normal incidence on spherical surfaces, and for the refraction on inclined plane surfaces; this problem is also present in folded or ring cavities of dye lasers and titanium-sapphire lasers.

As long as it is possible to factorize the field in two orthogonal directions, the beam is said to be *simply astigmatic*. In this case, the ellipse of equal intensity and the ellipse of equal phase, drawn on the transverse beam profile, have their axes of symmetry coincident, and the propagation can be determined with little difficulty over what was treated in the previous sections.

Starting again from Eq. (6.2.17), a solution for the amplitude of the field of a simply astigmatic mode of fundamental order is given by

$$\Psi(x',y',z)=\sqrt{\frac{q_{xo}q_{yo}}{q_x q_y}}\,e^{i\frac{nk_o}{2}\left(\frac{x'^2}{q_x}+\frac{y'^2}{q_y}\right)}. \tag{6.6.1}$$

where x', y' and z are the coordinates of a system of Cartesian axes respectively coincident with the aforementioned symmetry axes and with the propagation axis, and

$$\begin{aligned} q_x(z)&=q_{xo}+(z-z_{xo})=-iz_{Rx}+z-z_{xo},\\ q_y(z)&=q_{yo}+(z-z_{yo})=-iz_{Ry}+z-z_{yo}. \end{aligned} \tag{6.6.2}$$

Therefore, the beam is defined by the two Rayleigh distances z_{Rx} and z_{Ry} and the corresponding waist positions z_{xo} and z_{yo} along the z-axis, while the dependence from z is implicitly contained in the variables q_x and q_y.

Wanting, instead, to make the Gouy's phase explicit, we have that

$$\Psi(x',y',z)=\sqrt{\frac{z_{Rx}z_{Ry}}{|q_x q_y|}}\,e^{-i\phi_x-i\phi_y+i\frac{nk_o}{2}\left(\frac{x'^2}{q_x}+\frac{y'^2}{q_y}\right)}, \tag{6.6.3}$$

where the phase shift $\phi_x+\phi_y$ is given by

$$\phi_x+\phi_y=\arctan\left(\frac{z-z_{xo}}{z_{Rx}}\right)+\arctan\left(\frac{z-z_{yo}}{z_{Ry}}\right)=\arctan\left(\frac{\Im m(q_x q_y)}{\Re e(q_x q_y)}\right). \tag{6.6.4}$$

This solution can be rotated by an angle φ around the z-axis, whereby, with a new pair of Cartesian coordinates x and y such that

$$\begin{pmatrix}x\\y\end{pmatrix}=\begin{pmatrix}\cos\varphi & \sin\varphi\\ -\sin\varphi & \cos\varphi\end{pmatrix}\begin{pmatrix}x'\\y'\end{pmatrix}, \tag{6.6.5}$$

Eq. (6.6.2) becomes bit more of a general expression:

$$\Psi(x,y,z)=\sqrt{\frac{q_{xo}q_{yo}}{q_x q_y}}\times e^{i\frac{nk_o}{2}\left[\left(\frac{\cos^2\varphi}{q_x}+\frac{\sin^2\varphi}{q_y}\right)x^2+\left(\frac{\sin^2\varphi}{q_x}+\frac{\cos^2\varphi}{q_y}\right)y^2+\sin 2\varphi\left(\frac{1}{q_y}-\frac{1}{q_x}\right)xy\right]}. \tag{6.6.6}$$

J.A. Arnaud and H. Kogelnik have shown that this expression remains a solution to the paraxial equation, even for a complex angle $\varphi = \beta + i\alpha$, with which it describes a *generally astigmatic* Gaussian beam capable of representing the propagation of a fundamental mode in a non-orthogonal optical system [Arnaud and Kogelnik 1969].[3] This happens thanks to the property that $\sin^2\varphi + \cos^2\varphi = 1$, even for complex arguments.[4] Eq. (6.6.6) can be rewritten as

$$\Psi(x,y,z)=\sqrt{\frac{q_{xo}q_{yo}}{q_x q_y}}\, e^{i\frac{nk_o}{2}\left[Q_x x^2+Q_y y^2+2Q_{xy}xy\right]}, \tag{6.6.7}$$

where

$$Q_x=\frac{\cos^2\varphi}{q_x}+\frac{\sin^2\varphi}{q_y},\quad Q_y=\frac{\sin^2\varphi}{q_x}+\frac{\cos^2\varphi}{q_y},$$
$$Q_{xy}=\sin\varphi\cos\varphi\left(\frac{1}{q_y}-\frac{1}{q_x}\right)=\frac{1}{2}\tan 2\varphi\left(Q_y-Q_x\right), \tag{6.6.8}$$

which, in turn, can be imagined as the components of a symmetric matrix

$$\mathbf{Q}=\begin{pmatrix} Q_x & Q_{xy}\\ Q_{xy} & Q_y\end{pmatrix}=\begin{pmatrix}\cos\varphi & \sin\varphi\\ -\sin\varphi & \cos\varphi\end{pmatrix}\begin{pmatrix}1/q_x & 0\\ 0 & 1/q_y\end{pmatrix}\begin{pmatrix}\cos\varphi & -\sin\varphi\\ \sin\varphi & \cos\varphi\end{pmatrix}, \tag{6.6.9}$$

---

3 The complex representation of the field they use in their work is the conjugate of that of this text.

4 Suppose, in fact, that $E(x,y,z)$ is a solution of the paraxial equation (6.1.3) and R is a linear operator transforming the x, y in x', y' transverse coordinates, leaving z unchanged. Then, we have that

$$R\nabla_\perp^\dagger \nabla_\perp E=-ikR\frac{\partial}{\partial z}E\rightarrow R\nabla_\perp^\dagger R^{-1}R\nabla_\perp R^{-1}RE=-ik\frac{\partial}{\partial z}RE,$$

where the symbol † means the transposed conjugate. If, therefore, $R^\dagger = R^{-1}$, we have that RE is the solution to the paraxial equation in the system x', y', z. In other words, it is sufficient that R is a unitary transformation.

which corresponds to the rotation of a diagonal matrix expressed in the system x', y' of the axes of equi-intensity ellipses and equi-phase ellipses or hyperbolas.

If we now consider the case of a complex rotation φ, taking, for x, y, a pair of real Cartesian axes, the axes x', y' diagonalizing the matrix are complex. Such a transformation can be decomposed into a real rotation β and an imaginary rotation $i\alpha$, which, acting in the same plane, that is, around the z-axis, commute between them. For the real rotation, we go back again to that shown above, and therefore we will limit ourselves in the following only to imaginary rotations. Since $\cos(i\alpha) = \cosh\alpha$ and $\sin(i\alpha) = i\sinh\alpha$, we have then

$$\begin{aligned} Q_x &= \rho_x \cosh^2\alpha - \rho_y \sinh^2\alpha + i\left(\sigma_x \cosh^2\alpha - \sigma_y \sinh^2\alpha\right), \\ Q_y &= \rho_y \cosh^2\alpha - \rho_x \sinh^2\alpha + i\left(\sigma_y \cosh^2\alpha - \sigma_x \sinh^2\alpha\right), \\ Q_{xy} &= \left(\sigma_x - \sigma_y\right)\sinh\alpha\cosh\alpha \ + i\left(\rho_y - \rho_x\right)\sinh\alpha\cosh\alpha. \end{aligned} \tag{6.6.10}$$

To obtain this, in Eq. (6.6.8), the following substitutions were made:

$$\begin{aligned} &\varphi = i\alpha \\ &1/q_j = \rho_j + i\sigma_j, \quad j = x, y, \end{aligned} \tag{6.6.11}$$

where ρ_j and σ_j are, respectively, the terms of curvature and extinction of the beam. Their dependence on the coordinate z is deduced from Eq. (6.6.2), obtaining

$$\begin{aligned} \rho_j(z) &= \frac{z - z_{jo}}{\left(z - z_{jo}\right)^2 + z_{Rj}^2}, \\ \sigma_j(z) &= \frac{z_{Rj}}{\left(z - z_{jo}\right)^2 + z_{Rj}^2}, \end{aligned} \tag{6.6.12}$$

where z_{jo} are the z-coordinates for the waists, while z_{Rj} are the Rayleigh distances. Observing the expression of Q_{xy} in Eq. (6.6.10), it is noted that the effect of an imaginary rotation is to mix curvature and extinction, so it takes two distinct real rotations to diagonalize the real part or the imaginary part of the matrix $\mathbb{Q}$

$$\begin{pmatrix} x_P \\ y_P \end{pmatrix} = \begin{pmatrix} \cos\varphi_P & \sin\varphi_P \\ -\sin\varphi_P & \cos\varphi_P \end{pmatrix}\begin{pmatrix} x \\ y \end{pmatrix}, \quad \begin{pmatrix} x_I \\ y_I \end{pmatrix} = \begin{pmatrix} \cos\varphi_I & \sin\varphi_I \\ -\sin\varphi_I & \cos\varphi_I \end{pmatrix}\begin{pmatrix} x \\ y \end{pmatrix}. \tag{6.6.13}$$

The rotation angles $\varphi_P(z)$ and $\varphi_I(z)$, respectively, of the axial systems of the equi-phase and equi-intensity curves are given by[5]

$$\tan 2\varphi_P = -\frac{\sigma_x - \sigma_y}{\rho_x - \rho_y}\tanh 2\alpha, \quad \tan 2\varphi_I = \frac{\rho_x - \rho_y}{\sigma_x - \sigma_y}\tanh 2\alpha. \tag{6.6.14}$$

With reference to these two systems, the amplitude of the field can be rewritten as

$$\Psi = \sqrt{\frac{q_{xo}q_{yo}}{q_x q_y}}\, e^{i\frac{nk_o}{2}\left(\frac{x_P{}^2}{R_{xP}}+\frac{y_P{}^2}{R_{yP}}\right)-\left(\frac{x_I{}^2}{w_{xI}{}^2}+\frac{y_I{}^2}{w_{yI}{}^2}\right)}, \tag{6.6.15}$$

where the semi-axes of the equi-intensity curves at $1/e^2$, $w_\pm = w_{xI}$, w_{yI}, are given by

$$\frac{\lambda}{\pi w_\pm{}^2} = \frac{1}{2}\left[\sigma_x + \sigma_y \pm \sqrt{\left(\sigma_x - \sigma_y\right)^2 \cosh^2 2\alpha + \left(\rho_x - \rho_y\right)^2 \sinh^2 2\alpha}\right] \tag{6.6.16}$$

and the principal radii of curvature $R_\pm = R_{xP}$, R_{yP} of the wavefront are given by

$$\frac{1}{R_\pm} = \frac{1}{2}\left[\rho_x + \rho_y \pm \sqrt{\left(\rho_x - \rho_y\right)^2 \cosh^2 2\alpha + \left(\sigma_x - \sigma_y\right)^2 \sinh^2 2\alpha}\right]. \tag{6.6.17}$$

These expressions can be derived from Eqs. (6.6.10), separately for the real and imaginary parts, remembering, for example, that the trace and the determinant of a matrix are preserved in a rotation.

More details of the behavior of a generally astigmatic beam are described in the work by Arnaud and Kogelnik cited above. In particular, they note that the half-axes of the equi-intensity and equi-phase curves rotate by varying the position along the *z*-axis and never coincide, except in the special case $\alpha = 0$ of simple astigmatism or in the stigmatic case where $q_x = q_y$. They also note that the field confinement condition, for which the principal semi-axes w_x and w_y are real, imposes a limit on the angle α of the imaginary rotation

$$\cosh 2\alpha \le \left|\frac{q_x - q_y{}^*}{q_x - q_y}\right|; \tag{6.6.18}$$

this condition is independent of *z*, so that a confined beam remains as such in its propagation in free space.

---

[5] For a real symmetric matrix $\begin{pmatrix} a & c \\ c & b \end{pmatrix}$, the angle φ of rotation that diagonalizes it, defined as in Eq. (6.8), is given by $\tan 2\varphi = 2c/(b-a)$.

6.6.2 ABCD law for the propagation in a non-axial system

Y. Suematsu and H. Fukinuki extended the ABCD law to Gaussian beams propagating in a non-axial optical system described by 4x4 matrices. Their publication [Suematsu and Fukinuki 1968] is not readily available, and also contains a rather complicated formalism. A further demonstration was given by J.A. Arnaud, who also discusses the case of higher-order modes [Arnaud 1970]. Here, I present a simple derivation for the fundamental mode.

We have already seen how to express the complex curvature $1/q$ of the beam as a 2x2 matrix; furthermore, in §2.7, we have seen that the 4x4 matrices of a non-axial optical system can be considered as the composition of 4 sub-matrices 2x2 $\mathfrak{A}$, $\mathfrak{B}$, $\mathfrak{C}$, $\mathfrak{D}$. For our purpose, we could imagine using an expression like Eq. (6.5.3); however, in the case of matrices, there are two problems, the first being that, unlike the scalar quantities, the matrices do not commute between them, the second being that the ratio between two matrices can be defined as the product between the matrix at the numerator and the inverse matrix of that at the denominator, but in what order?

An answer to these questions has already been presented in §2.7.5, where we have specifically studied how a wavefront is transformed from an optical system using its 4x4 paraxial matrix; in practice, it only remains to verify that the ABCD law (2.7. 67) introduced there, which I show again here, is also valid for the complex curvatures Q of a Gaussian beam:

$$\mathfrak{Q}_2 = \left(\mathfrak{D}\mathfrak{Q}_1 + \mathfrak{C}\right)\left(\mathfrak{B}\mathfrak{Q}_1 + \mathfrak{A}\right)^{-1}. \qquad (6.6.19)$$

For this purpose, it is sufficient to prove that it is valid for each simple element that composes the optical system.

First, Eq. (6.6.19) is still compatible with the case of a simple rotation $\mathfrak{R}$, so $\mathfrak{A} = \mathfrak{D} = \mathfrak{R}$, $\mathfrak{B} = \mathfrak{C} = 0$, and therefore

$$\mathfrak{Q}_2 = \left(\mathfrak{R}\mathfrak{Q}_1\right)\left(\mathfrak{R}\right)^{-1}. \qquad (6.6.20)$$

For the propagation in a homogeneous space for a distance d, we have $\mathfrak{A} = \mathfrak{D} = \mathfrak{I}$, $\mathfrak{B} = d\,\mathfrak{I}$, $\mathfrak{C} = 0$, where $\mathfrak{I}$ is the identity matrix, and thus we have that

$$\mathfrak{Q}_2 = \mathfrak{Q}_1\left(d\mathfrak{Q}_1 + \mathfrak{I}\right)^{-1}. \qquad (6.6.21)$$

In this expression, we apply a rotation $\mathfrak{R}$ for which

$$\mathfrak{R}\mathfrak{Q}_2\mathfrak{R}^{-1} = \mathfrak{R}\mathfrak{Q}_1\mathfrak{R}^{-1}\mathfrak{R}\left(d\mathfrak{Q}_1 + \mathfrak{I}\right)^{-1}\mathfrak{R}^{-1} = \mathfrak{R}\mathfrak{Q}_1\mathfrak{R}^{-1}\left(d\mathfrak{R}\mathfrak{Q}_1\mathfrak{R}^{-1} + \mathfrak{I}\right)^{-1},$$

and, if $\mathfrak{R}$ is the complex rotation of Eq. (6.6.9) that diagonalizes $\mathfrak{Q}$, we get

$$\begin{pmatrix} 1/q_{2x} & 0 \\ 0 & 1/q_{2y} \end{pmatrix} = \begin{pmatrix} 1/q_{1x} & 0 \\ 0 & 1/q_{1y} \end{pmatrix} \begin{pmatrix} d/q_{1x}+1 & 0 \\ 0 & d/q_{1y}+1 \end{pmatrix}^{-1}$$
$$= \begin{pmatrix} 1/(q_{1x}+d) & 0 \\ 0 & 1/(q_{1y}+d) \end{pmatrix}.$$

This expression is equivalent to Eqs. (6.6.2) and therefore Eq. (6.6.19) is also valid for translation.

As for locally induced transformations from refractive or reflective surfaces, we will rely on the same transformation rules that a wavefront undergoes with these surfaces in Geometrical Optics, as we have already done with Eq. (6.6.5.1). In essence, this procedure neglects the contribution of diffraction to the propagation between the transverse planes to the principal ray that lie immediately before and after the refraction or reflection surface. Therefore, for the real part of the complex curvature, i.e., for the wavefront curvature, the ABCD law applied to such surfaces has already been demonstrated in §2.7.5. It remains to verify its validity for the imaginary part.

Consider the case of a plane dioptre. Setting the curvature coefficients in Eq. (2.7.52) to zero, we obtain the matrix representing it in its reference system of the tangential and sagittal coordinates for the principal ray

$$\begin{pmatrix} \dfrac{\cos\vartheta_R}{\cos\vartheta_I} & 0 & 0 & 0 \\ 0 & 1 & 0 & 0 \\ 0 & 0 & \dfrac{n_I\cos\vartheta_I}{n_R\cos\vartheta_R} & 0 \\ 0 & 0 & 0 & \dfrac{n_I}{n_R} \end{pmatrix}, \tag{6.6.22}$$

where the subscripts I and R respectively indicate the incident beam and the refracted one and the other symbols correspond to those already described in §2.7. This represents the anamorphic transformation of the plane dioptre in which it can be noted, in particular, how the angular magnification is the reciprocal of the transverse one multiplied by n_I/n_R in accordance with Eq. (2.6.18). Applying Eq. (6.6.19), we have that

$$\mathbf{Q}_2 = \frac{n_I}{n_R}\begin{pmatrix} \dfrac{\cos\vartheta_I{}^2}{\cos\vartheta_R{}^2}Q_{1xx} & \dfrac{\cos\vartheta_I}{\cos\vartheta_R}Q_{1xy} \\ \dfrac{\cos\vartheta_I}{\cos\vartheta_R}Q_{1xy} & Q_{1yy} \end{pmatrix}. \tag{6.6.23}$$

Due to the fact that the only non-zero sub-matrices, $\mathfrak{A}$ and $\mathfrak{D}$, of the matrix (6.6.22) are diagonal, this transformation does not mix the components of the matrix $\mathfrak{Q}$; furthermore, the real and imaginary parts of the individual components Q are transformed in the same way. So, we can see that the coefficients w_x, w_y, w_{xy} of equal intensity ellipses undergo the same anamorphic transformation of Geometrical Optics, just as expected. In particular, the sagittal component Q_{yy} is only multiplied by the ratio between the indices, as we have for a stigmatic beam incident orthogonally over a plane interface (see §6.5.4). Instead, the tangential component Q_{xx} is transformed as

$$Q_{2xx} = \frac{n_{\rm I}}{n_{\rm R}} \frac{\cos \vartheta_{\rm I}{}^2}{\cos \vartheta_{\rm R}{}^2} \left(\frac{1}{R_{1xx}} + i \frac{\lambda}{\pi w_{1xx}{}^2} \right); \tag{6.6.24}$$

its imaginary component is transformed in accordance with the transversal enlargement of w_{xx}, since $\lambda_2 = \lambda_1 n_{\rm I}/n_{\rm R}$ and $w_{2xx} = w_{1xx}\cos\vartheta_{\rm R}/\cos\vartheta_{\rm I}$, while the real component is transformed in accordance with the longitudinal magnification of R_{xx}. Similar considerations can also be made for Q_{xy}, for which the coefficient of transformation is the geometric mean of the coefficients of Q_{xx} and Q_{yy}. Therefore, if we consider the transformation rule in Geometrical Optics of the parameters R and w of the equi-phase and equi-intensity curves, then Eq. (6.6.23) remains fully justified.

In the case of a curved dioptre, its 4x4 matrix is given by Eq. (2.7.52), which can be factorized as

$$\begin{pmatrix} \cos\vartheta_{\rm R} & 0 & 0 & 0 \\ 0 & 1 & 0 & 0 \\ 0 & 0 & \dfrac{1}{n_{\rm R}\cos\vartheta_{\rm R}} & 0 \\ 0 & 0 & 0 & \dfrac{1}{n_{\rm R}} \end{pmatrix} \begin{pmatrix} 1 & 0 & 0 & 0 \\ 0 & 1 & 0 & 0 \\ \delta C_{\rm PP} & \delta C_{\rm PS} & 1 & 0 \\ \delta C_{\rm PS} & \delta C_{\rm SS} & 0 & 1 \end{pmatrix} \begin{pmatrix} \dfrac{1}{\cos\vartheta_{\rm I}} & 0 & 0 & 0 \\ 0 & 1 & 0 & 0 \\ 0 & 0 & n_{\rm I}\cos\vartheta_{\rm I} & 0 \\ 0 & 0 & 0 & n_{\rm I} \end{pmatrix},$$

where the diagonal matrices at the extremes are the factorization of Eq. (6.6.22) and

$$\delta = n_{\rm I} \cos\vartheta_{\rm I} - n_{\rm R} \cos\vartheta_{\rm R}$$

is the coefficient that determines the power of the dioptre: if the index jump was null, then δ would be null and the curvature of the interface would not have any effect. The transformation associated with the dioptre can then be decomposed into a succession of three operations: $\mathfrak{Q}_1 \rightarrow \mathfrak{Q}_2 \rightarrow \mathfrak{Q}_3 \rightarrow \mathfrak{Q}_4$. Together, the two extreme matrices represent the effect of anamorphism due to the refractive index jump and the change of direction, both on the beam section and on the wavefront curvature, which we have just examined for a plane dioptre. Finally, the matrix in the center represents the contribution of the curvature; the transformation (6.6.19) applied to this matrix is simply

$$\mathfrak{Q}_3 = \begin{pmatrix} Q_{2xx} & Q_{2xy} \\ Q_{2xy} & Q_{2yy} \end{pmatrix} + \delta \begin{pmatrix} C_{\mathrm{PP}} & C_{\mathrm{PS}} \\ C_{\mathrm{PS}} & C_{\mathrm{SS}} \end{pmatrix}, \tag{6.6.25}$$

which has the same meaning as Eq. (6.6.5.2) of the stigmatic case. Since the surface curvature matrix is real, it only intervenes on the wavefront curvature.

In conclusion, the transformation (6.6.19) is also valid for dioptres.

It remains to verify the behavior of a reflective surface, which, in Geometrical Optics, is given by Eq. (2.7.54); applying Eq. (6.6.19), we now directly have that

$$\mathfrak{Q}_2 = \begin{pmatrix} Q_{1xx} & -Q_{1xy} \\ -Q_{1xy} & Q_{1yy} \end{pmatrix} + 2 \begin{pmatrix} \dfrac{1}{\cos\vartheta_{\mathrm{I}}} C_{\mathrm{PP}} & -C_{\mathrm{PS}} \\ -C_{\mathrm{PS}} & \cos\vartheta_{\mathrm{I}} C_{\mathrm{SS}} \end{pmatrix}, \tag{6.6.26}$$

in accordance with Eqs. (2.7.64). Note that, for the definition of Eqs. (2.7.32), we chose to take, for the reflection, $\cos\vartheta_{\mathrm{R}} = -\cos\vartheta_{\mathrm{I}}$. Even in this case, as in the previous, the matrix $\mathfrak{Q}$ remains symmetrical.

In conclusion, even in the case of a non-axial optical system, represented with a succession of rotations, refractions or reflections from interfaces, and propagation in homogeneous spaces, the propagation (of the fundamental mode) of a generally astigmatic Gaussian beam can be derived from the knowledge of the 4x4 matrix of the system by applying the ABCD law (6.6.19).

However, it remains to establish the phase gained by the wave between the initial and the final planes. For this purpose, we assume that the amplitude of the electric field on the first plane is that of a fundamental Gaussian mode expressed by Eq. (6.6.7)

$$E_1(x,y) = E_{\mathrm{o}}\, e^{i\frac{n_1 k_{\mathrm{o}}}{2}\left[Q_{1x}x^2 + Q_{1y}y^2 + 2Q_{1xy}xy\right]}, \tag{6.6.27}$$

where E_{o} is the on-axis amplitude of the mode, and we take into account Collins' integral (4.2.26) applied to it

$$\begin{aligned} E_2(x_2,y_2) = -i\sqrt{\frac{n_1}{n_2}}\frac{n_1 k_{\mathrm{o}}}{2\pi\sqrt{|\mathfrak{B}|}} e^{ik_{\mathrm{o}}L_{\mathrm{o}}} E_{\mathrm{o}} \iint_{-\infty}^{\infty} dx_1\, dy_1\, e^{i\frac{n_1 k_{\mathrm{o}}}{2}\rho_1^{\mathrm{T}}\mathfrak{Q}_1\rho_1} \\ \times e^{i\frac{k_{\mathrm{o}}}{2}\left(n_1\rho_1^{\mathrm{T}}\mathfrak{B}^{-1}\mathfrak{A}\rho_1 - 2n_1\rho_1^{\mathrm{T}}\mathfrak{B}^{-1}\rho_2 + n_2\rho_2^{\mathrm{T}}\mathfrak{D}\mathfrak{B}^{-1}\rho_2\right)}, \end{aligned} \tag{6.6.28}$$

and $\rho_1^{\mathrm{T}} = (x_1, y_1)$, $\rho_2^{\mathrm{T}} = (x_2, y_2)$. Following Arnaud (1970) on an impervious climb, we can reformulate it as

$$E_2(x_2, y_2) = -i\frac{n_1 k_o}{2\pi\sqrt{|\mathfrak{B}|}} e^{ik_o L_o} E_o e^{i\frac{k_o}{2} n_2 \rho_2^{\mathrm{T}} \mathfrak{D}\mathfrak{B}^{-1}\rho_2} \times \iint_{-\infty}^{\infty} e^{-\rho_1^{\mathrm{T}} \mathfrak{M} \rho_1 - 2\rho_1^{\mathrm{T}} s} dx_1\, dy_1 , \tag{6.6.29}$$

where

$$\mathfrak{M} = -i\frac{n_1 k_o}{2}\left(\mathfrak{Q}_1 + \mathfrak{B}^{-1}\mathfrak{A}\right), \quad s = i\frac{n_1 k_o}{2}\mathfrak{B}^{-1}\rho_2 . \tag{6.6.30}$$

Given that $\mathfrak{M}$ is a square matrix, and assuming that it is non-singular, the argument of the exponential in the integral can now be rewritten as

$$-\rho_1^{\mathrm{T}}\mathfrak{M}\rho_1 - 2\rho_1^{\mathrm{T}} s = -\left(\rho_1^{\mathrm{T}} + s^{\mathrm{T}}\mathfrak{M}^{-1}\right)\mathfrak{M}\left(\rho_1 + \mathfrak{M}^{-1}s^{\mathrm{T}}\right) + s^{\mathrm{T}}\mathfrak{M}^{-1}s . \tag{6.6.31}$$

Since, then, $\mathfrak{M}$ is a symmetric matrix, because both $\mathfrak{Q}_1$ and $\mathfrak{B}^{-1}\mathfrak{A}$ are symmetric, for Eq. (6.6.31), and with a translation of variables, the integral of Eq. (6.6.29) becomes

$$e^{s^{\mathrm{T}}\mathfrak{M}^{-1}s} \iint_{-\infty}^{\infty} e^{-\rho_1'^{\mathrm{T}}\mathfrak{M}\rho_1'} dx_1'\, dy_1' = \frac{\pi}{\sqrt{|\mathfrak{M}|}} e^{s^{\mathrm{T}}\mathfrak{M}^{-1}s} , \tag{6.6.32}$$

with the condition that $\rho_1'^{\mathrm{T}}\,\Re e(\mathfrak{M})\rho_1'$ is defined as positive. For Eq. (6.6.30.a), this is ensured by the fact that $\mathfrak{Q}_1$ expresses a spatially confined Gaussian beam. In addition, the double bar on $\mathfrak{M}$ represents the *determinant* of the matrix. Finally, for the amplitude of the fundamental mode on the final plane, we have that

$$E_2(x_2, y_2) = \frac{-i}{\sqrt{|\mathfrak{M}|}}\sqrt{\frac{n_1}{n_2}}\frac{n_1 k_o}{2\sqrt{|\mathfrak{B}|}} e^{ik_o L_o} E_o \times e^{i\frac{n_2 k_o}{2}\rho_2^{\mathrm{T}}\left[\mathfrak{D}\mathfrak{B}^{-1} - \frac{n_1}{n_2}\mathfrak{B}^{-1\mathrm{T}}\left(\mathfrak{Q}_1 + \mathfrak{B}^{-1}\mathfrak{A}\right)^{-1}\mathfrak{B}^{-1}\right]\rho_2} . \tag{6.6.33}$$

Thanks to Eqs. (2.7.71.c) and (2.7.72.b), the term in square brackets may, in turn, be reformulated as

$$\mathfrak{D}\mathfrak{B}^{-1} - \frac{n_1}{n_2}\mathfrak{B}^{-1\mathrm{T}}\left(\mathfrak{B}\mathfrak{Q}_1 + \mathfrak{A}\right)^{-1} = \mathfrak{B}^{-1\mathrm{T}}\left[\mathfrak{B}^{\mathrm{T}}\mathfrak{D}\mathfrak{B}^{-1}\left(\mathfrak{B}\mathfrak{Q}_1 + \mathfrak{A}\right) - \frac{n_1}{n_2}\mathfrak{I}\right]\left(\mathfrak{B}\mathfrak{Q}_1 + \mathfrak{A}\right)^{-1}$$
$$= \mathfrak{B}^{-1\mathrm{T}}\left[\mathfrak{D}^{\mathrm{T}}\mathfrak{B}\mathfrak{Q}_1 + \mathfrak{D}^{\mathrm{T}}\mathfrak{A} - \frac{n_1}{n_2}\mathfrak{I}\right]\left(\mathfrak{B}\mathfrak{Q}_1 + \mathfrak{A}\right)^{-1} = \left(\mathfrak{D}\mathfrak{Q}_1 + \mathfrak{C}\right)\left(\mathfrak{B}\mathfrak{Q}_1 + \mathfrak{A}\right)^{-1} = \mathfrak{Q}_2, \tag{6.6.34}$$

where $\mathfrak{Q}_2$ is the complex curvature of the mode on the final plane. Arnaud has thus shown the generalization of ABCD law to non-orthogonal optical systems. Moreover, for the property of determinants, $|\mathfrak{B}||\mathfrak{M}|=|\mathfrak{B}\mathfrak{M}|$ and the field on the output plane is given by

$$E_2(x_2,y_2)=\frac{1}{\sqrt{|\mathfrak{B}\mathfrak{Q}_1+\mathfrak{A}|}}\sqrt{\frac{n_1}{n_2}}E_o e^{ik_oL_o} e^{i\frac{n_2k_o}{2}\rho_2{}^{\mathrm{T}}\mathfrak{Q}_2\rho_2}, \tag{6.6.35}$$

where the root sign is chosen to make the result coincide with the starting field within the limit of a null path in a homogeneous medium. The first fraction of this expression is generally a complex number whose phase is given by

$$\phi=-\frac{1}{2}\arctan\left(\frac{\Im m|\mathfrak{B}\mathfrak{Q}_1+\mathfrak{A}|}{\Re e|\mathfrak{B}\mathfrak{Q}_1+\mathfrak{A}|}\right). \tag{6.6.36}$$

Siegman called it *Gouy's generalized phase*, which is added to the phase k_oL_o earned along the principal ray.

When the optical system is orthogonal so that it can be described by two distinct ABCD arrays for two orthogonal directions x, y, and the incoming beam is simply astigmatic so that $\mathfrak{Q}_1$ is diagonal along the same directions, the output beam remains simply astigmatic. The determinant under the root in Eq. (6.6.35) then becomes the product $(B_xQ_x + A_x)(B_yQ_y + A_y)$. Therefore, the Gouy's phase is the sum of two contributions

$$\phi=\phi_x+\phi_y,\quad \phi_j=-\frac{1}{2}\arctan\left(\frac{\Im m\left(B_jQ_{1j}+A_j\right)}{\Re e\left(B_jQ_{1j}+A_j\right)}\right),\quad j=x,y. \tag{6.6.37}$$

Finally, as a particular case of application of Eq. (6.6.35), we consider a fundamental mode with its own waist on the initial plane and assume that the system is simply the empty space, travelled for a distance z. In this case,

$$\frac{1}{\sqrt{|\mathfrak{B}\mathfrak{Q}_1+\mathfrak{A}|}}=\frac{1}{1+\dfrac{iz}{z_{\mathrm{R}}}},$$

whose phase coincides with that of Gouy given by Eq. (6.2.15).

The propagation of transverse modes beyond the fundamental can still be described by the product of the solution of the wave equation of the fundamental mode for a function of the transverse coordinates, but their treatment is dramatically more complicated [Arnaud 1970].

6.7 Bessel's waves

In the physics of waves, diffraction is a universally present phenomenon, such as in acoustic and electromagnetic waves and also in quantum mechanics, in which it is the basis of the uncertainty relationships defined by Heisenberg. Generally, for an initially spatially delimited wave, the diffraction results in a dispersion and a shape variation during its propagation in space. However, there is a class of solutions of the wave equation

$$\nabla^2 \boldsymbol{E}(x,y,z,t) - \frac{1}{\upsilon^2}\frac{\partial^2}{\partial t^2}\boldsymbol{E}(x,y,z,t) = 0 \tag{6.7.1}$$

that maintains its intensity distribution in the plane, say, (x, y), regardless of z, namely,

$$I(x,y,z) = I(x,y,0). \tag{6.7.2}$$

These solutions are said to be *diffraction-free*. An example of this type is a plane wave, infinitely extended. As we have seen by studying interference, another example is the superposition of two plane waves of equal frequency: taking as the z-axis the bisector of the directions of their vectors $\boldsymbol{k}_1$ and $\boldsymbol{k}_2$, the intensity is modulated sinusoidally in the direction of $\boldsymbol{k}_1 - \boldsymbol{k}_2$. A less trivial example, which has been identified by J. Durnin, is constituted by the superposition of infinite plane waves of equal amplitude and frequency, whose wave vectors, with module k, have an assigned projection $\beta = k_z$ on the z-axis, as shown in Fig. 6.10(a) [Durnin 1987; Durnin et al 1987]. These vectors are therefore arranged on a circle, as shown in Fig. 6.10(b), defined by

$${k_x}^2 + {k_y}^2 = \alpha^2 = k^2 - \beta^2, \tag{6.7.3}$$

which lies on a sphere of radius k in the space of the wave vectors. In other words, the propagation direction of each plane wave is inclined with respect to the z-axis of a given angle ϑ so that it is $\alpha = k\sin\vartheta$ and $\beta = k\cos\vartheta$. We also assume that all waves have a zero phase in the origin of the coordinates at $t = 0$. The resultant field is then defined as the integral over the angular variable φ that runs through this circle

$$\boldsymbol{E}(x,y,z,t) = e^{i\beta z - i\omega t}\frac{1}{2\pi}\int_{-\pi}^{\pi} \boldsymbol{E}_\varphi e^{i\alpha(x\cos\varphi + y\sin\varphi)} d\varphi,$$

where $\boldsymbol{E}_\varphi/(2\pi)$ is the amplitude (density) contribution of a plane wave whose wave vector is $\boldsymbol{k}_\varphi$. Let us also suppose that the polarization is defined by the complex amplitude $\boldsymbol{E}_\varphi = (E_{x\varphi}, E_{y\varphi}, E_{z\varphi})$ of each plane wave, it is linear and it lies on the

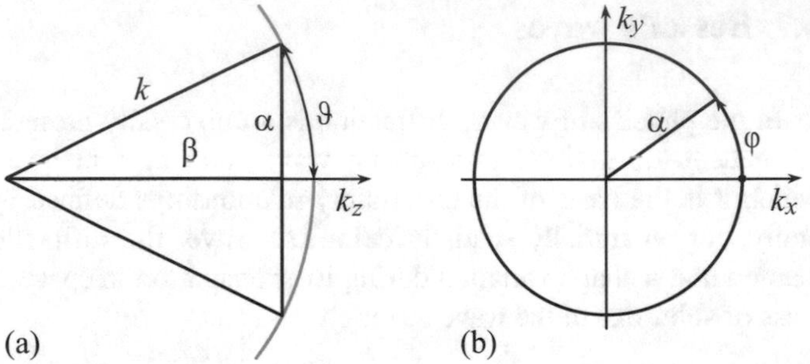

Fig. 6.10 Representation of the wave vectors of the plane waves that make a Bessel's beam

yz-plane, so it is $E_{x\varphi} = 0$. In particular, we assume the component $E_{y\varphi} = E_0$ constant, and therefore, for the orthogonality between $\boldsymbol{E}_\varphi$ and $\boldsymbol{k}_\varphi$ it follows that $E_{z\varphi} = -(\alpha/\beta)\sin\varphi\, E_{y\varphi}$. In the paraxial approximation, with $\vartheta << \pi/2$, we can ignore this contribution to the field and consider only its component y, so that, integrating, we have that

$$E(x,y,z,t) = \frac{1}{2\pi} E_o e^{i\beta z - i\omega t} \int_{-\pi}^{\pi} e^{i\alpha r \cos\varphi'} d\varphi' = E_o e^{i(\beta z - \omega t)} J_o(\alpha r), \quad (6.7.4)$$

where $r = \sqrt{x^2 + y^2}$ is the distance from the z-axis; furthermore, φ' is the difference between φ and the azimuthal angle φ'' corresponding to the couple (x, y), and, lastly E represents the y component of the field.

The Bessel's function J_o of order 0 constitutes a transverse modulation of the field, dependent on the sole distance r from the z-axis, with a maximum at $r = 0$ and radial oscillation whose spatial frequency is approximately $\alpha/(2\pi)$. The dependence on z and t is instead given by the exponential, from which it is seen that the phase velocity is ω/β. Because $\beta = k\cos\vartheta$ is given by the projection on the z-axis of the propagation velocity of each plane component, this speed is $c/(n\cos\vartheta)$, where n is the refractive index of the medium, and can therefore be varied with their angle ϑ of inclination: it can thus become much larger than c. The Bessel's wave intensity profile consists of a central peak surrounded by a series of concentric rings of decreasing intensity. The HWHM width of the central peak is given by

$$r_{\mathrm{HWHM}} \cong \frac{1.126}{\alpha} \cong 0.1792 \frac{\lambda}{\sin\vartheta}, \quad (6.7.5)$$

which, for relatively large angles, may be on the order of λ.

This profile is constant along the z direction, and is therefore free from diffraction effects in the sense specified above. It may, at first sight, appear well spatially localized, and therefore seems to violate the principle of uncertainty. However, the rate with which the rings' intensity decreases (Fig. 6.11) is only proportional to their radius, so each ring carries substantially the same energy as the central peak. In other words, as in the case of a single plane wave, the energy carried by the

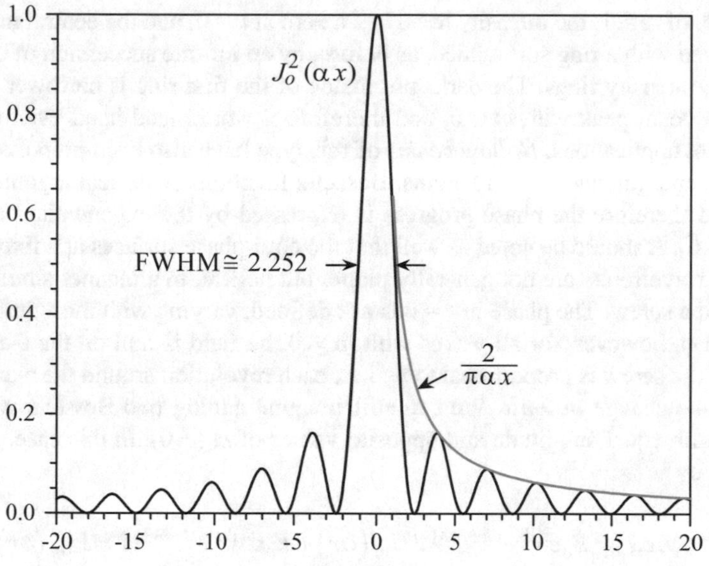

Fig. 6.11 A normalized intensity profile of a Bessel's wave along the *x*-axis. The gray curve indicates the asymptotic trend of the peaks

Bessel's wave is infinite, and therefore such a wave is not achievable experimentally, if not in approximate manner by the application of an opening function whose profile can take various shapes [see references quoted in Santarsiero (1996)]. One way to obtain an approximate Bessel's wave was already proposed in 1952 by J.H. McLeod, who had identified its *diffraction free* properties [McLeod 1953]. It is achieved by sending a plane wave toward an *axicon*, that is, among the various possibilities, a glass cone with a very large vertex angle. More recently, other authors have replicated the use of an axicon [Scott and McArdle 1992; Arlt and Dholakia 2000] or of diffractive elements [Vasara et al 1989] to produce approximate Bessel's beams or Bessel-Gauss beams of various order. In particular, zero-order Bessel's beams have found interest in the generation of second harmonics in non-linear crystals.

A generalization of the Bessel's wave can ultimately be obtained by associating each phase component with a phase term that varies periodically with the azimuthal angle φ. In this case, the integral of Eq. (6.7.4) is replaced with

$$\begin{aligned} E(x,y,z,t) &= \frac{1}{2\pi} E_o e^{i\beta z - i\omega t + im\varphi} \int_{-\pi}^{\pi} e^{+im\phi' + i\alpha r\cos\varphi'} d\varphi' \\ &= E_0 e^{i(\beta z - \omega t + m\varphi)}\, i^m J_m(\alpha r), \end{aligned} \tag{6.7.6}$$

where m is an integer number and J_m is the Bessel's function of order m. The polar coordinates (r, φ) still correspond to the Cartesian couple (x, y). In the case, for example, of $m = 1$, the intensity trend has a zero at $r = 0$, and the central maximum is replaced with a ring surrounded, as before, by an infinite succession of other decreasing intensity rings. The dark spot inside of the first ring is narrower than the Bessel's beam peak with $m = 0$, and therefore it would lend itself even better in alignment applications. *Hollow beams* of this type have also been proposed for the trapping and guidance of cold atoms. Bessel's functions (with real arguments) are real, and therefore the phase progress is expressed by the exponential present in Eq. (6.7.6). It should be noted as well that the equi-phase surfaces at a fixed time t, i.e., the wavefronts, are not generally plane, but helical, in a manner similar to the thread of a screw. The phase in $r = 0$ is not defined, varying with the azimuthal coordinate φ, however, for all waves with $m \neq 0$ the field is null on the z-axis. The pitch of the screw is proportional to m, i.e., each revolution around the z-axis causes a phase change of $2m\pi$. We can still imagine adding two Bessel's waves together with equal amplitude and opposite values of m ($\neq 0$). In this case, the field is given by

$$\begin{aligned} E(x,y,z,t) &= E_0 e^{i(\beta z-\omega t+m\varphi)}\, i^m \mathrm{J}_m(\alpha r) + E_0 e^{i(\beta z-\omega t-m\varphi)}\, i^{-m} \mathrm{J}_{-m}(\alpha r) \\ &= 2E_0 e^{i(\beta z-\omega t)}\, i^m \cos(m\varphi)\, \mathrm{J}_m(\alpha r), \end{aligned} \tag{6.7.7}$$

where we exploited the fact that $\mathrm{J}_{-m}(z) = (-)^m\, \mathrm{J}_m(z)$. The wavefronts return to be plane, but the intensity is modulated angularly, whereby each ring is broken into $2m$ arcs.

6.8 Bessel-Gauss beams

6.8.1 Zero-order Bessel-Gauss beams

A particular aperture function that can be applied to the Bessel's waves is that of a Gaussian function. In this way, one obtains Bessel-Gauss beams that can also be considered as the superposition of TEM_{00} Gaussian beams instead of plane waves, in the same manner as in the previous case with respect to the distribution of vectors $\boldsymbol{k}$ corresponding to their propagation axes [Gori et al 1987]. The main difference with Bessel's beams is that the intensity of Bessel-Gauss beams rapidly decays away from their axis of symmetry; therefore, they are effectively localized in the direction transverse to this axis. Under appropriate conditions, their transverse intensity profile remains roughly equal to itself, but only for a finite distance called *Bessel's distance*, which plays a similar role to the Rayleigh distance of Gaussian beams. Outside of the Bessel's range, the beam widens progressively, taking a ring shape.

Consider, then, the superposition of Gaussian beams of equal amplitude and sizes, whose propagation axes form a cone disposed around the z-axis and all intersect at the point ($x = 0, y = 0, z = 0$); in addition, let ϑ be the half-aperture angle of the cone around the z-axis. The plane $z = 0$ will also be the one where the waists of all beams lie. The resulting field is then given by

$$\begin{aligned} \boldsymbol{E}(x,y,z,t) &= e^{i\beta z - i\omega t}\frac{1}{2\pi}\frac{1}{1+iz/z_R} \\ &\times \int_{-\pi}^{\pi} \boldsymbol{E}_\varphi\, e^{i\alpha(x\cos\varphi + y\sin\varphi) - \frac{(x - z\tan\vartheta\cos\varphi)^2 + (y - z\tan\vartheta\sin\varphi)^2}{w_o^2(1+iz/z_R)}}\, d\varphi, \end{aligned} \tag{6.8.1}$$

where $\boldsymbol{E}_\varphi/(2\pi)$ is the amplitude contribution of each Gaussian component. Moreover, z_R is the Rayleigh distance that, within the paraxial limit, is approximated with its projection on the z-axis, similarly to what was done in §6.6. Lastly, w_o is the beam waist radius, and the ellipticity of their sections on planes orthogonal to the z-axis was neglected. By rearranging the terms, Eq. (6.8.1) becomes

$$\begin{aligned} \boldsymbol{E}(x,y,z,t) &= e^{i\beta z - i\omega t}\frac{1}{2\pi}\frac{1}{1+iz/z_R}\, e^{-\frac{r^2 + (z\tan\vartheta)^2}{w_o^2(1+iz/z_R)}} \\ &\times \int_{-\pi}^{\pi} \boldsymbol{E}_\varphi\, e^{\left[i\alpha + \frac{2z\tan\vartheta}{w_o^2(1+iz/z_R)}\right](x\cos\varphi + y\sin\varphi)}\, d\varphi, \end{aligned} \tag{6.8.2}$$

where r is the distance from the z-axis of the points with transverse coordinate (x,y). By assuming, within the paraxial limit, that $\boldsymbol{E}_\varphi = \boldsymbol{E}_o$ constant and $\vartheta \approx \tan\vartheta$, similarly to the case of Eq. (6.7.4), we get

$$\boldsymbol{E}(x,y,z,t) = \boldsymbol{E}_o e^{i\beta z - i\omega t}\frac{e^{-\frac{r^2 + (z\vartheta)^2}{w_o^2(1+iz/z_R)}}}{1+iz/z_R} \mathrm{J_o}\left(\alpha r - \frac{2iz\vartheta r}{w_o^2(1+iz/z_R)}\right), \tag{6.8.3}$$

where the Bessel's function $\mathrm{J_o}$ now has a complex argument. In particular, on the plane $z = 0$, the field is simply described by the product of a Gaussian for the Bessel's function:

$$\boldsymbol{E}(x,y,z=0,t) = \boldsymbol{E}_o e^{-i\omega t} e^{-r^2/w_o^2} J_o(\alpha r). \tag{6.8.4}$$

The behavior of a Bessel-Gauss beam is determined by two competing effects. On the one hand, each pair of Gaussian components tends to separate progressively with increasing $|z|$. On the other hand, the spot radius of each Gaussian component increases with respect to $|z|$, thus tending to prolong or even maintain the

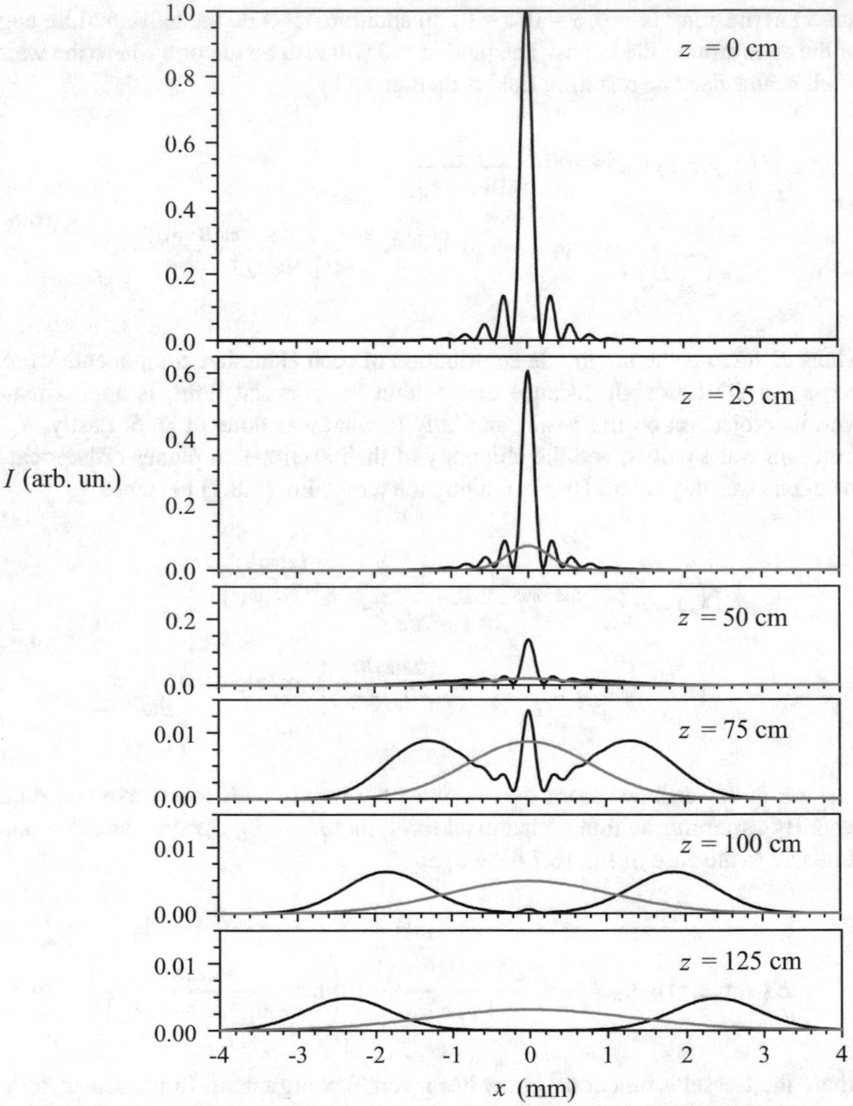

Fig. 6.12 Bessel-Gauss beam intensity profiles at different distances z, in the case of $\lambda = 1$ μm, $\vartheta = 2$ mrad, $w_0 = 1$ mm. These values correspond to a divergence ϑ_G of the Gaussian components equal to $(1/\pi)$ mrad, a ratio $\vartheta/\vartheta_G = 2\pi$ and a Bessel's distance $D = 50$ cm. The three figures at the bottom have the vertical scale magnified by 20 times compared to the other. The gray curves represent the trend of a Gaussian beam initially of the same width as the central peak of the Bessel-Gauss beam

overlapping of the various beams. In the far field zone, this overlap is determined by the ratio between the angular aperture of the cone of the wave vectors and the angle of divergence ϑ_G of each Gaussian beam:

$$\frac{\vartheta}{\vartheta_G} = \frac{\pi w_o}{\lambda} \arcsin\left(\frac{\alpha}{k}\right) \cong \frac{\alpha w_o}{2}, \tag{6.8.5}$$

where the approximation is valid for small angles ϑ. When this ratio is greater than 1, in the far field region, the shape of the Bessel-Gauss beam is that of a ring surrounding a dark region around the z-axis.

Instead, in the neighborhood of $z = 0$, the various components overlap, interfering with each other and giving rise to a central peak that reduces in height by maintaining the same width, as long as the tails of the various Gaussian components overlap appreciably (Fig. 6.12). This takes place indicatively up to the Bessel's distance D, defined in such a way that the center of each component has moved away from the z-axis of a distance equal to w_o, whereby

$$D = \frac{w_o}{\vartheta}. \tag{6.8.6}$$

This distance is a conservative estimate of the overlap distance, as the radius w of the beam components also grows with z.

When the ratio of Eq. (6.8.5) is less than 1, all Gaussian components continue to overlap and interfere, even in the far field zone. However, in this case, the Bessel-Gauss beam does not approach the case of a diffraction free beam; indeed, on the plane $z = 0$, the radius at half-height of the function $J_0(\alpha r)$ is about $2.4/\alpha$, while, for Eq. (6.8.5), $w_o < 2/\alpha$. This means that the subsequent oscillations of the function J_o are strongly damped, so that the beam profile is very similar to that of a Gaussian beam and, as such, widens progressively with increasing z.

In conclusion, a zero-order Bessel-Gauss beam represents a paraxial wave generalization, which has the extreme limits of a zero-order Bessel's beam for $w_o \to \infty$, and a Gaussian TEM_{00} beam for $\vartheta \to 0$.

6.8.2 Generalized Bessel-Gauss beams

Similarly to the Bessel's waves, higher order Bessel-Gauss beams can be obtained by adding to the various Gaussian components a periodic phase modulation, of the type $\exp(im\varphi)$, as the azimuthal angle φ changes: these are called *Bessel-Gauss beams of order m.*

An additional increase in complexity is introducing a radial distance between the center of the Gaussian beams and the z-axis at the plane of the waists (Fig. 6.13). This type of overlap produces what are called *generalized Bessel-Gauss beams*, which constitute a class of beams that contains those previously discussed as special cases [Bagini et al 1996]. Assuming, again within the paraxial limit, that $\boldsymbol{E}_\varphi = \boldsymbol{E}_0 e^{im\varphi}$ and $\vartheta \approx \tan\vartheta$, the resulting field is now

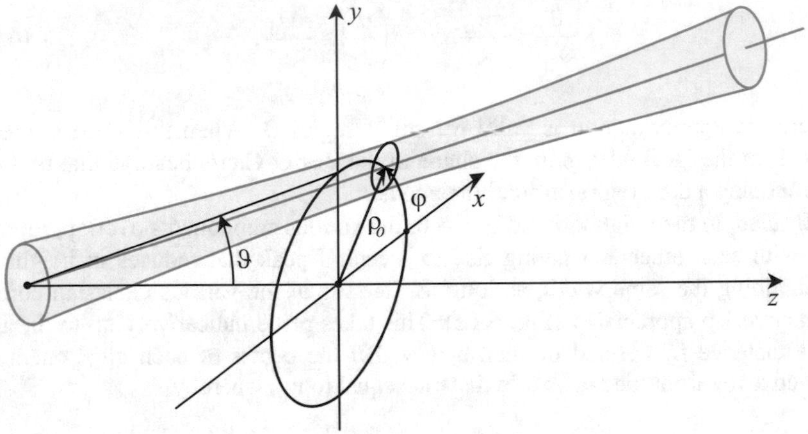

Fig. 6.13 Representation of one of the Gaussian components of a generalized Bessel-Gauss beam. ϑ is the half-aperture angle of the cone of the axes of the individual components and φ is their azimuthal angle; lastly, ρ_o is the radius that these axes trace on the waist plane, which is common to all components and on which the origin of the coordinates is fixed

$$E(x,y,z,t) = e^{i\beta z - i\omega t}\frac{1}{2\pi}\frac{1}{1+iz/z_R}\int_{-\pi}^{\pi} E_0\, e^{i\alpha(x\cos\varphi + y\sin\varphi)+im\varphi} \times e^{-\frac{[x-(\rho_o+z\vartheta)\cos\varphi]^2+[y-(\rho_o+z\vartheta)\sin\varphi]^2}{w_o^2(1+iz/z_R)}}\, d\varphi, \tag{6.8.7}$$

from which, rearranging the terms and expressing the field according to radial coordinates r and azimuthal coordinate φ, we have that

$$E(r,\varphi,z,t) = E_0 e^{i\beta z - i\omega t}\frac{e^{-\frac{r^2+(\rho_o+z\vartheta)^2}{w_o^2(1+iz/z_R)}+im\varphi}}{1+iz/z_R}\, i^m J_m\left[\alpha r - i\frac{2(\rho_o+z\vartheta)}{w_o^2(1+iz/z_R)}r\right]. \tag{6.8.8}$$

Specifically, for $\rho_o = 0$, Eq. (6.8.8) describes a Bessel-Gauss beam of order m.

But, in the case of taking that $\vartheta = 0$, and therefore $\alpha = 0$, the argument of the function J_m is purely imaginary, and the Bessel's function can be replaced with the modified Bessel's function. The resulting beam is then called a *modified Bessel-Gauss beam*. Since the Gaussian components are now parallel to each other, it can be generated simply by spatially filtering a Gaussian beam over its waist plane. This can be achieved by means of a lamina whose transmission is varied radially in amplitude and, in the case of $m \neq 0$, even in phase, in the azimuthal direction. For $m = 0$, with increasing parameter ρ_o, the modified beam takes a form that is transformed from Gaussian to flattened, and finally annular.

6.8.3 Propagation of a Bessel-Gauss beam through an axial optical system

Resuming the analysis of Santarsiero (1996), we can now easily determine the propagation of a generalized Bessel-Gauss beam through an axial optical system. It is enough to consider the effects that the system has on the individual Gaussian beam components, which, in this case, have all their propagation axis corresponding to a meridian ray. From Eq. (6.5.44), we therefore get

$$E_2(x_2,y_2,z_2,t)=E_{o1}\frac{-iz_{R1}}{(Aq_1+B)}\sqrt{\frac{n_1}{n_2}}$$
$$\times e^{-i\omega t+in_1k_{oz}z_1+ik_oL_o-i\frac{n_2k_o}{2}(C\rho_1\rho_2+B\vartheta_1\vartheta_2)-\frac{r_2^2+\rho_2^2}{w_{o2}^2(1+iz_2/z_{R2})}} \tag{6.8.9}$$
$$\times\frac{1}{2\pi}\int_{-\pi}^{\pi}d\varphi\, e^{im\varphi+\left[in_2k_o\vartheta_2+\frac{2\rho_2}{w_{o2}^2(1+iz_2/z_{R2})}\right](x_2\cos\varphi+y_2\sin\varphi)},$$

where the field is propagated between the initial plane 1 and the final plane 2 at which the ABCD matrix of the system is determined. These planes do not necessarily coincide with the waist planes of the Gaussian components, so z_j, with $j = 1$, 2, are the distances in the object space and in the image space from the corresponding waist planes, z_{Rj} are the Rayleigh distances, and w_{oj} are the waist radius. Besides, ρ_j are the radii of the circles that the axes of the component beams trace on planes 1 and 2, while ϑ_j are the half-aperture angles. Lastly, n_j are the refractive indices of the initial and final spaces, while k_o is the wave vector module in the vacuum and L_o is the optical path between planes 1 and 2. The transformation between the initial and final variables is given by the relations (6.5.37) for ρ_j and ϑ_j, by Eqs. (6.5.43) for the variables q_j, z_j and z_{Rj} and by Eq. (6.2.8) for w_{oj}.

By solving the integral of Eq. (6.8.9) and expressing the field according to the radial and azimuthal coordinates r and φ corresponding to the pair x_2, y_2, we get

$$E_2(r,\varphi,z_2,t)=$$
$$E_{o1}\frac{-iz_{R1}}{(Aq_1+B)}\sqrt{\frac{n_1}{n_2}}e^{-i\omega t+in_1k_{oz}z_1+ik_oL_o-i\frac{n_2k_o}{2}(C\rho_1\rho_2+B\vartheta_1\vartheta_2)} \tag{6.8.10}$$
$$\times e^{im\varphi-\frac{r^2+\rho_2^2}{w_{o2}^2(1+iz_2/z_{R2})}}i^m J_m\left[n_2k_o\vartheta_2 r-i\frac{2\rho_2 r}{w_{o2}^2(1+iz_2/z_{R2})}\right].$$

Consider, in particular, the propagation only for a distance z along the beam axis; therefore, the ABCD matrix is simply

$$\begin{pmatrix} A & B \\ C & D \end{pmatrix} = \begin{pmatrix} 1 & z \\ 0 & 1 \end{pmatrix}.$$

We also take the starting plane coincident with the waist plane of the incident beam. In this case,

$$q_1 = -iz_R = -i\pi w_o^2/\lambda, \; z_2 = z, \; z_{R2} = z_R, \quad w_{o2} = w_o, \; \vartheta_2 = \vartheta_1, \; \rho_2 = \rho_1 + \vartheta_1 z,$$

and, with these substitutions, we again find Eq. (6.8.8).

Suppose, instead, that we calculate the transformation of the generalized Bessel-Gauss beam by taking both the initial and final planes coinciding with the waist planes in their respective object and image spaces. Let A'B'C'D' be the matrix of the system between these planes:

$$\begin{pmatrix} A' & B' \\ C' & D' \end{pmatrix} = \begin{pmatrix} 1 & z_2 \\ 0 & 1 \end{pmatrix} \begin{pmatrix} A & B \\ C & D \end{pmatrix} \begin{pmatrix} 1 & z_1 \\ 0 & 1 \end{pmatrix} = \begin{pmatrix} A + Cz_2 & Az_1 + B + Cz_1 z_2 + Dz_2 \\ C & Cz_1 + D \end{pmatrix}. \tag{6.8.11}$$

Then, Eq. (6.8.10) becomes

$$E_2(r,\varphi,0,t) = E_{o2}\, e^{-i\omega t - \frac{r^2 + \rho_{o2}^2}{w_{o2}^2} + im\varphi}\, i^m \mathrm{J}_m\left[n_2 k_o \vartheta_2 r - i\frac{2\rho_{o2} r}{w_{o2}^2} \right], \tag{6.8.12}$$

where

$$E_{o2} = E_{o1} \frac{-iz_{R1}}{(iA' z_{R1} + B')} \sqrt{\frac{n_2}{n_1}}\, e^{ik_o L_o' - i\frac{n_2 k_o}{2}(C' \rho_{o1} \rho_{o2} + B' \vartheta_1 \vartheta_2)}, \tag{6.8.13}$$

with $L_o' = L_o + n_1 z_1 + n_2 z_2$. Eq. (6.8.12) has the same shape as Eq. (6.8.8) for $z = 0$, and thus the initial beam is transformed into another generalized Bessel-Gauss beam. Taking the origin of z coordinate on the waist plane in the object space, the new beam is therefore described by

$$E_2(r,\varphi,z,t) = E_{o2} \frac{e^{i\beta_2 z - i\omega t - \frac{r^2 + (\rho_{o2} + \vartheta_2 z)^2}{w_{o2}^2 (1 + iz/z_{R2})} + im\varphi}}{1 + iz/z_{R2}}\, i^m \mathrm{J}_m\left[\alpha_2 r - i\frac{2(\rho_{o2} + \vartheta_2 z) r}{w_{o2}^2 (1 + iz/z_{R2})} \right], \tag{6.8.14}$$

where we took that $\alpha_2 = n_2 k_o \vartheta_2$.

A particularly interesting case is when the *waist plane of the incident beam coincides with the object focal plane* of an optical system. Consequently, the emerging beam has its own waist plane on the image focal plane and the matrix to be applied has the shape of Eq. (2.7.14), whereby $A = D = 0$, $B = f_o$ and $C = -1/f_i$.

Applying the transformations (6.5.37), Eq. (6.8.14) becomes

$$E_2(r,\varphi,z,t) = E_{o1}\frac{-iz_{R1}}{f_o}\sqrt{\frac{n_1}{n_2}} \times e^{ik_o L_o + in_1 k_o \vartheta_1 \rho_{o1}} \frac{e^{i\beta_2 z - i\omega t - \frac{r^2 + (f_o\vartheta_1 - \rho_{o1}z/f_i)^2}{w_{o2}^2(1+iz/z_{R2})} + im\varphi}}{1+iz/z_{R2}} \times i^m J_m\left[-\frac{n_2 k_o \rho_{o1} r}{f_i} - i\frac{2(f_o\vartheta_1 - \rho_{o1}z/f_i)r}{w_{o2}^2(1+iz/z_{R2})}\right]. \quad (6.8.15)$$

where L_o is the optical path on the axis between the two focal planes. From the comparison between this equation and Eq. (6.8.8), it can be noted that the variables ρ and ϑ have exchanged their role. This can be easily understood by considering the behavior of the principal rays of the Gaussian beam components. In particular,

if, on the focal plane object, the incident beam is a modified Bessel-Gauss beam, that is, with $\vartheta_1 = 0$, on the image focal plane, it becomes an ordinary Bessel-Gauss beam, with $\rho_2 = 0$.

6.9 Resonant cavities

The Gaussian beams owe their popularity to the fact that they are a good representation of the laser beams generated within a class of resonant cavities. They typically form the solution for the electromagnetic field that is generated within such cavities containing an active medium in the presence of *feedback* that the optical system brings back on the active medium. This feedback is generally associated with the boundary conditions given by the cavity mirrors. Gaussian beams typically also give a good representation of the field within such passive optical cavities, i.e., without an internal active medium, injected externally by a laser beam.

The study of laser cavities is very complex, and I will only give you some vague notions here. For further details, see the monumental books of Anthony E. Siegman, *Lasers* [Siegman 1986] and Yu Anan'ev, *Laser resonator and the beam resonance problem* [Anan'ev 1992]. See also Solimeno et al (1986), Chap. VII.

In the optical region of the spectrum, the resonant cavities, with the exception of those of the semiconductor lasers that require a separate discussion, differ considerably from the resonant cavities that are used in the microwave region. These latter are *closed* cavities with metal walls of high reflectivity and with dimensions on the order of a few wavelengths for the radiation of interest. In a closed cavity, the walls require the cancellation of the field on their surface, and the field is divided into a set of modes with well-defined frequency. With relatively small cavities with respect to the wavelength, the number of modes to be taken into consideration is sufficiently limited. Instead, in the optical spatial region and non-microscopic cavities, for the Rayleigh-Jeans formula, the number of resonant modes included within a frequency range $\Delta\nu$ in a closed cavity of generic form is given by

$$\Delta N = \frac{8\pi}{c^3}\nu^2 V \Delta\nu\,, \tag{6.9.1}$$

where V is its volume and c is the speed of light. Taking, as a model, a He-Ne laser, with its typical volume and its frequency, the number of modes in its active band given by this formula would be exorbitant, with a spacing on the order of 1 Hz [Solimeno et al 1986]. Therefore, taking into account the presence of losses, the spectrum generated would essentially be continuous and incoherent as that of a spectral lamp.

To get out of this situation, A.M. Prokhorov and A.L. Schawlow and C.H. Townes proposed the use of open cavities consisting of a pair of plane mirrors [Prokhorov 1958; Schawlow and Townes 1958]. In this configuration, the modes that are called *transverse*, especially those of high order, undergo heavy losses and laser action only develops on a drastically reduced number of propagation modes around the optical cavity axis. Following this indication, in 1960, T.H. Maiman and R.J. Collins et al managed to get the first laser radiation in ruby bars pumped by flash lamps [Maiman 1960a 1960b; Collins et al 1960].

A further step was made by A.G. Fox and T. Li in their article in which they began to lay the theoretical bases for open resonant cavities with plane and concave mirrors, introducing the concept of diffraction losses [Fox and Li 1961]. This work was shortly followed by that of G.D. Boyd and J.P. Gordon and Boyd and H. Kogelnik, who presented a classification of resonant cavities with two mirrors depending on their radius of curvature [Boyd and Gordon 1961; Boyd and Kogelnik 1962]. The generalization to more complex optical systems was then described by S.A. Collins, and by H. Kogelnik and T. Li [Collins 1964; Kogelnik 1965b; Kogelnik and Li 1966a 1966b]. It was initially believed that only the resonators with low diffraction losses, namely, resonators called *stable*, could sustain laser action. In 1965, A.S. Siegman was the first to recognize the benefits of those referred to as *unstable* resonators, using a simple geometric analysis [Siegman 1965]. Finally, L.A. Weinstein completed the theory by collecting it in one of his important monographs [Weinstein 1966].

6.9.1 Resonant modes

As Anan'ev (1992) notes, a resonant cavity has a set of oscillating modes of the field, each characterized by a spatial structure and a frequency. In the absence of an external source, these modes are defined as a field distribution whose amplitude generally decays over time, while its shape remains constant. Therefore, the frequency is a complex value that, in its imaginary part, represents the field decay due to the losses that each mode undergoes. The field in the cavity may generally be represented by a wave that, in its propagation, is reflected by the mirrors and at least partially returns back upon itself. The discrete nature of the frequencies is therefore due to the fact that the cavity forces the field to a periodic behavior with a period equal to the time that the radiation takes to complete a lap in the cavity itself.

The calculation of these modes can then be traced back to find the spatiotemporal eigenfunctions of a propagation operator that represents the completion of a round trip in the cavity, between a starting plane and the same plane of arrival. For this purpose, in Paraxial Optics, the Huygens-Fresnel diffraction integral can be used, for which as a starting plane, one chooses one of the mirrors and the integration is carried out on its surface. For an optical system defined by an ABCD matrix, we can use the equivalent Collins' integral. In both cases, it is necessary to consider the effect of the finite size of the mirrors and other elements that make up the optical system. Moreover, as yet noted by Anan'ev (1992), it should be remembered that these integrals refer to strictly monochromatic waves with angular frequency ω, and, when the temporal part $\exp(-i\omega t)$ is made explicit, they produce a result that describes the wave on the plane of arrival for the very same moment at which it is defined on the starting plane, and not its propagation over time. In a case in which the frequencies are complex for the presence of losses, the treatment becomes subtly more complicated.

Here, I am only dealing with the simplest case of an ideal, losses-free resonator with unlimited-sized mirrors, thus neglecting the diffraction due to their apertures. However, this is often a good approximation, with high reflectivity mirrors and/or laser in appropriate dynamical conditions. Moreover, even in an unstable cavity, with high losses, it is possible to obtain, in a stable manner, a wave of constant intensity (CW, *continuous wave*), in which the active medium gain compensates the losses and the frequency of the modes can be considered real.

Applying the ABCD matrix corresponding to a closed path in a cavity to the propagation of a Gaussian beam, from Eq. (6.6.19), we can find the complex curvature of the field on the starting/arrival plane imposing equality between $\mathfrak{Q}_2$ and $\mathfrak{Q}_1$. The simplest case consists of a non-astigmatic linear cavity formed by two spherical mirrors with radius of curvature R_1 and R_2 separated by a distance L. For both, R is *positive* if the mirror is concave. For the symmetry of the result, we can imagine replacing each curved mirror with a lens placed next to a plane mirror (Fig. 6.14). By choosing the first of these fictitious planes as departure and arrival, the system is paraxially represented by the 2x2 ABCD matrix

Fig. 6.14
(a) scheme of a linear resonant cavity with two spherical mirrors.
(b) representation of the same cavity as consisting of two plane mirrors and two lenses. The focal distance of each lens is equal to the radius of curvature of the corresponding mirror

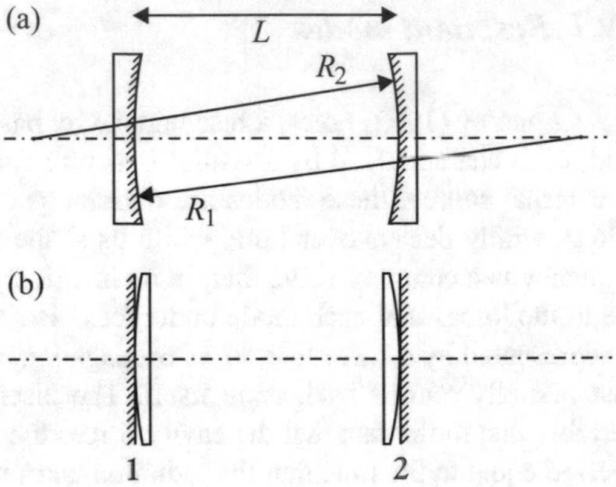

$$\begin{pmatrix} A & B \\ C & D \end{pmatrix} = \begin{pmatrix} 1 & 0 \\ -1/R_1 & 1 \end{pmatrix}\begin{pmatrix} 1 & L \\ 0 & 1 \end{pmatrix}\begin{pmatrix} 1 & 0 \\ -1/R_2 & 1 \end{pmatrix}\begin{pmatrix} 1 & 0 \\ -1/R_2 & 1 \end{pmatrix}\begin{pmatrix} 1 & L \\ 0 & 1 \end{pmatrix}\begin{pmatrix} 1 & 0 \\ -1/R_1 & 1 \end{pmatrix} =$$

$$\begin{pmatrix} g_2 & L \\ \left(\frac{1}{L}\right)(g_1g_2-1) & g_1 \end{pmatrix}\begin{pmatrix} g_1 & L \\ \left(\frac{1}{L}\right)(g_1g_2-1) & g_2 \end{pmatrix} = \begin{pmatrix} 2g_2g_1-1 & 2Lg_2 \\ \left(\frac{2}{L}\right)(g_1g_2-1)g_1 & 2g_2g_1-1 \end{pmatrix}, \tag{6.9.2}$$

where

$$g_1 = 1-\frac{L}{R_1}, \quad g_2 = 1-\frac{L}{R_2}. \tag{6.9.3}$$

By applying the ABCD law (6.5.3) to the condition $q' = q$ of "self-reproduction" of the field, we have that

$$Cq^2 + q(D-A) - B = 0\,, \tag{6.9.4}$$

which in the case of Eq. (6.9.2), simply becomes

$$\frac{1}{q} = \frac{1}{R} + i\frac{\lambda}{\pi w^2} = \pm\sqrt{\frac{C}{B}} = \pm\frac{1}{L}\sqrt{\frac{g_1}{g_2}(g_1g_2-1)}\,, \tag{6.9.5}$$

where, by definition, the radius of curvature R of the wavefront and the spot radius w should be real values.[6] On the other hand, g_1 and g_2 are real geometric parame-

[6] This limitation can be overcome under more general conditions of the optical system [Anan'ev 1992].

ters, and therefore the root can only be imaginary and the real part $1/R$ of the complex curvature on the dummy plane 1 is null, whereas, for the corresponding spot radius on mirror 1, it is

$$w_1^2 = \pm\frac{\lambda L}{\pi}\sqrt{\frac{g_2}{g_1(1-g_1g_2)}} = \pm\frac{\lambda R_1}{\pi}\sqrt{\frac{L(R_2-L)}{(R_1-L)(R_1+R_2-L)}}\,. \tag{6.9.6}$$

This equation would imply 4 values for w. On the other hand, for the two solutions indicated by the sign $\pm$, only the positive one corresponds to a Gaussian beam, while the other corresponds to an amplitude that diverges as the distance from the axis increases. Moreover, choosing a positive or negative sign of w can only change the phase of transverse field modes.

The complex field curvature on mirror 1 of Fig. 6.14(a) is obtained simply by applying the first matrix, the one to the right in Eq. (6.9.2), to the solution found for q on plane 1, i.e.,

$$\frac{1}{q_1} = \frac{1}{q} - \frac{1}{R_1}\,. \tag{6.9.7}$$

A similar discussion also applies to the field on mirror 2. In other words, since $1/q$ is imaginary, in a linear cavity, the wavefront on the mirrors follows the curvature of the mirrors themselves. Taking, as a positive direction, that from mirror 1 to mirror 2, the radii of curvature R_{w} of the beam on these mirrors are, respectively,

$$R_{\mathrm{w}1} = -R_1, \quad R_{\mathrm{w}2} = R_2\,. \tag{6.9.8}$$

In addition, the spot radius on the second mirror is obtained from Eq. (6.9.5) simply by exchanging indexes 1 and 2. In particular, we have that

$$\frac{w_2^2}{w_1^2} = \frac{g_1}{g_2}\,. \tag{6.9.9}$$

In conclusion, from Eq. (6.9.5), one finds that a Gaussian beam is a solution for the field in the cavity only when the following condition is verified:

$$0 \le g_1g_2 \le 1\,. \tag{6.9.10}$$

This is described in the diagram of Boyd-Kogelnik (Fig. 6.15), in which each linear resonator with two mirrors is represented by a point of coordinates g_1, g_2. If this point lies in the "butterfly" region of the 1st and 3rd quadrants, the cavity is said to be "stable", otherwise it is called "unstable". When the cavity is represented by a point well within the butterfly zone, the fundamental mode is strictly confined around the principal ray and, with reasonably large mirrors, the diffraction losses

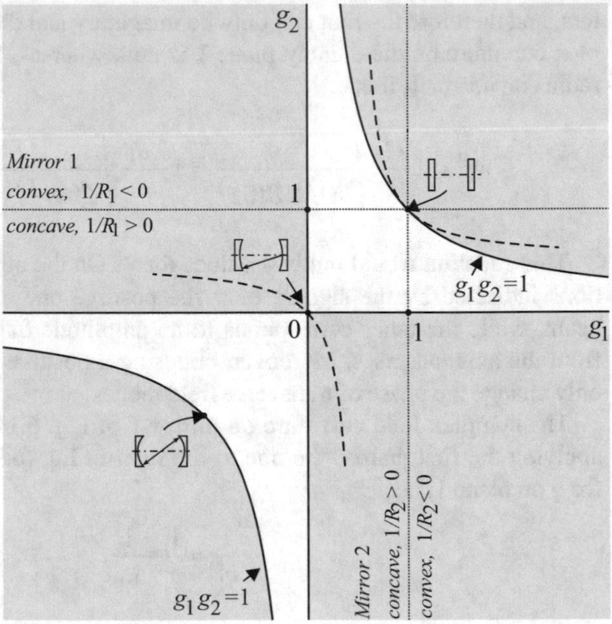

Fig. 6.15 Boyd-Kogelnik's diagram. Each point represents a linear cavity consisting of two spherical mirrors with radius R_1 and R_2. The "butterfly" region admits a Gaussian beam as a field solution within the cavity. The outer regions in gray are characterized by high losses. The two dotted hyperboles correspond to *confocal cavities*, for which $R_1 + R_2 = 2L$

due to their finished dimension are generally very low. Instead, with unstable cavities, the field tends to expand beyond the size of at least one of the mirrors and such losses are large. The symmetrical cavities with flat mirrors, confocal or concentric, are at the edge between the zones of stability and instability and are rather critical to align. The most common passive cavities are precisely the confocal cavities, represented in the center of the diagram. The condition $g_1 = g_2 = 0$ implies that the radii of curvature of the mirrors must be identical, otherwise the condition of confocality $R_1 + R_2 = 2L$ places them in the unstable area.

For a stable cavity, knowing the spot size w from the second Eq. (6.9.6) and the curvature radius R_w of the beam on the mirrors from Eq. (6.9.7), Eq. (6.2.12) allows us to determine the waist radius and the position of the mirrors in relation to the waist plane

$$w_o{}^2 = \frac{\lambda}{\pi}\frac{\sqrt{L(R_1 - L)(R_2 - L)(R_1 + R_2 - L)}}{R_1 + R_2 - 2L},$$
$$z_1 = -L\frac{R_2 - L}{R_1 + R_2 - 2L}, \quad z_2 = L\frac{R_1 - L}{R_1 + R_2 - 2L}. \tag{6.9.11}$$

In particular, for a cavity with equal radii $R_1 = R_2 = R_s$, it is

$$w_o{}^2 = \frac{\lambda}{2\pi}\sqrt{L(2R_s - L)}, \quad z_1 = -\frac{L}{2}, \quad z_2 = \frac{L}{2}. \tag{6.9.12}$$

In general, with cavities consisting of more than two mirrors, from the knowledge of the ABCD matrix corresponding to a round trip in the cavity, the resonant modes can be found by Eq. (6.5.3), imposing $q_1 = q_2$:

$$\frac{1}{q} = \frac{1}{2B}\left[D - A \pm \sqrt{(D+A)^2 - 4\,\mathrm{sgn}(D+A)}\right], \tag{6.9.13}$$

where the relationship (2.7.8) was used (with equal refractive indexes for the initial and final spaces) for the determinant of the matrix $AD - BC = 1$. Moreover, the sign sgn(D+A) was introduced for convenience, as we shall see shortly. Even in this case, there is a "stable" cavity, which admits a Gaussian beam as a field solution in thc cavity, when the term under the root is negative.

Otherwise, the cavity is called "unstable" and two real solutions are obtained for $1/q_\pm = 1/R_\pm$ that correspond to two distinct waves in Geometrical Optics [Anan'ev 1992]. In other words, there are two *real* points O_+ and O_- (whose distance from the reference plane is R_+ and R_-, respectively) along the principal ray that are reproduced in the same position by the optical system, in the sense that each is the image of itself after a lap into the cavity. Unstable cavities are always associated with an angular magnification $M_\pm \neq 1$. If we consider a ray of one of these waves, which therefore passes for one of two points O, on the reference plane considered at the beginning and end of a turn, its heights x_1 and x_2 are related to the relationship

$$x_2 = Ax_1 + B\vartheta_1 = \left(A + B\frac{1}{R}\right)x_1 .$$

Therefore, the magnification M is

$$M_\pm = \frac{x_2}{x_1} = \frac{1}{2}\left[D + A \pm \sqrt{(D+A)^2 - 4\,\mathrm{sgn}(D+A)}\right]. \tag{6.9.14}$$

It is easy to see that the two magnifications are mutually reciprocal, with $|M_+| > 1$ and $|M_-| < 1$, thanks to the term sgn. The second solution corresponds to a wave that contracts and its presence may induce us to believe that there may be a confined wave in the cavity. However, this is an unstable solution in that, iterating a wave with its center shifted slightly from O_-, it is increasingly displaced by O_-, that is, O_- is an unstable equilibrium point, while O_+ is stable. On the other hand, it is found that any Gaussian beam injected into one of these cavities (even one with a radius of curvature equal to R_- on the initial plane), even if it contracts at first, would end up expanding in the subsequent round trips, occupying the entire useful surface of at least one of the mirrors and overflowing from it.

Unstable cavities are used in high power lasers, in which the useful volume of the active medium is very large, due to the fact that the field is not strictly confined around the principal ray. For such expansion, a stable cavity would have a large Fresnel's number and the laser action would be on many transversal modes with an essentially unstable emission. Conversely, an unstable cavity leaves room for only fundamental modes, while the transverse ones have much larger losses and do not activate. The wave generated then exits from the cavity, bypassing the edge of a mirror, or is transmitted by a special mirror with a reflectivity profile decreasing towards the edge. Essentially, in the volume of the active medium, the nearest part to the principal ray acts as a *master oscillator*, while the outer one acts as an *amplifier*. The field profile is not of easy solution and depends on the finite size of mirrors and their specific features.

On the other hand, the nonlinear cavities are usually astigmatic, and the optical system is rather represented by a 4x4 matrix $\mathfrak{ABCD}$. In the simplest cases, this matrix can be diagonalized on two orthogonal transverse directions and reduced to two independent 2x2 matrices for which the cavity can be traced back to the analysis of stability described above. If the cavity is stable in both directions, its modes can be represented by simply astigmatic Gaussian beams. Generally, assuming that the cavity is stable, when the principal ray completing a closed loop in the cavity is not planar, there is a rotation on the plane of the transverse coordinates, and the modes become generally astigmatic. In this case, it is necessary to solve the most complicated Eq. (6.6.19), imposing $\mathfrak{Q}_2 = \mathfrak{Q}_1$. It can be brutally resolved iteratively, obtaining at least the form of the principal mode. The calculation of transverse modes is much more complicated and also involves the analysis of the polarization rotation. The description of this case is reported by J.A. Arnaud in a difficult work [Arnaud 1970].

6.9.2 Frequency of modes

Once the form of resonant modes has been established, their frequency remains to be found. This is achieved by imposing that, in a cavity, the phase accumulated by each mode in its propagation is a whole multiple of 2π. The simplest calculation is for a linear spherical stable cavity with two spherical mirrors, where we already know that the modes are typically those of Gauss-Hermite or those of Gauss-Laguerre described in §6.3. If, therefore, we consider the ideal case of a cavity with negligible losses and ignore the phase-out of the mirrors, the phase accumulated in a cavity is

$$\delta = 2kL - 2(\Sigma+1)\left[\arctan\left(\frac{z_2}{z_R}\right) - \arctan\left(\frac{z_1}{z_R}\right)\right], \qquad (6.9.15)$$

where $\Sigma = n + m$ for the Gauss-Hermite modes and $\Sigma = 2p + l$ for Gauss-Laguerre modes, and z_1 and z_2 are the position of the mirror vertices relative to the beam waist, with $z_2 - z_1 = L$. In other words, even the Gouy's phase difference between the mirrors is doubled in completing a turn in the cavity.

On the other hand, from Eq. (6.2.11), the first Eq. (6.9.6) and Eqs. (6.9.8), with some steps, we have that

$$-\frac{z_1}{z_R} = (1-g_1)\sqrt{\frac{g_2}{g_1(1-g_1g_2)}}, \quad \frac{z_2}{z_R} = (1-g_2)\sqrt{\frac{g_1}{g_2(1-g_1g_2)}}.$$

Taking that $\alpha = -\arctan(z_1/z_R)$, $\beta = \arctan(z_2/z_R)$, the term in the square bracket of Eq. (6.9.15) becomes the sum $\sigma = \alpha + \beta$. Recalling, then, that $\tan(\alpha+\beta) = (\tan\alpha + \tan\beta)/(1 - \tan\alpha\tan\beta)$, and that $\cos^2\sigma = 1/(1+\tan^2\sigma)$, through a laborious calculation, we find the Gouy phase for the fundamental mode in a cavity round trip

$$\theta = 2\left[\arctan\left(\frac{z_2}{z_R}\right) - \arctan\left(\frac{z_1}{z_R}\right)\right] = 2\arccos\left(s\sqrt{g_1g_2}\right), \tag{6.9.16}$$

where s is the sign of g_1, g_2, since, in the region of stability, $0 \leq g_1g_2 \leq 1$, and therefore g_1 and g_2 agree in the sign. If then s is positive, θ is between 0 and $\pi/2$, while, if s is negative, θ is between $\pi/2$ and π [Anan'ev 1992]. Finally, the condition $\delta = 2\pi q$, with q integer, implies that the frequencies of the resonant modes are

$$\nu_{q\Sigma} = \frac{\upsilon}{2L}\left[q + \frac{1}{\pi}(\Sigma+1)\arccos\left(s\sqrt{g_1g_2}\right)\right], \tag{6.9.17}$$

where υ is the velocity of the radiation inside of the cavity. Resonant transverse modes in the cavity for the same longitudinal order q are thus resolved in a regular sequence of frequencies with step

$$\Delta\nu_\perp = \frac{\upsilon}{2L}\frac{1}{\pi}\arccos\left(s\sqrt{g_1g_2}\right), \tag{6.9.18}$$

however, degeneration is not completely eliminated, as modes with the same value of Σ have the same frequency. These degenerate modes (Σ+1 transverse modes for each frequency) can then be combined by giving a profile that preserves its shape in propagation.

In their pioneering 1964 article, D. Herriott, H. Kogelnik and R. Kompfner, showed that, with special cavities, a laser beam injected off of the axis followed a closed path that formed a series of spots on the mirrors arranged along an ellipse [Herriott et al 1964]. In fact, for particular distances between mirrors, for which the Gouy's phase for a round trip is

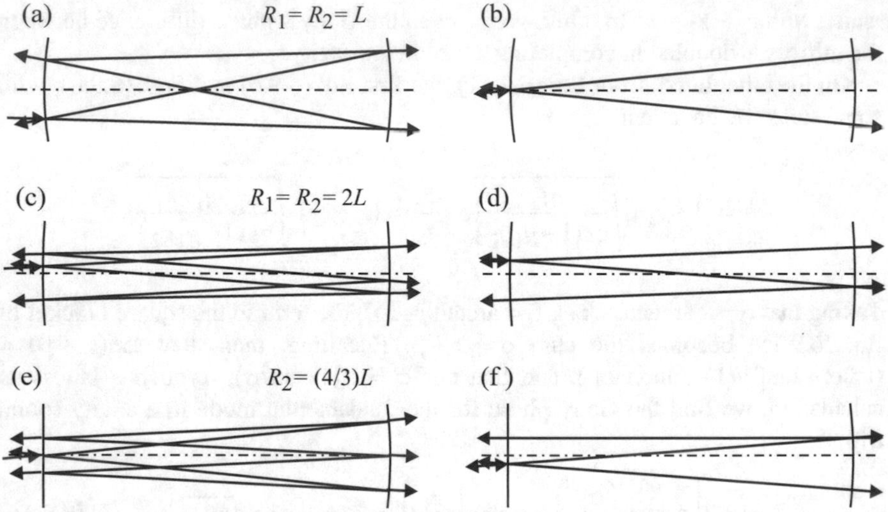

Fig. 6.16 Examples of cavities with mixed modes. (a), (b): confocal cavities, for which the focal points of the two mirrors coincide and $j = 2$. (c), (d): cavities with $j = 3$, in which the focal point of each mirror is at the vertex of the other mirror. (e), (f): cavities with $j = 3$, but the first mirror is plane. The path configuration examples on the right is such that the ray returns to itself with reflections perpendicular to the mirrors. For (a) and (c), the path shown is generic, while (e) represents a particular symmetrical case

$$\theta = 2\arccos\left(s\sqrt{g_1 g_2}\right) = 2\pi\frac{h}{j}\text{, with } h \text{ and } j \text{ integers,} \tag{6.9.19}$$

there is another type of degeneration that mixes transverse modes of different longitudinal orders. In other words, the frequency of modes with pairs of longitudinal and transverse orders

$$q_o, \Sigma_o;\;\; q_o - h, \Sigma_o + j;\;\; q_o - 2h, \Sigma_o + 2j;\;\; \ldots\;\; q_o - lh, \Sigma_o + lj;\;\; \ldots \tag{6.9.20}$$

with l integer, have the same frequency. In these cases, the field of any combination of these modes will reproduce itself after j round trips into the cavity (Fig. 6.16). These combinations are now called *mixed modes*, and they have the characteristic that their shape varies along the path, as they are constituted by elementary modes with different phase velocities.

In the case of cavities without axial symmetry, M.M. Popov and, with greater generality, J.A. Arnaud have shown that the frequencies of transverse resonant modes for an empty cavity are given by [Popov 1968a 1968b; Arnaud 1970]

$$\nu_{qnm} = \frac{c}{L_o}\left[q + \frac{1}{2\pi}\left(m + \frac{1}{2}\right)\theta' + \frac{1}{2\pi}\left(n + \frac{1}{2}\right)\theta''\right], \tag{6.9.21}$$

where L_o is the optical path on the principal ray in a complete revolution of the cavity, q, m, n are integer numbers, where q describes the longitudinal order and m, n the transverse one of the mode, and, lastly, θ' and θ'' are two contributions to the Gouy's phase, respectively associated with the index m and n. The degeneration of the frequencies that we see in Eq. (6.9.17) is then removed.

A further removal of degeneration occurs with a non-planar cavity for which, in addition to astigmatism, there is also a rotation of the image, in the sense that, having taken a ray on a meridian plane at the start, after one revolution, it returns on a different meridian plane. With metallic mirrors and almost normal incidences, this also typically involves a rotation of the polarization plane. With dielectric mirrors and elements that alter polarization, things are more complicated, but the final state of polarization can be determined step by step along the principal ray.

As an example, Arnaud considers the case of an optically axial cavity and no elements that alter polarization, but with planar mirrors arranged to produce a rotation of the transverse axes of an angle Ω. In other words, the optical cavity system can now be described by a matrix $\mathfrak{ABCD}$, with $\mathfrak{A} = A\mathfrak{I}$, etc., multiplied by a rotation matrix for an angle Ω. His resonant frequencies are given by

$$\nu_{qnm\pm} = \frac{c}{L_o}\left[q + \frac{1}{2\pi}(m+n+1)\theta + \frac{1}{2\pi}(m-n\pm 1)\Omega \right] . \tag{6.9.22}$$

The term ± 1 implies that right and left circular polarizations now have different resonance frequencies, in other words, degeneration of the polarization state is also removed for the fundamental mode.

The ring laser cavity, as in the dye laser or Ti:Sapphire lasers, are constructed to produce radiation with a single direction of propagation. This is achieved with a non-planar path on which there is a Faraday rotator, various laminae and interfaces at the Brewster angle, and elements that compensate for the astigmatism. The wave polarization that circulates in the cavity is linear and is determined by the orientation of the laminae of the Lyot's filter (see §7.11.7.2) at the Brewster's angle, which act concordantly as a polarizer. The Faraday's rotator compensates for the rotation of the polarization plane due to the non-planarity of the path for a single direction of travel, for which the laminae do not produce any loss for reflection. Instead, it doubles the rotation for the wave that propagates in the other direction, which is then extinguished by those laminae.

6.9.3 *Multipass cavity*

To explain the behavior of cavities that support the mixed modes described by Herriott et al (1964), we can use an argument from Geometrical Optics [Dingjan 2003]. Let us now search the eigenvalues and eigenvectors of the ABCD matrix (6.9.2):

$$\lambda\begin{pmatrix}\rho\\ \vartheta\end{pmatrix}=\begin{pmatrix}A & B\\ C & D\end{pmatrix}\begin{pmatrix}\rho\\ \vartheta\end{pmatrix}=\begin{pmatrix}2g_2g_1-1 & 2Lg_2\\ \left(\frac{2}{L}\right)(g_1g_2-1)g_1 & 2g_2g_1-1\end{pmatrix}\begin{pmatrix}\rho\\ \vartheta\end{pmatrix}. \tag{6.9.23}$$

We thus find two eigenvalues

$$\lambda_\pm = 2g_2g_1-1\pm 2i\sqrt{g_1g_2\left(1-g_2g_1\right)}, \tag{6.9.24}$$

where the relationship $0 \le g_1g_2 \le 1$ is exploited, valid for a stable cavity. Within these limits, $|\lambda_\pm|=1$, which is to say that the module of the magnification is unitary, and we can write

$$\lambda_\pm = e^{\pm i\theta}, \tag{6.9.25}$$

where $\cos\theta = 2g_1g_2-1$. So, the phase θ is just the Gouy's phase for a round trip given by Eq. (6.9.16). The two complex eigenvalues correspond to two complex eigenvectors $\boldsymbol{r}_+$ and $\boldsymbol{r}_-$ expressible with the relation

$$\boldsymbol{r}_\pm=\begin{pmatrix}\rho_o\\ \vartheta_\pm\end{pmatrix},\quad g_2\vartheta_\pm=\pm i\sqrt{g_1g_2\left(1-g_1g_2\right)}\frac{\rho_o}{L}, \tag{6.9.26}$$

where ρ_o is a real distance and ϑ is an imaginary angle. These two eigenvectors are each other's complex conjugate and do not correspond to real rays, but can be considered as a mathematical artifice useful for the calculation. For $g_1, g_2 \neq 0$, any real ray $\boldsymbol{r}_0$ can be expressed as a linear combination of these two eigenvectors

$$\boldsymbol{r}_0 = c_+\boldsymbol{r}_+ + c_-\boldsymbol{r}_-, \tag{6.9.27}$$

where c_+ and c_- are appropriate complex numbers. $\boldsymbol{r}_0$ being real, they are each the complex conjugate of the other, too. Propagating this ray into the cavity, after j turns, $\boldsymbol{r}_0$ becomes

$$\boldsymbol{r}_j = c_+e^{+ij\theta}\boldsymbol{r}_+ + c_-e^{-ij\theta}\boldsymbol{r}_- = \boldsymbol{r}_0\cos(j\theta)+\boldsymbol{s}_o\sin(j\theta), \tag{6.9.28}$$

where the vector

$$\boldsymbol{s}_o = i\left(c_+\boldsymbol{r}_+ - c_-\boldsymbol{r}_-\right) \tag{6.9.29}$$

is also real, because $c_-\boldsymbol{r}_-$ is the complex conjugate of $c_+\boldsymbol{r}_+$. All of this reasoning was made by implying the propagation of a ray on a meridian plane containing the optical cavity axis. For the cylindrical symmetry of the system, we can easily ex-

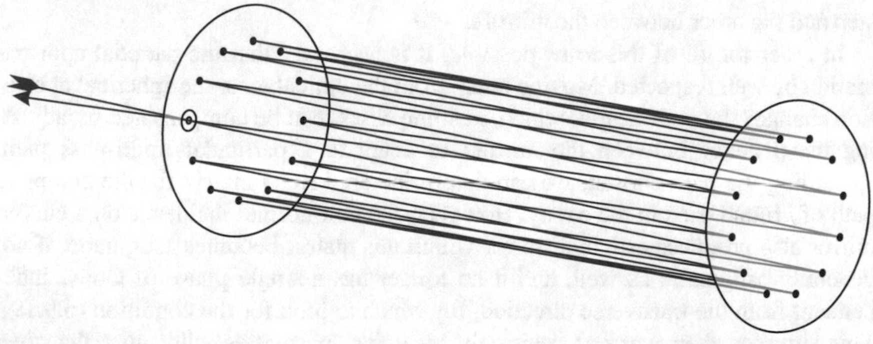

Fig. 6.17 Schematic of the path of the rays in a symmetric multipath cavity. The incident beam enters the cavity through a small opening so as to be reflected, forming a series of spots arranged in a circle on the mirrors, and then exit through the same opening. The radius of curvature of the mirrors is equal to approximately 29.348 times their distance L and corresponds to a Gouy's phase of $2\pi/12$ for a turn in the cavity. Within the cell, the total distance is therefore 24 L

tend it to a skew ray in three-dimensional space. Therefore, on plane 1 (Fig. 6.14), for which the ABCD matrix (6.9.2) is referred, in subsequent cycles in the cavity, the transverse coordinates of that ray can be expressed as

$$x_j = A\cos(j\theta + \alpha), \quad y_j = B\cos(j\theta + \beta), \tag{6.9.30}$$

where the constant A, α, B, β are a function of the corresponding values of $\boldsymbol{r}_{ox}$, $\boldsymbol{s}_{ox}$, and $\boldsymbol{r}_{oy}$, $\boldsymbol{s}_{oy}$, which are the projection of $\boldsymbol{r}_o$ and $\boldsymbol{s}_o$ on their respective meridian planes xz and yz. In general, the coordinate points x_j, y_j lie on an ellipse, filling it gradually with the growth of j.

When, however, the cavity is such that $j\theta = h2\pi$, with integer h, we have that, after j turns in the cavity, the ray returns to itself [Vaughan 1989]. This is the same condition (6.9.19) that causes the degeneration between the pairs of longitudinal and transverse modes described above. The two treatises can therefore be considered equivalent, and the Gouy's phase represents the conjunction ring.

In these cavities, *each* ray can be taken as the principal ray of a Gaussian beam that follows an off-axis path that reproduces itself after j round trips. Therefore, such cavities are also called multipass. Fig. 6.16 gives some examples with $j = 2$ and 3, while Fig. 6.17 describes the case of a Herriott's multipass cavity with $j = 12$. The latter is not conceived as a resonant cavity, as the laser beam enters and exits through a small aperture in a mirror without causing interference. The subsequent reflections here serve just to have, in a small space, a long path in an absorbent medium placed inside of the cavity.

In particular, in these cavities, if the beam is adapted to the same shape as the axial TEM_{00} mode, it keeps such shape even if it is off of the axis, and the mirrors form stains of the same size. More generally, they possess mixed modes that, while reproducing themselves after j turns in the cavity, change shape between a step and the other between the mirrors.

In order for all of this to be possible, it is necessary that the paraxial approximation be well respected. Moving away from the optical axis, the spherical aberration changes the path length, which, in some cases, can be compensated by adjusting the distance between the mirrors to adapt to a particular multi-pass path. Regarding the path chosen, we can determine an ABCD matrix for the complete path of j round trips in the cavity. However, the non-normal incidence on a curved mirror also produces astigmatism for which this matrix becomes astigmatic, if not generally astigmatic as well, and it no longer has a single phase of Gouy, independent from the transverse direction, for which to look for the condition (6.9.19). This situation then worsens decisively by going to cavities with more than two mirrors, for which the degeneration of the paths that reproduce themselves disappears.

6.9.4 Confocal cavity

Symmetric confocal cavities are made up of two identical spherical mirrors at a distance equal to their radius of curvature. They are represented by the point $g_1=g_2=0$ on the Boyd-Kogelnik diagram and are the simplest configuration for mixed modes, with $j = 2$, for which the field is self-reproduced with two turns in the cavity. Therefore, the frequency position of their transverse modes becomes degenerate with that of longitudinal modes.

If, however, the distance between the centers of curvature of the mirrors deviates a little from L, the degeneracy is broken and the frequencies of the transverse modes are crowded around the frequency of the fundamental modes. If, instead, the mirrors are not identical, while remaining in the confocal condition, the cavity is unstable.

Even compared to the other cavity with mixed modes, the confocal cavities represent a particular case. Indeed, following the same reasoning as for the ordinary cavity with two mirrors and a single turn in the cavity, for Eqs. (6.9.6) and (6.9.11), with $g_1=g_2=0$, the beam size on the mirrors is indeterminate, as is the position of the waist. Therefore, imposing that the wavefront curvature coincides with that of the mirrors, the only fixed quantity is the product $w_1 w_2 = \lambda L/\pi$. In other words, the beam size on a mirror is inversely proportional to that on the other mirror. This indetermination makes symmetric confocal cavities unsuitable as laser cavities, while they are often used as passive cavities for ease of use.

Indeed, these cavities were already proposed in 1956 by Pierre Connes as an alternative to the Fabry-Perot with plane mirrors and were called *spherical Fabry-Perots*, much less sensitive to the tilting defects of the mirrors [Connes 1956 1958]. Their fundamental merit is that, in paraxial approximation, each ray entering the cavity is reproduced after two turns. In fact, from Eq. (6.9.2), it is easy to see that the matrix corresponding to two turns in the cavity is reduced to the identity matrix. In Geometrical Optics, this means that all of the rays that start from every point inside of the cavity after four reflections on the mirrors converge at an image point coincident with the object point. Therefore, in paraxial approximation, all optical paths that complete two revolutions for any ray entering the cavity are equal, and hence the resonance conditions of the cavity are independent of both the position and the inclination of a beam that comes inside. This makes a symmetric confocal cavity much brighter than the Fabry-Perot with plane mirrors. As a counterpart, one loses the ability to vary the FSR of the cavity by changing the distance between the mirrors.

It is still possible to tune the cavity by varying, for example, the pressure of the gas contained inside, but also with small displacements of the mirrors produced by a piezoelectric motion

With relatively large apertures, spherical aberration becomes significant and, with a distant monochromatic source on the axis, circular interference rings generated by this aberration are observed. These are similar to those of the plane Fabry-Perot, in the sense that they become denser away from the axis. Approaching the mirrors a little bit, one can get a spacing such as to give an approximately linear dispersion with wavelength within a spectral order [Bradley and Mitchell 1968]. However, for mirrors with high reflection, these rings tend to become asymmetrical and widen, losing fineness.

With symmetrical confocal cavities used off-axis with monochromatic laser radiation, two emerging beams are observed at both the output and input sides, coinciding with the four reflections on the mirrors. The resonance condition occurs with an integer number q of wavelengths on the $4L$ path, so FSR = $c/(4L)$. When q is odd, bringing all of the paths to the axis, there is destructive interference between the two beams emerging from each mirror. There remains only the contribution of reflection from the input mirror. Only resonances with q even survive, and the separation between the transmission peaks returns to $c/(2L)$. In order for this to happen, however, the cavity must be aligned very precisely and, generally, there is a weak trace of intermediate modes. The width of the peaks, however, remains the same in the two situations: even if, in the off-axis condition, the FSR is halved, the transmission losses from the mirrors double, as four reflections should be considered instead of two. In P. Connes' original proposal, half of the surface of both mirrors had to be completely reflective in order to have partial transmission only for the input beam and for one of the beams on the rear mirror, but this solution has generally been abandoned, as such mirrors need to be specially built.

Symmetric confocal cavities also simplify the care for their *mode matching* with an incident laser beam. Generally, with resonating cavities, both the beam size, the wavefront curvature and the direction of propagation on the input mirror should be adjusted. Instead, within the paraxial approximation, any TEM_{00} laser beam finds a path that is self-reproduced in a symmetric confocal cavity.

In conclusion, confocal cavities are used as *optical spectrum analyzers* upon which a mirror is mounted on a piezoelectric transducer. By varying the voltage on the actuator, they are very useful for analyzing the radiation of a laser source due to their ability to have transverse degenerate modes. In fact, even if the cavity is not used on axis, its spectrum with monochromatic radiation is free from secondary peaks within the spectral range $c/4L$. Instead, the other cavities, as soon as they are slightly misaligned, are afflicted by the peaks corresponding to their transverse modes, which are distributed everywhere.

Bibliographical references

Abramowitz M. and Stegun I.A., *Handbook of Mathematical Functions*, Dover Publications Inc., New York (1972).

Allen L. M., Beijersbergen W., Spreeuw R.J.C., and Woerdman J.P., *Orbital angular momentum of light and the transformation of Laguerre-Gaussian laser modes*, Phys. Rev A **45**, 8185-8189 (1992).

Anan'ev Y.A., *Laser resonator and the beam resonance problem*, Hadam Hilger, Bristol, IOP publishing Ltd (1992).

Arlt J. and Dholakia K., *Generation of high-order Bessel beam by use of an axicon*, Opt. Comm. **177**, 297-301 (2000).

Arnaud J.A., *Nonorthogonal Optical Waveguides and Resonators*, The Bell System Technical Journal **49**, 2311-2348 (November 1970).

Arnaud J.A. and Kogelnik H., *Gaussian Light Beams with General Astigmatism*, Appl. Optics **8**, 1687-1693 (1969).

Bagini V., Frezza F., Santarsiero M., Schettini G., and Schirripa Spagnolo G., *Generalized Bessel-Gauss Beams*, J. of Mod. Optics **43,** 1155-1166 (1996).

Boyd G.D. and Gordon J.P., *Confocal Multimode Resonator for Millimeter Through Optical Wavelength Masers*, Bell. Syst. Tech. J. **40**, 489-508 (1961).

Boyd G.D. and Kogelnik H., *Generalized Confocal Resonator Theory*, Bell. Syst. Tech. J. **41**, 1347-1369 (1962).

Bradley D.J. and Mitchell C.J., *Characteristics of the defocused spherical Fabry-Perot interferometer as a quasi-linear dispersion instrument for high resolution spectroscopy of pulsed laser sources*, Phil. Trans. R. Soc. A **263**, 209-223 (1968).

Collins Stuart A. Jr., *Analysis of Optical Resonators Involving Focusing Elements*, Appl. Opt. **3**, 1263-1275 (1964). *Lens-System Diffraction Integral Written in Terms of Matrix Optics*, J. Opt. Soc. Am. **60**, 1168-1177 (1970).

Collins R.J., Nelson D.F., Schawlow A.L., Bond W., Garrett C.G.B., and Kaiser W., *Coherence, Narrowing, Directionality, and Relaxation Oscillations in the Light Emission from Ruby*, Phys. Rev. Lett. **5**, 303-305 (1960).

Connes P., *Augmentation du produit luminosité × résolution des interféromètres par l'emploi d'une différence de marche indépendante de l'incidence*, Revue Opt. **35**, 37-43 (1956). *L'étalon Fabry-Perot sphérique*, J. Phys. Radium **19**, 262-269 (1958).

Dingjan J., *Multi-mode optical resonators and wave chaos*, Thesis, Leiden University (2003), accessibile da http://iop.hqu.edu.cn/uploadfile/20111229162543941.pdf.

Durnin J., *Exact solution for nondiffracting beams. I. The scalar theory*, J. Opt. Soc. Am. A **4**, 651-654 (1987).

Durnin J., Miceli J.J. Jr, and Eberly J.H., *Diffraction-Free Beams*, Phys. Rev. Lett. **58**, 1499-1451 (1987).

Fox A.G. and Li T., *Resonant Modes in a Maser Interferometer*, Bell. Syst. Tech. J. **40**, 453-488 (1961).

Gori F., Guattari G. and Padovani C., *Bessel-Gauss Beams*, Opt. Comm. **64**, 491-495

(1987).

Guenther R., *Modern Optics*, John Wiley & Sons, New York (1990).

Herriott D., Kogelnik H., and Kompfner R., *Off-axis paths in spherical mirror interferometers*, Applied Optics **3**, 523-526 (1964).

Iizuka K., *Engineering Optics*, Springer-Verlag, Berlino (1987).

Kogelnik H., *On the Propagation of Gaussian Beams of Light Through Lenslike Media Including those with a Loss or Gain Variation*, Appl. Optics **4**, 1562-1569 (1965a). *Imaging of Optical Modes - Resonators with Internal Lenses*, Bell. Syst. Tech. J. **44**, 455-494 (1965b).

Kogelnik H. and Li T., *Laser Beams and Resonators*, Appl. Opt. 5, 1550-1567 (1966a). Proc. IEEE **54**, 1312-1320 (1966b).

Maiman T.H., *Stimulated optical radiation in ruby*, Nature **187**, 493-494 (1960a). *Optical and microwave-optical experiments in ruby*, Phys. Rev. Lett. **4**, 564-566 (1960b).

McLeod J.H., *The axicon: a new type of optical element*, J. Opt. Soc. Am. **44**, 592-597 (1953).

Piccirillo B., Slussarenko S., Marrucci L., and Santamato E., *The orbital angular momentum of light: genesis and evolution of the concept and of the associated photonic technology*. Rivista del Nuovo Cimento **36**, 501-554 (2013).

Popov M.M., *Resonators for lasers with unfolded directions of principal curvatures*, Opt. Spectrosc. **25**, 231-217 (1968); *Resonators for lasers with rotated directions of principal curvatures*, Opt. Spectrosc. **25**, 170-171 (1968).

Prokhorov A.M., *Molecular amplifier and generator for submillimeter waves*, Sov. Phys. JETP **7**, 1140-1141 (1958).

Santarsiero M., *Propagation of generalized Bessel-Gauss beams through ABCD optical Systems*, Opt. Comm. **132**, 1-7 (1996).

Schawlow A.L. and Townes C.H., *Infrared and optical maser*, Phys. Rev. **112**, 1940-1949 (1958).

Scott G. and McArdle N., *Efficient generation of nearly diffraction-free beams using an axicon*, Opt. Engineering **31**, 2640-2643 (1992).

Siegman A.E., *Unstable optical resonator for laser application*, Proc. IEEE **53**, 277-287 (1965). *Lasers*, University Science Books, Mill Valley, California (1986).

Solimeno S., Crosignani B., Di Porto P., *Guiding, Diffraction, and Confinement of Optical Radiation*, Academic Press, inc., Orlando (1986).

Suematsu Y. and Fukinuki H., *Matrix theory of light beam waveguides*, Bull. Tokyo Inst. Technol. **88**, 33-47 (1968).

Vasara A., Turunen J., and Friberg A.T., *Realization of general nondiffracting beams with computer-generated holograms*, J. Opt. Soc. Am. A **6**, 1748-1754 (1989).

Vaughan J.M., *The Fabry-Perot Interferometer*, A. Hilger, Bristol (1989).

Weinstein L.A., *Otkrytyye Rezonatory i otkrytyye volnovody* (Moskow: Sovetskoye Radio, 1966); *Open resonators and open waveguide*, Golem Press (1969).

Chapter 7
Light propagation in anisotropic media

Introduction

So far, the interaction between light and matter has been treated very succinctly, approximating the polarization **P** and the magnetization **M** with linear functions of the fields **E** and **B**. In this chapter, we will stick with this assumption, however, we will devote more attention to optical materials and, in particular, to the study of their anisotropic properties. The field of this study is nevertheless very vast and of great applicative importance, for the numerous effects that are encountered there.

There are many materials whose optical properties depend on the direction of propagation and the polarization of light. In other words, they have a polarization and magnetization that depend peculiarly on the direction of the fields **E** and **B**. The anisotropy can come from an ordered and regular arrangement of atoms, as in the crystals, or from the presence of ordered and regular structures of dimensions smaller than the wavelength, from the form called *right-handed* or *left-handed* of the molecules dissolved in a liquid, from application of an electric or magnetic field, and more. Several optical phenomena have arisen from this, in the following order: *birefringence*, *optical activity*, *Pockels* and *Kerr effects*, *Faraday effect*, etc.

One common characteristic of all of these effects is, as we will see, the removal of the degeneration in the polarization of the electromagnetic wave: for each direction of energy propagation, there are generally two eigenvectors in the polarization that correspond to two wave vectors $\boldsymbol{k}$ that are different in magnitude and direction.

In addition to the effects mentioned above, for which there is no absorption, the anisotropy also presents itself with a different transparency or reflection for two polarization components. Depending on the case, this is called dichroism or even pleochroism, because, with natural illumination, the transmitted or reflected radiation changes intensity, and also color, according to the polarization and the direction of propagation with respect to the anisotropy axes of the material.

Anisotropic materials are widely used to manipulate radiation at will, in particular, that of laser beams, and constitute devices such as polarizers, birefringent plates, and optical isolators. An extraordinary application synthesis is thus represented by liquid crystal displays, which, with appropriate ingenious means, also allow for a three-dimensional stereoscopic vision.

In this chapter, we will first look at a brief review of crystallography, and then we will consider the effects of anisotropy in linear optics. The main effect is to break the symmetry in polarizations and, in particular, to separate the directions of the wave vector and the Poynting's vector. As mentioned above, in general, for an

G. Giusfredi, *Physical Optics*, UNITEXT for Physics,
https://doi.org/10.1007/978-3-030-25279-3_7

assigned direction of one or the other of these vectors, two polarization eigenvectors can be determined with different values of the propagation speed. This has non-trivial consequences, which, in general, need a relativistic treatment to frame their phenomenology in a consistent way, including the non-reciprocal effects induced by magnetic fields and the effects of (uniform) motion of the medium. Finally, we will study how to exploit this anisotropy with various devices so as to manipulate the polarization.

7.1 Crystallography

Crystalline materials are an object of study and application that are very important in optics, and, in particular, in non-linear optics. In order to understand their behavior, it is necessary to have at least an idea of how the crystals are made and of their symmetry properties. There are two ways to address the issue. One is pragmatic: we stack the atoms according to the rules of chemical bonding and see what comes out; the other is analytical and abstract and aims to study their symmetries in space more directly. Both are good and, in reality, each cannot do without the other in a thorough study. Here, we will see only a nod to the second method of this vast subject, with an attempt to summarize the essential aspects that will be necessary to continue. One concise text that describes the way in which the atoms are stacked in the crystalline structure, their symmetry properties, and the X-ray diffraction is that of Hammond (1992)

A crystal is defined as a solid composed of atoms arranged according to an ordered and repetitive pattern. This means that we can identify a small group of atoms (a *base* or *cluster*) that are internally arranged according to a certain "motif", which is repeated, without variations, in a compact and orderly manner throughout the space. Consider, then, a single point of this motif and imagine its repetition in the space formed by the corresponding points of all of the repeated bases: in this way, we have a grid of points, which we will call *lattice points*.

The *crystal structure* is then formed by the lattice and the base. A crystal can have an infinite number of possible structures, which, in theory, can be classified among 230 *space groups*, 32 *crystal classes*, 14 *Frankenheim-Bravais lattices*, and 7 *crystal systems*. This list of classifications forms a hierarchy, in order, from the most specific to the most generic. Now, let us look at it starting from the lattices.

The condition to be respected is that *all of the lattice points are equivalent*, in particular, the neighborhood of each point must be the same for all points. To help the view, we can join each point with its neighbors to form cells with the same shape and orientation. Be aware that the lattice must remain the same if we choose different ways to connect the points: the same lattice can be described by cells of different shapes, but the one with the greatest symmetry is normally chosen. The *symmetry operations* are those that bring the crystal back onto itself with the same arrangement of the lattice points.

7.1.1 Two-dimensional lattices

As an example, let us see what happens in a two-dimensional space. Using only cells constructed by joining the reticular points, a plane can be filled with identical cells in the form of a parallelogram (square, rectangle, rhombus, ...) or of a hexagon. Any other geometric shape would not be appropriate: they could be triangles, but they would have different orientations, and, taken in pairs, we would have a parallelogram again. For the hexagonal lattice to be acceptable, it must also have a reticular point at its center, otherwise the points at the vertices would not all be equivalent.

In this way, the hexagonal lattice is equivalent to a rhombic lattice, in which the cells are still parallelograms. In total, in the plane, we have 5 types of lattice that can be rendered with the 5 *unit cells* on the left in Fig. 7.1.

When calculating how many lattice points are associated with a cell, it is taken into account that, in the plane, each vertex of a cell also belongs to three other adjacent cells, and therefore it counts for 1/4. The first, second, fourth, and fifth cells of Fig. 7.1 have only one point and are called *primitive cells*: the letter *p* in their name indicates this. The third cell is rectangular, with a dot in the center (the letter *c* indicates this), and therefore has two points in total. It is equivalent (in the sense that it leads to the same lattice) to a primitive cell shaped like a rhombus, whose diagonals are the sides of the rectangular cell; however, it is preferable to its primitive, since it shows a greater symmetry that is also present in the lattice.

In conclusion, the unit cells of Fig. 7.1 are all in the form of a parallelogram: their distinction derives from the symmetry operations that their lattices underlie (see the graphs on the right in Fig. 7.1). Thus, centering on the reticular points, the lattice of the first cell has, as its symmetry operation, only the *inversion*, also common to the other, and a 180° rotation axis. In addition to these operations, the lattices of the second and third cells also have two *lines of reflection* orthogonal to each other. For the fourth cell, at the reticular points, there is a *fourfold rotation axis*, that is, of 90°, and 4 reflection lines at 45° to each other. Finally, the fifth cell has a *sixfold* axis, that is, of 60°, and 6 lines of reflection at the reticular points. In the five cells, other elements of symmetry are then distributed regularly between the reticular points.

7.1.2 Three-dimensional lattices

In three-dimensional space, we have a similar situation, but with a greater number of possibilities. The first systematic work of classification of three-dimensional lattices was done by M.L. Frankenheim, who, in 1835, proposed that there were 15 lattices in all. However, two of his lattices were the same: in 1848, A. Bravais noticed this error and established 14 of them that now bear his name.

The *unit cells* of the 14 lattices of Frankenheim-Bravais are drawn in Fig. 7.2.

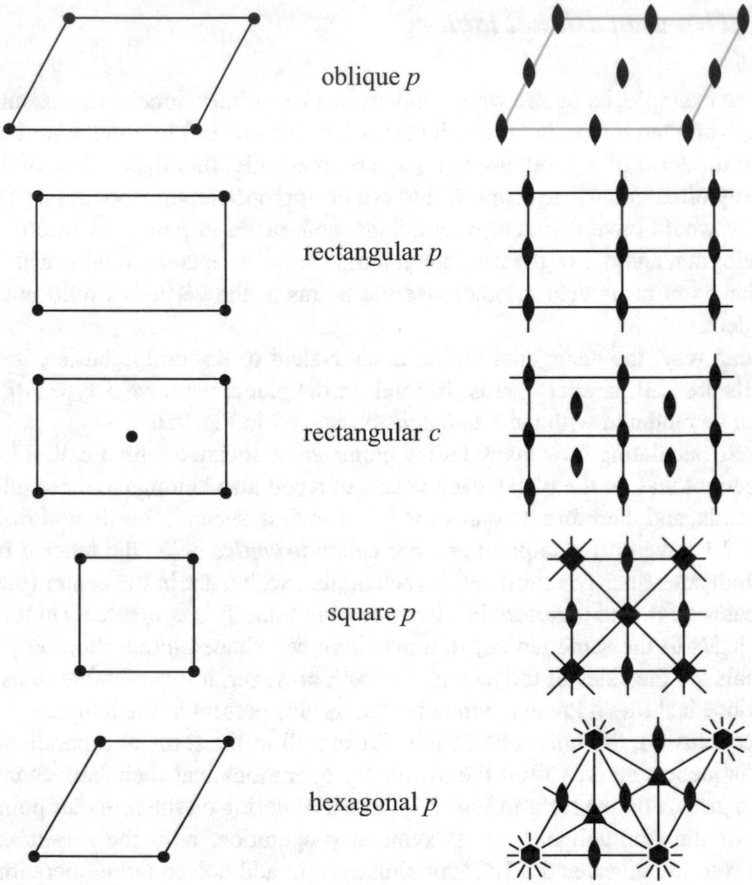

Fig. 7.1 Unity cells of the five plane lattices (on the left) and the symmetry elements present (on the right). The marked segments represent specular symmetry lines, the lens-shaped symbols the twofold axes, the triangles the threefold axes, the squares the fourfold axes and the hexagons the sixfold axes

In the list, there are 7 primitive unit cells. The others have reticular points centered on the faces or in the center of the cell: they are not primitive, but, as in the two-dimensional case, they highlight the symmetry of the lattice that would otherwise be hidden. These cells are conventionally distinguished by a letter: *P* for primitive, *I* for body-centered (*Innenzentrierte* in German), *F* for face-centered, *C* for base-centered, *R* for rhombic. All of these cells have the shape of parallelepipeds, which we can imagine as the bricks that make up the lattice itself. Indeed, for each unit cell, in each of the vertices, we have three intersecting edges defining the direction of the *crystallographic axes*.

We can then define three characteristic vectors, $\boldsymbol{a}$, $\boldsymbol{b}$, $\boldsymbol{c}$, oriented along these axes and having, as a module, the length a, b, c of the three corresponding edges that start from a vertex chosen as the origin of the axes.

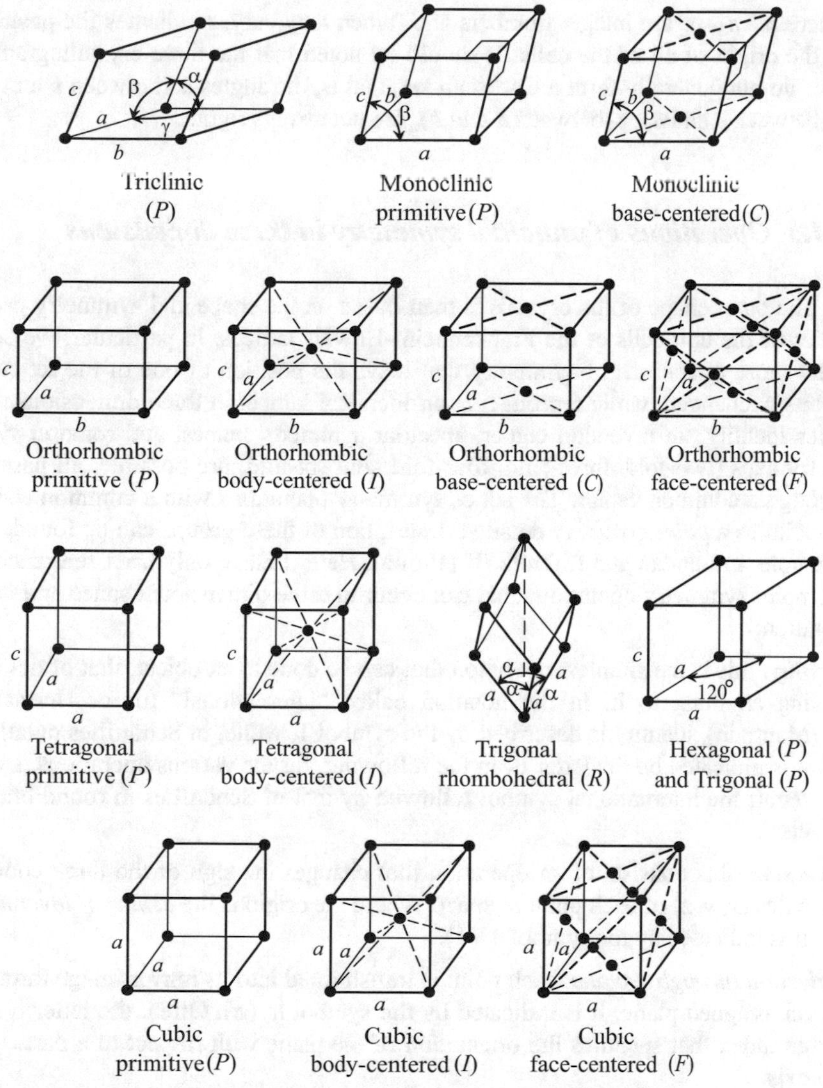

Fig. 7.2 The 14 Frankenheim-Bravais lattices represented by parallelepipeds. The lengths of the sides of the unit cells are indicated with *a*, *b*, *c*, while α, β, γ are the angles between the crystallographic axes. When the angle is not indicated or deducible from another, it is implied that it is right

The shape of each of the unit cells in Fig. 7.2 can then be defined by these three vectors and the lattice corresponding to each unit cell can be constructed with a simple translation operation:

$$\boldsymbol{t} = n_a\boldsymbol{a} + n_b\boldsymbol{b} + n_c\boldsymbol{c}\,, \tag{7.1.1}$$

where n_a, n_b, n_c are integer numbers and, when they vary, ***t*** indicates the position of the origin of all of the cells. It should be noted that the three crystallographic axes do not generally form a Cartesian set, that is, the angles α (between ***b*** and ***c***), β (between ***c*** and ***a***), γ (between ***a*** and ***b***), are not always right.

7.1.3 Operations of punctual symmetry in three dimensions

The outer shape of the crystals is then based on the shape and symmetry properties of the unit cells of the Frankenheim-Bravais lattices. In particular, we consider those operations of symmetry that leave the position of one of the reticular points unchanged, while reproducing an identical lattice: in three dimensions, besides identity, an inversion center, specular symmetry planes, and rotation symmetry axes (two-fold, three-fold, four-fold, and six-fold) are possible, all passing through a common center. The set of symmetry operations with a common center constitutes a *point group*. A detailed description of these groups can be found, for example, in Landau and Lifchitz III (1966a). Here, I make only brief reference to the point symmetry operations that can occur in three-dimensional space and their notation:

identity: this is the simplest operation that can be done to an object, that of not doing anything to it. In the notation called "international" (or of Hermann-Mauguin), identity is described by the symbol 1, while, in Schönflies notation, it is indicated by the letter E. In the following, for the various operations, I will report the international symbol followed by that of Schönflies in round brackets.

inversion: this consists of an operation that changes the sign of the three coordinates (x, y, z) of each point in space, where the origin is the *center of inversion*. It is indicated by the symbol $\bar{1}$ (I).

reflection through a plane: each point is transformed into its mirror image through an assigned plane. It is indicated by the symbol m (σ). Often, the letter σ has an index that specifies the orientation of the plane with respect to a particular axis.

proper rotation: if a rotation for an angle of 360°/n (where n is an integer) around an axis is a symmetry operation for a given object, then this object has an axis of symmetry for rotation of order n, which is indicated by the symbol n (C_n).

rotation-inversion and *rotation-reflection*: these are composed operations, called *improper*; although they can be put in relation, they are realized and described in different ways between the international approach and that of Schönflies. In the first case, a rotation is executed first, of order n, followed immediately by an inversion: the corresponding symbol is $\bar{n} = \bar{1}n$ (IC_n). In the second case, the rotation is instead immediately followed by a reflection, σ_h, through a plane perpendicular to the axis: its Schönflies symbol is $S_n = \sigma_h C_n$. The correspondences between the various operations must be searched case by case and they may appear strange. In particular, we have that $\bar{1} = S_2 = I$ is the inversion, $\bar{2} = m\ (\sigma)$, $\bar{3} = S_6{}^5$, $\bar{6} = S_3{}^5$.

On the basis of point symmetry operations, we cannot distinguish whether a lattice is face-centered or body-centered, for example, all of the cubic lattices, *P*, *I* and *F*, have the same point groups, the same for orthorhombic lattices, *P*, *I* , *C* and *F*, and so on. However, the distinction between these cells contributes to building the 230 groups of spatial symmetry mentioned above.

7.1.4 Crystal systems

Altogether, the Frankenheim-Bravais lattices can be organized into 7 *crystal systems*. They are substantially distinct from the shape of the unity cells, i.e., according to the lengths *a*, *b*, *c*, and the angles α, β, γ, (with one exception that we will see at the end). For each system, we have a maximum point symmetry (called *holohedral*) and a minimal point symmetry compatible with the shape of the lattice: we will return to this argument shortly. As before, in their description, I indicate the symbols that are now used to describe the *group* of point symmetry: the first one is the International symbol and the second one, in the round brackets, is that of Schönflies. For both, I use italic characters to distinguish groups from individual symmetry operations.

Triclinic: this is the most general case and is represented by a primitive cell with the three sides *a*, *b*, *c* different from each other and with any angles α, β, γ. It does not have any rotation symmetry: the only operations of symmetry of the lattice are the identity, which forms the group *1* (C_1), as a minimum symmetry, and the inversion, with its group $\bar{1}$ (C_i), as holohedry.

Monoclinic: this derives from the triclinic in the particular case in which one of the sides of the cell is perpendicular to the other two. In this way, the lattice possesses a twofold rotation axis, with the group *2* (C_2), that is, it is symmetrical for rotations of 180°, or has a plane of reflection, with the group *m* (C_{1h}), or both, *2/m* (C_{2h}), in the holohedral case, with m perpendicular to the axis. It is represented by two cells, a primitive, *P*, and one, *C*, with (only) two opposite centered faces, parallel to the axis.

Orthorhombic: this derives from the previous one by imposing that all of the angles are now right, $\alpha = \beta = \gamma = 90°$. Now, the lattice has at least two twofold axes or two specular planes normal to each other: in both cases, a third twofold axis automatically derives from them. The corresponding groups are *222* (D_2) and *mm2* (C_{2v}). The holohedry is given by the group *mmm* (D_{2h}) with three twofold axes and three mirror planes.

Tetragonal: similar to the orthorhombic case, but now two of the sides are of equal length, $a = b \neq c$. The minimum symmetry is given by a single proper fourfold axis, with its group *4* (C_4), or of rotation-inversion, with the group $\bar{4}$ (S_4). The holohedry is given by the group *4/mmm* (D_{4h}), which is obtained by transforming one of the twofold axes of the orthorhombic holohedry into a fourfold axis.

Cubic: the unitary cells of cubic crystals ultimately have all equal sides, $a = b = c$, and all right angles, $\alpha = \beta = \gamma = 90°$. However, even if it is somewhat surprising, the minimum symmetry needed is given by four threefold axes, along the diagonals of the cell, which transform each side into another (the angles between two diagonals, with neighboring vertices, are approximately of 109°28′). The corresponding group is that of the tetrahedron *23* (T). The holohedry is that of the octahedron with the group *m3m* (O_h).

The last two crystal systems, in the classification that is most useful to us, are distinguished by the presence of a single threefold symmetry axis or, alternately, of a sixfold symmetry axis (proper or of rotation-inversion) possessed by the crystal and not simply by its lattice, rather than the rhombic or hexagonal shape of the unit cells of the Frankenheim-Bravais lattices:

Trigonal: this system is defined as the one that has a single proper ternary axis, 3, or of rotation-inversion, $\bar{3}$. There are two possible lattices associated with this system: rhombic *R* or hexagonal *P*. The rhombic unit cell is a primitive cell, with equal sides $a = b = c$ and equal angles $\alpha = \beta = \gamma$, while the hexagonal cell, also primitive, has sides $a = b \neq c$ and angles $\alpha = \beta = 90°$, $\gamma = 120°$. The rhombic lattice can also be expressed by a non-primitive hexagonal cell, with three reticular points, two of which are placed inside of it in the positions $(2/3, 1/3, 1/3)$ and $(1/3, 2/3, 2/3)$, or $(2/3, 1/3, 2/3)$ and $(1/3, 2/3, 1/3)$, which are the normalized coordinates in the oblique system of the crystallographic axes $\boldsymbol{a}$, $\boldsymbol{b}$, $\boldsymbol{c}$ of the hexagonal cell. The punctual groups that report the essential symmetry of this crystalline system are *3* (C_3) and $\bar{3}$ (C_{3i}), while its holohedral group is $\bar{3}\,m$ (D_{3d}). It should be noted that a crystal of the cubic system passes to the trigonal system when, even with an infinitesimal perturbation, three of its four ternary axes are removed.

Hexagonal: the latter system requires a sixfold axis, proper or of rotation-inversion, and is associated with the hexagonal lattice *P* (which is therefore "distributed" between the hexagonal and trigonal systems). Its minimum symmetry is that of the groups *6* (C_6) and $\bar{6}$ (C_{3h}), while its holohedral group is *6/mmm* (D_{6h}).

7.1.5 Crystal classes

In describing the various crystal systems, I implicitly introduced some of the classes in which crystals are classified. In total, we have 32 classes, or groups of point symmetry, divided between the 7 systems: the groups mentioned above respectively correspond to the minimum and maximum symmetry compatible with each system. Overall, both the symmetries of the lattice and of the base that is repeated contribute to the symmetry of the crystal: let us now try to understand how to combine them. From a geometric point of view alone, we could assign a base to the reticular points with a group of any specific point symmetry. However, physically, to be compatible with a given crystalline system, it happens that the base must have at least the minimum symmetry properties required for that system. In fact, the bases interact with each other with electromagnetic forces that, if they do not have the symmetry properties of the lattice, end up distorting it, perhaps only a little, but enough to no longer have the exact original symmetry. The opposite is also true: if the lattice is "less" symmetrical than the base, the interaction between the bases, even if only slight, reduces the "excess" symmetry of the bases themselves. The way the bases are associated with reticular points is also important. On the one hand, to simplify things, we can always choose that the center of the base (i.e., the invariant point of the punctual group of the base) coincide with the reticular points. On the other hand, the various elements of symmetry, planes, axes, etc., of the base must be aligned with those of the lattice, otherwise this symmetry would be lost. In conclusion, the same crystalline system is divided into various crystalline classes by the symmetry of the base. In other words, the 32 classes correspond only to groups of punctual symmetry compatible with the construction of a lattice.

The 32 crystalline classes were originally identified by J.F.C. Hessel in 1830 from the symmetry properties of the faces of the crystals. Independently and much later, Bravais (in 1849) and Alex Gadolin (in 1867) came to the same conclusion. These classes are listed in Table 7.1 with the indication of some optical properties that will be explained later.

7.1.6 Space groups

Proceeding to define the crystalline structure in even greater detail, it is necessary to combine the symmetry operations of the point symmetry groups with the translational symmetries of the Frankenheim-Bravais lattices. Contributions to this study came from numerous researchers: the first steps were made by L. Sohncke, who introduced the notion of screw axes and glide planes as new elements of symmetry, followed by E.S. v.Fedorov in 1890, and, independently, A. Schönflies in 1891 and W. Barlow in 1894. A total of 230 distinct spatial symmetry groups have been obtained, each of which is defined as the set of symmetry operations that transform a periodic object (a crystal) into itself. The study of these groups would require a separate course, and, for this see, for example, Burns and Glazer (1978), which describes the symmetry properties of the crystals in great detail and provides the basis for the reading of the *International Tables for Crystallography*. But there are some aspects that are important for optics and that we need to know. All of the operations of a spatial group can be described by the Seitz's operator $\{\mathfrak{R}|\boldsymbol{t}\}$, consisting of a point operation $\mathfrak{R}$ followed by a translation $\boldsymbol{t}$.

System	International	Schönflies	Examples	χ_{ij}	χ_{ijk}	χ_{ijkl}
Triclinic	*1*	C_1		6*	18	81
	$\overline{1}$	$S_2\,(C_i)$		6	0	81
Monoclinic	*m*	C_{1h}		4†	10	41
	2	C_2	sucrose	4*	8	41
	2/m	C_{2h}		4	0	41
Orthorhombic	*2mm*	C_{2v}	KTP, LBO, $KNbO_3$	3†	5	21
	222	$D_2\,(V)$	$MgSO_4{\cdot}7HO_2$	3*	3	21
	2/m 2/m 2/m	$D_{2h}\,(V_h)$		3	0	21
Tetragonal	*4*	C_4		2*	4	21
	$\overline{4}$	S_4		2†	4	21
	4/m	C_{4h}	$LiYF_4$	2	0	11
	4mm	C_{4v}		2†	3	11
	$\overline{4}\ 2m$	$D_{2d}\,(V_d)$	KDP, ADP	2†	2	11
	422	D_4	TeO_2	2*	1	11
	4/m 2/m 2/m	D_{4h}	MgF_2, TiO_2, $BaTiO_3$	2	0	11
Trigonal	*3*	C_3	BBO	2*	6	27
	$\overline{3}$	$S_6\,(C_{3i})$		2	0	27
	3m	C_{3v}	$LiTaO_3$	2†	4	14
	32	D_3	quartz	2*	2	14
	$\overline{3}\ 2/m$	D_{3d}	calcite, sapphire	2	0	14
Hexagonal	$\overline{6}$ (*3/m*)	C_{3h}		2	2	19
	6	C_6		2*	4	19
	6/m	C_{6h}	α-$LiIO_3$	2	0	19
	62m	D_{3h}		2	1	10
	6mm	C_{6v}	CdSe, CdS, AlN	2†	3	10
	622	D_6		2*	1	10
	6/m 2/m 2/m	D_{6h}		2	0	10
Cubic	*23*	T	$NaClO_3$, BGO	1*	1	7
	$2/m\,\overline{3}$	T_h	Y_2O_3	1	0	7
	$\overline{4}\ 3m$	T_d	β-ZnS, GaAs, CdTe	1	1	4
	432	O		1*	0	4
	$4/m\ \overline{3}\ 2/m$	O_h	CaF_2, NaCl, LiF, diamond	1	0	4

Table 7.1 Classifications and optical properties of crystals [Bhagavantam 1966]. The classification system is based on the unit cell, and the symbols refer to the symmetry groups. The last three columns indicate the number of independent components for linear and nonlinear susceptibility. The presence of optical activity is indicated by an asterisk for 11 enantiomorphic classes and by a † for 7 non enantiomorphic classes, which we will discuss in §7.9.8

Screw axes: $2_1, 3_1, 3_2, 4_1, 4_2, 4_3, 6_1, 6_2, 6_3, 6_4, 6_5$		the symbol has the form n_q, which is interpreted as follows: a proper rotation of 360°/n and a translation τ that is a fraction q/n of a lattice step along the rotation axis.
Glide planes :		translation τ:
a, b, c	axial	1/2 of the corresponding side .
n	diagonal	1/2 of the diagonal of a face, or 1/2 of the main diagonals (tetragonal and cubic).
d	diamond	as n, but 1/4.

Table 7.2. Non-symmorphic symmetry operations

The effect of this operator on a generic position $\boldsymbol{r}$ in space is given by the expression

$$\{\mathfrak{R}|\boldsymbol{t}\}\boldsymbol{r} = \mathfrak{R}\boldsymbol{r} + \boldsymbol{t}, \tag{7.1.2}$$

which is an inhomogeneous linear transformation. One can easily verify that the set of operations defined by this operation forms a true group in the mathematical sense. In particular, we have an identity element given by $\{1\,|\,\mathbf{0}\}$ (in the international notation). Moreover, for a crystal, *all of* the reticular points can be obtained with the translations $\mathbf{t}_n = n_1\boldsymbol{a}_1 + n_2\boldsymbol{a}_2 + n_3\boldsymbol{a}_3$, with n_1, n_2, n_3 integer, and where $\boldsymbol{a}_1$, $\boldsymbol{a}_2$, $\boldsymbol{a}_3$ are a base of primitive vectors of the corresponding Frankenheim-Bravais lattice (the expression for $\mathbf{t}_n$ becomes a little more complicated when the unit, non-primitive, lattice cells are used). Then, the set of operations $\{1\,|\,\mathbf{t}_n\}$ constitutes a sub-group of the spatial group of that crystal.

The point symmetry operations can also be expressed in this form as $\{\mathrm{R}_m\,|\,\mathbf{0}\}$, where R_m is one of the operations of a point group. The combination of the translational groups of the various lattices with the punctual groups compatible with these gives rise to 73 distinct spatial groups, called symmorphic. For their indication, the symbol of the punctual group is simply preceded by the letter P for primitive, A, B, or C for base-centered (according to the concerned axis), F for face-centered, I for body-centered, R for rhombic, of the corresponding lattice.

With these combinations, we have not exhausted all of the possible crystal structures: there are other operations that can leave a crystal unchanged. These new symmetry operations are called *screw axis* and *glide plane*. The former derive from the combination of a proper rotation around an axis with a translation τ, parallel to such an axis, equal to a rational fraction of the lattice step along that axis. The latter derive from a reflection through a plane followed by a translation τ along that plane, still equal to a rational fraction of the lattice step corresponding to the unit cell along that direction.

A list of the new symmetry operations we need to add to our set is given in Table 7.2.

If we also consider these operations, we obtain another 157 spatial groups, which are called *non-symmorphic*. The international symbol that specifies them is given, as above, by a letter specifying the type of lattice, followed by the symbol of the point group in which the number of some axis of rotation has an index or the letter m is replaced by one of the letters from Table 7.2.

The symbol of the spatial group represents the elements of symmetry essential to characterize the crystalline structure; however, as in the case of plane lattices, there are many other elements of symmetry distributed regularly in the lattice and, in particular, there may also be screw axes and glide planes in the structure of symmorphic space groups.

It should be noted that the punctual group corresponding to a given non-symmorphic spatial group (i.e., the one obtained by canceling $\boldsymbol{\tau}$) is not a subgroup of it! One can then ask what sense there is in continuing to refer to it for its symmetry properties. The fact is that $\boldsymbol{\tau}$ is only a fraction of the size of a unit cell, which, in turn, is generally much smaller than the wavelength of an optical radiation. Therefore, these translations do not produce appreciable optical anisotropies. Yet, there is an important exception: a helical configuration of atoms, that is, a screw axis, can give optical activity, whereby the plane of polarization of light rotates when it crosses the medium. However, the discourse concerning this effect is broader, and we can essentially have optical activity for all of those spatial groups in which there are no inversion centers, or in which there are not enough elements of symmetry for a mutual cancellation of the contributions that generate it. The point groups for which one may have optical activity are indicated in Table 7.1 by a sign in the fifth column.

7.2 The dielectric tensor

In a crystal, the atoms are arranged in an ordered manner, so that the symmetry properties of the microscopic structure are reflected on the macroscopic variables, instead of being canceled by the averaging process. In the context of a linear theory, we will here consider the case of very small external fields with respect to those present within the crystal. Macroscopically, only the average value of the position of the elementary charges will emerge, associated with a harmonic potential and subject to the perturbations of the external field. A simple model for illustrating the anisotropy of the crystals is shown in Fig. 7.3. A bound electron is drawn there as a mass attached to a set of elastic springs, to represent an anisotropic harmonic potential. In general, the "springs" have different hardness in the various directions of displacement of the electron from its equilibrium position in the crystalline lattice.

Therefore, the displacement of the electron under the influence of an external field **E** depends on the direction of the field, as well as on its strength. In particular, the direction of this displacement does not necessarily coincide with the direc-

Fig. 7.3
"Mechanical" representation of an electron in the crystalline lattice

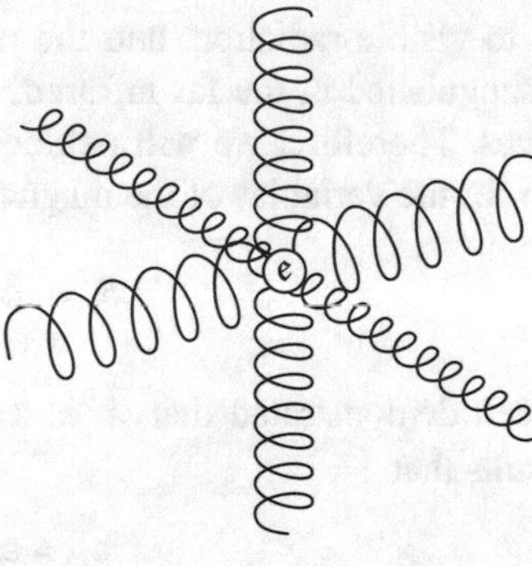

tion of **E** and generally depends on the frequency of the applied field. This is also true for the resulting polarization **P**. Moreover, the magnetization **M** also behaves in an anisotropic manner with respect to the magnetic field **B**, however, the effects of the magnetic type are generally very small and can be neglected.

We therefore consider the case of a monochromatic wave with angular frequency ω:

$$\mathbf{E} = \Re e\left[\boldsymbol{E}(\boldsymbol{r})e^{-i\omega t}\right]. \tag{7.2.1}$$

The induced polarization oscillates with the same frequency (the medium is assumed linear), and its relation to the electric field can now be expressed in a tensor form that generalizes that of Eq. (1.3.27):

$$\boldsymbol{P} = \varepsilon_0 \overset{\leftrightarrow}{\chi} \boldsymbol{E}, \tag{7.2.2}$$

where $\overset{\leftrightarrow}{\chi}$ is the *electric susceptibility tensor*:

$$\overset{\leftrightarrow}{\chi} = \begin{pmatrix} \chi_{11} & \chi_{12} & \chi_{13} \\ \chi_{21} & \chi_{22} & \chi_{23} \\ \chi_{31} & \chi_{32} & \chi_{33} \end{pmatrix}. \tag{7.2.3}$$

The corresponding displacement field $\boldsymbol{D}$ is given by

$$\boldsymbol{D} = \varepsilon_0\left(\mathbf{I} + \overset{\leftrightarrow}{\chi}\right)\boldsymbol{E} = \overset{\leftrightarrow}{\varepsilon}\boldsymbol{E}, \tag{7.2.4}$$

where $\mathbf{I}$ is the unity tensor and

$$\overset{\leftrightarrow}{\varepsilon} = \varepsilon_0\left(\mathbf{I} + \overset{\leftrightarrow}{\chi}\right) \tag{7.2.5}$$

is the *dielectric tensor*. There are also magnetic crystals, but they are generally not transparent to visible radiation, and the magnetic anisotropies, due to magnetization, are extinguished in the far infrared, at frequencies higher than those of centimeter waves. Therefore, we will neglect such anisotropies, maintaining a scalar relationship for the variables of the magnetic field, with

$$\boldsymbol{B} = \mu \boldsymbol{H} . \tag{7.2.6}$$

It has been demonstrated that $\vec{\vec{\varepsilon}}$ is a symmetrical tensor and, for its components, it results that

$$\varepsilon_{ij} = \varepsilon_{ji} . \tag{7.2.7}$$

This property follows from the *symmetry principle of the generalized kinetic coefficients*, that is, it follows from the symmetry properties, at the microscopic level, of the potential and kinetic energy of the elementary charges that make up the medium [Landau and Lifchitz III 1966, §127; Hopf and Stegeman 1985, Chap. 2]. Moreover, if the medium is transparent, it results that, for the conservation of energy, $\vec{\vec{\varepsilon}}$ is a Hermitian tensor [Landau and Lifchitz II 1966, §61 and 76]:

$$\varepsilon_{ij} = \varepsilon^*_{ji} . \tag{7.2.8}$$

Therefore, the comparison of Eqs. (7.2.7) and (7.2.8) shows that all of the elements ε_{ij} (and χ_{ij}) are real and, because of these symmetries, the maximum number of independent elements of the tensor is reduced from 9 to 6.

We can therefore also extend Eq. (1.5.15) for the energy density to the (electrically) anisotropic and transparent media:

$$\overline{w} = \frac{1}{4} \Re e \left(\boldsymbol{E} \bullet \boldsymbol{D}^* + \boldsymbol{H} \bullet \boldsymbol{B}^* \right) = \frac{1}{4} \left(\varepsilon_{ij} \, E_i^* E_j + \mu |\boldsymbol{H}|^2 \right), \tag{7.2.9}$$

where the bar on w indicates the time average and where the summation on the repeated indices was implied.

As we will see later, certain magnetic anisotropies, such as those responsible for optical activity or the Faraday effect, can be combined in the dielectric tensor, rather than in μ. With this inclusion, $\vec{\vec{\varepsilon}}$ remains Hermitian, but not symmetrical.

Excluding the latter case, for electrically anisotropic and transparent media, $\vec{\vec{\chi}}$ and $\vec{\vec{\varepsilon}}$ are therefore real and symmetrical tensors. Therefore, we can find a set of Cartesian axes, called *principal axes*, in which these tensors can be represented with a diagonal matrix:

$$\vec{\vec{\chi}} = \begin{pmatrix} \chi_{11} & 0 & 0 \\ 0 & \chi_{22} & 0 \\ 0 & 0 & \chi_{33} \end{pmatrix}, \qquad \vec{\vec{\varepsilon}} = \begin{pmatrix} \varepsilon_{11} & 0 & 0 \\ 0 & \varepsilon_{22} & 0 \\ 0 & 0 & \varepsilon_{33} \end{pmatrix}. \tag{7.2.10}$$

These three χ_{ii} are known as the *principal susceptibilities*, and the three corresponding quantities $\varepsilon_{ii} = 1 + \chi_{ii}$ are called *principal dielectric constants*. In the following, I will indicate, for brevity, with ε_1, ε_2, ε_3, the diagonal elements of the diagonalized matrix of $\vec{\vec{\varepsilon}}$.

Of some interest is a geometric construction associated with these tensors. Consider, for example, the electric part of the energy density of Eq. (7.2.9). Since, in our case, all ε_{ij} are real, let us consider a field described by amplitudes E_i and D_i, also real:

$$\begin{aligned}\overline{w}_e &= \frac{1}{4}\boldsymbol{E}_0 \cdot \boldsymbol{D}_o \\ &= \frac{1}{4}\left(\varepsilon_{xx}E_x{}^2 + \varepsilon_{yy}E_y{}^2 + \varepsilon_{zz}E_z{}^2 + 2\varepsilon_{xy}E_xE_y + 2\varepsilon_{yz}E_yE_z + 2\varepsilon_{zx}E_zE_x\right).\end{aligned} \tag{7.2.11}$$

This expression, at constant energy, describes an ellipsoid that, in the axis system that diagonalizes $\vec{\vec{\chi}}$ and $\vec{\vec{\varepsilon}}$, is simplified into

$$\overline{w}_e = \frac{1}{4}\left(\varepsilon_1 E_1{}^2 + \varepsilon_2 E_2{}^2 + \varepsilon_3 E_3{}^2\right), \tag{7.2.12}$$

where the indices 1, 2 and 3 here indicate the principal axes, which are then also the principal axes of the ellipsoid. Expressed in the form

$$\varepsilon_1 x^2 + \varepsilon_2 y^2 + \varepsilon_3 z^2 = C\,, \tag{7.2.13}$$

with C = constant, Eq. (7.2.12) becomes the equation of the surface of an ellipsoid that is called an *energy ellipsoid* of the dielectric tensor. Its symmetry properties will be useful in classifying the optical behavior of the various crystals. In the literature, we find several ellipsoids in the study of anisotropic media, each of which is indicated by various names, generating not a little confusion. The energy ellipsoid corresponds to what Fresnel simply calls an *ellipsoide*, Sommerfeld (1949) calls a *Fresnel's ellipsoid* and Born and Wolf (1980) call a *ray ellipsoid*. Contrastingly, Feynman (1969) calls an *energy ellipsoid* the one built with the components of $\varepsilon_0\vec{\vec{\chi}}$, which expresses the energy contribution generated by the polarization of the medium.

The energy density can also be expressed in terms of the field $\boldsymbol{D}$, and Eq. (7.2.12) becomes

$$\overline{w}_e = \frac{1}{4}\left(\frac{1}{\varepsilon_1}D_1{}^2 + \frac{1}{\varepsilon_2}D_2{}^2 + \frac{1}{\varepsilon_3}D_3{}^2\right). \tag{7.2.14}$$

This expression also describes a surface that is called an *ellipsoid of wave normals*, or an *ellipsoid of the indices* or *optical indicatrix*.

The ellipse obtained by intersecting this last surface with a plane passing through the origin and orthogonal to the direction of the wave vector, has the remarkable property of having the principal axes oriented along two linear polarization directions for the field ***D*** corresponding to two different refraction indexes. In particular, these elliptical sections become circular for two specific directions of the wave vector, which are called *optical axes*, and which we will discuss below. Even the Fresnel's *ellipsoide* has similar properties with respect to the *Poynting's vector*, the field ***E*** and the *radial axes*.

Fresnel referred to a *surface d'élasticité* to mean a surface of 4th degree that is the "reciprocal" of the ellipsoid of normals, and it is from this and from the previously mentioned *ray ellipsoid* that he derived all of the formulas for propagation in anisotropic media. Apart from the descriptive use of their symmetries, their application in some geometrical theorems and their historical importance, all of these ellipsoids are a difficult starting point for the calculation of propagation, and we will follow another path here; however, the symmetry or, in some cases, the surprising simplicity of some expressions derives from their underlying essentiality.

7.3 Crystal classes and principal axes

We have seen that, for a fixed angular frequency ω, it is always possible to find a system of axes in which the dielectric tensor is diagonal, with a number of distinct and non-zero elements of the dielectric tensor that is reduced from six to three. In other words, three of the six degrees of freedom of the tensor contain information on how the principal axis system is oriented. In some crystalline classes, this datum can be obtained from the symmetry properties of the crystals themselves. However, in triclinic and monoclinic crystals, the forces acting on the various elementary charges, which can be associated with different oscillation modes depending on their position in the lattice, are diagonalized in different axis systems. Therefore, overall, the main axis system has an orientation that depends on the frequency and is not directly associated with the crystal axes. In such crystals, all calculations involving more than one frequency, such as those in nonlinear optics, must be made judiciously. In all other crystalline classes, the diagonalized force system is determined by the crystal axes, or is degenerate.

In total, we can distinguish 5 cases in the number of independent elements, that is, degrees of freedom that are left by the symmetry of the crystal, which must be considered in the dielectric tensor.

One independent element

The ellipsoid of energy, described in the previous section, obviously rotates together with the crystal associated with it and must necessarily share its properties of symmetry by rotation. The crystals of the cubic class, especially those of the subclasses O and O_h, are symmetrical for rotations of 90° with respect to any of their three fourfold axes, and the force constants are equal in these three directions. But those of the subclasses T, T_h and T_d also have 4 threefold axes, each corresponding to one of the diagonals of the cubic cell. For all of these cases, the only ellipsoid with all of these symmetries is the sphere, and therefore

$$\varepsilon_1 = \varepsilon_2 = \varepsilon_3 = \varepsilon \,. \tag{7.3.1}$$

In this case, all axes are optically degenerate and any set of three orthogonal axes can be taken as a set of principal axes. In conclusion, all of the crystals of the cubic class are optically isotropic (two subclasses, however, have optical activity). Be aware that this isotropy does not necessarily apply to non-linear susceptibilities!

Two independent elements

The crystals of the hexagonal, trigonal and tetragonal classes essentially have *one* rotation axis (with or without reflection), respectively, sixfold, threefold and fourfold. The only ellipsoids with these symmetries are those of rotation around *that* axis. In our mechanical model, two force constants are equal, and the third, whose force is parallel to the rotation axis, is different. In this case, the diagonal elements of our dielectric tensor are reordered as follows:

$$\varepsilon_1 = \varepsilon_2 \neq \varepsilon_3 \,. \tag{7.3.2}$$

Thus, the principal axes 1 and 2 are degenerate and their direction is arbitrary on a plane orthogonal to axis 3, which is instead determined by the orientation of the crystal. This type of crystal is also called uniaxial, since it has only one optical axis. This terminology will become clear later.

Three independent elements

The crystals of the orthorhombic class possess three characteristic axes, orthogonal to each other, for which the operations of symmetry are essentially only rotations of 180° around one or all three axes, or reflections on planes normal to these axes. Now, the ellipsoids must necessarily have their main axes in correspondence with the crystal axes, but the three constants of force are now different. In this case, as in the following two, the main dielectric constants are all different from each other and are usually reordered as follows:

$$\varepsilon_1 < \varepsilon_2 < \varepsilon_3 \,. \tag{7.3.3}$$

These crystals are called *biaxial*, since they have 2 *optical axes*.

Four independent elements

The crystals of the monoclinic class are also optically biaxial. They have only one symmetry axis of rotation for 180° and/or reflection on a plane normal to it. Only one of the principal axes of the ellipsoid must now coincide with this characteristic axis of the crystal, while the direction of the other two around it can change with frequency.

Six independent elements

Finally, the crystals of the triclinic class have only the identity, or possibly also the inversion, as symmetry operations. They are therefore biaxial crystals for which the ellipsoid can have any orientation, and the direction of the three principal axes now depends on the frequency.

7.4 Propagation modes for the field D

7.4.1 Wave equation and eigenvectors

Let us now determine the *eigenvectors of the field*: in particular, we want to look for those plane waves that, by propagating with a well-defined direction of the wave vector $\boldsymbol{k}$, retain the initial polarization.

The fact that $\bar{\bar{\varepsilon}}$ is real suggests that the eigenvectors have a linear polarization. We can therefore look for a solution to Maxwell's equations with this form and verify their congruence. We therefore consider a wave expressed by

$$\mathbf{E} = \Re e\left[\boldsymbol{E}_{\mathrm{o}} e^{i(\boldsymbol{k}\cdot\boldsymbol{r}-\omega t)}\right] \tag{7.4.1}$$

and similar equations for all other fields.

As a first step, we replace these expressions in the Maxwell's Eqs. (1.2.6). For $\rho_m = 0, \boldsymbol{j}_m = 0$, and with $\boldsymbol{B} = \mu\boldsymbol{H}$, where μ is a constant real scalar, we get

$$\begin{cases} \boldsymbol{k}\cdot\boldsymbol{D}_{\mathrm{o}} = 0, \\ \boldsymbol{k}\cdot\boldsymbol{H}_{\mathrm{o}} = 0, \\ \boldsymbol{k}\times\boldsymbol{E}_{\mathrm{o}} = +\omega\mu\,\boldsymbol{H}_{\mathrm{o}}, \\ \boldsymbol{k}\times\boldsymbol{H}_{\mathrm{o}} = -\omega\,\boldsymbol{D}_{\mathrm{o}}. \end{cases} \tag{7.4.2}$$

From the first, second and fourth equations, one can immediately observe that $\boldsymbol{k}$, $\boldsymbol{D}_{\mathrm{o}}$, $\boldsymbol{H}_{\mathrm{o}}$ (and $\boldsymbol{B}_{\mathrm{o}}$) must form a set of three mutually orthogonal vectors. The third equation tells us only that $\boldsymbol{E}_{\mathrm{o}}$ must be perpendicular to $\boldsymbol{H}_{\mathrm{o}}$ and then lie in the plane determined by $\boldsymbol{k}$ and $\boldsymbol{D}_{\mathrm{o}}$.

In turn, the Poynting's vector defines a further direction $\mathbf{s}$ in which the wave energy is propagated: this fact is also valid for the anisotropic media, since it derives directly from the macroscopic Maxwell equations, without any other hypothesis on the media. So, we have that

$$I\mathbf{s} = \bar{\mathbf{S}} = \frac{1}{2}\Re e\left(\boldsymbol{E}_{\mathrm{o}}\times\boldsymbol{H}_{\mathrm{o}}^{*}\right), \tag{7.4.3}$$

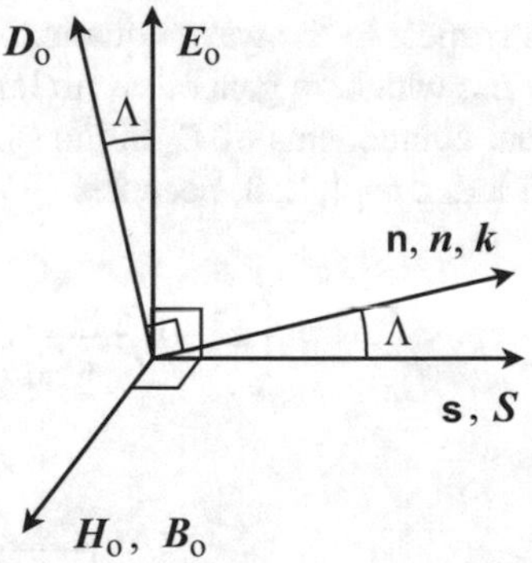

Fig. 7.4
Mutual orientation of the vectors

where **s** is the versor of **S** and I is the intensity of the wave. Because we are looking for linearly polarized solutions, we can take E_0 and H_o as real, whereby

$$I\mathbf{s} = \frac{1}{2}\boldsymbol{E}_0 \times \boldsymbol{H}_o. \tag{7.4.4}$$

Thus, $\boldsymbol{E}_0$, $\boldsymbol{H}_o$, **s** also form a set of mutually orthogonal vectors. It follows, in particular, that the angle Λ between $\boldsymbol{k}$ and **s** is the same as that between $\boldsymbol{D}_o$ and $\boldsymbol{E}_0$ (Fig. 7.4). Therefore, in general, the direction of the energy flow does not coincide with the direction of $\boldsymbol{k}$. We will see the consequences of this fact later.

Indicating the versor of $\boldsymbol{H}_o$ and $\boldsymbol{B}_o$ with **h**, and the versors of $\boldsymbol{D}_o$ and $\boldsymbol{E}_0$ with **d** and **e**, respectively, for Eq. (7.4.2), we also have the relations

$$\begin{aligned} \mathbf{n} &= \mathbf{d}\times\mathbf{h}, \quad \mathbf{d} = -\mathbf{n}\times\mathbf{h}, \quad \mathbf{h} = \mathbf{n}\times\mathbf{d}, \\ \mathbf{s} &= \mathbf{e}\times\mathbf{h}, \quad \mathbf{e} = -\mathbf{s}\times\mathbf{h}, \quad \mathbf{h} = \mathbf{s}\times\mathbf{e}. \end{aligned} \tag{7.4.5}$$

As a second step of our verification, we apply Eq. (7.4.1), and the similar ones of our plane wave, to the wave Eq. (1.4.3), which is valid for a *magnetically* isotropic and linear material. With the substitutions of Eqs. (1.4.13), we have that

$$-k^2\boldsymbol{E}_0 + (\boldsymbol{k}\boldsymbol{\cdot}\boldsymbol{E}_0)\boldsymbol{k} + \varepsilon_0\mu\omega^2\boldsymbol{E}_0 = -\mu\omega^2\boldsymbol{P}_o, \tag{7.4.6}$$

which, with Eq. (7.2.2), inverting the signs, becomes

$$k^2\boldsymbol{E}_0 - (\boldsymbol{k}\boldsymbol{\cdot}\boldsymbol{E}_0)\boldsymbol{k} - \varepsilon_0\mu\omega^2\boldsymbol{E}_0 = \varepsilon_0\mu\omega^2\vec{\vec{\chi}}\,\boldsymbol{E}_0, \tag{7.4.7}$$

that is,

$$k^2\boldsymbol{E}_0 - (\boldsymbol{k}\boldsymbol{\cdot}\boldsymbol{E}_0)\boldsymbol{k} = \mu\omega^2\vec{\vec{\varepsilon}}\,\boldsymbol{E}_0. \tag{7.4.8}$$

With respect to the wave equations for isotropic media, we must now keep the term $\boldsymbol{k} \cdot \boldsymbol{E}_o$, which, in general, is no longer null. Eq. (7.4.8), written in terms of the individual components of $\boldsymbol{E}_o$ in the system of the principal axes, with an equation for each index $j = 1, 2, 3$, becomes

$$\left(k^2 - n_j^2 \frac{\omega^2}{c^2}\right) E_j - (\boldsymbol{k} \cdot \boldsymbol{E}_o) k_j = 0, \tag{7.4.9}$$

where

$$n_j = c\sqrt{\mu \varepsilon_j} = \sqrt{\frac{\mu}{\mu_o}(1+\chi_j)}, \quad \text{with } j = 1, 2, 3, \tag{7.4.10}$$

are the refractive indices associated with the principal axes. For the hypotheses made on the medium, n_1, n_2, n_3 are real and positive.

Let us now introduce some quantities, related to the vector $\boldsymbol{k}$, which will be useful later. In particular, the wave vector will be conveniently factored as

$$\boldsymbol{k} = \frac{\omega}{c} n \mathbf{n} = \frac{\omega}{c} \boldsymbol{n} = \frac{\omega}{\upsilon} \mathbf{n}, \tag{7.4.11}$$

where $\mathbf{n}$ is the versor of $\boldsymbol{k}$ and n is the index of refraction associated with the propagation mode along $\boldsymbol{k}$. We then have a new vector $\boldsymbol{n} = n\mathbf{n}$. Lastly, $\upsilon = c/n$ is the *phase velocity* with which the wave propagates.

Eigenvalue equation

In the next section, the calculations will be continued from Eq. (7.4.9), however, it is good to note that Eq. (7.4.8) can be expressed in terms of the vector $\boldsymbol{D}_o$ in the form of an *eigenvalue equation*. Indeed, in the case that we are dealing with, the dielectric tensor is invertible (none of the ε_j is null), so we have

$$\vec{\vec{\varepsilon}}^{-1} \boldsymbol{D}_o - \mathbf{n}\left(\mathbf{n} \cdot \vec{\vec{\varepsilon}}^{-1} \boldsymbol{D}_o\right) = \frac{\mu c^2}{n^2} \boldsymbol{D}_o,$$

that is,

$$\frac{1}{\mu c^2} (\mathcal{I} - \mathbf{nn}) \cdot \vec{\vec{\varepsilon}}^{-1} \boldsymbol{D}_o = \frac{1}{n^2} \boldsymbol{D}_o, \tag{7.4.12}$$

where $\mathcal{I}$ is the unit tensor and $\mathbf{nn}$ is a dyad built with the versor $\mathbf{n}$.

In particular, $\mathcal{I} - \mathbf{nn}$ constitutes a projection operator in the subspace orthogonal to $\mathbf{n}$. Finally, from Eq. (7.4.12), we can immediately observe that the possible solutions for the sought after plane waves, with a direction of $\mathbf{n}$ fixed, correspond to the values of $\boldsymbol{D}_o$ that are the eigenvectors of the operator

$$\frac{1}{\mu c^2}(\mathbf{I}-\mathbf{nn})\cdot\vec{\vec{\varepsilon}}^{-1} \tag{7.4.13}$$

with eigenvalues $1/n^2$. The term *eigenvector* attributed to the solutions sought is therefore justified. However, the operator of the expression (7.4.13) is not symmetric, so we cannot directly deduce the orthogonality of the eigenvectors. This will be done later with simple considerations concerning Eq. (7.4.8). In the following, I will also indicate, with the term *eigenvalue*, the solutions for k, n, etc., which correspond to the eigenvalues $1/n^2$ of the operator (7.4.13).

Constraints imposed by prefixing the direction of the field instead of the wave vector

In solving Eq. (7.4.9), we will put the emphasis on determining the field once the propagation direction is fixed. On the other hand, if you fix the direction of the vector $\boldsymbol{E}_0$ (or of $\boldsymbol{D}_o$) in the medium, the direction of the vector $\boldsymbol{D}_o$ (or of $\boldsymbol{E}_0$) remains unequivocally determined, as does, except for 180°, the directions of the vectors $\boldsymbol{H}_o$, $\mathbf{n}$, $\mathbf{s}$. The value of the phase velocity υ is equally determined. But be aware: this does *not* mean that the field cannot propagate in other directions, but only that these are the only directions in which the assigned polarization is maintained! In fact, if you fix the direction of the vector $\boldsymbol{E}_0$ ($\boldsymbol{D}_o$), and therefore its polarization, that of the vector $\boldsymbol{D}_o$ ($\boldsymbol{E}_0$) is uniquely defined by the tensor relationship (7.2.4), which, in the system of the principal axes, has the simple form

$$D_x = \varepsilon_1 E_x, \quad D_y = \varepsilon_2 E_y, \quad D_z = \varepsilon_3 E_z\,. \tag{7.4.14}$$

Therefore, the plane ($\boldsymbol{E}_0$, $\boldsymbol{D}_o$), which is orthogonal to $\boldsymbol{H}_o$, is also defined. The direction of the latter is then determined, except for 180°, and so is the direction of the versors $\mathbf{n}$ and $\mathbf{s}$. In the special cases in which the directions of $\boldsymbol{E}_0$ and $\boldsymbol{D}_o$ coincide (when all of the ε are different, this occurs only in correspondence with the principal axes), the direction of $\boldsymbol{H}_o$ remains indeterminate on the normal plane to these vectors, while $\mathbf{n}$ and $\mathbf{s}$ coincide, completing the orthogonal set

The value of k, and thus those of n and υ, can be immediately obtained by multiplying the wave Eq. (7.4.8) scalarly for $\boldsymbol{D}_o$:

$$k^2 = \frac{\mu\omega^2 \boldsymbol{D}_o{}^2}{\boldsymbol{E}_0\cdot\boldsymbol{D}_o}, \quad n^2 = \frac{\mu c^2 \boldsymbol{D}_o{}^2}{\boldsymbol{E}_0\cdot\boldsymbol{D}_o}, \quad \upsilon^2 = \frac{\boldsymbol{E}_0\cdot\boldsymbol{D}_o}{\mu\boldsymbol{D}_o{}^2}\,. \tag{7.4.15}$$

The angle Λ can also be calculated easily through the scalar product between $\boldsymbol{E}_0$ and $\boldsymbol{D}_o$. If e_x, e_y, e_z are the cosine directors (i.e., the components of the versor) of $\boldsymbol{E}_0$ with respect to the principal axes, we have that

$$\Lambda = \arccos \frac{n_1^2 e_x^2 + n_2^2 e_y^2 + n_3^2 e_z^2}{\sqrt{n_1^4 e_x^2 + n_2^4 e_y^2 + n_3^4 e_z^2}}, \tag{7.4.16.a}$$

which, for a medium magnetically isotropic, coincides with the angle between the Poynting's vector and the wave vector. Similarly, it is also

$$\Lambda = \arccos \frac{n_1^{-2} d_x^2 + n_2^{-2} d_y^2 + n_3^{-2} d_z^2}{\sqrt{n_1^{-4} d_x^2 + n_2^{-4} d_y^2 + n_3^{-4} d_z^2}}. \tag{7.4.16.b}$$

We also have that

$$\upsilon = \frac{\sqrt{\boldsymbol{E}_o \cdot \boldsymbol{D}_o}}{\sqrt{\mu}\sqrt{\boldsymbol{D}_o^2}} = c \frac{\sqrt{n_1^2 e_x^2 + n_2^2 e_y^2 + n_3^2 e_z^2}}{\sqrt{n_1^4 e_x^2 + n_2^4 e_y^2 + n_3^4 e_z^2}}. \tag{7.4.17}$$

Moreover, from Eq. (7.4.15.c), we have another way to express the ellipsoid of the normals of Eq. (7.2.14):

$$\frac{c^2}{n^2} = \frac{1}{\mu \boldsymbol{D}_o^2} \boldsymbol{E}_o \cdot \boldsymbol{D}_o = \frac{d_x^2}{\mu\varepsilon_1} + \frac{d_y^2}{\mu\varepsilon_2} + \frac{d_z^2}{\mu\varepsilon_3}. \tag{7.4.18}$$

In conclusion, we can therefore state that, in a crystal, the electric vector ($\boldsymbol{E}_o$ or $\boldsymbol{D}_o$) is the *principal vector*. This is basically natural, since it is this vector that determines the electrical polarization of the medium, whose excitation constitutes the main contribution to the process of propagation of electromagnetic waves in material media.

7.4.2 Fresnel's equations

In order for Eqs. (7.4.9) to give a non-zero result for $\boldsymbol{E}_o$, it is necessary that their determinant be zero. This implies that a certain relationship must be satisfied between the vector $\boldsymbol{k}$, the angular frequency ω and the principal indices n_j. With several steps in the calculation of the determinant, it is finally found that

$$\begin{aligned} & k^2\left(k_x^2 n_1^2 + k_y^2 n_2^2 + k_z^2 n_3^2\right) \\ & -\frac{\omega^2}{c^2}\left[k_x^2 n_1^2\left(n_2^2 + n_3^2\right) + k_y^2 n_2^2\left(n_3^2 + n_1^2\right) + k_z^2 n_3^2\left(n_1^2 + n_2^2\right)\right] \\ & + \frac{\omega^4}{c^4} n_1^2 n_2^2 n_3^2 = 0. \end{aligned} \tag{7.4.19.a}$$

This equation is one of the forms in which the *Fresnel's equation for wave vectors* can be written. For convenience, I report here some equivalent forms that will serve us later, or that have a practical interest anyway. Replacing Eq. (7.4.11) in (7.4.19.a), we find an expression in n^2:

$$\begin{aligned} & n^4\left(\mathsf{n}_x^2 n_1^2+\mathsf{n}_y^2 n_2^2+\mathsf{n}_z^2 n_3^2\right) \\ & -n^2\left[\mathsf{n}_x^2 n_1^2\left(n_2^2+n_3^2\right)+\mathsf{n}_y^2 n_2^2\left(n_3^2+n_1^2\right)+\mathsf{n}_z^2 n_3^2\left(n_1^2+n_2^2\right)\right] \\ & +n_1^2 n_2^2 n_3^2=0, \end{aligned} \tag{7.4.19.b}$$

where n_x, n_y, n_z are the three components of the versor $\mathbf{n}$. Collecting the terms in n_x^2, n_y^2, n_z^2, this second equation can be rewritten as

$$\begin{aligned} & \mathsf{n}_x^2 n_1^2\left(n^2-n_2^2\right)\left(n^2-n_3^2\right)+\mathsf{n}_y^2 n_2^2\left(n^2-n_3^2\right)\left(n^2-n_1^2\right) \\ & +\mathsf{n}_z^2 n_3^2\left(n^2-n_1^2\right)\left(n^2-n_2^2\right)=0, \end{aligned} \tag{7.4.19.c}$$

(Eq. A of Fresnel), that is,

$$\frac{\mathsf{n}_x^2 n_1^2}{n^2-n_1^2}+\frac{\mathsf{n}_y^2 n_2^2}{n^2-n_2^2}+\frac{\mathsf{n}_z^2 n_3^2}{n^2-n_3^2}=0 . \tag{7.4.19.d}$$

Adding member to member $\mathsf{n}_x^2+\mathsf{n}_y^2+\mathsf{n}_z^2=1$, this equation can be placed in a form known for its elegance:

$$\frac{n_x^2}{n^2-n_1^2}+\frac{n_y^2}{n^2-n_2^2}+\frac{n_z^2}{n^2-n_3^2}=1, \tag{7.4.19.e}$$

where n_x, n_y, n_z are the components of the vector $\boldsymbol{n}=n\mathbf{n}$ that should *not* be confused with the values n_1, n_2, n_3 of the refractive index associated with the principal axes! Lastly, from Eq. (7.4.19.d), we also get

$$\frac{\mathsf{n}_x^2}{\frac{1}{n^2}-\frac{1}{n_1^2}}+\frac{\mathsf{n}_y^2}{\frac{1}{n^2}-\frac{1}{n_2^2}}+\frac{\mathsf{n}_z^2}{\frac{1}{n^2}-\frac{1}{n_3^2}}=0 . \tag{7.4.19.f}$$

Eq. (7.4.19.c) is quadratic in n^2, so that, once the direction of $\mathbf{n}$ is fixed, there are two possible roots for n^2. This equation also helps us to delimit the range of existence of these roots. In fact, it has the form

$$F\left(n^2\right)=0 \tag{7.4.20}$$

and, alternatively placing n_1, n_2, n_3 as arguments of F instead of n, we find the following equations:

$$F\left(n_1^2\right)=\mathrm{n}_x^2\, n_1^2\left(n_1^2-n_2^2\right)\left(n_1^2-n_3^2\right)\geq 0,$$
$$F\left(n_2^2\right)=\mathrm{n}_y^2\, n_2^2\left(n_2^2-n_3^2\right)\left(n_2^2-n_1^2\right)\leq 0,$$
$$F\left(n_3^2\right)=\mathrm{n}_z^2\, n_3^2\left(n_3^2-n_1^2\right)\left(n_3^2-n_2^2\right)\geq 0,$$

where the inequalities follow from having rearranged the numbering of the principal axes so that

$$n_1\leq n_2\leq n_3\,. \tag{7.4.21}$$

Therefore, the function $F(n^2)$ changes sign twice: the first between $n_1{}^2$ and $n_2{}^2$ and the second between $n_2{}^2$ and $n_3{}^2$. In other words, the equation $F(n^2) = 0$ has two real and positive roots n'^2 and n''^2 such that

$$n_1{}^2\leq n'^2\leq n_2{}^2\leq n''^2\leq n_3{}^2\,. \tag{7.4.22}$$

In conclusion, along the direction $\mathbf{n}$, there are two waves that propagate with two wave vectors of generally different modulus,

$$k'=\frac{\omega}{c}n',\quad k''=\frac{\omega}{c}n''\,. \tag{7.4.23}$$

In certain special cases, k' and k'' can coincide.

7.4.3 The surface of the wave vectors

Eq. (7.4.19.a) describes a surface in the space of vectors $\boldsymbol{k}$, which is therefore called the *surface of the wave vectors*. It is biquadratic in k and, indeed, consists of two concentric "ovals". To see what it looks like, let us consider one of the principal planes, for example, the xy-plane, so that $k_z = 0$. Rearranging the terms of Eq. (7.4.19.a), we find

$$\left[\left(n_3\frac{\omega}{c}\right)^2-k_x^2-k_y^2\right]\left[n_1{}^2k_x^2+n_2{}^2k_y^2-\left(n_1n_2\frac{\omega}{c}\right)^2\right]=0\,. \tag{7.4.24}$$

Since the product must cancel, one or both factors must be equal to zero. By

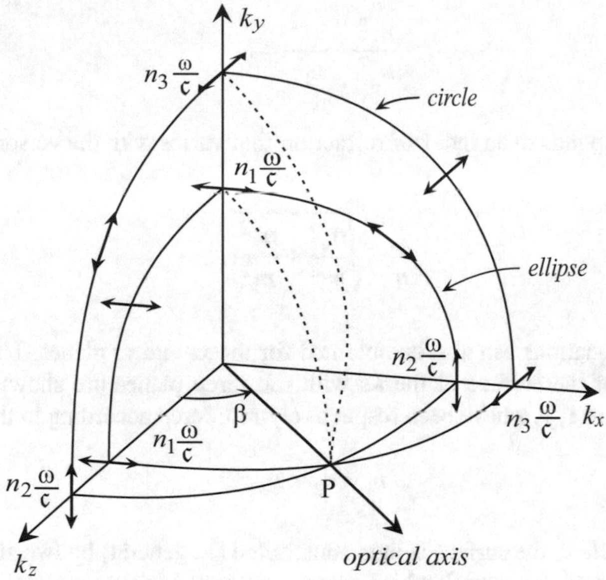

Fig. 7.5 Surface of wave vectors: here, an octant is represented through its intersections with the principal planes. By convention, the *y*-axis is coincident with the principal axis with which the intermediate index n_2 is associated, the *z*-axis with the highest index, n_3, and the *x*-axis with the lowest index, n_1. The optical axis is then on the *xz*-plane on the intersection of the circle, with the eigenvalue of index n_2 and the ellipse, with an index that varies between n_1 and n_3. The double arrows indicate the direction of the eigenvectors for the electric field

canceling the first factor, we find the equation of a circle:

$$k_x^2 + k_y^2 = \left(n_3 \frac{\omega}{c} \right)^2 , \tag{7.4.25}$$

which is equivalent to a constant index of refraction

$$n = n_3 , \tag{7.4.26}$$

while the second factor gives the equation of an ellipse

$$\left(\frac{c}{n_2 \omega} \right)^2 k_x^2 + \left(\frac{c}{n_1 \omega} \right)^2 k_y^2 = 1 , \tag{7.4.27}$$

that is,

$$\frac{n_x{}^2}{n_2{}^2}+\frac{n_y{}^2}{n_1{}^2}=1\,, \tag{7.4.28}$$

which corresponds to an index of refraction that varies with the versor **n** according to the law

$$\frac{1}{n}=\sqrt{\frac{\mathrm{n}_x{}^2}{n_2{}^2}+\frac{\mathrm{n}_y{}^2}{n_1{}^2}}\,. \tag{7.4.29}$$

Similar equations can also be obtained for the *xz* and *yz* planes. The various intersections of the surface of the ***k***s with the three planes are shown in Fig. 7.5, where the axes *x*, *y*, z have been respectively reordered according to the inequality

$$n_1 < n_2 < n_3\,. \tag{7.4.30}$$

In each octant, the surface is thus constituted (in general) by two sheets, one internal and the other external, which correspond to the two possible eigenvalues for *k*. They are associated with two corresponding field eigenvectors, with distinct polarizations (Fig. 7.5). We can now proceed with their determination.

7.4.4 Eigenvectors

Let us first consider the case in which one of the components of ***k*** is null, for example, when, as above, $k_z = 0$. Rearranging the terms, Eq. (7.4.8) with Eq. (7.4.11) becomes

$$\begin{cases} n_x n_y E_y &= \left(n_y{}^2 - n_1{}^2\right) E_x, \\ n_x n_y E_x &= \left(n_x{}^2 - n_1{}^2\right) E_y, \\ \quad 0 &= \left(n^2 - n_3{}^2\right) E_z, \end{cases} \tag{7.4.31}$$

where n_x, n_y are components of the vector ***n***. From the last of the Eqs. (7.4.31) it is observed that it can be $E_z \neq 0$ only for an eigenvalue of *n* that verifies Eq. (7.4.26). From the first two, it is instead found that

$$E_x = \frac{n_x n_y}{n_y{}^2 - n_1{}^2} E_y, \quad E_x = \frac{n_x{}^2 - n_2{}^2}{n_x n_y} E_y,$$

which, as expected, can both be verified only for an eigenvalue of *n* that verifies

Eq. (7.4.29), or if both the E_x and E_y fields are null. In general, for indices n_1, n_2, n_3 different from each other, the circle and the ellipse do not coincide, apart from their eventual intersection (the point P in Fig. 7.5) that will be discussed shortly. The points of the circle on the *xy*-plane have a field $\boldsymbol{E}$ with a linear polarization oriented along the z-axis and a refractive index $n = n_3$ as an eigenvector. On the contrary, the points of the ellipse on the *xy*-plane have a field $\boldsymbol{E}$ with linear polarization parallel to the same plane. With a few calculations, it is found that the relationship between E_x and E_y is

$$\frac{E_x}{E_y} = -\frac{\mathsf{n}_y\, n_2^{\,2}}{\mathsf{n}_x\, n_1^{\,2}}, \tag{7.4.32}$$

and therefore this eigenvector has the field $\boldsymbol{E}$ oriented tangentially to the ellipse. The corresponding index of refraction is given by Eq. (7.4.29). Similar relationships can also be found for the intersections of the surface of the $\boldsymbol{k}$s with the *xz* and *yz* planes. We will see later that these tangential properties of the eigenvalues of $\boldsymbol{E}$ with the surface of the $\boldsymbol{k}$s are general for all $\boldsymbol{k}$ directions.

Now, let us look at the more general case, in which all of the components of $\boldsymbol{k}$ in the system of the principal axes are different from zero. Dividing each of the Eqs. (7.4.9) for the corresponding k_j, one finds that

$$\frac{n^2 - n_1^{\,2}}{n_x} E_x = \frac{n^2 - n_2^{\,2}}{n_y} E_y = \frac{n^2 - n_3^{\,2}}{n_z} E_z, \tag{7.4.33}$$

and therefore, depending on the versor n, the eigenvectors $\boldsymbol{E}_o$ are oriented according to the vectors (not normalized):

$$\tilde{\boldsymbol{V}}_E^{(i)} = \left(\frac{\mathsf{n}_x}{n^{(i)2} - n_1^{\,2}}, \frac{\mathsf{n}_y}{n^{(i)2} - n_2^{\,2}}, \frac{\mathsf{n}_z}{n^{(i)2} - n_3^{\,2}} \right), \quad \text{with } (i) = {}', {}''$$

or, to avoid the divergences arising when n tends to one of the principal axis directions, these others, which are less elegant but more effective:

$$\boldsymbol{V}_E^{(i)} = \left[\left(n^{(i)2} - n_2^{\,2}\right)\left(n^{(i)2} - n_3^{\,2}\right)\mathsf{n}_x, \left(n^{(i)2} - n_1^{\,2}\right)\left(n^{(i)2} - n_3^{\,2}\right)\mathsf{n}_y, \left(n^{(i)2} - n_1^{\,2}\right)\left(n^{(i)2} - n_1^{\,2}\right)\mathsf{n}_z\right], \tag{7.4.34}$$

again, with $(i) = {}', {}''$ and not normalized. Recalling the relation $D_j = \varepsilon_j E_j$, the corresponding vector for the direction of $\boldsymbol{D}$ is

$$\boldsymbol{V}_D(i) = \left[\left(\frac{1}{n^{(i)2}} - \frac{1}{n_2{}^2}\right)\left(\frac{1}{n^{(i)2}} - \frac{1}{n_3{}^2}\right)\mathsf{n}_x, \left(\frac{1}{n^{(i)2}} - \frac{1}{n_1{}^2}\right)\left(\frac{1}{n^{(i)2}} - \frac{1}{n_3{}^2}\right)\mathsf{n}_y, \left(\frac{1}{n^{(i)2}} - \frac{1}{n_1{}^2}\right)\left(\frac{1}{n^{(i)2}} - \frac{1}{n_2{}^2}\right)\mathsf{n}_z\right], \tag{7.4.35}$$

again, with $(i) = {}', {}''$ and not normalized. In both expressions, n' and n'' are the two eigenvalues of the refractive index to be obtained through Eq. (7.4.19.b).

In conclusion, the procedure is as follows: knowing the direction n of $\boldsymbol{k}$ and the three principal indices, from Eq. (7.4.19.b), we calculate the two possible values of n; these are then replaced in Eqs. (7.4.34) and (7.4.35), obtaining two vectors that are finally normalized so as to obtain the versors of the eigenvectors of $\boldsymbol{E}_\mathrm{o}$ and $\boldsymbol{D}_\mathrm{o}$.

Let us now analyze the angular relationship between the eigenvectors corresponding to the eigenvalues n' and n'' of the refractive index. Applying Eq. (7.4.8), with the transformation (7.4.11), for the two solutions, we have that

$$\boldsymbol{E}'_\mathrm{o} - (\mathsf{n}\bullet\boldsymbol{E}'_\mathrm{o})\mathsf{n} = \mu\frac{c^2}{n'^2}\boldsymbol{D}'_\mathrm{o},$$
$$\boldsymbol{E}''_\mathrm{o} - (\mathsf{n}\bullet\boldsymbol{E}''_\mathrm{o})\mathsf{n} = \mu\frac{c^2}{n''^2}\boldsymbol{D}''_\mathrm{o}.$$

We then scalarly multiply the first equation for $\boldsymbol{D}''_\mathrm{o}$, the second for $\boldsymbol{D}'_\mathrm{o}$ and we make the difference in the results. Since $\boldsymbol{D}'_\mathrm{o} \bullet \mathsf{n} = \boldsymbol{D}''_\mathrm{o} \bullet \mathsf{n} = 0$, we get:

$$\boldsymbol{E}'_\mathrm{o}\bullet\boldsymbol{D}''_\mathrm{o} - \boldsymbol{E}''_\mathrm{o}\bullet\boldsymbol{D}'_\mathrm{o} = \mu c^2\left(\frac{1}{n'^2} - \frac{1}{n''^2}\right)\boldsymbol{D}'_\mathrm{o}\bullet\boldsymbol{D}''_\mathrm{o}.$$

Moreover, since

$$\boldsymbol{E}'_\mathrm{o}\bullet\boldsymbol{D}''_\mathrm{o} = \sum_j \varepsilon_j E'_j E''_j = \boldsymbol{E}''_\mathrm{o}\bullet\boldsymbol{D}'_\mathrm{o},$$

it results that

$$\left(\frac{1}{n'^2} - \frac{1}{n''^2}\right)\boldsymbol{D}'_\mathrm{o}\bullet\boldsymbol{D}''_\mathrm{o} = 0.$$

Therefore, when $n' \neq n''$, it is $\boldsymbol{D}'_\mathrm{o}\bullet\boldsymbol{D}''_\mathrm{o} = 0$. In conclusion, the two plane waves admitted by the wave equation for a given direction n of the vector $\boldsymbol{k}$ have eigenvectors $\boldsymbol{D}'_\mathrm{o}$ and $\boldsymbol{D}''_\mathrm{o}$ orthogonal to each other. Instead, the corresponding fields $\boldsymbol{E}'_\mathrm{o}$ and $\boldsymbol{E}''_\mathrm{o}$ are *not* generally orthogonal to each other. On the other hand, this non-orthogonality is not required, because the versors s' and s'' corresponding to these eigenvectors are generally different from each other. This is a direct consequence of the non-parallelism between $\boldsymbol{D}$ and $\boldsymbol{E}$ established by Eqs. (7.4.14).

On the other hand, when the propagation direction is such that this parallelism is present, the eigenvectors are orthogonal in both the $\boldsymbol{D}$ and $\boldsymbol{E}$ eigenvector pairs. For example, when $\boldsymbol{k}$ lies on the *xy*-plane of a uniaxial crystal, the eigenvectors have components

$$\left(D'_{ox}, D'_{oy}, 0\right) = \varepsilon\left(E'_{ox}, E'_{oy}, 0\right), \quad \left(0, 0, D''_{oz}\right) = \varepsilon_3\left(0, 0, E''_{oz}\right),$$

where $\varepsilon = \varepsilon_1 = \varepsilon_2$.

7.4.5 Optical axes

Observing Fig. 7.5, we see that, in one of the three principal planes, where a circle and an ellipse intersect, the two sheets of the surface of the wave vectors $\boldsymbol{k}$ touch themselves at a singular point P, forming a sort of "navel". Normally, it is expected that two surfaces intersect along a line, but the nature of the surface of the $\boldsymbol{k}$s is such that (in general, for indices n_1, n_2, n_3 that are all different) there are only 4 points of contact. With reference to Fig. 7.5, they are all on the *xz*-plane, one for each quadrant, and are diagonally arranged with each other (Fig. 7.6). In the neighborhood of each of these points, the separation between the two surfaces grows linearly with the distance on the *xz*-plane, while, for the symmetry of reflection around this plane, it grows quadratically in the *y* direction. The two internal and external surfaces therefore have 4 edges aligned in this last direction, which are progressively blunted with the growth of $|k_y|$.

The existence of only 4 points of contact, whereby the two eigenvalues n' and n'' coincide, immediately follows from the relation (7.4.22). It is indeed necessary that $n' = n'' = n_2$, and, in this case, Eq. (7.4.19.c) is reduced to

$$\mathsf{n}_y^2\, n_2^2 \left(n_2^2 - n_3^2\right)\left(n_2^2 - n_1^2\right) = 0\,. \tag{7.4.36}$$

In the case in which all of the indices are different from each other, this is only possible if $\mathsf{n}_y = 0$, i.e., only if the sought after point is on the *xz*-plane, and therefore corresponds to one of the 4 intersections already identified. Since each of these points defines a direction for which the two eigenvalues of k are equal, along this direction, called the *optical axis*, the field eigenvectors are degenerate and any polarization is preserved in crossing the medium.

In the general case in which the three principal indices are different, there are *two* optical axes and the crystal is called *biaxial*. These axes are shown in Fig. 7.6(a). In the conditions of Fig. 7.5, the angle β subtended between the optical axes and the *z*-axis can be easily obtained from the calculation of the intersection between the circle and the ellipse on the *xz*-plane, obtaining

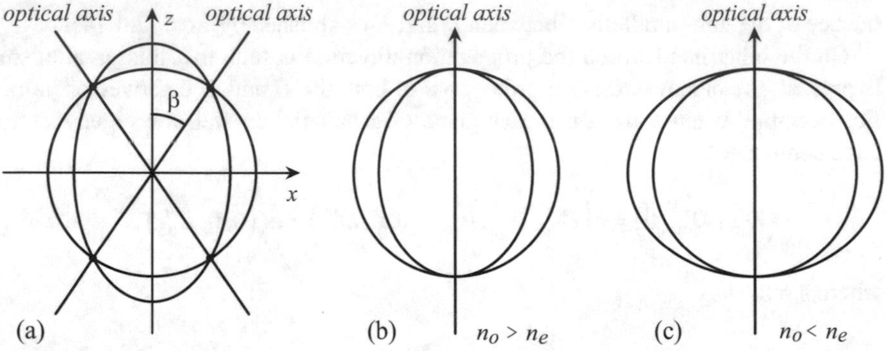

Fig. 7.6 Intersections of the wave vector surface with the *xz*-plane. (a): biaxial crystal; (b): negative uniaxial crystal; (c): positive uniaxial crystal. In both of these last two cases, the optical axis is conventionally taken as the *z*-axis; for case (b), this involves a different order of the principal indices: $n_o = n_1 = n_2 > n_3 = n_e$

$$\tan\beta = \frac{\mathsf{n}_x}{\mathsf{n}_z} = \pm\sqrt{\frac{n_1^{-2} - n_2^{-2}}{n_2^{-2} - n_3^{-2}}} = \pm\frac{n_3}{n_1}\sqrt{\frac{n_2^2 - n_1^2}{n_3^2 - n_2^2}} \quad \text{for } n_1 \le n_2 < n_3\,. \qquad (7.4.37)$$

Thanks to a theorem that uses the ellipsoid of normals [Born and Wolf 1980], it then turns out that

> the planes defined by the wave vector and the eigenvectors of the field ***D***, that is, the planes ($\mathbf{n}$,$\mathbf{d}'$) and ($\mathbf{n}$,$\mathbf{d}''$), are the internal and external bisector planes of the angle between the planes ($\mathbf{n}$,$\mathbf{N}_1$) and ($\mathbf{n}$,$\mathbf{N}_2$), where, here, $\mathbf{N}_1$ and $\mathbf{N}_2$ are the versors of the optical axes.

In many crystals, it happens that two of the principal indices are equal. In this case, there is only one optical axis ($\beta = 0$), which coincides with the principal axis with different index, and the crystal is called *uniaxial*. For such crystals, the surface of the ***k***s is much simpler and consists of a sphere and a revolution ellipsoid, whose axis coincides with the optical axis of the crystal, as shown in Figs. 7.6 (b) and (c).

In the uniaxial crystals, the index of refraction that corresponds to the two equal elements of susceptibility, $\chi_1 = \chi_2$, is called the *ordinary index*, n_o, and the other index, which corresponds to χ_3, is called the *extraordinary index*, n_e. If $n_o < n_e$, the crystal is said to be *positive*, while, if $n_o > n_e$, the crystal is said to be *negative*. Retaking the calculation of Eq. (7.4.19.c), we indeed have

$$\left[\mathsf{n}_\perp^2 n_o^2 \left(n^2 - n_e^2\right) + \mathsf{n}_z^2 n_e^2 \left(n^2 - n_o^2\right)\right]\left(n^2 - n_o^2\right) = 0\,, \qquad (7.4.38)$$

where $\mathsf{n}_\perp{}^2 = \mathsf{n}_x{}^2 + \mathsf{n}_y{}^2$. This equation is solved for

$$n_x{}^2 + n_y{}^2 + n_z{}^2 = n_o{}^2\,, \qquad (7.4.39)$$

OPTICALLY ISOTROPIC CRYSTALS (CUBIC)			n
Sodium chloride, NaCl	cubic, m3m		1.544[a]
Lithium fluoride, LiF	cubic, m3m		1.392[a]
Diamond, C	cubic, m3m		2.417[a]
Fluorite, CaF_2	cubic, m3m		1.434[a]
Spinel, $MgAl_2O_4$	cubic, (Fd$\bar{3}$m) m$\bar{3}$m		1.715[a]
Zinc selenide , ZnSe	cubic, $\bar{4}$3m	@4500nm	2.431[a]

POSITIVE UNIAXIAL CRYSTALS			n_o	n_e
Ice			1.309	1.313[a]
Sellaite, MgF_2	tetragonal, 4/mmm		1.378	1.390[a]
Quartz	trigonal, 32		1.544	1.553[a]
Zircon, $ZrSiO_4$	tetragonal		1.926	1.985[c]
Rutile, TiO_2	tetragonal, 4/mmm		2.616	2.903[a]
Yttrium orthovanadate, YVO_4	tetragonal, 4/m 2/m 2/m	@633nm	1.993	2.215[c]

NEGATIVE UNIAXIAL CRYSTALS			n_o	n_e
Beryl (synthetic) $Be_3Al_2(SiO_3)_6$	hexagonal, 6/m 2/m 2/m		1.563	1.560[c]
Calcite, $CaCO_3$	trigonal, $\bar{3}$m		1.658	1.486[a]
β-BBO, β-Ba $(BO_2)_2$	trigonal, (R3c), 3m		1.670	1.552[a]
α-BBO, β-Ba $(BO_2)_2$	trigonal, (R$\bar{3}$c), $\bar{3}$m		1.673	1.551[d]
Tourmaline	trigonal		1.669	1.638[a]
Sapphire	trigonal, $\bar{3}$m		1.768	1.760[a]
Lithium niobate , $LiNbO_3$	trigonal, 3m	@ 1300nm	2.220	2.146[a]
congruent melt		@ 1064nm	2.232	2.156
		@ 532 nm	2.324	2.235

BIAXIAL CRYSTALS				n_1	n_2	n_3
Gypsum, $CaSO_4 \cdot 2H_2O$		monocline, 2/m		1.520	1.523	1.530[b]
Mica (muscovite)		monocline, 2/m		1.552	1.582	1.588[b]
Topaz, $Al_2SiO_4\,(F,OH)_2$		orthorhombic, 2/m 2/m 2/m		1.619	1.620	1.627[b,c]
$KYb(WO_4)_2$	(C2/c)	monocline, 2/m	@633nm	2.021	2.065	2.112[e]
KTA, $KTiOAsO_4$	($Pna2_1$)	orthorhombic, mm2	@1064nm	1.782	1.790	1.868[f]
			@633 nm	1.808	1.814	1.905
KTP, $KTiOPO_4$	($Pna2_1$)	orthorhombic, mm2	@1064nm	1.740	1.748	1.830[a]
			@ 532 nm	1.780	1.790	1.888
LBO, LiB_3O_5	($Pna2_1$)	orthorhombic, mm2	@1064nm	1.565	1.591	1.605[a]
RTA, $RbTiOAsO_4$	($Pna2_1$)	orthorhombic, mm2	@1064nm	1.804	1.811	1.885[f]

Table 7.3 Indices of refraction of some crystals. Where not indicated, the corresponding wavelength is 589.3 nm. [a] Bass et al ed. (1995), *Handbook of Optics* II; [b] Fowles (1968); [c] Medenbach and Shannon (1997); [d] www.agoptics.com; [e] Pujol et al (2002); [f] Nikogosyan (1997)

which is the equation of a sphere, and for

$$n_x^2 n_o^2 + n_y^2 n_o^2 + n_z^2 n_e^2 = n_o^2 n_e^2 , \tag{7.4.40}$$

which is the equation of a rotation ellipsoid. From Eq. (7.4.38), we also find that, for a direction of the wave vector inclined at an angle θ with the optical axis, the two eigenvalues for n are

$$\begin{cases} n' = n_o, \\ n'' = \left(\dfrac{1}{n_e^2} \sin^2\theta + \dfrac{1}{n_o^2} \cos^2\theta \right)^{-\frac{1}{2}} . \end{cases} \tag{7.4.41}$$

Lastly, if all three indices are equal, the surface of the ***k***s degenerates into a single sphere and the crystal is not birefringent, but rather optically isotropic. Table 7.3 gives some examples of this classification.

7.4.6 The surface of phase velocity

The wave vector ***k*** is associated with the vector of the *phase velocity* $\boldsymbol{\upsilon}$ by the relations

$$\boldsymbol{\upsilon} = \frac{\omega}{k}\mathbf{n}, \quad \boldsymbol{k} = \frac{\omega}{\upsilon}\mathbf{n} , \tag{7.4.42}$$

which, transcribed with respect to the refractive index, become:

$$\boldsymbol{\upsilon} = \frac{c}{n}\mathbf{n}, \quad \boldsymbol{n} = \frac{c}{\upsilon}\mathbf{n} . \tag{7.4.43}$$

Replacing n with c/υ in Eq. (7.4.19.b), we find a biquadratic equation in υ, which, in turn, defines a surface (in the velocity space), which is called the *phase velocity surface*:

$$\begin{aligned} &\upsilon^4 n_1^2 n_2^2 n_3^2 \\ &-\upsilon^2 c^2 \left[\mathrm{n}_x^2 n_1^2 \left(n_2^2 + n_3^2 \right) + \mathrm{n}_y^2 n_2^2 \left(n_3^2 + n_1^2 \right) + \mathrm{n}_z^2 n_3^2 \left(n_1^2 + n_2^2 \right) \right] \\ &+ c^4 \left(\mathrm{n}_x^2 n_1^2 + \mathrm{n}_y^2 n_2^2 + \mathrm{n}_z^2 n_3^2 \right) = 0. \end{aligned} \tag{7.4.44}$$

Even this surface is double, and consists of two concentric "ovaloids": they provide the two possible values of phase velocities for each direction $\mathbf{n}$ of propagation of the wavefront.

The intersections of the surface of the phase velocity with the principal planes consist of circles and fourth-degree "oval" curves (which are the reciprocal of the ellipses of the surface of the ***k***s): for example, for the *xy*-plane, the two equations for the intersections are:

$$\upsilon^2 = \upsilon_x{}^2 + \upsilon_y{}^2 = \frac{c^2}{n_3{}^2}, \qquad \frac{\upsilon_x{}^2}{n_2{}^2} + \frac{\upsilon_y{}^2}{n_1{}^2} = \frac{\upsilon^4}{c^2}. \tag{7.4.45}$$

Also, the surface of the phase velocity has 4 points of auto-intersection, and therefore has two optical axes, which, however, coincide with those already found for the surface of the ***k***s. Finally, the eigenvalues Eq. (7.4.12) can be reformulated in terms of υ:

$$\frac{1}{\mu}(\mathbf{I} - \mathbf{nn}) \cdot \overline{\overline{\varepsilon}}^{-1} \boldsymbol{D}_\mathrm{o} = \upsilon^2 \boldsymbol{D}_\mathrm{o} \tag{7.4.46}$$

and the new operator on $\boldsymbol{D}_\mathrm{o}$ has eigenvalues equal to υ^2.

The solutions for υ^2 of Eq. (7.4.44) take a simple form that expresses the normal versor $\mathbf{n}$ in terms of the angles θ_1 and θ_2, which it does with the optical axes of the crystal [Born and Wolf 1980]; these will be useful for the study of the interference produced by birefringent laminae. By applying this substitution, the solutions for the phase velocity turn out to be

$$\upsilon'^2 = \frac{1}{2}\left[\upsilon_1{}^2 + \upsilon_3{}^2 + \left(\upsilon_1{}^2 - \upsilon_3{}^2\right)\cos(\theta_1 - \theta_2)\right], \qquad \upsilon''^2 = \frac{1}{2}\left[\upsilon_1{}^2 + \upsilon_3{}^2 + \left(\upsilon_1{}^2 - \upsilon_3{}^2\right)\cos(\theta_1 + \theta_2)\right], \tag{7.4.47}$$

where $\upsilon_j = c/n_j$, with $j = 1, 2, 3$. The phase velocity $\upsilon_2 = c/n_2$ does not appear explicitly in them; however, it is implicitly contained in the definition of the angles θ_1 and θ_2. The cosine of these angles is obtained from the scalar product of $\mathbf{n}$ with the versors $(\pm\sin\beta, 0, \cos\beta)^\mathrm{T}$ of the optical axes, whereby

$$\cos\theta_1 = \mathrm{n}_x \sin\beta + \mathrm{n}_z \cos\beta, \qquad \cos\theta_2 = -\mathrm{n}_x \sin\beta + \mathrm{n}_z \cos\beta, \tag{7.4.48}$$

where the angle β is given by Eq. (7.4.37). The solutions given by Eqs. (7.4.47) can be laboriously traced back to Eq. (7.4.44) developing the equation $(\upsilon - \upsilon')(\upsilon - \upsilon'') = 0$.

7.5 Propagation modes for the field E

7.5.1 *Group and energy velocity*

The phase velocity of a plane wave was introduced above with Eq. (7.4.42), and it corresponds to the speed at which the wave front propagates in the direction of $\boldsymbol{k}$. By *group velocity,* we instead mean the speed with which a pulse, that is, a wave packet or train, is propagated. As we shall see shortly, this velocity characterizes the speed and direction in which energy is propagated, even in anisotropic transparent media. The group velocity is mathematically defined by

$$\boldsymbol{u} = \frac{\partial \omega}{\partial \boldsymbol{k}}, \tag{7.5.1}$$

where the gradient operation in the $\boldsymbol{k}$ space has been indicated concisely. In order for this equation to become useful, it is necessary to explicitly determine the relationship existing between $\boldsymbol{k}$ and ω. One such relation for the anisotropic media is precisely the Fresnel's equation (7.4.19.a), which has the form

$$F\left(k_x, k_y, k_z, \omega\right) = 0\,. \tag{7.5.2}$$

For greater precision, the dependence on ω of the principal indices n_1, n_2, n_3 should also be made explicit in F and, in the case of triclinic or monoclinic crystals, even the variation on ω of the orientation of the principal axis system should be considered. In order not to complicate things too much, let us be content to consider only the non-dispersive media, or at least such that their dispersion can be neglected. On the other hand, when there is a strong dispersion, the absorption is generally not negligible and the equations that have allowed us to reach Eq. (7.4.19.a) would no longer be applicable.

Now, consider Eq. (7.5.1): for the constraint set on F, to a variation of $\boldsymbol{k}$, for example, a variation ∂k_x of the sole component k_x must correspond to a variation $\partial\omega$ such that

$$\frac{\partial F}{\partial k_x}\delta k_x = -\frac{\partial F}{\partial \omega}\delta\omega\,,$$

which yields

$$\boldsymbol{u} = \frac{\partial \omega}{\partial \boldsymbol{k}} = -\frac{\partial F}{\partial \boldsymbol{k}} \bigg/ \frac{\partial F}{\partial \omega}\,. \tag{7.5.3}$$

Therefore, the group velocity is proportional to $\partial F/\partial \boldsymbol{k}$. Furthermore, by definition of a partial derivative, $\partial F/\partial \boldsymbol{k}$ must be calculated at ω constant, and therefore the vector $\boldsymbol{u}$ is perpendicular to the surface of the wave vectors,

since this is, in turn, defined by the expression

$$F(\boldsymbol{k}, \omega = \text{cost.}) = 0.$$

This justifies that, in Fig. 7.5, the electric field is always represented tangent to the surface of the $\boldsymbol{k}$s.

The explicit calculation of $\boldsymbol{u}$ is now easily executable by replacing F in Eq. (7.5.3) with the left side of Eq. (7.4.19.a). The expression that is obtained is a little complicated, and I leave it for exercise (for the solution, see §7.5.3).

Let us now look for a relationship between phase and group velocity. Resolving Eq. (7.4.42) for ω, we have

$$\omega = k\,\upsilon(\boldsymbol{k}),$$

where $\upsilon(\boldsymbol{k})$ can be obtained by solving Eq. (7.4.44). Thus, remembering that $\partial k/\partial \boldsymbol{k} = \boldsymbol{k}/k = \mathbf{n}$, from the first equality of Eq. (7.5.3), we find

$$\boldsymbol{u} = \frac{\partial(k\upsilon)}{\partial \boldsymbol{k}} = \boldsymbol{\upsilon} + k\frac{\partial \upsilon}{\partial \boldsymbol{k}}.$$

Therefore, the phase velocity differs from the group velocity for the vector $k\partial\upsilon/\partial\boldsymbol{k}$. Its projection in the direction $\mathbf{n}$ has the module

$$\mathbf{n}k \bullet \frac{\partial \upsilon}{\partial \boldsymbol{k}} = \boldsymbol{k} \bullet \frac{\partial \upsilon}{\partial \boldsymbol{k}} = k\frac{\partial \upsilon}{\partial k},$$

where, in the last passage, the gradient of υ was projected in the direction of $\boldsymbol{k}$. Observing Eq. (7.4.44), we see that there is no explicit dependence between υ and k, because we are considering a non-dispersive medium. Thus, $\partial\upsilon/\partial k = 0$, and consequently

$$\mathbf{n} \bullet \boldsymbol{u} = \upsilon, \tag{7.5.4}$$

that is:

the projection of the group velocity in the direction $\mathbf{n}$, normal to the wave front, is equal to the phase velocity (Fig. 7.7).

On the other hand, for Eqs. (7.4.16) and (7.4.17),

$$u = \frac{\upsilon}{\cos\Lambda} = \frac{\mathrm{c}}{\sqrt{n_1^2 \mathrm{e}_x^2 + n_2^2 \mathrm{e}_y^2 + n_3^2 \mathrm{e}_z^2}}, \tag{7.5.5}$$

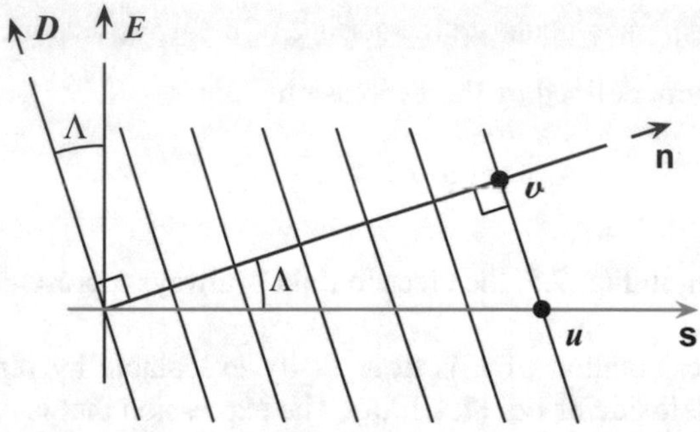

Fig. 7.7 Wavefronts in an anisotropic medium. The phase velocity $\boldsymbol{\upsilon}$ is the projection of the group velocity $\boldsymbol{u}$ in the direction of the wave vector

so this group velocity still remains lower than c.

It now remains to be shown that $\boldsymbol{u}$ coincides with the velocity of energy propagation, defined by

$$\boldsymbol{\upsilon}_{\mathrm{E}} = \frac{\boldsymbol{S}_{\mathrm{o}}}{\overline{w}}, \tag{7.5.6}$$

where $\overline{w}$ and $\boldsymbol{S}_{\mathrm{o}}$ are given by Eqs. (7.2.9) and (7.4.3), respectively. Consider the last two Eqs. (7.4.2) and suppose that $\boldsymbol{k}$ is varied by a quantity $\delta\boldsymbol{k}$. If $\delta\omega$, $\delta\boldsymbol{E}_{\mathrm{o}}$, and $\delta\boldsymbol{H}_{\mathrm{o}}$ are the corresponding changes in ω, $\boldsymbol{E}_{\mathrm{o}}$, and $\boldsymbol{H}_{\mathrm{o}}$, we have that

$$\begin{aligned}
\delta\boldsymbol{k}\times\boldsymbol{E}_{\mathrm{o}} + \boldsymbol{k}\times\delta\boldsymbol{E}_{\mathrm{o}} &= \delta\omega\mu\,\boldsymbol{H}_{\mathrm{o}} + \omega\mu\,\delta\boldsymbol{H}_{\mathrm{o}},\\
\delta\boldsymbol{k}\times\boldsymbol{H}_{\mathrm{o}} + \boldsymbol{k}\times\delta\boldsymbol{H}_{\mathrm{o}} &= -\delta\omega\,\varepsilon_{ij}E_j - \omega\varepsilon_{ij}\,\delta E_j.
\end{aligned}$$

With a scalar multiplication of the first Eq. for $\boldsymbol{H}_{\mathrm{o}}^*$ and of the second for $\boldsymbol{E}_{\mathrm{o}}^*$, and then subtracting the results, we finally have that

$$\begin{aligned}
&2\delta\boldsymbol{k}\cdot\Re e\left(\boldsymbol{E}_{\mathrm{o}}\times\boldsymbol{H}_{\mathrm{o}}^*\right) - \delta\omega\left(E_i^*\varepsilon_{ij}E_j + \mu\left|\boldsymbol{H}_{\mathrm{o}}\right|^2\right) =\\
&= \delta\boldsymbol{H}_{\mathrm{o}}\cdot\left(\omega\mu\,\boldsymbol{H}_{\mathrm{o}}^* - \boldsymbol{k}\times\boldsymbol{E}_{\mathrm{o}}^*\right) + \delta\boldsymbol{E}_{\mathrm{o}}\cdot\left(\varepsilon_{ji}^*E_i^* + \boldsymbol{k}\times\boldsymbol{H}_{\mathrm{o}}^*\right),
\end{aligned}$$

where, in the last parenthesis, the symmetry property (hermitianity) of $\vec{\vec{\varepsilon}}$ has been exploited. For the last two Eqs. (7.4.2), the second side of this equation is null, so that ultimately

$$\boldsymbol{u} = \frac{\partial\omega}{\partial\boldsymbol{k}} = \frac{2\Re e\left(\boldsymbol{E}_{\mathrm{o}}\times\boldsymbol{H}_{\mathrm{o}}^*\right)}{\varepsilon_{ij}E_i^*E_j + \mu\left|\boldsymbol{H}_{\mathrm{o}}\right|^2} = \boldsymbol{\upsilon}_{\mathrm{E}}, \tag{7.5.7}$$

which is what we wanted to prove.

In conclusion, the vector $\boldsymbol{u}$ is oriented along the unit vector **s** and coincides with the speed with which the intersections of wave fronts with a straight line parallel to **s** are moving (Fig. 7.7), and

$$u = \frac{\upsilon}{\cos\Lambda}, \tag{7.5.8}$$

where Λ is the angle subtended between **n** and **s,** and also between $\boldsymbol{D}_\mathrm{o}$ and $\boldsymbol{E}_\mathrm{o}$, given by Eqs. (7.4.16). Therefore, in the non-dispersive anisotropic crystals, the group velocity is greater than the phase velocity.

So far, we have considered the case of plane waves, which is infinitely extended, while, in practical cases, we have to deal with light beams of limited size. We have seen that, for the isotropic media, within the limits of the validity of Geometrical Optics, we can introduce the concept of a "ray of light" and justify it with the help of the Poynting's theorem, as a (localized) path of the energy. This concept remains valid even in the case of anisotropic media: the reasoning laid out in §2.2.1 can, in fact, easily be extended to them. In the anisotropic media, we have, as a remarkable consequence, that, in general, the rays are not perpendicular to the wavefronts. However, the laws of refraction and propagation in anisotropic media can be implemented in the ray tracking programs. Therefore, applying the constancy of the optical paths, even when an optical system contains anisotropic media, the wavefronts can be reconstructed.

7.5.2 Inversion theorem

The propagation of light in anisotropic media presents a remarkable reciprocity property that derives from the splitting, along two different directions of the phase velocity and the group velocity, so that, in addition to asking ourselves what the field's eigenvectors for a fixed wave vector direction are, we can also ask what (if any) the field's eigenvectors for a fixed direction of the Poynting vector are. We have discussed the answers to the first question in the previous sections. There is a notable theorem, called the *inversion theorem*, which allows us to duplicate the previously found relationships automatically, enabling us to respond quickly to the second question.

Consider again Eqs. (7.4.2) and, in particular, the third. Making the cross product of both sides of this equation for the versor **s**, we have that

$$\mathbf{s}\times\left(\mathbf{n}\times\boldsymbol{E}_\mathrm{o}\right)=\upsilon\mu\,\mathbf{s}\times\boldsymbol{H}_\mathrm{o},$$

where Eq. (7.4.42.b) was used to replace $\boldsymbol{k}$. Proceeding, we have that

$$\mathbf{s}\times \boldsymbol{H}_o = -\frac{1}{\upsilon\mu}\left[(\mathbf{s}\cdot\mathbf{n})\boldsymbol{E}_0 - (\mathbf{s}\cdot\boldsymbol{E}_0)\mathbf{n}\right] = -\frac{1}{\upsilon\mu}(\mathbf{s}\cdot\mathbf{n})\boldsymbol{E}_0 .$$

Since $\mathbf{s}\cdot\mathbf{n} = \cos\vartheta = \upsilon/u$, we finally get

$$\mathbf{s}\times \boldsymbol{H}_o = -\frac{1}{u\mu}\boldsymbol{E}_0 .$$

A similar calculation can also be made for the fourth of the Eqs. (7.4.2). Thus, we have two sets of equivalent equations:

$$\begin{cases} \mathbf{n}\cdot\boldsymbol{D}_o = 0, \\ \mathbf{n}\cdot\boldsymbol{B}_o = 0, \\ \mathbf{n}\times\boldsymbol{E}_0 = +\upsilon\boldsymbol{B}_o, \\ \mathbf{n}\times\boldsymbol{B}_o = -\upsilon\mu\boldsymbol{D}_o, \end{cases} \qquad \begin{cases} \mathbf{s}\cdot\boldsymbol{E}_0 = 0, \\ \mathbf{s}\cdot\boldsymbol{H}_o = 0, \\ \mathbf{s}\times\boldsymbol{D}_o = +\frac{1}{u}\boldsymbol{H}_o, \\ \mathbf{s}\times\boldsymbol{H}_o = -\frac{1}{u\mu}\boldsymbol{E}_0, \end{cases} \tag{7.5.9}$$

with $\quad \boldsymbol{E}_0 = \vec{\vec{\varepsilon}}^{-1}\boldsymbol{D}_o \quad$ and $\quad \boldsymbol{D}_o = \vec{\vec{\varepsilon}}\boldsymbol{E}_0 .$

Apart from an exchange of variables, we see that the Eqs. (7.5.9) on the right have the same structure as those on the left. On the other hand, the latter are formally equivalent to Eqs. (7.4.2), from which we have obtained all of the results of the previous sections. It is therefore evident that these results can be duplicated starting from the Eqs. (7.5.9) on the right or simply by applying the following substitutions to each of the relations in §7.4:

$$\updownarrow \begin{matrix} \mathbf{n} & \boldsymbol{D} & \boldsymbol{B} & \vec{\vec{\varepsilon}} & \mu & c & \omega & n_1 & n_2 & n_3 & \upsilon & n & k \\ \mathbf{s} & \boldsymbol{E} & \boldsymbol{H} & \vec{\vec{\varepsilon}}^{-1} & \frac{1}{\mu} & \frac{1}{c} & \omega & \frac{1}{n_1} & \frac{1}{n_2} & \frac{1}{n_3} & \frac{1}{u} & \frac{1}{n_r} & \frac{1}{k_r} \end{matrix} . \tag{7.5.10}$$

The variables $n_r = c/u$ and $k_r = \omega/u$ are, respectively, the *ray refractive index* and the *ray wave vector*, which are introduced here for completeness. However, it is probably better to avoid using them, replacing them with their equivalent expression, respectively, c/u and ω/u.

7.5.3 Eigenvectors and eigenvalues

With the help of the inversion theorem, we can directly transcribe the relations for the propagation modes for the field **E** in the direction of the Poynting vector, starting from those obtained for the propagation of the field **D** in the direction of ***k***. So, without the need to repeat the discussions above, let us now look at some of these new results.

For Eq. (7.4.12), the eigenvalue wave equation is

$$\mu\left(\mathbf{I}-\mathbf{ss}\right)\bullet\vec{\varepsilon}\,\boldsymbol{E}_{\mathrm{o}}=\frac{1}{u^2}\boldsymbol{E}_{\mathrm{o}},\tag{7.5.11}$$

and its eigenvectors (for the field $\boldsymbol{E}_{\mathrm{o}}$) are oriented according to the two non-normalized vectors

$$\begin{aligned}V_E^{(j)}=\Bigg[&\left(n_2{}^2-\frac{c^2}{u^{(j)2}}\right)\left(n_3{}^2-\frac{c^2}{u^{(j)2}}\right)\mathrm{s}_x,\left(n_1{}^2-\frac{c^2}{u^{(j)2}}\right)\left(n_3{}^2-\frac{c^2}{u^{(j)2}}\right)\mathrm{s}_y,\\&\left(n_1{}^2-\frac{c^2}{u^{(j)2}}\right)\left(n_2{}^2-\frac{c^2}{u^{(j)2}}\right)\mathrm{s}_z\Bigg]\quad\text{with }(j)=\;',\;'',\end{aligned}\tag{7.5.12}$$

while those corresponding to the field $\boldsymbol{D}_{\mathrm{o}}$ are

$$\begin{aligned}V_D^{(j)}=\Bigg[&\left(\frac{c^2}{u^{(j)2}}-\frac{1}{n_2{}^2}\right)\left(\frac{c^2}{u^{(j)2}}-\frac{1}{n_3{}^2}\right)\mathrm{s}_x,\left(\frac{c^2}{u^{(j)2}}-\frac{1}{n_1{}^2}\right)\left(\frac{c^2}{u^{(j)2}}-\frac{1}{n_3{}^2}\right)\mathrm{s}_y,\\&\left(\frac{c^2}{u^{(j)2}}-\frac{1}{n_1{}^2}\right)\left(\frac{c^2}{u^{(j)2}}-\frac{1}{n_2{}^2}\right)\mathrm{s}_z\Bigg]\quad\text{with }(j)=\;',\;''.\end{aligned}\tag{7.5.13}$$

u', and u'' are the group velocity solutions for the Fresnel's equation (Eq. D of Fresnel):

$$\begin{aligned}&u^4\left(\mathrm{s}_x^2\,n_2^2n_3^2+\mathrm{s}_y{}^2\,n_3^2n_1^2+\mathrm{s}_z^2\,n_1^2n_2^2\right)\\&-u^2c^2\left[\mathrm{s}_x^2\left(n_2^2+n_3^2\right)+\mathrm{s}_y{}^2\left(n_3^2+n_1^2\right)+\mathrm{s}_z^2\left(n_1^2+n_2^2\right)\right]+c^4=0.\end{aligned}\tag{7.5.14}$$

Contrary to what is obtained in §7.4.5, the eigenvectors for the field $\boldsymbol{E}_{\mathrm{o}}$ are now the ones orthogonal to each other, while the corresponding values for $\boldsymbol{D}_{\mathrm{o}}$ are generally not orthogonal.

7.5.4 The surface of group velocity

Eq. (7.5.14) is of the second degree in u^2 and has two real and positive roots u'^2, u''^2 for each direction of the versor **s**, such that

$$\frac{c}{n_1} \geq u' \geq \frac{c}{n_2} \geq u'' \geq \frac{c}{n_3}, \tag{7.5.15}$$

where the principal axes here are also ordered according to the inequalities (7.4.21). Similarly to what has been seen for the phase velocity, Eq. (7.5.14), when **s** changes, also describes a double surface in the velocity space, which is called the *group velocity surface*, or *ray surface*.

Eq. (7.5.14), however, looks more like the Fresnel's equation for n, i.e., Eq. (7.4.19.b), than that for υ. Therefore, the surface of the group velocity is similar to that of the wave vectors, and its intersections with the principal planes consist of a circle and an ellipse. For example, in the xy-plane, there is a circular intersection of radius

$$u = \frac{c}{n_3}, \tag{7.5.16}$$

while the other intersection is described by the equation of an ellipse:

$$n_2{}^2 u_x{}^2 + n_1{}^2 u_y{}^2 = c^2 . \tag{7.5.17}$$

Even the surface of the group velocity generally has 4 auto-intersection points and two corresponding *ray axes*. With the principal axes ordered in the usual way, the ray axes are in the xz-plane and the angle γ that each of them makes with the z-axis is given by

$$\tan\gamma = \sqrt{\frac{n_2{}^2 - n_1{}^2}{n_3{}^2 - n_2{}^2}} . \tag{7.5.18}$$

In biaxial crystals, the ray axes do not coincide with the optical axes (Fig. 7.8), while they do coincide in the uniaxial crystals, where $\gamma = \beta = 0$. Indeed, from the comparison with Eq. (7.4.37), one has that

$$\tan\gamma = \frac{n_1}{n_3} \tan\beta . \tag{7.5.19}$$

Therefore, $\gamma < \beta$ (with $n_1 < n_3$), however, the difference between γ and β is generally small in common biaxial crystals (for mica, it is about 25′).

For a *uniaxial crystal*, from Eq. (7.5.14), with various steps, we have that

$$\left[\left(u_x^2 + u_y^2 + u_z{}^2\right) n_o^2 - c^2\right]\left[u_x^2\, n_e^2 + u_y^2\, n_e^2 + u_z{}^2 n_o{}^2 - c^2\right] = 0. \tag{7.5.20}$$

This equation is solved for

$$\left(u_x^2 + u_y^2 + u_z{}^2\right) n_o^2 = c^2 , \tag{7.5.21}$$

which expresses a sphere, and for

$$u_x^2\, n_e^2 + u_y^2\, n_e^2 + u_z{}^2 n_o{}^2 = c^2 , \tag{7.5.22}$$

which is the equation of an ellipsoid. It can also be obtained by applying the inversion theorem to Eq. (7.4.40). Among these equations, the indices n_e and n_o are exchanged between the axial component along z and the transverse ones on the xy-plane. Therefore, the shape of this ellipsoid is "inverted" with respect to that of the corresponding ellipsoid of the wave vectors: if the first is prolate, the second is oblate, and vice versa.

7.6 Relations between the surfaces of phase velocity, group velocity and wave vectors

Both the phase and group velocity surfaces can be drawn in the same space, i.e., the velocity space. They have various points of contact: in particular, their circular intersections with the principal planes coincide, as can be verified by comparing Eq. (7.4.45.a) with Eq. (7.5.16). The two surfaces have contact points, in correspondence with the three principal axes, including for the ends of the non-circular intersections, as can be seen from the comparison between Eqs. (7.4.45.b)

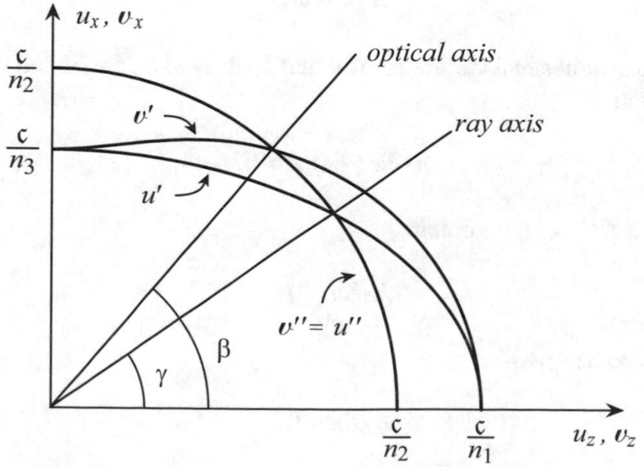

Fig. 7.8 Phase and group velocities in the xz-plane and optical axes

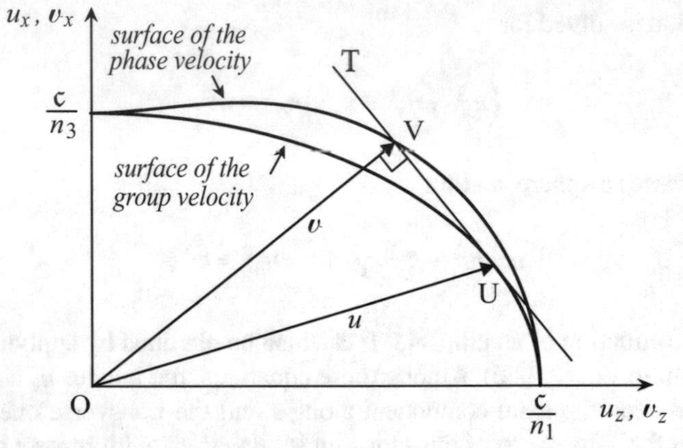

Fig. 7.9 Graphical construction of the group and phase velocity surfaces

and (7.5.17) (Fig. 7.8).

There are other interesting relationships between these surfaces, as well as between them and that of the wave vectors. We have already observed (see §7.5) that the direction **s** of $\boldsymbol{u}$ is perpendicular to the surface of the wave vectors, and this can be expressed by the equation (valid for ω fixed)

$$\mathbf{s} \cdot \delta \boldsymbol{k} = 0 \quad \text{and therefore also} \quad \boldsymbol{u} \cdot \delta \boldsymbol{k} = 0\,. \tag{7.6.1}$$

It is easy to show that a similar relationship between **n** and the group velocity surface also applies. We have seen that $\boldsymbol{\upsilon}$ is the projection of $\boldsymbol{u}$ in the direction of **n**, as expressed by Eq. (7.5.4); the latter can be rewritten by multiplying both sides by k, as

$$\boldsymbol{u} \cdot \boldsymbol{k} = \omega\,.$$

Thus, the infinitesimal variations of $\boldsymbol{u}$ and $\boldsymbol{k}$, at fixed ω, are linked to each other by the relation

$$\boldsymbol{u} \cdot \delta \boldsymbol{k} + \delta \boldsymbol{u} \cdot \boldsymbol{k} = 0\,.$$

Finally, for Eq. (7.6.1), we obtain

$$\boldsymbol{k} \cdot \delta \boldsymbol{u} = 0\,,$$

which can also be rewritten as

$$\boldsymbol{\upsilon} \cdot \delta \boldsymbol{u} = 0\,. \tag{7.6.2}$$

From Eqs. (7.5.4) and (7.6.2), we can therefore state that

the plane tangent to the surface of the group velocity is perpendicular to the direction **n** of the wave vector, and, in this direction, its distance from the origin is equal to the phase velocity. Furthermore, the field $\boldsymbol{D}$ is also tangent to the surface of the group velocity.

Limited to one of the principal planes, a graphical representation of this fact is given in Fig. 7.9.

The surfaces of the phase velocity and the group velocity can thus be drawn, each knowing the other. In fact, by indicating the origin of the axis system with O, the surface of the $\boldsymbol{v}$s corresponds to the set of points V thus determined: at each point of the surface of the $\boldsymbol{u}$s, draw a tangent plane T and, on this one, identify the point V such that the segment $\overline{\mathrm{OV}}$ is normal to T. Contrastingly, the surface of the $\boldsymbol{u}$s is given by the envelope of the T-planes passing through each point V of the surface of the $\boldsymbol{v}$s and normal to the segment $\overline{\mathrm{OV}}$.

7.7 Refraction at the interface with an anisotropic medium

In the previous sections, we analyzed the propagation modes of the field *inside* of the anisotropic materials: we now have to study what happens at the surface of these materials, which is one of the most important aspects from the phenomenological and applicative points of view. If one observes an object through an anisotropic material, the object itself appears split into two images, which rotate as the material rotates. The first to record this behavior, with a calcite crystal, was Rasmus Bartolin, in 1669, who recognized a special case of refraction in this effect and called it *double refraction*. Since then, optically anisotropic materials have been called *birefringent*. This effect was added to the difficulties already evident to the supporters of the corpuscular theory to explain the diffraction or the coloration of the thin laminae. Newton attributed the refraction to an attraction by the transparent bodies to the corpuscles of the lumen and suggested that these corpuscles might not be spherical, but that they would have a "pole": according to its orientation, the corpuscles would have divided in the two directions of refraction. In 1690, Huygens succeeded in giving a more satisfactory explanation of this phenomenon with his theory of wavelets. He supposed that, in the crystal, the wave motion is split into two waves, one composed of spherical wavelets for which the propagation speed is the same in all directions, while the other is composed of ellipsoidal wavelets for which the propagation speed changes according to the direction with respect to the crystal axes. The first is an *ordinary wave*, which obeys Snell's law, while the other is an *extraordinary wave*, for which Snell's law *would be* violated (calcite, among other things, is a uniaxial crystal). However, Huygens himself admitted that he could not explain why, if the first calcite crystal was followed by a second, the four rays that resulted from it were modulated in intensity as the relative orientation between the two crystals varied. Newton considered this a serious failure of the wave theory, but even he had to leave the field. In 1808,

Ètienne Louis Malus noticed that the images of the Sun reflected from a window and split by a calcite crystal varied in intensity in a complementary way between them with the rotation of the crystal, and therefore a reflection was also able to break the rotational symmetry around the direction of propagation. His conception of light was still that of Newton's theory, and he resumed the concept of a *polarization* of the luminous corpuscles.[1] Finally, in 1817, Young suggested that light could be made up of transverse waves, and, on this basis, in 1824, Fresnel presented an extensive theory of crystal optics.

Even for the anisotropic dielectric media, the problem of reflection and refraction can be solved from the boundary conditions that the field must obey on the interface. Well, the arguments used for the "kinematic condition" in §1.7.1 remain valid for birefringent media. In particular, the invariance condition for these media of the relative phase between the various waves, incident, reflexed and refracted, implies that the ***k***-vectors of the waves must all lie on the plane of incidence and their projections on the interface must all be the same. For generality, consider the case in which both media separated from the interface are birefringent and assume that the incident wave has a polarization corresponding to one of two possible eigenvectors for its direction of propagation. The space-time dependence to which the various waves are subjected implies that

$$k_{\mathrm{I}} \sin \vartheta_{\mathrm{I}} = k'_{\mathrm{R}} \sin \vartheta'_{\mathrm{R}} = k''_{\mathrm{R}} \sin \vartheta''_{\mathrm{R}} = k'_{\mathrm{T}} \sin \vartheta'_{\mathrm{T}} = k''_{\mathrm{T}} \sin \vartheta''_{\mathrm{T}}, \tag{7.7.1}$$

where k_{I} is one of the two possible eigenvalues for the assigned direction of the incident wave vector, corresponding to a given polarization eigenvector, and the apices indicate the two possible modes of propagation in the two media for the reflected wave (R) and for the transmitted wave (T), corresponding to the eigenvalue k_{I} of the incident wave (I). The angles ϑ are taken on the plane of incidence from the normal to the interface. It is important to note that the *incidence plane* thus defined is *the one on which* $\boldsymbol{k}_{\mathrm{I}}$ *lies*. A separate similar equation must be written for the other propagation eigenvector of the incident wave. Therefore,

Snell's law is still valid for both refracted waves,

but in the sense that the ratio of the sines of the angles of incidence and refraction is equal to the ratio of the refractive indices, which now depend on the direction of the normal to the wavefronts. Indeed, from Eq. (7.7.1), we get

$$\frac{\sin \vartheta_{\mathrm{I}}}{\sin \vartheta'_{\mathrm{T}}} = \frac{n'_{\mathrm{T}}(\vartheta'_{\mathrm{T}})}{n_{\mathrm{I}}(\vartheta_{\mathrm{I}})}; \quad \frac{\sin \vartheta_{\mathrm{I}}}{\sin \vartheta''_{\mathrm{T}}} = \frac{n''_{\mathrm{T}}(\vartheta''_{\mathrm{I}})}{n_{\mathrm{I}}(\vartheta_{\mathrm{I}})}. \tag{7.7.2}$$

---

[1] It should be noted that the choice of the polarization plane made by Malus was the exact opposite of what we do today with the *vibration plane* of the electric field. Many authors prefer to use this second term to avoid confusion, but, in this text, I adopted the current use of calling the polarization plane the vibration plane of the electric field.

Basically, these relationships alone are now insufficient to solve the problem. To complicate matters, there is also the fact that the ray direction (of the Poynting's vector) of the refracted and reflected rays does not generally coincide with that of the respective wave vectors and, in general, does not even lie on the plane of incidence.

7.7.1 Graphical constructions

One way to determine the direction of the reflected and refracted waves is shown, with some graphical license, in Fig. 7.10: let us consider the space of the wave vectors and, in this space, we trace a plane S, passing through the origin O and "parallel" to the interface, which divides this space in two. In each half-space, we trace the surfaces of the wave vectors, taking into account the orientations of the principal axes in each medium. Let us then consider the intersections of these surfaces with a plane, the one of the figure, "parallel" to the plane of incidence and still passing through the origin. Then, we draw the vector $\boldsymbol{k}_{\mathrm{I}}$ of the incident wave and its projection $\boldsymbol{k}_{\mathrm{S}}$ on the plane S of the interface, both with the arrow towards O: here, $\boldsymbol{k}_{\mathrm{I}}$ corresponds to one of the two eigenvectors. If we then duplicate $\boldsymbol{k}_{\mathrm{S}}$ from O, and we pull a line P perpendicular to S from its extreme, the intersection of P with the surfaces of the $\boldsymbol{k}$s finally determines the wave vectors of the reflected and refracted waves.

Then, to solve the second problem of finding the direction of the rays, at least conceptually, one can resort to the graphic construction introduced by Huygens for the case of the uniaxial crystals. Indeed, it can be extended to the biaxial ones. These were discovered more than a century later by Brewster, who studied the optical properties of over 150 different crystals and was the first to empirically establish the relationship between symmetry and birefringence. The Huygens' construc-

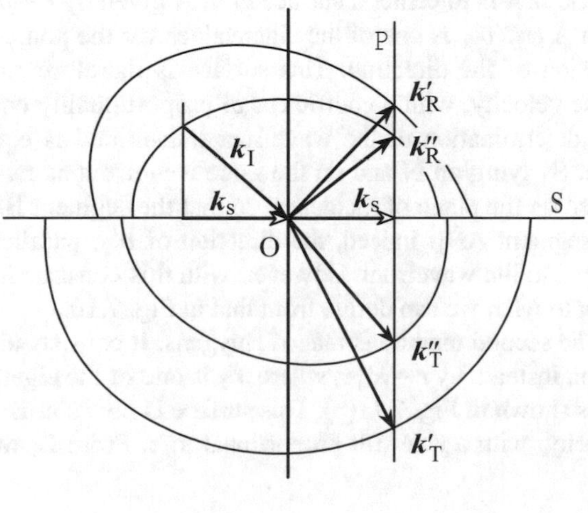

Fig. 7.10 Refraction and reflection between two anisotropic media. Here, $\boldsymbol{k}_{\mathrm{I}}$ corresponds to one of the polarization eigenvectors. Overall, for a generic incident polarization, we have 4 reflected and 4 transmitted waves

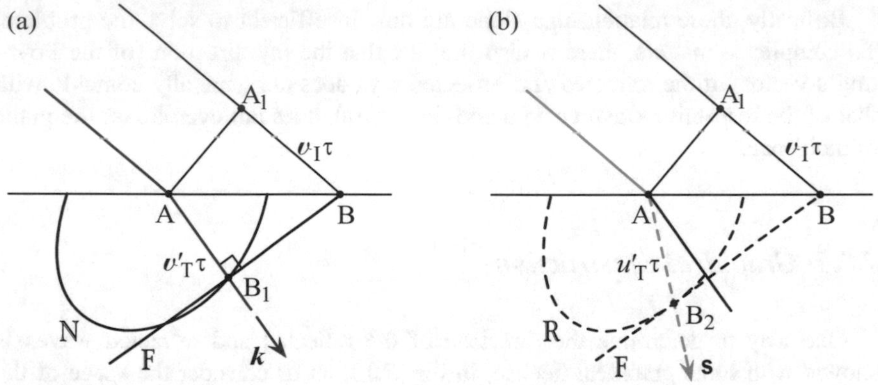

Fig. 7.11 Graphical methods for determining the direction of $\boldsymbol{k}$ and $\mathbf{s}$. In (a), the wavefront F intersects the surface N in B_1 obtained from the surface of the phase velocity. In (b), we have the construction of Huygens-Fresnel (see Fig. 10 of the *Second mémoire sur la double réfraction* by Fresnel), where F is tangent to the surface R in B_2 obtained with the group velocity. The point B_1 is on the incidence plane, while the point B_2 is usually out of this plane

tion was an important success for the wave theory, but it went into crisis with Brewster's discovery, which was initially used as an argument against this theory. Finally, the enigma was recomposed by the optical crystallographic studies of Fresnel.

Consider the case of a plane wave that is incident on a face of a birefringent crystal (Fig. 7.11) and consider two wave fronts that intersect the interface at points A and B in the plane of incidence. The distance ($\overline{A_1B}$) between the two fronts in the first (isotropic) medium is given by $v_I\tau$, where v_I is the phase velocity module and τ is the time required for the first front to reach the position of the second one. We can follow two methods to determine the direction of the second front within the crystal.

The first is to center a surface N in A given by $r = v'_T\tau$, where r is the distance from A and v'_T is one of the eigenvalues for the phase velocity in the crystal as a function of the direction. This surface is therefore similar to the surface of the phase velocity, with a coefficient of proportionality equal to τ, as in Fig. 7.11(a). The determination of the wavefront transmitted is equivalent to searching for a point B_1 lying on N and on the same wavefront as B. Finally, this point is determined on the plane of incidence, so that the segment BB_1 forms a right angle with the segment AB_1: indeed, the direction of $\boldsymbol{v}'_T$, parallel to AB_1, must, in fact, be normal to the wavefront. However, with this construction, we have not added anything to what we can derive from that in Fig. 7.10.

The second method is that of Huygens. It consists in centering a surface R in A given, instead, by $r = u'_T\tau$, where u'_T is one of the eigenvalues of the group velocity, as shown in Fig. 7.11(b). This surface is now similar to the surface of the group velocity, with a size still proportional to τ. From B, we now trace a plane that is normal to the plane of incidence and tangent to the surface R. Thanks to the properties of the group velocity surface (see §7.6), this plane constitutes the sought after wavefront and the point of contact B_2 determines the ray direction of the refracted wave on the AB_2 segment. Since the surface R is arranged in a more or less oblique way, B_2 does not generally lie on the plane of incidence, and therefore neither does the refracted ray.

It is then to be noted that the Huygens wavelets are those built with the group velocity! Indeed, the width of the surface R varies proportionally to τ, and therefore to the distance between A and B. By similarity, as the position of A changes, while B is fixed, all surfaces R_A remain tangent to the wavefront passing through B: in other words, this front is the envelope of the "wavelets" R_A. Recalling the shape of the surface of the group velocity, we find justification for Huygens' notation for the case of a uniaxial crystal, in which the waves are of two types: one spherical, *ordinary*, and one ellipsoidal, *extraordinary*. Instead, for a biaxial crystal, they have a more complicated shape, and both of the refracted waves can be considered extraordinary. The construction of Fig. 7.11(b) still applies in both cases. The same type of reasoning can also be conducted for reflection within an anisotropic medium.

When a wave has already undergone a birefringence process by entering an anisotropic medium, there is no birefringence at the output interface towards an isotropic medium. In fact, when entering the medium, the two waves that are generated have amplitudes of the field that are eigenvectors with defined polarization. The direction in which each of these waves leaves is unique, in that the external medium is here supposed to be isotropic, and, for it, there is a single solution to the Snell law for the incident wave vector. The two outgoing waves therefore conserve the orthogonal polarizations[2] that they had acquired by entering the anisotropic medium

Calcite crystals were the first birefringence polarizers to be used: observing a scene through such a crystal, it appears doubled in two images with orthogonal polarizations between them. This explains Huygens' experience with two calcite crystals in succession. Typically, by entering the second crystal, each image is still projected onto two orthogonal polarizations and, in all, four images are observed at the output. However, if the crystals are oriented to each other so that the polarizations that are generated in the second crystal are the same as those that emerge from the first, the transmitted images are reduced to two.

If, then, the birefringent medium is a lamina with plane and parallel faces immersed in an isotropic medium, we have that the two emerging waves both keep the same direction of the incident wave. Indeed, the projection of the wave vector of each of them within the lamina on the input face is equal to that of the incident wave. If the output face is parallel to the first, the projection on this second face is also the same for both waves, which then emerge parallel to the incident wave.

---

2 On this, however, it is necessary to consider the various geometric factors concerning the change of direction of the wave at the exit from the crystal.

7.7.2 Analytical method

The graphical constructions of the previous section, however attractive, are not very useful in a concrete problem, yet they can show us the way to an analytical solution. But, instead of looking for a single, very complicated expression that can replace Snell's equation, I prefer to indicate an algorithm that achieves the same purpose. For generality, suppose we have two anisotropic media separated by a plane interface.

1) First, we define three distinct Cartesian reference systems. For the first, $\mathbb{S}_1$, we choose it so that its axes coincide with the principal axes of the first medium, let us say, X_1, Y_1, Z_1, and, similarly, for the second, $\mathbb{S}_2$, we choose it with axes X_2, Y_2, Z_2 coinciding with the principal axes of the second medium. Then, we choose the third system, $\mathbb{S}_i$, with axes x, y, z such that the interface lies on the xy-plane.

2) Let us take the same origin for the three systems on a point O of the interface and define two unit matrices $\mathfrak{M}_1$ and $\mathfrak{M}_2$ for the transformation from one system to another:

$$\begin{pmatrix} r_{X1} \\ r_{Y1} \\ r_{Z1} \end{pmatrix} = \mathfrak{M}_1 \begin{pmatrix} r_x \\ r_y \\ r_z \end{pmatrix}, \quad \begin{pmatrix} r_{X2} \\ r_{Y2} \\ r_{Z2} \end{pmatrix} = \mathfrak{M}_2 \begin{pmatrix} r_x \\ r_y \\ r_z \end{pmatrix}, \tag{7.7.3}$$

where, for the same point, r_{j1} and r_{j2}, with j=X,Y,Z, indicate the coordinates expressed in the first and second systems and r_x, r_y, r_r the corresponding coordinates in the system $\mathbb{S}_i$ of the interface.

3) We establish the angular frequency ω of an incident wave in the first medium and the versor $\mathbf{n}_I$ of its wave vector with coordinates n_{Ix}, n_{Iy}, n_{Iz} in the system $\mathbb{S}_i$. From these, we obtain the coordinates with respect to the principal axes

$$\begin{pmatrix} \mathsf{n}_{IX1} \\ \mathsf{n}_{IY1} \\ \mathsf{n}_{IZ1} \end{pmatrix} = \mathfrak{M}_1 \begin{pmatrix} \mathsf{n}_{Ix} \\ \mathsf{n}_{Iy} \\ \mathsf{n}_{Iz} \end{pmatrix}. \tag{7.7.4}$$

Using these cosine directors, we then solve the Eq. (7.4.19.b) of Fresnel for n' and n'' to determine the eigenvalues $k_I' = \omega n'/c$ and $k_I'' = \omega n''/c$ of the wave vector. Of these, we choose one that we will now indicate, without the apex, as k_I. Applying n_{IX1}, n_{IY1}, n_{IZ1} and k_I to Eqs. (7.4.34) and (7.4.35), normalizing them, we respectively determine the versors $\mathbf{e}_I$ and $\mathbf{d}_I$ of the fields $\boldsymbol{E}_o$ and $\boldsymbol{D}_o$ of the incident wave expressed in the system $\mathbb{S}_1$. Finally, from one of the Eqs. (7.4.5), we have that $\mathbf{h}_I = \mathbf{n}_I \times \mathbf{d}_I$.

In the case in which the direction $\mathbf{s}_I$ of the Poynting's vector is defined, one proceeds as above with Eq. (7.7.4), with $\mathbf{s}$ instead of $\mathbf{n}$, transforming its coordinates from the $\mathbb{S}_\mathrm{i}$ system to those of $\mathbb{S}_1$. With them and Eq. (7.5.14), we determine which ones are the corresponding eigenvalues for the group velocity. Of these, we choose one, u_I, and apply it to Eqs. (7.5.12) and (7.5.13), together with the director cosines $\mathsf{s}_{\mathrm{I}X1}$, $\mathsf{s}_{\mathrm{I}Y1}$, $\mathsf{s}_{\mathrm{I}Z1}$ of $\mathbf{s}_\mathrm{I}$. Normalizing the resulting vectors, we obtain the versors $\mathbf{e}_\mathrm{I}$, $\mathbf{d}_\mathrm{I}$ and, finally, $\mathbf{h}_\mathrm{I} = \mathbf{s}_\mathrm{I} \times \mathbf{e}_\mathrm{I}$. The versor $\mathbf{n}_\mathrm{I}$ is obtained from the first of the Eqs. (7.4.5), for which $\mathbf{n}_\mathrm{I} = \mathbf{d}_\mathrm{I} \times \mathbf{h}_\mathrm{I}$. Having chosen the eigenvalue u_I of the group velocity, the eigenvalue of the phase velocity $v_\mathrm{I} = u_\mathrm{I}\, \mathbf{e}_\mathrm{I} \bullet \mathbf{d}_\mathrm{I}$ is also determined, and therefore $k_\mathrm{I} = \omega / v_\mathrm{I}$.

The cosine directors of $\mathbf{e}_\mathrm{I}$, $\mathbf{d}_\mathrm{I}$, $\mathbf{h}_\mathrm{I}$ and, in the second case, also those of $\mathbf{n}_\mathrm{I}$, are then converted into the $\mathbb{S}_\mathrm{i}$ system by the inverse transformation of (7.7.3).

4) For the refracted waves, we now solve the system:

$$\begin{cases} \begin{pmatrix} k_{\mathrm{T}X2} \\ k_{\mathrm{T}Y2} \\ k_{\mathrm{T}Z2} \end{pmatrix} = \mathfrak{M}_2 \begin{pmatrix} k_\mathrm{I} \mathsf{n}_{\mathrm{I}x} \\ k_\mathrm{I} \mathsf{n}_{\mathrm{I}y} \\ k_z \end{pmatrix}, \\ \text{Fresnel's equation (4.19.a) for } k_{\mathrm{TX2}}, k_{\mathrm{TY2}}, k_{\mathrm{TZ2}}, \end{cases} \tag{7.7.5}$$

where k_z is our unknown and $\mathsf{n}_{\mathrm{I}x}$, $\mathsf{n}_{\mathrm{I}y}$ are the cosine directors in the system of the interface. The system (7.7.5) is of fourth degree, and we will have to choose the solution to k_z that is consistent with the propagation towards the second medium. From this calculation, we derive the unknown components k_z of the two possible wave vectors $\boldsymbol{k}_{\mathrm{T}'}$ and $\boldsymbol{k}_{\mathrm{T}''}$ in the system of the interface. Then, we derive the corresponding components in the system of the second medium, as well as the direction cosines of the versors $\mathbf{n}_{\mathrm{T}'}$ and $\mathbf{n}_{\mathrm{T}''}$.

5) With the help of Eqs. (7.4.34) and (7.4.35), we then determine the components in the system $\mathbb{S}_2$ of the versors for the electric field, $\mathbf{e}_{\mathrm{T}'}$ and $\mathbf{e}_{\mathrm{T}''}$, $\mathbf{d}_{\mathrm{T}'}$ and $\mathbf{d}_{\mathrm{T}''}$, corresponding, respectively, to $\mathbf{n}_{\mathrm{T}'}$ and $\mathbf{n}_{\mathrm{T}''}$. From these, we can find the versors for the magnetic field

$$\mathbf{h}_{\mathrm{T}'} = \mathbf{n}_{\mathrm{T}'} \times \mathbf{d}_{\mathrm{T}'}, \quad \mathbf{h}_{\mathrm{T}''} = \mathbf{n}_{\mathrm{T}''} \times \mathbf{d}_{\mathrm{T}''}. \tag{7.7.6}$$

6) Finally, we calculate the two possible versors for the Poynting vector

$$\mathbf{s}_{\mathrm{T}'} = \mathbf{e}_{\mathrm{T}'} \times \mathbf{h}_{\mathrm{T}'}, \quad \mathbf{s}_{\mathrm{T}''} = \mathbf{e}_{\mathrm{T}''} \times \mathbf{h}_{\mathrm{T}''}, \tag{7.7.7}$$

which give us the ray directions of the refracted waves.

7) Applying the inverse transformation to Eq. (7.7.3.b), we can finally derive the components of all of these versors in the $\mathbb{S}_\mathrm{i}$ system of the interface.

For the calculation of the reflected waves, we proceed in the same way. It is understood that we will be able to meet all of the complications of the isotropic case multiplied by two: for example, we can have total reflection for the eigenvalues of k_z that would turn out to be imaginary.

The calculation of the field amplitudes and intensities of the refracted and reflected waves is also complicated. However, we can also imagine a resolution algorithm for these quantities:

8) With the choice of k_I or u_I, the polarization of the incident wave, that is, the linear polarization, was also defined, with the electric field oriented in the direction of $\mathbf{e}_\mathrm{I}$, which is preserved in the propagation. For a generic initial polarization, it is necessary to repeat the calculation for the other possible value of the group velocity or of the wave vector module and then apply the principle of superposition; the elegant separation in TE and TM waves of the isotropic case is not allowed here. Then, we fix the amplitude of the electric field as E_I, linearly polarized along $\mathbf{e}_\mathrm{I}$. From the previous calculation, we already have the versors for the fields of the incident wave, of the two reflected waves and of the two refracted waves. Given that the amplitude of the magnetic field is linked to that of the corresponding electric field, we have four unknowns in all. From the last of the Eqs. (7.5.9), multiplying vectorially for the versor $\mathbf{s}$, we indeed have that

$$\boldsymbol{H}_j = \frac{1}{u_j\mu}\mathbf{s}_j \times \boldsymbol{E}_j, \text{ where } j = \mathrm{I}, \mathrm{T}', \mathrm{T}'', \mathrm{R}', \mathrm{R}'' \text{ and } u_j = \frac{\omega}{k_j \mathbf{e}_j \cdot \mathbf{d}_j}. \quad (7.7.8)$$

9) As in the isotropic case, we then apply the boundary conditions for the continuity of the electrical and magnetic field components parallel to the interface: we still have 4 linear equations that allow us to calculate the four unknown amplitudes $E_{\mathrm{R}'}$, $E_{\mathrm{R}''}$, $E_{\mathrm{T}'}$, $E_{\mathrm{T}''}$ as a function of E_I:

$$\begin{aligned} E_\mathrm{I}\mathrm{e}_{\mathrm{I}x} + E_{\mathrm{R}'}\mathrm{e}_{\mathrm{R}'x} + E_{\mathrm{R}''}\mathrm{e}_{\mathrm{R}''x} &= E_{\mathrm{T}'}\mathrm{e}_{\mathrm{T}'x} + E_{\mathrm{T}''}\mathrm{e}_{\mathrm{T}''x}, \\ E_\mathrm{I}\mathrm{e}_{\mathrm{I}y} + E_{\mathrm{R}'}\mathrm{e}_{\mathrm{R}'y} + E_{\mathrm{R}''}\mathrm{e}_{\mathrm{R}''y} &= E_{\mathrm{T}'}\mathrm{e}_{\mathrm{T}'y} + E_{\mathrm{T}''}\mathrm{e}_{\mathrm{T}''y}, \\ n_{\mathrm{rI}}\mathrm{h}_{\mathrm{I}x}E_\mathrm{I} + n_{\mathrm{rR}'}\mathrm{h}_{\mathrm{R}'x}E_{\mathrm{R}'} + n_{\mathrm{rR}''}\mathrm{h}_{\mathrm{R}''x}E_{\mathrm{R}''} &= n_{\mathrm{rT}'}\mathrm{h}_{\mathrm{T}'x}E_{\mathrm{T}'} + n_{\mathrm{rT}''}\mathrm{h}_{\mathrm{T}''x}E_{\mathrm{T}''}, \\ n_{\mathrm{rI}}\mathrm{h}_{\mathrm{I}y}E_\mathrm{I} + n_{\mathrm{rR}'}\mathrm{h}_{\mathrm{R}'y}E_{\mathrm{R}'} + n_{\mathrm{rR}''}\mathrm{h}_{\mathrm{R}''y}E_{\mathrm{R}''} &= n_{\mathrm{rT}'}\mathrm{h}_{\mathrm{T}'y}E_{\mathrm{T}'} + n_{\mathrm{rT}''}\mathrm{h}_{\mathrm{T}''y}E_{\mathrm{T}''}, \end{aligned} \quad (7.7.9)$$

where I used the ray index definition $n_{\mathrm{r}j} = \mathrm{c}/u_j$.

10) Having determined the amplitudes of the electric field, we calculate those of the magnetic field according to Eqs. (7.7.8), for which

$$H_j = \frac{1}{u_j\mu}E_j, \quad (7.7.10)$$

and the intensity of the various waves

$$I_j = \frac{1}{u_j \mu} \left| E_j \right|^2 . \tag{7.7.11}$$

To calculate the reflectivity and transmissivity associated with each wave, that is, the ratio between the reflected and transmitted powers and the incident one, it is necessary to take into account the finite transverse dimension of the waves. This causes another geometric factor to intervene:

$$\begin{aligned} R' &= \frac{s_{R'z}\, I_{R'}}{s_{Iz}\, I_I}, \quad R'' = \frac{s_{R''z}\, I_{R''}}{s_{Iz}\, I_I}, \\ T' &= \frac{s_{T'z}\, I_{T'}}{s_{Iz}\, I_I}, \quad T'' = \frac{s_{T''z}\, I_{T''}}{s_{Iz}\, I_I}. \end{aligned} \tag{7.7.12}$$

7.7.3 Refraction with uniaxial crystals

The procedure of the previous section is simplified considering the most common refraction case from a uniaxial crystal, immersed in an isotropic medium. As we have seen before, the surface of wave vectors now consists of a sphere and an ellipsoid defined by Eqs. (7.4.39) and (7.4.40), which are tangent between them in correspondence with the optical axis of the crystal. For an external incident wave, we have two refracted waves as above. One follows the ordinary laws of refraction and is called an *ordinary wave*, while the other is called an *extraordinary wave*. Indeed, the ordinary wave corresponds to an electric field oriented perpendicular to the optical axis of the crystal, and therefore the phase velocity is independent of the direction of the wave vector, which lies on the sphere. Instead, the electric field of the extraordinary wave oscillates on a plane containing the optical axis; therefore, the phase velocity depends on the direction of the corresponding wave vector, which then lies on the ellipsoid. The two waves are generally not TE or TM with respect to the plane of incidence, except in the case in which the optical axis itself lies on this plane or is perpendicular to it. Let us now look at some special cases.

Consider the case of a lamina of birefringent material, with its optical axis aligned on the incidence plane and inclined at an angle Θ with the normal at its faces in the manner indicated in Fig. 7.12. Assuming that $n_I = 1$ for the index of the first medium, the law of refraction for the ordinary wave is simply

$$\begin{cases} \sin \vartheta_{To} = \dfrac{1}{n} \sin \vartheta_I, \\ n = n_o, \end{cases} \tag{7.7.13}$$

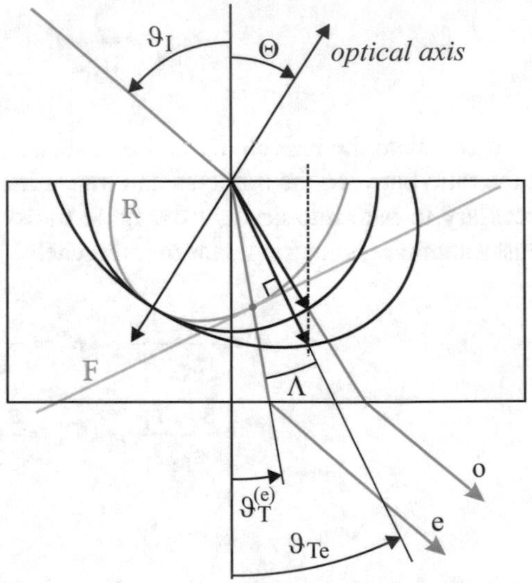

Fig. 7.12
Birefringence with a uniaxial lamina with $n_o < n_e$. This is a case in which the optical axis lies on the incidence plane, inclined at the angle Θ. The circle and the ellipse in black represent the intersection of the surface of the wave vectors with the plane of incidence, while the ellipse in gray represents the intersection of the ray surface R. The extraordinary ray passes through the point of tangency of this surface with wavefront F

while, for the extraordinary wave, we have to solve the system

$$\begin{cases} \sin\vartheta_{Te} = \dfrac{1}{n}\sin\vartheta_I, \\ \dfrac{1}{n^2} = \dfrac{1}{n_e^2}\sin^2(\vartheta_{Te}+\Theta) + \dfrac{1}{n_o^2}\cos^2(\vartheta_{Te}+\Theta), \end{cases} \tag{7.7.14}$$

where the second equation is derived from Eq. (7.4.41.b). The direction of the versor **s** of the Poynting vector of the ordinary wave coincides with that of $\boldsymbol{k}$, which is given by ϑ_{Te}. The direction **s** of the extraordinary wave still lies, in this case, on the plane of incidence, but its inclination with the normal to the surface is

$$\begin{aligned} \vartheta^{(e)}{}_T &= \vartheta_{Te} + \Lambda \quad \text{for } n_o > n_e, \\ \vartheta^{(e)}{}_T &= \vartheta_{Te} - \Lambda \quad \text{for } n_o < n_e, \end{aligned} \tag{7.7.15}$$

where the angle Λ between the directions of **n** and **s** is given by Eq. (7.4.16.b), from which

$$\Lambda = \arccos\frac{n_o^{-2}\sin^2(\vartheta_{Te}+\Theta) + n_e^{-2}\cos^2(\vartheta_{Te}+\Theta)}{\sqrt{n_o^{-4}\sin^2(\vartheta_{Te}+\Theta) + n_e^{-4}\cos^2(\vartheta_{Te}+\Theta)}}. \tag{7.7.16}$$

The signs + and – in Eqs. (7.7.15) can be deduced by looking at Fig. 7.6. Finally, the ordinary wave is a TE wave here, while the extraordinary wave is TM.

In particular, it may also happen that the extraordinary ray is inclined "backwards". Moreover, in the case in which $\vartheta_I = 0$, $\vartheta_{Te} = 0$; however, if the optical axis is inclined with respect to the normal to the surface of the lamina, the extraordinary ray is inclined at an angle Λ from this same normal.

More frequently, the uniaxial birefringent laminae are used with their optical axis parallel to their faces. Consider a reference system in which the z-axis coincides with the direction of the optical axis and the y-axis with the normal at the surface. In addition, let us indicate, with φ, the angle between the plane of incidence and the optical axis, measured on the xz-plane and with ϑ_{Te} as the angle of refraction on the plane of incidence. In this case, the intersection of the plane of incidence with the surface of the $\mathbf{k}$s is still given by a circle of radius $n_o\omega/c$ and by an ellipse expressed for the index of refraction by

$$\begin{aligned} &n_\perp^2\left(n_o^2\sin^2\varphi + n_e^2\cos^2\varphi\right) + n_y^2 n_o^2 = n_o^2 n_e^2, \\ &\text{with } n_\perp = n\sin\vartheta_{Te}, \quad n_y = n\cos\vartheta_{Te}, \end{aligned} \tag{7.7.17}$$

where Eq. (7.4.40) was used. Knowing the angles, Eqs. (7.7.17) can be used to determine n.

In conclusion, when the optical axis is parallel to the interface, the refraction angle for the extraordinary wave is obtained by solving the system

$$\begin{cases} \sin\vartheta_{Te} = \dfrac{1}{n}\sin\vartheta_I, \\ \dfrac{1}{n^2} = \dfrac{n_i^2}{n_o^2 n_e^2}\sin^2\vartheta_{Te} + \dfrac{1}{n_e^2}\cos^2\vartheta_{Te}, \end{cases} \tag{7.7.18}$$

where

$$n_i^2 = n_o^2\sin^2\varphi + n_e^2\cos^2\varphi\,. \tag{7.7.19}$$

In general, the direction of $\mathbf{s}$ is now not on the plane of incidence, and its determination is complicated, just as it is complicated to determine the wave polarization.

As a further simplification, we consider the case in which φ is null. The solution to the problem for the extraordinary wave is then similar to that with the optical axis orthogonal to the interface, and we have that

$$\begin{cases} \sin\vartheta_{Te} = \dfrac{1}{n}\sin\vartheta_I, \\ \dfrac{1}{n^2} = \dfrac{1}{n_o^2}\sin^2\vartheta_{Te} + \dfrac{1}{n_e^2}\cos^2\vartheta_{Te}, \end{cases} \tag{7.7.20}$$

where n_e and n_o are exchanged with respect to Eqs. (7.7.14). The ordinary wave is TE, while the extraordinary wave is TM.

If, instead, $\varphi = \pi/2$, we have that

$$\begin{cases} \sin \vartheta_{\text{Te}} = \dfrac{1}{n} \sin \vartheta_{\text{I}}, \\ n = n_{\text{e}}. \end{cases} \qquad (7.7.21)$$

In this case, the extraordinary wave behaves angularly like the ordinary one, but with index n_e. Its polarization is TE, while the ordinary wave is TM.

In the last case, we still consider a lamina with the optical axis parallel to its faces, but an angle of incidence ϑ_I null, while φ can vary. In this important case, ϑ_{To} and ϑ_{Te} are null and, for Eq. (7.4.16.b), the angle between the versors **s** and **n** is also null, and therefore there is no distinct refraction. The two waves, polarized with the electric field, one oscillating in the orthogonal direction and the other parallel to the optical axis, proceed superimposed, but with a different phase velocity. As we will study below, this situation is used to make birefringent laminas that modulate the polarization of radiation continuously by varying φ.

In all of the cases discussed above, if the second face of the lamina is parallel to the first, the two ordinary and extraordinary waves emerge with wave vectors parallel to that of the incident wave and between them. However, in general, there is a translation of the rays due to the different refraction on the faces.

7.7.4 Conical refraction

For biaxial crystals, there is a singularity condition: if the direction **n** of the wave vector $\boldsymbol{k}$ coincides with one of the optical axes, the direction **s** of the Poynting's vector is indeterminate. Likewise, if **s** coincides with one of the radial optical axes, the direction of the wave vector is undetermined. Indeed, in the first case, we have that the eigenvalues for the index of refraction are degenerate and any direction be taken for the field $\boldsymbol{D}_o$, however orthogonal to **n**, corresponds to an eigenvector and is preserved in the propagation. On the other hand, **s** lies on the plane of **n** and $\boldsymbol{D}_o$, at an angle θ with **n**, which is generally not null (it is zero for $\boldsymbol{D}_o$ parallel to the y-axis), and therefore varies with the direction of $\boldsymbol{D}_o$. The same reasoning also applies in the second case, exchanging $\boldsymbol{E}_o$ with $\boldsymbol{D}_o$ and **s** with **n**.

These singularities are particularly evident in the case of refraction, illuminating a biaxial plate immersed in air with a non-polarized collimated beam. When the direction of incidence is such that the refracted waves have their wave vectors $\boldsymbol{k}$ coinciding with an optical axis, they are equal in module and there is a decomposition of the incident beam in a continuous spectrum of directions **s** due to all of the possible directions of $\boldsymbol{D}_o$ around **n**. The refracted beam then takes the form of a cone and the beam emerging from the second face, taken parallel to the first, has the shape of a tube: if, in fact, it is intercepted by a screen, it produces a bright ring. This consequence of the Fresnel's equations was identified by William Rowan Hamilton in 1833, and was experimentally confirmed two months later by Humphrey Lloyd, whom Hamilton had asked to verify his prediction [Lloyd 1833; Hamilton 1833, 1837].

Let us now look at a geometrical demonstration of this effect taken from Born and Wolf (1980). Consider a linear wave, linearly polarized, which propagates in the direction n_a of one of the two optical axes of the crystal. Fig. 7.13 shows a geometric construction useful for this discussion: the surface of the group velocity is plotted in gray, and the plane N is orthogonal to the optical axis and intersects it at the point V on the surface of the phase velocity. The distance $\upsilon_a = \overline{OV}$ is therefore equal to the phase velocity on the optical axis.

In the system of the principal axes, n_a has components (n_{ax}, 0, n_{ay}) and the field $\boldsymbol{D}$ must satisfy the equation

$$\mathsf{n}_{ax} D_x + \mathsf{n}_{az} D_z = 0 .$$

We then have that, for the corresponding electric field,

$$\mathsf{n}_{ax} \varepsilon_1 E_x + \mathsf{n}_{az} \varepsilon_3 E_z = 0 .$$

Therefore, the field $\boldsymbol{E}$ must be perpendicular to the vector $\boldsymbol{w}$ = ($\mathsf{n}_{ax}\varepsilon_1$, 0, $\mathsf{n}_{az}\varepsilon_3$), hence it must lie on a plane parallel to the y-axis (and perpendicular to $\boldsymbol{w}$) indicated by S. The vector $\boldsymbol{w}$ is inclined with respect to the z-axis of an angle φ given by

$$\tan\varphi = \frac{\mathsf{n}_{ax}\varepsilon_1}{\mathsf{n}_{az}\varepsilon_3} = \frac{\mathsf{n}_{ax} n_1^2}{\mathsf{n}_{az} n_3^2} . \tag{7.7.22}$$

When $\boldsymbol{D}$ is aligned on the zx-plane, the group velocity vector $\boldsymbol{u}$ must also lie on this plane, and its vertex coincides with the point U. It is necessarily a point of tangency between the plane N and the (external) surface of the group velocity, according to the rule indicated in Fig. 7.9. Moreover, the segment OU, being perpendicular to the direction of $\boldsymbol{E}$ and lying on the zx-plane, is perpendicular to the plane S. In particular, the angle $\chi = \mathrm{U\hat{O}V}$ is given by

$$\tan\chi = \tan(\beta - \varphi) = \frac{\dfrac{\mathsf{n}_{ax}}{\mathsf{n}_{az}} - \dfrac{\mathsf{n}_{ax} n_1^2}{\mathsf{n}_{az} n_3^2}}{1 + \dfrac{\mathsf{n}_{ax}^2 n_1^2}{\mathsf{n}_{az}^2 n_3^2}} = \frac{1}{n_3 n_1}\sqrt{\left(n_2^2 - n_1^2\right)\left(n_3^2 - n_2^2\right)} , \tag{7.7.23}$$

where Eq. (7.4.37) was used.

By rotating $\boldsymbol{D}$, $\boldsymbol{E}$ and $\boldsymbol{u}$ also rotate, because they must lie on the plane defined by $\boldsymbol{D}$ and n_a. In particular, $\boldsymbol{u}$ draws a curve C on the surface of the group velocity: it also lies on the N plane, because, for the particular direction chosen for the ver-

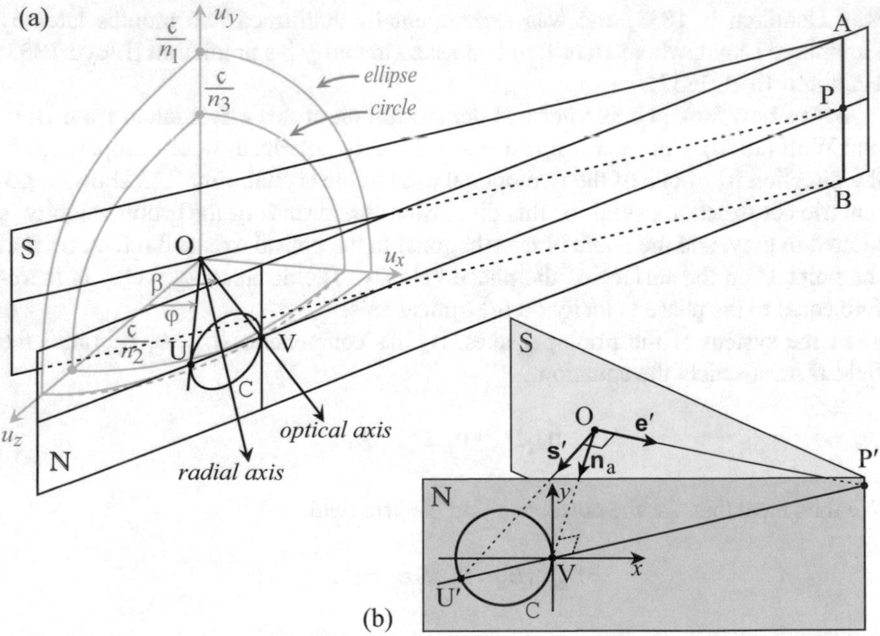

Fig. 7.13 Study of conical refraction. (a) an octant of the surface of the group velocity is shown in gray, in an axonometric projection. The dashed line indicates the phase velocity on the *zx*-plane. The circle C in black, with diameter UV , indicates the tangent curve with this surface of the plane N perpendicular to the optical axis OV, and surrounds one of the "navels" of the group velocity surface. The plane S passes through the origin O and is perpendicular to the segment OU. (b) diagram of the geometric construction of the curve C

sor **n** of ***k***, the phase velocity does not change. This means that, in the *velocity space*,

> when the phase velocity **n**υ is aligned with an optical axis of a biaxial crystal, the plane perpendicular to the direction of **n**, with distance from the origin equal to υ, is tangent to the surface of the group velocity along a continuous line.

On the other hand, the versor **e′** of ***E*** must remain on the plane S along a direction OP′, where P′ lies on the line AB of intersection between plane S and plane N, while the versor **s′** of ***u***, which lies on segment OU′, remains perpendicular to **e′**. Therefore, the two triangles VOP′ and U′OP′ are similar to each other and to the triangle U′OV. It follows that

$$\overline{\mathrm{U'V}}\;\overline{\mathrm{VP'}} = \overline{\mathrm{OV}}^2 ,$$

which can be rewritten as

$$\left(x^2+y^2\right)\left(a^2+Y^2\right)=\upsilon_a{}^4,\quad \frac{Y}{a}=\frac{y}{x},$$

where x and y are the Cartesian coordinates of U′ on the plane N, with origin in V, $Y=\overline{\mathrm{PP'}}$ and $a=\overline{\mathrm{VP}}=\upsilon_a/\tan\chi$. Therefore, the curve C is defined by the equation of a circle

$$\left(x+\upsilon_a\frac{\tan\chi}{2}\right)^2+y^2=\upsilon_a{}^2\frac{\tan^2\chi}{4}.$$

The directions assumed by **s** then form the surface of a cone (with an elliptical base, since the vertex does not lie on the axis of circle C). Therefore, when a collimated beam reaches the surface of a biaxial crystal, with such an inclination that the refracted wave vector is accurately aligned with an optical axis, it is refracted by opening itself in a cone inside the crystal. This phenomenon is called *internal conical refraction*.

A similar effect occurs when a wave propagates with the Poynting's vector aligned with a ray axis. Let us call W the intersection point of this axis with the group velocity surface. The singularity of W implies that, on it, there are infinite tangent planes to the surface. In particular, for each direction **e** of the electric field, transverse to the radial axis, there are two tangent lines to the surface that pass from W. On each of them, the corresponding value of the phase velocity is given by the intersection with an orthogonal line passing through the origin O, according to the rule in Fig. 7.9. By varying **e**, a closed curve is again obtained on the surface of the phase velocity. Therefore, in this case, there are the directions of the phase velocity that form a cone, around a "navel" of the phase velocity surface. The angular opening ψ of this cone is obtained by applying the inversion theorem to Eq. (7.7.23):

$$\tan\psi=\frac{1}{n_2{}^2}\sqrt{\left(n_2{}^2-n_1{}^2\right)\left(n_3{}^2-n_2{}^2\right)}=\frac{n_1n_2}{n_2{}^2}\tan\chi\,. \tag{7.7.24}$$

When a beam, which propagates in the crystal with these properties, reaches the surface, it is refracted by opening itself like a cone out of the crystal. This phenomenon is called *external conical refraction*. One way to generate this effect is to focus a light beam on the crystal's entrance face, with an angular opening large enough to cover, inside of the crystal, the wave vector cone mentioned above, and to place a small aperture on the output face in order to select those waves whose Poynting's vector is (almost) parallel to a radial axis. However, the opening produces diffraction.

Hamilton's prediction and Lloyd's experimental confirmation constituted a substantial experimental proof of the wave theory of light developed by Fresnel. Conical refraction can be observed by placing a screen in the path of radiation; however, a careful observation reveals that, instead of a single ring, there are two

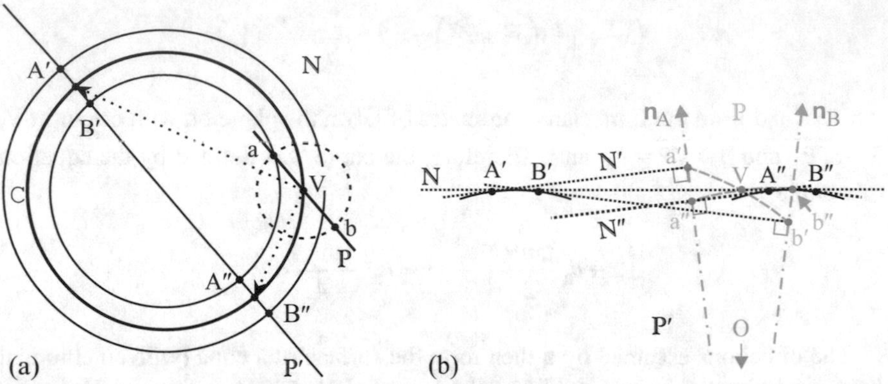

Fig. 7.14 Study of the double ring formation in internal conical refraction. (a): the plane of the figure coincides with the plane N of tangency to the surface of the group velocity on the circle C of Fig. 7.13. The point V is the intersection of N with one of the optical axes, which is perpendicular to N. The radiation beam is supposed to be made up of a spectrum of plane waves whose versors **n** are directed towards points of N contained within the dotted circle around V. The line ab identifies a plane P, containing the optical axis and perpendicular to N, on which the versor **n** of a plane wave component of the beam also lies. (b): the plane of the figure coincides with the plane P′ identified in (a) by the line A′B″ and parallel to P. It shows, in black, the section of the *group velocity* surface with P′. The projection on P′ of the section of the *phase velocity* surface with the plane P is drawn in gray with the dashed lines a′b′ and a″b″, while the dotted lines indicate the tangent planes to the surface of the group velocity (and perpendicular to P′)

concentric ones separated by a dark ring. This strange behavior was noted by J.C. Poggendorff and by von Haidinger [Poggendorff 1839; Haidinger 1855], but it remained misunderstood for a long time, until W. Voigt succeeded in giving a qualitative explanation [Voigt 1905]. A quantitative interpretation was achieved only recently with an analytic theory based on diffraction [Berry et al 2006].

In the case of internal conical refraction, the observation of the ring requires a collimated input beam of finite diameter, but this implies that its spectrum is extended in the space of wave vectors. It is therefore necessary to consider the plane wave components that are not aligned with the optical axis.

In Fig. 7.14(a), the plane N orthogonal to the optical axis whose versor is $\mathbf{n}_a$ is represented. As in Fig. 7.13, it crosses the axis at the point V on the surface of the phase velocity, and is tangent to the surface of the group velocity on the circle C. Near this circle, the outer surface of the group velocity is convex. The curvature of the surface is zero parallel to the circle C, and is therefore maximum in its radial direction.

Indeed, for a plane wave component, with a direction $\mathbf{n}_A$ slightly away from the optical axis, it happens that the degeneration of the eigenvalues is removed. Two distinct phase velocity eigenvalues exist for that value of $\mathbf{n}_A$, represented by the points a′ and a″ in Fig. 7.14(b), with specific polarizations of the field $\boldsymbol{D}$ orthogonal to each other. These correspond to two specific values of the versor $\mathbf{s}$ of the Poynting's vector, which are identified by the points of tangency A′ and A″

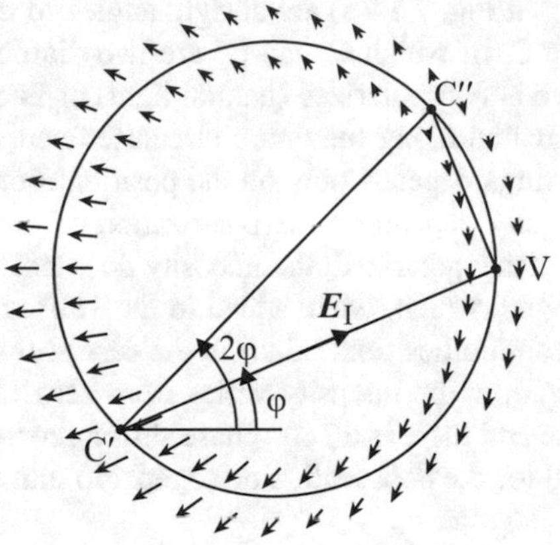

Fig. 7.15
Representation of the electric field in the two rings of the conical refraction, for a linear polarization of the incident field. At each point along the rings, the polarization is linear and the direction of the arrows indicates its direction. The length of the arrows indicates the width of the field, which is maximum in C′, where the polarization coincides with that of the incident wave, and is zero in C″, where the polarization is orthogonal. The phase of the field is constant along each ring

with the surface of the group velocity of two planes N′ and N″ orthogonal to $\mathbf{n}_A$ in which N splits itself.

Let us consider the change of direction of $\mathbf{n}_A$ as a rotation around the origin O on a plane P containing V and O, that is, radially along the line ab in Fig. 7.14(a). Then, as shown in Fig. 7.14(b), the tangent points on the surface of the group velocity move on the arcs A′B′ and A″B″ on a plane P′ parallel to P in two diametrically opposite positions of the circle C. This happens in correspondence with the arcs a′b′ and a″b″ on the phase velocity surface and with polarizations for the field $\boldsymbol{E}$ that is (almost) orthogonal. One of these positions is inside of the circle and the other is outside.

Turning, instead, $\Delta\mathbf{n} = \mathbf{n}_A - \mathbf{n}_a$, around the optical axis, with constant module r, we obtain that A′ and A″ travel along two circles one inside and the other outside C. This is represented by two circular "contour lines" concentric to the circle C in Fig. 7.14(a). On the other hand, the content of radiant energy in the spectrum of the wave vectors, and therefore in that of the phase velocities, corresponding to a radial interval δr, is proportional to $r\delta r$. It is distributed on directions of the Poynting's vector that lie in two concentric rings, of thickness $\delta r' \approx \delta r$, around the circle C on the surface of the group velocity. Therefore, this contribution is zeroed only at C, for which $r = 0$. Returning to the space of ordinary physical coordinates, the transformation of this spectrum of refracted plane waves corresponds to a conic beam. When it is intercepted by a screen, two rings separated by a dark ring appear instead of a single ring.

The behavior of the polarization along the circumference of the ring is very intriguing. At refraction, each incident plane wave component is divided into two linearly polarized waves in an orthogonal manner in the field $\boldsymbol{D}$, with amplitudes that depend on the initial polarization and the inclination of the wave with the optical axis. In particular, the direction of $\boldsymbol{D}$ is always contained on the plane defined by $\mathbf{n}$ and $\mathbf{s}$ and the polarization of the rings varies along the circumference in the manner indicated in Fig. 7.15, in which the arrows indicate how the polarization rotates by 180° in one revolution. It should be noted that the segments VC′ and

VC″ in Fig. 7.14(a) are at right angles to each other, since V is on the circumference C, of which C′ and C″ are two diametrically opposite points. If the incident wave is non-polarized (natural light) or is circularly polarized, the rings appear illuminated along the entire circumference. While the polarization direction along the rings depends only on the position along the circle C, the phase and the local intensity depend on the polarization of the incident wave. In particular, if this is circularly polarized, the intensity does not vary, but the phase changes by 180° in a round, which, when added to the 180° rotation of the arrows, does not produce discontinuities. Instead, as in the case shown in Fig. 7.15, with linear incident polarization, the intensity varies from zero to a maximum: the rings appear incomplete and there is a 180° phase shift in correspondence with the cancellation of the field for the polarization orthogonal to that incident.

7.8 Interference with birefringent plates

Many practical applications of anisotropic media exploit the phase difference that occurs between orthogonal linear polarizations in the path through them. The birefringent material is generally cut so as to have the input and output faces parallel to each other, with the optical axis oriented in various ways.

Here and in the discussion that follows, the phase is determined along a *path* that is always perpendicular to the wavefronts (and therefore not along a ray, associated with the flow of energy). The phase variation is calculated in proportion to the length of the path and the phase velocity. On the other hand, a ray is assimilated into a thin collimated beam, which is the place where there is constructive

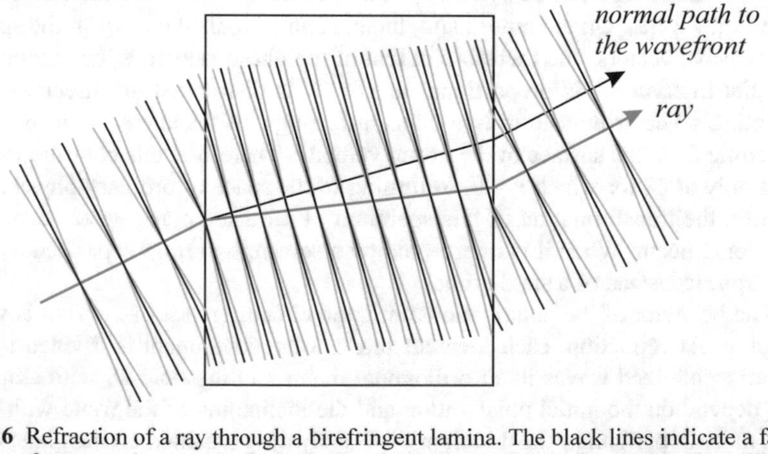

Fig. 7.16 Refraction of a ray through a birefringent lamina. The black lines indicate a family of wavefronts, equi-spaced in the optical path, corresponding to the carrier of a collimated beam. Similarly, the gray lines indicate the wavefronts of two plane wave components in the beam's spatial spectrum. The ray is the place of the intersection points of these wavefronts, where there is constructive interference between the waves of the spectrum

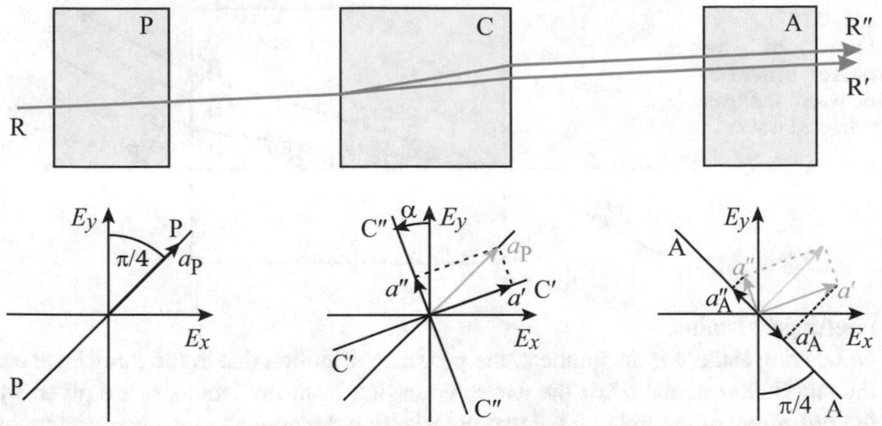

Fig. 7.17 Study of the interference produced by a birefringent lamina. The lamina C is placed between two crossed polarizers, P and A, with their transmission axis, PP and AA, at right angles to each other. A linearly polarized plane wave emerges with amplitude a_P from P. In the refraction at the entrance face of C, this wave is split into two waves. As the lamina has parallel faces, these two waves emerge with parallel directions and with orthogonal polarizations along the axes C′C′ and C″C″. Their amplitudes a' and a'' are the projection of a_P on these axes, except for a different phase factor introduced by the birefringent lamina. These amplitudes are then, in turn, projected on the direction AA of the analyzer A

interference between the plane waves that make up its spatial spectrum (Fig. 7.16); we can take its phase to be equal to that of the plane wave carrier around which the spectrum develops.

By themselves, orthogonal polarized waves do not give rise to local variations in intensity, and therefore do not produce visible fringes. However, interference fringes can be observed if a birefringent plate is placed between two polarizers (Fig. 7.17). The former ensures that the lamina is illuminated with polarized radiation, even when the source is natural. The second projects, along its transmission direction, the wave field corresponding to the two eigenvalues of the phase velocity. In this way, the two waves produce visible interference fringes.

The procedure is analogous to that of two-wave interference, in which the paths are separated and then recombined by means of mirrors and semi-reflecting laminae. Here, the separation is produced by the birefringent medium that decomposes the radiation in two orthogonal polarizations, and the recombination is generated by the second polarizer. Among other things, the fringes produced in this way are much more stable than those produced ordinarily, as the phase differences are produced within a solid body and are small, since they are proportional to the generally small difference between the two eigenvalues of the refractive index.

Suppose here that the transmission axes of the two polarizers are crossed, as in Fig. 7.17 [Sommerfeld 1949]. Other orientations are possible, but this configuration produces the best results in terms of visibility of the fringes, and it will be the only one analyzed here. Let us consider, in the meantime, a single plane wave that crosses the system, represented in the figure by a "ray" that doubles through the

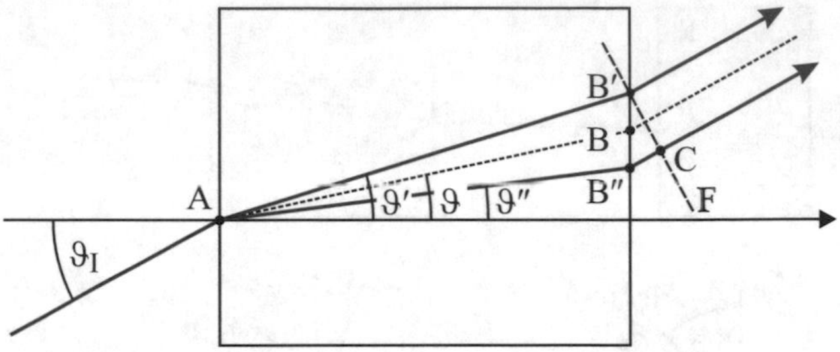

Fig. 7.18 Study of the phase difference between the two refracted waves

birefringent lamina.

Leaving aside, for the moment, the phase contribution due to the birefringence, the amplitudes a' and a'' of the waves emerging from the lamina C are given by the projection of the field on the two polarization directions of the eigenvectors of $\boldsymbol{D}$ in the lamina, which depend on the direction of the wave, for which

$$a' = a_P \cos\left(\frac{\pi}{4} - \alpha\right), \quad a'' = a_P \cos\left(\frac{\pi}{4} + \alpha\right),$$

where a_P is the module of the wave amplitude that passes through the polarizer P and α is the angle between the field eigenvectors and the axes of the reference system indicated in Fig. 7.17. In turn, these amplitudes are projected with the opposite sign on the transmission direction AA of the analyzer, whereby

$$a''_A = -a'_A = a_P \cos\left(\frac{\pi}{4} - \alpha\right)\cos\left(\frac{\pi}{4} + \alpha\right) = \frac{a_P}{2}\cos 2\alpha . \tag{7.8.1}$$

In addition to these values for the amplitude, it is necessary to consider the phase difference $\Delta\phi$ that is produced for the two emerging waves. Let us consider, as in Fig. 7.18, a path *normal to the wavefront* that splits when entering into a birefringent lamina at the point A. The two paths that are generated continue up to points B′ and B″ on the second face of the lamina, parallel to the first, and then continue, returning parallel to each other and orthogonal to the wavefront F, which intersects the two paths in B′ and C. The phase difference should therefore be calculated between A and F. For a lamina immersed in air, with a refractive index of 1, we have that

$$\Delta\phi = \frac{2\pi}{\lambda_o}\left(n''\overline{AB''} + \overline{B''C} - n'\overline{AB'}\right)$$
$$= \frac{2\pi}{\lambda_o} h\left[\frac{n''}{\cos\vartheta''} + (\tan\vartheta' - \tan\vartheta'')\sin\vartheta_I - \frac{n'}{\cos\vartheta'}\right],$$

where h is the thickness of the lamina. By expressing the tangents as the ratio of sine and cosine and applying the law of sines to each refracted path, we obtain

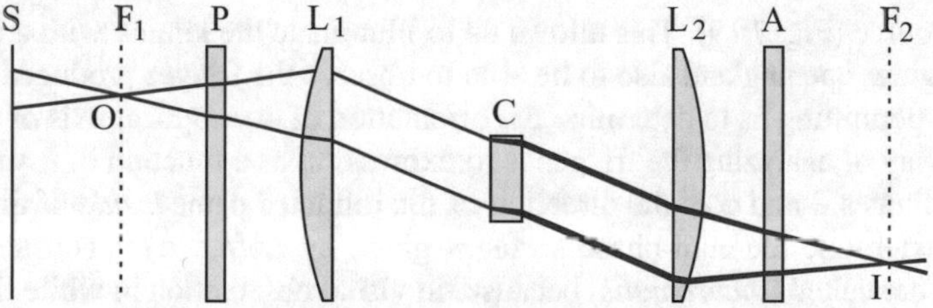

Fig. 7.19 Optical system for producing interference fringes in convergent polarized light. P is the polarizer and A is the analyzer. The points of the focal plane F_1 of the lens L_1 act as object points O, whose image I is on the focal plane F_2 of the lens L_2. The birefringent lamina is thus illuminated by a spectrum of plane waves, each of which originates from an object point O and is split into two plane waves that are then focused on the image point I. The transform of this spectrum is formed on the plane F_2, and the fringes observed there are produced in a manner similar to that used to visualize Haidinger's fringes and the Fraunhofer diffraction

$$\Delta\phi = \frac{2\pi}{\lambda_o} h\left(n'' \cos\vartheta'' - n' \cos\vartheta'\right). \tag{7.8.2}$$

Since the difference $\eta = n'' - n'$ is generally small with respect to the average value n of the index of refraction, we can use an approximate expression at the first order in η for $\Delta\phi$ [Born and Wolf 1980]:

$$\Delta\phi \cong \frac{2\pi}{\lambda_o} h\left(n'' - n'\right)\frac{d}{dn}\left(n\cos\vartheta\right) = \frac{2\pi}{\lambda_o} h\left(n'' - n'\right)\left(\cos\vartheta - n\sin\vartheta\frac{d\vartheta}{dn}\right), \tag{7.8.3}$$

where ϑ is the average value between ϑ' and ϑ''. On the other hand, differentiating the law of sines $\sin\vartheta_I = n\sin\vartheta$, with ϑ_I fixed, we have

$$0 = \sin\vartheta + n\cos\vartheta\frac{d\vartheta}{dn}.$$

Resolving for $d\vartheta/dn$ and replacing the result in Eq. (7.8.3), we get

$$\Delta\phi \cong \frac{2\pi}{\lambda_o}\frac{h}{\cos\vartheta}\left(n'' - n'\right). \tag{7.8.4}$$

With respect to the displacement produced by multiple reflections in an ordinary lamina, here, the cosine of the *inner angle* appears in the denominator. In this approximation, the phase difference is simply proportional to the distance $\overline{AB}$ and to the difference between the indices. However, in addition to ϑ, the indices of refraction also depend here on the conditions of incidence.

Typically, the interference fringes are observed by placing the lamina within a telescopic system, formed by two lenses, which receives the radiation of an extended source (Fig. 7.19). This allows us to illuminate the lamina with a relatively large angular opening and also to be able to observe the fringes produced by small crystals, permitting us to determine the orientation of their optical axis or axes.

One way of analyzing the fringes is to express $\Delta\phi$ as a function of h and the polar coordinates ϑ and φ of the direction of the refracted paths *inside* of the lamina, and to determine the equi-phase surfaces given by $\Delta\phi(h,\vartheta,\varphi)$ = constant. These surfaces are called *isochromatic*, because, in visual observation in white light, they are associated with a particular color, due to the contribution of the various wavelengths, on which $\Delta\phi$ depends. The second lens of the optical system of Fig. 7.19 allows us, in paraxial approximation, to transform the angular coordinates into spatial coordinates on its image focal plane. If, on the other hand, we reconsider Fig. 7.18, $\Delta\phi$ can also be expressed as a function of the *vector* $\overrightarrow{\mathrm{AB}}$, to which the variables ϑ and φ can be traced.

For Eq. (7.8.1), the intensity observed on the plane F_2 is therefore given by

$$I = I_o \cos^2 2\alpha \sin^2 \frac{\Delta\phi}{2}, \tag{7.8.5}$$

where I_o is a constant. This intensity is canceled in correspondence with the values of α equal to $\pi/4$, $3\pi/4$, $5\pi/4$ and $7\pi/4$, i.e., when the eigenvectors of the field $\boldsymbol{D}$ are aligned with the transmission axes of the polarizers. The lines where the intensity is canceled are called *isogyre.* Contrastingly, if the polarizers have their axes parallel between them, the complementary situation is obtained, in the sense that the figure obtained is the negative of the previous one and each color is replaced by its complementary with respect to the white light.

7.8.1 Interference with a uniaxial crystal

Following Born and Wolf (1980), in the case of a uniaxial crystal, from Eqs. (7.4.41), we find the Laplace's law

$$\frac{1}{n'^2} - \frac{1}{n''^2} = \left(\frac{1}{n_o{}^2} - \frac{1}{n_e{}^2}\right) \sin^2\theta, \tag{7.8.6}$$

where θ is the angle between the direction of the wave vector and the optical axis. Since the difference between ordinary and extraordinary indices is generally small compared to their value, we can approximate this equation with

$$n'' - n' = (n_e - n_o)\sin^2\theta$$

and the phase difference between the two paths becomes

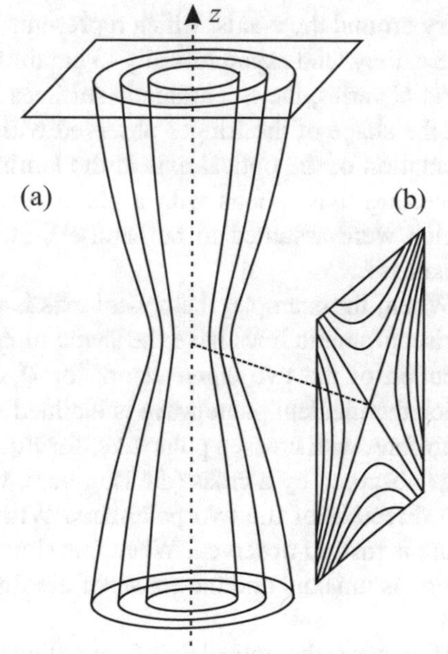

Fig. 7.20 Representation of the isochromatic surfaces. The shape of the fringes is obtained by cutting these surfaces with a plane placed at a distance from the center equal to the thickness of the lamina. (a) When the lamina is cut perpendicular to the optical axis *z*, the fringes appear circular with the center on the axis. (b) If, instead, the optical axis is parallel to the lamina, the fringes have the form of equilateral hyperbola

$$\Delta\phi \cong \frac{2\pi}{\lambda_o}\frac{h}{\cos\vartheta}(n_e - n_o)\sin^2\theta\,. \tag{7.8.7}$$

The isochromatic surfaces are then expressed by the equation

$$\rho\sin^2\theta = C, \quad \text{with } C \text{ constant}, \tag{7.8.8}$$

where $\rho = h/\cos\vartheta$ it is the distance traveled along the path through the lamina, i.e., the distance between points A and B in Fig. 7.18. We can imagine a space like the wave vector space, represented by a Cartesian reference frame such that its *z*-axis is parallel to the optical axis of the lamina, but with distances that are the same as the physical space. We will use this space only to represent each of the plane waves that cross the lamina by means of a straight line orthogonal to the wavefront and coming from the origin. In this space, we can then express ρ and θ by

$$\rho^2 = x^2 + y^2 + z^2, \quad \sin^2\theta = \frac{x^2 + y^2}{x^2 + y^2 + z^2}\,.$$

Therefore, in this space, the equi-phase surfaces are given by

$$\left(x^2 + y^2\right)^2 = C^2\left(x^2 + y^2 + z^2\right). \tag{7.8.9}$$

The appearance of these surfaces is shown in Fig. 7.20. They have rotation symmetry around the z-axis, which represents the optical axis of the lamina, and, for $z \to \pm\infty$, they tend asymptotically to paraboloids of rotation.

As C varies, the isochromatic surfaces form a series of shells around the z-axis and the shape of the fringes observed with the system in Fig. 7.19 depends on the orientation of the optical axis in the lamina. In other words, this form is obtained by cutting these shells with a plane surface parallel to the faces of the lamina (which were assumed to be parallel), at a distance from the origin equal to its thickness h.

When, for example, the optical axis is perpendicular to the sides of the lamina, the isochromatic lines have the shape of concentric circles. On the other hand, the direction of the two eigenvectors for $\boldsymbol{D}$ varies with the azimuthal angle φ, with which the incident plane wave is inclined with respect to the lamina, so that $\alpha = \varphi$. Therefore, with crossed polarizers, for Eq. (7.8.5), the fringes that are observed are rings obscured by a cross (the isogyres), with axes corresponding to the transmission direction of the two polarizers. With aligned polarizers, the complementary figure is instead observed. When this situation occurs, the material constituting the lamina is uniaxial and the center of the figure indicates the direction of the optical axis.

If, instead, the optical axis is parallel to the faces of the lamina, the fringes that are observed are equilateral hyperbola whose asymptotes are oriented at 45° with the optical axis. This can be easily understood (Born and Wolf 1980) by assuming that $x, z \ll y = h$ and developing Eq. (7.8.9) to the first order in $z^2/(x^2+y^2)$, for which we have that

$$C^2 = \frac{\left(x^2+y^2\right)^2}{\left(x^2+y^2+z^2\right)} = \frac{x^2+y^2}{\left(1+\dfrac{z^2}{x^2+y^2}\right)} \simeq \left(x^2+y^2\right)\left(1-\frac{z^2}{x^2+y^2}\right) = x^2+y^2-z^2,$$

whereby

$$x^2 - z^2 = C^2 - h^2 .$$

In this case, the directions of the field eigenvectors are independent of the direction of the considered plane wave, but it depends on the angle φ between the optical axis and the transmission direction of the analyzer for which $\alpha = \varphi - \pi/4$. The previous dark cross is not observed, but, rather, rotating the lamina, the fringe contrast instead varies cyclically with period $\pi/2$, passing from a maximum for $\varphi = \pi/4$ to a dark field for $\varphi = \pi/2$ and to a new maximum for $\varphi = 3\pi/4$, and so on. As above, with aligned polarizers, the complementary figures are observed.

The uniaxial birefringent laminae, with the optical axis parallel to the faces, are the most commonly used in optical systems; therefore, when using convergent radiation, this fringe structure must be remembered. On the other hand, with collimated radiation, we can vary the phase displacement introduced by the lamina, tilting and rotating it appropriately.

7.8.2 Interference with a biaxial crystal

For a biaxial crystal, Eq. (7.8.6) is replaced by the Biot's law

$$\frac{1}{n'^2}-\frac{1}{n''^2}=\left(\frac{1}{{n_1}^2}-\frac{1}{{n_3}^2}\right)\sin\theta_1\sin\theta_2\,, \tag{7.8.10}$$

which is obtained from Eqs. (7.4.47) [Born and Wolf 1980]. Since the difference between the indices is small compared to their value, this expression can be approximated as

$$n''-n'=\left(n_3-n_1\right)\sin\theta_1\sin\theta_2 \tag{7.8.11}$$

and, from Eq. (7.8.4), the phase difference now turns out to be

$$\Delta\phi\cong\frac{2\pi}{\lambda_o}\rho\left(n_3-n_1\right)\sin\theta_1\sin\theta_2\,, \tag{7.8.12}$$

where $\rho = h/\cos\vartheta$ and ϑ is, as before, the average angle of the two paths with respect to the normal to the faces of the lamina. The isochromatic surfaces are now expressed by

$$\rho\sin\theta_1\sin\theta_2=C,\quad\text{with } C \text{ constant}\,. \tag{7.8.13}$$

When the propagation direction is close to one of the optical axes, for example,

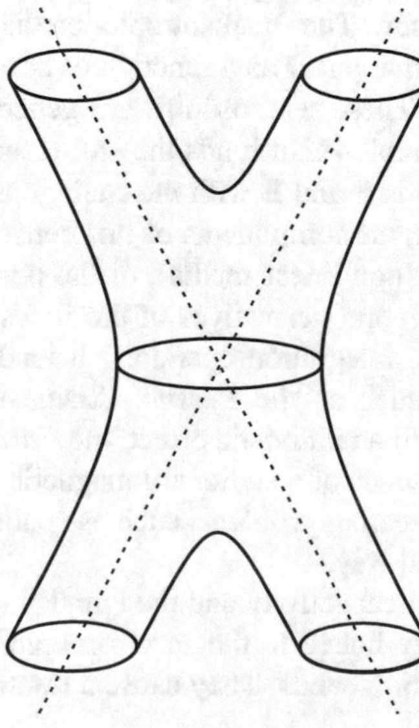

Fig. 7.21
Representation of an isochromatic surface for a biaxial crystal. It has the shape of two cylindrical surfaces, which are joined at the origin and whose axes correspond to the optical axes

when $\theta_1 \approx 0$, for which $\theta_2 \approx 2\beta$, Eq. (7.8.13) becomes

$$\rho \sin\theta_1 = \frac{C}{\sin 2\beta}, \tag{7.8.14}$$

but $\rho \sin\theta_1$ is the distance from the considered optical axis, so the isochromatic surfaces are cylindrical around this axis. The same happens when, vice versa, $\theta_2 \approx 0$, and the overall appearance is that of two cylinders that intersect each other as shown in Fig. 7.21.

As in the case of interference from uniaxial laminae, the intensity of the fringes is then modulated by the isogyre lines. For example, if the lamina is cut so that the principal z-axis is normal to the faces, and the polarizers are crossed, the isochromatic fringes appear to surround two "eyes" centered on the two optical axes and are interrupted by 4 dark bands passing through these eyes: see, for example, the *cover* of Born and Wolf (1980).

7.9 Bianisotropy

In electromagnetism, the materials are phenomenologically described by constitutive relations that connect the fields **E** and **B** with the fields **D** and **H**. According to this connection, the materials can be indicated as linear or non-linear; conductors or non-conductors; dispersive or non-dispersive; homogeneous or non-homogeneous; isotropic, anisotropic or *bianisotropic* [Kong 1972]. An isotropic medium is characterized by a permittivity and a permeability of both scalars, while, for an anisotropic medium, permittivity and/or permeability are represented by a tensor. The bianisotropic media are characterized by constitutive relationships that put **D** as a function of both **E** and **B**, and put **H** as a function of both **E** and **B**. These relationships are generally represented in a covariant form by a tensor of rank 4 that binds the 4x4 tensor of the electromagnetic field constituted by the fields **E** and **B** with the analogous tensor constituted by the fields **D** and **H**. In general, the components of this constitutive tensor can, in turn, be a function of the fields (non-linear media), of the position (inhomogeneous media), of the spatial or temporal derivatives of the fields (dispersive media) and also contain derivation and integration operators. It lends itself to explaining various optical phenomena, such as the *Fresnel-Fizeau effect* of dragging for moving media, by linking it to a relativistic effect, the *Faraday's effect* of rotation of the polarization in the presence of a stationary magnetic field, and the *optical activity*. Lastly, it allows for treating problems such as spatial and temporal inversions and *reciprocity* in a natural way.

The optical activity and the Faraday effect are "essentially dispersive" phenomena closely linked to the spectroscopic behavior of the medium, and therefore to its absorption bands. They cause a rotation of the polarization plane of an incident wave whose entity is dependent on the wave frequency and has a specific trend for each substance. This happens because of spectral regions where there is a peculiar form of different absorption for the two circular polarization components, which is called *circular dichroism* or *Cotton's effect*. Optical activity, in particular, is used as a spectroscopic investigation tool on organic molecules. In the following, we will limit ourselves to considering only the case of linear, homogeneous, and non-absorbent media.

As we will see shortly, the constitutive tensor can be decomposed into four 3x3 tensors. One is the dielectric tensor $\vec{\vec{\varepsilon}}$, which we have encountered above; one is the magnetic tensor $\left(\vec{\vec{\mu}}\right)^{-1}$; in addition, we have the "out diagonal" tensors $\vec{\vec{\gamma}}$ and $-\vec{\vec{\gamma}}^{\dagger}$, where the sign † indicates the transposed conjugate, whose imaginary part is responsible for the *optical activity* of the medium.

7.9.1 Optical activity

Some materials have the property of rotating the linear polarization of a wave that passes through them, and it is said that they possess *optical activity*. This is true for some crystalline media, such as quartz, when the wave propagates in the direction of the optical axis, and for unordered media, such as a solution of sugar in water. This effect has produced valuable information on the physics of radiation, the structure of molecules and the nature of life.

Optical activity[3] was noted for the first time in 1811 by Arago, and independently by Biot in 1812, in a quartz crystal, so, in particular conditions, the light transmitted and analyzed with a polarizer appeared variously colored. Biot showed that the generation of colors occurs because the polarization rotates around the direction of propagation, and this rotation depends on the color itself. Biot also discovered that quartz exists in two forms that produce opposite rotations. In fact, *looking towards the source* through an optically active medium, the rotation of the polarization may appear clockwise or counterclockwise; such a medium is therefore called *dextrorotary* or *levorotary*, respectively; this duplicity is typical of optically active inorganic crystals. Biot's main discovery was that optical activity is also present in gases and liquids, including the essences of turpentine and lemon and solutions of sugar and camphor. In particular, he discovered that the organic substances produced by living organisms always present themselves in the same form, levorotary or dextrorotary, according to their composition. For this reason, optical activity is often referred to as *natural*. Furthermore, the rotation appeared to be proportional to the thickness crossed and to the concentration of the optically active substance in the solution. This fact immediately found a wonderful application in quantitatively determining the sugar content in wine musts.

Fresnel gave an interpretation of the optical activity in terms of a different

---

[3] An historical account is given in Applequist (1987).

propagation velocity for the circular polarization components, and assumed that this difference was caused, for example, by a helical arrangement of the molecules of the optically active medium.

A few decades later, Louis Pasteur discovered that the optically active substances in solutions possess two molecular forms, two stereoisomers, each of which is a mirror image of the other, suggesting that helices and irregular tetrahedra were possible shapes. Each of these forms gives rise to rotation, in the opposite direction to that of the other; thus, if both are present in a solution in the same proportion, the optical activity is canceled out. In particular, he discovered that, when such a natural substance is artificially synthesized, both forms are obtained; therefore, living organisms must possess an ancestral property that produces only one of the two possibilities.

A non-crystalline medium is optically active only when it is *chiral*, that is, only when the substance of which it is composed is not congruent to its mirror image. In other words, it must *not* possess, among its operations of symmetry, either the *inversion*, or the *reflection*, or an axis of *rotation-inversion* or *rotation-reflection*. When this happens, the substance is called *enantiomorph*. This also applies, in particular, to molecules [Brand 1987]. A molecule is therefore called *chiral* when it is not superimposable to its mirror image. The stereoisomers that are each the mirror image of the other are called *enantiomers*. A complex molecule can have many distinct stereoisomers: when two of these are not specular to each other, they are called diastereomers. Chiral is synonymous with *dissymmetric*, while the term *asymmetric* means the lack of all symmetry operations, except identity. Therefore, an asymmetric molecule is also chiral, but a chiral molecule is not necessarily asymmetric, since it can possess, for example, an axis of rotation symmetry.

In liquids, optical activity is therefore associated with the shape of the molecules that compose them and manifests itself as a rotating power generated by a pure *circular birefringence*, in which the field eigenvectors are circularly polarized. In crystalline solids, optical activity may be due to the sole dissymmetric arrangement of atoms characteristic of the spatial group to which the crystal belongs. In fact, there are optically active soluble crystals, whose solution is not so. When ordinary birefringence is also present, the optical behavior of the medium is more complex and the eigenvectors for the field generally correspond to two elliptic polarizations whose ellipticity depends on the direction of propagation. Since, in the crystals, excluding those of the cubic classes, the linear birefringence is generally much stronger than the circular one, the optical activity manifests itself only as a small ellipticity of the polarization eigenvectors, except near the optical axes. Therefore, the crystal shows a rotating power only near these axes, for which the linear birefringence is canceled.

In the previous sections, we have assumed that the medium is anisotropic and also invariant under the translations. In this case, the elements of punctual symmetry give an adequate representation of a crystalline medium. However, as we saw in §7.1.6, some crystals have a more complex structure characterized by an arrangement of atoms such as to form a helix with a constant pitch around a screw axis. Therefore, in these cases, the spatial group to which this structure belongs must also be considered. In §7.9.8, we will take a detailed look at which point groups can have optical activity.

Optical activity is a particularly difficult phenomenon to explain, and its quantitative determination requires complex calculations of quantum mechanics on the structure of the medium [Mason 1968]. Also, at the classical level, various models have been proposed, including a helical motion of electrons along the molecules, and coupled oscillators in which each atom represents an oscillator, whose polarization influences the field of the others in a dissymmetric way, making the relationship between electric field and polarization *non-local*. The same also applies to magnetic field and magnetization.

The conductor solenoids have been taken as a model for the optical activity, and their behavior in response to microwaves has been studied experimentally by Tinoco [Tinoco and Freeman 1957]. This case is instructive for the way in which the magnitude of the effect, as the propagation direction changes, differs from what can be expected from an apparently reasonable discourse.

In the classic study of optical activity, the constitutive relations between the fields **D**, **H** and **E**, **B** must be reviewed. So far, we have limited ourselves to considering the sole polarization **P** by a *local* and *linear* relationship with the electric field. Usually, magnetization is neglected, since it produces generally small effects. However, in the case of an optically active medium, the cumulative effect of magnetization on relatively large propagation distances produces significant consequences. In the media in which the molecules are oriented, the electric quadrupole must also be considered [Nakano and Kimura 1969].

7.9.2 Constitutive equations

To interpret optical activity, at least four distinct constitutive relationships between the fields have been proposed in the literature, among which it is not easy to orientate oneself [Lakhtakia 1990; Lakhtakia et al 1988]. Fortunately, they substantially coincide when one considers the case of monochromatic waves, to which we can always return in Linear Optics. However, we shall see that the various constitutive equations possess definitions of permittivity and permeability that are different from each other.

After the unification of electromagnetism with optics, in 1892, Paul Drude proposed a model in which a single electron was constrained to move along a helix while it was elastically bound to a position of equilibrium on this curve [Condon 1937]. In 1900, he imagined that the polarization of the medium depended not only on **E**, but also on its rotor $\nabla\times\mathbf{E}$. This concept was revived in 1915 by Max Born, who criticized Drude's approach as *ad hoc*, seeking an interpretation based on a model with coupled oscillators arranged dissymetrically [Born 1915, 1935; Caldwell and Eyring 1971]. This model was well suited to the case of organic molecules in which carbon has the almost unique property of forming tetrahedral molecules, such as the CHFClBr dissymmetric molecule, which is optically active, and long molecular chains in which the structure of tetrahedra is present. In 1959, F.I. Fedorov reconsidered the constitutive equations of Drude-Born for the case of a crystalline medium [Fedorov 1959a, b; Bokut and Fedorov 1959]:

$$\begin{aligned} \mathbf{D} &= \vec{\vec{\varepsilon}}\mathbf{E} + \boldsymbol{\alpha}\nabla\times\mathbf{E}, \\ \mathbf{B} &= \vec{\vec{\mu}}\mathbf{H} + \boldsymbol{\beta}\nabla\times\mathbf{H}, \end{aligned} \tag{7.9.1}$$

where $\vec{\vec{\varepsilon}}$ and $\vec{\vec{\mu}}$ are symmetric tensors that respectively express the permittivity and the permeability of the medium, while $\boldsymbol{\alpha}$ and $\boldsymbol{\beta}$ are pseudo-tensors that express the gyrotropic effects. We note, in particular, that **D** also depends here on the spatial derivatives of the electric field, and it is therefore not only *locally* bound to **E**; the same is also true between **B** and **H**. Recalling the atomistic structure of matter, this fact derives from the influence of the charges near the point considered. In order that these effects do not cancel each other out, we need an asymmetry in the arrangement of atoms [Sommerfeld 1949].

Eqs. (7.9.1) are more symmetrical than those of Born, who completely neglected a contribution of anisotropy for the magnetic field; however, Fedorov did not impose any relation between $\boldsymbol{\alpha}$ and $\boldsymbol{\beta}$. This asymmetry in the equations for electric and magnetic fields has been challenged by Silverman, as it imposes a modification of the definition of a Poynting vector and of the boundary conditions, resulting in complicated expressions and different predictions for the reflection from an optically active medium [Silverman 1986]. Silverman brings this discrepancy back to the fact that Maxwell's equations are *a priori* invariant with respect to a transformation of duality between the fields [Jackson 1974, p. 252], while Eqs. (7.9.1) violate this invariance. These equations must therefore be replaced with

$$\begin{aligned} \mathbf{D} &= \vec{\vec{\varepsilon}}\left(\mathbf{E} + \mathbf{g}\nabla\times\mathbf{E}\right), \\ \mathbf{B} &= \vec{\vec{\mu}}\left(\mathbf{H} + \mathbf{g}^{\mathrm{T}}\nabla\times\mathbf{H}\right), \end{aligned} \tag{7.9.2}$$

where the gyrotropic effects are represented by $\mathbf{g}$, which is a pseudo-tensor of the second order and $\mathbf{g}^{\mathrm{T}}$ is its transpose.[4] The expression for energy conservation takes the form

$$\nabla\bullet\mathbf{S} + \partial u/\partial t + \mathbf{E}\bullet\mathbf{J} = 0\,, \tag{7.9.3}$$

with standard expressions for the Poynting vector and the energy density

[4] The experimental test to discriminate between the two theories consists in a difference in behavior in the reflection: it has been researched in vain, due to the smallness of the sought after effects [Silverman et al 1988; Silverman and Badoz 1990].

$$\mathbf{S}=\mathbf{E}\times\mathbf{H},\quad u=\mathbf{D}\cdot\vec{\vec{\varepsilon}}^{-1}\mathbf{D}+\mathbf{B}\cdot\vec{\vec{\mu}}^{-1}\mathbf{B}\,. \tag{7.9.4}$$

A second mechanism for explaining the optical activity is to consider a contribution to the polarization, coming from the variation of the magnetic field, and to the magnetization, from the variation of the electric field. In an ordinary birefringent medium, we have seen that, for a generic plane wave, the magnetization is parallel to the fields **B** and **H** and orthogonal to the fields **E** and **D**, while the polarization **P** is perpendicular to **B** and **H**.

In this situation, the motion of the charges is linear and the electric and magnetic dipoles are orthogonal to each other; therefore, **M** and **P** are perpendicular. Hence, in Eq. (7.9.1), $\nabla\times\mathbf{M}$ is parallel to **P** and adds only a small contribution that can be incorporated into **P** without changing its direction. Instead, in the optically active media, according to the single oscillator model [Condon et al 1937], the induced currents are constrained to a curved path [Caldwell and Eyring 1971]. Thus, **M** has a component orthogonal to **B**, induced by the currents associated with the variation of **P**, and therefore of **E**, and, in turn, **P** has a component orthogonal to **E** produced by the polarization resulting from the currents induced by the variation of **B**.

A simplified treatment is obtained when the ordinary anisotropy is absent, as in the case of a sugar solution, whereby $\nabla\cdot\mathbf{E}=0$. Here, I follow, with some adaptation, that which was reported in the important review article of Condon of 1937. [Condon 1937]. On average, on all of the possible orientations that the molecules take, we can assume that

$$\begin{aligned}\mathbf{P}&=\chi_e\varepsilon_0\mathbf{E}-g_C\frac{\partial}{\partial t}\mathbf{H},\\ \mu_0\mathbf{M}&=\chi_m\mu_0\mathbf{H}+g_C\frac{\partial}{\partial t}\mathbf{E},\end{aligned} \tag{7.9.5}$$

where χ_e and χ_m are, respectively, the electric and magnetic susceptibility taken here to be scalars and g_C is a *pseudo-scalar*, since **P** and **E** change sign for a parity transformation while **M** and **H** do not. The terms in g_C give only a small contribution to polarization and magnetization. The average operation on all possible orientations also cancels out the contribution of the electric quadrupole [Nakano and Kimura 1969; Caldwell and Eyring 1971].

The fields **D** and **B** can still be defined by Eqs. (1.2.4-5) as

$$\begin{aligned}\mathbf{D}&=\varepsilon_0\mathbf{E}+\mathbf{P}=\varepsilon\mathbf{E}-g_C\frac{\partial}{\partial t}\mathbf{H},\\ \mathbf{B}&=\mu_0\left(\mathbf{H}+\mathbf{M}\right)=\mu\mathbf{H}+g_C\frac{\partial}{\partial t}\mathbf{E}.\end{aligned} \tag{7.9.6}$$

where $\varepsilon=\varepsilon_0(1+\chi_e)$, $\mu=\mu_0(1+\chi_m)$. Eqs. (7.9.2) can be transformed into Eqs. (7.9.6) with some parameter rescaling [Lakhtakia et al 1988]. Indeed, through the

Maxwell's equations for the rotors of the fields, Eqs. (7.9.2) become

$$\begin{aligned}\mathbf{D}+\vec{\vec{\varepsilon}}\,\mathfrak{g}\vec{\vec{\mu}}\,\mathfrak{g}^{\mathrm{T}}\frac{\partial^2}{\partial t^2}\mathbf{D}&=\vec{\vec{\varepsilon}}\mathbf{E}-\vec{\vec{\varepsilon}}\,\mathfrak{g}\vec{\vec{\mu}}\frac{\partial}{\partial t}\mathbf{H},\\ \mathbf{B}+\vec{\vec{\mu}}\,\mathfrak{g}^{\mathrm{T}}\vec{\vec{\varepsilon}}\,\mathfrak{g}\frac{\partial^2}{\partial t^2}\mathbf{B}&=\vec{\vec{\mu}}\mathbf{H}+\vec{\vec{\mu}}\,\mathfrak{g}^{\mathrm{T}}\vec{\vec{\varepsilon}}\frac{\partial}{\partial t}\mathbf{E},\end{aligned}\tag{7.9.7}$$

and, for an isotropic medium and a monochromatic wave with an angular frequency ω, Eq. (7.9.7) coincides with Eq. (7.9.6), taking that

$$\vec{\vec{\varepsilon}}\to\varepsilon_{\mathrm{L}},\vec{\vec{\mu}}\to\mu_{\mathrm{L}},\vec{\vec{\varepsilon}}\,\mathfrak{g}\vec{\vec{\mu}}\to\beta,h=\left(1-\omega^2\varepsilon_{\mathrm{L}}^{-1}\mu_{\mathrm{L}}^{-1}\beta^2\right)^{-1},\varepsilon_{\mathrm{L}}h=\varepsilon,\mu_{\mathrm{L}}h=\mu,\beta h=g_{\mathrm{C}}.\tag{7.9.8}$$

In terms of field **B**, from Eqs. (7.9.5-6), we have that

$$\begin{aligned}\mathbf{P}&=\chi_e\varepsilon_o\mathbf{E}-\frac{g_{\mathrm{C}}}{\mu}\frac{\partial}{\partial t}\mathbf{B}+\frac{g_{\mathrm{C}}^2}{\mu}\frac{\partial^2}{\partial t^2}\mathbf{E}, &\quad \mathbf{D}&=\varepsilon\mathbf{E}-\frac{g_{\mathrm{C}}}{\mu}\frac{\partial}{\partial t}\mathbf{B}+\frac{g_{\mathrm{C}}^2}{\mu}\frac{\partial^2}{\partial t^2}\mathbf{E},\\ \mathbf{M}&=\frac{\chi_m}{\mu}\mathbf{B}+\frac{g_{\mathrm{C}}}{\mu}\frac{\partial}{\partial t}\mathbf{E}, &\quad \mathbf{H}&=\frac{1}{\mu}\mathbf{B}-\frac{g_{\mathrm{C}}}{\mu}\frac{\partial}{\partial t}\mathbf{E}.\end{aligned}\tag{7.9.9}$$

A further impulse to the theory of optical activity was given by the proposal of B.D.H. Tellegen of a non-reciprocal device, the *gyrator*, to be added to the classical linear elements of an electric network, such as resistors, capacitors, inductors, and transformers, [Tellegen 1948]. To this end, he considered an isotropic medium whose constitutive equations were

$$\begin{aligned}\mathbf{D}&=\varepsilon\mathbf{E}-\zeta\mathbf{H},\\ \mathbf{B}&=\mu\mathbf{H}+\zeta\mathbf{E}.\end{aligned}\tag{7.9.10}$$

For a harmonic field, these equations are equivalent to those of Condon, taking that $\zeta=i\omega g/\mathfrak{c}$.

A noteworthy set of equations, which resemble those of Tellegen, but are given for a generic bianisotropic medium, has been presented by E.J. Post, who deduced them from the observation that the constitutive equations must generally be invariant in form under the Lorentz transformations of the fields [Post 1962]. This means that the constitutive relationship must be expressed as

$$\mathfrak{G}^{\eta\iota}=\frac{1}{2}\mathfrak{K}^{\eta\iota\kappa\lambda}\mathfrak{F}_{\kappa\lambda},\tag{7.9.11}$$

where $\mathfrak{F}$ is the tensor of the electromagnetic field constituted by the fields **E** and **B** defined by Eq. (1.1.35) and $\mathfrak{G}$ is the tensor that brings together the fields **D** and **H**, defined by Eq. (1.2.8), while $\mathfrak{K}$ is a rank 4 tensor that describes the properties of the medium. $\mathfrak{K}$ can contain differential and integral operators as its elements if the medium is dispersive and depends on the coordinates if the medium is inhomogeneous, or on the fields if the medium is non-linear [Kong 1972].

In terms of $\mathfrak{F}$ and $\mathfrak{G}$, the Maxwell's equations become the Minkowski's Eqs. (1.2.10). The fraction is added to compensate for the double summation on λ and η. Since $\mathfrak{F}$ and $\mathfrak{G}$ are anti-symmetric tensors, the number of independent components of $\mathfrak{K}$ is reduced by the rules of symmetry to which it must underlie for the pairs η, ι and κ, λ, i.e.,

$$\begin{aligned} \mathfrak{K}^{\eta\iota\kappa\lambda} &= -\mathfrak{K}^{\eta\iota\lambda\kappa}, \\ \mathfrak{K}^{\eta\iota\kappa\lambda} &= -\mathfrak{K}^{\iota\eta\kappa\lambda}, \end{aligned} \tag{7.9.12}$$

to which, for media that are *without sources* and are *non-absorbent*, it is necessary to add

$$\mathfrak{K}^{\eta\iota\kappa\lambda} = \mathfrak{K}^{\kappa\lambda\eta\iota}, \tag{7.9.13}$$

which derives from energy conservation considerations. In all, the independent elements of $\mathfrak{K}$ are only 21 out of 256 of the tensor.

Wanting also to include the case of dispersive media, *in the absence of absorption*, passing to the complex representation of the fields in the space of the temporal frequencies, Eq. (7.9.13) is replaced by

$$\mathfrak{K}^{\eta\iota\kappa\lambda} = \mathfrak{K}^{\kappa\lambda\eta\iota *}. \tag{7.9.14}$$

To these relations, Post (1962) also adds the condition

$$\mathfrak{K}^{0123} + \mathfrak{K}^{0231} + \mathfrak{K}^{0312} + \mathfrak{K}^{2301} + \mathfrak{K}^{3102} + \mathfrak{K}^{1203} = 0, \tag{7.9.15}$$

which is known as the *Post constraint* [Lakhtakia 2004].

For a monochromatic wave (with angular frequency ω), the tensor $\mathfrak{K}$ can then be reduced to four 3x3 tensors in the relation between the amplitudes of the fields **E** and **B** and of the fields **D** and **H**, according to Table 7.4. Eq. (7.9.14) expresses the Hermitian symmetry of this matrix when the medium is non-dissipative. Therefore, in the complex representation of the fields, with $\mathbf{E} = \mathcal{Re}(\boldsymbol{E})$, $\boldsymbol{E} = \boldsymbol{E}_0(\boldsymbol{r})e^{-i\omega t}$ and the same for the other variables, the constitutive relations can be rewritten as [Kong 1972]

$$\begin{pmatrix} c\boldsymbol{D} \\ \boldsymbol{H} \end{pmatrix} = \mathfrak{C} \begin{pmatrix} \boldsymbol{E} \\ c\boldsymbol{B} \end{pmatrix}, \tag{7.9.16}$$

where the tensor $\mathfrak{C}$ is given by

$\mathfrak{K}^{\eta\iota\kappa\lambda}$		01 $-E_1$	02 $-E_2$	03 $-E_3$	23 cB_1	31 cB_2	12 cB_3
01	cD_1	$-\mathfrak{E}_{11}$	$-\mathfrak{E}_{12}$	$-\mathfrak{E}_{13}$	$\mathfrak{J}_{11}$	$\mathfrak{J}_{12}$	$\mathfrak{J}_{13}$
02	cD_2	$-\mathfrak{E}_{21}$	$-\mathfrak{E}_{22}$	$-\mathfrak{E}_{23}$	$\mathfrak{J}_{21}$	$\mathfrak{J}_{22}$	$\mathfrak{J}_{23}$
03	cD_3	$-\mathfrak{E}_{31}$	$-\mathfrak{E}_{32}$	$-\mathfrak{E}_{33}$	$\mathfrak{J}_{31}$	$\mathfrak{J}_{32}$	$\mathfrak{J}_{33}$
23	H_1	$-\mathfrak{L}_{11}$	$-\mathfrak{L}_{12}$	$-\mathfrak{L}_{13}$	$\mathfrak{M}_{11}$	$\mathfrak{M}_{12}$	$\mathfrak{M}_{13}$
31	H_2	$-\mathfrak{L}_{21}$	$-\mathfrak{L}_{22}$	$-\mathfrak{L}_{23}$	$\mathfrak{M}_{21}$	$\mathfrak{M}_{22}$	$\mathfrak{M}_{23}$
12	H_3	$-\mathfrak{L}_{31}$	$-\mathfrak{L}_{32}$	$-\mathfrak{L}_{33}$	$\mathfrak{M}_{31}$	$\mathfrak{M}_{32}$	$\mathfrak{M}_{33}$

Table 7.4 Decomposition of the contravariant tensor $\mathfrak{K}^{\eta\iota\kappa\lambda}$ into four tensors

$$\mathfrak{C} = \begin{pmatrix} \mathfrak{E} & \mathfrak{J} \\ \mathfrak{L} & \mathfrak{M} \end{pmatrix}, \tag{7.9.17}$$

where, from Eq. (7.9.14),

$$\mathfrak{E} = \mathfrak{E}^{\dagger}, \quad \mathfrak{M} = \mathfrak{M}^{\dagger}, \quad \mathfrak{L} = -\mathfrak{J}^{\dagger}. \tag{7.9.18}$$

$\mathfrak{E}$ and $\mathfrak{M}$ are therefore Hermitian tensors, while $\mathfrak{J}$ and $\mathfrak{L}$ are pseudo-tensors, since they connect the vector $\boldsymbol{E}$ with the pseudo-vector $\boldsymbol{H}$ and the pseudo-vector $\boldsymbol{B}$ with the vector $\boldsymbol{D}$. Therefore,

$$\begin{aligned} c\boldsymbol{D} &= \mathfrak{E}\boldsymbol{E} + c\mathfrak{J}\boldsymbol{B}, \\ \boldsymbol{H} &= -\mathfrak{J}^{\dagger}\boldsymbol{E} + c\mathfrak{M}\boldsymbol{B}. \end{aligned} \tag{7.9.19}$$

Finally, the condition of Post (7.9.15) here translates into

$$\Re e\left(\mathfrak{J}_{11} + \mathfrak{J}_{22} + \mathfrak{J}_{33}\right) = 0\,. \tag{7.9.20}$$

Eq. (7.9.16) can be converted into the form

$$\begin{pmatrix} \boldsymbol{D} \\ \boldsymbol{B} \end{pmatrix} = \mathfrak{C}_{\mathrm{EH}} \begin{pmatrix} \boldsymbol{E} \\ \boldsymbol{H} \end{pmatrix}, \tag{7.9.21}$$

where the tensor $\mathfrak{C}_{\mathrm{EH}}$ is given by

$$\mathfrak{C}_{\mathrm{EH}} = \begin{pmatrix} \vec{\vec{\varepsilon}} & \vec{\vec{\gamma}} \\ \vec{\vec{\gamma}}^{\dagger} & \vec{\vec{\mu}} \end{pmatrix} = \frac{1}{c} \begin{pmatrix} \mathfrak{E} + \mathfrak{J}\mathfrak{N}^{-1}\mathfrak{J}^{\dagger} & \mathfrak{J}\mathfrak{N}^{-1} \\ \mathfrak{N}^{-1}\mathfrak{J}^{\dagger} & \mathfrak{N}^{-1} \end{pmatrix}, \tag{7.9.22}$$

where $\vec{\vec{\varepsilon}}$ and $\vec{\vec{\mu}}$ are, respectively, the permittivity and permeability tensors introduced above. We should note that the permittivity is defined differently in the two representations of the constitutive equations. This transformation also keeps the tensors along the diagonal Hermitian.

The real part of $\mathfrak{E}$ and $\mathfrak{N}$, and of $\vec{\vec{\varepsilon}}$ and $\vec{\vec{\mu}}$, is associated with the birefringence, and their imaginary part with the Faraday effect. Instead, the real part of $\mathfrak{J}$ and $\vec{\vec{\gamma}}$ represents the Fresnel-Fizeau effect and the imaginary one the optical activity.

7.9.3 Maxwell's equations

For monochromatic waves and in regions of space without sources, it is possible to treat a bianisotropic medium as an equivalent anisotropic medium [Cheng and Kong 1968], in which the permittivity and permeability tensors also contain differential operators of the spatial coordinates, bringing back, in particular, the equations of Post and Kong to Eqs. (7.9.2) for optically active media.

In the absence of charges and currents, for a monochromatic wave, with the substitution $\partial/\partial t \to -i\omega$, the macroscopic Maxwell's Eqs. (1.2.6) become

$$\begin{cases} \nabla \cdot \boldsymbol{D} = 0, & \nabla \cdot \boldsymbol{B} = 0, \\ \nabla \times \boldsymbol{E} = i\omega \boldsymbol{B}, & \nabla \times \boldsymbol{H} = -i\omega \boldsymbol{D}, \end{cases} \tag{7.9.23}$$

and, applying the constitutive relations (7.9.19), we have that

$$\begin{cases} \nabla \cdot (\mathfrak{E}\boldsymbol{E} + c\mathfrak{J}\boldsymbol{B}) = 0, \\ \nabla \cdot \boldsymbol{B} = 0, \\ \nabla \times \boldsymbol{E} = i\omega \boldsymbol{B}, \\ c\nabla \times \left(-\mathfrak{J}^{\dagger}\boldsymbol{E} + c\mathfrak{N}\boldsymbol{B} \right) = -i\omega \left(\mathfrak{E}\boldsymbol{E} + c\mathfrak{J}\boldsymbol{B} \right). \end{cases} \tag{7.9.24}$$

On the other hand, by solving the equations (7.9.19) for the fields $\boldsymbol{E}$ and $\boldsymbol{H}$, we have that

$$\begin{aligned} \boldsymbol{E} &= c\mathfrak{E}^{-1}\boldsymbol{D} - c\mathfrak{E}^{-1}\mathfrak{J}\boldsymbol{B}, \\ \boldsymbol{H} &= -c\mathfrak{J}^{\dagger}\mathfrak{E}^{-1}\boldsymbol{D} + c\mathfrak{J}^{\dagger}\mathfrak{E}^{-1}\mathfrak{J}\boldsymbol{B} + c\mathfrak{M}\boldsymbol{B}. \end{aligned} \tag{7.9.25}$$

Using the third and fourth Maxwell Eqs. (7.9.23), we can rewrite them as

$$\boldsymbol{D} = \frac{1}{c}\mathfrak{E}\boldsymbol{E} - \frac{i}{\omega}\mathfrak{J}\nabla \times \boldsymbol{E}, \quad \boldsymbol{H} + \frac{i}{\omega}c\mathfrak{J}^{\dagger}\mathfrak{E}^{-1}\nabla \times \boldsymbol{H} = +c\mathfrak{J}^{\dagger}\mathfrak{E}^{-1}\mathfrak{J}\boldsymbol{B} + c\mathfrak{M}\boldsymbol{B}. \tag{7.9.26}$$

Therefore, we can bring these equations to the form of those of an anisotropic medium

$$\boldsymbol{D} = \vec{\vec{\varepsilon}}'\boldsymbol{E}, \quad \boldsymbol{B} = \vec{\vec{\mu}}'\boldsymbol{H}, \tag{7.9.27}$$

where

$$\vec{\vec{\varepsilon}}' = \frac{1}{c}\mathfrak{E}\left(\mathfrak{I} - i\frac{c}{\omega}\mathfrak{E}^{-1}\mathfrak{J}\vec{\vec{\nabla}}\right), \quad \vec{\vec{\mu}}' = \frac{1}{c}\left(\mathfrak{M} + \mathfrak{J}^{\dagger}\mathfrak{E}\mathfrak{J}\right)^{-1}\left(\mathfrak{I} + i\frac{c}{\omega}\mathfrak{J}^{\dagger}\mathfrak{E}^{-1}\vec{\vec{\nabla}}\right), \tag{7.9.28}$$

where, again, $\vec{\vec{\nabla}}$ is the anti-symmetric curl operator such that, for each vector $\boldsymbol{A}$, it is $\vec{\vec{\nabla}}\boldsymbol{A} = \nabla \times \boldsymbol{A}$. For the pseudo-tensor $\mathfrak{g}$ of Eq. (7.9.2), we therefore have the correspondence

$$\mathfrak{g} = -i\frac{c}{\omega}\mathfrak{E}^{-1}\mathfrak{J}. \tag{7.9.29}$$

If, instead, we apply Eqs. (7.9.25) to Eqs. (7.9.23), we obtain the Maxwell equations expressed in the fields $\boldsymbol{D}$ and $\boldsymbol{B}$ alone:

$$\begin{cases} \nabla \cdot \boldsymbol{D} & = 0, \\ \nabla \cdot \boldsymbol{B} & = 0, \\ \nabla \times \left(c\mathfrak{E}^{-1}\boldsymbol{D} - c\mathfrak{E}^{-1}\mathfrak{J}\boldsymbol{B}\right) & = i\omega\boldsymbol{B}, \\ \nabla \times \left(-c\mathfrak{J}^{\dagger}\mathfrak{E}^{-1}\boldsymbol{D} + c\mathfrak{J}^{\dagger}\mathfrak{E}^{-1}\mathfrak{J}\boldsymbol{B} + c\mathfrak{M}\boldsymbol{B}\right) & = -i\omega\boldsymbol{D}. \end{cases} \tag{7.9.30}$$

These equations are useful in dealing with plane waves, as the amplitudes of these fields remain perpendicular to the direction of the wave vector.

There is yet another way to write the Maxwell equations, defining new fields $\boldsymbol{D}$ and $\boldsymbol{H}$. If we take that

$$\begin{aligned} \boldsymbol{D}' &= \boldsymbol{D} + i\frac{1}{\omega}\nabla \times \mathfrak{J}^{\dagger}\boldsymbol{E}, \\ \boldsymbol{H}' &= \boldsymbol{H} + \mathfrak{J}^{\dagger}\boldsymbol{E}, \end{aligned} \tag{7.9.31}$$

Eqs. (7.9.23) are still formally the same, with $\boldsymbol{D}'$ and $\boldsymbol{H}'$ replacing $\boldsymbol{D}$ and $\boldsymbol{H}$, since, in the first Maxwell equation, the divergence of the curl added to $\boldsymbol{D}$ is zero, while, in the fourth equation, there is only a displacement of a term from the first to the second member. Using the third Maxwell's equation as well, we have that

$$\boldsymbol{D} = \overleftrightarrow{\varepsilon}''\boldsymbol{E}, \quad \boldsymbol{B} = \overleftrightarrow{\mu}''\boldsymbol{H}, \tag{7.9.32}$$

where now

$$\overleftrightarrow{\varepsilon}'' = \frac{1}{c}\mathfrak{E} + \frac{i}{\omega}\left(\overleftrightarrow{\nabla}\mathfrak{J}^{\dagger} - \mathfrak{J}\overleftrightarrow{\nabla}\right), \quad \overleftrightarrow{\mu}'' = \frac{1}{c}\mathfrak{M}^{-1}. \tag{7.9.33}$$

which constitute a different definition of permittivity and susceptibility. We will continue the discussion on this point in §7.9.8.

7.9.4 Symmetries and reciprocity

7.9.4.1 Spatial inversion

There are two approaches that are both useful for understanding *spatial inversion*. The first is to consider the physical system unchanged and to invert the spatial coordinates, while the second consists in considering a system that is the mirror image of the original one (plus a 180° rotation around the reflection axis, i.e., normal to the mirror). Let us indicate, with $\mathcal{P}$, the spatial inversion operator such that

$$\mathcal{P}\boldsymbol{r} = -\boldsymbol{r}, \quad \mathcal{P}\nabla = -\nabla. \tag{7.9.34}$$

To maintain the invariance of the Maxwell's equations under the spatial inversion, it is necessary that the fields be transformed in the following way:

$$\begin{aligned} &\mathcal{P}\mathbf{E}(\boldsymbol{r}) = -\mathbf{E}(-\boldsymbol{r}), \quad \mathcal{P}\mathbf{D}(\boldsymbol{r}) = -\mathbf{D}(-\boldsymbol{r}),\\ &\mathcal{P}\mathbf{B}(\boldsymbol{r}) = \mathbf{B}(-\boldsymbol{r}), \quad \mathcal{P}\mathbf{H}(\boldsymbol{r}) = \mathbf{H}(-\boldsymbol{r}). \end{aligned} \tag{7.9.35}$$

From Eq. (7.9.16), we can find the transformation properties for the representative tensors of the medium

$$\begin{aligned} &\mathcal{P}\mathfrak{E}(\boldsymbol{r}) = \mathfrak{E}(-\boldsymbol{r}),\\ &\mathcal{P}\mathfrak{M}(\boldsymbol{r}) = \mathfrak{M}(-\boldsymbol{r}),\\ &\mathcal{P}\mathfrak{J}(\boldsymbol{r}) = -\mathfrak{J}(-\boldsymbol{r}). \end{aligned} \tag{7.9.36}$$

7.9.4.2 Time reversal

Also, for the temporal inversion, we can consider two points of view, that of inverting the temporal coordinate, like watching a film that runs backwards in time, or considering a physical system in which the speeds and the rotation directions are reversed. The invariance of Maxwell's equations requires that

$$\begin{aligned} \mathcal{T}\mathbf{E}(t) &= \mathbf{E}(-t), \quad \mathcal{T}\mathbf{D}(t) = \mathbf{D}(-t), \\ \mathcal{T}\mathbf{B}(t) &= -\mathbf{B}(-t), \quad \mathcal{T}\mathbf{H}(t) = -\mathbf{H}(-t), \end{aligned} \tag{7.9.37}$$

where $\mathcal{T}$ represents the time reversal operator. For the representative tensors of the medium, we have that

$$\begin{aligned} \mathcal{T}\boldsymbol{\mathfrak{E}}(t) &= \boldsymbol{\mathfrak{E}}(-t), \\ \mathcal{T}\boldsymbol{\mathfrak{M}}(t) &= \boldsymbol{\mathfrak{M}}(-t), \\ \mathcal{T}\boldsymbol{\mathfrak{J}}(t) &= -\boldsymbol{\mathfrak{J}}(-t). \end{aligned} \tag{7.9.38}$$

In the complex representation of the fields, for a monochromatic plane wave, associated with the transformation $t \to -t$, we must also consider that $\boldsymbol{k} \to -\boldsymbol{k}$. For example, for a wave expressed by

$$\mathbf{E}(t,\boldsymbol{r}) = \frac{1}{2}\boldsymbol{E}_0 e^{i\boldsymbol{k}\cdot\boldsymbol{r}-i\omega t} + \frac{1}{2}\boldsymbol{E}_0^* e^{-i\boldsymbol{k}\cdot\boldsymbol{r}+i\omega t},$$

we have that

$$\mathcal{T}\mathbf{E}(t,\boldsymbol{r}) = \frac{1}{2}\boldsymbol{E}_0^* e^{i\boldsymbol{k}\cdot\boldsymbol{r}-i\omega t} + \frac{1}{2}\boldsymbol{E}_0 e^{-i\boldsymbol{k}\cdot\boldsymbol{r}+i\omega t},$$

where the first and second terms were exchanged between them, so that

$$\mathcal{T}\boldsymbol{E}_0 e^{i\boldsymbol{k}\cdot\boldsymbol{r}-i\omega t} = \boldsymbol{E}_0^* e^{i\boldsymbol{k}\cdot\boldsymbol{r}-i\omega t}. \tag{7.9.39}$$

For the constitutive tensors, Eqs. (7.9.38) are replaced by the following:

$$\begin{aligned} \mathcal{T}\boldsymbol{\mathfrak{E}}(t) &= \boldsymbol{\mathfrak{E}}^*(-t), \\ \mathcal{T}\boldsymbol{\mathfrak{M}}(t) &= \boldsymbol{\mathfrak{M}}^*(-t), \\ \mathcal{T}\boldsymbol{\mathfrak{J}}(t) &= -\boldsymbol{\mathfrak{J}}^*(-t). \end{aligned} \tag{7.9.40}$$

7.9.4.3 Reciprocity

By the concept of *reciprocity* [Potton 2004], we mean that, for two sources A and B (both at rest in a given inertial reference frame), the response of B to signals emitted by A is the same as the response of A to the signals emitted by B. For example, in the vacuum, antenna and receiver can exchange their roles; in Geometrical Optics, the rays that start from an object point O reach its image point I, and *vice versa*. Under ordinary conditions, the attenuation of a signal from A to B is the same as that from B to A, whether this occurs due to a progressive absorption of the medium or due to the diaphragms of the optical system. The interposed medium may, however, make the response *non-reciprocal*.

In electromagnetism, the *reciprocity theorem*, presented by Hendrik Lorentz in 1896, states that, under some conditions for the medium, the relationship between an oscillating current, placed in A, and the electric field generated by it, measured in B, does not change if the current is in B and the measure in A. When the currents are located in a finite volume, this theorem is expressed with the equation (see *www.wikipedia.org* under the entry *reciprocity*)

$$\int \boldsymbol{J}_1 \cdot \boldsymbol{E}_2 dV = \int \boldsymbol{J}_2 \cdot \boldsymbol{E}_1 dV\,, \tag{7.9.41}$$

where $\boldsymbol{E}_j$ is the electric field generated by the current $\boldsymbol{J}_j$, with $j = 1, 2$, oscillating harmonically at an angular frequency ω.

In order for the Lorentz reciprocity theorem to be valid, the medium must have certain conditions of symmetry; in other words, it is necessary that the constitutive tensors of Eq. (9.7.17) satisfy the relations

$$\mathfrak{E} = \mathfrak{E}^{\mathrm{T}}, \quad \mathfrak{M} = \mathfrak{M}^{\mathrm{T}}, \quad \mathfrak{L} = \mathfrak{J}^{\mathrm{T}}\,, \tag{7.9.42}$$

that result immediately in contrast to Eqs. (7.9.18). The comparison of these equations allows us to determine which contributions to the constitutive tensor, separated into real and imaginary parts, give reciprocal and non-reciprocal effects. Indeed, the behavior of the sign of these parts, which accompanies the transposition of the tensor according to Eqs. (7.9.18), can be in agreement or in disagreement

	reciprocal effects	non-reciprocal effects
$\mathfrak{E}$	$\mathcal{R}e(\mathfrak{E})$ *linear dielectric birefringence*	$\mathcal{I}m(\mathfrak{E})$ *Faraday's dielectric effect*
$\mathfrak{M}$	$\mathcal{R}e(\mathfrak{M})$ *linear magnetic birefringence*	$\mathcal{I}m(\mathfrak{M})$ *Faraday's magnetic effect*
$\mathfrak{J}, \mathfrak{L}$	$\mathcal{I}m(\mathfrak{J}) = -\mathcal{I}m(\mathfrak{L}^{\dagger})$ *optical activity*	$\mathcal{R}e(\mathfrak{J}) = -\mathcal{R}e(\mathfrak{L}^{\dagger})$ *Fresnel-Fizeau effect*

Table 7.5 Reciprocal and non-reciprocal contributions of the constitutive tensor

with the Eqs. (7.9.42). This is summarized in Table 7.5. Post notes, in particular, that those parts, real or imaginary, that change sign following the temporal inversion correspond to non-reciprocal effects.

7.9.5 Fresnel-Fizeau effect

When a dielectric isotropic medium is in motion, it becomes bianisotropic and "dragging" effects of radiation can be observed. In the past, these were associated with the drag effects of aether, but, when at least the motion is uniform, they are "easily" explained in relativistic terms, essentially applying the appropriate Lorentz transformation to the field calculated in the reference system in which the medium is at rest. On the other hand, it is instructive to see how the constitutive tensor $\mathfrak{C}$ is transformed for a medium in uniform motion. When the electromagnetic field is represented by a column vector, as in Eq. (7.9.16), with 6 components, the Lorentz's transformation from one inertial reference system to another in uniform relative motion along the z-axis with velocity V with respect to the first, is represented by the 6x6 matrix

$$\mathcal{L}_z = \begin{pmatrix} \gamma & 0 & 0 & 0 & -\gamma\beta & 0 \\ 0 & \gamma & 0 & \gamma\beta & 0 & 0 \\ 0 & 0 & 1 & 0 & 0 & 0 \\ 0 & \gamma\beta & 0 & \gamma & 0 & 0 \\ -\gamma\beta & 0 & 0 & 0 & \gamma & 0 \\ 0 & 0 & 0 & 0 & 0 & 1 \end{pmatrix}, \text{ with } \gamma = \frac{1}{\sqrt{1-\beta^2}} \text{ and } \beta = \frac{V}{c}, \quad (7.9.43)$$

which derives from Eqs. (1.1.37). The inverse transformation is obtained simply by exchanging β with $-\beta$. More generally, the transformation for a translation velocity $\boldsymbol{V}$ in an arbitrary direction, between two reference systems oriented so that $\boldsymbol{V}$ is represented in the two systems with the same components, can be calculated by applying a rotation matrix $\mathcal{R}$ for which $\mathcal{L} = \mathcal{R}^{\mathrm{T}}\mathcal{L}_z\mathcal{R}$, where

$$\mathcal{R} = \begin{pmatrix} \boldsymbol{r} & \mathbf{0} \\ \mathbf{0} & \boldsymbol{r} \end{pmatrix}, \quad (7.9.44)$$

where $\boldsymbol{r}$ is any matrix that rotates the vector $\boldsymbol{V}$ carrying it on the z-axis, for example,

$$\boldsymbol{r} = \begin{pmatrix} -v_x v_z / v_T & -v_y v_z / v_T & v_T \\ v_y / v_T & -v_x / v_T & 0 \\ v_x & v_y & v_z \end{pmatrix}, \quad \text{with } v_T = \sqrt{v_x{}^2 + v_y{}^2}, \quad (7.9.45)$$

where v_x, v_y, v_z are the components of the *versor* of $\boldsymbol{V}$.

The constitutive tensor of an isotropic medium in motion with velocity V along the z-axis can be obtained from the transformation $\mathfrak{C} = \mathscr{L}_z^{-1}\mathfrak{C}_0\mathscr{L}_z$, where $\mathfrak{C}_0$ is the constitutive tensor of the medium at rest. The succession of operators must be understood as follows: $\mathscr{L}_z$ transforms the fields from the initial reference system to that of the medium (in motion with speed V) in its rest system, while $\mathscr{L}_z^{-1}$ leads them back to the initial reference system. As a result, we have that

$$\mathfrak{C} = \gamma^2 \begin{pmatrix} c\varepsilon-\beta^2(c\mu)^{-1} & 0 & 0 & 0 & -\beta c\varepsilon+\beta(c\mu)^{-1} & 0 \\ 0 & c\varepsilon-\beta^2(c\mu)^{-1} & 0 & \beta c\varepsilon-\beta(c\mu)^{-1} & 0 & 0 \\ 0 & 0 & \gamma^{-2}c\varepsilon & 0 & 0 & 0 \\ 0 & -\beta c\varepsilon+\beta(c\mu)^{-1} & 0 & (c\mu)^{-1}-\beta^2 c\varepsilon & 0 & 0 \\ \beta c\varepsilon-\beta(c\mu)^{-1} & 0 & 0 & 0 & (c\mu)^{-1}-\beta^2 c\varepsilon & 0 \\ 0 & 0 & 0 & 0 & 0 & \gamma^{-2}(c\mu)^{-1} \end{pmatrix}, \tag{7.9.46}$$

where ε and μ are, respectively, the scalar permittivity and permeability of the medium at rest. For non-relativistic speeds, we can replace γ with 1 and neglect terms in β^2. With this approximation, for an isotropic moving medium with non-relativistic velocity $\boldsymbol{V}$ (generically oriented), the diagonal constitutive tensors of Eq. (7.9.17) can still be represented by scalar permittivity and permeability, while the pseudo-tensor $\mathfrak{J}$ is given by

$$\mathfrak{J} = \begin{pmatrix} 0 & -V_z\varepsilon\alpha & V_y\varepsilon\alpha \\ V_z\varepsilon\alpha & 0 & -V_x\varepsilon\alpha \\ -V_y\varepsilon\alpha & V_x\varepsilon\alpha & 0 \end{pmatrix}, \tag{7.9.47}$$

where $\alpha = 1-1/n^2$ is the *Fresnel drag coefficient*. This tensor is anti-symmetric, with $-\mathfrak{J}^\dagger=\mathfrak{J}$, as a consequence of the rotation symmetry of the medium around the direction of motion, and it expresses the fact that the radiation is dragged by the motion in proportion to α, as we shall see shortly.

7.9.6 Faraday's effect

One of the proofs of the existing connection between light radiation and electromagnetism was Faraday's discovery that a magnetic field applied to a dielectric medium is capable of rotating the polarization of a wave that passes through it. Indeed, an external magnetic field modifies the dielectric tensor $\mathfrak{E}$ or the magnetic tensor $\mathfrak{N}$ by means of an imaginary contribution; they therefore become Hermitian tensors. In the first case, the rotation of the polarization is called the *dielectric Faraday's effect*, while, in the second case, it is called the *magnetic Faraday's effect*. For example, when a magnetic field is applied to an isotropic dielectric medium along the *z*-axis, the tensors $\mathfrak{E}$ and $\mathfrak{N}$ take the form

$$\mathfrak{E} = c\begin{pmatrix} \varepsilon & i\varepsilon_I & 0 \\ -i\varepsilon_I & \varepsilon & 0 \\ 0 & 0 & \varepsilon_3 \end{pmatrix}, \quad \mathfrak{N} = \begin{pmatrix} \eta & i\eta_I & 0 \\ -i\eta_I & \eta & 0 \\ 0 & 0 & \eta_3 \end{pmatrix}. \tag{7.9.48}$$

The Faraday effect could be ascribed to the family of non-linear phenomena, since the coefficients of these tensors depend on the magnetic field; on the other hand, in ordinary applications, this field is constant over time and is applied "externally", and therefore "does not belong" to the wave taken into consideration. The peculiarity of the Faraday effect derives from the fact that the magnetic field that causes it is a pseudo-vector and, as we will see now, it behaves in a way that is clearly distinct from the optical activity for the counter-propagating waves.

7.9.7 *Phase velocity and rotatory power*

For a monochromatic plane wave with angular frequency ω and wave vector $\boldsymbol{k}$, the first two divergence Maxwell's equations (7.9.30) assure us that the fields $\boldsymbol{D}$ and $\boldsymbol{B}$ are orthogonal to $\boldsymbol{k}$, while the two equations for the rotors become

$$\begin{cases} c\boldsymbol{k}\times\left(\mathfrak{E}^{-1}\boldsymbol{D}_o - \mathfrak{E}^{-1}\mathfrak{J}\boldsymbol{B}_o\right) & = \omega\boldsymbol{B}_o, \\ c\boldsymbol{k}\times\left(-\mathfrak{J}^{\dagger}\mathfrak{E}^{-1}\boldsymbol{D}_o + \mathfrak{J}^{\dagger}\mathfrak{E}^{-1}\mathfrak{J}\boldsymbol{B}_o + \mathfrak{N}\boldsymbol{B}_o\right) & = -\omega\boldsymbol{D}_o, \end{cases} \tag{7.9.49}$$

where $\boldsymbol{D}_o$ and $\boldsymbol{B}_o$ are the amplitudes of the plane wave fields. By choosing a Cartesian reference such that $\boldsymbol{k}$ is oriented along the *z*-axis, these equations reduce to a homogeneous system of four linear equations with four unknowns, D_x, D_y, B_x, B_y, since D_z=0 and B_z=0. By requiring that the determinant be zero, we can find the solutions for the phase velocity of the wave. We consider here some simple special cases.

7.9.7.1 Fresnel-Fizeau effect

An *isotropic medium in uniform motion* with velocity V along the *z*-axis is represented by the matrix (7.9.46), so that, at non-relativistic velocity, the constitutive tensors are given by

$$\mathfrak{E} = c\varepsilon\mathfrak{I}, \quad \mathfrak{N} = \frac{1}{c\mu}\mathfrak{I}, \quad \mathfrak{J} = \begin{pmatrix} 0 & -V\varepsilon\alpha & 0 \\ V\varepsilon\alpha & 0 & 0 \\ 0 & 0 & 0 \end{pmatrix}, \tag{7.9.50}$$

where, again, $\alpha = 1 - 1/\left(c^2\varepsilon\mu\right)$ is the drag coefficient of Fresnel. For a wave with its wave vector also oriented along the z-axis, we have

$$\boldsymbol{k}\times = \begin{pmatrix} 0 & -k & 0 \\ k & 0 & 0 \\ 0 & 0 & 0 \end{pmatrix}. \tag{7.9.51}$$

Ignoring the terms in V^2, Eqs. (7.9.49) become

$$\begin{cases} 0 & -D_y & -\varepsilon(\upsilon - V\alpha)B_x & +0 & = 0, \\ D_x & +0 & +0 & -\varepsilon(\upsilon - V\alpha)B_y & = 0, \\ \mu(\upsilon - V\alpha)D_x & +0 & +0 & -B_y & = 0, \\ 0 & +\mu(\upsilon - V\alpha)D_y & +B_x & +0 & = 0, \end{cases} \tag{7.9.52}$$

where $\upsilon = \omega/k$ is the phase velocity. The non-trivial solutions of this system are given by the values of υ that cancel out the determinant:

$$\varepsilon^2\mu^2(\upsilon - V\alpha)^4 - 2\varepsilon\mu(\upsilon - V\alpha)^2 + 1,$$

from which we immediately find that

$$\upsilon = V\alpha \pm \frac{1}{\sqrt{\varepsilon\mu}}, \tag{7.9.53}$$

whose solutions are counted twice, since, in this case, they are degenerate with respect to the polarization. The sign ± indicates the two directions of travel of the wave in the medium. The phase velocity is therefore greater in absolute value when the wave propagates in the same direction as the medium; therefore, between two fixed positions along the z-axis, the phase displacement suffered in crossing it is less than when the wave propagates in the opposite direction.

7.9.7.2 Optical activity

For an *optically active isotropic medium*, the constitutive tensors are diagonal and are represented by

$$\mathfrak{E} = c\varepsilon\left(1-g^2\frac{\mu}{\varepsilon}\right)\mathfrak{I},\quad \mathfrak{N} = \frac{1}{c\mu}\mathfrak{I},\quad \mathfrak{J} = ig\,, \tag{7.9.54}$$

where g is a pseudo-scalar. In the first expression, the definition of permittivity of Eq. (7.9.27) was used. The correspondences with the tensor $\mathfrak{g}$ of Eq. (7.9.29) and with the pseudo-scalar g_C of Eq. (7.9.5) are, respectively,

$$\mathfrak{g} \to \frac{g}{\omega\varepsilon\left(1-g^2\,\mu/\varepsilon\right)},\quad g_C = \frac{g\mu}{\omega}\,.$$

Eqs. (7.9.49) here become

$$\begin{cases} \boldsymbol{k}\times\left(\boldsymbol{D}_\text{o} - ig\boldsymbol{B}_\text{o}\right) - \omega\varepsilon\, p\boldsymbol{B}_\text{o} = 0, \\ \boldsymbol{k}\times\left[ig\left(\mu/\varepsilon\right)\boldsymbol{D}_\text{o} + \boldsymbol{B}_\text{o}\right] + \omega\mu\, p\boldsymbol{D}_\text{o} = 0, \end{cases} \tag{7.9.55}$$

where $p = 1 - g^2\,\mu/\varepsilon$. More explicitly,

$$\begin{cases} 0 & -D_y & -\varepsilon\upsilon pB_x & +igB_y & =0, \\ D_x & +0 & -igB_x & -\varepsilon\upsilon pB_y & =0, \\ \mu\upsilon pD_x & -ig\left(\mu/\varepsilon\right)D_y & +0 & -B_y & =0, \\ ig\left(\mu/\varepsilon\right)D_x & +\mu\upsilon pD_y & +B_x & +0 & =0, \end{cases} \tag{7.9.56}$$

where υ is still the phase velocity. The determinant is now

$$\varepsilon^2\mu^2p^4\upsilon^4 - 2\varepsilon\mu\, p^2\upsilon^2\left(1+g^2\,\mu/\varepsilon\right) + p^2\,. \tag{7.9.57}$$

By setting it to 0, we find the solutions

$$\upsilon_{rs} = r\frac{1}{\sqrt{\varepsilon\mu}\left(1+sg\sqrt{\mu/\varepsilon}\right)}, \tag{7.9.58}$$

where $r = +1, -1$ is the sign of the direction of the wave, and $s = +1, -1$ is the sign that, applied to g, indicates the symmetry breaking between two polarization autovectors. Resolving the second two Eqs. (7.9.56) for B_x and B_y, and replacing the result in the first two, we find that

$$2ig\mu\upsilon pD_x + \left[\varepsilon\mu\left(\upsilon p\right)^2 - p\right]D_y = 0\,. \tag{7.9.59}$$

Then, replacing the solutions of υ in Eq. (7.9.56), we find the corresponding polarization eigenvectors

$$D_y^{(rs)} = irsD_x^{(rs)}, \quad \text{with } r = +1, -1, \quad s = +1, -1\,. \tag{7.9.60}$$

Here, the apex (rs) identifies the eigenvector; therefore, the various solutions for the phase velocity each have a polarization eigenvector, respectively,

$$\begin{aligned} \upsilon_{++} &\rightarrow \frac{1}{\sqrt{2}}\begin{pmatrix}1\\ i\end{pmatrix} = \mathbf{L}, \quad \upsilon_{+-} \rightarrow \frac{1}{\sqrt{2}}\begin{pmatrix}1\\ -i\end{pmatrix} = \mathbf{R},\\ \upsilon_{-+} &\rightarrow \frac{1}{\sqrt{2}}\begin{pmatrix}1\\ -i\end{pmatrix} = \mathbf{R}, \quad \upsilon_{--} \rightarrow \frac{1}{\sqrt{2}}\begin{pmatrix}1\\ i\end{pmatrix} = \mathbf{L}, \end{aligned} \tag{7.9.61}$$

where $\mathbf{L}$ and $\mathbf{R}$ are the versors already defined in Eqs. (1.6.5). As expected, the eigenvectors represent the two opposite circular polarizations. On the other hand, by solving Eqs. (7.9.56) for field $\boldsymbol{B}$ as well, we have that

$$B_x^{(rs)} = -r\sqrt{\frac{\mu}{\varepsilon}}D_y^{(rs)}, \; B_y^{(rs)} = r\sqrt{\frac{\mu}{\varepsilon}}D_x^{(rs)}, \tag{7.9.62}$$

and therefore we still have that $\boldsymbol{B}$ is orthogonal to $\boldsymbol{D}$. We note immediately that, for the same index s, and therefore the same absolute value of the phase velocity, the forward and backward waves exchange the versors $\mathbf{L}$ and $\mathbf{R}$.[5] Therefore, in the standard notation of the polarization, for the optical activity, we have that the indices of refraction of the waves $\mathbf{L}$ and $\mathbf{R}$ are

$$\begin{aligned} n_{\mathbf{L}} &= n_{++} = n_{-+} = c\sqrt{\varepsilon\mu} + gc\mu\,,\\ n_{\mathbf{R}} &= n_{+-} = n_{--} = c\sqrt{\varepsilon\mu} - gc\mu\,, \end{aligned} \tag{7.9.63}$$

where $n_{rs} = c/|\upsilon_{rs}|$.

7.9.7.3 Faraday's effect

For an *isotropic medium* immersed in a static magnetic field, of amplitude B, oriented along the z-axis, the constitutive tensors are represented by

---

[5] Much like a clock seen from behind appears to proceed counterclockwise.

$$\mathfrak{E} = \mathrm{c}\begin{pmatrix} \varepsilon & i\varepsilon_{\mathrm{I}} & 0 \\ -i\varepsilon_{\mathrm{I}} & \varepsilon & 0 \\ 0 & 0 & \varepsilon_3 \end{pmatrix}, \quad \mathfrak{N} = \frac{1}{\mathrm{c}\mu}\mathfrak{I}, \quad \mathfrak{J} = \mathfrak{O}, \tag{7.9.64}$$

for the case of the dielectric Faraday's effect, and from

$$\mathfrak{E} = \mathrm{c}\varepsilon\mathfrak{I}, \quad \mathfrak{N} = \frac{1}{\mathrm{c}}\begin{pmatrix} \eta & i\eta_{\mathrm{I}} & 0 \\ -i\eta_{\mathrm{I}} & \eta & 0 \\ 0 & 0 & \eta_3 \end{pmatrix}, \quad \mathfrak{J} = \mathfrak{O}, \tag{7.9.65}$$

for the case of the magnetic Faraday's effect.

In the first case, Eqs. (7.9.49) become

$$\begin{cases} D_x & +0 & +i\varepsilon_{\mathrm{I}}\upsilon B_x & -\varepsilon\upsilon B_y & = 0, \\ 0 & +D_y & +\varepsilon\upsilon B_x & +i\varepsilon_{\mathrm{I}}\upsilon B_y & = 0, \\ \mu\upsilon D_x & +0 & +0 & -B_y & = 0, \\ 0 & \mu\upsilon D_y & +B_x & +0 & = 0, \end{cases} \tag{7.9.66}$$

whose determinant is

$$\mu^2\upsilon^4\left(\varepsilon^2 - \varepsilon_{\mathrm{I}}{}^2\right) - 2\varepsilon\mu\upsilon^2 + 1, \tag{7.9.67}$$

which is canceled for

$$\upsilon_{rs} = r\sqrt{\frac{1}{\mu\left(\varepsilon - s\varepsilon_{\mathrm{I}}\right)}}, \tag{7.9.68}$$

where, again, $r = +1, -1$ is the sign of the travel direction of the wave, and $s = +1, -1$ is the sign that, applied to ε_{I}, indicates the symmetry breaking between two polarization eigenvectors. The corresponding polarization eigenvectors are

$$D_y = isD_x, \quad B_x = -r\sqrt{\frac{\mu}{\varepsilon - s\varepsilon_{\mathrm{I}}}}D_y, \quad B_y = r\sqrt{\frac{\mu}{\varepsilon - s\varepsilon_{\mathrm{I}}}}D_x, \tag{7.9.69}$$

with $r = +1, -1, \quad s = +1, -1,$

with the correspondences

$$\upsilon_{++} = -\upsilon_{-+} \to \mathbf{L}, \quad \upsilon_{+-} = -\upsilon_{--} \to \mathbf{R}. \tag{7.9.70}$$

In the case of the magnetic Faraday's effect, we instead have

$$\begin{cases} 0 & -D_y & -\upsilon\varepsilon B_x & 0 & =0, \\ D_x & +0 & +0 & -\upsilon\varepsilon B_y & =0, \\ \upsilon D_x & +0 & +i\eta_I B_x & -\eta B_y & =0, \\ +0 & +\upsilon D_y & +\eta B_x & +i\eta_I B_y & =0, \end{cases} \tag{7.9.71}$$

whose determinant is

$$\varepsilon^2\upsilon^4 - 2\varepsilon\eta\upsilon^2 + \eta^2 - \eta_I{}^2, \tag{7.9.72}$$

which is canceled for

$$\upsilon_{rs} = r\sqrt{\frac{\eta + s\eta_I}{\varepsilon}}. \tag{7.9.73}$$

The corresponding polarization eigenvectors are now

$$D_y^{(rs)} = -isD_x^{(rs)}, \quad B_x^{(rs)} = -\frac{r}{\sqrt{\varepsilon(\eta+s\eta_I)}}D_y^{(rs)}, \quad B_y^{(rs)} = \frac{r}{\sqrt{\varepsilon(\eta+s\eta_I)}}D_x^{(rs)},$$
$$\text{with } r = +1, -1, \quad s = +1, -1, \tag{7.9.74}$$

with the correspondences

$$\upsilon_{++} = -\upsilon_{-+} \to \mathbf{R}, \quad \upsilon_{+-} = -\upsilon_{--} \to \mathbf{L}. \tag{7.9.75}$$

From these equations, we can deduce the essential difference between optical activity and the Faraday effect. Indeed, in the second case, for the polarized waves **L** or **R**, the phase velocity remains the same in module regardless of the travel direction of the wave, while, for the optical activity, the phase velocities exchange these versors. In other words, in the standard notation of the polarization, for the Faraday effect, the phase velocity of the circular forward wave **L** (**R**) is the same as the circular backward wave **R** (**L**), while, for the optical activity, the phase velocity for the forward wave **L** (**R**) is the same as the backward wave **L** (**R**).

7.9.7.4 Rotatory power

An incident wave that propagates along the *z*-axis, linearly polarized and with amplitude E_o, can be decomposed according to Eq. (1.6.8) in its circular components with amplitude

$$E_L(0)=\frac{1}{\sqrt{2}}E_o e^{-i\alpha}, \quad E_R(0)=\frac{1}{\sqrt{2}}E_o e^{+i\alpha}, \tag{7.9.76}$$

where α is the angle that the direction of the electric field has with the x-axis, measured counterclockwise looking from positive to negative values of z. After having traversed a distance d in a *Faraday's medium*, the two amplitudes become

$$\begin{aligned} E_L(d) &= \frac{1}{\sqrt{2}}E_o e^{-i\alpha+i\omega d/\upsilon_L} = \frac{1}{\sqrt{2}}E_o e^{-i\alpha+in_L(\omega/c)|d|}, \\ E_R(d) &= \frac{1}{\sqrt{2}}E_o e^{+i\alpha+i\omega d/\upsilon_R} = \frac{1}{\sqrt{2}}E_o e^{-i\alpha+in_R(\omega/c)|d|}, \end{aligned} \tag{7.9.77}$$

where υ_L and υ_R are the phase velocities associated, respectively, with the versors **L** and **R**, and the corresponding indices of refraction are $n_L = c/|\upsilon_L|$ and $n_R = c/|\upsilon_R|$. The second equality follows from the fact that, in changing its direction of travel, d also changes sign. By expressing the field in the base of the linear polarizations, it is found that the angle of oscillation of the electric field, with respect to the x-axis, is equal to half of the difference between the phases of the two terms of Eqs. (7.9.77) [Huard 1997]. Therefore, the wave has undergone a rotation of its polarization plane equal to

$$\varphi = \frac{\pi}{\lambda_o}(n_R - n_L)|d|. \tag{7.9.78}$$

In other words, the rotation angle φ, defined as being done for the angle α above, does not depend on the direction of travel.

Instead, in the case of *optical activity*, applying the relations (7.9.61) and (7.9.63) and proceeding in the calculation of the rotation in a similar manner to what was done above, but by distinguishing between the two directions of travel, the rotations φ_+ and φ_- of the polarization plane for a wave that propagates forward or backward, respectively, along the z-axis are

$$\varphi_+ = -\omega\mu g|d|, \quad \varphi_- = \omega\mu g|d|, \tag{7.9.79}$$

where, again, the angles φ are measured with the same rule mentioned for the initial angle α. Formally, in both cases, the rotation angle is

$$\varphi = -\omega\mu gd \tag{7.9.80}$$

Crystals	λ (nm)	β (degree/mm)
Quartz	650	17
	589	21.7
	400	49
Cinnabar (HgS)	589	32.5
$NaClO_3$	589	3.1
$AgGaS_2$	505	430
	495	600
	485	950
Selenium	1000	30
	750	180
Tellurium	10000	40
	6000	15
Liquids		β (degree/cm)
Turpentine ($C_{10}H_6$)	589	−3.7
Aqueous solutions		β (degree/cm) @1g/cc
Dextrose (d-glucose)	589	+5.25
Levulose (l-glucose)	589	−5.14
Lactose	589	+5.24
Maltose	589	+13.85
Sucrose	589	+6.64

Tab. 7.6 Specific rotation of some optically active substances at room temperature. The solutions are indicated for a concentration of 1 g of solute per cm^3

and depends on the sign of d. Therefore, if the wave is reflected back from an ordinary mirror, it takes back the original oscillation direction. An effect observed by Fresnel should be noted here: if the incident radiation is not monochromatic, the different spectral components generally undergo a different rotation due to the dispersion, i.e., to the dependence of φ on ω. If, with a initially polarized *non-monochromatic* wave, the medium traversed is sufficiently long, the transmitted radiation may appear to be completely depolarized when analyzed by a polarizer. However, the radiation reflected back from the mirror returns to its original polarization state.

In conclusion, optical activity is a *reciprocal* effect.

The specific rotatory power of an optically active substance is usually expressed in degrees by

$$\beta = \frac{180°}{\lambda_o}\left(n_{\mathbf{R}} - n_{\mathbf{L}}\right), \tag{7.9.81}$$

for which the rotation is given by $\varphi = \beta d$ (see Table 7.6). Here, the two indices of

Substance	λ (nm)	V [degree/(kOested·cm)]
water	589	0.218
acetone	589	0.185
fluorite	589	0.015
diamond	589	0.20
sodium chloride	589	0.598
carbon disulfide (CS_2)	589	0.705
sphalerite (ZnS)	589	3.75
fused silica (SiO_2)	633	0.21(c)
	589	0.277
Schott glass BK-7(c)	633	0.23
Schott glass SF-57(c)	633	1.15
Schott glass SFS-6(a)	1300	0.30
	800	0.85
	600	1.72
	366	8.17
Kigre glass M-32	1064	1.67
	532	5.65
Hoya glass FR-5(b)	633	4.18
terbium Gallium garnet (TGG)(b)	694	5.88
	633	7.67
	514	12.5
	468	16.7

Table 7.7 Verdet constant of some substances at room temperature expressed in Gaussian units (Eq. (7.9.83)). In SI, these values must be multiplied by $4\pi 10^{-4}$ to obtain V in units of degree/[(A/m)m]. Assuming $\mu = \mu_0$, with $B = \mu_0 H$ and $\varphi = V'B|d|$, the same values of the table apply to $V' = V/\mu_0$ in units of degree/(kGauss·cm). (a) Robinson (1964). (b) Villaverde et al (1978). (c) Williams et al (1991)

refraction are assigned to the helicity **R** or **L** of the wave, and their difference is generally very small, for example, for quartz, with $\lambda_o = 589$ nm, it is 7.1×10^{-5}. Nevertheless, the rotation produced is in proportion to d/λ_o and even reaches considerable values; the great sensitivity that is obtained by measuring it is comparable to the interferential methods and is the basis of the *Polarization Spectroscopy* with which the structure of optically active substances is studied.

In the case of the dielectric Faraday effect, the rotations of the polarization plane of the waves forward and backward, at the first order in $\varepsilon_I/\varepsilon$, are both given by

$$\varphi = \omega\sqrt{\mu\varepsilon}\,\frac{\varepsilon_I}{\varepsilon}|d|\,, \tag{7.9.82}$$

while, for the magnetic Faraday effect, at the first order in η_I/η, we have that

$$\varphi = -\omega\sqrt{\frac{\varepsilon}{\eta}\frac{\eta_I}{\eta}}|d|. \tag{7.9.83}$$

In both cases, the rotation is independent of the sign of d: if the wave is reflected back from an ordinary mirror, the rotation of the polarization plane doubles, and therefore the Faraday effect is *non-reciprocal*.

The magnitude of the Faraday effect is proportional to the distance traveled and the applied magnetic field, if this is relatively weak. This proportionality is expressed by the equation

$$\varphi = VH|d|, \tag{7.9.84}$$

where V is the *Verdet's constant* that is also called *rotary power*. For ferromagnetic materials, rotation depends not on $\boldsymbol{H}$ but on magnetization, which tends to become saturated by strong magnetic fields, on the order of Tesla, and consequently also the rotation stops. The Faraday effect is essentially dispersive, and therefore the Verdet constant varies a lot with frequency and changes in sign corresponding to the electromagnetic resonances of the material.

The value of the Verdet constant is generally very small (see Table 7.7), and fields in the order of kGauss are required to have rotations on the order of degrees/cm. Very large rotations can be obtained with thin films of ferromagnetic materials such as iron, nickel or cobalt. For example, with an iron layer with a thickness of 0.1 μm and a field of 10000 Gauss, a rotation of about 2° is obtained for $\lambda_o = 589$ nm.

7.9.8 Optically active crystals

For a non-absorbent and stationary medium, i.e., one not in motion in the considered inertial system, the matrix of the pseudo-tensor $\mathfrak{J}$ has purely imaginary components. If the medium is crystalline, its symmetry properties determine which components can be non-zero, and thus generate optical activity. In particular, for crystals belonging to classes with a center of symmetry, all elements of $\mathfrak{J}$ are zero. On the other hand, other operations of punctual symmetry, i.e., the inversion, a rotation-inversion axis, a reflection plane, change the parity of a structure, and all of the punctual groups with these symmetries are non-enantiomorphic. That leaves 11 enantiomorph point groups: C_1 (*1*), C_2 (*2*), C_2 (*222*), C_4 (*4*), D_4 (*422*), C_3 (*3*), D_3 (*32*), C_6 (*6*), D_6 (*622*), T (*23*) and O (*432*), indicated by an asterisk in Table 7.1. Crystals that have a structure from these point groups can belong to a space group containing a screw axis with a unique helicity, and therefore *can* have optical activity.

However, in crystals belonging to four non-enantiomorphic classes, C_{1h} (*m*), C_{2v} (*2mm*), S_4 (*4*) and D_{2d} (*42m*), optical activity is possible even if they have a

$$C_1: \begin{pmatrix} \mathfrak{J}_{11} & \mathfrak{J}_{12} & \mathfrak{J}_{13} \\ \mathfrak{J}_{21} & \mathfrak{J}_{22} & \mathfrak{J}_{23} \\ \mathfrak{J}_{31} & \mathfrak{J}_{32} & \mathfrak{J}_{33} \end{pmatrix} \quad C_{1h}: \begin{pmatrix} 0 & 0 & \mathfrak{J}_{13} \\ 0 & 0 & \mathfrak{J}_{23} \\ \mathfrak{J}_{31} & \mathfrak{J}_{32} & 0 \end{pmatrix} \quad C_2: \begin{pmatrix} \mathfrak{J}_{11} & \mathfrak{J}_{12} & 0 \\ \mathfrak{J}_{21} & \mathfrak{J}_{22} & 0 \\ 0 & 0 & \mathfrak{J}_{33} \end{pmatrix}$$

$$D_2: \begin{pmatrix} \mathfrak{J}_{11} & 0 & 0 \\ 0 & \mathfrak{J}_{22} & 0 \\ 0 & 0 & \mathfrak{J}_{33} \end{pmatrix} \quad C_{2v}: \begin{pmatrix} 0 & \mathfrak{J}_{12} & 0 \\ \mathfrak{J}_{21} & 0 & 0 \\ 0 & 0 & 0 \end{pmatrix} \quad D_3, D_4, D_6: \begin{pmatrix} \mathfrak{J}_{11} & 0 & 0 \\ 0 & \mathfrak{J}_{11} & 0 \\ 0 & 0 & \mathfrak{J}_{33} \end{pmatrix}$$

$$C_{3v}, C_{4v}, C_{6v}: \begin{pmatrix} 0 & \mathfrak{J}_{12} & 0 \\ -\mathfrak{J}_{12} & 0 & 0 \\ 0 & 0 & 0 \end{pmatrix} \quad C_3, C_4, C_6: \begin{pmatrix} \mathfrak{J}_{11} & \mathfrak{J}_{12} & 0 \\ -\mathfrak{J}_{12} & \mathfrak{J}_{11} & 0 \\ 0 & 0 & \mathfrak{J}_{33} \end{pmatrix} \quad D_{2d}: \begin{pmatrix} \mathfrak{J}_{11} & 0 & 0 \\ 0 & -\mathfrak{J}_{11} & 0 \\ 0 & 0 & 0 \end{pmatrix}$$

$$S_4: \begin{pmatrix} \mathfrak{J}_{11} & \mathfrak{J}_{12} & 0 \\ \mathfrak{J}_{12} & -\mathfrak{J}_{11} & 0 \\ 0 & 0 & \mathfrak{J}_{33} \end{pmatrix} \quad T, O: \begin{pmatrix} \mathfrak{J}_{11} & 0 & 0 \\ 0 & \mathfrak{J}_{11} & 0 \\ 0 & 0 & \mathfrak{J}_{11} \end{pmatrix}$$

Tab. 7.8 Non-zero elements of the pseudo-tensor $\mathfrak{J}$ for the various crystalline classes referring to the principal axes [Post 1990]

symmetry plane or a rotation-inversion axis. In this case, due to the presence of a reflection plane, the substance has two screw axes with opposite helicity. These axes are not necessarily parallel, so that a circular birefringence of opposite sign is observed along their directions. Post (1990) and Fedorov (1959a, b) identify 3 other classes, C_{3v}, C_{4v}, and C_{6v}, for which the tensor $\mathfrak{J}$ can have non-zero components. Altogether, we have, for $\mathfrak{J}$, the 18 classes from Table 7.8.

The crystals of the classes S_4, D_{2d}, C_{3v}, C_{4v}, and C_{6v} are uniaxial, but, along their optical axis, the circular birefringence is canceled, so there is no rotary effect.[6] This also happens for the biaxial crystals of the C_{2v} class, because the optical axes are on one of the symmetry planes. The class C_{1h} remains the only non-enantiomorphic for which rotary power can be observed along its two optical axes [Landau and Lifchitz VIII 1966, §83].

---

6 M.V. Hobden, in his paper *Optical activity in a non-enantiomorphous crystal: $AgGaS_2$*, [Hobden 1968], reports that silver tiogallate crystals belonging to the class D_{2d} are accidentally isotropic at 497.4 nm. Therefore, at this wavelength, they have a rotating power of 522°/ mm along the dyadic axes with rotation in the opposite direction.

On the other hand, Post also notes that the components indicated in these matrices are the only ones that may not be null by virtue of the symmetry of the crystal, but that there may be other physical reasons that some are canceled.

Propagation in an optically active crystalline medium, such as quartz, is more difficult to analyze than in the case of linear birefringence alone. From the Maxwell's Eqs. (7.9.24), for a monochromatic plane wave of angular frequency ω and wave vector $\boldsymbol{k}$, we obtain the wave equation

$$\frac{c^2}{\omega^2}\boldsymbol{k}\times\mathfrak{M}\boldsymbol{k}\times\boldsymbol{E}_o+\mathfrak{E}\boldsymbol{E}_o+\frac{c}{\omega}\mathfrak{J}\boldsymbol{k}\times\boldsymbol{E}_o-\frac{c}{\omega}\boldsymbol{k}\times\mathfrak{J}^{\dagger}\boldsymbol{E}_o=0\,. \tag{7.9.85}$$

Generally, in the crystals, ordinary birefringence prevails over the circular one, thus, neglecting the term in $\mathfrak{J}^{\dagger}\mathfrak{M}\mathfrak{J}$ in Eq. (7.9.22), we can assume that $\mathfrak{E}=c\vec{\vec{\varepsilon}}$. Moreover, the crystals generally used in optics are non-magnetic, so we can also assume that $\mathfrak{M}$ is a simple scalar. Therefore, the constitutive tensors are given by

$$\mathfrak{E}=c\begin{pmatrix}\varepsilon_1 & 0 & 0\\ 0 & \varepsilon_2 & 0\\ 0 & 0 & \varepsilon_3\end{pmatrix},\quad \mathfrak{M}=\frac{1}{c\mu}\mathfrak{I},\quad \mathfrak{J}=\begin{pmatrix}\mathfrak{J}_{11} & \mathfrak{J}_{12} & \mathfrak{J}_{13}\\ \mathfrak{J}_{21} & \mathfrak{J}_{22} & \mathfrak{J}_{23}\\ \mathfrak{J}_{31} & \mathfrak{J}_{32} & \mathfrak{J}_{33}\end{pmatrix}, \tag{7.9.86}$$

where $\mathfrak{J}$ is one of the pseudo-tensors represented in Table 7.8, and we can write the wave equation as

$$k^2\boldsymbol{E}_{\mathrm{o}}-(\boldsymbol{k}\cdot\boldsymbol{E}_{\mathrm{o}})\boldsymbol{k}-\mu\omega^2\left(\vec{\vec{\varepsilon}}+\frac{1}{\omega}\mathfrak{J}\boldsymbol{k}\times-\frac{1}{\omega}\boldsymbol{k}\times\mathfrak{J}^{\dagger}\right)\boldsymbol{E}_{\mathrm{o}}=0\,, \tag{7.9.87}$$

where the operator

$$\vec{\vec{\varepsilon}}''=\vec{\vec{\varepsilon}}+\frac{1}{\omega}\mathfrak{J}\boldsymbol{k}\times-\frac{1}{\omega}\boldsymbol{k}\times\mathfrak{J}^{\dagger} \tag{7.9.88}$$

replaces $\vec{\vec{\varepsilon}}$ in Eq. (7.4.8) for the sole linear birefringence. Here, the operator $\boldsymbol{k}\times$ is described in general form by the matrix

$$\boldsymbol{k}\times=\begin{pmatrix}0 & -k_z & k_y\\ k_z & 0 & -k_x\\ -k_y & k_x & 0\end{pmatrix}. \tag{7.9.89}$$

Given that, with the sole optical activity, the tensor $\mathfrak{J}$, expressed by one of the matrices of Table 7.8, is imaginary, we can write it as $\mathfrak{J}=iJ$, where J is a real tensor. Therefore, the operator $\boldsymbol{k}\times\mathfrak{J}^{\dagger}-\mathfrak{J}\boldsymbol{k}\times$ is

$$\mathfrak{J}\boldsymbol{k}\times-\boldsymbol{k}\times\mathfrak{J}^{\dagger}=\boldsymbol{J}\boldsymbol{k}\times+\boldsymbol{k}\times\boldsymbol{J}^{\mathrm{T}}=$$
$$i\begin{pmatrix} 0 & \begin{matrix}-k_z\left(J_{11}+J_{22}\right)\\ +k_xJ_{13}+k_yJ_{23}\end{matrix} & \begin{matrix}k_y\left(J_{11}+J_{33}\right)\\ -k_xJ_{12}-k_zJ_{32}\end{matrix} \\ \begin{matrix}k_z\left(J_{11}+J_{22}\right)\\ -k_yJ_{23}-k_xJ_{13}\end{matrix} & 0 & \begin{matrix}-k_x\left(J_{22}+J_{33}\right)\\ +k_yJ_{21}+k_zJ_{31}\end{matrix} \\ \begin{matrix}-k_y\left(J_{11}+J_{33}\right)\\ +k_zJ_{32}+k_xJ_{12}\end{matrix} & \begin{matrix}k_x\left(J_{22}+J_{33}\right)\\ -k_zJ_{31}-k_yJ_{21}\end{matrix} & 0 \end{pmatrix}, \tag{7.9.90}$$

which is an anti-symmetric tensor (non-null elements have been written on two lines for compactness). It can be rewritten in the form

$$\mathfrak{J}\boldsymbol{k}\times-\boldsymbol{k}\times\mathfrak{J}^{\dagger}=i\left(\boldsymbol{G}\boldsymbol{k}\right)\times, \tag{7.9.91}$$

where $\boldsymbol{G}$ is the gyrotropic tensor given by

$$\boldsymbol{G}=\mathfrak{I}Tr\left(\boldsymbol{J}\right)-\boldsymbol{J}^{\mathrm{T}} \tag{7.9.92}$$

and Eq. (7.9.88) becomes

$$\overleftrightarrow{\varepsilon}''=\overleftrightarrow{\varepsilon}+\frac{i}{\omega}\left(\boldsymbol{G}\boldsymbol{k}\right)\times, \tag{7.9.93}$$

reconciling Post's theory with that of Born, Landau and Lifchitz [Peterson 1975], for which the constitutive equations of optical activity were

$$\boldsymbol{D}=\overleftrightarrow{\varepsilon}\boldsymbol{E}+\frac{i}{\omega}\left(\boldsymbol{G}\boldsymbol{k}\right)\times\boldsymbol{E},\quad \boldsymbol{B}=\mu\boldsymbol{H}, \tag{7.9.94}$$

and the wave equation was, instead,

$$k^2\boldsymbol{E}_{\mathrm{o}}-\left(\boldsymbol{k}\bullet\boldsymbol{E}_{\mathrm{o}}\right)\boldsymbol{k}-\mu\omega^2\overleftrightarrow{\varepsilon}\boldsymbol{E}_{\mathrm{o}}-i\mu\omega\left(\boldsymbol{G}\boldsymbol{k}\right)\times\boldsymbol{E}_{\mathrm{o}}=0\,. \tag{7.9.95}$$

This reconciliation is obtained from the transformation (9.7.31), but, as we mentioned earlier, Silverman has made clear that it is not innocent with respect to the definitions of a Poynting vector and energy density.

It should be noted that, while the gyrotropic tensor depends only on the anisotropies of the medium, the tensor $\left(\boldsymbol{G}\boldsymbol{k}\right)\times$ depends explicitly on the wave vector. As we stated earlier, for a uniaxial crystal, the rotatory power manifests itself essentially only along the optical axis, i.e., along the axis of symmetry that coincides with the principal axis 3, alias z, of this matrix. Therefore, when the wave vector is directed along the optical axis, the matrix (7.9.89) is simplified into

$$\mathfrak{J}\boldsymbol{k}\times-\boldsymbol{k}\times\mathfrak{J}^{\dagger}=\begin{pmatrix}0 & -k_z\left(\mathfrak{J}_{11}+\mathfrak{J}_{22}\right) & -k_z\mathfrak{J}_{32}\\ k_z\left(\mathfrak{J}_{11}+\mathfrak{J}_{22}\right) & 0 & k_z\mathfrak{J}_{31}\\ k_z\mathfrak{J}_{32} & -k_z\mathfrak{J}_{31} & 0\end{pmatrix} \tag{7.9.96}$$

and, comparing this equation with the Post matrices from Table 7.8, we can justify the disappearance of the rotatory power for the classes C_{2v}, C_{3v}, C_{4v}, C_{6v}, D_{2d}, S_4.

7.9.8.1 Optical activity of Quartz

For the particular case of a crystal of the class D_3 such as quartz, we have that

$$\vec{\vec{\varepsilon}}''=\begin{pmatrix}\varepsilon & -i2g_3\frac{k_z}{\omega} & i2g_1\frac{k_y}{\omega}\\ i2g_3\frac{k_z}{\omega} & \varepsilon & -i2g_1\frac{k_x}{\omega}\\ -i2g_1\frac{k_y}{\omega} & i2g_1\frac{k_x}{\omega} & \varepsilon_3\end{pmatrix}. \tag{7.9.97}$$

where $g_1 = g_2$ and g_3 are the diagonal, real components of the gyrotropic tensor $\boldsymbol{G}$

$$i2g_1=\mathfrak{J}_{11}+\mathfrak{J}_{33},\quad ig_3=\mathfrak{J}_{11}. \tag{7.9.98}$$

The tensor of Eq. (7.9.97) is Hermitian and resembles the permittivity of the Faraday effect, inducing similar effects of circular birefringence. However, it must be remembered that the representation of the optical activity with this tensor is only a mathematical simplification of convenience, and not a physical coincidence. The wave equation (7.9.87) becomes

$$\begin{cases}\left(k^2-k_xk_x-\mu\omega^2\varepsilon\right)E_x+\left(-k_xk_y+2i\mu\omega g_3k_z\right)E_y+\left(-k_xk_z-2i\mu\omega g_1k_y\right)E_z=0,\\ \left(-k_yk_x-2i\mu\omega g_3k_z\right)E_x+\left(k^2-k_yk_y-\mu\omega^2\varepsilon\right)E_y+\left(-k_yk_z+2i\mu\omega g_1k_x\right)E_z=0,\\ \left(-k_zk_x+2i\mu\omega g_1k_y\right)E_x+\left(-k_zk_y-2i\mu\omega g_1k_x\right)E_y+\left(k^2-k_zk_z-\mu\omega^2\varepsilon_3\right)E_z=0.\end{cases}$$

By canceling its determinant, we obtain the Fresnel's equation

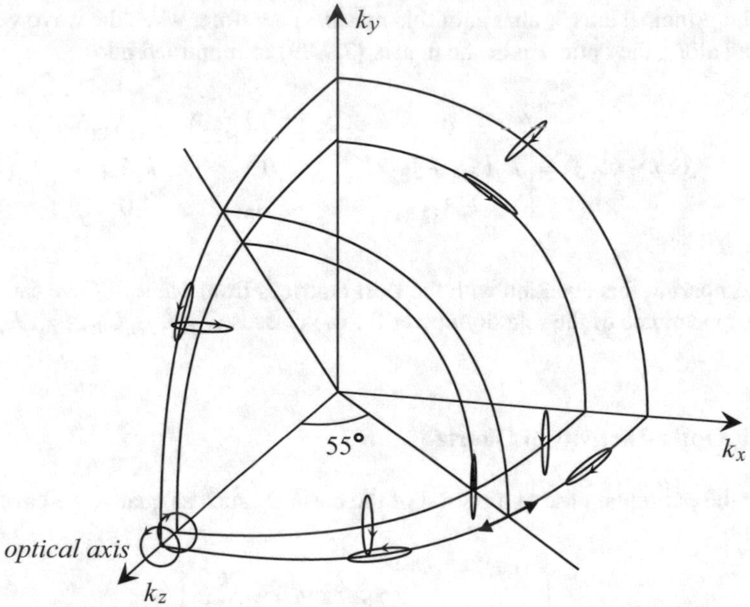

Fig. 7.22 Surface of wave vectors for quartz. The direction of rotation indicated with increasing time, for the eigenvectors of circular or elliptical polarization, is that of the dextrorotary crystals. It is inverted for the levorotary crystals. Birefringence, both circular and linear, is represented here amplified

$$k^4\left[\left(\varepsilon \mathsf{n}_\perp{}^2+\varepsilon_3\mathsf{n}_z{}^2\right)+4\mu\left(g_1-g_3\right)^2\mathsf{n}_\perp{}^2\mathsf{n}_z{}^2\right] \\ -\mu\omega^2k^2\left[\varepsilon^2\mathsf{n}_\perp{}^2+\varepsilon\varepsilon_3\left(1+\mathsf{n}_z{}^2\right)+4\varepsilon\mu g_1{}^2\mathsf{n}_\perp{}^2+4\varepsilon_3\mu g_3{}^2\mathsf{n}_z{}^2\right]+\mu^2\omega^4\varepsilon^2\varepsilon_3=0, \tag{7.9.99}$$

which defines the surface shape of wave vectors for quartz, where $\mathsf{n}_\perp{}^2 = \mathsf{n}_x{}^2+\mathsf{n}_y{}^2$ and n_x, n_y, n_z are the components of the versor of $\boldsymbol{k}$.

In various texts, g_1 is taken to be null. However, this assumption is not correct, as both experimental measures and theoretical calculations [Zhong et al 1993] show that $g_1 \approx -0.5\, g_3$, regardless of frequency, according to a symmetry argument [Jerphagnon and Chemla 1976]. It states that quartz consists of the union of SiO_4 tetrahedral groups; these tetrahedra locally possess a specular symmetry plane that cancels out the trace of the gyrotropic tensor $\boldsymbol{G}$, and that of $\mathfrak{J}$, too.

The optical activity eliminates the point of contact along the optical axis between the ellipsoidal surface of the extraordinary wave and the spherical one of the ordinary wave, removing the degeneration of the eigenvalues along that axis (Fig. 7.22). In the direction of the optical axis, the polarization eigenvectors are now circular. Along the z-axis, the solutions for n are

$$n^2 = \varepsilon\mu c^2 \pm 2c^2\mu g_3\sqrt{\varepsilon\mu}\,, \tag{7.9.100}$$

where the terms in $g_3{}^2$ have been neglected. Applying the binomial approximation to the root, the eigenvalues for the refractive index are

$$n_L \cong c\sqrt{\mu\varepsilon} + c\mu g_3, \quad n_R \cong c\sqrt{\mu\varepsilon} - c\mu g_3, \tag{7.9.101}$$

where we again find Eqs. (7.9.63). For quartz, at 589nm, $\Delta n \cong 3.36 \cdot 10^{-3}$. In the direction different from that of the optical axis, the eigenvectors are elliptic. However, ellipticity is reversed around 55° from this axis and has the opposite sign in a perpendicular direction. Indeed, taking that $g_1 = -0.5\, g_3$, $k_y = 0$, and looking for a solution for $E_y = 0$, it is found that, for the direction given by

$$\mathrm{n}_z{}^2 = \frac{\varepsilon}{2\varepsilon_3 + \varepsilon}, \quad \mathrm{n}_x{}^2 = \frac{2\varepsilon_3}{2\varepsilon_3 + \varepsilon}, \tag{7.9.102}$$

the two solutions to Eq. (7.9.99) for the index of refraction and the electric field are

$$n'^2 = \frac{1}{3} c^2 \mu (2\varepsilon_3 + \varepsilon), \quad E'_x = -\frac{\mathrm{n}_x}{2\mathrm{n}_z} E'_z, \quad E'_y = 0,$$

$$n''^2 = \frac{c^2 \mu\varepsilon (2\varepsilon_3 + \varepsilon)}{2\varepsilon_3 + \varepsilon + 24\mu g_1{}^2}, \quad E''_x = \frac{\mathrm{n}_x}{\mathrm{n}_z} E''_z = -4ig_1 \mathrm{n}_z \sqrt{\frac{\mu}{\varepsilon}} \sqrt{\frac{2\varepsilon_3 + \varepsilon}{2\varepsilon_3 + \varepsilon + 24\mu g_1{}^2}} E''_y, \tag{7.9.103}$$

while, in the direction orthogonal to the optical axis, say, along the x-axis, with $\mu g_1{}^2 << |\varepsilon_3 - \varepsilon|$, the solutions are

$$n'^2 = c^2 \mu\varepsilon \left\{ 1 - 4 \frac{\mu g_1{}^2}{\varepsilon_3 - \varepsilon} \right\}, \quad E'_z = -2i \frac{\sqrt{\varepsilon\mu} g_1}{\varepsilon_3 - \varepsilon} E'_y, \quad E'_x = 0, \tag{7.9.104.a}$$

$$n''^2 = c^2 \mu\varepsilon_3 \left\{ 1 + 4 \frac{\mu g_1{}^2}{\varepsilon_3 - \varepsilon} \right\}, \quad E''_y = -2i \frac{\sqrt{\varepsilon_3 \mu} g_1}{\varepsilon_3 - \varepsilon} E''_z, \quad E''_x = 0, \tag{7.9.104.b}$$

where, for quartz, at 589 nm, $2\sqrt{\varepsilon\mu}\,|g_1| / (\varepsilon_3 - \varepsilon) \cong 1 \cdot 10^{-3}$.

7.10 Form birefringence

Birefringence in crystals is due to their microscopic properties at the molecular level. However, it may also be present in mesoscopic structures of isotropic material, but with an anisotropic shape and dimensions much larger than the atomic ones, but, in any case, less than the wavelength. These structures can be constituted by particles of elongated dielectric material, aligned parallel to each other and with a refractive index different from that of the medium in which they are dispersed. This type of birefringence is referred to as *form birefringence* and is common in nature in refined structures of biological origin.[7] If the particles show absorption, we instead have dichroism, which is exploited to produce polarizers.

Modern techniques now allow for reproducing this situation artificially. One example is that of the columnar structures that normally form if appropriate precautions are not taken in the deposition of thin dielectric films. Vice versa, these structures may be desired to obtain various birefringent optical devices by deposition [Hodgkinson and Wu 1997]. The lithographic techniques employed in the realization of electronic circuits then allow for engraving the surface of a substrate and producing birefringent laminae for far infrared. Finally, in the realization of opto-electronic devices, such as detectors and semiconductor lasers, the optical anisotropy induced by their structure must be considered.

The form birefringence then has various applications in other fields of research. For example, it is exploited in biology to study viruses and bacteria, or in fluid dynamics. When particles of non-spherical shape are dispersed in a fluid, they are randomly oriented and there is no birefringence. However, in the presence of a velocity gradient, they tend to align themselves; observing the medium, for example, between two crossed polarizers, one can have an indication of the state of motion of the fluid around obstacles.

To illustrate the nature of form birefringence and the effect it has on a wave, let us consider the case of a medium consisting of a series of alternating parallel layers of two isotropic materials with different values ε_1 and ε_2 of electrical permittivity and the same magnetic permeability μ. We also assume that the thickness t_1 and t_2 of these layers is much smaller than the wavelength, but sufficiently large that the medium can still be treated with the macroscopic Maxwell's equations. For the boundary conditions on the interfaces, there is continuity of the component $\mathbf{E}_{\|}$ of the electric field parallel to the interfaces. For the corresponding component of the field $\mathbf{D}$ in the two media, we then have that [Born and Wolf 1980]

$$\mathbf{D}_{\|1} = \varepsilon_1 \mathbf{E}_{\|}, \quad \mathbf{D}_{\|2} = \varepsilon_2 \mathbf{E}_{\|} .$$

Therefore, if the thicknesses of the layers are small with respect to the wavelength, we can assume that the field is uniform inside of them. Furthermore, we can make the approximation that the waves reflected by the various interfaces destructively interfere, so that the wave propagates in the original direction without attenuation. Integrating on the volume, the mean value of the susceptibility for the tangential component of the field is then

---

[7] See, for example, Roberts et al (2009), *A biological quarter-wave retarder with excellent achromaticity in the visible wavelength region.*

$$\varepsilon_{\parallel} = \frac{\mathbf{D}_{\parallel}}{\mathbf{E}_{\parallel}} = \frac{t_1\varepsilon_1 + t_2\varepsilon_2}{t_1 + t_2} = f_1\varepsilon_1 + f_2\varepsilon_2 ,$$

where $f_1 = t_1/(t_1 + t_2)$, $f_2 = t_2/(t_1 + t_2)$. On the other hand, for the continuity of the component $\mathbf{D}_{\perp}$ perpendicular to the interfaces, we have that

$$\mathbf{E}_{\perp 1} = \frac{\mathbf{D}_{\perp}}{\varepsilon_1}, \quad \mathbf{E}_{\perp 2} = \frac{\mathbf{D}_{\perp}}{\varepsilon_2},$$

and therefore

$$\varepsilon_{\perp} = \frac{\mathbf{D}_{\perp}}{\mathbf{E}_{\perp}} = \frac{t_1 + t_2}{t_1/\varepsilon_1 + t_2/\varepsilon_2} = \frac{\varepsilon_1\varepsilon_2}{f_1\varepsilon_2 + f_2\varepsilon_1}.$$

These two susceptibilities are different, so that the set of layers behaves like a uniaxial medium whose optical axis is perpendicular to the interfaces with ordinary index $n_o = c\sqrt{\mu\varepsilon_{\parallel}}$ and extraordinary index $n_e = c\sqrt{\mu\varepsilon_{\perp}}$. Moreover, the birefringence of this medium is negative, since

$$n_e^2 - n_o^2 = -c^2\mu \frac{f_1 f_2 \left(\varepsilon_1 - \varepsilon_2\right)^2}{f_1\varepsilon_2 + f_2\varepsilon_1} \leq 0 .$$

In general, with other structures, the calculation becomes more complex and the medium is biaxial. An interesting case occurs with ellipsoidal particles (crystallites) aligned with susceptibility ε_c and dispersed in a medium with susceptibility ε_v. If f is the volume fraction of these particles, for $f << 1$, the main susceptibilities of the material turns out to be [Hodgkinson and Wu 1997; Wiener 1912; Bragg and Pippard 1953]

$$\varepsilon_j = \varepsilon_v + \frac{f\left(\varepsilon_c - \varepsilon_v\right)}{1 + \left(\varepsilon_c - \varepsilon_v\right)L_j/\varepsilon_v}, \quad j = 1, 2, 3 ,$$

where L_1, L_2, L_3 depend on the shape of the particles, with the property $L_1 + L_2 + L_3 = 1$. If the particles are cylindrical in shape, we can take that $L_1 = L_2 = 1/2$, $L_3 = 0$ and the medium is positive uniaxial, with the optical axis parallel to that of the cylinders.

7.11 Devices of manipulation and analysis of the polarization

The peculiar property of electromagnetic waves to possess a transverse nature doubles the amount of information that a wave can carry. This allows for countless applications, and practically all coherent optical apparatuses have at least one element based on polarization. Below, we will analyze the optical properties of the elementary devices that modify the polarization and their combinations. The main categories of these devices are: polarizers, phase delay laminae, and rotators (for optical activity or the Faraday effect). These components are, in most cases, made of birefringent materials, but also of dichroic materials and diffractive elements. In nature, there is a wide variety of materials with interesting optical properties, but the limitations due to transparency, mechanical properties, ease of processing and cost reduce the number of those used to a few units. Modern technology now makes it possible to create special materials, both as a composition and as a form, for specific applications, especially in fiber optic and planar guidance devices, but also deposited in the form of thin anisotropic films.

The devices that are used are mainly constituted by laminae with plane and parallel faces, so it is relatively simple and natural to treat the case of the propagation of plane waves, while, in principle, the propagation of various types of waves can be derived from their spectrum in plane waves.

7.11.1 Calculation of the transformation of pure polarization states by means of Jones vectors and matrices

There are various methods for calculating the propagation through a succession of elements that modify the state of polarization. The formalism introduced by R.C. Jones makes use of the homonymous vectors that we have encountered in §1.6 [Jones 1941a, b, 1942]; it adapts well to the case of completely polarized radiation and, in particular, of coherent radiation. Each element, polarizer, phase shifter, rotator, etc., is characterized by a 2x2 Jones' matrix $\mathfrak{M}$, which represents a linear operator on the state of polarization $\boldsymbol{J}$

$$\boldsymbol{J}' = \mathfrak{M}\boldsymbol{J} . \tag{7.11.1}$$

The coefficients of $\mathfrak{M}$ are assigned according to the base of orthonormal versors chosen for the Jones' vectors. This base is generally defined by two orthogonal linear polarizations, say, **X** and **Y**, but it can also be a base of circular polarizations, **L** and **R**, or a base consisting of two elliptic polarizations.

The coefficients of $\boldsymbol{J}$ change from a base $\mathfrak{r}$ to a base $\mathfrak{s}$ according to a unit transformation $\mathfrak{R}$

$$\boldsymbol{J}_{\mathfrak{s}} = \mathfrak{R}_{\mathfrak{s}}^{\mathfrak{r}} \boldsymbol{J}_{\mathfrak{r}} , \tag{7.11.2}$$

where the $\mathfrak{r}$ and $\mathfrak{s}$ labels indicate the direction of the transformation and the base [Lang 1987]. For example, between two ℓ and $\mathscr{L}$ bases of linear polarization such that the second is obtained by rotating the first of an angle φ, we have that

$$\mathfrak{R}^{\ell}_{\mathcal{L}} = \begin{pmatrix} \cos\varphi & \sin\varphi \\ -\sin\varphi & \cos\varphi \end{pmatrix}^{\ell}_{\mathcal{L}} . \quad (7.11.3)$$

Instead, to go from a base of linear polarization ℓ to a circular one c defined between them by the versors

$$\mathbf{P}_x = \begin{pmatrix} 1 \\ 0 \end{pmatrix}_{\ell}, \quad \mathbf{P}_y = \begin{pmatrix} 0 \\ 1 \end{pmatrix}_{\ell}, \quad \mathbf{L} = \frac{1}{\sqrt{2}}\begin{pmatrix} 1 \\ i \end{pmatrix}_{\ell}, \quad \mathbf{R} = \frac{1}{\sqrt{2}}\begin{pmatrix} 1 \\ -i \end{pmatrix}_{\ell}, \quad (7.11.4)$$

we apply the transformation (1.6.7) from ℓ to c

$$\mathfrak{R}^{\ell}_{c} = \frac{1}{\sqrt{2}}\begin{pmatrix} 1 & -i \\ 1 & i \end{pmatrix}^{\ell}_{c} . \quad (7.11.5)$$

The general rule, according to which one changes the elements of a matrix from a base r to a base s, is

$$\mathfrak{M}_{s} = \mathfrak{R}^{r}_{s}\mathfrak{M}_{r}\mathfrak{R}^{s}_{r}, \text{ where } \mathfrak{R}^{r}_{s} = \mathfrak{R}^{s\dagger}_{r}, \quad (7.11.6)$$

where the sign † indicates the transposed conjugate. In matrices that act between elements expressed in the same base, for the sake of brevity, the double index is contracted into one.

Consider a base $\mathcal{L}$ rotated by an angle φ with respect to the base ℓ and the respective circular bases c and $\mathcal{C}$ linked to the former by the transformation (7.11.5). The matrix of the transformation from c to $\mathcal{C}$ is obtained from the succession of transformations between the bases $c \to \ell \to \mathcal{L} \to \mathcal{C}$,

$$\mathfrak{R}_{\varphi}{}^{c}_{\mathcal{C}} = \frac{1}{2}\begin{pmatrix} 1 & -i \\ 1 & i \end{pmatrix}^{\mathcal{L}}_{\mathcal{C}} \begin{pmatrix} \cos\varphi & \sin\varphi \\ -\sin\varphi & \cos\varphi \end{pmatrix}^{\ell}_{\mathcal{L}} \begin{pmatrix} 1 & 1 \\ i & -i \end{pmatrix}^{c}_{\ell} = \begin{pmatrix} e^{i\varphi} & 0 \\ 0 & e^{-i\varphi} \end{pmatrix}^{c}_{\mathcal{C}} . \quad (7.11.7)$$

An elementary polarization device is identified by its eigenvalues a_1 and a_2 and polarization eigenvectors V_1 and V_2. For ordinary devices, these eigenvectors are orthogonal; therefore, in the chosen base r, their versors are

$$\mathbf{V}_1 = \begin{pmatrix} u \\ v \end{pmatrix}_{r}, \quad \mathbf{V}_2 = \begin{pmatrix} -v^* \\ u^* \end{pmatrix}_{r}, \quad V_1 \cdot V_2^* = 0, \quad \text{with } uu^* + vv^* = 1, \quad (7.11.8)$$

hence they form a new orthogonal base a. To convert the elements of the Jones vectors from the base a to the base r, one applies the unitary transformation

$$\mathfrak{R}_r^a = \begin{pmatrix} u & -v^* \\ v & u^* \end{pmatrix}_r^a . \tag{7.11.9}$$

On the base $\boldsymbol{a}$, the matrix of the device is given by

$$\mathfrak{M}_a = \begin{pmatrix} a_1 & 0 \\ 0 & a_2 \end{pmatrix}_a \tag{7.11.10}$$

and the corresponding matrix expressed in the base $\boldsymbol{r}$, for Eq. (7.11.6), turns out to be

$$\mathfrak{M}_r = \begin{pmatrix} a_1uu^* + a_2vv^* & (a_1 - a_2)uv^* \\ (a_1 - a_2)u^*v & a_1vv^* + a_2uu^* \end{pmatrix}_r . \tag{7.11.11}$$

The operator of a complex system, consisting of N elements, is represented by the product of the matrices of the individual elements

$$\mathfrak{M} = \mathfrak{M}_N \cdots \mathfrak{M}_2\mathfrak{M}_1 , \tag{7.11.12}$$

where it should be remembered that the order of this product is such that 1 is the first element encountered by the wave, 2 is the second and N is the last. All matrices must be expressed in the same base; otherwise, the transformation (7.11.6) must be applied to them.

When the eigenvalues of a matrix both have modulo 1, i.e., when there is neither attenuation nor gain, this matrix is unitary. This is the case of phase delay laminae and rotators, since they introduce only a phase difference between two orthogonal polarizations, without altering the intensity of the wave passing through them. These devices have the property of maintaining the orthogonality between the components of orthogonal polarization of the incident wave, while modifying both. Since the product of unitary matrices is still a unitary matrix, this property also extends to a succession of such elements. On the other hand, the matrices do not, in general, commute with each other, and thus, for a given incident wave, the effect that is obtained is usually different if the order of the elements is changed.

Instead, a polarizer operates a projection of the polarization of the incident wave along its axis.

The method of the matrices applies well to the case of uniform media. But, in the case of media whose properties change along the path, one needs to use a particular technique of integration, which makes use of matrices $\mathfrak{N}$ that define the properties of the medium for an infinitesimal path, so that [Huard 1997]

$$\mathfrak{M} = \exp \int_0^z \mathfrak{N} dz , \tag{7.11.13}$$

where the integral of a matrix is interpreted as the integral of its elements and the exponential of a matrix is interpreted as the polynomial development of $e^x = 1 + x + x^2/2 + x^3/6 + \cdots$, where x is replaced by the matrix

7.11.2 Calculation of the propagation of partially polarized radiation through the Stokes parameters and Mueller matrices

When the polarization state of radiation is not pure, but is partial or natural, Jones' vectors are not suitable to describe it. As we saw in §1.6.7, we must now treat polarization using the coherence matrix (1.6.15) or, equivalently, the Stokes parameters (1.6.17), which contain information related only to the intensity of the field, not to the amplitude.

The concept of pure and non-pure states is proper to Quantum Mechanics and, indeed, the manipulation of polarization is a modern research topic, for which the various optical devices that we deal with here, although they are macroscopic objects, are, for all purposes, quantum operators on the polarization state. Therefore, the mathematical tools that we are going to use have a close quantum correspondence.

Under certain conditions, it is possible to assimilate the Stokes parameters into a four-component vector and then calculate the transformation through an optical device by means of *Mueller's matrices*. These matrices depend only on how the device is constructed, and we can therefore determine their relationship with those of Jones.

Consider a monochromatic wave with a pure polarization state described by a Jones' vector

$$\boldsymbol{J} = \begin{pmatrix} a_x \\ a_y \end{pmatrix}, \tag{7.11.14}$$

where a_x and a_y are complex amplitudes. The corresponding coherency matrix is given by

$$\mathfrak{J} = \boldsymbol{J}\boldsymbol{J}^\dagger = \begin{pmatrix} a_x \\ a_y \end{pmatrix} \begin{pmatrix} a_x^* & a_y^* \end{pmatrix} = \begin{pmatrix} |a_x|^2 & a_x a_y^* \\ a_y a_x^* & |a_y|^2 \end{pmatrix}, \tag{7.11.15}$$

in which its *dyad* structure is made evident.[8] The sign † indicates the transposed conjugate. Since this matrix is Hermitian, it can be decomposed with real coefficients on a basis given by the Pauli's matrices [Huard 1997]

[8] If, in particular, we have chosen a normalized vector $\boldsymbol{J}$, this dyad can also be considered as a projection operator of any state on the polarization state described by $\boldsymbol{J}$.

$$\sigma_0 = \begin{pmatrix} 1 & 0 \\ 0 & 1 \end{pmatrix}, \quad \sigma_1 = \begin{pmatrix} 1 & 0 \\ 0 & -1 \end{pmatrix}, \quad \sigma_2 = \begin{pmatrix} 0 & 1 \\ 1 & 0 \end{pmatrix}, \quad \sigma_3 = \begin{pmatrix} 0 & -i \\ i & 0 \end{pmatrix}, \tag{7.11.16}$$

for which

$$\mathfrak{J} = \frac{1}{2}\left(S_0\sigma_0 + S_1\sigma_1 + S_2\sigma_2 + S_3\sigma_3\right), \tag{7.11.17}$$

where one can easily see, for comparison with Eqs. (1.6.17), that the coefficients S_i are proportional to the Stokes' parameters corresponding to the matrix $\mathfrak{J}$. On the other hand, the Jones' vector of a wave that passes through a device described by a matrix $\mathfrak{M}$ is transformed according to the rule $\boldsymbol{J}' = \mathfrak{M}\boldsymbol{J}$, and therefore its coherence matrix is transformed as

$$\mathfrak{J}' = \mathfrak{M}\boldsymbol{J}\boldsymbol{J}^\dagger\mathfrak{M}^\dagger = \mathfrak{M}\mathfrak{J}\mathfrak{M}^\dagger . \tag{7.11.18}$$

Developing the coherence matrices through Eq. (7.11.17) and multiplying by $\sigma_j\dagger$, we obtain the rule of transformation of the Stokes' parameters

$$S_j' = \sum_{j=0}^{3} M_{ij}\, S_i , \tag{7.11.19}$$

where M is the Mueller's matrix, which has 4x4 components given by

$$M_{ij} = \frac{1}{2}\mathrm{Tr}\left(\mathfrak{M}\sigma_i\mathfrak{M}^\dagger\sigma_j{}^\dagger\right), \tag{7.11.20}$$

where Tr indicates the trace. This result was obtained for a pure state of polarization; however, it also extends to the more general cases of partially polarized or natural radiation. Indeed, these situations can always be traced to an incoherent superposition of pure states for which the coherency matrix (7.11.15) is replaced by Eq. (1.6.15)

$$\mathfrak{J} = \begin{pmatrix} \langle \mathcal{E}_x\mathcal{E}_x{}^* \rangle & \langle \mathcal{E}_x\mathcal{E}_y{}^* \rangle \\ \langle \mathcal{E}_y\mathcal{E}_x{}^* \rangle & \langle \mathcal{E}_y\mathcal{E}_y{}^* \rangle \end{pmatrix}, \tag{7.11.21}$$

where $\mathcal{E}_x(t)$, $\mathcal{E}_y(t)$, indicate the analytical functions of the x and y components of the field and where the triangular brackets indicate the temporal average over a sufficiently long interval with respect to the inverse of the spectral band $\Delta\nu$ of the radiation.

Given that, for a stationary medium, Jones' matrices are invariant over time, Eq. (7.11.18) for the transformation of the coherence matrix is also valid introduc-

Device	Jones' matrix	Mueller's matrix
Polarizer, transmission axis along x and along y	$\begin{pmatrix} 1 & 0 \\ 0 & 0 \end{pmatrix}$ x ; $\begin{pmatrix} 0 & 0 \\ 0 & 1 \end{pmatrix}$ y	$\frac{1}{2}\begin{pmatrix} 1 & 1 & 0 & 0 \\ 1 & 1 & 0 & 0 \\ 0 & 0 & 0 & 0 \\ 0 & 0 & 0 & 0 \end{pmatrix}$ x ; $\frac{1}{2}\begin{pmatrix} 1 & -1 & 0 & 0 \\ -1 & 1 & 0 & 0 \\ 0 & 0 & 0 & 0 \\ 0 & 0 & 0 & 0 \end{pmatrix}$ y
Polarizer, transmission axis rotated by an angle φ from the x-axis	$\begin{pmatrix} \cos^2\phi & -\cos\phi\sin\phi \\ -\cos\phi\sin\phi & \sin^2\phi \end{pmatrix}$	$\frac{1}{2}\begin{pmatrix} 1 & \cos 2\phi & \sin 2\phi & 0 \\ \cos 2\phi & \cos^2 2\phi & \cos 2\phi \sin 2\phi & 0 \\ \sin 2\phi & \cos 2\phi \sin 2\phi & \sin^2 2\phi & 0 \\ 0 & 0 & 0 & 0 \end{pmatrix}$
Polarizer, transmission axis at +45° and −45° from x	$\frac{1}{2}\begin{pmatrix} 1 & 1 \\ 1 & 1 \end{pmatrix}$ +45° ; $\frac{1}{2}\begin{pmatrix} 1 & -1 \\ -1 & 1 \end{pmatrix}$ −45°	$\frac{1}{2}\begin{pmatrix} 1 & 0 & 1 & 0 \\ 0 & 0 & 0 & 0 \\ 1 & 0 & 1 & 0 \\ 0 & 0 & 0 & 0 \end{pmatrix}$ +45° ; $\frac{1}{2}\begin{pmatrix} 1 & 0 & -1 & 0 \\ 0 & 0 & 0 & 0 \\ -1 & 0 & 1 & 0 \\ 0 & 0 & 0 & 0 \end{pmatrix}$ −45°
Birefringent lamina, neutral axes along x and y	$\begin{pmatrix} 1 & 0 \\ 0 & e^{i\varphi} \end{pmatrix}$	$\begin{pmatrix} 1 & 0 & 0 & 0 \\ 0 & 1 & 0 & 0 \\ 0 & 0 & \cos\varphi & \sin\varphi \\ 0 & 0 & -\sin\varphi & \cos\varphi \end{pmatrix}$
Birefringent laminae, lambda/4 and lambda/2, neutral axes along x and y	$\begin{pmatrix} 1 & 0 \\ 0 & i \end{pmatrix}$ $\lambda/4$; $\begin{pmatrix} 1 & 0 \\ 0 & -1 \end{pmatrix}$ $\lambda/2$	$\begin{pmatrix} 1 & 0 & 0 & 0 \\ 0 & 1 & 0 & 0 \\ 0 & 0 & 0 & 1 \\ 0 & 0 & -1 & 0 \end{pmatrix}$ $\lambda/4$; $\begin{pmatrix} 1 & 0 & 0 & 0 \\ 0 & 1 & 0 & 0 \\ 0 & 0 & -1 & 0 \\ 0 & 0 & 0 & -1 \end{pmatrix}$ $\lambda/2$
Lambda/4 lamina, neutral axes at ±45° from x	$\frac{1+i}{2}\begin{pmatrix} 1 & i \\ i & 1 \end{pmatrix}$ +45° ; $\frac{1-i}{2}\begin{pmatrix} 1 & -i \\ -i & 1 \end{pmatrix}$ −45°	$\begin{pmatrix} 1 & 0 & 0 & 0 \\ 0 & 0 & 0 & -1 \\ 0 & 0 & 1 & 0 \\ 0 & 1 & 0 & 0 \end{pmatrix}$ +45° ; $\begin{pmatrix} 1 & 0 & 0 & 0 \\ 0 & 0 & 0 & 1 \\ 0 & 0 & 1 & 0 \\ 0 & -1 & 0 & 0 \end{pmatrix}$ −45°
Lambda/2 lamina, neutral axes rotated by an angle φ	$\begin{pmatrix} \cos 2\phi & -\sin 2\phi \\ -\sin 2\phi & -\cos 2\phi \end{pmatrix}$	$\begin{pmatrix} 1 & 0 & 0 & 0 \\ 0 & \cos 4\phi & \sin 4\phi & 0 \\ 0 & \sin 4\phi & -\cos 4\phi & 0 \\ 0 & 0 & 0 & -1 \end{pmatrix}$
Rotator by an angle φ, by optical activity or the Faraday effect	$\begin{pmatrix} \cos\phi & -\sin\phi \\ \sin\phi & \cos\phi \end{pmatrix}$	$\begin{pmatrix} 1 & 0 & 0 & 0 \\ 0 & \cos 2\phi & -\sin 2\phi & 0 \\ 0 & \sin 2\phi & \cos 2\phi & 0 \\ 0 & 0 & 0 & 1 \end{pmatrix}$
Depolarizer	not defined	$\begin{pmatrix} 1 & 0 & 0 & 0 \\ 0 & 0 & 0 & 0 \\ 0 & 0 & 0 & 0 \\ 0 & 0 & 0 & 0 \end{pmatrix}$

Table 7.9 Matrices of Jones and Mueller

ing this temporal average; therefore, the transformation (7.11.19) is also valid.

Some matrices of Mueller and Jones are shown in Table 7.9 for the most common optical devices. In particular, for a rotatory device of an angle φ, by optical activity or Faraday's effect, the Jones' matrix is,

$$\mathfrak{M}_\varphi = \begin{pmatrix} \cos\varphi & -\sin\varphi \\ \sin\varphi & \cos\varphi \end{pmatrix} \tag{7.11.22}$$

and the corresponding 4x4 matrix is obtained from Eq. (7.11.20). It turns out that

$$M_\varphi = \begin{pmatrix} 1 & 0 & 0 & 0 \\ 0 & \cos 2\varphi & -\sin 2\varphi & 0 \\ 0 & \sin 2\varphi & \cos 2\varphi & 0 \\ 0 & 0 & 0 & 1 \end{pmatrix}. \tag{7.11.23}$$

The matrix of a device rotated by an angle φ with respect to the base in which the matrix is known can be obtained by imagining placing this non-rotated object between two polarization rotators, for an angle −φ for the one preceding it and for an angle +φ for the one following it. In other words, the matrices $\mathfrak{A}_\varphi$ of Jones and A_φ of Mueller of the rotated device are given by

$$\mathfrak{A}_\varphi = \mathfrak{M}_\varphi \mathfrak{A}_0 \mathfrak{M}_\varphi{}^\dagger \quad \text{and} \quad A_\varphi = M_\varphi A_0 M_\varphi{}^\dagger , \tag{7.11.24}$$

where $\mathfrak{A}_0$ and A_0 are the respective matrices of the non-rotated device.

7.11.3 Polarizers

Polarizers are devices that *project* the polarization of the incident wave along a particular direction. In other words, they split the wave field into two orthogonal linear polarizations, one of which is transmitted while the other is not. This is the prototype of a *measurement* operation in Quantum Mechanics, projecting the state of polarization in one of the possible eigenstates. More generally, there are devices that split the field into two elliptical orthogonal polarizations, of which the linear and circular polarizers are two particular cases. These are made in many different forms and use one of five physical phenomena: dichroism, diffraction, reflection, diffusion, and birefringence, or a combination of these, for which there is an asymmetry of behavior on the polarization. There are essentially two types of polarizers. The first selects a single polarization component by absorbing (or dispersing) the other, while the other type spatially separates the two polarization components. In the first case, there is a loss of information, as the amplitude of the extinct wave is lost and its energy is transformed into heat, while, in the second case, there is no loss.

Linear polarizers are of particular importance, since, in practice, they are almost always necessary as a discriminating element, even in circular or elliptical polarization devices. In some applications, devices that produce a high degree of polarization purity of the transmitted radiation are required, while, with laser beams, the ability to withstand high intensities is often required.

In order to fully characterize a polarizer, several parameters must be specified, including the *extinction ratio*, *transmission*, *angular acceptance*, *ratio L/A between length and aperture width*, *spectral band* and *power damage threshold*.

Consider the case of a linear polarizer used to analyze a linearly polarized plane wave. The imperfections of the devices, the absorption, and the parasitic reflections cause the Jones' matrix of a real polarizer to be given by

$$\begin{pmatrix} T_{\mathrm{p}}^{1/2} e^{i\phi_{\mathrm{p}}} & 0 \\ 0 & T_{\mathrm{s}}^{1/2} e^{i\phi_{\mathrm{s}}} \end{pmatrix}, \tag{7.11.25}$$

where T_{p} is the transmission coefficient for the polarization component parallel to the axis of the polarizer and T_{s} is that for the perpendicular component, while ϕ_{p} and ϕ_{s} are the respective phases in the propagation through the device.[9]

Depending on the angle θ between the direction of the main axis of the polarizer and the direction of oscillation of the electric field of the incident wave, the transmission is given by

$$T = \left(T_{\mathrm{p}} - T_{\mathrm{s}}\right)\cos^2\theta + T_{\mathrm{s}}\,. \tag{7.11.26}$$

In the ideal case, we should have $T_{\mathrm{s}} = 0$, $T_{\mathrm{p}} = 1$, but, in general, we have that $0 < T_{\mathrm{s}} < T_{\mathrm{p}} < 1$ and the extinction ratio is defined as

$$\tau = \frac{T_{\mathrm{p}}}{T_{\mathrm{s}}}\,. \tag{7.11.27}$$

7.11.3.1 Dichroic polarizers

The first dichroic polarizers were made with tourmaline crystals, cut into the shape of a foil, with their optical axis parallel to the faces. In this way, the polarization component with the electric field oscillating in a direction perpendicular to the axis is strongly absorbed. The degree of extinction that is obtained is strongly dependent on the wavelength and the transmitted component is also partially ab-

---

9 More generally, for example, in the presence of optical activity, this matrix is diagonalized with eigenvectors of elliptic polarization, but we avoid this complication here.

sorbed.

At the end of the 19th century, Hertz used a grid of parallel conducting wires to demonstrate the transverse nature of radio waves. Currently, tungsten wire polarizers are available in the terahertz region. Different types of dichroic polarizer are based on this same principle. The most widespread ones are made of plastic sheets called *polaroids*. They can also have dimensions in meters and have a high degree of extinction, over 1:1000, maintaining a limited absorption for the component to be transmitted, with a useful spectral band covering the entire visible spectrum. Polaroids are made of polyvinyl alcohol molecules that are aligned through the stretching process that occurs at the time of lamination. These foils are then immersed in an iodine-rich ink, which binds to the long molecular chains. Conduction electrons associated with iodine can move along the chain, which behaves like a conducting wire. A wave polarized with the electric field parallel to the direction of the chains sets these electrons in motion, which then dissipate the energy in heat.

A new type of dichroic polarizer consists of a glass inside of which ellipsoidal nanoparticles are spread, all aligned in the same direction; they reach a particularly high degree of extinction (up to 10^5) and have a damage threshold for power greater than that of the plastic polarizers. The *wire polarizers* are now also available in the mid-infrared: they have a lower degree of extinction ($\approx$200), but allow for the separation of two orthogonal polarizations with good efficiency, since they reflect the electric field component parallel to the wires and transmit the other one. The wires have a very small pitch (2700 lines/mm) and are holographically engraved in a metal layer deposited on a transparent substrate. Observed visually, these polarizers have the appearance of a diffraction grating; however, for the infrared, their pitch is much smaller than the half-wavelength for which the diffracted orders are evanescent, and the reflected wave is only that at the zero order. On the other hand, the reflection is of a metallic type, due to the motion of the electrons near the surface of the wires, whose transverse dimension is much lower than the wavelength. Therefore, the electrons are constrained to move mainly along the wires and the grating behaves like a metal mirror only for the component of the electric field parallel to them. These polarizers are very delicate and, *alas*, cannot be cleaned in any mechanical way, or even remotely touched.

7.11.3.2 Polarizers by dielectric reflection

One frequently used polarization element is a simple *surface at the Brewster's angle*, for which, according to Eqs. (1.7.16-17) and (1.7.19-20), the transmitted wave has an extinction ratio equal to

$$\tau = \frac{T_{\parallel}}{T_{\perp}} = \frac{1}{4}\left[\frac{n_{\mathrm{T}}}{n_{\mathrm{I}}} + \frac{n_{\mathrm{I}}}{n_{\mathrm{T}}}\right]^2 . \tag{7.11.28}$$

The reflected wave, on the other hand, is, in principle, completely polarized, but with a rather low energy efficiency in general. However, in the mid-infrared, for a single surface of germanium at the Brewster angle, the reflection of the TE wave reaches 78%.

These surfaces are frequently used in laser cavities: despite their low extinction ratio, they introduce losses large enough to suppress the laser emission component with the electric field perpendicular to the incidence plane, while essentially completely transmitting the other, whereby the laser radiation is polarized with the electric field parallel to the incidence plane. The degree of extinction can be increased by placing various laminae in succession, with plane and parallel faces, inclined at the Brewster's angle. This solution is used when it is necessary to polarize high power radiation, like that of CO_2 lasers. For example, with six plates of zinc selenide, for $\lambda_o \approx 10$ μm, there is an extinction ratio of about 3700. However, the plates must be sufficiently thick and spaced so as to avoid multiple reflections between the faces and unwanted interference effects.

The interference is, vice versa, exploited to enhance the selective reflection of a polarization component in *polarizing beam splitters* consisting of a dielectric multilayer deposited on the hypotenuse of a right angle isosceles prism. A second identical prism is glued to the first to form a cubic prism whose internal diagonal face serves as a polarizing element. Instead, the external faces of the cube have an anti-reflection treatment to attenuate the amount of parasitic reflections that they introduce. The operation of these polarizers has already been described in the chapter on interference. They decompose the incident radiation into two beams orthogonal to each other, with an energy efficiency close to 100%, and have an extinction ratio on the order of 1000. They can also be made with a relatively large opening and an angular acceptance of more than ten degrees.

7.11.3.3 Polarizers by birefringence

Birefringent materials possess the property of decomposing the radiation by refraction into two components of orthogonal polarization, and therefore constitute a natural polarization device. The extent of the angular separation between the ordinary and the extraordinary waves depends on the orientation of the optical axes of the crystal and is generally small. However, by appropriately choosing the materials and the geometry of the device, the two waves can be well separated, either by refraction alone or by total reflection of one of the two.

The materials used are generally uniaxial and must have good transparency in the spectral region of use, as well as an appreciable difference between the indices of refraction. This limits the choice between a few crystalline materials as a function of the wavelength, such as magnesium fluoride (130-190 nm), calcite (220-3000 nm), rutile (500-4000 nm), alpha-BBO (189-3500 nm), and some others. The extinction ratio and the energy efficiency of birefringent polarizers are excellent; on the other hand, the need for them to be made of clear, defect-free crystals

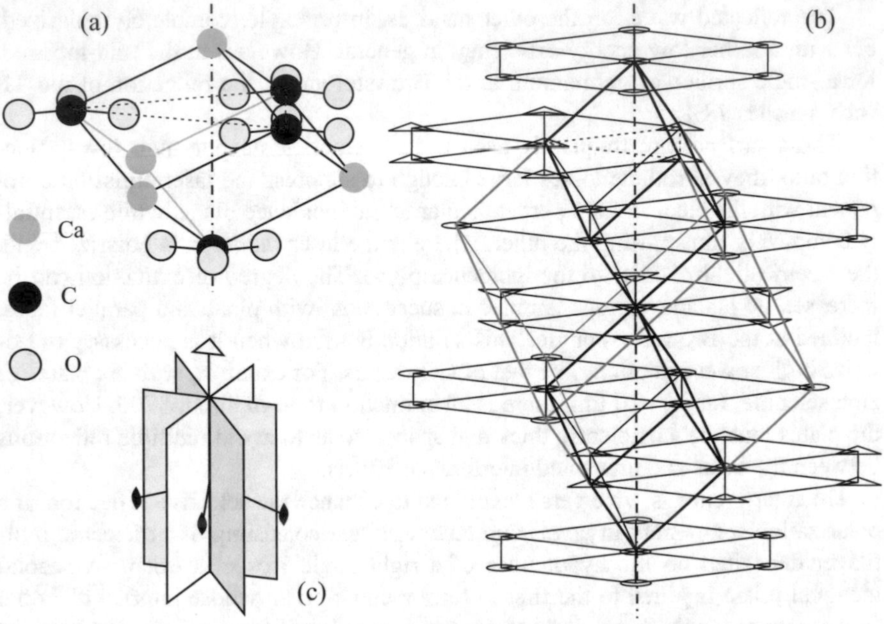

Fig. 7.23 Arrangement of atoms in calcite. (a) primitive cell consisting of a rhombohedrum with edges all of equal length: the optical axis passes diagonally into the cell between a calcium atom and a carbon atom. (b) unitary cell, consisting of the union of 8 primitive cells, each inverted with respect to the adjacent ones; two of these are indicated along the diagonal of the unit cell. The figure is vertically expanded to highlight the alternate-plane structure of carbonate groups and calcium atoms; the triangles represent the carbonate groups and the circles the calcium atoms. (c) crystal symmetry elements, consisting of a three-fold rotation axis (3 or C_3) and a three-fold rotation-inversion axis ($\overline{3}$), which is equivalent to a six-fold rotation-reflection axis (S_6) along the same direction, three two-fold axes, orthogonal to the three-fold axis, and three mirror planes with translation (*glide plane*)

means that their cost is high and/or that their apertures are limited to a few cm.

Among these materials, calcite is the most common. It consists of calcium carbonate ($CaCO_3$), which is a very common substance, generally present in nature in polycrystalline form, such as marble and limestone.[10] Clear crystals of calcite with useful dimensions in optics are very rare. Samples were discovered in Iceland, hence the name *Iceland spar*. It was thanks to them that, in 1669, Rasmus Bartholin discovered the phenomenon of birefringence. Unlike other crystals used in optics, calcite cannot be artificially grown with any kind of efficiency, and its crystals currently come from mines in Mexico, Africa, and Siberia. Recently, instead of calcite, the tendency as arisen to use α-BBO, which is the high-temperature phase of barium borate; it has the advantage of being artificially grown and has a greater transparency range.

10 Calcium carbonate has two other polymorphic crystalline forms: aragonite and vaterite (hexagonal).

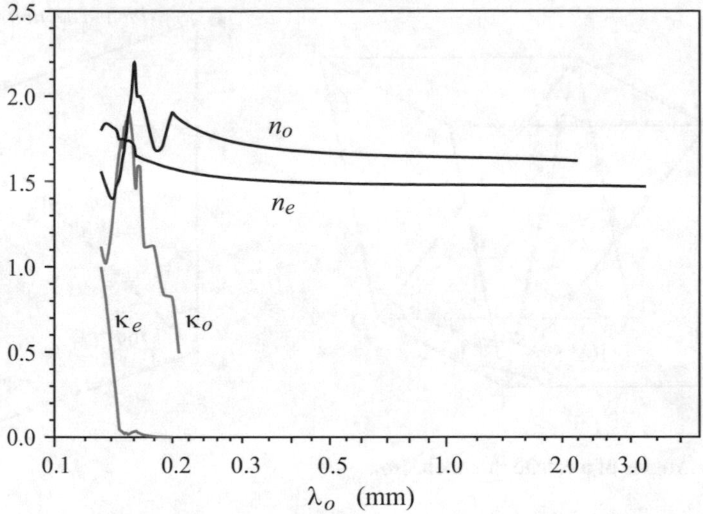

Fig. 7.24 Calcite. Trend of the refraction index, real part n and imaginary part κ, for the ordinary ray o and the extraordinary ray e as the wavelength varies. Values interpolated from the table of J.M. Bennett, *Polarizer*, Chap.3 of Bass et al ed. (1995), *Handbook of Optics*

The calcite crystals belong to the trigonal system ($\overline{3}$ 2/m) and to the space group $R\overline{3}c$ thanks to the CO_3 grouping that forms an equilateral triangle of oxygen atoms with the carbon atom in the center (Fig. 7.23). These plane groups lie on mutually parallel planes (with opposite orientation between contiguous planes) and in an alternating fashion with the calcium atoms at the vertices of an elementary cell with all identical edges. This cell has a three-fold rotation axis along a diagonal perpendicular to the CO_3 groups. For this symmetry, calcite has a single optical axis and, on the other hand, the planar form of the carbonate groups produces a strong index difference between the polarized waves parallel and perpendicular to that axis.

The transparency region of the calcite is between 0.214 and 3.3 μm for the extraordinary ray and between 0.23 and 2.2 μm for the ordinary ray (Fig 7.24). It presents dichroism at the extremes of these regions and, in the infrared, it finds application as a dichroic polarizer in the form of laminae with the optical axis parallel to the faces.

A calcite crystal can be flaked along its crystalline planes to form a rhombohedral prism (Fig. 7.25) that only has two "obtuse" vertices, diagonally opposite, whose three edges between them form angles all of 101° 55', while the other vertices have two angles of 78° 5' and one of 101° 55'. The axis that intersects an obtuse vertex at equal angles with its three edges corresponds to the direction of the optical axis. Imagine, then, cutting such a prism so that all of the edges are of the same length. Therefore, the line that joins the two obtuse vertices is parallel to the optical axis. The planes defined by two opposite parallel edges contain the optical axis, and these are called *principal sections*. It can be shown that they are perpen-

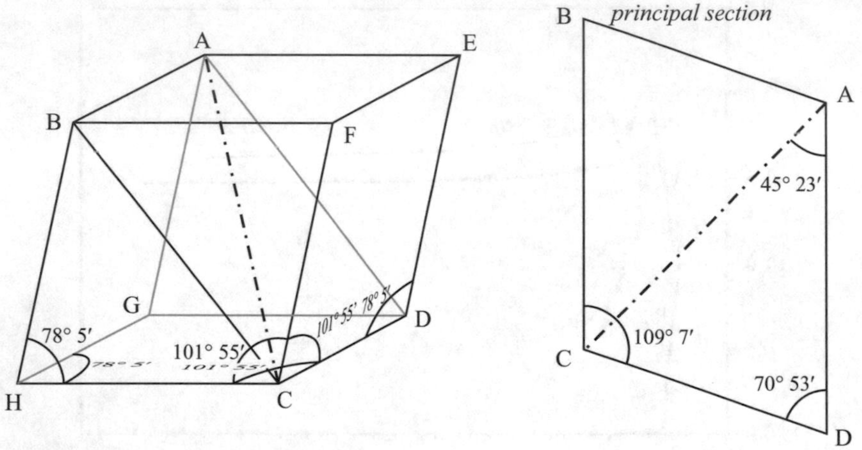

Fig. 7.25 Angles of a calcite rhombohedron

dicular to the prism faces that they cut diagonally.

Nicol's Prism. The first of the birefringent polarizers was created by William Nicol in 1828. The Nicol's prism is based on total internal reflection of a polarization component and is constituted by a calcite crystal of elongated shape with its faces coincident to its cleavage planes. It is cut diagonally in a direction perpendicular to a principal section and then joined by gluing it to the *Canada balsam* (Fig. 7.26). The cut is performed between the obtuse vertices of the entry and exit faces. The principal sections are inclined with respect to the lateral surfaces of the prism: the one that contains the axis of the polarizer, in addition to the optical axis, crosses the input and output faces diagonally.

The index of refraction of the balsam ($n \approx 1.55$) is intermediate between those

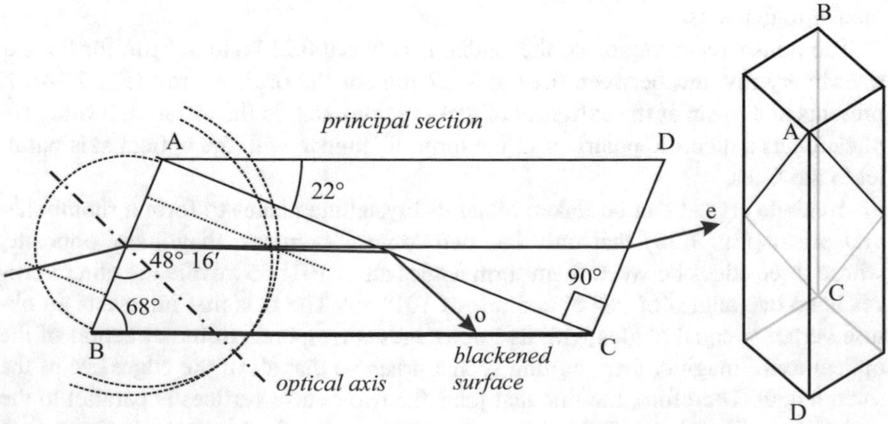

Fig. 7.26 Nicol's prism. The graphic construction for determining the refraction on the entrance face is shown

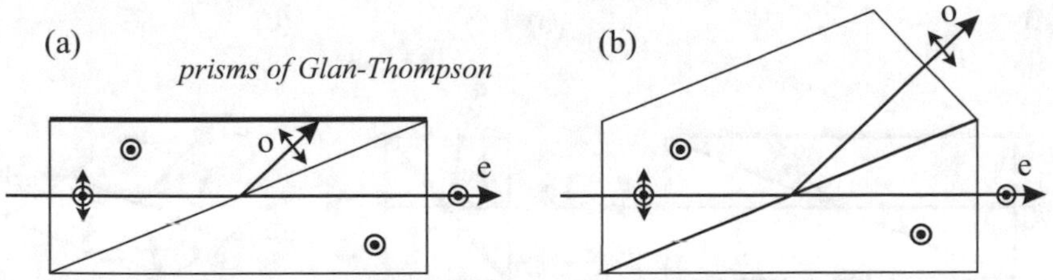

Fig. 7.27 Glan-Thompson prisms. Version (b) is used as a beam divider. The symbols ⊙ and ↕ indicate the directions of the optical axis and the electric field vibration plane, respectively perpendicular and parallel to the plane of the drawing

of the ordinary and extraordinary rays, and therefore the extraordinary ray (the one whose electric field oscillates parallel to the principal section) is transmitted; instead, the ordinary ray undergoes a total reflection and goes toward a strongly absorbing surface. However, the extraordinary ray is also partially reflected, so that its transmission is noticeably less than unity.

The second part of the prism only serves to bring the transmitted rays in the original direction. In principle, the calcite of the second prism could be replaced with a more economical medium, but the difference in thermal expansion would produce a shear stress on the balsam and induce birefringence, with the risk of a separation between the two prisms.

With respect to the axis of the polarizer, the angular acceptance in the incidence plane on the input face varies between about –9.7°, below which the extraordinary ray is also totally reflected, and + 18.8°, beyond which the ordinary ray starts to be transmitted. In order to make these angular deviations more symmetrical, the input faces of Nicol's prism are generally cut at an angle of 68°. However, there are different cutting variants [Bennett, in Bass et al ed. 1995, *Handbook of Optics*] of this type of polarizer.

Glan's Prisms. Although the Nicol's prism has a good angle of acceptance, one of its defects is to produce a translation of the transmitted light beam, which is particularly annoying for the analysis of the polarization. In fact, a rotation of the prism around its own axis produces a circular motion of the output beam. Currently, the calcite polarizers are cut to form a parallelepiped, with the input and output faces parallel to the optical axis, which is also parallel to two side surfaces of the prism. These polarizers are called Glan's prisms, and there are various types of them. Their functioning remains based on total reflection; however, by normal incidence, on the entrance face, there is no birefringence, i.e., separation, between ordinary and extraordinary rays. Compared to Nicol's prism, they have various advantages: since the optical axis is perpendicular to the longitudinal axis of the prism, there is the maximum difference in index between ordinary and extraordinary rays; this allows for obtaining a greater angular field and/or a shorter length/aperture ratio (L/A). Furthermore, the polarization that is obtained is more uniform. Finally, if the input and output faces are well cut, there is no appreciable translation of the beam while the prism is rotated around its axis (a deflection or a

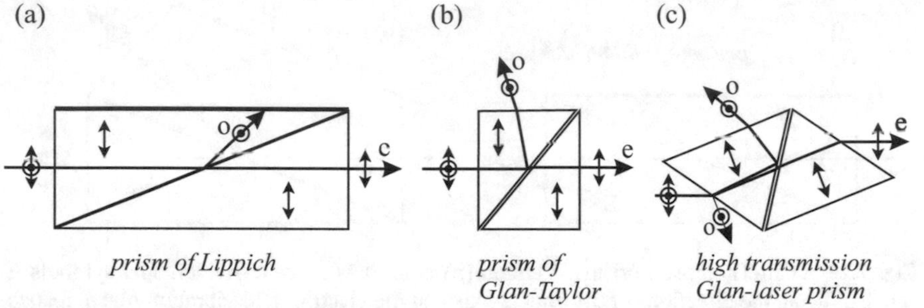

Fig 7.28. Prisms of the Lippich type

residual translation place a limit to be considered for the quality of these polarizers).

Depending on the type, the cut on which the total reflection is produced is still perpendicular to a principal section (*Lippich* and *Glan-Taylor prisms*), or it contains the optical axis (*Glan-Thompson* and *Glan-Foucault prisms*), or at 45° from the optical axis (*Frank-Ritter prisms*). These polarizers are then distinguished by the fact that the prisms that compose them are *cemented* or *spaced in air*.

Currently, the Canada balsam is replaced by other polymers, such as *n*-butilmethacrylate, or by very viscous substances that are more transparent in the ultraviolet, such as the silicone oil DC-200, gedamine (butyl alcohol solution of urea formaldehyde), or glucose. The cemented prisms are optically better, because they produce less loss on the extraordinary ray, both by reflection and diffusion, due to the residual roughness on the cut, especially using cements with an index close to that of the extraordinary ray.

The *Glan-Thompson* prisms (*Glazebrook*) (Fig. 7.27) reach a high degree of extinction, over 100,000, and can have a high angular opening, depending on the cutting angle, that is, the L/A ratio: 42° with L/A = 4, 26° with L/A = 3, and 15° with L/A = 2.5. These values increase by decreasing the wavelength.

The *Lippich's prism* has its own optical axis on the incidence plane on the cut, as shown in Fig. 7.28(a). In this way, the losses by reflection for the extraordinary ray are lower in the presence of a discrepancy between the extraordinary index of the calcite and the cement index. The angular aperture of the Lippich's prism is smaller, but this is generally not a problem with laser beams. Moreover, this type of prism would be preferable when using the polarizer as a separator for the two polarization components, with the ordinary wave that exits laterally. However, Lippich prisms are not a standard product on the market.

With high power or an ultraviolet laser, up to the limit of calcite transmission, around 214 nm, cement or resin can no longer be used, and it is necessary to use prisms spaced in air, which, however, have a rather small angular aperture. Among these, there is the *Glan-Foucault prism*, which is similar to that of Glan-Thompson, but has the disadvantage of presenting substantial multiple reflections, and therefore also interference in correspondence with the air spacing between the two prisms that compose it. Much better performance is obtained with the *Glan-*

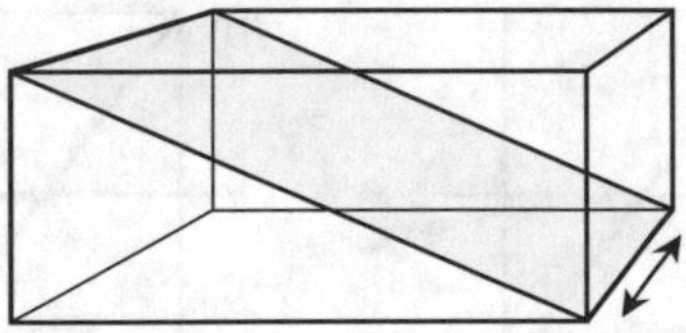

Fig. 7.29
Prism of Bertrand - Feussner

Taylor prism, indicated in Fig. 7.28(b), which is the air-spaced analog of Lippich's prism, since the reflection of the ordinary ray on the cutting faces is much smaller. Commonly, these prisms are made with the cut at about 39.7° with respect to the entrance face: this angle is equal to the angle of incidence of the principal ray. They have a total angular aperture of about 8.5° and an L/A ratio of 0.83.

A higher transmission version of the Glan-Taylor prism, called the *Glan-laser*, has a smaller cutting angle for approaching the Brewster's angle for the extraordinary ray, which is 33.9°. Consequently, these prisms have an asymmetrical angular aperture, and they are essentially used only with collimated laser beams. A further improvement of the transmission occurs with the prisms called *High-Transmission Glan-laser* (Fig. 7.28(c)), which have the principal ray with incidence angle close to the Brewster's angle for the extraordinary ray, either on the input and output faces and at internal interfaces. This is made possible by taking advantage of the deviation of the ordinary ray by birefringence on the entrance face.

Generally, even the entrance and exit faces of the prisms involve parasitic reflections that are attenuated through an anti-reflective treatment. Particular care must then be devoted to avoiding diffuse radiation of the ordinary beam deflected by the polarizer, especially when a high degree of extinction is desired.

Feussner's Prism. One particular type of polarizer is constituted by the *Feussner's prism* (Fig. 7.29), in which a birefringent thin lamina is cemented between two glass prisms. The operation is similar to that of the previous prisms, but the deflected ray by total reflection is now the extraordinary ray. In Feussner's original proposal, the optical axis of the lamina is perpendicular to its faces, while Bertrand later proposed orienting the optical axis parallel to them. Feussner-Bertrand prisms are made with birefringent laminae of mica or sodium nitrate.

Rochon, Sénarmont and Wollaston Prisms. These constitute another category of birefringent polarizers, with a larger cutting angle, and therefore work on the refraction of ordinary and / or extraordinary rays. Unlike previous prisms, in which the two crystals that compose them have parallel optical axes, these prisms are characterized by having these optical axes perpendicular to each other.

In the *Rochon's prism*, shown in Fig. 7.30(a), the first crystal has the optical axis perpendicular to the input face, while the second crystal has the optical axis parallel to both the output face and the plane of the cut. In this way, the ordinary ray is transmitted without deflection, while the extraordinary ray is refracted in the passage to the second crystal. Sometimes, the first crystal is replaced by crystalline quartz, which, however, has the defect of presenting optical activity. A better

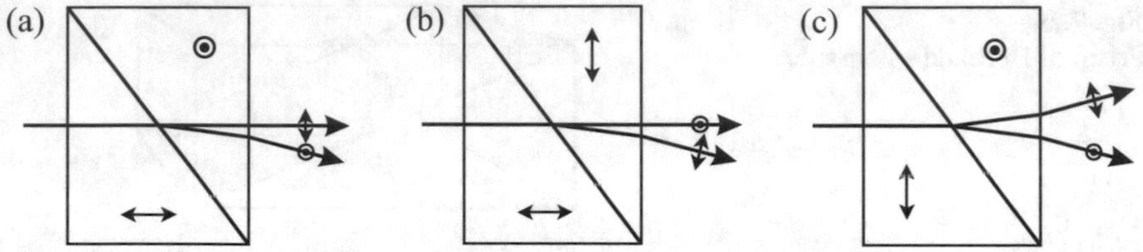

Fig. 7.30 Prisms of (a) Rochon, (b) Sénarmont, (c) Wollaston

choice is to replace the first crystal with a glass and, depending on its index, the deflected ray can alternatively be the extraordinary ray or the ordinary ray. The two prisms are generally cemented together; however, prisms consisting of crystals of the same material, quartz or magnesium fluoride joined optically, are also used. The prisms made with MgF_2 are used both in ultraviolet, down to 130 nm, and in the infrared, up to 7 μm.

The *Sénarmont prism* (Fig. 7.30(b)) is similar to that of Rochon, but has the optical axis of the second crystal perpendicular to the edge between the output face and the cut. Since the refractive index of the extraordinary ray does not come near its minimum value, the deflection of this ray is smaller.

The *Wollaston's prism* (Fig. 7.30(c)) is instead made with two crystals whose optical axes are both parallel to the input and output faces, remaining perpendicular to each other. In this way, at the cut, both rays are deflected, but in opposite directions: the ordinary ray in the first crystal becomes extraordinary in the second crystal, and vice versa. The deviation of each ray is about the same (but not the same), and the total deviation is about twice that of the Rochon prism.

All of these prisms, working by refraction, suffer from chromatism for the deviated polarization component.

7.11.4 Wave plate

The second important class of devices used to manipulate the polarization is composed of simple laminae of birefringent material with plane and parallel faces, usually uniaxial, whose optical axis lies parallel to the faces (Fig. 7.31). These are called *wave plates* or *retarders*. From Eq. (7.8.7), we have that, for small differences between the indices n_e, n_o with respect to their average value, the phase difference due to the difference in optical path between the extraordinary and ordinary rays that cross one of these uniaxial laminae is

$$\phi \cong \frac{2\pi}{\lambda_o} \frac{h}{\cos\vartheta}(n_e - n_o)\sin^2\theta = m\lambda_o, \tag{7.11.29}$$

where the indices n_e and n_o are those associated with the principal axes of the crystal (see §7.4.5), h is the thickness of the lamina, m is the *order* of the lamina,

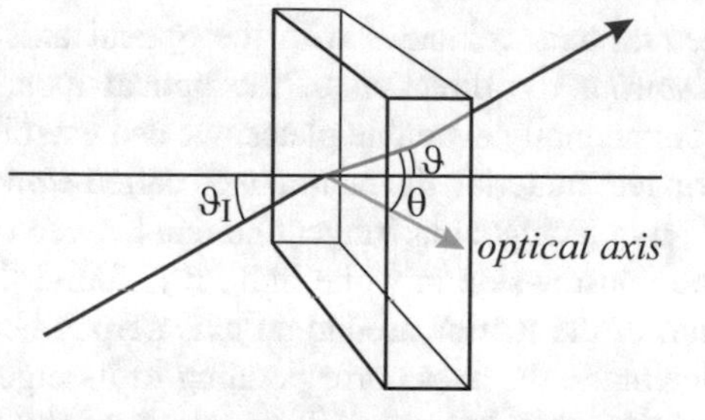

Fig. 7.31
Generic wave plate with plane parallel faces. The optical axis here has any direction. ϑ indicates the *internal* incidence angle, while θ is the angle between the average direction of ordinary and extraordinary rays with the optical axis

and, lastly, ϑ and θ are angles *inside* of the lamina. They are, respectively, the angle of incidence and the angle between the average direction of the ordinary and extraordinary rays with the optical axis. Under ordinary conditions, when the lamina is cut with its optical axis parallel to the faces, the difference in path, and therefore in phase, is indicated by the equi-phase hyperboloid lines of Fig. 7.20(b). These laminae are characterized by the phase difference expressed at normal incidence

$$\phi = 2\pi \frac{h}{\lambda_o}(n_e - n_o) \tag{7.11.30}$$

and are typically specified for a given wavelength, such as the *quarter-wave*, "λ/4", the *half-wave*, "λ/2", or the *full-wave*, "λ". In the three cases, the phase difference respectively amounts to

$$\phi = 2\pi\left(m + \frac{1}{2} \pm \frac{1}{4}\right), \quad \phi = 2\pi\left(m + \frac{1}{2}\right), \quad \phi = 2\pi m\,, \tag{7.11.31}$$

where m is an integer here. Neglecting the Fresnel's reflection losses from the faces and the typically dichroic internal absorption, the Jones matrix of a wave plate expressed in the base of its eigenvectors is

$$\mathfrak{M}_\varphi = \begin{pmatrix} 1 & 0 \\ 0 & e^{i\phi} \end{pmatrix} = e^{i\delta}\begin{pmatrix} e^{-i\delta} & 0 \\ 0 & e^{i\delta} \end{pmatrix}, \text{ with } \phi = 2\delta. \tag{7.11.32}$$

The second expression is useful in some calculations. The phase term in front of the matrix is generally irrelevant and is kept only for consistency with the first definition.

The direction of the eigenvectors corresponds to that of the *neutral axes* of the lamina, which are often called the *fast axis* and the *slow axis*, in reference to the phase velocity of the corresponding eigenvectors whose electric field oscillates in that direction. The *neutral* designation derives from the fact that, for an incident wave linearly polarized along one of these axes, the polarization is not modified. Typically, with a lamina of uniaxial material and the axis parallel to the faces, one neutral axis coincides with the optical axis and the other is perpendicular to it. *Knowing* the direction of the optical axis, which is generally indicated in the commercially available plates, we can establish that, for a *positive* (*negative*) birefringent material, the neutral axis called *slow* (*fast*) corresponds to the optical axis.

By placing a bircfringent lamina between two crossed polarizers, and observing the transmission in white light, it is found that this is canceled every 90° of rotation of the lamina around an axis perpendicular to the faces. In this way, we can determine the axes corresponding to its eigenvectors. The test of crossed polarizers does not, however tell us which of the two directions is the fast or the slow one.

The meaning of the matrix (7.11.32) is as follows. The E'_x and E'_y components of the field of a monochromatic wave transmitted by the lamina at its exit are

$$E'_x = E_{ox} e^{i(kz - i\omega t)} \cdot e^{i\phi_x} \cdot 1,$$
$$E'_y = E_{oy} e^{i(kz - i\omega t)} \cdot e^{i\phi_x} \cdot e^{i\phi},$$

where E_{ox} and E_{oy} are the input amplitudes in front to the lamina, z is the axial distance traveled from the lamina exit and ϕ_x is the phase shift corresponding to the optical path in the lamina for the component aligned along the x axis, while, for the y component, we have that $\phi_y = \phi_x + \phi$. Consider, for simplicity, the case of a linear input polarization at 45° from the axes, with $E_{ox} = E_{oy} = \mathrm{E_0}$ real, and we also omit the phase term common to the two components. For a given value of the position, for example, $z = 0$, the real field components oscillate in time as

$$\begin{aligned} \mathrm{E}_x &= \mathrm{E}_o \cos(\omega t), \\ \mathrm{E}_y &= \mathrm{E}_o \cos(\omega t - \phi). \end{aligned} \tag{7.11.33}$$

In other words, with ϕ *positive*, the E_y component oscillates in *delay* with respect to the component E_x, and vice versa with ϕ negative. Even with a generic incident polarization, with ϕ *positive*, the lamina *delays* the y component with respect to the x component, while it *anticipates* it with ϕ *negative*. Therefore, in Eq. (7.11.31), the x-axis corresponds to the fast axis in the first case and to the slow axis in the second. Often, however, the phase shift that appears in the Jones' matrix of a lamina is expressed by module π, and therefore the sign of the overall phase shift, for which it is not possible to attribute the fast or slow axes label, is not generally known.

The birefringent plates also have a strong dependence of the phase shift on the wavelength, especially if they are of a high order. This dependence derives not only from the explicit term $1/\lambda_o$, but also from the dependence of the indices n_e and n_o on λ_o.

When the plate that is available does not have the appropriate phase shift, it can be obtained by inclining it appropriately. Consider the most typical case of a lamina with the optical axis parallel to the faces and an incident plane wave (Fig.

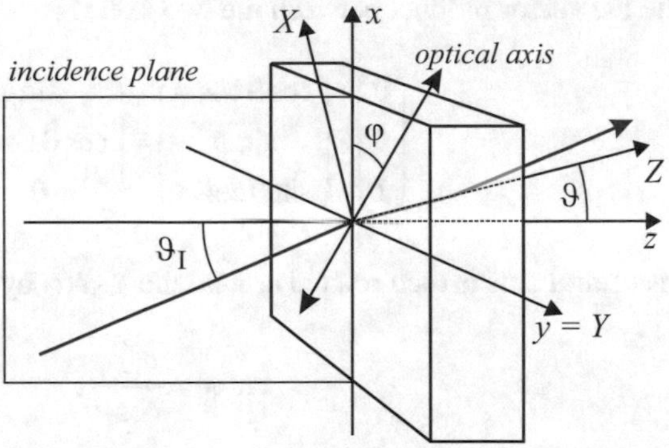

Fig. 7.32
Wave plate with plane and parallel faces with the optical axis parallel to them. The angle ϑ_I indicates the angle of incidence, while φ is the azimuthal angle between the optical axis and the plane of incidence

7.32). Let us fix two reference frames with the $y = Y$ axes in common, perpendicular to the incidence plane and parallel to the faces. We also establish that the z-axis of the first frame is perpendicular to the faces of the lamina, while the Z-axis of the second frame is oriented in the (average) propagation direction of the rays (ordinary and extraordinary) inside of the lamina. The two systems are then rotated between them at an angle equal to that of internal incidence ϑ given by

$$\sin\vartheta = \frac{1}{n}\sin\vartheta_I, \tag{7.11.34}$$

where ϑ_I is the external incidence angle and n is the average refractive index between n_e and n_o. If the optical axis lies on the xy-plane of the faces with an azimuthal angle φ with respect to the incidence plane, i.e., from the x-axis, its direction cosines in the system X, Y, Z are obtained by the rotation

$$\begin{pmatrix} \cos\vartheta & 0 & -\sin\vartheta \\ 0 & 1 & 0 \\ \sin\vartheta & 0 & \cos\vartheta \end{pmatrix}\begin{pmatrix} \cos\varphi \\ \sin\varphi \\ 0 \end{pmatrix} = \begin{pmatrix} \cos\vartheta\cos\varphi \\ \sin\varphi \\ \sin\vartheta\cos\varphi \end{pmatrix}.$$

So, for the angle θ between the propagation direction and the optical axis, we have that

$$\cos\theta = \sin\vartheta\cos\varphi, \tag{7.11.35}$$

and, finally, for Eq. (7.11.29), the phase difference between the two rays is

$$\phi \cong \frac{2\pi}{\lambda_o} h\left(n_e - n_o\right)\frac{1-\sin^2\vartheta\cos^2\varphi}{\cos\vartheta}. \tag{7.11.36}$$

Moreover, the neutral axis corresponding to the ordinary ray must be perpendicular to both the propagation direction and the optical axis, and therefore parallel to the vector product between the two axes:

$$\begin{pmatrix}0\\0\\1\end{pmatrix}\times\begin{pmatrix}\cos\vartheta\cos\varphi\\ \sin\varphi\\ \sin\vartheta\cos\varphi\end{pmatrix}=\begin{pmatrix}-\sin\varphi\\ \cos\vartheta\cos\varphi\\ 0\end{pmatrix}.$$

This neutral axis is then rotated around the Y-axis by an angle φ' given by

$$\tan\varphi' = \frac{\tan\varphi}{\cos\vartheta}. \tag{7.11.37}$$

The same also applies to the other neutral axis with respect to the incidence plane.

Thus, by rotating the lamina at an angle ϑ_I around the optical axis, with respect to the situation with normal incidence, we have that $\varphi = \pi/2$ and, from Eq. (7.11.36),

$$\phi \cong \frac{2\pi}{\lambda_o} h(n_e - n_o)\frac{1}{\cos\vartheta} = \frac{2\pi}{\lambda_o} h(n_e - n_o)\frac{1}{\sqrt{1-\left(n^{-1}\sin\vartheta_I\right)^2}}. \tag{7.11.38}$$

Indeed, with this rotation, the refraction indices of ordinary and extraordinary rays do not change, and the phase shift varies because of the increase of the path through the lamina, expressed by the term $\cos\vartheta$ at the denominator. Therefore, for $n_e > n_o$ ($n_e < n_o$), the phase shift increases (decreases) as ϑ grows .

Instead, by rotating the lamina around the other neutral axis, perpendicular to the optical axis, we have that $\varphi = 0$ and, from Eq. (7.11.36),

$$\phi \cong \frac{2\pi}{\lambda_o} h(n_e - n_o)\cos\vartheta = \frac{2\pi}{\lambda_o} h(n_e - n_o)\sqrt{1-\left(n^{-1}\sin\vartheta_I\right)^2}, \tag{7.11.39}$$

from which we can see that the increase in path is more than compensated by the variation of the n' and n'' indices of refraction of the ordinary and extraordinary rays as a function of ϑ (see §7.8). Therefore, the variation of the phase is in opposite sign to the previous case.

Also, the temperature has an effect on the phase delay introduced by the lamina, because its variations ΔT change both the thickness h and the difference $\eta = n_e - n_o$ between the indices. For normal incidence and an assigned wavelength, we have that

$$\left.\frac{\Delta\phi}{\phi}\right|_{\lambda_o} = \left(\frac{1}{h}\frac{\partial h}{\partial T} + \frac{1}{\eta}\frac{\partial \eta}{\partial T}\right)\Delta T. \tag{7.11.40}$$

Typically, the first term is positive because the material expands as the tempera-

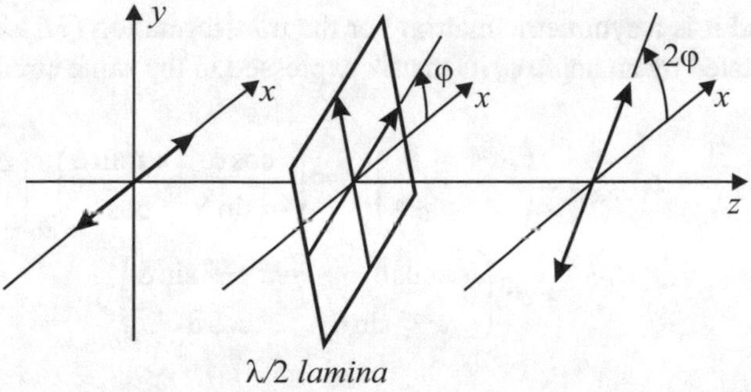

Fig. 7.33 Rotation of the polarization by a lamina λ/2

ture increases, and the fact that the dilatation in a crystal is generally not isotropic must be considered, while the second term is negative because the density decreases for the same reason. Therefore, the second term, which is approximately proportional to the increase in volume, generally prevails over the first and $\Delta\phi/\Delta T$ is generally negative.

7.11.4.1 Effect of a wave plate on polarization

The importance of wave plates is due to the potential that they offer to modify the polarization state of a polarized wave at will without (at least in principle) attenuating it. Indeed, the Jones' matrices that represent them, as in Eq. (7.11.32), are unitary, and therefore indicate a complex rotation in the space of polarizations. For partially polarized radiation, things get complicated and, indeed, this freedom is reduced, to the point of canceling out for the totally non-polarized radiation.

Consider the case of a fully polarized wave so that a lamina can be represented by a Jones' matrix. Of all of the wave plates, the most commonly used laminae are λ/4, λ/2, 3λ/4 and λ, which are respectively described by the matrices

$$\mathfrak{M}_{\pi/2} = \begin{pmatrix} 1 & 0 \\ 0 & i \end{pmatrix}, \mathfrak{M}_{\pi} = \begin{pmatrix} 1 & 0 \\ 0 & -1 \end{pmatrix}, \mathfrak{M}_{3\pi/2} = \begin{pmatrix} 1 & 0 \\ 0 & -i \end{pmatrix}, \mathfrak{M}_{2\pi} = \begin{pmatrix} 1 & 0 \\ 0 & 1 \end{pmatrix}, \quad (7.11.41)$$

for an assigned wavelength equal to λ. The lamina 3λ/4 is essentially equivalent to a lamina λ/4, in that a 90° rotation transforms the matrix of one into that of the other, apart from an overall phase factor. The same is true between a lamina λ/2 and one −λ/2, where the difference is a minus sign.

In a circular base referred to a Cartesian frame with the z-axis perpendicular to the faces of the lamina and the x- and y-axes chosen along the neutral axes, the matrix representing the lamina is

$$\mathfrak{M}_{2\delta c} = \frac{1}{2}\begin{pmatrix} 1 & -i \\ 1 & i \end{pmatrix}_c^{\ell} e^{i\delta} \begin{pmatrix} e^{-i\delta} & 0 \\ 0 & e^{i\delta} \end{pmatrix}_{\ell} \begin{pmatrix} 1 & 1 \\ i & -i \end{pmatrix}_{\ell}^{c} = e^{i\delta} \begin{pmatrix} \cos\delta & -i\sin\delta \\ -i\sin\delta & \cos\delta \end{pmatrix}_c, \quad (7.11.42)$$

and it is a symmetric matrix. For the transformation (7.11.7), if the lamina is then rotated by an angle φ, its matrix expressed in the same circular base is

$$\begin{aligned}\mathfrak{M}_{2\delta,\varphi c} &= \begin{pmatrix} e^{-i\varphi} & 0 \\ 0 & e^{i\varphi} \end{pmatrix}_c^{\mathscr{C}} e^{i\delta} \begin{pmatrix} \cos\delta & -i\sin\delta \\ -i\sin\delta & \cos\delta \end{pmatrix}_{\mathscr{C}} \begin{pmatrix} e^{i\varphi} & 0 \\ 0 & e^{-i\varphi} \end{pmatrix}_{\mathscr{C}}^{c} \\ &= e^{i\delta} \begin{pmatrix} \cos\delta & -ie^{-i2\varphi}\sin\delta \\ -ie^{i2\varphi}\sin\delta & \cos\delta \end{pmatrix}_c .\end{aligned} \tag{7.11.43}$$

In the linear basis, the matrix of the rotated lamina is given by the expression

$$\mathfrak{M}_{2\delta,\varphi\ell} = e^{i\delta} \begin{pmatrix} \cos\delta - i\cos 2\varphi \sin\delta & -i\sin 2\varphi \sin\delta \\ -i\sin 2\varphi \sin\delta & \cos\delta + i\cos 2\varphi \sin\delta \end{pmatrix}_\ell . \tag{7.11.44}$$

Lamina λ. A lamina λ, that is, of an integer order, has no effect on the polarization of a monochromatic wave with that wavelength, whatever its polarization and whatever the angle φ. Indeed, its Jones' matrix coincides with identity. On the other hand, spectrally, these laminae are not neutral, since their phase shift coincides with an integer multiple of 2π only for a discrete set of wavelengths. This property is used to construct spectral filters, as we will see later.

Lamina λ/2. This type of lamina, on the other hand, has the effect of reflecting the polarization direction with respect to the neutral axes (Fig. 7.33). Consider a linearly polarized wave at the angle α from the neutral axis oriented along the *x*-axis. Its Jones' vector is transformed as follows:

$$\boldsymbol{J}' = \begin{pmatrix} 1 & 0 \\ 0 & -1 \end{pmatrix} \begin{pmatrix} \cos\alpha \\ \sin\alpha \end{pmatrix} = \begin{pmatrix} \cos\alpha \\ -\sin\alpha \end{pmatrix} . \tag{7.11.45}$$

This effect coincides with rotation of the polarization. In fact, the Jones matrix of such a plate rotated by an angle φ from the *x*-axis, counterclockwise looking toward the source, is

$$\mathfrak{M}_{\pi,\varphi} = \begin{pmatrix} \cos\varphi & -\sin\varphi \\ \sin\varphi & \cos\varphi \end{pmatrix}_\ell^{\mathscr{L}} \begin{pmatrix} 1 & 0 \\ 0 & -1 \end{pmatrix}_{\mathscr{L}} \begin{pmatrix} \cos\varphi & \sin\varphi \\ -\sin\varphi & \cos\varphi \end{pmatrix}_{\mathscr{L}}^{\ell} = \begin{pmatrix} \cos 2\varphi & \sin 2\varphi \\ \sin 2\varphi & -\cos 2\varphi \end{pmatrix}_\ell . \tag{7.11.46}$$

For an incident wave polarized along the *x*-axis, the wave transmitted by the rotated lamina is represented by the vector

$$\boldsymbol{J}' = \begin{pmatrix} \cos 2\varphi \\ -\sin 2\varphi \end{pmatrix}, \tag{7.11.47}$$

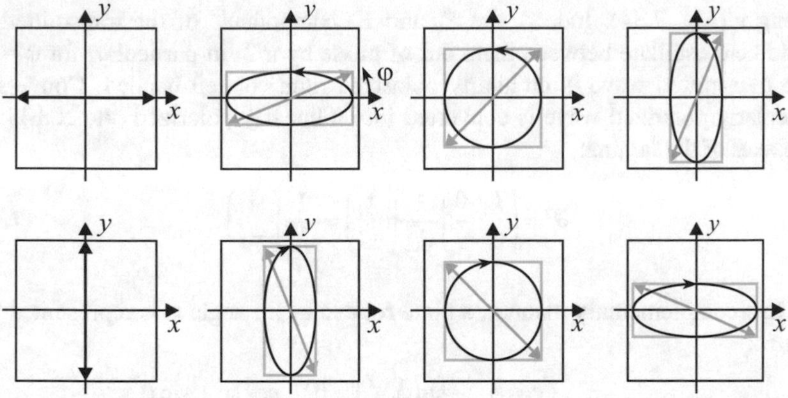

Fig. 7.34 Polarization transmitted by a lamina λ/4, represented by the matrix $\mathfrak{M}_{\pi/2}$ of Eq. (7.11.31), at different azimuthal angles φ of the incident linear polarization, indicated by the arrows in gray. The helicity of the transmitted polarization is indicated by the arrows on the ellipses. For a lamina represented by the matrix $\mathfrak{M}_{3\pi/2}$, the helicity is opposite

and its polarization is then rotated by an angle 2φ. It should be noted that, if a lamina λ/2 is inserted into a paraxial optical system, only the polarization is rotated and not the image. A circularly polarized wave reverses its helicity going through a plate λ/2:

$$\boldsymbol{J}' = \begin{pmatrix} 1 & 0 \\ 0 & -1 \end{pmatrix} \begin{pmatrix} 1 \\ i \end{pmatrix} = \begin{pmatrix} 1 \\ -i \end{pmatrix}, \tag{7.11.48}$$

and the same happens by rotating the lamina:

$$\boldsymbol{J}' = \begin{pmatrix} \cos 2\varphi & -\sin 2\varphi \\ -\sin 2\varphi & -\cos 2\varphi \end{pmatrix} \begin{pmatrix} 1 \\ i \end{pmatrix} = e^{-i2\varphi} \begin{pmatrix} 1 \\ -i \end{pmatrix}. \tag{7.11.49}$$

It should be noted that the phase expressed in the exponential varies in an opposite way for the two circular polarizations. Similarly, an elliptically polarized wave changes its helicity and the axes of the ellipse are reflected with respect to the neutral axes of the lamina, i.e., they are rotated as they are for linear polarization.

Lamina λ/4. Lastly, a lamina λ/4 has the effect of varying the helicity of the polarization. For example, consider an incident wave polarized at an angle φ from the x-axis. The Jones' vector of the transmitted wave is

$$\boldsymbol{J}' = \begin{pmatrix} 1 & 0 \\ 0 & \pm i \end{pmatrix} \begin{pmatrix} \cos\varphi \\ \sin\varphi \end{pmatrix} = \begin{pmatrix} \cos\varphi \\ \pm i \sin\varphi \end{pmatrix}. \tag{7.11.50}$$

Therefore, the transmitted polarization is elliptic with the axes of the ellipse coinciding with the neutral axes of the lamina, which are here oriented along x and along y (Fig. 7.34). Indeed, the E_x and E_y components of the transmitted wave field here oscillate between them out of phase by $\pi/2$. In particular, for $\varphi = \pm 45°$, the transmitted wave is circularly polarized right- or left-handed. Conversely, a circularly polarized wave is converted into a linearly polarized one at ±45° from the axes of the lamina:

$$\boldsymbol{J}' = \begin{pmatrix} 1 & 0 \\ 0 & i \end{pmatrix} \frac{1}{\sqrt{2}} \begin{pmatrix} 1 \\ \pm i \end{pmatrix} = \frac{1}{\sqrt{2}} \begin{pmatrix} 1 \\ \mp 1 \end{pmatrix}. \tag{7.11.51}$$

In a complementary manner, a plate rotated by an angle φ is represented by the matrix

$$\begin{aligned} \mathfrak{M}_{\pi/2,\varphi} &= \begin{pmatrix} \cos\varphi & -\sin\varphi \\ \sin\varphi & \cos\varphi \end{pmatrix} \begin{pmatrix} 1 & 0 \\ 0 & i \end{pmatrix} \begin{pmatrix} \cos\varphi & \sin\varphi \\ -\sin\varphi & \cos\varphi \end{pmatrix} \\ &= \frac{1-i}{2} \begin{pmatrix} i+\cos 2\varphi & \sin 2\varphi \\ \sin 2\varphi & i-\cos 2\varphi \end{pmatrix}, \end{aligned} \tag{7.11.52}$$

and, for a linear incident polarization along the x-axis, we have that

$$\boldsymbol{J}' = \frac{1-i}{2} \begin{pmatrix} i+\cos 2\varphi \\ \sin 2\varphi \end{pmatrix}. \tag{7.11.53}$$

By changing φ, the helicity varies periodically with a period of 180°. For $\varphi = 45°$, we have that

$$\mathfrak{M}_{\pi/2,45°} = \frac{1+i}{2} \begin{pmatrix} 1 & -i \\ -i & 1 \end{pmatrix}. \tag{7.11.54}$$

Variable wave plate by inclination. Nowadays, "variable" zero-order wave plates are commercially available, inserted into optical assemblies that allow them to be inclined and rotated. These are Berek's compensator laminae (see §7.11.4.6) made of uniaxial crystals with the optical axis perpendicular to the face and exploiting the dependence of the phase shift from the angle of incidence on the lamina. Therefore, they are useful only with collimated radiation, in particular, with laser beams.

7.11.4.2 Materials used for the realization of wave plates

The material of the laminae must be clear and transparent and generally have a low birefringence; indeed, low order laminae made of materials with greater birefringence would be too thin to be made. For example, a $\lambda/4$ plate of calcite should be less than 1 μm thick.

In the visible region, the most commonly used is crystalline quartz and, in the

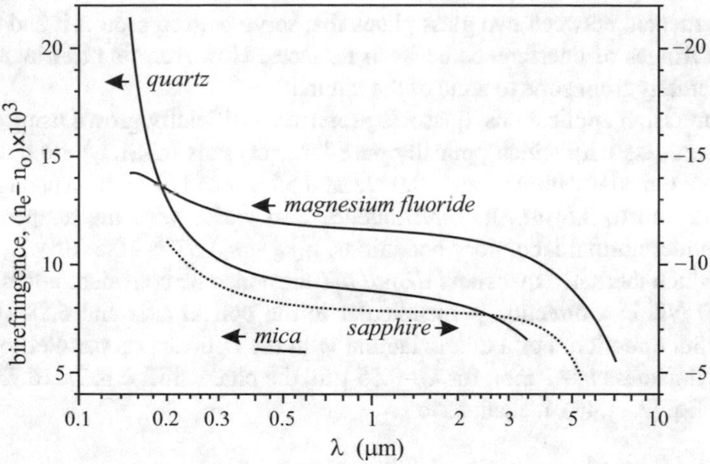

Fig. 7.35 Birefringence of some materials. The dotted curve for the sapphire refers to the scale on the right, while the others refer to the scale on the left. The birefringence of the quartz is drawn by the function $n_e - n_o = A_0 + A_1\lambda_o^2/(\lambda_o^2 + A_2) + A_3\lambda_o^2/(\lambda_o^2 + A_4)$, with λ_o in μm, where the parameters $A_0 = 0.00091038$, $A_1 = 0.00792684$, $A_2 = 0.0138030$, $A_3 = 0.0184746$, $A_4 = 108$, were fit by the author using known measurements between 150 nm and 3.5 μm. [Chandrasekharan and Damany 1968; Smartt and Steel 1959; Shields and Ellis 1956; Maillard 1971]. The curves for magnesium fluoride [Dodge 1984] and for sapphire are calculated from the Sellmeier equations reported in the *Handbook of Optics*, while the curve for mica, which is a biaxial crystal, refers to the difference $n_3 - n_2$ between two of the main indices, measured by Medhat and El-Zaiat (1997)

near infrared, magnesium fluoride, sapphire and mica (Fig. 7.35). Large, low-cost birefringent sheets are made by extruding a transparent plastic polymer, including polyvinyl alcohol, cellophane, and Mylar: the extrusion aligns the molecules of the material, making it anisotropic.

Mica is an aluminum and potassium silicate mineral with formula $KAl_2(AlSi_3O_{10})(F,OH)_2$, or $(KF)_2(Al_2O_3)_3(SiO_2)_6(H_2O)$, that forms pseudo-hexagonal biaxial monoclinic crystals with spatial group C 2/*m*. The most common variety is muscovite, which can easily be broken down into thin, flexible, large area sheets. The crystallographic axes *a* and *b* (see Fig. 7.2) lie on the cleavage plane that coincides with the base of the pseudo-hexagonal prism, while the axis *c* is almost perpendicular to the base and forms an angle β = 95.5° with the axis *a*. The axis *b* coincides with the two-fold axis of symmetry, while the plane *ac* corresponds to the plane of symmetry of the class. The principal axis *x* is almost parallel to the axis *c* and is almost perpendicular to the cleavage plane, the *y*-axis is almost parallel to this plane and to the axis *a*, and the *z*-axis coincides with the axis *b* [Sommerfeld 1949]. From one sample to another, the corresponding main indices vary within the intervals n_1 = 1.552-1.570, n_2 = 1.582-1.607, n_3 = 1.587-1.611, for λ_o = 689 nm. Therefore, the optical axes lie on the *zx*-plane about twenty degrees from the principal axis *x* (see Fig. 7.5).

Muscovite has the advantage of also being available with foils of several cm of diameter at a limited cost, as it does not require any optical processing. It is normally cemented between two glass plates that serve both to protect it and to eliminate the fringes of interference between its faces. However, its birefringence varies moderately from zone to zone of the lamina.

For precision applications, quartz is preferred, artificially grown using a hydrothermal process with which optically pure large crystals (α-SiO_2) are obtained. It is positive uniaxial with $n_e - n_o = 0.0092$ at 0.55 μm and has a transparency range from 0.25 μm to 2.5 μm. Its birefringence also varies with the temperature for which, under normal laboratory conditions, $d(n_e - n_o)/dT = -0.8\cdot 10^{-6}$/K, in addition to which thermal expansion $(1/L)\cdot dL/dT$ must also be considered. It is equal to $12.38\cdot 10^{-6}$/K in a direction perpendicular to the optical axis and $6.88\cdot 10^{-6}$/K in the parallel direction. For a quartz lamina with the optical axis parallel to the faces, with thickness $h = 1$ mm, for λ_o=0.55 μm, the phase shift ϕ is $2\pi\cdot 16.73$ radians and, for Eq. (7.11.40), it changes to

$$\Delta\phi \cong 2\pi\cdot 16.73\cdot\left(1.238\cdot 10^{-5} - 8.70\cdot 10^{-5}\right)\mathrm{K}^{-1}\Delta T \cong -2\pi\cdot 1.25\cdot 10^{-3}\mathrm{K}^{-1} dT \,. \tag{7.11.55}$$

Quartz also has optical activity, which becomes evident with a rotating power within a small angular range around the direction of the optical axis. In other directions, for a general wave passing through a quartz plate, this activity causes only a slight, essentially inappreciable, oscillation of the principal axes of the polarization ellipse as the wavelength or the thickness of the plate varies.

7.11.4.3 Multi-order wave plate

With the exception of mica, the crystalline laminae of reasonable thickness to be handled are, however, of a relatively high order, for example, a 1 mm thick quartz lamina has an order m of about 17 at 530 nm. The *multiple order* plates have a succession of wavelengths for which they behave like laminae $\lambda/4$, $\lambda/2$, and so on. In other words, as the wavelength varies, the phase shift induced by the lamina passes cyclically from the values in module 0, $\pi/2$, π, $3\pi/2$. Replacing $\nu = c/\lambda_o$ in Eq. (7.11.30), the step of this periodicity is approximately constant as a function of frequency:

$$\varphi = 2\pi \frac{h\nu}{c}\left(n_e - n_o\right), \tag{7.11.56}$$

from which we find that the period of this cycle is given by

$$\Delta\nu = \frac{c}{h\left(n_e - n_o\right)} \,. \tag{7.11.57}$$

For example, for a 1 cm thick quartz plate at 550nm, $\Delta\nu \cong 3.26$ THz and, for a sheet of calcite with the same thickness, $\Delta\nu \cong 173$ GHz. By placing a birefringent lamina between two polarizers with parallel or crossed axes, the transmitted radiation shows an oscillation of its intensity with a frequency step halved with respect to this value. However, this step varies according to the dependence of $\Delta\nu$ on the refractive indices, and hence on the wavelength. By measuring the position of the maxima and minima, various researchers have determined the birefringence of various materials in large spectral regions [Bennett 1995].

This periodicity with wavelength is useful for filtering the spectrum, excluding various components. In particular, the thickness can be fixed so that the lamina behaves differently for two distinct wavelengths. For example, a *double wavelength plate* is used to separate the pump beam, i.e., the carrier, at 1064 nm, from the second harmonic at 532 nm, generated in a non-linear duplicating crystal. Indeed, neglecting the dependence of the birefringence on the wavelength, the thickness is such that, for the second harmonic, the lamina is $\lambda/2$, while it is λ for the carrier. By exploiting the non-constancy of quartz birefringence and by appropriately choosing the thickness, double-effect plates are marketed for various other combinations, such as for the fundamental and third or fourth harmonics, at particular wavelengths.

7.11.4.4 Zero-order wave plate

On the other hand, the *zero-order* plates are formed by joining together two laminae of the same material, of slightly different thickness and with the optical axes rotated 90° to each other; in this way, the orders of each lamina are subtracted and the phase shift is obtained by placing h equal to the thickness difference in Eq. (7.11.30). For example, for a given wavelength, it is possible to obtain laminae of total order $m = 1/4$ or $1/2$ that behave like a thin lamina of that order. These composite plates have several advantages: their phase shift is much less sensitive to wavelength, inclination, and temperature variations. At the same time, at least with the natural radiation of a spectral lamp, they are sufficiently thick so as not to have interference fringes such as those afflicting the thin mica foils with the same phase shift. However, with laser radiation, it is good practice to apply an anti-reflective treatment to the external sides of the plate.

7.11.4.5 Achromatic wave plate

In some applications, for example, in Ellipsometry, the phase shift dependence from the wavelength constitutes a problem. In principle, the dependence on the term λ_o in the denominator of Eq. (7.11.30) could be compensated by a proportionality with λ_o of the difference of the indices of refraction in the numerator. However, there are no optically valid materials with this dependency. In fact, good achromatic laminae $\lambda/4$ consist of a Fresnel's rhombus. However, by joining laminae of different materials, it is possible to obtain achromatic plates that maintain an almost constant phase shift, for example, $\pi/2$, i.e., $\lambda/4$ plates, over the entire visible spectrum [Pancharatnam 1955]. The combination of one lamina of MgF_2 and one of KDP has a maximum deviation of only ±0.4% from 400 to 700nm [Bekers 1971]. The marketed achromatic laminae are now formed through the combination of several quartz and MgF_2 laminae cemented together. They have very good performance, but are sensitive to alignment. In low power applications in the visible and near infrared, there are achromatic plates consisting of a birefringent polymer multilayer inserted between two isotropic transparent sheets of support and protection.

7.11.4.6 Compensators

One popular, but uncomfortable, polarization analysis device is the *Babinet's compensator*, proposed by J. Babinet in 1837 [Babinet 1849]. It consists of two birefringent plates cut into wedges and superposed so as to collectively form a single plate of variable thickness with plane and parallel faces. Indeed, the two plates can slide between them controlled by a micrometer. One wedge has the optical axis parallel and the other has it perpendicular to the edge. This compensator looks like the Brewster prism, but the cutting angle with respect to the input and output faces is much smaller and the material used, typically quartz, has a weak birefringence, so there is no appreciable separation of the ordinary and extraordinary rays.

Since the optical axes of the two wedges are orthogonal to each other, the displacement that is obtained varies linearly with the position in the direction orthogonal to the edge:

$$\phi = \frac{2\pi}{\lambda_o}(n_e - n_o)(h_1 - h_2) = \frac{2\pi}{\lambda_o}(n_e - n_o)\,2x\tan\alpha\,, \tag{7.11.58}$$

where α is the angle between the faces of each wedge and the origin for the x-axis is taken on the line where the two thicknesses are equal. By placing the compensator between two crossed polarizers and observing an extended source through this system, there is a sinusoidal modulation of the intensity (localized on the compensator), with the maximum contrast when the axes of the polarizers are rotated by 45° with respect to the neutral axes of the compensator. In white light, only the fringe corresponding to $\phi = 0$ is black (or white, if the two polarizers are parallel), while the others are colored. The x position of these fringes can be adjusted by turning the micrometer and taking the "black" fringe as the reference "zero".

A generic wave to be analyzed is made incident on the compensator without the first polarizer. In the system of axes of the compensator, except for a common phase, such a wave can be defined as the vector

$$\begin{pmatrix} E_x \\ E_y e^{i\delta} \end{pmatrix},$$

Fig. 7.36
Soleil's compensator

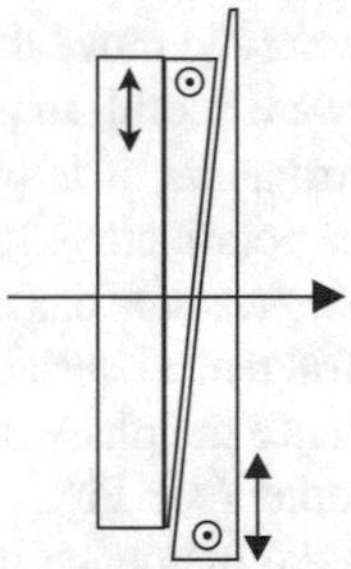

where E_x and E_y are its amplitudes taken as real and δ is the phase shift between the two polarization components. Proceeding along the x-axis, the overall phase shift $\phi + \delta$ of the wave transmitted by the compensator then passes through a succession of values ..., -4π, -2π, 0, 2π, 4π, ..., for which the transmitted wave is linearly polarized in a direction defined by the vector $(E_x, E_y)^T$. By adjusting the second polarizer in a direction perpendicular to it, for which the contrast has a maximum, a series of fringes can be observed whose intensity minima are shifted with respect to the previous calibration. By turning the micrometer, the fringes are returned to their original position. The phase shift δ of the wave in the compensator axis system is therefore deducted from the movement made. The intensity ratio between the two x and y components of the field can be determined by the orientation of the analyzer [Born and Wolf 1980]. Also, in the positions along the x-axis for which the phase shift $\phi + \delta$ is ..., -3π, $-\pi$, π, 3π, ..., the polarization of the transmitted wave is linear with the corresponding Jones' vector given by $(E_x, -E_y)^T$ (which is not generally perpendicular to the previous one). Therefore, even when orienting the analysis polarizer in an orthogonal direction to this vector, fringes with maximum contrast are observed. However, in white light, the fringes at $\pm\pi$ are not entirely "black", and this situation can be distinguished from the previous one.

The instrument that is generally marketed, however, is the *Soleil's compensator* [Soleil 1845, 1847] (Fig. 7.36), which is much more practical, consisting of two superimposed wedges, whose optical axes *are parallel to each other* and to the edges. In this way, the phase shift is constant on the opening and can be linearly changed by shifting the wedges between them. One of the wedges is cemented to a plate of the same birefringent material, with the optical axis rotated by 90°, and a thickness such as to compensate the phase shift of the two wedges in a given position. The result is a low-order variable wave plate. The use of this tool is similar to that of Babinet, except that, in the position corresponding to a null delay, with crossed polarizers, the entire field is dark. Therefore, one determines the translational positions that correspond to the extinction of the radiation transmitted by the polarizer analysis. This compensator has a delay resolution, expressed in distance, on the order of 0.001λ. The advantage of the Soleil compensator compared to that of Babinet is that the phase shift determination can be made with a detector. On the other hand, the use of a CCD camera can allow for the measurement of the fringes' translation in the Babinet compensator without the movement of the translator, but it is still necessary to move the analysis polarizer.

These compensators are useful, in principle, in polarimeters and ellipsometers. However, in these instruments, it is preferable to use *rotating* $\lambda/4$ (or even $\lambda/3$) plates, together with the polarizers [Chipman 1995; Azzam 1995].

Also, the *Berek's compensator* has a uniform phase shift. It consists of a simple sheet of uniaxial birefringent material with the optical axis perpendicular to the faces. By normal incidence the phase shift introduced is null, but it varies quadratically by tilting the lamina (see Fig. 7.31), with one neutral axis parallel and the other perpendicular to the incidence plane. Indeed, in Eq. (7.11.29), the angle θ now coincides with the internal angle of incidence ϑ, so that

$$\phi \cong 2\pi \frac{h}{\lambda_o}\left(n_e - n_o\right)\frac{\sin^2 \vartheta_I}{n^2 \cos\vartheta}, \qquad (7.11.59)$$

where n is the mean refractive index between n_e and n_o. This situation can occur involuntarily with vacuum windows of MgF_2 or sapphire, made with the optical axis normal to the faces.

Small corrections of the bias ellipticity can be performed with isotropic glass laminae, subjecting them to a uniform lateral compression along a transverse direction. On the other hand, an assembly of the optics that is too tight or an effort due to thermal expansion in cemented optics can alter the purity of the polarization.

7.11.4.7 Rotator of polarization by circular birefringence

Quartz has the peculiarity of also being optically active and can be grown in both levorotary and dextrorotary crystals. Its circular birefringence manifests itself when the propagation direction is close to the direction of the optical axis. Therefore, a quartz plate, cut with the faces perpendicular to this axis, has the property of rotating the polarization plane of a linearly polarized wave. The angle of rotation depends on the thickness of the lamina and on the wavelength. It is 21.7° per mm of thickness at 589 nm (see Tab. 7.6). However, the angular aperture for this effect is limited to a few degrees and, at 5° of inclination from the optical axis, the ordinary birefringence already prevails.

7.11.5 Depolarizers

While there are devices capable of effectively and uniformly polarizing natural radiation, there is no equally effective way of doing the opposite.

Yet, a depolarizing device can be useful, for example, in the detection of a signal: if the detector is, to some extent, sensitive to the polarization and if the polarization of the incident wave fluctuates due to unwanted causes, the revealed intensity will bear the sign of these fluctuations, adding noise to the signal. Ideally, a depolarizer should be represented by a Mueller matrix such as the one at the bottom of Table 7.9, with which the intensity of the wave is conserved, while all of the other Stokes coefficients are zeroed. A depolarizer should randomly alter, both spatially and temporally, the phase between any two orthogonal components of the wave field. One way to do this is to use a diffuser, possibly rotating. However, it has the defect of spreading the radiation spatially, subtracting it from the original direction. A less destructive, and also more economic, way to preserve the spatial coherence of the wave is to use a thin sheet of crumpled and then re-extended *cellophane*. In this manner, small areas of varying orientation are formed on the sheet, for each of which the produced phase shift is different. To be effective, the incident wave must be large enough to illuminate many of these areas. Currently, there are depolarizers made of a thin film of liquid crystal polymer sandwiched between two glass plates. A pattern of random varying retardation and fast-axis orientation is imprinted in the polymer. The retardation also depends on the wavelength, making these devices very effective with broadband radiation.

If the radiation has an extended spectral band, a good device is the *Lyot's pseudo-depolarizer*. It consists of two uniaxial birefringent laminae of high order and different thickness, rotated so that the neutral axes of the second lamina are at 45° from those of the first. The overall Mueller's matrix is

$$\begin{pmatrix} 1 & 0 & 0 & 0 \\ 0 & \cos\delta_2 & \sin\delta_2\sin\delta_1 & -\sin\delta_2\cos\delta_1 \\ 0 & 0 & \cos\delta_1 & \sin\delta_1 \\ 0 & \sin\delta_2 & -\cos\delta_2\sin\delta_1 & \cos\delta_2\cos\delta_1 \end{pmatrix}. \tag{7.11.60}$$

Applied to a generic Stokes vector, the polarization of the transmitted radiation is represented by

$$\begin{aligned} S_0' &= S_0, \\ S_1' &= \cos\delta_2 S_1 + \\ &\quad 0.5\left[\cos(\delta_2-\delta_1)-\cos(\delta_2+\delta_1)\right]S_2 - 0.5\left[\sin(\delta_2-\delta_1)+\sin(\delta_2+\delta_1)\right]S_3, \\ S_2' &= \cos\delta_1 S_2 + \sin\delta_1 S_3, \\ S_3' &= \sin\delta_2 S_1 + \\ &\quad 0.5\left[\sin(\delta_2-\delta_1)-\sin(\delta_2+\delta_1)\right]S_2 + 0.5\left[\cos(\delta_2-\delta_1)+\cos(\delta_2+\delta_1)\right]S_3. \end{aligned} \tag{7.11.61}$$

On the other hand, the phase shifts δ_1 and δ_2 depend on the wavelength according to the relation

$$\delta_1 = \Delta n 2\pi\sigma d_1, \quad \delta_1 = \Delta n 2\pi\sigma d_1, \tag{7.11.62}$$

where Δn is the birefringence of the medium, i.e., the difference between the refractive indices of the ordinary and extraordinary rays, d_1 and d_2 are the thicknesses of the two laminae and $\sigma = 1/\lambda_o$ is the spatial frequency, i.e., the wave number. Suppose that the incident radiation has a Gaussian power spectrum expressed by

$$p(\sigma) = p_o e^{-(\sigma-\sigma_o)^2/\Delta\sigma^2}, \tag{7.11.63}$$

where σ_o is the central wave number and $\Delta\sigma$ is the bandwidth. The Stokes parameters for the overall power of the transmitted wave are then given by the integral of the expressions (7.11.61) weighted with the Gaussian function of Eq. (7.11.63). These integrals are the sum of terms like

$$\begin{aligned} &\int_{-\infty}^{\infty} \sin(2h\sigma)e^{-(\sigma-\sigma_o)^2/\Delta\sigma^2}d\sigma = \Delta\sigma\sqrt{\pi}\sin(2h\sigma_o)e^{-\Delta\sigma^2h^2}, \\ &\int_{-\infty}^{\infty} \cos(2h\sigma)e^{-(\sigma-\sigma_o)^2/\Delta\sigma^2}d\sigma = \Delta\sigma\sqrt{\pi}\cos(2h\sigma_o)e^{-\Delta\sigma^2h^2}, \end{aligned} \tag{7.11.64}$$

where h takes the value of $\pi\Delta n d_1$, $\pi\Delta n d_2$, $\pi\Delta n(d_1 \pm d_2)$. In order that, for a generic polarization input, the output polarization vanish, it is necessary that $h >> 1/\Delta\sigma = l_c$, where l_c is the coherence length of the wave, that is, d_1, d_2, $|d_1 \pm d_2| >> 1/(\pi\Delta n\Delta\sigma)$.

This device does not destroy the information on the initial polarization, but hides it in the space of temporal frequencies. Indeed, with a plane wave incident perpendicular to the laminae, the transmitted polarization is periodically altered as the wave number changes and can be restored by an identical device with the neutral axes rotated by 90°.

With monochromatic radiation, or nearly so, the Lyot depolarizer is not adequate. If the polarization of the wave is linear, that is, if its opposite circular components have the same intensity, the *Cornu's pseudo-depolarizer* can be used, which exploits optical activity. It consists of two wedges of crystalline quartz, one levorotary and the other dextrorotary, joined together so as to form a single element with the optical axes perpendicular to the input and output faces. Its construction is similar to the Wollaston prism. Indeed, at the diagonal face of union between the two crystals, it separates the two circular components of the incident radiation, deflecting them in opposite directions along the transverse direction, say, x, orthogonal to the edge of the wedges, but this deviation is very small and has no practical application as a polarizer. The two waves therefore remain overlapped, but, locally, they add up, with different phases giving rise to a linear polarization that rotates with the position along the x-axis, forming "polarization fringes" parallel to the edge of the wedges. For a greater degree of depolarization, which involves all of the transverse directions, two of these depolarizers must be used, rotated 90° to each other. In order for Cornu's prism to be effective, the incident wave must be sufficiently large to cover many of these fringes. Moreover, at least with quartz, the angular acceptance, in order to neglect the ordinary birefringence, is only a few degrees, beyond which the rotation effect of the polarization disappears.

A pseudo-depolarizer can also be made with a prism similar to the Babinet compensator, but with the two wedges glued together. In this case, the optical axes of the two wedges are parallel to the faces and orthogonal to each other. When the incident wave is circularly polarized, a modulation of the transmitted polarization is obtained along the transverse direction perpendicular to the edge of the wedges. If their cutting angle is large enough, the modulation period can be made much smaller than the beam width, and the transmitted wave may appear unpolarized.

7.11.6 Combination of birefringent plates rotated between them

The case of two birefringent plates with different phase shift, with the neutral axes rotated from one another by whatever angle φ, is rather complicated and is not amenable to the behavior of a single birefringent lamina. I consider here only the case in which the two laminae are equal. We begin to study this combination as a transformation from the Cartesian frame ℓ of the eigenvectors of the first lamina to the analogous one $\mathscr{L}$ of the second lamina. For a rotation φ of the second lamina taken to be positive in the counterclockwise direction looking toward the source, the overall matrix of the transformation is given by

$$\mathfrak{M}^{\ell}_{\mathscr{L}} = e^{i2\delta}\begin{pmatrix} e^{-i\delta} & 0 \\ 0 & e^{i\delta}\end{pmatrix}_{\mathscr{L}}\begin{pmatrix} \cos\varphi & \sin\varphi \\ -\sin\varphi & \cos\varphi\end{pmatrix}^{\ell}_{\mathscr{L}}\begin{pmatrix} e^{-i\delta} & 0 \\ 0 & e^{i\delta}\end{pmatrix}_{\ell} =$$

$$= e^{i2\delta}\begin{pmatrix} e^{-i2\delta}\cos\varphi & \sin\varphi \\ -\sin\varphi & e^{i2\delta}\cos\varphi\end{pmatrix}^{\ell}_{\mathscr{L}}.$$

Now, we look for two rotations, one before and one after the combination, to diagonalize the resulting matrix. It turns out that these two rotations must be equal, with both, for $\cos 2\delta > 0$, in the opposite direction to the rotation of the second lamina with respect to the first one. The angle φ' of each is defined, module 90°, by

$$\tan 2\varphi' = -\frac{\tan\varphi}{\cos 2\delta}. \tag{7.11.65}$$

Indeed, including these two rotations, between the first frame ℓ' and the last frame $\mathscr{L}'$, the matrix of the device is

Fig. 7.37
Evans Compensator

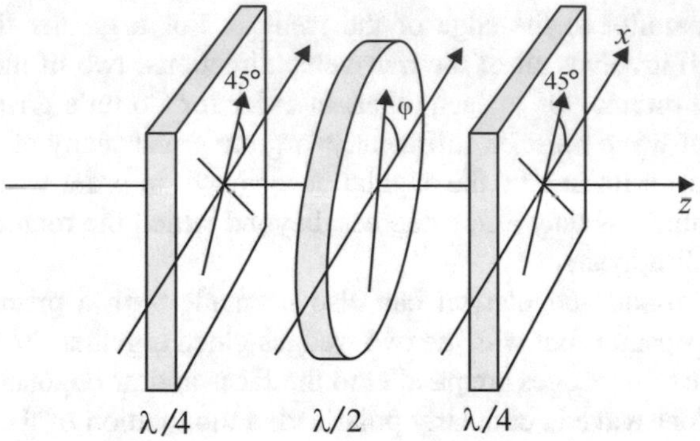

$$\mathfrak{M}_{\mathscr{L}'}^{\ell'} = e^{i2\delta}\begin{pmatrix} \cos\varphi' & \sin\varphi' \\ -\sin\varphi' & \cos\varphi' \end{pmatrix}_{\mathscr{L}'}^{\mathscr{L}} \mathfrak{M}_{\mathscr{L}}^{\ell} \begin{pmatrix} \cos\varphi' & \sin\varphi' \\ -\sin\varphi' & \cos\varphi' \end{pmatrix}_{\ell}^{\ell'}$$

$$= e^{i2\delta}\begin{pmatrix} -\dfrac{\sin\varphi}{\sin 2\varphi'} - i\cos\varphi\sin 2\delta & 0 \\ 0 & -\dfrac{\sin\varphi}{\sin 2\varphi'} + i\cos\varphi\sin 2\delta \end{pmatrix}_{\mathscr{L}'}^{\ell'}, \tag{7.11.66}$$

where Eq. (7.11.65) has been applied. The overall rotation from the first to the last Cartesian frame is $\varphi - 2\varphi'$. Only in the case in which $\delta = 0 \pm m\pi$ with m integer are the first and last frames equal. In other cases, the combination of two laminae is similar to a single birefringent lamina plus a rotation.

An interesting case is the combination of two $\lambda/4$ plates in series that may rotate between them. This device provides a simple compensator, in fact, the phase shift is zero when their fast axes are crossed and $\lambda/2$ when they are parallel. For Eq. (7.11.65), with $\delta = \pi/4$ and $0 \le \varphi \le 90°$, one possible solution is $\varphi' = 45°$, and Eq. (7.11.66) becomes

$$\mathfrak{M}_{\mathscr{L}'}^{\ell'} = \begin{pmatrix} e^{-i\varphi} & 0 \\ 0 & -e^{i\varphi} \end{pmatrix}_{\mathscr{L}'}^{\ell'}. \tag{7.11.67}$$

This device can be easily interpreted as follows: the first lamina transforms the linear polarization components at 45° from its neutral axes into circular. The subsequent rotation introduces a phase shift between these components, which are turned back in linear polarization from the second lamina at 45° from its neutral axes. Therefore, holding the first lamina and rotating the second one, this device behaves as a variable lamina, followed by a rotator of an angle equal to the rotation of the second lamina ∓90°. This fact makes it generally inconvenient to apply.

A device that functions as a variable compensator without the rotation of neutral axes has been proposed by J.W. Evans (Fig 7.37) to improve the performance of the Lyot's filter that we will study in the next section [Evans 1949]. It consists of two $\lambda/4$ plates with fast axes that are parallel or crossed, between which a rotatable $\lambda/2$ lamina is placed. In short, the first $\lambda/4$ lamina transforms the linear polarization components at 45° from its neutral axes into opposite circular polarizations. The $\lambda/2$ lamina, rotating, varies the phase between these circular components. Finally, the second $\lambda/4$ lamina re-transforms them in linear form, but with the phase shift introduced by the $\lambda/2$ lamina. Its matrix, expressed in the frame where the $\lambda/4$ plates both have their neutral axes rotated by 45°, is

$$\mathfrak{M}_{2\varphi} = \frac{1+i}{2}\begin{pmatrix} 1 & -i \\ -i & 1 \end{pmatrix}\begin{pmatrix} \cos 2\varphi & -\sin 2\varphi \\ -\sin 2\varphi & -\cos 2\varphi \end{pmatrix}\frac{1+i}{2}\begin{pmatrix} 1 & -i \\ -i & 1 \end{pmatrix} = i\begin{pmatrix} e^{i2\varphi} & 0 \\ 0 & -e^{-i2\varphi} \end{pmatrix}. \tag{7.11.68}$$

If, instead, the $\lambda/4$ laminae have their fast axes crossed, a rotation of 90° is added to the phase shift.

7.11.7 *Spectral filters by birefringence*

The dependence of the phase shift on the wavelength is exploited to build very efficient spectral filters, with which one can obtain spectral bands of less than 0.1 nm, two orders of magnitude lower than those of the interference filters. These filters can be interpreted as two-wave interferometers, in which the difference in path is given by birefringence.

The simplest example is given by a birefringent lamina between two polarizers whose transmission axes, which we take to be the x-axis, are parallel to each other and at 45° from the neutral axes of the lamina. The power transmission for the x component of the field is obtained from Eq. (7.11.44), with $\varphi = 45°$,

$$T = \cos^2(\delta), \quad \text{where } \delta = \pi\sigma\Delta n\, h = \phi/2\,, \tag{7.11.69}$$

where h is the thickness of the lamina, $\sigma = 1/\lambda_o$ is the wave number and Δn it is the difference between the indices of the extraordinary ray and the ordinary ray. The y component of the field is instead extinguished by the second polarizer. This filter is used to discriminate between two spectral lines of wavelength λ_1 and λ_2, extinguishing the first and transmitting the second [Strumia 1973]. For this purpose, it is necessary to determine the thickness h to be assigned to the lamina so that it behaves like a half-wave plate for the first line and (possibly) as a full-wave plate for the second line. Therefore, the following system must be solved:

$$\begin{cases} (m+l+0.5)\lambda_1 = h\Delta n(\lambda_1), \\ m\lambda_2 = h\Delta n(\lambda_2), \end{cases}$$

where m is the order of interference of the second line and $m + l + 0.5$ that of the first line. By solving, we get

$$\begin{cases} m = (l+0.5)\dfrac{\Delta n(\lambda_2)\lambda_1}{\Delta n(\lambda_1)\lambda_2 - \Delta n(\lambda_2)\lambda_1}, \\ h = (m+l+0.5)\dfrac{\lambda_1}{\Delta n(\lambda_1)}. \end{cases} \tag{7.11.70}$$

When the two lines are close with respect to their wavelength, we can request that the thickness of the lamina be such as to be able to assume that $l = 0$ for $\lambda_1 < \lambda_2$, or $l = -1$ for $\lambda_1 > \lambda_2$. In this case, with $\Delta n(\lambda_1) \cong \Delta n(\lambda_2)$, we have that

$$\begin{cases} m = \dfrac{\lambda_1}{2|\lambda_2 - \lambda_1|}, \\ h = (m \pm 0.5)\dfrac{\lambda_1}{\Delta n}, \quad + \text{ for } \lambda_1 < \lambda_2, \quad - \text{ for } \lambda_1 > \lambda_2. \end{cases} \tag{7.11.71}$$

With the thickness h so determined, the difference of order between the two wavelengths is 0.5. However, from the first equation, it also turns out that m is not generally an integer. Nevertheless, when m is large, we can approximate it with the nearest integer and insert it into the second equation to obtain the thickness that produces the extinction at the wavelength λ_1, even if the other is not at the maximum of the transmission. If the thickness h is very large, it cannot be calculated with the precision necessary to have the extinction at λ_1; in fact, Δn is not known with much accuracy (the uncertainty for the quartz is on the order of 0.1%). However, with thicknesses of 1 cm or more, we can exploit the dependence of birefringence with temperature (see §7.11.4.2). In this case, it is necessary to thermostat the filter, and we can also extinguish one or the other of the two wavelengths by varying the temperature. For example, suppose that λ_1 and λ_2 correspond to the lines D1 and D2 of sodium. Using quartz as the material, the thickness required for the filter lamina is 31.94 mm, corresponding to the order $m = 494$. By varying the temperature by about 8 °C, the transmission of the filter varies from one line to the other.

7.11.7.1 Lyot's filters

For the astrophysical observation of the solar corona, in 1933, B. Lyot and then independently, in 1937, Y. Öhman, proposed and used a composite filter consisting of a series of polarizers alternated with a series of birefringent lamina (Fig. 7.38) [Lyot 1944; Öhman 1938]. The thickness doubles from one lamina to the next, while the neutral axes remain parallel to each other and at 45° from the axes

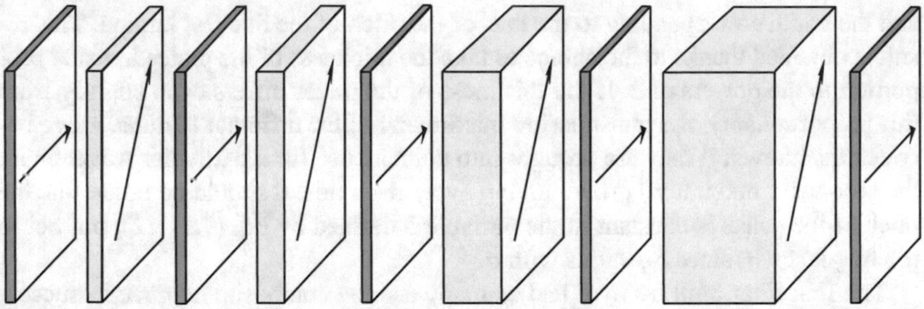

Fig. 7.38 Lyot's filter. The neutral axes of the plates (L) are rotated at 45° from the transmission axes of the polarizers (P)

of transmission of the polarizers, all of them also parallel to each other, along a direction that we take here to be the *x*-axis.

For the *j*-th polarizer-lamina pair, the power transmission for the *x* component of the field entering the next pair is now given by

$$T_j = \cos^2\left(2^{j-1}\delta\right), \quad \text{where } \delta = \pi\sigma\,\Delta n\, h\,, \tag{7.11.72}$$

where h is the thickness of the first lamina. The y component of the field is instead extinguished by the following polarizer. The transmission for the whole sequence of N pairs of polarizer-lamina + final polarizer, for the x component of the input field, is therefore given by

$$T = \prod_{j=1}^{N} T_j = \frac{1}{2^{2N}}\left[\prod_{j=0}^{N-1} e^{i2^j\delta}\left(1+e^{-i2^j2\delta}\right)\right]^2 . \tag{7.11.73}$$

Since

$$\sum_{j=0}^{N-1} 2^j = 2^N - 1\,, \text{ while } \prod_{j=0}^{N-1}\left(1+a^{2^j}\right) = 1+a+a^2+a^3+\cdots+a^{2^N-1}$$

is the sum of the first 2^N-1 integer powers of a, we have that

$$T = \frac{1}{2^{2N}} e^{i2\delta\left(2^N-1\right)} \frac{\left(1-e^{-2^N i2\delta}\right)^2}{\left(1-e^{-i2\delta}\right)^2} = \left[\frac{1}{2^N}\frac{\sin\left(2^N\delta\right)}{\sin(\delta)}\right]^2 . \tag{7.11.74}$$

The result is a succession of main peaks with the pitch of the thinner lamina and the width corresponding to the half of the pitch of the thickest lamina. This result is obtained thanks to the choice to take the thickness of the plates in exact proportion to the powers of 2. If the thickness of the plates differs even slightly from this proportionality, the transmission maximums of the different laminae move between them: even if they are brought into coincidence for a particular wavelength, the secondary maximums grow moving away from this. It should be noted that the pitch of the peaks is constant in the variable δ defined by Eq. (7.11.72), but not in the frequency σ, since Δn varies with σ.

The first filter built by Lyot had as many as nine composite laminae in succession, reaching a bandwidth of about 0.1 nm, much lower than that obtained with the interferential filters, maintaining a great brightness, much higher than that of a monochromator. This filter had various measures to increase the angular field of the instrument. In fact, the phase shift also depends on the incidence conditions of the radiation on the laminae.

Let us return to the case of a single sheet of uniaxial material with the optical axis parallel to the faces. Instead of using Eq. (7.11.36), let us start from Eq. (7.8.2) for the phase shift, applying a different approximation valid for small angles of incidence. Limiting ourselves to considering the terms up to the second order in ϑ, we have that

$$\Delta(\vartheta_I) = \frac{2\pi}{\lambda_o} h(n'' - n') - \frac{\pi}{\lambda_o} h\left(n''\vartheta''^2 - n'\vartheta'^2\right). \tag{7.11.75}$$

On the other hand, for Eqs. (7.4.41) with a uniaxial crystal, we have that $n' = n_o$, and, for small angles ϑ, n'' can be approximated by

$$n'' = n_e \left[1 + \frac{1}{2} n_e^2 \vartheta''^2 \cos^2\varphi \left(\frac{1}{n_e^2} - \frac{1}{n_o^2} \right) \right], \tag{7.11.76}$$

where Eq. (7.11.35) was used. We can also assume that

$$\vartheta'' \cong \vartheta_I / n'' \cong \vartheta_I / n_e, \quad \vartheta' \cong \vartheta_I / n' = \vartheta_I / n_o .$$

Therefore, taking only the terms up to ϑ^2, with various steps, we get

$$\Delta(\vartheta_I) = \frac{2\pi}{\lambda_o} h(n_e - n_o) \left[1 - \frac{1}{2n_o} \vartheta_I^2 \left(\frac{1}{n_o} \cos^2\varphi - \frac{1}{n_e} \sin^2\varphi \right) \right], \tag{7.11.77}$$

where φ is the angle between the plane of incidence and the optical axis (Fig. 7.32). With convergent monochromatic radiation, the filter field appears to be crossed by isochromatic fringes such as those in Fig. 7.20. These are families of hyperboles with asymptotes arranged at the angles $\pm\varphi_a$ from the optical axis such that $\tan^2 \varphi_a = n_e/n_o$. In the approximation of small birefringence with respect to the refractive index, they are equilateral hyperboles with 45 ° asymptotes from the optical axis. The dependence of the phase shift on the angle of incidence ϑ is expressed by Eq. (7.11.77) and is stronger for $\varphi = 0°$ and for $\varphi = 90°$, also having opposite signs for these two values. In the previous example of the sodium doublet, for a variation of the phase difference of less than 0.2π radians, the angular field for ϑ_I is limited to only 1.8°.

To alleviate this dependency, Lyot proposed various remedies. The first considers a device consisting of two thick birefringent plates, between which it is placed a $\lambda/2$ lamina of order zero that is as achromatic as possible. The two plates are made from the same piece worked and cut in two pieces of the same thickness, equal to half of that necessary in the previous case, and arranged with the optical axes at 90 ° to each other, which we take here to be x and y axes. The neutral axes of the $\lambda/2$ lamina are instead placed at 45° from those of the thick plates. As this lamina, so oriented, swaps the amplitudes of the fields between the x and y axes, the increase in phase shift that results from the inclination of the wave in a plate is largely compensated by the decrease in phase shift that one has with the other plate. Adding the two displacements together, we indeed have that

$$\Delta(\vartheta_I) = \frac{2\pi}{\lambda_o} h(n_e - n_o) - \frac{\pi}{\lambda_o} h(n_e - n_o)^2 \frac{1}{2n_o^2 n_e} \vartheta_I^2 , \qquad (7.11.78)$$

where h is the overall thickness of the two plates. The isochromatic fringes are now circles, and, with the parameters of the previous example, the angular field becomes approximately 33°. For astrophysical applications, this solution requires the use of a $\lambda/2$ achromatic compact lamina, with almost constant phase shift over the visible spectrum, something that was difficult to achieve in Lyot's time with existing materials.

To obviate this drawback, Lyot proposed, as a second remedy, a composite plate of two materials such as quartz and calcite, one positive and the other negative, with the optical axes parallel to each other. In this way, their phase shifts are partially erased. However, by appropriately calculating the relative thickness of the two crystals, it is possible to make the isochromatic fringes circular, eliminating the dependence on φ from the term to the second order in ϑ_I in the phase shift of the composite plate. This ratio between the thicknesses is calculated for a specific wavelength for which the maximum filter transmission is desired. For the other maxima, the dependence on φ is not exactly canceled, due to the different relative dispersion of the two materials. However, for the quartz-calcite combination, the spectral band, understood in this sense, extends over the visible spectrum and the angular field is only slightly less than in the previous case.

With these measures, the superiority of the birefringence filters is evident with respect to the interferential filters, which have much smaller fields of view, especially when their transmission band is very narrow.

The usefulness of a filter is greatly increased if it can be made tunable. This can be achieved by changing the temperature and, indeed, a very selective Lyot filter must be carefully thermo-stabilized in an oil bath with an index close to that of the optical components, which also serves to reduce the parasitic reflections on the faces of the laminae. However, the tunability is limited to a few tens of nm. For a larger scan, the thickness of the laminae should be varied with the same number of Soleil compensators. This issue was taken up by Evans, who has proposed a more elegant method for obtaining a continuously tunable filter. If, indeed, in the first type of Lyot filter, instead of the $\lambda/2$ lamina, we replace the Evans compensator, this time with the fast axes of the two $\lambda/4$ plates crossed and at 45° from those of the filter plates, we obtain, at the same time, a large angular field and the tunability given by the rotation of the $\lambda/2$ lamina. Each of the laminae of the Lyot filter in Fig. 7.38 is then divided into two with an Evans compensator in the middle. The angle of rotation of each $\lambda/2$ lamina must double in the following element. In this way, the whole filter becomes tunable.

7.11.7.2 Lyot's filters in laser cavities

With well-collimated laser beams, the dependence of the phase shift on the incidence conditions instead becomes an opportunity. First in the dye lasers, and then in the titanium-sapphire lasers, Coherent introduced a tunable Lyot's filter consisting of three quartz plates inclined at the Brewster's angle [Bloom 1974]: their rotation around an axis perpendicular to their faces allows us to vary the wavelength emitted by the laser. The inclination at the Brewster's angle has a dual function: to reduce, practically to zero, the losses by reflection for the polarization component of the field to which the laser operates and act as a polarizer attenuating the other component below threshold. The three plates are sufficiently thin, with a thickness such that the ordinary and extraordinary beams remain overlapped in the dimension of their diameter. To increase the fineness of the device, the thickness is quadrupled from one sheet to the next, without the secondary peaks going above the threshold. Given that the reflection at the Brewster's angle does not constitute an ideal polarizer, other secondary peaks appear that correspond to wavelengths for which the sum of the thicknesses of the laminae results in a phase shift equal to an integer multiple of 2π.

Consider one of these laminae. Since the approximation of a small incidence angle is not valid here, we instead take advantage of the small difference between the quartz refractive indices and use Eq. (7.11.36). At the Brewster's angle, we can assume that $\tan\vartheta = 1/n$, where n is the mean index between n_o and n_e, while ϑ is the internal angle, and therefore

$$\phi \cong \frac{2\pi}{\lambda_o} h\left(n_e - n_o\right) \frac{n^2 + \sin^2\varphi}{n\sqrt{1+n^2}}, \tag{7.11.79}$$

where I recall that φ is the angle between the optical axis and the plane of incidence. On the other hand, for Eq. (7.11.37), the neutral axes are rotated from the plane of incidence of an angle given by

$$\tan\varphi' = \sqrt{1+\frac{1}{n^2}}\tan\varphi\,. \tag{7.11.80}$$

For quartz, assuming that $\varphi' = 45°$ for the maximum modulation, $n_e - n_o = 0.00886$, $n = 1.542$, $\varphi \cong 40°$. Furthermore, with $\lambda_o = 860$ nm, $h = 1.97$ mm, the order of interference is about 20, which corresponds to a step for the wavelength between an order and the next of about 45 nm. With this order, the wavelength can be changed by about 26 nm by rotating the lamina by 5°.

7.11.7.3 Solc's filters

Solc's filters are made of N identical plates, whose neutral axes are oriented in a pre-established sequence. The filtering effect is obtained by placing only two polarizers, one at the beginning and one at the end. In Solc's *folded filter* the laminae have their axis alternately at the angles $+\varphi$ and $-\varphi$ with respect to the x-axis. Contrastingly, in Solc's *fan filter* one axis is progressively rotated with respect to the x-axis according to the sequence φ, 3φ, 5φ, ... , $(2N-1)\varphi$.

For the folded filter (Fig. 7.39), we consider the matrix of a pair of laminae referred to the Cartesian axes x, y

$$\mathfrak{M} = e^{i2\delta} \times \begin{pmatrix} A & B \\ C & D \end{pmatrix},$$

with $\begin{pmatrix} A & B \\ C & D \end{pmatrix} =$

$$= \begin{pmatrix} 1-2(\sin\delta\cos 2\varphi + i\cos\delta)\sin\delta\cos 2\varphi & -\sin^2\delta\sin 4\varphi \\ \sin^2\delta\sin 4\varphi & 1-2(\sin\delta\cos 2\varphi - i\cos\delta)\sin\delta\cos 2\varphi \end{pmatrix}.$$

Now, consider $M = N/2$ pairs of plates in succession. The overall matrix is obtained from Eq. (3.4.31)

$$\begin{pmatrix} A & B \\ C & D \end{pmatrix}^M = \begin{pmatrix} A - \dfrac{\sin(M-1)\alpha}{\sin M\alpha} & B \\ C & D - \dfrac{\sin(M-1)\alpha}{\sin M\alpha} \end{pmatrix} \frac{\sin M\alpha}{\sin\alpha}, \tag{7.11.81}$$

where $(A + D)/2$ is a real number between -1 and 1 and

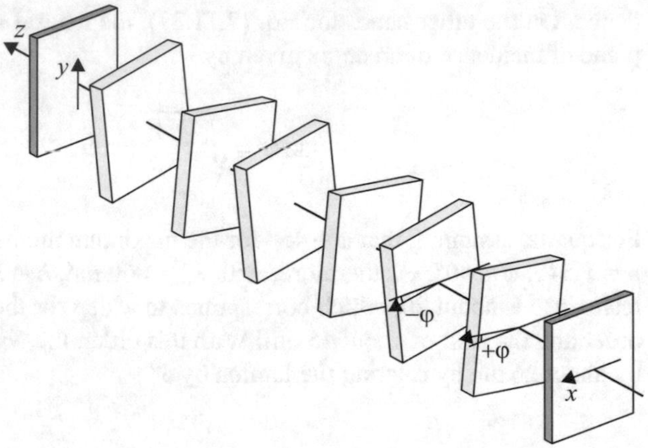

Fig. 7.39 Solc's Folded filter. N equal birefringent plates (in light gray) are alternately rotated by an angle $\varphi = \pi/(4N)$ clockwise and counter-clockwise. The series of plates is inserted between two polarizers (dark gray) whose transmission axes are crossed between them

$$\cos\alpha = \frac{A+D}{2} = 1 - 2\sin^2\delta\cos^2 2\varphi\,. \tag{7.11.82}$$

If the set of plates is disposed between two crossed polarizers, with the transmission axis of the first along the x-axis and the axis of the second along the y-axis, the transmitted power (for the x component of the input field, with power P_x) is simply

$$P_{\mathrm{T}} = P_x\left(\sin^2\delta\sin 4\varphi\frac{\sin M\alpha}{\sin\alpha}\right)^2. \tag{7.11.83}$$

Here, Yariv and Yeh (1984) suggest the change of variable $\alpha = \pi - 2\beta$, with which Eqs. (7.11.82-83) become, respectively,

$$\cos\beta = \sin\delta\cos 2\varphi\,, \tag{7.11.84}$$

$$P_{\mathrm{T}} = P_x\left(\tan 2\varphi\cos\beta\frac{\sin N\beta}{\sin\beta}\right)^2. \tag{7.11.85}$$

As an alternative, in the fan filter (Fig. 7.40), the matrix corresponding to the succession of laminae can be written as

$$\begin{aligned}\mathfrak{M} &= \mathfrak{R}_{-(2N-1)\varphi}\mathfrak{M}_{2\delta}\mathfrak{R}_{(2N-1)\varphi}\cdots\mathfrak{R}_{-3\varphi}\mathfrak{M}_{2\delta}\mathfrak{R}_{3\varphi}\mathfrak{R}_{-\varphi}\mathfrak{M}_{2\delta}\mathfrak{R}_{\varphi}\\ &= \mathfrak{R}_{-(2N+1)\varphi}\left(\mathfrak{R}_{2\varphi}\mathfrak{M}_{2\delta}\right)^N\mathfrak{R}_{\varphi}\,,\end{aligned}$$

where

$$\mathfrak{R}_{m\varphi} = \begin{pmatrix} \cos m\varphi & \sin m\varphi \\ -\sin m\varphi & \cos m\varphi \end{pmatrix}, \quad \mathfrak{M}_{2\delta} = e^{i\delta} \begin{pmatrix} e^{-i\delta} & 0 \\ 0 & e^{i\delta} \end{pmatrix}.$$

Therefore,

$$\left(\mathfrak{R}_{2\varphi}\mathfrak{M}_{2\delta}\right)^N = e^{iN\delta} \begin{pmatrix} e^{-i\delta}\cos 2\varphi - \dfrac{\sin(N-1)\beta}{\sin N\beta} & e^{i\delta}\sin 2\varphi \\ -e^{-i\delta}\sin 2\varphi & e^{i\delta}\cos 2\varphi - \dfrac{\sin(N-1)\beta}{\sin N\beta} \end{pmatrix} \frac{\sin N\beta}{\sin\beta},$$

where

$$\cos\beta = \cos\delta\cos 2\varphi. \tag{7.11.86}$$

We also take the rotation φ such that

$$\varphi = \frac{\pi}{4N}. \tag{7.11.87}$$

Therefore, the last lamina is rotated at the angle $(2N-1)\varphi = \pi/2 - \varphi$. In this way, after lengthy calculations, the matrix of the succession is

$$\mathfrak{M} = e^{iN\delta} \begin{pmatrix} \cos\delta\sin 2\varphi \dfrac{\sin N\beta}{\sin\beta} & -\cos N\beta - i\sin\delta \dfrac{\sin N\beta}{\sin\beta} \\ \cos N\beta - i\sin\delta \dfrac{\sin N\beta}{\sin\beta} & \cos\delta\sin 2\varphi \dfrac{\sin N\beta}{\sin\beta} \end{pmatrix}. \tag{7.11.88}$$

The fan filter is placed between parallel polarizers with their transmission axis along the *x*-axis. Applying Eq. (7.11.86), the transmitted power is still given by

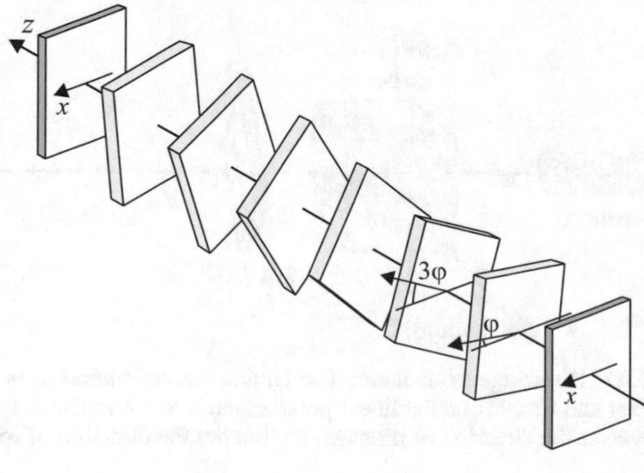

Fig. 7.40 Solc's fan filter. The N birefringent laminae are now progressively rotated by an angle $\varphi = \pi/(4N)$. The series of laminae is inserted between two polarizers whose transmission axes are parallel to each other

Eq. (7.11.85). The meaning of the parameter β is, however, different, and is expressed by Eq. (7.11.84) for the folded filter and (7.11.86) for the fan filter.

Also, for the folded filter, the angle φ is assigned with Eq. (7.11.87) and the laminae are taken to be thick so as to act as λ/2 laminae for the wavelength for which maximum transmission is desired. Instead, in the fan filter, this maximum is obtained with full wave plates.

The advantage of Solc's filters compared to those of Lyot is that only two polarizers are sufficient. However, their performance was considered inferior and the secondary peaks are more pronounced.

7.11.8 Optical isolators

In various applications, particularly with laser beams, it is necessary to prevent the return of radiation to the source. For example, in saturation spectroscopy, a mirror is used so that a beam travels backwards through the cell containing the gas to be analyzed, but is then deflected toward a detector. A similar situation occurs with the use of "double-pass" acousto-optical modulators used to produce a wave diffracted by the acoustic wave in the modulator material. The diffracted wave frequency is shifted as in the Doppler effect by an amount equal to twice that of the acoustic wave (typically, several tens or hundreds of MHz).

For this type of application, the optical system used as an isolator typically consists of a λ/4 lamina and a polarizing cube (Fig. 7.41). The cube is arranged to transmit (reflect) the maximum power of the beam, linearly polarizing the incident radiation along the direction of its axis. After the cube, there is a lamina λ/4 whose optical axes are at 45° from the transmission/reflection axes of the cube. In this way, the outgoing radiation is circularly polarized right or left. After passing through the cell, whose medium is supposed to be non-birefringent, the wave is reflected back by the mirror, and its helicity is inverted, too. The λ/4 lamina retransforms it into linear, but, this time, with the polarization direction rotated by

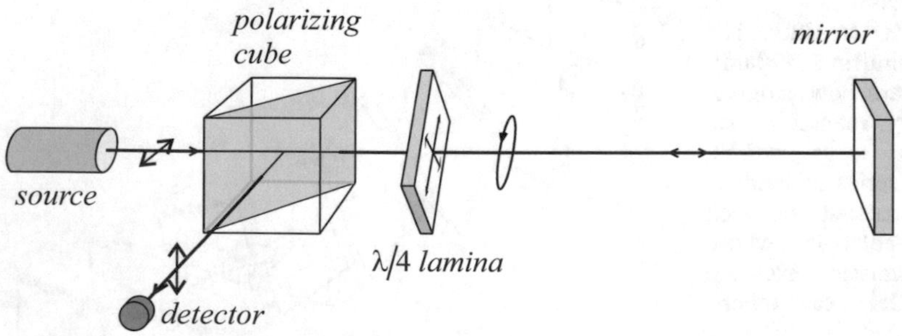

Fig. 7.41 Birefringence isolator. The lamina has its neutral axes at 45° from those of the polarizer and transforms the linear polarization into a circular one, and vice versa. The mirror reverses the direction of propagation, but not the direction of rotation of the field

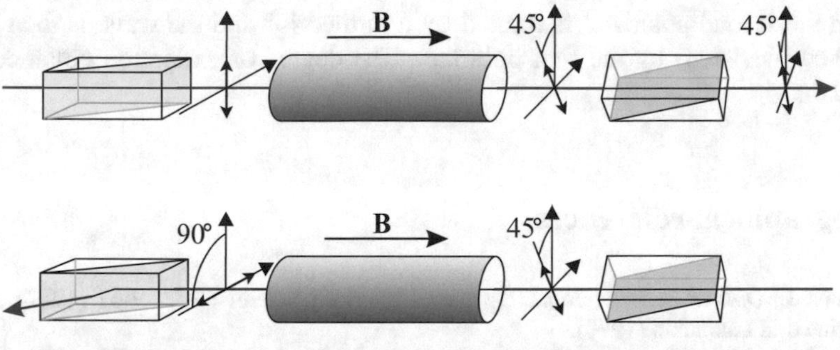

Fig. 7.42 Faraday's isolator. The magnetic medium is immersed in a static magnetic field **B**. The forward wave, after the first polarizer, has a vertical polarization that is rotated 45° by the magnetic medium and is ultimately transmitted by the second polarizer. The polarization of the backward wave, after the second polarizer, is at 45° from the vertical and is further rotated by the magnetic medium, becoming horizontal. This wave is ultimately deviated by the first polarizer

90° with respect to the going, and it is therefore reflected (transmitted) by the polarizing cube in the direction of the detector. In fact, the lamina $\lambda/4$ behaves overall in the two passages as a $\lambda/2$ lamina. This is easily understood by taking the lamina axes as the base directions of Jones's vectors: indeed the phase difference between the two components of the field along these directions is added in the two steps.

In this case, the polarizing cube + $\lambda/4$ lamina system behaves like a circulator. However, this works only if there are no other elements that alter the circular polarization in the path between the lamina and the mirror, such as another birefringent lamina or a polarizer, even partial. So, only if the polarization that reaches the mirror is circular.

With semiconductor lasers, it is especially important that any spurious reflection not return to the laser. Although this return could be much weaker by tens of dB than the emitted wave, the laser is disturbed in its emission spectrum. An almost ideal optical isolator, that is, with minimal losses, which functions for any polarization state of the reflected wave, can be realized only if it contains a *non-reciprocal* optical element (see §7.9.4 and following). This is possible thanks to the Faraday effect, in which the medium is made circularly birefringent by a magnetic field, which is a *pseudo*-vector.

A *Faraday's optical isolator* is composed of a transparent optical medium, generally a glass with a high Verdet constant, immersed in a longitudinal magnetic field (Fig. 7.42). Such a medium is placed between two polarizers, typically of calcite, whose optical axes of transmission are rotated from each other by 45°. The rotation of the polarization plane induced by the magnetic medium is calibrated to be 45° for an assigned wavelength, which is sometimes tunable by means of a ring that moves the medium in the field of the magnet. In this way, the forward wave, which has passed through the first polarizer, is also transmitted by the second polarizer. Instead, the polarization of the wave traveling backwards, after passing through the second polarizer, is rotated by a further 45° and the wave is then extinguished (deviated) by the first polarizer. The degree of extinction obtained is over 30 dB.

Bibliographical references

Applequist J., *Optical activity: Biot's bequest*, American Scientist 75, 59-67 (1987). Reprinted in Lakhtakia (1990).

Azzam R.M.A., *Ellipsometry*, in *Handbook of Optics*, Bass M. et al ed., Mc Graw-Hill, New York (1995).

Babinet J., *Sur le sens des vibrations dans les rayons polarisés*, Comptes Rendus Acad. Sci. Paris **29**, 514-515 (1849).

Bass M. et al ed., *Handbook of Optics*, McGraw-Hill, Inc., New York (1995).

Bekers J.M., *Achromatic linear retarders*, Appl. Opt. **10**, 973-975 (1971).

Bennett J.M., *Polarizer*, in *Handbook of Optics*, Bass M. et al ed. , Mc Graw-Hill, New York (1995).

Berry M.V., Jeffrey M.R., and Lunney J.G., *Conical diffraction: observations and theory*, Proc. R. Soc. A **462**, 1629-1642 (2006).

Bhagavantam S., *Crystal Symmetry and Physical Properties*, Academic Press, New York (1966).

Bloom A.L., *Modes of a laser resonator containing tilted birefringent plates*, J. Opt. Soc. Am. **64**, 447-452 (1974).

Bokut B.V. and Fedorov F.I., *On the theory of Optical activity in crystals. III. General equation of normal*, Opt. and Spectr. **6**, 342-344 (1959).

Born M., *Über die natürliche optische aktivität von flüssigkeiten und gasen*, Zeitschrift für Physik **16**, 251-258 (1915). *On the theory of optical activity*, Proc. Royal Soc. A **150**, 84-105 (1935). Reprinted in Lakhtakia (1990).

Born M. and Wolf E., *Principles of Optics*, Pergamon Press, Paris (1980).

Bragg W.L. and Pippard A.B., *The Form Birefringence of Macromolecules*, Acta Cryst. **6**, 865-867 (1953).

Brand D.J., *Molecular structure and chirality*, J. Chemical Education **64**, 1035-1038 (1987). Reprinted in Lakhtakia (1990).

Burns G., Glazer A.M., *Space Groups for Solid State Scientists*, Academic Press, New York, (1978).

Caldwell D.J. and Eyring H., *The theory of optical activity*, John Wiley & Sons, Inc., New York (1971).

Chandrasekharan V., Damany H., *Birefringence of Sapphire, Magnesium Fluoride, and Quartz in the Vacuum Ultraviolet, and Retardation Plates*, Appl. Opt. **7**, 939-941 (1968).

Cheng D.K. and Kong J.A., *Time harmonic fields in source free bianisotropic media*, J. Appl. Phys. **39**, 5792-5796 (1968).

Chipman R.A., *Polarimetry*, in *Handbook of Optics*, Bass M. et al ed., Mc Graw-Hill, New York (1995).

Condon E.U., *Theories of Optical Rotatory Power*, Rev. Mod. Phys. **9**, 432-457 (1937). Reprinted in Lakhtakia (1990).

Condon E.U., Altar W., and Eyring H., *One-Electron Rotatory Power*, J. Chem. Phys. **5**, 753-775 (1937).

Dodge M.J., *Refractive properties of magnesium fluoride*, Appl. Opt. **23**, 1980-1985

(1984).

Evans J.W., *The birefringent filter*, J. Opt. Soc. Am. **39**, 229-242 (1949).

Fedorov F.I., *On the theory of Optical activity in crystals. I. The law of conservation of energy and the Optical activity tensor*, Optics and Spectroscopy **6**, 49-53 (1959a), Reprinted in Lakhtakia (1990). *On the theory of Optical activity in crystals. II. Crystal of cubic symmetry and planar classes of central symmetry*, Opt. and Spectr. **6**, 237-240 (1959b).

Feynman R.P., Leighton R.B., Sands M., *The Feynman Lectures on Physics, Vol II*, Addison-Wesley publishing company, Reading MA (1969).

Fowles G.R., *Introduction to Modern Optics*, Holt, Rinehart, and Winston, New York (1968).

Guenther R., *Modern Optics*, John Wiley & Sons, New York (1990).

Hamilton W.R., *On some results of the view of a characteristic function in optics*, British Association Report, Cambridge 360-370 (1833). *Third supplement to an essay on the theory of systems of rays*, Trans. Roy. Irish Acad. **17**, 1-144 (1837).

Hammond C., *Introduction to crystallography*, Oxford University Press, Royal Microscopical Society (1992).

Hermann C.H. and Mauguin C., *International Tables for X-ray Crystallography*, Vol I, edited by K. Lonsdale. Birmingham: Kynoch Press (1952). See also the last edition at it.iucr.org: *International Tables for Crystallography, Vol. A: Space-group symmetry*, edited by Th. Hahn, International Union of Crystallography (2006).

Hobden M.V., *Optical activity in a non-enantiomorphous crystal: AgGaS2*, Acta Cryst. A **24**, 676-680 (1968).

Hodgkinson I.J. and Wu Q.H., *Birefringent thin films and polarizing elements*, World Scientific Publ. Co. Pte. Ltd., Singapore (1997).

Hopf F.A. and Stegeman G.I., *Applied Classical Electrodynamics, Vol. I: Linear Optics*, John Wiley & Sons, New York (1985).

Huard S., *Polarization of Light*, John Wiley & Sons, New York, Masson, Paris (1997).

Jackson J.D., *Classical Electrodynamics*, John Wiley & Sons, New York (1974).

Jenkins F.A. and White H.E., *Fundamental of Optics*, McGraw-Hill (1957).

Jerphagnon J. and Chemla D.S., *Optical activity of crystals*, J. Chem. Phys. **65**, 1522-1529 (1976).

Jones R.C., *A new calculus for the treatment of optical system, Part I*, J. Opt. Soc. Am. **31**, 488-493 (1941a); ... *Part II*, J. Opt. Soc. Am. **31**, 500-503 (1941b); ... *Part III*, J. Opt. Soc. Am. **32**, 486-493 (1942).

Kaminow I.P., *An introduction to electrooptic devices*, Academic Press (1974).

Kong J.A., *Theorems of bianisotropic media*, Proc. IEEE **60**, 1036-1046 (1972). Reprinted in Lakhtakia (1990).

Lakhtakia A., ed., *Selected Papers on Natural Optical Activity*, SPIE Milestone Series **15**, B.J. Thompson gen. ed., SPIE, Bellingham (1990). *On the genesis of Post constraint in modern electromagnetism*, Optik, Intern. J. for Light and Electron Optics **115**, 151-158 (2004).

Lakhtakia A., Varadan V.V., and Varadan V.K., *Field equations, Huygens's principle, integral equations, and theorems for radiation and scattering of electromagnetic waves in isotropic chiral media*, J. Opt. Soc. Am. A **5**, 175-184 (1988). Reprinted in Lakhtakia (1990).

Landau L. and Lifchitz E., (II), *Théorie du Champ*, MIR, Mosca (1966a). (III), *Physique Statistique*, MIR, Mosca (1966b). (VIII), *Électrodynamique des Milieux Continus*, MIR, Mosca (1966c).

Landsberg G.S., *Ottica*, MIR, Mosca (1979).

Lang S., *Linear Algebra*, Springer-Verlag, New York (1987).

H. Lloyd, *On the phænomena presented by light in its passage along the axes of biaxal crystals*, The London and Edinburgh Philosophical Magazine **2**, 112-120 (1833).

Lyot Bernard, *Le filtre monochromatique polarizant et ses applications en physique solaire*, Annales d'Astrophysique **7**, 1-49 (1944).

Maillard J.P., *Direct measurement of the birefringence of quartz at 3.39 and 3.50* μ, Opt. Comm. **4**, 175-177 (1971).

Mason S.F., *Optical activity and molecular dissymmetry*, Contemp. Phys. **9**, 239 (1968).

Medenbach G. and Shannon R.D., *Refractive indices and optical dispersion of 103 synthetic and mineral oxides and silicates measured by a small-prism technique*, JOSA B **14**, 3299-3318 (1997).

Medhat M. and El-Zaiat S.Y., *Interferometric determination of the birefringence dispersion of anisotropic materials*, Optics Comm. **141**, 145-149 (1997).

Nakano H. and Kimura H., *Quantum statistical-mechanical theory of optical activity*, J. Phys. Soc. Japan **27**, 519-535 (1969). Reprinted in Lakhtakia (1990).

Nikogosyan D.N., *Properties of Optical and Laser-Related Materials. A Handbook*, John Wiley & Sons, New York (1997).

Öhman Y., *A New Monochromator*, Nature **141**, n°3560, 157-158, and n°3563, 291-291 (1938).

Pancharatnam, S. , *Achromatic combinations of birefringent plates*, Proceedings of the Indian Academy of Sciences **41**, 130 (1955)

Peterson R.M., *Comparison of two theories of optical activity*, Am. J. of Physics **43**, 969-972 (1975).

Poggendorff J.C., *Ueber die konische refraction*, Ann. der Physik **129**, 461-462 (1839).

Post E.J., *Formal structure of electromagnetics*, (North-Holland, Amsterdam, 1962). Partially reprinted in Lakhtakia (1990).

Potton R.J., *Reciprocity in optics*, Reports on Progress in Physics **67**, 717-754 (2004).

Pujol M.C. et al, *Growth, optical characterization, and laser operation of a stoichiometric crystal $KYb(WO_4)_2$*, Phys. Rev. B **65**, 165121 (2002).

Roberts N.W., Chiou T.-H., Marshall N.J. and Cronin T.W., *A biological quarter-wave retarder with excellent achromaticity in the visible wavelength region*, Nature Photonics **3**, 641-644 (2009).

Robinson C.C., *The Faraday Rotation of Diamagnetic Glasses from 0.334 μ to 1.9* μ, Appl. Optics **3**, 1163-1166 (1964).

Shields J.H. and Ellis J.W., *Dispersion of Birefringence of Quartz in the Near Infrared*, J. Opt. Soc. Am. **46**, 363-365 (1956).

Silverman M.P., *Reflection and refraction at the surface of a chiral medium: comparison of gyrotropic constitutive relations invariant or noninvariant under a duality transformation*, J. Opt. Soc. Am. A **3**, 830-837 (1986). Reprinted in Lakhtakia (1990).

Silverman M.P. and Badoz J., *Light reflection from a naturally active birefringent medium*, J. Opt. Soc. Am. A **7**, 1163-1173 (1990).

Silverman M.P., Ritchie N., Cushman G.M., and Fischer B. , *Experimental configurations using optical phase modulation to measure chiral asymmetries in light specularly reflected from a naturally gyrotropic medium*, J. Opt. Soc. Am. A **5**, 1852-1862 (1988).

Sivoukhine D., *Optique*, MIR, Mosca (1984).

Smartt R.N. and Steel W.H., *Birefringence of Quartz and Calcite*, J. Opt. Soc. Am. **49**, 710-712 (1959).

Soleil J.B., *Nouvel appareil propre ä la mesure des deviations dans les experiences de polarisation rotatoire*, C. R. Acad. Sci. (Paris) **21**, 426-430 (1845); *Note sur un perfectionnement apporté au pointage du saccharimètre*, C. R. Acad. Sci. (Paris) **24**, 973-975 (1847).

Sommerfeld A., *Optics*, Academic Press, New York (1949).

Strumia F., *Appunti di conduzione elettrica nei gas*, Cap. IV - *I filtri*. Università degli Studi di Pisa, (a.a. 1973-1974).

Tellegen B.D.H., *The gyrator, a new electric network element*, Phillips Research Report **3**, 81-101 (1948). Reprinted in Lakhtakia (1990).

Tinoco I. and Freeman M., *The optical activity of oriented copper helices, I, experimental*, J. Physical Chemistry **61**, 1196-1200 (1957). Reprinted in Lakhtakia (1990).

Van Kranendonk J. and Sipe J.E., *Foundation of the macroscopic electromagnetic theory of dielectric media*, in *Progress in Optics XV*, 245-356, E. Wolf ed., North-Holland Publishing Company, Amsterdam (1977).

Villaverde A.B. et al, *Terbium gallium garnet Verdet constant measurements with pulsed magnetic field*, J. Phys. C: Solid State Physics **11**, L495-498 (1978).

Voigt W., *Bemerkung zur theorie der konischen refraktion*, Phys. Z. **6**, 672-673 (1905). *Ueber die Wellen flaeche zweiachsiger aktiver kristalle und ueber ihre konische refraktion*, Phys. Z. **6**, 787-790 (1905).

von Haidinger W., *Die konische Refraction am Diopsid, nebst Bemerkungen über einige Erscheinungen der konischen Refraction am Arragonit*, Annalen Physik **172**, 469-487 (1855).

Wiener O., *Allgemeine Sätze über die Dielektrizitätskonstanten der Mischkörper*, Abh. Sächs. Ges. Akad. Wiss., Math. Phys. Kl. **6**, 574-584 (1912).

Williams P.A. et al, *Temperature dependence of the Verdet constant in several diamagnetic glasses*, Appl. Optics **30**, 1176-1178 (1991).

Wood E.A., *Crystals and Light*, Dover Publications, Inc., New York (1977).

Yariv A. and Yeh P., *Optical waves in crystals*, John Wiley, New Jersey (1984).

Zhong H., Levine Z.H., Allan D.C., Wilkins J.W., *Band-theoretic calculations of the optical-activity tenor of α-quartz and trigonal Se*, Phys. Rev. B **48**, 1384-1402 (1993).

Appendix A
Conventions on electromagnetism

In this text, the following conventions are used:

a) The system of units used is the International System (SI). For comparison, in Table A.1, the essential formulas expressed in SI and in the "Gaussian" system are reported, while, from Table A.2, one can obtain the conversion between these two systems and, in Table A.3, the equivalence between the units of measurement. Many *theoretical* physicists insist on the use of systems such as CGS (founded on only three dimensions, with units of cm, gram, second) that has two different systems of unity, electrostatic (esu) and electromagnetic (emu), or Gaussian, which uses a combination of both. In this way, one obtains a greater formal elegance of the equations, which reaches its maximum with the Heaviside's system. This use by theorists is, unfortunately, ambiguous to the experimental, in which tools are used that are built by engineers, who instead use the SI.

b) In the complex abbreviated notation of fields, the term $+i(\boldsymbol{k}\cdot\boldsymbol{r}-\omega t)$ is used: in comparison with texts that use the (totally) opposite convention, it is sufficient to replace i with $-i$ in all of the formulas.

System	Gaussian	SI
ε_0	1	$\frac{10^7}{4\pi c^2}\frac{C^2}{Ns^2}$ $\left[\frac{Q^2T^2}{ML^3}\right]$
μ_0	1	$\frac{4\pi}{10^7}\frac{N}{A^2}$ $\left[\frac{ML}{Q^2}\right]$
D, H	$\mathbf{D}=\mathbf{E}+4\pi\mathbf{P}$, $\mathbf{H}=\mathbf{B}-4\pi\mathbf{M}$	$\mathbf{D}=\varepsilon_0\mathbf{E}+\mathbf{P}$, $\mathbf{H}=\frac{1}{\mu_0}\mathbf{B}-\mathbf{M}$
Macroscopic Maxwell's equations	$\nabla\cdot\mathbf{D}=4\pi\rho$, $\nabla\times\mathbf{E}=-\frac{1}{c}\frac{\partial\mathbf{B}}{\partial t}$ $\nabla\cdot\mathbf{B}=0$, $\nabla\times\mathbf{H}=\frac{4\pi}{c}\mathbf{J}+\frac{1}{c}\frac{\partial\mathbf{D}}{\partial t}$	$\nabla\cdot\mathbf{D}=\rho$, $\nabla\times\mathbf{E}=-\frac{\partial\mathbf{B}}{\partial t}$ $\nabla\cdot\mathbf{B}=0$, $\nabla\times\mathbf{H}=\mathbf{J}+\frac{\partial\mathbf{D}}{\partial t}$
Lorentz's force	$q\left(\mathbf{E}+\frac{\boldsymbol{v}}{c}\times\mathbf{B}\right)$	$q(\mathbf{E}+\boldsymbol{v}\times\mathbf{B})$
Constitutive equations	$\mathbf{P}=\chi_e\mathbf{E}$, $\mathbf{M}=\chi_m\mathbf{H}$ $\varepsilon=1+4\pi\chi_e$, $\mu=1+4\pi\chi_m$	$\mathbf{P}=\varepsilon_0\chi_e\mathbf{E}$, $\mathbf{M}=\mu_0\chi_m\mathbf{H}$ $\varepsilon=\varepsilon_0(1+\chi_e)$, $\mu=\mu_0(1+\chi_m)$

Table A.1 Comparison between the International System (SI) and the Gaussian System (noting that the latter, in turn, differs from the Electrostatic and Electromagnetic Systems)

G. Giusfredi, *Physical Optics*, UNITEXT for Physics,
https://doi.org/10.1007/978-3-030-25279-3

Quantity	Gaussian System	SI
Light speed	c	$c = 1/\sqrt{\varepsilon_0 \mu_0}$
Electrical permittivity	ε	$\varepsilon/\varepsilon_0$
Magnetic permeability	μ	μ/μ_0
Susceptibility	χ_e, χ_m	$\chi_e/(4\pi),\ \chi_m/(4\pi)$
Electric field, potential, voltage	$\mathbf{E}, \Phi, V$	$\sqrt{4\pi\varepsilon_0} \times \mathbf{E}, \Phi, V$
Electric displacement	$\mathbf{D}$	$\sqrt{\frac{4\pi}{\varepsilon_0}}\mathbf{D}$
Polarization	$\mathbf{P}$	$\frac{1}{\sqrt{4\pi\varepsilon_0}}\mathbf{P}$
Vector potential	$\mathbf{A}$	$\sqrt{\frac{4\pi}{\mu_0}}\mathbf{A}$
Magnetic field (induction)	$\mathbf{B}$	$\sqrt{\frac{4\pi}{\mu_0}}\mathbf{B}$
Magnetic field (displacement)	$\mathbf{H}$	$\sqrt{4\pi\mu_0}\mathbf{H}$
Magnetization	$\mathbf{M}$	$\sqrt{\frac{\mu_0}{4\pi}}\mathbf{M}$
Density of charge, current density, charge, current	$\rho, \mathbf{J}, q, I$	$\frac{1}{\sqrt{4\pi\varepsilon_0}} \times \rho, \mathbf{J}, q, I$
Conductivity	σ	$\frac{\sigma}{4\pi\varepsilon_0}$
Resistance, impedance, inductance	R, Z, L	$4\pi\varepsilon_0 \times R, Z, L$
Capacity	C	$\frac{C}{4\pi\varepsilon_0}$

Table A.2 Conversion table for symbols and formulas. To convert every formula expressed in the Gaussian System into a similar formula in the International System, it is sufficient to substitute the symbols of the first column with the expressions of the second column. Some symbols that follow the same substitution are gathered together. The inverse transformation is obtained by inverting the expressions, and also by applying the substitutions $\varepsilon_0 \to 1/(4\pi)$ and $\mu_0 \to 4\pi/c^2$. When an expression explicitly contains a unit of measurement, it follows the same rule of substitution as the corresponding physical quantity

Quantity	Gaussian System	SI
Electric charge	1 statCoulomb (esu) 1 Franklin	$\frac{1}{\text{ç}}\times 10^{-9}$ Coulomb
Electric current	1 statAmpere (esu)	$\frac{1}{\text{ç}}\times 10^{-9}$ Ampere
Electric potential	1 statVolt (esu)	$\text{ç}\times 10^{2}$ Volt
Electric field **E**	$1\frac{\text{statVolt}}{\text{cm}}$	$\text{ç}\times 10^{4}\frac{\text{Volt}}{\text{m}}$
Electric displacement **D**	$1\frac{\text{statCoulomb}}{\text{cm}^2}$	$\frac{1}{\text{ç}4\pi}\times 10^{-5}\frac{\text{Coulomb}}{\text{m}^2}$
Polarization **P**	$1\frac{\text{statCoulomb}}{\text{cm}^2}$	$\frac{1}{\text{ç}}\times 10\ ^{5}\frac{\text{Coulomb}}{\text{m}^2}$
Magnetic induction **B**	1 Gauss (emu)	$10^{-4}\frac{\text{Weber}}{\text{m}} = 10^{-4}$ Tesla
Magnetic field **H**	1 Oersted (emu)	$\frac{1}{4\pi}\times 10^{3}\frac{\text{Ampere-spira}}{\text{m}}$
Magnetization **M**	$1\frac{\text{emu}}{\text{cm}^3}$	$10^{3}\frac{\text{Ampere-spira}}{\text{m}}$
Capacity	1 cm	$\frac{1}{\text{ç}^2}\times 10^{-11}$ Farad
Inductance	$1\frac{\text{s}^2}{\text{cm}}$	$\text{ç}^2\times 10^{11}$ Henry
Resistance	$1\frac{\text{s}}{\text{cm}}$	$\text{ç}^2\times 10^{11}$ Ohm

Table A.3 Equivalences between electromagnetic units of measurement. The symbol ç is introduced for brevity and is 2.99792458, which are the figures of the speed of light, often approximated as 3. In carrying out the transformation, it is necessary to establish the precise physical entity to which the measure corresponds. For example, in the Gaussian system, **B** and **H** have the same dimensions, even if they are indicated with different names, Gauss and Oersted, while, in SI, they have different dimensions. Moreover, in the transformation for field **D**, an additional factor 4π appears to divide with respect to the transformation for the electric charge. There is a greater risk of confusion in regard to what is called *magnetic polarization* $\mathbf{P}_m$, which, in the Gaussian system, is equal to the magnetization **M**, while, in SI, it is $\mathbf{P}_m = \mu_0\mathbf{M}$, and therefore has different dimensions

[Moskowitz B.M., *Fundamental physical constant and conversion factors*, Global Earth Physics, a handbook of physical constants, ed. The American Geophysical Union (1995)].

Appendix B
Mathematical relations

B.1 Vector formulas

$$
\begin{aligned}
\boldsymbol{a}(\boldsymbol{b}\times\boldsymbol{c}) &= \boldsymbol{b}\cdot(\boldsymbol{c}\times\boldsymbol{a}) = \boldsymbol{c}\cdot(\boldsymbol{a}\times\boldsymbol{b}) \\
\boldsymbol{a}\times(\boldsymbol{b}\times\boldsymbol{c}) &= (\boldsymbol{a}\cdot\boldsymbol{c})\boldsymbol{b} - (\boldsymbol{a}\cdot\boldsymbol{b})\boldsymbol{c} \\
(\boldsymbol{a}\times\boldsymbol{b})\cdot(\boldsymbol{c}\times\boldsymbol{d}) &= (\boldsymbol{a}\cdot\boldsymbol{c})(\boldsymbol{b}\cdot\boldsymbol{d}) - (\boldsymbol{a}\cdot\boldsymbol{d})(\boldsymbol{b}\cdot\boldsymbol{c}) \\
\nabla\times\nabla\psi &= 0 \qquad \textit{(the rotor of a gradient is zero)} \\
\nabla\cdot(\nabla\times\boldsymbol{a}) &= 0 \qquad \textit{(the divergence of a rotor is zero)} \\
\nabla\times(\nabla\times\boldsymbol{a}) &= \nabla(\nabla\cdot\boldsymbol{a}) - \nabla^2\boldsymbol{a} \\
\nabla\cdot(\psi\boldsymbol{a}) &= \boldsymbol{a}\cdot\nabla\psi + \psi\nabla\cdot\boldsymbol{a} \\
\nabla\times(\psi\boldsymbol{a}) &= \nabla\psi\times\boldsymbol{a} + \psi\nabla\times\boldsymbol{a} \\
(\boldsymbol{a}\cdot\nabla)(\psi\boldsymbol{b}) &= \boldsymbol{b}(\boldsymbol{a}\cdot\nabla)\psi + \psi(\boldsymbol{a}\cdot\nabla)\boldsymbol{b} \\
\nabla(\boldsymbol{a}\cdot\boldsymbol{b}) &= (\boldsymbol{a}\cdot\nabla)\boldsymbol{b} + (\boldsymbol{b}\cdot\nabla)\boldsymbol{a} + \boldsymbol{a}\times(\nabla\times\boldsymbol{b}) + \boldsymbol{b}\times(\nabla\times\boldsymbol{a}) \\
\nabla\cdot(\boldsymbol{a}\times\boldsymbol{b}) &= \boldsymbol{b}\cdot(\nabla\times\boldsymbol{a}) - \boldsymbol{a}\cdot(\nabla\times\boldsymbol{b}) \\
\nabla\times(\boldsymbol{a}\times\boldsymbol{b}) &= \boldsymbol{a}(\nabla\cdot\boldsymbol{b}) - \boldsymbol{b}(\nabla\cdot\boldsymbol{a}) + (\boldsymbol{b}\cdot\nabla)\boldsymbol{a} - (\boldsymbol{a}\cdot\nabla)\boldsymbol{b} \\
\nabla^2(\psi\boldsymbol{a}) &= 2(\nabla\psi\cdot\nabla)\boldsymbol{a} + \boldsymbol{a}\nabla^2\psi + \psi\nabla^2\boldsymbol{a}
\end{aligned}
\tag{B.1}
$$

B.2 Integral theorems

In the following expressions, φ, ψ, and $\boldsymbol{A}$ are "regular" scalar and vector functions, defined in a three-dimensional volume V, delimited by the closed surface S. Integrals have dV as an element of volume and dS as a surface element; furthermore, **n** is a unitary vector normal to dS and is oriented outside of the volume. The circle in the integral symbol indicates that the surface is closed.

$$
\begin{aligned}
\int_V \nabla\cdot\boldsymbol{A}\,dV &= \oint_S \boldsymbol{A}\cdot\mathbf{n}\,dS \qquad \textit{divergence theorem} \\
\int_V \nabla\psi\,dV &= \oint_S \psi\mathbf{n}\,dS \\
\int_V \nabla\times\boldsymbol{A}\,dV &= \oint_S \boldsymbol{A}\times\mathbf{n}\,dS \\
\int_V \left(\phi\nabla^2\psi + \nabla\phi\cdot\nabla\psi\right)dV &= \oint_S \phi\mathbf{n}\cdot\nabla\psi\,dS \qquad \textit{Green's first identity} \\
\int_V \left(\phi\nabla^2\psi - \psi\nabla^2\phi\right)dV &= \oint_S (\phi\nabla\psi - \psi\nabla\phi)\cdot\mathbf{n}\,dS \qquad \textit{Green's theorem}
\end{aligned}
\tag{B.2}
$$

G. Giusfredi, *Physical Optics*, UNITEXT for Physics,
https://doi.org/10.1007/978-3-030-25279-3

In the two expressions that now follow, C is a closed line that delimits an open surface S, with line element dl. The versor **n** is normal to the surface element dS and is oriented towards the direction from which the circuit C is seen traveling counterclockwise. The circle in the integral indicates that the circuit is closed.

$$\int_S (\nabla \times \boldsymbol{A}) \cdot \mathbf{n} dS = \oint_C \boldsymbol{A} \cdot \boldsymbol{dl} \qquad \textit{Stokes' theorem}$$
$$\int_S \mathbf{n} \cdot \nabla \psi dS = \oint_C \psi \boldsymbol{dl} \tag{B.3}$$

B.3 Bessel's functions

The most common integral definition is

$$\mathrm{J}_n(x) = \frac{1}{\pi} \int_0^{\pi} \cos(x \sin\vartheta - n\vartheta) d\vartheta, \quad n \in \mathbb{Z}, \tag{B.4}$$

from which other equivalent definitions can be derived:

$$\mathrm{J}_n(x) = \frac{i^{+n}}{2\pi} \int_0^{2\pi} e^{-ix\cos\vartheta + in\vartheta} d\vartheta, \tag{B.5}$$

$$\mathrm{J}_n(x) = \frac{i^{-n}}{\pi} \int_0^{\pi} e^{ix\cos\vartheta} \cos(n\vartheta) d\vartheta. \tag{B.6}$$

In particular,

$$\mathrm{J}_1(x) = \frac{x}{\pi} \int_{-1}^{1} e^{ixu} \sqrt{1-u^2} du. \tag{B.7}$$

Eqs. (2), (3) and (4) can also be written by replacing i with $-i$, since $J_n(x)$ is real. There is also the recursive relationship

$$\frac{d}{dx}\left[x^{n+1} \mathrm{J}_{n+1}(x)\right] = x^{n+1} \mathrm{J}_n(x). \tag{B.8}$$

Series expansion:

$$\mathrm{J}_n(x) = \sum_{m=0}^{\infty} \frac{(-1)^m}{m!(m+n)!} \left(\frac{x}{2}\right)^{n+2m}. \tag{B.9}$$

Values in $x = 0$:

$$\mathrm{J}_0(0) = 1, \quad \mathrm{J}_n(0) = 0 \quad \text{per } n \neq 0. \tag{B.10}$$

Change of sign:

$$\mathrm{J}_n(-x) = (-1)^n \mathrm{J}_n(x), \quad \mathrm{J}_{-n}(x) = (-1)^n \mathrm{J}_n(x). \tag{B.11}$$

Recursive formulas:

$$\mathrm{J}_{n+1}(x) = \frac{2n}{x}\mathrm{J}_n(x) - \mathrm{J}_{n-1}(x), \tag{B.12}$$

$$2\mathrm{J}'_n(x) = \mathrm{J}_{n-1}(x) - \mathrm{J}_{n+1}(x). \tag{B.13}$$

B.4 Fresnel's integrals

These are defined by the expressions

$$\mathrm{C}(z) = \int_0^z \cos\left(\frac{\pi}{2}t^2\right)dt, \quad \mathrm{S}(z) = \int_0^z \sin\left(\frac{\pi}{2}t^2\right)dt, \tag{B.14}$$

that is, by

$$\begin{aligned} \mathrm{C}(z) &= \frac{1}{2} + f(z)\sin\left(\frac{\pi}{2}z^2\right) - g(z)\cos\left(\frac{\pi}{2}z^2\right), \\ \mathrm{S}(z) &= \frac{1}{2} - f(z)\cos\left(\frac{\pi}{2}z^2\right) - g(z)\sin\left(\frac{\pi}{2}z^2\right), \end{aligned} \tag{B.15}$$

also valid for $z \in \mathbb{C}$. For a real argument x such that $0 \le x \le \infty$, the auxiliary functions f and g are approximable with the fractions [Abramowitz and Stegun 1972]

$$\begin{aligned} f(x) &= \frac{1 + 0.926x}{2 + 1.792x + 3.104x^2} + \varepsilon(x), \quad |\varepsilon(x)| \le 2\times 10^{-3}, \\ g(x) &= \frac{1}{2 + 4.142x + 3.492x^2 + 6.670x^3} + \varepsilon(x), \quad |\varepsilon(x)| \le 2\times 10^{-3}. \end{aligned} \tag{B.16}$$

Symmetry relations:

$$\begin{aligned} &\mathrm{C}(-z) = -\mathrm{C}(z), \quad \mathrm{S}(-z) = -\mathrm{S}(z), \\ &\mathrm{C}(iz) = i\mathrm{C}(z), \quad \mathrm{S}(iz) = -i\mathrm{S}(z). \end{aligned} \tag{B.17}$$

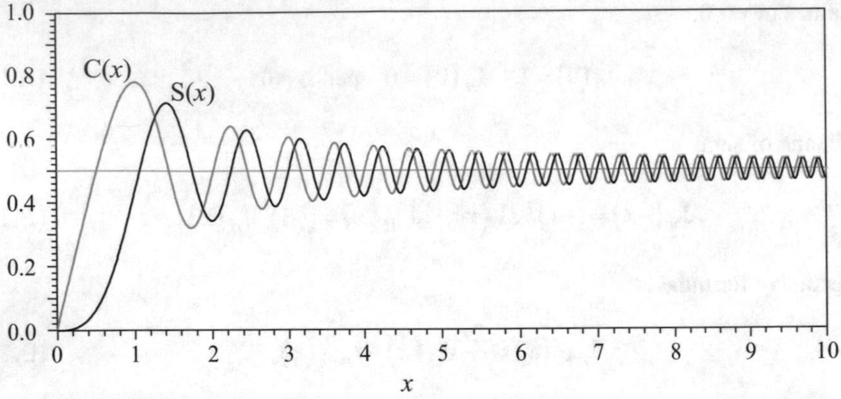

Fig. B.1 Fresnel's integrals

Relationship with the error function erf:

$$C(z)+iS(z)=\frac{1+i}{2}\operatorname{erf}\left[\frac{\sqrt{\pi}}{2}(1-i)z\right]. \tag{B.18}$$

B.5 Error function

This is defined as

$$\operatorname{erf}(z)=\frac{2}{\sqrt{\pi}}\int_0^z e^{-t^2}dt, \tag{B.19}$$

which, for real arguments, we can approximate as [Abramowitz and Stegun 1972]

$$\operatorname{erf}(x)=1-\left(a_1t+a_2t^2+a_3t^3\right)e^{-x^2}+\varepsilon(x), \tag{B.20}$$

with

$$t=\frac{1}{1+px}, \quad |\varepsilon(x)|\le 2.5\times 10^{-5}$$

$$a_1=0.3480242, \quad a_2=-0.0958798, \quad a_3=0.7478556, \quad p=0.47047\,.$$

[Abramowitz M. and Stegun I.A., *Handbook of Mathematical Functions* (Dover Publications Inc., New York, 1972)].

Appendix C
The founding fathers of Optics

Pythagòras of Samo	~580 – ~493 B.C.
Empedoklês of Agrigento	~490 – 435 B.C.
Eukléidēs of Alexandria	~330 – 260 B.C.
Archimédēs of Siracusa	287 – 212 B.C.
Heron of Alexandria	
Kleomédēs	~50 A.D.
Kláudios Ptolemâios	~90 – ~168
Galeno of Pergamo	129 – 200
Abu Ysuf Yacub ibn Is-Haq (Alkindi)	– ~873
Ibn al-Haythan al-Hazin (Alhazen)	~965 – 1039
Robert Grosseteste	1175 – 1253
Erazm Ciolek (Vitellione o Witelo)	~1210 – ~1285
Roger Bacon	1214 – 1294
Pierre le Pèlerin of Maricourt	~1269
Leonardo da Vinci	1452 – 1519
Francesco Maurolico	1494 – 1574
Robert Norman	~1590
Giovan Battista della Porta	1535 – 1605
William Gilbert	1544 – 1603

1564 ------------------------------- 1642	Galileo Galilei
1571 ------------------- 1620	Giovanfrancesco Sagredo
1571 ----------------------- 1630	Johannes Kepler
1580 ------------------ 1626	Willebrord Snel van Royen
1586--------------------------- 1650	Niccolò Cabeo
1596 --------------------- 1650	René Descartes
1601 ------------------------- 1665	Pierre de Fermat

G. Giusfredi, *Physical Optics*, UNITEXT for Physics,
https://doi.org/10.1007/978-3-030-25279-3

1602	1686	Otto von Guericke
1608	1647	Evangelista Torricelli
1618	1663	Francesco Maria Grimaldi
1620	1682	Jean Picard
1625	1698	Erasmo Bartholin
1625	1712	Gian Domenico Cassini
1627	1691	Robert Boyle
1629	1695	Christiaan Huygens
1635	1703	Robert Hooke
1642	1727	Isaac Newton
1644	1710	Olaf Römer
~1666	1736	Stephen Gray
1692	1762	James Bradley
1692	1761	Pieter van Musschenbroek
1698	1739	Charles François Dufay
1706	1790	Benjamin Franklin
1707	1783	Leonard Euler
1711	1787	Giuseppe Ruggero Boscovich
1716	1781	Giambattista Beccaria
1724	1802	Franz Aepinus
1724	1793	John Michell
1731	1810	Henry Cavendish
1733	1804	Joseph Priestly
1736	1813	Giuseppe Ludovico Lagrange
1736	1806	Charles-Augustin Coulomb
1737	1798	Luigi Galvani
1738	1822	William Herschel
1745	1827	Alessandro Volta
1749	1827	Pierre Simon de Laplace
1765	1833	Joseph Nicéphore Niepce
1766	1828	William Hyde Wollaston
1768	1830	Jean Baptiste Joseph Fourier

William Nicol	1768	1857
Thomas Young	1773	1829
Jean Baptiste Biot	1774	1862
Étienne Louis Malus	1775	1812
André Marie Ampère	1775	1836
Johann Wilhelm Ritter	1776	1810
Hans Christian Oersted	1777	1851
Karl Friedrich Gauss	1777	1855
Siméon Denis Poisson	1781	1840
David Brewster	1781	1868
Jean Dominique François Arago	1786	1853
Giovanni Battista Amici	1786	1863
Josef Fraunhofer	1787	1826
Jan Evangelista Purkyne	1787	1869
Louis Jacques Mandé Daguerre	1787	1851
Augustin Jean Fresnel	1788	1827
Georg Simon Ohm	1789	1854
Augustin Louis Cauchy	1789	1857
Felix Savart	1791	1841
Michael Faraday	1791	1867
George Green	1793	1841
Wilhelm Karl Haidinger	1795	1871
Joseph Henry	1799	1878
William Henry Fox Talbot	1800	1877
George Airy	1801	1892
Christian Doppler	1803	1853
Wilhelm Eduard Weber	1804	1891
William Rowan Hamilton	1805	1865
Josef Max Petzval	1807	1891
Rudolph Kohlrausch	1809	1858
Armand Hypolite Louis Fizeau	1819	1896

George Gabriel Stokes	1819 ---------------------------------1903
Jean Bernard Léon Foucault	1819 -------------------1868
Hermann von Helmholtz	1821 ------------------------------ 1894
Ludwig Philipp von Seidel	1821 ------------------------------1896
Gustav Robert Kirchhoff	1824 -------------------------- 1887
John Kerr	1824 ----------------------------------1907
William Thomson (*Lord Kelvin*)	1824 ---------------------------------- 1908
George Jonstone Stoney	1826 ---------------------------------- 1911
James Clerk Maxwell	1831 ------------------- 1879
Ernst Abbe	1840 ----------------------------1905
John William Strutt (*Lord Rayleigh*)	1842 --------------------------------1919
Wilhelm Konrad Röntgen	1845 --------------------------------- 1923
Woldemar Voigt	1850 -----------------------------1919
Oliver Heaviside	1850 ------------------------------- 1925
John Henry Poynting	1852 --------------------------1914
Albert Abraham Michelson	1852 ---------------------------------1931
Hendrik Antoon Lorentz	1853--------------------------------1928
Joseph John Thomson	1856 ----------------------------------1940
Heinrich Rudolf Hertz	1857 --------------- 1894
Max Plank	1858--------------------------------------1947
Paul Karl Ludwig Drude	1863 ---------------- 1906
Jean Baptiste Gaspard Gustave Alfred Perot	1863 ------------------------- 1925
Carl Alwin Pockels	1865--------------------- 1913
Pieter Zeeman	1865---------------------------------1943
Marie Paul Auguste Charles Fabry	1867--------------------------------1945
Arnold Johannes Wilhelm Sommerfeld	1868 ---------------------------------- 1951
Albert Einstein	1879--------------------------------- 1955
Frederik Zernike	1885 ----------------------------------1966
Arthur Compton	1892 ----------------------------- 1962

Index of authors

G. Giusfredi, *Physical Optics*, UNITEXT for Physics,
https://doi.org/10.1007/978-3-030-25279-3

Analytical index

G. Giusfredi, *Physical Optics*, UNITEXT for Physics,
https://doi.org/10.1007/978-3-030-25279-3

Printed in the United States
By Bookmasters

Printed in the United States
By Bookmasters